Theoretical Materials Science

Tracing the Electronic Origins of Materials Behavior

Theoretical Materials Science

Tracing the Electronic Origins of Materials Behavior

A. Gonis
Chemistry and Materials Science
Lawrence Livermore National Laboratory
Livermore, CA 94550

CAMBRIDGE
UNIVERSITY PRESS

University Printing House, Cambridge CB2 8BS, United Kingdom

One Liberty Plaza, 20th Floor, New York, NY 10006, USA

477 Williamstown Road, Port Melbourne, VIC 3207, Australia

314-321, 3rd Floor, Plot 3, Splendor Forum, Jasola District Centre, New Delhi - 110025, India

79 Anson Road, #06-04/06, Singapore 079906

Cambridge University Press is part of the University of Cambridge.

It furthers the University's mission by disseminating knowledge in the pursuit of education, learning and research at the highest international levels of excellence.

www.cambridge.org
Information on this title: www.cambridge.org/9781558995406

CODEN: MRSPDH

A catalogue record for this publication is available from the British Library

Library of Congress Cataloging in Publication data
Ttheoretical materials science: Tracing the electronic origins of materials behavior /
editor, A. Gonis
p.cm.
ISBN 1-55899-540-4
1. Materials science. 2. Solid state electronics. I. Gonis, A. II. Title.
TA403.6 .G63 2000
620.1´1—dc21 00-062470

ISBN 978-1-558-99540-6 Hardback

For

Deena,
Ifigenia, and
Antigone

PREFACE

When penetrating into a special field like the *ab initio* electron theory of solids, it is useful to have at least a rough idea of how this field is surrounded or embedded in the general scheme of scientific inquiry. The following remarks aim to give an impression of the width of materials problems that can be treated nowadays and that should be treatable in the future, of the different kinds of electron theory used in such treatments, and of the many links between quantum chemistry, solid state theory, materials science, microelectronics, and even biology that such treatments often entail.

Modern materials science deals not only with such traditional subjects as steel and metallic materials, ceramics, refractory phases, and compound systems, but also with such sophisticated subjects as conducting and insulating materials for electrical engineering, microelectronics, micromechanics and their combination, powders, sintertechnology, solid foams, hydrogen storage, high-temperature superconductors (HTSCs), and many others, too numerous to give a complete list. Nanoscience is developing which includes such esoteric subjects as self-organized quantum dots and carbon tubes, and even touches biology and pharmacy. In the future, 'traditional' and 'exotic' materials science, microelectronics, and quantum chemistry are expected to come closer together. An increasingly deeper understanding of the properties of the materials just mentioned on different levels, (micro-, meso-, macroscopic), will be followed/accompanied by the development of sophisticated computational tools of high predictive power.

Along with the diversity of materials types and problems, we have a hierarchy of solid-state theories available that proceed gradually from a macroscopic to an increasingly refined microscopic level of description. These cover the whole field with phenomenological solid-state theory, e.g., for heterogeneous materials like polycrystals or compound materials, and different sorts of microscopic theories. Within the latter, electron theory is found only at the last, deepest level. Although at the very end all properties of a material follow in principle from the Schrödinger equation and the Pauli principle, one should distinguish between electronic properties, which are determined directly by the electronic structure in a theoretically and computationally performable manner, and other properties, whose quantum-mechanical aspects are more hidden and cannot so easily be traced back to their fundamental origins. Examples are the heat capacity on the one hand and the ensemble (the 'sauerkraut') of the dislocations in a real solid on the other.

Electron theory began in the late twenties, essentially hand-in-hand with the developments in the quantum theory of matter. Since that time, the task has been to explain, and even predict, the properties of finite, extended and semi-infinite electron systems based 'only' on the solution of a corresponding Schrödinger equation under the constraints of the Pauli principle, without calling in or invoking any additional forces, principles, etc. Even within the simplifying Born-Oppenheimer approximation (nuclei at rest, so that the electron-phonon coupling requires an extension beyond this approximation) the general task consists of two partial, but by no means trivial elements, namely the solution of the many-electron problem and that of the many-center problem. With methods addressing these tasks in good approximation, electron theory deals with a number of diverse phenomena and systems. These include,

• the properties of atoms, molecules, small clusters, including fullerenes, carbon tubes, quantum dots, and also chemical reactions as processes in time, e.g., femtosecond dynamics at conical intersections,

• the mechanical, thermal, electric, magnetic, and optical properties of 'ordinary' solids as functions of temperature, pressure, (non-hydrodynamic) stress, magnetic and electric fields, incident light or particles, where by 'ordinary' systems or phenomena is meant,
a. metals, semiconductors, insulators, ferro- and antiferromagnets (e.g. magnetism of molecules, clusters, semiconductors, oxides, rare earths, actinides. also magnetic excitations), superconductors, ferroelectrics,
b. ordered and disordered bulk binary, ternary, quaternary, ..., alloys,
c. planar surfaces without and with physi- or chemisorbed adatoms, kinks, surfactants, surface chemical and enzymatic processes, corrosion, catalysis,
d. interfaces, including stacking faults, grain boundaries, semiconductor heterointerfaces, Schottky barriers, electric contacts, artificial multilayers, friction,
e. smoothly and sharply curved surfaces of large clusters, voids, edges, corners,
f. dislocations, Bloch walls, vortices,
g. scanning and atomic force microscopy, nanohardness among others,

• quasicrystals which, with their peculiarities, are an example of extraordinary solids, such as,
a. systems with special spin structure (non-collinear magnetism), and spin polarized transport, femtosecond spin-dynamics in thin magnetic films,
b. systems with coexistence of strong magnetism (ferromagnetism, antiferromagnetism) and superconductivity, e.g., quaternary boroncarbides,

• strongly correlated electron systems, such as,
a. transition metal oxides (metal-insulator transition and its coupling to para-, and ferromagnetic transitions), HTSCs, spin-dependent transport effects like giant, colossal, and tunneling magnetoresistance with three different physical mechanisms,
b. mixed valence compounds, UBe_{13}, or intermediate valence systems, SmS,
c. the Kondo effect and heavy fermions in compounds with lanthanides and actinides, where the phenomenon of Fermi surface melting appears (the discontinuity in momentum space vanishes),
d. the fractional quantum Hall effect in semiconductor systems,
e. Fermi to non-Fermi liquid transitions, and quantum critical points,

• localization due to disorder (Anderson localization) and the combination of such disorder effects with particle-particle interactions,

• persistent currents in mesoscopic systems,

• quantum dots (with phenomena such as self-organization and localization delocalization transitions), single-electron electronics (Coulomb blockade),

• quantum chaos.

There are also situations in real solids so complicated that they cannot be treated from first principles. Ensembles of dislocations have already been mentioned. A possibly simpler example is the Peierls force. Other complicated subjects are: structure of dislocation cores, plasticity, crack propagation, ductile and brittle fracture, hydrogen and helium embrittlement, defect migration and accumulation at semiconductor heterointerfaces, crystal and amorphous ice growth (treated with long time simulations). Examples where the nuclei appear as a probe of the electronic structure are: the nuclear spin-lattice relaxation time, the Knight shift, the coefficient of inner conversion, the Mößbauer effect. They show up in nuclear magnetic resonance measurements and hyperfine interactions. The gyromagnetic factor is an example of properties which require careful relativistic treatment.

Concerning the above-mentioned case of strong electron correlation, one should note that its peculiar interaction effects cannot be described by an effective field. Strong correlation causes novel states to materialize with unusual/anomalous phenomena, such as those mentioned previously. These result from the interplay between the Pauli and Coulomb repulsion, local properties, and lattice periodicity, and are so complicated that an *ab initio* treatment is so far not possible. In such cases, the use of model Hamiltonians, such as the Hubbard model and the periodic Anderson model, allows one to gain good insight into the competition between localized electrons (that move coherently because of strong on-site repulsion) and Bloch (itinerant) electrons in a lattice. The combination of such methodology with *ab initio* calculations leads to renormalized band structures that provide an explanation of a number of effects, such as the so-called heavy Fermion effect.

The present book fits into the broad picture described above by focusing on extended, weakly correlated systems that are known to be accurately described by the density-functional formalism of Hohenberg, Kohn, and Sham. This methodology is accurate for such quantities as electron total energies and densities. Also successful treatments of such systems can be obtained through quasiparticle theory (Dyson equation), which is good for band structures, including Fermi surfaces of metals and the band gaps of semiconductors. Thus, the many-electron problem is reduced to an effective one-particle theory. For the remaining multi-center problem such methods are available as that of Korringa-Kohn-Rostoker (KKR) or Green-function method, the method of augmented plane waves

(APW), its full-potential generalization (FLAPW), of linear muffin-tin orbitals (LMTO), tight-binding (TB), and (for disordered alloys) the coherent potential approximation (CPA). These have been successfully applied to metals and semiconductors, magnetic and non-magnetic solids, ordered and disordered bulk materials, point defects in otherwise ideal hosts, and to surfaces and interfaces. The many-electron and multi-center problems for non-metallic systems can be treated also with the incremental method, which successfully applies the highly accurate quantum chemical method (e.g., the coupled cluster method) to extended systems like ionic solids, semiconductors, and polymers.

This book summarizes and presents in a unifying way the vast development of the KKR or multiple scattering method since its inauguration by Korringa, Kohn, and Rostoker. Decisive contributions to this development are due to the author, and his parties and collaborators. Especially the extension of the original version of the KKR theory, restricted to so-called muffin-tin potentials, to a space-filling, full potential method was a great breakthrough.

Dealing with solid state physics and materials science at the level of the electronic structure means that one has in mind the calculation of typical quantities that characterize a solid or material, such as,

a. lattice structure and lattice constants,
b. elastic constants, compressibilities,
c. binding or atomization energies,
d. heat capacity, Debye temperature,
e. melting points, phase diagrams, band structure, Fermi surface, band gaps,
f. plasmon, phonon, magnon dispersions and corresponding dampings,
g. sound velocity, heat expansion,
h. electrical, thermal conductivity, thermopower,
i. fracture strength, surface energy, work function,
j. mechanical properties: Peierls force, plasticity, brittle or ductile behavior, hardness,
k. electrical properties: residual resistivity, dislocation resistivity, Schottky barriers, ferroelectrics,
l. magnetic properties: susceptibility, Bloch walls, Weiss domains,
m. optical properties: reflectivity, refractivity, absorption, X-ray spectra, photoelectron spectra, Bremsstrahlung spectra, inelastic scattering of electrons, neutrons, etc.

Ab initio theory aims at the treatment of such "simple" things like

a. the variation of lattice structure from bcc to fcc to hcp when going through a row in the periodic table,
b. the differences in Debye temperature across materials types,
c. the existence of structural phase transitions in many alloys as a function of temperature,
d. the isoelectronic phase transition α to β in Ce,

e. the electronic origins of the difference in behavior between, say, Cu, which is a metal, and Si, which is a semiconductor,
f. the pressure needed for a semiconductor to become a metal is high in some cases but low in others,
g. the differences in the work function between different surface orientations and/or different materials,
h. the Curie temperature for the para-ferromagnetic transition is high in some cases and low in others.

Such queries lie squarely within the realm of the electronic theory of solids, and some are explicitly treated in the present volume within DFT and the KKR method. These methods and their numerous applications are explained in detail. This gives the reader an excellent guide to penetrate into this particular field of materials science and enables him/her to contribute to this field with the means currently at our disposal. This expresses implicitly the hope, or better the certainty, that in the future a more sophisticated materials science will emerge and explain more complicated systems and phenomena. The present book by A. Gonis contributes to this evolution.

Paul Ziesche
Dresden
December, 1999

[illegible] questions [illegible] [illegible] [illegible] [illegible] [illegible] theory of [illegible], and [illegible] [illegible] [illegible] in the [illegible] [illegible] [illegible] and [illegible] [illegible] [illegible] methods and their numerous applications are explained in detail. [illegible] [illegible] [illegible] [illegible] [illegible] [illegible] into this [illegible] [illegible] [illegible] [illegible] [illegible] [illegible] [illegible] [illegible] [illegible] to the [illegible] [illegible] [illegible] [illegible] a [illegible] [illegible]. The [illegible]

[illegible]

[illegible] Zhang

[illegible] de

[illegible] 1998

AUTHOR'S PREFACE

In addition to providing the "glue" that holds atoms (ions) together to form a solid, electrons are responsible for most of the spectrum of materials behavior. Electrical and heat conductivity, magnetism, hardness, and ductile vs. brittle behavior are familiar properties of materials that are intimately related to electronic structure. Furthermore, electrons interacting via phonons can produce truly spectacular and singular effects, such as superconductivity.

The primary goal of a materials scientist is a predictive understanding of materials properties, and that requires a clear picture of the role played by electrons in determining materials behavior. Only then can one hope to design and build new materials with desirable physical, chemical, and engineering characteristics. Present-day research into this subject is carried out on the basis of quantum mechanics, through solution of the so-called single-particle Schrödinger equation that describes of the behavior of electrons in a solid.

This book is an attempt to describe one formal approach to solving the Schrödinger equation developed within the framework of multiple scattering theory (MST). Multiple-scattering theory is the mathematical framework that forms the basis for the so-called Korringa-Kohn-Rostoker (KKR) method or Green-function method in the study of the electronic structure of solids. We show how MST can be used to solve the single-particle Schrödinger equation to obtain the corresponding electronic Green function (or wave function) for a number of physical systems, such as bulk periodic solids, disordered alloys, and surfaces and interfaces. It is further shown how these Green functions can be made to yield information about physical properties of interest such equilibrium volumes, heats of alloy formation, and transport coefficients. Thus, one aim of the book is to complement many other works based on different approaches that also study the electronic structure of matter.

The developments in this book rest on the formal methodology of density functional theory (DFT) and the local-density approximation (LDA) [or the local-spin-density approximation (LSDA)]. Density functional theory underlies the vast majority of modern studies of electronic properties of matter because it provides a means to construct an effective single-particle potential that determines electronic motion through the corresponding Schrödinger equation. At the same time, first-principles multiple-scattering theory provides a very convenient way to solve this equation under the various boundary conditions that are associated with different geometric arrangements of atoms, e.g., surfaces and interfaces, impurities, and disordered materials. The specific features of MST that facilitate its application to such a diverse range of systems are commented upon below and are also exploited throughout the developments in the book.

The book is confined mostly to well-established developments within MST, although some important recent advances are also discussed. For example, MST has traditionally been used with respect to "muffin-tin" potentials (spherically symmetric cell potentials confined inside non-overlapping spheres) and has been implemented in terms of the so-called Korringa-Kohn-Rostoker (KKR) structure constants. In developments over the last 12 years or so, it has been established that MST can be used in connection with arbitrarily-shaped, contiguous cell potentials that fill the volume of a material. It has also been shown how the theory can be formulated in terms of structure constants that are of considerably shorter range than their KKR counterparts. These so-called "screened" structure constants can substantially simplify the application of the method and speed up its numerical convergence. Both of these advances are explicitly discussed and references are given to original papers.

In addition to MST based directly on the Schrödinger equation, so-called first-principles or *ab initio* MST, phenomenological tight-binding (TB) theory is also developed to some extent. This is done for a variety of reasons. The application of TB theory to electronic structure carries a number of formal similarities to *ab initio* MST that can shed light on the structure of the overall formalism. In addition, TB methodology can lead to computational algorithms that execute much more rapidly than corresponding *ab initio* methods, allowing one to get at least a glimpse of the underlying physics and boost motivation to develop more rigorous and accurate computational procedures. For example, current attempts to bridge the electronic and atomic behavior in materials, such as those based on quantum molecular dynamics, can be carried out quite efficiently within a TB scheme, and so-called TB molecular dynamics is currently widely used in such studies. Furthermore, TB affords a transparent way to test new computational methodologies in terms of simple model systems and identify formal advantages or limitations in a given approach. Finally, TB provides a formal framework for the derivations of formulae that can often be easily transcribed to an *ab initio* formalism. Many attempts to treat the many-body aspects of the electron gas in a solid are quite often initially implemented within a TB scheme.

At the same time, I have not included derivations of the so-called TB linear muffin-tin orbital (TB-LMTO), or screened LMTO formalism. This methodology strikes a balance between the rigor and accuracy of *ab initio* formalisms, such as first-principles MST, and the less accurate but considerably more efficient phenomenological TB schemes. The LMTO formalism in general is part and parcel of electronic structure literature, and the screened LMTO approach is explained well in recent articles and books. However, references to this computational approach are given along with results of its implementation when it is warranted.

The book is written primarily for graduate students beginning their career in theoretical/computational condensed matter physics and materials science. As such, it assumes a level of knowledge commensurate with the material usually covered in graduate-level courses in quantum mechanics and solid-state physics. More esoteric topics, such as the theory of scattering of a single particle from a spatially bounded potential or a collection of such potentials, and of single-particle Green functions, are developed in the body of the text and in appendices to levels that hopefully suffice for understanding their application to the electronic structure of solids. I also hope that the material will be found useful by seasoned scientists who may not be familiar with the specific formalism of MST and wish to gain an understanding and appreciation of the methodology.

In setting out to write a book for the readership just described, one must come to grips with a number of basic issues, such as rigor of presentation and completeness of material. Reflecting my own proclivities, and from my understanding of the expectations of students embarking on a new study, I have attempted to provide mathematical justifications for computable formulae derived in the text. My hope is that the student will derive sufficient familiarity with the basic formalism so that he or she can read scientific literature on the subject and follow derivations with a measure of comfort. At the same time, it has been obvious from the outset that a "complete" exposition of the theoretical study of the electronic structure of materials and its relation to materials properties hardly can be defined, let alone contained in a single volume. Consequently, for the most part I have chosen to develop a number of specific topics at sufficient depth to illustrate the methodology, which I hope facilitate its application to the study of phenomena not directly treated here.

Why choose to develop the formalism in terms of MST within a Green-function framework? The easy answer is that this approach is extremely useful and versatile, particularly in the treatment of boundary conditions associated with the geometry of a system, or its chemical composition. The point is that MST allows the treatment of the proper boundary conditions for a particular structure, in contrast to the necessity in other techniques of imposing artificial or analog boundary conditions that cannot be justified *a priori*. A pointed example in this connection is the practice commonly employed in many computational schemes of treating a surface as a sufficiently "thick" slab or repeating slabs or supercells in contrast to the proper treatment of a *semi-infinite* solid afforded by multiple-scattering theory. In this respect, a Green function formalism is indispensable.

Yet, even though Green functions are discussed in almost all graduate books on formal condensed matter physics, their computational use in the determination of electronic structure and its relation to materials properties usually is not given detailed exposure. In such works, Green functions are presented along with other methods based on the wave function, so that the reader can obtain a general knowledge of the various methodologies that have become available for carrying out such studies. However, the inclusion of many different methods mitigates against a thorough exposition of any one of them. This seems to compound the point of view that Green functions are somewhat esoteric quantities whose substance and computational characteristics are difficult to grasp. At least within MST, this book makes an attempt to illustrate that Green functions provide a versatile, powerful, and unifying approach to the study of the electronic structure of materials that is characterized by both conceptual appeal and often (although not always) computational ease. Also, Green functions are easy to manipulate in a formal sense, making them a convenient basis to develop formal and computational methods.

On the other hand, insistence on rigor exacts a price in terms of computational efficiency. A number of materials properties, e.g., the equilibrium structure, volume, and elastic constants of elemental solids or compounds, and surface magnetism, have also been studied extensively by methods that are based on the calculation of the wave function rather than the Green function. On occasion, the results of such calculations are used to illustrate the role of electronic structure in these studies.

The Green function formulation of MST is developed within what is commonly referred to as the *single-particle* picture, at the zero of temperature, and neglecting relativistic effects. Therefore, the book makes no attempt to cover the theory of Green functions as applied throughout the field of solid-state physics, but only in the restricted sense just mentioned. In the single-particle picture, the immensely complex problem presented by a very large (infinite) number of electrons interacting with the external potential of the nuclear charges and with one another is replaced by the problem of a single electron moving in an effective external potential field. This field is constructed so that, at least in principle, it reflects the effects of interparticle interactions and the statistical nature of the electrons, i.e., the requirements of the exclusion principle. In spite of its evidently restrictive nature, this methodology has been found extremely useful in tracing the electronic origins of materials behavior. Once again, I hope that the exposition of the subject matter in the body of the text will motivate not only further study, but also efforts to alleviate some of the approximations underlying present treatments.

At this point it may be useful to briefly consider the material discussed in the various parts of the book. Within a multiple-scattering framework, the study of materials properties using a Green-function approach has developed primarily on the basis of two procedures, which are performed iteratively until self-consistency is obtained:

1. The calculation of an effective potential for the one-electron Schrödinger equation in terms of the electron density (separately for spin-up and spin-down electrons, if necessary). This potential is most often obtained using the local density approximation (LDA) [local spin-density approximation (LSDA) in the case of spin] of density functional theory (DFT).

2. The calculation of the on-the-energy-shell scattering-path matrix (traditionally referred to as the scattering-path operator), which describes the

propagation of an electron in the field (assumed to be static) of the nuclei and the other electrons. From this quantity, the one-electron Green function can be calculated, leading to the charge density upon which DFT-LDA calculation of (1) above is based.

Part I contains a discussion of the basic principles and formal aspects that enter calculation of the Green functions of multiple-scattering theory (MST) within LDA. It briefly reviews the historical development of DFT and LDA and their connection to prior methodology, and provides a derivation of a number of key formulae that enter the calculation of the electronic structure of solids. This includes the treatment of both the Schrödinger and Poisson equations in solids using the formalism of MST. An introduction to Green functions is also found in this part. The concept of single-particle Green functions is used in essentially all subsequent developments in the book so that at the end the reader initially unfamiliar with the concept may obtain a working familiarity with it as well as an understanding of the basic concepts involved in the study of electronic structure. This part also contains a discussion of the concepts of strain and stress within quantum mechanics and the derivation of the so-called "force theorem". However, because the material on strain and stress lies somewhat outside the mainstream of electronic-structure studies, it can be skipped in a first reading as indicated by an asterisk.

Part II presents some applications of the formalism of Part I to the calculation of materials properties. Calculation of the electronic structure of periodic solids and some of their physical properties, such as equilibrium volumes and equilibrium structures, along with an introduction to the treatment of impurities is presented.

A study of electronic structure must invariably come to grips with the treatment of the electron-electron (Coulomb) interaction among the electrons. This interaction is most often responsible for the difference between results obtained in the LDA (which provides an approximate treatment) and experimental findings. This is not to diminish the importance and significance of the large number of successful calculations based on the LDA, but to point out that there is indeed a realm of phenomena, e.g., excitation spectra, that fall outside its purview. For this reason, many attempts have been made toward correcting the LDA to account for the electron-electron interaction in a more accurate way. This part provides a discussion of the basic effects of electron-electron correlations and discusses some of the methodology that has been introduced in attempts for a more accurate treatment than that of the LDA. A brief discussion of band magnetism is also provided. At the same time, the material on correlations is not directly germane to the developments in the rest of the book and can be skipped in a first reading.

The study of impurities in Part II paves the way for Part III, in which methods are developed for the study of the structural and phase stability of concentrated, substitutionally disordered alloys. We give a brief history of the development of methods to study substitutional alloys, culminating with the development of the coherent-potential approximation. It is in this material that the power of MST and the Green-function formalism solidly manifests itself. In dealing properly with random systems, it has been found necessary to rely exclusively on the Green function concept, and it is within such a framework that most advances in the field have been made.

Part IV contains details of the calculation of the electronic structure of surfaces and interfaces. Some highlights of the development of general surface science are mentioned, as well as the development of computational methods for the study of the electronic structure of semi-infinite systems. Also, a number of elementary properties of surface structure and electronic behavior are summarized, before proceeding with a rigorous treatment based on MST. The emphasis here is the treatment of the true semi-infinite solid, rather than surrogate systems consisting of slabs, repeating slabs, or supercells. Thus, the boundary conditions attending the presence of a two-dimensional defect such as a surface or an interface in a three-dimensional solid are properly taken into account. Here, the matrix structure of the scattering-path operator is particularly helpful, and two important methodologies designed for this treatment — the layered KKR method and the real-space MST method — are discussed at some length.

As in the case of bulk solids, the electronic structure of the surface can be used to study the physical properties of a surface or interface. Some examples of this use are presented here, although for the reasons already stated above, not all are derived from the use of MST.

Transport properties are discussed within the formalism of linear response in Part V. After a brief review of classical transport theory, the Kubo formula is derived and is used to discuss the transport properties of solids. Also, a number of exact relations between transport coefficients are derived. Particular emphasis is placed on the electronic transport of substitutionally disordered alloys, and the results of numerical applications of the transport formalism to alloys within the coherent-potential approximation are given.

Part VI contains an exposition of some fundamental properties of lattice vibrations (phonons) in solids, and of the interaction of radiation (photons) with matter. These developments rely heavily on the TB formalism. The electron-phonon interaction within the conceptual framework of MST is discussed and it is shown how, at least in principle, one can envision a fully self-consistent MST in which atomic displacements are taken into account directly to define a potential for electronic motion. The propagation of light through matter and the formation of photonic band structure are also discussed.

Part VII, the last part, contains a formal, albeit brief, presentation of Green-function theory for interacting many-particle systems. Thus, it goes considerably beyond the formalism of single-particle Green functions, on which the previous developments are based. Here, the equation of motion for the many-particle Green functions is derived, and its use is illustrated through application to electronic transport and superconductivity.

The book concludes with a number of appendices that contain explanatory or supplementary material.

A few words about references. Those selected reflect my intention of not to burden the reader with a plethora of literature, but rather to direct him or her to that literature containing seminal developments of a formal nature, or illustrative calculations in which these developments are applied. I also have restricted the references, as much as possible, to those that have withstood the test of time. My apologies to those colleagues whose work might have been expected to be referenced but is left out here. Not to apply a very strict selection process to references would yield an unmanageably large number of them. Still, references to some recent works do appear when in my opinion they are necessary. All in all, I must shoulder the blame for an omission that detracts from understanding, or an inclusion that does not particularly aid it.

There are many who are owed gratitude for contributing directly or indirectly to the final form of this volume. James Garland, my thesis advisor of old and now friend, read parts of the manuscript and made invaluable comments for improvement. Parts of the manuscript were also read by X. -G. Zhang, Patrice Turchi, and Thomas Schulthess, and the final form and structure of the book was improved greatly because of their comments. Dimitri Papaconstantopoulos, Per Söderlind, Julie Staunton, and Peter Dederichs provided figures for inclusion in the text. Words can only do poor justice in describing the immense benefit I received from interacting through many years with luminaries in the use of Green functions and MST in the study of electronic structure such as Sam Faulkner, Bill Butler, Balazs Györffy, Malcolm Stocks, Paul Ziesche, and Peter Dederichs, among others. It was mostly they who conceived and developed the formalism discussed in the following pages, and put it to use. Needless to say, any errors, omissions, or other defects still remaining in the text are entirely my responsibility.

I am extremely grateful to the editorial assistance provided by MRS. My contacts with Gail Oare and Mike Driver have been a delight and a reliable source of encouragement during the time preceding the publication of the book. Last, but certainly not least, I thank my family for their understanding during the long nights over the seven years or so that it took to complete the book, during which I had to leave their midst to devote myself to its writing.

In writing the book, I had the opportunity to learn a great deal about much of its subject matter, and I hope that this learning process may continue. I strongly encourage comments on content, structure, and errors and omissions that may catch the reader's attention. If only one student finds the book of value in pursuing a career in theoretical and computational condensed matter physics, the effort of writing it will have been well worth it.

A. Gonis
Pleasanton, CA
January, 2000

Contents

Part I

Concepts and Formalism

Chapter 1

Density-functional theory

1.1 General Comments

This chapter summarizes density-functional theory (DFT), and some formal developments that preceded it and led to it. Density-functional theory is that formal construct which allows the description of electronic systems, or of many-particle systems in general, whether atoms, molecules, or solids, in terms of the electron density, $n(\mathbf{r})$, i.e., the number of electrons per unit volume at point $\mathbf{r}$ in the material. The theory provides an account of the ground state of the inhomogeneous electron gas, where the density varies as a function of position, and allows the self-consistent determination of this varying density. Most importantly, it also allows the construction of a potential to be used in the Schrödinger equation whose solutions determine the particle density. In practice, it is often a good approximation to use the theory in the *local-density approximation* (LDA), in which one applies relations involving the constant density of the homogeneous electron gas to the density at $\mathbf{r}$ of the inhomogeneous system. This local assumption is clearly an approximation. It becomes more and more accurate when the electron density, $n(\mathbf{r})$, or the potential that affects the electron distribution, varies relatively little over a characteristic wavelength of the electrons in the system. This approximation has been found to be very useful in studying the electronic properties of solids and underlies the developments presented in this book.

Density-functional theory and its local approximation are based on the use of the charge density. This allows one to reduce a problem with $3N$ degrees of freedom (where N is the number of particles) to one with only three degrees of freedom, quite a significant simplification. Furthermore, in contrast to the wave function, the density is observable. Thus, the theory can be compared directly to the results of experimental observation.

For example, the charge density obtained through DFT can be compared directly to experimental data obtained in X-ray scattering from an atomic or molecular gas, or a solid. In the latter case, the Fourier components of the density are given essentially by the intensity of the scattered radiation at the Bragg peaks. Clearly, it has merit to attempt a description of an electron gas in terms of a directly measurable quantity, such as $n(\mathbf{r})$.

We begin with some basic notions and formal concepts that preceded and prepared the ground for the introduction of DFT. In particular, we discuss the virial theorem in the presence of Coulomb forces, the uniform electron gas in the absence as well as the presence of an external potential, and the Thomas-Fermi model. We attempt to make clear that the electron density plays a central role in the description of an electron system, using a discussion of the Hartree-Fock method and the Hellmann-Feynman theorem. Finally, we present an extended summary of DFT and the local-density approximation. The material presented in that discussion should suffice for the applications of DFT that will be made in this book. In particular, we develop those formal aspects that will be useful to us in the next chapter, and in some later chapters, in which we discuss the application of LDA to the calculation of the electronic structure of solid materials within a Green-function, multiple-scattering-theory approach. The reader interested in more details of applications of the method as well as its mathematical foundation may consult such excellent references as *Theory of the Inhomogeneous Electron Gas*, edited by Lundqvist and March [1], *Density-functional theory* by Dreitzler and Gross [2], and *Density-functional theory of Atoms and Molecules* by Parr and Yang [3].

1.2 Virial Theorem

The virial theorem assumes identical forms in both classical and quantum mechanics, when time averages are replaced by quantum-mechanical expectation values. As such, it provides a convenient point of departure for the development of quantum-mechanical methodology to study the electronic structure of materials, when this structure is described in terms of classical variables such as the electron density.

1.2.1 The virial theorem in classical mechanics

The virial theorem of classical mechanics can be expressed by [4]

$$\overline{T} = -\frac{1}{2}\sum_i \overline{\mathbf{F}_i \cdot \mathbf{r}_i}, \tag{1.1}$$

where the bars denote time averages. Here, T is the kinetic energy of a system of point particles labeled by the index i, the vector $\mathbf{r}_i$ denotes the position of the ith particle, and $\mathbf{F}_i$ is the total force acting on the particle, including external forces and forces of constraint[1]. The quantity under the bar on the right-hand side of the equation above is known as the virial. If the forces are derivable from a potential,

$$\mathbf{F}_i = -\nabla_i U, \tag{1.2}$$

then Eq. (1.1) takes the form

$$\overline{T} = \frac{1}{2}\sum_i \overline{\nabla_i U \cdot \mathbf{r}_i}, \tag{1.3}$$

where the operator ∇_i acts only on U. Now, consider the case of motion under a central force with the potential being a power-law function of the length of the vector $\mathbf{r}$,

$$U(r) = ar^{n+1}, \tag{1.4}$$

so that the force varies as r^n. In this case, Eq. (1.3) applied to a single particle takes the form

$$\begin{aligned} \overline{T} &= \frac{1}{2}\overline{\frac{\partial U}{\partial r}r} \\ &= \frac{n+1}{2}\overline{U}. \end{aligned} \tag{1.5}$$

For the further special case of the inverse square law, e.g., gravitational or Coulomb forces, the virial theorem takes the well-known form,

$$\overline{T} = -\frac{1}{2}\overline{U}, \tag{1.6}$$

or

$$2\overline{T} + \overline{U} = 0. \tag{1.7}$$

It is interesting to examine the last relation a little more closely in connection with a closed system of particles, i.e., a system that does not interact with its environment. If we set

$$\overline{T} + \overline{U} = \overline{E}, \tag{1.8}$$

where $\overline{E}$ is the total energy of the system of particles, we see that

$$\overline{T} = -\overline{E}, \tag{1.9}$$

[1]The forces $\mathbf{F}_i$ can include friction which, however, does not contribute to the sum on the right-hand side of Eq. (1.1). On the other hand, in the presence of frictional forces energy must be continually supplied to the system to maintain the motion; otherwise the motion will die down and all time averages will vanish when taken over sufficiently long periods.

and

$$\overline{U} = 2\overline{E}. \tag{1.10}$$

Furthermore, it is important to keep in mind that the kinetic and potential energies are those associated with all particles in the system. For example, consider a collection of positive and negative point particles with masses m_{i+} and m_{i-}, and charges q_{i+} and q_{i-}, respectively. Then the kinetic energy of the system is equal to

$$\begin{aligned} T &= T^+ + T^- \\ &= \sum_i \frac{p_{i+}^2}{2m_{i+}} + \sum_i \frac{p_{i-}^2}{2m_{i-}}, \end{aligned} \tag{1.11}$$

while the potential energy is given by

$$\begin{aligned} U &= U^{++} + U^{+-} + U^{--} \\ &= \frac{1}{2}\sum_{i,j\neq i} \frac{q_{i+}q_{j+}}{|\mathbf{R}_i - \mathbf{R}_j|} + \frac{1}{2}\sum_{i,j} \frac{q_{i+}q_{j-}}{|\mathbf{R}_i - \mathbf{r}_j|} + \frac{1}{2}\sum_{i,j\neq i} \frac{q_{i-}q_{j-}}{|\mathbf{r}_i - \mathbf{r}_j|}. \end{aligned} \tag{1.12}$$

Here, the $\mathbf{R}_i$ ($\mathbf{r}_i$) denote the positions of the positive (negative) charges. For a closed system, defined above, Eq. (1.7) holds for the values of T and U given in the last two equations.

Let us now consider the virial relation, Eq. (1.1), in connection with such a closed system. The kinetic energy is that given by Eq. (1.11). The forces, $\mathbf{F}_i$, are clearly of two kinds, those acting on the positive charges, $\mathbf{F}_{i+}$, and those acting on the negative charges, $\mathbf{F}_{i-}$. Thus, the virial relation takes the form,

$$\overline{T} = -\frac{1}{2}\sum_i \overline{\mathbf{F}_{i-}\cdot\mathbf{r}_i} - \frac{1}{2}\sum_i \overline{\mathbf{F}_{i+}\cdot\mathbf{R}_i}. \tag{1.13}$$

In addition, each of the two types of force is given by a gradient with respect to the appropriate coordinates of the total potential, U,

$$\mathbf{F}_{i+} = -\nabla_{\mathbf{R}_i} U, \tag{1.14}$$

and

$$\mathbf{F}_{i-} = -\nabla_{\mathbf{r}_i} U. \tag{1.15}$$

In view of Eq. (1.10), and using the fact that U is independent of time, Eq. (1.13) takes the form,

$$\overline{T} = -\sum_i \overline{\nabla_{\mathbf{r}_i} E\cdot\mathbf{r}_i} - \sum_i \overline{\nabla_{\mathbf{R}_i} E\cdot\mathbf{R}_i}. \tag{1.16}$$

Here, the energy E is ascribed its dependence on the nuclear and electronic coordinates, and the operator $\nabla_{\mathbf{R}_I}$ acts only on that quantity. Therefore, for a closed system characterized by Coulomb forces, the derivative of the potential (the force) can be replaced by the corresponding derivative of the energy. This result is derived and used again in subsequent discussion.

As an example, the force on the positive charge at position $\mathbf{R}_i$ is given by the expressions,

$$\begin{aligned}\mathbf{F}_{i+} &= -\nabla_{\mathbf{R}_i} E \\ &= -\nabla_{\mathbf{R}_i}\left[\sum_{j\neq i}\frac{q_{i+}q_{j+}}{|\mathbf{R}_i-\mathbf{R}_j|}+\sum_j\frac{q_{i+}q_{j-}}{|\mathbf{R}_i-\mathbf{r}_j|}\right].\end{aligned} \tag{1.17}$$

An analogous expression holds for the negative charges.

Now, consider the negative charges to move in the field of the positive charges which are constrained to remain stationary under the action of an externally applied force. Clearly, we now have $T^+ = 0$. Then, the virial relation can be written in the form,

$$\overline{T} - \frac{1}{2}\sum_i \overline{\nabla_{\mathbf{r}_i}\left[U^{--}+U^{+-}\right]\cdot\mathbf{r}_i} = \frac{1}{2}\sum_i \overline{\nabla_{\mathbf{R}_i}E\cdot\mathbf{R}_i}. \tag{1.18}$$

The expression on the right-hand side of the equation can be interpreted as the virial of the *external* forces used to keep the positive charges in position. (If, as is often the practice, one considers the reaction forces internal to the system, a negative sign must be introduced on the right-hand side of the last equation.) In this form, the virial theorem plays a crucial role in the study of the energetics of solid materials.

Finally, we note that in the case in which the system is characterized by a particle density distribution, $n(\mathbf{r})$, and the forces are derivable from a potential, Eq. (1.1) can be written in the form (where explicit indication of time averages is suppressed),

$$T = \frac{1}{2}\int d^3r n(\mathbf{r})\mathbf{r}\cdot\nabla_{\mathbf{r}}U(\mathbf{r}). \tag{1.19}$$

The last expression is significant for three principal reasons. First, as is shown below, this purely classical result carries over intact into a quantum mechanical description of a system of particles. Second, the force on a particle, given by the gradient of the potential or of the total energy, retains its form in quantum mechanics, when time averages are interpreted as the expectation values of the corresponding quantities. And, third, it depends explicitly on the electron density which is the central quantity of interest in studies based on DFT.

1.2.2 The virial theorem in quantum mechanics

The presence of the time average in Eq. (1.1) signifies the statistical nature of the virial theorem in classical mechanics. We now show that a quantum mechanical system held together by Coulomb forces also satisfies Eq. (1.7). In this case, the role of the time average of the kinetic and potential energy is played by the expectation values of the corresponding operators [5] with respect to the states of the medium. That Eq. (1.17) also holds in quantum mechanics is subsequently shown.

We consider a system of N (point) particles of masses m_i and charges e_i interacting via the Coulomb potential. In the absence of an external potential, the exact many-body wave function describing the states of the system satisfies the Schrödinger equation

$$\left[\sum_{i=1}^{N}\left(-\frac{\hbar^2}{2m_i}\right)\nabla_i^2 + \frac{1}{2}\sum_{i=1}^{N}\sum_{\substack{k=1 \\ k\neq i}}^{N}\frac{e_i e_k}{r_{ik}}\right]\Psi = E\Psi, \tag{1.20}$$

with $\Psi(\mathbf{r}_1, \mathbf{r}_2, \cdots, \mathbf{r}_N)$ being normalized according to the relation

$$\int \mathrm{d}^3 r_1 \int \mathrm{d}^3 r_2 \cdots \int \mathrm{d}^3 r_n \Psi^* \Psi = 1, \tag{1.21}$$

and with the integrations being carried out over the coordinates of all N particles in the system. (We suppress the explicit dependence of the wave function on spin.) The expectation values of the kinetic and potential energies of the system in the state associated with Ψ in principle can be obtained from the formulae

$$T = -\frac{\hbar^2}{2}\sum_{i}^{N}\frac{1}{m_i}\int \mathrm{d}^3 r_1 \int \mathrm{d}^3 r_2 \cdots \int \mathrm{d}^3 r_n \Psi^* \nabla_i^2 \Psi, \tag{1.22}$$

and

$$U = \frac{1}{2}\sum_{i=1}^{N}\sum_{\substack{k=1 \\ k\neq i}}^{N} e_i e_k \int \mathrm{d}^3 r_1 \int \mathrm{d}^3 r_2 \cdots \int \mathrm{d}^3 r_n \Psi^* \frac{1}{r_{ik}} \Psi. \tag{1.23}$$

Now, a scale transformation[2]

$$\mathbf{r}_i' = \lambda \mathbf{r}_i, \tag{1.24}$$

keeps the orthonormality relation, Eq. (1.21), intact and transforms the wave function according to

$$\Psi(\mathbf{r}_1, \mathbf{r}_2, \cdots, \mathbf{r}_N) = \lambda^{3N/2}\Psi(\lambda\mathbf{r}_1, \lambda\mathbf{r}_2, \cdots, \lambda\mathbf{r}_N). \tag{1.25}$$

[2]Under the scale transformation, $\mathbf{r}' \to \lambda\mathbf{r}$, a volume element transforms as $\mathrm{d}\tau' \to \mathrm{d}\tau/\lambda^3$, and the particle density as $n(\mathbf{r}) \to n(\lambda\mathbf{r})/\lambda^3$. Then the definition, $n_\lambda = \lambda^3 n(\lambda\mathbf{r})$ preserves normalization since, $\int n_\lambda(\mathbf{r})\mathrm{d}\tau = \int \lambda^3 n(\lambda\mathbf{r})\frac{\mathrm{d}\tau}{\lambda^3}$. Scale transformations are discussed in more detail in Chapter 5.

We also have the relations,

$$\nabla_i^2 = \lambda^2 \nabla_i'^2 \quad \text{and} \quad \frac{1}{r_{ik}} = \lambda \frac{1}{r'_{ik}}, \tag{1.26}$$

and, consequently, in the primed variables, Eqs. (1.22) and (1.23) become

$$T' \to \lambda^2 T \quad \text{and} \quad U' \to \lambda U. \tag{1.27}$$

The total energy of the system is the sum of the kinetic and potential energies, and for an arbitrary value of λ can be written as

$$E(\lambda) = \lambda^2 T + \lambda U. \tag{1.28}$$

This energy is a minimum for the true wave function of the system, i.e., when $\lambda = 1$, at which point we must have

$$\frac{\partial E(\lambda)}{\partial \lambda} = 0. \tag{1.29}$$

But, from Eq. (1.28) we obtain

$$\frac{\partial E(\lambda)}{\partial \lambda} = 2\lambda T + U, \tag{1.30}$$

so that for $\lambda = 1$ we have

$$2T + U = 0. \tag{1.31}$$

This is the quantum mechanical virial theorem that was to be proven. In addition to the scaling argument used above, Slater [6] has given a direct derivation of the virial theorem in quantum mechanics in terms of the many-body wave function. For purposes of comparison, we will also use a wave function approach in the derivation of the virial theorem within the local density approximation later in this chapter.

Having shown that the virial theorem takes the same form in classical and quantum mechanics, it is important to point out that this form is strictly associated with free systems, i.e., systems without external constraints. If external forces of constraint, $\mathbf{F}_i^{\text{ext}}$, are present, they might make a non-vanishing contribution to the virial. If these forces are electrostatic in nature, then the virial theorem (in both classical and quantum mechanics) takes the form

$$T = -\frac{1}{2}U - \frac{1}{2}\sum_i \mathbf{F}_i^{\text{ext}} \cdot \mathbf{r}_i. \tag{1.32}$$

This expression is of relevance [6] to molecular systems and solid materials. In the Born-Oppenheimer treatment of a molecule or solid, the nuclei are assumed to be at rest. This can be accomplished by means of external forces which balance the internal force, $-\nabla_{\mathbf{R}}E$, on the nucleus at position $\mathbf{R}$. Here, E is the total energy of the system consisting of electrons and nuclei. As the external force is opposite to the internal force on the nuclei, Eq. (1.18) shows that the virial of Eq. (1.32) takes the form

$$T = -\frac{1}{2}U + \sum_{\mathbf{R}} \nabla_{\mathbf{R}}E \cdot \mathbf{R}. \tag{1.33}$$

It is important to emphasize that the energy, E, in the last expression is the sum of the kinetic energy of the electrons (the nuclei being at rest) and of the potential energy of the entire system, including all interparticle Coulomb interactions. Thus in general, the virial theorem does not give the kinetic energy as equal to -1/2 the potential energy if only a part of an otherwise closed system, e.g., the electrons only, is considered. This would be the case if the contribution of external forces to the virial vanished, which can happen [6] in either of two cases: When the nuclei are infinitely far apart so that internuclear interactions go to zero, or, more importantly, when the system is in an equilibrium configuration so that $\nabla_{\mathbf{R}}E$ equals zero.

The virial theorem figures prominently in some of our subsequent discussion. For the moment, we point out that the theorem is exact, i.e., it holds for the exact wave function Ψ, and need not hold in connection with an approximation to that wave function. Therefore, it is somewhat remarkable that the theorem holds in the case of approximate theories, such as the Thomas-Fermi model[3].

1.3 Uniform Electron Gas

In this section, we summarize a few basic notions appropriate to a uniform, or homogeneous, noninteracting electron gas. The analogues of such relations are used in the local-density approximation to treat the interacting, inhomogeneous electron cloud.

1.3.1 Free electrons

We consider N electrons distributed uniformly inside a cubic box of side L (volume $V = L^3$) with a constant density

$$n_0 = \frac{N}{V}. \tag{1.34}$$

---

[3] As is discussed below, the reason for this behavior is that the virial is independent of exchange and correlation and hence is unaffected by approximations in treating those effects.

In connection with this or any density, n, we can define an effective interparticle separation, r_s, as the radius of a sphere which contains the volume occupied by a single electron at that density. Thus, we have

$$\frac{1}{n_0} = \frac{4\pi}{3} r_s^3, \quad \text{and} \quad r_s = \left(\frac{3}{4\pi n_0}\right)^{1/3}. \tag{1.35}$$

It follows that high (low) values of r_s correspond to low (high) electron density. In the case of nonuniform density, an effective r_s can be defined in terms of the density at a point $\mathbf{r}$; in that case r_s is also a function of $\mathbf{r}$.[4]

In order to keep the system neutral, we assume the presence of a uniform background of positive, stationary charge. This simplified construction is commonly referred to as the *jellium model.* We neglect the interaction among the electrons, and for the moment, we also neglect the influence of any external potential (such as that produced by the uniform background of positive charge). The effect of an external potential is considered in the next subsection. That of the electron-electron interaction will be addressed in subsequent sections. For the system at hand, we now derive expressions for the electron energy in the ground state, the average kinetic energy, and the pressure.

In a cube of side L (in the limit taken to become arbitrarily large) and assuming periodic boundary conditions, the electron wave functions are plane waves

$$\psi_{\mathbf{k}}(\mathbf{r}) = \frac{1}{L^{3/2}} e^{i(k_x x + k_y y + k_z z)}, \tag{1.36}$$

and the corresponding energies are given by[5]

$$E = \frac{\hbar^2 k^2}{2m} = \frac{\hbar^2 \pi^2}{2mL^2}(n_x^2 + n_y^2 + n_z^2), \tag{1.37}$$

where n_x, n_y, and n_z are integers ($= 0, 1, 2, 3, \cdots$). Each state, characterized by a specific set of values (n_x, n_y, n_z) can be occupied by two electrons of opposite spin. Clearly, plane-wave eigenstates are consistent with a constant charge density.

We consider the case of n large enough that the numbers n_x, n_y, n_z can be treated as quasi-continuous. In the space determined by the points (n_x, n_y, n_z) we define a radius vector, $\mathbf{n}$, length, n,[6] can be obtained from

---

[4] Often, one defines the alternative quantity, $\frac{1}{n_0} = \frac{4\pi}{3} r_0^3$, and the corresponding dimensionless quantity, $r_s = r_0/A_B$, where $a_B istheso-calledBohrradius, a_0 = 0.5292$Å.

[5] This is also the form of the kinetic energy, $\frac{\hbar^2 k^2}{2m}$, of a particle of mass m confined inside a cubic box of side L and with infinitely high potential walls. In this case, the wavefunctions are standing waves, and the individual components of the momentum vector, k_α, are quantized, $k_\alpha = n_\alpha \frac{\pi}{L}$, which leads to Eq. (1.37).

[6] Use of n denotes distance rather than density in this one point in the discussion.

the expression

$$n^2 = n_x^2 + n_y^2 + n_z^2. \tag{1.38}$$

Then, a spherical shell between n and $n + \mathrm{d}n$ confined to the first octant of the Cartesian coordinate system (n_x, n_y, n_z) (all of n_x, n_y, and n_z positive) will contain

$$\frac{1}{8} 4\pi n^2 \mathrm{d}n = \frac{\pi}{2} n^2 \mathrm{d}n \tag{1.39}$$

distinct points. With each point occupied by two electrons, the number of electrons in that same strip is given by

$$\mathrm{d}N = \pi n^2 \mathrm{d}n. \tag{1.40}$$

Now, we obtain for the change in energy of the electrons that lie in this shell of width $\mathrm{d}n$ at n,

$$\mathrm{d}E = \frac{\hbar^2 \pi^2}{mL^2} n \mathrm{d}n, \tag{1.41}$$

where Eqs. (1.37) and (1.38) have been used. In view of Eqs. (1.41) and (1.40), the number of electrons between energies E and $E + \mathrm{d}E$ is

$$\begin{aligned} \mathrm{d}N &= \pi \sqrt{\frac{2mL^2E}{\hbar^2\pi^2}} \cdot \frac{mL^2}{\hbar^2\pi^2} \mathrm{d}E \\ &= \alpha \sqrt{E} \mathrm{d}E \end{aligned} \tag{1.42}$$

where

$$\alpha = \sqrt{2m} \frac{mL^3}{\pi^2 \hbar^3}. \tag{1.43}$$

The expressions just derived allow a straightforward determination of the energy of the highest occupied level in the ground state of the system (where all states below a certain level are occupied, and all states above are empty). Denoting the energy of the highest occupied state by μ, we have the relation

$$N = \alpha \int_0^{\mu} \sqrt{E} \mathrm{d}E = \frac{2}{3} \alpha \mu^{3/2} \tag{1.44}$$

which, in view of the definition (1.34) of the density of the electron gas and of Eq. (1.43) leads to the expression

$$n_0 = \frac{1}{3\pi^2} \left(\frac{2m\mu}{\hbar^2} \right)^{3/2}. \tag{1.45}$$

We also have

$$\mu = \left(\frac{\hbar^2}{2m} \right) (3\pi^2 n_0)^{2/3}. \tag{1.46}$$

It is often convenient to define an effective Fermi wave vector by the relation,

$$k_{\mathrm{F}}^3 = 3\pi^2 n_0. \tag{1.47}$$

The energy μ of the highest occupied level is commonly referred to as the *Fermi energy*, or the *chemical potential* of the electron gas. We justify the latter designation later on in the discussion, when we show that μ can also be expressed as the variation of the energy with respect to the number of electrons. (So that μ is the energy difference resulting from changing the number of electrons by one.) For the moment, let us note that μ in the last equation is written solely in terms of the density of the electron gas. Using the density as the basic variable to express quantities of physical interest is the cornerstone of density-functional theory.

1.3.2 Average energy and pressure

As simple illustrations of the use that can be made of the energy relations established above, we derive formulae for the average energy and the pressure of a gas of noninteracting, uniformly distributed electrons. We will see again that these quantities also can be expressed in terms of the density. The average energy is given by

$$\overline{E} = \frac{\int E \mathrm{d}N}{\int \mathrm{d}N} \tag{1.48}$$

or, according to Eq. (1.42),

$$\overline{E} = \frac{\int E\sqrt{E}\mathrm{d}E}{\int \sqrt{E}\mathrm{d}E} = \frac{3}{5}\mu = \frac{3}{5}\left(\frac{\hbar^2}{2m}\right)(3\pi^2 n_0)^{2/3}. \tag{1.49}$$

Equation (1.46) has been used to arrive at the last line, which gives the energy in terms of the density. We see that the average energy for the noninteracting system equals 3/5 the energy of the highest occupied level. Note also that the total energy of the gas (at $T = 0$) is given by

$$E = N\overline{E} = \frac{3}{5}N\mu. \tag{1.50}$$

To obtain an expression for the pressure of the electron gas, we begin with the relation,

$$\mathrm{d}W = -p\mathrm{d}V \tag{1.51}$$

connecting the amount of ***positive*** work, $\mathrm{d}W$, which must be done against a pressure, p, to ***decrease*** the volume of the system by an amount $\mathrm{d}V$. This work appears as an increase[7] in the internal energy, E, of the gas,

$$\mathrm{d}W = \mathrm{d}E. \tag{1.52}$$

Using Eq. (1.37), we have

$$\frac{\mathrm{d}E}{E} = -\frac{2}{3}\frac{\mathrm{d}V}{V}, \tag{1.53}$$

so that from Eqs. (1.46) and (1.50) we obtain the result,

$$p = -\frac{\mathrm{d}E}{\mathrm{d}V} = \frac{2}{3}\frac{E}{V} = \frac{2}{5}n_0\mu = \frac{2}{5}\left(\frac{\hbar^2}{2m}\right)(3\pi^2)^{2/3}n_0^{5/3}. \tag{1.54}$$

Thus, the pressure is specified in terms of the density alone. Following these rather simple considerations, we now turn to a discussion of the effects of an external potential on the energetics of a noninteracting electron gas.

1.3.3 Electron gas in an external potential

When a gas of noninteracting electrons is immersed in a constant external potential, v, the modifications that must be made to the previous discussion are rather obvious. For example, Eq. (1.46) for the Fermi energy becomes

$$\mu = \frac{\hbar^2}{2m}(3\pi^2 n_0)^{2/3} + v. \tag{1.55}$$

In our treatment of the inhomogeneous electron cloud it is somewhat more convenient to work in terms of momentum rather than the energy. Introducing the Fermi momentum through the relation

$$\mu = \frac{p_F^2}{2m}, \tag{1.56}$$

we can cast Eq. (1.45) in the form

$$n_0 = \frac{p_F^3}{3\pi^2\hbar^3} = \frac{8\pi}{3h^3}p_F^3. \tag{1.57}$$

It is often convenient to define the Fermi wave number, k_F, through the relation

$$k_F = p_F/\hbar. \tag{1.58}$$

In order to develop a treatment of the inhomogeneous electron gas, we must ascribe a spatial dependence to both the charge density and the potential.

[7]In the case considered, in which the energy is entirely kinetic, increasing or decreasing the volume leads to a corresponding lowering or raising of the energy, as follows easily from Eq. (1.37).

The first step in this development is to assume that the relation in Eq. (1.57) holds between the density and momentum values at position $\mathbf{r}$,

$$n(\mathbf{r}) = \frac{8\pi}{3h^3} p_{\mathrm{F}}^3(\mathbf{r}), \tag{1.59}$$

where $p_{\mathrm{F}}(\mathbf{r})$ is the maximum momentum value at $\mathbf{r}$. Next, the classical relation for the energy acquires a position dependence

$$\mu = \frac{p_{\mathrm{F}}^2(\mathbf{r})}{2m} + v(\mathbf{r}), \tag{1.60}$$

where all electrons in the system are assumed to move in a common external potential, $v(\mathbf{r})$. It is important to realize that although the right-hand side of Eq. (1.60) is a sum of a kinetic and a potential energy both of which vary with position in the system, the left-hand side is a constant independent of position. For if the maximum energy varied from point to point, then the electron distribution could readjust itself to lower the energy of the system. But this would contradict the assumption made in this discussion that the system is in its ground state (state of lowest energy). Eliminating $p_{\mathrm{F}}(\mathbf{r})$ through the use of Eq. (1.59), we can write Eq. (1.60) in the form

$$\begin{aligned} \mu &= \frac{1}{2m}(3\pi^2\hbar^3)^{2/3}\,[n(\mathbf{r})]^{2/3} + v(\mathbf{r}) \\ &= \frac{1}{2m}\left(\frac{3h^3}{8\pi}\right)[n(\mathbf{r})]^{2/3} + v(\mathbf{r}). \end{aligned} \tag{1.61}$$

This equation is of central importance in the Thomas-Fermi theory of the electron cloud which is developed in the next section. Before turning to that discussion, however, we derive an expression for the kinetic and total energy of an interacting electron system by means of a variational principle. The kinetic energy plays a major role in the following discussion. Also, this derivation allows the unambiguous identification of μ as the chemical potential of the electron gas.

1.3.4 Variational formalism

We wish to obtain an expression for the total energy of the electron gas in terms of the electron density, $n(\mathbf{r})$. The total energy is taken to be the sum of the kinetic and potential energies, the latter including the interaction of the electrons with the positive background (nuclear charges), as well as the repulsive interaction between the electrons. At this point we will deal only with classical expressions, neglecting quantum mechanical effects. These effects, such as exchange and correlation, are discussed in later sections of this chapter.

To obtain an expression for the kinetic energy in terms of the electron density, we define the probability density, $I_r(p)$, for finding an electron at $\mathbf{r}$ with momentum $\mathbf{p}$.[8] Then the probability of an electron at position $\mathbf{r}$ with momentum between p and $p + dp$ is given by the relation,

$$\begin{aligned} I_r(p) &= \frac{4\pi p^2 \mathrm{d}p}{\frac{4}{3}\pi p_{\mathrm{F}}^3(\mathbf{r})} \quad \text{for} \quad p < p_{\mathrm{F}}^3(\mathbf{r}) \\ &= 0 \qquad\qquad\qquad \text{otherwise.} \end{aligned} \tag{1.62}$$

Since there are $n(\mathbf{r})$ electrons per unit volume at $\mathbf{r}$, the kinetic energy, $t(\mathbf{r})$, at that point is given by the integral

$$\begin{aligned} t(\mathbf{r}) &= \int_0^{p_{\mathrm{F}}(\mathbf{r})} n(\mathbf{r}) \frac{p^2}{2m} \frac{3p^2}{p_{\mathrm{F}}^3(\mathbf{r})} \mathrm{d}p \\ &= c_k \left[n(\mathbf{r})\right]^{5/3}, \quad \text{with} \quad c_k = \frac{3\hbar^2}{10m}\left(3\pi^2\right)^{2/3}, \end{aligned} \tag{1.63}$$

where Eq. (1.59) has been used to eliminate $p(\mathbf{r})$ in favor of the density, $n(\mathbf{r})$. The total kinetic energy is obtained through integration of $t(\mathbf{r})$ over the entire electron cloud,

$$T = c_k \int [n(\mathbf{r})]^{5/3} \, \mathrm{d}^3 r. \tag{1.64}$$

This equation shows that within the present formalism the kinetic energy of the system can be obtained solely from a knowledge of the electron density.

We now turn to a consideration of the potential energy, U. This consists of two terms, the attractive interaction of the electrons with the nuclei, U_{eN}, and the electron-electron repulsion, U_{ee}, $U = U_{\mathrm{eN}} + U_{\mathrm{ee}}$. The first term has the form,

$$U_{\mathrm{eN}} = \int n(\mathbf{r}) U_{\mathrm{N}}(\mathbf{r}) \mathrm{d}^3 r, \tag{1.65}$$

where U_{N} is the potential energy felt by a single electron at $\mathbf{r}$ due to the presence of the positively charged nuclei. The interaction of the cloud with itself is taken to have the classical form

$$U_{\mathrm{ee}} = \frac{\mathrm{e}^2}{2} \int \frac{n(\mathbf{r})n(\mathbf{r}')}{|\mathbf{r} - \mathbf{r}'|} \mathrm{d}^3 r \mathrm{d}^3 r', \tag{1.66}$$

where the factor of $1/2$ is used to correct for the double-counting effect of integrating $\mathbf{r}$ and $\mathbf{r}'$ independently over the same volume. (We note that

[8] It should be kept in mind that even though probability distribution functions can be meaningfully defined, in quantum mechanics the variables position and momentum do not commute and hence are not simultaneously measurable.

the energy of the system also includes an internuclear repulsion term U_{NN}. This term does not depend on the electron charge so that it need not be considered explicitly at present. However, this term must be taken into account in the final expression for the total energy of a system, such as a solid.) Now, from the expression,

$$\begin{aligned} E &= T + U \\ &= T + U_{\mathrm{eN}} + U_{\mathrm{ee}}, \end{aligned} \tag{1.67}$$

we obtain for the total energy of the electrons (but not of the entire system since we are leaving out internuclear repulsion),

$$\begin{aligned} E_{\mathrm{el}} &= c_k \int [n(\mathbf{r})]^{5/3}\, \mathrm{d}^3r + \int n(\mathbf{r}) U_{\mathrm{N}}(\mathbf{r}) \mathrm{d}^3r \\ &+ \frac{\mathrm{e}^2}{2} \int \frac{n(\mathbf{r})n(\mathbf{r}')}{|\mathbf{r}-\mathbf{r}'|} \mathrm{d}^3r \mathrm{d}^3r', \end{aligned} \tag{1.68}$$

an expression which is determined completely through the knowledge of the electron density and the nuclear potential, U_{N}.

The last equation expresses the energy as a functional of the electron density, and yields a value for *any* input $n(\mathbf{r})$. A variational principle is obtained through the imposition of the condition that E_{el} be minimized with respect to the correct electron density, subject to the conservation of particles in the system

$$N = \int n(\mathbf{r}) \mathrm{d}^3r. \tag{1.69}$$

This condition can be incorporated into the variational principle through the introduction of a Lagrange multiplier, μ. Thus, instead of $\delta E = 0$, we write

$$\delta(E_{\mathrm{el}} - \lambda N) = 0. \tag{1.70}$$

Using the expression for E_{el} given in Eq. (1.68), and the rules for functional differentiation, Appendix A, we have

$$\begin{aligned} \lambda &= \frac{5}{3} c_k [n(\mathbf{r})]^{2/3} + U_{\mathrm{N}}(\mathbf{r}) + \mathrm{e}^2 \int \frac{n(\mathbf{r})}{|\mathbf{r}-\mathbf{r}'|} \mathrm{d}^3r' \\ &= \frac{5}{3} c_k [n(\mathbf{r})]^{2/3} + U_{\mathrm{N}}(\mathbf{r}) + U_{\mathrm{e}}(\mathbf{r}), \end{aligned} \tag{1.71}$$

where $U_{\mathrm{e}}(\mathbf{r})$ is the potential energy felt by a single electron at $\mathbf{r}$ due to the presence of the other electrons in the system. With c_k given by Eq. (1.63), the last equation is identical to Eq. (1.61), provided that $v(\mathbf{r})$ in that equation is taken to be the sum of the nuclear and electronic potentials,

$v(\mathbf{r}) \equiv U(\mathbf{r}) = U_{\mathrm{N}}(\mathbf{r}) + U_{\mathrm{e}}(\mathbf{r})$. Thus, we have $\lambda = \mu$ and, furthermore, from Eq. (1.70) we see that

$$\mu = \frac{\delta E}{\delta N} \tag{1.72}$$

so that μ is the chemical potential of the electron cloud: *it denotes the minimum energy needed to add to or subtract an electron from the system, (to order N^{-1})* [8].

We have now set up the formal apparatus necessary for a discussion of the Thomas-Fermi theory, which is taken up in the next section.

1.4 Thomas-Fermi Model

Equation (1.61) can be solved for $n(\mathbf{r})$ to provide one relation connecting a spatially varying charge density with a varying potential. It is to be noted that this relation does not give an exact description of the charge density; rather it is an approximation whose validity increases for potentials that vary slowly compared to a characteristic electronic wavelength. In that case, we approximate locally the correct charge density by that of a free electron gas. In this approximation, knowledge of the potential determines the charge density at position $\mathbf{r}$. However, the potential in general is not given, but must be determined through a solution of the Poisson equation

$$\nabla^2 \Phi(\mathbf{r}) = 4\pi e n(\mathbf{r}), \tag{1.73}$$

where $\rho(\mathbf{r}) = -en(\mathbf{r})$ is the charge density of the electron cloud, $U = -e\Phi$, and the differential equation must be solved under the appropriate boundary conditions,

$$\Phi = \frac{Ze}{r} \quad \text{for} \quad r \to 0 \tag{1.74}$$

in the vicinity of the nuclear charge Ze, and

$$\Phi = \frac{ze}{r} \quad \text{for} \quad r \geq R, \tag{1.75}$$

where R is the radius of a positive ion of charge ze and is still to be determined. Because there can be no singularity of the charge density at R, both the potential and the electric field must be continuous there. This allows us to write the boundary condition in Eq. (1.75) in the form,

$$\Phi(R) = \frac{ze}{R} \quad \text{and} \quad \left(\frac{\mathrm{d}\Phi}{\mathrm{d}r}\right)_R = -\frac{ze}{R^2}. \tag{1.76}$$

The solution of the classical Poisson equation coupled with the quantum mechanical Eq. (1.61) allows a self-consistent description of electronic motion in a field that includes the potential due to electrons on themselves. This self-consistent procedure constitutes the Thomas-Fermi model [9-11].

As an example of the use of the Thomas-Fermi model, we will derive expressions for the potential and the charge density for an atom with nuclear charge Z. For an atom or a positively charged ion[9], Eq. (1.73) is assumed to hold over all space except at the origin where there is a point nuclear charge. This charge provides one boundary condition on the problem and allows one to estimate the "radius" R of the neutral atom, by demanding that the electronic charge inside a sphere of that radius must equal Z. One more boundary condition is provided by the expected behavior of a Coulomb potential at large distances; the potential there must approach that corresponding to the net charge of the system taken to be a point at the origin. Thus, for the case of the neutral atom, the potential at large distances must decrease faster than r^{-1}.

We can use Eqs. (1.61) and (1.73) to eliminate $n(\mathbf{r})$ and establish a relation between the energy of an electron and its momentum that explicitly involves the electrostatic potential. In any volume inside an atom (or ion), the momentum $p(\mathbf{r})$ is connected to its energy by the relation,

$$E = \frac{p(\mathbf{r})}{2m} - \mathrm{e}\Phi(\mathbf{r}). \tag{1.77}$$

For the electron to remain bound inside the atom, this energy must evidently be smaller than the potential energy, $-\mathrm{e}\Phi(\mathbf{r})$, at the surface. Therefore, at a distance r from the nucleus, and assuming spherical symmetry from now on, the energy of an electron cannot exceed a momentum larger than a maximum value given by,

$$\frac{p_{\max}^2}{2m} = e[\Phi(r) - \Phi(R)]. \tag{1.78}$$

Now, quantum statistics couples this maximum value of momentum with the electron density, in the manner indicated in Eq. (1.59),

$$n = 2 \cdot \frac{4\pi}{3} p_{\max}^3/(2\pi\hbar)^3. \tag{1.79}$$

Comparing the last two expressions, we arrive at the other fundamental equation of the problem,

$$n(r) = \frac{1}{3\pi^2\hbar^3}\{2me[\Phi(r) - \Phi(R)]\}^{3/2}. \tag{1.80}$$

We now introduce the dimensionless function

$$\phi(r) = \frac{r}{Z\mathrm{e}}[\Phi(r) - \Phi(R)], \tag{1.81}$$

[9] The Thomas-Fermi model does not yield stable solutions for negatively charged ions.

along with the dimensionless variable

$$x = r/a \quad \text{with} \quad a = \left(\frac{9\pi^2}{128Z}\right)^{1/3} \frac{\hbar^2}{me^2} = 0.88534Z^{-1/3}\frac{\hbar^2}{me^2}, \tag{1.82}$$

and obtain the universal differential equation,

$$\frac{d^2\phi}{dx^2} = \frac{\phi^{3/2}}{\sqrt{x}}. \tag{1.83}$$

In these terms, the boundary conditions, Eqs. (1.74) and (1.76) take the respective forms[10],

$$\phi(0) = 1 \tag{1.84}$$

and[11]
for $X = R/a$,

$$\phi(X) = 0; \quad X\phi'(X) = -\frac{z}{Z}. \tag{1.85}$$

The first interesting feature of Eq. (1.83) is that it is independent of Z. Thus, a single numerical solution of thisequation subject to the boundary condition that ϕ should approach zero faster than $1/X$ as $X \to \infty$ yields a universal solution for *all* atoms. From Eqs. (1.73) and (1.74) we infer the behavior near the nucleus

$$n(r \to 0) \approx \frac{1}{r^{3/2}} \to \infty, \tag{1.86}$$

[10]It can easily be checked that with these boundary conditions, all $Z - z$ electrons are enclosed inside a sphere of radius R. To see this, we note that from Eqs. (1.80) and (1.82), we have

$$4\pi \int_0^R n(r) r^2 dr = Z \int_0^X dx \sqrt{x}\phi(x)^{3/2}.$$

Using the differential equation, Eq. (1.83), to replace $\phi^{3/2}$ by ϕ'', the last expression is shown to be equal to

$$Z \int_0^X dx x\phi'' = Z[x\phi' - \phi]_0^X = Z\{\phi(0) + X\phi'(X)\}.$$

With the boundary conditions (1.84) and (1.85), this equals

$$Z\left(1 - \frac{z}{Z}\right) = Z - z.$$

This is indeed the number of electrons inside R.

[11]To derive Eq. (1.85), differentiate Eq. (1.81),

$$\frac{d\phi(r)}{dr}|_R = \frac{\phi(r) - \phi(R)}{Ze}|_R + \frac{r\phi'(r)}{Ze}|_R$$

, note that the first term to the right of the equals sign vanishes, and use Eq. (1.76) for $\phi'(r)$.

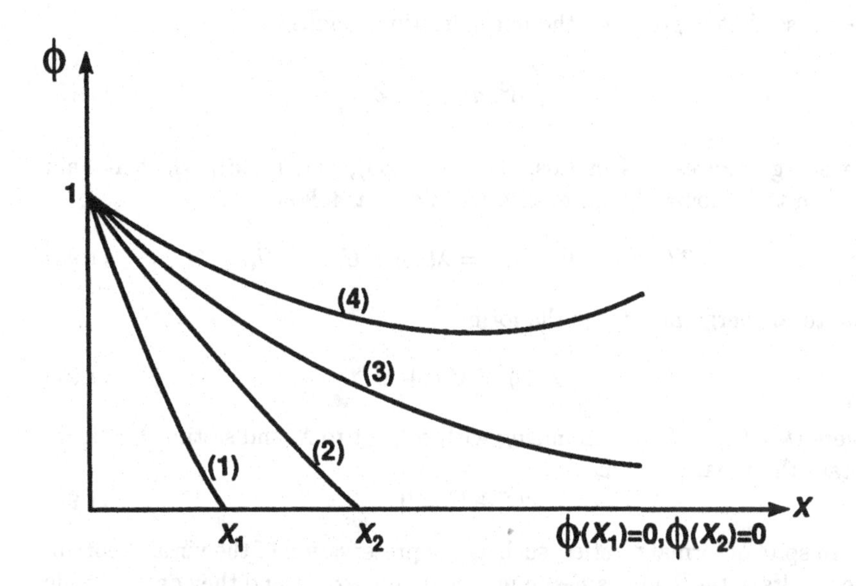

Figure 1.1: **Schematic representation of the solutions to the equations of the Thomas-Fermi model.**

while at large distances we can verify the analytic result $\Phi \simeq 144/X^3$, so that

$$n(r \to \infty) \approx \frac{1}{r^6}. \tag{1.87}$$

Figure 1.1 contains a schematic representation of the possible solutions of Eq. (1.83). Lines (1) and (2) lead to finite radii and correspond to positive ions. Line (3) is associated with a neutral atom and leads to an infinite radius. Solutions represented by line (4) have no direct physical significance, but could be useful for the description of atoms in crystals under boundary conditions that would allow the existence of negatively charged ions. It is interesting to note that the solution $144/X^3$ satisfies the differential equation exactly (and not only in the asymptotic limit), but has a singularity at $x = 0$. On the other hand, the practical utility of this solution is rather low, since even when $x = 100$ it differs by about 40% from the exact solution corresponding to neutral atoms, line (3) in the figure.

It is not difficult to prove that the virial theorem holds for the Thomas-Fermi atom. Under the scaling transformation (see footnote 2)

$$n_\lambda(\mathbf{r}) = \lambda^3 n(\lambda \mathbf{r}), \tag{1.88}$$

chosen so that it preserves the normalization condition

$$\int d^3r n_\lambda(\mathbf{r}) = Z \tag{1.89}$$

the energy expressions in Eqs. (1.64), (1.65), and (1.66), which remain valid in the Thomas-Fermi model, transform as follows,

$$T(\lambda) = \lambda^2 T; \quad U_{\mathrm{eN}} = \lambda U_{\mathrm{eN}}; \quad U_{\mathrm{ee}} = \lambda U_{\mathrm{ee}}. \tag{1.90}$$

The total energy now takes the form

$$E(\lambda) = \lambda^2 T + \lambda U \tag{1.91}$$

where $U = U_{\mathrm{eN}} + U_{\mathrm{ee}}$. Minimizing with respect to λ, and setting $\lambda = 1$, we obtain the virial theorem,

$$2T + U = 0. \tag{1.92}$$

In spite of formal niceties, such as the preservation of the virial theorem, the results of the Thomas-Fermi model are not exact, and they only provide rough agreement with realistic atomic charge densities. The model is of a statistical nature and its validity increases with an increasing number of electrons in the system. It is based on the assumption that one can treat an inhomogeneous gas by locally applying relations appropriate to a homogeneous electron cloud. Clearly, this aspect of the model provides a poor description of electronic states near a nucleus where the potential becomes singular. It is also invalid at large distances where the potential approaches the Fermi energy and the corresponding wavelength becomes very long. Also, the Thomas-Fermi model is incapable of yielding molecular binding at all [12-14]. At the same time, in the intermediate region the model often gives results that are surprisingly accurate.

The model also suffers from the serious drawback that it neglects both exchange and correlation effects. The former is a consequence of the Pauli exclusion principle, which prevents two electrons with the same spin from occupying the same state. The latter derives from the fact that electronic motion in a cloud is correlated, in the sense that the interaction between two electrons depends on the position of *both* particles. Thus, the potential $V(\mathbf{r})$ seen by a single electron varies as the other electrons move, and cannot be given exactly by the static charge distribution, $n(\mathbf{r})$. Both exchange [15-17] and correlation [18] effects can be introduced into the Thomas-Fermi model within various schemes. And the model often does provide a good starting point for calculations based on more accurate theoretical methods. However, it remained for the advent of density-functional theory to yield, at least in principle, a precise description of the many-body effects in an electron cloud and, through the LDA, to provide prescriptions for determining the potential in the calculation of the electronic structure. This theory is presented to lengths sufficient for our purposes later in this chapter. For more details, the reader is referred to a number of reviews in the literature [1, 3].

1.5 Elementary Quantum Mechanics

In order to set the stage for our development of density-functional theory, in this section we review certain basic notions of the wave-function description of quantum mechanical systems. One important feature which begins to emerge in this discussion is that the total energy, and indeed all properties, of the ground state of an electron cloud can be discussed entirely in terms of the electron density, $n(\mathbf{r})$. The implications of this statement in the calculation of the electronic structure and properties of materials will be presented at length in following chapters.

We consider a system of N electrons interacting by time-independent forces, in the presence of a background of positive (nuclear) charge. Such a system can be completely described by means of the Schrödinger equation, Eq. (1.20), modified to include the interaction of the electrons with the nuclei. In the Born-Oppenheimer approximation, i.e., nuclei fixed in position, and neglecting relativistic effects, we have,

$$\hat{H}\Psi = E\Psi, \tag{1.93}$$

where E is the electronic energy, $\Psi = \Psi(\mathbf{x}_1, \mathbf{x}_2, \cdots, \mathbf{x}_N,)$ is the true many-body wave function corresponding to the Hamiltonian operator $\hat{H}$ (operator symbols will be designated by a hat). In order to simplify our notation, we will be using atomic units in which the unit of length is the first Bohr radius of the hydrogen atom, $a_0 = 0.5292$Å, the unit of mass is the mass of the electron, $m = 1$, and the unit of charge is the electronic charge, e = 1. We also set $\hbar = 1$. In these units, the energy is given in Rydbergs, 1 Ry=13.6 eV, while the Hamiltonian operator of Eq. (1.93) takes the form

$$\hat{H} = \sum_{i=1}^{N}\left(-\frac{1}{2}\nabla_i^2\right) + \sum_{i}^{N} v(\mathbf{r}_i) + \sum_{i<j}^{N}\frac{1}{r_{ij}} \tag{1.94}$$

in which, [see also Eq. (1.65)],

$$v(\mathbf{r}_i) = -\sum_{\alpha}\frac{Z_\alpha}{r_{i\alpha}} \tag{1.95}$$

is the potential acting on an electron at $\mathbf{r}_i$ due to the presence of the nuclear charges Z_α located at positions $\mathbf{R}_\alpha$, with $r_{i\alpha} = |\mathbf{r}_i - \mathbf{R}_\alpha|$. In the expression for Ψ the "vectors" $\mathbf{x}_i$ include both positional and spin coordinates of the electron at $\mathbf{r}_i$. As it stands, Eq. (1.94) has the form of Eq. (1.67) as the sum of a kinetic energy term, and terms arising from the interactions of the electrons with the nuclei and with one another. In the presence of additional fields, the appropriate extra terms should be added. For future reference, we display explicitly the three terms in Eq. (1.94). We have

$$\hat{T} = \sum_i^N \left(-\frac{1}{2}\nabla_i^2\right) \tag{1.96}$$

for the kinetic energy operator, while

$$\hat{U}_{\text{eN}} = \sum_i^N v(\mathbf{r}_i) \tag{1.97}$$

is the operator describing the attractive interaction of the electrons and nuclei, and

$$\hat{U}_{\text{ee}} = \sum_{i<j}^N \frac{1}{r_{ij}} \tag{1.98}$$

describes the electron-electron repulsion. In order to obtain the total energy of the system consisting of electrons and nuclei, we must add to E in Eq. (1.67) a scalar[12] representing the repulsive energy of the nuclear charges,

$$U_{\text{NN}} = \sum_{\alpha<\beta} \frac{Z_\alpha Z_\beta}{R_{\alpha\beta}} \tag{1.99}$$

with $R_{\alpha\beta} = |\mathbf{R}_\alpha - \mathbf{R}_\beta|$. Thus, for the total energy of the system, we have

$$E_{\text{tot}} = E + U_{\text{NN}}. \tag{1.100}$$

One may solve Eq. (1.93) in the absence of U_{NN} and add the nuclear repulsion at the end, or include U_{NN} from the beginning.

The Schrödinger equation (1.93) must be solved under appropriate bound ary conditions. The solutions, Ψ, must be physically meaningful, so they must be square integrable over all space to have an interpretation as probability densities.[13] Specifically, the quantity

$$|\Psi(\mathbf{r}_1, s_1, \mathbf{r}_2, s_2, \cdots, \mathbf{r}_N, s_N)|^2 \mathrm{d}^3r_1 \mathrm{d}^3r_2 \cdots \mathrm{d}^3r_N \tag{1.101}$$

[12] As is standard practice, we designate the expectation value of an operator with the operator symbol but without the hat.

[13] It is to be noted that the Schrödinger equation describes also scattering states that are not square integrable.

represents the probability that the electrons in the system have position coordinates between $(\mathbf{r}_1, \mathbf{r}_2 \cdots \mathbf{r}_N)$ and $(\mathbf{r}_1 + \mathrm{d}\mathbf{r}_1, \mathbf{r}_2 + \mathrm{d}\mathbf{r}_2 \cdots \mathbf{r}_N + \mathrm{d}\mathbf{r}_N)$, where $\mathrm{d}\mathbf{r}_i$ is a point on the surface of the volume element $\mathrm{d}^3 r_i$, and spin values equal to $s_1, s_2, \cdots, s_N$. Because electrons obey Fermi-Dirac statistics, the wave function must be antisymmetric under exchange of the coordinates and spins of any two electrons in the system

$$\Psi(\cdots \mathbf{x}_i \cdots \mathbf{x}_j \cdots) = -\Psi(\cdots \mathbf{x}_j \cdots \mathbf{x}_i \cdots). \tag{1.102}$$

In general, there are many solutions, Ψ_n, to the Schrödinger Eq. (1.93) corresponding to different values of the energy. To the energy eigenvalues E_n there correspond eigenfunctions, Ψ_n, which can be taken to be orthonormal

$$\langle \Psi_n | \Psi_l \rangle = \sum_{s_1} \sum_{s_2} \cdots \sum_{s_N} \int \mathrm{d}^3 r_1 \mathrm{d}^3 r_2 \cdots \mathrm{d}^3 r_N \Psi_n{}^* \Psi_l = \delta_{nl}. \tag{1.103}$$

This expression defines the symbol $\langle A|B \rangle$ as an inner product consisting of an integral over the continuous spatial coordinates, $\{\mathbf{r}_i\}$, and a summation over the discrete spin variables $\{s_i\}$. The expectation value of an operator, $\hat{A}$, evaluated in a system described by a wave function Ψ, which in general can be a linear combination of eigenfunctions, is given by the expression

$$A = \langle \hat{A} \rangle = \frac{\langle \Psi | A | \Psi \rangle}{\langle \Psi | \Psi \rangle}. \tag{1.104}$$

For example, the expectation values of the kinetic and the potential energies are given by the formulae (assuming that Ψ is normalized to unity)

$$T \equiv T[\Psi] = \langle \hat{T} \rangle = \langle \Psi | \hat{T} | \Psi \rangle, \tag{1.105}$$

and

$$U \equiv U[\Psi] = \langle \hat{U} \rangle = \langle \Psi | \hat{U} | \Psi \rangle. \tag{1.106}$$

The expectation values of an operator are obtained as integrals over the wave function associated with the states of a given system. This type of relation is denoted by the square brackets in the last two equations and is an example of a *functional*: T and U are functionals of the wave function Ψ. (For an elementary account of functionals and some of their basic properties see Appendix A.)

As a simple example of Eq. (1.104), we derive a well-known relation between the expectation values of the energy of an electronic system. We denote the (non-degenerate) ground state of the system, i.e., the eigenstate with the lowest energy eigenvalue, by Ψ_0, and the corresponding energy by E_0. We will now show that for a system represented by a wave function Ψ, we have

$$\langle E \rangle \equiv E[\Psi] = \frac{\langle \Psi | \hat{H} | \Psi \rangle}{\langle \Psi | \Psi \rangle} \geq E_0. \tag{1.107}$$

This implies that the expectation value of the Hamiltonian, i.e., the value of the energy computed as an average over many measurements and calculated from a wave function that represents a physical system, forms an upper bound to the ground-state energy, E_0, or, in functional notation,

$$E[\Psi_0] = E_0 = \min_{\Psi} E[\Psi]. \tag{1.108}$$

The formal proof of this statement is elementary. The wave function, Ψ, is a linear combination of eigenstates (statistical mixture), $\Psi = \sum_n c_n \Psi_n$, so that Eq. (1.107) becomes

$$E[\Psi] = \frac{\sum_n |c_n|^2 E_n}{\sum_n |c_n|^2}. \tag{1.109}$$

Since E_n for $n \neq 0$ lies higher than E_0, the minimum principle expressed by Eq. (1.108) follows.

This minimum principle affords at least a formal means for obtaining Ψ_0 and E_0 through a variational procedure. First, we note that every eigenfunction Ψ_n is an extremum of $E[\Psi]$, so that

$$\delta E[\Psi]_{\Psi_n} = 0, \tag{1.110}$$

where δ denotes a variation of the functional $E[\Psi]$ under "small" variations of the wave function, Ψ. Therefore, a function that satisfies Eq. (1.110) also satisfies the Schrödinger equation (1.93), and vice versa. This is easy to see if we impose a normalization condition on the wave function. With E the Lagrange multiplier associated with the normalization condition, we write Eq. (1.110) in the form,

$$\delta[\langle\Psi|\hat{H}|\Psi\rangle - E\langle\Psi|\Psi\rangle] = 0. \tag{1.111}$$

Formal variation of Ψ (or Ψ^*) yields[14] Eq. (1.93).

In practical applications of Eq. (1.111), one determines Ψ as a function of E and adjusts E until normalization is achieved. For an approximate $\tilde{\Psi}_0$, this procedure leads to an $\tilde{E}_0$ which is an upper bound to the true ground-state energy, E_0. One can get arbitrarily close to E_0 by adjusting $\tilde{\Psi}_0$, a procedure which is often used in present-day calculations of the electronic structure of matter within the context of density-functional theory.

1.5.1 Slater determinants

Proceeding further can be made somewhat easier if we retain a mental picture of the form of the many-body wave function $\Psi(\mathbf{x}_1, \mathbf{x}_2, \cdots, \mathbf{x}_N)$. We

[14] Since Ψ is complex, Ψ and Ψ^* can be varied independently.

can obtain at least an approximate form for Ψ by demanding that the wave function satisfy the proper symmetry relations under particle interchange associated with Fermi-Dirac statistics. Denoting by $\hat{P}_{ij}$ the operator that interchanges the coordinates and spins of particles i and j, we have

$$\begin{aligned} \hat{P}_{ij}\Psi(\cdots \mathbf{x}_i \cdots \mathbf{x}_j \cdots) &= \Psi(\cdots \mathbf{x}_j \cdots \mathbf{x}_i \cdots) \\ &= -\Psi(\cdots \mathbf{x}_i \cdots \mathbf{x}_j \cdots). \end{aligned} \tag{1.112}$$

As was suggested by Slater, a wave function satisfying Eq. (1.112) can be constructed in terms of an orthonormal set of one-electron wave functions, $\phi_j(\mathbf{x}_i)$, where

$$\langle \phi_j | \phi_i \rangle = \sum_{s_1} \int \mathrm{d}^3 r_1 \phi_j^*(\mathbf{r}_1, s_1) \phi_i(\mathbf{r}_1, s_1) = \delta_{ij}. \tag{1.113}$$

Strictly speaking, the ϕ's are "spin orbitals" which can be written in the form

$$\phi_j(\mathbf{r}, s) \equiv \begin{pmatrix} \phi_j^1(\mathbf{r}) \\ \phi_j^2(\mathbf{r}) \end{pmatrix} = \phi_j^1(\mathbf{r})\alpha + \phi_j^2(\mathbf{r})\beta \tag{1.114}$$

with α and β being spinors for up, $\begin{pmatrix} 1 \\ 0 \end{pmatrix}$, and down, $\begin{pmatrix} 0 \\ 1 \end{pmatrix}$, spin directions. The functions of position in Eq. (1.114), $\phi_j^1(\mathbf{r})$ and $\phi_j^2(\mathbf{r})$ may be different from one-another, allowing the possibility of different spins being associated with different wave functions.

Using these spin orbitals, we can construct an N-body determinant of the form

$$\begin{aligned} \Psi_\nu^{(N)} &\equiv \frac{1}{\sqrt{N!}} \begin{vmatrix} \phi_1(1) & \phi_1(2) & \phi_1(3)\cdots & \phi_1(N) \\ \phi_2(1) & \phi_2(2) & \phi_2(3)\cdots & \phi_2(N) \\ \vdots & \vdots & \ddots & \vdots \\ \phi_N(1) & \phi_N(2) & \phi_N(3)\cdots & \phi_N(N) \end{vmatrix} \\ &\equiv |\phi_1 \phi_2 \phi_3 \cdots \phi_N|. \end{aligned} \tag{1.115}$$

These $\Psi_\nu^{(N)}$ are known as "Slater determinants". As the interchange of any two columns of a determinant produces a change in sign, these determinantal wave functions are antisymmetric with respect to the interchange of two particles, and thus satisfy Eq. (1.112). For orthonormal spin orbitals, ϕ_j, we have

$$langle\Psi_\nu^{(N)}|\Psi_\nu^{(N)}\rangle = 1. \tag{1.116}$$

In the method proposed by Slater, one represents a general, many-body wave function as a linear combination of determinantal wave functions[15]

$$\Psi = \sum_{\nu} C_\nu \Psi_\nu^{(N)}, \tag{1.117}$$

with the C_ν being c-numbers. The antisymmetry of Ψ follows from that of the $\Psi_\nu^{(N)}$'s. In the following discussion we suppress the superscript (N), it being understood that the discussion is concerned with systems of N particles, unless stated otherwise. Some useful properties of Slater determinants can be easily established [17] and are summarized below.

It can be shown [17] that if $\hat{F}$ is a sum of one-body operators,

$$\hat{F} = \sum_{i=1}^{N} \hat{f}(i) \tag{1.118}$$

and $\hat{G}$ is a sum of two-body operators

$$\hat{G} = \sum_{j>i}^{N} \sum_{i=1}^{N} \hat{g}(i,j) \tag{1.119}$$

then[16]

$$\begin{aligned} \langle \Psi_\nu | \hat{F} | \Psi_\nu \rangle &= \sum_n \nu_n \langle \phi_n | \hat{f} | \phi_n \rangle \\ &= \sum_n \nu_n \left[\sum_{s_1} \int d^3 r \phi_n^*(\mathbf{r}_1, s_1) \hat{f}(1) \phi_n(\mathbf{r}_1, s_1) \right], \end{aligned} \tag{1.120}$$

and

$$\begin{aligned} \langle \Psi_\nu | \hat{G} | \Psi_\nu \rangle &= \frac{1}{2} \sum_n \sum_m \nu_n \nu_m \left[\sum_{s_1} \sum_{s_2} \int d^3 r_1 \int d^3 r_2 \right. \\ &\times \left. \phi_n^*(1) \phi_m^*(2) \hat{g}(1,2) [1 - \hat{P}_{12}] \phi_m(2) \phi_n(1) \right]. \end{aligned} \tag{1.121}$$

In these expressions, we have defined the *occupation numbers*

$$\nu_n = \begin{cases} 1 & \text{if } \phi_n \text{ occurs in } \Psi_\nu. \\ 0 & \text{otherwise.} \end{cases} \tag{1.122}$$

[15] In quantum chemistry such an expression is referred to as a configuration interaction (CI).

[16] As always, the brackets denote integration over the positions of the particles and summations over all spins.

As a simple illustration of these results, let $\hat{F}$ represent the one-body operators in the many-body Hamiltonian, Eq. (1.94),

$$\hat{F} = \sum_{i=1}^{N} \left(-\frac{1}{2}\nabla_i^2\right) + \sum_{i=1}^{N} v(\mathbf{r}_i), \tag{1.123}$$

and $\hat{G}$ the repulsive electron-electron interaction,

$$\hat{G} = \frac{1}{2}\sum_{j=1}^{N}\sum_{i=1}^{N}\frac{1}{r_{ij}}. \tag{1.124}$$

Then, the expectation value of the Hamiltonian (1.94) with respect to a wave function that consists of a *single* determinant takes the form,

$$\begin{aligned} \langle\hat{H}\rangle &= E = \sum_{j=1}^{N}\left\{\int d^3r_1\phi_j^*(\mathbf{r}_1)[-\frac{1}{2}\nabla^2 + v(\mathbf{r}_1)]\phi_j(\mathbf{r}_1)\right. \\ &+ \frac{1}{2}\sum_{i=1}^{N}\left\{\int d^3r_1\int d^3r_2|\phi_j(\mathbf{r}_1)|^2\frac{1}{r_{12}}|\phi_i(\mathbf{r}_2)|^2\right. \\ &- \left.\int d^3r_1\int d^3r_2\phi_j^*(\mathbf{r}_1)\phi_i^*(\mathbf{r}_2)\frac{1}{r_{12}}\phi_j(\mathbf{r}_2)\phi_i(\mathbf{r}_1)\right\}, \end{aligned} \tag{1.125}$$

where a summation over spin indices is to be understood. We note that the self-interaction (SI) terms, $i = j$, cancel out exactly in the last two terms above. Denoting, respectively, the three integrals on the right-hand side of the last equation by H_i, J_{ij}, and K_{ij}, we can write,

$$E = \sum_{i=1}^{N} H_i + \frac{1}{2}\sum_{i,j}^{N}(J_{ij} - K_{ij}). \tag{1.126}$$

These integrals are real, and $J_{ij} > K_{ij} > 0$. The J_{ij} are commonly referred to as *Coulomb integrals*, and the K_{ij} as *exchange integrals*. Because,

$$J_{ii} = K_{ii} \tag{1.127}$$

the double summation in Eq. (1.126) can include the $i = j$ terms.

It is clear that J_{ii} is the classical interaction of a charge distribution with itself,

$$J_{ii} = \frac{1}{2}\int d^3r_1\int d^3r_2\frac{n(\mathbf{r}_1)n(\mathbf{r}_2)}{|\mathbf{r}_1 - \mathbf{r}_2|}. \tag{1.128}$$

For future reference, we point out that J_{ii} is a purely classical expression that neglects both exchange and correlation effects. Exchange can be treated through the introduction of integrals such as K_{ij} above, an approach that can lead to extremely cumbersome numerics. Both exchange and correlation are included, at least in principle, in a self-interaction of the form

$$U_{\rm ee}[n_2] = \frac{1}{2}\int {\rm d}^3r_1 \int {\rm d}^3r_2 \frac{n_2(\mathbf{r}_1,\mathbf{r}_2)}{|\mathbf{r}_1-\mathbf{r}_2|}, \tag{1.129}$$

where $n_2(\mathbf{r}_1,\mathbf{r}_2)$ is the exact *two-particle* density giving the number of particles per unit volume simultaneously at $\mathbf{r}_1$ and $\mathbf{r}_2$.

Of central interest is also the form assumed by the single-particle density when the wave function is a single Slater determinant. For orthonormal single-particle spin orbitals, (associated with the states of an electron in the absence of interactions with the other electrons) we have[17]

$$n(\mathbf{r}) = \sum_{i=1}^{N/2}\sum_{s} |\phi_i(\mathbf{r},s)|^2, \tag{1.130}$$

since a single particle at $\mathbf{r}$ has the probability $|\phi_i(\mathbf{r},s)|^2$ of being found in state ϕ_i with spin s. This form of the density is of great use in the development of density-functional theory. As the next step toward the presentation of that theory, we discuss the Hartree-Fock method.

1.5.2 Hartree-Fock equations

The Hartree-Fock method is aimed at the determination of a single Slater determinant that satisfies the variational principle of Eq. (1.111). We require that the expression for the energy given in Eq. (1.125) be stationary under variation of the ϕ_j, subject to the preservation of the orthonormality of these wave functions. Using the methods of Appendix A, we vary E in Eq. (1.125) and obtain the following set of equations:

$$\begin{aligned}[-\frac{1}{2}\nabla_1^2 &+ v(\mathbf{r}_1) + \sum_\ell \int \phi^*_{\ell\alpha}(\mathbf{r}_2)\phi_{\ell\alpha}(\mathbf{r}_2)\left(\frac{1}{r_{12}}\right){\rm d}^3r_2]\phi_{j\beta}(\mathbf{r}_2)\\ &- \sum_\ell \int \phi^*_{\ell\alpha}(\mathbf{r}_2)\phi_{j\alpha}(\mathbf{r}_1)\left(\frac{1}{r_{12}}\right)\phi_{\ell\beta}(\mathbf{r}_1){\rm d}^3r_2\\ &= \epsilon_j\phi_{j\beta}(\mathbf{r}_1),\end{aligned} \tag{1.131}$$

where a summation over repeated spin indices is to be understood. Equation (1.131) constitutes one form of the *Hartree-Fock equations* [20, 21]. For a system of N electrons, and hence of N functions ϕ_j, there are $2N$ coupled integrodifferential equations, the factor of two arising due to the presence

[17] This can also be shown through the use of Eq. (1.120) applied to the number operator in second-quantized notation [19].

of spin. The same equations, but without the exchange term (last integral) are known as the *Hartree equations.*

Certain features of these equations are worth emphasizing. First, if the last term is omitted, the equations have almost the form of an eigenvalue equation, e.g., the Schrödinger equation, with the single-particle potential consisting of the external (nuclear) contribution, $v(\mathbf{r}_1)$, and the Hartree term [the first integral in Eq. (1.131)]. This term represents the electrostatic interaction of an electron in state j with the field of all other electrons, and has the familiar form

$$U_{ee}(\mathbf{r}) = \int n(\mathbf{r}_2) \frac{1}{|\mathbf{r} - \mathbf{r}_2|} d^3 r_2. \tag{1.132}$$

Second, the inclusion of the exchange term arises because of the Pauli exclusion principle. Two electrons with parallel spins are less likely to be found in the vicinity of one another than if their spins were antiparallel, and thus the electrostatic interaction given by the Hartree term, which neglects this effect, is an overestimate. The exchange term has a sign opposite to that of the Hartree term and results in a reduction of the electron-electron interaction energy[18]. Within the approximation of using a wave function consisting of a single determinant[19], the Hartree-Fock equations provide an exact treatment of the Pauli exclusion principle. We note that these equations contain no contribution from terms with $\ell = j$ because an electron does not act on itself[20]. Note further that the exchange term is of a purely quantum mechanical origin, and does not have the form of an ordinary (classical) potential. Finally, it is important to realize that the Hartree-Fock equations completely neglect *correlation* effects, i.e., the fact that the Coulomb interaction between two electrons depends on the instantaneous position of both particles. The motion of the electrons in the system is correlated in the sense that the particles move in a manner greatly influenced by their mutual interactions; a proper description of the repulsive interaction of two electrons can only be given in terms of the two-particle density matrix as is indicated in Eq. (1.129). In any case, because in the Hartree and Hartree-Fock schemes the interaction is determined self-consistently, they are both referred to as *self-consistent field methods.*

It is of interest to cast the Hartree-Fock equations in a somewhat more general form that explicitly shows the preservation of the normalization of the spin orbitals. Straightforward application of Eq. (1.111) leads to the

---

[18] If the Pauli principle were to place electrons with parallel spin an infinite distance apart, then the exchange term would be absent, and in the Hartree term one would only consider the case in which α and β denote opposite spin orientations.

[19] The exact wave function can be written as a linear combination of Slater determinants, but is not itself necessarily a Slater determinant.

[20] In fact, if two ϕ_j's were identical, the determinantal wave function would vanish.

set of equations

$$\hat{F}\phi_i(\mathbf{x}) = \sum_{j=1}^{N} \epsilon_{ij}\phi_j(\mathbf{x}). \tag{1.133}$$

Here, the operator $\hat{F}$ is defined by

$$\hat{F} = -\frac{1}{2}\nabla^2 + v + \hat{\jmath} - \hat{k}, \tag{1.134}$$

where the meaning of the first two terms on the right side is obvious, and the Coulomb operator $\hat{\jmath}$ and exchange operator $\hat{k}$ are defined by means of their action on an arbitrary function, $f(\mathbf{x})$,

$$\hat{\jmath}(\mathbf{x}_1)f(\mathbf{x}_1) = \sum_{k=1}^{N} \int \phi_k^*(\mathbf{x}_2)\phi_k(\mathbf{x}_2)\frac{1}{r_{12}}f(\mathbf{x}_1)\mathrm{d}^3x_2 \tag{1.135}$$

and

$$\hat{k}(\mathbf{x}_1)f(\mathbf{x}_1) = \sum_{k=1}^{N} \int \phi_k^*(\mathbf{x}_2)f(\mathbf{x}_2)\frac{1}{r_{12}}\phi_k(\mathbf{x}_1)\mathrm{d}^3x_2. \tag{1.136}$$

With f replaced appropriately by one of the spin orbital functions, Eq. (1.133) yields Eq. (1.131). The right-hand side of Eq. (1.133) contains the Lagrange multiplier matrix, ϵ_{ij}, associated with the constraint that the ϕ_j be orthogonal. This matrix is in general complex, but

$$\epsilon_{ij}^* = \epsilon_{ji}, \tag{1.137}$$

so that $\underline{\epsilon}$ is Hermitian, where an underline denotes a matrix quantity.

It is also of interest to examine the significance of the quantities ϵ_j that occur in Eq. (1.131). Multiplying Eq. (1.133) by ϕ_i^* and integrating (along with a sum over spins), we obtain a formula for the "orbital energies",

$$\epsilon_i \equiv \epsilon_{ii} = H_i + \sum_{j=1}^{N}(J_{ij} - K_{ij}), \tag{1.138}$$

where the quantities J_{ij} and K_{ij} are defined by Eq. (1.125). Summing Eq. (1.138) over i and comparing with Eq. (1.126), we find for the energy in the Hartree-Fock method,

$$E_{\mathrm{HF}} = \sum_{i=1}^{N}\epsilon_i - U_{\mathrm{ee}}, \tag{1.139}$$

where the symbol U_{ee} denotes the total electron-electron interaction

$$U_{\mathrm{ee}} = \frac{1}{2}\sum_{i,j=1}^{N}(J_{ij} - K_{ij}). \tag{1.140}$$

The total energy of the system is obtained upon addition of the nuclear-nuclear repulsion to E_{HF}

$$\begin{aligned} E_{\text{tot}}^{\text{HF}} &= E_{\text{HF}} + U_{\text{NN}} \\ &= \sum_{i=1}^{N} \epsilon_i - U_{\text{ee}} + U_{\text{NN}} \\ &= \sum_{i=1}^{N} H_i + U_{\text{ee}} + U_{\text{NN}}. \end{aligned} \tag{1.141}$$

It is important to point out that neither E_{HF} nor $E_{\text{tot}}^{\text{HF}}$ is equal to the sum of orbital energies.

1.5.3 Hartree-Fock theory for free electrons

As a simple illustration of the Hartree-Fock method, we consider a cloud of free, non-interacting electrons. We describe this system in terms of plane-wave states (with δ-function normalization),

$$\phi_{\mathbf{k}}(\mathbf{r}) = \mathrm{e}^{\mathrm{i}\mathbf{k}\cdot\mathbf{r}} \times \text{spin function}, \tag{1.142}$$

with each vector $\mathbf{k}$ with magnitude less than k_{F} occurring twice, once in association with spin up and once with spin down (k_{F} is the magnitude of the momentum of the highest occupied state).

We can easily show that plane waves are indeed solutions to the Hartree-Fock equations for free electrons. We begin by noting that the density associated with free electrons and which determines the Hartree potential, Eq. (1.65), is uniform. But this potential cancels exactly against the similar term that arises from the density of the positively charged background, which is also taken to be uniformly distributed. In the Hartree-Fock equations, Eq. (1.131), only the exchange term remains (in addition to the kinetic energy), and for free electrons this term can be evaluated through the use of the Fourier transform of the Coulomb interaction

$$\frac{1}{|\mathbf{r}-\mathbf{r}'|} = 4\pi \sum_{q} \frac{1}{q^2} \mathrm{e}^{\mathrm{i}\mathbf{q}\cdot(\mathbf{r}-\mathbf{r}')} \to 4\pi \int \frac{\mathrm{d}^3 q}{(2\pi)^3} \frac{1}{q^2} \mathrm{e}^{\mathrm{i}\mathbf{q}\cdot(\mathbf{r}-\mathbf{r}')}. \tag{1.143}$$

If the last expression is used in the exchange term of Eq. (1.131) with the ϕ_i taken to be plane waves, the left hand side reduces to the form,

$$\epsilon(\mathbf{k})\phi_j, \tag{1.144}$$

where

$$\epsilon(\mathbf{k}) = \frac{k^2}{2} - 4\pi \int \frac{\mathrm{d}^3 k'}{(2\pi)^3} \frac{1}{|\mathbf{k}-\mathbf{k}'|}$$

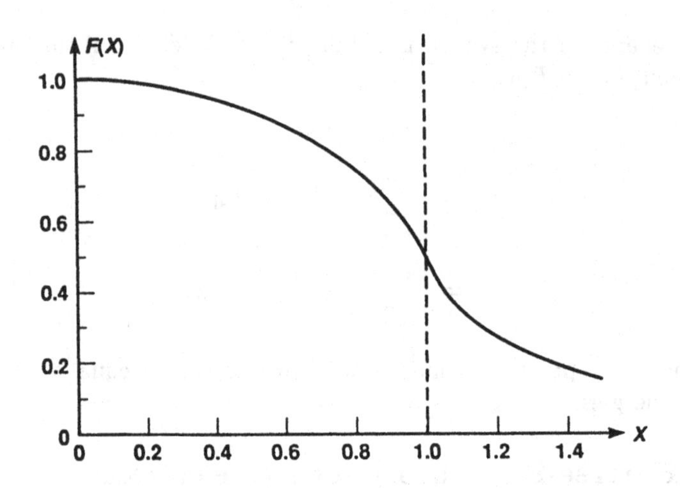

Figure 1.2: **A plot of the function $F(x)$ defined in the text.**

$$= \frac{k^2}{2} - \frac{2}{\pi} k_{\mathrm{F}} F\left(\frac{k}{k_{\mathrm{F}}}\right), \tag{1.145}$$

and

$$F(x) = \frac{1}{2} + \frac{1-x^2}{4x} \ln\left|\frac{1+x}{1-x}\right|. \tag{1.146}$$

It follows that plane waves solve the Hartree-Fock Eqs. (1.131), and that the energy levels with wave vector $\mathbf{k}$ are given by Eqs. (1.144) to (1.146). The function $F(x)$ and the energy $\epsilon(\mathbf{k})$ are plotted in Figs. 1.2 and 1.3, respectively. $F(x)$ has a divergence in slope at $x = 1$ which, being logarithmic, is not apparent in the figure and cannot be observed through changes in scale. For x large, $F(x)$ behaves as $1/3x^3$. The function plotted in Fig. 1.3 is taken to be

$$\epsilon(\mathbf{k})/\mu = \left[x^2 - 0.663\left(\frac{r_s}{a_0}\right) F(x)\right] \tag{1.147}$$

with $r_s/a_0 = 4$. The dashed line in the figure depicts the energy of free electrons, which is proportional to k^2. The introduction of exchange both depresses and widens the band. This increase in width, from 1 in the free-electron case to 2.23 is, however, not confirmed experimentally (presumably because the free-electron case is not realized in practice).

The band energy, $\epsilon(\mathbf{k})$, of the model system described by plane-wave states consists of a free-particle contribution and a term due to exchange. In order to obtain the total energy, we must sum over all $\mathbf{k}$ vectors with

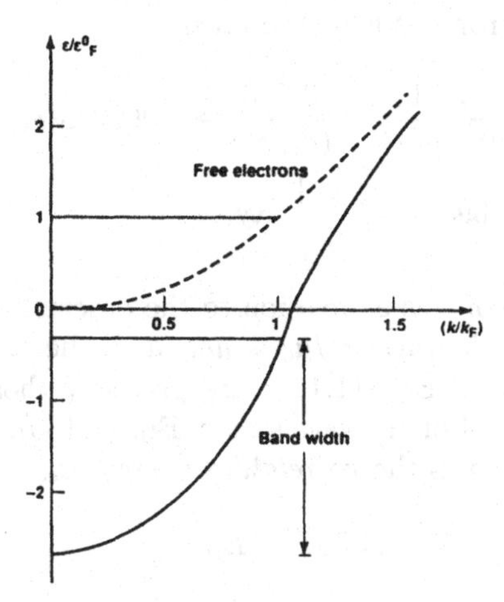

Figure 1.3: **A plot of the Hartree-Fock energy of free electrons for a particular choice of the electron density.**

magnitude less than k_F. This yields the expression

$$E = 2\sum_{k<k_F} \frac{k^2}{2} - \frac{k_F}{\pi}\sum_{k<k_F}\left[1 + \frac{k_F^2 - k^2}{2kk_F}\ln\left|\frac{k_F + k}{k_F - k}\right|\right]. \tag{1.148}$$

The first term is that given by Eq. (1.50). The second term can be evaluated through a transformation into an integral and gives for the average energy per particle

$$E/N = \left[\frac{3}{5}\mu - \frac{3}{4}\frac{k_F}{\pi}\right]. \tag{1.149}$$

This result can also be written in the form

$$\begin{aligned} \frac{E}{N} \equiv \frac{E_{HF}}{N} &= \frac{1}{2a_0}\left[\frac{3}{5}(\mu a_0)^2 - \frac{3}{2\pi}(k_F a_0)\right] \\ &= \left[\frac{2.21}{(r_s/a_0)^2} - \frac{0.916}{(r_s/a_0)}\right], \end{aligned} \tag{1.150}$$

where the units are Rydbergs, and r_s is the effective interparticle separation defined in Eq. (1.35). In metals, one has $2 \leq r_s/a_0 \leq 6$ so that the second term in the last equation is quite comparable in size to the first. Thus, electron-electron interactions cannot be ignored in a free-electron description of the electronic structure of metals. It can be shown [22] that the leading terms in a high density, i.e., small r_s/a_0, expansion of the ground state energy of an electron gas have the form

$$\begin{aligned} \frac{E}{N} &= \left[\frac{2.21}{(r_s/a_0)^2} - \frac{0.916}{(r_s/a_0)} + 0.0622\ln(r_s/a_0)\right. \\ &- \left. 0.096 + O(\frac{r_s}{a_0})\right] \text{ Ry.} \end{aligned} \tag{1.151}$$

However, the relevance of this expression to the description of electrons in metals is dubious since the ratio r_s/a_0 is *not* small there.

The first two terms in Eq. (1.151) are precisely those obtained in a Hartree-Fock treatment of plane-wave states, Eq. (1.150). The other terms are commonly referred to as the *correlation energy*, $E_{\rm cor}$,

$$E_{\rm cor} = E - E_{\rm HF}. \tag{1.152}$$

where E is the exact energy. By definition, the correlation measures the error made when making a rather crude first-order approximation, and has no physical significance *per se*. We see that by definition the Hartree-Fock approximation does not include *any* correlation effects[21].

As was mentioned above, the exchange term in the Hartree-Fock equation, Eq. (1.131), does not have the form of an ordinary potential. However, as suggested by Slater [17], one may attempt an approximate representation of the exchange term in terms of a single particle potential along the following lines. We note that the second term inside the brackets in Eqs. (1.149) and (1.150) represents the *average* change in the energy of a gas of free electrons due to exchange

$$\Delta E^{\rm exchg} = -\frac{3k_{\rm F}}{4\pi} = -\frac{0.916}{(r_s/a_0)} \text{ Ry.} \tag{1.153}$$

Slater's suggestion was to replace the exchange term in Eq. (1.131) by a potential energy equal to twice the term in Eq. (1.153), with $k_{\rm F}$ evaluated at the local density. The factor of two accounts for the spin of the electron. This leads to the introduction of a single-particle potential

$$U^{\rm exchg}(\mathbf{r}) = -2.95\left[a_0^3 n(\mathbf{r})\right]^{1/3} \text{ Ry.} \tag{1.154}$$

It is easy to see that this potential favors high densities, (becoming more negative there), and roughly mimics the dependence of the exchange term on the density of the electron gas. This term has been used in the performance of calculations, often in connection with a factor α that can be used as an adjustable parameter. This approach is commonly referred to as the

[21]Correlation effects are also *not* included in the classical self-interaction of a charge distribution which tends to keep electrons far apart.

$X\alpha$ approximation. However, in spite of its occasional success for specifically chosen values of α, this procedure is largely ad-hoc and has limited predictive power.

In addition to its neglect of correlation effects, the free-particle expression for the energy, Eq. (1.145), contains some other disturbing features. As seen in Fig. 1.3, the derivative $\partial/\partial k$ has a logarithmic infinity at $k = k_F$. This is indeed a disturbing feature, since the derivative is proportional to the velocity of electrons at the Fermi energy and figures prominently in the determination of metallic properties. For example, this feature leads to an electronic heat capacity that varies at low temperature as $T/|\ln T|$ rather than T [23]. This singularity is caused by the use of the long range, bare Coulomb potential and can be traced to the divergence of the Fourier transform $4\pi/k^2$ of $1/r$ at $k = 0$. Use of a Coulomb potential of the form $e^{-k_0 r}/r$ that reflects the effects of screening of any excess charge by the electron cloud would eliminate this unphysical behavior. *Screening effects* are extremely important in determining the behavior of charge-carrying disturbances in a metal. The treatment of these effects is indeed a *bona fide* and difficult many-body problem.

1.5.4 Koopmans' theorem

The Hartree-Fock equations, Eq. (1.131), lead to the determination of the "single-particle" energies, ϵ_i, defined in Eq. (1.138). This designation seems natural in light of the derivation of the Hartree-Fock equations. However, the liberal use of these energies as the energy levels of the many-particle system is not justified. Some insight into the nature of these orbital energies is afforded by Koopmans' theorem [24, 25].

Consider the difference in energy between a system of N electrons and the system that results when one of the electrons, say an electron in orbital ϕ_j, is removed. This difference can be easily evaluated if we assume that all occupied spin orbitals, ϕ_ℓ for $\ell \neq j$, still remaining are identical to those of the original N-electron system. This approximation of unchanged wave functions can be reasonably valid in a solid for states in a wide band. Electrons in such states are spread out throughout the volume of the system and correlation effects are fairly small. From Eq. (1.125), the difference in energy that results from the removal of an electron in spin-orbital j is

$$\begin{aligned}
\Delta E &= \int \phi_j^*(\mathbf{r}) \left[-\frac{1}{2}\nabla^2 + v(\mathbf{r}) \right] \phi_j(\mathbf{r}) \mathrm{d}^3 r \\
&= \sum_\ell \int\int |\phi_\ell(\mathbf{r}_1)|^2 |\phi_j(\mathbf{r}_2)|^2 \frac{1}{r_{12}} \mathrm{d}^3 r_1 \mathrm{d}^3 r_2 \\
&- \sum_\ell \int\int \phi_j^*(\mathbf{r}_1)\phi_\ell^*(\mathbf{r}_2) \frac{1}{r_1 2} \phi_j(\mathbf{r}_1)\phi_\ell(\mathbf{r}_2) \mathrm{d}^3 r_1 \mathrm{d}^3, \qquad (1.155)
\end{aligned}$$

where no factor of $\frac{1}{2}$ appears before the exchange and correlation terms because ϕ_j, the deleted state, occurs in both summations. Assuming that the ϕ_j are solutions of Eq. (1.131), we obtain

$$\Delta E = \epsilon_j. \tag{1.156}$$

Thus, the energy required to remove an electron from state j, the *ionization energy* associated with that state, equals the energy of the state. It follows, in the absence of system relaxation, i.e., of the orbitals being unaffected by the removal of the electron, that the energy required to remove an electron from state j and place it in state i is $\epsilon_i - \epsilon_j$. Thus, it is reasonable to think of the energies ϵ_j as single-particle energies, particularly in solids where the energy bands are determined from self-consistent field equations akin to the Hartree-Fock equations. At the same time, in spite of the physical insight afforded by Koopmans' theorem, Eq. (1.156) can be grossly incorrect. The main drawback of the derivation of that equation is the assumption that the remaining, occupied orbitals do not change upon removal of an electron. Because of correlation effects, system relaxation or rearrangement of levels can occur, and its neglect can lead to serious errors both in the case of atoms and molecules, and in that of solid materials.

The Hartree-Fock method has been used successfully in a number of calculations of the electronic structure of atoms and molecules. By and large, it has been found to provide a great improvement over the Thomas-Fermi model. At the same time, the Hartree-Fock method completely neglects correlation effects, which can be very important in the determination of electronic structure. Also, the method is rather difficult computationally to apply to the calculation of the electronic structure of solids. As is discussed in latter sections, the local density approximation to density-functional theory allows a treatment of both exchange and correlation, and is computationally much easier to use in the study of solid materials.

1.6 Hellmann-Feynman Theorem

We saw above that the virial theorem for particles interacting through purely Coulomb forces takes the same form in classical and quantum mechanics. Similarly, a classical expression is obtained for the force acting on a nucleus due to the presence of other nuclei and the electrons in the system. This is the content of the famous *electrostatics theorem* of Feynman [26]. In this section, we rederive[22] this theorem in a way that emphasizes the role of the electron density in describing the properties of a system of interacting charges.

[22] For overall clarity and insight, Feynman's original derivation is unsurpassed.

1.6.1 Electron density

The number of electrons per unit volume at point $\mathbf{r}$ in an electron gas in a given state is the electron density, $n(\mathbf{r})$, for that state. The exact single-particle electron density is given in terms of the many-body wave function, introduced in Eq. (1.93), by the expression,

$$n(\mathbf{r}) = N \int \cdots \int |\Psi(\mathbf{x}_1, \mathbf{x}_2, \cdots, \mathbf{x}_N)|^2 \mathrm{d}s_1 \mathrm{d}^3x_2 \mathrm{d}^3x_3 \cdots \mathrm{d}x_N, \qquad (1.157)$$

where $\mathrm{d}s_1$ implies a summation over the spin indices of the electron labeled by 1, and d^3x_i denotes an integration over the spatial coordinates and a summation over spin orientation of the ith electron. It follows that $n(\mathbf{r})$ is a simple function of the three variables, x, y, and z. Given the normalization property of Ψ, the density integrates to the total number of electrons

$$\int n(\mathbf{r}) \mathrm{d}^3r = N. \qquad (1.158)$$

For future reference, we also exhibit the form of the two-particle correlation function,

$$\begin{aligned} n_2(\mathbf{r}_1, \mathbf{r}_2) &= (N-1) \int \cdots \int |\Psi(\mathbf{x}_1, \mathbf{x}_2, \cdots, \mathbf{x}_N)|^2 \\ &\times \mathrm{d}s_1 \mathrm{d}s_2 \mathrm{d}^3x_3 \cdots \mathrm{d}x_N, \end{aligned} \qquad (1.159)$$

Let us use the electron density to express certain elementary results of quantum mechanics that will be useful in later developments. Under a perturbation, $\Delta U = \sum_i \delta u(\mathbf{r}_i)$, that changes the many-body wave function for a *fixed* number N of electrons in a state k from Ψ_k^0 to $\Psi_k^0 + \delta\Psi_k^1$, the energy change to first order in ΔU is given in first-order perturbation theory by the formula

$$\Delta E_k^{(1)} = \int \Psi_k^{0*} \Delta U \Psi_k^0 \mathrm{d}^3x_1 \cdots \mathrm{d}^3x_N = \int n_k(\mathbf{r}_1) \Delta u(\mathbf{r}_1) \mathrm{d}^3r_1 \qquad (1.160)$$

while the perturbed wave function is given, also to first order, by

$$|\Psi_k\rangle = |\Psi_k^0\rangle + \sum_{j \neq k} |\Psi_j\rangle \frac{\langle \Psi_j^0 | \Delta U | \Psi_k^0 \rangle}{E_k^0 - E_j^0}, \qquad (1.161)$$

where E_j^0 is the unperturbed energy of the electron gas in state j. The corresponding change in the density to first order in ΔU is then given by the expression

$$\Delta n_k(\mathbf{r}_1) = N \int \int \cdots \int \left(\Psi_k^* \Psi_k - \Psi_k^{0*} \Psi_k^0 \right) \mathrm{d}s_1 \mathrm{d}^3x_2 \mathrm{d}^3x_3 \cdots \mathrm{d}^3x_N$$

$$= 2N\Re\sum_{j\neq k}|\Psi_j\rangle\frac{\langle\Psi_j^0|\Delta U|\Psi_k^0\rangle}{E_k^0 - E_j^0}\int\cdots$$

$$\cdots\int\Psi_k^{0*}\Psi_j \mathrm{d}s_1\mathrm{d}^3x_2\mathrm{d}^3x_3\cdots\mathrm{d}^3x_N$$

$$\equiv \int\frac{\delta n_k(\mathbf{r}_1)}{\delta u(\mathbf{r}_2)}\Delta u(\mathbf{r}_2)\mathrm{d}^3r_2, \tag{1.162}$$

where $\Re$ denotes the real part of a complex quantity, and the functional derivative (see Appendix A) $\delta n/\delta u$ is defined by

$$\frac{\delta n_k(\mathbf{r}_1)}{\delta u(\mathbf{r}_2)} = \frac{\delta n_k(\mathbf{r}_2)}{\delta u(\mathbf{r}_1)} = 2N^2\sum_{j\neq k}$$

$$\times \quad \frac{\left[\int\cdots\int\Psi_j^{0*}\Psi_k \mathrm{d}s_2\mathrm{d}^3x_1\cdots\mathrm{d}^3x_N\right]\left[\int\cdots\int\Psi_k^{0*}\Psi_j^0\mathrm{d}s_1\mathrm{d}^3x_2\cdots\mathrm{d}^3x_N\right]}{(E_k^0 - E_j^0)}. \tag{1.163}$$

The change in density of the N-electron system produced by small changes in the potential is an example of a *linear response function.* We note that this function is symmetric in the indices 1 and 2, as is indicated in Eq. (1.163). This is expected behavior (at least for free electrons) as the response of the density at point 2 due to a change in potential at point 1 should be equal to the response at 1 due to a change at 2. Because of the orthonormality of the many-body wave functions associated with different states, we have

$$\int\frac{\delta n_k(\mathbf{r}_1)}{\delta u(\mathbf{r}_2)}\mathrm{d}^3r_1 = 0. \tag{1.164}$$

1.6.2 Hellmann-Feynman theorem

Consider a Hamiltonian, $\hat{H}(\lambda)$, depending on a parameter, λ, and the corresponding many-body wave function, $\Psi(\lambda)$. It follows from the first-order perturbation theory result, Eq. (1.160), that

$$\frac{\mathrm{d}E}{\mathrm{d}\lambda} = \frac{\langle\Psi|\partial\hat{H}(\lambda)/\partial\lambda|\Psi\rangle}{\langle\Psi|\Psi\rangle}, \tag{1.165}$$

a result known as the *differential Hellmann-Feynman theorem.* Integration of this formula from λ_1 to λ_2 yields the *integral Hellmann-Feynman theorem,*

$$E(\lambda_2) - E(\lambda_1) = \frac{\langle\Psi_2|\hat{H}(\lambda_2) - \hat{H}(\lambda_1)|\Psi_1\rangle}{\langle\Psi_2|\Psi_1\rangle}, \tag{1.166}$$

an expression which can also be put in the more general form

$$E_A - E_B = \frac{\langle \Psi_B | \hat{H}_A - \hat{H}_B | \Psi_A \rangle}{\langle \Psi_B | \Psi_A \rangle}. \tag{1.167}$$

The intermediate result

$$E(\lambda_2) - E(\lambda_1) = \int_{\lambda_1}^{\lambda_2} \frac{\langle \Psi_2 | \partial \hat{H}(\partial \lambda) | \Psi_1 \rangle}{\langle \Psi_2 | \Psi_1 \rangle} \mathrm{d}\lambda \tag{1.168}$$

is referred to as the *integrated Hellmann-Feynman theorem* [27]. In Eq. (1.167), $\hat{H}_A$ and $\hat{H}_B$, can be different Hamiltonians, not necessarily related by a parameter λ, acting on the same N-electron system, and Ψ_A and Ψ_B are the associated eigenfunctions, whose scalar product is assumed to be nonzero.

It is interesting to give a general derivation of Eq. (1.167). We begin with the relations

$$\hat{H}_A \Psi_A = E_A \Psi_A \quad \text{and} \quad \hat{H}_B \Psi_B = E_B \Psi_B, \tag{1.169}$$

multiply the first by Ψ_B^* and the second with Ψ_A^*, integrate, take the complex conjugate of the second result and subtract it from the first. The result is

$$(E_A - E_B)\langle \Psi_B | \Psi_A \rangle = \langle \Psi_B | \hat{H}_A - \hat{H}_B | \Psi_A \rangle, \tag{1.170}$$

from which Eq. (1.167) follows, provided that $\langle \Psi_B | \Psi_A \rangle \neq 0$. This procedure affords an alternative derivation of Eq. (1.165) for the case in which $\hat{H}_A$ and $\hat{H}_B$ are related through a parameter λ and differ infinitesimally from one another.

It is instructive also to emphasize a slightly different derivation of the Hellmann-Feynman theorem, Eq. (1.165). Consider the Schrödinger equation,

$$H(\lambda)\psi_\nu(\lambda) = E_\nu \psi_\nu(\lambda), \tag{1.171}$$

where the wave functions $\psi_\nu(\lambda)$ can be assumed to be normalized,

$$\langle \psi_\nu(\lambda) | \psi_\nu(\lambda) \rangle = 1. \tag{1.172}$$

Differentiating E_ν with respect to λ yields the set of equations

$$\begin{aligned} \frac{\mathrm{d}E_\nu}{\mathrm{d}\lambda} &= \langle \psi_\nu(\lambda) | \frac{\partial H}{\partial \lambda} | \psi_\nu(\lambda) \rangle + \langle \frac{\partial \psi_\nu(\lambda)}{\partial \lambda} | H | \psi_\nu(\lambda) \rangle \\ &+ \langle \psi_\nu(\lambda) | H | \frac{\partial \psi_\nu(\lambda)}{\partial \lambda} \rangle \\ &= 2\Re \langle \frac{\partial \psi_\nu}{\partial \lambda} | H | \psi_\nu \rangle + \langle \psi_\nu(\lambda) | \frac{\partial H}{\partial \lambda} | \psi_\nu(\lambda) \rangle, \end{aligned} \tag{1.173}$$

where, because of the normalization condition on the wave functions, we have

$$\langle\frac{\partial\psi_\nu(\lambda)}{\partial\lambda}|H|\psi_\nu(\lambda)\rangle + \langle\psi_\nu(\lambda)|H|\frac{\partial\psi_\nu(\lambda)}{\partial\lambda}\rangle = E_\nu\frac{\partial}{\partial\lambda}\langle\psi_\nu(\lambda)|\psi_\nu(\lambda)\rangle = 0. \tag{1.174}$$

The Hellmann-Feynman theorem now follows in the form,

$$\frac{\mathrm{d}E_\nu}{\mathrm{d}\lambda} = \langle\psi_\nu|\frac{\partial H}{\partial\lambda}|\psi_\nu\rangle = \frac{\partial\bar{H}}{\partial\lambda}. \tag{1.175}$$

As a simple illustration of the formulae just derived, let us calculate the force exerted on a nucleus α due to the presence of the other nuclei and the electron gas in a collection of atoms, e.g. a solid. We assume the presence of only Coulomb forces, so that the Hamiltonian of the system has the form of Eq. (1.94). Then, the only terms in the full Hamiltonian (including inter-nuclear repulsion) that depend on the nuclear positions are U_{eN}, Eq. (1.65), and U_{NN}, Eq. (1.99). Let λ be identified with the x-coordinate, X_α, of nucleus α. Then, Eq. (1.165) yields the expression,

$$\begin{aligned}\frac{\partial E_{\mathrm{tot}}}{\partial\lambda} &= -\sum_{\beta\neq\alpha}\frac{Z_\alpha Z_\beta}{R^3_{\alpha\beta}}(X_\alpha - X_\beta) - Z_\alpha\int n(\mathbf{r}_1)\frac{x_1 - X_\alpha}{r^3_{1\alpha}}\mathrm{d}^3r_1 \\ &= -\nabla_\alpha\left[U_{\mathrm{eN}} + U_{\mathrm{NN}}\right].\end{aligned} \tag{1.176}$$

Even though the system is described by the true many-body quantum mechanical Hamiltonian, Eq. (1.176) is purely classical. It shows that the force on a nucleus is what would have been computed from classical electrostatics, given a knowledge of nuclear positions and of the electronic charge density [28]. This is Feynman's electrostatics theorem, or the Hellmann-Feynman theorem[23].

As a further application of Eq. (1.165), let us calculate the formula obtained if one replaces Z_α by λZ_α everywhere it appears in $\hat{H}$, which is taken to include U_{NN}, and then compute $E_{\mathrm{tot}}(1) - E_{\mathrm{tot}}(0)$ for the ground state of the system. We note that the ground state energy of N electrons in the absence of any nuclei can be set equal to zero, $E_{\mathrm{tot}}(0) = 0$. Hence, for a ground state we have[24]

$$E_{\mathrm{tot}} = \sum_{\alpha<\beta}\frac{Z_\alpha Z_\beta}{R_{\alpha\beta}} - \sum_\alpha\int_0^1\mathrm{d}\lambda\int\frac{n(\mathbf{r}_1,\lambda)}{r_{1\alpha}}\mathrm{d}^3r_1, \tag{1.177}$$

where $n(\mathbf{r}_1,\lambda)$ is the electronic density obtained from the wave function $\Psi(\mathbf{x}_1,\mathbf{x}_2,\cdots,\mathbf{x}_N;\lambda)$ associated with the scaled nuclear charges. However,

[23] Feynman derived this result independently but following Hellmann's publication.

[24] More generally, the lower limit in the first integral should be replaced by a quantity $\lambda_{\mathrm{cr}} > 0$ because of binding considerations. For example, a nucleus with $Z < 0.9111...$ cannot bind 2 electrons (see ...)

it is to be emphasized that the electrostatics theorem holds for any state of the system whereas Eq. (1.177) is valid only for the ground state.

It is important to mention that the electrostatics theorem holds for the *exact* charge density. Applications of the theorem in connection with an approximate $n(\mathbf{r})$ are not a priori justified, and may lead to considerable inaccuracies. At the same time, approximate charge densities often satisfy some of the formulae discussed above. For example, Eq. (1.165) holds for the single-determinant wave function of the Hartree-Fock method.

It may have become apparent in the course of the discussion up to this point that many physical properties of the electron gas, such as the energy and internuclear forces, can be expressed in terms of the electronic charge density. In fact density-functional theory (DFT) asserts that the exact charge density suffices for the determination of *all* ground-state properties of an electron cloud. This theory is presented in the next section.

1.7 Density-Functional Theory

In this section we present the basic features of density-functional theory. This theory accomplishes the remarkable task of replacing the complicated many-body wave function, and the associated Schrödinger equation, with the *single-electron* density and its associated calculational scheme. This is remarkable indeed given that the single-particle density is an ordinary function of position and a much simpler quantity than the many-body wave function.

The idea of using the density to describe the properties of an N-electron system has a rather long history [9, 10, 29, 30]. As we saw above, it is the basis for the Thomas-Fermi model. In all cases a *statistical* model was suggested based on the notion that [9] "electrons are distributed uniformly in the six-dimensional space for the motion of an electron at the rate of two for each h^3 of volume", and these electrons are under the influence of an external field that "is itself determined by the nuclear charges and this distribution of electrons". Implementation of these assumptions leads directly to the Thomas-Fermi model of the atom. It is important to keep in mind that as originally stated these ideas could only lead to "models" rather than "theories" for the treatment of an N-electron system. As assumptions, they lacked rigorous justification. In addition, it was not clear to what theory they were approximations and consequently they did not provide a clear direction for improvement.

The situation was altered dramatically in 1964. That year, a paper by Hohenberg and Kohn [31] provided a rigorous justification for the use of the single-particle density, $n(\mathbf{r})$, in the *exact* description of the many-particle system. Among other characteristics, *Density-functional theory* (DFT), as the ensuing formalism has come to be known, clarified the status of earlier approaches, such as the Thomas-Fermi model and the Xα method. It can be shown that many such models can be derived as approximations or simplifications to DFT.

We begin our exposition of DFT with a discussion of the two important theorems proved by Hohenberg and Kohn [31]. In this discussion, we follow the lines of development in Parr and Yang [3], to which the reader is referred for further details. For a review with applications, see the article by Jones and Gunnarsson [32] and reference [1].

1.7.1 Theorems of Hohenberg and Kohn

Consider an N-electron system under the influence of an external potential, $v(\mathbf{r})$, which is not necessarily restricted to Coulomb interactions. The first theorem of Hohenberg and Kohn states that *the external potential, $v(\mathbf{r})$, acting on a fully interacting many-particle system in its ground state is determined, within an additive constant, by the electron density, $n(\mathbf{r})$*. The proof of this theorem is simple indeed.

Let the Hamiltonian describing the N-electron cloud have the form of Eq. (1.94)

$$\hat{H} = \sum_{i=1}^{N}\left(-\frac{1}{2}\nabla_i^2\right) + \sum_{i}^{N} v(\mathbf{r}_i) + \sum_{i<j}^{N} \frac{1}{r_{ij}}, \tag{1.178}$$

and consider the density $n(\mathbf{r})$ for the non-degenerate ground state of the system. The density determines the number of particles through Eq. (1.158). Now, let there be two external potentials, $v(\mathbf{r})$ and $v'(\mathbf{r})$, differing by more than a constant and each giving the same ground-state density $n(\mathbf{r})$. Let also $\hat{H}$ and $\hat{H}'$ be the corresponding many-body Hamiltonians, Ψ and Ψ', the associated ground-state wave functions. Take Ψ' to be a trial wave function for the system described by $\hat{H}$ (which includes $v(\mathbf{r})$) and use the minimum principle in Eq. (1.107) to obtain the result

$$\begin{aligned} E_0 < \langle\Psi'|\hat{H}|\Psi'\rangle &= \langle\Psi'|\hat{H}'|\Psi'\rangle + \langle\Psi'|\hat{H} - \hat{H}'|\Psi'\rangle \\ &= E_0' + \int n(\mathbf{r})\,[v(\mathbf{r}) - v'(\mathbf{r})]\,\mathrm{d}^3r, \end{aligned} \tag{1.179}$$

where E_0 and E_0' denote, respectively, the energy of the N-electron system (excluding nuclear-nuclear repulsion), under the influence of v and v'. The integral $\int n(\mathbf{r})v(\mathbf{r})\mathrm{d}^3r$ arises from the second term in Eq. (1.178) and describes the interaction of the charge distribution with the external field. Similarly, taking Ψ as a trial wave function for the system described by $\hat{H}'$, we have

$$\begin{aligned} E_0' < \langle\Psi|\hat{H}'|\Psi\rangle &= \langle\Psi|\hat{H}|\Psi\rangle + \langle\Psi|\hat{H}' - \hat{H}|\Psi\rangle \\ &= E_0 - \int n(\mathbf{r})\,[v(\mathbf{r}) - v'(\mathbf{r})]\,\mathrm{d}^3r. \end{aligned} \tag{1.180}$$

Adding the last two equations, we obtain

$$E_0 + E_0' < E_0' + E_0, \tag{1.181}$$

which is clearly a contradiction. (Note that there is *no* equals sign in either Eq. (1.179) or Eq. (1.180), and hence the contradictory statement that a quantity is smaller than itself.) It follows that there cannot be two different v's, differing by more than a constant, that give the same density $n(\mathbf{r})$ for the ground state of the system.

In retrospect, it may not be all that difficult to accept the theorem of Hohenberg and Kohn. Given the form of the Hamiltonian (1.94), the external potential $v(\mathbf{r})$ completely fixes the Hamiltonian (by specifying the second term), so that N and $v(\mathbf{r})$, specified within a constant, determine all ground-state properties. (Although the theorem was proven explicitly for the case of a non-degenerate ground state, degeneracy presents no special difficulty.) For example, when $v(\mathbf{r})$ is taken to represent the field set up by the nuclei in an atom, molecule, or solid, it becomes clear that it, along with the number of electrons, determines *all* electronic properties. Hence, $n(\mathbf{r})$ determines $v(\mathbf{r})$ as well as N, Eq. (1.158), and hence all properties of the ground state, and of excited states as well.

Among the properties of the electron cloud, the total energy and the distinct contributions to it are of particular interest. Thus, the kinetic energy, $T[n]$, the potential energy, $V[n]$, and the total energy[25], $E[n]$, are all determined as functionals of the density $n(\mathbf{r})$. For example, instead of Eq. (1.68) obtained within the Thomas-Fermi model, we have

$$\begin{aligned} E_v[n] &= T[n] + U_{\text{ext}}[n] + U_{\text{ee}}[n] \\ &= \int n(\mathbf{r})v(\mathbf{r})\mathrm{d}^3r + F_{\text{HK}}[n], \end{aligned} \tag{1.182}$$

where the subscript v indicates explicitly the presence of an external potential, and where

$$F_{\text{HK}} = T[n] + U_{\text{ee}}[n] \tag{1.183}$$

is the exact functional representing the kinetic energy and the mutual interaction among the electrons. This functional contains *all* many-body effects in the ground state of the system. To amplify its content somewhat, let us write

$$U_{\text{ee}} = J[n] + \text{ (nonclassical terms)}, \tag{1.184}$$

where

$$J[n] = \frac{1}{2}\int \frac{n(\mathbf{r}_1)n(\mathbf{r}_2)}{|\mathbf{r}_1 - \mathbf{r}_2|}\mathrm{d}^3r_1\mathrm{d}^3r_2 \tag{1.185}$$

[25]Unless there is danger of confusion we will not use the subscript "tot" to designate the total energy. Furthermore, when considering the total energy of the entire system of electrons and nuclei, the inter-nuclear repulsion should be taken into account even though it may not be exhibited explicitly in the formulae.

is the classical repulsive interaction of a charge distribution with itself. As its name indicates, the nonclassical terms contain all contributions arising from the nature of the electrons as indistinguishable particles obeying Fermi-Dirac statistics. Formally, this term includes the difference between $J[n]$ and $U_{ee}[n_2]$ defined in Eq. (1.129). This nonclassical term is of extreme importance. Although its exact form is not known, it provides the major contribution to the "exchange-correlation energy" which will be discussed in the next subsection. Improvements on the treatment of the inhomogeneous electron gas within DFT consist mostly of attempt s to construct a more accurate form to describe the exchange-correlation effects in an electron cloud.

The formal foundation of DFT becomes complete with the second theorem of Hohenberg and Kohn. This theorem provides a variational principle for the energy, and can be stated as follows: *For a trial density $\tilde{n}(\mathbf{r})$ such that $\tilde{n}(\mathbf{r}) \geq 0$ and $\int \tilde{n}(\mathbf{r})\mathrm{d}^3r = N$,*

$$E_0 \leq E_v[\tilde{n}] \tag{1.186}$$

where $E_v[\tilde{n}]$ is the energy functional of Eq. (1.182). This theorem is analogous to the corresponding theorem for wave functions, Eq. (1.107). Its proof rests on Hohenberg and Kohn's first theorem according to which $\tilde{n}$ determines its own $\tilde{v}$ and wave function $\tilde{\Psi}$. If we take this latter quantity as a trial wave function for a system in its ground state described by a Hamiltonian $\hat{H}$, and with external potential v, we obtain

$$\begin{aligned} \langle\tilde{\Psi}|\hat{H}|\tilde{\Psi}\rangle &= \int \tilde{n}(\mathbf{r})v(\mathbf{r})\mathrm{d}^3r + F_{\mathrm{HK}}[\tilde{n}] \\ &= E_v[\tilde{n}] \geq E_v[n] = E_0. \end{aligned} \tag{1.187}$$

Here, n is the correct density for the ground state so that $E_v[n] = E_0$ by definition. This proves the theorem.

In addition to the energy, the chemical potential is an important quantity within DFT. The chemical potential arises from an application of the variational principle implied in Eq. (1.186) under the condition $\int n(\mathbf{r})\mathrm{d}^3r = N$. Provided that $E_v[n]$ is differentiable, we have

$$\delta\tilde{E}[n(\mathbf{r})] \equiv \delta\left\{E_v[n] - \mu\left[\int n(\mathbf{r})\mathrm{d}^3r - N\right]\right\} = 0 \tag{1.188}$$

which in view of Eq. (1.182) yields the Euler-Lagrange equation,

$$\mu = \frac{\delta E_v[n]}{\delta n(\mathbf{r})} = v(\mathbf{r}) + \frac{\delta F_{\mathrm{HK}}[n]}{\delta n(\mathbf{r})}. \tag{1.189}$$

Let us emphasize certain features of the formalism developed thus far. We note that the functional $F_{\mathrm{HK}}[n]$ defined in Eq. (1.183) is a *universal functional* of n because it is defined independently of the external potential, $v(\mathbf{r})$. This functional, in principle, provides an exact description of the ground-state properties of the interacting electron gas. Thus, given an explicit form of $F_{\text{HK}}[n]$, the method is applicable to any system.

However, the exact form of F_{HK} has proven illusive, and exceedingly hard to achieve. One approximate but very successful approach has been to use the form of $F_{\text{HK}}[n]$ appropriate to a uniform electron gas with the constant density of the uniform system replaced by the local, varying density at position $\mathbf{r}$ of the interacting electron cloud. This local approximation is discussed in the remainder of this chapter, and is used in the developments presented in following chapters. At the same time, the existence of an exact theory has given impetus to a great deal of ongoing work toward the development of more accurate forms of $F_{\text{HK}}[n]$.

1.7.2 Kohn-Sham method

The Hohenberg and Kohn theorems yield the appealing result that the energy of the ground state of an interacting many-particle system can be obtained as the minimum of the energy functional[26] (with the internuclear repulsion added on at a convenient moment)

$$E[n] = \int n(\mathbf{r})v(\mathbf{r})\mathrm{d}^3r + F[n] \tag{1.190}$$

where

$$F[n] = T[n] + U_{\text{ee}}[n]. \tag{1.191}$$

The ground state electron density $n(\mathbf{r})$ uniquely determines $v(\mathbf{r})$ and satisfies the Euler-Lagrange equation

$$\mu = v(\mathbf{r}) + \frac{\delta F[n]}{\delta n(\mathbf{r})}, \tag{1.192}$$

where μ is the chemical potential, which enters the discussion as the Lagrange multiplier associated with the constraint

$$\int n(\mathbf{r})\mathrm{d}^3r = N. \tag{1.193}$$

Although these equations in principle provide an *exact* treatment of the interacting electron cloud, their numerical implementation is far from obvious. One approach would be to attempt the d irect determination of computable expressions for the various functionals, based on some approximation scheme. This is, in fact, the approach upon which the Thomas-Fermi method is based. Equation (1.68) is expressed in terms of quadratures involving functions of the density. However this direct use of the DFT

[26] We dispense with the subscript HK on functional quantities for now on.

equations has turned out to be greatly limited in accuracy, and provides no guidance for ways to improve it. An alternative, indirect method for the numerical implementation of DFT has been suggested by Kohn and Sham [33].

In order to illustrate the role played by the spin functions in the method of Kohn and Sham, let us recall a basic property of determinantal wave functions. For a wave function consisting of a single determinant, the density assumes the form given in Eq. (1.130) (we will often suppress spin variables)

$$n(\mathbf{r}) = \sum_i^N |\phi_i(\mathbf{r})|^2. \tag{1.194}$$

Furthermore, as in Eq. (1.125), the kinetic energy is given *exactly* in the form, (with spin indices summed)

$$\begin{aligned} T_s &= \sum_i^N \langle\phi_i| - \frac{1}{2}\nabla^2|\phi_i\rangle \\ &= \sum_i^N \int d^3r \phi_i^*(\mathbf{r}) \left(-\frac{1}{2}\nabla^2\right) \phi_i(\mathbf{r}). \end{aligned} \tag{1.195}$$

The subscript s helps distinguish the kinetic energy defined in terms of wave functions that are single determinants from the exact functional in Eq. (1.191). Although approximate as far as the real problem of the interacting electron cloud is concerned, Eqs. (1.194) and (1.195) are exact with respect to a wave function that is a single-determinant. In addition, the spin orbitals, $\phi_j(\mathbf{r}, s)$, provide a unique decomposition of the density $n(\mathbf{r})$, Eq. (1.130), and also yield a unique value for T_s.

Now, the spin orbitals that enter the construction of a Slater determinant are the solutions (eigenstates) of a single-particle Hamiltonian, $\hat{H}_s$ (hence the subscript s). Kohn and Sham introduced a *reference system*, of noninteracting electrons described by the Hamiltonian

$$\hat{H}_s = \sum_{i=1}^N \left(-\frac{1}{2}\nabla_i^2\right) + \sum_{i=1}^N v_s(\mathbf{r}), \tag{1.196}$$

which does not include the electron-electron repulsion, and for which *the ground-state electron density, as determined by Eq. (1.194), is exactly* n. Thus, the spin orbitals $\phi_j(\mathbf{r}, s)$ are the N lowest eigenstates of the one-electron Hamiltonian,

$$\hat{h}_s\phi_j = \left[-\frac{1}{2}\nabla_i^2 + v_s(\mathbf{r})\right] \phi_j = \epsilon_j\phi_j. \tag{1.197}$$

This construction sets up a noninteracting system described by the Hamiltonian (1.197) for which the kinetic energy is given *exactly* by T_s, and the density is given *exactly* by Eq. (1.194). However, it should be born in mind that the quantity $T_s[n]$, although uniquely determined for any density, is by no means the exact kinetic-energy functional, $T[n]$, of Eq. (1.191).

Kohn and Sham proposed a method in which *the kinetic energy component of the total energy is given exactly by $T_s[n]$*. Any difference between the exact kinetic energy, $T[n]$, and $T_s[n]$, along with any difference in the potential energy due to correlation effects are to be treated as a separate, and hopefully small, contribution to the energy. Therefore, we write

$$F[n] = T_s[n] + J[n] + E_{xc}[n] \tag{1.198}$$

where $J[n]$ is defined in Eq. (1.128) and is the classical interaction of a charge with itself. The last term, $E_{xc}[n]$, contains the difference $T[n]-T_s[n]$, and any difference between the exact self-interaction of the charge, Eq. (1.129),

$$U_{ee}[n_2] = \frac{1}{2}\int \mathrm{d}^3r_1 \int \mathrm{d}^3r_2 \frac{n_2(\mathbf{r}_1, \mathbf{r}_2)}{|\mathbf{r}_1 - \mathbf{r}_2|}, \tag{1.199}$$

and its classical approximation, $J[n]$. Thus, we have

$$E_{xc}[n] = T[n] - T_s[n] + U_{ee}[n_2] - J[n]. \tag{1.200}$$

This quantity is commonly referred to as the *exchange-correlation energy*. With thedefinition of $F[n]$ given in Eq. (1.198), the Euler-Lagrange Eq. (1.192) becomes

$$\mu = v_{\text{eff}} + \frac{\delta T_s[n]}{\delta n(\mathbf{r})}, \tag{1.201}$$

where the Kohn-Sham *effective potential* is defined by the expression

$$\begin{aligned} v_{\text{eff}}(\mathbf{r}) &= v(\mathbf{r}) + \frac{\delta J[n]}{\delta n(\mathbf{r})} + \frac{\delta E_{xc}[n]}{\delta n(\mathbf{r})} \\ &= v(\mathbf{r}) + \int \frac{n(\mathbf{r}')}{|\mathbf{r} - \mathbf{r}'|}\mathrm{d}^3r' + v_{xc}(\mathbf{r}), \end{aligned} \tag{1.202}$$

and where

$$v_{xc}(\mathbf{r}) = \frac{\delta E_{xc}[n]}{\delta n(\mathbf{r})}, \tag{1.203}$$

is the *exchange-correlation potential*. We recall that $v(\mathbf{r})$ in Eq. (1.202) is the externally applied potential, such as that due to the presence of the nuclear charges. We now note that *Eq. (1.201) with the constraint (1.193) is precisely the same equation as one would obtain from an application of density-functional theory to a system of non-interacting electrons moving under the influence of an external potential* $v_s(\mathbf{r}) = v_{\text{eff}}(\mathbf{r})$. Consequently, we introduce an effective single-particle Hamiltonian

$$\hat{h}_{\text{eff}} = -\frac{1}{2}\nabla^2 + v_{\text{eff}}(\mathbf{r}) \tag{1.204}$$

whose eigenfunctions are the spin orbitals $\phi_j(\mathbf{r})$,

$$\begin{aligned}\left[-\frac{1}{2}\nabla^2 + v_{\text{eff}}(\mathbf{r})\right]\phi_j(\mathbf{r}) &= \left[\frac{1}{2}\nabla^2 + \nu(\mathbf{r}) + \int \frac{n(\mathbf{r}')}{|\mathbf{r}-\mathbf{r}'|}\mathrm{d}^3r'\right. \\ &+ \left.\nu_{\text{xc}}(\mathbf{r})\right]\phi_j(\mathbf{r}) = \epsilon_j\phi_j(\mathbf{r}). \end{aligned} \tag{1.205}$$

There are N such equations for N electrons. The spin orbitals so obtained can be used to define the ground-state charge density through Eq. (1.130),

$$n(\mathbf{r}) = \sum_i |\phi_i(\mathbf{r})|^2. \tag{1.206}$$

Since $v_{\text{eff}}(\mathbf{r})$ depends on $v(\mathbf{r})$ through Eq. (1.203), Eqs. (1.202), (1.205), and (1.206) must be solved in a *self-consistent* manner: One begins with a guess for $n(\mathbf{r})$ and uses it to construct $v_{\text{eff}}(\mathbf{r})$ from Eq. (1.202). This v_{eff} is used in Eq. (1.205) to determine a set of spin orbitals, $\phi_j(\mathbf{r})$, which yield a new charge density through Eq. (1.206). The process is repeated until the difference between the input and output charge densities (and potentials) falls below a predetermined value. Finally the total energy is computed from Eq. (1.190). An alternative expression for the total energy is given in the next subsection, Eq. (1.211).

Equations (1.202), (1.203), (1.205), and (1.206) are the famous Kohn-Sham equations. What they accomplish is remarkable indeed: *In principle, they allow an exact treatment of the ground-state of the many-body problem in terms of a theory that is of the independent-particle form.* However, it should be kept in mind that in reality an exact treatment is not possible because $E_{\text{xc}}[n]$ is not known exactly.

1.7.3 Total energy

Using the quantities defined in Eqs. (1.202) and (1.203), the energy functional of Eq. (1.190) can be written in the form (to which the nuclear repulsion, U_{NN}, should be added),

$$\begin{aligned} E[n] &= T_s[n] + \int v(\mathbf{r})n(\mathbf{r})\mathrm{d}^3r + J[n] + E_{\text{xc}}[n] \\ &= \sum_i^N \sum_s \int \phi_i^*(\mathbf{r})\left[-\frac{1}{2}\nabla_i^2\right]\phi_i(\mathbf{r}) + \int v(\mathbf{r})n(\mathbf{r})\mathrm{d}^3r \\ &+ J[n] + E_{\text{xc}}[n], \end{aligned} \tag{1.207}$$

where $v(\mathbf{r})$ is the external potential (such as that due to the presence of the nuclei), and with the electron density being given by the familiar expression

$$n(\mathbf{r}) = \sum_{i=1}^{N} |\phi_i(\mathbf{r})|^2. \tag{1.208}$$

These equations express the energy in terms of N spin orbitals. Multiplying Eq. (1.205) by $\phi_i^*(\mathbf{r})$, integrating and summing over all occupied states, we obtain

$$\begin{aligned}\sum_{i=1}^{N} \langle\phi_i| \quad - \quad \frac{1}{2}\nabla^2 + v_{\text{eff}}(\mathbf{r})|\phi_i\rangle &= \sum_{i=1}^{N} \epsilon_i \\ &= T_s[n] + \int v_{\text{eff}}(\mathbf{r})n(\mathbf{r})\mathrm{d}^3r,\end{aligned} \tag{1.209}$$

which yields the following expression for the expectation value of the kinetic energy operator

$$\begin{aligned}T_s[n] &= \sum_{i=1}^{N} \langle\phi_i| - \frac{1}{2}\nabla^2|\phi_i\rangle \\ &= \sum_{i=1}^{N} \epsilon_i - \int n(\mathbf{r})v_{\text{eff}}(\mathbf{r})\mathrm{d}^3r \\ &= \sum_{i=1}^{N} \epsilon_i - \int n(\mathbf{r})v(\mathbf{r})\mathrm{d}^3r - \int n(\mathbf{r})v_{\text{xc}}(\mathbf{r})\mathrm{d}^3r.\end{aligned} \tag{1.210}$$

It now follows that the total energy can be written in the form

$$\begin{aligned}E &= \sum_i \epsilon_i - \frac{1}{2}\int \mathrm{d}^3r n(\mathbf{r}) \left[\int \frac{n(\mathbf{r}')}{|\mathbf{r}-\mathbf{r}'|}\mathrm{d}^3r' - \sum_n \frac{Z_n}{|\mathbf{r}-\mathbf{R}_n|}\right] \\ &- \frac{1}{2}\sum_n Z_n \left[\int \frac{n(\mathbf{r})}{|\mathbf{r}-\mathbf{R}_n|}\mathrm{d}^3r - \sum_{n'>n}^{\prime} \frac{Z_{n'}}{|\mathbf{R}_{n'}-\mathbf{R}_n|}\right] \\ &+ \int n(\mathbf{r})\nu_{rmxc}(\mathbf{r})\mathrm{d}^3r.\end{aligned} \tag{1.211}$$

We note that as in Hartree-Fock theory, the total energy is *not* the sum of orbital energies.

It is useful to compare the method of Kohn and Sham to both the Thomas-Fermi model and the Hartree-Fock method. The Kohn-Sham Eqs.

(1.202), (1.203), (1.205), and (1.206) yield N self-consistent equations for solution rather than the single Euler-Lagrange Eq. (1.192). By contrast, the Thomas-Fermi method provided a single equation for the single-particle density. However, even though the Kohn-Sham equations are much more complicated numerically than the Thomas-Fermi model, they are immensely more accurate. In addition, the method of Kohn and Sham in principle allows the exact treatment of exchange and correlation through the term E_{xc}. This term remains intact in the general theory, entering Eqs. (1.201) and (1.205) in exactly the same way.

We can also compare the method of Kohn and Sham to the Hartree-Fock and Hartree (Hartree-Fock without exchange) methods. All three are based on the self-consistent determination of single-particle orbitals for the description of an N-electron system. However, the Kohn-Sham treatment allows in principle for an exact treatment of exchange and correlation effects whereas the Hartree-Fock method neglects correlation completely[27], and the Hartree method neglects both exchange and correlation. Even if such effects are incorporated within these methods, they remain ad hoc. Furthermore, the Kohn-Sham method provides at least a virtually unique approach to improvement through the introduction of more accurate approximations to the unknown functional, E_{xc}. Attempts at obtaining more accurate descriptions of exchange and correlation are continuously under way. And last, but certainly not least, the Kohn-Sham equations are much easier to implement numerically under various approximations for E_{xc} than the strictly non-local Hartree-Fock equations. The Kohn-Sham equations are, in fact, no more difficult computationally than the Hartree equations[28]. It is to be noted that the Hartree-Fock equations are not an approximation to the Kohn-Sham equations because these two approaches handle exchange and correlation within incompatible schemes. (Recall the explicitly non-local character of the exchange term in the Hartree-Fock method; non-locality enters the Kohn-Sham equations implicitly through the exchange-correlation functional.)

1.7.4 Local density approximation (LDA)

The numerical implementation of the Kohn-Sham equations requires an explicit treatment of the largely unsettled and unknown exchange-correlation term. The accurate determination of E_{xc} continues to provide the greatest challenge to DFT. At the same time, reasonable approximation schemes can be invoked that allow the numerical application of the Kohn-Sham

[27]Incorporation of correlations into the Hartree-Fock method can be a very difficult numerical task.

[28]Which, incidentally, would also converge in a self-consistent iterative scheme such as those based on DFT (see further discussion).

method. The ***local density approximation (LDA)*** is the most widely used such scheme.

The idea, introduced by Kohn and Sham, rests on the local approximation that also underlies the Thomas-Fermi method. Once the kinetic energy, $T_s[n]$, is treated exactly, Kohn and Sham suggested the use of the uniform-electron-gas formula for handling the unknown parts of the energy functional. Thus, one applies uniform-electron-gas results to infinitesimal volumes in the interacting system that contain $n(\mathbf{r})\mathrm{d}^3r$ electrons, and sums these local contributions over all space. The resulting scheme is called the ***local density approximation (LDA)*** to density-functional theory.

Along the lines just described, we begin by defining the exchange-correlation energy as an integral,

$$E_{\mathrm{xc}}^{\mathrm{LDA}}[n] = \int n(\mathbf{r})\epsilon_{\mathrm{xc}}[n(\mathbf{r})]\mathrm{d}^3r, \tag{1.212}$$

where $\epsilon_{\mathrm{xc}}[n(\mathbf{r})]$ is the exchange and correlation energy per particle of a uniform electron gas of uniform density $n = n(\mathbf{r})$. Now, the exchange-correlation potential of Eq. (1.203) takes the form,

$$v_{\mathrm{xc}}^{\mathrm{LDA}} = \frac{\delta E_{\mathrm{xc}}^{\mathrm{LDA}}}{\delta n(\mathbf{r})} = \epsilon_{\mathrm{xc}}[n(\mathbf{r})] + n(\mathbf{r})\frac{\mathrm{d}\epsilon_{\mathrm{xc}}^{\mathrm{LDA}}[n]}{\mathrm{d}n}\Big|_{n=n(\mathbf{r})} \tag{1.213}$$

and the Kohn-Sham equations for the spin orbitals, Eq. (1.205), assumes the form,

$$\left[-\frac{1}{2}\nabla^2 + v(\mathbf{r}) + \int \frac{n(\mathbf{r})}{|\mathbf{r}-\mathbf{r}'|}\mathrm{d}^3r' + v_{\mathrm{xc}}^{\mathrm{LDA}}(\mathbf{r})\right]\phi_i = \epsilon_i\phi_i. \tag{1.214}$$

The self-consistent solution of these equations incorporated into the scheme of Eqs. (1.202), (1.205), and (1.206) defines the Kohn and Sham local density approximation, commonly referred to as the LDA.

1.7.5 Exchange-correlation functional

Much work has gone into the determination of the function $\epsilon_{xc}[n]$, which can be divided into separate exchange and correlation parts,

$$\epsilon_{\mathrm{xc}}[n] = \epsilon_{\mathrm{x}}[n] + \epsilon_{\mathrm{c}}[n]. \tag{1.215}$$

is known from Dirac's work [15],

$$\epsilon_{\mathrm{x}}[n] = -C_{\mathrm{x}}[n(\mathbf{r})]^{1/3}, \quad C_{\mathrm{x}} = \frac{3}{4}\left(\frac{3}{\pi}\right)^{1/3}, \tag{1.216}$$

or,

$$\epsilon_{\mathrm{x}}[n] = -\frac{0.5582}{r_s}, \tag{1.217}$$

where r_s is defined in Eq. (1.35). The coefficient 0.5582 can be compared with the value of 0.916 derived for the case of plane waves in Eq. (1.151). For a concise discussion of $\epsilon_{\rm x}[n]$ the reader is referred to Parr and Yang [3].

Neglecting correlation effects, Slater [17] had suggested the use of the so-called Xα-method and an exchange contribution of the form

$$v_{{\rm X}\alpha}(\mathbf{r}) = -\frac{3}{2}\alpha\left\{\frac{3}{\pi}n(\mathbf{r})\right\}^{1/3}. \tag{1.218}$$

The spin orbitals are now determined by the equations

$$\left[-\frac{1}{2}\nabla^2 + v(\mathbf{r}) + \int\frac{n(\mathbf{r}')}{|\mathbf{r}-\mathbf{r}'|}{\rm d}^3r' + v_{{\rm X}\alpha}(\mathbf{r})\right]\phi_j(\mathbf{r}) = \epsilon_j\phi_j, \tag{1.219}$$

where α has come to be viewed [3] as an adjustable parameter. For atoms and molecules, a value of $\alpha \simeq 0.75$ often seems to yield reasonable results.

The determination of the correlation contribution, $\epsilon_{\rm c}$, in Eq. (1.215) is a more complicated task. Values of $\epsilon_{\rm c}$ have been suggested by a number of authors, and we will quote the most commonly used ones. In the following expressions, $\epsilon_{\rm c}^0$ and $\epsilon_{\rm c}^1$ refer, respectively, to the correlation energy of a spin compensated (no net spin polarization) and to a polarized uniform electron gas. For derivations of the various results and the values of parameters that appear in the following discussion, the reader is referred to the original papers.

Using a random-phase analysis, von Barth and Hedin [37] suggested the form

$$\epsilon_{\rm c}^{\rm BH}[n,\zeta] = \epsilon_{\rm c}^0(r_s) + \left[\epsilon_{\rm c}^1(r_s) - \epsilon_{\rm c}^0(r_s)\right]f(\zeta), \tag{1.220}$$

where

$$f(\zeta) = \frac{1}{2}(2^{2/3}-1)^{-1}\{(1+\zeta)^{4/3} + (1-\zeta)^{4/3} - 2\}. \tag{1.221}$$

Here, ζ is the relative magnetization, $\zeta = (n_+ - n_-)/n$, where $n_\pm$ denotes the densities of electrons with up (+) and down (-) spins. They also suggested the use of analytic expressions for $\epsilon_{\rm c}^0$ and $\epsilon_{\rm c}^1$ proposed earlier by Hedin and Lundqvist [38]

$$\epsilon_{\rm c}^{0,{\rm HL}} = -\frac{1}{2}c_0F\left(\frac{r_s}{r_0}\right) \quad \epsilon_{\rm c}^{1,{\rm HL}} = -\frac{1}{2}c_1F\left(\frac{r_s}{r_1}\right), \tag{1.222}$$

where

$$F(Z) = (1+Z^3)\ln\left(1+\frac{1}{Z}\right) + \frac{Z}{2} - Z^3 - \frac{1}{3} \tag{1.223}$$

and

$$c_0 = 0.0504, \quad c_1 = 0.0254, \quad r_0 = 30, \quad \text{and} \quad r_1 = 75. \tag{1.224}$$

The factor of $\frac{1}{2}$ which appears in Eq. (1.222) is due to the use of atomic units, in contrast to the Rydberg units used in the original papers. One problem with Eqs. (1.220) and (1.222) is that they fail to reproduce the exact high- and low-density limits for the uniform electron gas [22]. Nevertheless, $\epsilon^{0,\mathrm{LH}}$ has been used extensively in the performance of electronic structure calculations of solids in the past, and is also of wide use currently.

The most accurate values of the correlation energy of the uniform electron gas, including both ϵ_c^0 and ϵ_c^1, have been obtained through the work of Ceperley and Alder [39]. These authors calculated the total energy for the uniform electron gas in spin-compensated and ferromagnetic (spin-polarized) states for various values of r_s using the quantum Monte Carlo method. From this, they subtracted the kinetic and exchange parts to obtain the correlation energy. Based on these results, Vosko, Wilk, and Nusair [40] used a Pade-approximant interpolation scheme to obtain an analytic expression for ϵ_c that both ϵ_c^0 and ϵ_c^1,

$$\epsilon_c(r_s) = \frac{A}{2}\left\{\ln\frac{x}{X(x)} + \frac{2b}{Q}\arctan\frac{Q}{2x+b} - \frac{bx_0}{X(x_0)}\right.$$
$$\times \left.\left[\ln\frac{(x-x_0)^2}{X(x)} + \frac{2(b+2x_0)}{Q}\arctan\frac{Q}{2x+b}\right]\right\}. \quad (1.225)$$

Here, $x = r_s^{1/2}$, $X(x) = x^2 + bx + c$, and $Q = (4c - b^2)^{1/2}$. For ϵ_c^0, the spin-compensated case, $A = 0.0621814$, $x_0 = -0.409286$, $b = 13.0720$, and $c = 42.7198$. For ϵ_c^1, the ferromagnetic case, $A = \frac{1}{2}(0.0621814)$, $x_0 = -0.743294$, $b = 20.1231$, and $c = 101.578$. These expressions do conform to both the high- and low-density limits, and are commonly accepted as the most accurate values available for the correlation energy of a uniform electron gas.

We close this discussion by quoting an expression for the total energy within the LDA. In the local approximation, Eq. (1.137) takes the form

$$E = T + U + E_{\mathrm{xc}} \quad (1.226)$$

where

$$T = \sum_i \int \phi_i^*(\mathbf{r})[-\frac{1}{2}\nabla^2]\phi_i(\mathbf{r})\mathrm{d}^3r \quad (1.227)$$

is the kinetic energy, and

$$U(\mathbf{r}) = -\sum_\alpha \frac{Z_\alpha n(\mathbf{r})}{|\mathbf{r}-\mathbf{R}_\alpha|} + \frac{1}{2}\int\frac{n(\mathbf{r})n(\mathbf{r}')}{|\mathbf{r}-\mathbf{r}'|}\mathrm{d}^3r\mathrm{d}^3r' + \sum_{\alpha<\beta}\frac{Z_\alpha Z_\beta}{|\mathbf{R}_\alpha - \mathbf{R}_\beta|} \quad (1.228)$$

is the Coulomb potential energy. Using the expression for $n(\mathbf{r})$ given by

Eq. (1.130), we can also write

$$\begin{aligned} E &= \sum_i^N \int \phi_i^*(\mathbf{r})[-\frac{1}{2}\nabla^2 + \sum_\alpha \frac{Z_\alpha}{|\mathbf{r}-\mathbf{R}_\alpha|}]\phi_i(\mathbf{r})\mathrm{d}^3r \\ &+ \frac{1}{2}\int \frac{n(\mathbf{r})n(\mathbf{r}')}{|\mathbf{r}-\mathbf{r}'|}\mathrm{d}^3r\mathrm{d}^3r' + \sum_{\alpha<\beta} \frac{Z_\alpha Z_\beta}{|\mathbf{R}_\alpha - \mathbf{R}_\beta|} + E_{\mathrm{xc}}. \end{aligned} \tag{1.229}$$

It is often convenient also to express E_{xc} as an integral over the electron density

$$E_{\mathrm{xc}} = \int n(\mathbf{r})\epsilon_{\mathrm{xc}}(n(\mathbf{r}))\mathrm{d}^3r. \tag{1.230}$$

In this case, we have (suppressing the superscript LDA from now on)

$$v_{\mathrm{xc}} = \frac{\mathrm{d}}{\mathrm{d}n}\left[n\epsilon_{\mathrm{xc}}[n]\right]|_{n=n(\mathbf{r})}. \tag{1.231}$$

Using Eq. (1.210), Eq. (1.229) assumes the form of Eq. (1.211) but with the exchange and correlation part calculated in the LDA.

1.8 Spin-Polarized DFT

The discussion of DFT within the LDA given above can be extended to treat spin-polarized systems and determine the magnetization density, $\mathrm{m}(\mathbf{r})$, as a ground-state variable. Such a treatment may be required in the study of (most) atoms and of ferromagnetic solids, such as Ni, Co, and Fe, that exhibit a net magnetic moment.

The formulation of DFT in the presence of spin proceeds along lines analogous to those followed in our discussion of nonmagnetic systems. In addition to the external potential, $\nu(\mathbf{r})$, we introduce an external magnetic field, $\mathbf{H}(0, 0, H([\mathbf{r}]))$, which we take to point along the z-direction. The Hamiltonian of the system now takes the form[29],

$$H = T + U + \int \mathrm{d}^3r v(\mathbf{r})n(\mathbf{r}) + \mu_{\mathrm{B}} \int \mathrm{d}^3r \mathbf{H}(\mathbf{r})m(\mathbf{r}), \tag{1.232}$$

where μ_{B} is the Bohr magneton, the density $n(\mathbf{r}) = n_+(\mathbf{r}) + n_-(\mathbf{r})$ is the sum of the partial densities, $n_+(\mathbf{r})$ and $n_-(\mathbf{r})$, associated with the up (+) and down (-) electron spins in the system, and $m(\mathbf{r}) = n_+(\mathbf{r}) - n_-(\mathbf{r})$ is

[29]The expression for the Hamiltonian is designed to emphasize the presence of spin and neglects the orbital contributions to the magnetic moment. Thus, it would be inappropriate to use in the study of most atoms and for many solids.

the magnetization density. The partial spin densities are given explicitly by the expressions,

$$n_{\pm}(\mathbf{r}) = \langle\Psi|\sum_i \delta(\mathbf{r}-\mathbf{r}_i)\delta_{\pm,\sigma_i^z}|\Psi\rangle, \tag{1.233}$$

where σ_i^z is the z-component of the spin operator for the ith electron. The form of the Hamiltonian in Eq. (1.232) allows us to consider both the particle density and the magnetization density on an equal footing as ground-state variables. In analogy with Eq. (1.107), we obtain the generalization of the energy functional, E[n(**r**)],

$$\begin{aligned} E[n(\mathbf{r}), m(\mathbf{r})] &= \min_{\Psi} \frac{\langle\Psi|H|\Psi\rangle}{\langle\Psi|\Psi\rangle} \\ &= \min_{\Psi} E[\Psi], \end{aligned} \tag{1.234}$$

for all wave functions that satisfy the condition,

$$n_{\pm}(\mathbf{r}) = \frac{\langle\Psi|\sum_i \delta(\mathbf{r}-\mathbf{r}_i)\delta_{\pm,\sigma_i^z}|\Psi\rangle}{\langle\Psi|\Psi\rangle}. \tag{1.235}$$

We now have,

$$E[n_+(\mathbf{r}), n_-(\mathbf{r})] \geq E[n_+^0(\mathbf{r}), n_-^0(\mathbf{r})] = E_0, \tag{1.236}$$

where $n_{\pm}^0(\mathbf{r})$ denotes the spin densities in the ground state.

Analogous with the paramagnetic case, we define single-particle wave functions, $\phi_{i\pm}(\mathbf{r})$, in terms of which we can write

$$n_{\pm}(\mathbf{r}) = \sum_i |\phi_{i\pm}(\mathbf{r})|^2. \tag{1.237}$$

The corresponding Kohn-Sham equations now take the form,

$$\begin{aligned} \left[-\frac{1}{2}\nabla^2 + v_{\text{eff}}(\mathbf{r})\right]\phi_j(\mathbf{r}) &= \left[\frac{1}{2}\nabla^2 + v(\mathbf{r})\right. \\ &+ \left.\int \frac{n(\mathbf{r}')}{|\mathbf{r}-\mathbf{r}'|}\mathrm{d}^3r' + v_{\text{xc}}^{\pm}(\mathbf{r}) \pm \mu_{\text{B}}H(\mathbf{r})\right]\phi_j(\mathbf{r}) \\ &= \epsilon_j\phi_j(\mathbf{r}), \end{aligned} \tag{1.238}$$

which should be compared to Eq. (1.205) obtained in the absence of spin polarization. We note the term proportional to the external magnetic field and that now $v_{\text{xc}}^{\pm}$ depends on spin,

$$v_{\text{xc}}^{\pm}(\mathbf{r}) = \frac{\delta E_{\text{xc}}(n_+, n_-)}{\delta n_{\pm}}. \tag{1.239}$$

The exchange interaction, $v_{xc}^{+} - v_{xc}^{-}$, is responsible for the onset of magnetism in the system since it allows electrons to gain energy by aligning their spins. As in the paramagnetic case treated within the LDA, here we can also approximate the exchange energy through a 1/3 law in analogy with Eq. (1.216),

$$v_x^{\pm}(\mathbf{r}) = -\left(\frac{6}{\pi} n_{\pm}(\mathbf{r})\right)^{\frac{1}{3}}. \tag{1.240}$$

This use of the local spin-density is referred to as the *local spin-density approximation* (LSDA). More accurate approximation schemes can be employed through consideration of correlation effects in a spin-polarized electron gas [41].

1.8.1 Exchange hole

We now consider the effect of spin polarization on the spatial distribution of electrons. In the presence of spin, the Hartree-Fock equations have the form

$$\begin{aligned}\left(-\frac{1}{2}\nabla^2 + v(\mathbf{r}) + \int d^3r' \frac{1}{|\mathbf{r}-\mathbf{r}'|} \sum_{ns'} |\phi_{ns'}(\mathbf{r}')|^2\right) \phi_{ms}(\mathbf{r}) \\ - \int d^3r' \frac{1}{|\mathbf{r}-\mathbf{r}'|} \sum_{ns'} \phi_{ns'}^{*}(\mathbf{r}')\phi_{ms}(\mathbf{r}')\phi_{ns'}(\mathbf{r}) = \epsilon_{ms}\phi_{ms}(\mathbf{r}).\end{aligned} \tag{1.241}$$

The Hartree and exchange terms can be written formally as

$$\int d^3r' \frac{1}{|\mathbf{r}-\mathbf{r}'|} \left[n(\mathbf{r}') - n_x^{ms}(\mathbf{r},\mathbf{r}')\right] \phi_{ms}(\mathbf{r}). \tag{1.242}$$

The second term in the last equation is the so-called *exchange density*,

$$n_x^{ms}(\mathbf{r},\mathbf{r}') = \frac{\sum_n \phi_{ns}^{*}(\mathbf{r}')\phi_{ms}(\mathbf{r}')\phi_{ns}(\mathbf{r})\phi_{ms}^{*}(\mathbf{r})}{\phi_{ms}(\mathbf{r})\phi_{ms}^{*}(\mathbf{r})}. \tag{1.243}$$

For $\mathbf{r} = \mathbf{r}'$, we have

$$n_x^{ms}(\mathbf{r},\mathbf{r}) \equiv n_s(\mathbf{r}), \tag{1.244}$$

and, in the absence of polarization $n_s(\mathbf{r}) = \frac{1}{2}n(\mathbf{r})$. Also,

$$\int d^3r' n_x^{ms}(\mathbf{r},\mathbf{r}') = 1, \tag{1.245}$$

which implies that in the neighborhood of any one electron there is an absence of exactly one electron. Equivalently, we say that an electron is

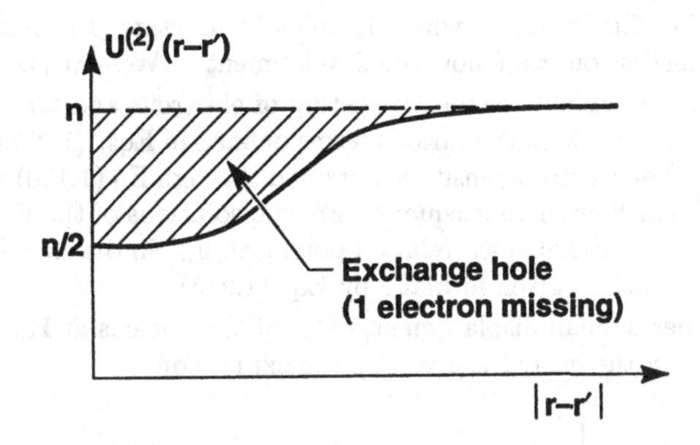

Figure 1.4: **Schematic representation of the Fermi hole.**

surrounded at short distances by an *exchange hole* or *Fermi hole.* In fact, the pair correlation function,

$$g(\mathbf{r},\mathbf{r}') = n(\mathbf{r}) - n_x(\mathbf{r},\mathbf{r}'), \tag{1.246}$$

is normalized to $N-1$, which also implies the absence of one electron. For homogeneous, the pair correlation function, $U^{(}2)$, which now depends only on the difference $\mathbf{r}-\mathbf{r}'$, has the form depicted in Fig. 1.4.

1.9 Force and Virial Theorems, and Pressure

As was done above for a system of free electrons and the Thomas-Fermi model, we examine the forms of the Hellmann-Feynman and virial theorems within the local density approximation to density-functional theory. We also derive a general expression for the pressure within the LDA. The expressions derived below will be useful when we discuss the self-consistent calculation of electronic structure and its role in determining materials properties[30].

[30]Material designated by an asterisk could be skipped on a first reading.

1.9.1 Hellmann-Feynman theorem in the LDA (*)

The electrostatic theorem is satisfied in any exact treatment of particles interacting via Coulomb forces. It also holds in some approximate treatments of Coulomb systems, such as the Thomas-Fermi model, as we saw above. In this section, we examine the form of the theorem within the LDA, and discuss the conditions under which it holds in its standard form. In this part of the discussion, we follow the development in Weinert [42].

The total energy of a system consisting of electrons and nuclei has the form of Eq. (1.226), with the various terms defined in Eqs. (1.227), (1.228), and (1.230). The electron density is obtained through Eq.(1.130) where the orbitals, ϕ_i, which enter that equation are the solutions of the Kohn-Sham Eq. (1.214). The exchange-correlation potential, v_{xc}, in the effective Kohn-Sham Hamiltonian is given formally by Eq. (1.213).

Now, under a small displacement, $\delta\mathbf{R}_\nu$, of the nucleus at $\mathbf{R}_\nu$, the first-order change in the energy is given by the expression

$$\begin{aligned}
\delta E &= \delta\mathbf{R}_\nu \cdot \left[-Z_\nu \frac{\partial}{\partial\mathbf{R}_\nu} \int \frac{n(\mathbf{r})}{|\mathbf{r}-\mathbf{R}_\nu|} + Z_\nu \frac{\partial}{\partial\mathbf{R}_\nu} \sum_{\alpha\neq\nu} \frac{Z_\alpha}{|\mathbf{R}_\alpha - \mathbf{R}_\nu|} \right] \\
&+ \delta\mathbf{R}_\nu \sum_i \int \mathrm{d}^3 r \left[\left(\frac{\partial}{\partial\mathbf{R}_\nu} \phi_i^*(\mathbf{r}) \right) \left(-\frac{1}{2}\nabla^2 \right) \phi_i(\mathbf{r}) \right. \\
&+ \left. \phi_i^*(\mathbf{r}) \left(-\frac{1}{2}\nabla^2 \right) \left(\frac{\partial}{\partial\mathbf{R}_\nu} \phi_i(\mathbf{r}) \right) \right] \\
&+ \delta\mathbf{R}_\nu \cdot \int \mathrm{d}^3 r \left(\frac{\partial}{\partial\mathbf{R}_\nu} n(\mathbf{r}) \right) \left[\int \mathrm{d}^3 r \frac{n(\mathbf{r})}{|\mathbf{r}-\mathbf{r}'|} \right. \\
&- \left. \sum_\alpha \frac{Z_\alpha}{|\mathbf{r}-\mathbf{R}_\alpha|} + v_{\mathrm{xc}} \right] . \qquad (1.247)
\end{aligned}$$

From Eq. (1.130) we can also write the change in the electron density brought about by the displacement of the nucleus at $\mathbf{R}_\nu$ in terms of the wave function ϕ_i:

$$\begin{aligned}
\frac{\partial}{\partial\mathbf{R}_\nu} n(\mathbf{r}) &= \sum_i \frac{\partial}{\partial\mathbf{R}_\nu} \phi_i^*(\mathbf{r}) \phi_i(\mathbf{r}) \\
&= \sum_i \left[\left(\frac{\partial}{\partial\mathbf{R}_\nu} \phi_i^*(\mathbf{r}) \right) \phi_i(\mathbf{r}) + \phi_i^*(\mathbf{r}) \frac{\partial}{\partial\mathbf{R}_\nu} \phi_i(\mathbf{r}) \right] .
\end{aligned} \qquad (1.248)$$

Substituting this result into Eq. (1.247) and using the definition of the single-particle Hamiltonian, Eq. (1.204), we can write

$$\begin{aligned}\delta E &= Z_\nu \delta \mathbf{R}_\nu \cdot \frac{\partial}{\partial \mathbf{R}_\nu}\left[\sum_{\alpha\neq\nu}\frac{Z_\alpha}{|\mathbf{R}_\alpha - \mathbf{R}|_\nu} - \int \mathrm{d}^3 r \frac{n(\mathbf{r})}{|\mathbf{r}-\mathbf{R}_\nu|}\right] \\ &+ \delta\mathbf{R}_\nu \cdot \sum_i \int \mathrm{d}^3 r \left[\left(\frac{\partial \phi_i^*(\mathbf{r})}{\partial \mathbf{R}_\nu}\hat{H}\phi_i(\mathbf{r})\right) + \phi_i^*(\mathbf{r})\hat{H}\frac{\partial\phi_i(\mathbf{r})}{\partial \mathbf{R}_\nu}\right].\end{aligned} \tag{1.249}$$

We now examine the last term in Eq. (1.249). Adding and subtracting the term $\left(\frac{\partial \phi_i(\mathbf{r})}{\partial \mathbf{R}_\nu}\hat{H}\phi_i^*(\mathbf{r})\right)$, and noting that all terms other than those in ∇^2 vanish, we can write,

$$\begin{aligned} I &= \int \mathrm{d}^3 r \left[\left(\frac{\partial \phi_i^*(\mathbf{r})}{\partial \mathbf{R}_\nu}\hat{H}\phi_i(\mathbf{r})\right) + \phi_i^*(\mathbf{r})\hat{H}\frac{\partial\phi_i(\mathbf{r})}{\partial \mathbf{R}_\nu}\right] \\ &= \int \mathrm{d}^3 r \left[\left(\frac{\partial \phi_i^*(\mathbf{r})}{\partial \mathbf{R}_\nu}\hat{H}\phi_i(\mathbf{r})\right) + \frac{\partial\phi_i(\mathbf{r})}{\partial \mathbf{R}_\nu}\hat{H}\phi_i^*(\mathbf{r})\right] \\ &+ \int \mathrm{d}^3 r \left[\phi_i^*(\mathbf{r})(-\frac{1}{2}\nabla^2)\frac{\partial\phi_i(\mathbf{r})}{\partial \mathbf{R}_\nu} \right.\\ &- \left.\frac{\partial\phi_i(\mathbf{r})}{\partial \mathbf{R}_\nu}(-\frac{1}{2}\nabla^2)\phi_i^*(\mathbf{r})\right]\end{aligned} \tag{1.250}$$

Using Green's second theorem, the last integral in this expression can be converted to a surface integral, with the surface of integration at infinity. Assuming that the wave functions, ϕ_i, are associated with bound electrons, and thus vanish at infinity, we can set the surface integrals equal to zero. Further, adding and subtracting terms proportional to the eigenvalue, ϵ_i, we obtain

$$\begin{aligned} I &= \int \mathrm{d}^3 r \left[\frac{\partial \phi_i^*(\mathbf{r})}{\partial \mathbf{R}_\nu}\left(\hat{H}-\epsilon_i\right)\phi_i(\mathbf{r}) + \frac{\partial\phi_i(\mathbf{r})}{\partial \mathbf{R}_\nu}\left(\hat{H}-\epsilon_i\right)\phi_i^*(\mathbf{r})\right] \\ &+ \epsilon_i \frac{\partial}{\partial \mathbf{R}_\nu}\int \mathrm{d}^3 \phi_i^*(\mathbf{r})\phi_i(\mathbf{r}).\end{aligned} \tag{1.251}$$

The last term, the derivative of a constant (the number of electrons), vanishes and the remaining term becomes

$$I = 2\Re \int \mathrm{d}^3 r \frac{\partial \phi_i^*(\mathbf{r})}{\partial \mathbf{R}_\nu}\left(\hat{H}-\epsilon_i\right)\phi_i(\mathbf{r}). \tag{1.252}$$

Hence, the change in the total energy is given to first order in the nuclear displacement by the expression

$$\delta E = Z_\nu \delta \mathbf{R}_\nu \cdot \frac{\partial}{\partial \mathbf{R}_\nu}\left[\sum_{\alpha\neq\nu}\frac{Z_\alpha}{|\mathbf{R}_\alpha - \mathbf{R}_\nu|} - \int \mathrm{d}^3 r \frac{n(\mathbf{r})}{|\mathbf{r}-\mathbf{R}_\nu|}\right]$$

$$+ \quad \delta\mathbf{R}_\nu \cdot 2\sum_i \Re\langle\frac{\partial\phi_i^*}{\partial\mathbf{R}_\nu}\left(\hat{H} - \epsilon_i\right)\phi_i\rangle. \tag{1.253}$$

So that the force on the νth nucleus takes the form

$$\begin{aligned}\mathbf{F}_\nu &= -\frac{\delta E}{\delta\mathbf{R}_\nu} \\ &= -Z_\nu\delta\mathbf{R}_\nu \cdot \frac{\partial}{\partial\mathbf{R}_\nu}\left[\sum_{\alpha\neq\nu}\frac{Z_\alpha}{|\mathbf{R}_\alpha - \mathbf{R}_\nu|} - \int \mathrm{d}^3r\frac{n(\mathbf{r})}{|\mathbf{r} - \mathbf{R}_\nu|}\right] \\ &+ \delta\mathbf{R}_\nu \cdot 2\sum_i \Re\langle\frac{\partial\phi_i^*}{\partial\mathbf{R}_\nu}\left(\hat{H} - \epsilon_i\right)\phi_i\rangle.\end{aligned} \tag{1.254}$$

The first term gives just the standard form of the Hellmann-Feynman theorem

$$\begin{aligned}\mathbf{F}_\nu^{\mathrm{HF}} &= -Z_\nu\delta\mathbf{R}_\nu \cdot \frac{\partial}{\partial\mathbf{R}_\nu}\left[\sum_{\alpha\neq\nu}\frac{Z_\alpha}{|\mathbf{R}_\alpha - \mathbf{R}_\nu|} - \int \mathrm{d}^3r\frac{n(\mathbf{r})}{|\mathbf{r} - \mathbf{R}_\nu|}\right] \\ &= -Z_\nu\left[\sum_{\alpha\neq\nu}\frac{Z_\alpha(\mathbf{R}_\alpha - \mathbf{R}_\nu)}{|\mathbf{R}_\alpha - \mathbf{R}_\nu|} - \int \mathrm{d}^3r\frac{n(\mathbf{r})(\mathbf{r} - \mathbf{R}_\nu)}{|\mathbf{r} - \mathbf{R}_\nu|^3}\right] \\ &= Z_\nu\nabla_{\mathbf{r}}U_{\mathrm{ext}}(\mathbf{r})|_{\mathbf{r}=\mathbf{R}_\nu},\end{aligned} \tag{1.255}$$

where

$$U(\mathbf{R})_\nu = -\left[\sum_{\alpha\neq\nu}\frac{Z_\alpha}{|\mathbf{R}_\mathbf{r}|} - \int \mathrm{d}^3r'\frac{n(\mathbf{r})}{|\mathbf{r}' - \mathbf{r}|}\right]|_{\mathbf{r}=\mathbf{R}_\nu} \tag{1.256}$$

is the so-called *Madelung potential.* The second term in Eq. (1.254) is a correction due to the possible presence of ϵ_i which are not the exact eigenvalues of $\hat{H}$, as may be the case when an incomplete basis is used to represent the wave function. Such effects are considered in more detail in a following chapter. Thus, provided that the correct LDA eigenvalues are used, the Hellmann-Feynman theorem assumes its standard form, Eq. (1.255), and has the same interpretation: *the force on a nucleus is given as the derivative of the classical electrostatic potential at the nucleus.*

1.9.2 Virial theorem in the LDA (*)

Next, we derive an expression for the virial theorem within the LDA. We start by applying the operator $\mathbf{r}\cdot\nabla$ on the left of Eq. (1.214) and multiplying subsequently by ϕ_i^* to obtain

$$\begin{aligned} - & \phi_i^*(\mathbf{r})(\mathbf{r}\cdot\nabla)\left[-\frac{1}{2}\nabla^2\phi_i(\mathbf{r})\right] + \phi_i^*(\mathbf{r})\left[v(\mathbf{r}) + v_{\mathrm{xc}}(\mathbf{r})\right](\mathbf{r}\cdot\nabla)\phi_i(\mathbf{r}) \\ + & \phi_i^*(\mathbf{r})(\mathbf{r}\cdot\nabla)\left[v(\mathbf{r}) + v_{\mathrm{xc}}(\mathbf{r})\right]\phi_i(\mathbf{r}) = \epsilon_i\phi_i^*(\mathbf{r})(\mathbf{r}\cdot\nabla\phi(\mathbf{r})). \end{aligned} \tag{1.257}$$

We now take the complex conjugate of Eq. (1.214), apply $\mathbf{r}\cdot\nabla$ on the right, multiply by ϕ_i and subtract the resulting equation from Eq. (1.257), and obtain the expression

$$\begin{aligned} \left[\frac{1}{2}\nabla^2\phi_i^*(\mathbf{r})\right](\mathbf{r}\cdot\nabla)\phi_i(\mathbf{r}) & - \phi_i^*(\mathbf{r})(\mathbf{r}\cdot\nabla)\left[\frac{1}{2}\nabla^2\phi_i(\mathbf{r})\right] \\ & + |\phi(\mathbf{r})|^2(\mathbf{r}\cdot\nabla)\left[v(\mathbf{r}) + v_{\mathrm{xc}}(\mathbf{r})\right] = 0. \end{aligned} \tag{1.258}$$

The first line of this equation can be simplified by means of the identity

$$\begin{aligned} \nabla\cdot\mathbf{A} &= \nabla\cdot\left[\phi_i^*(\mathbf{r})(\mathbf{r}\cdot\nabla)(\nabla\phi_i(\mathbf{r})) - (\nabla\phi_i^*(\mathbf{r}))(\mathbf{r}\cdot\nabla)\phi_i(\mathbf{r})\right] \\ &= \phi_i^*(\mathbf{r})(\mathbf{r}\cdot\nabla)\nabla^2\phi_i(\mathbf{r}) - \nabla^2\phi_i^*(\mathbf{r})(\mathbf{r}\cdot\nabla)\phi_i(\mathbf{r}) \\ &+ \phi_i^*(\mathbf{r})\nabla^2\phi_i(\mathbf{r}) - \nabla\phi_i^*(\mathbf{r})\cdot\nabla\phi_i(\mathbf{r}) \end{aligned} \tag{1.259}$$

where the term $[\nabla\phi_i^*(\mathbf{r})](\mathbf{r}\cdot\nabla)\nabla\phi_i(\mathbf{r})$ cancels out. This expression can be substituted into Eq. (1.258), and the result summed over all i and integrated over all space. By use of the divergence theorem, the term designated as $\nabla\cdot\mathbf{A}$ in the last equation leads to

$$\int_V \nabla\cdot\mathbf{A}\mathrm{d}^3r = \int_S \mathbf{A}\cdot\mathbf{n}\mathrm{d}S. \tag{1.260}$$

Also, by use of Green's first identity, we have

$$\int_V \nabla\phi_i^*(\mathbf{r})\cdot\nabla\phi_i(\mathbf{r}) = -\int_V \phi_i^*(\mathbf{r})\nabla^2\phi_i(\mathbf{r})\mathrm{d}^3r + \int_S \phi_i^*(\mathbf{r})\frac{\partial\phi_i(\mathbf{r})}{\partial n}\mathrm{d}S. \tag{1.261}$$

In all surface integrals, the surfaces of integration can be chosen at infinity. If the wave functions are those of bound states, then they vanish at sufficiently large distances and all surface integrals vanish accordingly[31]. Then, in view of Eq. (1.130), and of the definition of the kinetic energy, Eq. (1.195), the remaining volume integrals yield the expression [compare with the classical expression in Eq. (1.19)],

$$\begin{aligned} 2T &= \sum_i \int \phi_i^*(\mathbf{r})(-\nabla^2)\phi_i(\mathbf{r}) \\ &= \int n(\mathbf{r})(\mathbf{r}\cdot\nabla)\left[v_c(\mathbf{r}) + v_{\mathrm{xc}}(\mathbf{r})\right]\mathrm{d}^3r. \end{aligned} \tag{1.262}$$

[31]Note that the scaling arguments used in previous sections do not require the imposition of conditions associated with the behavior of the wave function at large distances and are thus of more general character.

Because the quantity $v_c + v_{xc}$ is the potential felt by an electron, the integral in the last expression can be interpreted as the average of the virial. Thus, the last expression is within a minus sign of the usual expression of the virial theorem. In fact, Slater [7] has shown that in connection with the $X\alpha$-method, and for systems containing both electrons and nuclei (molecules, solids), this equation readily yields the standard form of the virial, Eq. (1.33). In order to derive the form of the theorem within the LDA, we note that from the Hellmann-Feynman theorem, we have

$$\begin{aligned}\nabla_{\mathbf{R}} E &= \nabla_{\mathbf{R}}(U_{ee} + U_{eN}) \\ &= Z \int d^3r' \nabla' \frac{1}{|\mathbf{r} - \mathbf{r}'|} + Z^2 \sum_{\mathbf{R}' \neq \mathbf{R}} \frac{\mathbf{R}' - \mathbf{R}}{|\mathbf{R}' - \mathbf{R}|^3}. \end{aligned} \quad (1.263)$$

Thus, after some algebra we can show that

$$\sum_{\mathbf{R}} \mathbf{R} \cdot \nabla_{\mathbf{R}} E = -\int d^3r n(\mathbf{r})(\mathbf{r} \cdot \nabla) v_c(\mathbf{r}) - (U_{ee} + U_{eN}). \quad (1.264)$$

Finally, using Eq. (1.262), we arrive at the form of the virial theorem in the LDA,

$$2T = -(U_{ee} + U_{eN}) - \sum_{\mathbf{R}} \mathbf{R} \cdot \nabla_{\mathbf{R}} E + \int d^3r n(\mathbf{r})(\mathbf{r} \cdot \nabla) v_{xc}(\mathbf{r}). \quad (1.265)$$

This form clearly exhibits the contribution of the exchange and correlation part of the potential. (The reader is urged to fill in the steps between Eqs. (1.262) and (1.265) because they provide good practice in manipulating the various quantities involved.)

The last term in Eq. (1.265) can be simplified further. Consider the divergence theorem in terms of the vector $\mathbf{r}n(\mathbf{r})v_{xc}(\mathbf{r})$,

$$\int d^3r \nabla \cdot [\mathbf{r}n(\mathbf{r})v_{xc}(\mathbf{r})] = \int_S \mathbf{r}n(\mathbf{r})v_{xc}(\mathbf{r}) \cdot dS. \quad (1.266)$$

Taking the surface S sufficiently far away, where at least for bound systems, the density becomes negligible, we can set the surface integral equal to zero. We then have,

$$\begin{aligned}\int d^3r n(\mathbf{r})\mathbf{r} \cdot \nabla v_{xc}(\mathbf{r}) &= -3 \int d^3r n(\mathbf{r}) v_{xc}(\mathbf{r}) \\ &- \int d^3r v_{xc}(\mathbf{r}) \mathbf{r} \cdot \nabla n(\mathbf{r}) \\ &= -3 \int d^3r n(\mathbf{r}) v_{xc}(\mathbf{r}) \\ &- \int d^3r \left[\frac{\partial}{\partial n} n\epsilon_{xc}\right]_{n=n(\mathbf{r})}, \end{aligned} \quad (1.267)$$

where Eq. (1.231) has been used. We now note that,

$$\begin{aligned}\int\left[\frac{\partial}{\partial n}n\epsilon_{\rm xc}\right]\sum_i x_i\frac{\partial n}{\partial x_i}\Pi_i{\rm d}x_i &= \int\sum_i x_i\frac{\partial n\epsilon_{\rm xc}}{\partial x_i}\Pi_i{\rm d}x_i \\ &\equiv \int \mathbf{r}\cdot\nabla\left[n\epsilon_{\rm xc}\right]{\rm d}^3r. \end{aligned} \tag{1.268}$$

Using again the divergence theorem and setting the surface term equal to zero, leads to

$$\int {\rm d}^3rn(\mathbf{r})(\mathbf{r}\cdot\nabla)v_{\rm xc}(\mathbf{r}). = 3\int {\rm d}^3n(\mathbf{r})\left[\epsilon_{\rm xc}(\mathbf{r})-v_{\rm xc}(\mathbf{r})\right], \tag{1.269}$$

so that Eq. (1.265) can be cast in the form

$$2T = -(U_{\rm ee}+U_{\rm eN}) - \sum_{\mathbf{R}}\mathbf{R}\cdot\nabla_{\mathbf{R}}E + 3\int {\rm d}^3n(\mathbf{r})\left[\epsilon_{\rm xc}(\mathbf{r})-v_{\rm xc}(\mathbf{r})\right]. \tag{1.270}$$

We can now see that in the Xα method the last integral cancels out, so that we arrive at the result obtained by Slater [7]. Finally, we note that direct derivations of the Hellmann-Feynman theorem based on the wave function, as those just given, even though intuitively very appealing, are restricted to wave functions of bound states (that vanish at infinity). By contrast, the derivations based on scaling arguments do not carry that restriction.

1.9.3 Pressure within the LDA

An expression for the pressure within the LDA can be derived in a straightforward manner. We start with the thermodynamic relation

$$P = -\left(\frac{\partial E}{\partial V}\right)_S, \tag{1.271}$$

where S denotes the entropy of a system. Now, consider the scaling transformation,

$$\mathbf{R}\to\alpha\mathbf{R}^0. \tag{1.272}$$

We have,

$$\begin{aligned} 3PV &= -3V\left(\frac{\partial E}{\partial V}\right)_S = -\alpha\left(\frac{\partial E}{\partial\alpha}\right)_S \\ &= -\alpha\sum_{\mathbf{R}}\left(\frac{\partial E}{\partial\mathbf{R}}\right)\left(\frac{\partial\mathbf{R}}{\partial\alpha}\right)_S = -\sum_{\mathbf{R}}\mathbf{R}\cdot\nabla_{\mathbf{R}}E. \end{aligned} \tag{1.273}$$

So that from Eq. (1.270) we have

$$3PV = 2T + U - 3\int {\rm d}^3rn(\mathbf{r})\left[\epsilon_{\rm xc}(\mathbf{r})-v_{\rm xc}(\mathbf{r})\right]. \tag{1.274}$$

1.9.4 Summary of DFT and LDA

The basis of density-functional theory is Hohenberg and Kohn's proof that the ground-state energy of and the external potential acting on an electron gas are unique functionals of the electronic charge density, and that the energy functional assumes its minimum value for the correct ground-state density. Within the Kohn-Sham (KS) approach, one introduces single-particle functions (spin functions) $\phi(\mathbf{r})$ which are used to construct the density and expectation values of operators, in particular the kinetic-energy operator. There follows a short summary of the general procedure used in implementing the KS approach within the local density approximation (LDA).

For fixed nuclear positions, the most general form of the total energy of the system (including electrons and nuclei) can be written in the form,

$$E_{\mathrm{tot}} = E_{\mathrm{NN}} + E_{\mathrm{eN}} + E_{\mathrm{ee}} + T + E_{\mathrm{xc}}[n]. \tag{1.275}$$

Here,

$$E_{\mathrm{NN}} = \sum_{\alpha<\beta} \frac{Z_\alpha Z_\beta}{R_{\alpha\beta}} \tag{1.276}$$

is the direct interaction energy of the nuclei [denoted by U_{NN} in Eq. (1.99)],

$$E_{\mathrm{eN}} = \int n(\mathbf{r}) \sum_\alpha v(\mathbf{r} - \mathbf{R}_\alpha) \mathrm{d}^3 r \tag{1.277}$$

is the energy of interaction of the electrons with the nuclei, [denoted by U_{eN} in Eq. (1.97)],

$$E_{\mathrm{ee}} = \int \mathrm{d}^3 r \int \mathrm{d}^3 r' \frac{n(\mathbf{r}) n(\mathbf{r}')}{|\mathbf{r} - \mathbf{r}'|} \tag{1.278}$$

is the repulsive interaction of the electrons, [denoted by U_{ee} in Eq. (1.98)],

$$T = \sum_i \langle \phi_i | -\frac{1}{2} \nabla^2 | \phi_i \rangle \tag{1.279}$$

is the expectation value of the kinetic energy operator for the electrons, Eq. (1.195), and $E_{\mathrm{xc}}[n]$ is the exchange and correlation functional, Eq. (1.200). The sum of the nuclear and electron potentials acting on an electron,

$$V_{\mathrm{H}}(\mathbf{r}) = \int \frac{n(\mathbf{r}')}{|\mathbf{r} - \mathbf{r}'|} \mathrm{d}^3 r' + \sum_\alpha \frac{Z_\alpha}{|\mathbf{r} - \mathbf{R}_\alpha|}, \tag{1.280}$$

is also called the Hartree potential. The ground state electron density is given by the sum over occupied levels,

$$n(\mathbf{r}) = \sum_{i,s} |\phi_i(\mathbf{r}, s)|^2. \tag{1.281}$$

In the local density approximation (LDA) it is assumed that $E_{xc}[n]$ is a local function of the electron charge density,

$$E_{xc}[n] = \int n(\mathbf{r})\epsilon_{xc}[n(\mathbf{r})]\mathrm{d}^3r, \tag{1.282}$$

as in Eq. (1.212), where $\epsilon_{xc}[n(\mathbf{r})]$ is taken to be the exchange-correlation energy of a *uniform* electron gas of density $n(\mathbf{r})$. Within the LDA, the wave functions, $\phi_i(\mathbf{r})$, (with spin indices suppressed), are determined as the eigenstates of an effective, self-consistent, single-electron Hamiltonian,

$$H[n]\phi_i(\mathbf{r}) = \epsilon_i\phi_i(\mathbf{r}), \tag{1.283}$$

where the effective Hamiltonian is given by Eq. (1.214),

$$H = \left[-\frac{1}{2}\nabla^2 + v(\mathbf{r}) + \int \frac{n(\mathbf{r})}{|\mathbf{r}-\mathbf{r}'|}\mathrm{d}^3r' + v_{xc}^{\mathrm{LDA}}(\mathbf{r})\right], \tag{1.284}$$

with

$$v(\mathbf{r}) = \sum_\alpha \frac{Z_\alpha}{|\mathbf{r}-\mathbf{R}_\alpha|}, \tag{1.285}$$

being the nuclear potential at $\mathbf{r}$, and with

$$v_{xc}^{\mathrm{LDA}}(\mathbf{r}) = \frac{\delta E_{xc}^{\mathrm{LDA}}}{\delta n(\mathbf{r})} = \epsilon_{xc}(n(\mathbf{r})) + n(\mathbf{r})\frac{\mathrm{d}\epsilon_{xc}(n)}{\mathrm{d}n}\Big|_{n=n(\mathbf{r})} \tag{1.286}$$

being the exchange-correlation potential, Eq. (1.213).

In addition to the developments presented in this chapter, other approaches have been considered to the formulation of the exchange correlation potential. The study of the electron gas has a long history going back to the work of Wigner [43]. Attempts at improvement have been made by Perdew and Zunger [44] within a self-interaction correction to the density functionals for many electron systems. Exchange-correlation potentials for spin-polarized systems have been considered by U. von Barth and L. Hedin [37], and the scalar relativistic case has been considered by L. F. Mattheiss [45] and Coelling and Harmon [46].

Finally, we should keep in mind that DFT and the LDA make statements *only* about the density and the energy of the ground state. They are not meant to provide information about excited states or about subsidiary quantities which may be used in the calculation. Such a set of quantities are the spin orbitals introduced by Kohn and Sham[32]. Even though these wave functions are often interpreted as single-particle states of the interacting electron system and used (often successfully) in the construction of the

[32] For the possible use of DFT to study the excitation spectrum of the electron gas the reader is referred to the literature cited, e.g. reference [2].

band structure, they are not guaranteed to be accurate. Only their sum of squares, leading to the density, is of fundamental significance in the theory which can be held accountable for any lack of accuracy in it. Thus, the theory can be expected to yield the Fermi energy, but not necessarily the Fermi surface or the band structure of a solid.

Bibliography

[1] *Theory of the Inhomogeneous Electron Gas*, S. Lundqvist and N. H. March (eds.), Plenum, New York (1983).

[2] R. M. Dreitzler and E. K. U. Gross, *Density-functional theory*, Springer-Verlag, Berlin, Heidelberg, New York (1990).

[3] Robert G. Parr and Weitao Yang, *Density-Functional Theory of Atoms and Molecules*, Oxford, New York (1989).

[4] Herbert Goldstein *Classical Mechanics*, Addison Wesley, Reading, Massachusetts (1950).

[5] Siegfried Flügge, *Practical Quantum Mechanics*, Springer-Verlag, Berlin, (1971).

[6] J. C. Slater, *Quantum Theory of Molecules and Solids*, Vol. 1, Mc Graw-Hill, New York (1963), p. 29-34.

[7] J. C. Slater, J. Chem. Phys. **1**, 687 (1933).

[8] Alexander L. Fetter and John Dirk Walecka, *Quantum Theory of Many Particle Systems*, McGraw-Hill, New York (1971).

[9] L. H. Thomas, Proc. Camb. Phil. Soc. **23**, 542 (1926).

[10] E. Fermi, Z. Phys. **48**, 73 (1928).

[11] Harry J. Lipkin, *Quantum Mechanics*, North-Holland, (1988).

[12] E. Teller, Rev. Mod. Phys. **34**, 627 (1962).

[13] N. L. Balazs, Phys. Rev. **156**, 42 (1967).

[14] E. H. Lieb and B. Simon, Phys. Rev. Lett. **31**, 681 (1973); Adv. in Mat. **23**, 22 (1977).

[15] P. A. M. Dirac, Proc. Camb. Phil. Soc. **26**, 376 (1930).

[16] E. P. Wigner and F. Seitz, Phys. Rev. **43**, 804 (1933).

[17] J. C. Slater, Phys. Rev. **81**, 385 (1951); **82**, 538 (1951); **91**, 528 (1953).

[18] P. Gombas, *Die Statistische Theorie des Atoms und ihre Anwendungen* Springer-Verlag, Vienna (1949).

[19] John Avery, *Creation and Annihilation Operators*, Mc Graw Hill, New York (1976).

[20] V. Fock, Z. Phys. **61**, 126 (1930).

[21] V. Fock, Z. Phys. **2**, 795 (1930).

[22] M. Gell-Mann and K. Brueckner, Phys. Rev. **106**, 364 (1957). The first calculation of the coefficient of the logarithmic term was reparted by W. Macke, Z. Natur. **5a**, 192 (1950).

[23] Neil W. Ashcroft and N. David Mermin, *Solid State Physics*, Saunders College, Philadelphia (1976).

[24] T. Koopmans, Physica **1**,104 1934.

[25] Joseph Callaway *Quantum Theory of the Solid State*, Academic Press, Inc., New York (1974).

[26] R. P. Feynman, Phys. Rev. **56**, 340 (1939). The Hellmann-Feynman theorem was first reported in P. Göttinger, Z. Phys. **73**, 169 (1932).

[27] S. T. Epstein, A. C. Hurley, R. E. Wyatt, and R. G. Parr, J. Chem. Phys. **47**, 1275 (1967).

[28] *The Force Concept in Chemistry*, B. M. Deb (ed.), Van Nostrand Reinhold, New York (1981).

[29] E. Fermi, Z. Phys. **48**, 73 (1928).

[30] N. H. March, *Self-Consistent Fields in Atoms*, Oxford: Pergamon, New York (1975).

[31] P. Hohenberg and W. Kohn, Phys. Rev **136**, B864 (1964).

[32] R. O. Jones and O. Gunnarsson, *The Density Functional Formalism; Its Application and Prospects*, Rev. Mod. Phys. **61**, 689 (1989).

[33] W. Kohn and L. Sham, Phys. Rev. **140**, A1133 (1956).

[34] W. J. Carr and A. A. Maradudin, Phys. Rev. A**133**,371 (1964).

[35] W. J. Carr, Phys. Rev. **122**, 1437 (1961).

[36] P. Nozieres and D. Pines, *The Theory of Quantum Liquids*, Benjamin, New York (1966)

[37] V. von Barth and L. Hedin, J. Phys. C**5**, 1629 (1972).

[38] L. Hedin and B. I. Lundquist, , J. Phys. C**4**, 2064 (1971).

[39] D. M. Ceperley and B. J. Alder, Phys. Rev. Lett. **45**, 566 (1980).

[40] S. J. Vosko, L. Wilk and M. Nusair, Can. J. Phys. **58**, 1200 (1980).

[41] O. Gunnarsson and B. I. Lundqvist, Phys. Rev. B**13**, 4274 (1976).

[42] M. Weinert, private communication.

[43] E. Wigner, Trans. Faraday Soc. **34**, 678 (1938).

[44] J. P. Perdew and A. Zunger, Phys. Rev. B**23**, 5048 (1981).

[45] L. F. Mattheiss, Phys. Rev. **151**, 450 (1966).

[46] D. D. Koelling and B. N. Harmon, J. Phys. C**10**, 3107 (1977).

Chapter 2

Introduction to Green Functions

2.1 General Comments

So far we have developed methodology for treating the quantum many-body (many-electron) problem within the Born-Oppenheimer approximation and DFT within the formulation by Kohn and Sham and the LDA. We now have to deal with a characteristic feature of solids, namely the multicenter problem, so-called because of the presence of the force centers associated with the nuclei in the material. The methodology developed in this book to do this is based on the use of Green functions within the formalism of multiple-scattering theory (MST)[1]. In this chapter, we develop some basic notions about Green functions that will be sufficient for our immediate purposes, particularly with respect to MST, which is developed in the following chapter. A more detailed review of formal Green function theory is given in the last part of this book.

One way to think about Green functions is as the means of converting a (unit) stimulus into a response. For example, consider the solution of the Poisson equation, Eqs. (A.13) to (A.16). Equation (A.15) shows that the Green function is the functional derivative (somewhat loosely speaking, the rate of change) of the potential (the response) with respect to the charge

[1]Green functions can be used in connection with other formal methods for studying the electronic structure of solids such as that of the linear muffin-tin orbitals (LMTO's). In addition, alternative methodologies, not based on the Green function but the wave function, such as the augmented plane wave (APW) method and its refinements, the linearized APW (LAPW) and the full-potential linearized APW (FLAPW), and the pseudopotential method have been used quite successfully in the study of the properties of periodic solids.

density (the stimulus). In fact, Eq. (A.16) reveals explicitly that the Green function, $G(\mathbf{x}-\mathbf{x}')$, is the potential at $\mathbf{x}$ generated by a unit stimulus at $\mathbf{x}'$, represented by the delta function. Because of the connection they provide between stimulus and response, Green functions are a powerful tool in the solution of a great number of equations of mathematical physics.

In what follows, we first define Green functions within the context of vector spaces and linear operators. This discussion illustrates some very important features of Green functions, e.g., that the Green function directly embodies the boundary conditions placed on the solutions of a particular equation. We will also make contact with the application of Green functions in classical theories of ordinary and partial differential equations. The reader unfamiliar with the concepts of vector spaces and linear operators may profit by reading the material in Appendix B before proceeding further. Also, elementary treatments of Green functions can be found in a number of books on quantum mechanics and condensed matter physics [1-5].

2.2 Definition of Green Functions

Given an operator $\hat{L}$ with an inverse $\hat{L}^{-1}$ in a vector space V, we define the *Green function* or *Green operator* or *propagator* associated with $\hat{L}$ by the relation

$$\hat{G} \equiv \hat{L}^{-1}. \tag{2.1}$$

Thus, the solution of the equation,

$$\hat{L}|n\rangle = |m\rangle \tag{2.2}$$

can be obtained formally in terms of $\hat{G}$,

$$|n\rangle = \hat{G}|m\rangle. \tag{2.3}$$

It is to be emphasized that $\hat{G}$ exists only if the inverse operator $\hat{L}^{-1}$ exists. In particular, if $\hat{L}$ possesses eigenvalues that are equal to zero, $\hat{G}$ does not exist. As we see in the following discussion, it is in fact often necessary to consider inverses of singular operators. In such cases, the Green function must be defined in terms of a properly chosen limiting process.

The particular form of $\hat{G}$ depends on the basis chosen. In general, we have

$$\hat{G} = \sum_{\alpha\beta} |n\rangle G_{\alpha\beta} \langle\beta|, \tag{2.4}$$

with

$$G_{\alpha\beta} = \langle\alpha|\hat{G}|\beta\rangle, \tag{2.5}$$

where $G_{\alpha\beta}$ are the elements of the matrix representing $\hat{G}$. The expansion in Eq. (2.4) is called the *spectral representation* of the Green function. In a continuous basis, such as that afforded by the coordinate representation, Eq. (2.4) takes the form, (refer to Section 2.1 of Appendix B),

$$\begin{aligned} n(\mathbf{x}) &= \langle \mathbf{x}|n\rangle \\ &= \langle \mathbf{x}|\hat{G}|m\rangle \\ &= \int \mathrm{d}^3x' \langle \mathbf{x}|\hat{G}|\mathbf{x}'\rangle\langle \mathbf{x}'|m\rangle \\ &= \int \mathrm{d}^3x' G(\mathbf{x},\mathbf{x}')m(\mathbf{x}'), \end{aligned} \tag{2.6}$$

from which it follows that $G(\mathbf{x},\mathbf{x}')$ through its dependence on $\mathbf{x}$ satisfies the same boundary conditions as $n(\mathbf{x})$.

Let us establish the relation between the formal definition of the Green function given above in terms of operators in a vector space and the use of Green functions in connection with second-order partial differential equations. In the particular case of the Schrödinger equation,

$$\hat{H}\Psi = E\Psi, \tag{2.7}$$

it is customary to define the Green function as the inverse of the operator $(z - \hat{H})$, where z is a generally complex energy parameter. Thus, we have the formal relation,

$$(z - \hat{H})\hat{G} = \hat{I}. \tag{2.8}$$

Pre- and post-multiplying the last expression by the vectors $\langle \mathbf{x}|$ and $|\mathbf{x}'\rangle$, and using the property that for any function of coordinates (the position operator is usually denoted by $\hat{X}$)

$$\langle \mathbf{x}|F(\hat{X})|\mathbf{x}'\rangle = F(\mathbf{x})\delta(\mathbf{x}-\mathbf{x}'), \tag{2.9}$$

we obtain

$$\int \mathrm{d}^3x'' \langle \mathbf{x}| \left[z - \hat{H}(\mathbf{x}'')\right] |\mathbf{x}''\rangle\langle \mathbf{x}''|\hat{G}|\mathbf{x}'\rangle = \langle \mathbf{x}|\mathbf{x}'\rangle = \delta(\mathbf{x}-\mathbf{x}'), \tag{2.10}$$

or (in units of $\frac{\hbar^2}{2m} = 1$)

$$\int \mathrm{d}^3x'' \left[z + \nabla_{\mathbf{x}''} - v(\mathbf{x}'')\right] \delta(\mathbf{x}-\mathbf{x}'')G(\mathbf{x}'',\mathbf{x}) = \delta(\mathbf{x}-\mathbf{x}'). \tag{2.11}$$

From this follows the well-known equation for the Green function,

$$\left[z + \nabla_{\mathbf{x}} - v(\mathbf{x})\right] G(\mathbf{x},\mathbf{x}') = \delta(\mathbf{x}-\mathbf{x}'). \tag{2.12}$$

We note the formal similarity between the last expression and Eq. (A.16). In general, if $\hat{L}$ is a second order partial differential operator and $\hat{G}$ is the associated Green function, we have

$$L(\mathbf{x})G(\mathbf{x},\mathbf{x}') = \delta(\mathbf{x}-\mathbf{x}'). \tag{2.13}$$

We should also note the use of the generally complex energy parameter, z, in the equation for the Green function, Eq. (2.12). Since $\hat{H}$ is Hermitian its eigenvalues are real and $(z - \hat{H})$ is nonzero everywhere away from the real axis. It is in fact an analytic function of z everywhere on the upper or lower complex plane. On the real axis, however, $(z - \hat{H})$ vanishes at the eigenvalues of the Hamiltonian and, strictly speaking, $\hat{G} = (z - \hat{H})^{-1}$ does not exist there. It is then customary to define $\hat{G}(E)$ as the limit in which the imaginary part of the energy goes to zero,

$$\hat{G}(E) = \lim_{\epsilon \to 0} \hat{G}(E + i\epsilon), \tag{2.14}$$

where ϵ is an infinitesimal. An implementation of this procedure is illustrated explicitly in a later section.

2.2.1 Spectral representation

If the set $\{|n\rangle\}$ forms a biorthonormal basis (see Appendix B) determined by the eigenvectors of an operator $\hat{L}$, $\hat{L}|n\rangle = \lambda_n|n\rangle$, then the spectral representation of $\hat{G}$, Eq. (B.43), takes the form,

$$\hat{G} = \hat{L}^{-1} = \sum_n |n\rangle \frac{1}{\lambda_n} \langle n|. \tag{2.15}$$

In any other basis, $\{|\alpha\rangle\}$, we have

$$\langle \alpha|\hat{G}|\beta\rangle = \sum_n \langle \alpha|n\rangle \frac{1}{\lambda_n} \langle n|\beta\rangle. \tag{2.16}$$

Let us again choose $\hat{L} = (z - \hat{H})$, and let $\{|\alpha\rangle\}$ denote the coordinate representation. If $\{\phi_n(\mathbf{x})\}$ is the set of all eigenstates[2] of $\hat{H}$, $\hat{H}\phi_n(\mathbf{x}) = E_n\phi_n(\mathbf{x})$, then Eq. (2.16) yields

$$\begin{aligned} \langle \mathbf{x}|\hat{G}|\mathbf{x}'\rangle &= G(z;\mathbf{x},\mathbf{x}') \\ &= \sum_n \frac{\langle \mathbf{x}|n\rangle\langle n|\mathbf{x}'\rangle}{z - E_n} \\ &= \sum_n \frac{\phi_n(\mathbf{x})\phi_n^*(\mathbf{x}')}{z - E_n}, \end{aligned} \tag{2.17}$$

from which it is immediately evident that the poles of $\hat{G}$ coincide with the eigenvalues of the Hamiltonian. It is also evident that the Green function embodies the boundary conditions imposed on the wave functions $\phi_n(\mathbf{x})$.

[2] Care should be taken in evaluating the Green function in the case of scattering states in the continuum. For the moment, we will neglect such consideration in order to set down some important fundamental properties of Green functions.

Although the formal expression for the Green function can be rather simple, the actual evaluation of the summations or integrations involved can be a rather arduous task. Occasionally, and mostly for simple model systems, closed expressions can be found by analytic means. More often, however, the Green function must be evaluated numerically by an iterative process, or term by term in a (hopefully rapidly convergent) series expansion. Methods for evaluating the Green function in various manifestations of the Schrödinger equation are outlined in the remainder of this chapter and in subsequent chapters.

2.2.2 Density

We are now in a position to establish one of the most important and useful properties of one-particle Green functions, namely, that they lead directly to the single-particle density, the key quantity discussed in Chapter 1..

Using the limiting process indicated in Eq. (2.14) in connection with Eq. (2.17), and using the property

$$\lim_{\epsilon \to 0} \frac{1}{x + \mathrm{i}\epsilon} = \frac{P}{x} - \mathrm{i}\pi\delta(x), \tag{2.18}$$

where P denotes the Cauchy principal value, we obtain

$$G(\mathbf{x}, \mathbf{x}') = \sum_n \phi_n(\mathbf{x})\phi_n^*(\mathbf{x}') \left[\frac{P}{E - E_n} - \mathrm{i}\pi\delta(E - E_n) \right]. \tag{2.19}$$

We now define the energy-dependent single-particle density by the expression

$$n(\mathbf{x}, E) = \sum_n \phi_n(\mathbf{x})\phi_n^*(\mathbf{x})\delta(E - E_n), \tag{2.20}$$

which in view of Eq. (2.19) can be written in the form

$$n(\mathbf{x}, E) = -\frac{\Im}{\pi} G(E; \mathbf{x}, \mathbf{x}). \tag{2.21}$$

Here and subsequently, the symbol $\Im$ denotes the imaginary part of a complex quantity. An integral over energy now leads directly to the single-particle density

$$\begin{aligned} n(\mathbf{x}) &= -\frac{1}{\pi} \int \mathrm{d}E \Im G(E; \mathbf{x}, \mathbf{x}) \\ &= \sum_n \phi_n(\mathbf{x})\phi_n^*(\mathbf{x}), \end{aligned} \tag{2.22}$$

where the variable $\mathbf{x}$ can be interpreted to represent space and spin coordinates. This expression is identical to that in Eq. (1.130) that deal with bound states, (with spin indices suppressed). Thus, the Green function provides an immediate connection with theories such as DFT and the LDA which rely on the explicit use of the density as the basic variable in the study of electronic structure. This relationship holds also in the case of scattering states in the continuum. Coupled with a method such as multiple-scattering theory, which allows the direct evaluation of the Green function, one has a powerful tool indeed for studying the electronic structure and properties of a wide spectrum of physical systems including bulk materials, surfaces and interfaces, impurities, and substitutionally disordered alloys. It is particularly noteworthy that defining the charge density in terms of the Green function, rather than the wave function, allows the applicat ion of DFT and the LDA to systems, such as substitutionally disordered alloys, for which the wave function is often not a very useful quantity.

Finally, integrating Eq. (2.21) over space (and summing over spin indices included in the definition of $\mathbf{x}$), leads to the density of states,

$$\begin{aligned} n(E) &= -\frac{1}{\pi\Omega}\int d^3x \Im G(E;\mathbf{x},\mathbf{x}) \\ &= -\frac{1}{\pi\Omega}\mathrm{Tr}\hat{G}, \end{aligned} \tag{2.23}$$

where [3], Ω is the volume of the system.

2.3 Evaluation of Green Functions

The particular form of the Green function describing a physical system depends on the system as well as on the representation chosen. In many cases, the starting point in the discussion is the evaluation of $\hat{G}$ in the coordinate representation. Quite often, such evaluations are complicated due to the fact that the operator $\hat{G}^{-1}$ has zero eigenvalues and, strictly speaking, the Green function does not exist. It is then customary to consider the energy as a complex parameter and obtain $\hat{G}$ in the limit $\Im E \to 0$.

In order to illustrate these concepts and the complications involved in the evaluation of the Green function, we consider a one-dimensional problem in which $\hat{G}$ exists and can be calculated by contour integration. In a following subsection, we consider the evaluation of the free-particle propagator, i.e., the Green function describing particle propagation in free space.

[3] When no confusion can arise, the dependence of the Green function or other quantities on the energy or on various parameters will be suppressed. Often, we will also refrain from distinguishing between operators and their specific representations such as that of direct space.

2.3.1 Simple example

We consider the linear operator

$$\hat{L} = 1 - b^2 \frac{\mathrm{d}^2}{\mathrm{d}x^2}, \tag{2.24}$$

where $b^2 > 0$ is a real number. We seek the solution of the equation,

$$\hat{L}|\phi\rangle = |\psi\rangle, \tag{2.25}$$

or

$$\left[1 - b^2 \frac{\mathrm{d}^2}{\mathrm{d}x^2}\right] \phi(x) = \psi(x), \tag{2.26}$$

for a given $\psi(\mathbf{x})$ under the boundary conditions,

$$\phi(\pm\infty) = \psi(\pm\infty) = 0. \tag{2.27}$$

Formally, the solution of Eq. (2.26) is given in terms of the Green function,

$$\phi(x) = \hat{G}\psi(x), \tag{2.28}$$

so that we must evaluate $\hat{G}$ in the coordinate representation subject to the boundary conditions in Eq. (2.27). To this end, we note that the eigenfunctions (eigenvectors) of $\hat{L}$ are the plane waves, $(1/\sqrt{2\pi})\mathrm{e}^{\mathrm{i}kx}$, with corresponding eigenvalues $1 + b^2k^2$ which are always positive. Because there are no vanishing eigenvalues, the evaluation of the Green function is straightforward. From Eq. (2.17), we obtain

$$G(x, x'; b) = \frac{1}{2\pi} \int_{-\infty}^{\infty} \mathrm{d}k \frac{\mathrm{e}^{\mathrm{i}k(x-x')}}{1 + b^2k^2}. \tag{2.29}$$

This integral can be evaluated by contour integration in the complex $z = bk$ plane. We write,

$$G(x, x'; b) = \frac{1}{2b\pi} \int_{-\infty}^{\infty} \mathrm{d}z \frac{\mathrm{e}^{\mathrm{i}\frac{z}{b}(x-x')}}{(z+\mathrm{i})(z-\mathrm{i})}, \tag{2.30}$$

and choose the contour such that $G(x, x'; b)$ satisfies the boundary conditions, Eq. (2.27). For $x > x'$, we close the contour in the upper half plane, as indicated in Fig. 2.1, which leads to the vanishing of the contribution of the semicircular arc as $|z| \to \infty$. Using the residue theorem, we include the contribution of the pole at i and obtain

$$G(x, x'; b) = \frac{1}{2b}\mathrm{e}^{-\frac{1}{b}(x-x')}, \quad \text{for} \quad x > x'. \tag{2.31}$$

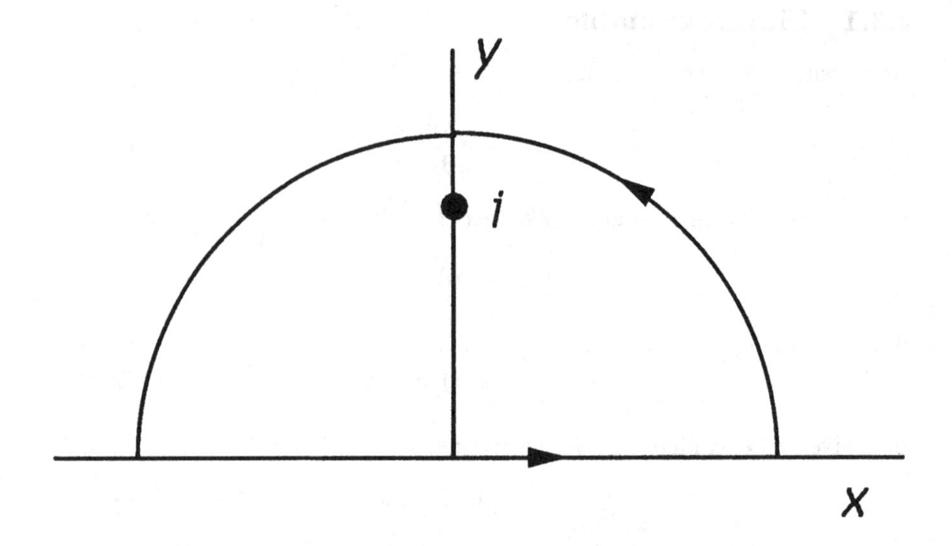

Figure 2.1: **Contour of integration for the evaluation of the Green function.**

Similarly, for $x < x'$, we complete the contour in the lower half plane. Including the contribution of the pole at$-$i, we obtain,

$$G(x, x'; b) = \frac{1}{2b} e^{-\frac{1}{b}(x'-x)}, \quad \text{for} \quad x' > x. \tag{2.32}$$

The last two expressions can be combined into a single one,

$$G(x, x'; b) = \frac{1}{2b} e^{-\frac{1}{b}|x'-x|}, \quad \text{for} \quad x' \neq x, \tag{2.33}$$

an expression which by construction satisfies the boundary conditions given in Eq. (2.27). (The reader is asked to verify that the Green function in Eq. (2.33) satisfies the defining equation $\left[1 - b^2 \frac{\mathrm{d}^2}{\mathrm{d}x^2}\right] G(x, x'; b) = \delta(x - x')$.) Having obtained an expression for $G(x, x'; b)$, we can solve Eq. (2.26), by means of Eq. (2.28). In the coordinate representation, we have,

$$\phi(x) = \int_{-\infty}^{\infty} G(x, x'; b)\psi(x')\mathrm{d}x'. \tag{2.34}$$

Once again, the reader is invited to verify directly that the function $\phi(x)$ satisfies the boundary conditions in Eq. (2.27).

2.3.2 Scattering analogy

We now turn to the evaluation of the Green function in the coordinate representation associated with the differential operator

$$\hat{H}_0 = -\nabla^2 - k_0^2, \tag{2.35}$$

where $k_0^2 = 2mE/\hbar^2$, and seek solutions that are finite at infinity. We recognize $\hat{H}_0$ as the Hamiltonian operator for free-particle motion whose eigenfunctions are the plane wave states, $1/(2\pi)^{3/2}e^{i\mathbf{k}\cdot\mathbf{x}}$ (where in one dimension, we have[4] $1/(2\pi)^{3/2}e^{ikx}$), and whose eigenvalues are given by the real numbers $k^2 - k_0^2$. From Eq. (2.17) we obtain the formal expression, (setting $\hbar^2/2m = 1$)

$$\begin{aligned} G_0(\mathbf{x}',\mathbf{x}) &= \langle \mathbf{x}'|\hat{G}_0|\mathbf{x}\rangle \\ &= \sum_{\mathbf{k}} \frac{\langle \mathbf{x}'|\mathbf{k}\rangle\langle \mathbf{k}|\mathbf{x}\rangle}{E - E(\mathbf{k})} \\ &= \frac{1}{(2\pi)^3}\int \mathrm{d}^3k \frac{e^{i\mathbf{k}\cdot(\mathbf{x}'-\mathbf{x})}}{k_0^2 - k^2}, \end{aligned} \tag{2.36}$$

where $E = k^2$ and the region of integration extends over the whole of k-space. The subscript 0 in $\hat{G}_0$ denotes a *free-particle Green function* or *propagator*.

Now, in contrast to the example discussed in the previous subsection, the denominator $k_0^2 - k^2$ can vanish at $k = \pm k_0$ and the Green function cannot be defined there. In practical terms, the integral in Eq. (2.36) cannot be evaluated as it stands due to the singularities of the integrand. In order to define a Green function for free particles, we must use a limiting process.

In order to develop this limiting process, it is convenient to consider the more general problem represented by the operator $\hat{H} = \hat{H}_0 + \hat{H}_1$, where $\hat{H}_1$ is a perturbation introduced into free space. We now seek solutions of the equation,

$$(\hat{H} - E)|n\rangle = 0, \tag{2.37}$$

or,

$$\hat{H}_1|n\rangle = (E - \hat{H}_0)|n\rangle, \tag{2.38}$$

so that

$$|n\rangle = (E - \hat{H}_0)^{-1}\hat{H}_1|n\rangle. \tag{2.39}$$

As suggested by Sommerfeld, the last equation can be interpreted as describing the scattering of particles (or waves) by a potential field represented

[4] The symbols **x** and **r** will both be used to denote position vectors when no confusion can arise.

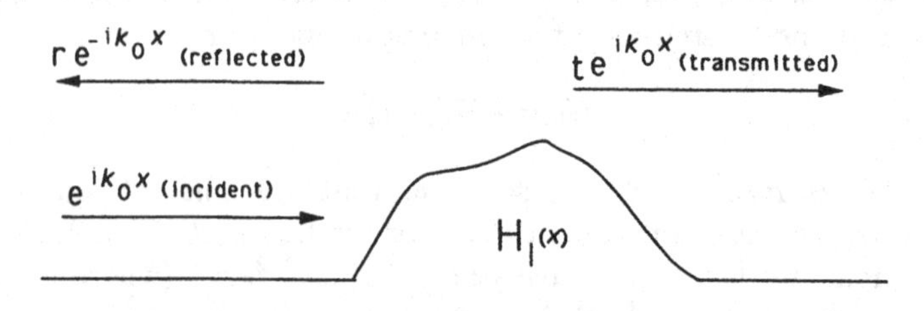

Figure 2.2: **Incident, transmitted, and reflected waves at an one-dimensional potential barrier.**

by $\hat{H}_1$, which we take to be spatially bounded. A schematic representation of this *scattering analogy* in the case of one dimension is shown in Fig. 2.2. Here, an incident plane wave, $\langle x|k_0\rangle = \mathrm{e}^{ik_0x}$, impinges on a potential field of finite spatial extent, giving rise to a transmitted wave, $\simeq t\mathrm{e}^{ik_0x}$, traveling in the same direction as the incident wave, and to a reflected wave, $\simeq r\mathrm{e}^{-ik_0x}$, traveling in the opposite direction. We have imposed the *extra* condition that the transmitted and reflected waves must be proportional to e^{ik_0x} and to e^{-ik_0x}, respectively. As we demonstrate below, this condition, called the *Sommerfeld radiation condition*, can be met by allowing the incident energy to become complex and taking the limit of the complex part going to zero. It is to be emphasized that this condition is imposed in addition to the boundary conditions placed on the wave function or the Green function at infinity. Its purpose is to fix the contours of integration in the complex plane for evaluating Eq. (2.36) (as done in more direct fashion in the example of the previous section).

Continuing our discussion of the one-dimensional case, we write Eq. (2.39) in the coordinate representation,

$$\begin{aligned}
\langle x|n\rangle &= \int\int \mathrm{d}x'\mathrm{d}x''\langle x|(E-\hat{H}_0)^{-1}|x'\rangle\langle x'|\hat{H}_1|x''\rangle\langle x''|n\rangle \\
&= \int\int \mathrm{d}x'\mathrm{d}x'' G_0(x,x')H_1(x')\delta(x'-x'')\langle x''|n\rangle \\
&= \int \mathrm{d}x' G_0(x,x')H_1(x')\langle x'|n\rangle,
\end{aligned} \tag{2.40}$$

where Eq. (2.9) has been used. When the perturbation vanishes, $H_1(x) = 0$, the solution of the last equation is the incoming wave, $\langle x|n\rangle_0 = \mathrm{e}^{ik_0x}$,

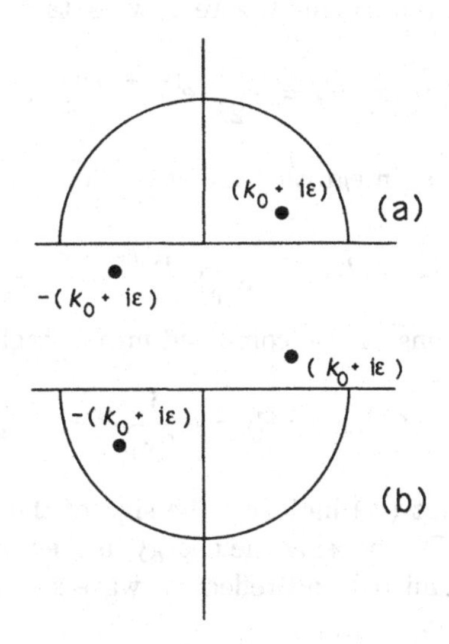

Figure 2.3: **Contours of integration for the evaluation of the one-dimensional free-particle propagator.**

and accordingly the complete solution which is the sum of the homogeneous solution and a particular solution can be written in the form,

$$\langle x|n\rangle = \langle x|n\rangle_0 + \int dx' G_0(x,x')H_1(x')\langle x'|n\rangle. \tag{2.41}$$

We now demonstrate by construction that if the energy, E, is allowed to possess an infinitesimal positive (negative) imaginary part, the corresponding Green functions lead to solutions of Eq. (2.41) in agreement with the Sommerfeld radiation condition.

2.3.3 Free-particle propagator in one dimension

In Eq. (2.36) we put $E = E + i\epsilon'$ or, equivalently, $k_0 = k_0 + i\epsilon$ and obtain the expression

$$G_0(x,x';E) = \frac{1}{2\pi}\lim_{\epsilon\to 0}\int_{-\infty}^{\infty}\frac{e^{ik(x-x')}}{(k+k_0+i\epsilon)(k-k_0-i\epsilon)}dk. \tag{2.42}$$

Here, the poles of the integrand have been shifted off the real axis to the positions $k_0 + i\epsilon$ and $-k_0 - i\epsilon$. For $x > x'$, we close the contour in the upper half-plane, Fig. 2.3a so that the integral along the semicircular arc vanishes and, using the residue theorem, we obtain

$$G_0(x,x';E) = -\frac{\mathrm{i}}{2k_0}\mathrm{e}^{\mathrm{i}k_0(x-x')}, \quad x > x'. \tag{2.43}$$

Similarly, for $x' > x$, integrating along the lower contour, Fig. 2.3b, we find,

$$G_0(x,x';E) = -\frac{\mathrm{i}}{2k_0}\mathrm{e}^{-\mathrm{i}k_0(x-x')}, \quad x < x'. \tag{2.44}$$

The last two equations can be combined into a single expression,

$$G_0^+(x,x';E) = -\frac{\mathrm{i}}{2k_0}\mathrm{e}^{\mathrm{i}k_0|x-x'|}, \tag{2.45}$$

where the superscript (+) indicates the sign of the imaginary part of the energy parameter, E. Choosing the energy in the upper half-plane leads to outgoing, i.e., transmitted and reflected, wave solutions of Eq. (2.45), as we now demonstrate.

To see that $G_0^+(x,x';E)$ satisfies the appropriate boundary conditions for outgoing waves, i.e., the radiation condition, we substitute Eq. (2.45) into Eq. (2.41) and obtain,

$$\langle x|n\rangle = \langle x|n\rangle_0 - \frac{\mathrm{i}}{2k_0}\int \mathrm{d}x'\mathrm{e}^{\mathrm{i}k_0|x-x'|}H_1(x')\langle x'|n\rangle, \tag{2.46}$$

where the integral extends over all space (effectively over the range of $H_1(x)$). For x large and positive, all contributions to the integral arise from $x' < x$ and we can write,

$$\begin{aligned}\langle x|n\rangle &= \langle x|n\rangle_0 - \frac{\mathrm{i}}{2k_0}\mathrm{e}^{\mathrm{i}k_0x}\int \mathrm{d}x'\mathrm{e}^{-\mathrm{i}k_0x'}H_1(x')\langle x'|n\rangle \\ &= \langle x|n\rangle_0 + t\mathrm{e}^{\mathrm{i}k_0x}.\end{aligned} \tag{2.47}$$

Similarly, for x large and negative, we have

$$\begin{aligned}\langle x|n\rangle &= \langle x|n\rangle_0 - \frac{\mathrm{i}}{2k_0}\mathrm{e}^{-\mathrm{i}k_0x}\int \mathrm{d}x'\mathrm{e}^{\mathrm{i}k_0x'}H_1(x')\langle x'|n\rangle \\ &= \langle x|n\rangle_0 + r\mathrm{e}^{-\mathrm{i}k_0x}.\end{aligned} \tag{2.48}$$

The last two expressions contain the implicit formal definition of the transmission and reflection coefficients, t and r, respectively. With these definitions, these expressions are consistent with the form required by the radiation condition, i.e., the existence of transmitted and reflected waves.

The Green functions defined in the limit in which the real axis is approached from the upper half-plane are examples of *retarded* or *causal* Green functions. Fourier transforming these functions into an explicit time dependence [1] shows that they describe a response *following* a stimulus, or that they preserve *causality*. Green functions and associated state vectors, obtained in the limit in which the real axis is approached from the lower half of the complex-energy plane, can also be defined. Although lacking direct physical significance, such quantities are of great formal value in the development of scattering theory.

2.3.4 G_0 in three dimensions - Coordinate representation

From Eq. (2.36) we obtain the expression,

$$G_0^+(\mathbf{r},\mathbf{r}';E) = \frac{1}{(2\pi)^3}\int d^3k \frac{e^{i\mathbf{k}\cdot(\mathbf{r}-\mathbf{r}')}}{k_0^2 - k^2 + i\epsilon}. \tag{2.49}$$

If we set $\boldsymbol{\rho} = \mathbf{r} - \mathbf{r}'$, and take $\boldsymbol{\rho}$ along the z-axis, we have

$$\mathbf{k}\cdot\boldsymbol{\rho} = k\rho\cos\theta, \quad \text{and} \quad d^3k = k^2\sin\theta d\theta d\phi dk, \tag{2.50}$$

where θ and ϕ are the angular coordinates of $\mathbf{k}$, and Eq. (2.50) becomes

$$G_0^+(\mathbf{r},\mathbf{r}';E) = \frac{1}{(2\pi)^3}\int_0^\infty k^2 dk \int_0^\pi \sin\theta d\theta \int_0^{2\pi} d\phi \frac{e^{ik\rho\cos\theta}}{k_0^2 - k^2 + i\epsilon}. \tag{2.51}$$

The angular integrals are elementary and yield the expression,

$$G_0^+(\mathbf{r},\mathbf{r}';E) = \frac{1}{(2\pi)^3}\frac{1}{2i\rho}\int_{-\infty}^\infty \frac{kdk\left[e^{ik\rho} - e^{-ik\rho}\right]}{k_0^2 - k^2 + i\epsilon}. \tag{2.52}$$

Proceeding as in the one-dimensional case, we replace the denominator in the integrand by $(k_0 - k - i\epsilon)(k_0 + k + i\epsilon)$ and use the contours in Fig. 2.3. For the first (second) term in the square brackets in Eq. (2.52) we close the contour in the upper (lower) half plane. Adding the results of these integrations, we obtain the final expression for the three-dimensional free-particle propagator in the coordinate representation,

$$G_0^+(\mathbf{r},\mathbf{r}';E) = -\frac{1}{4\pi}\frac{e^{ik_0|\mathbf{r}-\mathbf{r}'|}}{|\mathbf{r}-\mathbf{r}'|}. \tag{2.53}$$

One important property of the free-particle Green function is its dependence on the difference $\mathbf{r} - \mathbf{r}'$, rather than on $\mathbf{r}$ and $\mathbf{r}'$ separately. This behavior reflects the invariance of free space under simultaneous translations of $\mathbf{r}$ and $\mathbf{r}'$. It is easily shown that $G_0^+(\mathbf{r},\mathbf{r}';E)$ satisfies Eq. (2.12) in the form,

$$(\nabla^2 + E)G_0^+(\mathbf{r},\mathbf{r}';E) = \delta(\mathbf{r} - \mathbf{r}'). \tag{2.54}$$

2.3.5 G_0 in the angular momentum representation

The angular momentum representation is extremely useful in the study of condensed matter physics; it is indispensable in connection with multiple scattering theory. As a first step in using this representation in later discussion, we now obtain the form of the free-particle propagator as a summation over angular momentum eigenstates. A brief discussion of the spherical harmonics on which the angular momentum representation is based is presented in Appendix C.

Our starting point is again Eq. (2.36),

$$\langle \mathbf{r}|\hat{G}_0|\mathbf{r}'\rangle = \frac{1}{(2\pi)^3}\int_{-\infty}^{\infty} \mathrm{d}^3k \frac{\mathrm{e}^{\mathrm{i}\mathbf{k}\cdot(\mathbf{r}-\mathbf{r}')}}{k_0^2 - k^2}. \tag{2.55}$$

With the help of Bauer's identity,

$$\mathrm{e}^{\mathrm{i}\mathbf{k}\cdot\mathbf{r}} = 4\pi \sum_{\ell=0}^{\infty}\sum_{m=-\ell}^{\ell} \mathrm{i}^{\ell} j_\ell(kr) Y_{\ell m}(\hat{r}) Y^*_{\ell m}(\hat{k}), \tag{2.56}$$

which expresses a plane wave in terms of Bessel functions, $j_\ell(kr)$, and spherical harmonics, $Y_{\ell m}$, Eq. (2.55) can be written in the form

$$\begin{aligned} G_0^+(\mathbf{r},\mathbf{r}';E) &= \frac{1}{2\pi}\int k^2\mathrm{d}k \int \mathrm{d}\Omega_k \left[4\pi \sum_{\ell,m} \mathrm{i}^{\ell} j_\ell(kr) Y_{\ell m}(\hat{r}) Y^*_{\ell m}(\hat{k})\right] \\ &\times \frac{1}{k_0^2 - k^2}\left[4\pi \sum_{\ell' m'} (-\mathrm{i})^{\ell'} j_{\ell'}(kr') Y_{\ell' m'}(\hat{r}') Y^*_{\ell' m'}(\hat{k})\right]. \end{aligned} \tag{2.57}$$

Integrating over the angles of $\mathbf{k}$ and using the orthonormality condition for spherical harmonics, Eq. (C.12), we obtain the expression,

$$G_0^+(\mathbf{r},\mathbf{r}';E) = \sum_{\ell,m} G_\ell(r,r';E) Y_{\ell m}(\hat{r}) Y^*_{\ell m}(\hat{r}'), \tag{2.58}$$

where

$$\begin{aligned} G_\ell(r,r';E) &= \frac{2}{\pi}\int_0^{\infty} \frac{j_\ell(kr) j_\ell(kr')}{k_0^2 - k^2} k^2\mathrm{d}k \\ &= \frac{1}{\pi}\lim_{\epsilon\to 0}\int_{-\infty}^{\infty} \frac{j_\ell(kr) j_\ell(kr')}{k_0^2 - k^2 + \mathrm{i}\epsilon} k^2\mathrm{d}k, \end{aligned} \tag{2.59}$$

where the last line follows from the fact that the integrand is even in k.

The last integral can be evaluated by contour integration in the complex k-plane. We use Eqs. (C.27) through (C.29) to find the identity,

$$j_\ell(kr) = \frac{1}{2}\left[h_\ell(kr) + h_\ell^*(kr)\right], \tag{2.60}$$

where $h_\ell(kr) \equiv h_\ell^+(kr)$ is a Hankel function of the first kind with the asymptotic behavior,

$$h_\ell(z) \to \mathrm{i}^{-\ell-1}\frac{\mathrm{e}^{\mathrm{i}z}}{z}, \quad \text{as} \quad z \to \infty. \tag{2.61}$$

Because

$$j_\ell(z) \to \frac{1}{2}\sin\left(z - \frac{\ell\pi}{2}\right), \quad \text{as} \quad z \to \infty, \tag{2.62}$$

the first product, $j_\ell(kr)h_\ell(kr')$ in Eq. (2.59), can be evaluated for $r' > r$ by closing the contour in the upper half plane. In this case the behavior of the Hankel function dominates with increasing distance and overwhelms the exponentially growing part of the Bessel function. This yields the contribution

$$-\frac{1}{2}\mathrm{i}k_0 h_\ell(k_0 r')j_\ell(k_0 r), \quad r' > r. \tag{2.63}$$

The term $j_\ell(kr)h_\ell^*(kr')$ can be evaluated by closing the contour in the lower half plane and yields the contribution

$$-\frac{1}{2}\mathrm{i}k_0 h_\ell^*(-k_0 r')j_\ell(k_0 r), \quad r' > r. \tag{2.64}$$

Since

$$j_\ell(-x) = (-\mathrm{i})^\ell j_\ell(x) \quad \text{and} \quad h_\ell^*(-x) = (-1)^\ell h_\ell(x), \tag{2.65}$$

the last two expressions can be combined into a single term

$$-\mathrm{i}k_0 h_\ell(k_0 r')j_\ell(k_0 r), \quad r' > r. \tag{2.66}$$

A similar contribution results for $r > r'$ leading to the expression

$$G_0^+(\mathbf{r} - \mathbf{r}') = -\mathrm{i}k_0 \sum_{\ell,m} h_\ell(k_0 r_>)j_\ell(k_0 r_<)Y_{\ell m}(\hat{r})Y_{\ell m}^*(\hat{r}'), \tag{2.67}$$

where $r_>$ ($r_<$) denotes the larger (smaller) of the lengths r and r'. We note that the free-particle Green function is diagonal in the angular momentum representation reflecting the invariance of free space under rotations.

If we introduce the symbol $A_L(k\mathbf{r}) = \alpha_\ell(kr)Y_{\ell m}(\hat{r})$, where $\alpha_\ell(x)$ denotes any of the spherical Bessel, Hankel, or Neumann functions, and note that

$A_L(\mathbf{r})$ can be arranged in the form of a bra or ket vector indexed by L, we can cast the last expression in the convenient form,

$$\begin{aligned} G_0^+(k, \mathbf{r} - \mathbf{r}') &= -ik \sum_L J_L(k\mathbf{r}_<) H_L^*(k\mathbf{r}_>) \\ &= -ik \langle J(k\mathbf{r}_<) | H(k\mathbf{r}_>) \rangle. \end{aligned} \tag{2.68}$$

It will often be convenient to absorb the factor $-ik$ into the definition of the Hankel function and not display it explicitly. Also, it is often convenient to employ linear combinations of spherical harmonics which are real (see Appendix C), in which case it is no longer necessary to display the complex conjugates of the spherical functions.

At this point we must emphasize the restriction on the lengths of vectors which enter the expansion of the free-particle propagator in the last expression. We note that the length of the vector argument of the Hankel function (the irregular solution) must be larger than that of the Bessel function (the regular solution), otherwise the expansion is inherently divergent.

2.3.6 Free-particle density of states

The particle density and the density of states for free particles can be calculated immediately once the corresponding Green function has been obtained. From Eq. (2.21) the particle density at energy E is given by

$$n_0(\mathbf{r}, E) = -\frac{1}{\pi} \Im G_0(\mathbf{r}, \mathbf{r}) = \frac{1}{\pi} k \sum_L j_\ell^2(kr) Y_{\ell m}(\hat{r}) Y_{\ell m}^*(\hat{r}). \tag{2.69}$$

Using the well-known result $\sum_m Y_{\ell m}(\hat{r}) Y_{\ell m}^*(\hat{r}) = \frac{2\ell+1}{4\pi}$, and recalling the sum rule $\sum_\ell (2\ell + 1) j_\ell^2(z) = 1$, we obtain

$$n_0(\mathbf{r}, E) = \frac{k}{4\pi}. \tag{2.70}$$

We see that the particle density is independent of position and, along with the density of states, varies as the square root of the energy.

2.4 Iterative Expansions

In this section, we discuss certain standard manipulations designed to facilitate the determination of the Green function as well as the wave function in the presence of a perturbing potential $\hat{H}_1 \equiv \hat{V}$. The methods described below form the basis for much of the discussion of multiple scattering theory (MST) in the following chapters.

2.4.1 Iterative expansions for the Green function

For a given Hamiltonian $\hat{H} = H_0 + V$, the Green function is defined by Eq. (2.8),

$$\hat{G}(E) = (E - \hat{H})^{-1}. \tag{2.71}$$

This can also be written in the form (suppressing the explicit indication of the dependence on the energy variable),

$$\hat{G} = \hat{G}_0 + \hat{G}_0\hat{V}\hat{G}, \tag{2.72}$$

where $\hat{G}_0 = (E - \hat{H}_0)^{-1}$ is the free-particle propagator. Equation (2.72) is the *Dyson* equation for $\hat{G}$. One way of attempting to solve this equation is by means of successive iterations starting from $\hat{G}_0$. Thus, to zeroth order, we set

$$\hat{G}^{(0)} = \hat{G}_0. \tag{2.73}$$

When $\hat{G}^{(0)}$ is substituted into the right hand side of Eq. (2.72), there results the first approximation to the Green function,

$$\hat{G}^{(1)} = \hat{G}_0 + \hat{G}_0\hat{V}\hat{G}_0. \tag{2.74}$$

Proceeding in this manner, we obtain the ***Born series*** expansion for $\hat{G}$,

$$\hat{G} = \hat{G}_0 + \hat{G}_0\hat{V}\hat{G}_0 + \hat{G}_0\hat{V}\hat{G}_0\hat{V}\hat{G}_0 + \cdots. \tag{2.75}$$

Provided that the expansion converges, which we will assume is the case, the last equation provides a fairly general method for evaluating the Green function. As we have seen, the boundary conditions of the problem can be readily incorporated into the Green function, through the appropriate free-particle propagator. Of course, the specific form of $\hat{G}$ depends on the particular representation chosen. For example, in the coordinate representation, we obtain the ***integral equation***,

$$\begin{aligned} G(\mathbf{r},\mathbf{r}') &= G_0(\mathbf{r},\mathbf{r}') \\ &+ \int \mathrm{d}^3r_1 G_0(\mathbf{r},\mathbf{r}_1)V(\mathbf{r}_1)G_0(\mathbf{r}_1,\mathbf{r}') \\ &+ \int \mathrm{d}^3r_1 \int \mathrm{d}^3r_2 G_0(\mathbf{r},\mathbf{r}_1)V(\mathbf{r}_1)G_0(\mathbf{r}_1,\mathbf{r}_2) \\ &\times V(\mathbf{r}_2)G_0(\mathbf{r}_2,\mathbf{r}') + \cdots. \end{aligned} \tag{2.76}$$

In many applications, the free-particle Hamiltonian is simply taken to be the kinetic energy operator, ∇^2, with $\hat{V}$ describing the potential characterizing the system, e.g., the potential of the nuclei and the electrons acting at a given point inside a material.

2.4.2 Iterative expansions of the wave function

Like the Green function, the solution of the Schrödinger equation $(E - \hat{H})|n\rangle = 0$ can also be treated by iterative means. In general, we have

$$(E - \hat{H})|n\rangle = (E - \hat{H}_0 - \hat{V})|n\rangle = 0, \tag{2.77}$$

or

$$(E - \hat{H}_0)|n\rangle = \hat{V}|n\rangle, \tag{2.78}$$

from which we obtain the equation

$$|n\rangle = \hat{G}_0\hat{V}|n\rangle. \tag{2.79}$$

As was pointed out in relation to Eq. (2.41), the complete solution of the last equation is the sum of the homogeneous solution and a particular solution,

$$|n\rangle = |n\rangle_0 + \hat{G}_0\hat{V}|n\rangle. \tag{2.80}$$

Again the boundary conditions imposed on the solutions of Eq. (2.80) can be incorporated through the use of the proper Green function for free motion. For example, if the scattered (outgoing) solutions of Eq. (2.80) are desired, we have

$$|n^+\rangle = |n\rangle_0 + \hat{G}_0^+\hat{V}|n^+\rangle. \tag{2.81}$$

This is the *Lippmann-Schwinger* equation of scattering theory. It will be made the starting point in the development of MST in the following chapter.

The Lippmann-Schwinger equation describes the evolution of the unperturbed state $|n\rangle_0$ to the scattered state $|n^+\rangle$, at the same energy as $|n\rangle_0$. In the coordinate representation, Eq. (2.81) takes the form,

$$\langle \mathbf{r}|n^+\rangle = \langle \mathbf{r}|n\rangle_0 + \int \mathrm{d}^3r' G_0^+(\mathbf{r}-\mathbf{r}')V(\mathbf{r}')\langle \mathbf{r}'|n^+\rangle, \tag{2.82}$$

where $G_0^+(\mathbf{r}-\mathbf{r}')$ is given by Eq. (2.53).

The Lippmann-Schwinger equation can be treated formally by an iterative process. Using an analogous procedure to that used with the Green function, a series of approximations to $|n^+\rangle$ beginning with $|n\rangle_0$ yields the *Born series* expansion of Eq. (2.82),

$$\begin{aligned}\langle \mathbf{r}|n^+\rangle &= \langle \mathbf{r}|n\rangle_0 + \int \mathrm{d}^3r' G_0^+(\mathbf{r}-\mathbf{r}')V(\mathbf{r}')\langle \mathbf{r}'|n\rangle_0 \\ &+ \int \mathrm{d}^3r' \int \mathrm{d}^3r'' G_0^+(\mathbf{r}-\mathbf{r}')V(\mathbf{r}')G_0^+(\mathbf{r}'-\mathbf{r}'') \\ &\times V(\mathbf{r}'')\langle \mathbf{r}''|n\rangle_0 + \cdots. \end{aligned}\tag{2.83}$$

Provided that this series converges, its truncation after the nth term constitutes the nth *Born approximation* to the wave function $\langle \mathbf{r}|n^+\rangle$.

2.4.3 Transition matrix

The Born series for the wave function can be written in the (abstract) form

$$|n^+\rangle = \left[1 + \left(\hat{G}_0^+ + \hat{G}_0^+\hat{V}\hat{G}_0^+ + \cdots\right)\hat{V}\right]|n\rangle_0. \tag{2.84}$$

This in turn can be cast into a form involving the full Green function for the system

$$|n^+\rangle = |n\rangle_0 + \hat{G}\hat{V}|n\rangle_0. \tag{2.85}$$

This equation appears to provide a simple solution for $|n^+\rangle$ in terms of the known state $|n\rangle_0$. This simplicity is, however, illusory because the Green function of the system, $\hat{G}$, is unknown and still to be determined. In addition to Eq. (2.85), we can also write

$$|n^+\rangle = |n\rangle_0 + \hat{G}_0\hat{t}|n\rangle_0, \tag{2.86}$$

through the introduction of the *transition* operator or *t-matrix*,

$$\begin{aligned} \hat{t} &= \hat{V} + \hat{V}\hat{G}_0\hat{V} + \hat{V}\hat{G}_0\hat{V}\hat{G}_0\hat{V} + \cdots \\ &= \hat{V} + \hat{V}\hat{G}_0^+\hat{t} \\ &= \left(1 - \hat{V}\hat{G}_0\right)^{-1}\hat{V}. \end{aligned} \tag{2.87}$$

The t-matrix, also often loosely called the scattering matrix, will be of central importance in the development of multiple scattering theory in the next chapter. It follows from the last two expressions that the t-matrix satisfies the relations

$$\hat{V}|n^+\rangle = \hat{t}|n\rangle_0. \tag{2.88}$$

Also, from Eq. (2.85) and (2.86) we have

$$\hat{V}\hat{G} = \hat{t}\hat{G}_0. \tag{2.89}$$

Using these expressions, we see that the Green function, Eq. (2.72), can be written in the form,

$$\hat{G} = \hat{G}_0 + \hat{G}_0\hat{t}\hat{G}_0. \tag{2.90}$$

In the coordinate representation, the equation for the t-matrix takes the form,

$$t(\mathbf{r},\mathbf{r}') = V(\mathbf{r})\left[\delta(\mathbf{r}-\mathbf{r}') + \int d^3r'' G_0(\mathbf{r}-\mathbf{r}'')t(\mathbf{r}'',\mathbf{r}')\right]. \tag{2.91}$$

The t-matrix describes the transition from the unperturbed or incident state $|n\rangle_0$ to the scattered out state $|n^+\rangle$. Because the scattered and the incident states are at the same energy, the t-matrix is said to be *on the energy shell* or on-shell. The on-shell t-matrix provides a *complete* description of the scattering of a particle (or wave) from a spatially bounded field of force and is a central quantity upon which the multiple scattering description of the electronic structure can be based.

2.5 Optical Theorem

There are two useful and important relations that in MST retain essentially the same form as they have in operator space. To derive the first, we note that from the general definition, Eq. (2.71), and the hermiticity of the Hamiltonian, we obtain

$$\Im G = \frac{1}{2\mathrm{i}}\left[\hat{G} - \hat{G}^{\dagger}\right] = G^{\dagger}\left[-\Im z\right]G, \tag{2.92}$$

where z is a complex energy. Therefore, we have the general property, *the reality condition*, governing the imaginary part of the Green function,

$$\Im \hat{G}(z) \leq 0 \ \text{ for } \ \Im z \geq 0, \tag{2.93}$$

with an analogous expression holding for $\Im z \leq 0$. (The reader is invited to prove these relations.) Second, the imaginary part of the t-matrix becomes,

$$\begin{aligned} \hat{t} - \hat{t}^{\dagger} &= \left(\hat{V}^{-1} - \hat{G}_0^{\dagger}\right)^{-1}\left[\hat{V}^{-1} - \hat{G}_0^{\dagger} - \hat{V}^{-1} + \hat{G}_0\right]\left(\hat{V}^{-1} - \hat{G}_0\right)^{-1} \\ &= \hat{t}^{\dagger}\left(\hat{G}_0 - \hat{G}_0^{\dagger}\right)\hat{t}, \end{aligned} \tag{2.94}$$

or,

$$\Im \hat{t} = \hat{t}^{\dagger}\Im[\hat{G}_0]\hat{t}. \tag{2.95}$$

This result is known as the *optical theorem* of scattering theory. More detailed discussions of t-matrices is given as necessary in following chapters and can also be found [6] in standard treatises on scattering theory.

Bibliography

[1] A. Bohm, *Quantum Mechanics*, Springer Verlag, New York (1979).

[2] C. Kittel, *Quantum Theory of Solids*, John Wiley and Sons, New York (1963).

[3] G. Rickayzen, *Green Functions and Condensed Matter*, Academic Press, New York (1980).

[4] A. Gonis, *Green Functions for Ordered and Disordered Systems*, North Holland, Amsterdam (1992).

[5] K. Eik and W. Glasser, in *Die Methoden der Greenshen Funktionen in der Festkörperphysik*, (Akademie-Verlag, Berlin, 1979).

[6] Charles J. Joachain, *Quantum Collision Theory*, North Holland, New York and Amsterdam, (1983).

Chapter 3

Multiple Scattering Theory

3.1 General Comments

Multiple scattering theory (MST) is a generalization of the theory of s-cattering of a single particle by an external, spatially bounded potential. The single-scatterer formalism is presented in some detail in Appendix D, and the reader may wish to peruse that material before proceeding further. Some formal aspects of general scattering theory are presented in Appendix E. The study of the scattering of a single particle by a number of non-overlapping, spatially bounded potential cells requires a generalization of these formal methods. The resulting approach is particularly useful for the study of the electronic structure and related properties of materials.

There are three interrelated features that distinguish MST and set it apart from virtually any other method devised for the solution of second-order partial differential equations associated with complex systems: generality, versatility, and power.

Although it is based on the physical picture of a propagating wave being scattered by obstacles or imperfections contained in the medium of propagation, MST is much more general, because it is applicable to essentially every manifestation of the wave equation, including the Schrödinger equation, the Dirac equation, the Laplace equation, the Poisson equation, the Debye-Hückel equation, and the vector wave equation. The spectrum of applicability of MST includes most, if not all, of the linear partial differential equations of mathematical physics applied to complex systems, such as a solid, within both classical and quantum mechanics.

Another aspect of the generality of MST is connected with the types of physical systems to which it is applicable. With reference to the calculation of the electronic structure, MST can be used to treat the Schrödinger equation associated with pure, translationally invariant elemental solids, ordered alloys and compounds, substitutional or interstitial impurities, substitutionally disordered alloys, extended defects such as surfaces and interfaces in ordered or disordered systems, stacking faults, and many others. It's worth the risk of overstating the case to emphasize the desirable property of MST that allows the treatment of diverse crystal structures without the need to impose artificial boundary conditions, such as slabs, repeating slabs, and supercells to represent single impurities, surfaces, interfaces, and other defects.

The versatility of MST refers to its applicability in the study of diverse physical phenomena and properties. This is not surprising given the generality of MST mentioned above. A method that can be used with virtually every manifestation of the wave equation could be expected to be of use in connection with the physical properties described by these equations. In addition, MST can be used to study properties not described by a pure form of the wave equation but by derived expressions. For the physical properties of solid materials, MST can be used within the LDA to determine static properties, such as equilibrium volumes, as well as dynamic quantities such as the ac conductivity of substitutionally disordered alloys. Classical applications vary from the determination of the index of refraction of a heterogeneous medium, studied by Lord Rayleigh in the founding paper of MST [1], to the dielectric response of chains of disordered polarizable spheres [2].

Finally, the power of MST refers to its ability to yield directly the scattering matrix or, equivalently, the Green function for a given system or physical process, and to provide unique and well-defined approaches to convergence in determining these quantities. The importance of determining the Green function directly, and under the boundary conditions imposed by the problem at hand, can hardly be overstated, at least in connection with the single-particle Schrödinger equation. *All single-particle properties of a system are encoded in the analytic structure of the corresponding Green function.* We have already seen that the Green function leads directly to the particle (charge) density, the basic variable for studying electronic systems within density functional theory and its local approximation, the LDA. When expressed in the angular momentum representation, the convergence of the solutions of the Schrödinger equation can be tested uniquely by means of increasing the number of angular momentum states used in an expansion. Furthermore, the Green function retains its conceptual integrity and practical usefulness even when the wave function becomes of little practical value, as in the case of concentrated, substitutionally disordered alloys.

Having pointed out the advantages of MST, it is only fair to point out its drawbacks. These are associated with the computational characteristics of MST-based methods in studying condensed matter, and in the single-particle nature of the theory.

Even though it is formally applicable to them, multiple scattering theory is by no means the method of choice for the treatment of *all* of the physical systems and properties enumerated above. It provides by no means the fastest computational method for the determination of the electronic structure in general, and the approach to convergence, in spite of its uniqueness, can become very sluggish. Like all methods based on the single-particle framework, it must be strained to yield information about collective excitations, such as phonon spectra, and is virtually silent in predicting *bona-fide* many-particle properties such as superconductivity (except to the extent that single-particle properties, e.g., the form of the density of states at the Fermi energy, can be used in semi-phenomenological models to study many-body effects).

On the other hand, accepting the limitations inherent in the LDA, multiple scattering theory provides a *general* method of solution of virtually every problem associated with the single-particle wave equation. At the very least, it allows one the ability to construct models and derive essentially exact expressions from which efficient computational techniques can be obtained through approximation schemes. Quite often it yields directly computational schemes that are both efficient as well as transparent.

One purpose of the present work is to illustrate the utility of MST by explicitly exhibiting the equations to be solved in the study of diverse physical systems and physical properties. Such an exposition should by no means lull one into a belief that MST can be brought to bear upon any and all problems in the realm of materials properties. Rather, it is hoped that it will provide a differentiating measure as to where MST could and should be used, and increase the motivation for the search for generalizations, modifications, or alterations that would extend its realm of practicality.

3.2 Basic Tenet of MST

There are a number of different physical pictures that can be associated with multiple-scattering theory. As mentioned above, its most common interpretation is one of waves propagating through a given medium, being scattered by the impurities and imperfections contained in it. The basic notion here is that the *interaction* between the propagating wave and the scatterers can be treated separately from the *propagation* of the wave between scattering events. The interaction depends only on the nature of the scatterer and is fully described by the single-scatterer t-matrix. The propagation, on the other hand, is only a function of the medium itself. Thus, in ordinary scattering theory, a scattering process is viewed as consisting of three separate processes: the propagation of the wave to the scatterer, the scattering event, and the propagation of the wave to the detector (or to another scattering center).

Perhaps the single most important concept in the scattering picture of MST is that the single-cell (single-site) t-matrix suffices to represent the scattered wave at *any* point outside the scattering region, regardless of the shape of the region, or of how close to the potential one takes the observation point to be. This result [3] provides the resolution to two large obstacles to an intuitive understanding of MST. First, it clarifies the validity of MST in treating collections of scatterers of arbitrary shape and separated by arbitrarily small distances, even scattering potentials with common boundaries that fill all space. In such systems, each scattering region acts independently as if it were in isolation rather than in the vicinity of another scatterer.

The second obstacle is the apparent contradiction contained in the statement that a wave propagates freely (as if it were in free space) between scattering events even in cases in which the wave or particle is manifestly always under the influence of the potential, with no free space available for propagation to take place. This is clearly the case when MST is applied to the study of, say, periodic materials of infinite extent described by space-filling cells. The original days of MST were darkened by precisely this objection to it. The idea is not altogether unrelated to more recent notions, only recently clarified, that the scattering from a given potential cell could not be decoupled from that of an adjacent cell unless the spheres bounding the two cells did not overlap. The argument here is that it is only outside the bounding sphere that the scattered wave function assumes its asymptotic value and is fully describable in terms of the t-matrix. However, the fact that the cell t-matrix describes the scattering *everywhere* outside a given cell [3], including points in free space that lie inside a bounding sphere in the so-called moon region, suffices to dispel this apparent contradiction. In fact, MST applied to space-filling cells is formally *no* different, with the exception of some technical details for enhancing convergence, from MST applied to collections of scatterers any two of which are separated by non-overlapping spheres.

In addition to scattered waves, there is one other, very important interpretation of MST. This interpretation, which is more analytic (rather than pictorial) than the scattering picture, is based on the properties of solutions of linear second-order partial differential equations. As is well known, a solution to such an equation is continuous with continuous derivative across any arbitrary boundary in the system. We refer to the discussion in Appendix D in which the t-matrix is obtained by demanding that the wave function should be continuous with continuous derivative across a surface (usually taken to be a sphere) that separates the potential region from free space. MST carries this construction to its ultimate utility. ***The solution of the Schrödinger equation for a collection of potential cells is obtained by matching the solutions of individual cells across cell boundaries.*** Now, no mention need be made of the cell t-matrix, asymptotic behavior, or scattering. It suffices only that the matching across cell boundaries be effected to obtain the global solution. Thus the objection to MST based on the lack of free space between contiguous cells is not justified.

In this chapter, we discuss both approaches to MST, the scattering picture and the algebraic one. Before turning to formal considerations, we summarize briefly the historical development of the method, although this material is mostly of historical interest and can be skipped in a first reading.

3.3 Historical Development of MST (*)

Multiple scattering theory was born in 1892, the year in which Lord Rayleigh [1] published his famous paper on the electrical conductivity of a periodic array of spheres in an otherwise uniform matrix. Rayleigh's formalism was generalized to the frequency-dependent or dynamic conductivity by Kasterin [4], who used it to study the scattering of sound by an assembly of spheres. Korringa, Kronig, and Smit [5] applied Kasterin's theory to the problem of sound reflecting from a porous surface. All these were, of course, classical applications. Then in 194 7, Korringa published his classic paper [6] that showed how multiple scattering theory could be applied to the solution of the Schrödinger equation.

In 1951 Harrison [7] showed that for the case of one-dimensional systems, Korringa's formalism was equivalent to a technique that had been developed by Saxon and Hutner [8]. Three years later, Kohn and Rostoker [9] rederived MST within a variational formalism, showed that their results were equivalent to those of Korringa, and used the method to calculate the electronic structure of elemental lithium solid. Since the publication of that paper, the method has come to be known as the Korringa-Kohn-Rostoker (KKR) method for the calculation of electronic structure. Because of its explicit reliance on the Green function, the method is also referred to as the Green-function method, or the KKR Green-function method.

Soon after the publication of these first papers on MST, the method gained a degree of popularity because of its rapid convergence. It led to systems of linear equations much smaller than competing techniques such as the augmented plane wave (APW) technique introduced in the thirties by Slater [10]. In the early 1960s, the method was further developed by Ham and Siegel [11], Treusch and Sandrock [12], Faulkner, Davis, and Joy [13], and Johnson [14], who used MST to treat a finite cluster of atoms rather than infinite solids.

Initial applications of MST to the study of electronic structure were made with respect to potentials of the muffin-tin (MT) form, i.e., spherically symmetric potentials separated by non-overlapping spheres. These early calculations proved to be very successful, especially in the case of close-packed metals and alloys, culminating in the book by Moruzzi, Janak, and Williams [15] containing the results of calculated electronic structures and related properties of metallic elemental solids.

In addition to pure or ordered materials, MST in connection with MT potentials was also successfully applied to the study of the electronic properties of substitutionally disordered alloys. These applications were made on the basis of the coherent potential approximation (CPA) [16, 17] of Soven and Taylor, the resulting technique coming to be known as the KKR-CPA. The technique was developed and refined by a number of workers [18-22] such as Faulkner, Soven, Shiba, Stocks, Temmerman, Györffy, and Bansil. The marriage of the KKR Green-function method and the CPA led to a very elegant and successful description of the single-particle properties of nonmagnetic metallic alloys.

The use of spherically symmetric MT potentials in applications of the KKR or the KKR-CPA method continued until the mid 1970s. The method had obviously reached a rather satisfactory level of both conceptual as well as computational development. It afforded a very clear picture of the scattering events that take place inside a material: An electron wave propagates freely to a scattering cell, scatters, reaches its "asymptotic" form before encountering another scatterer, and procedes via free propagation to the next scattering event. Here was a computationally convenient and obvious separation of the potential and the structural aspects of the problem of solving the Schrödinger equation in a solid, the first being embodied in the cell t-matrices, the latter in the structure constants of the underlying lattice.

However, it was realized that the use of spherically symmetric potentials had a number of limiting features. For example, it provided a poor description of the electronic structure in systems with low rotational symmetry, such as surfaces and interfaces, and of materials with highly unidirectional charge distributions, such as covalently bonded semiconductors. At the same time other methods, notably the APW, were being generalized to treat the potential inside the Wigner-Seitz cell without approximations to its shape. The full-potential linearized APW (FLAPW) method [23, 24] has been applied quite successfully to the treatment of materials forming open, non-close-packed structures, and to surfaces and interfaces, the latter treatments often being based on the use of analog boundary conditions, e.g., slabs or repeating slabs, to simulate surface or interface regions. In addition to the FLAPW, full-potential techniques based on the method of linear MT orbitals (LMTO) [25] have also been developed [26], relying on the use of a slab or supercell geometry.

The flurry of activity in developing full-potential methods not directly based on the Green function failed to produce computational techniques that covered the needed spectrum of applications. For example, the study of random alloys through the use of the CPA in connection with any method other than one based on Green functions has still not been developed, and even the treatment of low-dimensional defects such as surfaces or interfaces has raised questions, at least at the conceptual level, in connection with the spurious effects of the interactions between repeating slabs or supercells.

Full-potential electronic structure methods based directly on MST and the use of Green functions within the formal construct of KKR were slow in becoming available. The reasons for this delay are not altogether clear, but in retrospect some educated guesses can be made. To begin with, the existence of methods such as the LMTO (faster than MST in its implementation but sometimes not as accurate), and the FLAPW (very accurate for periodic solids but also very sluggish), may have played a role in diminishing the desirability of further development of full-cell electronic structure codes. Also, the perception that had developed (as the result of numerous applications) that MST is rigorously valid *only* in the case of MT potentials may have played a role. As shown in Appendix D, the derivation of the single-cell t-matrix relies on the existence of a bounding sphere to separate the cell from the free space beyond. One can see explicitly that *outside* such a sphere the wave function attains the form that is completely described in terms of the t-matrix. If one were to bring two non-spherical scatterers close enough so that their bounding spheres overlapped, (more precisely, if the bounding sphere of one cell overlapped the potential region of another cell), their individual scattering events would be inextricably coupled, rendering impossible any description of the scattering in terms of individual cell t-matrices. In short, the scattering from one cell would begin before that in the adjacent cell had been completed. It was then conjectured [27-29] that in the case of space-filling potential cells one would have to introduce so-called near-field corrections (NFCs) into the equations of MST that describe scattering from assemblies of MT potentials.

Had the conjecture of the existence of NFCs turned out to be valid, it would have meant a great setback in the development of methods for the study of the electronic structure of solids. As has already been mentioned, Green functions, particularly those arising within MST, can be used to study systems for which the wave function, on which essentially all alternative techniques for electronic structure calculations are based, may not be a useful concept. Substitutionally disordered alloys are a case in point. Furthermore, as is discussed in later chapters of this book, Green functions allow the exact treatment of the Schrödinger equation associated with systems such as surfaces and interfaces, without the necessity of imposing artificial boundary conditions such as slabs, repeating slabs, or other superstructures. The discussion of the application of MST to such systems illustrates the hindrance that would result from the presence of N-FCs. The corresponding developments are possible because, in fact, NFCs do *not* exist.

Efforts to generalize MST to space-filling potential cells have given rise to a great body of work. Early attempts include the proposal of Bross and Anthony [30] and of Beleznay [31] to treat the potential in the interstitial region, i.e., outside the MT spheres inscribed inside a Wigner-Seitz cell, in a perturbative way, while Keller [32] suggested the use of MST with an ever increasing number of empty spheres (spheres not containing nuclei) to represent the scattering in the interstitial region. Evans and Keller [33] and John et. al. [34] treated the case of non-spherical MT scatterers, i.e., scatterers of general shape separated by non overlapping spheres. These early efforts contributed significantly toward the conceptual development of MST and in the appreciation of the difficulties inherent in its generalization to other than MT potentials.

The first formulation of MST in connection with space-filling cells was made in 1972 by Williams and Morgan [35]. A second paper [36] by the same authors appeared in 1974. Morgan [37] provided numerical tests of the convergence in angular momentum expansions of the solutions of the Poisson equation that seemed to show a reasonably good, although by no means perfect, convergence through $L = 4$. These papers sparked a controversy that lasted nearly 20 years.

In response to the papers by Williams and Morgan, Ziesche [28, 29], Faulkner [27], and Brown and Ciftan [38-42] pointed out that straightforward application to space-filling cells of the MST equations in the MT form would lead to serious divergences in the angular momentum expansions. It was shown that the derivations provided by Williams and Morgan violated the $r_</r_>$ conditions for the expansion of the free-particle propagator, Eq. (2.67). These comments were in agreement with Morgan's observations [37] that use of full-cell potentials or charge densities seemed to lead to divergent behavior in the solutions of the Poisson equation. The divergences seemed to be of the "near-field" type, i.e., to depend on the potential of a cell that was inside the bounding sphere of an adjacent cell. Thus, the near-field corrections of MST seemed to raise their unpleasant head in other manifestations of the wave equation. Some additional papers of that era by Scheire [43], Keister [44], and Fereira [45] should also be mentioned. These works purported to show rigorously that MST was indeed applicable in its MT form to potentials of arbitrary, even space-filling shape. However, invariably these papers did not succeed in providing fully satisfactory proofs of such applicability.

The controversy that followed the publication of the papers by Williams and Morgan [35-37] flared up considerably in the 1980s. In their work, Brown and Ciftan [38-42] suggested that in order to account for near-field effects one should consider the potential inside the bounding sphere when determining the cell t-matrix (or the cell sine and cosine functions discussed in Appendix D). They also objected to the procedure for determining the cell t-matrix suggested by Williams and Morgan, suggesting that it is inherently divergent. Faulkner [46, 47, 48] disagreed with this statement but agreed on the need to treat NFCs. He suggested a modification of the structure constants of a lattice to account for such effects. These modified structure constants would contain information about the potential in adjacent cells (the near field). The rather disturbing feature of this phase in the discourse was that the question of the validity of MST had permeated down to the level of the single scatterer. At this point the entire field of second-order partial differential equations as developed within the angular-momentum representation, was being seriously challenged. To make matters even more disconcerting, Badralexe and Freeman [49] provided formal arguments to show that MST was not correct even in the case of MT potentials!

The three camps, represented by Brown and Ciftan, Faulkner, and Badralexe and Freeman argued their points with tenacity and some measure of conviction. What seemed to emerge from the various arguments was that MST was by no means the powerful mathematical tool that it had given promise to be, but was riddled with difficult questions of convergence or formal validity even in its simplest applications.

Fortunately, the dust began to settle in the mid 1980s. In 1986, a paper by Gonis [50] showed formally that the MT form of the full t-matrix for a two-scatterer assembly properly described convex scatterers of arbitrary shape, even if they shared common boundaries. This was accomplished by bypassing expansions in terms of the individual t-matrices and studying the summed form of the full t-matrix. First, the centers of the angular momentum expansions for the cell t-matrices were moved far enough away (an infinite distance in the limit of cells with common boundaries) so that spheres centered at the moved positions and bounding the cells no longer overlapped. Once the resulting MT expressions were summed, the centers were moved back into their original positions through a similarity transformation.

From that point onward, progress was rapid. In 1987, Zeller [51] rigorously showed that the MT form of MST was inherently convergent when applied to MT potentials. He also showed that NFCs could be overcome by using Borel summation techniques. In later work [52], he also showed through a perturbative treatment that no near-field corrections arise to third order in the strength of the potential. This countered the numerical results of Faulkner [46, 47], who had reported rather large effects in connection with the two-dimensional empty lattice (a shifted constant potential that extends over all space). Also, Molenaar [53] used the underlying idea in Gonis's work to derive MST for an arbitrary number of space-filling cells.

A further breakthrough was achieved when Nesbet [54, 55] proved that the formalism of Williams and Morgan and that of Brown and Ciftan leads to *identical* results in the converged limit for the cell t-matrix, or the sine and cosine functions (see Appendix D). He also pointed out that the sine and cosine functions describing the scattering are independent of the potential in the moon region. One immediate and very useful aspect of this result is the possibility of using potentials in the moon region that would speed up convergence in L expansions. Conclusions similar to those reached by Nesbet emerged from further work by Molenaar [56]. Thus, Zeller's work coupled with that of Nesbet removed any doubts as to the applicability of MST to MT potentials, or to the use of the t-matrix in describing the scattering from individual cells. These works opened the door for a formal and rigorous justification of the validity of MST in connection with space-filling cells.

The proofs were not long in coming [57-60]. In addition to formal work, very careful numerical investigations [61, 62], coupled with formal arguments, revealed the inherently *convergent* nature of MST. The discrepancies in the single-site wave functions observed by Brown and Ciftan in applications of the method of Williams and Morgan were traced to problems in the convergence of internal summations [3]. The divergences reported by Morgan could also be overcome [63] through the replacement of a single divergent sum by a conditionally convergent double sum. In addition to the calculations of the electronic structures of elemental Nb, Rh, and Zn based on full-cell potentials by Gonis, Zhang, and Nicholson [57, 58], Nicholson and Faulkner [64] also confirmed the numerical usefulness of full-cell MST in their study of elemental Nb. The works mentioned above, the controversies that ensued from them, and the formal and computational work that led to the final resolution of the problem have been reviewed extensively in a number of publications [3, 65].

In the remainder of this chapter we provide a fo rmal development of MST for the cases of MT and space-filling cell potentials. It is hoped that the formal arguments will clarify the nature of the problem encountered in early attempts to generalize MST to space-filling cells as one *purely of geometry*, connected with the question of expansions of the free-particle propagator. Furthermore, we derive explicit expressions for the t-matrix and the Green function for an assembly of scattering cells, as well as related properties such as the Bloch function and the Lloyd formula. The expressions derived in this chapter are extended further and are generalized in the chapters that follow to the treatment of substitutionally disordered alloys, surfaces and interfaces, and other systems.

3.4 Fundamental Expressions of MST

In this section, we derive a number of basic expressions of multiple scattering theory involving the wave function, the Green function, and the t-matrix for an assembly of scattering potential cells. The discussion in this section is phrased in terms of operators, without reference to a particular representation. Expressions in the coordinate and the angular momentum representation are then presented in subsequent sections. Also, this discussion is based on the scattering picture of a wave propagating through a medium and being scattered by the presence of local (spatially bounded) potential cells. An algebraic approach, based on the properties of solutions of second-order partial differential equations and the corresponding integral equations is given in a later section. We begin by assuming the presence of a finite number of scatterers and subsequently generalize to the limit of an infinite number. In much of the discussion in this section we follow the development in Braspenning [66].

3.4.1 Wave function

Let $|\phi\rangle$ denote a state vector associated with force-free motion, e.g., $|\mathbf{k}\rangle$ denoting a plane-wave state, incident upon a finite collection of spatially bounded potential cells. Ω_i denotes the volume of cell i, centered at position $\mathbf{R}_i$. The potential of the system, V, is taken to be a linear superposition of nonoverlapping individual cell potentials

$$\hat{V} = \sum_i \hat{v}_i. \tag{3.1}$$

We are interested in obtaining expressions for the outgoing wave function $|\psi^+\rangle$ that evolves from $|\phi\rangle$ by means of its interaction with the scattering assembly.

Our discussion in Chapter 2, Eqs. (2.81) and (2.86), has shown that $|\psi^+\rangle$ and $|\phi\rangle$ are connected by the Lippmann-Schinger equation,

$$\begin{aligned} |\psi^+\rangle &= |\phi\rangle + \hat{G}_0^+ \hat{V} |\psi^+\rangle \\ &= |\phi\rangle + \hat{G}_0^+ \hat{T} |\phi\rangle \\ &= \left[1 + \hat{G}_0^+ \hat{T}\right] |\phi\rangle, \end{aligned} \tag{3.2}$$

where $\hat{T}$ is the t-matrix operator associated with the entire scattering assembly. Note that the last line expresses $|\psi^+\rangle$ as a renormalized solution of the free-particle Schrödinger equation, $|\phi\rangle$, but with $\hat{T}$ still unknown and to be determined.

To make further progress, it is convenient to consider a decomposition of the system t-matrix analogous to that of the potential

$$\hat{T} = \sum_i \hat{T}^i, \tag{3.3}$$

where $\hat{T}^i$ describes the multiple scattering events *ending* at site (cell) i. It is important to realize that $\hat{T}^i$ is not equal to the individual cell t-matrix for the potential at i, although it is related to it through a multiple-scattering series. In order to exhibit this relation, we substitute Eq. (3.1) into the defining equation for the t-matrix, Eq. (2.87), and expand in powers of $\hat{V}$ to obtain the expression,

$$\hat{T} = \sum_i \hat{v}_i + \left(\sum_i \hat{v}_i\right) \hat{G}_0^+ \left(\sum_j \hat{v}_j\right) + \cdots. \tag{3.4}$$

We now use Eq. (2.87) in connection with the potential in cell i and obtain the single-cell t-matrix

$$\begin{aligned} \hat{t}^i &= \left(1 - \hat{v}_i \hat{G}_0^+\right)^{-1} \hat{v}_i \\ &\equiv \left(1 - \hat{v}_i \tilde{G}_0^+\right)^{-1} \hat{v}_i, \end{aligned} \tag{3.5}$$

where $\tilde{G}_0^+$ denotes the strictly cell-diagonal part of the free-particle propagator (in a coordinate representation, both arguments of $\tilde{G}_0^+$ would lie in the same cell). Using Eq. (3.5), we can write (3.4) in the form

$$\hat{T} = \sum_i \hat{t}^i + \left(\sum_i \hat{t}^i\right) \bar{G}_0^+ \left(\sum_j \hat{t}_j\right) + \cdots, \tag{3.6}$$

where $\bar{G}_0^+$ denotes the strictly off-diagonal part of $\hat{G}_0^+$, i.e., the part that connects only different cells. The use of the off-diagonal propagator is necessary because by definition the single-cell t-matrix, $\hat{t}^i$, contains all scattering information about the potential $\hat{v}_i$ and no further scattering can take place unless the wave (or particle) first propagates to a different scatterer. Clearly, we have the operator equation,

$$\hat{G}_0 = \tilde{G}_0^+ + \bar{G}_0^+. \tag{3.7}$$

Now, using Eq. (3.3), we can write Eq. (3.6) in the form

$$\sum_i \hat{T}^i = \sum_i \hat{t}^i \left[1 + \bar{G}_0^+ \sum_{j \neq i} \hat{T}^j\right] \tag{3.8}$$

from which we can extract the desired relation between $\hat{T}^i$ and $\hat{t}^i$,

$$\hat{T}^i = \hat{t}^i \left[1 + \bar{G}_0^+ \sum_{j \neq i} \hat{T}^j \right] . \tag{3.9}$$

Here, the exclusion in the sum emphasizes the fact that the off-diagonal part of the free-particle propagator connects only different cells.

The multiple scattering character of the last expression is to be noted. It consists of a single scatterer contribution described by $\hat{t}^i$, at site i, and a superposition of all scattering events ending at all *other* sites j and propagating toward site i via the free-particle Green function, $\bar{G}_0^+$, where the last scattering, expressed by $\hat{t}^i$, takes place. This description allows us to set up a picture of *incoming* and *outgoing* waves at each site i by defining an individual source term, $\hat{v}_i|\psi^+\rangle$, by means of the relation,

$$\hat{v}_i|\psi^+\rangle = \hat{t}^i|\psi_i^{\mathrm{inc}}\rangle, \tag{3.10}$$

where $|\psi_i^{\mathrm{inc}}\rangle$ is the incoming wave at site i. This relation is written in analogy with Eq. (2.88). As in that expression, the potential acting on the final, outgoing wave produces the same effect as the corresponding t-matrix acting on the incident wave.

There are now two very informative expressions that can be written in terms of $|\psi_i^{\mathrm{inc}}\rangle$. First, applying $\hat{v}_i$ on the left of Eq. (3.2) and using the expansion for the system t-matrix in Eqs. (3.6) and (3.9), we find

$$\hat{v}_i|\psi^+\rangle = \hat{t}^i \left[1 + \bar{G}_0^+ \sum_{j \neq i} \hat{T}^j \right] |\phi\rangle, \tag{3.11}$$

so that upon comparison with Eq. (3.10) we have

$$\hat{T}^i|\phi\rangle = \hat{v}^i|\psi^+\rangle = \hat{t}^i|\psi_i^{\mathrm{inc}}\rangle, \tag{3.12}$$

and

$$\begin{aligned} |\psi_i^{\mathrm{inc}}\rangle &= \left[1 + \bar{G}_0^+ \sum_{j} \hat{T}^j \right] |\phi\rangle \\ &= |\phi\rangle + \bar{G}_0^+ \sum_j \hat{t}^j |\psi_j^{\mathrm{inc}}\rangle. \end{aligned} \tag{3.13}$$

This expression *provides a connection between the incoming waves at different sites of the system.* As is shown below, in Eq. (3.19), it allows solutions for $|\psi_i^{\mathrm{inc}}\rangle$ even in the absence of an incoming wave $|\phi\rangle$. The condition that such solutions exist leads to a secular equation that determines the bound states of the scattering assembly and, in the case of solids, the electronic structure or *band structure* of the material.

The second important expression is derived by substituting Eq. (3.5) into Eq. (3.10), which leads to the modified Lippmann-Schwinger equation for $|\psi^+\rangle$,

$$|\psi^+\rangle = \left[1 + \tilde{G}_0^+ \hat{t}^i\right] |\psi_i^{\rm inc}\rangle, \tag{3.14}$$

in which $|\psi_i^{\rm inc}\rangle$ plays the role of the incident wave $|n\rangle_0$ in Eq. (2.81). Finally, by using Eq. (3.3) and the last expression, we can write Eq. (3.2) for $|\psi^+\rangle$ in the form

$$|\psi^+\rangle = |\phi\rangle + \sum_j \hat{G}_0^+ \hat{t}^j |\psi_j^{\rm inc}\rangle. \tag{3.15}$$

This expression exhibits clearly the multiple-scattering nature of the final, outgoing wave. It is to be noted that it involves the entire free-particle propagator, including cell-diagonal and cell-off-diagonal parts. Thus, the outgoing wave is built up from contributions of scattering events that take place at all sites (cells) in the system. As it stands, Eq. (3.15) constitutes a *multi-center* expansion for the total wave function, and shows that it is possible for such a wave to exist even in the absence of an incident wave, $|\phi\rangle$. Combined with Eq. (3.13), which describes the incoming wave at site i as a superposition of waves scattered at all other sites in the material, it is consistent with the interpretation of Eqs. (3.9) and (3.12).

In analogy with an incoming wave at site i, we can define an *outgoing* wave by means of the relation,

$$|\psi_i^{\rm out}\rangle = \tilde{G}_0^+ \hat{t}^i |\psi_i^{\rm inc}\rangle. \tag{3.16}$$

Then, in view of Eq. (3.7), Eq. (3.15), can be written in the form

$$|\psi^+\rangle = |\phi\rangle + \sum_{j\neq i} \bar{G}_0^+ \hat{t}^j |\psi_j^{\rm inc}\rangle + |\psi_i^{\rm out}\rangle. \tag{3.17}$$

This expression is a multicenter expansion for $|\psi^+\rangle$ in the neighborhood of cell i. At the same time, by virtue of Eq. (3.13), the first two terms can be combined into $|\psi_i^{\rm inc}\rangle$ so that $|\psi^+\rangle$ can be written as a *single-center* expansion,

$$|\psi^+\rangle = |\psi_i^{\rm inc}\rangle + |\psi_i^{\rm out}\rangle, \tag{3.18}$$

in the vicinity of any site i in the system. This one-center expansion and its equivalent multi-center expansion in Eq. (3.15) are connected by Eq. (3.13).

The multi-center and one-center expansions for the system wave function derived above are modified forms of the Lippmann-Schwinger equation and are valid at any point in the system. The equation that connects them, Eq. (3.13), can be written in the form,

$$\sum_j \left[\delta_{ij} - (1-\delta_{ij})\bar{G}_0^+ \hat{t}^j\right] |\psi_j^{\mathrm{inc}}\rangle = |\phi\rangle, \tag{3.19}$$

for the set of incoming waves $|\psi_i^{\mathrm{inc}}\rangle$. We see that the absence of an incident wave $|\phi\rangle$ does not preclude the existence of $|\psi_i^{\mathrm{inc}}\rangle$. A finite $|\psi_i^{\mathrm{inc}}\rangle$ can exist even in the absence of an overall incident wave at energies at which the determinant of the coefficients of $|\psi_i^{\mathrm{inc}}\rangle$ in Eq. (3.19) vanishes. Thus, the condition

$$\det \left| \left[\delta_{ij} - \bar{G}_0^+ \hat{t}^j\right] \right| = 0, \tag{3.20}$$

determines the bound states of a scattering assembly. When expressed in a representation such as that of angular momentum, the *secular equation*, Eq. (3.20), determines the expansion coefficients for the wave function and the band structure of the system under consideration.

3.4.2 Green function and t-matrix in MST

Expressions analogous to those for the wave functions (state vectors) of MST obtained in the previous subsection also exist for the Green function and the t-matrix of a scattering assembly. These two quantities are related by the general expressions, see Eqs. (2.72) and (2.90),

$$\begin{aligned} \hat{G} &= \hat{G}_0^+ + \hat{G}_0^+ \hat{V} \hat{G} \\ &= \hat{G}_0^+ + \hat{G}_0^+ \hat{T} \hat{G}_0^+ . \end{aligned} \tag{3.21}$$

It follows from this equation that the states of the scattering potentials can be discussed either in terms of the Green function or of the t-matrix. In fact, a description in terms of the t-matrix is often more convenient because the poles of the free-particle propagator have been largely (although not completely) factored out in such a description.

We saw in Chapter 2, Eq. (2.89), that the system Green function satisfies the relation

$$\hat{G}\hat{V} = \hat{G}_0^+ \hat{T}. \tag{3.22}$$

Now, the first line in Eq. (3.13) can be written in the form,

$$|\psi_i^{\mathrm{inc}}\rangle = \sum_j \left[\delta_{ij} + \hat{G}\hat{v}_j\right] |\phi\rangle, \tag{3.23}$$

whereas Eq. (3.16) for $|\psi_i^{\mathrm{out}}\rangle$ becomes

$$|\psi_i^{\mathrm{out}}\rangle = \tilde{G}\hat{v}_i |\psi_i^{\mathrm{inc}}\rangle. \tag{3.24}$$

Here, we have introduced a splitting of the system Green function into cell diagonal, $\tilde{G}$, and cell off-diagonal parts $\bar{G}$,

$$\hat{G} = \tilde{G} + \bar{G}, \tag{3.25}$$

analogous to the corresponding decomposition of the free-particle propagator in Eq. (3.7). It follows that the off-diagonal part, $\bar{G}$, accounts for the multiple scattering contributions in the system.

We now derive the *fundamental multiple-scattering expression* for the system t-matrix. Consider Eq. (3.9) which expresses the component $\hat{T}^i$ describing all multiple scattering events *ending* at site i. We wish to obtain that part of $\hat{T}$ associated with scattering events *beginning* at site j. In view of Eq. (3.22) we can define the quantity

$$\begin{aligned} \hat{T}^{ij} &= \hat{v}_j \delta_{ij} + \hat{v}_i \hat{G} \hat{v}_j \\ &= \hat{v}_j \delta_{ij} + \hat{v}_i \hat{G}_0^+ \hat{T}_j, \end{aligned} \tag{3.26}$$

where $\hat{T}_j$ is the component of the system t-matrix describing all scattering events that begin at site j. In analogy with Eq. (3.9), $\hat{T}_j$ is given by the expression

$$\begin{aligned} \hat{T}_j &= \hat{v}_j + \hat{T} \hat{G}_0^+ \hat{v}_j \\ &= \left(1 + \hat{T} \hat{G}_0^+\right) \hat{t}^j. \end{aligned} \tag{3.27}$$

Clearly, we have

$$\hat{T} = \sum_{ij} \hat{T}^{ij} = \sum_i \hat{T}^i = \sum_j \hat{T} Jj, \tag{3.28}$$

in terms of either the components describing the scattering that starts or that ends at a site. It follows that Eq. (3.26) can be written in the form,

$$\hat{T}^{ij} = \hat{v}_j \delta_{ij} + \hat{v}_i \hat{G}_0^+ \sum_k \hat{T}^{kj}, \tag{3.29}$$

or, as follows upon iteration and resummation in terms of individual cell t-matrices

$$\hat{T}^{ij} = \hat{t}^j \delta_{ij} + \hat{t}^i \hat{G}_0^+ \sum_k \hat{T}^{kj}. \tag{3.30}$$

The last expression is commonly referred to as the *equation of motion* for the system t-matrix. It is an equation for the matrix elements $\hat{T}^{ij}$ describing all scattering events that start at site j and end on site i. For the sake of clarity, we exhibit an alternative form of the equation of motion

$$\hat{T}^{ij} = \hat{t}^j \left[\delta_{ij} + \sum_{k \neq i} \bar{G}^{ik} \hat{T}^{kj} \right]. \tag{3.31}$$

This equation[1] will form the starting point of much of our development of MST and its applications.

It is interesting to point out that Eq. (3.31) can be solved formally in terms of the inverse of a matrix,

$$\hat{T}^{ij} = \left[\hat{M}^{-1}\right]^{ij}, \tag{3.32}$$

where the matrix $\underline{M}$ is defined by its elements,

$$\hat{M}^{ij} = \left[\hat{t}^i\right]^{-1}\delta_{ij} - \bar{G}^{ij}(1-\delta_{ij}). \tag{3.33}$$

The bound states of a scattering assembly are determined by the poles of the t-matrix (or the Green function), and thus occur at the zeroes of the determinant

$$\det|\underline{M}| = \det|\left[\hat{t}^i\right]^{-1}\delta_{ij} - \bar{G}^{ij}(1-\delta_{ij})| = 0. \tag{3.34}$$

When this equation is expressed in a specific representation, such as that of angular momentum, it leads to the eigenvalue spectrum, or the *band structure* of a scattering assembly.

3.5 Coordinate Representation

The operator equations[2] of the previous section can be readily expressed in the coordinate representation where they take the form of integral equations. Consider, for example, Eq. (3.31). With $|\mathbf{r}\rangle$ and $|\mathbf{r}'\rangle$ position vectors in cells i and j, respectively, we have

$$\langle\mathbf{r}|\hat{T}^{ij}|\mathbf{r}'\rangle \equiv T^{ij}(\mathbf{r},\mathbf{r}'). \tag{3.35}$$

Inserting the identity in the form $\int d^3r|\mathbf{r}\rangle\langle\mathbf{r}|$ between all products in Eq. (3.31), we obtain the expression,

$$\begin{aligned} T^{ij}(\mathbf{r},\mathbf{r}') &= t^i(\mathbf{r},\mathbf{r}')\delta_{ij} \\ &+ \sum_{k\neq i}\int_{\Omega_i} d^3r_1 \int_{\Omega_j} d^3r_2 t^i(\mathbf{r},\mathbf{r}_1)G_0^{ik}(\mathbf{r}_1,\mathbf{r}_2)T^{kj}(\mathbf{r}_2,\mathbf{r}'), \end{aligned} \tag{3.36}$$

which is the coordinate representation for the system t-matrix. Here, the subscripts in the integrals indicate that the ranges of the integration over

[1] We hope that the structure and meaning of the terms in this expression should by now be familiar.

[2] We now suppress explicit operator notation and do not distinguish between operators and their expressions in a specific representation.

$\mathbf{r}_1$ and $\mathbf{r}_2$ extend over the domains of the cells i and j , respectively. We realize that all quantities T^{ij} are evaluated at the same energy so that the equation of motion represents only the *on-the-energy-shell* part of the t-matrix operator[3]. All other expressions presented so far in operator form can be similarly converted into integral equations in the coordinate representations.

The integral equations obtained in the coordinate representation do not lend themselves easily to efficient numerical work. The angular momentum representation, which we discuss in the next section, converts these equations into matrix equations, which are much easier to manipulate and handle computationally.

3.6 Angular Momentum Representation

In the angular momentum representation, expansions are written in terms of the spherical harmonics, $Y_L(\hat{r})$, where L represents the pair (l,m) (see Appendix C). A simple example of such an expansion was encountered in Chapter 2 where it was shown that at energy $E = k^2$ the free-particle propagator can be written in the form, Eqs. (2.58), (2.67), and (2.68),

$$\begin{aligned} G_0(k;\mathbf{r},\mathbf{r}') &= \sum_L G_\ell(k;r,r')Y_L^*(\hat{r})Y_L(\hat{r}') \\ &= -ik\sum_L h_L(kr_>)j_L(kr_<)Y_L^*(\hat{r})Y_L(\hat{r}') \\ &= -ik\sum_L H_\ell(k\mathbf{r}_>)J_\ell(k\mathbf{r}_<), \end{aligned} \tag{3.37}$$

where real spherical harmonics are assumed. Here, we explicitly show the dependence of the various quantities on the energy variable, and use the notation $r_>$ and $r_<$ to denote the larger and smaller of the vectors $\mathbf{r}$ and $\mathbf{r}'$. It is important to keep in mind that in expanding G_0^+ in the manner indicated above, the larger vector should be associated with the irregular solution of the free-particle Schrödinger equation. This condition was noted when Eq. (2.67) was derived. If this condition is violated, the sum over L in Eq. (3.37) diverges. To see this, note the asymptotic expansions of the Bessel and Hankel functions in Eq. (C.34). For large ℓ, the product $h_\ell(x)j_\ell(y)$ behaves as $\frac{1}{x}\left(\frac{y}{x}\right)^\ell \frac{1}{2\ell+1}$, which summed over ℓ diverges when $y > x$.

Let us now consider the expansion of the free-particle propagator in the case in which the vectors $\mathbf{r}$ and $\mathbf{r}'$ are confined inside two adjacent,

[3]A proper t-matrix operator is defined when off-shell parts, connecting states at different energies, are included.

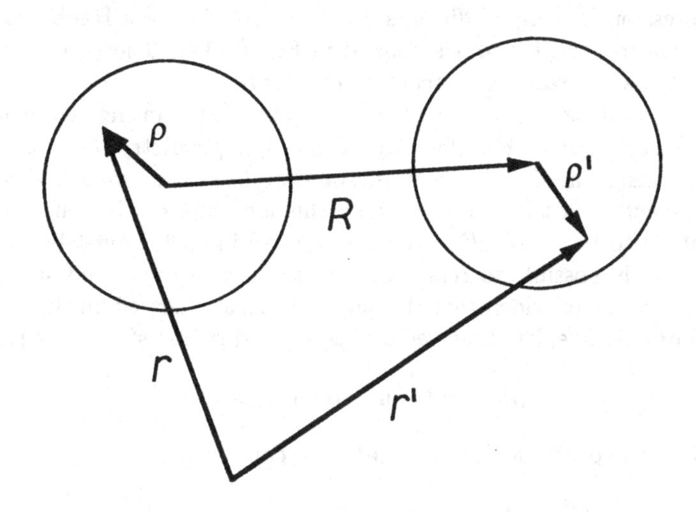

Figure 3.1: **Schematic representation of the MT geometry**

non-overlapping spheres, as is indicated schematically in Fig. 3.1. The geometric arrangement depicted in the figure is an example of the familiar and often used muffin-tin (MT) geometry. If $\boldsymbol{\rho}$ and$\boldsymbol{\rho}'$ are two vectors measured from the centers of and confined inside two non-overlapping spheres, and $\mathbf{R}$ is the vector connecting the centers of the spheres, then the following inequalities hold,

$$|\mathbf{R} - \boldsymbol{\rho}| > |\boldsymbol{\rho}'|, \;\; |\mathbf{R} + \boldsymbol{\rho}'| > |\boldsymbol{\rho}|, \;\; |\mathbf{R}| > |\boldsymbol{\rho}' - \boldsymbol{\rho}|. \tag{3.38}$$

Clearly, we also have

$$R > \rho, \quad \text{and} \quad R > \rho'. \tag{3.39}$$

Then, the vector difference $\mathbf{r} - \mathbf{r}'$ equals $\mathbf{R} + \boldsymbol{\rho} - \boldsymbol{\rho}'$, and using Eqs. (F.7) and (F.10), we can expand the free-particle propagator in the form,

$$\begin{aligned} -\frac{1}{4\pi}\frac{e^{ik|\mathbf{R}+\boldsymbol{\rho}-\boldsymbol{\rho}'|}}{|\mathbf{R}+\boldsymbol{\rho}-\boldsymbol{\rho}'|} &= -ik\sum_{L} H_L(k(\mathbf{R}+\boldsymbol{\rho}))J_L(k\boldsymbol{\rho}') \\ &= -ik\sum_{LL'} J_L(k\boldsymbol{\rho})G_{LL'}(\mathbf{R})J_{L'}(k\boldsymbol{\rho}'). \end{aligned} \tag{3.40}$$

Note that the conditions involving the lengths of the vectors associated with the regular (J_L) and irregular (H_L) solutions of the free-particle Schrödinger equation (also known as the Helmholtz equation) are satisfied at each step of the expansion. The quantities $G_{LL'}(\mathbf{R})$ occurring in the last expression are the coefficients for the expansion of a Hankel function about a shifted origin and are defined in Eq. (F.11). They are commonly referred to as the ***real-space structure constants.***

It is interesting to examine the double summation in the last expansion a little more closely. For the case of the MT geometry and for vectors ρ and ρ' inside muffin tin spheres, the double sum converges absolutely and irrespective of the order in which the sums are carried out [51]. For this feature to be valid, *all* of the inequalities in (3.38) must be satisfied. However, it is possible to relax some of these conditions and still obtain converged sums, provided that the summations are carried out in a specific order. For example, let three vectors, ρ, $\mathbf{R}$, and ρ', satisfy the inequalities,

$$|\mathbf{R}| > |\rho| \quad \text{and} \quad |\mathbf{R}+\rho| > |\rho'|. \tag{3.41}$$

Then, because of the second inequality we can write

$$-\frac{1}{4\pi}\frac{e^{ik|\mathbf{R}+\rho-\rho'|}}{|\mathbf{R}+\rho-\rho'|} = -ik\sum_{L} H_L(k(\mathbf{R}+\rho))J_L(k\rho'), \tag{3.42}$$

and because of the first, we have

$$\begin{aligned} & - \quad ik\sum_{L} H_L(k(\mathbf{R}+\rho))J_L(k\rho') = \\ & - \quad ik\sum_{L}\left[\sum_{L'} J_{L'}(k\rho)G_{L'L}(\mathbf{R})\right] J_L(k\rho'), \end{aligned} \tag{3.43}$$

so that

$$-\frac{1}{4\pi}\frac{e^{ik|\mathbf{R}+\rho-\rho'|}}{|\mathbf{R}+\rho-\rho'|} = -ik\sum_{L}\left[\sum_{L'} J_{L'}(k\rho)G_{L'L}(\mathbf{R})\right] J_L(k\rho'). \tag{3.44}$$

The presence of the brackets indicates that the sum over L' is to be performed *fully* before that over L. If this order is reversed, the double sum may diverge in the case in which $|\mathbf{R}-\rho'| < |\rho|$. To see this, interchange the order of summations to obtain the forma l result

$$\begin{aligned} & - \quad ik\sum_{L'} J_{L'}(k\rho)\sum_{L} G_{L'L}(k\mathbf{R})J_L(k\rho') = \\ & - \quad ik\sum_{L'} J_{L'}(k\rho)H_{L'}(k(\mathbf{R}-\rho')). \end{aligned} \tag{3.45}$$

Since the argument of the regular solution is larger than that of the irregular solution, this sum in general diverges.

However, the last discussion reveals a very useful feature of the expansions of the free-particle propagator. Namely, the possibility of converting a divergent single sum, Eq. (3.45), into a conditionally convergent double sum, Eq. (3.44). This method, which is related ***purely to the geometric character*** of the expansion, can be used to sum otherwise divergent series that can occur in applications of MST to space-filling potentials[4].

3.7 MST for MT Potentials

We now obtain expressions in the angular momentum representation for the t-matrix and the Green function associated with an assembly of MT scattering cells. Although applications of MST have been made almost exclusively in connection with spherically symmetric potentials, we do not impose that restriction in the following discussion. Thus, we consider a finite assembly of spatially bounded cells of arbitrary shape but arranged in space so that the cells are separated from one another by non-overlapping bounding spheres. A schematic representation of such a collection of scattering cells is shown in Fig. 3.2. Given any two cells i and j, we denote by $\mathbf{R}_{ij}$ the vector connecting the centers of the cells, taken to be also the centers of the bounding spheres. If $\mathbf{r}_i$ and $\mathbf{r}_j$ are vectors measured from the cell (sphere) centers and confined inside the cells i and j, respectively, then they satisfy the MT inequalities in (3.38). This simplifies considerably the determination of the t-matrix and the Green-function for the system of scatterers within MST.

3.7.1 t-matrix for MT potentials

We begin with the equation of motion, Eq. (3.36), which we write in the form,

$$\begin{aligned} T^{ij}(\mathbf{r},\mathbf{r}') &= t^i(\mathbf{r},\mathbf{r}')\delta_{ij} \\ &+ \sum_{k\neq i}\int_{\Omega_i} \mathrm{d}^3r_i \int_{\Omega_k} \mathrm{d}^3r_k t^i(\mathbf{r},\mathbf{r}_i)G_0^{ik}(\mathbf{r}_i,\mathbf{r}_k)T^{kj}(\mathbf{r}_k,\mathbf{r}'). \end{aligned} \tag{3.46}$$

In this expression, we insert the expansion of the free-particle propagator, Eq. (3.40), and obtain

$$T^{ij}(\mathbf{r},\mathbf{r}') = t^i(\mathbf{r},\mathbf{r}')\delta_{ij}$$

[4]Alternative uses of such conditional summations involve the introduction of modified structure constants that depend exclusively on geometry but in a such a way as to lead to converged results in the summations of MST. Detailed discussions of such constructions can be found in the literature [3].

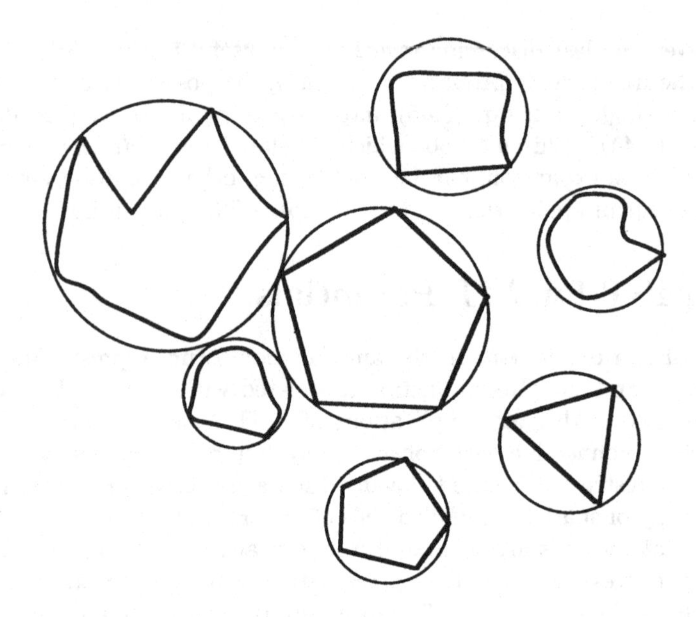

Figure 3.2: **A collection of cells satisfying the MT conditions.**

$$+ \sum_{k \neq i} \int_{\Omega_i} \mathrm{d}^3 r_i \int_{\Omega_k} \mathrm{d}^3 r_k t^i(\mathbf{r}, \mathbf{r}_i)$$

$$\times \left[\sum_{LL'} J_L(k\mathbf{r}_i) G_{LL'}(\mathbf{R}_{ik}) J_{L'}(k\mathbf{r}_k) \right] T^{kj}(\mathbf{r}_k, \mathbf{r}'), \quad (3.47)$$

where the prefactor, $-ik$, in the expansion of G_0 is absorbed into the definition of the structure constants, $G_{LL'}$. Performing the integrals over $\mathbf{r}_i$ and $\mathbf{r}_k$ replaces the corresponding integrations by summations over the angular momentum indices L and L'. In fact, the integral Eq. (3.46) can be converted into a matrix equation based on the following definition: For any quantity $A^{ij}(\mathbf{r}_i, \mathbf{r}_j)$, we define the matrix elements,

$$A_{LL'} = \int_{\Omega_i} \mathrm{d}^3 r_i \int_{\Omega_j} \mathrm{d}^3 r_j J_L(k\mathbf{r}_i) A^{ij}(\mathbf{r}_i, \mathbf{r}_j) J_{L'}(k\mathbf{r}_j), \quad (3.48)$$

or in matrix form,

$$\underline{A} = \int_{\Omega_i} \mathrm{d}^3 r_i \int_{\Omega_j} \mathrm{d}^3 r_j |J(k\mathbf{r}_i)\rangle A^{ij}(\mathbf{r}_i, \mathbf{r}_j) \langle J(k\mathbf{r}_j)|, \quad (3.49)$$

where Ω_i denotes the interior of cell i.

In the manner indicated in Appendix F, an underline denotes a matrix in angular momentum and/or site space, and $|J(k\mathbf{r}_i)\rangle$ and $\langle J(k\mathbf{r}_j)|$ denote row and column vectors, respectively, whose components are given by $J_L(k\mathbf{r}_i)$. For example, for the single-site t-matrix, we have

$$\underline{t}^i = \int_{\Omega_i} \mathrm{d}^3 r_1 \int_{\Omega_i} \mathrm{d}^3 r_2 |J(k\mathbf{r}_1)\rangle t^i(\mathbf{r}_1, \mathbf{r}_2) \langle J(k\mathbf{r}_2)|. \tag{3.50}$$

We now multiply Eq. (3.47) by $J_{L_1}(k\mathbf{r})$ on the left and $J_{L_2}(k\mathbf{r}')$ on the right, with $\mathbf{r}$ and $\mathbf{r}'$ confined in Ω_i and Ω_j, respectively, and integrate over the domains of these cells. Using the definition in Eq. (3.48), we obtain the expression,

$$T^{ij}_{L_1 L_2} = t^i_{L_1 L_2} \delta_{ij} + \sum_{k \neq i} \sum_{LL'} t^i_{L_1 L} G_{LL'}(\mathbf{R}_{ik}) T^{kj}_{L' L_2}. \tag{3.51}$$

In view of Eq. (3.49), this equation can be written in matrix notation

$$\underline{T}^{ij} = \underline{t}^i \left[\delta_{ij} + \sum_{k \neq i} \underline{G}^{ik} \underline{T}^{kj} \right]. \tag{3.52}$$

This expression is the *matrix* form of the t-matrix for a multiple-scattering system in the angular momentum representation. Compared to the corresponding expression in real (coordinate) space, it holds the distinct advantage of expressing integral equations in terms of matrix multiplications. There are a number of additional features of this expression worth emphasizing.

First, Eq. (3.52) is formally identical to the operator equation of motion, Eq. (3.31). Also, as already mentioned, it involves only the on-the-energy-shell t-matrix which, by itself, does not define an operator. However, the on-shell part of the t-matrix is all that is required to determine the single-particle electronic states and related properties of a system of scattering centers. Second, the exclusion in the summations over sites indicates that no two consecutive scattering events take place at the same site. This must be the case because the cell t-matrices describe the complete scattering from any given cell. Third, the structure constants $\underline{G}^{ik}$ are the cell (site) off-diagonal matrix elements of the free-particle propagator and are independent of potential. Their presence illustrates the complete separation of the *potential* aspects of the system, embodied in the cell t-matrices, from the *structural* aspects reflected in the structure constants. This separation not only makes the equations of MST appealing at the conceptual level, but has also important computational ramifications. For example, the structure constants associated with a given lattice structure need be calculated only once and used as necessary in connection with different potentials (as arise in a charge self-consistent calculation). Also, the fact that the cell t-matrices become rapidly negligible with increasing L and L' means that often the matrices in angular momentum space need be carried no further than a small value of L. For most metallic, close-packed systems very accurate results can be obtained for values of L as small as 2.

The separation of the structure from the potential, so easily effected in the MT case, must be justified separately in the case of space-filling cells. As is shown in the following sections, this justification can be obtained in a number of different ways, including the replacement of single, possibly divergent, sums by conditionally convergent double (and occasionally multiple) sums. However, in all cases the final expressions of MST can be written in forms that retain the separation of potential from structure and are often identical to the simple forms obtained within the MT geometry. It is only necessary to associate with these forms the proper summation procedures to guarantee convergence.

Finally, in order to conform with established notation, we designate the matrix elements $\underline{T}^{ij}$ by the symbol $\underline{\tau}^{ij}$. The quantity $\underline{\tau}$ is referred to as the *scattering path operator* (SPO) [67] and denotes the on-the-energy-shell elements of the system t-matrix. Because $\underline{\tau}$ is not an operator in the proper sense of the word, a possibly more appropriate term for it would be the scattering-path matrix. However, in our discussion we follow established practice hoping that no misunderstandings arise from it as to the nature of various quantities. Thus, the equation of motion, Eq. (3.52), can be written in the form,

$$\underline{\tau}^{ij} = \underline{t}^{i}\left[\delta_{ij} + \sum_{k\neq i} \underline{G}^{ik}\underline{\tau}^{kj}\right]. \tag{3.53}$$

Even though the equation of motion has been obtained for a finite cluster of scatterers, the condition that the system be finite has not been used explicitly in the derivations leading to Eq. (3.52) or (3.53). Thus, these equations apply equally well to systems with arbitrarily large (infinite) numbers of scattering cells. The equation of motion, Eq. (3.53), forms the starting point of much of the discussion of the electronic structure of different physical systems in later sections and chapters of the book.

Before leaving this section, let us note that like its operator counterpart, Eq. (3.53) has an immediate solution in terms of the inverse of a matrix

$$\underline{\tau}^{ij} = \left[\underline{M}^{-1}\right]^{ij}, \tag{3.54}$$

where $\underline{M}$ is a matrix in both site and angular momentum space defined by the elements

$$\begin{aligned} \underline{M}^{ij} &= \left[\underline{t}^{i}\right]^{-1}\delta_{ij} - \underline{G}^{ij}\left(1-\delta_{ij}\right) \\ &= \underline{m}^{i}\delta_{ij} - \underline{G}^{ij}\left(1-\delta_{ij}\right). \end{aligned} \tag{3.55}$$

In this expression it is assumed that the inverse of the single-cell t-matrix, $\underline{m} = \underline{t}^{-1}$, exists and converges sufficiently rapidly with increasing L. Note that this expression is formally identical to the corresponding Eq. (3.33) obtained in operator space. As we will see later, the matrix $\underline{M}$ can be put in a number of different forms that are distinguished by different convergence and variational properties.

3.7.2 Electronic structure of a periodic solid

Before proceeding with further consideration of Eq. (3.53), let us illustrate how this equation can be used to obtain the E vs. $\mathbf{k}$ dispersion relation, (the band structure), of a translationally invariant material characterized by a given[5] set of cell potentials, v_i. We consider the case in which identical scatterers are arranged on the sites of a simple Bravais lattice, so that $\underline{t}^i = \underline{t}$ for all i. Then the SPO, $\underline{\tau}$, is diagonal in reciprocal space, $\mathbf{k}$-space, and Eq. (3.53) can be solved by means of lattice Fourier transforms. Introducing the Fourier transforms,

$$\underline{\tau}(\mathbf{k}) = \frac{1}{N}\sum_{n,m} e^{i\mathbf{k}\cdot(\mathbf{R}_m - \mathbf{R}_n)}\underline{\tau}^{mn}, \tag{3.56}$$

$$\begin{aligned} \underline{\tau}^{mn} &= \frac{1}{N}\sum_{\mathbf{k}} e^{-i\mathbf{k}\cdot(\mathbf{R}_m - \mathbf{R}_n)}\underline{\tau}(\mathbf{k}) \\ &= \frac{1}{\Omega_{BZ}}\int_{BZ} d^3k e^{-i\mathbf{k}\cdot(\mathbf{R}_m - \mathbf{R}_n)}\underline{\tau}(\mathbf{k}), \end{aligned} \tag{3.57}$$

where BZ denotes the first Brillouin zone, and

$$\underline{G}(\mathbf{k}) = \frac{1}{N}\sum_{n,m} e^{i\mathbf{k}\cdot(\mathbf{R}_m - \mathbf{R}_n)}\underline{G}(\mathbf{R}_{mn}), \tag{3.58}$$

we obtain from Eq. (3.54) the solution

$$\underline{\tau}(\mathbf{k}) = [\underline{m} - \underline{G}(\mathbf{k})]^{-1}. \tag{3.59}$$

Both $\underline{t}$ and $\underline{G}(\mathbf{k})$ in the last expression are functions of the energy, E, but $\underline{t}$ depends only on the potential of a cell, whereas $\underline{G}(\mathbf{k})$ depends only on the structure of the lattice. Thus, Eq. (3.59) reflects the complete separation of the potential and the structural aspects of the multiple scattering assembly within the computationally convenient angular momentum representation. The quantities $\underline{G}(\mathbf{k})$ were introduced by Korringa [6] and Kohn

[5]In realistic calculations, the potential must be determined self-consistently through, say, the LDA of DFT.

and Rostoker [9], and are commonly referred to as the *KKR structure constants.* These structure constants need to be calculated only once for a given lattice, stored and used as it becomes necessary in connection with different MT potentials on the latice. Methods for the evaluation of the KKR structure constants, often also called the *structural Green function*[6], can be found in the literature [11].

Let us now return to the calculation of the electronic structure. It is given by the distribution in energy of the **k**-dependent eigenvalues of the Hamiltonian, which in turn are determined by the poles of the system t-matrix. Thus, at a given energy, E, the allowed eigenvalues, $E(\mathbf{k})$, can be obtained from the solutions of the so-called *secular equation,*[7]

$$\det\left|\underline{\tau}^{-1}(\mathbf{k})\right| = \det\left|\underline{m} - \underline{G}(\mathbf{k})\right| = 0, \tag{3.60}$$

where $\underline{m}$ depends on the energy. This equation is evidently the angular momentum representation of the abstract operator Eq. (3.34). Alternatively, Eq. (3.60) can be solved as a function of the energy for $\mathbf{k}(E)$ in the first Brillouin zone of the reciprocal lattice. These two alternative ways of searching for the roots of the secular equation are referred to as the "E-search" and "k-search" modes, respectively. Either one leads to the determination of the band structure, $E_n(\mathbf{k})$, associated with the various bands, n, of the periodic solid.

If the potentials are taken to be spherically symmetric, then the secular Eq. (3.60) can be written in the simpler forms

$$\det\left|\underline{m} - \underline{G}(\mathbf{k})\right| = 0 \quad \text{or} \quad \det\left|\cot\delta_\ell \underline{I} + \underline{B}(\mathbf{k})\right| = 0 \tag{3.61}$$

where δ_ℓ is the phase shift for the ℓth partial wave introduced in Appendix D, and $\underline{B} = -\underline{G} + ik\underline{I}$. A direct derivation of the second form of the secular equation is given in a later section.

It is to be noted that the secular Eq. (3.60) involves in principle infinite-dimensional matrices labeled by the angular momentum index, L. In realistic calculations, however, it is customary to truncate the values of L at some relatively small number, say $L = 2$ or $L = 4$. Such truncations have been found to be well justified in many cases of physical interest, such as that of metallic systems on close-packed lattices. In these cases the potentials are very nearly spherically symmetric, which justifies the MT approximation, and a truncation at low L values is not unreasonable. However, the MT approximation may not be very realistic in the case of systems with low rotational and translational symmetry e.g., surfaces and

[6]The term structural Green function is also used in the literature to denote a quantity closely related to the scattering-path operator.

[7]The secular equation can be written in a number of different forms, which in the case of MT potentials lead to identical results. However, in non-MT cases, these forms may exhibit different numerical behavior, as is discussed in following sections.

highly anisotropic charge distributions such as occur in covalently bonded materials. Thus, there arises the need to extend multiple scattering theory to arbitrarily shaped, space-filling cell potentials, a task which will be undertaken beginning with the next section.

We conclude this discussion with the derivation of a useful expression for the total t-matrix of a scattering assembly, i.e., the t-matrix describing the scattering ensemble as a single scattering cell, centered at the center of one of the individual cells. Consider the t-matrix, $T(\mathbf{r}, \mathbf{r}')$, associated with an N-scatterer system. According to Eq. (3.48), in the angular momentum representation we have,

$$\underline{T} = \int \mathrm{d}^3 r \int \mathrm{d}^3 r' |J(k\mathbf{r})\rangle T(\mathbf{r}, \mathbf{r}') \langle J(k\mathbf{r}')|. \tag{3.62}$$

We now break up the integrals into sums over the domain of each cell in the assembly and obtain the expression

$$\underline{T} = \sum_{i=1}^{N} \sum_{j=1}^{N} \int_{\Omega_i} \mathrm{d}^3 r_i \int_{\Omega_j} \mathrm{d}^3 r_j |J(k(\mathbf{R}_i + \mathbf{r}_i))\rangle T^{ij}(\mathbf{r}_i, \mathbf{r}_j) \langle J(k(\mathbf{R}_i + \mathbf{r}_i))|, \tag{3.63}$$

where $\mathbf{r}_i$ ($\mathbf{r}_j$) denotes the same point as $\mathbf{r}$ ($\mathbf{r}'$) but measured from $\mathbf{R}_i$ ($\mathbf{R}_j$), the center of cell I (j). Using the expansion of the Bessel function in Eq. (F.15), this last expression can be written in the form

$$\underline{T} = \sum_{i=0}^{N} \sum_{j=0}^{N} \underline{g}(\mathbf{R}_{i0}) \underline{\tau}^{ij} \underline{g}(\mathbf{R}_{j0}). \tag{3.64}$$

This expression has a straightforward interpretation. The full t-matrix is the sum of all intercell scattering events, represented by the SPO, referred to a common center by means of the elements, $\underline{g}(\mathbf{R})$, of the translation operator, given by Eq. (F.5).

3.7.3 Green function

We now use the framework of MST to derive expressions for the system Green function of an assembly of scatterers. This is a very important quantity since it forms the basis of a great deal of subsequent development, and we provide its derivation in some detail. In fact, because of its formal and computational importance we provide alternative derivations of the Green function in following sections.

We begin by considering the single-scatterer contribution to the Green function of a scattering assembly. From Eqs. (3.21) and (3.31) we have the general expansion

$$\begin{aligned}\hat{G} &= \hat{G}_0^+ + \hat{G}_0^+ \left[\sum_i \hat{t}^i + \sum_{ijk} \hat{t}^i \bar{G}^{ik} \hat{T}^{kj}\right] \hat{G}_0^+ \\ &= \hat{G}_0^+ + \hat{G}_0^+ \sum_i \hat{t}^i \hat{G}_0^+ + \hat{G}_0^+ \sum_i \hat{t}^i \hat{G}_0^+ \sum_j \hat{t}^j \hat{G}_0^+ + \cdots, \quad (3.65)\end{aligned}$$

where the summations are taken to exclude consecutive scatterings from the same site.

In the coordinate representation, the single-scatterer term, $\hat{G}_0^+ \hat{t}^i \hat{G}_0^+$, takes the form

$$\begin{aligned}\langle \mathbf{r}|G_0 t^i G_0|\mathbf{r}'\rangle &= \int \mathrm{d}^3 y \int \mathrm{d}^3 z G_0(\mathbf{y}-\mathbf{r}) t^i(\mathbf{y}-\mathbf{R}_i, \mathbf{z}-\mathbf{R}_i) G_0(\mathbf{r}'-\mathbf{z}) \\ &= \int \mathrm{d}^3 y' \int \mathrm{d}^3 z' G_0(\mathbf{R}_i - \mathbf{r} + \mathbf{y}') \\ &\times\ t^i(\mathbf{y}', \mathbf{z}') G_0(\mathbf{r}' - \mathbf{z}' - \mathbf{R}_i) \quad (3.66)\end{aligned}$$

where $\mathbf{R}_i$ labels the center of scattering cell (or MT) i, and $\mathbf{y}'$ and $\mathbf{z}'$ are the relative coordinates of the vectors $\mathbf{y}$ and $\mathbf{z}$ with respect to $\mathbf{R}_i$. Let us consider specifically the case in which both arguments of $G(\mathbf{r}, \mathbf{r}')$ lie outside all bounding spheres, in the interstitial region. Using Eq. (3.37), we have,

$$\begin{aligned}\langle \mathbf{r}|G_0 t^i G_0|\mathbf{r}'\rangle &= \int \mathrm{d}^3 y' \int \mathrm{d}^3 z' \sum_L (-\mathrm{i}k) H_L(\mathbf{R}_i - \mathbf{r}) J_L(-\mathbf{y}') \\ &\times\ t^i(\mathbf{y}', \mathbf{z}') \sum_{L'} (-\mathrm{i}k) J_{L'}(\mathbf{z}') H_{L'}(\mathbf{r}' - \mathbf{R}_i) \\ &= \sum_{LL'} (-\mathrm{i}k)^2 H_L(\mathbf{R}_i - \mathbf{r}) t^i_{LL'} H_{L'}(\mathbf{r}' - \mathbf{R}_i), \quad (3.67)\end{aligned}$$

where the definition of $t^i_{LL'}$, Eq. (3.50), has been used. Similarly, the two-scatterer term can be written in the form,

$$\begin{aligned}\langle \mathbf{r}|G_0 t^i G_0 t^j G_0|\mathbf{r}'\rangle &= \sum_{LL_1} \sum_{L_2 L'} (-\mathrm{i}k)^2 H_L(\mathbf{R}_i - \mathbf{r}) \\ &\times\ t^i_{LL_1} G_{L_1 L_2}(\mathbf{R}_{ij}) t^j_{L_2 L'} H_{L'}(\mathbf{r}' - \mathbf{R}_j). \quad (3.68)\end{aligned}$$

Proceeding in this manner, we obtain for $\mathbf{r}$ and $\mathbf{r}'$ both in the interstitial region the expression

$$\begin{aligned}G(\mathbf{r}, \mathbf{r}') &= G_0(\mathbf{r} - \mathbf{r}') \\ &+ \sum_{m,n} \sum_{L_1 L_2} (-\mathrm{i}k)^2 H_{L_1}(\mathbf{R}_m - \mathbf{r}) \tau^{mn}_{L_1 L_2}(E) H_{L_2}(\mathbf{r}' - \mathbf{R}_n) \quad (3.69)\end{aligned}$$

in terms of the scattering path operator [67].

The equation for the Green function can be written in a form exhibiting the solutions of the Schrödinger equation associated with the individual MT potentials. Again restricting the arguments of $G(\mathbf{r},\mathbf{r}')$ to lie outside the bounding spheres but inside a given Wigner-Seitz cell, say the one at the origin, we can cast Eq. (3.69) in the form [68],

$$\begin{aligned} G(\mathbf{r},\mathbf{r}') &= G_0(\mathbf{r},\mathbf{r}') + (-ik)^2\langle H(\mathbf{r})|\underline{\tau}^{00}|H(\mathbf{r}')\rangle \\ &+ (-ik)^2\sum_{n\neq 0}\langle H(\mathbf{r})|\underline{\tau}^{0n}\underline{G}(\mathbf{R}_n)|J(\mathbf{r}')\rangle \\ &+ (-ik)^2\sum_{m\neq 0}\langle J(\mathbf{r})|\underline{G}(\mathbf{R}_m)\underline{\tau}^{m0}|H(\mathbf{r}')\rangle \\ &+ (-ik)^2\sum_{n\neq 0}\sum_{m\neq 0}\langle J(\mathbf{r})|\underline{G}(\mathbf{R}_m)\underline{\tau}^{mn}\underline{G}(\mathbf{R}_n)|J(\mathbf{r}')\rangle, \end{aligned} \tag{3.70}$$

where the vector/matrix notation introduced in Appendix F has been used. The expressions

$$\underline{\tau}^{00} = \underline{t}^0 + \underline{t}^0\sum_{m\neq 0}\underline{G}^{0m}\underline{\tau}^{m0} \tag{3.71}$$

and

$$\underline{\tau}^{00} = \underline{t}^0 + \sum_{n\neq 0}\underline{\tau}^{0n}\underline{G}^{n0}\underline{t}^0, \tag{3.72}$$

now imply that

$$\begin{aligned} \sum_{m\neq 0}\underline{G}(\mathbf{R}_m)\underline{\tau}^{m0} &= \sum_{\mathbf{k}}\underline{G}(\mathbf{k})\underline{\tau}(\mathbf{k}) \\ &= \underline{m}^0\underline{\tau}^{00} - 1, \end{aligned} \tag{3.73}$$

$$\sum_{m\neq 0}\underline{\tau}^{0m}\underline{G}(\mathbf{R}_m) = \underline{\tau}^{00}\underline{m}^0 - 1, \tag{3.74}$$

and

$$\sum_{m\neq 0}\sum_{n\neq 0}\underline{G}(\mathbf{R}_m)\underline{\tau}^{mn}\underline{G}(\mathbf{R}_n) = \underline{m}^0\underline{\tau}^{00}\underline{m}^0 - \underline{m}^0, \tag{3.75}$$

allowing us to recast Eq. (3.70) in the form,

$$\begin{aligned} G(\mathbf{r},\mathbf{r}') &= \langle J(\mathbf{r})\underline{m}^0 - ikH(\mathbf{r})|\underline{\tau}^{00}|\underline{m}^0 J(\mathbf{r}') - ikH(\mathbf{r}')\rangle \\ &- \langle J(\mathbf{r})\underline{m}^0 - ikH(\mathbf{r})|J(\mathbf{r}')\rangle. \end{aligned} \tag{3.76}$$

Here, we used the expansion of the free-particle propagator, Eq. (3.37), and chose $\mathbf{r}'$ greater than $\mathbf{r}$. From the fact that the Green function is continuous with respect to either of its arguments, we can write an expression for the Green function even when the vectors $\mathbf{r}$ and $\mathbf{r}'$ lie inside the MT spheres

$$G(\mathbf{r},\mathbf{r}') = G^{(1)}(\mathbf{r},\mathbf{r}') + \langle Z(\mathbf{r})|\underline{\tau}^{00}|Z(\mathbf{r}')\rangle \tag{3.77}$$

where we have defined the "single-scatterer" part of the Green function corresponding to the cell at the origin

$$G^{(1)}(\mathbf{r},\mathbf{r}') = -\langle Z(\mathbf{r})|\tilde{J}(\mathbf{r}')\rangle. \tag{3.78}$$

Here, $|Z(\mathbf{r})\rangle$ and $|\tilde{J}(\mathbf{r})\rangle$ are those regular and irregular solutions, respectively, of the Schrödinger equation which at the radius of the MT sphere join smoothly to the corresponding external functions,

$$|Z(\mathbf{r})\rangle \to \underline{m}|J(\mathbf{r})\rangle - ik|H(\mathbf{r})\rangle \tag{3.79}$$

and

$$|\tilde{J}(\mathbf{r})\rangle \to |J(\mathbf{r})\rangle. \tag{3.80}$$

For real potentials and real energies the functions Z and $\tilde{J}$ can be chosen to be real.

Although Eq. (3.77) was derived for $\mathbf{r}$ and $\mathbf{r}'$ outside the bounding sphere (but inside the cell at the origin), it satisfies the defining equation for the Green function,

$$\left[-\nabla^2 - k^2 + V\right] G(\mathbf{r},\mathbf{r}') = \delta(\mathbf{r}-\mathbf{r}'), \tag{3.81}$$

and contains the proper boundary conditions so that it can readily be continued inside the sphere and the potential region. Thus, for $\mathbf{r}$ ($\mathbf{r}'$) inside no other WS cells than $n(m)$ and for $r' > r$, we can write

$$G(\mathbf{r},\mathbf{r}') = \langle Z^n(\mathbf{r})|\underline{\tau}^{mn}|Z^m(\mathbf{r}')\rangle - \langle Z^n(\mathbf{r})|\tilde{J}^n(\mathbf{r}')\rangle\delta_{nm}. \tag{3.82}$$

Once $G(\mathbf{r},\mathbf{r}')$ has been determined, all single-particle properties of the system under consideration can be calculated. For example, the electron number density at position $\mathbf{r}$ inside cell n and at a real energy E is given by the expression,

$$\begin{aligned} n^n(\mathbf{r};E) &= -\frac{1}{\pi}\Im G(\mathbf{r},\mathbf{r}') \\ &= -\frac{1}{\pi}\langle Z^n(\mathbf{r})|\Im\underline{\tau}^{nn}|Z(\mathbf{r})\rangle \\ &= -\frac{1}{\pi}\sum_L\sum_{L'} Z^n_L(\mathbf{r})\Im\tau^{nn}_{LL'}Z_{L'}(\mathbf{r}), \end{aligned} \tag{3.83}$$

where the charge density is given by the expression $\rho(\mathbf{r},E) = -en(\mathbf{r},E)$. This expression will be found to be valid for nonspherical as well as spherical potentials. [In calculations carried out at complex energies the single-scatterer term, $G^{(1)}$, in Eq. (3.77), acquires an imaginary part and must be taken into consideration.] The number density is given by the relation

$$n(\mathbf{r}) = \int \mathrm{d}En(\mathbf{r};E), \tag{3.84}$$

and the density of states (DOSs) at a real energy E is an integral over all space of the number density,

$$n(E) = \int \mathrm{d}^3rn(\mathbf{r};E). \tag{3.85}$$

We see that the number density and, consequently, the density of states (DOS) are determined directly in terms of the imaginary part of $\underline{\tau}^{nn}$.

These equations also indicate how MST can be used to carry out self-consistent calculations of the charge density and the potential in a solid. The density determined in Eq. (3.84) can be used to obtain a new value of the potential through the formalism of the LDA, which leads to a new value of the density. This iterative process continues until the density (or the ground state energy) converge to some preset criteria.

3.8 Alternative Derivation

We can gain further insight into the formalism of multiple scattering theory by considering a derivation of the secular equation of MST based directly on the properties of the wave function. This derivation demonstrates that the wave function determined within MST is properly continuous (satisfies the appropriate matching conditions) across any surface, such as that of a MT sphere.

The aim of multiple scattering theory is the solution of the single-particle Schrödinger equation whose integral representation is given by the Lippmann-Schwinger Eq. (2.81),

$$\psi(\mathbf{r}) = \chi(\mathbf{r}) + \int G_0(\mathbf{r},\mathbf{r}')V(\mathbf{r}')\psi(\mathbf{r}')\mathrm{d}^3r'. \tag{3.86}$$

We are interested in obtaining the stationary states of the potential V and thus set $\chi(\mathbf{r}) = 0$. Therefore, we have to solve the integral equation,

$$\psi(\mathbf{r}) = \int G_0(\mathbf{r},\mathbf{r}')V(\mathbf{r}')\psi(\mathbf{r}')\mathrm{d}^3r'. \tag{3.87}$$

We continue to work within the MT geometry in which the potential, (not necessarily spherically symmetric) is taken to vanish outside the spheres inscribed in the cells, n. Accordingly, Eq. (3.87) can be rewritten in the form,

$$\begin{aligned}\psi(\mathbf{r}+\mathbf{R}_n) &- \int_{\Omega_n} d^3r G_0(\mathbf{r}-\mathbf{r}')V_n(\mathbf{r}')\psi(\mathbf{r}+\mathbf{R}_n) \\ &= \sum_{m\neq n}\int_{\Omega_m} d^3r' G_0(\mathbf{r}-\mathbf{r}'+\mathbf{R}_{nm})V_n(\mathbf{r}')\psi(\mathbf{r}+\mathbf{R}_m),\end{aligned} \tag{3.88}$$

with $\mathbf{r}$ and $\mathbf{r}'$ now denoting relative cell coordinates (measured from the centers of the cells). In the spirit of Eq. (3.18), we write the wave function, ψ, in each cell n, as a *one-center expansion*

$$\begin{aligned}\psi(\mathbf{r}) &= \sum_L \gamma_L^n \psi_L^n(\mathbf{r}) \\ &= \langle \gamma^n | \psi^n(\mathbf{r})\rangle\end{aligned} \tag{3.89}$$

in terms of the basis functions $\psi_L^n(\mathbf{r})$ associated with the potential in the cell. The validity and convergence of this expansion of the wave function is discussed in the literature [3]. The following discussion demonstrates that MST provides a procedure for the determination of the coefficients γ_L^n in Eq. (3.89).

From Eqs. (D.84) and (D.85) the cell basis functions $\psi_L^n(\mathbf{r})$ satisfy the integral equation

$$\psi_L^n(\mathbf{r}) = \sum_{L'} J_{L'}(\mathbf{r})C_{LL'}^n + \int d^3r' G_0(\mathbf{r}-\mathbf{r}')V_n(\mathbf{r}')\psi_L^n(\mathbf{r}'), \tag{3.90}$$

and were obtained explicitly[8] in Eq. (D.94),

$$\psi_L^n(\mathbf{r}) = \sum_{L'} \left[C_{LL'}^n(r)J_{L'}(\mathbf{r}) - S_{LL'}^n(r)N_{L'}(\mathbf{r})\right], \tag{3.91}$$

where the phase functions $\underline{S}^n(\mathbf{r})$ and $\underline{C}^n(\mathbf{r})$ are given by Eqs. (D.92) and (D.93), respectively. In the interstitial region, these functions assume their asymptotic constant values, $\underline{C}^n$ and $\underline{S}^n$. The normalization of the solutions $\psi_L^n(\mathbf{r})$ is fixed by the requirement that

$$\lim_{r\to 0} \psi_L^n(\mathbf{r}) = J_L(\mathbf{r}), \tag{3.92}$$

[8] For the sake of completeness of presentation, in this discussion we employ the Neumann functions rather than the Hankel functions.

consistent with the boundary condition $C^n_{LL'}(0) = \delta_{LL'}$.

Thus for $\mathbf{r}$ in the interstitial region, $\psi^n_L(\mathbf{r})$ takes the asymptotic form

$$|\psi^n(\mathbf{r})\rangle = \underline{C}^n|J(\mathbf{r})\rangle - \underline{S}^n|N(\mathbf{r})\rangle. \tag{3.93}$$

At this point we need to specify the form of $G_0(\mathbf{r}-\mathbf{r}')$. Consistent with our search for stationary solutions, we use Neumann functions and write (for $r' > r$)

$$\begin{aligned} G_0(\mathbf{r}-\mathbf{r}') &= -(-ik)\sum_L J_L(\mathbf{r})N_L(\mathbf{r}') \\ &= -\langle J(\mathbf{r})|N(\mathbf{r}')\rangle, \end{aligned} \tag{3.94}$$

where the factor $-ik$ has been absorbed into the Neumann function in the second line of this equation. With the use of the single-center expansion, Eq. (D.92), of Eqs. (D.86) and (3.94), and for $\mathbf{r}$ in the interstitial region of cell n, the left-hand side of Eq. (3.88) takes the form

$$\begin{aligned} \langle J(\mathbf{r})|\underline{C}^n|\gamma^n\rangle &- \langle N(\mathbf{r})|\underline{S}^n|\gamma^n\rangle - \int_{\Omega_n} \mathrm{d}^3r' G_0(\mathbf{r}-\mathbf{r}')V_n(\mathbf{r}')\langle\psi^n(\mathbf{r}')|\gamma^n\rangle \\ &\equiv \langle J(\mathbf{r})|\underline{C}^n|\gamma^n\rangle, \end{aligned} \tag{3.95}$$

since the second and third terms on the left-hand side cancel. The integral equation now reads,

$$\langle J(\mathbf{r})|\underline{C}^n|\gamma^n\rangle = \sum_{m\neq n}\int_{\Omega_m} \mathrm{d}^3r' G_0(\mathbf{r}-\mathbf{r}'+\mathbf{R}_{nm})V_n(\mathbf{r}')\psi(\mathbf{r}+\mathbf{R}_m). \tag{3.96}$$

With the $n = m$ term included, the sum constitutes a *multicenter expansion* for the wave function and should be compared with Eq. (3.15). We now let $\mathbf{r}$ lie on the surface of the MT sphere in cell n, so that the vectors $\mathbf{r}$, $\mathbf{r}'$, and $\mathbf{R}_{mn}$ satisfy conditions (3.38) and we have

$$G_0(\mathbf{r}-\mathbf{r}'+\mathbf{R}_{nm}) = \langle J(\mathbf{r})|\underline{B}(\mathbf{R}_{mn})|J(-\mathbf{r}')\rangle, \tag{3.97}$$

where the coefficients matrix $\underline{B}(\mathbf{R}^{mn})$ is given by the expression,

$$B_{LL'} = -4\pi\sum_{L''} \mathrm{i}^{\ell-\ell'+\ell''} C(LL'L'')N_{L''}(\mathbf{R}_{mn}). \tag{3.98}$$

Thus, the $\underline{B}$'s are the real-space structure constants corresponding to standing waves and are simply related to the structure constants for outgoing waves [see Eq. (3.61)]. Using the expansion in Eq. (3.97), and Eq. (D.93), we obtain a consistency relation between the single-center expansion, Eq. (3.89), and the many-center expansion for $\psi(\mathbf{r})$,

$$\sum_m \left[\underline{C}^m\delta_{nm} + \underline{B}(\mathbf{R}_{nm})\underline{S}^m\right]|\gamma^m\rangle = 0 \tag{3.99}$$

or

$$\sum_j \left[\underline{m}^i \delta_{ij} + \underline{B}(\mathbf{R}_{ij})\right] |\gamma^j\rangle = 0. \tag{3.100}$$

Here, $\underline{m}^i = \underline{C}^i \underline{S}^{i^{-1}}$ is the appropriate generalization[9] of the quantity $\cot \delta_\ell$ to nonspherical potentials, and we have assumed that the inverse of the sine function exists, at least formally. We note that the last two equations give two different forms of the secular equation. For the case of spherically symmetric MT potentials, these two forms have identical mathematical properties. The homogeneous Eqs. (3.99) or (3.100) yield nontrivial solutions for the coefficients $|\gamma^m\rangle$ provided that the determinant of the coefficients vanishes,

$$\det \left|\underline{m}^i \delta_{ij} + \underline{B}(\mathbf{R}_{ij})\right| = 0. \tag{3.101}$$

This, of course, is the secular Eq. (3.61). For systems characterized by translational invariance, the use of lattice Fourier transforms yields the secular equation for non-spherical MT potentials,

$$\det \left|\underline{m} + \underline{B}(\mathbf{k})\right| = 0. \tag{3.102}$$

The analytic and convergence properties of MST in connection with spherically symmetric MT potentials are discussed in reference [3].

Before leaving this section, we comment on an important and often overlooked characteristic of the single-cell expansion in Eq. (D.92). The expansion in that equation can be shown to converge provided that $\mathbf{r}$ lies inside the MT sphere or inside a sphere that does not overlap an adjacent potential. Otherwise, the expansion may diverge. Therefore, in *all* cases, including that of MT potentials, the region of convergence of the single-cell expansion is defined by the region of validity of the expansion of G_0. Further discussion of the convergence properties of MST expansions and the various forms of the secular equation can be found in the literature [3].

3.9 MST for Space-Filling Cell Potentials(*)

The MT form of MST allows the development of a great deal of formalism and also brings out some very important characteristics of the theory. At the same time, the MT conditions impose a great restriction on the types of systems that can be treated. For example, systems that deviate substantially from spherical symmetry, such as two-dimentional defects, e.g., surfaces and interfaces, or are characterized by elongated, unidirectional

[9] We note that $\tilde{\underline{m}}$ is not identical to the (inverse) of the corresponding cell t-matrix $\tilde{\underline{t}}$. The difference arises because of the Neumann function used here rather than the Hankel function used in defining $\tilde{\underline{t}}$. For spherically symmetric potentials, we have $\tilde{\underline{m}} = \cot \delta_\ell$.

cells, e.g., covalently bonded materials, fall formally outside the perview of MST in its MT form.

In this section, we discuss the generalization of MST to the case of non-MT, space-filling potential cells. As we did in the case of MT potentials, we develop the formalism within the angular momentum representation. We show that MST is valid in the general case of nonoverlapping, space-filling scattering regions, retaining the basic features that characterize its application to MT potentials. In particular, it retains the separation of potential and structure from which MST derives much of its conceptual transparency and computational usefulness.

As part of our formal development, we derive expressions for the total t-matrix of an assembly of scatterers, the corresponding Green f unction, and the Bloch function of a translationally invariant material. We also derive the secular equation for the eigenenergies of a system of scattering potentials and for the band structure of a periodic system. These equations exhibit clearly the separation obtained within MST of the potential aspects, embodied in the single-cell phase functions and t-matrices, from the structure of the system, represented by the KKR structure constants. In these various derivations, we make every effort to treat explicitly the geometric subtleties associated with the application of MST to non-MT, space-filling cells. In addition, a derivation of full-cell MST that emphasizes the geometric aspects of the formalism is presented in Appendix G.

As we pointed out in our discussion of MT potentials, in deriving MST one can take either a scattered-wave approach, or view the problem as that of obtaining the solution of a second-order partial differential equation. Within the scattering formalism, introduced for electronic structure calculations by Korringa [6], one considers the wave function of a collection of potentials as being built up from contributions arising from the multiple scattering of waves from the potential cells comprising the system. The condition that the disturbance incident upon a given cell (scatterer) consists of the sum of the outgoing waves scattered by all cells, including the one in question, leads to the secular equation that determines the eigenvalues and eigenenergies of the system of scatterers. This formal approach to MST is particularly attractive because it endows the theory with a pictorial quality that greatly clarifies its content.

In the differential equation approach, one attempts to construct the wave function for a collection of potential cells from the knowledge of the corresponding wave function for each of the cells separately. This is done through the use of the continuity properties of the solutions of second-order partial differential equations. Thus, the individual cell wave functions are combined so that the final wave function is continuous with continuous derivatives across all cell boundaries. This matching of the wave function across cell boundaries can be effected either directly or within a variational formalism.

Kohn and Rostoker [9] used both approaches in their original derivation of MST for spherically symmetric MT potentials. The variational derivation of full-cell MST can be found in the literature [3].

3.9.1 Wave functions inside the MT sphere

In this section we provide a derivation of the secular equation of multiple scattering theory for the general case of arbitrarily shaped and even interpenetrating cells. The derivation is valid provided the following two restrictions are satisfied: First, it is assumed that the center of any one cell lies outside the sphere bounding any other cell. In other words, the shortest intercell vector that connects the origins of two cells is larger than any intracell vector. Second, it is assumed that there exists a finite neighborhood around the center of each cell that lies in the domain of the cell. Under these two conditions, which are easily satisfied for most realistic materials, the secular equation of MST assumes one of the common MT forms in all cases[10] of space-filling cells.

We begin with the Lippmann-Schwinger Eq. (2.81), with $\chi(\mathbf{r})$ explicitly set equal to zero,

$$\psi(\mathbf{r}) = \int G_0(\mathbf{r}-\mathbf{r}')V(\mathbf{r}')\psi(\mathbf{r}')d^3r', \tag{3.103}$$

which can also be written in the form,

$$\psi(\mathbf{r}) = \int G_0(\mathbf{r}-\mathbf{r}')\left(\nabla'^2 + E\right)\psi(\mathbf{r}')d^3r'. \tag{3.104}$$

Thus we are looking for the bound states of the potential, $V(\mathbf{r})$, i.e., for the states that exist even in the absence of an incident wave. Using the defining equation for the free-particle propagator,

$$(\nabla^2 + E)G_0(\mathbf{r}-\mathbf{r}') = \delta(\mathbf{r}-\mathbf{r}'), \tag{3.105}$$

we also have,

$$\psi(\mathbf{r}) = \int \left[\left(\nabla'^2 + E\right)G_0(\mathbf{r}-\mathbf{r}')\right]\psi(\mathbf{r}')d^3r', \tag{3.106}$$

which can be subtracted from Eq. (3.104) to yield the result,

$$0 = \int \left[G_0(\mathbf{r}-\mathbf{r}')\nabla'^2\psi(\mathbf{r}') - \nabla'^2 G_0(\mathbf{r}-\mathbf{r}')\psi(\mathbf{r}')\right]d^3r'$$

[10] With the exception of the variational form of the equation in which a modified version of the structure constants must be used as is shown below.

$$= \sum_n \int_{\Omega_n} \left[G_0(\mathbf{r}-\mathbf{r}'_n)\nabla'^2\psi(\mathbf{r}'_n) - \nabla'^2 G_0(\mathbf{r}-\mathbf{r}'_n)\psi(\mathbf{r}'_n) \right] d^3r'_n. \tag{3.107}$$

The second line of this equation follows upon partitioning the volume of integration into the volumes of individual cells, with a subsequent summation over cells. This partitioning is valid provided the total solution, $\Psi(\mathbf{r})$, is continuous with continuous derivatives across all cell boundaries which we assume to be the case[11]. Through the use of Green's theorem, this equation can be converted to a sum of surface integrals,

$$\sum_n \int_{S_n} [G_0(\mathbf{r}-\mathbf{r}'_n)\nabla'\psi(\mathbf{r}'_n) - \nabla' G_0(\mathbf{r}-\mathbf{r}'_n)\psi(\mathbf{r}'_n)]\, d^3r'_n = 0, \tag{3.108}$$

where now the symbol ∇' denotes the outward derivative over the surface, S_n, of the cell at n. Writing

$$\psi(\mathbf{r}) = \sum_L A_L^n \psi_L^n(\mathbf{r}), \tag{3.109}$$

for $\mathbf{r}$ in cell n, substituting in Eq. (3.108), and writing,

$$G_0(\mathbf{r}-\mathbf{r}'_n) = \sum_L G_L(\mathbf{r},\mathbf{r}'_n), \tag{3.110}$$

with

$$G_L(\mathbf{r},\mathbf{r}'_n) = J_L(\mathbf{r})H_L(\mathbf{r}'_n) \quad \text{for} \quad r'_n > r, \tag{3.111}$$

we can cast Eq. (3.108) in the form,

$$\begin{aligned} 0 &= \sum_n \int_{S_n} d^2r' \sum_{LL'} A_L^n \left[G_{L'}(\mathbf{r}-\mathbf{r}'_n)\nabla'\psi_L^n(\mathbf{r}'_n) \right. \\ &\quad - \left. \psi_L^n(\mathbf{r}'_n)\nabla' G_{L'}(\mathbf{r}-\mathbf{r}'_n) \right]. \end{aligned} \tag{3.112}$$

Now, consider $G_0(\mathbf{r}-\mathbf{r}')$ with $\mathbf{r}$ in cell p and $\mathbf{r}'$ on the *surface* of cell n. Upon restricting $\mathbf{r}$ to lie inside a spherical region surrounding the center of cell p, such as the MT sphere, we have,

$$G_L(\mathbf{r}_p,\mathbf{r}_n) = J_L(\mathbf{r}_p)H_L(\mathbf{r}'_n), \tag{3.113}$$

and Eq. (3.112) takes the form,

$$0 = \sum_{L'} J_L(\mathbf{r}_p) \sum_{Ln} \int_{S_n} d^2r' \sum_{LL'} [H_L(\mathbf{r}'_n)\nabla'\psi_L^n(\mathbf{r}'_n) - \psi_L^n(\mathbf{r}'_n)\nabla' H_L(\mathbf{r}'_n)]\, A_L^n. \tag{3.114}$$

[11]This will certainly be the case if the potential is analytic throughout space.

Finally, using the expression,

$$C^{pn}_{LL'} = \int_{S_n} d^2r' \sum_{LL'} [H_L(\mathbf{r}'_n)\nabla'\psi^n_L(\mathbf{r}'_n) - \psi^n_L(\mathbf{r}'_n)\nabla' H_L(\mathbf{r}'_n)]\, A^n_{L'}, \quad (3.115)$$

we obtain

$$\sum_{L'} J_L(\mathbf{r}_p) \sum_{Ln} C^{pn}_{LL'} A^n_L = 0. \quad (3.116)$$

The cell off-diagonal quantities, $C^{pn}_{LL'}$, are the generalizations of the corresponding single-cell cosine functions that were defined in Appendix D. In fact, we have,

$$C^{nn}_{LL'} = C^n_{LL'}. \quad (3.117)$$

The secular equation determining the coefficients A^n_L now follows from Eq. (3.116). Since $J_L(\mathbf{r}_p)$ does not vanish identically, we must have,

$$\sum_{Ln} C^{pn}_{LL'} A^n_L = 0, \quad (3.118)$$

which has non-trivial solutions for the A^n_L only if the determinant of the coefficients vanishes,

$$\det |C^{pn}_{LL'}| = 0. \quad (3.119)$$

We note that in this form, the secular equation is valid for all cases of nonoverlapping potential cells, *even for those in which the intracell vectors are not always smaller than intercell vectors that connect the origins of two cells.*

When compared with previously derived forms of the secular equation, e.g., Eq. (3.102), Eq. (3.119) looks somewhat strange. To see that it is equivalent to those equations, we use the condition $|\mathbf{R}_{pn}| > |\mathbf{r}'_n|$ (which now does become a restriction on the derivation), expand $H_{L'}(\mathbf{r}'_n)$ in the form,

$$H_{L'}(\mathbf{r}'_n) = \sum_L G_{L'L}(\mathbf{R}_{np}) J_L(\mathbf{r}'_p), \quad (3.120)$$

and use Eq. (3.108) to write,

$$C^{np}_{LL'} = C^n_{LL'}\delta_{np} - \sum_L G_{LL''}(\mathbf{R}_{np}) S^p_{L''L'}(1-\delta_{np}). \quad (3.121)$$

Provided that the inverse of $\underline{S}^p$ exists, the last equation and Eq. (3.119) lead readily to Eq. (3.102).

It is now important to note that even though the secular Eq. (3.119) uniquely determines the coefficients A^n_L, the use of these coefficients to obtain the wave function throughout the cell using the single-center expansion, Eq. (3.109), may lead to divergent results. This is because the secular equation was derived by matching wave functions inside a small sphere around the origin, and the single-center expansion converges only to the point that this matching is valid. However, these coefficients do lead to converged results when used with a multi-center expansion for the wave function. The reader who desires more detailed information about the convergence and other features of full-cell MST is referred to reference [3].

In what follows, we will tacitly assume that the MST expressions are convergent as written in the knowledge that even if an explicit form diverges, it can be made convergent through the replacement of divergent single sums with conditionally convergent double sums. This may require the use of appropriately defined modified structure constants describing free-electron propagation between nonspherical cells (see discussion following Eq. (G.29)). In any case, the corresponding t-matrices are defined only by the potential in a cell and the structure constants depend only on the geometry of the lattice. Thus, the separation of potential from structure, so useful in applications of MST within the MT approximation, is retained in applications to full-cell potentials. We now turn to the calculation of the Green function associated with an assembly of nonoverlapping, space-filling potential cells.

3.10 Green Functions

For the sake of completeness, but also to illustrate some of the versatility of MST, in this section we derive a number of equivalent expressions for the Green function for an assembly of scattering potential cells using a functional derivative formalism.

The formalism presented below is based on two observations. First, we note that from the Lippmann-Schwinger Eq. (3.103) we obtain the expression

$$\begin{aligned} \frac{\delta\Psi}{\delta V} &= G_0\Psi + G_0 V\frac{\delta\Psi}{\delta V} \\ &= [1 - G_0 V]^{-1} G_0 \Psi \\ &= G\Psi. \end{aligned} \tag{3.122}$$

In particular, we have

$$\frac{\delta|\psi^m\rangle}{\delta V_m} = G_m|\psi^m\rangle, \tag{3.123}$$

where G_m is the Green function associated with the potential V_m and which, according to Eq. (D.112), can be written in the form

$$G_m(\mathbf{r},\mathbf{r}') = \sum_L F_L^m(\mathbf{r})\psi_L^m(\mathbf{r}') \equiv \langle F^m(\mathbf{r})|\psi^m(\mathbf{r}')\rangle, \quad \text{for} \quad r > r', \tag{3.124}$$

in terms of the regular, $|\psi^m\rangle$, and irregular, $|F^m\rangle$, solutions of the Schrödinger equation for the potential in cell m. Using the expansion of Ψ in terms of basis functions, Eq. (3.109), we have

$$\begin{aligned}\frac{\delta\Psi(\mathbf{r})}{\delta V_m(\mathbf{r}')} &= \sum_{n,L}\frac{\delta\psi_L^m(\mathbf{r})}{\delta V_m(\mathbf{r}')}A_L^m + \sum_L \frac{\delta A_L^m}{\delta V_m(\mathbf{r}')}\psi_L^m(\mathbf{r}) \\ &= \sum_n \langle\frac{\delta A^n}{\delta V_m(\mathbf{r}')}|\psi^n(\mathbf{r}')\rangle + \langle A^m|\psi^m(\mathbf{r}')\rangle\langle F^m(br)|\psi^m(\mathbf{r}')\rangle,\end{aligned} \tag{3.125}$$

fot $r > r'$, and with both vectors $\mathbf{r}$ and $\mathbf{r}'$ in cell m.

The second relevant observation is that the matrix $\underline{C}^{nm}$, can be written in the form

$$\underline{C}^{nm} = [|H^n\rangle, \langle\psi^m|]_m, \tag{3.126}$$

where the brackets denote an integral of the Wronskian of the two functions, $H_L^n(\mathbf{r})$ and $\psi_L^m(\mathbf{r})$, centered about the origins in cells n and m, respectively, over the surface of cell m. The last equation provides a proper representation of the matrix $\underline{C}^{nm}$ for space-filling cell potentials. Now, we recall that the matrix $\underline{C}^{nm}$ and the vectors $|A^n\rangle$ satisfy the set of homogeneous linear equations,

$$\sum_m \underline{C}^{nm}|A^m\rangle = 0. \tag{3.127}$$

Varying Eq. (3.127) with respect to $V(\mathbf{r}')$ we obtain the relation

$$\frac{\delta\underline{C}^{nm}}{\delta V_m(\mathbf{r}')}|A^m\rangle + \sum_j \underline{C}^{nj}\frac{\delta|A^j\rangle}{\delta V_m(\mathbf{r}')} = 0 \tag{3.128}$$

so that

$$\frac{\delta|A^n\rangle}{\delta V_m(\mathbf{r}')} = \sum_j [\underline{C}^{-1}]^{nj}\frac{\delta\underline{M}^{jm}}{\delta V_m(\mathbf{r}')}|A^m\rangle. \tag{3.129}$$

But from Eq. (3.126) we have (for $\mathbf{r}'$ in cell m)

$$\frac{\delta\underline{C}^{nm}}{\delta V_m(\mathbf{r}')} = [|H^n\rangle, \frac{\delta}{\delta V_m(\mathbf{r}')}\langle\psi^m(\mathbf{r}')|]_m, \tag{3.130}$$

so that Eq. (3.129) can be written in the form,

$$\frac{\delta|A^n\rangle}{\delta V_m(\mathbf{r}')} = \sum_j [\underline{C}^{-1}]^{nj}\underline{N}^{jm}|\psi^m(\mathbf{r}')\rangle\langle\psi^m(\mathbf{r}')|A^m\rangle. \tag{3.131}$$

In the last expression the matrix $\underline{N}^{jm}$ is defined by its elements

$$\underline{N}^{nm} = [|H^n\rangle, \langle F^m|]_m \tag{3.132}$$

and is strictly site off-diagonal. This last property follows from the form of the irregular solution, Eq. (H.3), (see Appendix H),

$$|F^m\rangle = [\underline{C}^m]^{-1}|H^m\rangle. \tag{3.133}$$

In fact, assuming that intercell vectors that connect cell origins are longer than intracell vectors, we can write

$$\underline{N}^{nm} = \underline{G}(\mathbf{R}_{nm})[\underline{C}^m]^{-1}, \tag{3.134}$$

an expression that clearly displays the off-diagonal nature of the matrix $\underline{N}^{nm}$.

Now, using the expansion of the total wave function in terms of cell basis functions, Eq. (3.109), we can write Eq. (3.131) in the form

$$\frac{\delta|A^n\rangle}{\delta V_m(\mathbf{r}')} = \sum_j [\underline{C}^{-1}]^{nj}\underline{N}^{jm}|\psi^m(\mathbf{r}')\rangle\Psi(\mathbf{r}'). \tag{3.135}$$

An expression for the variation of the total wave function with respect to the potential in cell m can now be obtained through the use of Eq. (3.125),

$$\begin{aligned}\frac{\delta\Psi(\mathbf{r})}{\delta V_m(\mathbf{r}')} &= \sum_{nj}\langle\psi^n(\mathbf{r})|[\underline{C}^{-1}]^{nj}\underline{N}^{jm}|\psi^m(\mathbf{r}')\rangle\Psi(\mathbf{r}') \\ &+ \langle F^m(\mathbf{r})|\psi^m(\mathbf{r}')\rangle\Psi(\mathbf{r}') \quad \text{for } r > r'.\end{aligned} \tag{3.136}$$

Finally, comparing the last equation with Eq. (3.122) we obtain the expression for the Green function

$$\begin{aligned}G(\mathbf{r},\mathbf{r}') &= \langle\psi^n(\mathbf{r})|\tilde{\underline{\tau}}^{nm}|\psi^m(\mathbf{r}')\rangle \\ &+ \langle\psi^m(\mathbf{r})|F^m(\mathbf{r}')\rangle\delta_{nm}, \quad \text{for } r' > r\end{aligned} \tag{3.137}$$

where the vectors $\mathbf{r}$ and $\mathbf{r}'$ are confined in cells n and m, respectively, and the matrix $\tilde{\underline{\tau}}^{nm}$ is defined by the expression

$$\tilde{\underline{\tau}}^{nm} = \sum_j [\underline{C}^{-1}]^{nj}\underline{N}^{jm}. \tag{3.138}$$

The expression for the Green function in Eq. (3.137) has a familiar form, consisting of single-scatterer and multi-scatterer contributions. In the case of crystal lattices and for a choice of cell centers in which the intercell vectors are larger than intracell vectors, Eq. (3.137) can be readily converted to the form,

$$G(\mathbf{r},\mathbf{r}') = \langle\psi^n(\mathbf{r})|\hat{\underline{\tau}}^{nm}|\psi^m(\mathbf{r}')\rangle + \langle\psi^m(\mathbf{r})|F^m(\mathbf{r}')\rangle\delta_{nm}, \tag{3.139}$$

where $\underline{\hat{\tau}}^{nm}$ is the inverse of the matrix,

$$\underline{\hat{M}}^{nm} = \underline{S}^{n\dagger}\underline{C}^{n}\delta_{nm} - \underline{S}^{n\dagger}[\underline{G}(\mathbf{R}_{nm})\underline{S}^{m}](1-\delta_{nm}). \tag{3.140}$$

As is shown in the next section, this expression is formally equivalent to Eq. (3.77), and can also be put into a number of other, also equivalent, forms. However, Eq. (3.140) has been derived within a variational procedure and yields eigenvalues and eigenfunctions that are variational with respect to changes in the energy. Other forms that can be derived from Eq. (3.140) by purely algebraic means may not possess this property [3].

Even though Eqs. (3.139) and (3.137) do not involve individual t-matrices, and consequently tend to obscure somewhat the scattering aspects of an ensemble of cells, they also do *not* require the inversion of any phase functions, a feature which can be of great computational value. As long as the order of operations is kept as indicated by the brackets in Eq. (3.140), the expression for the Green function is guaranteed to converge. Furthermore, the product $\underline{GS}$ occurring in the last two expressions can be replaced by the Wronskian integral, Eq. (3.126). The entire double product, $\underline{\tilde{S}}[\underline{GS}]$, can also be evaluated in fully converged fashion in terms of $G_0(\mathbf{r},\mathbf{r}')$ rather than its expansion in angular momentum states.

3.10.1 Alternative expressions

For some applications, it is convenient to have expressions for the single-particle Green function explicitly in terms of the basis states $|\psi(\mathbf{r})\rangle$. Using some straightforward algebra, we can rewrite Eq. (3.139) in the forms,

$$\begin{aligned} \underline{G}(\mathbf{r},\mathbf{r}') &= \langle\psi^i(\mathbf{r})\,|\underline{\tilde{\tau}}^{ij}|\,\psi^j(-\mathbf{r}')\rangle \\ &- [\,\Theta(r-r')\langle Z^i(\mathbf{r}')\,|\,S^i(\mathbf{r})\rangle \\ &+ \Theta(r'-r)\langle Z^i(\mathbf{r})\,|\,S^i(\mathbf{r}')\rangle\,]\,\delta_{ij}, \end{aligned} \tag{3.141}$$

or

$$\begin{aligned} \underline{G}(\mathbf{r},\mathbf{r}') &= \langle\psi^i(\mathbf{r})\,|\underline{\hat{\tau}}^{ij}|\,\psi^j(-\mathbf{r}')\rangle \\ &- [\,\Theta(r-r')\langle\phi^i(\mathbf{r}')\,|\,F^i(\mathbf{r})\rangle \\ &+ \Theta(r'-r)\langle\phi^i(\mathbf{r})\,|\,F^i(\mathbf{r}')\rangle\,]\,\delta_{ij}. \end{aligned} \tag{3.142}$$

Here, $\Theta(x)$ is the Heaviside step function that equals one for positive values of its argument and vanishes otherwise.

In these expressions, we have defined the quantities

$$\underline{\tilde{\tau}} = [\underline{C} - \underline{GS}]^{-1}\underline{S}^{-1} \tag{3.143}$$

and

$$\underline{\hat{\tau}} = [\underline{S}^{\dagger}\underline{C} - \underline{S}^{\dagger}\underline{GS}]^{-1}. \tag{3.144}$$

The functions $|\psi^i\rangle$ and $|F^i\rangle$ are those regular and irregular solutions of the Schrödinger equation associated with the potential in cell i, which at the surface of the sphere bounding the cell join smoothly to the functions $\underline{C}^i|J\rangle + \underline{S}^i|H\rangle$ and $\underline{S}^{-1}|J\rangle$, respectively. Some of these alternative forms may be more convenient than others in certain applications. In any case, all of these forms lead to secular equations for the determination of the electronic structure. For example, in the case of translationally invariant materials, Eq. (3.144) leads to a secular equation of the form[12],

$$\det\left|[\underline{S}^{\dagger}\underline{C} - \underline{S}^{\dagger}\underline{G}\underline{S}]\right| = 0. \tag{3.145}$$

3.10.2 Bloch functions for periodic, space-filling cells

It is now fairly straightforward to derive an expression for the wave function of a scattering assembly through a diagonalization of the Green function. This can be easily demonstrated for the case of lattice periodic systems. Neglecting the (real) single-scatterer term for real values of the energy, we obtain after a Fourier transformation of Eq. (3.142),

$$G_{\mathbf{k}}(\mathbf{r},\mathbf{r}') = \langle Z(\mathbf{r})\,|\underline{\tau}(\mathbf{k})|\,Z(-\mathbf{r}')\rangle, \tag{3.146}$$

where $\mathbf{r}$ and $\mathbf{r}'$ are cell vectors. Near a singularity of the matrix $\underline{\tau}(\mathbf{k})$, i.e., where the secular Eq. (3.145) is satisfied, we can write,

$$G_{\mathbf{k}}(\mathbf{r},\mathbf{r}') = \langle \phi_{\mathbf{k}}(\mathbf{r})\left|[E - E(\mathbf{k})]^{-1}\right|\phi_{\mathbf{k}}(-\mathbf{r}')\rangle. \tag{3.147}$$

By definition, the $|\phi_{\mathbf{k}}(\mathbf{r})\rangle$ are the eigenstates of the system, which in this case are the Bloch functions of the material, and are given by the relation,

$$|\phi_{\mathbf{k}}(\mathbf{r})\rangle = \underline{C}(\mathbf{k})|\,Z(\mathbf{r})\rangle, \tag{3.148}$$

where $\underline{C}(\mathbf{k})$ is the matrix that diagonalizes $\underline{\tau}(\mathbf{k})$. Again we note that this equation has precisely the same form as in the case of MT potentials.

Now, an expression for the electronic cell number density can be obtained. By definition, at energy E, we have,

$$\begin{aligned} n(\mathbf{r},E) &= -\frac{1}{\pi}\Im \mathrm{Tr} G(\mathbf{r},\mathbf{r}') \\ &= -\frac{1}{\pi}\frac{1}{\Omega_{\mathrm{BZ}}}\int d^3k \langle Z(\mathbf{r})\,|\Im\underline{\tau}(\mathbf{k})|\,Z(-\mathbf{r})\rangle. \end{aligned} \tag{3.149}$$

Upon using Eq. (3.148) we obtain the well-known and general expression,

$$n(\mathbf{r},E) = \frac{1}{\Omega_{\mathrm{BZ}}}\int_{\mathrm{BZ}} d^3k \langle \phi_{\mathbf{k}}(\mathbf{r})\mid \phi_{\mathbf{k}}(-\mathbf{r})\rangle, \tag{3.150}$$

[12]Where the possible need for the use of modified structure constants, as discussed in Ref. 3 and after Eq. (G.29), should be kept in mind.

where the integration extends over the volume, Ω_{BZ}, of the first Brillouin zone of the reciprocal lattice. The number density follows from Eq. (3.150) upon integration over the energy, and the density of states is given by an integral over the coordinates, $\mathbf{r}$.

3.11 Lloyd Formula(*)

As is derived in Appendix D, Eq. (D.125), the change in the integrated density of states (total number of states with energy less than E) due to the existence of a single, spherically symmetric potential is given by

$$\Delta N(E) = \frac{1}{\pi}\Im \mathrm{Tr} \ln t_L = \frac{1}{\pi}\sum_{\ell}(2\ell+1)\delta_\ell(E). \tag{3.151}$$

We now generalize this result to an arbitrary number of space-filling scatterers.

Consider the operator function

$$\hat{L}(z) = \mathrm{Tr} \ln \left(z - \hat{H}\right). \tag{3.152}$$

Since the derivative of $\hat{L}$ with respect to the energy variable, z, is the Green function,

$$\frac{\mathrm{d}\hat{L}(z)}{\mathrm{d}z} = \mathrm{Tr}\left(z - \hat{H}\right)^{-1} = \hat{\mathrm{G}}(\mathrm{z}), \tag{3.153}$$

it is clear that the integrated density of states is given by

$$N(z) = -\frac{1}{\pi}\Im \hat{L}(z). \tag{3.154}$$

Now, $z - \hat{H} = z - \hat{H}_0 - \hat{V} = \hat{G}_0^{-1} - \hat{V}$, so that $\hat{L}(z)$ can be written in the form

$$\hat{L}(z) = \mathrm{Tr} \ln \left(1 - \hat{G}_0\hat{V}\right) - \mathrm{Tr} \ln \hat{G}_0. \tag{3.155}$$

Expanding the first term in powers of the potential, and using the fact that $\hat{V}$is the sum of non-overlapping potential cells, Eq. (3.1), $\hat{V} = \sum_n \hat{v}_n$, we have

$$\begin{aligned}\mathrm{Tr} \ln \left(1 - \hat{G}_0\hat{V}\right) &= \mathrm{Tr}\left\{\sum_n \hat{G}_0\hat{v}_n + \frac{1}{2}\sum_{nm} \hat{G}_0\hat{v}_n\hat{G}_0\hat{v}_m\right. \\ &+ \left.\frac{1}{3}\sum_{nmp} \hat{G}_0\hat{v}_n\hat{G}_0\hat{v}_m\hat{G}_0\hat{v}_p + \cdots\right\}. \end{aligned} \tag{3.156}$$

By collecting together the terms involving repeated scattering at the same site, we can rewrite this expression in terms of the individual cell t-matrices

$$\begin{aligned}\mathrm{Tr}\ln\left(1-\hat{G}_0\hat{V}\right) &= \mathrm{Tr}\left\{\ln\sum_n\left(1-\hat{G}_0\hat{v}_n\right)\right.\\ &+ \frac{1}{2}\sum_n\sum_{n\neq m}\hat{G}_0\hat{t}_n\hat{G}_0\hat{t}_m\\ &+ \left.\frac{1}{3}\sum_n\sum_{m\neq n}\sum_{p\neq m}\hat{G}_0\hat{t}_n\hat{G}_0\hat{t}_m\hat{G}_0\hat{t}_p+\cdots\right\}.\end{aligned} \tag{3.157}$$

In the case of MT potentials, this operator expression can be converted into the angular momentum representation through pre- and post-multiplication by $J_L(\mathbf{r})$ and $J_{L'}(\mathbf{r}')$, integrating over $\mathbf{r}$ and $\mathbf{r}'$, and using the expansion of the free-particle propagator in terms of the Bessel functions and the structure constants, Eq. (3.40),

$$-\frac{1}{4\pi}\frac{\mathrm{e}^{\mathrm{i}k|\mathbf{R}+\rho-\rho'|}}{|\mathbf{R}+\rho-\rho'|} = -\mathrm{i}k\sum_{LL'}J_L(k\rho)G_{LL'}(\mathbf{R})J_{L'}(k\rho'). \tag{3.158}$$

The same manipulations are possible in the case of arbitrarily shaped space-filling cells provided that the proper summation procedures are followed in evaluating sums over L (See Appendices F and G for details.) With this caveat in mind, we can write

$$\hat{L}(z) = \sum_n \mathrm{Tr}\ln\left(1-\hat{G}_0\hat{v}_n\right) + \mathrm{Tr}\ln\left[1-\underline{Gt}\right] - \mathrm{Tr}\ln\hat{G}_0, \tag{3.159}$$

where the form $1-\underline{Gt}$ is now expressed in the angular momentum representation. We note that $\mathrm{Tr}\underline{Gt}=0$ because the structure constant matrix contains no diagonal elements, and $\underline{t}$ is diagonal.

The first term on the right of the equals sign above gives the change in the integrated density of states caused by each scatterer in the absence of the others,

$$-\frac{1}{\pi}\Im\ln(1-G_0v_n) = N_n(E) - N_0(E). \tag{3.160}$$

This result, which is true in any representation, follows upon noticing the relation

$$\frac{\mathrm{d}}{\mathrm{d}z}(1-G_0v_n) = G_0t_nG_0 - G_0. \tag{3.161}$$

From this follows the generalization of the Lloyd formula (sometimes also called the Friedel sum rule) for the case of a non-spherical scatterer

$$N_n - N_0 = -\frac{1}{\pi}\Im\mathrm{Tr}\ln\underline{t}^n. \tag{3.162}$$

Now, combining Eqs. (3.159) and (3.162), and assuming the existence of the inverses of the cell t-matrices, we obtain the generalization of Lloyd's formula for the change in the integrated density of states caused by a scattering assembly

$$\begin{aligned} N(z) - N_0(z) &= -\frac{1}{\pi}\Im \mathrm{Tr} \ln \underline{M} \\ &= -\frac{1}{\pi}\left[\underline{m}^i \delta_{ij} - \underline{G}^{ij}(1-\delta_{ij})\right]. \end{aligned} \tag{3.163}$$

Various alternative formulations of the Friedel sum rule and the Lloyd formula have been given in the literature [69]. For spherically symmetric cell potentials, Lehmann [70] showed that the change in the integrated density of states can be written in the form,

$$N(E) - N^0(E) = -\frac{2}{\Omega\pi}\Im \ln \det||\cos\eta_L^i \delta_{LL'}\delta_{ii'} + \sin\eta_L^I (N_{LL'}^{ii'} - \mathrm{i}J_{LL'}^{ii'})|| \tag{3.164}$$

where

$$N_{LL'}^{ii'} = (1-\delta_{ii'})4\pi \sum_{L''} \mathrm{i}^{\ell-\ell'+\ell''} n_{L''}(\kappa \mathbf{R}_{ii'}) C(LL'L'') \tag{3.165}$$

and

$$J_{LL'}^{ii'} = 4\pi \sum_{L''} \mathrm{i}^{\ell-\ell'+\ell''} j_{L''}(\kappa \mathbf{R}_{ii'}) C(LL'L'') \tag{3.166}$$

and η_L^i denotes an element of the phase shift for the scatterer at site i, and ω denotes volume. This form avoids the use of the tangent of the phase shifts, or the inverse of the t-matrix, which may diverge for certain cases, such as at the resonances occurring in the scattering off atoms of the transition metal series.

3.12 Use of MST within LDA

As described above, MST allows the solution of the Schrödinger equation for any given value of the potential. In order to carry out the determination of the electronic structure of a solid, one uses MST in an iterative way. The density obtained at each step of the iteration process is used to generate a new potential using the functional forms of the LDA, and this potential is treated within MST to yield a new density. The process stops when the density (or the ground-state energy) falls within preassigned convergence limits.

3.13 Other Methods

Even though MST has a number of impressive features, it is not necessarily the method of choice for the treatment of a number of problems associated with the electronic structure of solid materials. For example, computational methods much faster than MST in specific cases have been developed that yield accurate single-particle solutions of the Schrödinger equation in a solid. These methods include [25, 71] that of linear muffin-tin orbitals (LMTOs), and its full-cell generalization [26] and that of augmented plane waves (APW), and its application to space-filling cells [23, 24]. Within the atomic sphere approximation (ASA), the LMTO method has even been extended [72] to the calculation of the Green function and hence can be applied to the study of substitutionally disordered alloys. However, it is difficult to conceive of any other method with the power and flexibility of MST which is formally unrelated to it or which does not involve the use of Green functions.

Lack of space, and the desire to emphasize a unified approach to the study of the electronic structure of solids, precludes anything but a cursory mention of these other techniques. The reader interested in further details is urged to review the literature cited. At the same time, we will have occasion to use results of methods other than MST to illustrate features of electronic structure calculations when it is convenient to do so, or when results of MST-based calculations are lacking.

Bibliography

[1] Lord Rayleigh, Phil. Mag. **34**, 481 (1892).

[2] For a recent compilation of applications of MST to classical and quantum systems, see Materials Research Society, Proceedings of Symposium V on *Applications of Multiple Scattering Theory to Materials Science*, W. H. Butler, P. H. Dederichs, A. Gonis, and R. L. Weaver (eds.), Materials Research Society, Vol. **253** (1992).

[3] A. Gonis and W. H. Butler, *Multiple Scattering Theory in Solids*, Springer Verlag, Berlin (1999).

[4] N. Kasterin, Koninklijke Akademie van Wetenschappen: Verslagen van de gewone vergaderingen der wis-en natuurkundige afdeeling, **VI**, pp. 460-480 (1898).

[5] J. Korringa, R. Kroning, and A. Smit, Physica **11**, 209 (1945).

[6] J. Korringa, Physica **13**, 392 (1947).

[7] R. J. Harrison, Phys. Rev. **84**, 377 (1951).

[8] D. Saxon and R. J. Hutner, Philips Res. Rep. **4**, 481 (1949).

[9] W. Kohn and N. Rostoker, Phys. Rev. **94**, 1111 (1954).

[10] J. C. Slater, Phys. Rev. **51**, 151 (1937).

[11] F. S. Ham and B. Segall, Phys. Rev. **124**, 1786 (1961).

[12] J. Treusch and R. Sandrock, Phys. Stat. Solidi **16**, 487 (1966).

[13] J. S. Faulkner, H. L. Davis, and H. W. Joy, Phys. Rev. **161**, 656 (1967).

[14] K. H. Johnson, J. Chem. Phys. **45**, 3085 (1966).

[15] V. L. Moruzzi, J. F. Janak, and A. R. Williams, *Calculated Electronic Properties of Metals*, Pergamon, New York (1978).

[16] P. Soven, Phys. Rev. **156**, 809 (1967).

[17] D. W. Taylor, Phys. Rev. **156**, 1017 (1967).

[18] P. Soven, Phys. Rev. B**2**, 4715 (1970).

[19] H. Shiba, Prog. Theor. Phys. **46**, 77 (1971).

[20] B. L. Györffy, Phys. Rev. B**5**, 2382 (1972).

[21] G. M. Stocks, W. M. Temerman, and B. L. Györffy, Phys. Rev. Lett. **41**, 339 (1978).

[22] A. Bansil, Phys. Rev. Lett. **41**, 1670 (1978).

[23] H. Krakauer, M. Posternak, and A. J. Freeman, Phys. Rev. B**19**, 1706 (1979).

[24] E. Wimmer, H. Krakauer, M. Weinert, and A. J. Freeman, Phys. Rev. B**24**, 864 (1981).

[25] H. L. Skriver, *The LMTO Method*, Springer Verlag, Berlin, Heidelberg, New York, (1984).

[26] M. S. Methfessel, *Multipole Green Functions for Electronic Structure Calculations*, Thesis, Katholike Universiteit te Nijmegen (1986), unpublished.

[27] J. S. Faulkner, Phys. Rev. B**19**, 6186 (1979).

[28] P. Ziesche, J. Phys. C**7**, 1085 (1974).

[29] P. Ziesche and G. Lehmann, *Ergebnisse in der Elektronentheorie der Metalle*, Akademie-Verlag, Berlin (1983), p. 151.

[30] H. Bross and K. H. Anthony, Phys. Stat. Sol. **22**, 667 (1967).

[31] F. Beleznay and M. J. Lawrence, J. Phys. C**1**, 1288 (1968).

[32] J. Keller, J. Phys. C**4**, L85 (1971).

[33] R. Evans and J. Keller, J. Phys. C**4**3155 (1971).

[34] W. John, G. Lehmann, and P. Ziesche, Phys. Stat. Sol. **53**, 287 (1972).

[35] A. R. Williams and J. van W. Morgan, J. Phys. C**5**, 1293 (1972).

[36] A. R. Williams and J. van W. Morgan, J. Phys. C**7**, 37 (1974).

[37] J. van W. Morgan, J. Phys. C**10**, 1181 (1977).

[38] R. G. Brown and M. Ciftan, Phys. Rev. B**27**, 4564 (1983).

[39] R. G. Brown and M. Ciftan, Phys. Rev. B**32**, 1343 (1985).

[40] R. G. Brown and M. Ciftan, Phys. Rev. B**33**, 7937 (1986).

[41] R. G. Brown and M. Ciftan, Phys. Rev. B**32**, 3454 (1985).

[42] R. G. Brown and M. Ciftan, Phys. Rev. B**39**, 10415 (1989).

[43] L. Sheire, Physica A **81**, 613 (1975).

[44] B. D. Keister, Am. J. Phys. **149**, 162 (1983).

[45] L. G. Fereira, A. Agostino, and D. Lida, Phys. Rev. B**14**, 354 (1976).

[46] J. S. Faulkner, Phys. Rev. B**34**, 5931 (1986).

[47] J. S. Faulkner, Phys. Rev. B**38**, 1686 (1988).

[48] J. S. Faulkner, Phys. Rev. B**32**, 1339 (1985).

[49] E. Badralexe and A. J. Freeman, Phys. Rev. B**36**, 1378 (1987); **36**, 1389 (1987); **36**, 1401 (1987); **38**, 10469 (1988).

[50] A. Gonis, Phys. Rev. B**33**, 5914 (1986).

[51] R. Zeller, J. Phys. C**20**, 2347 (1987).

[52] R. Zeller, Phys. Rev. B**38**, 5993 (1988).

[53] J. Molenaar, J. Phys. C**21**, 1455 (1988).

[54] R. K. Nesbet, Phys. Rev B**30**, 4230 (1984); **33**, 3027 (1986).

[55] R. K. Nesbet, Phys. Rev. B**41**, 4948 (1990).

[56] J. Molenaar, J. Phys. Condens. Matter **1**, 6559 (1989).

[57] A. Gonis, X. -G. Zhang, and D. M. Nicholson, Phys. Rev. B**38**, 3564 (1988).

[58] A. Gonis, X. -G. Zhang, and D. M. Nicholson, Phys. Rev. B**40**, 947 (1989).

[59] X. -G. Zhang and A. Gonis, Phys. Rev. B**39**, 10373 (1989).

[60] W. H. Butler, A. Gonis, and X. -G. Zhang, Phys. Rev. B**45**, 11527 (1992).

[61] W. H. Butler and R. K. Nesbet, Phys. Rev. B**42**, 1518 (1990).

[62] Chin-Yu Yeh, A. -B. Chen, D. M. Nicholson, and W. H. Butler, Phys. Rev. B**42**, 10976 (1990).

[63] A. Gonis, Erik C. Sowa, and P. A. Sterne, Phys. Rev. Lett. **66**, 2207 (1991).

[64] D. M. Nicholson and J. S. Faulkner, Phys. Rev. B39, 8187 (1989).

[65] A. Gonis, *Green Functions for Ordered and Disordered Systems*, North Holland, Amsterdam, New York (1992).

[66] P. J. Braspenning, *A Multiple Scattering Treatment of Dilute Metal Alloys*, Thesis, Vrije Universiteit, Amsterdam, The Netherlands (unpublished).

[67] B. L. Györffy and M. J. Stott, in *Band Structure Spectroscopy of Metals and Alloys*, D. J. Fabian and L. M. Watson (eds.), Academic Press, London (1972).

[68] W. H. Butler, Phys. Rev. B**14**, 468 (1976).

[69] W. John and P. Ziesche, phys. Stat. Sol. (b)**47**, K83 (1971).

[70] G. Lehmann, J. Phys. C**6**, 1881 (1973).

[71] Terry Loucks, *Augmented Plane Wave Method*, W. A. Benjamin, New York (1967).

[72] J. Kudrnovsky, I. Turek, V. Drchal, and M. Sob, Proceedings of NATO ASI on *Stability of Materials*, A. Gonis, P. E. A. Turchi, and J. Kudrnovsky (eds.), Plenum, New York, V. **355** (1996), p. 237.

Chapter 4

The Poisson Equation

4.1 General Comments

The discussion of density functional theory and its local approximation indicated the central role played by the electronic potential in the determination of the ground-state properties of the electronic gas in a solid [1-4]. It is this potential that enters the Schrödinger equation and the expression for the total energy. Realistic calculations of the electronic structure of a material are commonly based on the self-consistent calculation of the potential, which in turn requires the iterative solution of an interrelated system of Poisson and Schrödinger equations. Here, an input potential is used to generate a wave function and hence a charge density that in turn yields a new value of the potential for the next iteration. In addition to the potential, the determination of the total energy requires an accurate evaluation of the electrostatic interaction associated with the calculated charge. It is clear that in proceeding to self-consistency, the Schrödinger and the Poisson equations are of equal importance.

Having considered the application of multiple scattering theory in obtaining the solution to the one-particle Schrödinger equation in solid materials, we now turn our attention to the determination of the potential. As may be expected from the formal similarity between the Schrödinger and Poisson equations, we will see that the methods of multiple-scattering theory can be used essentially intact in the solution to the latter, and in the calculation of the electrostatic interaction in solids described by arbitrarily shaped, space-filling charges. In fact, the methods employed to treat the Poisson equation for such systems are directly analogous to the MST methods for the study of space-filling potentials. This is discussed at length in the literature [5], to which the reader is referred for further details.

4.2 Multipole Moments

In this chapter, we develop methods for the exact treatment of the Poisson equation associated with a charge density, $\rho(\mathbf{r})$,

$$\nabla^2 V = -4\pi\rho \tag{4.1}$$

or, equivalently, the evaluation of the integral

$$V(\mathbf{r}) = \int \frac{\rho(\mathbf{r}')}{|\mathbf{r}-\mathbf{r}'|} d^3r' \tag{4.2}$$

which is the solution of Eq. (4.1) in free space [6]. Also, we wish to consider the related problem of calculating the electrostatic energy of the charge $\rho(\mathbf{r})$,

$$\begin{aligned} E_e &= \int V(\mathbf{r})\rho(\mathbf{r})\mathrm{d}^3r \\ &= \int\int \frac{\rho(\mathbf{r})\rho(\mathbf{r}')}{|\mathbf{r}-\mathbf{r}'|}\mathrm{d}^3r\mathrm{d}^3r'. \end{aligned} \tag{4.3}$$

Over the years, a number of methods have been proposed for the evaluation of the integral in Eq. (4.2), most of them geared toward the special features of the system under consideration. Perhaps the simplest form of electrostatic interaction is that associated with equal numbers of positive and negative point charges arranged on the sites of a lattice. The self-potential of such an ionic crystal has been considered by Madelung [7]. A general procedure for obtaining the self-interaction of a translationally invariant material consisting of cell charges, $\rho^i(\mathbf{r})$, was subsequently given by Ewald [8], whose method has given rise to a number of specific applications [8-10], modifications, and generalizations [11-17]. Works based on Ewald's method have encompassed charge distributions of the muffin-tin (MT) form, i.e., spherically symmetric cell charges confined inside nonoverlapping spheres situated in a constant interstitial potential (often chosen to be zero), or more generally shaped, nonspherical arrangements of charge. The general approach in these works has been to consider the short-range Coulomb interactions arising from neutral cell charges, and to take account of the long-range interaction by lattice sums using Ewald's method. Also, approaches based on the use of finite differences [19] and on multipole expansions [20-26] of the charge $\rho(\mathbf{r})$ have been used.

Of the approaches based on multipole expansions, that of Morgan [26] must be singled out as the first attempt to solve the Poisson equation for a solid described by space-filling charges entirely in terms of the multipole moments of the charge. In this approach, one expresses the potential due to a spatially bounded charge distribution in the form

$$V(\mathbf{r}_0) = \sum_{l,m} \frac{4\pi}{2l+1} \frac{Y_{lm}(\hat{\mathbf{r}}_0)}{r_0^{l+1}} Q_{lm}, \tag{4.4}$$

where the multipole moments for the charge are given by the integral

$$Q_{lm} = \int r^l Y_{lm}^*(\hat{\mathbf{r}}) \rho(\mathbf{r}) d^3 r. \tag{4.5}$$

Here, $Y_{lm}(\hat{\mathbf{r}})$ is a spherical harmonic of order $L(= l, m)$, and an asterisk denotes complex conjugation. It is to be noted that Eq. (4.4) is rigorously valid only outside a sphere bounding the charge, e.g., for charges that satisfy the muffin-tin conditions. Thus, for a system of MT charges the electrostatic potential at point $\mathbf{r}_0$ inside cell j is given by the expression

$$V(\mathbf{r}_0) = \sum_{n \neq j} \sum_{l,m} \frac{4\pi}{2l+1} \frac{Y_{lm}(\hat{\mathbf{r}}_0)}{r_0^{l+1}} Q_{lm}^n + V_0^j(\mathbf{r}_0), \tag{4.6}$$

where the Q_{lm}^n are the multipole moments of the charge in cell n centered at point $\mathbf{R}_n$, and $\mathbf{r}_n = \mathbf{r} - \mathbf{R}_n$. The quantity $V_0^j(\mathbf{r}_0)$ denotes the potential inside cell j contributed by the charge in that cell, which can be obtained by a number of means [24]. Morgan's method used Eq. (4.6) even in cases in which the potential did not conform to the MT description, such as the space-filling, Wigner-Seitz cells in a solid. However, for points near the face of such a cell, Eq. (4.6) can diverge, an effect noted by Morgan. In the discussion that follows, we will formulate an expansion for the potential in terms of the multipole moments which is free of divergences and leads to well-defined values of the potential and electrostatic interaction everywhere in space.

4.3 Compared Schrödinger and Poisson Equations

In spite of the difficulties noted by Morgan, the idea of expressing the potential in terms of individual multipole moments is very appealing for a number of reasons. First, it would be in nearly perfect accord with the multiple-scattering theory approach to electronic structure, which also relies on the evaluation of cell multipole expansions, in this case the cell t-matrix to solve the Schrödinger equation. This similarity can be seen clearly upon comparison of the two equations. With the formal identification of $-4\pi\rho(\mathbf{r})$ in the Poisson equation with $V(\mathbf{r})\psi_L(\mathbf{r})$ in the Schrödinger equation, and the realization that the quantity $r^\ell Y_L(\hat{\mathbf{r}})$ is the zero-energy limit of $J_L(\mathbf{r})$

[27], we see that $Q_{\ell m}$ is the formal equivalent of the t-matrix. In fact, as $E \to 0$ ($k \to 0$), the free-particle propagator that appears in Eq. (D.84) reduces to the form $\frac{1}{|\mathbf{r}-\mathbf{r}'|}$ and the Poisson equation becomes in appearance equivalent to the Schrödinger equation. Of course, the source term in the Schrödinger equation is proportional to the wave function that results in a homogeneous equation still to be solved, whereas the integral form of the Poisson equation represents a formal solution of Eq. (4.1). Nevertheless, the similarity is great enough that the two equations can be attacked by formally similar means.

Second, a full-cell solution of the Poisson equation would be ideally suited to the treatment of point defects, either substitutional or interstitial, as is MST, obviating the need for using artificial constructions such as supercells in the treatment of the embedding problem. Third, it would conceivably apply to systems without full translational invariance, because of the possibility of summing Eq. (4.6) directly in real space in ma ny cases of physical interest, e.g. cubic systems. This last feature would be of great relevance to those applications of MST to the calculation of the electronic structure [27-31] that allow the treatment of systems with severely reduced or no translational invariance. It would also fit in well with applications of MST to potentials partitioned into space-filling cells. Provided that these features were made part of a computationally practical procedure, the use of multipole expansions in solving both the Schrödinger and Poisson equations would provide a unique and transparent solution of the self-consistent calculation of the electronic structure of solids, with essentially no restrictions on the shape of the cells or the underlying lattice structure.

Now, a serious problem in implementing such a procedure is the divergence of the multipole expansion of the MT form when applied to space-filling cell charges. The MT expression, Eq. (4.6), and the corresponding form for the electrostatic interaction diverge when applied to points that lie in the charge-free region inside a sphere bounding the charge, i. e., in the moon region of the cell.

This behavior is analogous to that obtained when the asymptotic form of the wave function is used inside the moon region of a cell. The divergence is analogous to that which may set in when a MT expression for the wave function is used to represent the wave function inside the bounding sphere circumscribing a cell. Conceivably, one could attempt a Padé-approximant-like continuation of the multipole expansion into the moon region. However, besides being conceptually unattractive, such a continuation would be computationally rather cumbersome, would have to be examined for convergence on a case-by-case basis, and would not provide a reliable measure of the possible errors introduced by its use. Obtaining the potential in the moon region through direct integration over the domains of adjacent cells might improve the convergence aspect of the calculation. However, that would also detract from the overall unifying features of the method, and might also be computationally sluggish in treating impurities, as one would have to treat the moon regions differently from the rest of the material. The method would come virtually to a full stop in the case of random alloys. As we show below, one way to resolve these convergence difficulties is to use a multipole expansion in which a single, divergent sum is replaced by a conditionally convergent double or multiple sum, just as was done in the previous chapter in considering MST for space-fillin cells.

In attempting an evaluation of the potential in a crystal, one may work either with the Poisson Eq. (4.1) or with the integral representation of its solution, Eq. (4.2). We will follow the latter approach. As mentioned above, the integral equation provides a proper solution of the differential equation and leads directly to expressions for the potential and the electrostatic interaction in terms of the multipole moments. The resulting expressions, which are generalizations of those corresponding to MT charges, are analytic and provide a transparent explanation for the divergences noted by Morgan [22], as well as a resolution of the difficulty. Moreover, these expressions are fairly straightforward to code, involving in addition to the multipole moments the well-known structure constants of the underlying lattice. These can be evaluated once and for all for a given, translationally invariant structure, and used repeatedly as necessary. Even in cases with reduced or no translational symmetry, these expressions provide a means of calculating the electrostatics of space-filling cells through direct real-space summations. Finally, they provide a point of comparison against which numerical procedures, e.g. Padé approximants, can be judged.

For convex-shaped cells bounded by planar surfaces, e.g. Wigner-Seitz cells, the methods presented below become particularly straightforward. They allow the calculation of the potential, as well as of the electrostatic interaction of a material, in terms of quadratures. However, the method can be extended to the treatment of concave cells. Thus, even in its simplest form, the method is characterized by all the desirable features quoted above and can be used readily in the performance of charge-self-consistent, total-energy calculations in solids.

4.4 Convex Polyhedral Cells

4.4.1 Mathematical preliminaries

The reason that Eq. (4.6) leads to divergent results is that the points $\mathbf{r}_0$ in the moon region of a cell may fail to satisfy the $r_</r_>$ criterion with respect to points $\mathbf{r}$ inside the cell in the expansion of the Green function, $G_0(\mathbf{r}_0 - \mathbf{r}) = 1/|\mathbf{r}_0 - \mathbf{r}|$. This criterion is rigorously satisfied in the formalism presented below. In order to see how this is accomplished, we begin with a review of those expansions of the Green function that are relevant to our discussion.

We begin with the expansion of the potential at $\mathbf{r}_0$ due to a unit point charge at point $\mathbf{r}$, i.e. the Green function $G_0(\mathbf{r}_0 - \mathbf{r})$ of electrostatics,

$$\frac{1}{|\mathbf{r} - \mathbf{r}_0|} = \sum_{l=0}^{\infty} \frac{r_<^l}{r_>^{l+1}} P_l(\cos\gamma), \tag{4.7}$$

where $r_<(r_>)$ is the smaller (larger) of $|\mathbf{r}_0|$ and $|\mathbf{r}|$, and γ is the angle between $\mathbf{r}_0$ and $\mathbf{r}$. We note that $G_0(\mathbf{r} - \mathbf{r}')$ satisfies the Laplace equation

$$\nabla^2 G_0(\mathbf{r} - \mathbf{r}') = -4\pi\delta(\mathbf{r} - \mathbf{r}'). \tag{4.8}$$

Upon using the expansion of Legendre polynomials in terms of spherical harmonics [6], we can also write

$$\frac{1}{|\mathbf{r} - \mathbf{r}_0|} = \sum_{l=0}^{\infty} \sum_{m=-l}^{l} \frac{r_<^l}{r_>^{l+1}} Y_{lm}(\hat{\mathbf{r}}_0) Y_{lm}^*(\hat{\mathbf{r}}), \tag{4.9}$$

where $\hat{\mathbf{r}}$ denotes a unit vector in the direction of $\mathbf{r}$. It can be easily established [27] that this expression constitutes the zero-energy limit of Eq. (2.68). Thus, Eq. (4.9) represents the potential in completely factorized form in the coordinates $\mathbf{r}_0$ and $\mathbf{r}$. Furthermore, for two vectors $\mathbf{r}$ and $\mathbf{R}$ such that $R > r$, we have the expansion [27]

$$\begin{aligned} \frac{\mathrm{i}^l}{|\mathbf{r} - \mathbf{R}|^{l+1}} Y_{lm}(\widehat{\mathbf{r} - \mathbf{R}}) &= 4\pi \sum_{l'} \sum_{m=-l'}^{l'} r^{l'} \mathrm{i}^{l'} Y_{l'm'}(\hat{\mathbf{r}}) \frac{(2l'' - 1)!!}{(2l - 1)!!(2l' + 1)!!} \\ &\times C(LL'L'') \frac{1}{R^{l''}} \left[\mathrm{i}^{l''} Y_{l''m''}(\hat{\mathbf{R}})\right]^*, \end{aligned} \tag{4.10}$$

where $l'' = l + l'$, $m'' = m' - m$, $C(LL'L'')$ is a Gaunt number (the integral of the product of three spherical harmonics), and for n odd the symbol $n!!$ denotes the product $1 \times 3 \times 5 \times \cdots \times n$. Equations (4.7) to (4.10) are the only ones needed to obtain the solution of Poisson's equation for any distribution of cell charges. As is well known, and as was quoted in the previous section, the use of these equations allows one to write the potential outside a sphere bounding a charge distribution in a form that involves the multipole moments of the charge density, Eq. (4.5). Crucial in the derivation of Eq. (4.4) and its multicell counterpart, Eq. (4.6), is that the "observation" point $\mathbf{r}_n$ lies outside a sphere circumscribing the cell at $\mathbf{R}_n$, so that $r_n > r$ for any intracell vector $\mathbf{r}$. Thus, at any point

$\mathbf{r}$ *outside all* bounding spheres in a collection of cells centered at sites $\mathbf{R}_n$, the potential can be written as the sum

$$V(\mathbf{r}) = \sum_n \sum_{l,m} \frac{4\pi}{2l+1} \frac{Y_{lm}(\widehat{\mathbf{r}_n})}{r_n^{l+1}} Q_{lm}^n. \tag{4.11}$$

This is essentially Eq. (4.6), but with the potential $V_0^j(\mathbf{r})$ also expressed as a sum over multipole moments. Thus, Eq. (4.11) allows one to express a collective property of an assembly of cells, the potential $V(\mathbf{r})$ at points outside all bounding spheres, as a linear combination of individual and independently determined cell quantities, the multipoles Q_{lm}^n. For MT potentials, Eq. (4.11) can be modified to the form of Eq. (4.6), in which form it holds for all points $\mathbf{r}$ in the material. These MT forms of the multipole expansion are particularly convenient for computational purposes. For example, in order to find the change in the potential at $\mathbf{r}$ due to the change in the charge distribution in the cell at $\mathbf{R}_n$, one simply replaces the original multiple moments Q_{lm}^n in Eqs. (4.6) or (4.11) by the ones appropriate to the new charge distribution. The solution of Poisson's equation for a solid under the MT approximation has been described in the literature [33], and has been used extensively [34] in the performance of electronic-structure calculations. The non-MT formulae are derived explicitly below.

4.4.2 Non-MT, space-filling cells of convex shape

A great number of physical situations, e.g., non-spherical charge distributions, potentials near surfaces, and interfaces and in the neighborhood of impurities, may not conform to the MT model. In such cases, Eq. (4.6) is no longer valid in the "moon" regions, i.e. between the charge distribution inside a cell and the circumscribing sphere, such as point P_2 in Fig. 4.1. It is now no longer evident *a priori* that a knowledge of the multipole moments suffices to determine the potential in this region. To address these problems, one requires a method for treating electrostatic interactions for complex systems consisting of space-filling cells. A number of attempts [25, 26] have been made in this direction. However, these attempts have not led to a satisfactory solution to the problem. Because Weinert's [25] approach utilizes only the multipole moments of the charge inside a sphere inscribed in the cell, it does not allow the straightforward replacement of the cell multipole moments Q_{lm}^n to treat impurities or other localized imperfections. Morgan's approach [26], based on the use of MT expansions to treat space-filling cell charges, does utilize the full-cell moments, but runs into severe convergence difficulties that effectively make it unreliable.

In this section, we provide the solution to this problem. We show rigorously that it is indeed formally and computationally possible to obtain

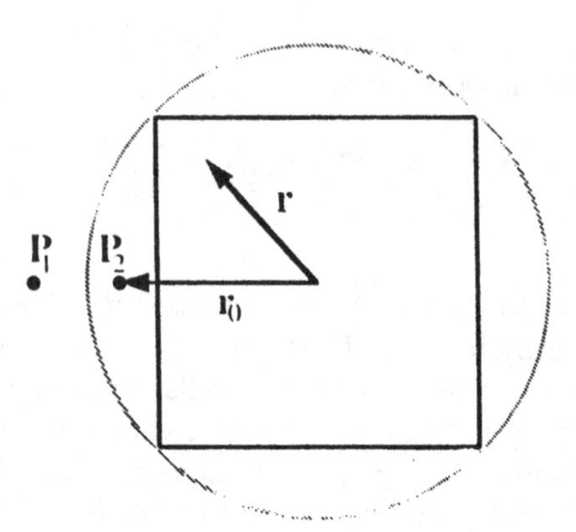

Figure 4.1: **A cubic cell of charge and its bounding sphere. The moon region is defined as the part of free space outside the cell but inside the bounding sphere.**

the intercell contributions to the potential and the electrostatic interaction of an assembly of space-filling cells in terms of the individual multipole moments of these cells. For the sake of clarity, we confine our discussion to cells of convex shape bounded by planar surfaces, such as Wigner-Seitz cells. The extension to more general, even interpenetrating cell charges has been reported in the literature [5]. For the case of convex cells, we derive explicit expressions for both the solution of Poisson's equation as well as the electrostatic interaction in a solid in terms of the cell multipole moments. We illustrate these formal expressions through the results of numerical calculations that are reported in the following section. These calculations also clarify the origin of the divergences noted by Morgan [26], and demonstrate the analytic process that can be used to circumvent such behavior. The resulting formulae are straightforward enough to be incorporated into most existing computer codes for the calculation of electronic structure.

Let us consider the potential outside a convex cell, but inside the sphere circumscribing the cell. As an example, we consider the unit cube shown schematically in Fig. 4.1. At any point P_1 outside the sphere bounding the cell, the potential is given by Eq. (4.4). We inquire as to whether the potential at P_2, inside the moon region, can similarly be expressed in terms of the multipole moments of the charge density, $\rho(\mathbf{r})$, inside the cell.

Let us denote by $\mathbf{r}_0$ the radius vector from the center of the cell to point P_2. We note that a straightforward expansion of the denominator in Eq. (4.2), aimed at recovering the form of Eq. (4.4), is not possible because $\mathbf{r}_0$ is not larger than all intracell vectors $\mathbf{r}$. (For example, r can exceed r_0 when $\mathbf{r}$ points to a corner of the cell.) Thus, Eq. (4.4) in general is not valid for $\mathbf{r}_0$ inside the moon region. In fact it leads to divergent behavior there, as is illustrated by the results of numerical calculations. However, an expansion[1] in terms of the cell multipole moments, Q^n_{lm}, can be obtained even in this case by means of the following procedure: We add and subtract a vector **b** in the denominator of the integrand in Eq. (4.2) such that for all **r** in the cell we have $b < |\mathbf{r} - (\mathbf{r}_0 + \mathbf{b})|$. Such a vector can always be constructed for *each planar face of a convex cell*, and for all **r** and $\mathbf{r}_0$. In fact, for each given face of the cell it suffices to choose any vector **b** normal to a face and pointing away from the cell. In view of Eq. (4.9), we can now write

$$V(\mathbf{r}_0) = \sum_{l,m} \frac{4\pi}{2l+1} b^l Y_{lm}(\widehat{-\mathbf{b}}) \int_\Omega \rho(\mathbf{r}) \frac{Y^*_{lm}[\mathbf{r} - \widehat{(\mathbf{r}_0 + \mathbf{b})}]}{|\mathbf{r} - (\mathbf{r}_0 + \mathbf{b})|^{l+1}} d^3r, \quad (4.12)$$

where the sum and the integral can now be safely interchanged because the sum converges for all values of the integrand. At this point the magnitude of **b** is still arbitrary. We now impose the further condition that **b** have a magnitude larger than the maximum distance between the bounding sphere and the face of the cell intersected by $\mathbf{r}_0$. In other words, **b** is such that for all $\mathbf{r}_0$ in the moon region adjacent to a face, $\mathbf{r}_0 + \mathbf{b}$ lies outside the bounding sphere. Then, **r** in Eq. (4.12) satisfies the condition $r < |\mathbf{r}_0 + \mathbf{b}|$, so that we can use Eq. (4.10) to expand once again and obtain the result

$$V(\mathbf{r}_0) = \sum_{l,m} \frac{4\pi}{2l+1} b^l Y_{lm}(\widehat{-\mathbf{b}}) \left[\sum_{l',m'} (-)^l S_{lm;l'm'}(\mathbf{r}_0 + \mathbf{b}) Q_{l',m'} \right], \quad (4.13)$$

where the structure constants **S** are defined by the expression

$$\begin{aligned} S_{lm;l'm'}(\mathbf{R}) &= 4\pi \frac{(2l+2l'-1)!!}{(2l-1)!!(2l'+1)!!} C(lm;l'm';l+l',m'-m) \\ &\times \frac{Y_{l+l',m'-m}(\hat{\mathbf{R}})}{R^{l+l'+1}} \\ &= \alpha(lm;l'm') Y_{l+l',m'-m}(\hat{\mathbf{R}})/R^{l+l'+1}. \end{aligned} \quad (4.14)$$

These 'real-space' structure constants are closely related to those used in the KKR [35, 36] and LMTO [27], band-structure methods, and can be readily calculated[2] for arbitrary values of l and m.

It is to be emphasized that Eq. (4.12) is valid for all convex-shaped cells, since it is always possible to choose **b** such that $b < |\mathbf{r} - (\mathbf{r}_0 + \mathbf{b})|$ and $r < |\mathbf{r}_0 + \mathbf{b}|$. Thus, Eq. (4.13) leads to the potential in the moon region, as well as everywhere else outside the cell, entirely in terms of the multipole

[1]The reader is urged to apply the methods of the present section to derive a form of MST involving the cell t-matrix that converges inside the moon region of a convex cell.

[2]They are essentially the zero energy limit of the KKR structure constants.

moments of the cell charge. The double summation indicated in Eq. (4.13) has been derived in such a way that, if the internal sum is carried out first, the double sum converges. This double summation is the foundation of the formalism in this section. Therefore, it may be instructive to gain a "pictorial" understanding of the derivation of Eq. (4.13), along with its convergence characteristics.

First, note that the single sum in Eq. (4.4), which generally diverges in the moon region, has been replaced by a double sum in Eq. (4.13). The brackets in Eq. (4.13) indicate that the internal sum must be performed first. This is because that sum was introduced after the outer sum, involving $\mathbf{b}$, and the order of summations must reflect that fact. Proceeding backwards from Eq. (4.13), with the internal sum carried out first, one readily obtains the integral formula, Eq. (4.2). Pictorially, it is as though point P_2 is moved outside the bounding sphere through a translation by the vector $\mathbf{b}$, in a region where the potential can be represented in terms of the cell multipole moments. This expression corresponds to the sum inside the brackets in Eq. (4.13), and by itself yields the (l, m)-component of the potential at $\mathbf{r}_0 + \mathbf{b}$. Next, the potential at $\mathbf{r}_0 + \mathbf{b}$ is analytically continued (moved back) to the original point $\mathbf{r}_0$ by means of the outer, or second sum. Alternatively, one may consider the cube as being displaced by $-\mathbf{b}$, causing all points in the moon region to fall outside the bounding sphere, and leading to the multipole expansion of the internal sum in Eq. (4.13). Then, the outer sum returns the cube back to its original position. Whether one invokes the passive or active interpretation of the double sum in Eq. (4.13), i.e. moving the points or moving the cell, the order of summations must be maintained as indicated to guarantee convergence. Also, it is to be noted that the Q_{lm}'s in Eq. (4.13) are the multipole moments of the cell about its original, undisplaced center. In spite of the points or the cell "moving around," no moments about displaced centers enter Eq. (4.13).

Equation (4.13) is manifestly different from Eq. (4.4), but reduces to that equation as $\mathbf{r}_0$ moves outside the bounding sphere. For such values of $\mathbf{r}_0$ the order of the sums in Eq. (4.13) can be reversed, causing it to collapse to the simpler form of Eq. (4.4). At the same time, in applying this formalism to realistic systems it may be convenient to use Eq. (4.13) for all points outside a cell, avoiding the cumbersome process of switching formulae on a point-by-point basis. Thus, the potential inside a given cell in a material is the sum of three terms: a term $V_0(\mathbf{r})$ arising from the charge in the cell itself, one term contributed by those cells whose bounding spheres do not intersect the cell in question, and a term contributed by a set $\{\delta\}$ of nearby cells whose bounding spheres do intersect the cell. The first term, V_0, can be found by direct integration or other means, the second is given by Eq. (4.4) for each cell not in the set $\{\delta\}$, and the third by an equation with a double sum, as in Eq. (4.13), for each of the cells in $\{\delta\}$. Explicitly, we have the expression

$$
\begin{aligned}
V(\mathbf{r}) = V_0(\mathbf{r}) \quad &+ \quad \sum_{n\notin\{\delta\}}\sum_{l,m}\frac{4\pi}{2l+1}\frac{Y_{lm}(\widehat{\mathbf{r}_n})}{r_n^{l+1}}Q_{lm}^n \\
&+ \quad \sum_{n\in\{\delta\}}\sum_{l,m}\frac{4\pi}{2l+1}R_n^l Y_{lm}(\widehat{\mathbf{R}_n}) \\
&\times \quad \left[\sum_{l',m'}(-)^l S_{lm;l'm'}(\mathbf{r}_n-\mathbf{R}_n)Q_{l',m'}^n\right], \qquad (4.15)
\end{aligned}
$$

where the symbol $\in$ ($\notin$) denotes the relation of belonging (not belonging) to a set, and we have set $-\mathbf{b}$ equal to the intercell vector $\mathbf{R}_n$. Now, we see that the divergences noted by Morgan arise from the use of the MT expressions, consisting of only the first two terms in Eq. (4.15) being used for all cells, including those in $\{\delta\}$. This violates the conditions for expanding the Green function in the moon region and can indeed lead to divergent results (see next section). We end the present section with the derivation of an expression for the Coulomb interaction of any cell in a material, say the one at the origin, with all other cells. Upon expanding the structure constants as prescribed in Eq. (4.10), we readily obtain the result

$$
\begin{aligned}
U = U_0 &+ \sum_{n\notin\{\delta\}}\sum_{l,m}\frac{4\pi}{2l+1}Q_{lm}^{0*}\sum_{l',m'}(-)^l S_{lm;l'm'}(\mathbf{R}_n)Q_{l'm'}^n \\
&+ \sum_{n\in\{\delta\}}\sum_{l,m}\frac{4\pi}{2l+1}R_n^l Y_{lm}(\widehat{\mathbf{R}_n}) \\
&\times \left[\sum_{l'm'}\sum_{l''m''}(-)^{l+l'}Q_{l'm'}^0\alpha(lm;l'm')S_{l+l',m'-m;l''m''}^*(2\mathbf{R}_n)Q_{l''m''}^{n*}\right].
\end{aligned}
\qquad (4.16)
$$

Even though they look rather involved, Eqs. (4.15) and (4.16) are not particularly difficult to use. More importantly, they lead to converged results, provided that the various sums contained in them are performed in the proper order as is explicitly indicated by the brackets.

4.5 Numerical Results for Convex Cells

In this section, we report the results of calculations based on the formalism of the previous section. We explicitly treat the case of constant charge densities, set equal to one in arbitrary units, confined inside cells of convex shape. As a prototype of such cells, we consider both $(1 \times 1 \times 1)$ unit cubes and $(1 \times 1 \times 2)$ prisms. In each case, the exact potential and electrostatic interaction were obtained analytically using an algebraic-manipulation program [37], and were compared to the results of the MT expansions and the modified expressions, Eqs. (4.13) and (4.16).

In the following figures, we plot the ratios $\Delta V/V$ and $\Delta E/E$, where ΔV (ΔE) denotes the difference in potential (energy) between the exact values, obtained analytically, of the potential (energy) and those computed through various multipole expansions. In all figures, a dotted line depicts the results of the MT-like expansions, while a solid line represents the results of the newly-derived expressions, Eqs. (4.13) and (4.16). The ratio $\Delta V/V$ vs. the value of the outer variable l in Eq. (4.13) for the case of the unit cube of unit charge density is shown in Fig. 4.2 at various distances Z along the z-axis away from the center of the cube. In each case, the value of the internal sum l' in Eq. (4.13) was truncated at 30. Outside the bounding sphere ($Z > \sqrt{3}/2$), both Eqs. (4.4) and (4.13) converge rapidly to the exact value of the potential, as is shown in panel (a) of the figure. In applying Eq. (4.13), a value of $b = 0.5$ was used. (All vectors $\mathbf{b}$ are chosen perpendicular to the cell face, i.e. along the z-axis.) On the other hand, inside the moon region, where it is invalid, Eq. (4.4) leads to an oscillatory behavior with increasing l as the observation point gets closer and closer to the face of the cube, as is indicated by the dotted curves in panels (b) and (c). In contrast to this behavior, Eq. (4.13) converges smoothly (solid curves) to the exact results for an appropriate choice of b and of internal summation cutoff ($l' \geq 30$). The results in panels (b) and (c) correspond to points just inside the bounding sphere and just outside the cube face, respectively. The latter position of the observation point clearly violates the MT condition rather severely. (A value of b was chosen such that $Z + b = 3/2$ lies outside the bounding sphere, which has radius $r^2 = 3/4$). For the case in panel (b), point P_2 lies $1/10$ inside the bounding sphere and the error produced by the use of Eq. (4.13) amounts to only 0.002% of the exact answer. Even when P_2 is moved to a position of only $1/200$ outside the cube face [panel (c)], violating the MT condition severely, the error only increases to 0.07%, for $l' \approx l'_{\text{max}}/2$, where l'_{max} is the upper limit of the range over which l' is summed (l'_{max}=30 for the calculations illustrated). If greater accuracy is desired, both sums should be increased, with the internal sum carried out to larger values of the angular momentum than the outer sum. The computational labor involved in performing these sums, even to values of l and l' that give the impression of being "large" is not excessive, especially after the various expansions have been coded for a particular cell shape. Once this is done, these expressions can be used repeatedly in different applications.

Results analogous to those for the $(1 \times 1 \times 1)$ cube are shown in Fig. 4.3 for the $(1 \times 1 \times 2)$ prism. Here, the observation point is taken outside

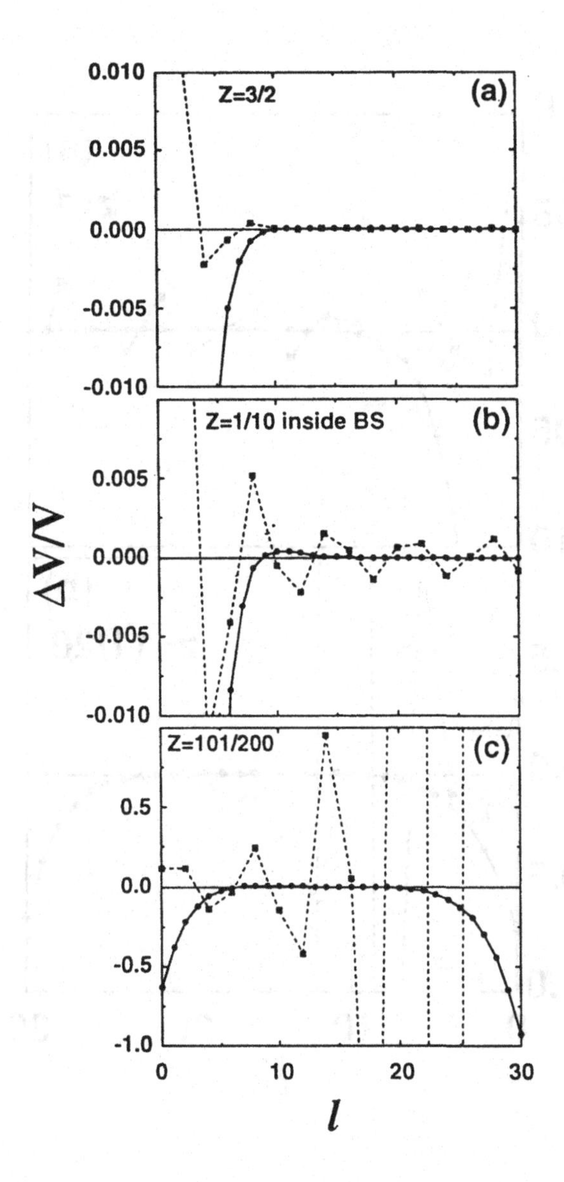

Figure 4.2: **Convergence of multipole expansions for a uniform charge confined inside a cubic cell.**

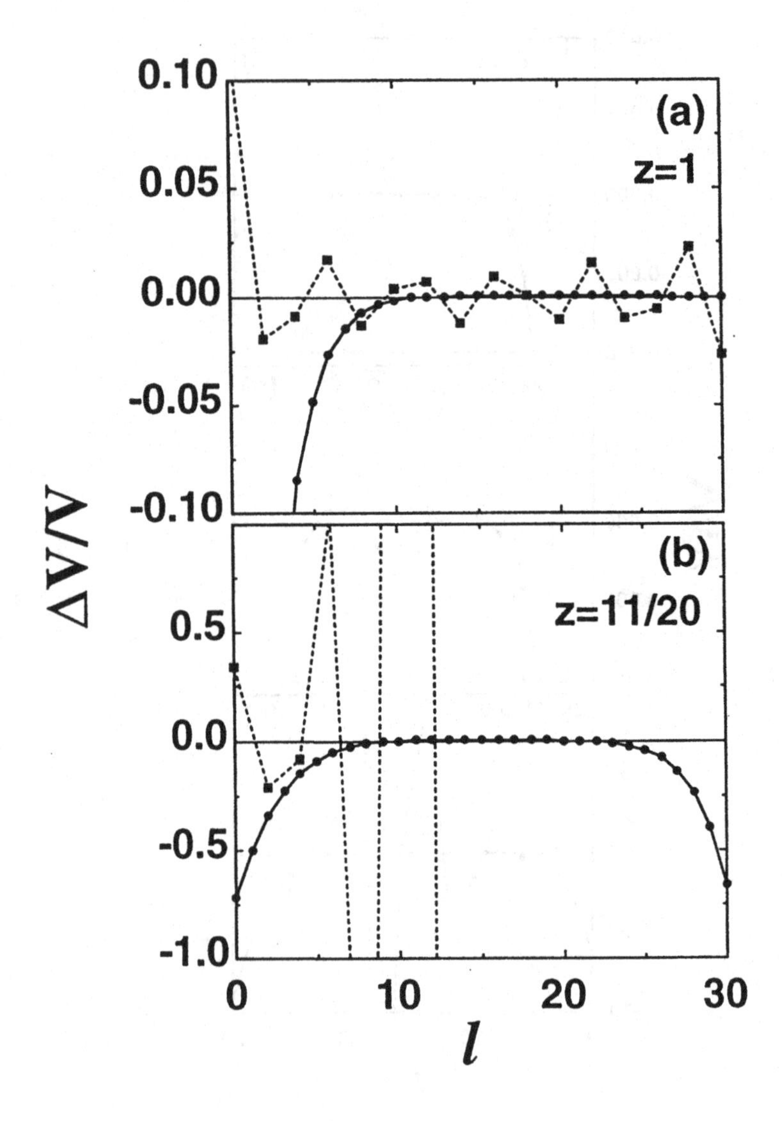

Figure 4.3: **Convergence of multipole expansions for a uniform charge confined inside a prism.**

a long face of the prism, so that the MT condition is violated even more severely than for a unit cube. The value of Z is indicated explicitly in the various panels, b was chosen equal to 2, and the internal sum in Eq. (4.13) was again truncated at 30. Only the results for points P_2 inside the moon region are shown, as both Eqs. (4.4) and (4.13) converge rapidly to the exact result outside the bounding sphere. As expected, in this case as the observation point moves closer to the prism, the failure of the MT expression (Eq. (4.4) becomes much more pronounced than in the case of the cube. As can be seen in panel (b), the results obtained through Eq. (4.4) (dotted lines) oscillate wildly even at relatively small values of $l \cong 10$ when $Z = 11/20$, i.e. when the observation point lies 1/20 above the long face of the prism. On the other hand, Eq. (4.13) converges rapidly to the exact result over a substantial range of the outer l-sum (solid curves). At the same time, as the maximum value of l used in the outer sum becomes close to $l'_{\rm max}$, even Eq. (4.13) begins to yield results that deviate from the exact ones, both for the cube and the prism unit cells, as is indicated in Figs. 4.2 and 4.3. This behavior can be readily understood on the basis of the convergence properties of the double sum in Eq. (4.13).

As already discussed following Eq. (4.13), the sums over l and l' are infinite in extent, and the internal sum over l', which corresponds to the second expansion of the denominator in Eq. (4.2), should be carried out first, while the outer sum over l, corresponding to the first expansion, should be carried out last. Under these conditions, Eq. (4.13) converges to the integral expression, Eq. (4.2). However, in practical applications both sums must be truncated. Under these circumstances, the internal sum must be carried out to values significantly larger than those of the outer sum in order to maintain the correct priority between them. In this way, the convergence of the outer sum is subjugated to that of the inner one. Thus, for a fixed value of b for l'_{max}, Eq. (4.13) will converge over a range of values of the outer sum before beginning to diverge, as l approaches l'. This effect arises because, as l approaches l', the distinction between inner and outer sum is obscured, and the expansion behaves more and more like Eq. (4.4), which diverges in the moon region. This behavior is clearly exhibited in Figs. 4.2 and 4.3 for high values of $l(\geq 20)$. On the other hand, provided that the inner sum is carried out to sufficiently large values of l' so that the outer sum converges up to some l, the two sums can be interchanged, because they are now both finite.

The convergence of the double sum in Eq. (4.13) also depends on the choice of $\mathbf{b}$. As is shown in Fig. 4.4, for small values of b the sum rapidly approaches the exact answer, but then quickly diverges as l gets larger. As b increases, the sum approaches the exact results more slowly, but matches it over a wider range of l. Practical applications then depend on a suitable choice of $\mathbf{b}$ and the maximum values of l and l'. As Fig. 4.4 indicates, an

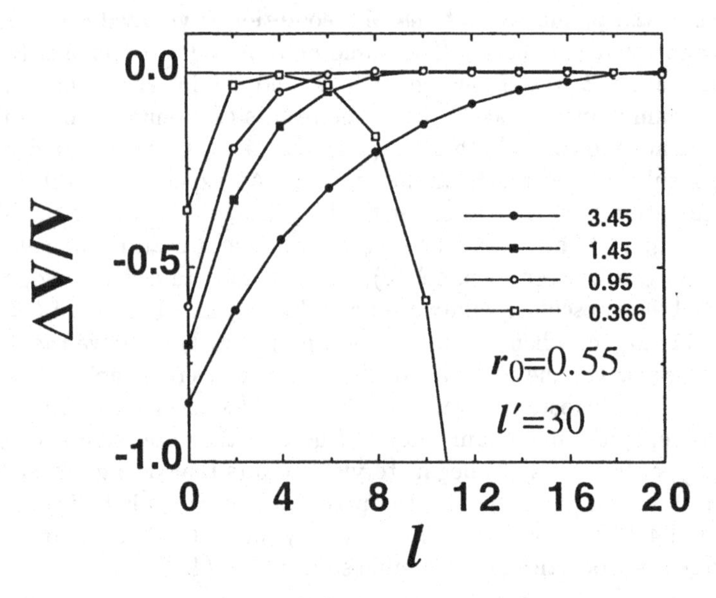

Figure 4.4: **Convergence of multipole expansions in the double sum as a function of b.**

intermediate b, approximately the size of the radius of the bounding sphere of a cell, leads to an acceptable range and rate of convergence.

At this point, it should be emphasized that the relatively large values of l at which the internal summation must be truncated are due to the expansion of a non-spherical shape into angular momentum eigenfunctions that are orthogonal on the surface of a sphere. This feature is particularly relevant to the performance of electronic structure calculations. In using the formalism presented here to calculate the electrostatic potential and energy in a solid, one need not necessarily calculate the charge density to correspondingly high values of l. It suffices to obtain only the first few elements in an angular momentum expansion, say $l = 6$ or 8, (corresponding to wave function expansions to $l = 3$ or 4) and use the spherical harmonic expansion of the cell to generate the higher multipole moments. The expansion of the cell shape can be calculated once and for all for each cell shape and used repeatedly in the calculation of the potential (and the energy) of a material.

The relative ratios $\Delta E/E$ of the electrostatic interaction between two cubes in contact along a face, and between two $(1 \times 1 \times 2)$ prisms touching along a long face, are shown in Figs. 4.5 and 4.6, respectively. In these calculations, l'_{max} was set equal to 30, and the values of b were chosen equal to 1 and 2 for the cube and prism, respectively. We see once again

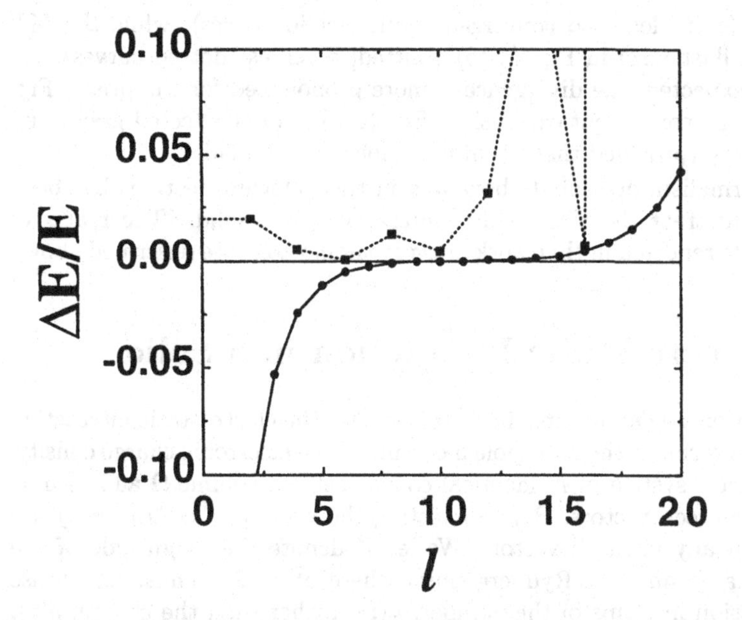

Figure 4.5: **Convergence of multipole expansions of the electrostatic interaction between two cubic cells.**

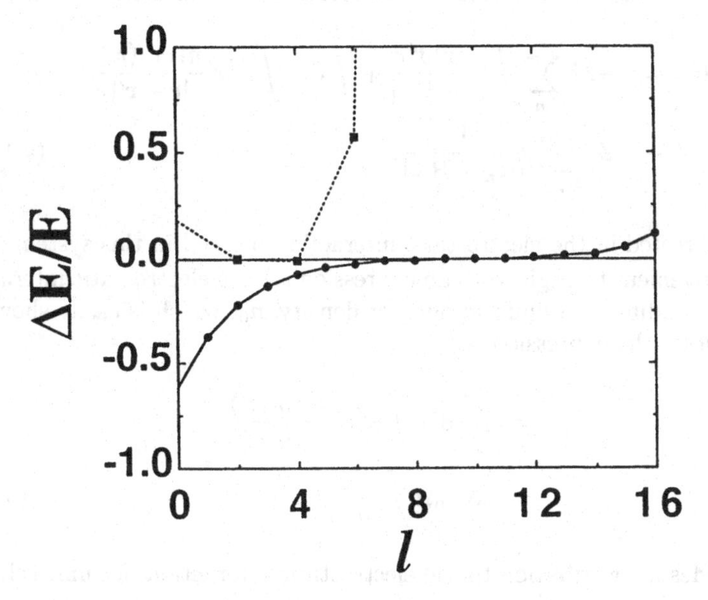

Figure 4.6: **Convergence of multipole expansions of the electrostatic interaction between two prisms.**

that Eq. (4.16) leads to converged results (solid curves), while the MT expression [last term in Eq. (4.16) omitted] diverges (dotted curves). As might be expected, the divergence is more pronounced for the prism, Fig. 4.6, while the convergent expression, Eq. (4.16), is not affected greatly by the geometry (provided that a suitable choice of b is made).

The formalism presented above and in the following sections has been used [40] to treat the electrostatic interaction in a solid. The r ates of convergence reported in that work are very similar to those described above.

4.6 Electrostatic Interaction in a Solid

In this section we derive explicit expressions for the electrostatic interaction in a solid in terms of the multipole moments of the electronic charge density. We consider a system of N identical convex cells of volume Ω and assume that the intercell vectors, $\mathbf{R}_{ij}$, connecting the centers of cells i and j are larger than any intracell vector. We let Z denote the magnitude of the nuclear charge and use Rydberg units where $\mathrm{e}^2 = 2$. Thus, we phrase our discussion in terms of the number, $n(\mathbf{r})$, rather than the charge, $\rho(\mathbf{r})$, density where $\rho(\mathbf{r}) = \mathrm{e}n(\mathbf{r})$.

The electrostatic interaction in a system of electrons and nuclei consists of electron-nuclear, electron-electron, and nuclear-nuclear contributions,

$$\begin{aligned} U &= -2Z\sum_n \int \frac{n(\mathbf{r})\mathrm{d}^3r}{|\mathbf{r}-\mathbf{R}_n|} + \int \mathrm{d}^3r \int \mathrm{d}^3r' \frac{n(\mathbf{r})n(\mathbf{r}')}{|\mathbf{r}-\mathbf{r}'|} \\ &+ Z^2 \sum_{n\neq m} \frac{1}{|\mathbf{R}_n - \mathbf{R}_m|}. \end{aligned} \tag{4.17}$$

We are interested in the electrostatic interaction per cell in this system.

It is convenient to begin with an expression for the electrostatic interaction corresponding to a uniform number density, n_0, which is easily shown to be given by the expression,

$$\begin{aligned} U &= \int \mathrm{d}^3r \int \mathrm{d}^3r' \frac{n(\mathbf{r})n(\mathbf{r}')}{|\mathbf{r}-\mathbf{r}'|} \\ &= Nn_0^2\Omega \int \frac{\mathrm{d}^3r}{r}. \end{aligned} \tag{4.18}$$

This provides a contribution to the electrostatic interaction per unit cell,

$$u^{(1)} = n_0^2\Omega \int \frac{\mathrm{d}^3r}{r}. \tag{4.19}$$

To this background, we now add the nuclei which make a contribution to the electrostatic interaction per unit cell, say the cell at $\mathbf{R}_0$, equal to

$$u^{(2)} = -2Zn_0 \int \frac{\mathrm{d}^3 r}{r} + Z^2 {\sum_{\mathbf{R}}}' \frac{1}{\mathbf{R}}, \tag{4.20}$$

where the sum runs over all intercell vectors other than the zero vector, as is noted by the prime on the summation symbol.

The sum of $u^{(1)}$ and $u^{(2)}$, excluding the internuclear contribution, can also be written in terms of the multipole moments, Q_L^0, of the uniform background charge,

$$\begin{aligned} u^{(3)} &= -\left\{ 2Zn_0 \int_\Omega \frac{\mathrm{d}^3 r}{r} - n_0^2 \int_\Omega \int_\Omega \frac{\mathrm{d}^3 r \mathrm{d}^3 r'}{|\mathbf{r} - \mathbf{r}'|} \right\} \\ &- \left\{ 2Z {\sum_{\mathbf{R}}}' \sum_L \frac{Q_L^0}{R^{\ell+1}} - 2 {\sum_{\mathbf{R}}}' \sum_{LL'} Q_L^0 \left[S_{LL'}(\mathbf{R}) Q_{L'}^0 \right] \right\}. \end{aligned} \tag{4.21}$$

The electrostatic interaction of the charge density (number density) $n(\mathbf{r})$ in the presence of the nuclei, expressed in multipole moments, Q_L, would be[3]

$$\begin{aligned} u^{(4)} &= -2Z \int_\Omega \frac{n(\mathbf{r}) \mathrm{d}^3 r}{r} + \int_\Omega \int_\Omega \frac{n(\mathbf{r}) n(\mathbf{r}')}{|\mathbf{r} - \mathbf{r}'|} \mathrm{d}^3 r \mathrm{d}^3 r' \\ &- \left\{ 2Z {\sum_{\mathbf{R}}}' \sum_L \frac{Q_L}{R^{\ell+1}} - 2 {\sum_{\mathbf{R}}}' \sum_{LL'} Q_L \left[S_{LL'}(\mathbf{R}) Q_{L'} \right] \right\}. \end{aligned} \tag{4.22}$$

With the definition $\Delta Q_L = Q_L - Q_L^0$, and choosing the uniform charge to be the average nuclear charge, $n_0 = Z/\Omega$, we can write

$$\begin{aligned} u &= -2Z \int_\Omega \frac{n(\mathbf{r}) - n_0}{r} \mathrm{d}^3 r + \int_\Omega \int_\Omega \frac{n(\mathbf{r}) n(\mathbf{r}') - n_0^2}{|\mathbf{r} - \mathbf{r}'|} \mathrm{d}^3 r \mathrm{d}^3 r' \\ &- 2Z {\sum_{\mathbf{R}}}' \sum_L \frac{\Delta Q_L}{R^{\ell+1}} + 2 {\sum_{\mathbf{R}}}' \sum_{LL'} \Delta Q_L \left[S_{LL'}(\mathbf{R}) \Delta Q_{L'} \right] \\ &+ 4 {\sum_{\mathbf{R}}}' \sum_{LL'} \Delta Q_L \left[S_{LL'}(\mathbf{R}) Q_{L'}^0 \right] \\ &- Z^2 A. \end{aligned} \tag{4.23}$$

Here, we have defined the quantity

$$A = \frac{1}{\Omega} \int \frac{\mathrm{d}^3 r}{r} - {\sum_{\mathbf{R}}}' \frac{1}{\mathbf{R}}, \tag{4.24}$$

[3]If necessary, modified structure constants could be used in these expressions in a manner analogous to that indicated in Eq. (4.13).

which for a given crystal structure can be evaluated by means of Ewald summations. Values of this constant corresponding to the sc, bcc, and fcc lattices are given in the literature [33]. We note that the expressions involving the difference in multipole moments do *not* contain the multipole expansion of the cell shape which has been incorporated into the constant A. Thus, the quantities ΔQ_L are determined solely in terms of the moments of the charge for $\ell > 0$.

Using Eq. (4.10), the expression for the electrostatic interaction may be further simplified to read

$$
\begin{aligned}
u &= -2Z\int_{\Omega}\frac{n(\mathbf{r})-n_0}{r}\mathrm{d}^3r + \int_{\Omega}\int_{\Omega}\frac{n(\mathbf{r})n(\mathbf{r}')-n_0^2}{|\mathbf{r}-\mathbf{r}'|}\mathrm{d}^3r\mathrm{d}^3r' \\
&- 2Z\sum_{\mathbf{R}}{}'\sum_{L}\frac{\Delta Q_L}{R^{\ell+1}} + 2\sum_{\mathbf{R}}{}'\sum_{LL'}\Delta Q_L\left[S_{LL'}(\mathbf{R})\Delta Q_{L'}\right] \\
&+ 4n_0\sum_{\mathbf{R}}{}'\sum_{L}\Delta Q_L\left[\int \mathrm{d}^3r\rho(\mathbf{r})H_L(\mathbf{r}-\mathbf{R})\right] \\
&- Z^2A.
\end{aligned}
\tag{4.25}
$$

where $H_L(\mathbf{r}-\mathbf{R})$ denotes the quantity on the left side of Eq. (4.10). For periodic systems with cubic symmetry, in which the charge possesses no moments between $\ell = 0$ and $\ell = 4$, and since the $\ell = 0$ contribution has been removed from ΔQ_L^0, the summations over $\mathbf{R}$ can be calculated directly in real space.

4.7 Analogy with Multiple Scattering Theory

In this section we provide an alternative treatment of the Poisson equation, which is directly related to the formalism of multiple scattering theory. In this approach, we combine the solutions (potentials) associated with individual cell charges so that the solution for a collection of cell charges is continuous in value and derivative across cell boundaries. In order to emphasize the formal relation to MST, we will use the notation and formal approach that was established in Chapter 3. However, it should be kept in mind that we are now dealing with the $E \to 0$ ($k \to 0$) limit of the various functions used in connection with MST. For example, in this limit the free-particle propagator, Eq. (D.128), is given by Eq. (4.9), whereas the expansion formula for the Hankel function, Eq. (F.10), takes the form of Eq. (4.10) [27]. Thus, in the following discussion we will use the symbols $J_L(\mathbf{r})$ and $H_L(\mathbf{r})$ to indicate the functions,

$$J_L(\mathbf{r}) = r^lY_L(\hat{\mathbf{r}}) \quad \text{and} \quad H_L(\mathbf{r}) = r^{-(l+1)}Y_L(\hat{\mathbf{r}}). \tag{4.26}$$

Because the formalism of this section is based on the knowledge of potential functions associated with individual, spatially bounded cell charge, we begin with a brief review [6] of the general form of these potentials.

4.7.1 Single cell charges

We consider a charge distribution, $\rho(\mathbf{r})$, which vanishes outside a bounding sphere of radius r_s. The potential, $v(\mathbf{r})$, associated with this charge is given by the integral

$$\begin{aligned} v(\mathbf{r}) &= \int \frac{\rho(\mathbf{r}')\mathrm{d}^3 r}{|\mathbf{r}-\mathbf{r}'|} \\ &= \sum_L \left\{ J_L(\mathbf{r}) \int_{r'>r} H_L^*(\mathbf{r}')\rho(\mathbf{r}')\mathrm{d}^3 r' \right. \\ &+ \left. H_L(\mathbf{r}) \int_{r'<r} J_L^*(\mathbf{r}')\rho(\mathbf{r}')\mathrm{d}^3 r' \right\} \end{aligned} \tag{4.27}$$

where we have used the expansion of the Green function given in Eq. (4.9), and the definition in Eq. (4.26). With obvious definitions, we can write

$$v(\mathbf{r}) = \sum_L \{J_L(\mathbf{r})C_L(r) + H_L(\mathbf{r})S_L(r)\} \tag{4.28}$$

or

$$v(\mathbf{r}) = \sum_L v_L(\mathbf{r}), \tag{4.29}$$

where

$$v_L(\mathbf{r}) = J_L(\mathbf{r})C_L(r) + H_L(\mathbf{r})S_L(r). \tag{4.30}$$

We note that outside the bounding sphere, $r > r_s$, we have

$$v_L(\mathbf{r}) = H_L(\mathbf{r})S_L \tag{4.31}$$

where

$$S_L \equiv Q_L = \int J_L^*(\mathbf{r}')\rho(\mathbf{r}')\mathrm{d}^3 r' \tag{4.32}$$

is the asymptotic value of the function defined implicitly in Eq. (4.27), and is precisely the multipole moment of order L, Eq. (4.5), associated with the charge $\rho(\mathbf{r})$. It follows that for $r > r_s$ the potential is given by the multipole expansion, Eq. (4.11).

4.7.2 Multiple scattering solution of the Poisson equation

We seek to obtain the potential, $V(\mathbf{r})$, associated with a collection of spatially bounded but nonoverlapping cell charges, $\rho^n(\mathbf{r})$, centered at positions specified by vectors $\mathbf{R}_n$. For the sake of clarity, we begin by assuming that the cells are far enough apart from one another as to satisfy the MT conditions, i.e., the sphere circumscribing any one cell does not intersect similar spheres around other cells[4]. Space-filling charges are considered in the following subsection.

We proceed in a manner typical of MST: We obtain solutions within each cell that contain a set of undetermined coefficients which are to be chosen so that these solutions are matched smoothly to solutions obtained for other cells. With $\mathbf{r}_n = \mathbf{r} - \mathbf{R}_n$, the solution inside cell n may be written in the form (where the vector/matrix notation of Appendix F is used to denote sums over angular momentum states),

$$\begin{aligned} V_{In}(\mathbf{r}) &= v_n(\mathbf{r}_n) + \sum_L c_L^n J_L(\mathbf{r}_n) \\ &= v_n(\mathbf{r}_n) + \langle J|c^n\rangle. \end{aligned} \tag{4.33}$$

Here, the function $v_n(\mathbf{r}_n)$ is a local solution of the Poisson equation associated with the charge $\rho^n(\mathbf{r}_n)$ and satisfies Eq. (4.1). The general form of these local solutions is given by Eqs. (4.28) to (4.33). We will assume that $v_n(\mathbf{r}_n)$ in specific cases can be obtained through the use of standard techniques, such as the LDA, so that the remaining problem is the determination of the coefficients, c_L^n, in Eq. (4.33).

The determination of these coefficients requires the use of two additional forms of the global solution, $V(\mathbf{r})$. Outside a sphere bounding the charge in cell n, the potential can be written either as a single-center or a multicenter expansion. In analogy with Eq. (3.89) we obtain the single-center expansion

$$\begin{aligned} V_{IIn}(\mathbf{r}) &= \sum_L [a_L^n J_L(\mathbf{r}_n) + b_L^n H_L(\mathbf{r}_n)] \\ &= \langle a^n|J(\mathbf{r}_n)\rangle + \langle b^n|H(\mathbf{r}_n)\rangle, \end{aligned} \tag{4.34}$$

while the multi-center expansion takes the form,

$$\begin{aligned} V_{II}(\mathbf{r}) &= \sum_n \sum_L b_L^n H_L(\mathbf{r}_n) \\ &= \langle b^n|H(\mathbf{r}_n)\rangle, \end{aligned} \tag{4.35}$$

which corresponds to Eq. (3.94).

[4]It suffices that a sphere bounding one cell does not intersect the charge densities in adjacent cells.

Equations (4.33), (4.34), and (4.35) contain three sets of coefficients, a_L^n, b_L^n, and c_L^n, which are still to be determined. The number of sets of unknowns can be reduced to two through the observation that b_L^n is related to the multipole moments of the charge in cell n,

$$b_L^n = \frac{4\pi Q_L^n}{2l+1}. \tag{4.36}$$

Furthermore, use of the fundamental condition of MST that the single-center and multicenter expansion be consistent leads to the relation

$$\begin{aligned} \langle a^n | J(\mathbf{r}_n) \rangle &= \sum_{n' \neq n} \langle b^{n'} | H(\mathbf{r}_n) \rangle \\ &= \sum_{n' \neq n} \langle b^{n'} | S^{n'n} | J(\mathbf{r}_n) \rangle, \end{aligned} \tag{4.37}$$

where the structure-constant matrix is given by Eq. (4.14) and its use here is justified because the cell charges conform to the MT geometry. Finally, the requirement that V_{In} and V_{IIn} be consistent leads to the following expression for the coefficients c_L^n:

$$c_L^n = \sum_{n' \neq n} \sum_{L'} b_{L'}^{n'} S_{L'L}^{n'n}. \tag{4.38}$$

The use of this expression in Eq. (4.33) and the use of Eq. (4.10) leads to the expression for the potential given in Eq. (4.6).

The generalization of the formalism given above to space-filling cells can be made in a number of ways, such as replacing the single sum over L in Eq. (4.38) by a double, or a multiple, conditionally convergent sum as was demonstrated in Section 3.4. Further details of the solution of the Poisson equation for space-filling charges can be found in the literature [5, 38].

4.8 MT Potentials and Charge Densities

Spherically symmetric MT potentials have been used and continue to be used in great numbers of calculations of the electronic structure, particularly those directed at metallic systems and based on multiple scattering theory and the use of Green functions. In close packed systems, the approximation of the cell potential by a spherically symmetric potential confined inside a sphere inscribed in the cell, the MT sphere of radius r_s, has been found to be quite useful at least in the determination of total energies and related quantities. Because of its practical importance, in this section we derive explicit expressions for the electrostatic energy and the potential associated with a collection of MT potentials and charge densities.

In the MT approximation, the electronic charge density in each unit cell is divided into a spherically symmetric part confined inside the MT sphere, and a constant, spatially averaged part, ρ_0, outside, in the interstitial region of the cell. For a cell of volume Ω, the interstitial volume is given by

$$\Omega_0 = \Omega - \frac{4}{3}\pi r_s^3, \tag{4.39}$$

and the constant charge density outside the MT is chosen to maintain the charge neutrality of the cell

$$\begin{aligned} \rho_0 &= \frac{Ze - 4\pi \int_0^{r_s} \rho_0(r) r^2 dr}{\Omega_0} \\ &= \frac{Z - Q_{\mathrm{MT}}}{\Omega_0}, \end{aligned} \tag{4.40}$$

where Ze is the nuclear charge, with e being the charge of the electron, and Q_{MT} the total charge inside the MT sphere. It follows that the cell charge outside the MT sphere is equal to

$$Z_{\mathrm{out}} = \rho_0 \Omega_0 = Ze - Q_{\mathrm{MT}}. \tag{4.41}$$

We note that the electronic charge density is simply related to the number density,

$$\rho(\mathbf{r}) = -en(\mathbf{r}). \tag{4.42}$$

In the following discussion, we use atomic units in which $e^2 = 2$.

We now seek an expression for the electrostatic energy of a system consisting of N identical neutral cells of volume Ω, each containing a nucleus of charge Ze and an electronic charge density of the MT form. We are interested in the limit as $N \rightarrow \infty$, which is appropriate for the case of a periodic solid.

An expression for the electrostatic energy can be obtained in terms of the cell multipole moments, following the formalism of the previous sections. Clearly, the interaction of the nuclei with the electrons and with one another would have to be added to the expression in Eq. (4.16). However, Eq. (4.16) is unduly complicated in the case of MT potentials. The result can be obtained in computationally more convenient form and by simpler means.

We can think of the charge distribution of the system as the superposition of the following four distinct components: A constant charge density, ρ_0, extending throughout space; the nuclear charges, Ze, located on the positions $\mathbf{R}$; the spherically symmetric MT charge densities, $\rho(r)$, centered also at $\mathbf{R}$; and a constant density, $-\rho_0$, inside each MT sphere. This last component corrects for the inclusion of ρ_0 inside the MT spheres that is contributed by the constant charge density. The electrostatic energy of the system can be found by combining the interactions as each component of charge is added to the system.

The electrostatic energy of the constant charge density and the nuclear charges is given by Eqs. (4.18) to (4.20). On this system of the nuclei and constant background charge, we now add the MT spherically symmetric charge densities inside the MT spheres. This yields the next contribution to the electrostatic energy per unit cell

$$\begin{aligned} u^{(3)} &= -8\pi Z \int_0^{r_s} rn(r)\mathrm{d}r + 2(4\pi)^2 \int_0^{r_s} rn(r)\mathrm{d}r \int_0^r r'^2 n(r')\mathrm{d}r' \\ &- 2ZQ_{\mathrm{MT}} {\sum_{\mathbf{R}}}' \frac{1}{R} + Q_{\mathrm{MT}}^2 {\sum_{\mathbf{R}}}' \frac{1}{R} + 2Q_{\mathrm{MT}} n_0 \left[\int \frac{\mathrm{d}^3 r}{r} - \int_0^R \frac{\mathrm{d}^3 r}{r} \right]. \end{aligned} \tag{4.43}$$

From the sum $u^{(1)} + u^{(2)} + u^{(3)}$, we must now subtract the effect of the charge ρ_0 that lies inside the MT sphere. This gives the final contribution,

$$\begin{aligned} u^{(4)} &= 8\pi Z \frac{{r_s}^2}{2} - 2(4\pi)^2 n_0^2 \frac{{r_s}^5}{15} + 2ZQ_{\mathrm{MT}}^0 {\sum_{\mathbf{R}}}' \frac{1}{R} \\ &+ {Q_{\mathrm{MT}}^0}^2 {\sum_{\mathbf{R}}}' \frac{1}{R} - 2Q_{\mathrm{MT}}^0 n_0 \left[\int \frac{\mathrm{d}^3 r}{r} - \int_0^{r_s} \frac{\mathrm{d}^3 r}{r} \right] \\ &- 2Q_{\mathrm{MT}} Q_{\mathrm{MT}}^0 {\sum_{\mathbf{R}}}' \frac{1}{R}, \end{aligned} \tag{4.44}$$

where

$$Q_{\mathrm{MT}}^0 = \frac{4}{3}\pi {r_s}^3 n_0 = \frac{Z_{\mathrm{out}}}{\Omega_0} \Omega_{\mathrm{MT}}, \tag{4.45}$$

is the total charge inside the MT sphere contributed by the constant charge density. The electrostatic energy per unit cell is now given by

$$\begin{aligned} u &= u^{(1)} + u^{(2)} + u^{(3)} + u^{(4)} \\ &= -8\pi Z \int_0^R rn(r)\mathrm{d}r + 2(4\pi)^2 \int_0^R rn(r)\mathrm{d}r \int_0^r r'^2 n(r')\mathrm{d}r' \\ &- \frac{1}{2} C \frac{Z_{\mathrm{out}}}{a}, \end{aligned} \tag{4.46}$$

where a is the lattice constant and C, whose value depends on the lattice structure, is given by

$$C = \frac{4\pi a^3}{\Omega_0^2} \left[\frac{\Omega^2 A}{2\pi a^3} - \frac{6\Omega + 4\Omega_0}{5} \left(\frac{R}{a} \right)^2 \right]. \tag{4.47}$$

In this expression, the constant A is defined by

$$\frac{A}{a} = \frac{1}{\Omega}\int \frac{d^3r}{r} - \sum_{\mathbf{R}}{}' \frac{1}{R}, \tag{4.48}$$

and can be evaluated using the Ewald method. Values for A for the canonical sc, bcc, and fcc lattices, $a = 1$, have been given by Coldwell-Horsfall and Maradudin [39].

The electrostatic potential (*not* including exchange and correlation) can be found as the functional derivative of u with respect to $n(\mathbf{r})$. Keeping in mind that in the MT approximation the potential is set to zero outside the MT sphere, the MT zero, we have

$$v(r) = \begin{cases} 0 \quad r > R \\ -\frac{2Z}{r} + 8\pi \int_0^r \left(\frac{r'^2}{r} - r'\right) n(r')dr' \\ +8\pi \int_0^R rn(r)dr + C\frac{Z_{out}}{a}, \quad r \leq R. \end{cases} \tag{4.49}$$

Bibliography

[1] P. Hohenberg and W. Kohn, Phys. Rev. **136B**, 864 (1964).

[2] W. Kohn and L. J. Sham, Phys. Rev. **140A**, 1133 (1965).

[3] Robert C. Parr and Weitao Yang, *Density-Functional theory of Atoms and Molecules*, Oxford University Press, New York (1989).

[4] R. M. Dreizler and E. K. U. Gross, *Density Functional Theory*, Springer Verlag, New York (1990).

[5] A. Gonis and W. H. Butler, *Multiple Scattering Theory in Solids*, Springer Verlag, Berlin (1999).

[6] J. D. Jackson, *Classical Electrodynamics*, John Wiley and Sons, New York (1975).

[7] E. Madelung, Physik Z.**19**, 524 (1918). See also, K. Fuchs, Proc. Roy. Soc. A**151**, 585 (1935).

[8] P. P. Ewald, Ann. Phys. **64**, 253 (1921).

[9] P. D. De Cicco, Phys. Rev. bf 153, 931 (1967).

[10] G. S. Painter, Phys. Rev. B **7**, 3520 (1973).

[11] N. Elyasher and D. D. Koelling, Phys. Rev. B **13**, 5362 (1976).

[12] B. R. A. Nijboer and F. W. de Wette, Physica **23** , 309 (1957).

[13] B. R. A. Nijboer and F. W. de Wette, Physica **24**, 422 (1958).

[14] F. W. de Wette and B. R. A. Nijboer , Physica **24**, 1105 (1958).

[15] F. G. Fumi and M. P. Tosi, Phys. Rev. **117**, 1466 (1960).

[16] W. E. Rudge, Phys. Rev. **181**, 1020 (1969).

[17] J. L. Birman, J. Phys. Chem. Solids **6**, 65 (1958).

[18] M. P. Tosi, Solid State Phys. **16**, 1 (1964).

[19] T. L. Loucks, *Augmented Plane Wave Method*, Benjamin, New York (1967).

[20] E. J. Baerends, D. E. Ellis, and P. Ross, Chem. Phys. **2**, 41 (1973).

[21] B. I. Dunlap, J. W. D. Connolly, and J. R. Sabin, J. Chem. Phys. **71**, 4993 (1979).

[22] J. W. Mintmire, Int. J. Quantum Chem. Symp. bf 13, 163 (1979).

[23] J. Harris and G. S. Painter, Phys. Rev. B **22**, 2614 (1980).

[24] G. S. Painter, Phys. Rev. B **23**, 1624 (1981).

[25] M. Weinert, J. Math. Phys. **22**, 2433 (1981).

[26] J. van W. Morgan, J. Phys. C **10**, 1181 (1977).

[27] H. L. Skriver, *The LMTO Method*, Springer Verlag, Berlin (1984).

[28] X. -G. Zhang and A. Gonis, Phys. Rev. Lett. **61**, 1161 (1989).

[29] X. -G. Zhang, A. Gonis, and James M. MacLaren, Phys. Rev. B **40**, 3694 (1989).

[30] X. -G. Zhang, a. Gonis, and D. M. Nicholson, Phys. Rev. B **40**, 947 (1989).

[31] R. K. Nesbet, Phys. Rev. B **41**, 4948 (1990).

[32] W. H. Butler and R. K. Nesbet, Phys. Rev. B **42**, 1518 (1990).

[33] J. F. Janak, Phys. Rev. B **9**, 3985 (1974).

[34] V. L. Moruzzi, J. F. Janak, and A. R. Williams, *Calculated Electronic Properties of Metals*, Pergamon Press, New York (1978).

[35] J. Korringa, Physica **13**, 392 (1947).

[36] W. Kohn and N. Rostoker, Phys. Rev. **94**, 1111 (1954).

[37] *Mathematica* 1.2, Wolfram Research, Inc., Champaign, IL, (1989).

[38] A. Gonis, Erik C. Sowa, and P. A. Sterne, Phys. Rev. Lett. **66**, 2207 (1991).

[39] R. A. Coldwell-Horsfall and A. A. Maradudin, J. Math. Phys. **1**, 395 (1960).

[40] L. Vitos, J. Kollár, and H. L. Skriver, Phys. Rev. B**49**, 16694 (1994).

Chapter 5

Strain and Stress(*)

5.1 General Comments

Our discussion of density functional theory (DFT) in Chapter 1 was aimed at the calculation of *global* properties, such as energy and pressure, of an interacting, many-body quantum mechanical system. At the same time, our discussion of the virial theorem and the Hellmann-Feynman theorem showed how classical concepts can often find direct analogies within quantum mechanics. We established the forms that both of these theorems take within the local density approximation to DFT.

In studying an interacting many-body system such as a solid, one may also adopt a *local* approach.[1] Now one is interested in the direct calculation of internal forces, or stresses, in the solid acting over regions of microscopic extent. At least in principle, there is quite a bit to be gained from the calculation of forces inside the solid. Quantities of a local nature can often be calculated within the total-energy scheme of DFT by means of a very large number, often prohibitively large, of calculations. On the other hand, in principle, a description based on forces can lead directly and by means of a single calculation to the determination of important physical quantities such as phonon dispersion curves [1], internal degrees of freedom [2], and surface reconstruction [3, 4], to name a few.

For the sake of completeness, we now turn our attention to local quantities such as forces and stresses in a material. In the following discussion, the virial theorem again plays an important role underlying the development in both the classical and quantum cases. However, the concepts presented here have not found wide application in actual calculations of materials properties. A conceivable reason for this discrepancy with respect to the

[1] As the following discussion attempts to make clear, the term local is not to imply the use of the LDA of DFT.

calculations of global properties such as energy may be the lack of variational principles in the present case of the kind on which density and energy calculations are based. These principles tend to lead to stable and easily monitored computational procedures. Because of this lack of general utility, this chapter could be skipped in a first reading.

In this chapter, we derive general relations for the stress, considered as an intrinsic property of the ground state of a quantum mechanical, interacting, many-body system. These relations supplement those based on the total energy, and can also lead to the determination of global physical properties, such as equilibrium lattice constants and pressures. We will see that, both in classical and quantum mechanics, the set of relations of stresses and forces to strain and deformation are essentially a generalization of the virial theorem, so that they offer a complete description of the equation of state of an infinite system [5-7]. To facilitate the following discussion, certain elements of classical elasticity theory are summarized in Appendix I.

Stress is an important concept in characterizing the state of condensed matter [8-11]. For the case of matter in which stress is homogeneous in volumes of macroscopic dimension, the equation of state is the relation between the stress and the internal variables, such as the density and the temperature [11][2].

Quite generally, *stress* is the force per unit area that is exerted on a given part of a system, e.g., a solid, by the surrounding material. We say that a body is in a state of stress if it is acted upon by external forces or, more generally, if one part of the body exerts forces upon another part [8, 9]. We can distinguish two different types of forces acting on a body in a state of stress: Forces acting directly in the interior of any given volume element inside the body, and those internal forces that arise from the presence of stress and are transmitted throughout the interior of the volume.

As in Chapter 1, we continue to employ the Born-Oppenheimer approximation. This approximation can lead to quite an accurate description of solid materials, since the dynamical vibrations of the nuclei are on such a slow time scale that the electrons can always be considered as being in their ground state for any instantaneous positions of the ions. Thus, it is meaningful to seek a description of *both* the dynamical vibrations as well as of strictly static properties in terms of a generalized equation of a solid in the electronic ground state (see also chapter on forces). As mentioned above, the same results can be obtained exclusively from extensive calculations of

[2]Consider the case of a liquid that can only sustain pressure, i.e., homogeneous, isotropic stress. The equation of state is then the relation of the pressure, p, to the density and temperature. Taking an ideal gas as a special case of a liquid, we have the well-known relation $pV = nRT$, where V is the volume, n the number of moles, T the absolute temperature, and R the gas constant.

the total energy as a function of distortion, with subsequent fittings and differentiations[3]. We will in fact discuss this procedure in later chapters. On the other hand, a method which in addition to the energy allows the *ab initio* calculation of stress can potentially increase dramatically the efficiency of computational work.

We begin with a generalization of the virial theorem that is particularly well suited for the discussion of strain and stress in quantum mechanics. We then prove the quantum-mechanical stress theorem, in terms of the wave-function, as well as in terms of the electron density. We also show explicitly the form taken by the stress theorem in the LDA. The following discussion makes use of concepts encountered in the classical theory of elasticity, such as the basic definitions of strain and stress, elastic tensors, etc., and knowledge of this material is assumed. Some relevant notions from classical elasticity theory are collected in Appendix I, and reference to that material will be made as needed in the discussion. The reader who may need to review such material may turn to that appendix, or consult some of the standard references given in the bibliography before proceeding further.

5.2 Generalization of the Virial Theorem

The virial theorem was discussed at some length in Chapter 1, within the context of both classical and quantum mechanics. It was shown there that the virial theorem can be used to derive relations among *global* quantities of a system of particles, such as the various contributions to the total energy of a system of electrons and nuclei, and the forces acting on the nuclear subsystem, Eq. (1.137).

The virial theorem, or an appropriate generalization of it, also plays an important role in the study of *local* physical properties, such as the calculation of forces acting on a given nucleus, or the stress across an interface. In fact, as we will see, ***the generalized virial theorem is exactly the expression for the total stress, in both classical and quantum mechanics.*** In this section, we provide a generalization of the classical virial theorem explicitly in terms of the stress tensor, a quantity that is by definition local in nature. It is this form of the theorem that will be exploited in our discussion of stress and related properties in quantum mechanical systems, beginning with the next section.

In analogy with the derivation of the ordinary virial theorem of classical mechanics [12], we consider a system of N particles at positions $\mathbf{r}_i$ and with momenta $\mathbf{p}_i$, and examine the time average of the time derivative of the

[3]Force is the (negative) derivative of the potential (or the energy) with respect to displacement, and stress is the derivative of the energy with respect to strain.

dyadic

$$\overleftrightarrow{G} = \sum_i \mathbf{p}_i \mathbf{r}_i. \tag{5.1}$$

Differentiation with respect to time, denoted by a dot over the symbol, yields the expression,

$$\dot{\overleftrightarrow{G}} = \sum_{i=1}^{N} [\mathbf{p}_i \dot{\mathbf{r}}_i + \dot{\mathbf{p}}_i \mathbf{r}_i]. \tag{5.2}$$

Using the definition

$$\overleftrightarrow{T} = \frac{1}{2} \sum_{i=1}^{N} m_i \mathbf{v}_i \mathbf{v}_i, \tag{5.3}$$

we can write Eq. (5.2) in the form,

$$\dot{\overleftrightarrow{G}} = 2\overleftrightarrow{T} + \sum_{i=1}^{N} \mathbf{F}_i \mathbf{r}_I, \tag{5.4}$$

where Newton's equations of motion have been used.

Now, the average of $\dot{\overleftrightarrow{G}}$ over a time τ is given by the expression,

$$\overline{\dot{\overleftrightarrow{G}}} = \frac{1}{\tau} \int_0^{\tau} \dot{\overleftrightarrow{G}} \mathrm{d}t = \frac{1}{\tau} \left[\overleftrightarrow{G}(\tau) - \overleftrightarrow{G}(0) \right]. \tag{5.5}$$

In the limit $\tau \to \infty$, the quantity in the brackets vanishes when the motion of the system is periodic, and also when $\overleftrightarrow{G}$ is bounded, i.e., when the motion is confined to a finite region of space and when the velocities are finite. (We recall from the last chapter the requirement that the wave function of a system vanishes at infinity in deriving the virial theorem in quantum mechanics.) In that case, Eq. (5.4) yields the expression,

$$2\overline{\overleftrightarrow{T}} = -\sum_{i=1}^{N} \overline{\mathbf{F}_i \mathbf{r}_i}. \tag{5.6}$$

This expression is the tensor generalization of the virial theorem quoted in Chapter 1, Eq. (1.1). In fact, Eq. (1.1) follows immediately upon taking the trace of the last expression (i.e., replacing the dyadic product $\mathbf{p}_i \mathbf{r}_i$ in Eq. (5.1) by the dot or scalar product $\mathbf{r}_i \cdot \mathbf{p}_i$).

Now, let $\mathbf{F}_i$ consist of two parts, an "internal" part, derivable from a potential, $\mathbf{F}_i^{\text{int}} = -\nabla_i U$, and an "external" part, $\mathbf{F}_i^{\text{ext}}$. Equation (5.6) can now be written in the form,

$$2\overline{\overleftrightarrow{T}} - \overline{\mathbf{r}_i \nabla_i U} = -\sum_{i=1}^{N} \overline{\mathbf{F}_i^{\text{ext}} \mathbf{r}_i}, \tag{5.7}$$

In the case of Coulomb potentials, we can use the relation

$$\frac{1}{2}\frac{[\mathbf{r}_i\nabla_i + \mathbf{r}_j\nabla_j]}{r_{ij}} = -\frac{1}{2}\frac{\mathbf{r}_{ij}\mathbf{r}_{ij}}{r_{ij}^3} \tag{5.8}$$

to write

$$\sum_i \mathbf{r}_i\nabla_i \sum_{j\neq i}\frac{1}{r_{ij}} = -\frac{1}{2}\sum_{i,j\neq i}\frac{\mathbf{r}_i\mathbf{r}_j}{r_{ij}^3} \equiv -\overleftrightarrow{U}, \tag{5.9}$$

so that Eq. (5.7) takes the form

$$2\overline{\overleftrightarrow{T}} + \overline{\overleftrightarrow{U}} = -\sum_{i=1}^{N}\overline{\mathbf{F}_i^{\text{ext}}\mathbf{r}_i}. \tag{5.10}$$

In the case of a closed system $\mathbf{F}_i^{\text{ext}} = 0$, and we obtain

$$2\overleftrightarrow{T} + \overleftrightarrow{U} = 0, \tag{5.11}$$

in direct analogy with Eq. (1.7).

As was the case with Eq. (1.32), Eq. (5.7) is a global relation, encompassing all particles in the system. Since we are interested in the study of systems with infinite extent, it is convenient to pass to the thermodynamic limit by dividing the last equation by the volume of the system. Consideration of this limit will also facilitate the application of the (generalized) virial theorem to the study of strictly local properties in the following sections. Defining $\overline{\overleftrightarrow{T}}_0$ and $\overline{\overleftrightarrow{U}}_0$ as the kinetic-energy and potential-energy tensors per unit volume, respectively, we obtain

$$2\overline{\overleftrightarrow{T}_0} + \overline{\overleftrightarrow{U}_0} = -\overline{\overleftrightarrow{\sigma}_0}. \tag{5.12}$$

It is easy to verify that the right-hand side of the last equation is indeed the stress tensor, given by the generalized virial on the left. Assuming that the external force is independent of the coordinates $\mathbf{r}_i$, we note that Eq. (5.7) yields the expression,

$$\text{Div}_i\overleftrightarrow{\sigma} = -\mathbf{F}_i^{\text{ext}}, \tag{5.13}$$

where $\mathbf{F}_i^{\text{ext}}$ is now defined as the external force density (from now on the superscript ext will mostly be omitted). This is the condition for static equilibrium, Eq. (I.49), applied to the particle at $\mathbf{r}_i$.

5.2.1 Force on a particle

We are interested in using analysis based on the concept of stress to determine quantities of a local nature, such as the force acting on a given particle, or a collection of particles assumed to form a closed system. To this end, it is convenient to define a total energy tensor,

$$\overleftrightarrow{E} = \overleftrightarrow{T} + \overleftrightarrow{U}, \tag{5.14}$$

so that in view of Eq. (5.11) we obtain the relations,

$$\overleftrightarrow{T} = -\overleftrightarrow{E} \tag{5.15}$$

and

$$\overleftrightarrow{U} = 2\overleftrightarrow{E}. \tag{5.16}$$

In the case in which external forces are acting on the system, their tensor virial must be included yielding Eq. (5.12). The last three expressions are the obvious generalizations of Eqs. (1.8), (1.9), and (1.10). It is also useful to define partial energy tensors associated with the particles confined inside a volume Ω_i of the system at hand. Thus, we set

$$\overleftrightarrow{T} = \frac{1}{2}\sum_i [m_i \mathbf{v}_i \mathbf{v}_i]_{\Omega_i}, \tag{5.17}$$

and

$$\overleftrightarrow{U} = \sum_i [\mathbf{r}_i \nabla_{\mathbf{r}_i} U]_{\Omega_i}, \tag{5.18}$$

where the subscript Ω_i indicates that the summations extend only over the particles in volume Ω_i.

We now consider a partition of the system into subsystems of volumes Ω_i, so that $V = \sum_i \Omega_i$, where V is the volume occupied by the system as a whole. If the system is closed, then we have

$$0 = \int_V \mathbf{F} \mathrm{d}^3 r = \sum_i \int_{\Omega_i} \mathbf{F}_i \mathrm{d}^3 r, \tag{5.19}$$

where $\mathbf{F}_i$ is the force density acting on the particles in Ω_i. Using Eq. (I.50), we can convert volume integrals to surface integrals and obtain an expression for the force acting on one subsystem,

$$-\mathbf{F}_{\Omega_i} = -\int_{\Omega_i} \mathbf{F}_i \mathrm{d}^3 r = \int_{\Omega_i} \mathrm{d}^3 r \mathrm{Div} \overleftrightarrow{\sigma}_i = \int\int_{S_i} \overleftrightarrow{\sigma}_i \cdot \mathrm{d}\mathbf{S}, \tag{5.20}$$

where S_i denotes the surface of region Ω_i. As follows from Eq. (5.12), $\overleftrightarrow{\sigma}_i$ is the stress associated with the particles in Ω_i, defined in terms of the kinetic and potential energy tensors for the subsystem by the relation,

$$-\overleftrightarrow{\sigma}_i = 2\overleftrightarrow{T}_i + \overleftrightarrow{U}_i. \tag{5.21}$$

We note that

$$\sum_i \int\int_{S_i} \overleftrightarrow{\sigma}_i \cdot d\mathbf{S} = \int\int_S \sum_i \overleftrightarrow{\sigma}_i \cdot d\mathbf{S} = 0. \tag{5.22}$$

This result can be seen when we note that the integrals over surfaces shared by the different subsystems cancel, leaving only the integral over the surface S of the system. Thus, Eq. (5.22) is consistent with Eq. (5.19).

To illustrate the results obtained above, and in order to prepare the ground for the discussion of quantum-mechanical systems, let us return to the collection of positively and negatively charged point particles considered in Section 1.2.1. We let the positively charged particles be kept fixed in position, say by externally applied forces, and consider the force acting on a volume Ω separated from the rest of the system by a surface S_Ω. In view of Eq. (I.48), this force is given by the expression,

$$\begin{aligned} -F_\Omega &= \int\int_{S_\Omega} \overleftrightarrow{\sigma} \cdot d\mathbf{S} \\ &= \int\int_{S_\Omega} \Big\{ \sum_i [m_i \mathbf{v}_i \mathbf{v}_i] - \sum_i \mathbf{r}_i \nabla_{\mathbf{r}_i} [U^{+-} + U^{--}] \\ &\quad - \sum_j \mathbf{R}_j \nabla_{\mathbf{R}_j} [U^{+-} + U^{++}] \Big\}_\Omega \cdot d\mathbf{S}, \end{aligned} \tag{5.23}$$

where the vanishing kinetic energy of the stationary positive particles has been omitted. Note that the subscript Ω is a reminder that the summations extend only to particles in Ω.

Now, in the limit in which the surface S_Ω encloses only a single, positively charged particle, the first two terms in the last expression vanish and $\mathbf{F}_\Omega$ reduces to Eq. (1.17), giving the force on a single particle. To see this, write $\mathbf{F}_\Omega$ as the volume integral of $\mathrm{Div}\overleftrightarrow{\sigma}$ and integrate over Ω. The term $\int \mathbf{R}_j \nabla^2 [U^{+-} + U^{++}] d^3r$ gives two volume integrals over delta functions whose arguments lie on opposite sides of the surface S_Ω[4] and therefore vanish identically. Only the term $\nabla_{\mathbf{R}_j}[U^{+-} + U^{++}]$ yields a non-vanishing contribution, leading to

$$\mathbf{F}_{i+} = -\nabla_{\mathbf{R}_i}[U^{+-} + U^{++}], \tag{5.24}$$

for the force on the particle, in agreement with Eq. (1.17).

As a second illustration, we now consider the case in which the positive particles are held in position by means of an isotropic stress, or pressure,

[4]Note that $\nabla^2[\frac{1}{|\mathbf{r}-\mathbf{r}'|}] = -\frac{1}{4\pi}\delta(\mathbf{r}-\mathbf{r}')$, because $\frac{1}{|\mathbf{r}-\mathbf{r}'|}$ is the Green function of the operator ∇^2. See Appendix A.

p. In view of Eq. (I.65), the trace of Eq. (5.7) yields the relation

$$2T + U = 3pV. \tag{5.25}$$

This relation agrees with Eq. (1.1) provided that we identify

$$\sum_i \mathbf{R}_i \cdot \nabla_{\mathbf{R}_i} E = -3pV. \tag{5.26}$$

This also agrees with Eq. (1.273). (The difference of a minus sign arises because here we are considering the action of the external force itself rather than the internal reaction of the system to it.)

Before leaving this section, let us consider some important features of the formalism presented above. One outstanding result of this formalism is the fact that volume integrals over the force density can be replaced by surface integrals over stress (force per unit area). This can be of computational significance since it bypasses the calculation of the derivatives of the potential in favor of the potential itself. We showed how this process does in fact lead to the force on a particle. Let us now consider the case in which the negative charges in the system are at equilibrium for a given distribution, not necessarily at equilibrium, of the positive charges. This means that the force on the negative charges vanishes identically, or that the integral over a surface containing *only* negative charges equals zero. Now, the force on a positive charge can be calculated by means of an integral over a surface enclosing a *finite* volume containing the positive charge and any number of negative charges. In this case, the kinetic and potential energy terms associated with the negative charges in Eq. (5.23) cancel identically in the integral over the surface leaving only the force on the positive charge. Clearly, this force will also vanish for any equilibrium configuration of the positive charges themselves.

We might gain further understanding of the results obtained in this section by imagining force lines, in the sense of elementary electrostatics, associated with the fields set up by the charges in the system in any stationary configuration. It is now easy to see that pairs (or clusters) of particles entirely inside or outside the surface enclosing a volume element make no contribution to the force on the volume. This force is determined solely by the field lines that cross the surface. This interpretation is consistent with Eq. (5.23), which shows that the force on a positive particle is due to the total field produced by all the charges in the system. The concept of field lines is explored further in sections 2.8 and 2.9, in which the Maxwell stress tensor is introduced.

Another important feature of the classical formalism developed here is that it carries essentially intact (under appropriate identification of parameters) into quantum mechanics. Since electrons move much faster than nuclei, in an electron system such as a solid it is usually an excellent approximation to consider the electrons as having reached equilibrium for *any* distribution of the nuclear charges. Then the integral of the stress tensor over a surface in the system yields *only* the force acting on the nuclear charges enclosed by the surface. The computational advantages that can derive from this approach are those mentioned above with respect to the system of positive and negative charges. Therefore, in this case there is a clear analogy between classical and quantum mechanics. This analogy was emphasized in Chapter 1 in connection with the virial theorem and the Hellmann-Feynman theorem. Consideration of stress within quantum mechanics makes this analogy even more evident. In fact, the virial theorem and the Hellmann-Feynman theorem can both be derived as simple cases of the more general *stress theorem* discussed in the next section.

5.3 Stress in Quantum Mechanics

In this section, we begin the formal discussion of stresses and forces within a quantum mechanical description of a many-body system such as a solid. The discussion is based directly on generalizations of the virial theorem expressed in a form that accommodates the tensor nature of the various quantities of interest. We establish a number of basic results involving the properties of the total stress, as well as local stress fields. These results are derived both using the wave function and using the charge density as the fundamental quantity, in the hope that development along different lines can help in clarifying the meaning of the underlying concepts. Some of the computational aspects of the formalism presented here are discussed in succeeding chapters.

The following discussion makes use of the close relation between the Hellmann-Feynman theorem and the virial theorem, both of which were discussed in Chapter 1. That the net force on a particle is given as the expectation value of the negative gradient of the potential was first shown by Ehrenfest [13], and subsequently derived by Pauli [14], Hellmann [15], and Feynman [16]. This result is known as the *Hellmann-Feynman theorem*, the *electrostatics theorem*, or the *force theorem*. This theorem was one of the cornerstones in establishing the correspondence between classical and quantum mechanics.

Closely related to the force theorem is the virial theorem, which establishes that the total pressure in a many-particle system is given by the kinetic energy and the virial of the potential, again in complete analogy with classical mechanics (see previous sections). The quantum mechanical virial theorem was proven first by Born, Heisenberg, and Jordan [17], and later by Finkelstein [18], Hylleraas [19], Fock [20], Pauli [14], and Slater [21]. Tensor generalizations of the virial theorem have also been considered by Schrödinger [22], Pauli [14], Feynman [23], and other authors [5, 6, 24-27]. Consideration of the stress field, rather than just the force, in a material provides a particularly satisfying and unified approach, as *both* the ordinary virial theorem and the force theorem become special cases of a more general result.

The consideration of stress and force fields at each point in space, as opposed to total stresses and forces, has additional formal and practical consequences for the theory of matter. Pauli [14] showed that the force field is uniquely defined in terms of the kinetic and potential energy operators. On the other hand, as already mentioned in previous sections, the stress field can be any tensor field whose divergence equals the vector force-density field, Eq. (5.13). Thus, it is unaffected by the addition of the curl of any arbitrary tensor field. This nonuniqueness can be used to advantage in practical applications, through an appropriate choice of gauge, e.g., the selection of a particular tensor field to represent the stress. Specific choices may facilitate computational applications of the formalism. We will explicitly consider the Maxwell stress tensor [28], which is particularly appropriate to particles interacting via Coulomb fields. An alternative form, applicable to arbitrary fields has been suggested by Kugler [25]. This form, however, has a number of disadvantages when compared to the Maxwell form, as has been discussed in the literature [5].

Although there exists a great formal similarity between the classical and quantum mechanical descriptions of stress and related quantities, points of departure exist as well. The *microscopic* fields in matter vary on the scale of atomic dimensions. Since the range of forces is of the same length scale, we may expect that the stress at point $\mathbf{r}$ is a *non-local* function of the state of matter at all points in some vicinity of $\mathbf{r}$, in contrast to the strictly local fields that are assumed in a classical treatment. It should be kept in mind that the stress theorem is derived in terms of volume integrals in connection with the presence of an isotropic strain, and thus gives only the *average* macroscopic stress of the system. The treatment of *inhomogeneous* stress requires additional considerations, such as assuming that the system consists of subsystems that are small but macroscopic so that the stress theorem applies to each one separately [5]. At the same time, a number of useful classical results find appropriate generalizations in quantum mechanics. For example, the quantum-mechanical expression for the force on a particle is still given in terms of surface integrals of the surface tensor. Thus, the macroscopic stresses on part of a system can be determined as surface integrals of the microscopic stress field. The corresponding expressions are complementary to the volume integrals used to derive the stress theorem.

A word of caution is in order before we begin our formal discussion of stress within quantum mechanics. In spite of the simplicity of many of the relations derived below, their numerical implementation can be very difficult, due to the need to calculate many-particle correlations. This is, of course, the same difficulty that arises in considering the ground state properties of an interacting system, discussed in Chapter 1. In fact, one approach to applying the stress relations derived below is to use density functional theory (DFT), and in particular its local density approximation (LDA) [6]. In the last section of this chapter, we give explicit expressions for stress in terms of the electron density within the LDA [5, 6]. Added numerical difficulties in applying the various force-stress relations derived in this chapter arise because of the lack of extremal properties, e.g., energy-minimum principles, which govern the calculation of the energy within DFT. The calculation of forces in a solid is discussed in more detail in a subsequent chapter.

5.3.1 Stress theorem: First derivation

To derive the total stress of a system in a state of equilibrium, i.e., a stationary system, we use the variational principle suggested by Eq. (I.79). We compute the expectation value of the Hamiltonian, $\langle\hat{H}\rangle$, for a state characterized by a uniform strain, a quantity related directly to the elastic potential, $W(\epsilon)$. Then, the internal stress is obtained as the negative variation of this expectation value with respect to the strain.

The presence of strain in a material can be represented in at least two different ways. For example, we may consider the strain transformation $\mathbf{r} \to (1 + \underline{\epsilon})\mathbf{r}$ applied to the wave function and compute the expectation value of the *unstrained* Hamiltonian. Or, we may keep the wave function unchanged and incorporate the strain into the Hamiltonian itself. Certainly, one should *not* consider the presence of strain in *both* the Hamiltonian and the wave function, because this leads to a double-counting of the effect. To see this, note that the effects of strain to first order (of interest to us here) can be calculated by means of first order perturbation theory. This, in turn, can be accomplished by calculating the change in the energy by varying either the wave function, or the Hamiltonian for a fixed wave function, but not both. The following derivation illustrates the interplay of these two different ways, and their complementary nature.

The derivation of the stress theorem given below is based on the formalism of Nielsen and Martin [5]. We consider only time-independent potentials, thus excluding the presence of magnetic fields.

Let us consider a general, many-body Hamiltonian of the form of Eq. (1.94), which we write in the form,

$$\hat{H} = \sum_i \frac{\hat{p}_i^2}{2m_i} + \hat{U}$$

$$= \sum_i \frac{\hat{p}_i^2}{2m_i} + \hat{U}_{\text{int}} + \hat{U}_{\text{ext}}, \tag{5.27}$$

where the summation over i goes over *all* particles in a system, $\mathbf{p}_i$ is the momentum of the ith particle, $\hat{U}_{\text{int}}$ is the potential internal to the system, and $\hat{U}_{\text{ext}}$ describes the influence of external forces. For an electron-nucleon system, (such as a solid treated in the Born-Oppenheimer approximation), the summation extends over all electrons, and $\hat{U}_{\text{int}}$ consists of all Coulomb interactions among the electrons, among the nuclei, and between the electrons and nuclei. These interactions are displayed explicitly in Eqs. (1.97) to (1.99). The eigenstates of the system are determined as the solutions of the Schrödinger equation, $\hat{H}\Psi = E\Psi$, Eq. (1.93), with E denoting an eigenenergy. The ground-state energy of the system is obtained as the minimum of $E = \langle\Psi|\hat{H}|\Psi\rangle$. This allows the use of a variational principle in connection with all allowed variations in Ψ in determining the ground-state energy. In the following discussion all expectation values of operators refer to the ground state, unless noted otherwise[5].

We begin by applying an infinitesimal, homogeneous strain to the ground state wave function, $\Psi(\mathbf{r}_1, \mathbf{r}_2, \cdots, \mathbf{r}_N) \equiv \Psi(\mathbf{r})$, by "stretching" each particle coordinate, $\mathbf{r}_i \to (1+\underline{\epsilon})\mathbf{r}_i$, where $\underline{\epsilon}$ is a symmetric, i.e., rotation-free, strain tensor, as exhibited in Eq. (I.10)[6]. Under such a transformation, the wave function becomes,

$$\Psi_\epsilon(\mathbf{r}) = \det(1+\underline{\epsilon})^{-\frac{1}{2}}\Psi(1+\underline{\epsilon}\mathbf{r}), \tag{5.28}$$

where, as shown below, the prefactor serves to preserve the normalization of the wave function (see also footnote 2 in Chapter 1). The expectation value of the Hamiltonian with respect to the strained state described by $\Psi_\epsilon(\mathbf{r})$ takes the form,

$$\langle\hat{H}\rangle \equiv \langle\Psi_\epsilon|\hat{H}|\Psi_\epsilon\rangle. \tag{5.29}$$

Under the change of variable, $\mathbf{r}_i \to (1+\underline{\epsilon})\mathbf{r}_i$, noting that under the strain transformation considered $\mathrm{d}^3r \to \det(1+\underline{\epsilon})\mathrm{d}^3r$ and $\nabla_{(1+\underline{\epsilon})\mathbf{r}} \simeq (1-\underline{\epsilon})\nabla_{\mathbf{r}}$, Eq. (5.29) can be written in the form

$$\begin{aligned}
\langle\Psi_\epsilon|\hat{H}|\Psi_\epsilon\rangle &\equiv [\det(1+\underline{\epsilon})]^{-1}\int \Psi^*((1+\underline{\epsilon})^{-1}\mathbf{r})\hat{H}(\mathbf{r})\Psi((1+\underline{\epsilon})^{-1}\mathbf{r})\mathrm{d}^3r \\
&= \int \Psi^*(\mathbf{r})\hat{H}((1+\underline{\epsilon})\mathbf{r})\Psi(\mathbf{r})\mathrm{d}^3r \\
&= \int \Psi^*(\mathbf{r})\left[\sum_i \frac{1}{2m_i}\left(p_i^2 - 2\sum_{\alpha\beta}\epsilon_{\alpha\beta}p_{i\alpha}p_{i\beta}\right.\right.
\end{aligned}$$

[5] Thus, both a direct approach, analogous to the classical Eq. (I.79), and a minimization procedure are used to derive the stress theorem.

[6] In Chapter 1, scaling was described in terms of a single parameter λ, which is formally equivalent to $1+\underline{\epsilon}$.

$$+ \sum_{\alpha\beta\gamma} \epsilon_{\alpha\beta}\epsilon_{\alpha\gamma}p_{i\beta}p_{i\gamma} \Bigg)$$
$$+ \quad U_{\text{int}}((1+\underline{\epsilon})\mathbf{r})\Big]\Psi(\mathbf{r})\mathrm{d}^3r. \tag{5.30}$$

There are now at least two different but equivalent ways in which we may proceed. First, we may view the difference in expectation values of the strained and unstrained Hamiltonians with respect to the unstrained wave function in terms of perturbation theory. We may then identify this difference with the increase in energy of the system in the presence of the perturbation, that is, with the elastic potential. Since we are interested in the internal stresses, irrespective of how they originate, we ignore the outside potential in taking the expectation value. Then the stress is given by the negative derivative of $\langle\Psi_\epsilon|\hat{H}|\Psi_\epsilon\rangle$ with respect to the strain, keeping terms only to first order. We find[7]

$$\overleftrightarrow{\sigma} = -\sum_i \langle\Psi|\frac{\mathbf{p}_i\mathbf{p}_i}{m_i} - \mathbf{r}_i\nabla_i(U_{\text{int}})|\Psi\rangle. \tag{5.31}$$

Equivalently, we may consider the *closed* system described by the Hamiltonian of Eq. (5.27), including the external potentials. Then we can use the variational principle to write

$$\frac{\delta\langle\Psi_\epsilon|\hat{H}|\Psi_\epsilon\rangle}{\delta\epsilon} = 0 = \sum_i \langle\Psi|\frac{\mathbf{p}_i\mathbf{p}_i}{m_i} - \mathbf{r}_i\nabla_i(U_{\text{int}} + U_{\text{ext}})|\Psi\rangle. \tag{5.32}$$

Noting that the term involving U_{ext} yields the internal stress by means of the relation,

$$\overleftrightarrow{\sigma} = -\sum_i \langle\Psi|\mathbf{r}_i\nabla_i(U_{\text{ext}})|\Psi\rangle, \tag{5.33}$$

we obtain Eq. (5.31).

Equation (5.31) is one form of the ***stress theorem***. We note that the expression for stress, Eq. (5.31), is identical to the generalized virial expression, Eq. (5.11). It expresses the total *macroscopic stress* in terms only of the *expectation values* of operators internal to the system. It is also important to note that this expression does not account for possible spatial variations of the stress field inside the system. Such *local* stress fields are considered in a later section. Furthermore, in spite its simple, "classical" form, Eq. (5.31) involves many-body correlations between the coordinates of any given particle and the gradient of the potential produced by all other particles. Thus, its evaluation is far from simple. Approximate schemes,

[7]Note that for Coulomb potentials we have $\overleftrightarrow{\sigma} = -[2\overleftrightarrow{T} - \overleftrightarrow{U}]$.

notably the LDA, that can provide a means of evaluating the stress are discussed below. Finally, we note that the expression for stress, Eq. (5.31), is identical to the classical generalized virial theorem, Eq. (5.12), for the stress.

There is another disadvantage to Eq. (5.31), which is easier to deal with. In its present form, it involves the coordinates of the particles explicitly, and is thus inconvenient for use in connection with large systems such as a solid. A more convenient form can be obtained in terms of the relative coordinates between the particles. In the case of two-body potentials of the form

$$U_{\text{int}} = \frac{1}{2} \sum_{i,j \neq i} U_{ij}(|\mathbf{r}_i - \mathbf{r}_j|), \tag{5.34}$$

we can write,

$$\overleftrightarrow{\sigma} = -\sum_i \langle \Psi | \frac{\mathbf{p}_i \mathbf{p}_i}{m_i} - \frac{1}{2} \sum_{j \neq i} \frac{(\mathbf{r}_i - \mathbf{r}_j)(\mathbf{r}_i - \mathbf{r}_j)}{|(\mathbf{r}_i - \mathbf{r}_j)|} \nabla_{(\mathbf{r}_i - \mathbf{r}_j)}(U_{\text{int}}) | \Psi \rangle. \tag{5.35}$$

This form is manifestly symmetric, i.e., torque free, and depends only on the relative positions of the particles. In macroscopic systems with a well-defined volume Ω, the average stress density is defined as

$$\overleftrightarrow{\sigma}_0 = \frac{\overleftrightarrow{\sigma}}{\Omega}, \tag{5.36}$$

which corresponds to the classical expression, Eq. (5.12).

5.3.2 Stress theorem: Second derivation

In addition to the scaling-argument derivation given above[8], the stress theorem can be derived by evaluating the commutator of the Hamiltonian with the generalized virial operator, $\sum_i \mathbf{r}_i \mathbf{p}_i$. This approach shows that, in a sense, the generalized virial operator is the generator of an infinitesimal scaling of the wave function. We begin by showing the equivalence of the scaling and commutation approaches, and then derive the stress theorem using the latter. The formal arguments presented below closely follow the development in reference [6].

It is convenient to express the scaling transformation in terms of a single, scalar parameter, λ, by writing, $\mathbf{r} = (1 + \lambda \underline{\epsilon})\mathbf{r}$. The correctly normalized

[8] At first glance the virial theorem, in either classical or quantum mechanics, does not seem to apply to free particles. This casts doubt on the validity of the derivations based on scaling arguments, since these arguments apparently apply to all systems, even when the force vanishes. This apparent contradiction can be removed through a limiting process, which in quantum mechanics consists of considering the system confined inside a box whose volume is allowed to increase beyond all bounds.

many-particle wave function in the presence of an isotropic scaling transformation of all particle coordinates takes the form of Eq. (5.28), which we now write in the form, (omitting reference to spin variables),

$$\Psi_\lambda(\mathbf{r}_1, \mathbf{r}_2, \cdots, \mathbf{r}_N) = \det(1 + \lambda\underline{\epsilon})^{N/2}\Psi(\mathbf{r}'_1, \mathbf{r}'_2, \cdots, \mathbf{r}'_N), \tag{5.37}$$

in which the presence of the N particles is displayed explicitly[9]. With this "stretched" wave function, we form the expectation value of the unchanged Hamiltonian,

$$\begin{aligned} \langle\Psi_\lambda|\hat{H}|\Psi_\lambda\rangle &= \det(1 + \lambda\underline{\epsilon})^N \int d^3r_1 \int d^3r_2 \cdots \int d^3r_N \\ &\times \Psi^*(\mathbf{r}'_1, \mathbf{r}'_2, \cdot, \mathbf{r}'_N)\hat{H}\Psi(\mathbf{r}'_1, \mathbf{r}'_2, \cdot, \mathbf{r}'_N), \end{aligned} \tag{5.38}$$

and consider the derivative with respect to λ at $\lambda = 0$, i.e., in the limit of vanishing strain ($\Psi_{\lambda=0}$ corresponds to the unstretched wave function). In view of Eqs. (I.31) and (I.32), we obtain,

$$\begin{aligned} \frac{\partial}{\partial\lambda}\langle\Psi_\lambda|\hat{H}|\Psi_\lambda\rangle\Big|_{\lambda=0} &= N\mathrm{Tr}\,\overleftrightarrow{\epsilon}\langle\Psi|\hat{H}|\Psi\rangle \\ &+ \mathrm{i}\langle\Psi| - \sum_i \mathbf{p}_i\overleftrightarrow{\epsilon}\mathbf{r}_i\hat{H} + \hat{H}\sum_i \mathbf{r}_i\overleftrightarrow{\epsilon}\mathbf{p}_i|\Psi\rangle. \end{aligned} \tag{5.39}$$

Then, with the use of Eq. (I.33), we can write,

$$\frac{\partial}{\partial\lambda}\langle\Psi_\lambda|\hat{H}|\Psi_\lambda\rangle\Big|_{\lambda=0} = \mathrm{i}\langle\Psi|\left[\hat{H}, \sum_i \mathbf{r}_i\overleftrightarrow{\epsilon}\mathbf{p}_i\right]|\Psi\rangle. \tag{5.40}$$

Since Ψ is an eigenfunction of $\hat{H}$, both sides of the last expression must vanish. The derivation of the stress theorem given in the previous subsection was based on setting the left-hand side equal to zero. To establish the equivalence of that derivation with the commutation approach used here, we set the right-hand side equal to zero and use the fact that $\overleftrightarrow{\epsilon}$ is a symmetric but otherwise arbitrary tensor to obtain the result

$$\mathrm{i}\langle\Psi|\left[\hat{H}, \sum_i \mathbf{r}_i\mathbf{p}_i\right]|\Psi\rangle = 0. \tag{5.41}$$

This relation shows that scaling arguments can be expressed in terms of the expectation value of the commutator of $\hat{H}$ and the generalized virial operator, suggesting a possible interpretation of this operator as the generator of infinitesimal strain transformations. Now, the derivation of the stress theorem starting with Eq. (5.41) is straightforward.

---

[9] In the discussion given in the previous subsection, the normalization factors canceled out through a change of variable.

Using the commutation rules for spatial coordinates and momentum operators, and commutator algebra, we obtain,

$$\mathrm{i}\left[\hat{H}, \sum_i \mathbf{r}_i \mathbf{p}_i\right] = \sum_i \mathbf{p}_i \nabla_{\mathbf{p}_i} \hat{T} - \sum_i \mathbf{r}_i \nabla_{\mathbf{r}_i} \hat{U}. \tag{5.42}$$

With the identification

$$\sum_i \mathbf{p}_i \nabla_{\mathbf{p}_i} \hat{T} = 2\overleftrightarrow{T}, \tag{5.43}$$

the generalized virial theorem reads,

$$2\overleftrightarrow{T} - \sum_i \mathbf{r}_i \nabla_{\mathbf{r}_i} \hat{U} = 0. \tag{5.44}$$

First, we note that this equation is formally identical to the classical expression, Eq. (5.11). For the case of Coulomb potentials, we have

$$2\overleftrightarrow{T} + \overleftrightarrow{U} = 0. \tag{5.45}$$

With the potential considered to be the sum of an external and an internal part, this equation leads directly to the stress theorem, Eq. (5.31).

5.4 Stress Fields and Microscopic Stress

Having discussed the total stress and force acting on the particles of a system, we turn to the consideration of *local, microscopic* stress and force fields [14, 16, 22, 24, 25]. Such fields are useful for the study of inhomogeneous systems, in which stress might be a function of position and, in addition, can be used to provide alternative derivations of Eq. (5.31), for the total stress.

Our previous consideration of stress (and force) was closely linked to the virial theorem, as is evidenced by Eqs. (5.12) and (5.31). That approach was motivated by the close connection between classical quantities and the expectation values of the corresponding quantum mechanical operators, which involve integrals over the volume of the system. In the treatment of local fields, we must work directly with the operators themselves, rather than with their expectation values.

Pauli [14] showed that one can obtain a *uniquely* defined force density field from the equation of motion of the momentum operator (so that Newton's second law carries over intact). *The stress field is then defined as the tensor whose divergence equals the force field.* However, this stress field is no longer uniquely defined, because the curl of any tensor field can be added to it without affecting the divergence. On the other hand, as already mentioned, this nonuniqueness can be turned to advantage and can facilitate application to specific problems in many-body systems.

The fields under consideration can be derived along the lines proposed by Nielsen and Martin [5]. We begin by deriving expressions for the force density by considering the equation of motion of the (symmetrized) momentum density operator for each particle i,

$$\begin{aligned} \mathbf{P}_i(\mathbf{r}) &= \frac{1}{2}\left[\mathbf{p}_i\delta(\mathbf{r}-\mathbf{r}_i)+\delta(\mathbf{r}-\mathbf{r}_i)\mathbf{p}_i\right] \\ &= \frac{1}{2}\left[\mathbf{p}_i, \delta(\mathbf{r}-\mathbf{r}_i)\right]_+, \end{aligned} \tag{5.46}$$

where $[]_+$ denotes the anticommutator. The time derivative[10] of $\mathbf{P}_i(\mathbf{r})$ can be interpreted as the *force density* operator acting on particle i along the αth direction,

$$\begin{aligned} f_{i\alpha}(\mathbf{r}) &= \frac{\partial P_{i\alpha}(\mathbf{r})}{\partial t} = \frac{1}{\mathrm{i}\hbar}\left[p_{i\alpha}(\mathbf{r}), \hat{H}\right] \\ &= -\frac{1}{4\pi}\left[p_{i\alpha}, \sum_\beta \left[p_{i\beta}, \nabla_\beta\delta(\mathbf{r}-\mathbf{r}_i)\right]_+\right]_+ \\ &- \nabla_{i\alpha}(U_{\text{int}}+U_{\text{ext}})\delta(\mathbf{r}-\mathbf{r}_i). \end{aligned} \tag{5.47}$$

It is to be emphasized that the quantity $f_{i\alpha}(\mathbf{r})$ is not the force itself but the force density. This interpretation implies that the force can be obtained by taking the expectation value of Eq. (5.47), which involves an integral over the volume of the system. In such an integral, the contribution of the kinetic energy term vanishes, at least for finite systems, since through the use of Green's theorem, the derivatives of the wave functions can be expressed in terms of the wave functions and derivatives at infinity. For bound systems, the wave function and its derivative can be taken to vanish in that limit. To see this, consider the single-particle, time-dependent Schrödinger equation,

$$\mathrm{i}\hbar\frac{\partial\Psi}{\partial t} = \left[-\nabla^2+U\right]\Psi, \tag{5.48}$$

and the expectation value of the rate of change of the momentum operator,

$$\begin{aligned} \langle\frac{\partial\mathbf{p}}{\partial t}\rangle &= -\mathrm{i}\hbar\int\frac{\partial\Psi^*}{\partial}\nabla\Psi\mathrm{d}^3r - \mathrm{i}\hbar\int\Psi^*\nabla\left(\frac{\partial\Psi}{\partial t}\right)\mathrm{d}^3r \\ &= -\mathrm{i}\hbar\int\left[\left(\nabla^2\Psi^*\right)\nabla\Psi - \Psi^*\nabla^2(\nabla\Psi)\right]\mathrm{d}^3r \\ &+ \int\left[U\Psi^*\nabla\Psi - \Psi^*\nabla(U\Psi)\right]\mathrm{d}^3r. \end{aligned} \tag{5.49}$$

[10] In the Heisenberg picture.

The integral containing Laplacians can be transformed to surface integrals using Eq. (J.33). These integrals can be taken to vanish at infinity, at least for bound states. The last integral gives

$$\langle\frac{\mathrm{d}\mathbf{p}}{\mathrm{d}t}\rangle = -\int \Psi^* \nabla V \Psi \mathrm{d}^3 r = -\langle\nabla U\rangle = -\mathbf{F}, \tag{5.50}$$

where $\mathbf{F}$ is the force on the particle. This result is *Ehrenfest's theorem*, the quantum mechanical analog of Newton's second law.

Let us now write $U = U_{\mathrm{int}} + U_{\mathrm{ext}}$. Taking the expectation value of Eq. (5.47) we obtain

$$\langle\frac{\partial \mathbf{p}_i(\mathbf{r})}{\partial t}\rangle = \frac{1}{\mathrm{i}\hbar}\langle\left[\mathbf{p}_i(\mathbf{r}), \hat{H}\right]\rangle = -\langle\nabla_i\left[U_{\mathrm{int}} + U_{\mathrm{ext}}\right]\rangle. \tag{5.51}$$

For stationary systems, $\langle\frac{\partial \mathbf{p}_i(\mathbf{r})}{\partial t}\rangle$ vanishes, so that the force associated with the internal motions of the particles and needed to balance any externally applied forces is given by the Hellmann-Feynman theorem

$$\mathbf{F}_i = -\langle\nabla_i U_{\mathrm{int}}\rangle, \tag{5.52}$$

which is seen from this derivation to be a direct consequence of Ehrenfest's theorem.

We now define the stress density, $\sigma_{\alpha\beta}(\mathbf{r})$, as a tensor field whose divergence is the vector force-density field. Therefore, the stress field due to the internal interactions satisfies the equation,

$$\begin{aligned}\left[\mathrm{Div}\overleftrightarrow{\sigma}(\mathbf{r})\right]_\alpha &= \sum_\beta \frac{\partial \sigma_{\alpha\beta}(\mathbf{r})}{\partial r_\beta} \\ &= \sum_i \langle\frac{\partial P_{i\alpha}(\mathbf{r})}{\partial t} + \nabla_{i\alpha} U_{\mathrm{ext}}\delta(\mathbf{r}-\mathbf{r}_i)\rangle, \end{aligned}\tag{5.53}$$

where the term involving U_{ext} is added in order to cancel the corresponding term in Eq. (5.47) and remove the contribution of the external stress.

Because of the ambiguity in the definition of the stress tensor already mentioned, it is possible to find alternative expressions for $\overleftrightarrow{\sigma}(\mathbf{r})$. For example, in view of Eq. (A.16), we see that the expression

$$\begin{aligned}\overleftrightarrow{\sigma}(\mathbf{r}) &= -\sum_i \left\{\frac{1}{4\pi}\left[\mathbf{p}_i, \left[\mathbf{p}_i, \delta(\mathbf{r}-\mathbf{r}_i)\right]_+\right]_+ \right. \\ &- \left. \frac{1}{4\pi}\nabla_{i\alpha}(U_{\mathrm{int}})\nabla_\beta\left(|\mathbf{r}-\mathbf{r}_i|^{-1}\right)\right\}\end{aligned}\tag{5.54}$$

satisfies Eq. (5.53). This form, which holds for arbitrary potentials U_{int}, was suggested by Kugler [25], and some of its properties have been discussed by Nielsen and Martin [5]. An alternative solution of Eq. (5.53), appropriate to Coulomb systems, is provided by combining the kinetic energy part of the last expression with the Maxwell stress tensor for the potential part. In the absence of time-dependent fields (magnetic fields) the Maxwell stress tensor takes the form,

$$\overleftrightarrow{M}(\mathbf{r}) = \frac{1}{4\pi}\left[\mathbf{E}(\mathbf{r})\mathbf{E}(\mathbf{r}) - \frac{1}{2}E^2(\mathbf{r})\overleftrightarrow{I}\right]. \tag{5.55}$$

The Maxwell stress tensor is discussed further in Appendix K, and in the next section. As pointed out in reference [5], there are a number of advantages in using the Maxwell stress tensor rather than Eq. (5.54) in the study of systems characterized by Coulomb interactions.

Regardless of the explic it form of the stress field used, the force on a volume element Ω can be found either by a volume or a surface integral. Furthermore, when convenient, we may treat separately the kinetic and potential parts of the stress tensor, so that we can write,

$$\mathbf{F}_\Omega = -\sum_I \frac{1}{4\pi}\int_S \langle[\mathbf{p}_i, [\mathbf{p}_i \cdot \hat{\mathbf{n}}, \delta(\mathbf{r}-\mathbf{r}_i)]]_+\rangle \mathrm{d}S - \sum_{i\in\Omega}\langle\nabla_i U_{\mathrm{int}}\rangle. \tag{5.56}$$

Note the presence of the brackets, indicating the expectation value with respect to the many-body wave function. The second term on the right is the result of a volume integral over a delta function and consequently the summation is restricted to particles confined in Ω. We note again the great correspondence with the classical expression, Eq. (5.23). The discussion following that equation is also pertinent in this case. Equation (5.56) can be viewed as a generalization of the Hellmann-Feynman theorem or force theorem, and also projects out an important feature of the Born-Oppenheimer approximation. Clearly no internal force is acting on the electrons, as they are always in equilibrium for any position of the nuclei. A force may, however, act on the nuclei themselves, due to internal interactions. This follows from the form of Eq. (5.56) by choosing the volume Ω to contain only a single nucleus. In that case, we obtain readily the force theorem,

$$\mathbf{F}_i = -\langle\nabla_i U_{\mathrm{int}}\rangle, \tag{5.57}$$

which is identical to the classical expression, Eq. (5.24).

Finally, it is also possible to derive Eq. (5.31) for the total stress, starting from the microscopic stress, e.g., Eq. (5.54). To see this, consider an infinite planar surface, Fig. 5.1, that divides space into left and right semi-infinite regions, denoted by Ω_L and Ω_R, respectively. Such a surface could be chosen as a plane perpendicular to the β-axis at r_β. The stress $S_{\alpha\beta}(r_\beta)$ is the component of the force in the α direction transmitted across

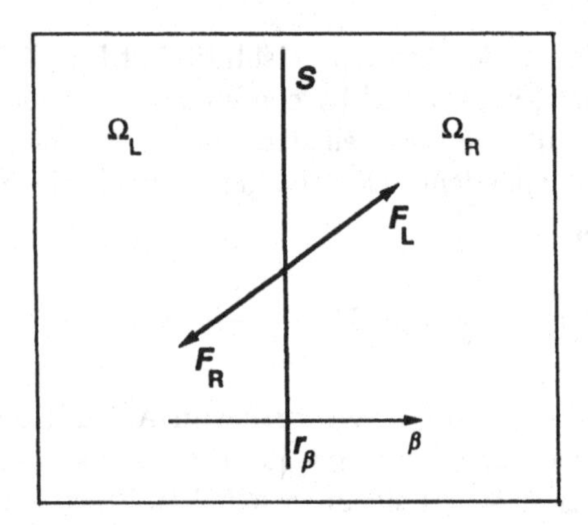

Figure 5.1: **An infinite planar surface, S, dividing space into two semi-infinite parts, Ω_L and Ω_R. $\mathbf{F}_L$ is the force exerted on Ω_R by Ω_L across the surface, and it is equal and opposite to $\mathbf{F}_R$.**

S, say from left to right. From Eq. (5.56), we obtain,

$$\begin{aligned} S_{\alpha\beta}(r_\beta) &= -\sum_i \frac{1}{4\pi}\langle \left[p_{i\alpha}, [p_{i\beta}, \delta(r_\beta - r_{i\beta})]_+\right]_+\rangle \\ &- \sum_i \langle \nabla_{i\alpha}(U_{\text{int}})\frac{1}{2}\text{sgn}(r_\beta - r_{i\beta})\rangle, \end{aligned} \tag{5.58}$$

where the sign function takes account of the change in sign of the force across the surface. The reason for the factor of $\frac{1}{2}$ is made clear below. The total stress is obtained by integrating the last expression over all r_β. From the property of delta functions,

$$\int \delta^{(n)}(x-a)f(x)\mathrm{d}x = (-1)^n f^{(n)}(a), \tag{5.59}$$

we can show that

$$\langle \left[p_{i\alpha}, [p_{i\beta}, \delta(r_\beta - r_{i\beta})]_+\right]_+\rangle = 4\langle p_{i\alpha}p_{i\beta}\rangle. \tag{5.60}$$

Also, (note the factor of 2), we have

$$\int_{-a}^{a} \text{sgn}(x-a)\mathrm{d}x = -2a, \tag{5.61}$$

so that

$$\sigma_{\alpha\beta} = \int S_{\alpha\beta}(r_\beta)\mathrm{d}r_\beta = -\sum_i \left[\frac{\langle p_{i\alpha}p_{i\beta}\rangle}{m_i} - r_{i\beta}\nabla_{i\alpha}\langle U_{\text{int}}\rangle\right], \tag{5.62}$$

which is Eq. (5.31).

As was the case with Eq. (5.35), we write Eq. (5.58) in a form appropriate for the limit of an infinite system. For the case of two-body potentials, the planar stress can be written in the form,

$$\begin{aligned} S_{\alpha\beta}(r_\beta) &= -\sum_i \frac{1}{4m_i}\langle \left[p_{i\alpha}, [p_{i\beta}, \delta(r_\beta - r_{i\beta})]_+\right]_+ \rangle \\ &- \sum_{i<j} \langle \nabla_{i\alpha} \left[U_{ij}(\mathbf{r}_i - \mathbf{r}_j)\right] \frac{1}{2} \left[\mathrm{sgn}(r_\beta - r_{i\beta}) \right. \\ &- \left. \mathrm{sgn}(r_\beta - r_{j\beta})\right] \rangle. \end{aligned} \tag{5.63}$$

This form also makes clear the significance of surface crossing terms. Particles that are confined entirely to either side of the surface, S, do not contribute to the stress (as the sgn terms cancel). The stress arises due to pairs of particles on opposite sides of the surface, whose "force-lines" intersect S. The concept of force lines is also useful in clarifying the meaning of the Maxwell stress tensor, and is discussed further in Appendix K.

5.5 Stress Theorem and Electron Density

In the previous sections, the stress theorem has been derived through scaling arguments applied to the wave function as well as by consideration of the generalized virial operator. As modern studies of electronic structure almost invariably are based on density functional theory (and its local approximation, the LDA), it is useful to derive the stress theorem using particle densities and correlation functions as the basic variables. Such a derivation will facilitate the discussion of stress in the next section, as well as the application of the results obtained to electronic structure studies carried out within a Green function approach.

The formal development presented below follows the lines of that in Ziesche et al. [6]. We first consider a cluster of a finite number of particles, and pass to the limit of infinite systems subsequently.

5.5.1 Density matrices

We consider a many-body system of interacting electrons and nuclei in the Born-Oppenheimer approximation, and the associated Hamiltonian, Eq. (1.94). Specifically, the kinetic energy operator includes only electronic motion, Eq. (1.96), and the potential energy of the system is given by the sum of the terms in Eqs. (1.97) and (1.99). It is convenient to define density operators for the different types of particles in the system. We introduce the nuclear charge density function,

$$\rho(\mathbf{r}) = \sum_{\alpha} Z_\alpha \delta(\mathbf{r} - \mathbf{R}_\alpha), \tag{5.64}$$

and the nuclear pair distribution function,

$$\rho_2(\mathbf{r}_1, \mathbf{r}_2) = \sum_{\alpha \neq \beta} Z_\alpha Z_\beta \delta(\mathbf{r}_1 - \mathbf{R}_\alpha)\delta(\mathbf{r}_2 - \mathbf{R}_\beta). \tag{5.65}$$

Similarly, for the electron subsystem, we define the *operators*,

$$\hat{n}(\mathbf{r}) = \sum_{i} \delta(\mathbf{r} - \mathbf{r}_i), \tag{5.66}$$

and

$$\hat{n}_2(\mathbf{r}_1, \mathbf{r}_2) = \sum_{i \neq j} \delta(\mathbf{r}_1 - \mathbf{r}_i)\delta(\mathbf{r}_2 - \mathbf{r}_j). \tag{5.67}$$

We note that the electrons are described by operators in contrast to the nuclei that are taken to behave "classically" and can be described by ordinary functions[11]. It is also convenient to define the *total* charge density operator,

$$\hat{\nu}(\mathbf{r}) = \rho(\mathbf{r}) - \hat{n}(\mathbf{r}), \tag{5.68}$$

and the *total* pair correlation operator

$$\hat{\nu}_2(\mathbf{r}_1, \mathbf{r}_2) = \rho_2(\mathbf{r}_1, \mathbf{r}_2) + \hat{n}_2(\mathbf{r}_1, \mathbf{r}_2) - [\rho(\mathbf{r}_1)\hat{n}(\mathbf{r}_2) + \rho(\mathbf{r}_2)\hat{n}(\mathbf{r}_1)]\,. \tag{5.69}$$

In terms of these density functions and operators, the various contributions to the potential in the many-body Hamiltonian, Eq. (1.94), can be written in the forms,

$$\hat{U}_{\mathrm{NN}} = \frac{1}{2}\int \mathrm{d}^3 r_1 \int \mathrm{d}^3 r_2 \rho_2(\mathbf{r}_1, \mathbf{r}_2)\frac{1}{r_{12}}, \tag{5.70}$$

$$\hat{U}_{\mathrm{Ne}} = -\frac{1}{2}\int \mathrm{d}^3 r_1 \int \mathrm{d}^3 r_2 \left[\rho_1(\mathbf{r})\hat{n}(\mathbf{r}_2) + \hat{n}(\mathbf{r}_1)\rho(\mathbf{r}_2)\right]\frac{1}{r_{12}}, \tag{5.71}$$

and

$$\hat{U}_{\mathrm{ee}} = \frac{1}{2}\int \mathrm{d}^3 r_1 \int \mathrm{d}^3 r_2 \hat{n}_2(\mathbf{r}_1, \mathbf{r}_2)\frac{1}{r_{12}}, \tag{5.72}$$

where $r_{12} = |\mathbf{r}_1 - \mathbf{r}_2|$. Using the total pair correlation operator, we can combine the last three expressions into a single potential energy operator,

$$\hat{U} = \frac{1}{2}\int \mathrm{d}^3 r_1 \int \mathrm{d}^3 r_2 \hat{\nu}_2(\mathbf{r}_1, \mathbf{r}_2)\frac{1}{r_{12}}. \tag{5.73}$$

Expressions for the kinetic energy are given below in subsection 5.5.3.

[11] Whether a classical or quantum mechanical description should be used with respect to nuclear motion depends on the particular problem under consideration.

5.5.2 Ground-state expectation values

The single-particle and pair distribution functions defined in Eqs. (1.157) and (1.159) are the expectation values with respect to the ground-state wave function of the operators defined in Eqs. (5.66) and (5.67),

$$n(\mathbf{r}) = \langle \hat{n}(\mathbf{r}) \rangle, \tag{5.74}$$

and

$$n_2(\mathbf{r}_1, \mathbf{r}_2) = \langle \hat{n}(\mathbf{r}_1, \mathbf{r}_2) \rangle. \tag{5.75}$$

Let us verify Eq. (5.74). Suppressing spin indices for the sake of clarity of presentation, we have

$$\begin{aligned}
\langle \hat{n}(\mathbf{r}) \rangle &= \int d^3r_1 \cdots d^3r_N \Psi^*(\mathbf{r}_1, \cdots \mathbf{r}_N) \sum_i \delta(\mathbf{r} - \mathbf{r}_i) \Psi(\mathbf{r}_1, \cdots \mathbf{r}_N) \\
&= \int d^3r_2 \cdots d^3r_N |\Psi(\mathbf{r}, \mathbf{r}_2, \cdots \mathbf{r}_N)|^2 \\
&+ \int d^3r_1, d^3r_3 \cdots d^3r_N |\Psi(\mathbf{r}_1, \mathbf{r}, \mathbf{r}_3, \cdots \mathbf{r}_N)|^2 + \cdots \\
&= N \int d^3r_2 \cdots d^3r_N |\Psi(\mathbf{r}, \mathbf{r}_2, \cdots \mathbf{r}_N)|^2 \\
&= n(\mathbf{r}).
\end{aligned} \tag{5.76}$$

The next to the last line follows from the equivalence of all electrons, and the last line from the definition of the single-particle density matrix, Eq. (1.157). We also have,

$$\nu(\mathbf{r}) = \langle \hat{\nu}(\mathbf{r}) \rangle = \rho(\mathbf{r}) - n(\mathbf{r}). \tag{5.77}$$

The expectation values of the kinetic and potential energy can be obtained in the manner indicated in Eqs. (1.22) and (1.23).

5.5.3 Kinetic energy (Ehrenfest) force density

The kinetic energy tensor field can be evaluated as the expectation value of the operator $\sum_i \mathbf{P}_i(\mathbf{r})\mathbf{P}_i(\mathbf{r})$ where $\mathbf{P}_i(\mathbf{r})$ is given by Eq. (5.46). We have,

$$\begin{aligned}
2\overset{\leftrightarrow}{t}(\mathbf{r}) &= \langle \sum_i \frac{1}{m} \mathbf{P}_i(\mathbf{r})\mathbf{P}_i(\mathbf{r}) \rangle \\
&= \int d^3r_1 \cdots d^3r_N \Psi^*(\mathbf{r}_1', \cdots, \mathbf{r}_N') \left[\sum_i \frac{1}{m} \mathbf{P}_i(\mathbf{r})\mathbf{P}_i(\mathbf{r}) \right] \\
&\times \; \Psi(\mathbf{r}_1, \cdots, \mathbf{r}_N)\Big|_{\mathbf{r}'=\mathbf{r}} \\
&= \frac{\tilde{P}\tilde{P}}{m} n(\mathbf{r}, \mathbf{r}')\Bigg|_{\mathbf{r}'=\mathbf{r}},
\end{aligned} \tag{5.78}$$

where

$$\tilde{\mathbf{P}} = \frac{1}{2}\left(p'^{*} + p\right) = \frac{\hbar}{2\mathrm{i}}(-\nabla' + \nabla). \tag{5.79}$$

Here, we have made use of the indistinguishability of the electrons and of the definition of the single-particle density matrix, Eq. (1.157). We note that $\mathbf{r}'$ should be set equal to $\mathbf{r}$ *after* the action of the operators $\mathbf{P}(\mathbf{r})$ has been taken into account. Taking the trace of Eq. (5.78), we obtain the expectation value of the kinetic energy operator,

$$\mathrm{Tr} \int \mathrm{d}^3 r \frac{\tilde{P}\tilde{P}}{2m} n(\mathbf{r},\mathbf{r}')\Big|_{\mathbf{r}'=\mathbf{r}} = \langle \hat{T} \rangle = T. \tag{5.80}$$

As we saw in the previous sections, the divergence of the kinetic energy tensor yields the force density on the electrons,

$$\mathrm{Div}\ 2\overleftrightarrow{t}(\mathbf{r}) = \mathbf{f}^{-}(\mathbf{r}). \tag{5.81}$$

It is customary to refer to $\mathbf{f}^{+}$ as the Hellmann-Feynman force density and $\mathbf{f}^{-}$ as the Ehrenfest force density.

5.5.4 Electrostatic force density

The *classical* electrostatic potential in the system is the solution of the Poisson equation,

$$\Phi(\mathbf{r}) = \int \mathrm{d}^3 r' \nu(\mathbf{r}') \frac{1}{|\mathbf{r}-\mathbf{r}'|} = \Phi^{+}(\mathbf{r}) + \Phi^{-}(\mathbf{r}) \tag{5.82}$$

which is a sum of a nuclear, Φ^{+}, and an electronic (Hartree), Φ^{-}, contribution. The corresponding electric field is given by the negative gradient of the potential,

$$\mathbf{E} = -\nabla\Phi(\mathbf{r}) = \mathbf{E}^{+}(\mathbf{r}) + \mathbf{E}^{-}(\mathbf{r}), \tag{5.83}$$

where

$$\mathbf{E}^{+}(\mathbf{r}) = \int \mathrm{d}^3 r_1 \rho(\mathbf{r}_1) \frac{\mathbf{r}-\mathbf{r}_1}{|\mathbf{r}-\mathbf{r}_1|^3} = \int \mathrm{d}^3 r_1 \rho(\mathbf{r}_1)\mathbf{E}_1(\mathbf{r}), \tag{5.84}$$

and

$$\mathbf{E}^{-}(\mathbf{r}) = \int \mathrm{d}^3 r_1 n(\mathbf{r}_1)\mathbf{E}_1(\mathbf{r}). \tag{5.85}$$

It follows that the total electrostatic force density at point $\mathbf{r}$ is given by,

$$\begin{aligned}
\mathbf{f}(\mathbf{r}) &= \frac{1}{2}\int \mathrm{d}^3 r_1 \int \mathrm{d}^3 r_2 \nu_2(\mathbf{r}_1,\mathbf{r}_2)\left[\delta(\mathbf{r}-\mathbf{r}_1)\mathbf{E}_2(\mathbf{r}) + \delta(\mathbf{r}-\mathbf{r}_2)\mathbf{E}_1(\mathbf{r})\right] \\
&= -\frac{1}{2}\mathrm{Div} \int \mathrm{d}^3 r_1 \int \mathrm{d}^3 r_2 \nu(\mathbf{r}_1,\mathbf{r}_2) \overleftrightarrow{M}_{12}(\mathbf{r}) \\
&= \mathrm{Div}\overleftrightarrow{M}(\mathbf{r}),
\end{aligned} \tag{5.86}$$

where $\overleftrightarrow{M}_{12}(\mathbf{r})$ is the Maxwell stress tensor (see Appendix K) associated with the presence of two charges at positions $\mathbf{r}_1$ and $\mathbf{r}_2$. The form of $\overleftrightarrow{M}_{12}(\mathbf{r})$ is obtained by subtracting the tensor self-interaction of the charges from the general expression for the Maxwell tensor,

$$\begin{aligned}
\overleftrightarrow{M}_{12}(\mathbf{r}) &= -\frac{1}{4\pi}\left\{[\mathbf{E}_1(\mathbf{r})+\mathbf{E}_2(\mathbf{r})][\mathbf{E}_1(\mathbf{r})+\mathbf{E}_2(\mathbf{r})] - \frac{1}{2}[\mathbf{E}_1(\mathbf{r})+\mathbf{E}_2(\mathbf{r})]^2\overleftrightarrow{I}\right. \\
&- [\mathbf{E}_1(\mathbf{r})\mathbf{E}_1(\mathbf{r})] - \frac{1}{2}\left[\mathbf{E}_1(\mathbf{r})\cdot\mathbf{E}_1(\mathbf{r})\overleftrightarrow{I}\right] - [\mathbf{E}_2(\mathbf{r})\mathbf{E}_2(\mathbf{r})] \\
&- \left.\frac{1}{2}\left[\mathbf{E}_2(\mathbf{r})\cdot\mathbf{E}_2(\mathbf{r})\overleftrightarrow{I}\right]\right\} \\
&= -\frac{1}{4\pi}\left[\mathbf{E}_1(\mathbf{r})\mathbf{E}_2(\mathbf{r}) + \mathbf{E}_2(\mathbf{r})\mathbf{E}_1(\mathbf{r}) - \mathbf{E}_1(\mathbf{r})\mathbf{E}_2(\mathbf{r})\overleftrightarrow{I}\right]. \qquad (5.87)
\end{aligned}$$

As shown in Appendix K, the divergence of the Maxwell stress tensor yields the *total* electrostatic force density, given here by Eq. (5.86),

$$\mathrm{Div}\overleftrightarrow{M} = -\mathbf{f}(\mathbf{r}). \qquad (5.88)$$

The stress field at point $\mathbf{r}$ is now obtained by adding the kinetic and potential energy contributions. Combining Eq. (5.78) with $\overleftrightarrow{M}(\mathbf{r})$ defined in Eq. (5.86), we have

$$-\overleftrightarrow{\sigma}(\mathbf{r}) = 2\overleftrightarrow{t}(\mathbf{r}) + \overleftrightarrow{M}(\mathbf{r}), \qquad (5.89)$$

for the total stress tensor field at $\mathbf{r}$. In view of Eqs. (5.81) and (5.88), the divergence of the stress tensor yields the force density on the nuclei at $\mathbf{r}$,

$$\mathrm{Div}\overleftrightarrow{\sigma}(\mathbf{r}) = \mathbf{f}^{+}(\mathbf{r}). \qquad (5.90)$$

Once again, we find that the local stress tensor is related only to the forces on the nuclei. Its divergence yields the force needed to keep the nuclei fixed in position, according to the Born-Oppenheimer approximation.

5.5.5 Thermodynamic limit

The considerations presented above for the case of finite clusters can be readily extended to infinite systems, i.e., $\Omega \to \infty$. In this case all quantities must be calculated per unit cell, in a manner analogous to that used to define quantities per unit volume in Eq. (5.12). However, care should be taken to take the limit in such a way as to guarantee the neutrality of the system, and the vanishing of macroscopic fields. Under these conditions, the relations derived above hold essentially unchanged, but with all extensive quantities divided by the number of cells N_c in the material so that $\Omega \to \infty$ and $N_c \to \infty$ while Ω/N_c remains finite [6].

5.6 Stress in the LDA

The stress relations derived in the previous sections carry over essentially intact into the formalism of density functional theory and the LDA (for a brief summary see the last section in Chapter 1). For example, the total stress within the LDA consists of the following contributions [5]:

(1) the kinetic stress

$$-\frac{1}{m}\sum_i \langle \phi_i | p_\alpha p_\beta | \phi_i \rangle, \tag{5.91}$$

which in a generalization of Eq. (1.262) can be written in terms of the electron density in the form,

$$-\frac{1}{2}\int \mathrm{d}^3 r r_\alpha \nabla_\beta \left[v_\mathrm{c}(\mathbf{r}) + v_\mathrm{xc}(\mathbf{r})\right] n(\mathbf{r}). \tag{5.92}$$

Here, the ϕ_i are the effective one-particle spin orbitals occurring in Eq. (1.283);

(2) the ion-electron potential stress,

$$-\frac{1}{2}\sum_I \int \mathrm{d}^3 n(\mathbf{r}) U'_\mathrm{Ne}(\mathbf{r}-\mathbf{R}_I)\frac{(\mathbf{r}-\mathbf{R}_I)_\alpha(\mathbf{r}-\mathbf{R}_I)_\beta}{|\mathbf{r}-\mathbf{R}_I|}, \tag{5.93}$$

where a prime denotes differentiation;

(3) the ion-ion potential stress,

$$-\frac{1}{2}\sum_{I,J\neq I} Z_I Z_J \frac{(\mathbf{R}_I-\mathbf{R}_J)_\alpha(\mathbf{R}_I-\mathbf{R}_J)_\beta}{|\mathbf{R}_I-\mathbf{R}_J|^3}; \tag{5.94}$$

(4) the Hartree stress

$$-\frac{1}{2}\int \mathrm{d}^3 r_2 n(\mathbf{r}_1) n(\mathbf{r}_2)\frac{(\mathbf{r}_1-\mathbf{r}_2)_\alpha(\mathbf{r}_1-\mathbf{r}_2)_\beta}{|\mathbf{r}_1-\mathbf{r}_2|^3}; \tag{5.95}$$

and

(5) the exchange and correlation stress,

$$\delta_{\alpha\beta}\int \mathrm{d}^3 r n(\mathbf{r})\left\{\epsilon_\mathrm{xc}[n(\mathbf{r})] - v_\mathrm{xc}^\mathrm{LDA}[n(\mathbf{r})]\right\}. \tag{5.96}$$

We note that the exchange-correlation term in the LDA is a local function of $\mathbf{r}$, and contributes only an isotropic pressure to the stress.

5.7 Summary of Stress Relations

In both classical and quantum mechanics, the generalized virial theorem leads to the stress theorem, which takes the form

$$-\overleftrightarrow{\sigma} = 2\overleftrightarrow{T} + \overleftrightarrow{U}. \tag{5.97}$$

Considering the nuclei in a solid as an independent subsystem external to the electronic system, we have (for the case of Coulomb potentials),

$$2\overleftrightarrow{T} - \overleftrightarrow{U} = -\sum_{\mathbf{R}} \mathbf{R}\nabla_{\mathbf{R}} E, \tag{5.98}$$

an expression which is valid in both classical and quantum mechanics. We have seen that the expressions for the total, macroscopic stress can also be obtained from those of microscopic stress fields by means of integration over the volume of the system.

Bibliography

[1] K. Kunc and R. M. Martin, Phys. Rev. Lett. **48**, 406 (1982); M. T. Yin and M. L. Cohen, Phys. Rev. B**25**, 4317 (1982).

[2] R. Biswas, R. M. Martin, R. J. Needs, and O. H. Nielsen, Phys. Rev. B**30**, 3210 1984

[3] J. E. Northrup and M. L. Cohen, J. Vac. Sci. Technol. **21**, 333 1982.

[4] O. H. Nielsen, R. M. Martin, D. J. Chadi, and K. Kunc, J. Vac. Sci. Technol. B**1**, 714 (1983).

[5] O. H. Nielsen and Richard M. Martin, Phys. Rev. B**32**, 3780; 3792 (1985).

[6] P. Ziesche, J. Gräfenstein, and O. H. Nielsen, Phys. Rev. B**37**, 8167 (1988).

[7] P. Ziesche, p. 85 in *Applications of Multiple Scattering Theory to Materials Science*, W. H. Butler, P. H. Dederichs, A. Gonis, and R. L. Weaver (eds.), Materials Research Society Proceedings, Vol. **253** (1992). See also, P. Ziesche, in *Density Functional Theory*, E. K. U. Gross and P. H. Dederichs (eds.), (Plenum Press, New York, 1995), p. 559.

[8] A. Sommerfeld, *Mechanics of Deformable Bodies*, Lectures on Theoretical Physics, Vol. **II**, Academic Press, New York (1964).

[9] J. F. Nye, *Physical Properties of Crystals*, Oxford University Press, Oxford (1958).

[10] L. Landau and I. Lifshitz, *Theory of Elasticity*, Pergamon, London, (1958).

[11] A. A. Wilson, *Thermodynamics and Statistical Mechanics*, Cambridge University Press, Cambridge (1957).

[12] Herbert Goldstein *Classical Mechanics*, Addison Wesley, Reading, Massachusetts (1950).

[13] P. Ehrenfest, Z. Phys. **45**, 455 (1927).

[14] W. Pauli, in *Handbuch der Physik, band XXIV, Teil 1*, Springer, Berlin, (1933), pp. 83-272; Vol. **V**, part 1 (1958).

[15] H. Hellmann, *Einfürung in die Quantenchemie*, Deuticke, Leipzig (1937), pp. 61 and 285.

[16] R. P. Feynman, Phys. Rev. **56**, 340 (1939).

[17] M. Born, W. Heisenberg, and P. Jordan, Z. Phys. **35**, 557 (1926).

[18] B. Finkelstein, Z. Phys. **50**, 293 (1928).

[19] E. Hylleraas, Z. Phys. **54**, 347 (1929).

[20] V. Fock, Z. Phys. **63**, 855 (1930).

[21] J. C. Slater, J. Chem. Phys. **1**, 687 (1933).

[22] E. Schrödinger, Ann. Phys. (Leipzig) **82**, 265 (1927).

[23] R. P. Feynman, Thesis, Massachusetts Institute of Technology (1939), unpublished.

[24] P. C. Martin and J. Schwinger, Phys. Rev. **115**, 1324 (1959).

[25] A. Kugler, Z. Phys. **198**, 236 (1967).

[26] N. O. Folland, Int. J. Quantum Chem. Symp. **15**, 369 (1981).

[27] A. G. McLellan, Am. J. Phys. **42**, 239 (1974); J. Phys. C**17**, 1 (1984).

[28] J. D. Jackson, *Classical Electrodynamics*, Wiley, New York (1962).

Part II

Periodic Solids and Impurities

Chapter 6

Electronic Structure of Periodic Solids

6.1 General Comments

When atoms combine to form a crystalline solid, the electrons in the sharp energy states of an isolated atom, particularly the most mobile ones farthest away from the nucleus, begin to interact with electrons on neighboring atoms. This causes the corresponding states to broaden and give rise to levels closely spaced in energy called *energy bands*. Many of these bands are separated from one another by energy regions in which no electronic states in the solid exist, at least in the bulk. These *energy gaps* are very important in determining the physical properties of crystalline solids. One of the aims of *band theory* is to calculate the *band structure* of crystalline materials, that is to say, the manner in which the energy, E, of an electron varies as a function of crystal momentum, $\mathbf{k}$, for each band in the solid. In this chapter, we describe how this can be done based on the formalism of MST previously discussed.

Band structure calculations for real materials are far from simple. Strictly speaking, one must solve the Schrödinger equation associated with a Hamiltonian that contains the kinetic energies and the electrostatic interactions of a very large number (Avogadro's number) of nuclei and their associated electrons. This is clearly an impossible task[1], prompting one to seek an approximate treatment, focusing on the calculation of quantities of interest. Band theory has evolved primarily around the so-called

[1] It is also an undesirable approach even if it could be implemented. An *exact* solution to such a complex problem would be of such complexity itself as to render very difficult, if not impossible, the extraction of useful information.

single-particle approximation in conjunction with the Born-Oppenheimer approximation. In the latter, one considers the nuclei as being at rest and determines the ground state of the electron cloud for an instantaneous distribution of nuclear positions. In the single-particle approximation, the many-body electron system is commonly treated within the local density approximation (LDA) to density functional theory (DFT), along the lines presented in Chapter 1. The solution to the much simplified Schrödinger equation associated with the effective LDA Hamiltonian yields "electron states" in the solid characterized by a band index, n, and the value of the wave vector, $\mathbf{k}$, in the first Brillouin zone (BZ) of the reciprocal lattice[2].

But even under such drastic approximations, complications exist. As a result of the interactions in the solid, electronic energy bands usually cannot be characterized simply by individual atomic levels. To complicate matters even further, these energy bands may overlap, and forbidden energy regions (gaps) for one band may correspond to allowed energies for another band. A schematic representation [1] of a typical band structure along a given direction inside the BZ of a periodic solid is shown in Fig. 6.1. This figure shows explicitly the presence of various bands along the energy axis, the existence of band gaps, and the possible overlap between different bands. In addition, one must calculate the potential in the solid in a *self-consistent* way in order to account for the interatomic interactions, and the interactions between the electrons and the ions and among the electrons themselves.

In this chapter, we discuss the basic features underlying the calculation of the electronic structure of periodic solid materials within a multiple-scattering theory, Green-function method. We begin by summarizing a number of important concepts about the form of the potential and the charge density in a periodic material, and review a number of conditions imposed by periodicity on the nature of the electronic states. Subsequently, we discuss a number of elementary notions about the density of states (DOS) and lattice Fourier transforms. Following this discussion, we present the basic elements of a self-consistent band structure method carried out within the LDA, both for the case of MT potentials as well as for the case of space-filling cells.

---

[2] Our discussion of the Kohn-Sham (KS) formalism in Chapter 1 shows that the individual particle states used there have no particular significance other than to reproduce the single-particle density. Hence, the commonly used assignment of single-particle states to the KS spin functions lacks rigorous justification. The discussion in the last chapter shows that a proper description of quasi-particle states requires an integral self-energy operator, that is in general energy dependent, complex, and non local rather than the static, real, and local exchange and correlation potential used in the LDA.

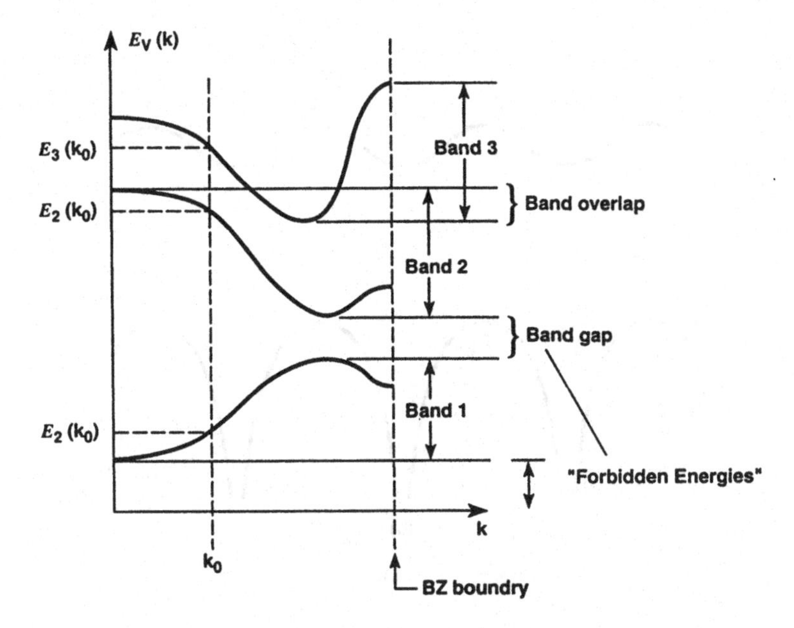

Figure 6.1: **Schematic representation of a band structure along a given direction inside the BZ.**

6.2 Electronic States in Crystals

In order to be specific, we phrase our discussion in terms of crystalline metallic materials that are characterized by the presence of electrons that are, in some sense, free to move. The high electrical conductivity of most metals, e.g., the alkalis and the noble metals such as silver and gold, offer experimental evidence that the concept of nearly free-moving electrons in these materials is not too distant from the truth. And yet the electrons in a material interact very strongly with one another and with the ions, forming a gas of anything but free particles. To a large extent, however, it is possible to treat the interactions in a solid in terms of an *effective* one-electron potential, along the general lines described in our discussion of the LDA in Chapter 1, which is sufficiently weak to justify the approximation of considering the particles as moving in nearly free space..

6.2.1 Potential in a metal

Since the negatively charged electrons tend to screen out the positive ionic potentials, an electron moving along the line that runs parallel to nuclear positions but at some distance from the nuclei themselves would experience a potential that may be represented roughly in the manner indicated in Fig.

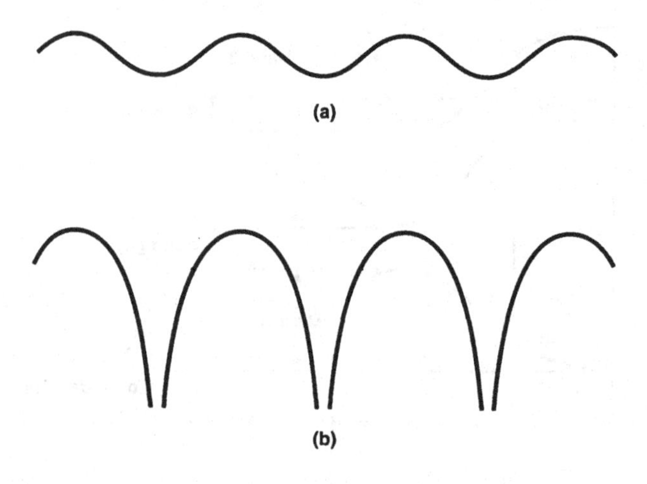

Figure 6.2: **Schematic representation of the crystal potential in a metal along a line parallel but not through nuclear positions (a), and along a line through nuclear positions, (b).**

6.2a. Apart from mild and periodic fluctuations, this potential is nearly a constant. Of course, the situation changes drastically when we consider the potential along a line through nuclear positions as is shown schematically in Fig. 6.2b. Here, the potential approaches $-\infty$ in the vicinity of each ion, indicating that the electron has penetrated the screening cloud of the other electrons and "sees" the bare nuclear charge. Now, apart from the regions very near the ions, the potential in most of the space inside a metal will look not unlike the simple picture shown in Fig. 6.2a. Thus, the strongly interacting electron gas can be treated, at least to first approximation, as a gas of independent particles moving in a constant potential that could be set equal to zero. In fact, in the *Sommerfeld model* one solves for the electronic states in a metal assuming that each electron moves in a box of constant (zero) potential that rises sharply at the surface of the material, (thus preventing the electrons from spilling out into the vacuum).

It should be pointed out that in the simple model for the potential described above, even in the Sommerfeld model, electron-electron interactions are not completely neglected. For instance, the screening of the ionic charge by the electron cloud is a part of that interaction which is not entirely ignored. It is this interaction that causes the potential far from the nucleus to be only mildly varying. On the other hand, if one electron happens to be on

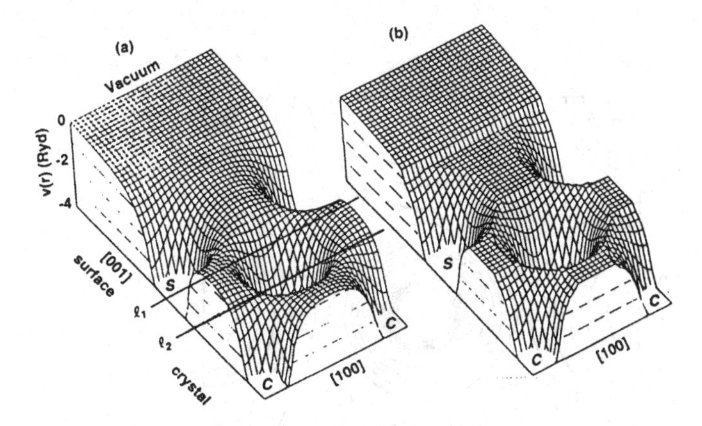

Figure 6.3: **Self-consistent one-electron potential near the (100) surface of a Ni crystal. Part (a) shows the full potential and part (b) its muffin-tin approximation [2].**

one side of a particular nucleus, a second electron is more likely to be on the other because of the repulsion between the two electrons. Such *correlation effects* are ignored completely in simple theories about the potential inside a metal. It is only the advance of quite sophisticated methods, such as DFT and the LDA discussed in Chapter 1, that has led to the construction of realistic potentials in which correlation and other effects are included at least in some approximation.

One property of the crystal potential seen by a single electron must be retained, however, in any theory. The crystal potential must retain any symmetry, translational or rotational, of the underlying lattice structure. In translationally invariant materials the potential, $V(\mathbf{r})$, experienced by an electron at position $\mathbf{r}$ must satisfy the condition,

$$V(\mathbf{r}+\mathbf{R}) = V(\mathbf{r}), \tag{6.1}$$

for all lattice vectors $\mathbf{R}$. Corresponding to such a *single-particle* potential, the electronic states in the material can be determined by the solutions of the Schrödinger equation (in units $\hbar = 1$, $m = \frac{1}{2}$, $e^2 = 2$)

$$\nabla^2\psi(\mathbf{r}) + [E - V(\mathbf{r})]\psi(\mathbf{r}) = 0. \tag{6.2}$$

Let us now consider a more realistic model of the crystal potential than that shown in Fig. 6.2. Figure 6.3a shows the self-consistently determined potential [2] near the (100) surface of a Ni crystal, computed using the method of augmented plane waves [3]. The main feature of the potential is its nearly spherical shape about the center of each cell. Also, we note that the variation of the potential along a line such as l_1 is not unlike

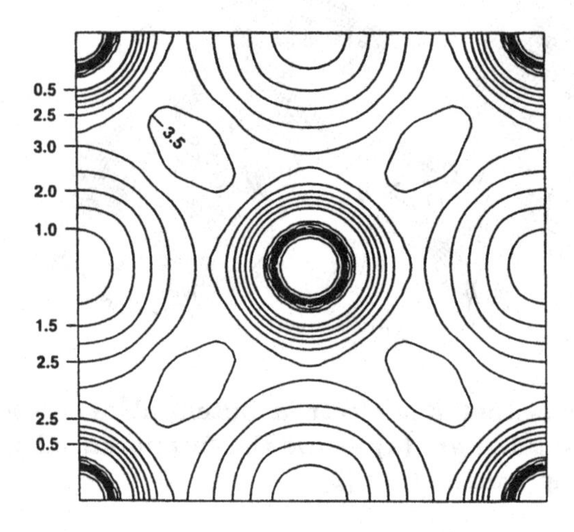

Figure 6.4: **Charge density contours on the (001) plane of elemental fcc Al.**

that depicted in Fig. 6.2a, while the variation along a line through nuclear positions, such as line l_2, would resemble closely that shown in Fig. 6.2b.

Given the form of the potential in Fig. 6.3a, it is easily seen that it can be represented quite accurately by its *muffin-tin* average, shown in part (b) of the figure. In the MT approximation the potential is spherically averaged inside the MT sphere (the sphere inscribed inside the WS cell of the lattice) and is set equal to a constant (the flat pieces in Fig. 6.3a) in the interstitial region outside. The value of the potential there is usually set equal to zero, the *muffin-tin zero.* In that case, the vacuum outside the crystal is represented by a positive potential step, also indicated in Fig. 6.3b.

6.2.2 Charge density

The valence charge density of fcc elemental Al in the (001) plane is shown [4] in Fig. 6.4. As is to be expected for a close-packed metallic system, the charge density exhibits a great deal of spherical symmetry. On the other hand, the charge density of Si, a semiconductor based on a zincblende structure, depicted in Figs. 6.5 [4] and 6.6 [5], can deviate substantially from its spherical average. Note, particularly in Fig. 6.6, the peak in the charge density midway along a bond (arrow A in the figure) signifying the

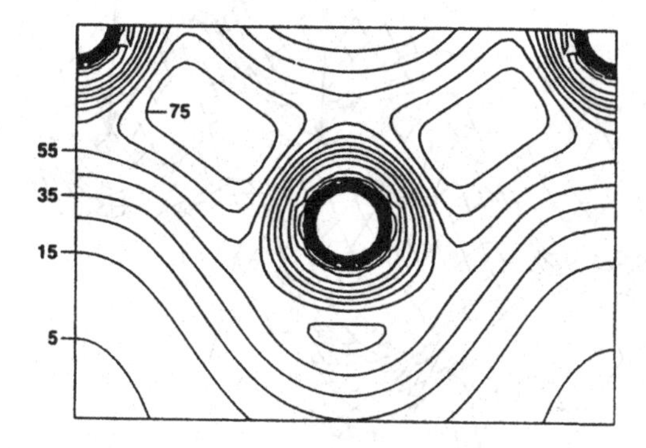

Figure 6.5: **Charge density contours on the (110) plane of elemental Si.**

covalent nature of the bond in this system. Contrast this with the charge density in Al that decreases rapidly and in spherical fashion away from the center of a cell. Note also the rather large regions nearly empty of charge (arrow B in Fig. 6.6). The treatment of such charge distributions is much more complicated numerically than that of the nearly spherical charge distribution in simple close-packed metals such as Al. A straightforward treatment within MST would require the use of full-cell potentials and charge densities[3]. But even within such a treatment, it may be expeditious to partition space into cells with cubic symmetry by inserting "empty" cells, i.e., cells that do not contain nuclei, in the open regions of the lattice. This partition would produce cells with higher symmetry and would facilitate the convergence of expansions in angular momentum eigenstates.

6.2.3 Bloch theorem

The periodicity of the potential is also reflected in the particle (charge) density, as is illustrated in the last three figures. The periodicity of the density implies that the crystal wave function can change only by a phase factor of modulus unity under translations by lattice vectors. As is well-

[3]It should be kept in mind that the use of full-cell potentials in no way overcomes the difficulties associated with the use of the LDA.

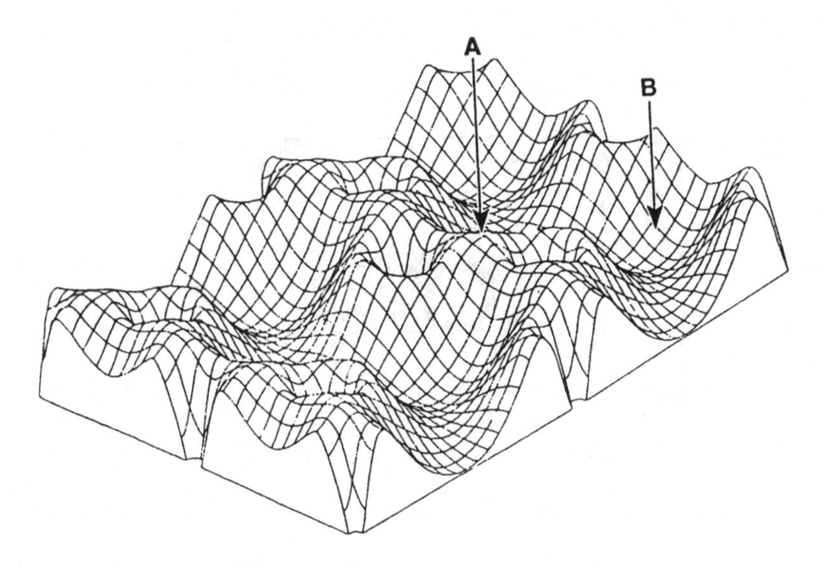

Figure 6.6: **Projected charge density on the (110) plane of elemental Si, calculated as if Si were a simple metal.**

known [6], the eigenfunctions of a periodic system must satisfy the *Bloch theorem*, relating the values of the wave function (the eigenfunctions) at two points differing by a direct lattice vector

$$\psi_{\mathbf{k}}(\mathbf{r}+\mathbf{R}) = e^{i\mathbf{k}\cdot\mathbf{R}}\psi_{\mathbf{k}}(\mathbf{r}), \tag{6.3}$$

where $\mathbf{k}$ is a vector in the BZ of the reciprocal lattice. It follows that the wave function associated with $\mathbf{k}$ must have the form

$$\psi_{\mathbf{k}}(\mathbf{r}) = e^{i\mathbf{k}\cdot\mathbf{r}}u_{\mathbf{k}}(\mathbf{r}), \tag{6.4}$$

where $u_{\mathbf{k}}(\mathbf{r})$ is periodic with the period of the lattice, $u_{\mathbf{k}}(\mathbf{r}+\mathbf{R}) = u_{\mathbf{k}}(\mathbf{r})$.

6.2.4 Density of states

We saw in Chapter 2, Eq. (2.20), that the particle density at energy E is given by the expression

$$n(\mathbf{r},E) = \sum_m \phi_m^*(\mathbf{r})\phi_m(\mathbf{r})\delta(E-E_m). \tag{6.5}$$

An integral over the energy yields the density, Eq. (2.22), the basic variable in LDA-based methods. Also of great importance is the *density of states*

(DOS), which results upon integration of $n(\mathbf{r}, E)$ over all space. Using the orthonormality of the functions $\phi_n(\mathbf{r})$, we have

$$\begin{aligned} n(E) &= \int \mathrm{d}^3 r n(\mathbf{r}, E) \\ &= \sum_m \delta(E - E_m). \end{aligned} \tag{6.6}$$

If we designate each eigenvalue, E_m, by the value of the $\mathbf{k}$ vector in the first BZ and the band index n, we obtain the *density of states function*

$$n(E) = \sum_n \sum_{\mathbf{k}} \delta[E - E_n(\mathbf{k})]. \tag{6.7}$$

The *partial* DOS associated with a particular band n is given by

$$\begin{aligned} n_n(E) &= \sum_{\mathbf{k}} \delta[E - E_n(\mathbf{k})] \\ &= \int \frac{\mathrm{d}^3 k}{(2\pi)^3} \delta[E - E_n(\mathbf{k})]. \end{aligned} \tag{6.8}$$

The density of states function has a number of important properties. In the process we also derive an alternative expression for evaluating the DOS for the case of translationally invariant materials. Because it denotes the *number* of states per unit energy per unit volume having energy E, $n(E)$ is never negative

$$n(E) \geq 0. \tag{6.9}$$

We note also that the partial DOS function, $n_n(E)$, can be defined in an equivalent way as the number of allowed wave vectors, $\Delta_n(E)$, in the nth band per unit energy range, ΔE, per unit volume, V,

$$n_n(E) = \Delta_n(E)/(V\Delta E). \tag{6.10}$$

Now, the number of allowed wave vectors in any volume of $\mathbf{k}$-space equals the ratio of that volume to the volume per allowed wave vector, $\Delta\mathbf{k} = \mathrm{d}^3 k = (2\pi)^3/V$. Thus,

$$\Delta_n = \frac{V}{(2\pi)^3} \int \mathrm{d}^3 k \Delta_n(\mathbf{k}), \tag{6.11}$$

where the function

$$\Delta_n(\mathbf{k}) = \begin{cases} 1 \text{ if } E \leq E_n(\mathbf{k}) \leq E + \Delta E \\ 0, \text{ otherwise,} \end{cases} \tag{6.12}$$

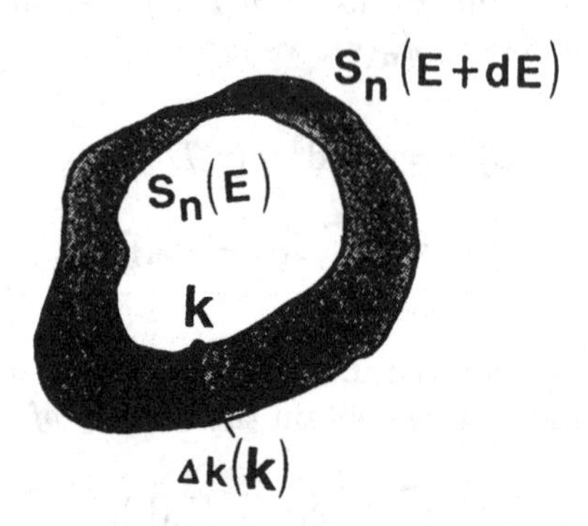

Figure 6.7: **Schematic two-dimensional diagram of a volume element contained between two constant energy surfaces, $S_n(E)$ and $S_n(E+\mathrm{d}E)$, and the distance $\Delta k(\mathbf{k})$ at a particular k. The shaded region represents the integral $\int \mathrm{d}S_n \Delta k(\mathbf{k})$.**

denotes the number of allowed **k** vectors in a given energy range in band n. From Eqs. (6.10) and (6.11), we obtain the expr ession,

$$n_n(E)\mathrm{d}E = \int \frac{\mathrm{d}^3k}{(2\pi)^3}\Delta_n(\mathbf{k}), \tag{6.13}$$

for the density of states function associated with band n.

As it stands, Eq. (6.13) is not in a computationally useful form. A useful and more revealing expression can be derived by noting that the product $\mathrm{d}^3k\Delta_n(\mathbf{k})$ denotes the infinitesimal volume of reciprocal space associated with band n near a particular **k** vector, i.e., the vector for which $E \leq E_n(\mathbf{k}) \leq E+\mathrm{d}E$. This volume element can be easily obtained, as indicated in Fig. 6.7. Let $S_n(E)$ denote a surface defined by the tips of all vectors **k** in the BZ for which $E_n(\mathbf{k}) = E$. Let also $\Delta k(\mathbf{k})$ be the perpendicular distance at **k** between the surface $S_n(E)$ and the similarly defined surface $S_n(E+\mathrm{d}E)$. Now, we have the relation

$$\mathrm{d}^3k\Delta_n(\mathbf{k}) = \mathrm{d}S_n(E)\Delta k(\mathbf{k}). \tag{6.14}$$

Finally, we note that

$$E + \mathrm{d}E = E + |\nabla E_n(\mathbf{k})|\Delta k(\mathbf{k}), \tag{6.15}$$

so that

$$\Delta k(\mathbf{k}) = \frac{\mathrm{d}E}{|\nabla E_n(\mathbf{k})|}, \tag{6.16}$$

and we can write the partial DOS as a surface integral

$$n_n(E) = \int_{S_n(E)} \frac{dS}{(2\pi)^3} \frac{1}{|\nabla E_n(\mathbf{k})|}. \tag{6.17}$$

This form reveals the relation between the partial density of states and the band structure of a periodic solid material. The total DOS is given as a sum over the band index of the expression for the partial density of states. As a simple illustration of these results, let us derive the expression for the DOS for a gas of non-interacting electrons, and compare it to the expression derived in Eq. (2.70) from the imaginary part of the free-particle propagator.

6.2.5 DOS for the free-electron gas

The energy of free electrons (or any other particles) of mass m is given in terms of the momentum $\hbar\mathbf{k}$ of the particle by the relation

$$E(\mathbf{k}) = \frac{\hbar^2 k^2}{2m}. \tag{6.18}$$

Thus, a surface of constant energy is a sphere in $\mathbf{k}$-space with origin at $\mathbf{k} = 0$ and a surface area $S_n = 4\pi k^2$. The integral over $S(E)$ in Eq. (6.17) can now be done immediately and yields the result,

$$\begin{aligned} n(E) &= \frac{4\pi k^2}{8\pi^3 \frac{\hbar^2 k}{m}} \\ &= \frac{mk}{2\pi\hbar^2}. \end{aligned} \tag{6.19}$$

In atomic units, $\hbar = 1$, $m = \frac{1}{2}$, this expression agrees with that in Eq. (2.70). A factor of two should be included in all expressions for the DOS exhibited above if closed systems are considered and if the presence of spin is to be taken into account.

Equation (6.17) yields the partial DOS for materials with well-defined dispersion relations, $E_n(\mathbf{k})$, such as a periodic solid. For such systems, in the converged limit it yields identical results with those obtained from the imaginary part of the Green function, discussed in Chapter 2. However, Eq. (6.17) becomes impractical in a number of cases, such as that of concentrated, substitutionally disordered alloys. As will be seen in a following chapter, such systems are often described by a non-Hermitian Hamiltonian (or self-energy) whose "eigenvalues" are complex quantities. In this case, $E_n(\mathbf{k})$ acquires an imaginary part leading to bands with a width, (more like ribbons rather than strings), proportional to the imaginary part of the eigenvalue. Now, Eq. (6.17) becomes rather ill-defined. Also, even in the case of periodic solids it is often computationally expedient to carry out calculations using complex values of the energy parameter,resulting in similar difficulties with the use of Eq. (6.17). In such cases, however, the Green function remains a well-defined quantity and its imaginary part always leads to the DOS function and the charge density.

6.3 Lattice Fourier Transforms

Many formal considerations that are valid for translationally invariant systems are also of great utility in the study of disordered materials. One important concept in this regard is that of *lattice Fourier transforms.*

As is well-known, the set of plane-wave functions

$$u_{\mathbf{k}}(\mathbf{r}) = \frac{1}{(2\pi)^{3/2}} e^{i\mathbf{k}\cdot\mathbf{r}}, \tag{6.20}$$

form a complete orthonormal basis in coordinate space in which one can expand any function, $f(\mathbf{r})$. The conditions of completeness and orthonormality are expressed by the relations,

$$\begin{aligned} \int u_{\mathbf{k}}^{*}(\mathbf{r}) u_{\mathbf{k}}(\mathbf{r}') d^3k &= \frac{1}{(2\pi)^3} \int e^{i\mathbf{k}\cdot(\mathbf{r}-\mathbf{r}')} d^3k \\ &= \delta(\mathbf{r}-\mathbf{r}'), \text{ (completeness)} \end{aligned} \tag{6.21}$$

and

$$\int u_{\mathbf{k}}^{*}(\mathbf{r}) u_{\mathbf{k}'}(\mathbf{r}) d^3r = \delta(\mathbf{k}-\mathbf{k}') \text{ (orthonormality).} \tag{6.22}$$

The integrations in the last two equations extend over all values of the *continuous* variables $\mathbf{r}$ and $\mathbf{k}$. In many physical applications, it is convenient to restrict $\mathbf{r}$ inside a box of volume V and to impose Born-von Karman periodic boundary conditions. In that case, the normalization factor $(2\pi)^{-3/2}$ is replaced by $V^{-3/2}$ and, since $\mathbf{k}$ becomes a discrete variable, the Dirac delta function, $\delta(\mathbf{k}-\mathbf{k}')$, in Eq. (6.21) is replaced by the Kronecker delta, $\delta_{\mathbf{k}\mathbf{k}'}$.

Analogous to expansions in plane-waves in continuous space, a function that is periodic with the period of a lattice can be expanded in plane waves involving only vectors in the reciprocal lattice. Any function, $f(\mathbf{r})$, with lattice periodicity can be expanded in the form

$$f(\mathbf{r}) = \sum_{\mathbf{K}} f_{\mathbf{K}} e^{i\mathbf{K}\cdot\mathbf{r}}, \tag{6.23}$$

where the sum extends over all reciprocal lattice vectors (recips), $\mathbf{K}$. We now obtain an expression for the expansion coefficients, $f_{\mathbf{K}}$.

Multiplying Eq. (6.23) by $e^{-i\mathbf{K}'\cdot\mathbf{r}}$ and integrating over all $\mathbf{r}$, we obtain the expression

$$\int d^3r f(\mathbf{r})e^{-i\mathbf{K}'\cdot\mathbf{r}} = \sum_{\mathbf{K}} f_{\mathbf{K}} \int d^3r e^{-i(\mathbf{K}'-\mathbf{K})\cdot\mathbf{r}}, \tag{6.24}$$

where we have assumed that the order of the summation and the integration can be interchanged. It follows from Eq. (6.22) that the right-hand side of Eq. (6.24) vanishes unless $\mathbf{K} = \mathbf{K}'$. Also, because of the assumed periodicity of $f(\mathbf{r})$, the integral over $\mathbf{r}$ can be restricted to the volume v of the unit cell of the direct lattice and we obtain

$$f_{\mathbf{K}} = \frac{1}{\Omega} \int_{\text{cell}} d^3r f(\mathbf{r})e^{-i\mathbf{K}\cdot\mathbf{r}}, \tag{6.25}$$

where Ω denotes the volume of a cell. The two expansion formulae, Eqs. (6.23) and (6.25), find exact analogies in the case of functions that are periodic in the reciprocal space of a lattice. For any function $\phi(\mathbf{k})$ such that $\phi(\mathbf{k}+\mathbf{K}) = \phi(\mathbf{k})$ for all $\mathbf{k}$ and all recips $\mathbf{K}$, we can write,

$$\phi(\mathbf{k}) = \sum_{\mathbf{R}} e^{i\mathbf{k}\cdot\mathbf{R}} \phi_{\mathbf{R}} \tag{6.26}$$

where

$$\phi_{\mathbf{R}} = \Omega \int \frac{d^3k}{(2\pi)^3} e^{-i\mathbf{k}\cdot\mathbf{R}} \phi(\mathbf{k}). \tag{6.27}$$

These expressions become evident when we recall that the reciprocal of the reciprocal lattice is the direct lattice, and that the volume of a primitive cell in the reciprocal lattice is $(2\pi)^3/v$.

Let us now consider a function that is periodic in real space in the sense of the Born-von Karman boundary conditions,

$$f(\mathbf{r} + N_i\mathbf{a}_i) = f(\mathbf{r}), \quad i = 1, 2, 3. \tag{6.28}$$

The three vectors $N_i\mathbf{a}_i$ generate a very large (and unphysical) lattice whose reciprocal lattice has the primitive vectors $\mathbf{b}_i/N_i$, where the $\mathbf{b}_i$ are the primitive reciprocal lattice vectors of the lattice determined by the $\mathbf{a}_i$. A vector in this unphysical reciprocal lattice is given as a linear combination of the $\mathbf{b}_i/N_i$ and the volume of the primitive cell in the direct lattice approaches the volume of the material. Thus, Eq. (6.25) now becomes

$$f_{\mathbf{K}} = \frac{1}{\Omega} \int_{\text{cell}} d^3r f(\mathbf{r})e^{-i\mathbf{K}\cdot\mathbf{r}}, \tag{6.29}$$

with Eq. (6.23) remaining unchanged. If in addition to the Born-von Karman periodicity, the function $f(\mathbf{r})$ is also periodic in the direct lattice defined by the $\mathbf{a}_i$, then Eq. (6.29) reduces to

$$f_{\mathbf{k}} = \frac{1}{\Omega} \int_{\text{cell}} \mathrm{d}^3 r e^{-i\mathbf{k}\cdot\mathbf{r}} f(\mathbf{r}) \tag{6.30}$$

where $\mathbf{k}$ is a vector in the first BZ of the reciprocal lattice. Under these conditions, Eq. (6.23) takes the form,

$$\begin{aligned} f(\mathbf{r}) &= \sum_{\mathbf{k}} f_{\mathbf{k}} e^{i\mathbf{k}\cdot\mathbf{r}} \\ &= \frac{1}{\Omega_{\text{BZ}}} \int_{\text{BZ}} \mathrm{d}^3 k f(\mathbf{k}) e^{i\mathbf{k}\cdot\mathbf{r}}. \end{aligned} \tag{6.31}$$

Equations (6.30) and (6.31) play a central role in many future developments.

6.3.1 Two identities of lattice Fourier transforms

The pair of Eqs. (6.26) and (6.27) [or (6.30) and (6.31)], are simply related, each being the Fourier transform of the other. Such Fourier transformations can be easily obtained by means of the identity,

$$\frac{1}{N} \sum_{\mathbf{R}} e^{i\mathbf{k}\cdot\mathbf{R}} = \delta_{\mathbf{k},0}. \tag{6.32}$$

Here, $\mathbf{R}$ runs through the N sites of a Bravais lattice, and $\mathbf{k}$ is any vector in the first BZ of the reciprocal lattice, consistent with the Born-von Karman periodic boundary conditions.

To prove the last identity, we note that the sum remains unchanged when $\mathbf{R}$ is replaced by $\mathbf{R} + \mathbf{R}_0$, for any $\mathbf{R}_0$ in the direct lattice,

$$\begin{aligned} \sum_{\mathbf{R}} e^{i\mathbf{k}\cdot\mathbf{R}} &= \sum_{\mathbf{R}} e^{i\mathbf{k}\cdot(\mathbf{R}+\mathbf{R}_0)} \\ &= e^{i\mathbf{k}\cdot\mathbf{R}_0} \sum_{\mathbf{R}} e^{i\mathbf{k}\cdot\mathbf{R}}. \end{aligned} \tag{6.33}$$

The last equality holds if the sum vanishes or if $e^{i\mathbf{k}\cdot\mathbf{R}_0} = 1$ for *all* vectors $\mathbf{R}_0$ which, in turn, holds only if $\mathbf{k}$ is a recip. But the only reciprocal lattice vector in the first BZ is the zero vector, $\mathbf{k} = 0$; thus, for all $\mathbf{k} \neq 0$ the sum in Eq. (6.33) vanishes, and is equal to N for $\mathbf{k} = 0$. This establishes the identity in Eq. (6.32). The reader can easily establish the related result,

$$\frac{1}{N} \sum_{\mathbf{k}} e^{i\mathbf{k}\cdot\mathbf{R}} = \delta_{\mathbf{R},0}, \tag{6.34}$$

which holds for all vectors $\mathbf{k}$ in the reciprocal lattice, and all vectors $\mathbf{R}$ in the direct lattice. We now illustrate the use of Eqs. (6.32) and (6.34) by means of some simple but very useful examples.

Let A_{ij} be a quantity depending on the position of sites $\mathbf{R}_i$ and $\mathbf{R}_j$ in the lattice. Quite generally, we can form the sum,

$$A(\mathbf{k},\mathbf{k}') = \frac{1}{N}\sum_{\mathbf{R}_i}\sum_{\mathbf{R}_j} A_{ij}\mathrm{e}^{\mathrm{i}(\mathbf{k}\cdot\mathbf{R}_i - \mathbf{k}'\cdot\mathbf{R}_j)}. \tag{6.35}$$

If A_{ij} is translationally invariant, i.e., depends only on the vector distance, $\mathbf{R}_i - \mathbf{R}_j$, and not on the particular positions $\mathbf{R}_i$ and $\mathbf{R}_j$,

$$\begin{aligned} A_{ij} &\equiv A(\mathbf{R}_i, \mathbf{R}_j) \\ &= A(\mathbf{R}_i - \mathbf{R}_j), \end{aligned} \tag{6.36}$$

then Eq. (6.35) takes a simple and useful form. We can now write,

$$\begin{aligned} A(\mathbf{k},\mathbf{k}') &= \frac{1}{N}\sum_{\mathbf{R}_i}\sum_{\mathbf{R}_j} A(\mathbf{R}_i - \mathbf{R}_j)\mathrm{e}^{\mathrm{i}\mathbf{k}\cdot(\mathbf{R}_i - \mathbf{R}_j) - \mathrm{i}\mathbf{R}_j\cdot(\mathbf{k}-\mathbf{k}')} \\ &= \frac{1}{N}\sum_{\mathbf{R}_j}\mathrm{e}^{-\mathrm{i}\mathbf{R}_j\cdot(\mathbf{k}-\mathbf{k}')}\left[\sum_{\mathbf{R}_i} A(\mathbf{R}_i - \mathbf{R}_j)\mathrm{e}^{\mathrm{i}\mathbf{k}\cdot(\mathbf{R}_i - \mathbf{R}_j)}\right]. \end{aligned} \tag{6.37}$$

Since the vector difference of lattice vectors runs through the same values as the vectors themselves, the term inside the brackets can be replaced by the quantity

$$A(\mathbf{k}) = \sum_{\mathbf{R}} A(\mathbf{R})\mathrm{e}^{\mathrm{i}\mathbf{k}\cdot\mathbf{R}}, \tag{6.38}$$

where the sum runs through the N vectors of the direct lattice. The sum over $\mathbf{R}_j$ in Eq. (6.36) can be evaluated by means of Eq. (6.32),

$$\frac{1}{N}\sum_{j}\mathrm{e}^{-\mathrm{i}\mathbf{R}_j(\mathbf{k}-\mathbf{k}')} = \delta_{\mathbf{k},\mathbf{k}'}, \tag{6.39}$$

so that Eq. (6.35) reduces to the form,

$$A(\mathbf{k},\mathbf{k}') = A(\mathbf{k})\delta_{\mathbf{k},\mathbf{k}'}, \tag{6.40}$$

for $\mathbf{k}$ a vector in the first BZ of the reciprocal lattice. This equation expresses the property that any quantity with the periodicity of the lattice is diagonal in the reciprocal space (crystal momentum) representation.

In many applications, it is more convenient to evaluate the Fourier transform $A(\mathbf{k})$ of a translationally invariant quantity and to obtain A_{ij} by means of the inverse transformation,

$$A_{ij} = \frac{1}{\Omega_{\mathrm{BZ}}}\int_{\mathrm{BZ}} \mathrm{d}^3k\,\mathrm{e}^{\mathrm{i}\mathbf{k}\cdot(\mathbf{R}_j - \mathbf{R}_i)}A(\mathbf{k}). \tag{6.41}$$

This equation can be obtained from Eq. (6.38) through the use of Eq. (6.32).

As a final example, consider the matrix product, $C_{ij} = \sum_k A_{ik}B_{kj}$, of two translationally invariant quantities, A_{ij} and B_{ij}. The use of Eqs. (6.32) and (6.38) allows us to show that

$$\begin{aligned} C(\mathbf{k},\mathbf{k}') &= C(\mathbf{k})\delta_{\mathbf{k},\mathbf{k}'} \\ &= A(\mathbf{k})B(\mathbf{k}). \end{aligned} \tag{6.42}$$

This result is the well-known convolution theorem of lattice Fourier transforms.

At this point, the reader may wish to glance back at Section 5.7.2 where the concepts of lattice Fourier transforms were used to solve the MST equations of motion for the case of a periodic solid. These notions will also be useful in our discussion of substitutionally disordered alloys, discussed later in this book.

6.4 Total Energy and Pressure

In Chapter 1, we derived a number of relations, both exact and within the LDA, involving the total energy and the pressure of an interacting electron cloud. In this subsection we summarize the application of these relations to the electronic structure of solid materials.

Within the LDA the total energy, E, of an interacting electron system can be written in the form

$$E = T + U + E_{\text{xc}}, \tag{6.43}$$

where, T, the expectation value of the kinetic energy operator is given by the expression, Eq. (1.195),

$$T = \sum_i^{\text{occ}} \int \phi_i^*(\mathbf{r})[-\nabla^2]\phi_i(\mathbf{r})\text{d}^3r, \tag{6.44}$$

where the $\phi_n(\mathbf{r})$, the Kohn-Sham functions, are the solutions of the one-electron effective Hamiltonian in Eq. (1.196), and the sum extending over all occupied electron states. We have also seen that the expectation value of the kinetic energy can be written in the general form, Eq. (1.210),

$$T = \sum_n^{\text{occ}} \epsilon_n - \int \text{d}^3r n(\mathbf{r})v_{\text{eff}}(\mathbf{r}). \tag{6.45}$$

The meaning of the various terms appearing in the last expression has been explained in connection with Eq. (1.210). The potential energy, U, in Eq.

(6.43) contains the interaction between the electrons and the nuclei, the electrons among themselves, and the nuclei among themselves, and can be written in the form, (with $e^2 = 2$),

$$\begin{aligned} U &= -2\sum_{\mathbf{R}_n} Z_n \int d^3r \frac{n(\mathbf{r})}{|\mathbf{r}-\mathbf{R}_n|} + \int d^3r \int d^3r' \frac{n(\mathbf{r})n(\mathbf{r}')}{|\mathbf{r}-\mathbf{r}'|} \\ &+ \sum_{\mathbf{R}_n \neq \mathbf{R}_m} \frac{Z_n Z_m}{|\mathbf{R}_n - \mathbf{R}_m|}, \end{aligned} \tag{6.46}$$

where Z_n is the nuclear charge (in atomic units) at position $\mathbf{R}_n$, and

$$n(\mathbf{r}) = \sum_n^{\mathrm{occ}} \phi_n^*(\mathbf{r})\phi_n(\mathbf{r}) \tag{6.47}$$

is the single-electron density at $\mathbf{r}$. The term E_{xc} contains the exchange and correlation contribution to the total energy. This term is *defined* as the difference between the exact total energy of the system and the quantity $T + U$ defined above. Throughout our discussion of the determination of the electronic structure of a solid, the exchange-correlation term will be treated in the LDA.

As we saw in Chapter 1, in the LDA one must solve an effective, one-electron Schrödinger equation of the form of Eq. (1.205),

$$[-\nabla^2 + v_{\mathrm{eff}}(\mathbf{r})]\phi_n(\mathbf{r}) = \epsilon_n\phi_n(\mathbf{r}), \tag{6.48}$$

where $v_{\mathrm{eff}}(\mathbf{r})$ is given by Eq. (1.202),

$$\begin{aligned} v_{\mathrm{eff}} &= v(\mathbf{r}) + \frac{\delta J[n]}{\delta n(\mathbf{r})} + \frac{\delta E_{\mathrm{xc}}[n]}{\delta n(\mathbf{r})} \\ &= v(\mathbf{r}) + \int \frac{n(\mathbf{r}')}{|\mathbf{r}-\mathbf{r}'|} d^3r' + v_{\mathrm{xc}}(\mathbf{r}), \end{aligned} \tag{6.49}$$

and

$$v_{\mathrm{xc}}(\mathbf{r}) = \frac{\delta E_{\mathrm{xc}}}{\delta n(\mathbf{r})}, \tag{6.50}$$

is the exchange-correlation potential, Eq. (1.203).

We also derived an expression for the pressure, Eq. (1.274),

$$3PV = 2T + U - 3\int d^3n(\mathbf{r})\left[\epsilon_{\mathrm{xc}}(\mathbf{r}) - v_{\mathrm{xc}}(\mathbf{r})\right]. \tag{6.51}$$

There, the quantity ϵ_{xc} is defined by the expression [see Eq. (1.213)],

$$E_{\mathrm{xc}} = \int d^3rn(\mathbf{r})\epsilon_{\mathrm{xc}}[n(\mathbf{r})]. \tag{6.52}$$

The expressions in Eqs. (6.43) to (6.50) are exact, and are applicable to *any* treatment of exchange and correlation. These equations are most commonly used within the context of the LDA, in which the exchange-correlation contribution to the energy is given in terms of the local electron density by Eq. (6.52). We note that evaluating the pressure using Eq. (6.51) may lead to numerically cumbersome procedures due to the large contributions to T and U made by the electrons in "core" states, i.e., states concentrated close to the nucleus with very little or no amplitude outside the MT sphere. However, we will see that both in the MT approximation and in the treatment of full cell potentials these terms can be made to cancel, which greatly facilitates the performance of numerical calculations. In addition, having two different expressions for the kinetic energy, Eqs. (6.44) and (6.45), allows a very stringent test on the precision of a calculation. It should also be noted that the kinetic energy expressions considered above provide only a part of the kinetic energy calculated within the LDA. Still to be added, is a contribution due to exchange and correlation, usually denoted by T_{xc}.

6.4.1 Total energy and pressure in the MT approximation

We recall that Eqs. (4.46) and (4.49) give, respectively, the expressions for the electrostatic energy per unit cell and the electrostatic potential acting on an electron at point $\mathbf{r}$. The effective one-electron potential is obtained by adding $v_{\mathrm{xc}}(\mathbf{r})$ to $v(\mathbf{r})$ of Eq. (4.49),

$$v(r) = \begin{cases} 0 \quad r > R \\ -\frac{2Z}{r} + 8\pi \int_0^r \left(\frac{r'^2}{r} - r' \right) n(r') \mathrm{d}r' \\ +8\pi \int_0^R r n(r) \mathrm{d}r + C \frac{Z_{\mathrm{out}}}{a} + v_{\mathrm{xc}}(\mathbf{r}) - v_{\mathrm{xc}}(n_0), \quad r \leq R, \end{cases} \tag{6.53}$$

where $v_{\mathrm{xc}}(n_0)$ is the exchange-correlation potential corresponding to the constant interstitial charge. It appears with a negative sign inside the MT sphere since the *entire* potential has been shifted so that the potential in the interstitial region can be set equal to zero. We note that the last equation is a special case of Eq. (6.49) appropriate to the case of MT potentials.

Now, using the relation, (for $r < R$)

$$(\mathbf{r} \cdot \nabla) v_{\mathrm{eff}}(\mathbf{r}) = r \frac{\mathrm{d}v_{\mathrm{eff}}(\mathbf{r})}{\mathrm{d}r} = \frac{2Z}{r} - 8\pi \int_0^R \mathrm{d}^3 r' r'^2 n(r') + (\mathbf{r} \cdot \nabla) v_{\mathrm{xc}}(\mathbf{r}), \tag{6.54}$$

we can write Eq. (4.46) in the form,

$$\begin{aligned} u &= -4\pi \int_0^{R^-} r^2 n(r)(\mathbf{r}\cdot\nabla)v_{\text{eff}}(\mathbf{r})\mathrm{d}r \\ &+ 4\pi \int_0^{R^-} r^2 n(r)(\mathbf{r}\cdot\nabla)v_{\text{xc}}(\mathbf{r})\mathrm{d}r - \frac{1}{2}C\frac{Z_{\text{out}}^2}{a}. \end{aligned} \tag{6.55}$$

The integrals in these expressions extend to just inside the MT sphere, as indicated by the upper limit R^-, and so they do not include contributions from the discontinuities of the potential there.

6.4.2 Kinetic energy of core states

It follows from Eq. (1.211) that the kinetic energy of the core states is given by the relation

$$t_{\text{c}} = \sum_{\text{c}} \epsilon_{\text{c}} - \int n_{\text{c}}(r)v_{\text{eff}}(r)\mathrm{d}r, \tag{6.56}$$

where $n_{\text{c}}(r)$ is the number density associated with the core states. At the same time, the virial theorem yields the expression, [Eq. (1.210) applied to core states]

$$2t_{\text{c}} = \int_\Omega n_{\text{c}}(r)(\mathbf{r}\cdot\nabla)v_{\text{eff}}\mathrm{d}^3r. \tag{6.57}$$

Used jointly, the last two relations provide a very stringent test of the precision of the calculation of the core states since it must be true that

$$\sum_{\text{c}} \epsilon_{\text{c}} - \int n_{\text{c}}(r)v_{\text{eff}}(r)\mathrm{d}r = \frac{1}{2}\int_\Omega n_{\text{c}}(r)(\mathbf{r}\cdot\nabla)v_{\text{eff}}\mathrm{d}^3r. \tag{6.58}$$

We note that due to the spherical symmetry of $v_{\text{eff}}(\mathbf{r})$, we also have

$$\begin{aligned} 2t_{\text{c}} &= \int_\Omega n_{\text{c}}(r)(\mathbf{r}\cdot\nabla)v_{\text{eff}}\mathrm{d}^3r \\ &= 4\pi \int_0^{R} r^2 n(r)\frac{\mathrm{d}v_{\text{eff}}(\mathbf{r})}{\mathrm{d}r}\mathrm{d}r. \end{aligned} \tag{6.59}$$

6.4.3 Pressure and total energy

The use of Eq. (6.59) allows a considerable simplification of the pressure expression in Eq. (6.51). Using subscripts c and v to denote the core and valence contributions, respectively, to various quantities, so that for example,

$$n(\mathbf{r}) = n_{\text{c}}(\mathbf{r}) + n_{\text{v}}(\mathbf{r}), \tag{6.60}$$

substituting Eq. (6.59) into Eq. (6.55) and canceling the core contribution, we find

$$\begin{aligned} 3P\Omega &= 2t_v - 4\pi\int_0^{R^-} r^3 n(r)\frac{dv_{\text{eff}}(\mathbf{r})}{dr}dr - \frac{1}{2}C\frac{Z_{\text{out}}^2}{a} \\ &+ 4\pi\int_0^{R^-} drr^3 n(r)\frac{dv_{\text{xc}}(r)}{dr} - 3\int_\Omega n(\mathbf{r})[\epsilon_{\text{xc}} - v_{\text{xc}}(\mathbf{r})]d^3r. \end{aligned} \tag{6.61}$$

Here, the valence contribution to the kinetic energy is given by

$$2t_v = \sum_v \epsilon_v - \int_\Omega n_v(\mathbf{r})v_{\text{eff}}(\mathbf{r})d^3r. \tag{6.62}$$

When we substitute this equation into Eq. (6.61), and integrate the term $\frac{dv_{xc}}{dr}$ by parts, we obtain

$$\begin{aligned} 3P\Omega &= 2\sum_v \epsilon_v - 4\pi\int_0^{R^-} drrn_v(r)\frac{d}{dr}[r^2 v_{\text{eff}}(r)] - \frac{1}{2}C\frac{Z_{\text{out}}^2}{a} \\ &- 3Z_{\text{out}}\left[\epsilon_{\text{xc}}(n_0) - v_{\text{xc}}(n_0)\right] - 4\pi R^3 n(R)\left[\epsilon_{\text{xc}}(R) - v_{\text{xc}}(R)\right]. \end{aligned} \tag{6.63}$$

The use of Eq. (6.59) also allows a simplification of the total energy per unit cell. From Eqs. (6.43) and (6.55), we find

$$\begin{aligned} e = E/N &= -2\pi\int_0^R drr^3 n_c(r)\frac{dv(r)}{dr} \\ &- 4\pi\int_0^{R^-} drr^3 n_v(r)\frac{dv(r)}{dr} + 4\pi\int_0^{R^-} drr^3 n_v(r)\frac{dv_{\text{xc}}(r)}{dr} \\ &- \frac{1}{2}C\frac{Z_{\text{out}}^2}{a} + \int_\Omega n(\mathbf{r})\epsilon_{\text{xc}}(\mathbf{r})d^3r. \end{aligned} \tag{6.64}$$

Finally, using Eq. (6.62) and integrating $\frac{dv_{xc}}{dr}$ by parts, we obtain [7]

$$\begin{aligned} e &= -2\pi\int_0^R drr^3 n_c(r)\frac{dv(r)}{dr} + \sum_v \epsilon_v \\ &- 4\pi\int_0^{R^-} drr^2 n_v(r)\frac{drv(r)}{dr} \\ &- \frac{1}{2}C\frac{Z_{\text{out}}^2}{a} + Z_{\text{out}}\epsilon_{\text{xc}}[n_0] \\ &+ 4\pi\int_0^R drn(r)[4\epsilon_{\text{xc}} - 3v_{\text{xc}}] - 4\pi R^3 n(R)\left[\epsilon_{\text{xc}}(R) - v_{\text{xc}}(R)\right]. \end{aligned} \tag{6.65}$$

One computational advantage of the last expression is that the core contribution to the energy is included in the first term, allowing a relatively easy numerical treatment.

6.5 Electronic Structure with Full-Cell Potentials

The formalism developed thus far has been used in a great number of calculations of the electronic structure of metals [8] in connection with potentials of the muffin tin form. The calculation of the electronic structure within MST in terms of space-filling cell potentials proceeds along the same lines as those followed in the MT approximation. The general scheme of a self-consistent calculation is that indicated in Fig. 7.1. For easy reference, we summarize the main steps of the calculation [9].

6.5.1 Total energy and potential

The total energy of a solid is written as the sum of the kinetic, potential, and exchange correlation terms, as in Eq. (6.43). In the LDA, these separate contributions are given in Eqs. (6.44), (6.46), and (6.52), with the density of the electrons given as a sum over the lowest occupied states, Eq. (6.47). The pressure is given by Eq. (6.51).

Now, the Coulomb potential at point $\mathbf{r}$ is given by the expression,

$$V_{\mathrm{c}}(\mathbf{r}) = \int \frac{n(\mathbf{r}')}{|\mathbf{r}-\mathbf{r}'|} \mathrm{d}^3 r' - \sum_n \frac{Z_n}{|\mathbf{r}-\mathbf{R}_n|}, \tag{6.66}$$

and, assuming N unit cells each of volume Ω in the crystal, the electrostatic energy per cell is given by

$$u = \int_\Omega n(\mathbf{r}) V_{\mathrm{c}}(\mathbf{r}) \mathrm{d}^3 r - \sum_n Z_n V_{\mathrm{M}}(\mathbf{R}_n). \tag{6.67}$$

Here, we have defined a generalized *Madelung potential,*

$$V_{\mathrm{M}}(\mathbf{R}_n) = \int \frac{n(\mathbf{r})\mathrm{d}^3 r}{|\mathbf{r}-\mathbf{R}_n|} - \sum_m{}' \frac{Z_m}{|\mathbf{R}_m-\mathbf{R}_n|}, \tag{6.68}$$

which is simply the Coulomb potential at $\mathbf{R}_n$ due to all charges in the material, except the nuclear charge at $\mathbf{R}_n$. The kinetic energy and the total energy are given by the general expressions in Eqs. (1.210) and (1.211).

It is convenient to cast the expression for the energy in a form that allows the exact cancellation of the Coulomb singularities in the potential.

Equation (1.211) can be written in the form (reverting to units where $e^2 = 1$)

$$\begin{aligned} E &= \sum_i \epsilon_i - \frac{1}{2}\Bigg[\int n(\mathbf{r})V_c(\mathbf{r})\mathrm{d}^3r \\ &+ \sum_n Z_n V_M(\mathbf{R}_n)\Bigg] - \int_\Omega n(\mathbf{r})v_{xc}(\mathbf{r})\mathrm{d}^3r + \int_\Omega n(\mathbf{r})\epsilon_{xc}(\mathbf{r})\mathrm{d}^3r. \end{aligned} \tag{6.69}$$

The core contributions cancel when the terms inside the square brackets are combined. To see this, expand both the cell charge density and the potential centered at $\mathbf{R}_n$ in spherical harmonics,

$$n(\mathbf{r}_n) = \sum_{\ell m} n_{\ell m}(r_n)Y_{\ell m}(\hat{r}_n), \tag{6.70}$$

and

$$V_c(\mathbf{r}_n) = \sum_{\ell m} V_{c,\ell m}(r_n)Y_{\ell m}(\hat{r}_n), \tag{6.71}$$

and consider the term inside the brackets centered at $\mathbf{R}_n$,

$$\begin{aligned} \int n(\mathbf{r})V_c(\mathbf{r})\mathrm{d}^3r + Z_n\int \frac{n(\mathbf{r}_n)\mathrm{d}^3r_n}{r_n} &= \int n(\mathbf{r}_n)\mathrm{d}^3r_n\left[V_c(\mathbf{r}_n) + \frac{Z_n}{r_n}\right] \\ &= \int \sum_{\ell m} n_{\ell m}(r_n)Y_{\ell m}(\hat{r}_n)\mathrm{d}^3r_n \\ &\times \left[\sum_{\ell' m'} V_{c,\ell' m'}(\hat{r}_n) + \frac{Z_n}{r_n}\right]. \end{aligned} \tag{6.72}$$

Now, the Coulomb singularity comes in only through the monopole term, $V_{c,00}(r_n)$, which can be written as the sum of a nuclear part and a smooth, nonsingular electronic contribution, $\hat{V}_{00}(r_n)$,

$$V_{c,00}(r_n) = -\sqrt{4\pi}\frac{Z_n}{r_n} + \hat{V}_{00}(r_n). \tag{6.73}$$

Clearly, the first term in this expression cancels against the second term inside the brackets in Eq. (6.72).

The various lattice sums appearing in the expression for the potential and the energy can be evaluated in a number of ways [10, 11], often geared to specific approaches taken to solve the Schrödinger equation. An expression for the energy per unit cell in terms of multipole moments was derived in Chapter 4. In this case, the lattice sums in connection with periodic solids can be evaluated by the use of lattice Fourier transforms of the structure constants, $S_{LL'}(\mathbf{k})$, defined in Eq. (4.14). These, in turn are obtained through Ewald-type methods, as is discussed in a later section of this chapter[4]. It should be pointed out that the electrostatic interaction between adjacent cells could be evaluated either through the use of conditional summations, or even double sums, Eqs. (4.15) and (4.16), or by direct integration over the domains of adjacent cells.

6.5.2 Band structure with space-filling cells

The band structure of a material described in terms of full-cell potentials is given by the solutions of the secular equation, Eq. (3.60), which can be written in the generic form

$$\det |\underline{m} - \underline{G}(\mathbf{k})| = 0. \tag{6.74}$$

As was discussed in Chapter 3, this equation can be put in various different forms, e.g., Eq. (3.119) or Eq. (6.74), which may possess different formal and computational characteristics. For example, Eq. (6.74) displays the slowest rate of convergence as it involves the inverse of the non-diagonal cell t-matrix. Eq. (6.75) has the advantage of converging relatively fast and of leading to a variational dependence of the energy with respect to the wave function. For the case of a periodic solid, the secular equation can be written in the form[5]

$$\det |\underline{\tilde{S}}\underline{C} - \underline{\tilde{S}}[\underline{G}(\mathbf{k})\underline{S}]| = 0. \tag{6.75}$$

Here, $\underline{S} \equiv \underline{S}^n$ and $\underline{C} \equiv \underline{C}^n$ are the sine and cosine matrices associated with the cell potential in the material, and a tilde denotes the transpose of a matrix. The brackets indicate that one of the two sums involving the structure constants should be carried to convergence before the other begins, and is valid when $R_{nm} > r$, for all vectors $\mathbf{r}$ inside a cell. If the last condition is violated, at least one of the two internal sums should be converted into a conditionally convergent double sum. In either case, one of the sums could be replaced with an integral over the surface of the adjacent cells, as discussed in detail in reference [12], to facilitate convergence.

Figure 6.8 shows the full-cell band structure [13] of elemental fcc Rh obtained through the use of square matrices. The evolution of the band structure is shown as a function of the maximum value of ℓ, from $\ell =$

---

[4]The lattice Fourier transform of $\underline{S}$ is the zero-energy limit of the KKR structure constants.

[5]Some forms of the secular equation may require the use of modified structure constants (similar to those that enter the discussion of the Poisson equation for space-filling charges) in order to lead to converged results in general.

1 to $\ell = 4$. It appears that there is essentially no change in the band structure after $\ell = 2$, with panels (c) and (d) being indistinguishable from one another, indicating a fairly rapid rate of convergence in this case. This rapid convergence may perhaps be expected due to the high rotational symmetry of the WS cell in an fcc lattice. Similar results were found in the case of elemental bcc Nb [14]. However, applications to crystal structures with low symmetry, e.g., the zincblende structure, should be made with care and with closely monitored tests of convergence.

6.5.3 Green function and Lloyd formula

Any of the forms of the MST Green function derived in Chapter 3 can be used in connection with space-filling cells, e.g., Eqs. (3.139), (3.141), or (3.142), as long as their use is consistent with the form of the secular equation. The imaginary part of the cell-diagonal element of the Green function yields the charge density, Eq. (3.149), and the imaginary part of the Green function in reciprocal space yields the Bloch function, (see discussion in Chapter 3). For both the cases of MT potentials and space-filling potential cells, the change in the integrated DOS brought about by the introduction of a potential into otherwise free space is given by the Lloyd formula, Eq. (3.163).

6.6 Comparison of MT and Full-Cell Calculations

The general procedure followed in a self-consistent determination of the electronic structure is the same within the MT approximation and the full-cell applications of MST. The use of MST allows the separation of the structural aspects of the system, embodied in the KKR structure constants of a lattice[6], from the potential aspects, reflected in the individual cell t-matrices, or the sine and cosine functions. Also, the use of the angular momentum representation leads in many cases, e.g., lattices with cubic symmetry, to rapidly convergent expansions in terms of low-dimensional, square matrices. The question of convergence becomes very important in the case of low-symmetry systems, where it may require the use of rectangular matrices (so as to facilitate the convergence of internal sums), and close monitoring. With this word of caution, a full-cell application of MST should make it possible to study reliably the electronic structure of impurities and other low-symmetry systems, such as surfaces, interfaces, and complex lattices. It should also allow accurate determination of elastic constants and

[6]Even if the modified structure constants must be used, they only depend on the geometry of the lattice and not on the potential.

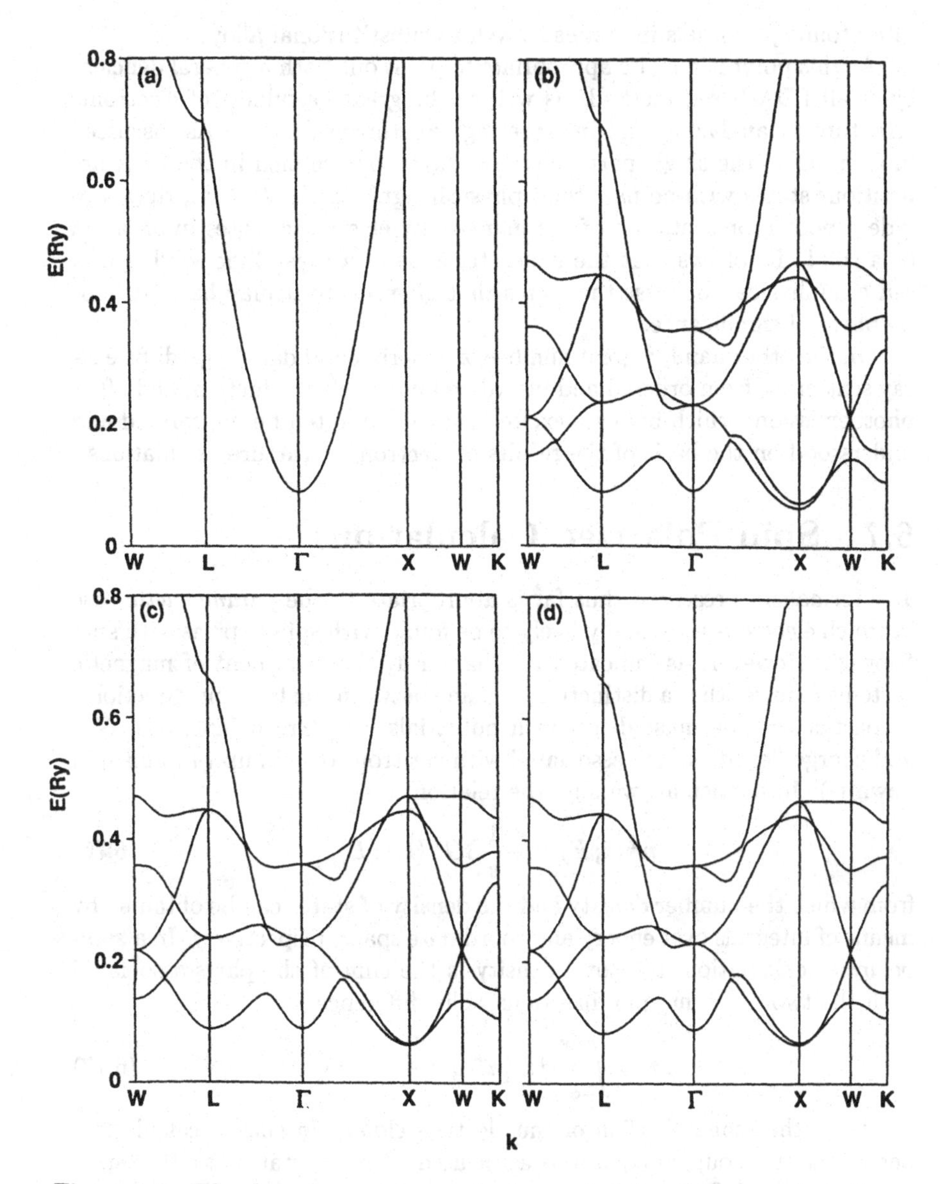

Figure 6.8: **The band structure of elemental fcc Rh as a function of ℓ. Panels (a), (b), (c), and (d) correspond to $\ell = 1$, 2, 3 and 4.**

interatomic potentials in metals as well as substitutional alloys.

At this point it may be appropriate to point out both a general deficiency of all LDA-based methods as well as the great usefulness of electronic structure in analyzing and interpreting experimental data. As discussed in Chapter 1, the single-particle states, $\phi_n(\mathbf{r})$, determined in the LDA are fictitious states with no particular physical significance. At best, they provide a poor representation of the many-body electronic states in an atom or a solid. It follows that the band structure of a crystalline solid is in a sense unphysical, because the states that give rise to it may have little or no physical significance.

On the other hand, a great number of experimental data, e.g., diffuse X-ray scattering from ordered and disordered materiel, conduction, and often photoemission (which involves excited states) can often be interpreted and understood on the basis of the results of electronic structure calculations.

6.7 Spin Polarized Calculations

The formalism presented thus far is appropriate to the *paramagnetic* case in which electrons are equally likely to be found with spin "up" as with spin "down". However, as indicated in Chapter 1, the treatment of magnetic systems requires that a distinction be made between the two spin directions. In particular, one must distinguish potentials, $v^{\pm}$, Green functions, $G^{\pm}$, and charge densities, $n^{\pm}$, associated with electrons of spin up (+) and spin down (-). In particular, we note the relation

$$n^{\pm}(\mathbf{r};E) = -\frac{1}{\pi}\Im G^{\pm}(\mathbf{r},\mathbf{r};E), \tag{6.76}$$

from which the number density and the density of states can be obtained by means of integrals over energy and coordinate space, respectively. In a *spin-polarized* calculation, the total density is the sum of the parts associated with the two different spin directions. The difference

$$M = \int_{-\infty}^{E_F} dE\,\left[n^{+}(E) - n^{-}(E)\right] \tag{6.77}$$

measures the magnetization of the electron cloud. In such a calculation, one solves two coupled equations associated with the states of either spin, with properly defined exchange-correlation potentials and effective single-particle Hamiltonians.

6.8 KKR Structure Constants(*)

The implementation of the MST equations requires the explicit use of the structure constants, $\underline{G}(\mathbf{R})$, defined in Eq. (F.11). The lattice Fourier transforms of these quantities are known as the KKR structure constants. These energy- and wave-vector dependent quantities need to be calculated only once for each lattice structure and used as necessary in subsequent calculations. It is important to realize that no separate calculations are needed for different lattice constants, because the KKR structure constants for different volumes are simply related, as is seen below. In this section we give a brief discussion of the evaluation of the KKR structure constants or, as they are also sometimes referred to, the structural Green function[7].

For the sake of clarity, we shall refer to the case of MT potentials when necessary to elucidate various notions in the discussion. The final expressions for the KKR structure constants are, of course, the same regardless of which approximation is used for the potential in the crystal[8]. From the defining equation of the free-particle propagator, Eq. (2.54), the Lippmann-Schwinger equation, Eq. (2.81), and the Bloch condition on the wave function, Eq. (6.3), we have

$$\psi_{\mathbf{k}}(\mathbf{r}) = \int \mathrm{d}^3r' G_{\mathbf{k}}(\mathbf{r}-\mathbf{r}')V(\mathbf{r}')\psi_{\mathbf{k}}(\mathbf{r}'), \tag{6.78}$$

where

$$\begin{aligned} G_{\mathbf{k}}(\mathbf{r}-\mathbf{r}') &= -\frac{1}{4\pi}\sum_{\mathbf{R}\neq 0}\frac{\mathrm{e}^{ik|\mathbf{r}-\mathbf{r}'+\mathbf{R}|}}{|\mathbf{r}-\mathbf{r}'+\mathbf{R}|}\mathrm{e}^{\mathrm{i}\mathbf{k}\cdot\mathbf{R}} \\ &= -\frac{1}{4\pi}\sum_{\mathbf{R}}\frac{\mathrm{e}^{ik|\mathbf{r}-\mathbf{r}'+\mathbf{R}|}}{|\mathbf{r}-\mathbf{r}'+\mathbf{R}|}\mathrm{e}^{\mathrm{i}\mathbf{k}\cdot\mathbf{R}} - G_0(\mathbf{r}-\mathbf{r}'). \end{aligned} \tag{6.79}$$

Because the free-particle propagator, $G_0(\mathbf{r}-\mathbf{r}')$, is periodic with respect to translations by direct lattice vectors, we can use the defining equation,

$$(\nabla^2 + E)G_0(\mathbf{r}-\mathbf{r}') = \delta(\mathbf{r}-\mathbf{r}'), \tag{6.80}$$

to write

$$\begin{aligned} -(\nabla^2+E)G_{\mathbf{k}}(\mathbf{r}-\mathbf{r}') &= \sum_{n\neq 0} -(\nabla^2+E)G_{\mathbf{k}}(\mathbf{r}-\mathbf{r}'-\mathbf{R}_n)\mathrm{e}^{\mathrm{i}\mathbf{k}\cdot\mathbf{R}_n} \\ &= \sum_{n\neq 0} -\delta(\mathbf{r}-\mathbf{r}'-\mathbf{R}_n)\mathrm{e}^{\mathrm{i}\mathbf{k}\cdot\mathbf{R}_n} \\ &= \sum_{n} -\delta(\mathbf{r}-\mathbf{r}'-\mathbf{R}_n)\mathrm{e}^{\mathrm{i}\mathbf{k}\cdot(\mathbf{r}-\mathbf{r}')} + \delta(\mathbf{r}-\mathbf{r}') \end{aligned}$$

[7] The term structural Green function is also used to denote various forms involving the scattering path operator.

[8] The possible need to use modified structure constants for some forms of the secular equation or in connection with non-cubic cells should be kept in mind.

$$
\begin{aligned}
&= -\frac{1}{v_c}\sum_n e^{i\mathbf{K}_n\cdot(\mathbf{r}-\mathbf{r}'-\mathbf{R}_n)}e^{i\mathbf{k}\cdot(\mathbf{r}-\mathbf{r}')} + \delta(\mathbf{r}-\mathbf{r}') \\
&= -\frac{1}{v_c}\sum_n e^{i(\mathbf{k}+\mathbf{K}_n)\cdot(\mathbf{r}-\mathbf{r}')} + \delta(\mathbf{r}-\mathbf{r}'), \qquad (6.81)
\end{aligned}
$$

so that

$$
G_{\mathbf{k}}(\mathbf{r}-\mathbf{r}') = \sum_n \frac{1}{v_c}\frac{e^{i\mathbf{k}_n\cdot(\mathbf{r}-\mathbf{r}')}}{\mathbf{k}_n^2 - E} - G_0(\mathbf{r}-\mathbf{r}'), \qquad (6.82)
$$

where

$$
\mathbf{k}_n = \mathbf{k} + \mathbf{K}_n \qquad (6.83)
$$

with $\mathbf{K}_n$ being a recip, and v_c the volume of a unit cell. The third line in Eq. (6.81) follows because of the presence of the delta function, and the last line from the definition of the reciprocal lattice vectors, $\mathbf{K}_n$. Now, in Eq. (6.82) we use Bauer's identity, Eq. (D.26),

$$
e^{i\mathbf{k}_n\cdot\mathbf{r}} = 4\pi\sum_L i^\ell j_\ell(k_n r) Y_L(\hat{r}) Y_L^*(\hat{k}_n), \qquad (6.84)
$$

the expansion of the free-particle propagator, Eq. (2.67),

$$
\begin{aligned}
G_0(\mathbf{r}-\mathbf{r}') &= -ik\sum_L J_L(k\mathbf{r}_<)H_L(k\mathbf{r}_>) \\
&= k\sum_L \left[J_L(k\mathbf{r}_<)N_L(k\mathbf{r}_>) - iJ_L(k\mathbf{r})J_L(k\mathbf{r})\right], \qquad (6.85)
\end{aligned}
$$

where the relation between the Bessel, Neumann, and Hankel functions, $h_\ell = j_\ell + in_\ell$ has been used, and compare with the expression

$$
G_0(\mathbf{r}-\mathbf{r}') = \sum_{LL'} i^{\ell-\ell'} J_L(k\mathbf{r}) G_{LL'}(E,\mathbf{k}) J_{L'}(k\mathbf{r}'), \qquad (6.86)
$$

which is valid in the MT case, and obtain the result

$$
G_{LL'}(E,\mathbf{k}) = -ik\delta_{LL'} - B_{LL'}(E,\mathbf{k}), \qquad (6.87)
$$

where

$$
\begin{aligned}
B_{LL'}(E,\mathbf{k}) &= -\frac{(4\pi)^2}{v_c}\sum_n \frac{[j_\ell(k_n r) j_{\ell'}(k_n r')] Y_L(\hat{k}_n) Y_{L'}(\hat{k}_n)}{j_\ell(kr) j_{\ell'}(kr')(k_n^2 - E)} \\
&- k\frac{n_\ell(kr)}{j_\ell(kr')}\delta_{LL'}. \qquad (6.88)
\end{aligned}
$$

The single summation over n in the last expression converges very slowly. The evaluation of the structure constants [15, 16] is consequently carried out using alternative methods based for the most part on the *Ewald identities*. With c denoting a contour in the complex ξ plane, we have

$$\frac{\mathrm{e}^{ikR}}{R} = \frac{2}{\sqrt{\pi}} \int_{0(c)}^{\infty} \mathrm{e}^{[-R^2\xi^2+k^2/4\xi^2]} \mathrm{d}\xi, \tag{6.89}$$

and

$$\sum_n \mathrm{e}^{-(\mathbf{r}-\mathbf{R}_n)^2\xi^2+\mathrm{i}\mathbf{k}\cdot(\mathbf{R}_n-\mathbf{r})} = \frac{\pi^{3/2}}{v_c\xi^2} \sum_n \mathrm{e}^{-k_n^2/4\xi^2+\mathrm{i}\mathbf{k}_n\cdot\mathbf{r}}. \tag{6.90}$$

These identities allow the evaluation of the structure constants through a partition into separate contributions in direct and reciprocal space. Starting with Eq. (6.79) and choosing the contour c to run along the real axis beyond $\xi = \sqrt{\eta}/2$, where η is called the *Ewald parameter*, we obtain

$$G_{\mathbf{k}}(\mathbf{R}) - G_0(\mathbf{R}) = G_1(\mathbf{R}) + G_2(\mathbf{R}), \tag{6.91}$$

where

$$G_1(\mathbf{R}) = -\frac{1}{2\pi^{3/2}} \int_{\sqrt{\eta}/2}^{\infty} \mathrm{d}\xi \sum_n \mathrm{e}^{\mathrm{i}\mathbf{k}\cdot\mathbf{R}_n-(\mathbf{R}_n-\mathbf{R})^2\xi^2+\frac{E}{4\xi^2}}, \tag{6.92}$$

is the summation in real space, and

$$G_2(\mathbf{R}) = -\frac{1}{v_c} \sum_n \frac{\mathrm{e}^{\mathrm{i}\mathbf{k}_n\cdot\mathbf{R}-(k_n^2/\eta)+E/\eta}}{\mathbf{k}_n^2 - E}, \tag{6.93}$$

is the summation in reciprocal space. The last two expressions can be expanded in spherical harmonics, leading to the result [15, 16],

$$B_{LL'} = 4\pi \sum_{L''} D_{L''} C(L'', L', L), \tag{6.94}$$

where $C(L'', L', L)$ is a Gaunt number (the integral of three spherical harmonics)

$$C(L'', L', L) = \int \mathrm{d}\hat{r} Y_{L''}^*(\hat{r}) Y_{L'}(\hat{r}) Y_L(\hat{r}), \tag{6.95}$$

and

$$D_L = D_L^{(1)} + D_L^{(2)} + D_0^{(3)}\delta_{L0}, \tag{6.96}$$

where

$$D_L^{(1)} = -\left(\frac{4\pi}{v_c}\right)\frac{1}{k^\ell}\exp\left(\frac{E}{\eta}\right)\sum_n \frac{k_n^\ell \mathrm{e}^{-k_n^2/\eta}}{k_n^2 - E} Y_L(\hat{k}_n), \tag{6.97}$$

$$D_L^{(2)} = \frac{(-2)^{\ell+1}\mathrm{i}^\ell k^{-\ell}}{\sqrt{\pi}} \sum_n R_n^\ell \mathrm{e}^{\mathrm{i}\mathbf{k}\cdot\mathbf{R}_n} Y_L(\hat{R}_n) \int_{\sqrt{\eta}/2}^{\infty} \xi^{2\ell} \mathrm{e}^{-\xi^2 R_n^2+\frac{E}{4\xi^2}} \mathrm{d}\xi, \tag{6.98}$$

and

$$D_0^{(3)} = -\frac{\sqrt{\eta}}{2\pi}\sum_{n=0}^{\infty}\frac{(E/\eta)^n}{n!(2n-1)}. \tag{6.99}$$

It is to be noted that the various sums in the last three expressions converge very rapidly, and that they become independent of η in the converged limit. The structure constants for a given structure are clearly independent of the potential. They can be stored as functions of wave vector and energy, and used as necessary in a given application. Finally, it follows from Eq. (6.82) that the ratio of the structure constants corresponding to two different lattice constants of a given structure is equal to the inverse ratio of the cell volumes, or

$$\frac{\underline{G}_1(E,\mathbf{k})}{\underline{G}_2(E,\mathbf{k})} = \left(\frac{a_2}{a_1}\right)^3. \tag{6.100}$$

6.9 Complex Lattices

Many elemental solids and ordered compounds form structures with more than one atom per unit cell. These are commonly referred as *complex lattices* and the calculation of their band structure can become quite involved. In this case, the matrix $\underline{M}$ in Eq. (3.126) becomes complex and its dimensions increase by a factor n_b, where n_b is the number of atoms in the unit cell. Correspondingly computational times, which for matrix inversions are proportional to the third power of the dimensionality of a matrix, also increase. However, apart from these technical considerations, the general features of MST remain the same.

Let the unit cell contain n_b atoms at positions $\mathbf{a}_j$ with respect to the center of the cell. The matrix $\underline{M}$ now has indices in $\mathbf{a}_j$ as well as in L, assuming the form

$$\underline{M}^{ij}_{LL'}(\mathbf{k}) = m^i_{LL'}\delta_{ij} - G_{LL'}(\mathbf{a}_{ij})(1-\delta_{ij}) - G^{ij}_{LL'}(\mathbf{k}). \tag{6.101}$$

Here, $m^i_{LL'}$ is the inverse scattering matrix for the ith potential in the cell, $G_{LL'}(\mathbf{a}_{ij})$ is the real-space structure constant connecting the atoms at positions $\mathbf{a}_i$ and $\mathbf{a}_j$ in the unit cell, and $G^{ij}_{LL'}(\mathbf{k})$ is the KKR structure constant between positions i and j. For example, for a complex lattice containing two atoms per unit cell, one at the origin of the cell and one at $\mathbf{a}$, we have

$$\underline{M} = \left[\begin{pmatrix} \underline{m}^1 & -\underline{G}(\mathbf{a}) \\ -\underline{G}(-\mathbf{a}) & \underline{m}^2 \end{pmatrix} - \begin{pmatrix} \underline{G}^{11}(\mathbf{k}) & \underline{G}^{12}(\mathbf{k}) \\ \underline{G}^{21}(\mathbf{k}) & \underline{G}^{22}(\mathbf{k}) \end{pmatrix}\right]. \tag{6.102}$$

Here, the first matrix inside the square brackets corresponds to the $\underline{M}$ matrix of the isolated diatomic basis. The structure constants indicate that this inverse matrix is repeated with the period of the underlying lattice. In particular, we have

$$\underline{G}^{ij}(\mathbf{k}) = \sum_{\mathbf{R}} e^{i\mathbf{k}\cdot\mathbf{R}} \underline{G}(\mathbf{R} + \mathbf{a}_j - \mathbf{a}_i). \tag{6.103}$$

The structure constants of a complex lattice can be evaluated in a generalization of the procedure used with simple lattices [16]. The secular equation for determining the band structure has the usual form

$$\det \underline{M} = 0, \tag{6.104}$$

and, analogous to its simple-lattice counterpart, can be put into a variety of forms with individual analytic and computational properties. Finally, the scattering path operator and the Green function can be obtained as straightforward generalizations of the expressions for simple lattices.

Bibliography

[1] P. H. Dederichs, in *Festkörperforschung für die Informationstechnik*, IFF-Ferienkurs, Forschungszentrum Jülich, GmbH, Institut für Festkörperforschung, Band 21, pp. 4.1-4.38.

[2] O. K. Andersen, p. 59 in *Canonical Description of the Band Structure of Metals*, in Proceedings of the International School of Physics, "Enrico Fermi", Course LXXXIX, Edited by F. Bassani and F. Fumi, Italian Physical Society (1985).

[3] O. Jepsen, J. Madsen, and O. K. Andersen, J. Mag. Magn. Mater. **15-18**, 867 (1980); Phys. Rev. B**26**, 2790 (1982)

[4] Prabhakar P. Singh, and A. Gonis, Phys. Rev. B**48**, 2139 (1993).

[5] Walter A. Harrison, *Electronic Structure and The Properties of Solids*, Dover, New York (1989).

[6] Neil W. Aschcroft and N. David Mermin, *Solid State Physics*, Saunders College, PA, (1976).

[7] J. F. Janak, Phys. Rev. B**9**, 3985 (1974).

[8] V. L. Moruzzi, J. F. Janak, and A. R. Williams, *Calculated Electronic Properties of Metals*, Pergamon Press, Inc. New York (1978).

[9] M. Weinert, E. Wimmer, and A. J. Freeman, Phys. Rev. B**26**, 4571 (1982).

[10] Terry Loucks, *Augmented Plane Wave Method*, W. A. Benjamin, Inc. New York, Amsterdam (1967).

[11] M. Weinert, J. Math. Phys. **22**, 2433 (1981).

[12] A. Gonis and W. H. Butler, *Multiple Scattering Theory in Solids*, Springer Verlag, Berlin (1999).

[13] A. Gonis, X. -G. Zhang, and D. M. Nicholson, Phys. Rev. B**40**, 947 (1989).

[14] A. Gonis, X. -G. Zhang, and D. M. Nicholson, Phys. Rev. B**38**, 3564 (1988).

[15] B. Segall, Phys. Rev. **105**, 108 (1957).

[16] F. S. Ham and B. Segall, Phys. rev. **124**, 1786 (1961).

Chapter 7

Calculation of Properties

7.1 General Comments

The discussion in the previous chapter laid the groundwork for the determination of the electronic structure of a material in its ground state (at zero temperature). That discussion was based on the knowledge of the crystal potential for fixed nuclear positions — the Born-Oppenheimer approximation. In a realistic study of materials properties it is necessary to follow a self-consistent procedure in determining the potential in the solid, as well as to consider the effects of elevated temperatures and nuclear displacements away from equilibrium. For example, the study of the electronic specific heat is directly tied into the response of the electron cloud to rising temperature. The study of elastic properties and thermal expansion within an electronic structure approach requires the treatment of the response of the electron cloud to nuclear displacements. Obviously, elevated temperatures are a basic cause of lattice vibrations and a study of the lattice specific heat — to be discussed in a subsequent chapter — must be based on the combined and interrelated effects of temperature and atomic displacement.

In this chapter, we provide an outline of the procedure used to study materials properties, such as those just mentioned, based on the determination of the electronic structure of a material. As we have done in previous chapters, we phrase the discussion in terms of multiple scattering theory. However, in order to emphasize the central role played by the electronic structure — rather than a particular method used in its determination — we will quote results obtained by other techniques. It is hoped that the reader will be motivated to study such other methods and identify those that best suit particular applications.

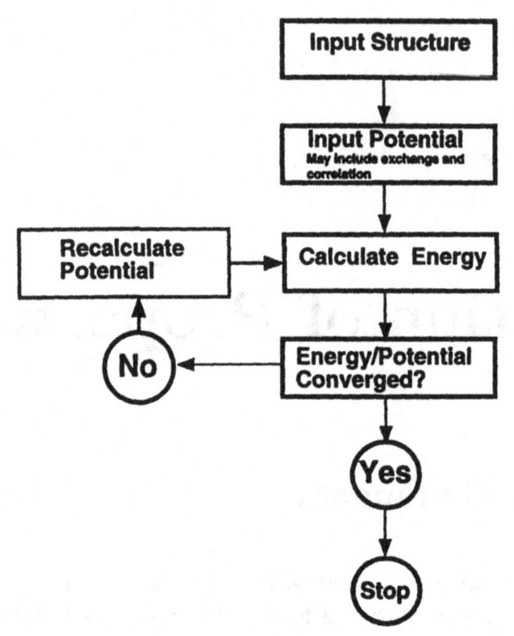

Figure 7.1: **Schematic representation of the self-consistency loop in an electronic structure calculation.**

7.2 Self-Consistent Calculations

For fixed nuclear positions and fixed (input) values for the charge density and hence for v_{eff}, the formalism of the previous chapter allows the determination of the total energy of the system. However, this energy is not necessarily a minimum, i.e., it does not correspond to the ground state of the system. The most common use of LDA-based methods is in connection with the identification of the structure of a material in its ground state, i.e., at $T = 0^{\circ}$K. In these calculations, the most stable structure is sought from a set of likely structures. The most stable structure is determined by minimizing the total energy of the system through a self-consistent, iterative process that terminates when the changes in the calculated values of various relevant parameters, e.g., energy, potential, fall within preascribed limits. This self consistent process is outlined in Fig. 7.1.

As is indicated in the figure, from the input charge density (and potential), one obtains the Green function [or the single-particle orbitals $\phi_n(\mathbf{r})$] which in turn leads to a new value for the charge density. This is used to calculate a new value of the potential, and completes the self-consistency loop. Therefore, it is necessary to have a computational procedure for the evaluation of the single-particle Green function (or the single particle s-tates). One approach for obtaining the Green function is based on MST, as described in Chapter 3.

In order to apply the formalism of MST to a self-consistent determination of the electronic structure, we consider the potential, $v_{\text{eff}}(\mathbf{r})$, to be the sum of individual cell potentials,

$$v_{\text{eff}} = \sum_n v_n(\mathbf{r}). \tag{7.1}$$

In the MT approximation, for example, each cell-potential $v_n(\mathbf{r})$ is taken to be spherically symmetric inside the MT sphere and to vanish in the interstitial region, as described in Eq. (6.53). The effect of each spherical potential on an incident wave is described by the cell scattering matrix, (or scattering amplitude), Eq. (D.62). These scattering matrices can be used in building the scattering matrix of the system, the scattering path operator in Eqs. (3.9), (3.52), or (3.53). For a periodic system, the use of lattice Fourier transforms allows the solution of the equation of motion in the form of Eq. (3.59). Integration of $\underline{\tau}(\mathbf{k})$ over the first BZ of a crystal leads directly to the site-diagonal element, $\underline{\tau}^{00}$, and hence to the associated Green function, Eq. (3.77), and to the charge density. This determines the output charge density, $n^{\text{out}}(\mathbf{r})$, for the current self-consistency loop. This new charge can be put into the MT form, and can be used to determine the new effective single-particle potential, $v_{\text{eff}}^{\text{out}}(r)$. If these output values for the density and the potential are within some predetermined tolerance of the input values, the self-consistency loop terminates, as indicated in Fig. 7.1. If not, the output values of the density (and the potential) become the input values for the next iteration[1].

Provided that self-consistency is reached, the value of the total energy corresponding to the value of the lattice parameter used in the calculation, $E(a)$, is obtained. The value of a at which the total energy exhibits a minimum (and where the pressure vanishes) determines[2] the equilibrium (ground) state of the system of electrons and nuclei.

In the following sections, we give illustrative examples of the self-consistent electronic structure calculations for metallic systems based on the KKR method within the MT approximation, as well as other techniques.

---

[1]The rate of convergence of the self-consistency loop is heavily dependent on the scheme used to mix the input and output charge densities.

[2]Although formally equivalent, calculations based on energy minimization or the vanishing of the pressure often yield slightly different values for equilibrium volumes. Because of the variational principles that govern the behavior of the energy with respect to density changes, calculations based on energy minimization are usually more reliable.

7.3 Application to Metallic Systems

7.3.1 Band structure and related properties

The solutions of the secular equation, Eq. (3.60), yield the *band structure* of a material, i.e., the variation of energy with wave vector inside the first BZ of the reciprocal lattice. The band structure is usually depicted as a set of curves, corresponding to the various bands, showing the dependence of the energy on **k** along various directions inside the zone. The band structure [1] of elemental Li, (atomic number 3), and of (non-magnetic) Ni (atomic number 28) are shown in Fig. 7.2. We note the presence of nearly parabolic, free-electron like bands at low energy and near the center of the zone, (the Γ point). These bands arise from the states with s symmetry in the atoms, and in the case of Li remain essentially free-electron like at energies above the Fermi level, denoted by E_F in the figure. On the other hand, *hybridization effects* that affect the mixing of symmetries in the solid, are very strong in the case of Ni. Because of such effects, the nearly free-electron bands in Ni exhibit gaps and a strong mixing with the narrow bands arising from electrons in atomic d states.

The density of states (DOS) for each metal are shown by the solid curves in the lower panels. These curves were calculated through the use of Eq. (6.17). Consistent with the free-electron like character of the band structure, the DOS in the case of Li is also essentially that of free electrons, displaying the characteristic $\sqrt{E}$ dependence on energy as expected from Eq. (2.70). On the other hand, the DOS for Ni is dominated by the contribution from d states. These states are much more localized around the nucleus than the s and p states and give rise to narrow bands and sharp peaks in the density of states.

Both DOS curves exhibit the presence of *van Hove singularities*. These arise when the gradient of $E_n(\mathbf{k})$ vanishes, causing the integrant in Eq. (6.17) to exhibit divergent behavior. In one-dimension, van Hove singularities are non-integrable, leading to DOS that become infinite at band edges [2]. In three dimensions, the singularities are integrable[3]. They lead to finite values of the DOS but to infinite values of the slope, as is illustrated in Fig. 7.3.

The dotted curve in the lower panel of Fig. 7.2 shows the variation with energy of the integrated density of states, (IDOS),

$$N(E) = \int_0^E n(E')\mathrm{d}E'. \tag{7.2}$$

[3]In three dimensions, it is unlikely to obtain divergence at a given point along all directions, which leads to integrable singularities.

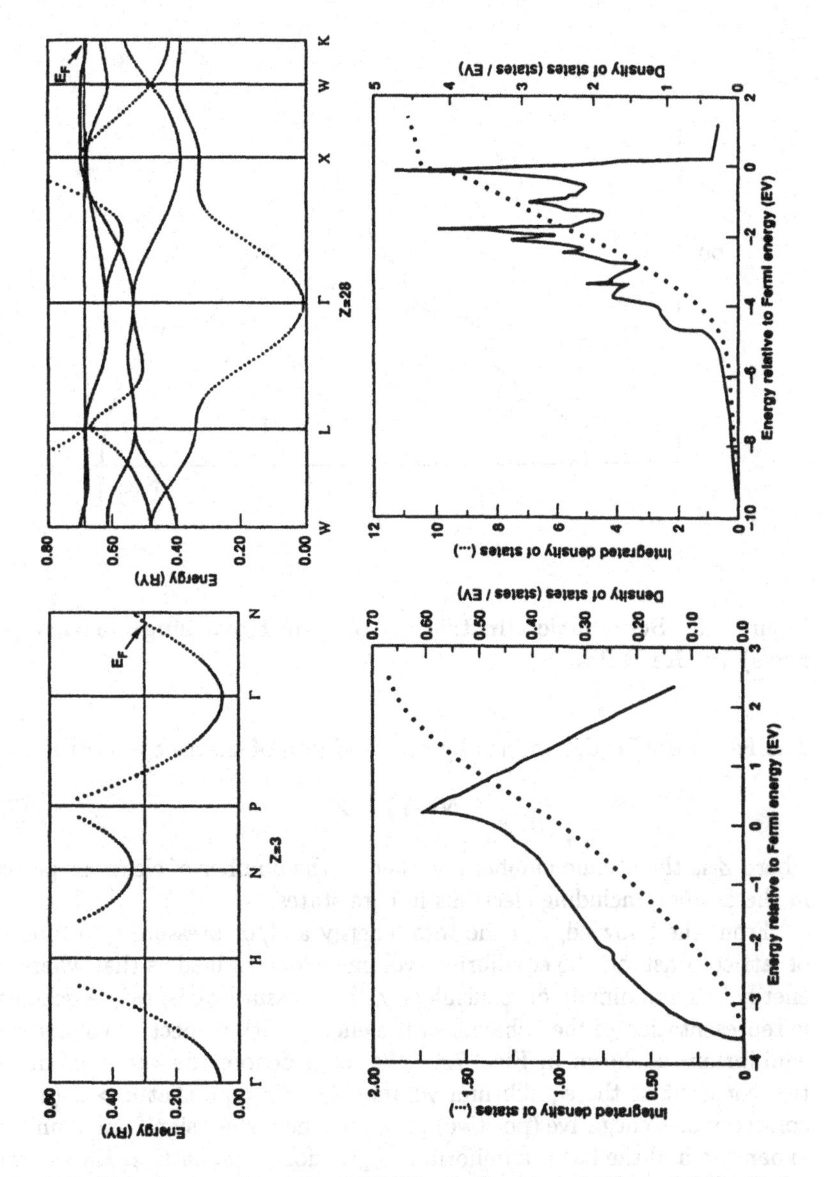

Figure 7.2: **Band structure and DOS of elemental Li and Ni. Calculations from reference [1].**

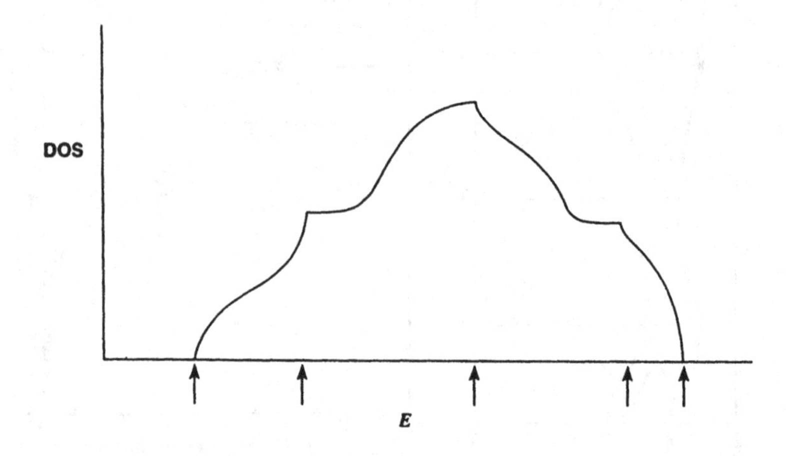

Figure 7.3: **Schematic illustration of van Hove singularities (arrows) in the DOS.**

The Fermi level is determined by the condition of charge conservation,

$$N(E_{\mathrm{F}}) = Z, \tag{7.3}$$

where Z is the atomic number and, hence, the number of electrons per cell in the system, including electrons in core states.

From the knowledge of the total energy and/or pressure as a function of lattice constant, the equilibrium volume is determined as that where the energy has a minimum or, equivalently, the pressure vanishes[4]. A schematic representation of the behavior of the energy with respect to volume near equilibrium is shown in Fig. 7.4. We let $\bar{a}$ denote the value of the lattice constant at the equilibrium volume V_0. We note that $a > \bar{a}$ ($a < \bar{a}$) corresponds to negative (positive) pressure, the former signifying a uniform expansion and the latter a uniform compression of the lattice. Energy minimization for different competing crystal structures can lead to predictions about the stable structure of a solid in its ground state. This is illustrated in Fig. 7.5, which shows the energy-volume curves for two different arrangements of an ordered Cu-Pt alloy containing equal numbers of Cu and Pt atoms. The results clearly indicate that the $L1_1$ configuration has the lowest energy which is in agreement with experiment.

[4]Because of numerical inaccuracies and other approximations, the minimum of the

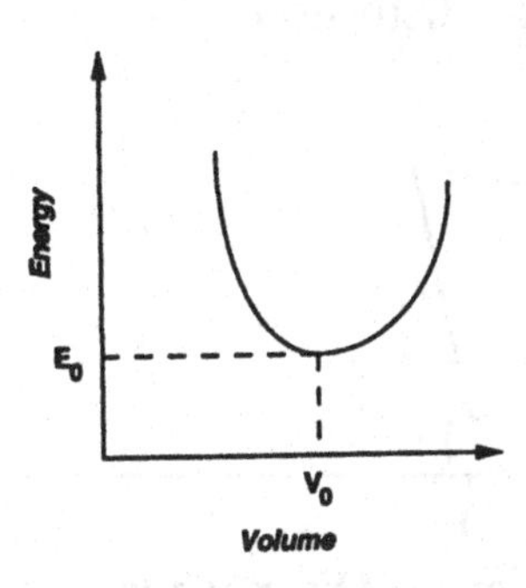

Figure 7.4: **Schematic representation of the variation of the energy with respect to volume near the equilibrium value of the lattice parameter of a crystalline material.**

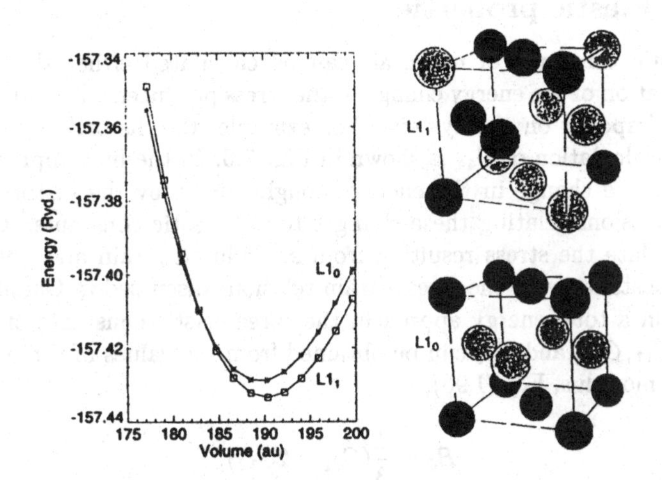

Figure 7.5: **Energy-volume curves for the $L1_0$ and $L1_1$ structures of CuPt. The crystal structures are also shown, with Cu atoms represented by filled circles and Pt atoms by open circles. We note the different stacking arrangement between the two structures. The calculations indicate that $L1_1$ has the lowest energy in agreement with experiment. Volume is expressed in atomic units (au), such that 1 au$\approx$ 0.529 Angströms**

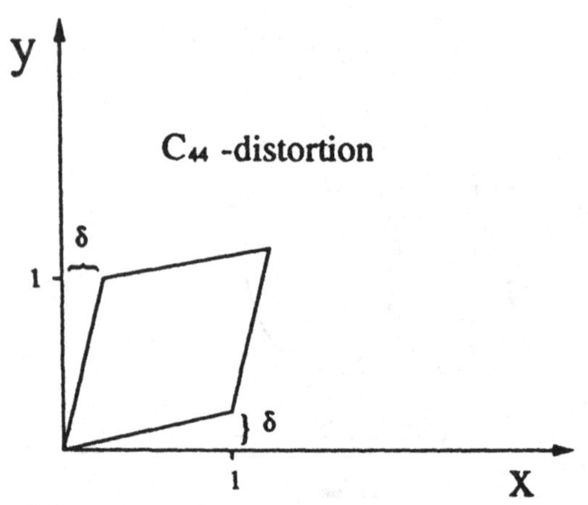

Figure 7.6: **The distortion in the x-y plane for the calculation of C_{44}. The transformation is volume conserving and it corresponds to a cell being compressed/elongated in the z-direction. From reference [8].**

7.3.2 Elastic properties

The elastic constants of materials can be calculated either through the determination of the energy change or the stress produced when an external strain is imposed on the system. For example, the distortion associated with the calculation of C_{44} is shown in Fig. 7.6. In the first approach, one calculates the change in the energy brought about by the distortion and uses expressions relating these changes to the elastic constants. One can also calculate the stress resulting from an induced strain and obtain the elastic constants from the stress-strain relations discussed in Chapter 5.

Within a total-energy approach, the three elastic constants of a cubic lattice, C_{11}, C_{12}, and C_{44} can be obtained from the values of three moduli, the bulk modulus, Eq. (I.90),

$$B_0 = \frac{1}{3}(C_{11} + 2C_{12}), \tag{7.4}$$

the tetragonal modulus,

$$C' = \frac{1}{2}(C_{11} - C_{12}), \tag{7.5}$$

and the trigonal shear modulus,

$$B_t = C_{44}. \tag{7.6}$$

---

energy may be somewhat displaced from the zero of the pressure.

Material	a (Bohr)	B_0	C'	C_{44}
Mo (cal)	5.914	2.864	1.743	2.214
(exp)	5.935	2.653	1.386	1.250
TiC (cal)	8.150	2.557	2.582	2.173
(exp)	8.161	2.417	2.035	1.780
MgO (cal)	7.985	1.603	0.670	1.030
(exp)	7.958	1.650	1.066	1.586

Table 7.1: **Experimentally assessed and calculated equilibrium lattice parameters, bulk moduli, and elastic constants for some elemental Mo, TiC, and MgO. See reference [3].**

The variation of the total energy with volume (lattice parameter), e.g., under isotropic (or hydrostatic) pressure, gives the bulk modulus, Eq. (5.8). Similar variations of the energy under tetragonal and shear distortions give the other two moduli allowing the determination of the elastic constants.

However, computing the changes in energy brought about by small distortions of the lattice places heavy demands on the precision of a band structure calculation. The changes in energy are typically of the order of milli Rydbergs (mRy), whereas the total energy of a solid may be of the order of several tens of thousands of Ry. The calculation of such small changes requires accurate **k**-space integration schemes, and often the treatment of the entire potential in a cell without approximation to its shape. In particular the MT approximation can fail rather badly in such applications.

As an illustration of the difficulties involved and the effects of the MT approximation, we summarize in Table 7.3.2 [3] the results of the equilibrium lattice parameter, the three moduli, and the elastic constants of Mo, TiC, and MgO, carried out within an application of the KKR method within a MT approximation to the potential. The results of the calculations are also compared to experimental data.

In all cases, the calculated equilibrium lattice parameters and bulk moduli correspond very closely to the experimental values. However, for the shear modulus, C', and the elastic constants, the agreement is less satisfactory. As indicated above, this may be understood on the basis of the types of distortions needed to calculate the various quantities. The calculation of lattice parameters and bulk moduli involves isotropic expansions or compressions that do not change the relative positions of the atoms on the lattice. In this case, the MT approximation can remain a viable procedure. The calculation of the shear modulus, on the other hand, involves distortions that significantly alter the shape of the unit cells in a material. The treatment of such distortions would require a full-cell treatment and cannot be addressed adequately within a MT approximation.

The increased accuracy that can be achieved with a proper treatment of the potential is indicated in Figs. 7.7 and 7.8. The figures show the elastic constants C_{44} and C′ for the elements in the 3d, 4d, and 5d transition series calculated within a full-potential LMTO approach. For a more detailed discussion of the calculations whose results are presented here, we refer to the literature [4, 5, 6].

7.4 Electronic Specific Heat

In this section, we discuss the behavior of the electron density as a function of temperature and derive an expression for the specific heat of the electron gas near $T = 0^o$K.

Let the electron density corresponding to a wave vector $\mathbf{k}$ in the ground state of the system be denoted by $n^0_{\mathbf{k}}$, and consider the deviations, $\delta n_{\mathbf{k}}$, caused by small temperature increases above zero. We write,

$$n_{\mathbf{k}} = n^0_{\mathbf{k}} + \delta n_{\mathbf{k}}. \tag{7.7}$$

The density describes the distribution of electrons in energy at a given temperature and is consequently given by the Fermi-Dirac distribution function,

$$n_{\mathbf{k}} = [1 + \exp[\epsilon_{\mathbf{k}} - E_F/k_B T]]^{-1}, \tag{7.8}$$

where $\epsilon_{\mathbf{k}}$ denotes the energy eigenvalue for the state $\mathbf{k}$, (given as a solution to the secular equation for the solid as discussed in the previous chapter), E_F is the Fermi energy, and k_B is Boltzmann's constant. Now, the group velocity of a particle in state $\mathbf{k}$ is given by the relation,

$$v_{\mathbf{k}} = \frac{1}{\hbar}\frac{\partial \epsilon_{\mathbf{k}}}{\partial \mathbf{k}}. \tag{7.9}$$

For an isotropic system, we can also write the velocity in the form,

$$v_{\mathbf{k}} = \frac{\hbar \mathbf{k}}{2m^*}. \tag{7.10}$$

In this expression, the quantity m^* is the *effective mass* characterizing a particle in state $\mathbf{k}$. This quantity is in general different than the free electron mass because the response of the particle to an external field is influenced by its interaction with all other particles. It is customary to use the term *quasi-particle* to describe an entity that behaves essentially as a single particle, e.g., its velocity can be written in the free-particle form of the last expression, but is characterized by values of internal quantities, e.g., mass, which may be different from those of free particles[5]. Treating a

[5]The notion of quasi-particle is discussed further in the next chapter

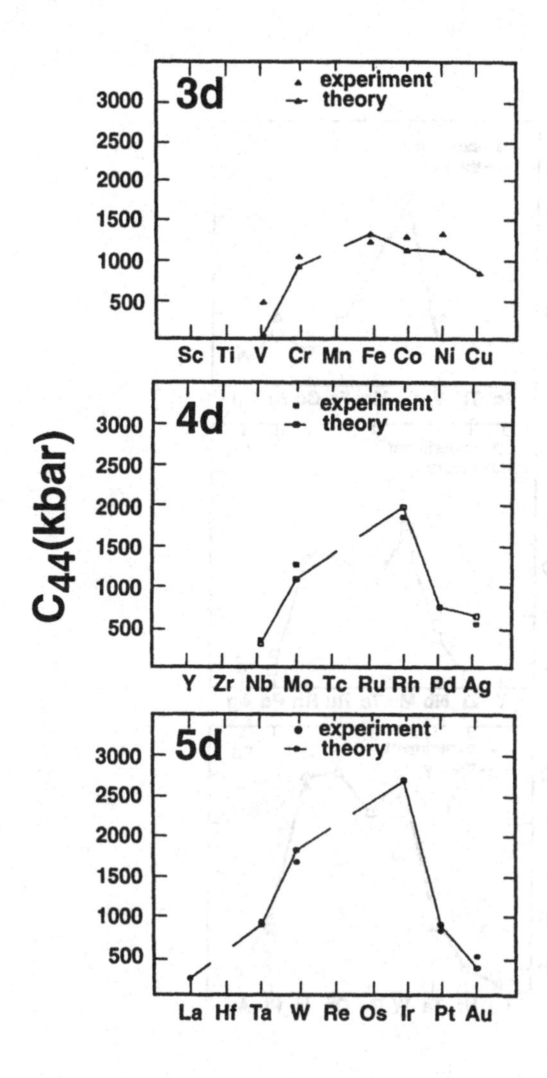

Figure 7.7: **Calculated values of C_{44} compared with experimental results for the elements in the 3d (upper panel), 4d (middle panel), and 5d transition metals. No data are available for the hexagonal structures. See reference [4].**

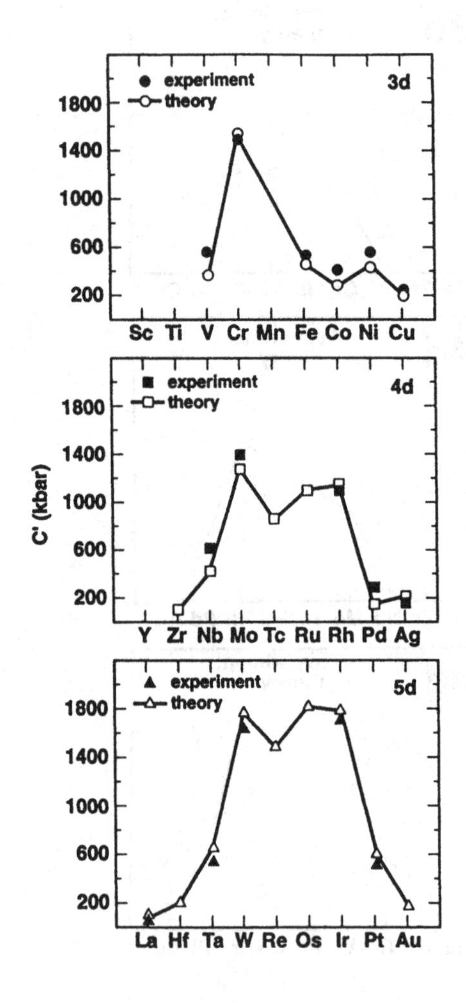

Figure 7.8: **Calculated values of C' compared with experimental results for the elements in the 3d (upper panel), 4d (middle panel), and 5d transition metals. For the hexagonal metals, an fcc structure is assumed in order to demonstrate trends [4].**

quasi-particle as an independent particle, we have,

$$n(E_{\mathrm{F}}) = \frac{\Omega k_{\mathrm{F}} m^*}{\pi^2 \hbar^2}, \tag{7.11}$$

where Ω denotes the volume of the system and k_{F} is the Fermi momentum. Now, the specific heat at constant volume is calculated from the relation,

$$\delta E = c_{\mathrm{v}} \delta T = \sum_{\mathbf{k}} \epsilon_{\mathbf{k}} \delta n_{\mathbf{k}}, \tag{7.12}$$

with $\delta n_{\mathbf{k}}$ being obtained from Eq. (7.8). Thus, we find,

$$c_{\mathrm{v}} = \frac{\pi^2 \mathrm{k_B}^2 T}{3\Omega} n(E_{\mathrm{F}}) = \frac{m^* k_{\mathrm{F}} \mathrm{k_B}^2 T}{3\hbar^2}. \tag{7.13}$$

We see that the electronic specific heat at low temperature varies linearly with T.

7.5 Thermal Expansion

Thermal expansion, referring to the temperature dependence of the volume of a material, is a consequence of the vibrations of the atoms (ions) around their equilibrium values as the temperature rises away from zero. Within the harmonic approximation, in which the energy of a displaced ion varies as the square of the displacement (Hook's law), no thermal expansion takes place because the average position of the vibrating ions is that of the equilibrium (ground state) configuration. The situation changes when unharmonic terms become important. Now, the energy (more precisely the free energy) of a system may be lower under expansion than under compression resulting in a finite, positive coefficient of thermal expansion.

These effects are illustrated [8] in Fig. 7.9 for the case of elemental Pu. It is shown here that the free energy of the metal in the harmonic approximation (dotted curve) is symmetric about the equilibrium volume[6] at $T = 0°\mathrm{K}$. Under these conditions the equilibrium positions of the ions remain on average the same at all temperatures as at vanishing temperature, and no expansion takes place. A more accurate calculation using [8] a full-potential LMTO approach (solid curve) reveals an asymmetry between the compressed and expanded systems. This behavior leads to a volume increase as the temperature rises and to a finite coefficient of thermal expansion. *For FCC Pu, this coefficient is negative.*

[6] The equilibrium volume for Pu predicated by LDA is considerably smaller than the experimental value because of the strong correlation effects of the f-electrons in the metal.

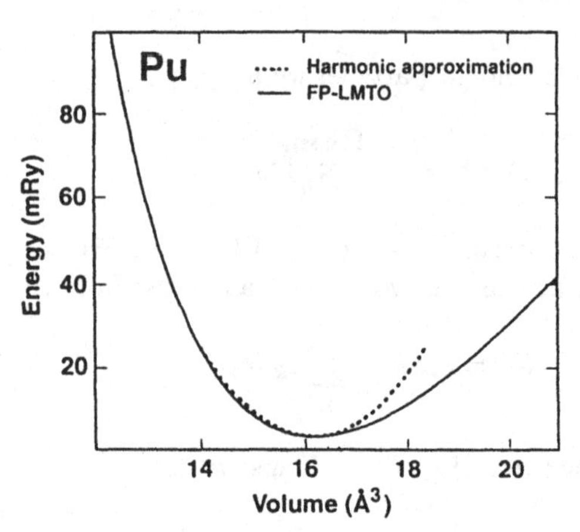

Figure 7.9: **The free energy of a vibrating atom in Pu metal as a function of atomic displacement (solid line). The dashed line indicates the free energy inthe harmonic approximation. See reference [4].**

To study the thermal expansion of a metallic system, we consider the Helmholtz free energy as a function of volume and temperature. This quantity is written in terms of contributions from both the electronic structure and the lattice vibrations (phonon) spectrum of the material. We can make fairly good progress in understanding thermal expansion by considering the free energy under two simplifying assumptions: First, we assume that the electronic structure is independent of temperature[7] and, second, we neglect the interaction of the electrons and the phonons, the so-called electron-phonon interaction[8]. Although these assumptions neglect some essential physics, they are found to be adequate at moderate temperatures. We now provide a brief description of the *Debye-Grüneisen* theory for the calculation of the coefficient of thermal expansion of a solid [7, 9, 10, 11].

It is convenient to express the volume of the system in terms of the Wigner-Seitz radius, r, which is the radius of a sphere whose volume equals that of the WS cell in the material. We write the Helmholtz free energy in the form,

$$H(r,T) = E_{\rm el}(r) + E_{\rm ph}(r,T) - T\left[S_{\rm el}(r,T) + S_{\rm ph}(r,T)\right], \qquad (7.14)$$

where $E_{\rm el}(r)$, $E_{\rm ph}(r,T)$, $S_{\rm el}(r,T)$, and $S_{\rm ph}(r,T)$ are, respectively, the contributions of the electrons and phonons to the energy and entropy of the

[7]The Fermi energy in a typical metal is of the order of tens of thousands of degrees Kelvin, while the region of interest in studying thermal expansions is of the order of the Debye temperature, a few hundred degrees. The concept of Debye temperature is introduced in Chapter 22.

[8]A discussion of the electron-phonon interaction can be found in Chapter 23.

system. Consistent with our assumptions above, $E_{\mathrm{el}}(r)$ is taken to be the total energy of the electron gas in the solid at zero temperature as determined in the calculation of the electronic structure. The electronic entropy involves the electron density and its behavior as the temperature increases and, because the electrons can be considered "cold" at low temperatures, will be neglected. This approximation has been found [9, 10, 11] to be viable for many metallic systems in the transition-metal and actinide series.

In the Debye model, the phonon contribution to the free energy takes the form,

$$E_{\mathrm{ph}}(r,T) = 3\mathrm{k_B}TD(\theta/T) + 9\mathrm{k}_B\theta/8, \tag{7.15}$$

where the Debye function is given by the expression,

$$D(x) = \frac{3}{x^3}\int_0^x \mathrm{d}y \frac{y^3}{\mathrm{e}^y - 1}, \tag{7.16}$$

and the last term in Eq. (7.15) describes the zero-point motion of the nuclei. In the Debye model, the phonon contribution to the entropy can be written in the form,

$$S_{\mathrm{ph}} = 3\mathrm{k_B}\left[\frac{4}{3}D(\theta/T) - \ln(1 - \mathrm{e}^{-\theta/T})\right]. \tag{7.17}$$

Now, for given values of r, θ, and T, we can write the Helmholtz free energy in the form,

$$H(r,T) = E_{\mathrm{el}}(r,T) - \mathrm{k_B}T\left[D(\theta/T) - 3\ln\left(1 - \mathrm{e}^{-\theta/T}\right) - \frac{9\theta}{8T}\right]. \tag{7.18}$$

The Debye temperature, θ, can be defined in terms of the cutoff frequency,

$$\omega_{\mathrm{D}} = \left(\frac{6\pi^2 v_m^3}{V}\right)^{1/3}, \tag{7.19}$$

where V is the volume of the WS cell in the material, and v_m is the mean velocity of sound. For approximately isotropic materials, this velocity can be obtained from the formula [12],

$$v_m = \left(\frac{2}{3v_{\mathrm{s}}^3} + \frac{1}{3v_{\mathrm{l}}^3}\right)^{-1/3}, \tag{7.20}$$

where v_{l} and v_{s} denote the longitudinal and shear velocities, respectively. These quantities, in turn, are given in terms of the bulk modulus, B, the Young's modulus, E, the shear modulus, G, and Poisson's ratio, ν, by the expressions [13],

$$v_{\mathrm{l}} = \left[\frac{G}{\rho}\left(\frac{4G - E}{3G - E}\right)\right]^{1/2}, \tag{7.21}$$

$$v_s = \left(\frac{G}{\rho}\right)^{1/2}, \tag{7.22}$$

$$G = \frac{E}{2(1+\nu)}, \tag{7.23}$$

and

$$B = \frac{E}{3(1-2\nu)}. \tag{7.24}$$

Using these expressions, the mean sound velocity, v_m, can be written in the form,

$$v_m = r\left(\frac{4\pi}{3}\right)^{1/2} F(\nu)\left(\frac{rB}{M}\right)^{1/2}, \tag{7.25}$$

where $M = \rho V$ is the atomic mass, and the function $F(\nu)$ is defined by the expression,

$$F(\nu) = \left[\frac{2}{3}\left(\frac{2}{1-2\nu}\right)^{3/2} + \frac{1}{3}\left(\frac{2}{1-\nu}\right)^{3/2}\right]^{-1/3}\left[\frac{3}{1+\nu}\right]^{1/2}. \tag{7.26}$$

An expression for the Debye temperature is obtained from the cutoff frequency,

$$\theta = \frac{h\omega_D}{2\pi k_B} = \frac{h}{k_B}\left(\frac{4\pi}{3}\right)^{-1/6} F(\nu)\left(\frac{rB}{M}\right)^{1/2}. \tag{7.27}$$

Let us denote by θ_0 the Debye temperature evaluated at the equilibrium radius, r_0. The unharmonicity in the crystal is introduced by the Grüneisen parameter, γ, defined by the expression,

$$\gamma = -\frac{\partial \ln \theta}{\partial \ln V}, \tag{7.28}$$

which through use of Eq. (7.27) can be expressed as

$$\gamma = -\frac{1}{6} - \frac{\partial \ln B}{2\partial \ln V}. \tag{7.29}$$

Using the definition of the bulk modulus,

$$B(V) = -V\frac{\partial P}{\partial V}, \tag{7.30}$$

the Grüneisen parameter can be expressed in the form,

$$\gamma = -\frac{2}{3} - \frac{V}{2}\frac{\partial^2/\partial V^2}{\partial P/\partial V}, \tag{7.31}$$

where the lattice pressure is obtained from the electronic energy,

$$P = -\frac{\partial E_{\rm el}(V)}{\partial V}, \tag{7.32}$$

and vanishes at the equilibrium volume as it should[9]. For a fixed temperature, γ is evaluated at the equilibrium volume and thus acquires a temperature dependence. The relation in Eq. (7.28) between θ and γ can be used to obtain the volume dependence of θ,

$$\theta(r,T) = \theta_0 \left(\frac{r_0}{r}\right)^{3\gamma}. \tag{7.33}$$

We are now in a position to determine the linear coefficient of thermal expansion, $\alpha(T)$, of a material. From the self-consistent determination of r_0 and θ at a given T, we obtain,

$$\alpha(T) = \frac{1}{r_0(T)} \frac{{\rm d}r_0(T)}{{\rm d}T}. \tag{7.34}$$

Bibliography

[1] V. L. Moruzzi, J. F. Janak, and A. R. Williams, *Calculated Electronic Properties of Metals*, Pergamon Press, Inc., New York (1978).

[2] A. Gonis, *Green Functions for Ordered and Disordered Solids*, (North Holland, Amsterdam, New York 1992).

[3] G. Y. Guo and W. M. Temerman, p.321 in *Applications of Multiple Scattering Theory to Materials Science*, Materials Research Society Proceedings, Vol. **253**, W. H. Butler, P. H. Dederichs, A. Gonis, and R. L. Weaver (eds.), (1992).

[4] Per Söderlind, Thesis, Uppsala University (1994), and references therein.

[5] R. F. S. Hearnon, Landolt-Börnstein, (eds.) K.-H Hellwege and A. M. Hellwege, *New Series*, Vol **11** Springer-Verlag, Berlin (1979).

[6] G. Simmons and H. Wang, *Single Crystal Elastic Constants and Calculated Aggregate Properties*, MIT Press, Cambridge, Massachusetts (1971).

[7] V. L. Moruzzi, J. F. Janak, and K. Schwartz, Phys. Rev. B **37**, 790 (1988).

[8] P. Söderlind, *Theoretical Studies of Elastic, Thermal, and Structural Properties of Metals*, Dissertation, Uppsala Univ., Sweden, April (1994).

[9] As commented previously, because of numerical uncertainties, the volume corresponding to zero pressure calculated directly from the pressure formula derived in Chapter 6 may differ from that obtained from the minimum of the energy.

[9] P. Söderlind and Börje Johansson, Thermochimica Acta **218**, 145 (1993)

[10] P. Söderlind, Lars Nortsröm, Lou Yongming, and Börje Johansson, Phys. Rev. B **42**, 4544 (1990).

[11] P. Söderlind, B. Johansson, L. Yongming, and L. Nordström, Intern. Journ. of Thermophysics, **12**, No. 4, 611 (1991).

[12] O. L. Anderson, in *Physical Acoustics*, W. P. Mason (ed.), Academic Press, New York (1965), Vol. III-B.

[13] S. P. Timoshenko and J. N. Goodier, *Theory of Elasticity*, 3rd ed. McGraw-Hill, Tokyo (1970).

Chapter 8

Correlation Effects(*)

8.1 General Comments

The DFT-based approach to electronic structure, implemented within the LDA, has contributed greatly to our understanding of the electronic structure of solids and its relation to materials behavior. Quantitative predictions of ground-state properties such as binding energies, equilibrium geometries, vibrational frequencies, surface energies, phase stability, and many others can now be made almost routinely. However, the LDA is not a fully satisfactory theory of electronic structure, failing on occasion to predict the correct lattice structure at zero temperature. The case of Fe is a well-known example of this failure. Furthermore, DFT provides a theoretical treatment of the ground state of the system and is largely silent about excited states. Predicting the band gap in semiconductors and insulators and the band width of simple metals are examples of excited-state properties that lie outside the purview not only of the LDA but of DFT itself. The same is true of phenomena such as relaxation shifts, lifetime broadening, and satellites in photoemission spectra. Attempts to obtain information about excited states within DFT have been made and can be found in the literature cited, e.g., [1].

The electron-electron interaction is the main culprit for the failure of the LDA for ground-state properties, as well as for the failure of DFT in general for excited-state properties of materials. The calculation of ground-state properties can often be improved through correction schemes applied to LDA, such as the generalized gradient approximation (GGA) and the LDA+U method, both of which are discussed in this chapter. In this discussion, we will also introduce the concept of quasi-particle that can be used to study a number of phenomena associated with the excitation part of the spectrum. At the same time, we will not delve in any detail into formal approaches geared toward the study of excited states. This forms the substance of many-body theory, a subject which lies outside the scope of this book.

8.2 Ground-State Properties

The framework of DFT-LDA underlies nearly all calculations of ground-state electronic properties. The corresponding formalism is based on the calculation of the single-particle density, $n(\mathbf{r})$, which can be compared directly with experiments. In this approach, the exact density of the interacting many-particle system is represented by the single-particle solutions, $\psi_i(\mathbf{r})$, of a (fictitious) non-interacting system,

$$n(\mathbf{r}) = \sum_i \psi_i^*(\mathbf{r})\psi_i(\mathbf{r}). \tag{8.1}$$

We recall from previous discussion that the wave functions $\psi_i(\mathbf{r})$ satisfy a single-particle Schrödinger equation of the form,

$$\left[-\nabla^2 + v_{\text{eff}}(\mathbf{r})\right]\psi_i(\mathbf{r}) = E_i\psi_i(\mathbf{r}), \tag{8.2}$$

where, as we have seen in Chapter 1, the effective one-electron potential, $v_{\text{eff}}(\mathbf{r})$, is given by the expression,

$$v_{\text{eff}}(\mathbf{r}) = v_{\text{ion}}(\mathbf{r}) + \int d^3r' \frac{n(\mathbf{r}')}{|\mathbf{r}-\mathbf{r}'|} + \mu_{\text{xc}}(\mathbf{r}). \tag{8.3}$$

Here, $v_{\text{ion}}(\mathbf{r})$ is the potential felt by an electron in the field of the ions, and

$$\mu_{\text{xc}}(\mathbf{r}) = \frac{\delta E_{\text{xc}}[n(\mathbf{r})]}{\delta n(\mathbf{r})} \tag{8.4}$$

is the exchange-correlation potential. The exchange and correlation energy, $E_{\text{xc}}[n]$, contains all exchange and correlation effects along with the difference in the kinetic and potential energies calculated within the LDA and their exact counterparts. The total energy of the system in its ground state takes the form,

$$\begin{aligned} E[n] &= T_s[n] + \int d^3r v_{\text{ion}}(\mathbf{r})n(\mathbf{r}) + \frac{1}{2}\int d^3r \int d^3r' \frac{n(\mathbf{r})n(\mathbf{r}')}{|\mathbf{r}-\mathbf{r}'|} \\ &+ E_{\text{xc}}[n], \end{aligned} \tag{8.5}$$

where $T_s[n]$ is the kinetic energy of the non-interacting system with density $n(\mathbf{r})$ represented by the wave functions, $\psi_i(\mathbf{r})$.

Up to this point, the formalism outlined above is exact. However, its implementation relies heavily on the unknown functional, $E_{xc}[n(\mathbf{r})]$, which can only be evaluated within an approximation. Let us consider the formal expansion,

$$E_{xc}[n] = \int d^3r \left\{ \epsilon_{xc}[n(\mathbf{r})]n(\mathbf{r}) + \epsilon_{xc}^{(2)}[n(\mathbf{r})]|\nabla^n(\mathbf{r})|^2 + \cdots \right\}, \tag{8.6}$$

where $\epsilon_{xc}[n(\mathbf{r})]$ is the exchange-correlation energy (per electron) of a uniform electron gas of density $n(\mathbf{r})$. Retaining only this term in the expansion leads directly to the local density approximation discussed in Chapter 1. The exchange-correlation potential, Eq. (8.4), now takes the form,

$$\mu_{xc}(\mathbf{r}) = \frac{d}{dn}\{\epsilon_{xc}[n]n\}. \tag{8.7}$$

The local density approximation to the exchange and correlation energy is exact in two limiting cases: for very slowly varying densities and for very high densities. In the LDA, the total energy takes the form (see also discussion in Chapter 1),

$$\begin{aligned} E[n] &= \sum_i^{\mathrm{occ}} E_i - \frac{1}{2}\int d^3r \int d^3r' \frac{n(\mathbf{r})n(\mathbf{r}')}{|\mathbf{r}-\mathbf{r}'|} \\ &- \int d^3r \mu_{xc}(\mathbf{r})n(\mathbf{r}) + E_{xc}[n(\mathbf{r})]. \end{aligned} \tag{8.8}$$

The LDA has been used extensively in the study of equilibrium properties. By varying the atomic geometries, equilibrium configurations and volumes can be determined, as discussed in the previous chapter. Equilibrium lattice constants are typically within 1% of experimental values. By displacing specific atoms from equilibrium, interatomic forces and vibrational frequencies can be obtained. Comparisons with experimental results usually bear out the viability of the LDA in these studies.

On the other hand, the validity of the LDA is severely tested in a number of calculations. Cohesive energies are typically overestimated by about 10 to 20%. For example, the average cohesive energy for diamond is 7.37 eV/atom, whereas the LDA predicts 8.63 eV/atom. This is to be compared with Monte Carlo calculations that yield 7.45 eV/atom. The difficulties encountered by the LDA can be traced to the neglect of the angular symmetries of orbitals in calculating $E_{xc}[n]$. This is a direct consequence of the local nature of this approximation. This defect is especially damaging when the hybridization between s, p, and d states in the material is large. Detailed discussions of this problem can be found in Jones and Gunnarson [2, 3]. The failures of the LDA also include occasional predictions of the incorrect equilibrium structure, such as the ground state of Fe (discussed below). One possible scheme for correcting the LDA, is to consider the next term in the expansion of E_{xc} in Eq. (8.6).

8.3 Generalized Gradient Approximation (GGA)

In their original work [4], Kohn and Sham suggested a gradient expansion of the type shown in Eq. (8.6) as a possible way of introducing non-local information into the exchange and correlation functional. The resulting approximation came to be known as the *gradient expansion approximation* (GEA).

Original applications of the GEA met with little success. In addition to inaccurate results, the GEA suffered from conceptual problems, such as the failure to satisfy fundamental sum rules. Subsequent work has produced considerable improvements to the effect that the *generalized gradient approximation* (GGA), as the approach came eventually to be called, can yield consistent and reliable results in many cases.

The approach followed in the GGA does not necessarily involve an expansion in powers of the gradient of the density [5, 6]. Instead, the exchange and correlation functional is written as the sum of two parts, $E_{xc} = E_x + E_c$, which, in turn, are written in the forms,

$$E_x = \int \mathrm{d}^3 r n(\mathbf{r}) \epsilon_x F(s) \tag{8.9}$$

$$E_c = \int \mathrm{d}^3 r n(\mathbf{r}) \left[\epsilon_c + H(t)\right]. \tag{8.10}$$

In general E_x, E_c, ϵ_x, ϵ_c, F, and H depend on the spin density, while s and t are two different scaled densities,

$$s = \frac{|\nabla n|}{2k_F n} \tag{8.11}$$

$$t = \frac{|\nabla n|}{2gk_s n}, \tag{8.12}$$

where k_F is the *Fermi wave vector* defined by the relation $k_F = (3\pi^2 n)^{1/3}$, and k_s is the local screening wave vector defined by

$$k_s = (4k_F/\pi)^{1/2} \tag{8.13}$$

Defining the quantity,

$$\zeta = \frac{n\uparrow - n\downarrow}{n\uparrow + n\downarrow}, \tag{8.14}$$

we have

$$g(\zeta) = \frac{1}{2}\left[(1+\zeta)^{2/3} + (1-\zeta)^{2/3}\right]. \tag{8.15}$$

It is clear that in the limits $F \to 1$ and $H \to 0$, we obtain the LDA, so that these quantities represent non-local effects. These functions can be chosen so that they possess a number of desirable features, such as the satisfaction of fundamental sum rules [7].

An example of the improved results obtained [9] through the use of the GGA compared with the LDA is indicated by the case of Fe, illustrated in Fig. 8.1. As is shown in this figure, the local spin density approximation (LSDA) used with a von Barth-Hedin [8] exchange and correlation functional predicts a paramagnetic fcc ground state for Fe. The GGA used either within a full-cell potential approach (such as the FP-LMTO used in the present calculation) or within the atomic sphere approximation (ASA) leads to the correct bcc magnetic structure for the ground state. However, the FP and the ASA calculations do give different results for the Fe moment, with the ASA leading to a rather poor description compared to the essentially exact value of $2.2\mu_B$ at the equilibrium volume obtained in the FP approach. In general, improvements in other quantities, such as atomic volumes and bulk moduli also result when the GGA is used in the determination of the electronic structure.

The version of the GGA used in the calculation reported here has been tested in a variety of systems. In general, it is found that the improvements are more pronounced when the GGA is used in connection with a full-potential approach rather than the ASA. Occasionally, an LDA-ASA calculation may give more accurate results than a GGA-ASA scheme possibly due to cancellation of errors in the former. More formal details of the GGA can be found, for example, in Söderlind [9].

8.4 LDA+U Method

The GGA is an attempt to correct for the effects of the local character of the LDA. Another deficiency of the LDA is connected with its treatment of the electron-electron interaction, i.e., of correlation effects. Although the LDA can boast a number of successful applications in the study of metallic systems, its record in the case of strongly correlated systems is much poorer. For example, systems that contain transition or lanthanide metal ions with partially filled d or f shells can be difficult to describe within the LDA, or the LSDA. In this case, one obtains as a result a partially filled d band with metallic-type electronic structure and itinerant d electrons, a result which is known to be incorrect in the case of late-transition metal oxides and rare-earth metal compounds. These systems contain well-localized d and f

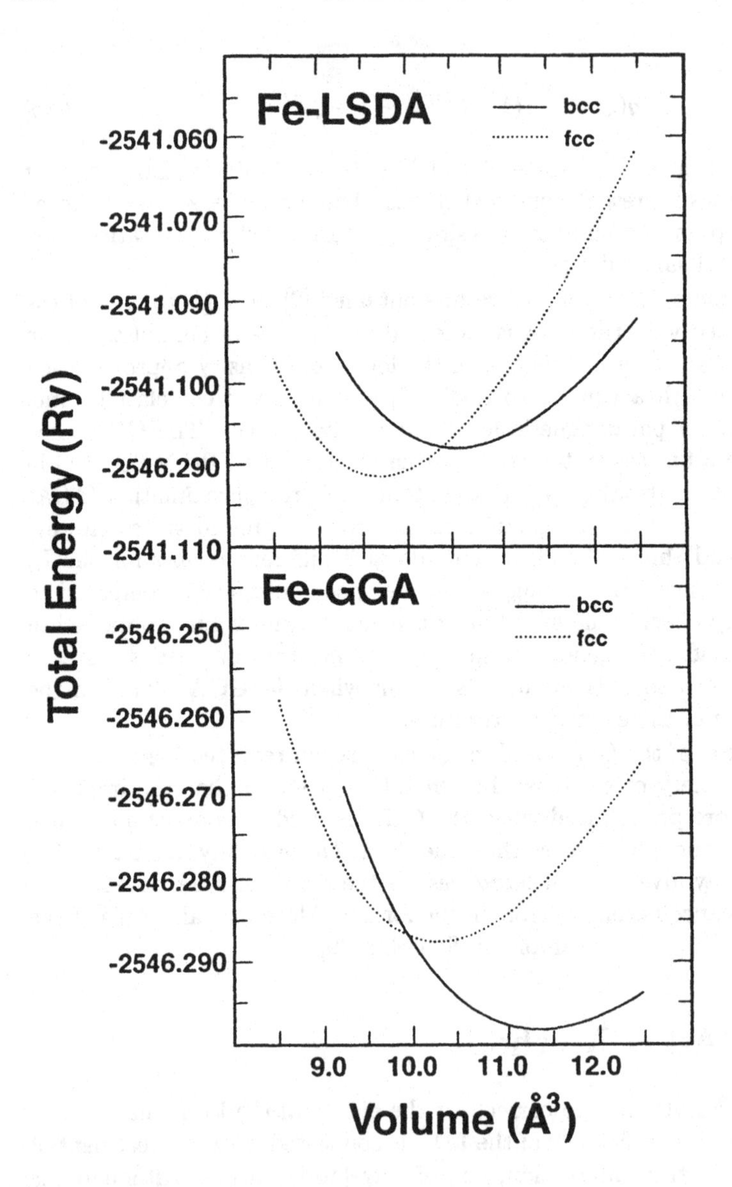

Figure 8.1: **The energy vs. volume curve for fcc and bcc Fe as obtained in the LDA (upper panel) and the GGA (lower panel). See reference [9].**

electrons with a large difference in energy between occupied and unoccupied bands.

There are a number of attempts to correct the LDA for its approximate treatment of correlation effects. In the so-called self-interaction-correction (SIC) method [10], one corrects for the fact that within the LDA an electron is allowed to interact with itself. This is evident from the form of the electrostatic interaction in the energy functional of the LDA, which explicitly contains the self interaction of the charge density. This methodology can give fairly accurate results for the localization of d and f states in transition and rare-earth ions, but yields eigenvalues that are usually in strong disagreement with experimental data.

In the so-called LDA+U method, briefly discussed here [11], the non-local and energy-dependent electron self-energy is approximated by a frequency-independent but non-local screened Coulomb potential. For the sake of easy reference, we refer to localized electrons as d electrons, in contrast to itinerant electrons in s and p bands.

Consider a d-ion system as containing a fluctuating number of d electrons. We assume that the LDA describes adequately the Coulomb energy of the d-d interactions in terms of the total number of d electrons $N = \sum_i n_i$, where n_i are d orbital occupancies, but not necessarily the orbital energies (eigenvalues). The expression for this energy takes the form $E = UN(N-1)/2$. Let us subtract this term from the LDA expression for the total energy functional and add a Hubbard-like term [neglecting for the moment exchange effects and the effects of non spherical charge distributions (multi-pole terms)] to obtain,

$$E = E_{\text{LDA}} - UN(N-1)/2 + \frac{1}{2}U\sum_{i\neq j} n_i n_j. \tag{8.16}$$

The orbital energies (eigenvalues) are obtained from this expression as derivatives with respect to n_i,

$$\epsilon_i = \frac{\partial E}{\partial n_i} = \epsilon_{\text{LDA}} + U(\frac{1}{2} - n_i). \tag{8.17}$$

Thus, within this approximation, the LDA-occupied orbitals ($n_i = 1$) are shifted by $-U/2$, while the unoccupied ($n_i = 0$) ones are shifted by $+U/2$. A formula similar to that is found for the orbital-dependent potential,

$$V_i(r) = \frac{\partial E}{\partial n_i(r)} = V_{\text{LDA}}(r) + U\left(\frac{1}{2} - n_i\right). \tag{8.18}$$

The relative shift of U between the occupied and unoccupied bands is in good agreement with the general behavior exhibited by so-called Mott-Hubbard insulators. At the same time, in order to carry out self-consistent calculations using this scheme, one needs a procedure to define an orbital basis set and take into account direct and exchange Coulomb interactions within a partially-filled d band. Methods for carrying out such calculations have been discussed in the literature [11], and the reader is referred to those works for details.

The quantity U to be used in a calculation can also be estimated from self-consistent calculations of the total energy, $E(n)$. It is obtained through the expression,

$$U = E(n_{\mathrm{d}} + 1) + E(n_{\mathrm{d}} - 1) - 2E(n_{\mathrm{d}}), \tag{8.19}$$

which is a measure of the energy difference that arises when electrons are promoted to or removed from a d-band state. Also, the site energy, ϵ_{d}, can be defined in terms of the removal energy, $E(n_{\mathrm{d}}) - E(n_{\mathrm{d}} - 1)$,

$$\epsilon_{\mathrm{d}} + U(n_{\mathrm{d}} - 1) \approx E(n_{\mathrm{d}}) - E(n_{\mathrm{d}} - 1). \tag{8.20}$$

In principle, these expressions should be used in connection with a single d state in the solid. On the other hand, the screening in metals is usually so effective that in practice one can alter the d occupation on every site.

In retrospect, the LDA+U methodology is a reasonable approximation to the treatment of correlations. At the same time, it is a corrective procedure that does not provide a suggestion for its improvement (apart from technical details having to do with such issues as the incorporation of multipole terms, for example). Also, it does not correct for the self-interaction terms, and its treatment of correlation effects is far from fully satisfactory. The need, therefore, remains for a method that will treat correlation effects systematically and rigorously.

8.5 Quasi-particles

Density functional theory provides a formally exact description of the ground state of an interacting electron system in an external potential, e.g., an atom or a solid. At the same time, DFT is equally remarkable for what it does not say as for what it does: namely, DFT is silent about phenomena connected with the excitation properties of an electron cloud. This is a basic feature of DFT and is not alleviated by corrections to LDA, such as those discussed above. For example, the single-particle, Kohn-Sham states determined within an LDA application of DFT cannot formally be used to interpret phenomena such as photoemission that involve the excitation spectrum of a system.

Photoemission and optical spectra in general provide information about the response of a system to incident radiation that ultimately moves the interacting particles away from their ground state. The description of such phenomena must, therefore, be sought beyond the confines of a ground-state theory. It is well understood that many features of excited states, such as screening and their decay, lead to characteristic phase shifts and line broadening that cannot be obtained within a ground-state theory. Furthermore, it is also known that measured band gaps in semiconductors and insulators, as well as the band width of simple metals such as Na, can be reproduced only through the use of many-body techniques that go beyond the realm of validity of DFT.

In studying excitation properties, it is convenient to consider the response of the rest of the system of particles when a particle is added (or removed) from the system. One usually refers to the removal of a particle as the creation of a *hole* in the system. An added particle or hole created in the system can often be studied in terms of concepts, e.g., mass, momentum, energy, familiar from the description of ordinary particles, but after assigning to such quantities "effective" values that may be very different from those usually ascribed to non-interacting single particles. Such entities are commonly referred as ***quasi-particles***. There is a Schrödinger-like equation describing quasi-particle behavior that has the general form [12]

$$\begin{aligned} &\left[-\nabla^2 + v_{\text{ion}}(\mathbf{r}) + v_{\text{Hartree}}(\mathbf{r})\right]\psi(\mathbf{r}) \\ &+ \int \mathrm{d}^3 r' \Sigma(\mathbf{r}, \mathbf{r}'; E)\psi(\mathbf{r}6/) = E\psi(\mathbf{r}), \end{aligned} \tag{8.21}$$

where $\Sigma(\mathbf{r}, \mathbf{r}'; E)$ is a non-local, energy-dependent *self-energy*[1] [12, 13, 14]. Here, $v_{\text{ion}}(\mathbf{r})$ represents the interaction of a particle with the external potential of the nuclei (ions) in the system, and $v_{\text{Hartree}}(\mathbf{r})$ is the bare (classical) interaction of a charge with itself. In spite of its rather innocent appearance as an eigenvalue equation, the last equation is extremely difficult to solve, primarily because of the energy dependence of the self-energy.

In addition to being non-local, the self-energy is complex and hence non-Hermitian. Consequently, the eigenergies of the last equation, the energies of the quasi-particle states, are also complex. The imaginary part of these energies determines the life-time of these quasi-particles. It can be shown [12] that Σ satisfies the exact expression,

$$\Sigma(1,2) = \mathrm{i}\int \mathrm{d}(3) \int \mathrm{d}(4) G(1,3) W(1,4) \Gamma(4,2;3), \tag{8.22}$$

where the notation (n) stands for $(\mathbf{r}_n, \sigma_n, t_n)$, G is the exact single-particle Green function, W is the dynamically screened Coulomb interaction between two particles, and Γ is the so-called vertex function. All these quantities depend on the quasi-particle spectrum and a self-consistent solution of the last equation is very difficult. It has been found useful to consider

[1] More precisely, Σ is the exchange-correlation part of the self-energy operator.

a simple approximation by setting $\Gamma = 1$, i.e., neglecting so-called vertex corrections. In this case, we obtain the *GW approximation* in which the self energy is given by the relation,

$$\Sigma(\mathbf{r},\mathbf{r}';E) = \frac{\mathrm{i}}{2\pi}\int \mathrm{d}\omega \mathrm{e}^{-\mathrm{i}\omega\delta} G(\mathbf{r},\mathbf{r}';E-\omega)W(\mathbf{r},\mathbf{r}';\omega), \tag{8.23}$$

where δ is an infinitesimal. In a manner analogous to Eq. (8.21), the Green function satisfies the equation,

$$\begin{aligned} [E + v_{\mathrm{ion}}(\mathbf{r}) + v_{\mathrm{Hartree}}(\mathbf{r})]\, G(\mathbf{r},\mathbf{r}') &= \delta(\mathbf{r}-\mathbf{r}') \\ + & \int \mathrm{d}^3r''\Sigma(\mathbf{r},\mathbf{r}'';\omega)G(\mathbf{r}'',\mathbf{r}';\omega). \end{aligned} \tag{8.24}$$

Knowledge of the Green function leads to the *spectral function*,

$$A(\mathbf{r},\mathbf{r}';E) = -\frac{1}{\pi}\Im G(\mathbf{r},\mathbf{r}';E). \tag{8.25}$$

Provided that non-diagonal elements can be neglected, the Fourier transform of this expression assumes the simple Lorenzian form,

$$A(\mathbf{k};E) = \frac{1}{\pi}\frac{\Gamma(\mathbf{k})}{[E-E(\mathbf{k})]^2 + \Gamma^2(\mathbf{k})}, \tag{8.26}$$

where $E(\mathbf{k})$ and $\Gamma(\mathbf{k})$ are the real and the imaginary parts of the eigenenergies associated with the solutions of Eq. (8.21). The quantity $\Gamma(\mathbf{k})$, the width of the Lorenzian, gives the life time of the quasi-particle state with energy $E(\mathbf{k})$.

Even though exact solutions of Eqs. (8.21) and (8.24) have not been given, approximate treatments often yield distinct improvements over the results of the LDA. A straightforward illustration is given below.

8.5.1 Band gap in semiconductors

Computational methods based on DFT and the LDA do a fairly accurate job of representing the ground-state properties of an interacting quantum system, such as the density and total energy. When used to study excitation properties, DFT-based methods are found to be much less accurate[2]. For example, the calculated widths of band gaps of semiconductors and insulators can be underestimated by more than a factor of two. When this effect was first noticed, it was unclear whether it was a consequence of the local character of the LDA or a more general limitation of DFT.

[2]We have already commented that DFT is essentially silent in a formal sense about the excitation spectrum of a system.

This question has been addressed in a number of published works. Godby et al. [15]used a fully non-local exchange-correlation potential to calculate the band structure of elemental Si. They found the result to be essentially the same as that of the LDA. Thus, DFT itself is inherently limited in describing excitation spectra. This description requires a proper treatment of the quasi-particle nature of the states in the valence and conduction bands.

Hubertsen and Louie [16] have used the GW approximation to study the band structure of Si, Ge, and diamond. The self-energy is evaluated as indicated in Eq. (8.23). The Green function in that expression is obtained in the form,

$$G(\mathbf{r},\mathbf{r}';E) = \sum_{n\mathbf{k}} \frac{\psi_{n\mathbf{k}}(\mathbf{r})\psi^*_{n\mathbf{k}}(\mathbf{r}')}{E - E_{n\mathbf{k}} - i\delta}, \tag{8.27}$$

where $\psi_{n\mathbf{k}}(\mathbf{r})$ is an eigenstate obtained in the LDA. The so-called screened Coulomb interaction takes the form,

$$W(\mathbf{r},\mathbf{r}';E) = \int d^3\mathbf{r}''\epsilon^{-1}(\mathbf{r},\mathbf{r}'';E)U(\mathbf{r}''-\mathbf{r}'), \tag{8.28}$$

where $U(\mathbf{r}''-\mathbf{r}')$ is the bare Coulomb repulsion and ϵ^{-1} is the full inverse dielectric function. This latter quantity is evaluated in two steps. First, a random-phase approximation (RPA) was used to determine the static, $\omega = 0$, dielectric matrix, and second, a generalized plasmon mode was used to extend the dielectric matrix to finite frequencies. These calculations resulted in band gaps whose width agrees well with experiments [16]. For example, the experimental band gaps in eV for Si, Ge and diamond are, respectively, 1.17, 0.74, and 5.5., as compared with 0.52, about 0.0, and 3.9 as obtained in the LDA. Quasi-particle calculations, on the other hand, yield 1.21, 0.75, and 5.6, respectively. The general conclusion from these calculations, and others like them, is that DFT and the LDA can be expected to yield accurate ground-state properties, but quasi-particle calculations are in general necessary for the study of excitation spectra.

8.6 Band Magnetism

One of the best known and most spectacular consequences of exchange and correlation effects in materials, as well as one of the most important technologically, is that associated with magnetic behavior. In this section, we provide a brief overview of a very broad and fertile field by considering the band magnetism of ferromagnetic and nearly ferromagnetic materials such as Fe and Ni. Lack of space prevents a discussion of other, equally important magnetic features, such as antiferromagnetism, spin-glass behavior, and the Kondo effect. Our focus on the ferromagnetism of transition metals is justified because the observed magnetic moments in these solids were among the first magnetic properties to be linked with theoretical electronic structure calculations at $T = 0°$K. Although the dependence of the magnetic properties of the pure transition metals at finite temperature is less well understood, certain features, e.g., spin-wave energies, have found fairly clear theoretical descriptions. Our aim in this section is to provide an interpretation of magnetic phenomena in transition metals based on the underlying electronic structure. The following discussion is based on a clear article by Edwards [17] to which the reader is referred for more details.

8.6.1 Stoner model

There is good reason to attempt to understand the magnetic behavior of metals such as Fe, Co, and Ni on the basis of electronic structure calculations. In fact, the magnetic moments in transition metals are due to d electrons whose spins are aligned by exchange interactions. This is attested to by the non-integral values of the moments whose saturation values are 2.2, 1.7, and 0.6 μ_B for Fe, Co, and Ni, respectively, the large electronic specific heat, associated with partly filled d-bands, and the observance of d-like parts of the Fermi surface by de Haas van-Alphen measurements [18]. Thus, the d electrons are itinerant rather than localized on atoms, as in the Heisenberg model of magnetic insulators, and a study within band theory may prove profitable. This remains true in spite of the large degree of hybridization that is known to exists among the s and p electron bands with the d bands in transition metals. Furthermore, the basic nature of the mechanism for moment formation can be understood even if we neglect the multiband nature of the d electrons.

Following up on the last comment, let us consider a single-band Hamiltonian of the tight-binding (TB) type having the form,

$$H = \sum_i \epsilon_i a^\dagger_{i\sigma} a_{i\sigma} + \sum_{ij \neq i} W_{ij} a^\dagger_{i\sigma} a_{j\sigma} + U \sum_i n_{i\uparrow} n_{i\downarrow}. \tag{8.29}$$

In this so-called *Hubbard* Hamiltonian, the ϵ_i denotes the site energy associated with site i, W_{ij} are "hopping integrals" describing electron transfer between sites i and j, and U is a parameter describing the Coulomb repulsion of two electrons on the same site. The $a^\dagger_{i\sigma}$ and $a_{i\sigma}$ create and destroy, respectively, an electron of spin σ on site i, and $n_{i\sigma}$ is the corresponding number operator. More details of the TB description of the electronic structure can be found in a number of texts [19, 20]. A brief review is also given in Chapter 10.

Let us choose $\epsilon_i = \epsilon$ to be independent of site index, and treat the Hamiltonian of Eq. (8.29) in the Hartree-Fock approximation. In this treatment, the parameter U is to be regarded as an effective interparticle potential describing not only the screening by s-p electrons but also the correlation of the d electrons. It is convenient to write the first two terms of the Hamiltonian in the Bloch representation. Defining the quantities,

$$W(\mathbf{k}) = \sum_j W_{ij} e^{i\mathbf{k}\cdot(\mathbf{R}_j - \mathbf{R}_i)}, \tag{8.30}$$

and

$$n_{\mathbf{k}\sigma} = \sum_i n_{i\sigma} e^{i\mathbf{k}\cdot\mathbf{R}_i}, \tag{8.31}$$

we have

$$H = \sum_{\mathbf{k}\sigma} [\epsilon + W(\mathbf{k})]\, n_{\mathbf{k}\sigma} + U \sum_i n_{i\uparrow} n_{i\downarrow}. \tag{8.32}$$

The exchange integral between Bloch states has the form,

$$T_x = \int\int d^3r_1 d^3r_2 \psi^*_{\mathbf{k}}(\mathbf{r}_1)\psi^*_{\mathbf{k}}(\mathbf{r}_2) H_{\text{int}} \psi_{\mathbf{k}}(\mathbf{r}_1)\psi_{\mathbf{k}}(\mathbf{r}_2), \tag{8.33}$$

where $H_{\text{int}} = U \sum_i n_{i\uparrow} n_{i\downarrow}$. For the case considered here, and writing a Bloch state in the form

$$\psi_{\mathbf{k}}(\mathbf{r}) = \frac{1}{\sqrt{N}} \sum_i e^{i\mathbf{k}\cdot\mathbf{R}_i} \phi(\mathbf{r} - \mathbf{R}_i), \tag{8.34}$$

where $\phi(\mathbf{r}) - \mathbf{R}_i$ is an "atomic orbital" centered on site i, the exchange integral equals U/N. The intra-atomic Coulomb interaction, U, is often referred to as an *exchange parameter*[3].

Now, in the Hartree-Fock approximation, we have,

$$\begin{aligned} U \sum_i n_{i\uparrow} n_{i\downarrow} &= U \sum_i n_{i\uparrow} \langle n_{i\downarrow} \rangle \\ &= \frac{1}{4} \frac{U}{N} \left[(n_{i\uparrow} + n_{i\downarrow})^2 - (n_{i\uparrow} - n_{i\downarrow})^2 \right] \\ &= -\frac{U}{4N} n^2 \zeta^2 + \text{constant}, \end{aligned} \tag{8.35}$$

where $n = n_{i\uparrow} + n_{i\downarrow}$ is the total number of electrons of either spin and $\zeta = n_{i\uparrow} - n_{i\downarrow}$ is the relative magnetization. Within a constant, the energy of the system now takes the form,

$$E(\zeta) = \sum_{\mathbf{k}\sigma} W(\mathbf{k}) n_{\mathbf{k}\sigma} - \frac{U}{4N} n \zeta^2. \tag{8.36}$$

[3] Some authors consider the intra-atomic exchange integral, denoted by J, between different orbitals in a multi-band calculation and write $U + 4J$ in place of U.

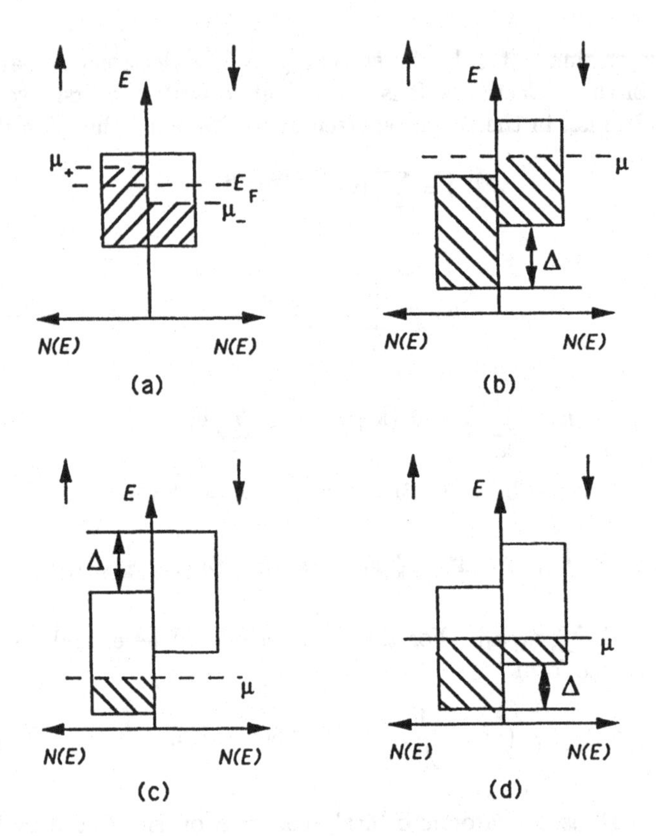

Figure 8.2: **Schematic representation of the DOSs in an itinerant ferromagnet. μ^+ and μ^- are the energy levels (chemical potentials) of the highest occupied spin-up and spin-down states. μ is the chemical potential of the ferromagnetic state.**

This rather simple expression for the energy was used by Slater [21] and Stoner [22] to obtain an understanding of itinerant electron magnetism in early studies of the phenomenon. A criterion for the occurrence of ferromagnetism at $T = 0°\mathrm{K}$ can be obtained by comparing the energy $E(\zeta)$ for small ζ to $E(0)$, by means of an expansion in ζ.

This expansion can be effected through reference to panel (a) of Fig. 8.2. We find,

$$E(\zeta) = E(0) + N\int_{E_{\mathrm{F}}}^{\mu^+} En(e)\mathrm{d}E - N\int_{\mu^-}^{E_{\mathrm{F}}} En(e)\mathrm{d}E - \frac{U}{4N}n^2\zeta^2, \quad (8.37)$$

where $n(E)$ is the density of states at energy E, E_{F} is the Fermi energy of the paramagnetic state, and μ^+ and μ^- are the Fermi energies of the spin up and spin down states in the magnetized material. The number of spins (per unit volume) turned around in going to the magnetized state is $n\zeta/2$ while the average energy increase for each spin reversal is equal to $n\zeta/2Nn(E_F)$. Thus,

$$E(\zeta) = E(0) + \frac{1}{4N}\frac{n^2\zeta^2}{n(E_F)}[1 - Un(E_F)] + O(\zeta^2), \tag{8.38}$$

and it follows that the energy of the magnetized state is lowered with respect to the paramagnetic state when

$$Un(E_F) \leq 1.0. \tag{8.39}$$

This relation is known as the *Stoner criterion* for ferromagnetism.

If the Stoner criterion is satisfied, the ground state of the system corresponds either to maximum spin alignment, panels (a) or (c) in Fig. 8.2, (corresponding to the band being more or less than half full), or to a state of partial alignment, panel (d), with ζ corresponding to the minimum value of $E(\zeta)$.

It follows from Eq. (8.35) that an up-spin electron sees an interaction energy $U\langle n_{i\downarrow}\rangle = n_\downarrow U/N$. Therefore, the difference in energy between an up-spin and a down-spin electron in states with the same value of $\mathbf{k}$ is equal to

$$\Delta = U(n_\uparrow - n_\downarrow)/N = Un\zeta/N. \tag{8.40}$$

This is the exchange splitting between the majority (spin-up) and minority (spin-down) bands illustrated schematically in Fig. 8.2. Because it is proportional to the magnetization, Δ corresponds to a molecular field of magnitude $Um/2\mu_B^2$, where m is the magnetic moment on an atom.

The simple argument given above provides an intuitive understanding of itinerant ferromagnetism in the transition metals. Realistic calculations based on a *spin-polarized* version of the LDA, (LSDA), lead to results that are qualitatively well described by the schematic diagrams in Fig. 8.2. Figures 8.3 and 8.4 show the exchange-split DOS for Ni and Fe, respectively, calculated by Moruzzi, Janak, and Williams [23]. The results for Fe correspond to the bcc structure[4].

8.6.2 Magnetic susceptibility

The term *susceptibility* is used to describe the response of a system to an external perturbation (stimulus). A general discussion of susceptibility within

[4]We saw above that the LSDA can lead to the incorrect structure for Fe in certain instances.

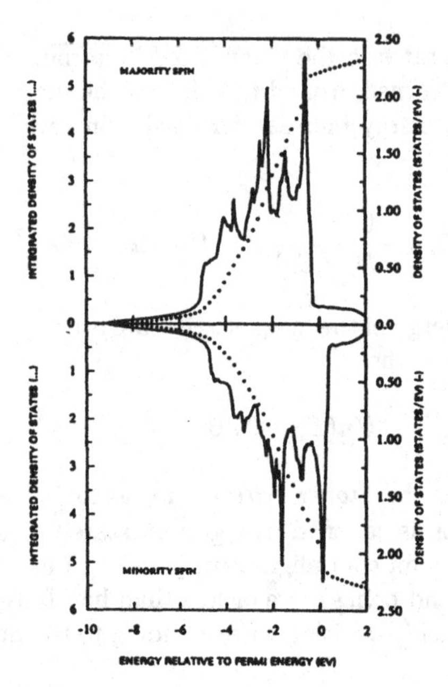

Figure 8.3: **Density of states for ferromagnetic Ni. The dots represent the integrated DOS. See reference [23].**

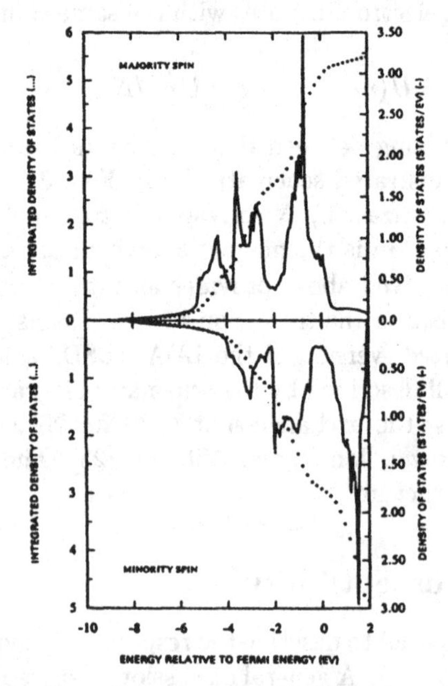

Figure 8.4: **Density of states for ferromagnetic Fe. See reference [23].**

linear response theory is reserved for Chapter 19. Here, we provide a preliminary account for the case of the magnetic susceptibility of an itinerant ferromagnet. Again, we follow Edwards [17].

We consider a paramagnetic metal in which correlations are important but in which the Stoner criterion is not satisfied. At site i of such a system we apply a time-dependent magnetic field of magnitude $h_i \exp(\mathrm{i}\omega t)$. The concept of molecular field introduced above allows us to express the total field at site i as a sum of the external field at the site and the field contributed by the *change* in the *local* moment at that site,

$$H_i \exp(\mathrm{i}\omega t) = (h_i + u_i \delta m_i) \exp(\mathrm{i}\omega t). \tag{8.41}$$

Here, $\delta m_i \exp(\mathrm{i}\omega t)$ is the change in the local moment and

$$u_i = U_i / 2\mu_{\mathrm{B}}^2. \tag{8.42}$$

We now introduce the *non-interacting* susceptibility, χ_{ij}^0, which relates the moment induced at site i to the field at site j,

$$\begin{aligned} \delta m_i &= \sum_j \chi_{ij}^0(\omega) H_j \\ &= \sum_j \chi_{ij}^0 \left[h_i(\omega) + u_j \delta m_j(\omega) \right]. \end{aligned} \tag{8.43}$$

If we consider the applied field as acting only on site j, we can define the physical or *interacting* susceptibility, $\chi_{ij}(\omega)$, through the first line of the equation above but with H_j replaced by h_j. We now have,

$$\chi_{ij}(\omega) = \chi_{ij}^0(\omega) + \sum_l \chi_{il}^0(\omega) u_l \chi_{lj}(\omega). \tag{8.44}$$

This expression is applicable to the transverse spin susceptibility of a paramagnet or a ferromagnet. The treatment of the longitudinal susceptibility is more complicated and must take into account the coupling between spin and density fluctuations.

In the case of a pure, translationally invariant material, the last expression can be solved by means of a lattice Fourier transform. Here, $u_l \equiv u$, and defining the quantities

$$\chi(\mathbf{q};\omega) = \sum_j \chi_{ij}(\omega) \mathrm{e}^{\mathrm{i}\mathbf{k}\cdot(\mathbf{R}_j - \mathbf{R}_i)} \tag{8.45}$$

we obtain

$$\chi(\mathbf{q};\omega) = \frac{\chi^0(\mathbf{q};\omega)}{1 - u\chi^0(\mathbf{q};\omega)}, \tag{8.46}$$

with $\chi^0(\mathbf{q};\omega)$ being defined in a manner analogous to that in Eq. (8.45).

The interacting susceptibility, $\chi(\mathbf{q};\omega)$, describes the response of the system to an applied field of the form $h\exp[-i(\mathbf{q}\cdot\mathbf{r}-\omega t)]$ with wave vector $\mathbf{q}$ and angular frequency ω. We note that if

$$u\chi^0(\mathbf{q};\omega) > 1.0, \tag{8.47}$$

then (for somewhat smaller values of u) $\chi(\mathbf{q};\omega)$ in Eq. (8.46) diverges. This implies that the paramagnetic state has become unstable against the formation of a spin-density wave of wave vector $\mathbf{q}$. In fact, Eq. (8.47) is a restatement of the Stoner criterion discussed above.

Bibliography

[1] R. M. Dreizler and E. K. U. Gross, *Density Functional Theory*, Springer-Verlag, Berlin, Heidelberg, New York (1009). See also, P. Ziesche, S. Kurth, and J. Perdew, *Density Functionals from LDA to GGA*, Comp. Mat. Science **11** 122 (1998).

[2] R. O. Jones and O. Gunnarsson, Phys. Rev. Lett. **55**, 107 (1985).

[3] O. Gunnarsson and R. O. Jones,Phys. Rev. B **31**, 7588 (1985).

[4] W. Kohn and L. Sham, Phys. Rev. **140**, A1133 (1965).

[5] J. P. Perdew, Phys. Rev. B**33**, 8822 (1986).

[6] J. P. Perdew, in *Electronic Structure of Solids*, P. Ziesche and H. Eschrig (eds.), Akademie Verlag, Berlin (1991), Vol. 11.

[7] J. P. Perdew and Y. Wang, Phys. Rev. B**33**, 8800 (1986).

[8] U. von Barth and L. Hedin, J. Phys. C**5**, 1629 (1972).

[9] P. Söderlind, Thesis, Uppsala University (1994), and references therein (unpublished).

[10] A. Svane and O. Gunnarsson, Phys. Rev. Lett. **65**, 1148 (1990).

[11] Vladimir I Anisimov, F. Aryasetiawan, and A. I. Lichtenstein, J. Phys. Condens. Matter, **9**, 767 (1997).

[12] Lars Hedin and Stig Landqvist, in *Solid State Physics*, Suppl.**23**, Academic Press, New York (1969).

[13] P.O. Gartland and B. J. Slagsvold, Solid State Commun. **25**, 489 (1978).

[14] N. O. Lang and W. Kohn, Phys. Rev. B**1**, 4555 (1970).

[15] R. W. Godby, M. Schlütter, and L. Sham, Phys. Rev. Lett.**56**, 2415 (1986).

[16] M. S. Hubertsen and S. G. Louie, Phys. Rev. Lett.**55**, 1418 (1985).

[17] D. M. Edwards, in *Electrons in Disordered Metals and at Metallic Surfaces*, P. Phariseau, B. L. Györfy, and L. Scheire (eds.), NATO-ASI, B**42**, Plenum, New York (1979), p. 355.

[18] A. V. Gold, J. Low Temp. Phys. **16**, 3 (1974).

[19] Neil W. Ashcroft and N. David Mermin, *Solid State Physics*, Saunders College, Philadelphia (1976).

[20] A. Gonis, *Green Functions for Ordered and Disordered Systems*, Elsevier, Amsterdam (1992).

[21] J. C. Slater, Phys. Rev. **49**, 537, 931 (1936).

[22] E. C. Stoner, Proc. Roy. Soc. A**165**, 372 (1938).

[23] V. L. Moruzzi, J. F. Janak, and A. R. Williams, *Calculated Electronic Properties of Metals*, Pergamon Press, New York (1978).

Chapter 9

Impurities

9.1 General Comments

The discussion in the previous chapter was concerned with the electronic structure of an infinite, translationally invariant solid. Although the treatment of such systems is extremely useful, providing a great deal of insight into the electronic origins of many physical phenomena, it is also limiting in many important respects. Real materials are most often, if not exclusively, characterized by the presence of imperfections or disorder of one form or another. For our purposes, we define *disorder* in a rather circuitous way as the absence of *order*. In turn, we ascribe to order the commonly perceived intuitive notion of the repetition of a basic structure in a regular manner. In the case of crystalline materials, perfect order often means the presence of a rigid, regular lattice of infinite extent whose points are "decorated" with a basic unit of atoms or molecules. With respect to such a system, the simplest type of disorder consists of a change, either chemical or structural, confined to a bounded region of space such as a unit cell. We call such a defect a *localized impurity* or a *point defect*.

The substitution of a host atom by an atom of a different kind in an otherwise pure, translationally invariant solid, the *host*, is an example of a *substitutional impurity*. An atom introduced into an off-lattice position in the interstitial region is referred to as an *interstitial impurity*. A *structural impurity* indicates the presence of a localized deviation from lattice periodicity, such as the displacement of an atom from its equilibrium position. It is fairly obvious that an impurity may exhibit a number of these different characteristics (for example, an interstitial impurity that causes the host atoms in its vicinity to move away from their equilibrium lattice positions).

These are some of the types of impurities studied in this and in subsequent chapters[1]. Our aim is to describe the effects of impurities on the electronic structure and related properties of metallic systems[2], within a Green function, multiple-scattering approach.

The treatment of impurities in metals has a long history and encompasses a number of different theoretical/computational models. The *jellium model* has been used in a number of calculations [1-8]. Here, a point defect is modeled as a spherical potential embedded in a uniform background of charge, with the rearrangement of the electrons around the impurity calculated within density functional theory (DFT).

Almbladt and von Barth [3] introduced the *spherical solid model* in which the effect of the near neighbors on an impurity is described by spherically averaging the potentials in each neighboring shell. The model has been used in a number of calculations by Manninen and Nieminen [9] and by Nieminen and Puska [10].

An often-used model is based on the treatment of a finite, usually small, number of atoms such as a site and its first few neighboring shells, surrounded by a spherical potential. Extensive calculations based on such a *cluster* model have been performed by Johnson and collaborators [11-13], by Müller, et al., [14], and Gunnarson and Jones [15], who used the method of linear muffin-tin orbitals (LMTO) [16]. Cluster calculations have also been performed by Seifert and Ziesche [17] on the basis of tight-binding trial functions, while Blaha and Callaway [18] employed a Gaussian basisset.

However, in a cluster of even large size most of the atoms are on the surface[3], and determining bulk properties on the basis of cluster calculations becomes unreliable. In addition, at real energies, the eigenvalue spectrum of such a system consists of a series of delta functions that further obscures the physical content of the method. Finally, cluster calculations cannot address phenomena that are tied in to characteristics of the Fermi surface.

Some of the undesirable features of the calculations based on finite (small) clusters can be alleviated by using methods that allow the treatment of thousands of atoms. One such method is based on the recursion formalism [19]. Calculations based on the recursion method in connection with a cluster embedded in a constant potential have been performed by Rijsenbrij and Lodder [20], Riess and Winter [21], and Lodder and Braspenning [22]. In order to improve the model, John and Keller [23] suggested embedding the cluster in an energy-dependent potential. However, calculations based on very large numbers of atoms yield numerical results which, although

[1] We will also be concerned with extended defects such as surfaces and interfaces. However, we will not explicitly address structurally disordered systems such as amorphous materials and liquids, or quasiperiodic systems such as quasicrystals for which no fully satisfactory treatments of the electronic structure have been developed thus far.

[2] Although we restrict the discussion to metallic systems, the methods developed here are in principle applicable to other types of materials as well.

[3] In a cubic arrangement, 600 atoms in a cluster of 1000 sites are on the surface.

possibly accurate and precise, tend to obscure the underlying physics. The validity and aesthetic quality of a physical theory increases with its ability to extract a maximum of physical insight from the simplest possible set of assumptions.

Perhaps the simplest cluster model is due to Friedel [24-27], and consists of a single, spherically symmetric potential embedded in the free-electron gas of the host conduction band. This model has been used [28] to investigate the residual resistivity of substitutional impurities in noble-metal hosts.

The effects associated with the finite size of the cluster or with the embedding potential can be circumvented to a large extent within a ***supercell*** calculation. Here, a large cell, containing the impurity at its "center" and a number of host atoms (which may be perturbed by the impurity potential), is repeated periodically on the points of a superlattice, which can then be handled by standard band-structure techniques [29, 30]. However, the usually large number of atoms in the unit cell of the superstructure often limits the applicability of the method. In addition, even in cases in which the unit cell contains as many as 20 or more atoms, the impurities in adjacent cells are still close enough to one another so that their interaction cannot be neglected *a priori*. Under these conditions, the results of supercell calculations lose much of their relevancy to the single-impurity problem.

Point defects in materials, such as substitutional and interstitial impurities, can be treated successfully within a Green function method that takes proper account of the boundary conditions imposed by the geometry. In such a treatment, one can properly calculate the effect that the impurity has on the surrounding host atoms, without introducing spurious effects associated with the finite size of a cluster, the embedding medium, or the impurity-impurity interaction. Consequently, a number of Green-function-based techniques have been developed to treat impurities in metals and semiconductors, often geared toward a particular type of system or a band-structure method.

Green functions based on a Wannier-function representation for the host crystal were originally used by Koster and Slater [31]. Seeger, et al., [32, 33], Hehl and Mann [34], and Mann [35] used a Wannier representation in the study of transport phenomena, such as magnetoresistance and Hall coefficients, in the presence of impurities, and to calculate the influence of impurities on the de Haas-van Alphen effect.

Tight-binding Green functions have also been used [36-44] to study impurities in a number of different physical systems, particularly semiconductors.

Of the Green function methods that have been used to treat impurities, we will be particularly interested in those based on MST, or the Korring a-Kohn-Rostoker (KKR) Green function method. This approach has been used in a number of applications, of which we cite a representative sample [45-81].

In developing the formalism, we proceed along lines formally similar to those we followed in the development of MST in previous chapters. We assume that we can solve "exactly" for the scattering properties and the Green function of the reference host medium, and thus we can determine the corresponding properties such as equilibrium lattice constants and charge densities. The effects of an impurity are then obtained by considering the further scattering produced by the impurity on waves already scattered by the presence of the host potential.

The presence of a localized impurity can be associated with a spatially bounded change, ΔV, in the one-electron potential in the system. However, the spatial extent of the perturbing potential is usually broader than that of the impurity due to the effect of the impurity on the surrounding host atoms. In metallic systems, impurities are usually screened out within a length typically equal to k_{F}^{-1}, where k_{F} is the Fermi wave vector. This eliminates long-range Coulomb fields and facilitates numerical treatment. It is then convenient to think of an *impurity cluster*, consisting of the impurity and the affected host atoms, as embedded in the host medium. Multiple scattering theory provides an ideal framework for the treatment of such a system.

It is also fairly obvious that in calculating the effects of impurities on the electronic structure of a material one should not treat *only* the scattering from the impurity cluster. An electron wave will scatter many times within the cluster *and* from the surrounding host atoms, necessitating the multiple-scattering treatment of the system as a whole. The formal framework for such a treatment is developed in this chapter. The foundation of the method developed below hinges on applying MST to calculate the additional scattering caused by the impurity cluster of an electron wave already scattered by the reference medium, i.e., the potential of the host. The concept of the reference medium is a very useful one, because it allows an almost straightforward extension of the MST formalism of Chapter 3, where the reference medium was taken to be free-space, to the case of scattering in a field of force. By making the formal analogy between free space and the reference medium, and of a collection of potentials in free space and an impurity cluster in a host material, we can transcribe essentially intact the formalism of Chapter 3 to the case at hand[4].

As a first step in the development of a formal treatment of scattering in a field of force, we discuss the scattering produced when two potentials act simultaneously in a given region of space.

[4]There are a few subtle differences that will be brought out as needed.

9.2 Two-Potential Scattering

We consider the scattering of a wave (or particle) induced by the presence of two potentials [say nuclear and screened Coulomb (interelectron) forces] acting simultaneously. Although the formalism will eventually be extended to infinite systems, we begin with the study of spatially bounded potentials that vanish outside a finite volume, Ω.

9.2.1 Distorted waves

The Hamiltonian of the system is taken to have the form, (we do not designate operators by hats trusting that context will help clarify the meaning of the various symbols),

$$H = H_0 + U + V, \tag{9.1}$$

where

$$H_0 = -\nabla^2. \tag{9.2}$$

We assume that the system described by the presence of only the potential U can be treated exactly, and we define it as the *reference system*. We now write

$$H = \bar{H}_0 + V, \tag{9.3}$$

where

$$\bar{H}_0 = H_0 + U \tag{9.4}$$

is the Hamiltonian of the reference system. We are interested in determining the scattering properties, i.e., the scattering wave function and the scattering matrix associated with the fully interacting system characterized by the potential $U + V$.

As a first guess, we may be tempted to use the Green function

$$\bar{G}_0(z) = (z - \bar{H}_0)^{-1} \tag{9.5}$$

of the reference system, which is assumed to be known, to write the scattered wave function in the form of a Lippmann-Schwinger equation (2.81)

$$|\psi_a^{\pm}(U+V)\rangle = |\phi_a(U)\rangle + \left(E_a - \bar{H}_0 \pm \mathrm{i}\epsilon\right)^{-1} V|\psi_a^{\pm}(U+V)\rangle, \tag{9.6}$$

where the free-particle solutions and Green function have been replaced by the corresponding quantities associated with the reference medium (see below). We will rigorously show that in fact the modified Lippmann-Schwinger equation exhibited above properly describes the state that evolves out of a specific solution of the reference medium in the presence of the perturbing potential. Also, we will show that the *additional* scattering caused by the presence of V in a wave already distorted by U is given by the expression

$$t_{ba}(V/U) = \langle \phi_b^-(U)|V|\psi_a^\pm(U+V)\rangle, \tag{9.7}$$

which is completely analogous to the definition of the t-matrix when the medium in which the scattering takes place is free space, $U = 0$. In the last expression, the functions $|\phi_a^-(U)\rangle$ are the solutions of the Lippmann-Schwinger equation for the potential U acting alone

$$|\phi_a^\pm(U)\rangle = |\chi_a\rangle + (E_a - H_0 \pm \mathrm{i}\epsilon)^{-1}\, U|\phi_a^\pm(U)\rangle, \tag{9.8}$$

where $|\chi_a\rangle$ is a free-particle state. We now proceed with the justification of the statements made above, beginning with the t-matrix.

The true scattering states of the *entire* potential $U+V$ are given as the solutions of the Lippmann-Schwinger equation

$$|\psi_a^\pm(U+V)\rangle = |\chi_a\rangle + (E_a - H_0 \pm \mathrm{i}\epsilon)^{-1}\,(U+V)|\psi_a^\pm(U+V)\rangle, \tag{9.9}$$

where

$$H_0|\chi_a\rangle = E_a|\chi_a\rangle. \tag{9.10}$$

Correspondingly, an application of Eq. (2.88) yields the total t-matrix,

$$\begin{aligned} t_{ba}(U+V) &= \langle \chi_b|U+V|\psi^+(U+V)\rangle \\ &= \langle \chi_b|t(U+V)|\chi_a\rangle, \end{aligned} \tag{9.11}$$

defined as a matrix element in terms of free-particle states.

It is possible to cast [82, 83] Eq. (9.11) in a form that separately exhibits the effects of the two potentials, $U+V$. We note that the scattering states $|\phi_a^\pm\rangle$ associated with U alone are obtained as the solutions of Eq. (9.8,) which can also be written in the "closed" form, (in which, however, the Green function is still unknown)

$$|\phi_a^\pm(U)\rangle = |\chi_a(U)\rangle + \left(E_a - \bar{H}_0 \pm \mathrm{i}\epsilon\right)^{-1} U|\chi_a\rangle. \tag{9.12}$$

Solving this expression for $|\chi_a\rangle$ in terms of $|\phi_a^-\rangle$,

$$|\chi_b\rangle = |\phi_b^-\rangle - \left[E_b - \bar{H}_0 - \mathrm{i}\epsilon\right]^{-1} U|\phi_b^-\rangle, \tag{9.13}$$

and noting that $\langle \chi_b| = |\chi_b\rangle^*$, we can write Eq. (9.11) (in somewhat condensed notation) in the form

$$\begin{aligned} \langle \chi_b|t|\chi_a\rangle &= \langle \chi_b|U|\psi_a^+\rangle + \langle \phi_b^-|V|\psi_a^+\rangle \\ &- \langle \chi_b|U\left[E_b - \bar{H}_0 + \mathrm{i}\epsilon\right]^{-1} V|\psi_a^+\rangle. \end{aligned} \tag{9.14}$$

The third term on the right may be transformed through the use of the Lippmann-Schwinger equation for $|\psi_a^+\rangle$. On the energy shell, $E_a = E_b = E$, we have

$$\begin{aligned} \langle\chi_b|U\left[E_b - \bar{H}_0 + i\epsilon\right]^{-1} V|\psi_a^+\rangle &= \langle\chi_b|U\left[E - \bar{H}_0 + i\epsilon\right]^{-1} V|\chi_a\rangle \\ + \quad \langle\chi_b|U\left[E - \bar{H}_0 + i\epsilon\right]^{-1} V\left[E - H + i\epsilon\right]^{-1} (U+V)|\chi_a\rangle, \end{aligned} \tag{9.15}$$

or,

$$\begin{aligned} \langle\chi_b|U\left[E_b - \bar{H}_0 + i\epsilon\right]^{-1} V|\psi_a^+\rangle &= \langle\phi_b^-|V|\chi_a\rangle - \langle\chi_b|V|\chi_a\rangle \\ + \quad \langle\chi_b|U\left[E - \bar{H}_0 + i\epsilon\right]^{-1} V\left[E - H + i\epsilon\right]^{-1} (U+V)|\chi_a\rangle. \end{aligned} \tag{9.16}$$

We now use the operator identity,

$$B^{-1}(B-A)A^{-1} = A^{-1} - B^{-1}, \tag{9.17}$$

with $A \equiv E - H + i\epsilon$ and $B \equiv E - \bar{H}_0 + i\epsilon$, and obtain,

$$\begin{aligned} &\langle\chi_b|U\left[E - \bar{H}_0 + i\epsilon\right]^{-1} V\left[E - H + i\epsilon\right]^{-1} (U+V)|\chi_a\rangle \\ = \quad &\langle\chi_b|U\left[E - H + i\epsilon\right]^{-1} (U+V)|\chi_a\rangle \\ - \quad &\langle\chi_b|U\left[E - \bar{H}_0 + i\epsilon\right]^{-1} (U+V)|\chi_a\rangle. \end{aligned} \tag{9.18}$$

With the use of the Lippmann-Schwinger equation for $|\psi_a^+\rangle$ and Eq. (9.13), the last expression can be written in the form,

$$\begin{aligned} &\langle\chi_b|U\left[E - \bar{H}_0 + i\epsilon\right]^{-1} V\left[E - H + i\epsilon\right]^{-1} (U+V)|\chi_a\rangle \\ = \quad &\langle\chi_b|U|\psi_a^+\rangle - \langle\phi_b^-|U+V|\chi_a\rangle + \langle\chi_b|V|\chi_a\rangle. \end{aligned} \tag{9.19}$$

Now, combining with Eq. (9.16), we find

$$\langle\chi_b|U\left[E - \bar{H}_0 + i\epsilon\right]^{-1} V|\psi_a^+\rangle = \langle\chi_b|U|\psi_a^+\rangle - \langle\phi_b^-|U|\chi_a\rangle. \tag{9.20}$$

So that Eq. (9.14) yields the final result,

$$\langle\chi_a|t|\chi_b\rangle = \langle\phi_b^-|U|\chi_a\rangle + \langle\phi_b^-|V|\psi_a^+\rangle. \tag{9.21}$$

Using the easily proven equality $\langle\phi_b^-|U|\chi_a\rangle = \langle\chi_a|U|\phi_b^+\rangle$, this can also be written in the form,

$$\langle\chi_a|t|\chi_b\rangle = \langle\chi_a|U|\phi_b^+\rangle + \langle\ phi_b^-|V|\psi_a^+\rangle. \tag{9.22}$$

or,

$$t_{ba}(U+V) = t_{ba}(U) + t_{ba}(V/U). \tag{9.23}$$

Here $t_{ba}(U)$ is the t-matrix associated with the potential U acting alone, and $t_{ba}(V/U)$ is a scattering-matrix-like quantity, defined with respect to reference states rather than free states,

$$t_{ba}(V/U) = \langle\phi_b^-|V|\psi_a^+\rangle, \tag{9.24}$$

which describes the additional scattering caused by V in a wave that has already been scattered by U.

9.2.2 Generalized Lippmann-Schwinger equation

The Lippmann-Schwinger equations, Eq. (2.81),

$$|\psi_a^{\pm}\rangle = |\chi_a\rangle + [E_a - H_0 \pm \mathrm{i}\epsilon]^{-1} V|\psi_a^{\pm}\rangle, \tag{9.25}$$

yield the scattered states $|\psi^+\rangle$ ($|\psi^-\rangle$) of the Schrödinger equation corresponding to a potential V, which evolve out of the incident, unperturbed state $|\chi_a\rangle$ at the same energy E_a as the scattered states. In Eq. (9.25), H_0 is the Hamiltonian of the unperturbed system, taken here to be that of force-free space. As was the case with the t-matrix in the previous subsection, the Lippmann-Schwinger equation can be generalized so that H_0 can denote the Hamiltonian associated with a reference system characterized by a potential U, with V being an additional perturbation. Then, the free-particle states and Green function in Eq. (9.25) become those of the reference system. As we now show, in this case, the Lippmann-Schwinger equations can be written in the form

$$\begin{aligned} |\psi_a^{\pm}(U+V)\rangle &= |\phi_a^{\pm}(U)\rangle + \left(E_a - \bar{H}_0 \pm \mathrm{i}\epsilon\right)^{-1} V|\psi_a^{\pm}(U+V)\rangle \\ &= |\phi_a^{\pm}(U)\rangle + \bar{G}_0(E_a)V|\psi_a^{\pm}(U+V)\rangle, \end{aligned} \tag{9.26}$$

where $\bar{G}_0$ is the Green function for the reference Hamiltonian and is defined in Eq. (9.5).

In order to derive Eq. (9.26), we begin with the general expression for the Green functions of the reference and perturbed system,

$$\bar{G}_0(z) = (z - H_0 - U)^{-1}, \tag{9.27}$$

and

$$G(z) = (z - H_0 - U - V)^{-1}, \tag{9.28}$$

where z is a complex energy parameter. Using the operator identity in Eq. (9.17), we can write

$$\begin{aligned} G &= \bar{G}_0 + GV\bar{G} \\ &= \bar{G} + \bar{G}VG. \end{aligned} \tag{9.29}$$

Then the Lippmann-Schwinger equation, Eq. (9.26) can be transformed in the following series of steps,

$$\begin{aligned} |\psi_a^{\pm}(U+V)\rangle &= |\chi_a\rangle + G(U+V)|\chi_a\rangle \\ &= |\chi_a\rangle + \bar{G}_0(U+V)|\chi_a\rangle + \bar{G}_0VG(U+V)|\chi_a\rangle \\ &= |\phi_a^{\pm}\rangle + \bar{G}_0V\left[|\chi_a\rangle + G(U+V)|\chi_a\rangle\right] \\ &= |\phi_a^{\pm}\rangle + \bar{G}_0V|\psi_a^{\pm}(U+V)\rangle. \end{aligned} \tag{9.30}$$

In deriving the third line above, we used the expression

$$|\phi_0^{\pm}(U)\rangle = |\chi_a\rangle + \bar{G}_0 U |\chi_a\rangle, \tag{9.31}$$

and the last line follows from the fact that the term in the square brackets equals $|\psi_a^{\pm}(U+V)\rangle$, according to Eq. (9.25).

Equation (9.30) provides the result that we set out to prove. It shows that the Lippmann-Schwinger equation can be used in the intuitive suggestive form in which the solutions corresponding to a perturbation V are given in terms of the solutions and Green functions of the reference system in which the perturbation is acting. It is left as an exercise for the reader to show that Eq. (9.30) can also be cast in the form

$$|\psi_a^{\pm}(U+V)\rangle = |\phi_a^{\pm}(U)\rangle + GV|\phi_a^{\pm}(U)\rangle. \tag{9.32}$$

in direct analogy with the "closed" solution form of the Lippmann-Schwinger equation given in Eq. (2.85).

9.2.3 Green function and t-matrix

Equation (9.29) provides the generalization of the Dyson Eq. (2.72) to the case in which two potentials act simultaneously. Iterating the expression in Eq. (9.29) yields the distorted-wave t-matrix defined in Eq. (9.23)

$$\begin{aligned} t(V/U) &= V + V\bar{G}_0 V + V\bar{G}_0 V\bar{G}_0 V + \cdots \\ &= \left(1 - V\bar{G}_0\right)^{-1} V. \end{aligned} \tag{9.33}$$

Therefore, we can write,

$$G(U+V) = \bar{G}_0(U) + \bar{G}_0(U) t(V/U) \bar{G}_0(U), \tag{9.34}$$

which is a direct generalization of Eq. (2.90). Similarly, we have

$$VG(U+V) = t(V/U)\bar{G}_0(U), \tag{9.35}$$

which is the corresponding generalization of Eq. (2.89). The reader is invited to show that a similar expression holds for the wave function

$$V|\psi_a^{\pm}(U+V)\rangle = t(V/U)|\phi_a^{\pm}(U)\rangle. \tag{9.36}$$

From the general expression, $G(U) = G_0 + G_0 T(U) G_0$, and Eq. (9.34), we easily deduce the relation

$$T(U+V) = T(U) + (1 + G_0 T(U))\, T(V/U)\, (1 + T(U) G_0). \tag{9.37}$$

Furthermore, Eq. (9.33) can be rearranged to yield the expression,

$$T(V/U) = T(V)\, [1 - G_0 T(U) G_0 T(V)]^{-1}. \tag{9.38}$$

Combining the last two expressions, we obtain the following relation for the total scattering matrix associated with two-potentials acting simultaneously,

$$\begin{aligned} T(V+U) &= T(U) + (1+T(U)G_0)\,T(V)\,[1-G_0T(U)G_0T(V)]^{-1} \\ &\times\ (1+G_0T(U))\,. \end{aligned} \tag{9.39}$$

This completes our generalization of some basic concepts of scattering theory to the case of two-potential scattering.

9.2.4 Change in the density of states

In this section we derive an expression for the change in the integrated density of states when a potential $V(\mathbf{r})$ is introduced into the region of a potential $U(\mathbf{r})$. From Eq. (D.119) and the general expression for the Green function,

$$G(z) = (z-H)^{-1}, \tag{9.40}$$

we have

$$\Delta N_U(E) = \frac{1}{\pi}\left[\Im \mathrm{Tr} \ln G(U) - \Im \mathrm{Tr} \ln G_0\right]. \tag{9.41}$$

Also,

$$\Delta N_{U+V}(E) = \frac{1}{\pi}\left[\Im \mathrm{Tr} \ln G(U+V) - \Im \mathrm{Tr} \ln G_0\right]. \tag{9.42}$$

Therefore, the additional change that results when $V(\mathbf{r})$ is "switched on", is given by the expression,

$$\Delta N_{V/U}(E) = \frac{1}{\pi}\left[\Im \mathrm{Tr} \ln G(U+V) - \Im \mathrm{Tr} \ln G(U)\right]. \tag{9.43}$$

We now relate this result to the distorted wave t-matrix, $t(V/U)$. Using Eq. (9.34), we can write the last equation in the form,

$$\begin{aligned} \Delta_{V/U}(E) &= \frac{1}{\pi}\Im \mathrm{tr} \ln \left\{1 + \left[1 - V\bar{G}_0(U)\right]^{-1} V\bar{G}_0(U)\right\} \\ &= \frac{1}{\pi}\Im \mathrm{tr} \ln \left[1 - V\bar{G}_0(U)\right]^{-1}. \end{aligned} \tag{9.44}$$

If the potential $V(\mathbf{r})$ is real, then $\Im \ln V = 0$ and we can write,

$$\begin{aligned} \Delta N_{V/U}(E) &= \frac{1}{\pi}\Im \mathrm{Tr} \ln \left[1 - V\bar{G}_0(U)\right]^{-1} V \\ &= \frac{1}{\pi}\Im \mathrm{Tr} \ln t(V/U). \end{aligned} \tag{9.45}$$

This is a generalization to two-potential scattering of the Lloyd formula[5], Eq. (D.126)

[5]Care should be taken in applying Eq. (9.45) to see that the wave functions are normalized consistently. An example of the effects of normalization is given at the end of the next section.

9.2.5 Spherical potentials

As an illustrative example of the foregoing formal results, we consider the case of two overlapping, spherically symmetric, spatially bounded potentials, $V_0(r)$ and $V_1(r)$, acting over the same region of space. We let $\delta_\ell^{(0)}$ and δ_ℓ denote the phase shifts of the ℓth partial wave corresponding to the reference potential $V_0(r)$ and to the total potential $V_0(r) + V_1(r)$. We also define the difference phase shift, $\delta_\ell^1 = \delta_\ell - \delta_\ell^{(0)}$. Using the definition of the t-matrix, Eq. (D.68), we have (with $k = \sqrt{E}$),

$$\begin{aligned} t_\ell &= -\frac{1}{k}\left[\mathrm{e}^{\mathrm{i}\delta_\ell}\sin\delta_\ell\right] = -\frac{1}{k}\left[\frac{\mathrm{e}^{2\mathrm{i}\delta_\ell}-1}{2\mathrm{i}}\right] \\ &= -\frac{1}{k}\left[\frac{\mathrm{e}^{2\mathrm{i}\delta_\ell^{(0)}}-1}{2\mathrm{i}} + \frac{\mathrm{e}^{2\mathrm{i}\delta_\ell}-\mathrm{e}^{2\mathrm{i}\delta_\ell^{(0)}}}{2\mathrm{i}}\right] \\ &= -\frac{1}{k}\mathrm{e}^{\mathrm{i}\delta_\ell^{(0)}}\sin\delta_\ell^{(0)} - \frac{1}{k}\mathrm{e}^{2\mathrm{i}\delta_\ell^{(0)}}\mathrm{e}^{\mathrm{i}\delta_\ell^{(1)}}\sin\delta_\ell^{(1)} \\ &= t_\ell^{(0)} + t_\ell^{(1)}. \end{aligned} \tag{9.46}$$

Here, $t_\ell^{(0)}$ is clearly the t-matrix corresponding to the potential $V_0(r)$, while $t_\ell^{(1)}$ is the extra scattering induced by $V_1(r)$ on a wave perturbed by the field of $V_0(r)$. Note the explicit appearance of the S-matrix element, $\mathrm{e}^{2\mathrm{i}\delta_\ell^{(0)}}$, associated with $V_0(r)$ in the second term of Eq. (9.46). This term is a reminder that the scattering described by the second term does not takes place in free space but in a region where a potential is acting.

It is also instructive to consider the explicit expressions for the t-matrix in the coordinate representation. We let $v_\ell(r)$ and $u_\ell(r)$ denote, respectively, the solutions of the radial Schcrödinger equation corresponding to potentials $V_0(r)$ and $V_0(r) + V_1(r)$. Then from Eq. (D.68) we can write

$$t_\ell = \int_0^\infty r^2 j_\ell(kr)V_0(r)u_\ell(r)\mathrm{d}r + \int_0^\infty r^2 j_\ell(kr)V_1(r)u_\ell(r)\mathrm{d}r, \tag{9.47}$$

whereas from Eq. (9.23) we obtain

$$t_\ell = \int_0^\infty r^2 j_\ell(kr)V_0(r)v_\ell(r)\mathrm{d}r + \int_0^\infty r^2 v_\ell(kr)V_1(r)u_\ell(r)\mathrm{d}r. \tag{9.48}$$

With the normalization of the wave functions as indicated in Eq. (D.66), we find

$$\begin{aligned} t_\ell^{(1)} &= -\frac{1}{k}\mathrm{e}^{2\mathrm{i}\delta_\ell^{(0)}}\mathrm{e}^{\mathrm{i}\delta_\ell^{(1)}}\sin\delta_\ell^{(1)} \\ &= -\frac{1}{k}\int_0^\infty r^2 v_\ell(kr)V_1(r)u_\ell(r)\mathrm{d}r. \end{aligned} \tag{9.49}$$

Equation (9.47) represents the complete scattering from the two potentials, $V_0(r)+V_1(r)$, and the two terms in it have no independent significance. On the other hand, the first term in Eq. (9.48) represents the scattering from $V_0(r)$ with $V_1(r)$ "switched off", while the second term gives the *additional* scattering produced by $V_1(r)$ in the field of $V_0(r)$.

These equations lead to an interesting result for the functional derivative of t with respect to the potential. Taking the limit, $V_1 \equiv \delta U \to 0$, we can set $u_\ell = v_\ell$ so that with $\delta t_\ell = t_\ell - t_\ell^{(0)} \equiv t_\ell^{(1)}$ we have

$$\frac{\delta t_\ell}{\delta U} = -\frac{1}{k}\int_0^\infty r^2 v_\ell(kr) v_\ell(kr)\mathrm{d}r. \tag{9.50}$$

The reader is invited to show that in the general case of non-spherical potentials, we have

$$\frac{\delta \underline{t}}{\delta U} = \int \mathrm{d}^3 r |\phi(\mathbf{r})\rangle\langle\phi(\mathbf{r})|, \tag{9.51}$$

where $|\phi(\mathbf{r})\rangle$ is a solution of the Schrödinger equation associated with a potential $U(\mathbf{r})$ expressed in the angular momentum representation.

It is always possible to choose solutions of the Schrödinger equation that are real. Such a solution is obtained in the present case through multiplication of the radial wave function in Eq. (D.66) by $\mathrm{e}^{-\mathrm{i}\delta_\ell}$, so that in the asymptotic region, we have,

$$\bar{R}_\ell(r) = j_\ell(kr)\cos\delta_\ell - n_\ell(kr)\sin\delta_\ell. \tag{9.52}$$

Using different normalizations leads to different explicit expressions for the distorted-wave t-matrix, $t(V/U)$. For example, introducing the functions,

$$\begin{aligned} \bar{\bar{R}}_\ell(r) &= \mathrm{e}^{-\mathrm{i}\delta_\ell^{(0)}} R_\ell(r) \\ &= \mathrm{e}^{\mathrm{i}(\delta_\ell - \delta_\ell^{(0)})}\bar{R}_\ell(r) \end{aligned} \tag{9.53}$$

we obtain from Eq. (9.49) the expression,

$$\begin{aligned} \bar{t}_\ell^{(1)} &= \mathrm{e}^{-\mathrm{i}\delta_\ell^{(0)}} t_\ell^{(1)} \mathrm{e}^{-\mathrm{i}\delta_\ell^{(0)}} \\ &= -\frac{1}{k}\int_0^\infty r^2 \bar{v}_\ell(kr) V_1(r) \bar{\bar{u}}_\ell(r)\mathrm{d}r \\ &= -\frac{1}{k}\int_0^\infty r^2 v_\ell^*(kr) V_1(r) u_\ell(r)\mathrm{d}r \\ &= -\frac{1}{k}\mathrm{e}^{\mathrm{i}\delta_\ell^{(1)}}\sin\delta_\ell^{(1)}. \end{aligned} \tag{9.54}$$

The t-matrix $\bar{t}_\ell^{(1)}$ is a natural extension of the t-matrix associated with a single potential, and describes the additional scattering due to $V_1(r)$.

As a final application, we consider the additional change in the density of states resulting when a perturbing potential is introduced into a force field. From Eq. (3.162) we obtain the change in the integrated density of states caused by the presence of $U = V_0(r)$, the reference potential,

$$\begin{aligned} \Delta N_U(E) &= \frac{1}{\pi}\Im \mathrm{Tr} \ln t(U) \\ &= \frac{1}{\pi}\sum_{\ell}(2\ell+1)\delta_\ell^{(0)}. \end{aligned} \tag{9.55}$$

The corresponding change in the presence of the entire potential $U + V = V_0(r) + V_1(r)$ takes the form,

$$\begin{aligned} \Delta N_{U+V}(E) &= \frac{1}{\pi}\Im \mathrm{Tr} \ln t(U+V) \\ &= \frac{1}{\pi}\sum_{\ell}(2\ell+1)\delta_\ell. \end{aligned} \tag{9.56}$$

It follows that the change in the integrated density of states introduced by $V_1(r)$, is given by the expression

$$\begin{aligned} \Delta N_{V/U}(E) &= \frac{1}{\pi}\sum_{\ell}(2\ell+1)\left(\delta_\ell - \delta_\ell^{(0)}\right) \\ &= \frac{1}{\pi}\sum_{\ell}(2\ell+1)\delta_\ell^{(1)} \\ &= \frac{1}{\pi}\Im \mathrm{Tr} \ln \bar{t}_\ell^{(1)}. \end{aligned} \tag{9.57}$$

We see that the t-matrix $\bar{t}_\ell^{(1)}$ properly describes the added scattering resulting from the addition of $V_1(r)$ into the field of $V_0(r)$.

9.3 Substitutional Impurities

The formalism of the previous section allows the calculation of the wave function, the t-matrix, and the Green function when a spatially bounded potential $U(\mathbf{r})$ is changed by an amount $V(\mathbf{r})$. We now extend this formalism to the case in which $V(\mathbf{r})$ denotes a localized perturbation not in a spatially bounded system but in a translationally invariant host material.

The term localized will be taken to mean any spatially bounded deviation from the periodic potential characterizing the host medium. Perhaps the simplest such perturbation is that which results from the substitution of a host atom in a metallic system by an atom of a different kind. Localized perturbations can also result when a host atom is displaced from its equilibrium lattice position, or a foreign or even host atom is embedded in the interstitial region of an otherwise pure metal. It is important that the host has metallic character so that screening by the conduction electrons can confine the perturbation to a finite region around the disturbance. Because of its relative formal simplicity, first we consider the case of substitutional impurities, deferring the discussion of atomic displacements and interstitial impurities until later.

In the simplest model of a substitutional impurity the change in the potential is confined to the impurity cell. However, a substitutional impurity will affect the host atoms in its vicinity in many and physically significant ways. Differences in valence between the host and the impurity atoms will, in general, give rise to charge rearrangement around the impurity site that may extend to the first few coordination cells around the impurity site. This rearrangement of charge may imply that the Wigner-Seitz cells defined by the lattice containing the impurity and some of its neighbors are no longer charge neutral. We say that an "effective" charge transfer has taken place to or from the impurity site. It is conceivable, of course, especially in metallic systems, that space can be repartitioned into cells of unequal volume so that each cell is neutral. However, *potential relaxation* where the potential on nearby cells is changed by the presence of the impurity will persist even under repartition into neutral cells. Also, the difference in shape and volume that must accompany such partitions can pose a difficult computational problem in a realistic, full-cell treatment of the electronic structure. In addition, the electrostatic interaction that results from such rearrangements of charge as well as any difference in size between the host and impurity ions, may cause the host atoms nearest the impurity to be displaced from their equilibrium lattice positions, an effect referred to as *lattice relaxation*. Thus, the model of a substitutional impurity whose perturbation is confined to the impurity cell can be a rather severe idealization. On the other hand, the treatment of this idealized problem can form the starting point for the development of more refined methods that would allow the self-consistent treatment of the charge and the potential in the region of a localized perturbation. This would contain the effects of the substitution as well as any accompanying relaxation effects.

In the following discussion, we confine our attention to an *impurity cluster*, C, consisting of a number of substitutional impurities and their "cloud" of perturbed host atoms in their vicinity. A schematic diagram of such a cluster is given in Fig. 9.1. Our aim is to evaluate the Green function associated with the perturbed system. In general, this Green function can be written in the form,

$$G = \bar{G}_0 + \bar{G}_0 \Delta V G, \tag{9.58}$$

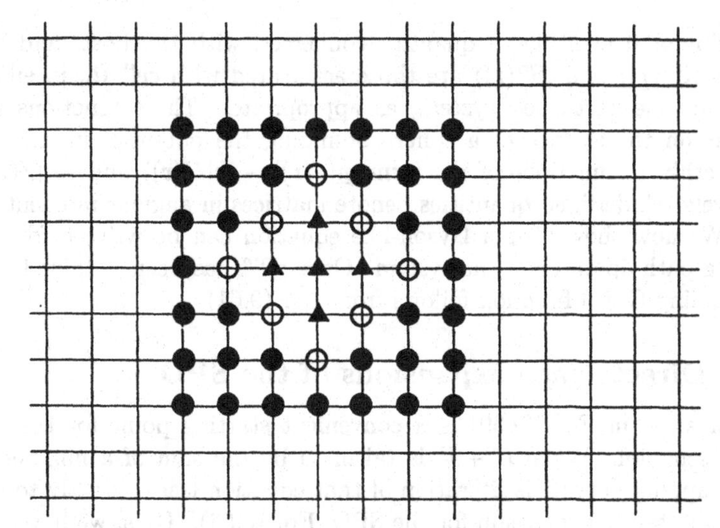

Figure 9.1: **Schematic representation of an impurity cluster consisting of substitutional impurities, (open triangles), and perturbed host atoms, (open circles), surrounded by the unperturbed host, (filled circles).**

where

$$\Delta V = \sum_n \Delta v_n \tag{9.59}$$

with Δv_n being the change in potential in cell Ω_n in the impurity cluster caused by the presence of the impurity or impurities. The Green function $\bar{G}_0$ is that of the unperturbed host material and is assumed known.

Although it is possible to work directly with the Dyson equation for the Green function Eq. (9.58), or with the wave function [84], it is somewhat more convenient within the formal lines followed in this book to develop the formalism in terms of the system t-matrix, or the scattering-path operator (SPO), defined in Eq. (3.53). We recall from Eq. (3.139) that for $\mathbf{r}$ in cell Ω_n and $\mathbf{r}'$ in cell Ω_m, the Green function for a scattering assembly is given in terms of the corresponding SPO and the regular and irregular solutions of the Schrödinger equation for the potentials in those cells. Thus, in the angular momentum representation we have the expressions,

$$\bar{G}_0(\mathbf{r}, \mathbf{r}') = \langle Z_0^n(\mathbf{r}) | \underline{\tau}_0^{nm} | Z_0^m(\mathbf{r}') \rangle - \langle Z_0^n(\mathbf{r}) | S_0^n(\mathbf{r}') \rangle \delta_{nm}, \tag{9.60}$$

and

$$\bar{G}(\mathbf{r}, \mathbf{r}') = \langle Z^n(\mathbf{r}) | \underline{\tau}^{nm} | Z^m(\mathbf{r}') \rangle - \langle Z^n(\mathbf{r}) | S^n(\mathbf{r}') \rangle \delta_{nm}, \tag{9.61}$$

for the Green functions of the unperturbed host material and the perturbed system containing the impurity cluster, respectively. In these equations, a subscript of 0 identifies a quantity connected with the host, and the functions $|Z^n(\mathbf{r})\rangle$ and $|S^n(\mathbf{r})\rangle$ are those associated with cell Ω_n in either the host or the perturbed system, as appropriate. These functions are such that on the surface of a sphere bounding the potential in the cell join smoothly to functions of the form $\underline{m}|J(\mathbf{r})\rangle - ik|H(\mathbf{r})\rangle$ and to $|J(\mathbf{r})\rangle$, respectively. Underlined quantities denote matrices in angular momentum space. We now show that a Dyson-like equation can be written for the scattering-path operator of the system. Once $\underline{\tau}^{nm}$ has been obtained, the corresponding Green function follows from Eq. (9.61).

9.3.1 Direct space expansions of the SPO

The expression in Eq. (9.39) is a convenient starting point for the following discussion. When $U + V$ is taken to be the sum of a number of non-overlapping potentials, iteration of that equation leads directly to the multiple-scattering expression for the SPO, Eq. (3.53). Thus, we have,

$$\underline{\tau}^{nm} = \underline{t}^n \left[\delta_{nm} + \sum_{k \neq n} \underline{G}^{nk} \underline{\tau}^{km} \right], \tag{9.62}$$

where $\underline{G}^{kn} = \underline{G}(\mathbf{R}_{kn})$ is a real-space structure constant. Before proceeding with formal manipulations of Eq. (9.62), a few words about the convergence of the summations over angular momentum indices that are inherent in this equation are in order. As we discussed in Chapter 3, in the case of MT potentials all summations over L converge absolutely, regardless of the order in which the sums are carried out. In the case of convex-shaped, space-filling cells, in which all the intercell vectors, $\mathbf{R}_{nm}$, are larger than any intracell vector, the sums also converge irrespective of order provided that each sum is carried to convergence before another summation is started. Finally, in the case in which intracell vectors may exceed in length vectors between the cells, or in the case of interpenetrating, concave cells, other summation procedures can be invoked [85] to guarantee convergence. In any case, the final, summed results are valid in all cases of non-overlapping potentials, regardless of cell shapes or the proximity between cells.

Now, the second term on the right-hand side of Eq. (9.62) can be treated formally as a perturbation leading to the following series of approximations for the site-diagonal part of the SPO,

$$\begin{aligned} \left[\underline{\tau}^{(1)}\right]^{00} &= \underline{t}^0 \\ \left[\underline{\tau}^{(2)}\right]^{00} &= \underline{t}^0 + \sum_{k \neq 0} \underline{t}^0 \underline{G}^{0k} \underline{t}^k \underline{G}^{k0} \underline{t}^0 \end{aligned}$$

$$\begin{aligned}[\underline{\tau}]^{00} &= \underline{t}^0 + \sum_{k\neq 0} \underline{t}^0 \underline{G}^{0k} \underline{t}^0 k \underline{G}^{k0} \underline{t}^0 \\ &+ \sum_{k\neq 0} \sum_{m\neq k,0} \underline{t}^0 \underline{G}^{0k} \underline{t}^k \underline{G}^{km} \underline{t}^m \underline{G}^{m0} \underline{t}^0 + \cdots . \end{aligned} \tag{9.63}$$

This series can be summed in the form,

$$\begin{aligned}\underline{\tau}^{00} &= \underline{t}^0 + \underline{t}^0 \underline{\Delta}^0 \underline{\tau}^{00} \\ &= \left(\underline{m}^0 - \underline{\Delta}^0\right)^{-1}, \end{aligned} \tag{9.64}$$

where $\underline{m} = \underline{t}^{-1}$, and the *renormalized interactor*, $\underline{\Delta}^0$, is given by the expression,

$$\begin{aligned}\underline{\Delta}^0 &= \sum_{n\neq 0} \underline{G}^{0n} \underline{t}^n \underline{G}^{n0} + \sum_{n\neq 0} \sum_{m\neq 0} \underline{G}^{0n} \underline{t}^n \underline{G}^{nm} \underline{t}^m \underline{G}^{m0} + \cdots \\ &= \sum_{n\neq 0} \sum_{m\neq 0} \underline{G}^{0n} \underline{\Gamma}^{(0)}_{nm} \underline{G}^{m0}. \end{aligned} \tag{9.65}$$

The renormalized interactor can be interpreted as the sum of all scattering paths that start and end on site 0 but avoid site 0 at all intermediate steps. Consequently, the quantity $\underline{\Gamma}^{(0)}_{nm}$ is the scattering path operator corresponding to the system that results when site 0 is removed from the original system, or equivalently is replaced by vacuum. Thus, $\underline{\Gamma}^{(0)}_{nm}$ is the minor of the matrix $\underline{\tau}^{nm}$ corresponding to site 0, and will be referred to as the *defective medium* scattering path operator.

We note that from Eq. (9.64) we have the relation,

$$\underline{\Delta}^0 = \underline{m}^0 - \left[\underline{\tau}^{00}\right]^{-1}. \tag{9.66}$$

This equation can often be used to evaluate the renormalized interactor. Consider the case of a translationally invariant material where $\underline{\tau}^{00}$ is given as an integral over the Brillouin zone, as indicated in Eq. (3.57),

$$\underline{\tau}^{00} = \frac{1}{\Omega_{\mathrm{BZ}}} \int_{\mathrm{BZ}} \underline{\tau}(\mathbf{k}) \mathrm{d}^3 k. \tag{9.67}$$

The calculation of the renormalized interactor in this case is of comparable difficulty to that of the SPO.

We now consider an application of Eq. (9.64) to the rather simple model in which the perturbation due to a substitutional impurity at site 0 is confined entirely to the impurity site, leaving the surrounding atoms undisturbed. First, we note that the renormalized interactor defined in Eq.

(9.65) depends on the nature of the material surrounding site 0, but has no explicit dependence on the t-matrix of the site itself. Denoting the inverse of the impurity t-matrix by $\underline{m}_{\mathrm{I}}^{0}$, and realizing that within the model used here the renormalized interactor is that corresponding to the undisturbed host material, we find the expression,

$$\begin{aligned} \underline{\tau}_{\mathrm{I}}^{00} &= \left(\underline{m}_{\mathrm{I}}^{0} - \underline{\Delta}^{0}\right)^{-1} \\ &= \left[\underline{m}_{\mathrm{I}}^{0} - \underline{m}^{0} + \left(\underline{\tau}^{00}\right)^{-1}\right]^{-1}. \end{aligned} \tag{9.68}$$

From this expression and Eq. (9.61) with $n = m = 0$, we readily obtain the cell-diagonal element of the Green function associated with the impurity site, (for $r' > r$),

$$G(\mathbf{r},\mathbf{r}';E) = \langle Z^{\mathrm{I}}(\mathbf{r};E)|\underline{\tau}_{\mathrm{I}}^{00}(E)|Z^{\mathrm{I}}(\mathbf{r}';E)\rangle - \langle Z^{\mathrm{I}}(\mathbf{r};E)|S^{\mathrm{I}}(\mathbf{r}';E)\rangle, \tag{9.69}$$

where it is hoped that the definition of the various symbols is clear, and the dependence of the various quantities on the energy is explicitly indicated.

Once the Green function has been determined, the number (and charge) density of the electronic system at energy E follows from the usual relation,

$$n(\mathbf{r};E) = -\frac{1}{\pi}\Im G(\mathbf{r},\mathbf{r};E). \tag{9.70}$$

From this the number density can be obtained as an integral over the energy,

$$n(\mathbf{r}) = -\frac{1}{\pi}\int \mathrm{d}E n(\mathbf{r};E), \tag{9.71}$$

while the impurity (local) density of states at energy E is given as an integral over the domain of the cell,

$$n(E) = -\frac{1}{\Omega_0}\int n(\mathbf{r};E)\mathrm{d}^3r. \tag{9.72}$$

It is important at this point to amplify the meaning of the term local DOS used above in connection with Eq. (9.72). The density of states function (and the density matrix in general) represent global properties of the system. Thus, both can be defined as direct integrals over the Brillouin zone of the reciprocal lattice of a translationally invariant material. Note that a single impurity does not affect the density of states in a global sense. This involves the average of the trace of the Green function over all sites of the system, and an impurity cell embedded in an infinite host makes a negligible contribution to this trace. However, Eqs. (9.71) and (9.72) define the density and the density of states in a strictly *local* sense, in terms of the Green function whose arguments are restricted to a specific

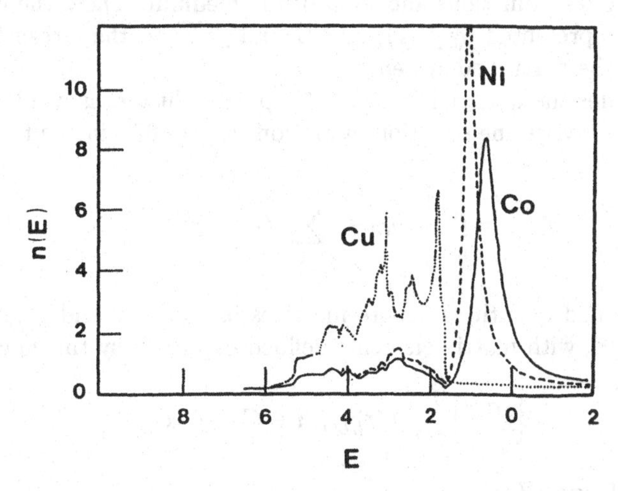

Figure 9.2: **Local DOS for Ni and Co single impurities embedded in Cu. (P. H. Dederichs, private communication.)**

region of space. Such local quantities can deviate substantially from their global counterparts, as is illustrated in Fig. 9.2. This figure shows the local DOS in states per atom per eV as a function of energy in eV for single Ni and Co impurities embedded in elemental Cu. The figure indicates clearly the high peaks by which the impurity manifests itself in the DOS. These peaks correspond to fairly sharp (in energy) resonant states of the impurity potential, which are broadened through the interaction with the conduction electrons in the host. We will have occasion to examine the presence and significance of such ***impurity states*** when we apply MST to the study of substitutionally disordered alloys.

9.3.2 Clusters of substitutional impurities

In our attempt to derive the relation between the SPO of the host and the corresponding quantity for a system with substitutional impurities, we next generalize the single-site results of the previous subsection to an impurity cluster. Figure 9.1 shows a schematic representation of such a cluster. We designate by C this entire cluster of atoms, consisting of substitutional impurities or host atoms that are perturbed by the impurity. We also use the symbol $\in$ to indicate that a site n belongs, $n \in C$, or does not belong, $n \notin C$, to the cluster C.

As in the single-impurity case, the following development is based on the two-potential scattering formalism presented above. Note that the potential of the reference medium corresponds to $U(\mathbf{r})$ in Eq. (9.34) with $\bar{G}(U)$ and

$\underline{\tau}$ being the Green functions and SPO of the medium. Then, the impurity potential is represented by $V(\mathbf{r})$, and G and $\underline{\tau}^{CC'}$ to the Green function and SPO of the perturbed system.

We consider the material, host plus impurity cluster, as a collection of clusters C and write the equation of motion, Eq. (9.62), in the form,

$$\underline{\tau}^{CC'} = \underline{t}^C \left[\delta_{CC'} + \sum_{C'' \neq n} \underline{G}^{CC''} \underline{\tau}^{C''C'} \right]. \tag{9.73}$$

Here, underlined quantities denote matrices in *both* site and angular momentum space, with matrix elements defined explicitly by the equations,

$$\left[\underline{\tau}^{CC'}\right]^{ij}_{LL'} = \tau^{ij}_{LL'}, \quad i \in C,\ j \in C', \tag{9.74}$$

$$\left[\underline{G}^{CC'}\right]^{ij}_{LL'} = G(\mathbf{R}_{ij})_{LL'}, \quad i \in C,\ j \in C',\ \ C \neq C', \tag{9.75}$$

and

$$\left\{ [\underline{t}]^{-1} \right\}^{ij}_{LL'} \equiv [\underline{m}^C]^{ij}_{LL'} = \begin{cases} m^i_{LL'} & i = j,\ i \in C \\ -G_{LL'}(\mathbf{R}_{ij}), & i \neq j,\ i, j \in C. \end{cases} \tag{9.76}$$

Now, Eq. (9.73) can be iterated with the second term on the right treated formally as a perturbation in exact analogy with the single-site Eq. (9.62). This yields the cluster-diagonal part of $\underline{\tau}^{CC}$, corresponding to cluster C,

$$\underline{\tau}^{CC} = \left[(\underline{m}^C)^{-1} - \underline{\Delta}^C\right]^{-1}, \tag{9.77}$$

where, the *cluster renormalized interactor*, $\underline{\Delta}^C$, describes the interaction of the cluster with the surrounding medium. The reader is invited to show that the site matrix elements of $\underline{\Delta}^C$ are given by the expression,

$$\begin{aligned} [\underline{\Delta}]^{ij} &= \sum_{k \notin C} \underline{G}(\mathbf{R}_{ik}) \underline{t}^k \underline{G}(\mathbf{R}_{kj}) \\ &+ \sum_{k \notin C} \sum_{n \notin C} \underline{G}(\mathbf{R}_{ik}) \underline{t}^k \underline{G}(\mathbf{R}_{kn}) \underline{t}^n \underline{G}(\mathbf{R}_{nj}) + \cdots \\ &= \sum_{k \notin C} \sum_{n \notin C} \underline{G}(\mathbf{R}_{ik}) \underline{\Gamma}^{(C)}_{kn} \underline{G}(\mathbf{R}_{nj}). \end{aligned} \tag{9.78}$$

Like the corresponding single-site quantity defined in Eq. (9.65), $\left[\underline{\Delta}^C\right]^{ij}$ corresponds to the sum of all paths that start at site i and end at site j of the cluster but avoid all sites in the cluster at intermediate steps. Thus, $\underline{\Delta}^C$ depends on the nature of the material surrounding cluster C, but is independent of the cluster potential itself. We can also identify $\underline{\Gamma}^{(C)}$ as the defective medium SPO for the system that results when the cluster C has been removed from the host material (or equivalently replaced by vacuum).

The properties of the renormalized interactor just mentioned allow us to obtain a useful and calculable expression for $\underline{\Delta}^C$ in the case of an impurity cluster embedded in an otherwise unperturbed host material. With a subscript I designating quantities associated with the impurity cluster, we have

$$\underline{\tau}_{\mathrm{I}}^{CC} = \left[\underline{m}_{\mathrm{I}}^C - \underline{\Delta}^C\right]^{-1}, \tag{9.79}$$

where $\left[\underline{m}_{\mathrm{I}}^C\right]_{LL'}^{ij}$ is given by Eq. (9.76), but with $m_{LL'}^i$ in that definition being replaced by the quantity associated with cell Ω_i in the impurity cluster. Now, since the renormalized interactor is independent of the potential in the cluster, it remains unchanged when every cell in the cluster is replaced with an unperturbed cell of the host material. Thus, the quantity $\underline{\tau}^{CC}$ for a host cluster is given by the expression,

$$\underline{\tau}^{CC} = \left[\underline{m}^C - \underline{\Delta}^C\right]^{-1}. \tag{9.80}$$

Here, $\underline{m}^C$ is defined through Eq. (9.76) with $\underline{m}^i$ denoting the inverse of the cell t-matrix of the host. From the last expression, we obtain the result,

$$\underline{\Delta}^C = \underline{m}^C - \left[\underline{\tau}^{CC}\right]^{-1}, \tag{9.81}$$

where, the elements of the host SPO are given as integrals over the BZ, as is indicated in Eq. (3.57). It follows that we can write,

$$\underline{\tau}_{\mathrm{I}}^{CC} = \left[\underline{m}_{\mathrm{I}}^C - \underline{m}^C + \left[\underline{\tau}^{CC}\right]^{-1}\right]^{-1}. \tag{9.82}$$

Since the real space structure constants in Eq. (9.76) do not depend on the occupation of the sites in the cluster, the quantity $\underline{m}_{\mathrm{I}}^C - \underline{m}^C$ is purely site diagonal,

$$\left[\underline{m}_{\mathrm{I}}^C - \underline{m}^C\right]^{ij} = \left(\underline{m}_{\mathrm{I}}^i - \underline{m}^i\right)\delta_{ij}. \tag{9.83}$$

Once $\underline{\tau}_i^{CC}$ has been determined, the Green function can be found through the general relation, Eq. (3.82). For example, with $\mathbf{r}$ in cell Ω_i and $\mathbf{r}'$ in cell Ω_j in the impurity cluster, and setting $r' > r$, we have

$$G(\mathbf{r},\mathbf{r}';E) = \langle Z^i(\mathbf{r};E)|\underline{\tau}_{\mathrm{I}}^{ij}(E)|Z^j(\mathbf{r}';E)\rangle - \langle Z^i(\mathbf{r};E)|S^i(\mathbf{r}';E)\rangle\delta_{ij}. \tag{9.84}$$

From this expression, related quantities such as the number density and the density of states follow the usual way, Eqs. (9.71) and (9.72). This completes the formal development needed to derive the Dyson equation between the SPO of the system with the impurity cluster and that for the unperturbed host.

9.4 Dyson Equation for the SPO

A Dyson equation for the SPO can be derived as follows: Let $\underline{\tau}^C$ denote the scattering-path operator for the perturbed system. Then,

$$\underline{\tau}^C = \left[\underline{\tau} - \delta\underline{m}^C\right]^{-1}, \tag{9.85}$$

where $\delta\underline{m}^C = \underline{m}^C - \underline{m}_I^C$. We note that $\underline{\tau}^C$ corresponds to the entire material, and in contrast to the cluster diagonal part, $\underline{\tau}^{CC}$, has site matrix elements for all sites in the system. From the last expression, we obtain

$$\begin{aligned} \underline{\tau}^C &= \underline{\tau} + \underline{\tau}\delta\underline{m}\underline{\tau} + \underline{\tau}\delta\underline{m}\underline{\tau}\delta\underline{m}\underline{\tau} + \cdots \\ &= \underline{\tau} + \underline{\tau}\delta\underline{m}\underline{\tau}^C \\ &= \left[1 - \underline{\tau}\delta\underline{m}\right]^{-1}\underline{\tau}. \end{aligned} \tag{9.86}$$

The second line of the last expression gives the Dyson-like equation for the scattering path operator, and the third line gives its formal solution. This equation can be put in a number of useful forms.

The quantity $\delta\underline{m}^C$ describes the perturbation in the system that results from the change in the cell (inverse) t-matrix. With the definition,

$$\begin{aligned} \underline{X}^{CC} &= \delta\underline{m}^C\left[1 - \underline{\tau}\delta\underline{m}^C\right]^{-1} \\ &= \left[1 - \delta\underline{m}^C\underline{\tau}\right]^{-1}\delta\underline{m}^C, \end{aligned} \tag{9.87}$$

we can cast Eq. (9.86) in the form,

$$\underline{\tau}^C = \underline{\tau} + \underline{\tau}\underline{X}^{CC}\underline{\tau}. \tag{9.88}$$

Equations (9.86), (9.87), and (9.88) should be compared, respectively, to Eqs. (2.74), (2.87), and (2.90). Clearly the quantities $\underline{\tau}$, $\underline{\tau}^C$, $\delta\underline{m}^C$, and $\underline{X}^{CC}$ in the former set of equations correspond to the quantities G_0, G, V, and t in the latter. Thus, $\underline{X}^{CC}$ describes the *additional* scattering of waves in a medium described by $\underline{\tau}$ that is caused by the presence of the impurity perturbation, $\delta\underline{m}^C$. Now, from Eq. (9.88) we obtain the site matrix elements of $\underline{\tau}^C$ with respect to any two sites in the lattice,

$$\left[\underline{\tau}^C\right]^{ij} = \underline{\tau}^{ij} + \sum_{k\in C}\sum_{n\in C}\underline{\tau}^{ik}\left[\underline{X}^C\right]^{kn}\underline{\tau}^{nj}. \tag{9.89}$$

The last expression can be used to obtain a useful form for the cluster defective medium scattering path operator, $\underline{\Gamma}^{(C)}$. First, we note that by definition $\underline{X}^{CC}$ is cluster diagonal. Then, in Eq. (9.89) we take the limit $\delta\underline{m}^C \to \infty$, which corresponds to replacing the cluster cells by vacuum (so that the t-matrix vanishes). In this limit, $\underline{\tau}^C \to \underline{\Gamma}^{(C)}$ and from Eq. (9.89), we have

$$\left[\underline{\Gamma}^{(C)}\right]^{ij} = \underline{\tau}^{ij} - \sum_{k\in C}\sum_{n\in C} \underline{\tau}^{ik}\left\{\left[\underline{\tau}^{CC}\right]^{-1}\right\}^{kn}\underline{\tau}^{nj}, \tag{9.90}$$

or, in matrix form,

$$\underline{\Gamma}^{(C)} = \underline{\tau} - \underline{\tau}\left[\underline{\tau}^{CC}\right]^{-1}\underline{\tau}. \tag{9.91}$$

Consistent with these expressions, the matrix elements of $\underline{\Gamma}^{(C)}$ between cluster sites vanish.

Having obtained the scattering path operator for the perturbed system, the Green function is given by Eq. (3.82). Related quantities, such as the number density and the DOS then follow from the usual relations.

9.5 Displaced Atoms

As another example of the formalism developed above, we consider the calculation of the Green function associated with a cluster of atoms, C, which have been displaced from their equilibrium lattice position. *Lattice relaxation* plays a major role in determining a number of physical properties of materials, e.g., in the formation of magnetic moments. Although no general treatment exists for ***structurally disordered*** materials such as liquids or amorphous solids, comparable in rigor to treatments of ordered materials, it is possible to treat a *finite* cluster of atoms displaced from their equilibrium lattice sites by embedding the cluster in an otherwise translationally invariant medium.

Figure 9.3 shows a schematic diagram of a number of atoms, j, displaced from their equilibrium positions by vectors, $\boldsymbol{\rho}_j$. Let us assume that the vectors $\boldsymbol{\rho}_j$ are smaller than the intercell vectors $\mathbf{R}_{ij}$ linking cell Ω_j to its nearest neighbors, thus satisfying the MT conditions between intercell and intracell vectors. This is a convenient condition that allows the performance of a number of formal manipulations, but it is not absolutely necessary. As we saw in Chapter 3, it is formally possible within MST to replace divergent single sums with conditionally convergent double (or multiple) sums that lead to well-defined results.

The Green function of the system, Eq. (9.61), is given in terms of the scattering path operator for the system, which in turn is given by Eq. (9.82),

$$\underline{\tau}_{\mathrm{I}}^{CC} = \left[\underline{m}_{\mathrm{I}}^{C} - \underline{\hat{\Delta}}^{C}\right]^{-1}. \tag{9.92}$$

Here, the matrix $\underline{m}_{\mathrm{I}}^{C}$ is given by an expression analogous to Eq. (9.76),

$$\left\{\left[\underline{t}\right]^{-1}\right\}^{ij}_{LL'} \equiv \left[\underline{m}^C\right]^{ij}_{LL'} = \begin{cases} m^i_{LL'} & i=j,\ i\in C \\ -G_{LL'}(\mathbf{R}_{ij}+\boldsymbol{\rho}_j), & i\neq j,\ i,j\in C, \end{cases} \tag{9.93}$$

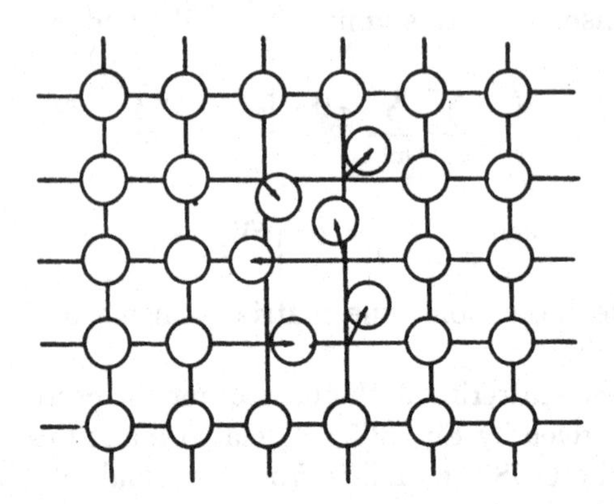

Figure 9.3: **Schematic representation of an impurity cluster consisting of displaced atoms.**

where i, j are sites in the cluster, C, and $m^i_{LL'}$ is the inverse cell t-matrix for the ***displaced*** cell Ω_i. This t-matrix in principle contains the effects of the displacement on the potential of the cell. However, we impose the condition that the cluster of the impurity cells (which can contain the displaced atoms and host atoms perturbed by them) is finite and is surrounded by host atoms unperturbed by the impurity. The structure constants in the last equation are evaluated at the true intercell distances in the impurity cluster,

$$\mathbf{R}_{ij} + \boldsymbol{\rho}_{ij} \equiv \mathbf{R}_j + \boldsymbol{\rho}_j - (\mathbf{R}_i + \boldsymbol{\rho}_i) \,. \tag{9.94}$$

Also, the renormalized interactor, $\hat{\underline{\Delta}}^C$, is distinguished by a hat to indicate that the sites in the cluster are displaced from their equilibrium (undisplaced) lattice positions. Therefore, $\hat{\underline{\Delta}}^C$ is given by an expression similar to that in Eq. (9.78), *but* in terms of the displaced positions of the sites i and j,

$$\left[\hat{\underline{\Delta}}^C\right]^{ij} = \sum_{k \notin C} \sum_{n \notin C} \underline{G}\left[\mathbf{R}_k - (\mathbf{R}_i + \boldsymbol{\rho}_i)\right] \underline{\Gamma}^{(C)}_{kn} \underline{G}\left[(\mathbf{R}_i + \boldsymbol{\rho}_i) - \mathbf{R}_n\right] . \tag{9.95}$$

The quantity $\underline{\Gamma}^{(C)}_{kn}$ is the SPO for the defective medium (the material from which C has been removed) and depends only on the host surrounding the cluster. Thus, it is given by Eq. (9.91).

As is the case with all multiple scattering expansions that involve products of structure constants and cell t-matrices, care should be taken to employ a convergent procedure in performing summations over the angular momentum index. With this in mind, we expand $\underline{G}(\mathbf{R}_{ik} + \boldsymbol{\rho}_i)$ in the form, Eq. (F.14),

$$\underline{G}(\mathbf{R}_{ik} + \boldsymbol{\rho}_i) = \underline{g}(-\boldsymbol{\rho}_i)\underline{G}(\mathbf{R}_{ik}), \tag{9.96}$$

and cast Eq. (9.95) in the form,

$$\left[\hat{\underline{\Delta}}^{C}\right]^{ij} = \underline{g}(\rho_i)\left[\underline{\Delta}^{C}\right]^{ij}\underline{g}(-\rho_j), \tag{9.97}$$

where $[\underline{\Delta}]^{ij}$ is the renormalized interactor corresponding to undisplaced sites in the cluster, and is given by Eq. (9.81). Now, Eq. (9.92) can be written as

$$\underline{\tau}_{\mathrm{I}}^{CC} = \left[\underline{m}_{\mathrm{I}}^{C} - \underline{g}^{C}\underline{\Delta}^{C}\underline{g}^{C\dagger}\right]^{-1}, \tag{9.98}$$

where the site-diagonal matrix $\underline{g}^{C}$ has matrix elements

$$\left[\underline{g}^{C}\right]^{ij} = \underline{g}(\rho_i)\delta_{ij}. \tag{9.99}$$

We note that since $\underline{g}^{C}\underline{g}^{C\dagger} = 1$, Eq. (F.23), we can also write

$$\underline{\tau}_{\mathrm{I}}^{CC} = \underline{g}^{C}\left[\underline{g}^{C\dagger}\underline{m}_{\mathrm{I}}^{C}\underline{g}^{C} - \underline{\Delta}^{C}\right]^{-1}\underline{g}^{C\dagger}. \tag{9.100}$$

This expression can be interpreted as follows. First, the inverse t-matrices, $\underline{m}^{i}$, centered at the displaced positions, are expanded in an angular momentum basis about the equilibrium lattice sites. The inverse in the last expression then gives the SPO connected with these undisplaced sites being occupied by the *displaced* t-matrices. To obtain the scattering path operator about the displaced sites, a final re-expansion is necessary, as is indicated by the presence of the pre- and post factors of $\underline{g}^{C}$ or $\underline{g}^{C\dagger}$. However, even though the last expression is formally exact, it is also rather cumbersome to use computationally. It can only be hoped that advances in computational technology will make the implementation of expressions such as this possible within the near term.

9.6 Interstitial Impurities

As a further example, we consider the case of interstitial impurities. This problem is physically significant encompassing absorption of hydrogen in metals, PdH_x, TiH_x, and interstitial impurities in semiconductors. In the metal case, hydrogen is known to profoundly affect the mechanical behavior of materials. It is largely responsible for *environmental embrittlement* effects, in which the ductility of the system is considerably reduced from that of samples free of hydrogen. Impurities in semiconductors can cause states to appear in the gap of the host material, thus affecting and often qualitatively altering its electronic properties.

Figure 9.4 shows a schematic representation of the geometric arrangement associated with an interstitial impurity. Again, for the sake of clarity

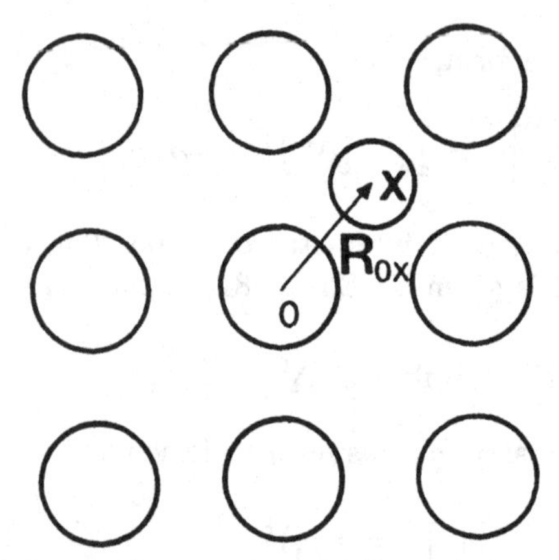

Figure 9.4: **Schematic representation of an interstitial impurity.**

of exposition, we assume that the impurity does not affect the host atoms in its vicinity. Thus, we neglect *both* potential and lattice relaxation effects. These can be taken into account in a formally straightforward manner using the results of the previous sections.

It is convenient to designate one of the sites surrounding the impurity, the closest one to it if such a site can be identified, as the reference site, labeled by 0 in the figure. Using the expansion properties of the structure constants, Eq. (F.14), and realizing that $\underline{\Gamma}^{(x)}$ is identical to the SPO, $\underline{\tau}$, for the host material without the impurity we can write the renormalized interactor for the impurity in the form,

$$\begin{aligned}
\underline{\Delta}^{x} &= \sum_{k}\sum_{n} \underline{G}^{xk}\underline{\tau}^{kn}\underline{G}^{nx} \\
&= \underline{G}^{x0}\underline{\tau}^{00}\underline{G}^{0x} + \underline{G}^{x0}\left[\sum_{k\neq 0}\underline{\tau}^{0k}\underline{G}^{k0}\right]\underline{g}^{0x} \\
&+ \underline{g}^{x0}\left[\sum_{k\neq 0}\underline{G}^{k0}\underline{\tau}^{k0}\right]\underline{G}^{0x} \\
&+ \underline{g}^{x0}\left[\sum_{k\neq 0}\sum_{n\neq 0}\underline{G}^{kk}\underline{\tau}^{kn}\underline{G}^{n0}\right]\underline{g}^{0x}. \qquad (9.101)
\end{aligned}$$

Now, we note the relations,

$$\begin{aligned}
\sum_{k\neq 0}\underline{\tau}^{0k}\underline{G}^{k0} &= \frac{1}{\Omega_{\rm BZ}}\int_{\rm BZ}\left(\underline{m}-\underline{G}(\mathbf{k})\right)^{-1}\,{\rm d}^3k \\
&= -1+\underline{\tau}\underline{m}, \qquad (9.102)
\end{aligned}$$

and

$$\sum_{k\neq 0}\sum_{n\neq 0}\underline{G}^{0k}\underline{\tau}^{kn}\underline{G}^{n0} = \underline{m}(\underline{\tau}^{00}-\underline{t})\underline{m} \equiv \underline{\theta}^{00}, \tag{9.103}$$

and we write

$$\underline{\Delta}^{x} = (\underline{G}^{x0}, \underline{g}^{x0}) \begin{pmatrix} \underline{\tau}^{00} & \underline{\tau}^{00}\underline{\Delta}^{00} \\ \underline{\Delta}^{00}\underline{\tau}^{00} & \underline{\theta}^{00} \end{pmatrix} \begin{pmatrix} \underline{G}^{0x} \\ \underline{g}^{0x} \end{pmatrix}. \tag{9.104}$$

It follows that the scattering path operator associated with the impurity site is given by the expression, Eq. (9.79),

$$\underline{\tau}_{\rm I}^{xx} = (\underline{m}^{x} - \underline{\Delta}^{x})^{-1}, \tag{9.105}$$

and the site-diagonal impurity Green function by the usual relation, with $(r' > r)$

$$G(\mathbf{r},\mathbf{r}') = \langle Z^{x}(\mathbf{r})|\underline{\tau}_{\rm I}^{xx}|Z^{x}(\mathbf{r}')\rangle - \langle Z^{x}(\mathbf{r})|S^{x}(\mathbf{r}')\rangle. \tag{9.106}$$

9.7 Screening of Impurities

One important physical aspect of scattering in metallic systems is the *screening* of an impurity charge by the conduction electrons. Let an impurity atom be introduced into a metal, and let ΔZ denote the difference between the charge on the impurity and the host atoms. The charge distribution in the host will distort so that this difference is completely screened out at large distances. It follows that the integral over all space over the distorted charge density, $\Delta\rho$, must obey the sum rule,

$$\int \Delta\rho(\mathbf{r})\mathrm{d}^3r = \Delta Z. \tag{9.107}$$

From the relation between the charge density and the Green function, we also have

$$\begin{aligned}\int \Delta\rho(\mathbf{r})\mathrm{d}^3r &= \int_{-\infty}^{E_{\rm F}} \mathrm{d}^3E \int \mathrm{d}^3r\,[\Im G(\mathbf{r},\mathbf{r}) - \Im G_0(\mathbf{r},\mathbf{r})] \\ &= \int_{-\infty}^{E_{\rm F}} \mathrm{d}^3E\Delta n(E) = N(E_{\rm F}) = \Delta Z,\end{aligned} \tag{9.108}$$

where $E_{\rm F}$ denotes the Fermi level of the host.

We have already seen that when the reference or host system is a single cell, the change in the integrated density of states, in the case of spherically symmetric potentials, is given by the Friedel sum rule,

$$\Delta n(E) = \frac{1}{\pi}\sum_{\ell}(2\ell+1)\frac{\mathrm{d}\delta_\ell^{(1)}}{\mathrm{d}E}, \tag{9.109}$$

where $\delta_\ell^{(1)} = \delta_\ell - \delta_\ell^{(0)}$ is the difference between the phase shifts of the impurity and the host cell. In the case of generally shaped potentials, $\Delta n(E)$ is obtained from the Lloyd formula, Eq. (3.162),

$$\Delta n(E) = \frac{1}{\pi}\Im \mathrm{Tr} \ln \frac{\mathrm{d}\Delta \underline{t}}{\mathrm{d}E}, \tag{9.110}$$

where $\Delta \underline{t} = \underline{t} - \underline{t}^0$ is the difference in the t-matrices of the impurity and reference potentials. However, when the substitution takes place in a metal, the effects of scattering from the host atoms must be considered. These *back-scattering effects*, can be calculated form the Lloyd formula, Eq. (3.163), and Eq. (9.86), relating the SPOs of the pure and perturbed systems. Thus, for the difference in the density of states brought about by the introduction of the impurity, and for an appropriate normalization of the wave functions, we find,

$$\begin{aligned}
\Delta N(E) &= \frac{1}{\pi}\Im rmTr\left[\ln \underline{\tau}^C - \ln \underline{\tau}\right] \\
&= \frac{1}{\pi}\Im \mathrm{Tr} \ln \left[1 - \delta \underline{m}\underline{\tau}\right]^{-1} \\
&= \frac{1}{\pi}\Im \mathrm{Tr}\left\{ \ln \underline{t}_{\mathrm{I}}[\delta \underline{m}^{-1} - \underline{\tau}]^{-1}\underline{t} - \ln \Delta \underline{t}\right\}.
\end{aligned} \tag{9.111}$$

The term in $\Delta \underline{t}$ describes the change in the integrated density of states that would have resulted if the reference system (the host) consisted of a single potential in free space. It is the term denoted by $\underline{t}_\ell^{(1)}$ in Eq. (9.54). The first term inside the curly brackets describes the back-scattering effects, which arise due to the presence of the host atoms in the material in which the impurity is embedded [86].

9.8 Distorted Bloch Waves

In addition to the Green-function formalism discussed in previous sections, the effects of impurities on the electronic structure of a solid can often be treated within a wave-function approach. The following treatment is most appropriate for the case of dilute substitutional impurities, or of isolated impurity clusters embedded in an otherwise host material.

We recall in Eq. (3.148) that the Bloch function in the vicinity of the cell centered at $\mathbf{R}_n$ takes the form (with the band index explicitly indicated)

$$\bar{\phi}_{\nu\mathbf{k}}(\mathbf{r}) = \underline{\bar{C}}(\nu\mathbf{k})|\bar{\phi}(\mathbf{r})\rangle, \tag{9.112}$$

where a bar over a symbol designates a quantity associated with the translationally invariant host material. Using the Bloch condition,

$$\bar{\phi}_{\nu\mathbf{k}}(\mathbf{r}+\mathbf{R}_n) = \mathrm{e}^{\mathrm{i}\mathbf{k}\cdot\mathbf{R}_n}\bar{\phi}_{\nu\mathbf{k}}(\mathbf{r}), \tag{9.113}$$

we can write

$$\bar{\phi}_{\nu\mathbf{k}}(\mathbf{r}+\mathbf{R}_n) = \underline{\bar{C}}^n(\nu\mathbf{k})|\bar{\phi}^n(\mathbf{r})\rangle, \tag{9.114}$$

where

$$\underline{\bar{C}}^n(\nu\mathbf{k}) = \mathrm{e}^{\mathrm{i}\mathbf{k}\cdot\mathbf{R}_n}\underline{\bar{C}}(\nu\mathbf{k}), \tag{9.115}$$

and $|\bar{\phi}^n(\mathbf{r})\rangle = |\bar{\phi}(\mathbf{r})\rangle$ in a translationally invariant system.

Now, let us introduce a change in potential, $\Delta v_{n'}(\mathbf{r})$, in a selected set of sites, $\{n'\}$. Using the two-potential scattering formalism of Section 9.2, we can write the Lippmann-Schwinger equation for the wave function at *any* site n in the form,

$$\begin{aligned}\phi_{\nu\mathbf{k}}(\mathbf{r}+\mathbf{R}_n) &= \bar{\phi}_{\nu\mathbf{k}}(\mathbf{r}+\mathbf{R}_n) \\ &+ \sum_{n'}\int_{\Omega_{n'}} \mathrm{d}^3r\bar{G}(\mathbf{r}+\mathbf{R}_n,\mathbf{r}'+\mathbf{R}_{n'})\Delta v_{n'}(\mathbf{r}')\phi_{\nu\mathbf{k}}(\mathbf{r}'+\mathbf{R}_{n'}),\end{aligned} \tag{9.116}$$

where $\bar{G}$ denotes the Green function of the unperturbed host. As discussed in Section 3.13, the host Green function can be written as

$$\bar{G}(\mathbf{r}+\mathbf{R}_n,\mathbf{r}'+\mathbf{R}_{n'}) = \bar{G}^n(\mathbf{r},\mathbf{r}')\delta_{nn'} + \langle\bar{\phi}^n(\mathbf{r})|\underline{\bar{\tau}}^{nn'}|\bar{\phi}^{n'}(\mathbf{r}')\rangle. \tag{9.117}$$

Here, the single-scatterer term, $\bar{G}^n(\mathbf{r},\mathbf{r}')$, has the form (with $r' > r$)

$$\bar{G}^n(\mathbf{r},\mathbf{r}') = \langle\bar{\phi}(\mathbf{r})|\bar{F}(\mathbf{r}')\rangle, \tag{9.118}$$

with $\bar{\phi}(\mathbf{r})$ and $\bar{F}(\mathbf{r})$ being the regular and irregular solutions of the Schrödinger equation for the potential at cell n. In the presence of an impurity in cell n the functions $\bar{\phi}(\mathbf{r})$ satisfy the Lippmann-Schwinger equation,

$$|\phi^n(\mathbf{r})\rangle = |\bar{\phi}^n(\mathbf{r})\rangle + \int_{\Omega_n}\mathrm{d}^3\bar{G}^n(\mathbf{r},\mathbf{r}')\Delta v_n(\mathbf{r}')|\phi^n(\mathbf{r})\rangle. \tag{9.119}$$

In the vicinity of site n, the perturbed wave function is a distorted Bloch wave that can be written in the form,

$$\phi_{\nu\mathbf{k}}(\mathbf{r}+\mathbf{R}_n) = \underline{C}^n(\nu\mathbf{k})|\phi^n_{\nu\mathbf{k}}(\mathbf{r})\rangle. \tag{9.120}$$

Using Eqs. (9.114), (9.116), (9.117), (9.118), and (9.119), we derive the following relation

$$\underline{C}^n = \underline{\bar{C}}^n + \sum_{n'} \underline{\bar{\tau}}^{nn'} \Delta \underline{t}^{n'} \underline{C}^{n'}, \tag{9.121}$$

where we have used the relation, Eq. (9.36),

$$\Delta v_n(\mathbf{r})|\phi^n_{\nu\mathbf{k}}(\mathbf{r})\rangle = \Delta t_n(\mathbf{r})|\bar{\phi}^n_{\nu\mathbf{k}}(\mathbf{r})\rangle. \tag{9.122}$$

From Eq. (9.121), we can write

$$\underline{C}^n(\nu\mathbf{k}) = \sum_{n'} \underline{D}^{nn'} \underline{\bar{C}}^{n'}(\nu\mathbf{k}), \tag{9.123}$$

where

$$\underline{D}^{nn'}(E) = \left\{ [1 - \underline{\bar{\tau}} \Delta \underline{t}]^{-1} \right\}_{nn'}. \tag{9.124}$$

In this expression, $\Delta \underline{t}$ is a site-diagonal matrix with elements equal to $\Delta \underline{t}_n$ associated with all perturbed sites n. The quantity $\underline{\bar{\tau}}$ is the scattering-path operator for the host material. The reader can verify that if the perturbation is confined to a single site, n, we have

$$\underline{C}^n(\nu\mathbf{k}) = [1 - \underline{\bar{\tau}}^{nn}(E) \Delta \underline{t}_n]^{-1} \underline{\bar{C}}^n(\nu\mathbf{k}), \tag{9.125}$$

where

$$\underline{\bar{\tau}}^{nn} = \left\{ [\underline{t} - \underline{tGt}]^{-1} \right\}_{nn}, \tag{9.126}$$

is the site-diagonal element of $\underline{\bar{\tau}}$ of the host medium.

9.8.1 Transition matrix

By definition, the t-matrix describing the scattering of a Bloch wave at an impurity site (or sites) is given by the relation

$$T_{\nu\mathbf{k},\nu'\mathbf{k}'} = \int \mathrm{d}^3 r \langle \bar{\phi}^n_{\nu\mathbf{k}}(\mathbf{r})|\Delta v_n(\mathbf{r})|\phi^n_{\nu'\mathbf{k}'}(\mathbf{r})\rangle, \tag{9.127}$$

where the integral extends over the region of the perturbation $\Delta v(\mathbf{r})$. With the definition

$$\Delta v(\mathbf{r}) = \sum_n \Delta v_n(\mathbf{r}), \tag{9.128}$$

and using Eqs. (9.112) and (9.120) we can write,

$$\begin{aligned} T_{\nu\mathbf{k},\nu'\mathbf{k}'} &= \sum_n \int \mathrm{d}^3 r \langle \bar{\phi}^n(\mathbf{r})|\underline{\bar{C}}^n(\nu\mathbf{k}) \Delta v_n(\mathbf{r}) \underline{C}^n(\nu'\mathbf{k}')|\bar{\phi}^n(\mathbf{r})\rangle \\ &= \sum_n \sum_L \sum_{L'} \sum_{L'\prime} \underline{\bar{C}}^n_{LL'}(\nu\mathbf{k}) \underline{C}^n_{L'L''}(\nu'\mathbf{k}') I^n_{LL''}, \end{aligned} \tag{9.129}$$

where

$$I^{n}_{LL''} = \int_{\Omega_n} \mathrm{d}^3 r \bar{\phi}^{n*}_{L}(\mathbf{r}) \Delta v_n(\mathbf{r}) \phi^{n}_{L''}(\mathbf{r}). \tag{9.130}$$

The formulae derived in this section become considerably simpler in the case of spherically symmetric potentials, as the reader can easily verify.

Bibliography

[1] R. Nieminen, M.Manninen, P. Hautojärvi, and J. Arponen, Solid State Commun. **16**, 831 (1975).

[2] Z. D. Popovic, M. J. Stott, J. P. Carbotte, and G. R. Piercy, Phys. Rev. B**13**, 590 1976

[3] C. O. Almbladt, and U. von Barth, Phys. Rev. B**13**, 3307 (1976).

[4] E. Zaremba, L. M. Sanders, H. B. Shore, and J. H. Rose, J. Phys. F**7**, 1763 (1977).

[5] M. Manninen, P. Hautojärvi, and R. Nieminen, Solid State Commun. **23**, 795 (1977).

[6] P. Jena and K. S. Singwi, Phys. Rev. B**17**, 1592 (1978); 3518 (1978).

[7] N. J. Norskov, Phys. Rev. B**20**, 446 (1979).

[8] A. Hintermann and M. Manninen, Phys. Rev. B**27**, 7262 (1983).

[9] M. Manninen and R. Nieminen, J. Phys. F**9**, 1333 (1979).

[10] R. Nieminen and M. Puska, J. Phys. F**10**, L123 (1980).

[11] K. H. Johnson, J. Chem. Phys. **45**, 3985 (1976).

[12] K. H. Johnson, in *Advances in Quantum Chemistry*, Vol. **1**, P. O. Löwdin (ed.), Academic Press, New York (1977), p. 143.

[13] K. H. Johnson, D. D. Vveddenski, and R. P. Messner, Phys. Rev. B**19**, 1519 (1979).

[14] Ch. Müller, G. Seifert, G. Lautenschläger, H. Wonn, P. Ziesche, and E. Mrosan, Phys. Stat. Sol. (b) **91**, 605 (1979).

[15] O. Gunnarson and R. O. Jones, Physica Scripta **21**, 394 (1980).

[16] H. L. Skriver, *The LMTO method*, Springer Verlag, Berlin (1984).

[17] G. Seifert and P. Ziesche, Proc. 15th Symp. Electronic Structure, P. Ziesche (ed.), TU Dresden (1985), p.109.

[18] P. Blaha and J. Callaway, Phys. Rev. B**33**, 1706 (1986).

[19] V. Heine, D. Bullet, R. Haydock, and M. J. Kelly, Solid State Phys. Vol. **35**, H. Ehrenreich, et al. (eds.), Academic Press, New York (1980).

[20] D. B. B. Rijsenbrij and A. Lodder, J. Phys. F**6**, 1053 (1976).

[21] G. Ries and H. Winter, J. Phys. F**9**, 1589 (1979); F**10**, 1 (1980).

[22] A. Lodder and P. J. Braspenning, J. Phys. F**10**, 2259 (1980); F**11**, 79 (1981).

[23] W. John and W. Keller, J. Phys. F**7**, L223 (1977).

[24] J. Friedel, Phil. Mag. **43**, 153 (1952).

[25] J. Friedel, J. Phys. **34**, 1190 (1956).

[26] J. Friedel, J. Phys. C**3**,285 (1958).

[27] J. Friedel, Nuovo Cimento, Suppl. **7**, 287 (1958).

[28] F. J. Blatt, Phys. Rev. **108**, 286 (1957).

[29] R. P. Gupta, R. W. Siegel, Phys. Rev. Lett. **39**, 1212 (1977).

[30] R. P. Gupta and R. W. Siegel, Phys. Rev. B**22**, 4572 (1980).

[31] G. F. Koster and J. C. Slater, Phys. Rev. **94**, 1392 (1954); 1165 (1954).

[32] A. Seeger, E. Mann, and K. Clausecker, Phys. Stat. Sol. (b) **24**, 721 (1967).

[33] K. Clausecker, E. Mann, and A. Seeger, Phys. Kond. Mat. **9**, 73 (1969).

[34] W. Hehl and E. Mann, Phys. Stat. Sol. (b) **48**, 271 (1971).

[35] E. Mann, Phys. Cond. Mat. **12**, 210 (1971).

[36] R. Riedinger, J. Phys. F**1**, 392 (1971).

[37] C. L. Cook and P. V. Smith, J. Phys. F**4**, 1344 (1974).

[38] J. C. Parlebas, J. Phys. F**4**, 1392 (1974).

[39] C. A. Shell and P. V. Smith, J. Phys. F**10**, 811 (1980).

[40] G. A. Baraff and M. Sclüter, Phys. Rev. Lett. **41**, 892 (1978).

[41] G. A. Baraff, and M. Sclüter, Phys. Rev B**19**, 4965 (1979).

[42] J. Bernholc, N. O. Lipari, and S. T. Pantelides, Phys. Rev. Lett. **41**, 895 (1978).

[43] J. Bernholc, N. O. Lipari, and S. T. Pantelides, Phys. Rev. B**21**, 3545 (1980).

[44] S. P. Singhal and J. Callaway, Phys. Rev. B**19**, 5049 (1979).

[45] T. H. Dupree, Ann. Phys. (NY) **15**, 63 (1961).

[46] J. Morgan, Proc. Roy. Soc. London **16**,365 (1966).

[47] J. L. Beeby, Proc. Roy. Soc. London **A302**, 113 (1967).

[48] R. Harris, J. Phys. C**3**, 172 (1970).

[49] P. T. Coleridge, N. A. W. Holzwarth, and M. J. G. Lee, Phys. Rev. B**10**, 1213 (1974).

[50] N. A. W. Holzwarth, Phys. Rev. **B11**, 3718 (1975).

[51] G. Lehmann, Phys. Stat. Sol. (b) **70**, 735 (1975).

[52] K. Terakura, J. Phys. Soc. (J) **40**, 450 (1976).

[53] K. Terakura, J. Phys. F**6**, 1385 (1976).

[54] K. Terakura, Physica **91**B, 162 (1977).

[55] H. Katayama-Yoshida, K. Terakura, and J. Kanamori, J. Phys. Soc. (J) **48**, 1504 (1980).

[56] H. Katayama-Yoshida, K. Terakura, and J. Kanamori, J. Phys. Soc. (J) **49**, 972 (1980).

[57] I. Mertig and E. Mrosan, J. Phys. F**10**, 417 (1980).

[58] I. Mertig, E. Mrosan, and R. Schöpke, J. Phys. F**12**, 1689 (1982).

[59] I. Mertig and E. Mrosan, J. Phys. F**12**, 1139 (1982).

[60] I. Mertig and E. Mrosan, J. Phys. F**12**, 3031 (1982).

[61] I. Mertig and E. Mrosan, J. Phys. F**13**, 373 (1983).

[62] R. Zeller and P. H. Dederichs, Phys. Rev. Lett. **42**, 1713 (1979).

[63] R. Podloucky, R. Zeller, and P. H. Dederichs, Phys. Rev. B**22**, 5777 (1980).

[64] P. J. Braspenning, R. Zeller, A. Lodder, and P. H. Dederichs, Phys. Rev. B**29**, 703 (1984).

[65] B. M. Klein and W. E. Picket, J. Less-Com. Met. **88**, 231 (1982).

[66] B. M. Klein and W. E. Picket, in *Electronic Structure and Properties of Hydrogen in Metals*, P. Jena and C. B. Satterthwaite (eds.), Plenum, New York (1983).

[67] B. M. Klein and W. E. Picket, J. Less-Com. Met. **103**, 185 (1984).

[68] B. M. Klein and W. E. Picket, Phys. Rev. B**29**, 1597 (1984).

[69] B. M. Klein and W. E. Picket, Phys. Rev. B**29**, 1588 (1984).

[70] B. M. Klein and W. E. Picket, Phys. Rev. B**31**, 156273 (1984).

[71] P. H. Dederichs, S. Blügel, R. Zeller, and A. Oswald, in the Proceedings of the 15th Symp. Electronic Structure, P. Ziesche (ed.), TU Dresden (1985).

[72] M. Akai, H. Akai, and J. Kanamori, J. Phys. Soc. (J), **54**, 4246 (1985).

[73] H. Akai, M. Akai, and J. Kanamori, J. Phys. Soc. (J), **54**, 4257 (1985).

[74] A. Oswald, R. Zeller, and P. H. Dederichs, Phys. Rev. Lett. **56**, 1419 (1986).

[75] C. König and E. Daniel, J. Phys. Lett. (Paris) **42**, L193 (1981).

[76] C. König, P. Leonard, and E. Daniel, J. Physique **43**, 1015 (1981).

[77] O. Gunnarson, O. Jepsen, and O. K. Andersen, Phys. Rev. B**27**, 7144 (1983).

[78] R. Zeller, in *Electronic Band Structure and its Applications*, M. Yussouff (ed.), Springer Verlag, Berlin (1986), p.106.

[79] B. Drittler, H. Ebert, R. Zeller, and P. H. Dederichs, Phys. Rev. B**39**, 6334 (1989).

[80] B. Drittler, M. Weinert, R. Zeller, and P. H. Dederichs, Solid State Commun. **79**, 31 (1991).

[81] P. H. Dederichs, T. Hoshimo, B. Drittler, K. Abraham, and R. Zeller, Physica B**172**, 203 (1991).

[82] M. Gell-Mann and M. L. Goldberger, Phys. Rev. **91**, 398 (1953).

[83] Charles J. Joachain, *Quantum Collision Theory*, North Holland, Amsterdam, New York, (1983).

[84] Ingrid Mertig, Eberhard Mrosan, and Paul Ziesche, *Multiple Scattering Theory of Point Defects in Metals*, Teubner Texte zur Physik, Band **11** Teubner, Leipzig, (1987).

[85] A. Gonis and W. H. Butler, *Multiple Scattering Theory in Solids*, Springer Verlag, Berlin (1999).

[86] P. J. Braspenning, *A Multiple Scattering Treatment of Dilute Metal Alloys*, Thesis, Vrije Universiteit, Amsterdam, The Nertherlands, unpublished.

[87] B. Ujfalussy, L. Szunyogh, P. Weinberger, and J. Kollar, in *Metallic Alloys: Experimental and Theoretical Perspectives*, J. S. Faulkner and R. G. Jordan (eds.), Proceedings of NATO Advanced Research Workshop, Boca Raton, FL. July 16-21, 1994, Kluwer Academic Publishers, Dordrecht (1994), Vol. 256, p. 301.

[88] L. Szunyogh, B. Ujfalussy, P. Weinberger, and J. Kollar, Phys. Rev. B**49**, 2721 (1994) .

[89] P. J. Braspenning and A. Lodder, Phys. Rev. B**49**, 10222 (1994) .

[90] O. K. Andersen and O. Jepsen, Phys. Rev. Lett. **53**, 2571 (1984)

[91] W. R. L. Lambrecht and O. K. Andersen, Surf. Sci. **178**, 256 (1986).

[92] B. Wenzien, J. Kudrnovsky, V. Drchal, and M. Sob, J. Phys.: Condens. Matter **1**, 9893 (1989).

[93] J. Kudrnovsky, P. Weinberger, and V. Drchal, Phys. Rev. B**44**, 6410 (1991).

[94] V. Drchal, J. Kudrnovsky, L. Udvardi, P. Weinberger, and A. Pasturel, Phys. Rev. B**45**, 1432 (1992).

[95] J. Kudrnovsky, I. Turek, V. Drchal, P. Weinberger, S. K. Bose, and A. Pasturel, Phys. Rev. B**47**, 16525 (1993).

[96] H. L. Skriver and N. M. Rosengaard, Phys. Rev. B**43**, 9538 (1991).

Part III

Substitutional Alloys

Chapter 10

Basic Features of Disorder

10.1 General Comments

With this chapter we begin our discussion of the electronic structure and related properties of substitutionally disordered alloys. After a few comments on the general types of disorder encountered in materials and a brief summary of the development of methods for the study, we present numerical examples of electronic densities of states for model systems. We also mention the simplest of approximation schemes that have been introduced to study disordered materials. The discussion in this and the next chapter is formulated in terms of a phenomenological tight-binding (TB) approximation to the Hamiltonian of a disordered system. Although a TB description of this kind[1] is of somewhat limited accuracy in describing realistic materials, it has a number of advantages when compared to first-principles methods. It allows a particularly clear identification of the various separate features characterizing a disordered material and of the methods that can be used in their study. Also, it allows the performance of model calculations that can be used to test the validity of a particular formalism or computational scheme. The importance of this latter feature can hardly be overstated. Many promising conceptual schemes have been abandoned when it was discovered that they gave mathematically and physically incorrect results when tested in connection with simple one-dimensional model systems. A summary of a number of theoretical methods, for which there is no room in the present volume, has been given in previous work [1].

It is convenient to think of *disorder* as random deviations from a reference *ordered* medium. A particularly useful picture of an ordered medium is provided by a material with the translational invariance of a Bravais lat-

[1] Recent developments of first-principles TB methods alleviate many of the drawbacks of the phenomenological approach.

tice. In this case, physical observables, e.g., the electron density at point $\mathbf{r}$, are unaffected by translations through lattice vectors, $\mathbf{R}$, so that

$$n(\mathbf{r}+\mathbf{R}) = n(\mathbf{r}). \tag{10.1}$$

It is easily seen that this condition is fulfilled by a charge density that is given by the square of the modulus of a Bloch state.

Let us consider a translationally invariant material, the *host*, into which a finite concentration of substitutional impurities has been introduced in random fashion. If we denote the impurity concentration by c, then a given site on the lattice has probability c of being occupied by an impurity[2], and $(1-c)$ by a host atom. We often write $A_{1-c}B_c$ to denote a substitutionally disordered *binary alloy*, consisting of atoms of type A and type B distributed randomly with probability $1-c$ and c, respectively, over the N sites of a lattice. Clearly, in a random alloy Eq. (10.1) no longer holds. In fact, the entire construct of Bloch's theorem and associated $\mathbf{k}$-selection rules fail, and the calculation of the electronic structure of a substitutionally disordered alloy presents a conceptually and computationally very difficult problem.

Because of the inherent difficulties just mentioned, it was only about the early 1970s that disordered materials began to receive attention from theoretical physicists comparable to that paid to ordered, translationally invariant systems. The reasons for this increased interest are many and varied, reflecting the growing realization of the technological importance of such systems in the modern world.

10.2 Types of Disorder

As already mentioned, disorder is defined with reference to an ideal, ordered state. The simplest example of such a state is provided by vacuum that possesses the highest degree of spatial order. Next to vacuum, the highest degree of spatial order is found in crystals. In a periodic, crystalline material, we can learn the properties of the entire, infinite material by studying only a part of very small spatial extent, i.e., the unit cell of the system under consideration.

There are several different ways in which a material can deviate from crystalline order. In possibly the simplest type of disorder, the material is still characterized by an underlying regular lattice, but atoms of two or more different kinds may be randomly distributed over the sites of this lattice. This type of disorder is known as *chemical disorder*. If the material contains atoms of type A, B, $C, \cdots$, distributed with concentrations c_A, c_B, $c_C, \cdots$, over the N sites of the lattice, we write $A_{c_A}B_{c_B}C_{c_C}\cdots$,

[2]For high enough concentrations, the distinction between host and impurity is lost, but it is convenient to use these terms for reference purposes.

where clearly $\sum_i c_i = 1$. For example, as shown earlier, a binary alloy is denoted by the general symbol, $A_{1-c}B_c$.

In *topologically disordered* materials, such as liquids or glasses, no ghost of an underlying lattice remains. The atoms or molecules are distributed through the available space under very broad restrictions of size or the constraints of chemical bonding. Clearly, a combination of substitutional and topological disorder is quite possible, and such systems as liquid alloys and spin glasses present many challenging and as yet generally unsolved problems.

Substitutional and topological disorder are probably the most prevalent but by no means exhaust the possible types of disorder. Thus *magnetic disorder*, observed in disordered Heisenberg ferromagnets, and *ice disorder* [2] provide interesting and important problems for study.

In this book, we concentrate on methods and materials for which conceptual and computational progress has been the most extensive. Thus, our discussion focuses on the study of the electronic structure and related properties of substitutionally disordered metallic alloys. Occasionally, we illustrate the methods discussed with the results of numerical calculations. Although these methods are in principle applicable to any material, computationally their application to metallic systems appears to be the easiest. The presence of mobile electrons in metallic systems, which effectively screen impurities and other imperfections, makes metallic systems amenable to fairly accurate treatment by methods that must, due to the complexity of the problem, be approximate at best.

10.3 Historical Development

Prior to the late 1960s, the study of disorder had proceeded very slowly, although a few important results were obtained in certain cases. In 1892 Lord Rayleigh [3], in a paper that can be regarded as having given birth to the formal framework of MST, addressed the question of the permitivity of a heterogeneous medium. In 1906, Einstein [4] considered the dielectric and optical properties of a suspension of small particles in a liquid. In 1945 Foldy [5] showed how the macroscopic index of refraction of a heterogeneous medium can be related to its averaged forward scattering amplitude. However, even this comparatively modern theory was primarily limited to the strong-scattering, low-concentration regime and was not self-consistent. In 1951, Lax [6, 7] introduced within a multiple-scattering theory the idea of an effective medium in which fluctuations are embedded, and the parameters of which are determined by an averaging similar to that introduced by Foldy [5], but with an explicit self-consistent equation for determining the index of refraction. His work forms the basis for the modern, self-consistent theories of the electronic (and other excitation) structure of disordered materials.

After the early 1970s, interest in and work on the problems posed by and associated with disordered systems increased dramatically. Technological and metallurgical advances have led to more precise, varied, and informative experimental studies of disordered materials. These in turn have led to a need for much more realistic theoretical models and descriptions of such systems.

The most commonly used and to date the most successful approach in the theoretical study of substitutionally disordered alloys is based on obtaining approximate expressions for the Green function of a disordered system. The single-particle Green functions give the thermodynamic properties of disordered systems, and in particular, the charge density and the density of states and spectral weight-functions for the corresponding elementary excitations; the two-particle Green functions give the transport properties. A great deal of theoretical effort has been expended toward the construction of such improved, general theories.

The simplest theories that have been introduced for the study of substitutional alloys are the rigid-band model (RBM) [8], the *virtual crystal approximation* (VCA) [9], and the *virtual bound-state model* [10]. A brief discussion of the RBM and the VCA is provided in Section 10.6.2. Foldy [5] proposed a fundamentally different approach; the restoration of translational invariance through the introduction of an effective medium. A second major step was taken by Lax [6, 7], who introduced a multiple-scattering formalism. This approach was used and extended by many investigators. Edwards [11] used diagrammatic perturbation theory to treat the dilute alloy limit, while Korringa [12] and Beeby [13] introduced the average t-matrix approximation (ATA). The ATA is the most satisfactory, non-self-consistent single-site theory of the electronic structure and the structure of other excitations in disordered systems, and is still in use today [14-16]. In 1967, Soven [17] and Taylor [18], making use of physical ideas first expressed by Anderson and McMillan [19], simultaneously proposed a greatly improved, self-consistent multiple-scattering theory, the *coherent potential approximation* (CPA).

The ATA and the CPA are very similar theories. In both, one determines an effective medium to represent the substitutionally disordered material. In the ATA the medium is assumed to possess scattering properties that are the averages of those of the pure constituents, while the CPA allows for a self-consistently determined t-matrix.

The CPA was derived by Soven [17] and Taylor [18] within the single-site approximation. That is, only the occupation of a single site was considered explicitly and averaged over, with the rest of the heterogeneous system replaced by a homogeneous effective medium. Many different techniques and approaches have been used to study disordered systems within a single-site approximation; they all yield the single-site CPA (SSCPA) as the unique best single-site approximation. As is discussed in the next two chapters, the CPA possesses many outstanding characteristics, in particular, analyticity, translational invariance, complete self-consistency, and practicability for computation. At the same time, it also possesses a number of limitations, such as the general inability to provide a treatment of statistical fluctuations in site occupancy in an alloy. The CPA is examined in some detail in the next two chapters, and is compared closely with the ATA.

10.3.1 Development of the CPA

The coherent potential approximation was formulated by Soven [17] and Taylor [18] as a single-site theory within the framework of multiple scattering theory. An early detailed exposition of the theory has been given by Velicky, *et al.* [20]. The theory has also been reviewed by several authors [21, 22, 23].

In the multiple scattering approach, one considers an electron (or a lattice wave in the case of lattice vibrations) propagating through a disordered material, subject to scattering at each site by an individual atomic scatterer. Such division of the total scattering potential into a sum of individual atomic scattering centers is assumed *a priori* and is essential to the multiple-scattering formulation of the theory in terms of individual cell t-matrices. In order to determine the t-matrix for scattering by a single cell (site), the rest of the material is replaced by an effective medium. In a single-site approximation, one then averages only over the possible chemical occupation of a single site (or scattering center). In the ATA, [15, 24] an effective scatterer described by the average t-matrix of the pure constituents is chosen. In the SSCPA, there is one step further: The medium t-matrix is chosen so that the additional scattering that occurs when a real cell is embedded in the medium vanishes on the average. This is a self-consistent procedure that determines the *self-energy* and the corresponding Green function in the CPA.

Independently of the work of Soven [17] and Taylor [18], other authors such as Onodera and Toyozawa [25] obtained the SSCPA as a formalism that interpolates smoothly between the correct weak- and strong-coupling limits for the exciton problem. Also, in 1968, Velicky *et al.* [20] showed that the CPA results from a self-consistent mean-field theory, Yonezawa [26] derived the SSCPA diagrammatically using a continued fraction cumulant expansion, and Leath [27] derived it using a self-consistent approach based on diagrammatic expansions. Finally, in 1972 Schwartz and Siggia [28] derived the SSCPA using a functional-derivative technique. In all of the above derivations, the excitations were assumed to propagate freely with disorder scattering treated as a perturbation to free motion; this approach constitutes the *propagator formalism*. The SSCPA can also be derived within the *locator formalism* [29-32], in which the excitations are considered as localized to lowest order with perturbation terms giving rise to itinerant behavior.

Additional and rather different techniques have been used to derive the SSCPA. Using the idea of Haydock, *et al.* [33], Mookerjee in a series of papers [34-37] proposed a new approach to disordered systems characterized by the introduction of an extended Hilbert space. The CPA (and various generalizations) can be obtained by appropriately choosing diagrams in this space. Ducastelle [38] and Lloyd and Best [39] have given derivations of the SSCPA based on a variational approach.

The fact that the SSCPA can be derived equally well within either a propagator or a locator formalism, as well as a number of different and diverse mathematical schemes, renders this approximation unique among single-site approximations for the treatment of substitutional disorder. The derivations of the CPA both within a propagator and a locator formalism are given in the next chapter in connection with TB Hamiltonians. First principles, MST derivations of the CPA, are then presented in the following chapter.

In this chapter, we discuss a number of general features of disordered alloys and introduce some basic formal approaches to the evaluation of the Green function of a disordered material. For clarity of presentation, we base our discussion on a TB, single-band Hamiltonian. We show representative samples of one-dimensional model calculations of the DOS for both random alloys and alloys exhibiting short range order. These calculations provide a particularly vivid representation of some important aspects of the electronic structure of substitutional alloys. And even though real, three-dimensional materials usually do not exhibit the structure in the DOS shown by one-dimensional systems, these model calculations serve a particularly useful purpose: They can be used as a testing ground for various approximation schemes whose validity can be assessed upon comparison with such numerical results.

10.4 Tight-Binding Hamiltonians

Within a tight-binding description [1, 40], the Hamiltonian of an electronic system can be written in the form

$$H = \sum_{i,\alpha,\beta} \epsilon_{i\alpha,i\beta} |i\alpha\rangle\langle i\beta| + \sum_{i,j\neq i} \sum_{\alpha\beta} W_{i\alpha,j\beta} |i\alpha\rangle\langle j\beta|, \tag{10.2}$$

where[3] the $|i\alpha\rangle$ are state vectors associated with "orbitals" α centered on site i, the $\epsilon_{i\alpha,i\beta}$ are the corresponding atomic on-site energies, and the site off-diagonal terms, $W_{i\alpha,j\beta}$, describe electron hopping from orbital β on site j to orbital α on site i. These terms vanish for $i = j$. The site-diagonal terms, $\epsilon_{i\alpha,i\beta}$, include inter-orbital electron hopping at the same site. Within appropriate definitions of the various parameters [1], the Hamiltonian of Eq. (10.2) can be made to describe different types of excitations, such as electrons, phonons, and magnons, in disordered materials. (See also the discussion of itinerant magnetism in Chapter 8.)

For our purposes, we use the simplest form of the tight-binding Hamiltonian, corresponding to only a single orbital (single-band model), and restrict most of our discussion to electron hopping between nearest neighbors (although this last condition is not necessary for the following development). This single-band Hamiltonian takes the form

$$H = \sum_i \epsilon_i |i\rangle\langle i| + \sum_{i,j\neq i} W_{ij} |i\rangle\langle j|. \tag{10.3}$$

The corresponding Green function takes the usual form

$$G(z) = (z - H)^{-1}. \tag{10.4}$$

Before examining ways for evaluating the Green function associated with a disordered material, we present a few examples of the DOS of model systems.

10.4.1 Numerical examples

We consider a substitutional, disordered binary alloy, $A_{1-c}B_c$, described by the Hamiltonian of Eq. (10.3). In such a system, both the site-diagonal and site off-diagonal terms, ϵ_i and W_{ij}, respectively, in principle take random values, being functions of the chemical occupation of sites i and j, as well as of the local environment of these sites. The variation of the ϵ_i is referred to as *diagonal disorder*, and that of the W_{ij} as *off-diagonal disorder* (ODD). In most applications, the dependence on local environment is neglected (it is in fact very difficult to treat, certainly within a strictly single-site theory), and only the dependence of various quantities on the occupation of the sites connected by the terms in the Hamiltonian is taken into account. Thus, we will assume that in the case of a binary alloy, $A_{1-c}B_c$, the site energies can assume the "values" ϵ_A or ϵ_B, with probability $1 - c$ and c, respectively. In the presence of ODD, the terms W_{ij} can take the values W_{ij}^{AA}, W_{ij}^{BB},

[3]Rigorously, the complex conjugate of the terms shown should also be included, but these terms can be taken into account in a straightforward way within the formalism and will not be indicated explicitly.

or $W_{ij}^{AB} = W_{ij}^{BA}$, depending on whether the sites i and j are occupied by two atoms of type A, of type B, or by an atom of type A and one of type B, respectively. The generalization of these concepts to multiband, multicomponent alloys is evidently straightforward.

We are interested in the evaluation of the configurationally averaged DOS and charge density of an alloy, which in turn are given in terms of the corresponding Green function. Clearly, the evaluation of the average of the Green function over all alloy configurations cannot be carried out exactly and must be obtained within an approximation. Approximate methods have been introduced both within analytic as well as computational schemes [1]. These are based on the general properties of Green functions and in particular the density of states (DOS) function.

The DOS of any configuration of a substitutionally disordered system is given by the expression, Eq. (2.23),

$$n(E) = -\frac{1}{\pi N}\Im \mathrm{Tr} G(E), \tag{10.5}$$

where N denotes the number of sites in the system, and Tr the sum over the site-diagonal elements of the Green function. Experimentally observed quantities are averages over all alloy configurations, so that we must calculate the corresponding quantity

$$n(E) = -\frac{1}{\pi N}\Im \mathrm{Tr}\langle G(E)\rangle, \tag{10.6}$$

where $\langle ... \rangle$ signifies an average over the possible distributions of the atoms over the sites in the lattice. If the sites are topologically and statistically equivalent, then

$$n(E) = -\frac{1}{\pi}\Im\langle G_{00}(E)\rangle, \tag{10.7}$$

where $G_{00}(E)$ is the site-diagonal element of the system Green function associated with site 0.

In an ordered material, $G_{00}(E)$ can be obtained by means of a lattice Fourier transform along the lines indicated in Eq. (3.57) (see also discussion below). Because of the lack of periodicity, the eigenvalue spectrum of a disordered system cannot be obtained by FTs, and other techniques must be used. In addition to approximate formal methods, to be discussed in following sections and in the next chapter, numerical techniques, some of which are in principle exact, have been developed [41-50], including the *negative eigenvalue counting theorem* [41], the *recursion method* [33, 42-44], *moment methods* [45-47], the *continued fraction method* [48], the *augmented space formalism* [35-37], and *position-space renormalization* techniques [50]. These methods can potentially yield an exact representation of the DOS, at

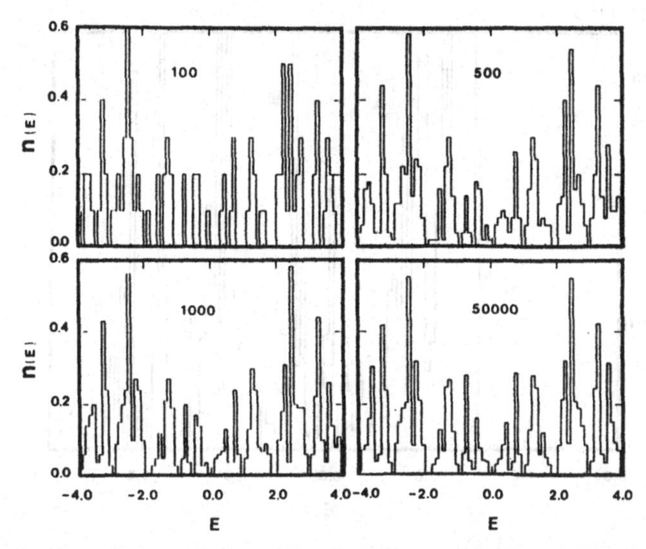

Figure 10.1: **The evolution and converges of the DOS of a one-dimensional alloy with increasing number of sites.**

least for simple model systems, to which the results of approximate analytic theories can be compared.

It may be thought that in order to obtain a proper statistical average of the DOS, one would need to average the results obtained for a large number of extended systems, using the appropriate statistical weight for each system. However, this strict procedure is not necessary. The DOS converges rapidly as the number of sites in a *single* system increases, and is said to possess the *self-averaging* property. For example in the case of a single-band, one-dimensional model alloy reliable densities of states can be obtained numerically by using a single chain with about 1,000 to 1,500 sites.

The self-averaging property of the DOS is illustrated in Fig. 10.1. The figure shows histograms of the DOS as a function of energy (in arbitrary units) for one-dimensional, single-band tight-binding alloys with only nearest-neighbor hopping. The panels show the results of the application of the negative eigenvalue theorem [41] to single chains of 100, 500, 1,000, and 50,000 atoms. The Hamiltonian parameters chosen in this model are $\epsilon_A = -\epsilon_B = 2$ and nearest-neighbor hopping $W = 1$, while the concentration was set at $c = 0.50$. These parameters correspond to a *scattering strength*

$$\begin{aligned} \delta &= \frac{|\epsilon_A - \epsilon_B|}{w} \\ &= 2, \end{aligned} \tag{10.8}$$

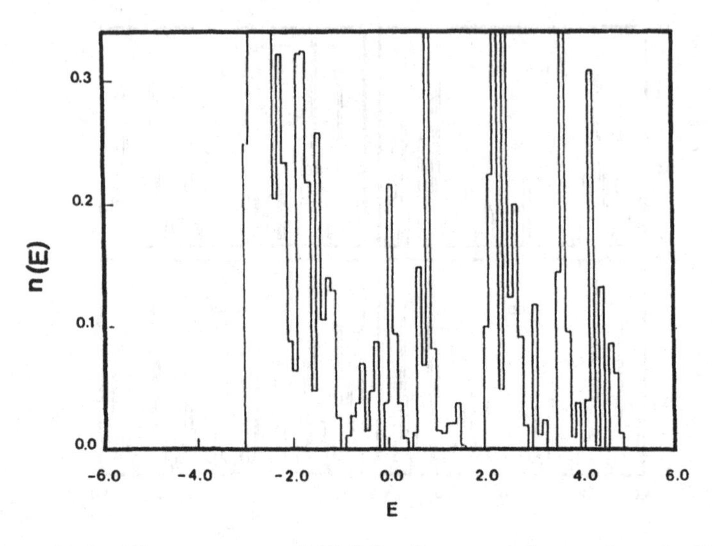

Figure 10.2: **The converged DOS of a one-dimensional alloy with ODD.**

where $w = 2W$ corresponds to one half the width of the band (see discussion following Eq. (10.25)). We see that there is little, if any, difference between the last two panels, indicating that the DOS has essentially converged for 1,000 (or more) sites, even in this relatively strong-scattering alloy. Similar convergence rates are observed in alloys with ODD, Fig. 10.2. For alloys characterized by weaker scattering, or for three-dimensional systems, the convergence is even faster.

The rich structure of the DOS curves in the last two figures should be noted. The alternating high peaks and gaps, where the DOS nearly vanishes, are quite characteristic of one-dimensional TB systems with only nearest-neighbor (short-range) hopping. The structure increases with increasing scattering strength, δ, being essentially the envelope of very many sharp peaks at each of the N eigenvalues of the alloy Hamiltonian. Consistent with the normalization used here, the DOS integrates to 1.0. In the case of Fig. 10.1, the DOS clearly exhibits the existence of two separate subbands, or partial DOSs associated with each of the chemical species in the alloy, each centered at the corresponding value of ϵ_i. In the *split-band limit* these partial DOSs are separated by a gap such as the one in the figure at $E = 0.0$. By contrast, due to the asymmetry in hopping the DOS of the alloy with ODD, Fig. 10.2, is no longer symmetric in energy and there is no longer a clear indication of subband formation.

Further studies reveal more interesting features of the DOS of disordered systems. They show that these DOSs are consistent with the *bounding limit* theorem which states that the eigenvalue spectrum of an alloy, and hence the DOS, must vanish in any energy region in which the corresponding spectra of all the pure constituents of the alloy themselves exhibit a gap. This theorem can be proved rigorously only for one-dimensional systems. Its validity in higher dimensions can be surmised on the following intuitive grounds. The DOS near the edges of the band consists of contributions from very large clusters of like atoms, which in a randomly disordered alloy, have an exponentially small probability of occurrence. One can argue that the partial DOS cannot exceed the limits of the DOS for the pure solids.

The behavior of the DOS of a disordered one-dimensional system near band edges or near a band gap is illustrated [51] in Fig. 10.3. The figure shows that near band edges the DOS decays exponentially, a behavior that can be connected to the exponential behavior of the probability of occurrence [52] of large clusters of like atoms. Thus, the structure of the edges of the DOS is a function of long-distance statistical fluctuations in the alloy, in contrast to the structure inside the band that is associated with fluctuations in compact clusters of sites (see discussion of cluster theories beginning with Section 11.8). This type of behavior is also expected in higher-dimensional systems, although because of the increasing coordination number and the corresponding decrease in the statistical weight of a given site, the structure in the DOS decreases substantially with increasing dimensionality. Because of this behavior, any analytic theory will be severely tested in attempting to account for *both* local and long-distance chemical correlation effects in disordered systems. Therefore, it has been found to be a sound procedure to test any new analytic technique against the numerical results for model one-dimensional systems. This testing allows one to assess many of the advantages and limitations of a particular computational scheme before applying it to realistic materials.

Finally, it is important to mention again that substitutionally disordered alloys often exhibit short-range order (SRO) effects whereby atoms of a given kind tend to surround themselves with like atoms, *clustering* or *phase separation*, or different atoms, *ordering*. The analytical study of SRO is more complicated than that of random alloys, and methods for its investigation have evolved more slowly than those directed at totally random systems. Some of these methods are discussed in following chapters and are illustrated with the results of numerical calculations.

10.5 Expressions for TB Green Functions

In this section, we provide brief derivations of explicit expressions for the Green functions of one- two- and three-dimensional systems described by a TB Hamiltonian of the form shown in Eq. (10.3). A complete discussion can be found in Economou [53] to which the reader is referred to for further details.

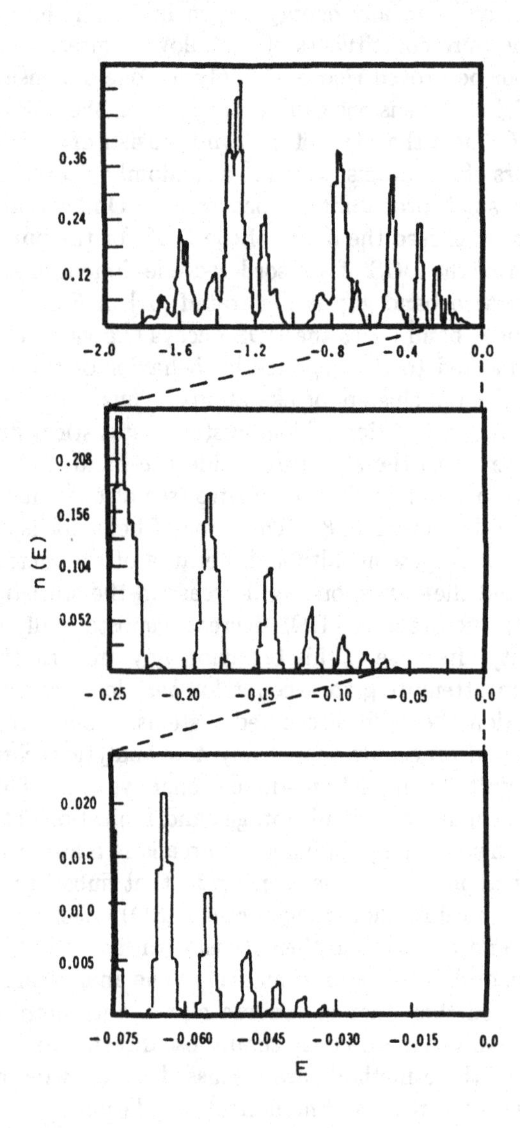

Figure 10.3: **Exponentially decaying DOS of one-dimensional disordered alloys near band edges. Note the different scales in the various panels. See reference [51].**

We begin by noting that the Hamiltonian given in Eq. (10.3) can be written in the form of a matrix

$$\mathbf{H} = \begin{pmatrix} \epsilon_1 & W_{12} & W_{13} & W_{14} & \cdots \\ W_{21} & \epsilon_2 & W_{23} & W_{24} & \cdots \\ \cdot & \cdot & \cdot & \cdot & \cdot \\ \cdot & \cdot & \cdot & \cdot & \cdot \\ \cdots & W_{i,i-1} & \epsilon_i & W_{i,i+1} & \cdots \\ \cdot & \cdot & \cdot & \cdot & \cdot \\ \cdot & \cdot & \cdot & \cdot & \cdot \end{pmatrix}. \tag{10.9}$$

The corresponding Green function takes the form

$$\mathbf{G}(E) = (\mathbf{E} - \mathbf{H})^{-1}. \tag{10.10}$$

Here, $\mathbf{E} \equiv E\mathbf{I}$, where $\mathbf{I}$ is the N-dimensional unit matrix, N being the number of sites. The scalar energy parameter, E, is assumed to have an infinitesimal imaginary part. From Eq. (10.10) we obtain (suppressing the dependence of various quantities on the energy)

$$(\mathbf{E} - \mathbf{H})\,\mathbf{G} = \mathbf{I}, \tag{10.11}$$

or,

$$(\mathbf{E} - \boldsymbol{\epsilon})\,\mathbf{G} = \mathbf{I} + \mathbf{W}\mathbf{G}, \tag{10.12}$$

where $\boldsymbol{\epsilon}$ is a site-diagonal matrix defined by the ϵ_i, and $\mathbf{W}$ is a purely off-diagonal one. The last equation can be written in terms of matrix elements in the form

$$(E - \epsilon_i)G_{ij} = \delta_{ij} + \sum_{k \neq i} W_{ik}G_{kj}. \tag{10.13}$$

The quantities G_{ij} are the matrix elements of the Green function in a site representation. It is convenient to cast Eq. (10.13) in the form

$$G_{ij} = g_i\delta_{ij} + g_i \sum_{k \neq i} W_{ik}G_{kj}, \tag{10.14}$$

where

$$g_i = (E - \epsilon_i)^{-1} \tag{10.15}$$

is called the *bare locator* for site i. Equation (10.14) is called the *locator equation of motion* for the Green function. This equation can be made the starting point in the discussion of disordered materials within a TB formalism. We note the formal similarity between this expression and the equation of motion for the scattering-path operator, Eq. (3.52). This similarity has important consequences: Not only does it often clarify the content of a certain approximation scheme, but it makes it possible to transcribe formal manipulations developed with respect to a TB formalism to the framework of first-principles MST. We also point out that in deriving these equations of motion no assumption is made as to the probability distribution of the Hamiltonian matrix elements. Therefore, these equations are valid in both the cases of ordered and disordered systems.

As an illustration of the use of Eq. (10.14), we now obtain expressions for the Green function and the DOS of ordered translationally invariant single-band materials in one, two, and three dimensions. In such systems, the site-diagonal elements of the Hamiltonian are independent of site index, $\epsilon_i = \epsilon_0$, and the off-diagonal elements, W_{ij}, depend only on the distance between sites i and j. Consequently, all bare locators are identical and independent of site index, $g_i = g$. It follows that the equation of motion, Eq. (10.14), can be solved through the use of lattice Fourier transforms. Multiplying both sides of that equation by $\mathrm{e}^{\mathrm{i}(\mathbf{k}\cdot\mathbf{R}_i - \mathbf{k}'\cdot\mathbf{R}_j)}$, summing over i and j and using Eq. (6.35) we obtain

$$G(E;\mathbf{k},\mathbf{k}') = g\delta_{\mathbf{k}\mathbf{k}'} + g\frac{1}{N}\sum_{ijk\neq i} W_{ik}G_{kj}\mathrm{e}^{\mathrm{i}(\mathbf{k}\cdot\mathbf{R}_i - \mathbf{k}'\cdot\mathbf{R}_j)}. \tag{10.16}$$

Since W_{ik} is translationally invariant, we introduce its Fourier transform in the usual way and we can write

$$G(E;\mathbf{k},\mathbf{k}') = g\delta_{\mathbf{k}\mathbf{k}'} + gW(\mathbf{k})G(E;\mathbf{k},\mathbf{k}'), \tag{10.17}$$

or,

$$\begin{aligned} G(E;\mathbf{k},\mathbf{k}') &= \left[g^{-1} - W(\mathbf{k})\right]^{-1}\delta_{\mathbf{k}\mathbf{k}'} \\ &= G(E;\mathbf{k})\delta_{\mathbf{k}\mathbf{k}'}. \end{aligned} \tag{10.18}$$

Thus, the Green function of a translationally invariant system is diagonal in a reciprocal-space representation. This expression for the Green function directly yields the band structure of a material as those values of the energy that correspond to the poles of $G(E;\mathbf{k})$.

Now, the density of states of a translationally invariant system is given by the expression,

$$n(E) = -\frac{1}{\pi}\Im G_{00}(E), \tag{10.19}$$

where G_{00} is given as an integral of $G(E;\mathbf{k})$ over the BZ of the reciprocal lattice,

$$G_{00}(E) = \frac{1}{\Omega_{\mathrm{BZ}}}\int_{\mathrm{BZ}} G(E;\mathbf{k})\mathrm{d}^3k. \tag{10.20}$$

In general, the site matrix elements of the Green function are given by the expression

$$G_{ij}(E) = \frac{1}{\Omega_{\mathrm{BZ}}}\int_{\mathrm{BZ}} G(E;\mathbf{k})\mathrm{e}^{-\mathrm{i}\mathbf{k}\cdot(\mathbf{R}_i - \mathbf{R}_j)}\mathrm{d}^3k. \tag{10.21}$$

Specific examples of the evaluation of Green function matrix elements are given in the next subsection.

In closing the present discussion we mention two fundamental properties of Green functions that are preserved in both ordered and disordered systems. First, it follows from the definition in Eq. (10.10) that the imaginary part of the Green function has the opposite sign from that of the energy. Thus, for energies in the upper half of the complex plane, $\Im z > 0$, we have

$$\Im G = -\Im z G^{\dagger} G < 0. \tag{10.22}$$

It follows also from the definition of the Green function that as $z \to \infty$,

$$G(z) \to \frac{1}{z}. \tag{10.23}$$

Both of these results can be exploited to advantage in the implementation of approximate methods such as the CPA in obtaining the electronic structure of disordered materials.

10.5.1 TB Green functions of ordered systems

First we consider the case of a one-dimensional ordered system of lattice constant a, characterized by site energies ϵ_0 and nearest-neighbor hopping W. The Green function matrix elements, $G_{ij}(z)$, corresponding to a complex energy parameter, z, are given by Eq. (10.10),

$$G_{ij}(z) = \frac{a}{2\pi} \int_{-\pi}^{\pi} \frac{\mathrm{e}^{-\mathrm{i}\mathbf{k}\cdot(\mathbf{R}_j - \mathbf{R}_i)}}{\mathrm{d}} kz - \epsilon_0 - 2W \cos ka, \tag{10.24}$$

where $2W \cos ka$ is the FT $W(\mathbf{k})$ in this case, and the BZ has been defined as the interval $[-\pi/a, \pi/a)$. This integral can be evaluated by means of contour integration [53]. With the definition $\alpha = z - \epsilon_0$ and for $i = j$, we find

$$G_{00}(z) = -\frac{\mathrm{i}}{\sqrt{4W^2 - \alpha^2}}, \tag{10.25}$$

where the positive square root is to be understood.

For z real and equal to the energy E the Green function has an imaginary part for values of E such that $\epsilon_0 - 2W < E < \epsilon_0 + 2W$. Thus, $2W = w$ is equal to one half the band width of this single-band, nearest neighbor model system. For such values of the energy, the density of states is given by, Eq. (10.19),

$$n(E) = \frac{\theta(4W - |\alpha|)}{\sqrt{4W^2 - \alpha^2}}, \tag{10.26}$$

where $\theta(x)$ is the Heaviside step function, $\theta(x) = 1$ if $x \geq 0$, and $\theta(x) = 0$ if $x < 0$. Both the real and imaginary parts of the site-diagonal Green

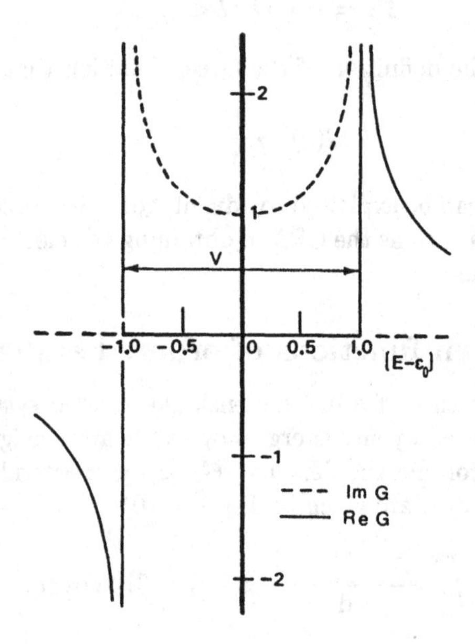

Figure 10.4: Imaginary part (dotted line) and real part (solid line) of $G_{00}(E)$ for a one-dimensional periodic material with nearest-neighbor hopping, $W = 0.5$ in arbitrary units. The band width V equals $4W$. See reference [53].

function are plotted in Fig. 10.4. We note the square-root singularities in $n(E)$ (dotted curve) at the band edges, $E = \epsilon_0 \pm 2W$, and the discontinuity of the real part there.

For a single-band material based on a two-dimensional square lattice of constant a and characterized by nearest-neighbor hopping, the FT of the hopping matrix takes the form

$$W(\mathbf{k}) = 2W(\cos ak_x + \cos ak_y), \tag{10.27}$$

and the corresponding BZ integral of Eq. (10.24) for the site-diagonal Green function yields [53]

$$G_{00}(z) = \frac{2}{\pi(z - \epsilon_0)} K(\lambda), \tag{10.28}$$

where $\lambda = 4W/(z - \epsilon_0)$ and $K(\lambda)$ is an elliptic integral of the first kind. The density of states can be obtained by taking the imaginary part of G_{00}, and it is plotted in Fig. 10.5. This figure clearly exhibits the discontinuity of the DOS (imaginary part of G_{00}) at the edges of the band, and the corresponding logarithmic singularity of the real part there. In contrast to this behavior, note the discontinuity of the real part of G_{00} and the logarithmic singularity of the imaginary part.

For a three-dimensional single-band material with nearest-neighbor hopping based on a cubic lattice, we have

$$W(\mathbf{k}) = 2W(\cos ak_x + \cos ak_y + \cos ak_z), \tag{10.29}$$

and the corresponding site-diagonal element of the Green function can be brought to the form [53]

$$G_{00} = \frac{1}{2\pi^2 W} \int_0^{\pi} \mathrm{d}\phi_x t K(t), \tag{10.30}$$

where

$$t = 4W/(z - \epsilon_0 - 2W\cos\phi_x). \tag{10.31}$$

The integral in Eq. (10.30) must be evaluated numerically. The real and imaginary parts of G_{00} for the simple cubic lattice are shown in Fig. 10.6.

The figure shows the presence of the van Hove singularities inside the band, where the real and imaginary parts of G_{00} are continuous but have discontinuous derivatives. The imaginary part of G_{00} approaches the band edges as $|\Delta E|^{1/2}$, where ΔE denotes the distance from the edge. The real part of G_{00} remains finite at the band edges but its derivative diverges as one approaches the band edges from outside the band.

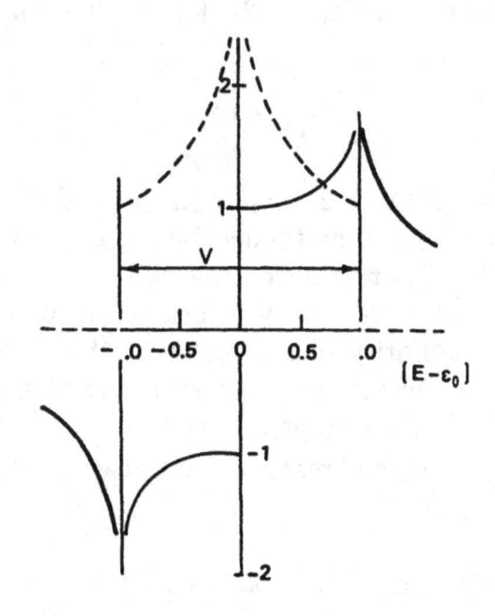

Figure 10.5: **Imaginary part (dotted line) and real part (solid line) of $G_{00}(E)$ for a two-dimensional periodic material based on a square lattice with nearest-neighbor hopping. The band width V equals $8W$. See reference [53].**

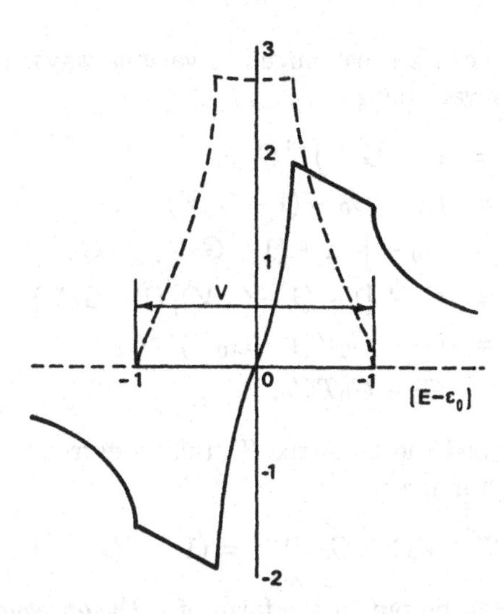

Figure 10.6: **Imaginary part (dotted line) and real part (solid line) of $G_{00}(E)$ for a three-dimensional periodic material based on a cubic lattice with nearest-neighbor hopping. The band width V equals $12W$. See reference [53].**

10.6 Standard Manipulations

The Green function can be written in various forms, some of which may be more appropriate than others in the discussion of specific problems. We consider the Hamiltonian H of a disordered system as the sum of the Hamiltonian H_0 of a pure, translationally invariant material and a perturbing part V, (not necessarily small),

$$H = H_0 + V. \tag{10.32}$$

(For the sake of convenience of presentation, we may consider H_0 as describing the invariant off-diagonal part, W, of an alloy with only diagonal disorder, with V being the random diagonal part.) The Green function $G = (z - H)^{-1}$, can be expressed in terms of the unperturbed Green function $G_0 = (z - H_0)^{-1}$, which is assumed to be known, and the perturbation, V,

$$\begin{aligned} G(z) &= (z - H_0 - V)^{-1} \\ &= \left[(z - H_0)(1 - (z - H_0)^{-1}V)\right]^{-1} \\ &= [1 - G_0(z)V]^{-1} G_0(z). \end{aligned} \tag{10.33}$$

The last expression can be rearranged in various ways. For example (suppressing the energy variable z),

$$\begin{aligned} G &= (1 - G_0V)^{-1}G_0 \\ &= G_0 - G_0 + (1 - G_0V)^{-1}G_0 \\ &= G_0 + \left[-1 + (1 - G_0V)^{-1}\right] G_0 \\ &= G_0 + [1 - (1 - G_0V)]\,(1 - G_0V)^{-1}G_0 \\ &= G_0 + G_0V(1 - G_0V)^{-1}G_0 \\ &= G_0 + G_0TG_0, \end{aligned} \tag{10.34}$$

where we have defined the t-matrix, T, (also referred to as the scattering matrix or transition matrix)

$$T = V(1 - G_0V)^{-1} = (1 - VG_0)^{-1}V. \tag{10.35}$$

Also, Eq. (10.33) can be put in the form of a *Dyson equation*,

$$G = G_0 + G_0VG, \tag{10.36}$$

whose iteration yields a perturbation-like expansion for G,

$$G = G_0 + G_0VG_0 + G_0VG_0VG_0 + \cdots. \tag{10.37}$$

This operator expression represents the propagation of an excitation in the invariant material represented by H_0 (or W), with scattering due to V taking place at the impurity sites. It should be noted that the last expression is meaningful only if the expansion in powers of V converges, which we will take to be the case without further scrutiny. Given this expression, various approximations can be introduced for the evaluation of $\langle G\rangle$, i.e., the average of the Green function over all configurations of a disordered material.

Quite generally, we seek approximations to the exact *self-energy* $\Sigma(z)$ which is such that the ensemble averaged Green function can be written in the form,

$$\langle G\rangle = G_0 + G_0\Sigma\langle G\rangle = (G_0^{-1} - \Sigma)^{-1}. \tag{10.38}$$

The exact self-energy is a function of the energy, z, and must be translationally invariant reflecting the corresponding property of $\langle G\rangle$,

$$\langle G\rangle(\mathbf{k}) = [z - W(\mathbf{k}) - \Sigma(z,\mathbf{k})]^{-1}. \tag{10.39}$$

In the following discussion, we will introduce several methods that lead to approximate evaluations of the self-energy Σ. Effectively, all of these methods attempt to replace the true disordered material by a translationally invariant medium characterized by an approximate self-energy, and then calculate the averaged Green function through Eq. (10.39).

10.6.1 Rigid band model

The simplest way to introduce order into the alloy problem is to *neglect entirely* the difference between the potentials of the alloy constituents, as is done in the *rigid band model* (RBM) [8]. In this model, the electronic states of a pure A metal are assumed to be *identical* to those of a pure B metal and to those of any $A-B$ alloy! Therefore, the self-energy is assumed to have the form

$$\Sigma \equiv V_A \equiv V_B. \tag{10.40}$$

Thus, the only difference between a pure system and an alloy is postulated to be the number of electrons per atom that determines the Fermi energy. For example, according to this model, Ni behaves differently from Cu because it contains one less electron per atom, and the Fermi energy in Ni lies lower (closer to the band bottom) than it does in Cu. Also, because of this difference in position, the nature of the electronic states at the Fermi energy in Ni is different from that in Cu. In fact, the model predicts that the addition of Ni to Cu causes the Fermi energy to move to lower energies relative to the band (which remains rigidly fixed in position).

The original applications of the RBM to CuNi alloys seemed to meet with some success in explaining the properties of these systems, especially magnetism. Increasing experimental evidence and computational sophistication, however, have shown that the RBM in general provides a wrong description of disordered materials. The model is in fact totally inappropriate for Cu-Ni alloys, due to the difference in the band structure of the two pure elemental solids as indicated in Figs. 10.7 and 10.8. Although the two band structures are similar, they are by no means identical. Furthermore, experimental studies (quoted in the next chapter) indicate that upon Ni additions to Cu, the Fermi level stays essentially fixed in position while the Ni structure of the DOS evolves progressively out of the one for Cu. This behavior is contrary to the predictions of the RBM.

The failure of the RBM is due to the fact that the electronic structure of a substitutionally disordered alloy is *qualitatively* different from that of any of the pure elements that go into the formation of the alloy. However, the model can give reasonable results in the *rigid band limit*, in which the potentials of the alloy constituents are nearly identical. This is a limiting case that is not often realized in practice.

10.6.2 Virtual crystal approximation

The next level in the approximation schemes for calculating the electronic states of a random alloy is to assume that the alloy potential is periodic and equal to the concentration average of the potentials of the pure species

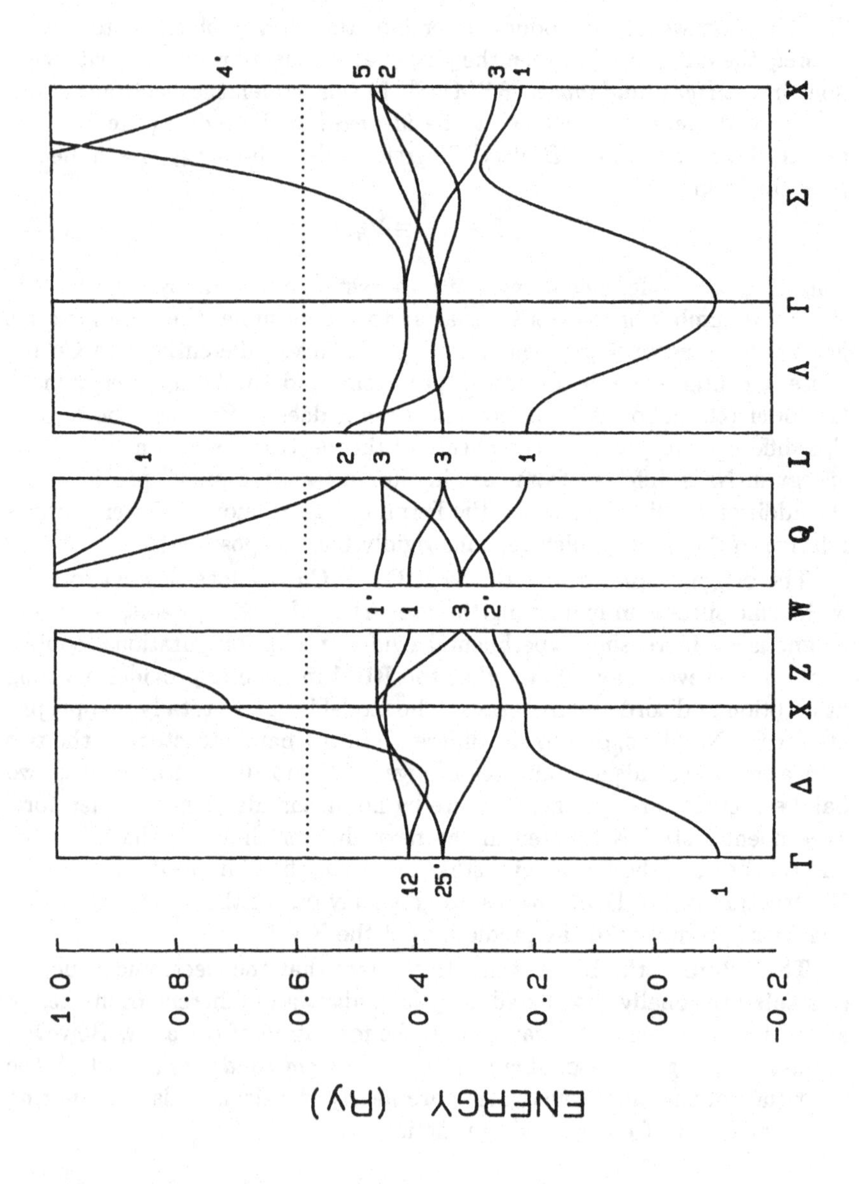

Figure 10.7: **The energy bands of elemental Cu along various directions in the BZ. The horizontal dashed line indicates the Fermi energy. (D. Papaconstantopoulos, private communication.)**

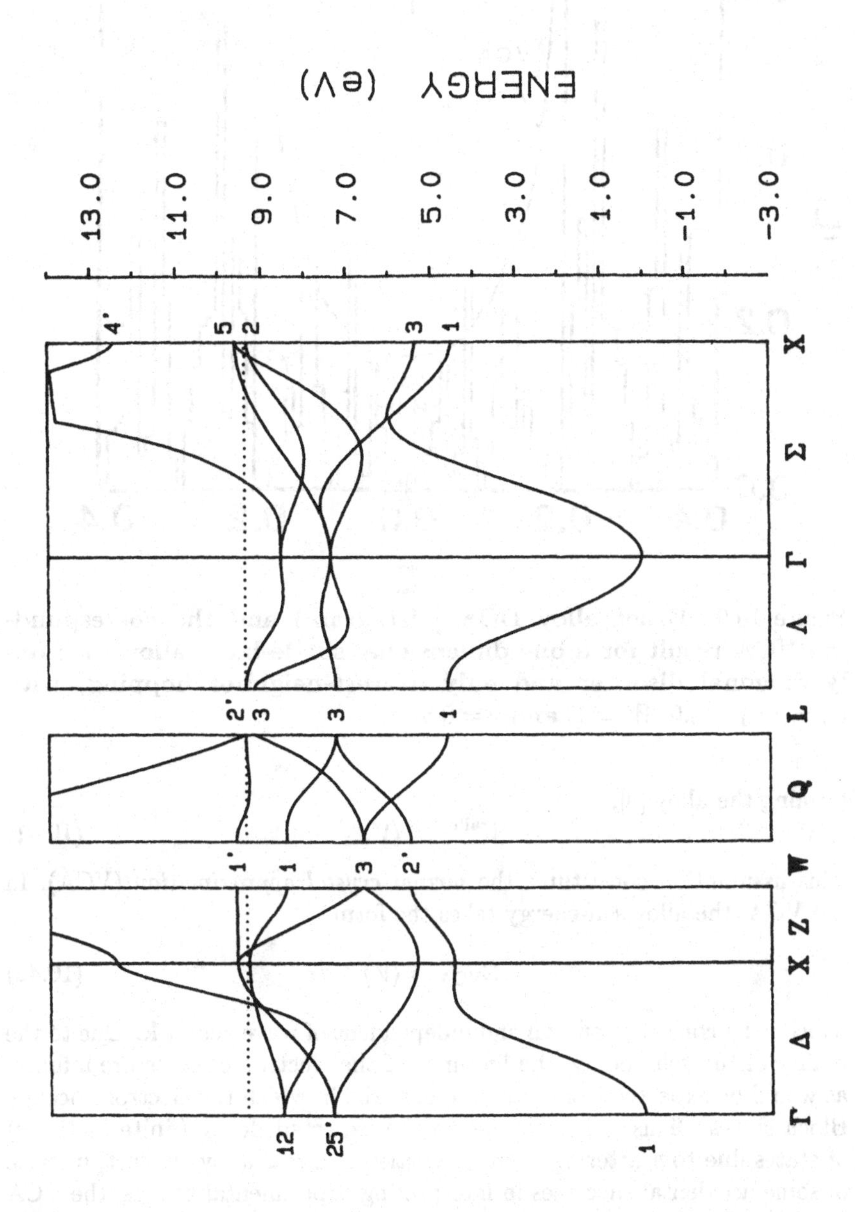

Figure 10.8: **The energy bands of elemental Ni along various directions in the BZ. The horizontal dashed line indicates the Fermi energy. This band structure should be compared closely with the one for Cu in the previous figure. (D. Papaconstantopoulos, private communication.)**

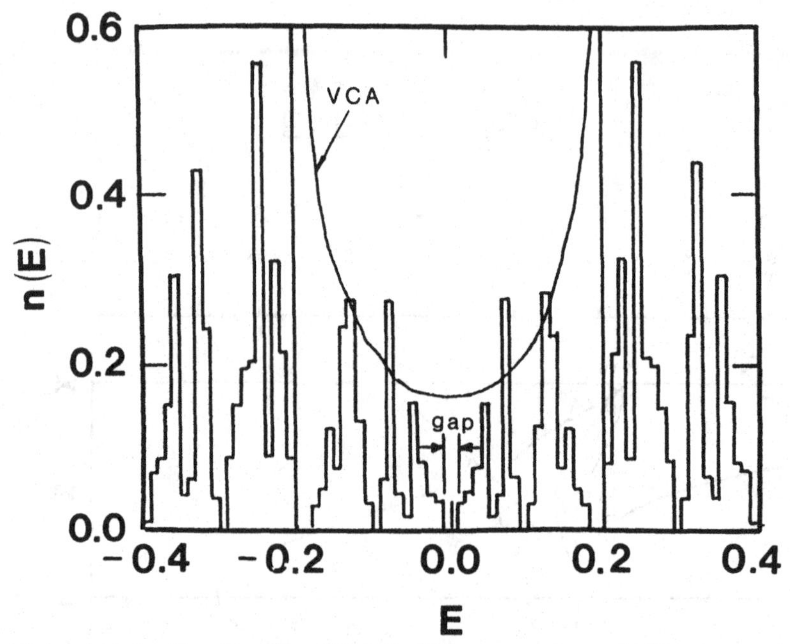

Figure 10.9: **Exact alloy DOS (histogram) and the corresponding VCA result for a one-dimensional single-band alloy with only diagonal disorder and only nearest-neighbor hopping, with $\epsilon_A = -\epsilon_B = 2.0$, $W = 1$, and $c = 0.5$.**

forming the alloy [9],

$$V^{\text{alloy}} = \langle V \rangle. \tag{10.41}$$

This assumption constitutes the ***virtual crystal approximation*** (VCA). In the VCA, the alloy self-energy takes the form

$$\Sigma_{\text{VCA}} = \langle V \rangle, \tag{10.42}$$

which is by construction real and independent of wave vector **k**. Due to the reality of the self-energy, the lifetimes of the electronic states are infinite, as would be expected from any pure material in which the electrons occupy Bloch states. This is in contrast to the expected decay (finite lifetimes) of states due to scattering from impurities in a real alloy. In fact, in spite of some accidental successes in interpreting experimental results, the VCA can be shown to be mathematically wrong. This can easily be illustrated by means of the DOS of model one-dimensional systems, like the one shown [1] in Fig. 10.9. The histogram in the figure shows the density of states of a one-dimensional, single-band TB alloy with only diagonal disorder and only nearest-neighbor hopping, obtained using the negative eigenvalue theorem [41]. In these exact results, one clearly distinguishes an A and B subband separated by a gap at $E = 0.0$, and each containing 0.5 electrons per atom. The VCA result, on the other hand, shows only a *single* band, corresponding to a pure material with a site energy

$$\epsilon^{\text{VCA}} = \langle \epsilon \rangle = c_A \epsilon_A + c_B \epsilon_B = 0.0. \tag{10.43}$$

Not only do the VCA results fail to reproduce the gap between the A and the B subbands, thus failing to satisfy the bounding limit theorem, but the VCA densities of states are ***qualitatively*** different from those of the alloy.

The situation depicted in Fig. 10.9 is an example of the ***split-band limit*** in which the alloy DOS consists of separate and well-defined subbands associated with each of the alloy constituents. This split-band behavior also exists in three-dimensional systems, and the VCA is equally inadequate in general to describe the electronic structure there.

As was the case with the RBM, the VCA also has a small range of validity. It gives reasonable results when the potentials of the pure systems do not differ greatly, in the so-called ***virtual-crystal limit***, and also often near the limits $c = 0.0$ or $c = 1.0$. It has also been found to be fairly successful in describing some properties of III-V and II-VI semiconductor alloys, e.g., (GaAl)As and (ZnCd)Se, particularly the states at conduction-band minima. The wave functions corresponding to those states are generally spread over a few hundred angstroms, effectively averaging over the alloy disorder. However, in spite of such glimpses of success, the VCA remains in general a rather unsatisfactory alloy theory.

10.6.3 Average t-matrix approximation

The ***average t-matrix approximation*** (ATA) is based on the expression for the t-matrix in Eq. (10.35), which can be written in the form,

$$T = V + VG_0T. \tag{10.44}$$

It follows from Eq. (10.35) that the configurational average of the Green function can be written in the form,

$$\langle G \rangle = G_0 + G_0 \langle T \rangle G_0, \tag{10.45}$$

so that the self-energy takes the general form

$$\Sigma = \langle T \rangle \left(1 + G_0 \langle T \rangle\right)^{-1}. \tag{10.46}$$

This is an exact expression, yielding the self-energy in terms of the configurational average of the t-matrix of the entire system.

Let us now consider a potential that is the sum of individual site contributions,

$$V = \sum_i V_i. \tag{10.47}$$

In this case, the t-matrix in Eq. (10.35) can be written in a site representation,

$$\begin{aligned} T_{ij} &= V_i\delta_{ij} + V_iG_0^{ij}V_j + V_i\sum_k G_0^{ik}V_kG_0^{kj}V_j + \cdots \\ &= t_i\delta_{ij} + t_i\sum_{k\neq i} G_0^{ik}T_{kj}, \end{aligned} \tag{10.48}$$

where an individual scatterer t-matrix is defined by the usual expression,

$$t_i = (1 - V_iG_0)^{-1}V_i, \tag{10.49}$$

in terms of the on-site potential, V_i. We note that no site-diagonal elements of the Green function occur in Eq. (10.48) reflecting the fact that the t-matrix, t_i, provides a full description of the scattering from the potential V_i. We also recognize that Eq. (10.48) has the same form as the equation of motion of the scattering matrix, Eq. (3.52), but having been derived here in terms of quantities appropriate to a TB description of a material. Now, the configurational average of the system t-matrix, takes the form,

$$\langle T_{ij}\rangle = \langle t_i\rangle\delta_{ij} + \langle t_i\sum_{k\neq i} G_0^{ik}T_{kj}\rangle. \tag{10.50}$$

Because of the correlations between sites occurring in the second term on the right-hand side, the exact configurational average in the last equation can be evaluated only approximately. One approximation scheme consists in neglecting statistical correlations between site occupancies and writing

$$\langle T_{ij}\rangle = \langle t_i\rangle\delta_{ij} + \langle t_i\rangle\sum_{k\neq i} G_0^{ik}\langle T_{kj}\rangle. \tag{10.51}$$

In this approximation, the site-diagonal element of the system t-matrix becomes,

$$\langle T_{ii}\rangle = \langle t_i\rangle\left(1 - \bar{G}_0\langle t_i\rangle\right)^{-1}, \tag{10.52}$$

leading to the approximate expression for the self-energy,

$$\Sigma_{ij} = \langle t_i\rangle\left(1 - \bar{G}_0\langle t_i\rangle\right)^{-1}\delta_{ij}. \tag{10.53}$$

In the last two expressions, $\bar{G}_0$ denotes the site off-diagonal part of G_0. This self-energy is clearly site-diagonal. The average of the site t-matrix, $\langle t_i\rangle$, equals the configurational average of the t-matrices of the pure constituents of the alloy,

$$\langle t\rangle = \sum_i c_it_i. \tag{10.54}$$

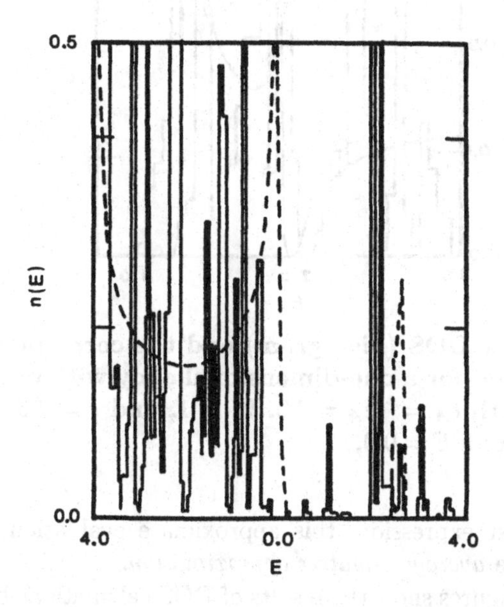

Figure 10.10: Exact DOS (histogram) and the corresponding ATA result (solid curve) for a one-dimensional alloy with only diagonal disorder, and with $\epsilon_A = -\epsilon_A = 2.0$, $W = 1$, and $c = 0.05$.

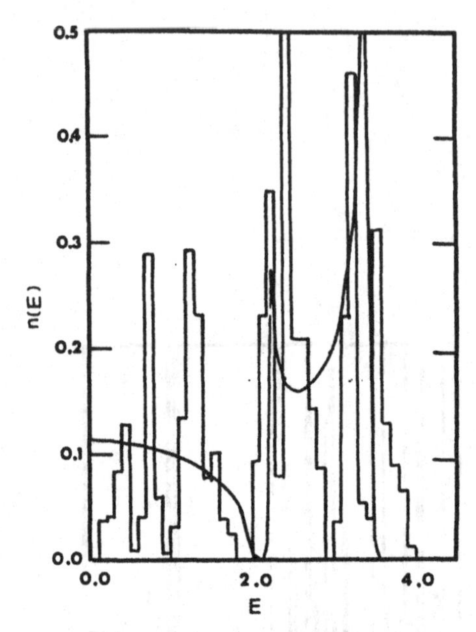

Figure 10.11: **Exact DOS (histogram) and the corresponding ATA result (solid curve) for a one-dimensional alloy with only diagonal disorder, and with $\epsilon_A = -\epsilon_A = 2.0$, $W = 1$, and $c = 0.5$. The DOS is symmetric about $E = 0.0$.**

Because of the last expression, this approximate evaluation of the self-energy is called the *average t-matrix approximation.*

The following figures show the results of DOS calculations based on the ATA. The DOS in Fig. 10.10 corresponds to a very dilute alloy (see figure caption) for which the ATA may be expected to provide relatively accurate results due to the smallness of the effects of intersite correlations at low concentrations. It is seen in this figure, however, that even at such concentrations, the exact DOS (histogram) contains a great deal of structure that is due to the statistical fluctuations in the local environment of a site. This structure is completely smoothed out in the ATA (solid curve). Furthermore, even though the ATA produces an impurity peak at $E \simeq 2.50$, it does so inaccurately, the exact peak being positioned at $E \approx 2.10$, and the peak also has the wrong weight. We also note that the ATA produces states well inside the gap region at $E \approx 0.0$ separating the host band from the impurity levels.

As the concentration increases, correlations between scatterers become more and more important, and the accuracy of the ATA is expected to decrease even further. This is shown in Fig. 10.11 which shows the exact DOS (histogram) and the corresponding ATA result for a concentrated alloy. The ATA results completely miss the gap at $E \approx 0.0$, thus violating the bounding limit theorem, and yield a rather poor representation of the rest of the band. Note in particular that the ATA yields a DOS that is narrower than the exact result. This feature is common to all single-site approximations, which account only for statistical fluctuations confined to a single site but neglect correlation effects between different sites. On the other hand, the ATA provides a much improved description of the DOS compared to the VCA, (compare with Fig. 10.9), and can be used as an interpolation scheme at all concentrations since the positions of the DOSs in the ATA follow that of the exact DOS. The ATA is discussed further and compared with the CPA in the following two chapters.

10.7 Propagator and Locator Expansions

The operator expression in Eq. (10.37) represents the propagation of an excitation in a translationally invariant material represented by H_0 (or W), with scattering taking place at the impurity sites described by V. In a site representation, this equation can also be put in the form

$$G_{ij} = G_0^{ij} + \sum_k G_0^{ik} t_k G_{kj}, \tag{10.55}$$

where t_i is the site t-matrix corresponding to the potential at site i. This expression for the Green function is commonly referred to as the ***propagator equation of motion.*** In this expression an excitation, e.g., an electron wave, is considered as propagating freely between sites, and scattered upon incidence on an impurity site.

An alternative approach to the propagator expansion is provided by the locator equation of motion, Eq. (10.14). This equation can be "solved" formally by an iterative process. In particular, we seek an expression for the site-diagonal part of G,

$$G_{ii} = g_i + g_i \sum_{k \neq i} W_{ik} G_{ki}. \tag{10.56}$$

Following a procedure formally identical to that used to treat Eq. (9.62), we can write

$$G_{ii} = (g_i^{-1} - \Delta_i)^{-1} = (z - \epsilon_i - \Delta_i)^{-1}, \tag{10.57}$$

where Δ_i is the renormalized interactor for site i. It is given by an expression identical to that in Eq. (9.65) when the quantities $\underline{t}_i$ and $\underline{G}_{ij}$ in that equation are replaced respectively by g_i and W_{ij}. This also allows the identification of the defective-medium Green function, corresponding to the defective-medium scattering-path operator, Eq. (9.65). We have,

$$\begin{aligned}\Delta_i &= \sum_{k\neq i} W_{ik} g_k W_{ki} + \sum_{k\neq i}\sum_{j\neq i,k} W_{ik} g_k W_{kj} g_j W_{ji} + \cdots \\ &= \sum_{k\neq i} W_{ik} \Gamma_{kj}^{(0)} W_{ji} \end{aligned} \tag{10.58}$$

where

$$\Gamma_{ij}^{(0)} = G_{ij} - G_{i0} G_{00}^{-1} G_{0i}. \tag{10.59}$$

The renormalized interactor and defective medium Green function associated with a cluster of impurities can be obtained in a straightforward way through the generalization of scalar quantities to matrices.

The renormalized interactor associated with a single impurity embedded in an otherwise translationally invariant host material can be easily evaluated. First, we note that Δ_i is independent of the chemical occupation of site i [see discussion following Eq. (9.67)] so that it has the same value when site i is occupied by a host atom or an impurity atom. In the former case, equating the expression in Eq. (10.57) with the site-diagonal element of the Green function obtained through a lattice Fourier transform, Eq. (10.20), we find

$$\Delta_0 = z - \epsilon_0 - G_{00}^{-1}, \tag{10.60}$$

where, ϵ_0 is the site energy in the host material. Thus, we obtain for the site-diagonal element of the Green function associated with the impurity site,

$$G_{00}^{\text{imp}} = \left(\epsilon_{\text{imp}} - \epsilon_0 + G_{00}^{-1}\right)^{-1}, \tag{10.61}$$

where ϵ_{imp} is the site energy at the impurity site.

A local mode solution, called an *impurity mode*, exists whenever ϵ_{imp} lies sufficiently outside the band of the host material. For $\Re z = E$ well outside the band, we obtain

$$G_{00}^{\text{imp}} \simeq \frac{1}{E - \epsilon_{\text{imp}}}. \tag{10.62}$$

Considering E as having an infinitesimal, positive imaginary part, we see that the local mode is a δ-function at the impurity energy. The existence of a local mode well outside the host band is illustrated schematically in Fig. 10.12. For a realistic representation of an impurity state, see Fig 9.2.

The considerations above apply when the impurity energy is well outside the host band. As ϵ_{imp} approaches the energies of the host material, the impurity states begin to hybridize with the host states, until the local mode is absorbed in the host band[4]. In materials containing small but finite

[4]Impurity states may also appear in the gap of the energy spectrum of semiconductors and insulators.

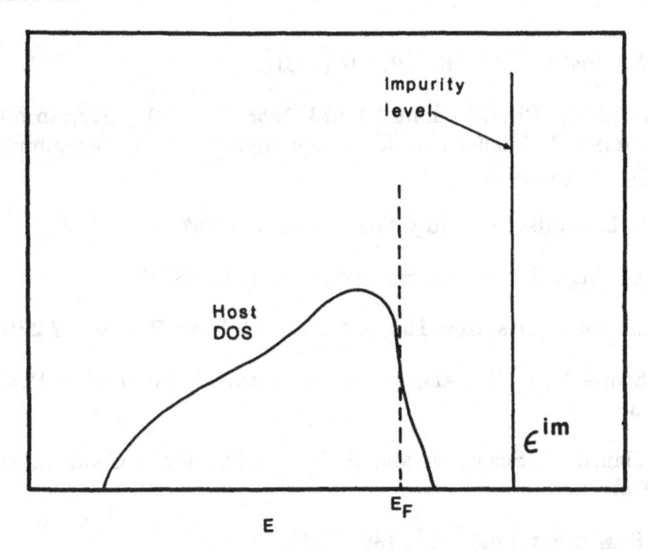

Figure 10.12: **Schematic representation of a local mode. The states below the Fermi energy, E_F, of the host are occupied, those above empty.**

concentrations of impurities, one often recognizes the formation of impurity bands that lie outside the band of host states. For sufficiently strong-scattering alloys, these impurity bands persist at all concentrations and give rise to separate subbands for the host and impurity states. This corresponds to the split-band regime, an example of which is given in Fig. 10.9.

Bibliography

[1] A. Gonis, *Green Functions for Ordered and Disordered Systems*, North Holland, Amsterdam, New York (1992).

[2] M. Ziman, *Models of Disorder*, Cambridge University Press, London (1979).

[3] Lord Rayleigh, Phil. Mag. Str. 5, **34**, 481 (1892).

[4] A. Einstein, *Investigations on the Theory of Brownian Movement*, (1906), Dover Reprint, New York (1956).

[5] L. L. Foldy, Phys. Rev. **67**, 107 (1945).

[6] M. Lax, Rev. Mod. Phys. **23**, 287 (1951).

[7] M. Lax, Phys. Rev. **85**, 621 (1952).

[8] J. Friedel, Phil. Mag. **43**, 153 (1952); Adv. Phys. **3**, 446 (1953).

[9] L. Nordheim, Ann. Phys. **9**, 607 (1931).

[10] J. Friedel, J. Phys. Radium **19**, 573 (1958); **23**, 501 (1962); in *Metallic Solid Solutions*, J. Friedel and A. Guinier (eds.), W. A. Benjamin, New York (1963), paper XIX.

[11] S. F. Edwards, Proc. Roy. Soc. (London) A**267**, 518 (1962).

[12] J. Korringa, J. Phys. Chem. Solids **7**, 252 (1958).

[13] J. L. Beeby, Phys. Rev. **135**, A 130 (1964); Proc. Roy. Soc. A**279**, 82 (1964).

[14] L. Schwartz, F. Brouers, A. Vedyayev, and H. Ehrenreich, Phys. Rev. B**4**, 3383 (1971).

[15] A. Bansil, L. Schwartz, and H. Ehrenreich, Phys. Cond. Matter **19**, 391 (1975).

[16] R. Bass, Phys. Lett. **46**A, 189 (1973).

[17] P. Soven, Phys. Rev. **156**, 809 (1967).

[18] D. W. Taylor, Phys. Rev. **156**, 1017 (1967).

[19] P. W. Anderson and W. L. McMillan, in *Proceedings of the International School of Physics, "Enrico Fermi"* course 37, W. Marshall (ed.), Academic Press Inc., New York (1967).

[20] B. Velicky, S. Kirkpatrick, and H. Ehrenreich, Phys. Rev. **175**, 747 (1968).

[21] F. Yonezawa and K. Morigaki, Prog. Theor. Phys. Suppl. **53**, 1 (1973).

[22] R. J. Elliott, J. A. Krumhansl, and P. L. Leath, Rev. Mod. Phys. **46**, 465 (1974).

[23] H. Ehrenreich and M. L. Schwartz, *Solid State Physics*, H. Ehrenreich, F. Seitz, and D. Turnbull (eds.), **31**, 149 (1976).

[24] L. Hodges, R. E. Watson, and H. Ehrenreich, Phys. Rev. B**5**, 3952 (1972).

[25] Y. Onodera and Y. Y. Yonezawa, J. Phys. Soc. (Japan) **24**, 341 (1968).

[26] F. Yonezawa, Progr. Theor. Phys. (Kyoto) **40**, 734 (1968).

[27] P. L. Leath, Phys. Rev. **171**, 725 (1968).

[28] L. Schwartz and E. Siggia, Phys. Rev. B**5**, 383 (1972).

[29] F. Ducastelle, J. Phys. C**4**, L 75 (1971).

[30] P. L. Leath, Phys. Rev. B**2**, 3078 (1970).

[31] T. Matsubara, Prog. Theor. Phys. Suppl. **46**, 326 (1970).

[32] F. Brouers, M. Cyrot, and F. Cyrot-Lackmann, Phys. Rev. B**7**, 4370 (1972).

[33] R. Haydock, V. Heine, and M. J. Kelly, J. Phys. C**5**, 2845 (1972).

[34] A. Mookerjee, J. Phys. C**6**, 1340 (1973); L 205 (1973).

[35] A. R. Bishop and A. Mookerjee, J. Phys. C**7**, 2165 (1973).

[36] A. Mookerjee, J. Phys. C**8**, 29 (1975).

[37] A. Mookerjee, J. Phys. C**8**, 1524 (1975).

[38] F. Ducastelle, J. Phys. C**8**, 3297 (1975).

[39] P. Lloyd and P. R. Best, J. Phys. C**8**, 3752 (1975).

[40] Neil W. Aschcroft and M. David Mermin, *Solid State Physics*, Holt, Reinhart, and Winston (1976).

[41] P. Dean, Rev. Mod. Phys. **44**, 127 (1972).
C**5**, 2845

[42] R. Haydock, V. Heine, and M. J. Kelly, J. Phys. C**8**, 2591 (1975).

[43] R. Haydock, *Solid State Physics*, H. Ehrenreich, F. Seitz, and D. Turnbull (eds.), **35**, 215 (1980), and references therein.

[44] M. J. Kelly, *Solid State Physics*, H. Ehrenreich, F. Seitz, and D. Turnbull (eds.), **35**, 295 (1980), and references therein.

[45] J. P. Gaspard and F. Cyrot-Lackmann, J. Phys. C**6**, 3077 (1973).

[46] M. C. Desjonqueres and F. Cyrot-Lackmann, J. Phys. F**7**, 61 (1976).

[47] A. Trias, M. Kiwi, and M. Weissmann, Phys. rev. B**18**, 1859 (1983).

[48] P. Turchi, F. Ducastelle, and G. Treglia, J. Phys. C**15**, 2891 (1982).

[49] A. Mookerjee, J. Phys. C**6**, L205 (1973); 1340 (1973); J. Phys. C**8**, 29 and 2943 (1975).

[50] M. Hwang, R. Podloucky, A. Gonis, and A. J. Freeman, Phys. Rev. B**23**, 765 (1985), and references therein.

[51] J. E. Gubernaitis and P. L. Taylor, J. Phys. C**6**, 1889 (1973).

[52] I. M. Lifshitz, Soviet Physics Usp. **7**, 549 (1965).

[53] E. N. Economou, *Green's Functions in Quantum Physics*, Springer Verlag, Heidelberg, Berlin, New York (1979).

Chapter 11

Single Site CPA and Clusters

11.1 General Comments

The attempt to regain translational invariance made within the formalism of the VCA and the ATA is a sound one, and the drawbacks of these approximations should not discourage further efforts in that direction. The methods failed primarily because they led to choices of a translationally invariant effective medium that were mathematically incorrect or not very accurate. Much more appropriate choices of such a medium can be made that can yield calculated physical quantities in close agreement with experiment. A particularly successful method for doing this is the *coherent potential approximation* (CPA).

The CPA was introduced simultaneously by Soven [1] in connection with the electronic structure and by Taylor [2] in connection with the lattice dynamics of substitutionally disordered alloys. However, ideas similar to those embodied in the CPA had been expressed previously by several authors, and following its introduction the theory has been developed and generalized in many respects in a number of publications (see preceding chapter). Also, the CPA has been discussed and reviewed in a number of publications [3-8], using a number of diverse formalisms [9-21].

The CPA belongs to the same class of mean-field theories as exemplified by the Weiss molecular-field theory of magnetism, or the Bragg-Williams approximation to the solution of the Ising problem in statistical mechanics. In such theories the properties of the entire material are determined from the behavior at a localized region, usually taken to be a single cell. Even though there have been several applications of the CPA based on many independent formalisms, the nature of the theory can perhaps be understood best in the context of the multiple-scattering description of disordered systems. In this description, we consider the propagation of an electron or a lattice wave through a disordered material as a succession of elementary scatterings at the random atomic (point) scatterers. Thus, at the outset a separation between independent scatterings from different sites is invoked. At the end, an average over all configurations of the disordered system consisting of these scattering centers must be attempted.

In order to render the problem of configurational averaging tractable, we may proceed along the following lines. First, we note that at least in the case of alloys based on a lattice with one atom per unit cell (monatomic material), the configurational average of any physical quantity over all alloy configurations is independent of position (site) in the lattice. We may then replace the material surrounding any single site by an otherwise translationally invariant medium[1] so constructed as to reflect the ensemble average over all alloy configurations. This medium can be chosen in some physically and intuitively reasonable manner or it can be determined in a self-consistent way. The former procedure is exemplified by the VCA and the ATA, the latter by the CPA. In the CPA, we assume that averages over the occupation of a site embedded in the effective medium should yield quantities indistinguishable from those associated with a site of the medium itself. We now make use of the fact that a translationally invariant medium produces no scattering of a wave, and impose the condition that the scattering from a real atom embedded in a CPA medium vanishes on the average. Equivalently, the site-diagonal part of the Green function of a real atom embedded in a CPA medium averaged over the possible occupations of a single site should equal the corresponding Green function of the medium itself. This self-consistency condition yields an effective Hamiltonian describing the medium, often referred to as the *coherent-potential Hamiltonian.*

In the following two sections, we derive the self-consistency condition of the CPA both within the propagator and the locator formalisms. We present both derivations of the CPA in some detail in order to clarify the meaning and content of the theory as much as possible. Alternative derivations of the CPA, e.g., based on diagrammatic expansions, have been reviewed in the literature [8].

For the sake of clarity, we consider explicitly binary alloys, $A_{1-c}B_c$, describable by a single-band TB Hamiltonian. The generalization to multicomponent, multiband alloys can be made in a straightforward way through the replacement of scalar quantities by appropriately defined matrices [8]. Within a single-band framework, we consider $N_A = (1-c)N$ and $N_B = cN$ atoms of type A and B, respectively, distributed over the N sites of a lat-

[1] In general, this medium has the symmetry of the underlying lattice.

tice, with Born-Von Karman periodic boundary conditions. Our aim is to obtain approximations to the average $\langle G\rangle$ of the single-particle Green function over the configurations of the system.

11.2 CPA in the Propagator Formalism

We begin with some general considerations. The one-electron Hamiltonian corresponding to a given alloy configuration is denoted by H, and it is usually taken to be the sum, $H = H_0 + V$, of a translationally invariant part, H_0, and a random part, V, Eq. (10.31). The corresponding Green function is given by the expression,

$$G(z) = (z - H)^{-1}. \tag{11.1}$$

We are interested in the configurational average, $\langle G(z)\rangle$, from which all single-particle properties (of the system described by H) can be obtained. We note that in contrast to the disordered Hamiltonian, the statistically averaged Green function has the full symmetry of the empty lattice. We also introduce an energy-dependent effective Hamiltonian, H_{eff}, characterizing the exact averaged material, through the definition,

$$\langle G(z)\rangle = (z - H_{\text{eff}})^{-1}, \tag{11.2}$$

or

$$H_{\text{eff}} = z - \langle G(z)\rangle^{-1}. \tag{11.3}$$

If the physical Hamiltonian, H, possesses a translationally invariant part, H_0, then H_{eff} is simply equal to $H_0 + \Sigma$, where Σ is the exact self-energy defined in Eq. (10.38). However, the definition of H_{eff} given in Eq. (11.3) is valid in all cases of disorder, including diagonal and off-diagonal disorder. The effective Hamiltonian possesses the full symmetry of the lattice, and is an analytic function of the complex energy, z, in both (upper and lower) complex energy half planes. However, H_{eff} is non-Hermitian (since the site-diagonal elements of the self-energy are complex in general), satisfying instead the *reality* condition,

$$H_{\text{eff}}(z^*) = H^{\dagger}_{\text{eff}}(z), \tag{11.4}$$

where z^* denotes the complex conjugate of z. These properties of H_{eff} follow directly from Eq. (11.3). Suppose now that an approximation, $K(z)$, to the exact effective Hamiltonian is known. Such an approximate Hamiltonian clearly should have the same analytic structure as H_{eff} and could be used in the definition of the averaged Green function through the relation,

$$\langle G\rangle = \bar{G} + \bar{G}(H_{\text{eff}} - K)\langle G\rangle \tag{11.5}$$

where $\bar{G}$ is the Green function associated with the approximate effective Hamiltonian, K,

$$\bar{G}(z) = (z - K)^{-1}. \tag{11.6}$$

From now on the Green function associated with an effective Hamiltonian (as opposed to a real one) will be designated by an overbar. In terms of our previous discussion, K could have been chosen in the VCA, the ATA, or in some other approximation. We have already seen that both the VCA and the ATA contain unsatisfactory features. Hence arises the motivation to search for a better approximation to H_{eff}.

In the language of MST, Eq. (11.5) is expressed in terms of the t-matrix. Using Eq. (10.34), $G = \bar{G} + \bar{G}T\bar{G}$, we have $T = (H_{\text{eff}} - K)(1 + \bar{G}T)$ and we obtain the relation,

$$\langle G \rangle = \bar{G} + \bar{G}\langle T \rangle \bar{G}, \tag{11.7}$$

between the configurational average of the Green function and the t-matrix. Use of this equation in Eq. (11.5) yields the expression,

$$\langle T \rangle = (H_{\text{eff}} - K) + (H_{\text{eff}} - K)\bar{G}\langle T \rangle, \tag{11.8}$$

which can be arranged in the form,

$$H_{\text{eff}} = K + \langle T \rangle (1 + \bar{G}\langle T \rangle)^{-1}. \tag{11.9}$$

From this point, there are two different paths: insert the value of $\langle T \rangle$ corresponding to a given $\bar{G}$, which is the process underlying the ATA, or, choose $\bar{G}$ in such a way that $\langle T \rangle$ vanishes, i.e.,

$$\langle T[\bar{G}] \rangle = 0, \tag{11.10}$$

which leads to the equality

$$H_{\text{eff}} = K. \tag{11.11}$$

This expression provides a ***self-consistent condition*** for K, where the notation $T[\bar{G}]$ denotes the functional dependence of T on K. Once a Hamiltonian K has been found that satisfies Eq. (11.10) within some approximation, the Hamiltonian H_{eff} is also determined to within the same approximation. The self-consistent approach given by Eq. (11.10), although more complicated than any non-self-consistent method, has several advantages over a non-self-consistent determination of the effective Hamiltonian. Among other properties, such a self-consistently determined Hamiltonian leads to physically meaningful results for all values of the alloy parameters, satisfies exact limits and fundamental sum rules, and preserves all analytic and symmetry properties of the exact H_{eff}. We now show how this self-consistent approach can be implemented within a single-site approximation.

The methods of multiple scattering theory can be used in the discussion of the disordered alloy problem through the partitioning of the random, perturbing Hamiltonian, $V = H - K$, into a sum of non-overlapping, local contributions,

$$V = H - K = \sum_n V_n, \tag{11.12}$$

where most often n refers to a single site. This decomposition is of sufficient generality to apply to most cases of physical interest. Thus, it is applicable to systems describable in terms of TB, muffin-tin (MT), or more general Hamiltonians. Note that due to the surrounding material, V_n is usually not the potential associated with a free atom; in fact, it generally depends on the chemical occupation of the sites surrounding site n, (see discussion of self-consistent electronic structure calculations above).

Upon using the Dyson equation, Eq. (2.72), in the form

$$G = \bar{G} + \bar{G}(H - K)G, \tag{11.13}$$

and Eq. (2.87), we obtain the expression,

$$\begin{aligned} T &= (H - K)(1 + \bar{G}T) \\ &= \sum_n T_n, \end{aligned} \tag{11.14}$$

which expresses the t-matrix of the entire system as a sum of contributions describing all scattering events that end (or start) on a single site. Using the definition of the scattering matrix, t, associated with a single scatterer, Eq. (10.49),

$$t_n = (1 - V_n\bar{G})^{-1}V_n, \tag{11.15}$$

we can write

$$T_n = t_n \left(1 + \bar{G} \sum_{m \neq n} T_m\right). \tag{11.16}$$

As discussed in connection with Eq. (3.9), T_n describes the total contribution of the scatterer at n to the scattering matrix. We see that Eq. (11.16) is the equation of motion for the columns of the operator T in a site representation. We recall from our discussion of MST, Chapter 3, that the quantity T_n describes the total scattering from an individual scatterer in the alloy as the combination of the scattering from an isolated cell, described by t_n, and an effective term arising from the scattering in the alloy from all sites other than the one in question. Equations (11.13) to (11.16) are exact and represent a closed form of the formalism of MST in the present, TB, representation.

Now, from Eq. (11.14) we obtain the exact expression for the configurational average of the t-matrix

$$\langle T \rangle = \sum_n \langle T_n \rangle, \tag{11.17}$$

where

$$\langle T_n \rangle = \langle t_n \left(1 + \bar{G} \sum_{m \neq n} T_m \right) \rangle. \tag{11.18}$$

The last equation can also be written in the suggestive form,

$$\langle T_n \rangle = \langle t_n \rangle \left(1 + \bar{G} \sum_{m \neq n} \langle T_m \rangle \right) + \langle t_n \bar{G} \sum_{m \neq n} (T_m - \langle T_m \rangle) \rangle, \tag{11.19}$$

in which the first term on the right-hand side describes the average scattered wave seen by the atom at site n, while the second describes fluctuations away from the average wave. Clearly, this second term is much more difficult to treat than the first. As a *first approximation to the evaluation of* T, we neglect the second term and obtain the closed form

$$\langle T_n \rangle = \langle t_n \rangle \left(1 + \bar{G} \sum_{m \neq n} \langle T_m \rangle \right) \tag{11.20}$$

for the averaged quantities $\langle T_n \rangle$. Since $\sum_{m \neq n} T_m = T - T_n$, Eq. (11.14), we have

$$\langle T_n \rangle = (1 + \langle t_n \rangle \bar{G})^{-1} \langle t_n \rangle (1 + \bar{G} \langle T \rangle). \tag{11.21}$$

Finally, substituting this expression in Eq. (11.9) we see that the effective Hamiltonian is given by the expression,

$$H_{\text{eff}} = K + \sum_n (1 + \langle T_n \rangle \bar{G})^{-1} \langle t_n \rangle, \tag{11.22}$$

or, equivalently,

$$H_{\text{eff}} = K + \sum_n \langle t_n \rangle (1 + \bar{G} \langle T_n \rangle)^{-1}. \tag{11.23}$$

Thus, H_{eff} is the Hamiltonian K of the underlying medium corrected for the effects arising by the average scattering of impurities in that medium.

The single-site quantity $\langle t_n \rangle (1 + \bar{G} \langle t_n \rangle)^{-1}$ can be interpreted as an effective scattering potential corresponding to the average scattering from the atom at site n. Thus, Eq. (11.23) is the form taken by the general expression, Eq. (11.9), within a single-site approximation. If a reasonable choice for $\langle t_n \rangle$ and, therefore, for K can be made, one obtains a corresponding approximation for H_{eff}. In the ATA, $\langle t_n \rangle$ is taken to be the average of the scattering matrices characterizing the pure constituents of an alloy. One can proceed further and determine K in a self-consistent way by imposing the condition that the second term in Eq. (11.23) vanishes. Within a single-site approximation, this corresponds to the imposition of the condition,

$$\langle t_n \rangle = 0, \tag{11.24}$$

which yields the relation

$$H_{\text{eff}} = K, \tag{11.25}$$

for the effective Hamiltonian. In physical terms, Eq. (11.24) means that the ***additional scattering produced by a real atom (of type A or B in a binary alloy) vanishes on the average.***

The procedure resulting in Eq. (11.25) constitutes the *coherent potential approximation* (CPA) to the exact effective Hamiltonian of a substitutionally disordered alloy. Because all averaged quantities possess the symmetry of the underlying lattice, it is sufficient in the case of monatomic systems to consider only one site, say the 0th site, in applying the self-consistency condition, Eq. (11.24). This condition determines an effective, energy-dependent, site-diagonal and generally complex potential, (often denoted by σ or σ_0), which is associated with each site in the lattice. For alloys with only diagonal disorder[2], the off-diagonal part of $H_{\text{eff}} = K$ is simply equal to the translationally invariant part, W, of the alloy Hamiltonian, H. Thus, K is translationally invariant,

$$K_{ij} = \sigma\delta_{ij} + W_{ij}(1 - \delta_{ij}), \tag{11.26}$$

with σ being an approximation to the exact self-energy, Σ. The explicit form of t_n in Eq. (11.24) is obtained from Eq. (11.15) with ϵ^{imp} and G_0 replaced, respectively, with $V_n = \epsilon_n - \sigma$ and $\bar{G} = (z - K)^{-1}$. We have,

$$t_n^\alpha = \frac{\epsilon_\alpha - \sigma}{1 - (\epsilon_\alpha - \sigma)\bar{G}_{00}} = \delta_\alpha/(1 - \delta_\alpha \bar{G}_{00}), \tag{11.27}$$

where $\delta_\alpha \equiv \epsilon_\alpha - \sigma$ is the impurity scattering potential for an atom of type α embedded in a CPA effective medium. This impurity potential determines the ***additional*** scattering caused by the presence of an atom (cell) of type α embedded in the medium. The equation determining the self-energy σ of that medium, Eq. (11.24), has the explicit forms,

$$(1 - c)\frac{\epsilon_A - \sigma}{1 - (\epsilon_A - \sigma)\bar{G}_{00}} + c\frac{\epsilon_B - \sigma}{1 - (\epsilon_B - \sigma)\bar{G}_{00}} = 0, \tag{11.28}$$

for a binary alloy or, more generally,

$$\sum_\alpha c_\alpha \delta_\alpha/(1 - \delta_\alpha \bar{G}_{00}) = 0, \tag{11.29}$$

[2]The case of off-diagonal disorder is treated in Section 11.

for a multicomponent substitutionally disordered material. These equations can be cast into alternative equivalent forms. We will use the term *CPA-algebra* to denote a set of algebraic manipulations that are based on and make explicit use of the CPA self-consistency condition. For a binary alloy, simple CPA algebra allows us to cast Eq. (11.28) in the form,

$$\sigma = (1-c)\epsilon_A + c\epsilon_B - (\epsilon_A - \sigma)\bar{G}_{00}(\epsilon_B - \sigma). \tag{11.30}$$

Also, for any multicomponent alloy, we have

$$\sum_\alpha c_\alpha (1 - \delta_\alpha \bar{G}_{00}) = 1. \tag{11.31}$$

The reader is urged to satisfy him/herself of the validity of the last two expressions. Because $\bar{G}_{00}$ is given in terms of σ by the BZ integral, see Eq. (10.20),

$$\bar{G}_{00} = \frac{1}{\Omega_{\mathrm{BZ}}} \int_{\mathrm{BZ}} \left[z - \sigma - W(\mathbf{k})\right]^{-1} \mathrm{d}^3 k, \tag{11.32}$$

the CPA self-consistency conditions are non-algebraic and in general must be solved by numerical means [8].

Before leaving this section, we summarize the two basic assumptions embodied in the CPA. First, we assumed that the atoms of the various species forming the alloy were randomly distributed over the sites of the lattice. This assumption neglects the possible effects of short-range order (SRO) which for some materials can be very important at certain compositions, pressure, and temperatures. Second, we assumed that the perturbing Hamiltonian could be written as a sum of individual atomic scatterers that act independently of the chemical composition of neighboring sites. Thus, multiple, correlated scattering processes that are represented by the second term on the right side of Eq. (11.20) were neglected. Although SRO may be rather unimportant in some cases, multiple scattering correlations are always present. The CPA as a single-site, mean-field theory cannot describe either the effects of SRO or correlated scattering. In further discussion, we present methods which have been introduced in order to take account of such effects.

11.3 CPA in the Locator Formalism

In this section, we present an alternative but formally equivalent derivation of the CPA self-consistency condition within the locator formalism. In this formalism, the equation of motion for the Green function takes the form, Eq. (10.56),

$$G_{ij} = g_i \delta_{ij} + g_i \sum_{k \neq i} W_{ik} G_{kj}. \tag{11.33}$$

The site-diagonal element of G can be written in the form, [see Eq. (10.57)],

$$G_{ii} = (z - \epsilon_i - \Delta_i)^{-1}, \tag{11.34}$$

which is an exact formal result for any given alloy configuration. The renormalized interactor Δ_i is defined by the expansion in Eq. (10.58). For alloys with only diagonal disorder, Δ_i denotes the interaction of site i with the surrounding medium and is independent of the occupation of site i, (depending, however, on the configuration of the rest of the material). Consistent with the assumptions underlying the CPA, we replace the rest of the material, not including site i, by an effective medium characterized by a site-diagonal self-energy, σ. We denote the corresponding renormalized interactor by $\bar{\Delta}_0$. This quantity is given by an expression analogous to that in Eq. (10.58) but with every g_i in that equation replaced by the *bare locator* for the effective medium at site 0,

$$\bar{g} = (z - \sigma)^{-1}. \tag{11.35}$$

We now obtain an approximation for the site-diagonal element, G_{00}^{α}, associated with a real atom of type α embedded in the CPA effective medium,

$$G_{00}^{\alpha} = (z - \epsilon_\alpha - \bar{\Delta}_0)^{-1}. \tag{11.36}$$

We note that G_{00}^{α} depends only on the occupation of site 0 (in addition to its dependence on the quantity $\bar{\Delta}_0$ and the energy). In order to determine $\bar{\Delta}_0$, and hence G_{00}^{α}, we make use of the independence of the effective renormalized interactor, $\bar{\Delta}_0$, of the chemical occupation of site 0, and the translational invariance of the CPA effective medium. Because of these features, the site-diagonal element of the CPA effective-medium Green function is given by the expression,

$$\bar{G}_{00} = (z - \sigma - \bar{\Delta}_0)^{-1}, \tag{11.37}$$

and together with Eq. (11.32),

$$\bar{G}_{00} = \frac{1}{\Omega_{BZ}} \int_{BZ} [z - \sigma - W(\mathbf{k})]^{-1} \mathrm{d}^3 k, \tag{11.38}$$

we have

$$\bar{G}_{00} = z - \sigma - \bar{G}_{00}^{-1}, \tag{11.39}$$

and

$$G_{00}^{\alpha} = (\sigma - \epsilon_\alpha + \bar{G}_{00}^{-1})^{-1}. \tag{11.40}$$

For a given self-energy, σ, Eq. (11.40) determines the site-diagonal element of the Green-function for a real atom α embedded in the host medium. For example, σ could be chosen within the VCA, the ATA, or some other approximation. In the CPA, the self-energy σ is determined self-consistently through the condition

$$\langle G_{00}^{\alpha} \rangle = \bar{G}_{00}. \tag{11.41}$$

It is also possible to solve directly for the renormalized interactor by casting the last expression in the form (which the reader is urged to verify)

$$\begin{aligned} \langle G_{00}^{\alpha} \rangle &= \langle (z - \epsilon_{\alpha} - \bar{\Delta}_0)^{-1} \rangle \\ &= \frac{1}{\Omega_{\mathrm{BZ}}} \int_{\mathrm{BZ}} \left(\langle G_{00}^{\alpha} \rangle^{-1} + \bar{\Delta}_0 - W(\mathbf{k}) \right)^{-1} \mathrm{d}^3 k. \end{aligned} \tag{11.42}$$

The reader is encouraged to explore alternative ways of casting the CPA self-consistency relation. We see from Eq. (11.36) that we can define species resolved DOSs, i.e., DOSs associated with each kind of species in the alloy,

$$n^{\alpha}(E) = -\frac{1}{\pi} \Im G_{00}^{\alpha}(E). \tag{11.43}$$

Thus, the CPA (as well as the ATA but not the VCA) allows us to study separately the electronic structure associated with each alloy species. Within the CPA (but not the ATA), these DOSs satisfy the relation,

$$\sum_{\alpha} c_{\alpha} n^{\alpha}(E) = \bar{n}(E), \tag{11.44}$$

where $\bar{n}(E)$ is obtained in terms of the imaginary part of the effective medium Green function, $\bar{G}_{00}$.

11.4 Applications

The CPA has been used to calculate the DOSs and related electronic properties both of model systems described by TB Hamiltonians, and of realistic materials; see references [5, 8] and literature cited therein. In this section we present a brief review of some of those results, beginning with calculations based on model, one-dimensional, single-band, TB Hamiltonians.

Figure 11.1 shows in histogram form the DOSs obtained through the use of the negative eigenvalue theorem (for references see previous chapter), compared with those obtained in the CPA, shown as a smooth curve. The alloy parameters are specified in the figure caption. The value of the scattering strength, Eq. (10.8), corresponds to the onset of split-band behavior in which the DOSs of each species (more generally the centers of gravity of the species subbands) are well separated in energy. In the figure, the two subbands are separated by a gap at $E = 0.0$. For this alloy, the VCA density of states is shown in Fig. 10.9, while the corresponding

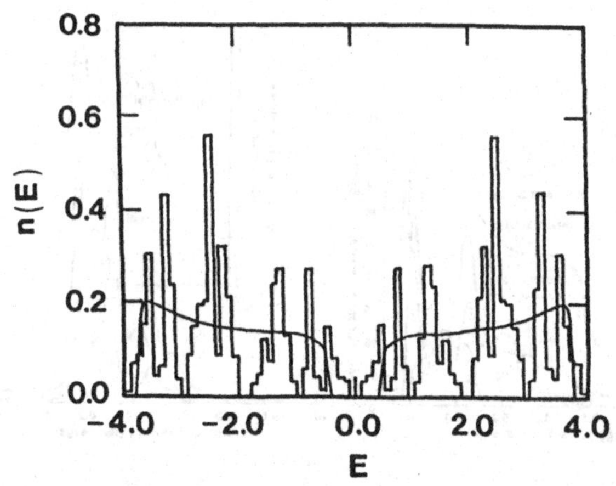

Figure 11.1: **Exact DOS histogram for a one-dimensional, single-band alloy with only nearest-neighbor hopping, $W = 1$, $c = 0.50$, and with $\epsilon_A = -\epsilon_B = 2.0$, compared with the corresponding results obtained in the CPA (solid curve). These values of the parameters correspond to a scattering strength $\delta = 2.0$.**

ATA results are shown in Fig. 10.10. Note that the exact results (histogram) exhibit a great deal of structure, with gaps and peaks interspersed throughout the band. This structure is due to the statistical fluctuations in the local environment of a site, and is particularly pronounced because of the one-dimensional character of the system. The statistical importance of these fluctuations and their effect on the DOSs diminishes with increasing coordination number and dimensionality. Consequently, the accuracy of the CPA increases with increasing dimensionality or coordination number [8].

First, we note that the DOS obtained in the VCA and the ATA, Figs. 10.9 and 10.10, respectively, fail rather badly in reproducing the main feature of the exact DOS, namely the separation into well-defined subbands. This separation is certainly reflected in the CPA results. At the same time, the CPA yields DOSs that are narrower than the exact results. This is caused by the failure of the CPA to account for long-range statistical fluctuations, which are responsible for the tails in the DOS. For the same reason, the CPA yields DOSs that do not have the exponentially decaying character of the exact results near the edges of the band. This last limitation is also characteristic of self-consistent cluster theories such as the molecular CPA (MCPA) [22], as well as non-self-consistent cluster methods discussed below. All such methods account for local statistical fluctuations while neglecting long-range fluctuations that determine the shape of the edges of the bands. Further discussion of the application of the CPA to model, one-dimensional systems can be found in books [8] and review

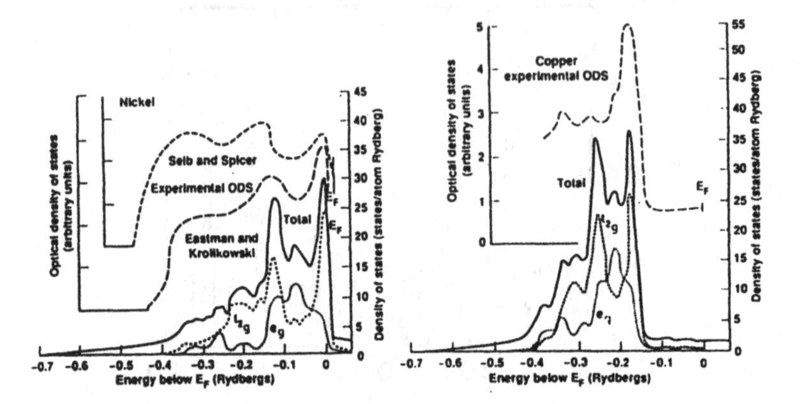

Figure 11.2: **Species resolved and total DOS's for two different concentrations of the Cu_cNi_{1-c} alloy system, compared to photoemission spectra designated as experimental optical densities of states (ODS). For photoemission results, see reference [89]. For CPA calculations, see reference [90].**

articles [5].

Figure 11.2 shows species resolved and total DOSs for Cu-Ni alloys calculated in the CPA, compared with photoemission spectra[3]. Note that as the Ni concentration increases, the Ni character of the DOS develops gradually just below the Fermi energy, which otherwise remains essentially fixed in position. This result is in direct contradiction to what one would expect in the rigid-band picture, in which the Fermi energy would move along the energy axis in accordance with the number of electrons per atom (see previous chapter). More discussion on these points, the use of parametrization schemes to obtain parameters for the TB representation of realistic alloys, and the role of model calculations can be found in the literature [8].

11.5 Advantages and Limitations of the CPA

The SSCPA possesses a large number of desirable properties that unambiguously classify it as the unique, fully self-consistent, best single-site theory of substitutionally disordered alloys. In justifying this statement we first examine the behavior of the CPA in several simple limits and compare it to that of the ATA and other approximate theories. Like the ATA and

[3]Photoemission spectra in general include the effects of matrix elements of transitions between electronic states and are not exact representations of the DOS. However, they often preserve important and pronounced features of the DOS that can be compared with results obtained in electronic structure calculations in which such matrix elements are not taken into account.

other single-site theories, the SSCPA is exact up to, but not including, order Z^{-1} to all orders in c, with leading errors in the CPA being on the order of $Z^{-1}c^2$, where Z is the coordination number of the lattice. However, unlike the ATA and other single-site theories, the CPA is exact in *both* the weak-scattering and the narrow-band limits. In the weak-scattering or virtual-crystal limit, it is exact to order δ^3 in the scattering strength of the alloy. In the narrow-band limit it is exact to order w^2, where w is the half width of the band[4]. By contrast, the ATA gives results which are exact only to order δ^2 in the weak scattering limit, and which are not in general correct in the narrow-band (split-band) limit. The lowest order corrections to $\langle G \rangle$ are of order t^4 in the CPA [1], while being of order t in the ATA [8].

In addition to being valid in all the limits discussed above, the SSCPA also provides a valid interpolation procedure between those limits. It yields results that are analytic functions of all alloy parameters, and which are properly symmetric with respect to the concentrations c and $1-c$ of the two types of species in a binary alloy. Most importantly, the CPA yields self-energies and Green functions, total and species resolved, which are analytic everywhere in the complex energy plane, except for points on the real energy axis corresponding to bound states or band states. The analytic properties [4, 7, 23-26] of the CPA are established in Appendix L.

The main limitations of the CPA are consequences of its single-site nature [8]. Some of these limitations have been pointed out above, in connection with the numerical applications of the CPA. Many of these limitations can be overcome within a cluster extension of the theory, and have provided a strong motivation for the search for such theories. A number of attempts to study the effects of SRO and of statistical fluctuations in general are presented in following sections in this chapter, and in some of the subsequent chapters.

11.6 Off-Diagonal Disorder

The term off-diagonal disorder (ODD) is used to describe the dependence of the transfer integrals or hopping terms of a TB Hamiltonian on the configuration of a disordered system. Up to this point in our discussion, the effects of ODD on the electronic structure of alloys have been ignored. This, in turn, allowed us to present a great deal of formalism in a fairly simple form. However, ODD can be expected to play a significant role in

[4]In TB alloys with only diagonal disorder all alloy constituents have the same band structure and the band width is uniquely defined. It is given by the union of the band widths of the pure systems, with the species resolved DOS centered at the particular site energies, e.g., at ϵ_α for a single-band system. In more general cases, the band width may be taken to be the energy difference between the bottom of the band and the Fermi energy.

determining the electronic properties of many susbstitutionally disordered alloys. In this section, we generalize the formalism of the CPA so as to take into account the effects of ODD.

In general, the off-diagonal elements of the Hamiltonian, W_{ij}, of an alloy depend not only on the chemical occupation of the sites i and j but also on the atomic configurations of the local environments of those sites. For computational but also formal reasons, the effects of the local environment have been neglected in the treatments that have been developed for the study of ODD, and we shall do the same in this book. Thus, we shall treat the case in which the off-diagonal elements of the Hamiltonian assume the "values" $W_{ij}^{\alpha\beta}$, when atoms of kinds α and β occupy sites i and j, respectively. For example, in a binary alloy, the hopping terms of a TB Hamiltonian can in principle assume the values, W_{ij}^{AA}, W_{ij}^{BB}, or $W_{ij}^{AB} = W_{ij}^{BA}$, depending on whether the sites i and j are occupied by twoatoms of type A, two atoms of type B, or by one atom of each species, respectively. The case in which no special relation is assumed to hold between the various values of the hopping terms is often referred to as *general off-diagonal disorder*. It is this general type of ODD that we treat in this book. This approach encompasses somewhat simplified models of ODD, which are characterized by special relations among the various species-dependent values. Even though simpler models such as the additive model

$$W_{ij}^{AB} = \frac{1}{2}\left(W_{ij}^{AA} + W_{ij}^{BB}\right), \tag{11.45}$$

or the geometric model

$$\left(W_{ij}^{AB}\right)^2 = W_{ij}^{AA} W_{ij}^{BB}, \tag{11.46}$$

are somewhat easier to treat than general ODD, such models are of restrictive validity. In addition, they can all be incorporated into the general treatment presented in the following pages.

The fundamental difficulty in treating ODD using the methods presented thus far becomes apparent when we consider a basic quantity of the theory such as the self-energy or the renormalized interactor in the presence of ODD. The renormalized interactor, Δ_0, associated with site 0 is now no longer independent of the chemical composition of this site. The dependence on the chemical species occupying site 0 arises from the corresponding dependence of the hopping terms: The first hop away from site zero when occupied by, say, an atom of type A in a binary alloy, is either W_{0j}^{AA} or W_{0j}^{AB}, depending on the occupation of site j. Thus, a straightforward application of the propagator or the locator formalism to alloys with general ODD is not possible.

There had been many attempts [27-56] in the literature to extend the CPA to alloys with ODD, all of which met with little or no success. It remained for Blackman, Esterling, and Berk (BEB) [57] to show how diagonal and off-diagonal disorder can be decoupled and treated on an equal footing within a single-site formalism. The formalism of BEB, originally based on a functional approach, has been rederived [58] within the locator picture, and has been shown to possess the proper analytic properties. It has also been extended to a cluster method, which allows the treatment of local environment effects in substitutionally disordered alloys.

11.7 Method of BEB

The self-consistent approximation of BEB takes exact account of the correlation between diagonal and off-diagonal disorder within a single-site theory. Like the original SSCPA, it is a completely self-consistent theory, with the correct analytic properties, which treats both diagonal and off-diagonal disorder on an equal footing. For the sake of clarity, we present the basic features of this formalism in connection with a binary, single-band alloy. The extension to multicomponent, multiband alloys is conceptually straightforward, although computationally more involved.

We consider a binary alloy, $A_{1-c}B_c$, describable by a single-band TB model Hamiltonian with matrix elements,

$$H_{ij} = \epsilon_i \delta_{ij} + W_{ij}(1 - \delta_{ij}), \tag{11.47}$$

in a site representation. The site-diagonal energies, ϵ_i, are assumed to take only the values ϵ_A or ϵ_B, depending on whether an atom of type A or B occupies site i. The transfer integrals (hopping terms) W_{ij} are allowed to assume three independent values, $W_{ij}^{AA} = \alpha_{ij}$, $W_{ij}^{BB} = \beta_{ij}$, or $W_{ij}^{AB} = W_{ij}^{BA} = \zeta_{ij}$, depending on the occupation of the sites i and j by two A atoms, two B atoms, or an A and a B atom, respectively. No restriction is placed on the range of W_{ij}. We are interested in obtaining an approximation to $\langle G \rangle$, the configurationally averaged Green function of the alloy.

Corresponding to any alloy configuration, the Green function matrix elements satisfy the equation of motion, Eq. (11.33),

$$G_{ij} = g_i \delta_{ij} + g_i \sum_{k \neq i} W_{ik} G_{kj}. \tag{11.48}$$

Note that in contrast to the case of only diagonal disorder, the values of g_i and W_{ik} on the right-hand side of the last equation are correlated with one another. For example, $g_i = g^A$ implies that W_{ik} is either equal to W_{ik}^{AA} or W_{ik}^{AB}. In order to decouple the two types of disorder, BEB proceeded along the following novel lines:

Introduce a site-occupation index, x_i, defined to be equal to 1(0) if the atom occupying site i is of type $A(B)$. Similarly, define the index $y_i = 1-x_i$, which equals 1(0) for an atom B(A) on site i. It is clear that these indices behave like projection operators satisfying the relations,

$$x_i + y_i = 1, \tag{11.49}$$

$$x_i x_i = x_i, \tag{11.50}$$

$$y_i y_i = y_i, \tag{11.51}$$

and

$$x_i y_i = y_i x_i = 0. \tag{11.52}$$

Now, multiply the equation of motion, Eq. (11.48), on the left and on the right by all possible combinations of x_is and y_is, and insert the expression in Eq. (11.49) between the W_{ik} and the G_{kj} on the right-hand side. For example, upon multiplication by x_i on the left and by x_j on the right one obtains the result,

$$x_i G_{ij} x_j = g_i x_i \delta_{ij} + g_i x_i \sum_{k \neq i} W_{ik}(x_k + y_k) G_{kj} x_j. \tag{11.53}$$

Similarly,

$$x_i G_{ij} y_j = g_i x_i \sum_{k \neq i} W_{ik}(x_k + y_k) G_{kj} y_j, \tag{11.54}$$

where the $g_i \delta_{ij}$ term on the right is missing in the last expression, because of Eq. (11.52) (a site cannot be occupied simultaneously by an atom of type A and one of type B). Let the indices μ and ν denote collectively xs and ys and note that the quantities appearing in the last two equations can be viewed as elements of 2×2 matrices in the space spanned by the occupation numbers x and y. For a particular such element of G_{ij}, we can write,

$$G_{ij}^{\mu\nu} = g_i^{\mu} \delta_{ij} \delta_{\mu\nu} + g_i^{\mu} \sum_{k \neq i} \sum_{\gamma} W_{ik}^{\mu\gamma} G_{kj}^{\gamma\nu}. \tag{11.55}$$

Here, we have defined the two-dimensional matrix locators, Green functions and transfer integrals, respectively, by the expressions,

$$\underline{g}_i = \begin{pmatrix} \frac{x_i}{z-\epsilon_A} & 0 \\ 0 & \frac{y_i}{z-\epsilon_B} \end{pmatrix} = \begin{pmatrix} x_i g^A & 0 \\ 0 & y_i g^B \end{pmatrix}, \tag{11.56}$$

$$\underline{G}_{ij} = \begin{pmatrix} x_i G_{ij} x_j & x_i G_{ij} y_j \\ y_i G_{ij} x_j & y_i G_{ij} y_j \end{pmatrix}, \tag{11.57}$$

and

$$\underline{W}_{ij} = \begin{pmatrix} W_{ij}^{AA} & W_{ij}^{AB} \\ W_{ij}^{BA} & W_{ij}^{BB} \end{pmatrix} = \begin{pmatrix} \alpha_{ij} & \zeta_{ij} \\ \zeta_{ij} & \beta_{ij} \end{pmatrix}. \tag{11.58}$$

We note that the matrices $\underline{g}_i$ are singular. In terms of the matrices defined in the last three expressions, the equation of motion, Eq. (11.48) can be written in the form

$$\underline{G}_{ij} = \underline{g}_i \delta_{ij} + \underline{g}_i \sum_{k \neq i} \underline{W}_{ik} \underline{G}_{kj}. \tag{11.59}$$

This last matrix expression contains a great simplifying feature when compared to its scalar counterpart, Eq. (11.48): Because the matrix $\underline{W}_{ij}$ does *not* involve occupation indices, it is only a function of the distance $|\mathbf{R}_i - \mathbf{R}_j|$ between sites i and j, and the disorder is confined strictly to the site-diagonal quantity, $\underline{g}_i$. Consequently, the methods developed for the treatment of diagonal disorder can be used directly in the study of Eq. (11.59).

Treating the last term in Eq. (11.59) formally as a perturbation, we find the expression,

$$\underline{G}_{ii} = \left(\underline{1} - \underline{g}_i \underline{\Delta}_i\right)^{-1} \underline{g}_i, \tag{11.60}$$

which is analogous to Eq. (11.42). However, in contrast to that equation, the use of the inverse of the site locators is avoided, since $\underline{g}_i$ is a singular matrix. In Eq. (11.60), the fully renormalized interactor,

$$\underline{\Delta}_i = \begin{pmatrix} \Delta_i^{AA} & \Delta_i^{AB} \\ \Delta_i^{BA} & \Delta_i^{BB} \end{pmatrix}, \tag{11.61}$$

is defined by means of the expansion,

$$\begin{aligned} \underline{\Delta}_i &= \sum_{j \neq i} \underline{W}_{ij} \underline{g}_j \underline{W}_{ji} + \sum_{j \neq i} \sum_{k \neq i,j} \underline{W}_{ij} \underline{g}_j \underline{W}_{jk} \underline{g}_k \underline{W}_{ki} + \cdots \\ &= \sum_{j,k \neq i} \underline{W}_{ij} \underline{\Gamma}_{jk}^{(i)} \underline{W}_{jk}, \end{aligned} \tag{11.62}$$

which is analogous to Eq. (10.58) and has an identical interpretation as the sum of all paths through the material that start and end on site i but avoid site i at intermediate steps. It is important to emphasize that the matrix renormalized interactor associated with site i is independent of the occupation of that site (in the following the subscript i will often be omitted). Note also that the terms of the type Δ^{AB} are unphysical since they correspond to the occupation of a given site by two different types of atoms. However, these terms do not appear in any real-space expansion of the Green function, as can be readily verified by formally expanding the inverse in Eq. (11.60), and thus, they do not lead to nonphysical results.

We are now ready to derive the CPA self-consistency condition for alloys with ODD. First, we note that because of Eqs. (11.50) and (11.51), the average of the site-diagonal Green function is diagonal in the two-dimensional space of the occupation indices,

$$\langle \underline{G}_{ii} \rangle = \begin{pmatrix} \langle G_{ii}^{AA} \rangle & 0 \\ 0 & \langle G_{ii}^{BB} \rangle \end{pmatrix}. \tag{11.63}$$

In the spirit of the CPA, we replace the material surrounding site i by a translationally invariant effective medium characterized by a two-dimensional self-energy matrix,

$$\underline{\sigma} = \begin{pmatrix} \sigma^{AA} & \sigma^{AB} \\ \sigma^{BA} & \sigma^{BB} \end{pmatrix}. \tag{11.64}$$

An expression for the CPA Green function, $\underline{\bar{G}}$, corresponding to this effective medium can be obtained through a Fourier transform,

$$\underline{\bar{G}}(\mathbf{k}) = [\underline{z} - \underline{\sigma} - \underline{W}(\mathbf{k})]^{-1}, \tag{11.65}$$

where $\underline{z}$ is proportional to the two-dimensional unit matrix, $\underline{\sigma}$ is defined by Eq. (11.64), and $\underline{W}(\mathbf{k})$ is defined in the usual way,

$$\underline{W}(\mathbf{k}) = \sum_{i \neq 0} \underline{W}_{0i} e^{i\mathbf{k}\cdot\mathbf{R}_{0i}}. \tag{11.66}$$

It follows that the site-diagonal element of the Green function associated with site 0 of the effective medium is given by the expression,

$$\underline{\bar{G}}_{00} = \frac{1}{\Omega_{\mathrm{BZ}}} \int_{\mathrm{BZ}} \bar{G}(\mathbf{k}) d^3k = (\underline{z} - \underline{\sigma} - \underline{\Delta})^{-1}, \tag{11.67}$$

which is analogous to Eq. (11.37) with the appropriate replacement of scalar quantities with matrices. Now, the self-consistency condition determining the matrix self-energy, $\underline{\sigma}$, within the matrix formulation of the CPA takes the form

$$\langle \underline{G} \rangle = \underline{\bar{G}}. \tag{11.68}$$

As was the case with the scalar form of the CPA, its matrix formulation can be cast in a number of alternative expressions, e.g., corresponding to Eq. (11.42), when matrices are employed to represent the various quantities in that equation. It can also be written in terms of a t-matrix within the propagator formalism [58][5]. Using the relation,

$$\underline{G} = \underline{\bar{G}} + \underline{\bar{G}t\bar{G}}, \tag{11.69}$$

[5]The reader is encouraged to derive explicit expressions for the t-matrix $\underline{t}$.

we can cast Eq. (11.68) in the form,

$$\langle \underline{t} \rangle = 0. \tag{11.70}$$

These expressions make clear the role of a matrix renormalized interactor and of the "unphysical" elements, Δ_i^{AB}. Like σ^{AB}, these elements serve to satisfy the matrix CPA self-consistency condition and do not appear in any real-space expansions of the Green function. As was pointed out by BEB, the physically significant Green function for alloys with ODD is the sum of all elements of the matrix Green function. The reason for this is simply that physical space is the sum over all configurational occupations of a single site. Thus, the physical, site-diagonal Green function is given by the sum of the elements

$$\langle G_{00}^{\mathrm{Ph}} \rangle = \langle G_{00}^{AA} \rangle + \langle G_{00}^{BB} \rangle. \tag{11.71}$$

It can be shown [57] that in the absence of ODD the physical Green function defined by the last expression becomes identical to that obtained in the non-matrix (scalar) formulation of the CPA. In momentum space (**k**-space) the physical Green function of the effective medium is given by the expression,

$$\left[\bar{G}(\mathbf{k})\right]^{\mathrm{Ph}} = \sum_{\alpha\beta} \left[\bar{G}(\mathbf{k})\right]_{\alpha\beta}, \quad \alpha, \beta = A \text{ or } B \text{ (for binary alloys)}. \tag{11.72}$$

From the last expression, one may define a physical effective medium through the relation,

$$H_{\mathrm{eff}}^{\mathrm{Ph}}(\mathbf{k}) = z - \left\{ \left[\bar{G}(\mathbf{k})\right]^{\mathrm{Ph}} \right\}^{-1}. \tag{11.73}$$

The physical effective medium defined by the last expression has a very interesting property: Introducing an impurity (say of type A or B) at a site in this medium results in a perturbation that is of infinite rather than localized extent. The formalism of BEB can be easily generalized to multicomponent, multiband alloys, in a straightforward way, as the reader is urged to verify.

11.7.1 Applications

The single-site approximation of BEB has been used to carry out numerical calculations for the DOS of model one-dimensional alloys with both diagonal and off-diagonal disorder. The matrix formulation of the CPA has also been used in calculations of realistic, three-dimensional multiband alloys. These calculations show that the matrix CPA indeed provides the most satisfactory treatment of ODD. A comparison of the BEB formalism with other methods introduced for the treatment of ODD, such as Shiba's [30]

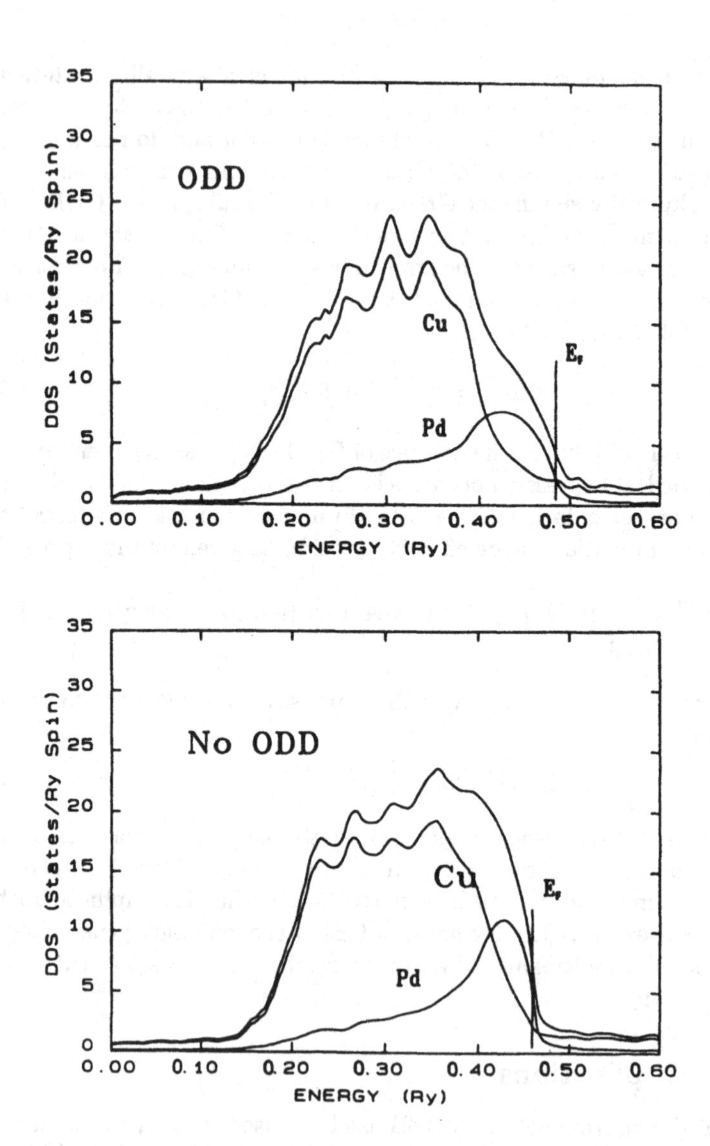

Figure 11.3: Densities of states for the $Cu_{90}Pd_{10}$ alloy calculated with the inclusion of off-diagonal disorder, upper frame, and in the presence of only diagonal disorder, lower frame [91].

approach to ODD of the geometric kind, has been given in the literature [8].

Figure 11.3 shows the DOS for the alloy $Cu_{90}Pd_{10}$ calculated both in the absence and the presence of ODD. For details of the calculation, the reader is referred to the work cited in the figure caption. These calculations reveal that CuPd alloys do indeed exhibit effects associated with ODD and the corresponding DOS, lower frame, is more closely related to the DOS obtained in more realistic applications of the CPA to muffin-tin potentials than those obtained in the absence of ODD.

11.8 Cluster Theories

As we mentioned above, in spite of its many desirable properties, the SSCPA does not provide a fully satisfactory theory for the calculation of the electronic structure of substitutionally disordered alloys. All the disadvantages of the CPA stem from its single-site nature. In particular, as a single-site theory, the CPA cannot account for the effects of correlated scattering from clusters of sites or for the presence of SRO.

There have been many attempts [59-91] to develop a multisite or cluster generalization of the SSCPA. A description of the best known such attempts, in particular of the molecular CPA (MCPA) [22-74] has been given elsewhere [8]. These attempts are quite varied, reflecting the many different approaches to the problem. The earliest multi-site theories were based on diagrammatic methods and attempted to retain full self-consistency [52-72]. These n-site theories, collectively called n-CPA, do improve upon the single site CPA, retaining many of the desirable properties of that approximation while also allowing for the treatment of local environment effects. However, close numerical studies [71] have shown that these theories can yield non-analytic results for certain ranges of relevant alloy parameters. Such results include the presence of non-unique, crossing roots, and of negative DOSs. Consequently, the n-CPA methods are unreliable as general computational tools and have been almost completely abandoned.

Among attempts to obtain a self-consistent cluster generalization of the SSCPA, we mention the molecular CPA (MCPA) [22, 73, 74], the self-consistent central-site approximation (SCCSA) [75-78] and boundary-site approximation (SCBSA) [79], the so-called truncated t-matrix (TTM) approximation [35-37], variationally derived cluster theories [82], theories based on the augmented space formalism [83], and the traveling cluster approximation (TCA) [84]. More recent attempts [85] include rather sophisticated uses of the augmented space formalism, but even these methods have failed to yield analytic results in reported applications. Of all self-consistent cluster methods, the MCPA appears to be the most satisfactory from a conceptual standpoint. However, the application of the MCPA to all but the simplest model systems seems to be prohibitively difficult. A detailed discussion of various cluster methods with comparisons and contrast is given in reference [8].

Upon reflection, it appears that there are two well-separated problems associated with the development of any cluster theory of disordered alloys. The first, and by far the more difficult problem, is that of determining an optimal choice of effective medium. As just discussed, the solution of this problem has not been accomplished in satisfactory fashion as of this writing. The second and easier problem is that of finding a computationally practical technique for the evaluation of the Green function associated with a cluster of sites embedded in a CPA effective medium. This problem has in principle been solved [86, 87] and the associated technique, the *embedded cluster method* (ECM) is discussed in the next section.

11.9 Embedded Cluster Method

The *embedded cluster method* (ECM) has proved to be perhaps the most successful non-self-consistent cluster theory for the study of substitutionally disordered alloys. The method is physically transparent, based on the treatment of a cluster of real sites embedded in an effective medium. Although there is a great deal of latitude in choosing this medium, e.g., the ATA or even the VCA can be used for this purpose, the one determined in the CPA has been found to be the most useful. A review of the method has been given in reference [8]. The reader may wish to review the material in Chapter 9 discussing the embedding of impurities in otherwise pure host materials before reading further.

The ECM consists of two main steps:

- Determine an optimal choice of a translationally invariant, analytic effective medium. There are strong indications from numerical calculations that the best such medium is that obtained in the SSCPA.
- Obtain the Green function of a cluster of "real" scatterers embedded in the medium determined in the first step.

For a given configuration J of the cluster, one can define the density of states associated with any site i in the cluster by means of the relation,

$$n_i^J(E) = -\frac{1}{\pi}\Im G_{ii}^J(E). \tag{11.74}$$

In the usual applications of the ECM, the cluster is taken as consisting of a central site, labeled 0, and the first few coordination cells around this site. The density of states of the material is then approximated by the average over cluster configurations of $n_0^J(E)$,

$$n(E) = n_0(E) = \sum_J P_J n_0^J(E), \tag{11.75}$$

where P_J is the probability of occurrence of configuration J. It follows that SRO effects can be incorporated into the average in the last expression through a proper definition of the probabilities P_J [8].

We are interested in obtaining an expression for the Green function of a system consisting of a cluster C, of real atoms, embedded in an effective medium. The Hamiltonian of the system can be separated into three parts: (i) The part confined to the cluster C; (ii) the part confined to the medium surrounding the cluster; and (iii) the part describing the coupling of the cluster with the surrounding medium. In a TB description, the matrix elements of the Hamiltonian are given by the following expressions:

(a) For i and j sites in the cluster C, we have

$$H_{ij} = \epsilon_i \delta_{ij} + W_{ij} \left(1 - \delta_{ij}\right). \tag{11.76}$$

Here, ϵ_i can assume any value consistent with the constituents of the alloy, e.g., ϵ_A or ϵ_B in the case of a $A_{1-c}B_c$ alloy, and W_{ij} are the configuration independent, translationally invariant hopping terms. The restriction to diagonally disordered alloys is made only for convenience of exposition. The extension of the ECM to alloys with ODD is fairly straightforward [8].

(b) For i and j sites in the medium surrounding the cluster C, we have

$$H_{ij} = \sigma_0 \delta_{ij} + \bar{W}_{ij} \left(1 - \delta_{ij}\right). \tag{11.77}$$

In most applications of the ECM, σ_0 is taken as the site-diagonal self-energy determined in the CPA, and $\bar{W}_{ij}$ is set equal to W_{ij}. In principle, however, other choices could be made for these parameters.

(c) For i a site in the cluster C and j a site in the medium surrounding cluster C, we have

$$H_{ij} = \tilde{W}_{ij}. \tag{11.78}$$

Clearly, no diagonal elements of this part of the Hamiltonian exist. Again, although other choices may be possible in general, one usually sets $\tilde{W} = W$. In the following discussion, we use the symbol W'_{ij} to denote any one of the three parameters, W_{ij}, $\bar{W}_{ij}$, or $\tilde{W}_{ij}$, with the proper symbol displayed as necessary.

The starting point for the evaluation of the matrix elements G_{ij} is the locator equation of motion, Eq. (11.48)

$$G_{ij} = g_i \left(\delta_{ij} + \sum_{k \neq i} W'_{ik} G_{kj} \right). \tag{11.79}$$

Here,

$$g_i = \begin{cases} (z - \epsilon_i)^{-1} & \text{if } i \text{ is in cluster } C, \\ (z - \sigma_0)^{-1} & \text{if } i \text{ is in the medium surrounding } C. \end{cases} \tag{11.80}$$

We wish to calculate the matrix elements G_{ij} for i and j both in the cluster C. For these elements the equation of motion, Eq. (11.79), can be written in the form,

$$G_{ij} = g_i \left(\delta_{ij} + \sum_{k \in C} W_{ik} G_{kj} + \sum_{k \notin C} \tilde{W}_{ik} G_{kj} \right), \tag{11.81}$$

where $k \in C$ ($k \notin C$) implies that site k is (is not) in cluster C.

The reader is invited to show (see Chapter 9) that by treating the last term on the right-hand side of the last equation formally as a perturbation, one can write

$$\sum_{k \notin C} \tilde{W}_{ik} G_{kj} = \sum_{k \in C} \Delta_{ik}^{(C)} G_{kj}, \tag{11.82}$$

where $\Delta_{ik}^{(C)}$ is given by

$$\Delta_{ij}^{(C)} = \sum_{k,l \in C} \tilde{W}_{ik} \bar{\Gamma}_{kl}^{(C)} \tilde{W}_{lj}. \tag{11.83}$$

It is clear from this expansion that $\bar{\Gamma}^{(C)}$ is the Green function for the effective medium with the cluster C removed. It is also clear, [see discussion following Eq. (10.58)], that $\Delta_{ik}^{(C)}$ arises from the hopping of electrons from site i in the cluster to site j in the cluster, touching at least one site outside the cluster but avoiding cluster sites at all intermediate steps. Substituting Eq. (11.82) into Eq. (11.81), we obtain

$$\left(g_i^{-1} - \Delta_{ii}^{(C)} \right) G_{ij} - \sum_{\substack{k \in C \\ k \neq i}} \left(W_{ik} + \Delta_{ik}^{(C)} \right) G_{kj} = \delta_{ij}, \tag{11.84}$$

or in matrix form

$$\underline{G}_{CC} = \left(\underline{g}_C^{-1} - \underline{W}_{CC} - \Delta_{CC}^{(C)} \right)^{-1}. \tag{11.85}$$

where $\underline{A}_{CC'}$ denotes a matrix with elements A_{ij} with the site i in cluster C and the site j in cluster C'. Sinc e $\underline{g}_C$ and $\underline{W}_{CC}$ are defined by the alloy Hamiltonian, a determination of $\underline{G}_{CC}$ requires only the determination of the *cluster renormalized interactors*, $\Delta_{ik}^{(C)}$.

Let us evaluate these interactors for the case $\tilde{W}_{ij} = W_{ij}$, in which case $\Delta_{ik}^{(C)}$ is independent of the configuration of the cluster. Replacing the quantities $\underline{g}_C$ and $\underline{W}_{CC}$ with the corresponding effective medium quantities, $\bar{\underline{g}}_C$ and $\bar{\underline{W}}_{CC}$, we obtain the expression

$$\bar{\underline{G}}_{CC} = \left(\bar{\underline{g}}_C^{-1} - \bar{\underline{W}}_{CC} - \bar{\Delta}_{CC}^{(C)} \right)^{-1}, \tag{11.86}$$

for the cluster-diagonal part of the effective-medium Green function. The translational invariance of this medium implies that the site matrix elements of $\bar{\underline{G}}_{CC}$ are given by the relations,

$$\left[\bar{\underline{G}}_{CC}\right]_{ij} = \frac{1}{\Omega_{\mathrm{BZ}}} \int_{\mathrm{BZ}} \left[z - \sigma_0 - \bar{W}(\mathbf{k})\right]^{-1} e^{i\mathbf{k}\cdot[\mathbf{R}_i - \mathbf{R}_j]} d^3k. \tag{11.87}$$

It now follows from Eq. (11.86) that the cluster renormalized interactor matrix is given by the expression

$$\Delta_{CC}^{(C)} = \bar{\underline{g}}_C^{-1} - \bar{\underline{W}}_{CC} - \bar{\underline{G}}_{CC}^{-1}. \tag{11.88}$$

Upon substituting Eq. (11.88) into Eq. (11.85) one obtains the expression

$$\underline{G}_{CC} = \left\{ \left(\underline{g}_C^{-1} - \underline{W}_{CC}\right) - \left(\bar{\underline{g}}_C^{-1} - \bar{\underline{W}}_{CC}\right) + \left[\bar{\underline{G}}_{CC}\right]^{-1} \right\}^{-1}, \tag{11.89}$$

for the cluster Green function corresponding to any given cluster configuration. Physical quantities, e.g., the densities of states, $n_i^J(E)$, can be calculated using Eq. (11.74) for any site i in the cluster and any cluster configuration, J.

Figure 11.4 shows the DOSs obtained in a seven-site application of the ECM to a one-dimensional, binary, random alloy characterized by the parameters, $\epsilon_A = -\epsilon_B = 2.0$, nearest neighbor hopping, $W = 1.0$, and $c = 0.5$. As expected, the ECM reproduces the structure in the exact DOS rather accurately, but fails in reproducing the behavior at band edges. This is due to the fact that the internal structure of the DOS results from statistical fluctuations inside a compact cluster, which are accounted for within the ECM, but the edges are shaped by the presence of long-distance fluctuations that are not taken into account in either the SSCPA or in the ECM.

The ECM has been used to calculate DOSs for both random and non-random, substitutionally-disordered alloys described by model, TB Hamiltonians [8], and also by more realistic MT potentials. Results of the latter applications are discussed in the next chapter. We conclude this chapter with a discussion of the main characteristics of a satisfactory cluster theory.

11.10 Features of Cluster Theories

One condition that is often imposed in discussions of disordered systems is that the self-energy should have a non vanishing **k** dependence, i.e., it should not be restricted to site-diagonal terms. Such a feature is certainly desirable, but one must be aware of pitfalls. A self-energy that has non site-diagonal elements poses difficulties in the treatment of the so-called embedding problem, i.e., the study of isolated impurities embedded in the effective

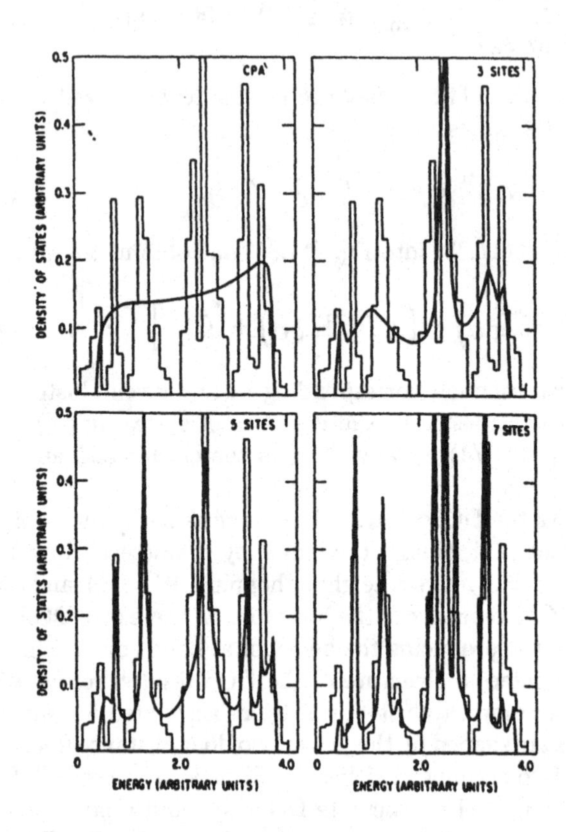

Figure 11.4: **Exact (histogram) and seven-site ECM DOSs for a one-dimensional, single-band alloy with only nearest-neighbor hopping, $W = 1$, $c = 0.50$, and with $\epsilon_A = -\epsilon_B = 2.0$. The DOSs are symmetric around the zero of energy.**

medium characterized by the self-energy. This is because there is no unique prescription for determining the elements of the Hamiltonian coupling the impurity to the host medium. Another, related, difficulty is connected with the fundamental features of scattering theory. The t-matrix is rigorously defined with respect to propagation in free space. The self-energy defines a medium that is characterized by a complex, energy-dependent potential, whose scattering properties may not be completely known. Thus, difficulties may arise in attempting to derive a theory within both the locator and propagator formalisms, and to reach full self-consistency in terms of a self-energy that has a non vanishing $\mathbf{k}$ dependence.

A satisfactory cluster theory of the electronic structure of substitutionally disordered alloys must meet a rather stringent set of criteria the most prominent of which are listed below. Even though many of these criteria are not mutually exclusive, they are mentioned individually for the sake of emphasizing their importance. Among others properties, a fully satisfactory alloy theory should:

- yield analytic, physically meaningful results, i.e., non-negative DOS's in real and reciprocal space;
- be derivable within both a locator (real space) and a propagator (reciprocal space) formalism;
- provide a unique prescription for the choice of physical parameters, such as the coupling of a cluster of real atoms to the surrounding embedding medium;
- become exact in all physical limits, i.e., as $c \to 0, 1$;
- be applicable to model as well as realistic Hamiltonians;
- be applicable to the treatment of various types of disorder, i.e., diagonal and ODD;
- become exact in the limit of large cluster sizes;
- allow the calculation of one- and two-particle properties within essentially the same formalism;
- provide a rigorous, self-consistent treatment of SRO effects; and
- be computationally feasible.

Some of these criteria are satisfied by cluster extensions of the CPA proposed in the literature, and a discussion can be found in reference [8].

Bibliography

[1] P. Soven, Phys. Rev. **156**, 809 (1967).

[2] D. W. Taylor, Phys. Rev. **156**, 1017 (1967).

[3] B. Velicky, S. Kirkpatrick, and H. Ehrenreich, Phys. Rev. **157**, 747 (1968).

[4] R. J. Elliott, J. A. Krumhansl, and P. L. Leath, Rev. Mod. Phys. **46**, 465 (1974).

[5] J. S. Faulkner, in *Progress in Materials Science*, J. W. Christian, P. Hassen, and T. B. Massalski (eds.), Pergamon Press, New York (1982). No. 1-2.

[6] B. L. Györffy and G. M. Stocks, in *Electrons in Disordered Metals and Metallic Surfaces*, P. Phariseau, B. L. Györffy, and L. Scheire (eds.), Plenum, New York (1978), p. 89.

[7] F. Yonezawa and K. Morigaki, Prog. Theor. Phys. Suppl. **53**, 1 (1973).

[8] D. A. Papaconstantopoulos, *Handbook of the Bandstructure of Elemental Solids*, Plenum, New York (1986). The tight-binding CPA for certain systems where charge transfer is not an issue gives reliable results comparable to those of the KKR-CPA. See, for example, P. M. Laufer and D. A. Papaconstantopoulos, Phys. Rev. B**35**, 9019 (1987).

[9] Y. Onodera and Y. Toyozawa, J. Phys. Soc. (Japan) **24**, 341 (1968).

[10] P. L. Leath, Phys. Rev. **171**, 725 (1968).

[11] F. Yonezawa, Prog. Theor. Phys. **39**, 1076 (1968); **40**, 734 (1968).

[12] R. N. Aiyer, R. J. Elliott, J. A. Krumhansl, and P. L. Leath, Phys. Rev. **181**, 1006 (1969).

[13] B. G. Nickel and J. A. Krumhansl, Phys. Rev. B**4**, 4354 (1971).

[14] P. L. Leath, Phys. Rev. B**5**, 1643 (1972).

[15] W. H. Butler, Phys. Lett. A**39**, 203 (1972).

[16] F. Brouers, M. Cyrot, and F. Cyrot-Lackmann, Phys. Rev. B**7**, 4370 (1973).

[17] F. Ducastelle, J. Phys. C**4**, L75 (1971).

[18] P. L. Leath, Phys. Rev. B**2**, 3078 (1970).

[19] T. Matsubara, Prog. Theor. Phys. Suppl. **46**, 326 (1970).

[20] F. Ducastelle, J. Phys. C**8**, 3297 (1975).

[21] P. Lloyd and P. R. Best, J. Phys. C**8**, 3752 (1975).

[22] M. Tsukada, J. Phys. Soc. (J) **32**, 1475 (1972).

[23] E. Müller-Hartmann, Solid State Comm. **12**, 1269 (1973).

[24] F. Ducastelle, J. Phys. C**7**, 1795 (1974).

[25] F. Ducastelle and G. Tréglia, J. Phys. F**10**, 2137 (1980).

[26] M. Plischke and D. Mattis, Phys. Rev. Lett. **27**, 42 (1971).

[27] M. M. Pant and S. K. Joshi, Phys. Rev B**2**, 1704 (1976).

[28] M. P. Das and S. K. Joshi, Can. J. Phys. **50**, 2856 (1972); Phys. Rev B**4**, 4363 (1971).

[29] P. Soven, Phys. Rev. B**2**, 4715 (1970).

[30] H. Shiba, Progr. Theor. Phys. **46**, 77 (1971).

[31] B. L. Györffy, Phys. Rev. B**5**, 2382 (1972).

[32] An-Ban Chen, Phys. Rev. B**7**, 2230 (1973).

[33] L. Schwartz and A. Bansil, Phys. Rev. B**10**, 3261 (1974).

[34] F. Brouers and J. van der Rest, J. Phys. F**2**, 1070 (1972).

[35] N. F. Berk and R. A. Tahir-Kheli, Physica **67**, 501 (1973).

[36] N. F. Berk, D. J. Shazeer, and R. A. Tahir-Kheli, Phys. Rev. B**8**, 2496 (1973).

[37] S. M. Bose and E-Ni Foo, J. Phys. (Paris) **35**, C4 95, (1974).

[38] S. Takeno, Progr. Theor. Phys. (Kyoto) **40**, 942 (1968).

[39] Y. Izyumov, Proc. Phys. Soc. (London) **87**, 505 (1966).

[40] N. F. Berk, Phys. Rev. B**1**, 1136 (1970).

[41] K. Niizeki, Progr. Theor. Phys. **53**, 74 (1973); **54** 1648 (1975).

[42] F. Brouers, Solid State Comm. **10**, 757 (1972).

[43] T. Kaplan and L. J. Gray, Phys. Rev. B**14**, 3462 (1976); J. Phys. C**9**, L303 (1976).

[44] A. Mookerjee, J. Phys. C**8**, 2943 (1975).

[45] E-Ni Foo, S. M. Bose, and A. Ausloos, Phys. Rev. B**7**, 2454 (1973).

[46] A. Bergman and V. Halperin, J. Phys. C**7**, 289 (1974).

[47] T. Horiguchi, C. C. Chen, and T. Morita, Solid State Comm. **12**, 1243 (1973); **13**, 975 (1973).

[48] C. Herscovici, Phys. Stat. Sol. (b) **72**, 689 (1975).

[49] J. A. Blackman, J. Phys. F**3**, L31 (1973).

[50] G. Kerker, Phys. Lett. **48**A, 345 (1975).

[51] D. M. Esterling, Phys. Rev. B**12**, 1596 (1975).

[52] H. Braeter, Th. Eifrig, and E. Kirstein, phys. stat. sol. (b), **59**, 693 (1973).

[53] Th. Eifrig, E. Kolley, and W. Kolley, phys. stat. sol (b), 225 (1975).

[54] W. John and J. Schreiber, phys. stat. sol. (b) **66**, 197 (1974).

[55] T. Saso, R. Yamamoto, and M. Doyama, Solid State Comm. **15**, 699 (1974).

[56] K. Bass and P. L. Leath, Phys. Rev. B**9**, 2769 (1974).

[57] J. A. Blackman, D. M. Esterling, and N. F. Berk, Phys. Rev. B**4**, 2412 (1971).

[58] A. Gonis and J. W. Garland, Phys. Rev. B**16**, 1495 (1977).

[59] F. Yonezawa, Prog. Theor. Phys. (Kyoto) **40**, 734 (1968).

[60] P. Leath, Phys. Rev. **171**, 725 (1968).

[61] F. Ducastelle, J. Phys. F**2**, 468 (1972).

[62] P. L. Leath, Phys. Rev. B**5**, 1643 (1972).

[63] J. Zittartz, Solid State Commun. **14**, 51 (1973); Z. Physik **267**, 243 (1974).

[64] F. Ducastelle, J. Phys. (Paris) **33**, C3 269 (1972).

[65] V. Chapek, phys. stat. sol. (b) **43**, 61 (1971).

[66] F. Cyrot-Lackmann and F. Ducastelle, Phys. Rev. Lett. **27**, 429 (1971).

[67] F. Cyrot-Lackmann and F. Cyrot, J. Phys. C**5**, L209 (1972).

[68] V. Shrivastava and S. K. Joshi, J. Phys. F**3**, L179 (1973).

[69] L. Schwartz and E. Siggia, Phys. Rev. B**5**, 383 (1972).

[70] L. Schwartz and H. Ehrenreich, Phys. Rev. B**6**, 2923 (1972).

[71] B. G. Nickel and W. H. Butler, Phys. Rev. Lett. **30**, 373 (1973).

[72] I. Mertsching, phys. stat. sol. (b) **59**, 227 (1973); **63**, 241 (1974).

[73] A. R. Bishop and A. Mookerjee, J. Phys. C**7**, 2165 (1973).

[74] P. Leath, J. Phys. C**6**, 1559 (1973).

[75] W. H. Butler, Phys. Rev. Lett. **39**A, 203 (1972).

[76] V. Chapek, phys. stat. sol. (b) **52**, 399 (1972).

[77] V. Kumar and S. K. Joshi, J. Phys. C**8**, L148 (1975).

[78] F. Brouers, F. Ducastelle, and J. van der Rest, J. Phys. F**3**, 1704 (1973).

[79] W. H. Butler, Phys. Rev. B**8**, 4499 (1973).

[80] U. Krey, Phys. Cond. Matt. **18**, 17 (1974).

[81] G. Szycholl and H. Kühl, Z. Phys. B**24**, 257 (1976).

[82] P. Lloyd and P. R. Best, J. Phys. C**8**, 3752 (1975).

[83] T. Kaplan, P. L. Leath, L. J. Grey, and H. W. Diehl, Phys. Rev. **21**, 4230 (1980).

[84] R. Mills and P. Ratanavararaksa, Phys. Rev. B**18**, 5219 (1978).

[85] S. S. Razee and R. Prasad, Phys. Rev. B**48**, 1349 (1993-II).

[86] A. Gonis and J. W. Garland, Phys. Rev. B**16**, 2424 (1977).

[87] C. W. Myles and J. D. Dow, Phys. Rev. B**19**, 4939 (1979).

[88] A. Gonis and A. J. Freeman, Phys. Rev. B**28**, 5487 (1983); Phys. Rev. B**31**, 2506 (1985).

[89] D. H. Seib and W. E. Speicer, Phys. Rev. B**2**, 1676 (1970).

[90] G. M. Stocks, R. W. Williams, and J. S. Faulkner, Phys. Rev. B**4**, 4390 (1971).

[91] D. Papaconstantopoulos, A. Gonis, and P. M. Laufer, Phys. Rev. B**40**, 12196 (1989).

Chapter 12

The KKR-CPA and Cluster Methods

12.1 General Comments

The modern era in the theoretical/computational study of substitutionally disordered alloys undoubtedly began with the introduction of the CPA [1]. As was mentioned in connection with the tight-binding (TB) approximation to the Hamiltonian in the previous chapter, the CPA has been found to be the best single-site theory for the treatment of substitutional disorder. In many cases it is a great improvement over simpler theories such as the rigid-band model (RBM), the virtual crystal approximation (VCA), and the average t-matrix approximation (ATA)[1]. In addition, the CPA possesses a large number of desirable properties such as analyticity, uniqueness, and satisfaction of various fundamental sum rules.

However, a study of disordered systems within the phenomenological TB framework suffers from several unsatisfactory features[2]. In spite of a number of successful applications of the TB formalism to many realistic materials, this approach lacks the generality, power, and conceptual integrity afforded by the use of more realistic potentials[3]. More importantly, a T-B approach requires the development of essentially new techniques for the treatment of distinct features of disorder, such as ODD, band hybridization, and many others. Such features are taken into account automatically

[1] A longer discussion and references to these methods can be found in the previous chapter.

[2] The formalism presented in this chapter applies virtually intact to treatments of disorder within a first-principles TB formalism.

[3] Recent applications of first-principles TB theory remove many of these unsatisfactory features.

and are treated on an equal footing in a formulation of the CPA within a first-principles, multiple scattering, Green function formalism. Because the first such applications of the CPA were carried out [2, 3, 4] in connection with muffin-tin (MT) potentials in an extension of the Green function formalism of Korringa, Kohn, and Rostoker (see Chapter 3), the method has come to be known as the KKR-CPA. We will also refer to it in this manner, but we will take the term to include the use of generally-shaped potentials in addition to ones of the MT form[4].

In developing the KKR-CPA it is possible, at least in a formal sense, to transcribe our entire discussion of the CPA given in terms of TB systems to alloys described by realistic potentials. This can be effected by means of correspondences between the parameters entering the respective descriptions [5]. A generalization of the CPA to realistic potential functions carried out through this sort of analogy might help bring out the unifying features of the formalism and save us a great deal of labor. At the same time, arguing by analogy has some serious pedagogical and practical drawbacks. It can easily obscure much of the underlying physical basis of the formalism, possibly hindering an intuitive understanding of the method. In addition, in such an approach important practical considerations, such as the accounting of certain types of ODD, may be hidden from view. Thus, risking a certain amount of repetition, in this chapter we provide a fairly detailed exposition of the KKR-CPA. We develop the formalism explicitly in a way that is appropriate to MT potentials (but not necessarily spherically symmetric ones), as well as to potentials of arbitrary shape, when due observance is paid to the conditions for convergence of angular momentum expansions.

In addition to the single-site KKR-CPA, we also briefly discuss one of its cluster extensions, namely the embedded cluster method (ECM) [6]. We recall from the previous chapter that the SSCPA, in spite of its many desirable properties, is not a fully satisfactory theory of substitutionally disordered alloys. Its most serious limitations are the inability to describe correlated scattering from clusters of sites, and providing a treatment of SRO effects in the alloy. A summary of attempts to develop a cluster theory, which in principle would alleviate these shortcomings, was given in the previous chapter and in review publications [5]. Many, if not all of these methods, were proposed in connection with TB Hamiltonians, and are usually of the single-band variety. In addition to perturbative methods based on the SSCPA, the only multi-site technique that has been finding increasing application in connection with realistic potentials, especially of the MT form, is the embedded cluster method [6] (see also previous chapter for more references). A description of the form of the ECM within a realistic

[4]As of this writing, the CPA has not yet been implemented in connection with generally shaped, space-filling cell potentials.

description of the potential of a disordered system is given following the development of the KKR-CPA.

We begin with a brief discussion of the VCA and ATA within a realistic description of the potential. This discussion can help to emphasize the similarities and contrasts between these methods and the CPA, possibly leading to a better understanding of these approaches to the study of substitutional disorder.

12.2 VCA and ATA

In addition to features particular to realistic potentials, such as those of the MT form, the treatment of disorder within a realistic description of the Hamiltonian presents many of the same conceptual problems that one encounters within a TB formalism. Not altogether surprising, the solutions to these problems can be attempted along formally similar lines within the two frameworks.

Since the performance of exact averages within a statistical ensemble characterizing a disordered material is not practically possible in general, we are forced into an approximate treatment. A particularly fruitful approach is to attempt a representation of the true disordered material by means of an appropriately chosen effective medium. Clearly, the simplest such medium which can be constructed is that obtained within the virtual crystal approximation (VCA).

In a realistic potential description of an alloy, such as that afforded by the MT approximation, the VCA consists of assigning to every cell in the material the *same* potential, $\bar{v}(\mathbf{r})$, which is the average over the concentrations, c_α,

$$\bar{v}(\mathbf{r}) = \sum_\alpha c_\alpha v_\alpha(\mathbf{r}), \tag{12.1}$$

of the cell potentials $v_\alpha(\mathbf{r})$ associated with species α in the material. Evidently, $\bar{v}(\mathbf{r})$ is a real potential whose scattering properties are fully described in terms of the single-cell scattering matrix, $t_{LL'}$, (see Appendix D). If we consider spherically symmetric MT potentials, then (assuming the definitions of the various parameters are familiar)

$$t_{LL'}^{\mathrm{VCA}} \equiv t_\ell^{\mathrm{VCA}} \delta_{\ell\ell'} \delta_{m0} = -\mathrm{e}^{\mathrm{i}\delta_\ell^{\mathrm{VCA}}} \sin \delta_\ell^{\mathrm{VCA}} / \sqrt{E}. \tag{12.2}$$

Because the VCA potential corresponds to a translationally invariant system, its band structure and related electronic properties can be readily determined by the methods of Chapter 6. Clearly, the potential function, $\bar{v}(\mathbf{r})$, and the electronic structure change smoothly with concentration, at least in the absence of structural transformations, approaching the corresponding quantities associated with the αth species in the alloy as the concentration c_α approaches 1.0. Thus, the VCA is a somewhat more sophisticated approximation than the rigid band model. However, as we had occasion to remark in our discussion of TB systems, the VCA is generally unreliable and is, in fact, mathematically incorrect. On the other hand, it can give fairly accurate predictions of electronic properties when the individual potentials $v_\alpha(\mathbf{r})$ are not very different from one another. And although the VCA often misses important features of the electronic structure, occasionally it can provide a realistic picture of alloying effects.

As was mentioned in the TB description of disorder, the next more sophisticated choice of an effective medium is afforded through the average t-matrix approximation (ATA). We recall that in the ATA, a disordered material is described in terms of an *averaged* scattering matrix,

$$\bar{\underline{t}} = \sum_\alpha c_\alpha \underline{t}^\alpha, \tag{12.3}$$

where $\underline{t}^\alpha$ is the cell scattering matrix associated with species α in the material. At first glance, it might appear that the ATA should offer no great improvement over the VCA. Closer scrutiny, however, indicates that the ATA should not be dismissed out of hand.

As an example, we consider the Argand diagrams (for a description see Section D.3) obtained in an application of the ATA to $Cu_{1-c}Ni_c$ alloys several of which are shown in Fig. 12.1. Argand plots in which the scattering amplitude lies inside the unitary circle represent inelastic scattering, so that the ATA scattering amplitude, $\bar{f}$, must have the form,

$$\begin{aligned} \bar{f} &= (1-c)f_{\mathrm{Cu}} + cf_{\mathrm{Ni}} \\ &= \frac{1}{2\mathrm{i}}\left(\eta_\ell \mathrm{e}^{2\mathrm{i}\delta_\ell} - 1\right), \end{aligned} \tag{12.4}$$

where η_ℓ represents the strength of the inelasticity. The presence of inelastic behavior is indeed unexpected, since the scattering amplitudes for both Cu and Ni lie on the unitarity circle. This inelasticity implies that although the ATA medium is translationally invariant, electronic states have finite lifetimes, as is expected in a substitutionally random system. Furthermore, the inelasticity exhibits the physically meaningful behavior of being small at the limits of dilute alloys, and increasing with concentration. Also, the ATA scattering amplitude is symmetric with respect to the two species of a binary alloy, as one should expect from a reasonable theory. It is interesting to mention that the inelastic behavior exhibited by the scattering amplitude in the ATA is a direct consequence of the disorder in the material. A wave propagating through the random alloy scatters at each site according to f_{Cu} or f_{Ni} with respective probabilities $1-c$ and c. The outgoing wave is attenuated due to destructive interference resulting from the random superposition of the two component waves. Thus, unlike the VCA, the

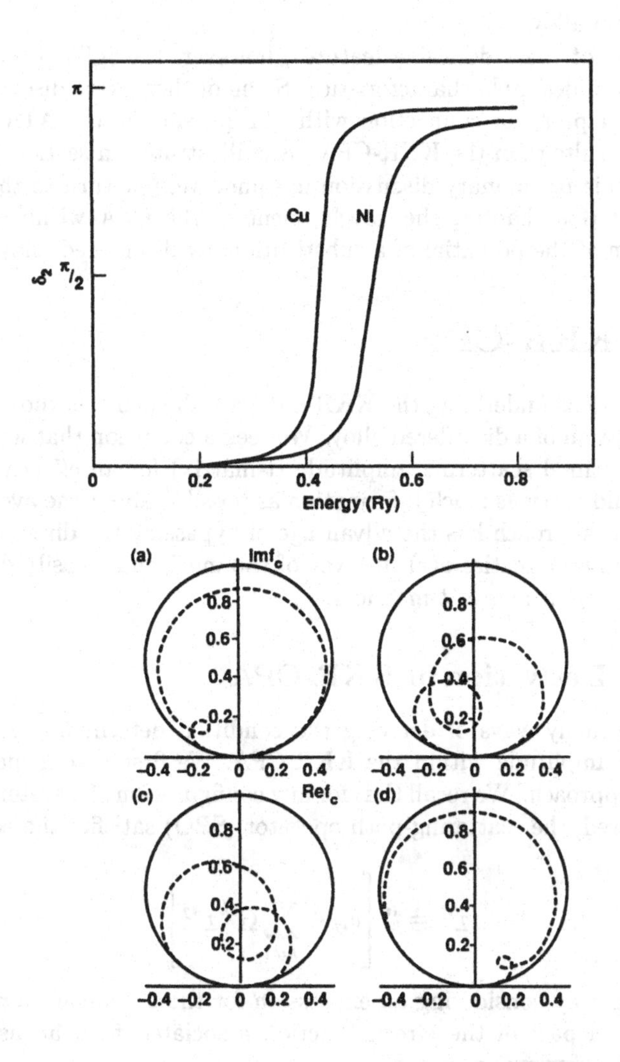

Figure 12.1: **The $\ell = 2$ phase shifts (upper panel) for Ni and Cu spherically symmetric MT potentials, and associated Argand plots for the ATA scattering amplitudes for several $Cu_{1-c}Ni_c$ alloys [2].**

ATA embodies at least some of the expected physical features of scattering in a random alloy.

In spite of these desirable features, however, the ATA does possess a number of undesirable characteristics. Some of them were discussed in the previous chapter. In connection with MT potentials, the ATA gives less accurate results than the KKR-CPA, as is illustrated in section 12.6.

With this preliminary discussion in mind, we now turn to the primary concern of this chapter, the development of the CPA within a realistic description of the potential of a substitutionally disordered alloy.

12.3 KKR-CPA

The basic ideas underlying the KKR-CPA are the same as those used in a TB description of a disordered alloy. We seek a condition that would determine an optimal scattering amplitude (t-matrix) for an effective medium which would carry as much information as possible about the averaged material. This approach has the advantage of bypassing the direct evaluation of an effective potential $\bar{v}(\mathbf{r})$ in favor of the much more easily determined scattering amplitudes or t-matrices.

12.3.1 Derivation of KKR-CPA

There are many ways of deriving the condition determining an effective scattering amplitude within the KKR-CPA. We first take a more or less intuitive approach. We recall that for any configuration of a system, ordered or disordered, the scattering-path operator (SPO) satisfies the equation of motion[5]

$$\underline{\tau}^{ij} = \underline{t}^{i} \left[\delta_{ij} + \sum_{k \neq i} \underline{G}^{ik} \underline{\tau}^{kj} \right] . \tag{12.5}$$

Let us begin by considering an expression for the ensemble average of the site-diagonal part of the Green function associated with an assembly of

[5]In the case of non-MT, space-filling cells, this equation must be carefully justified. In the case in which the length of intercell vectors exceeds that of vectors inside a cell, the summations over angular momentum indices converge provided each sum is carried to convergence completely, before another sum begins or, if necessary, through the use of modified structure constants (see Chapter 3). In fact, the use of the equation of motion is not really necessary, since as is discussed in Chapter 3, the Green function for an assembly can be derived directly in terms of $\underline{\tau}$, or its inverse $\underline{M}$. The formal approach that leads to Eq. (3.139) could be used to derive the KKR-CPA condition, and the equation of motion *per se* need not be employed. However, due to its intuitive appeal, we will continue to base much of our discussion explicitly on the equation of motion, in the knowledge that the various expressions derived from it can be justified on an independent basis for the case of space-filling cells.

scattering potential cells, and given in terms of the corresponding element of the SPO. In both the case of MT and space-filling potentials, we use Eq. (3.77) to obtain the expression (with energy variables suppressed)

$$\langle G(\mathbf{r},\mathbf{r}')\rangle = \langle\langle Z^0(\mathbf{r})|\underline{\tau}^{00}|Z^0(\mathbf{r}')\rangle\rangle + \langle G^{(1)}(\mathbf{r},\mathbf{r}')\rangle, \tag{12.6}$$

where $\mathbf{r}$ and $\mathbf{r}'$ are cell vectors in cell 0 and the outermost brackets denote an average over all configurations of a disordered alloy (at a given concentration). Further progress can now be made by introducing an approximate treatment for the evaluation of the exact configurational average indicated in this expression.

Consistent with the spirit of the CPA, we consider the true disordered material as being represented by a translationally invariant medium characterized by an effective scattering matrix $\bar{\underline{t}}$ and associated scattering-path operator $\bar{\underline{\tau}}$ (see Chapter 3). From Eq. (9.68), the site-diagonal part of $\bar{\underline{\tau}}$ has the form,

$$\bar{\underline{\tau}}^{00} = \left(\bar{\underline{m}} - \bar{\underline{\Delta}}\right)^{-1}, \tag{12.7}$$

where $\bar{\underline{m}} = \bar{\underline{t}}^{-1}$, and a bar over the symbol indicates a quantity associated with the effective medium. Since the renormalized interactor, $\bar{\underline{\Delta}}$, is independent of the occupation of site 0, we can obtain the SPO for an "impurity" atom of type α, embedded in the effective medium,

$$\bar{\underline{\tau}}^{00;\alpha} = \left(\bar{\underline{m}}^{\alpha} - \bar{\underline{\Delta}}\right)^{-1}. \tag{12.8}$$

Now, an approximation to the exact configurational average of the site-diagonal Green function in Eq. (12.6) can be obtained by replacing $\underline{\tau}^{00}$ with $\bar{\underline{\tau}}^{00;\alpha}$, and replacing the configurational average in that equation by an average over the occupations of a single site. This leads to the expression (with $r' > r$),

$$\langle G(\mathbf{r},\mathbf{r}')\rangle = \langle\langle Z^0(\mathbf{r})|\underline{\tau}^{00;\alpha}|Z^0(\mathbf{r}')\rangle\rangle + \langle Z^0(\mathbf{r})|S^0(\mathbf{r}')\rangle. \tag{12.9}$$

Exhibiting this average explicitly and writing $\underline{\tau}^{\alpha}$ for $\underline{\tau}^{00;\alpha}$, we have

$$\begin{aligned}\langle G(\mathbf{r},\mathbf{r}')\rangle &= \sum_{\alpha} c_{\alpha}\langle Z^{\alpha}(\mathbf{r})|\underline{\tau}^{\alpha}|Z^{\alpha}(\mathbf{r}')\rangle \\ &- \theta(r-r')\sum_{\alpha} c_{\alpha}\langle Z^{\alpha}(\mathbf{r}')|\tilde{J}(\mathbf{r})\rangle \\ &- \theta(r'-r)\sum_{\alpha} c_{\alpha}\langle Z^{\alpha}(\mathbf{r})|\tilde{J}(\mathbf{r}')\rangle,\end{aligned} \tag{12.10}$$

where $\underline{\tau}^{\alpha}$ is given by Eq. (12.8). Here, the $|Z^{\alpha}(\mathbf{r})\rangle$ are "vectors" whose components are indexed by angular momentum states, $L = (\ell, m)$. It now remains to specify the conditions determining $\underline{\bar{m}}$ or, equivalently, $\underline{\bar{\Delta}}$ which would then allow the evaluation of Eq. (12.10). Such a condition can be obtained as follows.

First, we note from Eq. (12.7) [see also Eq. (11.39)] that

$$\underline{\bar{\Delta}} = \underline{\bar{m}} - \left(\underline{\bar{\tau}}^{00}\right)^{-1}, \tag{12.11}$$

so that we can write

$$\begin{aligned} \underline{\tau}^{\alpha} &= \left[\underline{m}^{\alpha} - \underline{\bar{m}} + \left(\underline{\bar{\tau}}^{00}\right)^{-1}\right]^{-1} \\ &= \underline{D}^{\alpha}\underline{\bar{\tau}}^{00} \\ &= \underline{\bar{\tau}}^{00}\underline{\tilde{D}}^{\alpha}, \end{aligned} \tag{12.12}$$

where we have defined the matrix

$$\underline{D}^{\alpha} = \left[\underline{I} + \underline{\bar{\tau}}^{00}\left(\underline{m}^{\alpha} - \underline{\bar{m}}\right)\right]^{-1}, \tag{12.13}$$

and its transpose, $\underline{\tilde{D}}^{\alpha}$. We now impose the condition that *the single-site average of the site-diagonal matrix element of the scattering-path operator be equal to the corresponding effective-medium quantity*, and obtain the self-consistency condition of the KKR-CPA

$$\sum_{\alpha} c_{\alpha}\underline{\tau}^{\alpha} = \underline{\bar{\tau}}^{00}, \tag{12.14}$$

or, equivalently,

$$\sum_{\alpha} c_{\alpha}\underline{D}^{\alpha} = 1. \tag{12.15}$$

These expressions constitute two commonly used forms of the KKR-CPA condition. It is a matter of some straightforward algebra to show that this condition can also be written in the form,

$$\sum_{\alpha} c_{\alpha}\underline{X}^{\alpha} = 0, \tag{12.16}$$

where the quantity

$$\underline{X}^{\alpha} = \left(\underline{m}^{\alpha} - \underline{\bar{m}}\right)\left[\underline{I} - \underline{\bar{\tau}}^{00}\left(\underline{m}^{\alpha} - \underline{\bar{m}}\right)\right]^{-1}, \tag{12.17}$$

defines the additional scattering that results when an atom (cell) of type α is embedded in the CPA effective medium. The quantity $\underline{X}^{\alpha}$ is analogous to the t-matrix encountered in our discussion of TB systems in the two previous chapters. Thus, in both cases of TB and general potentials, the CPA condition implies that the scattering from an atom embedded in the

CPA medium vanishes on the average. Finally, in the case of binary alloys, the KKR-CPA self-consistency condition can be put into the useful form

$$\underline{\bar{m}} = c_A \underline{m}^A + c_B \underline{m}^B + \left(\underline{\bar{m}} - \underline{m}^A\right) \underline{\bar{\tau}}^{00} \left(\underline{\bar{m}} - \underline{m}^B\right). \tag{12.18}$$

It is instructive to note that the scattering-path matrix, $\underline{\tau}$, and the scattering matrix, $\underline{X}$, are connected by the equation,

$$\underline{\tau}^{nm} = \underline{\bar{\tau}}^{nm} + \sum_{kl} \underline{\bar{\tau}}^{nk} \underline{X}^{kl} \underline{\bar{\tau}}^{lm}, \tag{12.19}$$

while the scattering matrix satisfies the equation of motion,

$$\underline{X}^{nm} = \underline{X}^n \delta_{nm} + \underline{X}^n \sum_{k} \underline{\bar{\tau}}^{nk} \underline{X}^{km}. \tag{12.20}$$

The reader can easily verify the formal analogy of these expressions with expressions in the TB formulation of the CPA, and the fact that the TB expressions can often be transcribed directly to the first-principles framework discussed here. These equations lead immediately to the condition in Eq. (12.16).

Applications of these expressions to transition metal alloys are often made in connection with spherically symmetric MT potentials, and the angular momentum expressions are often carried out to values $\ell \leq 2$. In this case, the various matrix quantities appearing in the formulation of the KKR-CPA (with certain obvious exceptions, such as the KKR structure constants) become diagonal in angular momentum indices that greatly simplify the performance of numerical calculations. The KKR-CPA self-consistency condition can be expressed in a number of additional equivalent forms, as has been quoted in the literature [7].

In solving the KKR-CPA self-consistency equations, it is often convenient to eliminate either of the quantities $\underline{\bar{m}}$ or $\underline{\Delta}$ in favor of the other. For example, Eq. (12.11) can be written in terms of $\underline{\bar{m}}$ in the form

$$\sum_{\alpha} c_\alpha \left[\underline{m}^\alpha - \underline{\bar{m}} + \left(\underline{\bar{\tau}}^{00}\right)^{-1}\right]^{-1} = \underline{\bar{\tau}}^{00}. \tag{12.21}$$

Because $\underline{\bar{\tau}}^{00}$ is the scattering-path operator for a translationally invariant effective medium, it can be obtained as an integral over the Brillouin zone,

$$\begin{aligned} \underline{\bar{\tau}}^{00} &= \frac{1}{\Omega_{\mathrm{BZ}}} \int_{\mathrm{BZ}} \underline{\bar{\tau}}(\mathbf{k}) \mathrm{d}^3 k \\ &= \frac{1}{\Omega_{\mathrm{BZ}}} \int_{\mathrm{BZ}} \left[\underline{\bar{m}} - \underline{G}(\mathbf{k})\right]^{-1} \mathrm{d}^3 k, \end{aligned} \tag{12.22}$$

where $\underline{G}(\mathbf{k})$ denotes the KKR structure constants of the alloy. Now, Eqs. (12.21) and (12.22) involve only a single unknown matrix quantity, $\underline{\bar{m}}$. The solution of these equations can then be obtained by a number of numerical procedures that include direct iteration of the equations, or a Newton-Raphson process. The reader interested in the details of solving the KKR-CPA equations may consult some of the references given in the bibliography [4, 8, 9].

The KKR-CPA self-consistency condition in its various forms determines an effective medium characterized by a single-site t-matrix, $\underline{\bar{t}}$, which properly reflects many of the physical properties of the true disordered materials. For example, the effective t-matrix has a non-vanishing and non-trivial imaginary part even at real energies, and the corresponding scattering amplitude, $\underline{\bar{f}} = -\sqrt{E}\underline{\bar{t}}$, lies entirely inside the unitarity circle in an Argand plot, indicating the existence of inelastic scattering processes and finite lifetimes.

It should be emphasized that the t-matrices $\underline{t}^{\alpha}$ and corresponding scattering amplitudes entering the KKR-CPA expressions are not in general those determined in an independent calculation of single scatterers in vacuum or in bulk elemental solids. Instead, they must be determined in a charge self-consistent manner based on density functional theory, as discussed below. Thus, the charge self-consistent KKR-CPA is a first-principles theory of substitutionally disordered alloys in the same sense that the KKR Green-function method is a first-principles theory of the electronic structure of ordered materials. Furthermore, the KKR-CPA is quite general, allowing the treatment of many effects associated with disorder without the need to impose additional parameters describing individual features. For example, effects such as ODD that arise from differences in chemical species (but notably not those associated with lattice relaxation) and multiple and/or degenerate bands are automatically included in the formalism.

12.3.2 Charge density and DOS

Once the KKR-CPA equations have been solved and the effective-medium t-matrix, $\underline{\bar{t}}$, has been determined, the corresponding single-scatterer Green function associated with an individual species in the alloy and related properties follow. For example the number density inside a cell of type α is given by the expression

$$n^{\alpha}(\mathbf{r}) = -\frac{1}{\pi}\Im\int_{-\infty}^{E_F} G^{\alpha}(E;\mathbf{r},\mathbf{r})\mathrm{d}E, \tag{12.23}$$

whereas the corresponding DOS is given by

$$n^{\alpha}(E) = -\frac{1}{\pi}\Im\int_{\Omega} G^{\alpha}(E;\mathbf{r},\mathbf{r})\mathrm{d}^3 r, \tag{12.24}$$

where E_F denotes the Fermi energy, and G^α is given in terms of the SPO

$$\begin{aligned} G^\alpha(E;\mathbf{r},\mathbf{r}') &= \langle Z^\alpha(\mathbf{r})|\underline{\tau}^\alpha|Z^\alpha(\mathbf{r}')\rangle \\ &- \theta(r-r')\langle Z^\alpha(\mathbf{r}')|\tilde{J}^\alpha(\mathbf{r})\rangle \\ &- \theta(r'-r)\langle Z^\alpha(\mathbf{r})|\tilde{J}^\alpha(\mathbf{r}')\rangle. \end{aligned} \tag{12.25}$$

Since the wave functions can be chosen to be real at real energies, the only imaginary part that needs to enter the expressions for the charge density and DOS is that of the scattering-path operator.

The total charge densities and DOS are given by averages over the concentration, so that

$$\bar{n}(\mathbf{r}) = \sum_\alpha c_\alpha n^\alpha(\mathbf{r}), \tag{12.26}$$

and

$$\bar{n}(E) = \sum_\alpha c_\alpha n^\alpha(E). \tag{12.27}$$

Introducing the matrix quantities

$$\underline{F}^\alpha = \int_\Omega |Z^\alpha(\mathbf{r})\rangle\langle Z^\alpha(\mathbf{r})| \mathrm{d}^3r \tag{12.28}$$

and

$$\begin{aligned} \underline{\bar{F}} &= \int_\Omega \underline{\bar{F}}(\mathbf{r},\mathbf{r})\mathrm{d}^3r \\ &= \sum_\alpha c_\alpha \underline{F}^\alpha \underline{D}^\alpha, \end{aligned} \tag{12.29}$$

we can write

$$n^\alpha(\mathbf{r}) = -\frac{1}{\pi}\int_{-\infty}^{E_F} \Im \mathrm{Tr} \underline{F}^\alpha(E;\mathbf{r},\mathbf{r})\underline{D}^\alpha \underline{\bar{\tau}}^{00} \mathrm{d}E, \tag{12.30}$$

and

$$\bar{n}(E) = -\frac{1}{\pi}\Im \mathrm{Tr} \sum_\alpha c_\alpha \underline{F}^\alpha \underline{D}^\alpha \underline{\bar{\tau}}^{00}. \tag{12.31}$$

It is emphasized that the partial, species-resolved, quantities defined above satisfy the sum rules of Eqs. (12.26) and (12.27). Also, the reader can verify [10] that the integrated density of states in the KKR-CPA is given by the expression

$$\begin{aligned} \bar{N}(E) &= N_0(E) - \frac{1}{\pi}\Im \ln \det\left[\underline{\bar{m}}(E) - \underline{G}(E)\right] \\ &+ \sum_\alpha c_\alpha \Im \ln \det\left[\underline{I} + (\underline{m}^\alpha - \underline{\bar{m}})\,\underline{\bar{\tau}}^{00}\right]. \end{aligned} \tag{12.32}$$

The integrated density of states within the KKR-CPA can also be expressed as a Lloyd formula, which gives results equivalent to those obtained from expressions based on the Green function [11].

The site-diagonal Green function obtained in the KKR-CPA can be used in the study of many physical properties of substitutionally disordered alloys, such as energies of mixing, electronic transport, and others. The subject of alloy phase stability in particular is taken up in the next chapter.

12.4 Alternative Derivation

Although the derivation of the KKR-CPA given above has intuitive appeal, it also lacks a certain amount of rigor. Note, for example, an important difference between the derivations of the CPA in the TB formalism and in the MT or general potential cases discussed above: Whereas in the TB case one has an *independent* definition for the Green function of an effective medium, no such definition is given in the discussion of the KKR-CPA. Indeed, the site-diagonal Green function for the medium is defined to be the average of the Green functions corresponding to individual species of atoms embedded in the CPA medium.

The difficulty with defining an effective medium Green function in the case of the KKR-CPA is connected with the fact that cell potentials associated with different chemical species lead to different single-site wave functions, $|Z^{\alpha}(\mathbf{r})\rangle$. Furthemore, no wave function can be identified that corresponds to the effective potential of the CPA medium [11]; indeed, such a potential is not even determined within the CPA. Thus, the effective medium Green function cannot be written in terms of $\bar{\underline{\tau}}$ alone. On the other hand, the wave functions never enter the discussion in the phenomenological TB description of a system. Their effect is included in the values of the TB parameters such as the ϵ_i and the hopping integrals, W_{ij}. This makes it possible to determine $\bar{G}$ directly and write the CPA condition in the form, Eq. (11.41),

$$\langle G \rangle = \bar{G}. \tag{12.33}$$

The lack of an independent definition of the effective medium Green function in the KKR-CPA is more than just a formal inconvenience. For example, in a charge self-consistent calculation, one would expect the average of the charge density over chemical species to be equal to the independently determined charge of the effective medium. Also, in the study of $\mathbf{k}$-dependent properties, e.g., transport, it is necessary to have a proper description of the two-particle Green function. Such a description, as is seen in the next part of the book, requires the proper correlation between species-dependent scattering of an electron propagating through the effective medium and cannot be obtained unless the Green function of that medium is known.

In this section, we show that an effective-medium Green function can be defined within the KKR-CPA, and that the CPA condition can be written in the form of Eq. (12.33). We can obtain an indication of how such a construction can be made by recalling the approach used by Blackman, Esterling, and Berk to treat the effects of ODD (see previous chapter). After all, ODD arises because the wave functions and their overlap, which determines such quantities as band width, are different for individual atomic species. Thus, ODD heralds the existence of different band widths for the pure elemental solids formed by each species of a substitutional alloy. The difficulties associated with species-dependent wave functions is fully alleviated through the introduction of the configurational matrices in the formalism of BEB.

For the sake of clarity, we explicitly treat the case of binary alloys, $A_{1-c}B_c$. The results obtained below can, however, be immediately generalized to the case of multicomponent alloys. Our starting point is the equation of motion of the scattering-path operator, Eq. (12.5). Introducing occupation numbers, x_i and $y_i = 1 - x_i$, defined to be equal to 1 and 0 when site i is occupied by an atom of type A, and to be equal to 0 and 1 for an atom of type B, (see corresponding discussion in previous chapter), and multiplying Eq.(12.5) on the left and on the right by all possible combinations of these numbers, we can cast the equation of motion in the form,

$$\underline{\underline{\tau}}^{ij} = \underline{\underline{t}}^i \left[\delta_{ij} + \sum_{k \neq i} \underline{\underline{G}}^{ik} \underline{\underline{\tau}}^{kj} \right], \tag{12.34}$$

where double underlines denote matrices in angular momentum and configuration space. In particular, we have,

$$\underline{\underline{t}}^i = \begin{pmatrix} x_i \underline{t}^A & 0 \\ 0 & y_i \underline{t}^B \end{pmatrix}, \tag{12.35}$$

where

$$\underline{\underline{t}}^A = \begin{pmatrix} \underline{t}^A & 0 \\ 0 & 0 \end{pmatrix}, \tag{12.36}$$

and

$$\underline{\underline{t}}^B = \begin{pmatrix} 0 & 0 \\ 0 & \underline{t}^B \end{pmatrix}, \tag{12.37}$$

$$\underline{\underline{\tau}}^{ij} = \begin{pmatrix} x_i \underline{\tau}^{ij} x_j & x_i \underline{\tau}^{ij} y_j \\ y_i \underline{\tau}^{ij} x_j & y_i \underline{\tau}^{ij} y_j \end{pmatrix}, \tag{12.38}$$

and

$$\begin{aligned}\underline{\underline{G}}(\mathbf{R}_{ij}) &= \begin{pmatrix} \underline{G}(\mathbf{R}_{ij}) & \underline{G}(\mathbf{R}_{ij}) \\ \underline{G}(\mathbf{R}_{ij}) & \underline{G}(\mathbf{R}_{ij}) \end{pmatrix} \\ &= \underline{G}(\mathbf{R}_{ij}) \times \begin{pmatrix} 1 & 1 \\ 1 & 1 \end{pmatrix}. \end{aligned} \tag{12.39}$$

We note that like the case of ODD within a TB formalism the single-site t-matrix for cell n is a random variable, taking on values according to the occupation of the cell. However, unlike the TB case, all "hopping" terms are the same, $\underline{G}^{AA} = \underline{G}^{BB} = \underline{G}^{AB} = \underline{G}^{BA} = \underline{G}$. This matrix formulation, however, allows us to correlate exactly the species at a given site with the corresponding wave function, $|Z^n(\mathbf{r})\rangle$. To see this, we introduce the wavefunction matrices,

$$|\underline{Z}^n(\mathbf{r})\rangle = \begin{pmatrix} |Z^A(\mathbf{r})\rangle & 0 \\ 0 & |Z^B(\mathbf{r})\rangle \end{pmatrix}, \; |\underline{\tilde{J}}^n(\mathbf{r})\rangle = \begin{pmatrix} |\underline{\tilde{J}}^A(\mathbf{r})\rangle & 0 \\ 0 & |\underline{\tilde{J}}^B(\mathbf{r})\rangle \end{pmatrix}, \tag{12.40}$$

and their transposes, $\langle \underline{Z}^n(\mathbf{r})|$ and $\langle \underline{\tilde{J}}^n(\mathbf{r})|$, and note that the Green function for an arbitrary assembly can now be written as a matrix in configuration space,

$$\begin{aligned}\underline{G}(\mathbf{r},\mathbf{r}') &= \langle \underline{Z}^n(\mathbf{r})|\underline{\tau}^{nm}|\underline{Z}^n(\mathbf{r}')\rangle \\ &- \Big[\theta(r-r')\langle \underline{Z}^n(\mathbf{r}')|\underline{\underline{Q}}_n|\underline{\tilde{J}}^n(\mathbf{r})\rangle \\ &+ \theta(r'-r)\langle \underline{Z}^n(\mathbf{r})|\underline{\underline{Q}}_n|\underline{\tilde{J}}^n(\mathbf{r}')\rangle\Big]\,\delta_{nm}, \end{aligned} \tag{12.41}$$

where we have introduced the matrix $\underline{\underline{Q}}_n = \begin{pmatrix} x_n & 0 \\ 0 & y_n \end{pmatrix}$. In order to simplify the discussion we proceed by neglecting the single-scatterer term in the last expression for the Green function. For real energies, this term can be chosen to be real and does not contribute to the charge density. In any case, it can easily be brought back when necessary, e.g., when the Green function is evaluated in the complex plane.

We are now in a position to derive the KKR-CPA conditions within the configurational matrix formalism just outlined. Consistent with the spirit of the CPA, we replace the true, disordered single-site t-matrix (which is a matrix in configuration space) by an effective scattering matrix which is the same for all sites of the effective medium,

$$\underline{\underline{\bar{t}}} = \begin{pmatrix} \underline{\bar{t}}^{AA} & \underline{\bar{t}}^{AB} \\ \underline{\bar{t}}^{BA} & \underline{\bar{t}}^{BB} \end{pmatrix}. \tag{12.42}$$

The corresponding matrix equation of motion is then solved by a FT and the effective medium scattering-path operators are given by the expression,

$$\bar{\underline{\underline{\tau}}}^{ij} = \frac{1}{\Omega_{\mathrm{BZ}}} \int_{\mathrm{BZ}} \mathrm{d}^3k \left[\bar{\underline{\underline{t}}}^{-1} - \underline{\underline{G}}(\mathbf{k})\right]^{-1} \mathrm{e}^{\mathrm{i}\mathbf{k}\cdot\mathbf{R}_{ij}}. \tag{12.43}$$

Following familiar iteration procedures, we can write the site-diagonal element of the SPO in the form,

$$\bar{\underline{\underline{\tau}}}^{00} = \left(\bar{\underline{\underline{m}}} - \bar{\underline{\underline{\Delta}}}^{00}\right)^{-1}, \tag{12.44}$$

where

$$\bar{\underline{\underline{\Delta}}}^{00} \equiv \bar{\underline{\underline{\Delta}}} = \begin{pmatrix} \underline{\Delta}^{AA} & \underline{\Delta}^{AB} \\ \underline{\Delta}^{BA} & \underline{\Delta}^{BB} \end{pmatrix}, \tag{12.45}$$

is a matrix renormalized interactor, and $\bar{\underline{\underline{m}}} = \bar{\underline{\underline{t}}}^{-1}$. We note that

$$\bar{\underline{\underline{\Delta}}} = \bar{\underline{\underline{m}}} - \left(\bar{\underline{\underline{\tau}}}^{00}\right)^{-1}, \tag{12.46}$$

so that Eq. (12.43) can be written in terms of the renormalized interactor

$$\bar{\underline{\underline{\tau}}}^{ij} = \frac{1}{\Omega_{\mathrm{BZ}}} \int_{\mathrm{BZ}} \mathrm{d}^3k \left[\left(\bar{\underline{\underline{\tau}}}^{00}\right)^{-1} + \bar{\underline{\underline{\Delta}}} - \underline{\underline{G}}(\mathbf{k})\right]^{-1} \mathrm{e}^{\mathrm{i}\mathbf{k}\cdot\mathbf{R}_{ij}}. \tag{12.47}$$

Because the renormalized interactor depends only on the medium surrounding a site in the material, we can write the SPO associated with an atom (cell) of type α in the form,

$$\underline{\underline{\tau}}^{00;\alpha} = \left(\underline{\underline{I}} - \underline{\underline{t}}^{\alpha}\bar{\underline{\underline{\Delta}}}\right)^{-1} \underline{\underline{t}}^{\alpha} = \underline{\underline{\tau}}^{\alpha}. \tag{12.48}$$

In this expression care should be taken not to use the inverse of the cell t-matrix since that quantity is now singular, see e.g., Eq. (12.35).

Now, the KKR-CPA condition can be expressed either in terms of the SPO or the Green function. Following the latter approach, we set the single-site average of the site-diagonal part of the Green function, Eq. (12.41), written in terms of $\underline{\underline{\tau}}^{\alpha}$, to the corresponding element[6] of the effective medium Green function written in terms of $\bar{\underline{\underline{\tau}}}^{00}$. For $\mathbf{r}$ and $\mathbf{r}'$ in cell 0, we have

$$\langle \underline{G}(\mathbf{r}, \mathbf{r}') \rangle = \bar{\underline{G}}(\mathbf{r}, \mathbf{r}'). \tag{12.49}$$

Both sides of this expression are pre- and post-multiplied by the wave function matrices defined in Eq. (12.40). Since these matrices are generally

[6] The single-scatterer terms are identical on both sides of the equation.

non-vanishing and configuration independent, the last expression leads directly to a condition on the matrix SPO. Written in terms of the renormalized interactor, (the reader is urged to cast the following expression in a form involving the inverse of the effective cell t-matrix), this condition becomes

$$
\begin{aligned}
&\begin{pmatrix} c_A\left[\underline{m}^A-\underline{\Delta}^{AA}\right]^{-1} & 0 \\ 0 & c_A\left[\underline{m}^B-\underline{\Delta}^{BB}\right]^{-1} \end{pmatrix} \\
\equiv\ & \langle \underline{\underline{\tau}}^{00} \rangle = \begin{pmatrix} \underline{\bar{\tau}}^{AA} & 0 \\ 0 & \underline{\bar{\tau}}^{BB} \end{pmatrix} \\
=\ & \frac{1}{\Omega_{\mathrm{BZ}}}\int_{\mathrm{BZ}} \mathrm{d}^3k \left[\begin{pmatrix} \underline{\bar{\tau}}^{AA} & 0 \\ 0 & \underline{\bar{\tau}}^{BB} \end{pmatrix}^{-1} + \begin{pmatrix} \underline{\bar{\Delta}}^{AA} & \underline{\bar{\Delta}}^{AB} \\ \underline{\bar{\Delta}}^{BA} & \underline{\bar{\Delta}}^{BB} \end{pmatrix} \right. \\
-\ & \left. \begin{pmatrix} \underline{\bar{G}}^{AA}(\mathbf{k}) & \underline{\bar{G}}^{AB}(\mathbf{k}) \\ \underline{\bar{G}}^{BA}(\mathbf{k}) & \underline{\bar{G}}^{BB}(\mathbf{k}) \end{pmatrix} \right]^{-1}, \qquad (12.50)
\end{aligned}
$$

where we have used the fact that $\langle x \rangle = c_A$ and $\langle y \rangle = c_B$. Here, the site-diagonal part of the SPO is diagonal in configuration space reflecting the fact that a site cannot be occupied simultaneously by an atom of type A and one of type B. The fact that the renormalized interactor $\underline{\bar{\Delta}}^{AB}$ appears explicitly in the KKR-CPA conditions does not violate this requirement. This off-diagonal element of the renormalized interactor does not appear in any real-space expansion of the SPO (or the Green function), and its presence in the self-consistency condition is to guarantee the diagonality of $\underline{\underline{\tau}}^{00}$ in configuration space.

In the absence of ODD, i.e., when the structure constants have no configurational dependence, the matrix formulation of the KKR-CPA yields results identical to those obtained in the scalar version [13]. In that case, the renormalized interactor takes the form

$$
\underline{\underline{\bar{\Delta}}} = \underline{\bar{\Delta}} \begin{pmatrix} 1 & 1 \\ 1 & 1 \end{pmatrix}. \qquad (12.51)
$$

Finally, it is easily seen that the matrix formulation of the CPA can be generalized to n-component alloys in terms of n-dimensional matrices in configuration space. As was the case in a TB description of the Hamiltonian, physical quantities are obtained as sums over all elements of a configurational matrix.

The matrix formulation of the KKR-CPA leads to equivalent descriptions based on the Green function or the SPO. The self-consistency condition can also be put in the form

$$
\langle \underline{\underline{X}} \rangle = 0, \qquad (12.52)
$$

which the reader is urged to verify by deriving the form of the matrix $\underline{\underline{X}}$. Thus, various sum rules, e.g. on the partial charge densities, follow immediately from formal arguments. This matrix approach, although more difficult to implement computationally than the ordinary "scalar" formulation of the CPA, finds a number of useful applications, as is discussed in the next and following chapters.

12.4.1 Full-cell KKR-CPA

It is useful to derive expressions for the KKR-CPA based on a full-cell treatment of the cell potentials and charge densities, rather than the MT approximation. The basic reasons for such an extension are the same as for the corresponding extension of MST in the study of ordered solids. For example, the study of alloy surfaces and interfaces, the calculation of elastic constants, and the energetics associated with distortion, etc., cannot be undertaken reliably unless the CPA is implemented so that it can account properly for the shape of the potential and the charge distribution within a cell. In addition, thetreatment of so-called charge-transfer effects associated with the presence in the alloy of non-neutral cells also requires the use of a matrix formulation [14, 15].

It is possible to formulate a full-cell version of the KKR-CPA using the formalism already developed. All that is necessary is the use of the cell t-matrices that correspond to scattering from the non-spherical potentials associated with the atomic species in an alloy (with proper attention paid to questions of convergence of angular-momentum expansions). At the same time, the use of the t-matrix involves the inverse of the sine function, $S_{LL'}$, which can be the source of numerical instabilities. In this section, we provide an alternative formulation that avoids such inversions. This formulation requires the use of the matrix formalism discussed above. We also make use of the expression for the matrix $\underline{M}$ given in Eq. (3.140), and the associated expressions for the Green function.

In order to simplify our notation, we introduce the quantities

$$\underline{q}^n = \underline{\tilde{S}}^n \underline{C}^n \tag{12.53}$$

and[7]

$$\underline{Q}(\mathbf{R}_{nm}) = \underline{\tilde{S}}^n [\underline{G}(\mathbf{R}_{nm}) \underline{S}^m] \tag{12.54}$$

so that Eq. (3.140) can be written in the form

$$\underline{M}^{nm} = \underline{q}^n \delta_{nm} - \underline{Q}(\mathbf{R}_{nm})(1 - \delta_{nm}). \tag{12.55}$$

[7]The possible need for the use of modified structure constants in the double product should be kept in mind.

Multiplying Eq. (12.55) on the left and right by all possible combinations of occupation numbers, and collecting the results in a single matrix, we obtain the expression

$$\underline{\mathbf{M}}^{nm} = \underline{\mathbf{q}}^{n}\delta_{nm} - \underline{\mathbf{Q}}(\mathbf{R}_{nm})(1-\delta_{nm}). \tag{12.56}$$

where underlined bold-face symbols denote matrices in both angular momentum and configuration space. Thus, the site-diagonal matrices such as $\underline{\mathbf{q}}^n$ have matrix elements in configuration space, (the space spanned by the $\{x_\alpha\}$), of the form

$$\begin{aligned} \underline{\mathbf{q}}^{n,\alpha\beta} &= \underline{q}^{\alpha}\delta_{\alpha\beta} \\ &= \tilde{\underline{S}}^{\alpha}\underline{C}^{\alpha}\delta_{\alpha\beta} \end{aligned} \tag{12.57}$$

while site off-diagonal matrices have elements in configuration space given by the expression

$$\underline{\mathbf{Q}}^{\alpha\beta}(\mathbf{R}_{nm}) = \tilde{\underline{S}}^{\alpha}[\underline{G}(\mathbf{R}_{nm})\underline{S}^{\beta}]. \tag{12.58}$$

For the sake of clarity we quote the explicit values of these expressions for the case of binary alloys, $A_{1-c}B_c$. In this case, each matrix in angular momentum space, either site diagonal or off-diagonal, is augmented into a matrix of dimension twice that of the original matrix. Thus, we have

$$\underline{\mathbf{q}}^n = \begin{cases} \begin{pmatrix} \underline{q}^{A} & 0 \\ 0 & 0 \end{pmatrix} & \text{if an A atom occupies site } n \\ \begin{pmatrix} 0 & 0 \\ 0 & \underline{q}^{B} \end{pmatrix} & \text{if a B atom occupies site } n. \end{cases} \tag{12.59}$$

In general, we can write

$$\underline{\mathbf{q}}^n = \begin{pmatrix} x_n\underline{q}^{A} & 0 \\ 0 & y_n\underline{q}^{B} \end{pmatrix}. \tag{12.60}$$

Each matrix quantity in direct space $\underline{Q}^{nm}$ can be augmented to the two-dimensional form

$$\underline{\mathbf{Q}}^{nm} = \begin{pmatrix} x_n\underline{Q}^{AA}x_m & x_n\underline{Q}^{AB}y_m \\ y_n\underline{Q}^{BA}x_m & y_n\underline{Q}^{BB}y_m \end{pmatrix}. \tag{12.61}$$

At the same time, matrices in reciprocal space, such as $\underline{\tau}(\mathbf{k})$ and $\underline{Q}(\mathbf{k})$ take the form

$$\underline{\tau}(\mathbf{k}) = \begin{pmatrix} \underline{\tau}^{AA}(\mathbf{k}) & \underline{\tau}^{AB}(\mathbf{k}) \\ \underline{\tau}^{BA}(\mathbf{k}) & \underline{\tau}^{BB}(\mathbf{k}) \end{pmatrix}, \tag{12.62}$$

and

$$\underline{\mathbf{Q}}(\mathbf{k}) = \begin{pmatrix} \underline{Q}^{AA}(\mathbf{k}) & \underline{Q}^{AB}(\mathbf{k}) \\ \underline{Q}^{BA}(\mathbf{k}) & \underline{Q}^{BB}(\mathbf{k}) \end{pmatrix}. \tag{12.63}$$

We also introduce the matrix ket wave functions

$$|\underline{\psi}^n\rangle = \begin{pmatrix} |\psi^{\mathrm{A}}\rangle & 0 \\ 0 & |\psi^{\mathrm{B}}\rangle \end{pmatrix}. \tag{12.64}$$

and the corresponding bra quantities, $\langle\underline{\psi}^n|$. Similar definitions can be made of all other cell functions, such as $\langle \underline{Z}^m|$. Now, with the vectors $\mathbf{r}$ and $\mathbf{r}'$ in no cells other than Ω_n and Ω_m, respectively, we can define a matrix Green function for any configuration of the system in the form,

$$\mathbf{G}(\mathbf{r},\mathbf{r}') = \langle\underline{\psi}^n|\underline{\tau}^{nm}|\underline{\psi}^m\rangle + \langle\underline{Z}^n|\underline{\tilde{J}}^n\rangle\delta_{nm}. \tag{12.65}$$

The physical Green function is obtained as a sum over all the matrix elements of the matrix Green function in configuration space,

$$G(\mathbf{r},\mathbf{r}') = \sum_{\alpha\beta} \mathbf{G}^{\alpha\beta}(\mathbf{r},\mathbf{r}'), \tag{12.66}$$

where α and β run over all alloy species.

We are now in a position to derive the self-consistency condition of the KKR-CPA. Imposing the condition that the single-site average of the site-diagonal element of the matrix Green function, $\mathbf{G}(\mathbf{r},\mathbf{r}')$, be equal to the corresponding quantity associated with the effective medium, we obtain the result

$$\langle\mathbf{G}(\mathbf{r},\mathbf{r}')\rangle = \bar{\mathbf{G}}(\mathbf{r},\mathbf{r}') \tag{12.67}$$

where $\bar{\mathbf{G}}(\mathbf{r},\mathbf{r}')$ is given by an expression analogous to that in Eq. (12.65), but with $\bar{\underline{\tau}}^{nm}$ replacing $\underline{\tau}^{nm}$.

The self consistency condition, Eq. (12.67), can also be expressed in terms of the effective medium scattering-path operator,

$$\langle\underline{\tau}^{00}\rangle = \bar{\underline{\tau}}^{00}, \tag{12.68}$$

which is obtained when we consider the matrix scattering-path on both sides of Eq. (12.67).

For the sake of clarity, we display explicitly the form of the self-consistency condition, Eq. (12.68), for binary alloys

$$\begin{aligned} \bar{\underline{\tau}}^{00} &= \begin{pmatrix} c_{\mathrm{A}}[\underline{q}^{\mathrm{A}} - \bar{\underline{\Delta}}^{\mathrm{AA}}]^{-1} & 0 \\ 0 & c_{\mathrm{B}}[\underline{q}^{\mathrm{B}} - \bar{\underline{\Delta}}^{\mathrm{BB}}]^{-1} \end{pmatrix} = \begin{pmatrix} \bar{\underline{\tau}}^{\mathrm{AA}} & 0 \\ 0 & \bar{\underline{\tau}}^{\mathrm{BB}} \end{pmatrix} \\ &= \frac{1}{\Omega_{\mathrm{BZ}}}\int_{\mathrm{BZ}} \mathrm{d}^3k \left[\begin{pmatrix} \bar{\underline{\tau}}^{\mathrm{AA}} & 0 \\ 0 & \bar{\underline{\tau}}^{\mathrm{BB}} \end{pmatrix}^{-1} - \begin{pmatrix} \bar{\underline{\Delta}}^{\mathrm{AA}} & \bar{\underline{\Delta}}^{\mathrm{AB}} \\ \bar{\underline{\Delta}}^{\mathrm{BA}} & \bar{\underline{\Delta}}^{\mathrm{BB}} \end{pmatrix} \right. \\ &\quad \left. - \begin{pmatrix} \underline{Q}^{\mathrm{AA}}(\mathbf{k}) & \underline{Q}^{\mathrm{AB}}(\mathbf{k}) \\ \underline{Q}^{\mathrm{BA}}(\mathbf{k}) & \underline{Q}^{\mathrm{BB}}(\mathbf{k}) \end{pmatrix} \right]^{-1} \end{aligned} \tag{12.69}$$

Upon multiplication on the left and right by the matrix cell functions, Eq. (12.64), the last equation leads directly to a self-consistency condition in terms of Green functions. Note that in the case of alloys with n components, the self-consistency condition is expressed in terms of n-dimensional matrices.

The KKR-CPA self-consistency condition expressed in the form of Eq. (12.69) involves only products of the cell functions and can be expected to display better convergence characteristics than forms based on the t-matrix. Of course, in this form, the KKR-CPA formulation loses its attractive interpretation of waves propagating through a medium and being scattered at each site by the cell t-matrices. However, if this loss leads to more stable numerical properties it may well be worth it.

12.5 Bloch Spectral Function

The Bloch spectral function, $A^{\mathrm{B}}(\mathbf{k}, E)$, is defined as the trace of the imaginary part of the Green function in $\mathbf{k}$-space, Eq. (3.149),

$$A_{\mathrm{B}}(\mathbf{k}, E) = -\frac{1}{\pi} \Im \mathrm{Tr} \underline{G}(\mathbf{k}, E), \tag{12.70}$$

where the underline denotes a matrix in a given representation (excluding configuration space; see below). For a translationally invariant system, the Bloch spectral function is a series of δ functions

$$A^{\mathrm{B}}(\mathbf{k}, E) = \sum_{n} \delta(E - E_n(\mathbf{k})), \tag{12.71}$$

with $E_n(\mathbf{k})$ as the eigenstates of the system. These eigenstates form the band structure of the material as discussed in Chapter 6.

A Bloch spectral function can also be defined in the case of substitutionally disordered alloys within the formalism of the KKR-CPA. The defining equation is the same as that for ordered systems, but involves the Green function of the KKR-CPA effective medium. However, one of the effects of disorder is to broaden the δ-functions into essentially Lorenzian distributions whose half-width is related to the finite lifetime of the electron states. For example, such lifetimes can be observed experimentally in studies of the de Haas-van Alphen effect in dilute alloys where they show up as the so-called Dingle temperature. In concentrated alloys, the broadening can be observed using angle-resolved photoemission techniques. Using the formal machinery set up above, it is fairly straightforward to obtain an expression for $A^{\mathrm{B}}(\mathbf{k}, E)$ in alloys of arbitrary concentration.

In order to define the Bloch spectral function, it is necessary to have an expression for the Green function in reciprocal space, a quantity that conveniently results from the matrix formulation of the KKR-CPA. The physical Green function is the sum over all elements of the corresponding configurational matrix so that in the case of random substitutional alloys we can define

$$\begin{aligned} A^{\mathrm{B}}(\mathbf{k},E) &= -\frac{1}{\pi}\Im\sum_{\alpha\beta} A^{\mathrm{B},\alpha\beta}(\mathbf{k},E) - \frac{1}{\pi}\Im\sum_{\alpha\beta} G^{\alpha\beta}(\mathbf{k},E) \\ &= -\frac{1}{\pi}\Im\int \mathrm{d}^3r \sum_{\alpha\beta}\langle Z^{\alpha}(\mathbf{r},E)|\underline{\tau}^{00,\alpha\beta}(\mathbf{k},E)|Z^{\beta}(\mathbf{k},E)\rangle, \end{aligned} \tag{12.72}$$

where α and β range over the species in the alloy. It follows from the definition of the density of states that

$$n(E) = \frac{1}{\Omega_{\mathrm{BZ}}}\int_{\mathrm{BZ}} \mathrm{d}^3k A^{\mathrm{B}}(\mathbf{k},E). \tag{12.73}$$

It is interesting to note that Eq. (12.72) allows the identification of species-resolved spectral functions,

$$A^{\mathrm{B},\alpha\beta}(\mathbf{k},E) = \frac{1}{\pi}\Im\int \mathrm{d}^3r\langle Z^{\alpha}(\mathbf{r},E)|\underline{\tau}^{00,\alpha\beta}(\mathbf{k},E)|Z^{\beta}(\mathbf{k},E)\rangle. \tag{12.74}$$

These quantities are very useful in a number of formal studies of materials properties, in particular in studies of transport properties as discussed in the next part of the book.

12.6 Numerical Applications of KKR-CPA

Both the KKR-CPA and the ATA have been used extensively in the study of the electronic structure and properties of substitutionally disordered alloys [2, 3, 4], mostly within the MT approximation to the potential. In this section, we provide a few examples of such applications with a comparison of some of the most important features of these two approximations.

Figure 12.2 shows the total (solid line) and concentration-weighted, species-resolved densities of states for several Cu-Ni alloys, with concentrations as indicated in the legend. In each case the concentration-weighted sum of the species-resolved DOS equals the total DOS. Symmetry resolved DOSs corresponding to the irreducible representations on an fcc lattice are shown in Fig. 12.3. These DOSs, $n_s(E)$, satisfy the sum rule $\sum_s n_s(E) = n(E)$, where $n(E)$ is the total DOS for the alloy. Symmetry resolved DOSs corresponding to individual alloy species can also be defined and obey the obvious relation, $\sum_s n_s^{\alpha}(E) = n^{\alpha}(E)$.

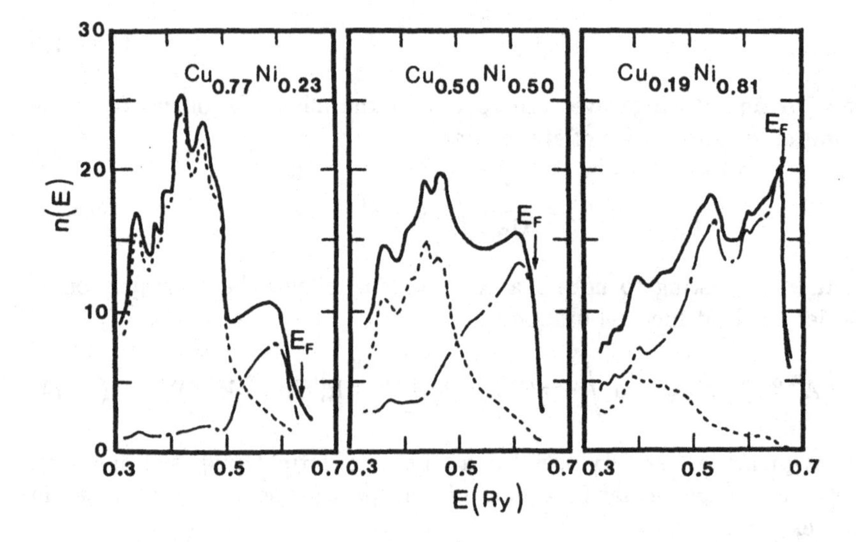

Figure 12.2: **Species resolved DOSs for Cu-Ni alloys calculated within the KKR-CPA. The dashed line is the Cu contribution, and the dot-dashed line the Ni contribution to the total DOS, (solid line) [11].**

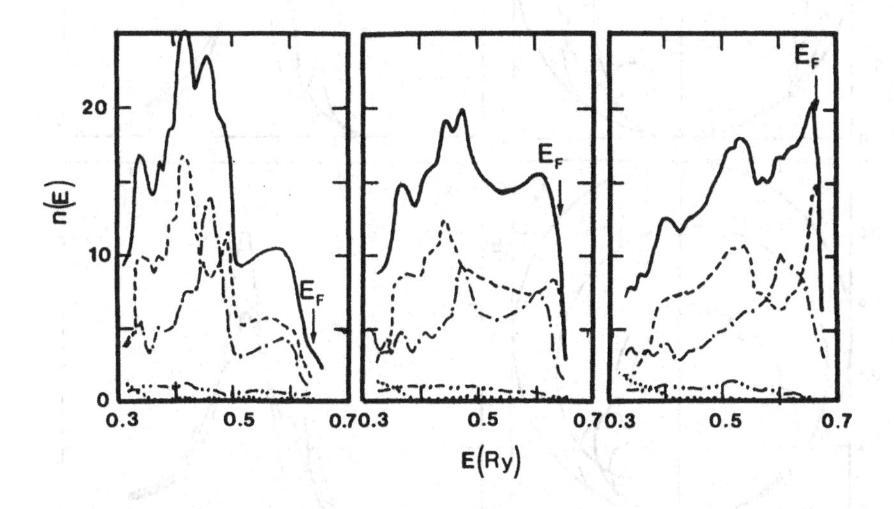

Figure 12.3: **Symmetry resolved DOSs for the alloys depicted in the previous figure. The dotted line indicates (a_{1g}), the dot-dot-dash line (t_{1u}), the dash line (t_{2g}), and the dot-dash line (e_g) symmetry [11].**

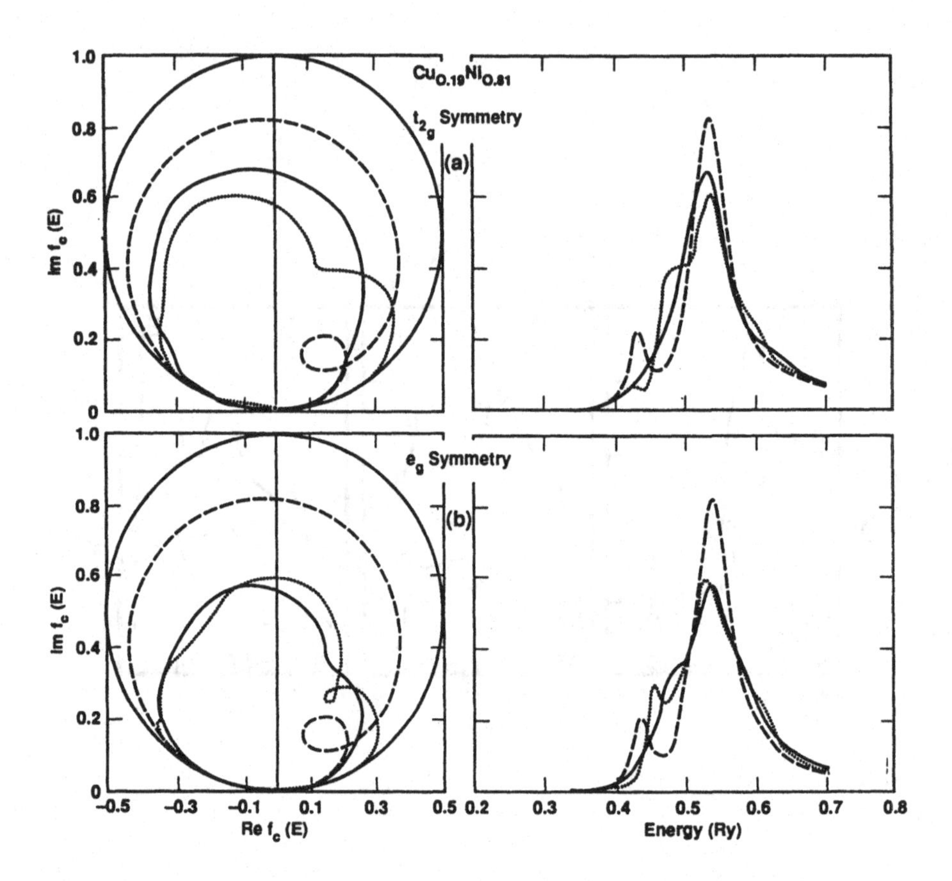

Figure 12.4: **The t_{2g} and e_g effective scattering amplitudes in the KKR-CPA (solid line) and the ATA (dashed line) and an approximate solution involving inverting a finite cluster in real space of the KKR-CPA (dotted line) for a Cu-Ni alloy [2].**

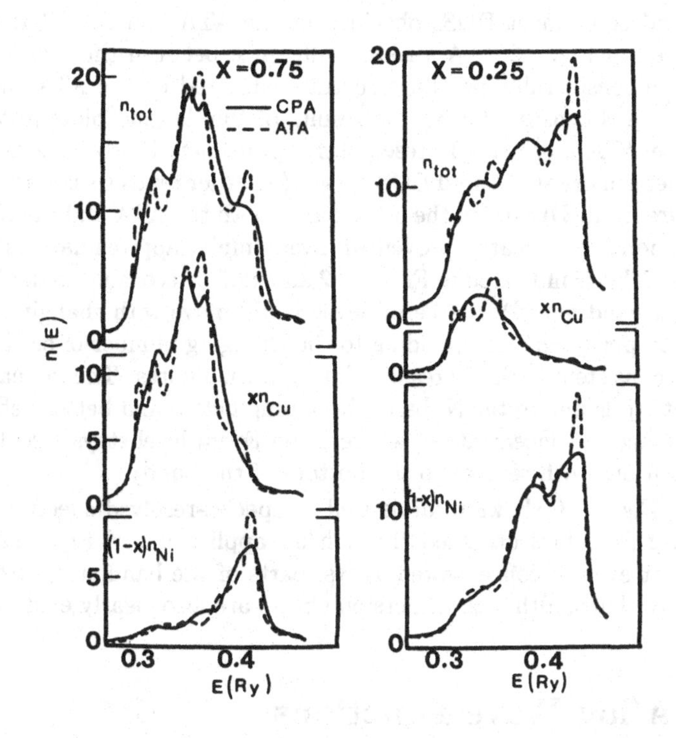

Figure 12.5: **Total and species-resolved DOSs in the ATA and the KKR-CPA for two Ni-Cu alloys [12].**

Figure 12.4 allows a comparison of the results obtained within the ATA and the CPA for the same alloy system. The results are shown in terms of Argand diagrams (Appendix D.3). We see that the ATA gives two resonances (the extra loop), whereas the solution of the KKR-CPA clearly contains a single resonance, [although the approximate solution of the C-PA (dotted line), obtained through an inversion of the equation of motion of the SPO for a finite cluster in real space, gives indications of an extra resonance). Furthermore, the ATA curves that lie closer to the circumference of the unitarity circle correspond to less absorptive scatterers than the KKR-CPA ones. More detailed examination of these results confirms that only the resonance shown by the CPA is realistic, emphasizing the importance of full self consistency in determining the electronic properties of substitutionally disordered alloys. Lack of self consistency can cause the appearance of spurious resonances in the scattering amplitudes, which can affect the results of calculations through their effect on spectral functions (and densities of states).

Total and component DOSs obtained in the ATA and the KKR-CPA are compared in Fig. 12.5. A striking difference between the two sets of results is the considerably greater structure exhibited by the ATA curves. This structure is consistent with the results of the Argand plots just discussed. The ATA more closely resembles a pure, nonabsorbtive scatterer and thus yields more structured DOS curves (disorder tends to smooth over the structure in the DOS). On the other hand, both the ATA and the KKR-CPA correspond to a great improvement over simpler approximations such as the VCA. This is indicated in Figs. 12.2 and 12.3. According to the VCA (or the rigid band model), the Fermi level would move with changing concentration to positions corresponding to the changing number of electrons. This is not consistent with the results shown in the figures. The increase in the DOS at E_F is due to the Ni bands becoming better and better defined with increasing Ni concentration, whereas the Fermi level stays essentially fixed in position (with respect to the bottom of the band).

Finally, Fig. 12.6 shows total as well as species-resolved spectral densities for Ag-Pd alloys obtained through an application of Eq. (12.74). We note that even in concentrated alloys, parts of the band structure can remain fairly sharp, although dispersion effects are also clearly evident.

12.7 Alloy Wave Functions

We have mentioned that it is not possible to ascribe the usual meaning of wave function to an alloy potential. In fact, no such potential is calculated within the CPA. However, a wave function can still be defined as a quantity that can be used to construct the Green function through its spectral representation, Eq. (2.17). We recall that the Green function for a pure, translationally invariant material can be written in terms of Bloch wave functions in the form, Eq. (2.17),

$$G(\mathbf{r},\mathbf{r}';E) = \sum_n \sum_{\mathbf{k}} \frac{\psi_{n\mathbf{k}}(\mathbf{r})\psi_{n\mathbf{k}}(\mathbf{r}')}{E - E_n(\mathbf{k})}, \tag{12.75}$$

where the summations extend over all bands, n, and wave vectors $\mathbf{k}$ in the first Brillouin zone of the lattice. Introducing cell vectors, $\mathbf{x}$ and $\mathbf{x}'$, measured from the center of a cell, and using Bloch's theorem, Eq. (6.4), we can write

$$G(\mathbf{x}+\mathbf{R},\mathbf{x}'+\mathbf{R}';E) = \sum_n \sum_{\mathbf{k}} \frac{\psi^*_{n\mathbf{k}}(\mathbf{x})\psi_{n\mathbf{k}}(\mathbf{x}')e^{i\mathbf{k}\cdot(\mathbf{R}-\mathbf{R}')}}{E - E_n(\mathbf{k})}. \tag{12.76}$$

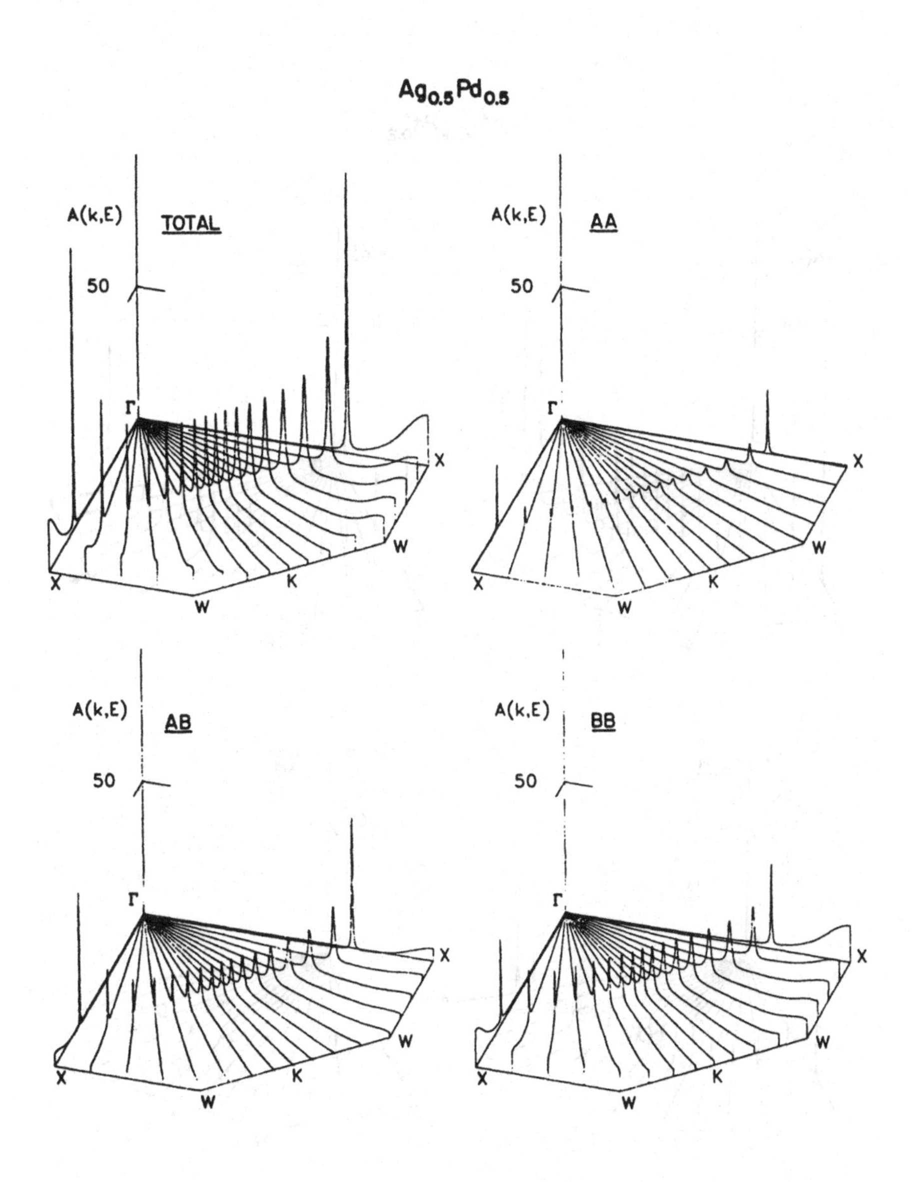

Figure 12.6: **Total and species-resolved Bloch spectral densities for $Ag_{0.5}Pd_{0.5}$ alloys as functions of energy along various direction-s in the first Brillouin zone, as obtained within the KKR-CPA approach [23].**

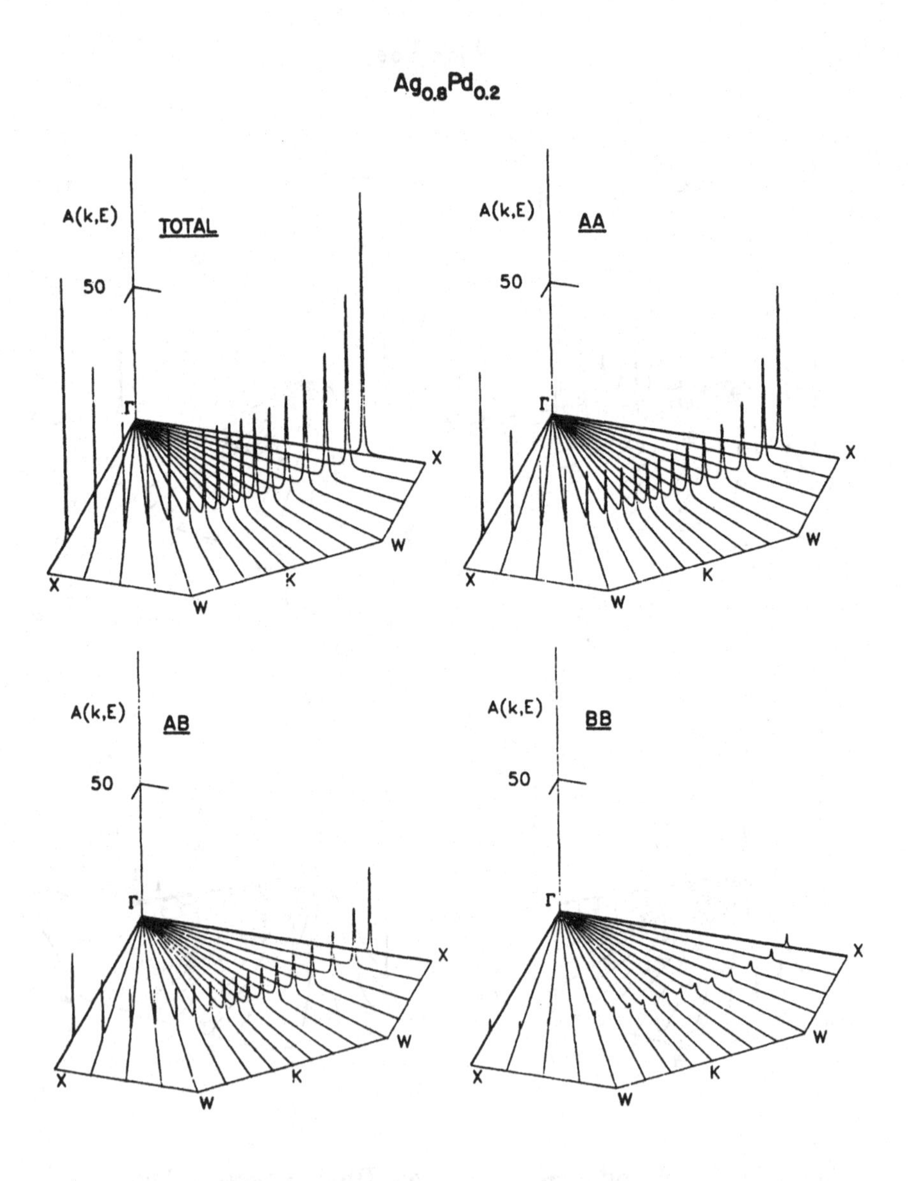

Figure 12.7: **Results analogous to those in the previous figure, but for $Ag_{0.8}Pd_{0.2}$ alloys [23].**

At the same time, we can use the properties of Fourier transformations to write

$$\underline{G}(\mathbf{x}+\mathbf{R},\mathbf{x}'+\mathbf{R}';E)=\sum_{\mathbf{k}}\langle\underline{Z}(\mathbf{x})|\underline{\underline{\tau}}(\mathbf{k})|\underline{Z}(\mathbf{x}')\rangle \mathrm{e}^{\mathrm{i}\mathbf{k}\cdot(\mathbf{R}-\mathbf{R}')}. \quad (12.77)$$

Now, let the matrix $\underline{\underline{C}}(\mathbf{k},E)$ diagonalize $\underline{\underline{\tau}}(\mathbf{k})$, in the form (e.g., near a pole of $\underline{\tau}(\mathbf{k})$)

$$\left[\underline{\underline{C}}(\mathbf{k},E)\underline{\underline{\tau}}(\mathbf{k})\underline{\underline{C}}^{\dagger}(\mathbf{k},E)\right]=\left[\underline{\underline{I}}E-\underline{\underline{E}}_{n}(\mathbf{k})\right]^{-1}\delta_{nm}. \quad (12.78)$$

Comparison with Eq. (12.76) allows us to define the formal quantity

$$|\underline{\psi}_{\mathbf{k}}(\mathbf{r})\rangle=\sum_{L}\underline{\underline{C}}^{\dagger}(\mathbf{k},E)|\underline{Z}(\mathbf{x})\rangle \mathrm{e}^{-\mathrm{i}\mathbf{k}\cdot\mathbf{R}}\text{ , with } \mathbf{r}=\mathbf{x}+\mathbf{R}. \quad (12.79)$$

This matrix quantity is one way of defining a "wave function" for the case of substitutionally disordered alloys.

An alternative approach in the definition of an alloy wave function has been taken by Faulkner and Stocks [11]. They showed that the site diagonal element of the SPO connecting a cell occupied by an atom of type α and one of type β is given by the expression[8]

$$\underline{\tau}^{nm,\alpha\beta}=\underline{D}^{\alpha}\underline{\bar{\tau}}^{nm}\underline{D}^{\beta}. \quad (12.80)$$

The corresponding elements of the physical Green function (the sum of the matrix elements in configuration space) is then given by the expression,

$$G(\mathbf{r},\mathbf{r}');E)=\mathrm{Tr}\underline{F}^{\mathrm{cc}}(\mathbf{x},\mathbf{x}';E)\underline{\bar{\tau}}^{nm}, \quad (12.81)$$

where the cell vectors $\mathbf{x}$ and $\mathbf{x}'$ are confined to cells Ω_n and Ω_m, respectively, and

$$\begin{aligned}\underline{F}^{\mathrm{cc}}(\mathbf{x},\mathbf{x}';E) &= c_A^2\underline{\tilde{D}}^{A}\underline{F}^{AA}(\mathbf{x},\mathbf{x}';E)\underline{D}^{A}\\ &+ c_Ac_B\underline{\tilde{D}}^{B}\underline{F}^{AB}(\mathbf{x},\mathbf{x}';E)\underline{D}^{A}\\ &+ c_Ac_B\underline{\tilde{D}}^{A}\underline{F}^{BA}(\mathbf{x},\mathbf{x}';E)\underline{D}^{B}\\ &+ c_B^2\underline{\tilde{D}}^{B}\underline{F}^{BB}(\mathbf{x},\mathbf{x}';E)\underline{D}^{B}. \end{aligned} \quad (12.82)$$

The quantities $\underline{F}^{\alpha\beta}(\mathbf{x},\mathbf{x}';E)$ are defined by an obvious extension of the expressions in Eq. (12.28),

$$\underline{F}^{\alpha\beta}=\int_{\Omega}|Z^{\alpha}(\mathbf{r})\rangle\langle Z^{\beta}(\mathbf{r})|\mathrm{d}^3r. \quad (12.83)$$

[8]The reader is urged to verify that this expression is identical to the $\alpha\beta$ element of $\underline{\underline{\tau}}^{ij}$ obtained in the configurational-matrix approach to the CPA.

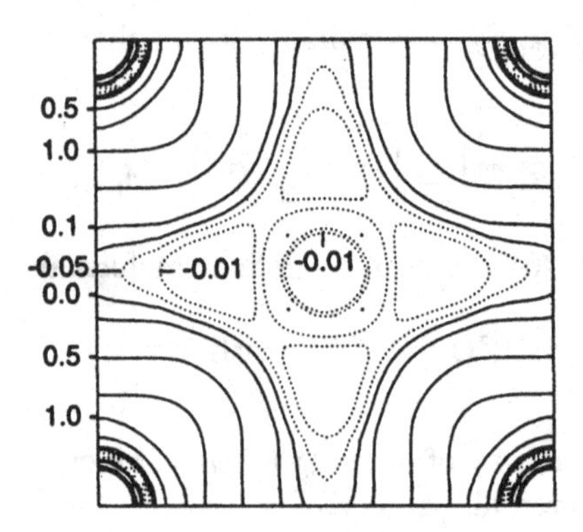

Figure 12.8: **The differences in the charge densities on the (001) planes between an Al cell (center) in an Al-Li alloy at equiatomic composition, and in the ordered LI_0 configuration. The alloy results were based on the wave function expressions quoted in the text. The contours are in units of 10^{-2} electrons/a.u.**

The matrix $\underline{F}^{cc}(\mathbf{x}, \mathbf{x}'; E)$ can now be calculated from an equation such as Eq. (12.29) provided that one defines the bra "wave function" (with an obvious analogous definition for the corresponding ket)

$$\begin{aligned} \langle Z^c(\mathbf{r}, \mathbf{E})| &= \left[c_A \langle Z^A(\mathbf{r}, \mathbf{E})|\underline{D}^A \right. \\ &+ \left. c_B \langle Z^A(\mathbf{r}, \mathbf{E})|\underline{D}^B \right]. \end{aligned} \tag{12.84}$$

The reader is urged to show that the definitions of the "wave functions" given in Eqs. (12.79) and (12.84) are equivalent, but is warned that the CPA algebra required is not trivial[9].

The last expression has been used [16] to construct charge density contours for Al-Li alloys, an example of which is shown in Fig. 12.8. The figure shows the difference in valence charge density in the (001) plane for an Al cell between the disordered alloy at equiatomic composition and the ordered LI_0 configuration. The charge densities for the disordered alloy were obtained on the basis of Eq. (12.84). In the figure, solid

[9] This demonstration becomes somewhat easier if all matrix quantities are taken to commute, which is rigorously valid in the case of single-band systems.

lines correspond to positive and dotted lines to negative charge differences, $\Delta n(\mathbf{r}) = n_{\mathrm{LI_0}}(\mathbf{r}) - n_{\mathrm{alloy}}(\mathbf{r})$. We note that the greatest changes in the charge density between the ordered and disordered configurations occur midway along nearest-neighbor Al-Al directions, and can be ascribed to the presence of the effective KKR-CPA "atoms" around a cell of given composition.

12.8 ECM

As was mentioned already, the CPA is a single-site theory that cannot account for the effects of local statistical fluctuations in disordered materials. The treatment of such effects is important in determining many materials properties, e.g., phase stability, magnetism, transport, and various methods for such treatment have been proposed [5]. Almost invariably, these methods rely on the study of site clusters, and can be divided into self-consistent and non-self-consistent approaches. However, even a cursory review of the various formalisms associated with self-consistency reveals that in most, if not all cases, these methods are computationally very difficult. In addition, conceptual problems also arise, such as uniqueness, and the identification of the coupling of impurities with the surrounding materials. The latter difficulty precludes a unique definition even of the Hamiltonian describing the impurity system. A number of cluster methods, both of the self-consistent and non-self-consistent variety are reviewed in reference [5].

A fairly simple, although non-self-consistent treatment of a cluster of impurity atoms embedded in a CPA effective medium is provided by the embedded cluster method (ECM). This method was discussed in the previous chapter in connection with TB systems, and was also used in Chapter 9 within the KKR (multiple-scattering) formalism to study the electronic structure of impurities embedded in an otherwise pure host material. With appropriate changes in the interpretation of the various parameters that enter the discussion, the formalism of Chapter 9 can be made to apply to the case of clusters embedded in an effective medium determined within the KKR-CPA.

12.8.1 Formal details

We identify the quantities, $\underline{m}$, $\underline{\Delta}^C$, and $\underline{\tau}^{CC}$, that occur in the discussion of clusters embedded in pure host materials, Eqs. (9.79), with the KKR-CPA quantities, $\bar{\underline{m}}$, $\bar{\underline{\Delta}}^C$, and $\bar{\underline{\tau}}^{CC}$, and the quantities describing impurities in the former case with those for real atoms in the KKR-CPA, and obtain the expressions,

$$\underline{\tau}^{CC} = \left[\underline{m}^C - \bar{\underline{\Delta}}^C\right]^{-1}$$

$$= \left[\underline{m}^C - \bar{\underline{m}}^C + \left(\bar{\underline{\tau}}^{CC}\right)^{-1}\right]^{-1}, \tag{12.85}$$

for the SPO associated with a cluster of "real" atoms embedded in the KKR-CPA medium. The renormalized interactor, $\bar{\underline{\Delta}}^C$, is given by an expression analogous to that in Eq. (9.81), but with host quantities in that equation replaced with the corresponding effective medium quantities. The effective medium SPO has matrix elements $\bar{\underline{\tau}}^{nm}$ given as integrals over the first Brillouin zone of the reciprocal lattice,

$$\begin{aligned} \bar{\underline{\tau}}^{nm} &= \frac{1}{\Omega_{\mathrm{BZ}}} \int_{\mathrm{BZ}} \bar{\underline{\tau}}(\mathbf{k}) e^{i\mathbf{k}\cdot\mathbf{R}_{nm}} d^3k \\ &= \frac{1}{\Omega_{\mathrm{BZ}}} \int_{\mathrm{BZ}} \left[\bar{\underline{m}} - \underline{G}(\mathbf{k})\right]^{-1} e^{i\mathbf{k}\cdot\mathbf{R}_{nm}} d^3k. \end{aligned} \tag{12.86}$$

Once the matrix elements of the SPO corresponding to all sites in a cluster have been determined, the Green function associated with any sites n and m in the cluster is given by the expression,

$$G(\mathbf{r},\mathbf{r}';E) = G^{(1)}(\mathbf{r},\mathbf{r}';E)\delta_{nm} + \langle Z^n(\mathbf{r})|\underline{\tau}^{nm}|Z^m(\mathbf{r}')\rangle, \tag{12.87}$$

where $\underline{\tau}^{nm}$ denotes the (n,m)th matrix element of $\underline{\tau}^{CC}$. Further discussion of the ECM, including the use of symmetry in its implementation, is given in reference [5].

12.8.2 Numerical results

The use of the ECM is illustrated in Fig. 12.9, which shows DOSs for the $Ag_{0.5}Pd_{0.5}$ alloy. The DOSs shown here are associated with the center of a 13-atom, nearest-neighbor cluster on an fcc lattice, embedded in a medium determined within the KKR-CPA. The left four frames in the figure show component DOS's for a Ag atom at the center of the cluster, and the right four frames for a Pd atom at the center. Frame (a) shows the s ($\ell = 0$) component DOS surrounded by 12 Ag atoms (solid curve), 10 Ag atoms and 2 Pd atoms (dash-dotted curve), 6 Ag and 6 Pd atoms (dashed curve), and 12 Pd atoms (dotted curve). In the configuration chosen for the case of 10 Ag and 2 Pd atoms, the two Pd atoms are second nearest neighbors, while the 6 Ag and 6 Pd configuration was chosen at random from among the many inequivalent configurations possible.

Frame (c) in the figure shows the p ($\ell = 1$) component DOS and frame (e) the d ($\ell = 2$) component for the same configurations as in frame (a), and with the same designation attached to the various curves. Frame (g) shows the total DOS at the center of the cluster, i.e., the sums of the DOS shown in the frames above it. Frames (b), (d), (f), and (h) show results analogous to those on the frames to their left but with the roles of Ag and Pd atoms interchanged.

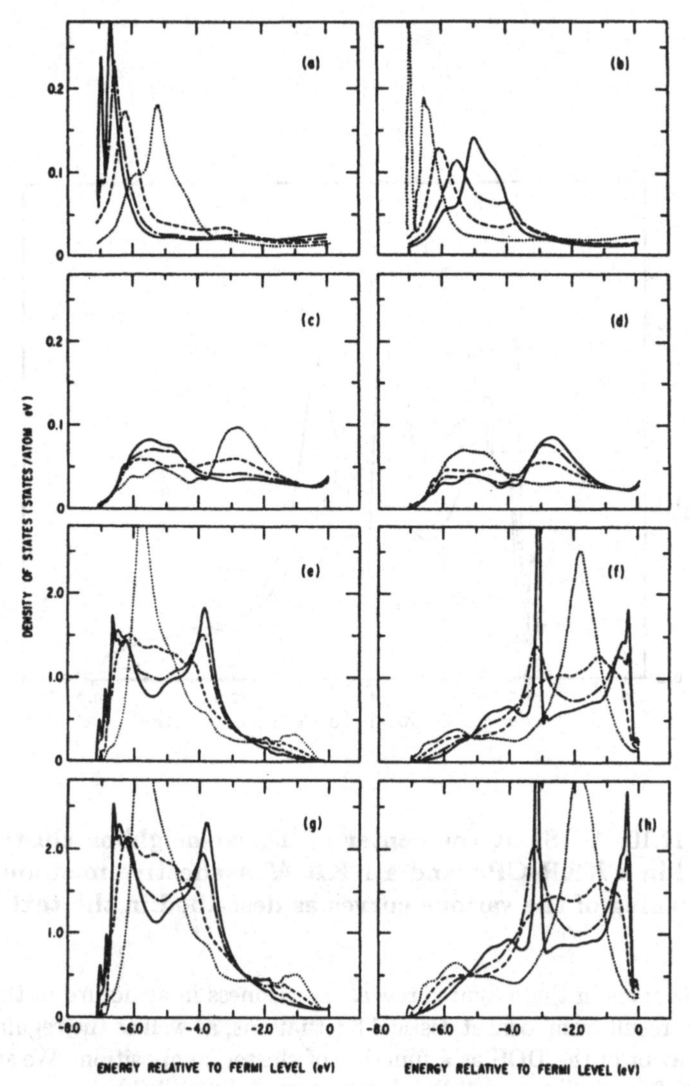

Figure 12.9: **DOSs associated with the center of a near-neighbor cluster in the $Ag_{0.5}Pd_{0.5}$ alloy for various cluster configurations. Ag (Pd) centered cluster DOSs are shown on the left (right) column of frames. Frame (a): s-component DOS for a Ag atom surrounded by 12 Ag atoms (solid curve), 10 Ag atoms and 2 Pd atoms (dash-dotted curve), 6 Ag and 6 Pd atoms (dashed curve), and 12 Pd atoms (dotted curve). Frames (c), (e), and (g) depict analogous results for p, d, and the total DOS, respectively. Frames (b), (d), (f), and (h) show results corresponding to those in frames (a), (c), (e), and (g), but with the roles of the Ag and Pd atoms interchanged [6].**

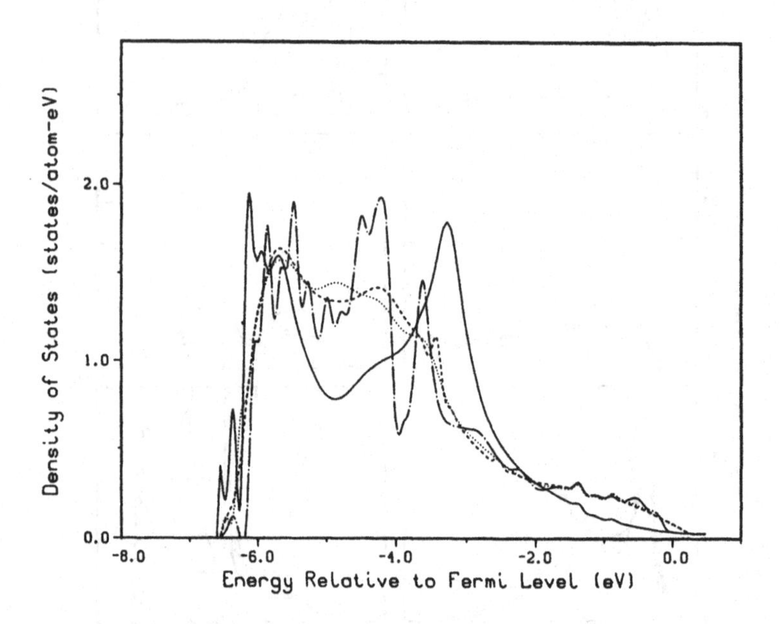

Figure 12.10: **DOSs at the center of a near-neighbor cluster embedded in a KKR-CPA and a KKR-ATA effective medium, with the meaning of the various curves as described in the text [6].**

The curves in these figures reveal the richness in structure in the DOS that can result from local statistical fluctuations, as well as the regularity in the behavior of the DOS as a function of cluster composition. We see that the DOS for an all-Ag or all-Pd cluster, frames (g) and (h), are very similar to those that are obtained for the corresponding bulk material [17]. They clearly display the two-peaked structure of transition-metal DOSs that is associated with the bonding (lower peak) and antibonding components of t_{2g} symmetry. On the other hand, an Ag atom surrounded by 12 Pd atoms, or a Pd atom surrounded by 12 Ag atoms, dotted curves in panels (g) and (h), display the characteristic impurity peak associated with the formation of a resonant level, somewhat broadened by disorder. The DOSs curves for intermediate cluster configurations move gradually from one to the other DOSs corresponding to these two extreme configurations.

Figure 12.10 shows DOSs for clusters embedded in a KKR-CPA and in a KKR-ATA effective medium. Here, the solid curve represents the d-band DOS at the center of an all-Ag cluster in the KKR-CPA medium, the dotted curve is the DOS for an Ag atom surrounded by 6 Ag and 6 Pd atoms also in a CPA medium, and the dashed curve is the Ag-site DOS determined in the KKR-CPA. The dash-dotted curve is the DOS for an Ag atom surrounded by 6 Ag and 6 Pd atoms embedded in a medium described by the ATA. We note that the SSCPA curve, (dashed line), is quite similar to the DOS for the Ag atom surrounded by 6 Ag and 6 Pd atoms (dotted curve). This is not an unexpected result, since the concentration of Ag atoms in the cluster is the same as that of the overall alloy concentration. In contrast, the DOS obtained for the same cluster concentration in the ATA is much more structured than the corresponding CPA curves. This structure is reminiscent of that exhibited by the ATA density of states in Fig. 12.5. The structure in both cases is spurious, and reflects the fact that the ATA represents the averaged material less accurately than the CPA.

12.9 DFT and KKR-CPA

The development of the KKR-CPA presented in the preceding sections was based on a given set of potentials, $v_\alpha(\mathbf{r})$, and corresponding charge densities, $n^\alpha(\mathbf{r})$, associated with each species α in an alloy. However, the study of the properties of alloy systems almost invariably depends on the knowledge of the energies of the various structures, and in particular the identification of the structure with the lowest energy, i.e., the ground state of the system. This is particularly true at T=0 K, where entropy terms do not arise and the free energy of a system reduces to the total energy. It is, therefore, necessary to have a procedure that yields the ground state energy of a substitutionally disordered alloy as a functional of the electron density, along the lines described for ordered systems in Chapter 6. Such a procedure can be constructed based on DFT, or its local density approximation, the LDA. In this section, we show how the KKR-CPA can be implemented within a DFT, self-consistent field mode (SCF-KKR-CPA).

We recall that DFT is based on the theorem of Hohenberg and Kohn (see Chapter 1), which states that the total energy of a quantum mechanical system is a *unique* functional of the electron density, $n(\mathbf{r})$, and that it is a minimum for the true density of the system. As we saw in Chapter 1, the corresponding minimization principle yields an effective, single-particle Hamiltonian with which is associated a set of single-particle-like Schrödinger equations and wave functions. These wave functions allow a straightforward determination of the electron density. And as we have seen, the determination of the ground-state density and total energy proceeds via an iterative process that leads to the consistency of the charge density and the potential. Such a process can be implemented within the KKR-CPA framework.

12.9.1 Charge self-consistency

For the sake of simplicity of formal expressions, we consider the case of generally-shaped, space-filling potential cells. We are interested in obtaining an expression for an effective potential $v^{\alpha}_{\text{eff}}(\mathbf{r})$, associated with a cell of type α, to be used in obtaining the scattering properties of the cell in the KKR-CPA medium. These scattering properties are determined through a solution of the corresponding single-particle Schrödinger equation, Eq. (1.205), for each species α in the alloy. It is consistent with the spirit of the CPA to write $v^{\alpha}_{\text{eff}}(\mathbf{r})$, the potential in a cell of type α, in the form

$$\begin{aligned} v^{\alpha}_{\text{eff}}(\mathbf{r}) &= v^{\alpha}(\mathbf{r}) + \int_{\Omega^{\alpha}} \frac{n^{\alpha}(\mathbf{r}')}{\mathbf{r}-\mathbf{r}'} \mathrm{d}^3 r' + \int_{R^3-\Omega^{\alpha}} \frac{\bar{n}(\mathbf{r}')}{\mathbf{r}-\mathbf{r}'} \mathrm{d}^3 r' \\ &+ v^{\alpha}_{\text{xc}}(\mathbf{r}), \end{aligned} \tag{12.88}$$

where

$$v^{\alpha}(\mathbf{r}) = \frac{Z_{\alpha}}{r} + \sum_{\mathbf{R}}{}' \frac{\bar{Z}}{|\mathbf{r}-\mathbf{R}|} \tag{12.89}$$

is the external potential acting on the electron cloud due to the presence of the nuclear charges. Note that in this expression for the potential, an electron in cell Ω_{α} interacts with the cell number $n^{\alpha}(\mathbf{r})$ and the nucleus in the cell, Z^{α}, but interacts with the concentration averages of these quantities,

$$\bar{n}(\mathbf{r}) = \sum_{\alpha} c_{\alpha} n^{\alpha}(\mathbf{r}) \tag{12.90}$$

and

$$\bar{Z} = \sum_{\alpha} c_{\alpha} Z_{\alpha} \tag{12.91}$$

outside the cell. We note that this is consistent with the domains of the spatial integrations, where the second integral extends over all space outside cell Ω^{α}. The term

$$v^{\alpha}_{\text{xc}}(\mathbf{r}) = \frac{\delta E^{\alpha}_{\text{xc}}}{\delta n^{\alpha}(\mathbf{r})} \tag{12.92}$$

is the exchange and correlation potential for a cell of type α. The first term on the right side of Eq. (12.89) gives the interaction of an electron in the cell with the cell nucleus, and the second term with the average nuclear charge outside. Because the system as a whole is neutral, we have the condition,

$$\sum_{\alpha} c_{\alpha} \int_{\Omega^{\alpha}} n^{\alpha}(\mathbf{r}) \mathrm{d}^3 r = \bar{Z}. \tag{12.93}$$

The expressions given above allow the following self-consistency procedure in determining the ground-state charge densities and potentials in the KKR-CPA: For a given $n^\alpha(\mathbf{r})$, one determines $v^\alpha_{\text{eff}}(\mathbf{r})$, and hence the t-matrix, $\underline{t}^\alpha$. The KKR-CPA self-consistency condition then leads to the Green functions $G^\alpha(\mathbf{r},\mathbf{r}')$ and, hence, to a new number density for cell α,

$$n^\alpha(\mathbf{r}) = -\frac{1}{\pi}\Im\int G^\alpha(E;\mathbf{r},\mathbf{r})\mathrm{d}E. \tag{12.94}$$

The process is repeated until the charge density stops changing within some preset tolerance [18, 19]. Schemes that combine the iterations for charge self-consistency with those for solving the KKR-CPA have also been developed [20], greatly improving computational times.

12.9.2 Total energy

Once the equations of the KKR-CPA have been solved in the self-consistent field mode (SCF-KKR-CPA), the corresponding ground-state energy of an alloy can be found in a fairly straightforward way (at least in principle). Let us recall Eq. (1.211), which gives the total energy of a system of electrons and nuclei, and note that the sum of orbital energies (band energies) can be written by definition in terms of the density of states in the form

$$\sum_i^N \epsilon_i = \int_{-\infty}^{E_F} En(E)\mathrm{d}E. \tag{12.95}$$

Thus, *quite generally*, i.e., in the case of periodic materials and random alloys, the total energy associated with a cell of type α is given by the expression[10]

$$\begin{aligned} E^\alpha &= \int_{-\infty}^{E_F} En^\alpha(E)\mathrm{d}E - \frac{1}{2}\int_{\Omega^\alpha}\mathrm{d}^3r n^\alpha(\mathbf{r})\left[\int\frac{n(\mathbf{r}')}{|\mathbf{r}-\mathbf{r}'|}\mathrm{d}^3r - \sum_n\frac{Z_n}{|\mathbf{r}-\mathbf{R}_n|}\right] \\ &- \frac{1}{2}Z_\alpha\left[\int\frac{n(\mathbf{r})}{|\mathbf{r}-\mathbf{R}_\alpha|}\mathrm{d}^3r - {\sum_{\beta\neq\alpha}}'\frac{Z_\beta}{|\mathbf{R}_\alpha-\mathbf{R}_\beta|}\right] \\ &+ E^\alpha_{\text{xc}}\left[n^\alpha(\mathbf{r})\right]. \end{aligned} \tag{12.96}$$

In order to obtain the corresponding expression in the KKR-CPA, we replace the electronic charge outside cell Ω^α by the average charge, $\bar{n}(\mathbf{r})$, and also replace all nuclear charges outside the cell by the average nuclear charge, $\bar{Z}$. Of course, inside the cell, we use the appropriate cell electronic

[10] The general expression for the total energy can be put in various forms as dictated by convenience of application within different computational schemes.

and nuclear charges. If the cells are not individually neutral for each alloy species, this procedure can lead to large inaccuracies in the calculation of the electrostatic interactions and energies in an alloy. In this case, the CPA medium may be inadequate for the study of many alloy properties such as phase stability and transport. It also provides a poor basis for the use of perturbation theory. Methods for treating charge transfer effects within the formal construct of the CPA have been discussed in the literature [14, 15]. Thus, in determining the potential and the total energy, all intercell interactions are written in terms of averages, whereas intracell interactions are calculated in terms of cell quantities. Thus, in the KKR-CPA Eq. (12.96) takes the form,

$$\begin{aligned} E^{\alpha}_{\mathrm{CPA}} &= \int_{-\infty}^{E_{\mathrm{F}}} E n^{\alpha}(E)\mathrm{d}E - \frac{1}{2}\int_{\Omega^{\alpha}} \mathrm{d}^3 r n^{\alpha}(\mathbf{r}) \\ &\times \left[\int_{\Omega^{\alpha}} \frac{n^{\alpha}(\mathbf{r}')}{|\mathbf{r}-\mathbf{r}'|}\mathrm{d}^3 r' + \int_{R^3-\Omega^{\alpha}} \frac{\bar{n}(\mathbf{r}')}{|\mathbf{r}-\mathbf{r}'|}\mathrm{d}^3 r' \right. \\ &- \left. \frac{Z_{\alpha}}{r} - \sum_{\mathbf{R}_n \neq \mathbf{R}_{\alpha}}{}' \frac{\bar{Z}}{|\mathbf{r}-\mathbf{R}_n|}\right] \\ &- \frac{1}{2}Z_{\alpha}\left[\int_{\Omega^{\alpha}} \frac{n^{\alpha}(\mathbf{r})}{r}\mathrm{d}^3 r + \int_{R^3-\Omega^{\alpha}} \frac{\bar{n}(\mathbf{r})}{r}\mathrm{d}^3 r \right. \\ &- \left. \sum_{\mathbf{R}_n \neq \mathbf{R}_{\alpha}} \frac{\bar{Z}}{|\mathbf{R}_n - \mathbf{R}_{\alpha}|}\right] + E^{\alpha}_{\mathrm{xc}}\left[n^{\alpha}(\mathbf{r})\right], \end{aligned} \tag{12.97}$$

where $n^{\alpha}(\mathbf{r})$ is the self-consistently determined charge within the KKR-CPA, and where we have chosen the origin at the center of the cell α, so that $\mathbf{R}_{\alpha} = \mathbf{0}$. This expression gives the total, averaged ground-state energy of a substitutionally disordered alloy within the KKR-CPA,

$$E^{\mathrm{tot}}_{\mathrm{CPA}} = \sum_{\alpha} c_{\alpha} E^{\alpha}_{\mathrm{CPA}}. \tag{12.98}$$

Using the variational property of the CPA discussed in Appendix M, one can show [21] that $E^{\mathrm{tot}}_{\mathrm{CPA}}$ is stationary with respect to small variations, δn, in the electron charge density. It follows that the change in the energy resulting from such changes is of second order, $(\delta n)^2$. The explicit forms taken by the expressions derived above for the potential and the energy in the case of MT potentials has been given by Johnson, et al. [21].

12.9.3 Application to Cu-Zn alloys

The SCF-KKR-CPA in connection with MT potentials has been used successfully to study [22] the electronic properties of a large number of transition metal alloys. Here, we give a brief illustration of the results obtained in such studies in the case of Cu-Zn alloys [21].

The study of the Cu-Zn system quoted here was based on an fcc lattice[11]. The $Cu_{1-c}Zn_c$ alloys form stable fcc structures for values of $c \leq 0.38$, and some metastable fcc alloys also exist up to $c = 0.55$. The equilibrium volume and charge density of an alloy can be obtained at each concentration by means of an energy minimization procedure analogous to that used in the case of pure elemental solids or ordered compounds. The equilibrium lattice parameter[12] $a_{\rm E}$ of the Cu-Zn alloy system as a function of concentration is shown in Fig. 12.11. In the calculations reported here, the maximum value of the angular momentum used in the MST expansions of the t-matrices and the structure constants, (see Chapter 3), was set equal to 2. The lattice parameter determined as the value corresponding to vanishing pressure is $a_{\rm p} = 6.933$ a. u., and is about 0.5% smaller than $a_{\rm E}$. Such discrepancies are common in electronic structure calculations, with the pressure leading to equilibrium lattice parameters about 0.5% smaller than the corresponding quantities obtained through energy minimization.

The most important feature revealed by this figure is the good (although not perfect) agreement with Vegard's law, which predicts a linear variation of lattice parameter with concentration and is verified by the experimental results. At the same time, the calculated equilibrium lattice constants are about 2% larger than the experimental ones.

More details of the application of the KKR-CPA to Cu-Zn alloys can be found in reference [21]. This alloy is considered further in the next chapter, where we study its ordering tendencies and phase diagram.

12.10 Screened KKR

As has probably become apparent, applications of MST rely heavily on the construction, storage, and implementation of the structure constants of the underlying lattice. Although these structure constants can be constructed once and for all for a particular lattice, their integration over the Brillouin zone of the reciprocal lattice of a given Bravais lattice is never free of difficulties. These difficulties are accentuated when **k**-space-dependent properties must be calculated such as the electronic conductivity, or the shape of the Fermi surface. For example, the treatment of the free-electron poles that manifest themselves in the Fourier transform of the structure

[11] Note that Zn crystallizes in the hcp structure.

[12] In these calculations, the minimum of the energy occurs at the value $a_{\rm E} = 6.963$ a.u., and lies within 0.5% of the experimental value. Although very good, this agreement is somewhat fortuitous because measurements are usually carried out at finite temperatures, whereas in the present study equilibrium volumes are by and large calculated at zero temperature.

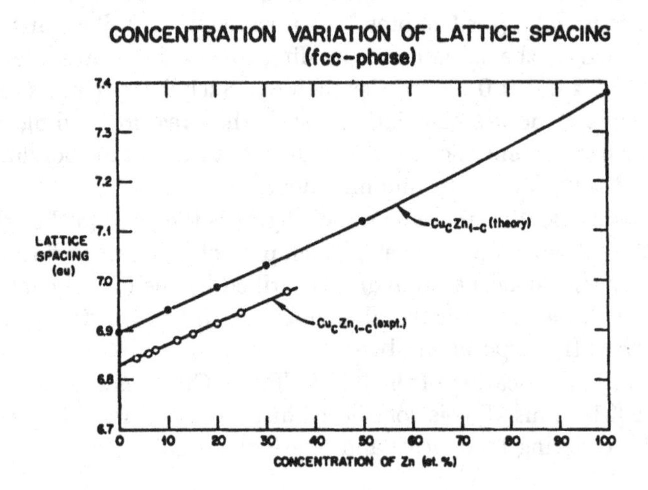

Figure 12.11: **The variation of the equilibrium lattice constant with Zn concentration for chemically random, fcc-based Cu-Zn alloys, determined through total-energy minimization [21].**

constants can be particularly cumbersome.

Therefore, it is sensible to search for the development of MST within a framework that would keep intact its advantageous properties while also reducing the difficulties associated with the infinite extent of the structure constants of the lattice. A reduction in the spatial extent of the structure constants would remove many of the difficulties just mentioned while also increasing the conceptual attraction of the theory. Such a framework has been proposed [24, 25, 26] that leads to the construction of *screened* structure constants whose extent need not reach beyond nearest neighbors in a lattice. In this regard, they are similar to the screened structure constants obtained through transformations derived within the linear muffin-tin orbitals (LMTO) method by Andersen and Jepsen [27]. The resulting formalism was further advanced by a number of investigators [28-33] and has been successfully applied not only to bulk materials but to surfaces as well. One important difference in the two sets of structure constants, those of the LMTO and the KKR methods, is pointed out below.

The reduction in spatial extent of the structure constants is accomplished through an application of MST to the treatment of impurities embedded in a translationally invariant host material, as described in Section 9.4 and in Section 12.3.1. Onebegins by calculating the SPO for a "material" that consists of a constant MT positive potential at every lattice site. The basic idea is to consider an impurity cluster consisting of real potential cells embedded in this medium and extending over the *entire* solid. Now, medium propagation takes place through the scattering path operator which, effectively describing propagation at negative energies, decays rapidly with distance. These rapidly decaying medium propagators play the role of the structure constants in the propagation of an electron wave through the medium that is being scattered by the difference in potential between the host and any impurities embedded in it.

Derivation of screened structure constants

The development of MST presented thus far is based on the treatment of the scattering of waves by potentials with free-particle propagation between scattering events. This free-particle propagation is described in reciprocal space by the KKR structure constants, and by their Fourier transforms in real space. These structure constants connect scattering potentials that can be arbitrarily far apart and thus are of infinite extent. This extended spatial dependence of the structure constants can result in slow rates of convergence of the MST expressions.

A reduction in the range of the structure constants can be achieved by taking advantage of the decaying properties of waves propagating at negative energies. The amplitudes of such waves is known to decrease exponentially with distance, and MST can be formulated in away that takes full advantage of this behavior. To see how this can be done, we recall the equation of motion of the scattering-path operator, Eq. (12.5),

$$\underline{\tau}^{ij} = \underline{t}^i \left[\delta_{ij} + \sum_k \underline{G}^{ik} \underline{\tau}^{kj} \right], \tag{12.99}$$

where $\underline{G}^{ik}$ is the element of the structure-constant matrix that denotes free-electron propagation from site k to site i in the lattice. We also recall the form of the SPO, Eq. (12.85), describing electron scattering when a cluster of impurity sites is embedded in an otherwise uniform medium characterized by an effective SPO denoted by $\bar{\underline{\tau}}^{ij}$,

$$\underline{X}^{ij} = \underline{x}^i \left[\delta_{ij} + \sum_k \bar{\underline{\tau}}^{ik} \underline{X}^{kj} \right]. \tag{12.100}$$

Here, $\underline{x}^I = [1 - \bar{\underline{\tau}}^{ii} \Delta \underline{m}^i]^{-1} \bar{\underline{\tau}}^{ii}$ denotes the extra scattering induced when the scattering potential at site i in the host material is replaced by an impurity potential. The quantity $\Delta \underline{m}^I = \underline{m} - \bar{\underline{m}}$ is the difference in the inverse site scattering matrices defined by the impurity potential, $\underline{m}$, and that of the medium, $\bar{\underline{m}}$. It is easily seen that in the last expression the role of the structure constants in Eq. (12.99) is played by the SPO, $\bar{\underline{\tau}}$, of the host medium. We note that the quantity $\underline{X}^{ij}$ has non-vanishing elements only when i and j are sites in the impurity cluster. The SPO of the entire system containing the impurity cluster is given by the expression, Eq. (12.20),

$$\underline{\tau}^{ij} = \bar{\underline{\tau}}^{ij} + \sum_{kl} \bar{\underline{\tau}}^{ik} \; unl X^{kl} \bar{\underline{\tau}}^{lj}. \tag{12.101}$$

We now think of the real system of scatterers as a superposition of two sets of scattering systems: a system that is characterized by an "effective" or "reference" inverse scattering matrix $\bar{\underline{m}}$, and one that is characterized by a cell scattering matrix $\Delta \underline{m}$. Thus, we consider that every cell in the lattice of effective or reference potentials corresponding to $\bar{\underline{m}}$, is replaced by a set of potentials corresponding to $\Delta \underline{m}$. This replacement restores the original system, but the freedom in choosing the reference potentials can be used to advantage in view of the last two expressions above, Eqs. (EqX2) and (12.101). The reader is urged to verify that the SPO of the system described as such a superposition is also given by the Fourier transform,

$$\underline{\tau}^{ij} = \frac{1}{\Omega_{\mathrm{BZ}}} \int \mathrm{d}\mathbf{k} \bar{\underline{\tau}}(\mathbf{k}) \left[1 - \Delta \underline{m} \right]^{-1}. \tag{12.102}$$

On possible choice for the reference system is that defined [87, 88] by a repulsive constant cell potential, say one of the muffin-tin form. In this

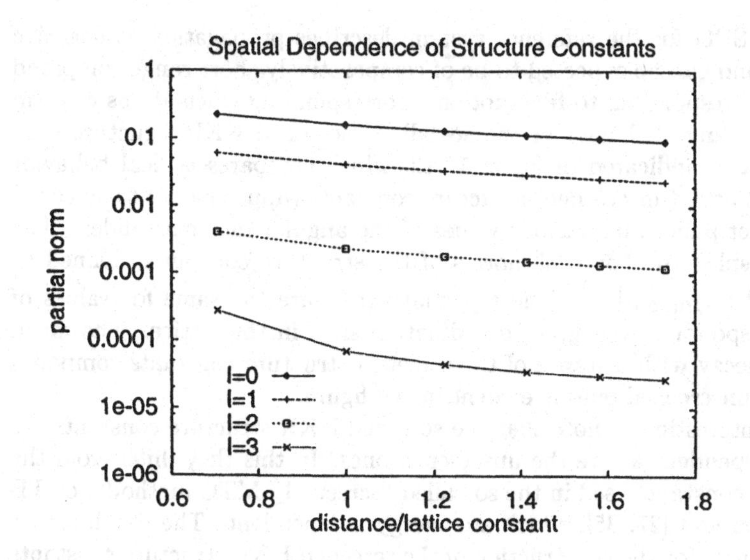

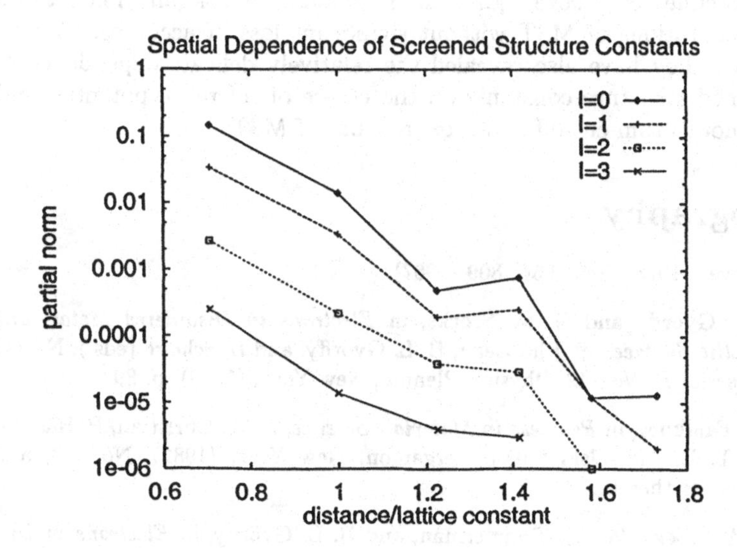

Figure 12.12: **A comparison of "screened structure constants" and their free-space counterparts [34].**

case, the SPO for the reference system describes propagation at negative energies and can be expected to be of comparatively short range compared to that corresponding to free motion. Corresponding calculations confirm this short-range behavior of the so-called "screened KKR structure constants", as is indicated in figure 12.12, which compares typical behavior of the ordinary (unscreened) structure constants (upper panel) to screened ones (lower panel), for various values of the angular-momentu index. The figures display the "partial" norms of the structure constants, defined as $\left[\sum_{mm'} \bar{\mathcal{I}}^{ij}(E)_{\ell m,\ell m'}\right]^{1/2}$. These partial norms are the same for values of R_{ij} corresponding to a given coordination shell in the lattice. The much sharper decay with distance of the screened structure constants compared with the unscreened ones is evident in the figure.

It is interesting to note that the screened KKR structure constants are energy dependent as are the unscreened ones. In this they differ from the structure constants used in the so-called "screened LMTO method", or TB LMTO method [27, 35], which are energy independent. The development of procedures for the construction of the screened KKR structure constants is currently under intensive consideration. As these efforts indicate, it is indeed possible to achieve significant reductions in computational times in implementations of MST without significant loss of accuracy. At the same time, they have also revealed the relatively delicate dependence of the screened structure constants on the choice of reference potential and angular-momentum cut-off in the expressions of MST.

Bibliography

[1] P. Soven, Phys. Rev. **156**, 809 (1967).

[2] B. L. Györffy and G. M. Stocks, in *Electrons in Disordered Metals and Metallic Surfaces*, P. Phariseau, B. L. Györffy, and L. Scheire (eds.), NATO ASI series B, Vol. 42, Physics, Plenum, New York, (1979), p. 89.

[3] J. S. Faulkner, in *Progress in Materials Science*, J. W. Christian, P. Haasen, and T. B. Massalski (eds.), Pergamon, New York, (1982), No. 1-2, and references therein.

[4] G. M. Stocks, W. M. Temmerman, and B. L. Györffy, in *Electrons in Disordered Metals and Metallic Surfaces*, P. Phariseau, B. L. Györffy, and L. Scheire (eds.), NATO ASI series B, Vol. 42, Physics, Plenum, New York, (1979), p. 193.

[5] A. Gonis, *Green Functions for Ordered and Disordered Systems*, North Holland, Amsterdam, (1992).

[6] A. Gonis, G. M. Stocks, W. H. Butler, and H. Winter, Phys. Rev. B**29**, 555 (1984).

[7] W. H. Butler, Phys. Rev. B**14**, 468 (1976).

[8] F. J. Pinski and G. M. Stocks, Phys. Rev. B**32**, 4204 (1985).

[9] W. M. Temmerman and Z. Szotek, Comp. Phys. Rep. **5**, 175 (1987).

[10] B. L. Györffy and G. M. Stocks, J. Phys. (Paris) **35**, C4-75 (1974).

[11] J. S. Faulkner and G. M. Stocks, Phys. Rev. B**21**, 3222 (1980).

[12] A. Bansil, in *Electrons in Disordered Metals and Metallic Surfaces*, P. Phariseau, B. L. Györffy, and L. Scheire (eds.), NATO ASI series B, Vol. 42, Physics, Plenum, New York, (s1979), p. 223.

[13] J. A. Blackman, D. M. Esterling, and N. F. Berk, Phys. Rev. B**4**, 2412 (1971).

[14] A. Gonis, P. E. A. Turchi, J. Kudrnovsky, V. Drchal, and I. Turek, J. Phys.: Condens. Matter **8**, 7869 (1996).

[15] A. Gonis, P. E. A. Turchi, J. Kudrnovsky, V. Drchal, and I. Turek, J. Phys.: Condens. Matter **8**, 7883 (1996).

[16] Prabhakar P. Singh and A. Gonis, Phys. Rev. B**48**, 2139 (1993).

[17] V. L. Moruzzi, J. F. Janak, and A. R. Williams, *Calculated Electronic Properties of Metals*, Pergamon Press, Inc., New York, (1978).

[18] H. Winter and G. M. Stocks, Phys. Rev. B**27**, 882 (1983).

[19] G. M. Stocks and H. Winter, Z. Phys. B**46**, 864 (1982).

[20] H. Akai, J. Phys. Cond. Matt. **1**, 8045 (1989).

[21] D. D. Johnson, D. M. Nicholson, F. J. Pinski, B. L. Györffy, and G. M. Stocks, Phys. Rev. B**41**, 9701 (1990).

[22] See Alloy section in the proceedings of the symposium on *Applications of Multiple Scattering Theory to Materials Science*, edited by W. H. Butler, P. H. Dederichs, A. Gonis, and R. L. Weaver, Materials Research Society Vol. **253**, (1992).

[23] A. Gonis, A. J. Freeman, and P. Weinberger, Phys. Rev. B**32**, 7720 (1985).

[24] B. Ujfalussy, L. Szunyogh, P. Weinberger, and J. Kollar, in *Metallic Alloys: Experimental and Theoretical Perspectives*, J. S. Faulkner and R. G. Jordan (eds.), Proceedings of NATO Advanced Research Workshop, Boca Raton, FL. July 16-21, 1994, Kluwer Academic Publishers, Dordrecht, (1994), Vol. 256, p. 301.

[25] L. Szunyogh, B. Ujfalussy, P. Weinberger, and J. Kollar, Phys. Rev. B**49**, 2721 (1994) .

[26] P. J. Braspenning and A. Lodder, Phys. Rev. B**49**, 10222 (1994) .

[27] O. K. Andersen and O. Jepsen, Phys. Rev. Lett. **53**, 2571 (1984)

[28] W. R. L. Lambrecht and O. K. Andersen, Surf. Sci. **178**, 256 (1986).

[29] B. Wenzien, J. Kudrnovsky, V. Drchal, and M. Sob, J. Phys.: Condens. Matter **1**, 9893 (1989).

[30] J. Kudrnovsky, P. Weinberger, and V. Drchal, Phys. Rev. B**44**, 6410 (1991).

[31] V. Drchal, J. Kudrnovsky, L. Udvardi, P. Weinberger, and A. Pasturel, Phys. Rev. B**45**, 1432 (1992).

[32] J. Kudrnovsky, I. Turek, V. Drchal, P. Weinberger, S. K. Bose, and A. Pasturel, Phys. Rev. B**47**, 16525 (1993).

[33] H. L. Skriver and N. M. Rosengaard, Phys. Rev. B**43**, 9538 (1991).

[34] N. Moghadam and G. M. Stocks, private communication.

[35] I. Turek, V. Drchal, J. Kudrnovský, MŠob, and P. Weinberger, *Electronic Structure of Disordered Alloys, Surfaces, and Interfaces*, Kluwer Academic, Boston (1997), and references therein.

Chapter 13

Alloy Phase Stability

13.1 General Comments

This chapter is concerned with the role of electronic structure calculations in the study of the phase stability and ordering tendencies of substitutional alloys. These properties of disordered materials lie squarely within the realm of studies concerned with the ground-state characteristics of extended systems. For example, consider the question of whether or not two given elements form a stable substitutional alloy at given ranges of concentration, temperature, and pressure. That is to say, is the energy (at finite temperature, the free energy) of the alloy lower when compared to the (average of the) energies of the two elemental solids? More specifically, what ordered structures of the alloy would be stable as the temperature approaches zero? How do these structures evolve as the temperature rises? And how can one understand and predict alloy phase characteristics on the basis of physically meaningful parameters[1]. In this chapter we explore the extent to which questions such as these can be addressed by means of electronic structure calculations.

Being able to provide answers, even approximate ones, through the performance of self-consistent electronic structure calculations can be very helpful in the study of materials properties. Such a process may provide a predictive understanding of the physical and mechanical properties of alloy systems and aid in the design of alloys and compounds with technologically desirable characteristics. It may even lead to the identification of precur-

[1]It is possible to go a long way, even construct alloy phase diagrams, in terms of parameters that have no clear physical meaning. However, such parameters often obscure the underlying physical mechanisms driving alloy phase stability, and in general do not shed light on the phenomena of phase formation in materials. A computational scheme, however accurate, is of no use to the alloy engineer unless it leads to a clear understanding of physical principles.

sor phenomena associated with failure in a mechanical or chemical context. One task facing the materials scientist is the development of concepts and computational algorithms that allow this study to take place.

The alloy designer must be able to address relatively accurately most of the questions posed above. The attempts to answer such questions have been based for the most part on the establishment of criteria that would allow one to predict which ordered state, among many competing ones, would actually occur under given external conditions. These criteria have often been based on phenomenological models, the great number [1-13] of which reflect the richness of the phenomena under study. These models vary from fairly simple ones [1-3] based on the relative sizes of the atoms (or ions) forming an alloy, to electronegativity (an atom's affinity for electrons) [4], to covalency and ionicity [6, 7], to structure maps [7-13] based on coordinates made up of combinations of atomic numbers.

However, in spite of many successful applications of phenomenological models, such as structure maps, these models suffer from a number of shortcomings. Chief among these is the lack of clarity of the physical reasons responsible for the phenomena under study. It is certainly useful to be able to predict with relative accuracy that a given binary system will form a specific ordered configuration at low temperatures, but one must also be able to understand why it does so. It may then be possible to understand more complex systems than the ones for which the model was originally constructed. Many such complex systems defy analysis to this date[2]. Furthermore, the coordinate space in which the structure maps are constructed contains no natural metric [14] so that two binary compounds may form in the same ordered structure but for widely different reasons. Finally, the variables entering the construction of structural maps, such as concentration and electronegativity, are not necessarily independent so that the effects on stability of each of them individually cannot be assessed.

The shortcomings outlined above are characteristic of most phenomenological models proposed for the study of alloy phase stability. Therefore, one hopes that these models can be either justified (or even replaced) by theories based on first-principles[3]. Here, the attempt would be to derive rather than conjecture a set of parameters or a computational procedure for the study of the phase boundaries of an alloy as functions of concen-

[2]Certain ordered configurations contain many hundreds of atoms per unit cell, a remarkable phenomenon that remains largely a mystery.

[3]The term first-principles has no unambiguous, universally accepted definition. Rather loosely, first-principles theories in the present context are based on a quantum-mechanical treatment of the physical Hamiltonian (not a parametrized model Hamiltonian) of an interacting many-particle system, and both the Hamiltonian and the treatment are free of parametrization or fitting procedures. At the same time, any treatment of an interacting many-particle system is bound to contain approximation schemes, so that a first-principles treatment is not necessarily an *exact* treatment.

tration, temperature, and pressure. These parameters and the associated computational process could possibly shed light on the underlying physical basis governing alloy stability. This, in turn, might guide attempts to understand fairly complex systems for which even a phenomenological study may be too complicated to carry out.

But, what are these parameters, and how does one determine them from first-principles? A particularly fruitful approach has been to extract them from self-consistent electronic structure calculations. In this chapter, we review a number of techniques that allow the determination of such sets of parameters. This subject has also been discussed at great length by Ducastelle [14], and the reader is encouraged to consult that work for more details.

13.2 Alloy Hamiltonians and the Ising Model

Complex physical phenomena are often understood in terms of models built upon one's understanding of processes simpler than the ones under consideration. Usually, a connection is made on the basis of geometric (spatial) features that are common to both sets of phenomena, even though the physics underlying them may be quite different in the two cases. Nevertheless, the similarity in features may lead to a formulation or a way of attack on a particular problem. In the case of alloys, a connection with the formal framework surrounding the Ising [15] model, i.e., equilibrium thermodynamics, has proved to be of great utility.

13.2.1 Ising Model

The Ising model [15] was introduced in an attempt to treat the ordering characteristics of magnetic systems, and has been discussed extensively in a number of textbooks [16, 17]. The model is built upon the physical picture of spin-like variables, σ_n, (interpreted as the z-component of a magnetic moment) associated with sites n in a lattice of N sites, which can be aligned either parallel (up) or antiparallel (down) to a given direction, usually taken as the positive direction of the z-axis of a Cartesian coordinate system. Thus, the spin variables are assigned the values,

$$\sigma_n = \begin{cases} 1, & \text{if the spin is up} \\ -1, & \text{if the spin is down.} \end{cases} \tag{13.1}$$

Clearly, we have

$$\sigma_n^2 = 1. \tag{13.2}$$

Within the original model, the energy of a system of spins in a configuration J, in which some of the spins point up and some down, is taken to be of the pairwise form,

$$H = \frac{1}{2}\sum_{ij} J_{ij}\sigma_i\sigma_j, \tag{13.3}$$

where the interactions or "exchange couplings" are taken to be negative, $J_{ij} < 0$, and thus favoring a ferromagnetic arrangement. (The energy of a pair of spins is lower when both point in the same direction than when they point in opposite directions.) In the usual applications of the Ising model, the coupling constants are assumed to be independent of configuration and concentration, i.e., of the fraction of spins pointing up (or down).

The expression for the Ising Hamiltonian, Eq. (13.3), leads to the energy, $E(\mathrm{J})$, of any configuration, J, of spins. Assuming that the system is in equilibrium, each configuration is assigned the usual statistical weight, $\exp[-E(\mathrm{J})/k_{\mathrm{B}}T]$, where k_{B} is the Boltzmann constant and T is the absolute temperature. Given these energies, one must decide in which configuration the system will tend to settle as the temperature is lowered toward zero. More generally, what structures are preferred as a function of temperature and other external constraints, such as pressure? The thermodynamics of the Ising model, which is concerned with questions of this type, constitutes a rich and elegant part of mathematical physics [14, 16, 17].

Before proceeding further, it is useful to note a number of relations satisfied by some basic parameters of the Ising model. Denoting the number of spins pointing up (down) by the symbol N_+ (N_-) we have

$$N_+ + N_- = N. \tag{13.4}$$

Also, the concentration of up or down spins is defined by the ratios,

$$c_\pm = \frac{N_\pm}{N}, \tag{13.5}$$

and obviously,

$$c_+ + c_- = 1. \tag{13.6}$$

We can view these expressions as defining a system in the *thermodynamic limit*, in which N_+, N_-, and N approach infinity but do so in a way which preserves the relations in Eqs. (13.5) and (13.6).

Let us now consider a system with *fixed* concentration. The set of all possible configurations consistent with a given value of c_+ constitutes a *canonical ensemble.* Within a canonical (C) ensemble, the free energy of a configuration is given by the expression [16, 17],

$$F(T, c_+) = -k_{\mathrm{B}}T \ln Z, \tag{13.7}$$

where the *partition function*, Z, is given by

$$Z = \sum_{\mathrm{J}} \exp\left[-E(\mathrm{J})/k_{\mathrm{B}}T\right]. \tag{13.8}$$

We note that summing over all configurations is equivalent to summing over all possible values of all spin variables, consistent with the concentration

$$\sum_{\mathrm{J}} \Leftrightarrow \sum_{\{\sigma_n\}}. \tag{13.9}$$

It is convenient to denote such sums over spin variables by the symbol Tr (for trace), in a notation that can be readily applied to quantum mechanical systems. Thus, Eq. (13.8) can be written in the form,

$$Z = \mathrm{Tr}\exp\left[-E(\mathrm{J})/k_{\mathrm{B}}T\right]. \tag{13.10}$$

We now see, however, that the evaluation of the trace within a canonical ensemble may cause severe counting problems since the constraint of fixed concentration induces interdependencies in the values of the spin variables[4]. Therefore, it is convenient to introduce a formalism that allows the trace to be obtained as a sum over independent spin variables.

For either finite or infinite systems, such a process is allowed within a *grand-canonical ensemble*, consisting of all possible configurations at all concentrations. However, we would still like to keep track of concentration, and the common way of doing this is to introduce a *Lagrange multiplier*, μ_α, and write the Hamiltonian in the form,

$$\tilde{H}(\{\sigma_n\}) = H(\{\sigma_n\}) - \sum_{\alpha=\pm} \mu^\alpha \sum_n \sigma_n^\alpha. \tag{13.11}$$

The *chemical potentials*, μ^α, differentiate between systems of different concentrations by assigning a concentration dependent part to the Hamiltonian. In the grand-canonical (GC) ensemble, the independent thermodynamic variables are the temperature T and the chemical potentials, μ^α. The *grand thermodynamic potential* or grand potential, $\tilde{F}$, is a generalizati on of the free-energy, F, and is given by the expression

$$\tilde{F}(T, \{\sigma_n\}) = -k_{\mathrm{B}}T\ln\tilde{Z}, \tag{13.12}$$

where

$$\tilde{Z} = \mathrm{Tr}\exp\left[-\tilde{H}(\{\sigma_n\})/k_{\mathrm{B}}T\right]. \tag{13.13}$$

In the thermodynamic limit, we have the relation

$$\tilde{F} = F - \sum_\alpha \mu^\alpha N_\alpha. \tag{13.14}$$

---

[4]Consider a system of three sites, with a fixed concentration $c_+ = 1/3$. If one spin is known to be pointing up, then the other two spins *must* be pointing down, and the spin variables associated with different sites are no longer independent.

Thus, the chemical potentials are connected with particle conservation according to the relations,

$$\begin{aligned} N_\alpha &= -\frac{\partial \tilde{F}}{\partial \mu^\alpha} \\ \mu^\alpha &= -\frac{\partial \tilde{F}}{\partial N_\alpha}. \end{aligned} \tag{13.15}$$

[See Eq. (1.72) describing the Fermi energy as a chemical potential.] It follows that the chemical potential μ^α is the energy difference that results when the number of particles of type α is changed by ± 1.

Here, it may be appropriate to raise a point that will play an important role in the following discussion. Namely, the presence of the chemical potential is an inherent feature of a GC averaging process. In fact, it is this presence that keeps track of concentration and guarantees that a treatment within a GC ensemble will give the same results as a treatment within a C ensemble for the concentration corresponding to the chemical potential. For example, when $\sum_n \sigma_n$ vanishes, and hence the term involving the chemical potential drops out, the corresponding GC average is entirely equivalent to an average for which $\sum_n \sigma_n = 0$ or, correspondingly, $c_+ = 0.5$.

As is well known [16, 17], the average of a quantity Q, denoted by $\langle Q \rangle$, within a C or a GC ensemble, is given in terms of the partition function. Within a canonical ensemble, we have

$$\langle Q \rangle = \frac{1}{Z} \sum_{\mathrm{J}} Q(\mathrm{J}) \exp\left[-H(\{\sigma_n\})/k_{\mathrm{B}}T\right] = \mathrm{Tr}\rho Q, \tag{13.16}$$

where

$$\rho(\{\sigma_n\}) = \frac{1}{Z} \exp\left[-H(\{\sigma_n\})/k_{\mathrm{B}}T\right] \tag{13.17}$$

is referred to as the *density matrix*. Corresponding averages within the grand-canonical ensemble can be defined with the quantities $\tilde{H}$, $\tilde{Z}$, and $\tilde{\rho}$ replacing H, Z, and ρ, respectively, in the expressions above. The thermodynamic behavior of the classical Ising model has been discussed extensively in a number of textbooks [16, 17], and the reader is referred there for further details.

13.2.2 Connection with alloy Hamiltonians

There is an obvious and compelling analogy between the classical Ising model and a binary substitutionally disordered alloy[5]: clearly, a given configuration J of up and down spins corresponds to a configuration of a binary

[5] Extensions to multicomponent systems are straightforward, and references to such systems will be made freely in the discussion.

alloy. Let us quantify this analogy. For a given binary alloy, A_cB_{1-c}, we introduce *occupation number* functions $p_n^\alpha(\text{J})$, $\alpha = A$ or B, defined over alloy configurations so that

$$p_n^\alpha(\text{J}) = \begin{cases} 1, & \text{if site n is occupied by an atom of type } A \\ 0, & \text{otherwise.} \end{cases} \tag{13.18}$$

The values of these functions are denoted simply by p_n^α, and it is customary to suppress the distinction between these values and the corresponding functions. By definition, the *occupation numbers*, p_n^α, satisfy the relation

$$\sum_\alpha p_n^\alpha = 1, \tag{13.19}$$

since we assume that every site in the alloy is occupied by an atom of some kind. It follows that for an s-components alloy there are $(s-1)$ independent occupation numbers associated with each site. For a binary, $A-B$, alloy, only one such number exists. Choosing p_n^A as the independent variable, we have

$$p_n^A = p_n, \quad p_n^B = (1-p_n). \tag{13.20}$$

In addition to the sum rule in Eq. (13.19), the occupation numbers obey the relations

$$p_n^\alpha p_n^\beta = p_n^\alpha \delta_{\alpha\beta}, \tag{13.21}$$

which simply says that a site cannot be occupied by more than one atom. In the case of binary alloys, we have

$$p_n^2 = p_n. \tag{13.22}$$

The concentration, c^α, of a given species, α, in an alloy is given by the obvious relation,

$$N_\alpha = \sum_n p_n^\alpha = Nc^\alpha, \quad \sum_\alpha c^\alpha = 1, \tag{13.23}$$

where N is the total number of sites in the lattice.

Clearly, the occupation numbers p_n^α could also be used to describe the Ising model or, alternatively, the spin variables, σ_n, could be used in the description of an alloy. Indeed, for a binary alloy, the two sets of variables are related,

$$p_n = \frac{1}{2}(1+\sigma_n). \tag{13.24}$$

This establishes a unique, one-to-one correspondence between the configurations of a magnetic Ising system and of a substitutionally disordered binary alloy[6]. The question, now, arises as to whether or not this analogy can be carried further and be used in the study of alloy stability.

The answer to this question is by no means obvious. It is certainly true that the Ising model and an alloy system can be characterized by configurational energies, and both exhibit transformations between various configurations under varying external conditions. Furthermore, thermodynamics tells us that phase transformations can be studied in terms of the partition function, Eq. (13.10), which involves a sum over such configurations. However, the physical mechanisms that underlie the two systems are vastly different. In the Ising model, a spin can reverse direction at the cost of an energy of the order of the coupling constant. Thus, the macroscopic properties of the model, e.g., its phase diagram, are connected directly to its microscopic properties. The situation is considerably different in an alloy. In an alloy, an atom of a given species cannot change into another (at least not through the expenditure of energy differences between different configurations). Thus, one must think of a physical alloy system as being confined to a single concentration. Furthermore, even at a given concentration two atoms may not be able to exchange positions, as may be dictated by purely configurational energetics, because of the presence of energy barriers that require a certain amount of *activation* energy in order for exchange to occur. In fact, atomic rearrangements in alloys happen only because of the presence of vacancies, voids, dislocations, and other imperfections that make motion possible. Thus, the information as to which alloy configuration has the lowest energy may be largely irrelevant to the study of phase stability, unless it comes coupled with information that the *kinetics* characterizing the alloy would allow transformations to take place. Kinetics may indeed prevent motion at an atomic scale and thus nullify predictions based on thermodynamic arguments[7].

Information on kinetics is very difficult to obtain, certainly within a first-principles electronic structure calculation. Because of such difficulties, the mechanisms driving alloy phase stability cannot be identified solely from the knowledge of configurational energies. In addition, a number of effects that are known to be important, e.g., lattice vibrations at elevated temperatures, are ordinarily not taken into account within electronic structure methods for the study of alloy phase stability. Finally, even if all these considerations were ignored, one still faces an important distinction between a magnetic Ising model system and a substitutional alloy. Namely, the energies characterizing the Ising model were chosen to be independent

[6]Multicomponent alloys can be associated with more general spin systems as is indicated often in the discussion.

[7]At the same time, it is equally untrue that knowledge of kinetics alone will allow determination of thermodynamic equilibrium properties. Thermodynamic behavior is an intricate function of the configurational and kinetics features of a system and its study should account for both of them.

of concentration, whereas the way two atoms interact does depend in a profound way on the concentration. It is this dependence, and other features, that proclaim a substitutional alloy as being a quantum mechanical rather than a classical system. Thus, the study of alloy phase stability carried out within equilibrium thermodynamics, the formal framework of the Ising model, is by no means a trivial, or even straightforward, undertaking.

Present-day, first-principles methods for the calculation of alloy phase stability and phase diagrams often contain the implicit assumption that kinetics would allow a phase transformation to occur in accordance with thermodynamic arguments[8]. Under these assumptions, it is customary to attempt a study of alloy phase stability based on the determination of the alloy partition function,

$$Z = \mathrm{Tr} \exp\left[-\beta E(\mathrm{J})\right], \tag{13.25}$$

where $\beta = 1/k_{\mathrm{B}}T$. In this chapter, we examine the most prominent developments, founded on physical and mathematical principles, that have been made within this approach to the study of alloy phase stability. Keeping in mind that the Ising model and alloy systems represent very different pieces of physical reality, it is encouraging that thermodynamic arguments often lead to successful predictions of alloy behavior. At the same time, we should not regard such instances as proofs of validity, but rather as motivation to continue the search for an even deeper understanding of alloy properties.

13.3 Correlation Functions

In this section, we establish a number of fundamental relations concerning averages of products of occupation numbers. In our discussion, we follow the development in Ducastelle [14], where the reader is referred for more details.

An n-point *correlation function* is defined by the general expression

$$\xi^{\alpha_1\alpha_2\cdots\alpha_n}_{i_1 i_2\cdots i_n} = \langle p^{\alpha_1}_{i_1} p^{\alpha_2}_{i_2} \cdots p^{\alpha_n}_{i_n} \rangle, \tag{13.26}$$

where the variables α_k range over the species of atoms forming an alloy, and the brackets denote an average over either a grand canonical (GC) or a canonical (C) ensemble. Therefore, we distinguish between GC and C correlation functions. Once again we draw attention to the fact that in the thermodynamic limit, a GC and a C average ultimately yield the same results, because the chemical potential in a GC average keeps track of the

[8] Care should be taken, however, not to identify mechanisms by which these transformations take place based solely on thermodynamic information.

concentration[9].

The simplest correlation function is that associated with a single site and is denoted by c_n^α,

$$c_n^\alpha = \langle p_n^\alpha \rangle = \text{Tr}\rho p_n^\alpha. \tag{13.27}$$

Similarly, the pair correlation function is given by the expression

$$\xi_{nm}^{\alpha\beta} = \langle p_n^\alpha p_m^\beta \rangle = \text{Tr}\rho p_n^\alpha p_m^\beta. \tag{13.28}$$

The last expressions on the right of the equals sign of these equations follow from Eq. (13.16), and are valid, with the appropriate choice of the density matrix, $\tilde{\rho}$ or ρ, either within a GC or a C ensemble. Within a GC ensemble, it is convenient [14] to introduce site-dependent chemical potentials μ_n^α, and in view of Eqs. (13.15) to write,

$$c_n^\alpha = \text{Tr}\tilde{\rho} p_n^\alpha = \frac{\partial \ln \tilde{Z}}{\partial(\beta\mu_n^\alpha)}. \tag{13.29}$$

We also have,

$$\begin{aligned} \frac{\partial^2 \ln \tilde{Z}}{\partial(\beta\mu_n^\alpha)\partial(\beta\mu_m^\beta)} &= \langle p_n^\alpha p_m^\beta \rangle - \langle p_n^\alpha \rangle \langle p_m^\beta \rangle \\ &= \langle (p_n^\alpha - c_n^\alpha)(p_m^\beta - c_m^\beta) \rangle \\ &= \langle p_n^\alpha p_m^\beta \rangle_c, \end{aligned} \tag{13.30}$$

where c_n^α is the one-point correlation function defined in Eq. (13.27), and the subscript c denotes a *cumulant*. The n-point cumulant expression is given by the general relation,

$$\langle p_{i_1}^{\alpha_1} p_{i_2}^{\alpha_2} \cdots p_{i_n}^{\alpha_n} \rangle_c = \frac{\partial^n \ln \tilde{Z}}{\partial(\beta\mu_{i_1}^{\alpha_1}) \cdots \partial(\beta\mu_{i_n}^{\alpha_n})}, \tag{13.31}$$

and is a linear combination of correlation functions associated with 1,2,...,n, sites. Clearly, the cumulant average of independent variables vanishes identically. From Eqs. (13.29) and (13.30) we can define the *susceptibility function*

$$\chi_{nm}^{\alpha\beta} = \frac{\partial c_n^\alpha}{\partial \mu_m^\beta} = \frac{\partial c_m^\beta}{\partial \mu_n^\alpha} = \langle p_n^\alpha p_m^\beta \rangle / k_B T, \tag{13.32}$$

[9]One should not interpret an average over a GC ensemble in which the chemical potential does not appear as an average over all configurations at all concentrations with equal weight. What in fact takes place in this case is an average over the configurations of the concentration at which $\sum_i \sigma_i = 0$, or for a concentration equal to 0.50. For example, for a binary alloy, a GC average with $\mu = 0.0$ corresponds only to a canonical average at $c = 0.50$.

which measures the concentration fluctuation (change in correlation function) at site n induced by a change in chemical potential at site m (and vice versa). Such susceptibility functions are special cases of the *fluctuation-dissipation* theorem, discussions of which can be found in the literature [18-21].

Now, the pair correlation function and the corresponding cumulants can be used to define the *short-range order* (SRO) parameter for pairs of sites, $\alpha_{ij}^{\alpha\beta}$. This parameter is a measure of the degree to which the pair correlation function deviates from the product of two point correlation functions, and can be defined by the relation,

$$\langle p_n^\alpha p_m^\beta \rangle_c = k_B T \chi_{nm}^{\alpha\beta} = -\langle p_n^\alpha \rangle \langle p_m^\beta \rangle \alpha_{nm}^{\alpha\beta}, \tag{13.33}$$

or

$$\langle p_n^\alpha p_m^\beta \rangle = \langle p_n^\alpha \rangle \langle p_m^\beta \rangle \left(1 - \alpha_{nm}^{\alpha\beta}\right). \tag{13.34}$$

These definitions of the $\alpha_{nm}^{\alpha\beta}$ coincide with the *Warren-Cowley* SRO parameters defined with respect to binary alloys. We note that at high temperatures[10] the statistical variables p_n^α are independent and the cumulant expression along with the SRO parameters, $\alpha_{nm}^{\alpha\beta}$, vanish. These quantities also vanish when the occupation numbers are statistically independent, as is the case in random alloys. In general, the pair correlation functions, $\langle p_n^\alpha p_m^\beta \rangle$, yield the conditional probability that an atom of type β is at site m when it is known that an atom of type α occupies site n, multiplied by c_n^α, the concentration of atoms α at n.

In the absence of *long-range order* (LRO) in the alloy [14], e.g., when $c_n^\alpha = c^\alpha$ is independent of site index, the Warren-Cowley SRO parameters depend only on the distance $\mathbf{R}_n - \mathbf{R}_m$, and also $\alpha_{nm}^{\alpha\beta} = \alpha_{nm}^{\beta\alpha}$. In addition, because $\sum_\alpha p_n^\alpha = 1$, there exist only $s(s-1)/2$ independent SRO parameters in an s-component alloy. This can be easily seen from the relations,

$$\sum_\beta \langle p_n^\alpha p_m^\beta \rangle = \langle p_n^\alpha \rangle = c_n^\alpha, \tag{13.35}$$

and

$$\sum_\alpha c^\alpha \alpha_{nm}^{\alpha\beta} = 0, \text{ or } \sum_\alpha \chi_{nm}^{\alpha\beta} = 0. \tag{13.36}$$

In the case of a binary alloy, $A_c B_{1-c}$, there exists only a single SRO parameter. Choosing that parameter to be $\alpha_{nm} = \alpha_{nm}^{AB}$, we find the identities,

$$\langle p_n p_m \rangle = c^2 + c(1-c)\alpha_{nm}$$

[10] The concept of high temperature is only meaningful in connection with the energy scales defined by the physical system under consideration. Thus, in the Ising model, high temperature implies that $k_B T >> J$. In the case of alloys, such a definition is not as meaningful since kinetics may prevent atoms from exchanging positions even at temperatures much larger than the energy differences that may result from such an exchange.

$$\begin{aligned}\langle p_n(1-p_m)\rangle &= c(1-c)(1-\alpha_{nm}) \\ \langle(1-p_n)(1-p_m)\rangle &= (1-c)^2 + c(1-c)\alpha_{nm}, \end{aligned} \tag{13.37}$$

where $c = c^A$ and $p_n = p_n^A$. Similarly, only a single susceptibility function is needed to describe the alloy

$$\chi_{nm} = \chi_{nm}^{AA} = \chi_{nm}^{BB} = -\chi_{nm}^{AB}. \tag{13.38}$$

Thus, we have,

$$k_B T \chi_{nm} = c(1-c)\alpha_{nm}. \tag{13.39}$$

We note that it is convenient to define $\alpha_{nn} = 1$, so that Eqs. (13.37) and (13.39) are satisfied also when $n = m$. For multicomponent alloys, we have

$$\alpha_{nm}^{\alpha\beta} = 1 - \delta_{\alpha\beta}/c^A. \tag{13.40}$$

We see that $\alpha_{nm} > 0$ (< 0) corresponds to a predominance of like (unlike) neighbors, with the number of AA or BB pairs being higher (lower) than in the completely random system.

Finally, in the case of binary alloys, the correlation functions defined above can also be written in terms of the spin variables σ_n, and the corresponding r-site correlation function

$$x_{\{\sigma_r\}} = \langle \sigma_{n_1}\sigma_{n_2}\cdots\sigma_{n_r}\rangle. \tag{13.41}$$

Using Eq. (13.24), we have

$$x_n = \langle\sigma_n\rangle = 2c_n - 1, \tag{13.42}$$

and

$$x_{\{n,m\}} = \langle\sigma_n\sigma_m\rangle = 4\langle p_n p_m\rangle - 2(c_n + c_m) + 1, \tag{13.43}$$

so that in the absence of LRO, we have

$$\langle\sigma_n\sigma_m\rangle_c = 4\langle p_n p_m\rangle = 4c(1-c)\alpha_{nm} = 4k_B T\chi_{nm}. \tag{13.44}$$

13.4 Concentration Waves

In analyzing experimental data, particularly from diffraction experiments, it is convenient to work in reciprocal space. To this end, we introduce the Fourier transforms,

$$\begin{aligned} p_{\mathbf{k}}^{\alpha} &= \frac{1}{N}\sum_n \mathrm{e}^{\mathrm{i}\mathbf{k}\cdot\mathbf{R}_n} p_n^{\alpha} \\ p_n^{\alpha} &= \sum_{\mathbf{k}} \mathrm{e}^{-\mathrm{i}\mathbf{k}\cdot\mathbf{R}_n} p_{\mathbf{k}}^{\alpha}, \end{aligned} \tag{13.45}$$

where $\mathbf{k}$ is a vector in the first Brillouin zone of the lattice. Because p_n^α is real, we have

$$(p_{\mathbf{k}}^\alpha)^* = p_{-\mathbf{k}}^\alpha, \tag{13.46}$$

where an asterisk denotes complex conjugate. Let us consider a few examples.

In a homogeneous system where $c_n^\alpha = c^\alpha$, the Fourier transform $c_{\mathbf{k}}^\alpha$ vanishes unless $\mathbf{k}$=0 [see Eq. (6.34)]. Thus, only concentration fluctuations, $\delta c_n^\alpha = c_n^\alpha - c^\alpha$, contribute to $c_{\mathbf{k}}^\alpha$ for $\mathbf{k} \neq 0$,

$$\begin{aligned} c_{\mathbf{k}}^\alpha &= c^\alpha \delta_{\mathbf{k}0} + \frac{1}{N}\sum_n \mathrm{e}^{-\mathrm{i}\mathbf{k}\cdot\mathbf{R}_n} \delta c_n^\alpha \\ c_n^\alpha &= c^\alpha + \sum_{\mathbf{k}} \mathrm{e}^{\mathrm{i}\mathbf{k}\cdot\mathbf{R}_n} \delta c_{\mathbf{k}}^\alpha. \end{aligned} \tag{13.47}$$

These equations show that the form of $c_{\mathbf{k}}^\alpha$ is intimately connected with the state of order of a substitutional alloy. A detailed analysis based on these *concentration waves* can be found in reference [14].

We can also define the Fourier transform of the SRO parameters, $\alpha_{nm}^{\alpha\beta}$. In the disordered state, these parameters depend only on the distance between the sites n and m, $\mathbf{R}_n - \mathbf{R}_m$, and we can write

$$\begin{aligned} \alpha^{\alpha\beta}(\mathbf{k}) &= \frac{1}{N}\sum_{nm} \mathrm{e}^{\mathrm{i}\mathbf{k}\cdot(\mathbf{R}_m - \mathbf{R}_n)} \alpha_{nm}^{\alpha\beta} \\ &= \sum_n \mathrm{e}^{\mathrm{i}\mathbf{k}\cdot\mathbf{R}_n} \alpha_{0n}^{\alpha\beta}. \end{aligned} \tag{13.48}$$

The Fourier transform of the correlation function, $\langle p_n^\alpha p_m^\beta \rangle_c$, is now given by the expression

$$\frac{1}{N}\sum_{nm} \mathrm{e}^{\mathrm{i}\mathbf{k}\cdot(\mathbf{R}_n - \mathbf{R}_m)} \langle p_n^\alpha p_m^\beta \rangle_c = -k_{\mathrm{B}} T \chi^{\alpha\beta}(\mathbf{k}) = -c^\alpha c^\beta \alpha^{\alpha\beta}(\mathbf{k}). \tag{13.49}$$

We note that because $p_{-\mathbf{k}}^\alpha = (p_{\mathbf{k}}^\alpha)^*$, the quantity $\chi^{\alpha\alpha}(\mathbf{k})$ is always positive. In the case of binary alloys, there is only one independent correlation function (or susceptibility), and we have

$$k_{\mathrm{B}} T \chi(\mathbf{k}) = c(1-c)\alpha(\mathbf{k}). \tag{13.50}$$

We also note an important sum rule satisfied by $\alpha(\mathbf{k})$. Writing the inverse Fourier transform of Eq. (13.48) in terms of integrals rather than sums over the first Brillouin zone, we have (with $n \equiv 0n$),

$$\alpha_n = \frac{\Omega}{(2\pi)^3} \int_{\mathrm{BZ}} \mathrm{d}^3 k \mathrm{e}^{-\mathrm{i}\mathbf{k}\cdot\mathbf{R}_n} \alpha(\mathbf{k}), \tag{13.51}$$

where Ω denotes the volume of the unit cell in the lattice. Because $\alpha_{nn} = 1$, we have the sum rule

$$\frac{\Omega}{(2\pi)^3} \int_{BZ} \alpha(\mathbf{k}) = 1. \tag{13.52}$$

Further discussion of SRO parameters and their properties can be found in reference [14]. The concept of concentration waves and its relation to alloy phase stability is discussed in Section 13.8.3.

13.5 Effective Cluster Interactions

Having decided to apply classical statistical thermodynamics to the study of alloy phase stability, how do we go about evaluating the partition function? At first glance, this requires knowledge of the energies of all alloy configurations. A direct experimental determination of these energies is certainly out of reach in a practical sense. On the theoretical side, the problem has been studied by phenomenological methods that include a judicious use of experimental information [22].

In attempts to discover the physical mechanisms underlying alloy phase stability, it has been found convenient to begin with the self-consistent determination of the electronic structure of the alloys under consideration. This, of course, leads to $E(\mathrm{J})$, and in principle allows the evaluation of the partition function. However, a complete calculation encompassing all alloy configurations is not feasible. An alternative way of evaluating the trace in Eq. (13.25) would be to convert it into a sum over all possible sets of occupation numbers, in the manner indicated in Eq. (13.9). It is now clear that an implementation of this approach requires the determination of interaction energies for an alloy, analogous to the spin interchange energies that enter the Ising model. However, the alloy Hamiltonian is given by an expression such as the one in Eq. (1.211) which, at first glance, has nothing to do with occupation numbers.

Furthermore, the original Ising Hamiltonian in Eq. (13.3) involves pairwise interactions that are independent of concentration. It has been found through numerical applications that neither of these assumptions is necessarily true in the case of alloy systems. In fact, if an Ising-like Hamiltonian is to represent the energy of an alloy, then it must be of the form,

$$H = \frac{1}{2}\sum_{ij} V_{ij}\sigma_i\sigma_j + \frac{1}{3}\sum_{ijk} V_{ijk}\sigma_i\sigma_j\sigma_k + \cdots, \tag{13.53}$$

involving interactions between pairs, triplets, and higher-order clusters of sites. The coefficients $V_{i_1 i_2 \cdots i_n}$ are commonly referred to as *effective cluster interactions* (ECIs) and are, on obvious physical grounds, dependent on concentration. As we see below, traditionally these ECIs have not been defined in a unique way. However, the use of such ECIs allows the calculation of the energy of any given configuration and also the trace in Eq. (13.25). The rest of this section is concerned with ways in which ECIs have been defined based on first-principles electronic structure calculations. Specific methods for determining these quantities are then presented in the following sections.

13.5.1 Electronic structure and ECIs

Traditionally, there have been two different approaches taken in determining ECIs from the results of electronic structure calculations. One approach is to extract these parameters froma study of a number of ordered structures at $T = 0$K. These concentration-independent parameters are then used to study the thermodynamic behavior of the system as the temperature is raised based on well-defined statistical methods for the treatment of the Ising model. The other approach is based on the study of the completely disordered material at high temperature and almost invariably involves an application of the CPA. Now, the thermodynamic behavior of the system is studied as the temperature is lowered. Both of these two different approaches make the reasonable assumption that the interactions are temperature independent, and since they begin from opposite ends of the temperature range, they can be thought of as complementary. In fact, the most complete information with a minimum of computational effort can be extracted from a study of both ordered and disordered alloys within the same formal framework [23].

However, the numerical values of these interactions are not the same in each of the two schemes, and neither is their interpretation. And it should be emphasized that, for reasons already mentioned, e.g., kinetics, neither scheme can be taken to provide rigorous understanding of the actual mechanisms underlying alloy transformations, although the ECIs determined within one scheme may be perceived as having clearer physical significance than those of the other. In any case, both schemes attempt a mapping of the true alloy Hamiltonian, Eq. (1.211), onto a generalized Ising form, Eq. (13.53). Thus, both involve an expansion of the energy in terms of orthonormal cluster functions, either within a GC or a C scheme.

Let us begin by considering an alloy associated with a *finite* number, N, of sites, and the space defined by all possible configurations, J, over all concentrations of this alloy. A function over this grand-canonical ensemble takes values corresponding to each configuration. Over this space, we define *cluster functions*, $\Phi_\alpha(\mathrm{J})$, which are products of spin variables (or equivalently of occupation numbers) and can be shown to be both orthonormal and complete in GC configuration space[11]. Thus the generalized Hamiltonian of Eq. (13.53) can be written in the form,

$$H(\mathtt{J}) = \sum_{\alpha s} K_{\alpha s} \Phi_{\alpha s}(\mathtt{J}), \tag{13.54}$$

where $K_{\alpha s}$ are the coefficients denoted by $V_{ij\cdots k}$ in Eq. (13.53), α denotes a particular cluster of sites, and s is the order of the cluster (the number of sites in it). Since the specification of the cluster sites also specifies the order, we will often use a single subscript for the cluster functions and their coefficients. These coefficients, $K_{\alpha s}$, are referred to as effective cluster interactions within the GC ensemble. These ECIs are given by the relation

$$K_{\alpha s} = \frac{1}{N} \sum_{\mathtt{J} \in \mathtt{D}_N} H(\mathtt{J}) \Phi_{\alpha s}(\mathtt{J}), \tag{13.55}$$

where $\mathtt{D}_N$ is the set of configurations in a GC ensemble over N sites.

A summation over configurations is equivalent to an averaging process. Thus, the configurational average of any function $f(\mathtt{J})$ of configuration over a GC ensemble takes the form,

$$\begin{aligned} \langle f \rangle &= \sum_{\mathtt{J} \in \mathtt{D}_N} P(\mathtt{J}) f(\mathtt{J}) \\ &= \sum_{\alpha} K_\alpha \xi_\alpha, \end{aligned} \tag{13.56}$$

where

$$\xi_\alpha = \langle \Phi_\alpha \rangle \tag{13.57}$$

is a *correlation function* for cluster α. For example, the GC average of the total energy of an alloy is given by the expression,

$$\langle E \rangle_{\mathrm{GC}} = \sum_{\alpha} K_\alpha \xi_\alpha. \tag{13.58}$$

It is to be noted that this GC average of the energy is not an experimentally attainable quantity, and the physical meaning of the expansion coefficients, K_α, is not altogether clear. In any case, a GC average as described here is only meaningful with respect to finite systems and can only yield concentration-independent interactions.

Expressions analogous to those derived above within the GC ensemble exist also in connection with a canonical ensemble. Such canonical or concentration-dependent interactions, however, can only be defined with

[11] It is hoped that the use of Greek letters to denote very different quantities such as types of atoms and clusters of sites will be clear from the content of the discussion.

respect to infinite systems, and they are the only kind of interactions that can be defined in this case. Denoting a canonical correlation function by ξ_α^c, we have in place of the last expression

$$\langle E\rangle_c = \sum_\alpha K_\alpha^c \xi_\alpha^c. \tag{13.59}$$

In contrast to the grand canonical average indicated in Eq. (13.58), experimental data reflect an average over configurations at a fixed concentration and calculated canonical averages can be compared directly with experimental results. We also note that in the thermodynamic limit $N \to \infty$ the number of configurations in a C ensemble is equal to that in the GC ensemble[12]. Furthermore, the coefficients K_α determined within a GC ensemble (for finite systems) are ***concentration independent***, whereas the coefficients K_α^c (for infinite systems) depend explicitly on concentration. Finally, these latter sets of coefficients appear in expansions within non-overlapping configuration spaces and one set (at one concentration) cannot be obtained from the other (at another concentration), in spite of attempts [24] in the literature to do so.

We have now laid the groundwork for the determination of ECIs from the results of electronic structure calculations. As we have demonstrated, there are two distinct kinds of ECIs, one independent of concentration, the other concentration dependent. Quantities of the former kind are determined on the basis of Eq. (13.58) based on a procedure first proposed and implemented by Connolly and Williams [25], and referred to as the Connolly-Williams method (CWM). This procedure carries with it the added approximation of an infinite system by one of strictly finite size. In spite of such approximations, the method can yield reliable results provided that the ECIs are essentially concentration-independent, and of rather short range.

In fact, the only "relation" between the two sets of interactions consists of the property that in the thermodynamic limit the concentration-independent interactions approach the concentration dependent ones corresponding to $\mu = 0.0$. On physical grounds, the connection between these two different types of interactions is even more tenuous. Concentration-independent interactions are obtained by "fitting" the energies of alloys at

[12]This may sound a bit surprising, but often infinite subsets of infinite sets contain as many elements as the sets themselves. Thus, the number of even integers is equal to the number of integers, although a ***finite***, continuous subset of the integers contains about twice as many (within ± 1) integers as even integers. Thus, investigations into the properties of infinite sets should not rely only on limiting procedures from the finite side. For example, the number of configurations of a $A_{0.5}B_{0.5}$ binary alloy based on a lattice with a ***finite*** number of sites becomes overwhelmingly larger than the number of configurations at any other concentration as the number of sites approaches infinity. However, this behavior is not true of infinite systems, for which all finite concentrations contain the same number of configurations.

different concentrations, each of which defines a different physical system (characterized by generally different properties such as Fermi level, elastic moduli, and possibly magnetic properties). These physical systems cannot be viewed within a single context any more than could alloys of different elements altogether.

Finally, the CWM suffers from a severe formal defect. The method is based on averages over a grand canonical ensemble that must be carried out in terms of a chemical potential. However, no chemical potential is used in the determination of the cluster interactions within the CWM, and the method is not thermodynamically justified. Furthermore, since the nature of the interactions describing a given alloy system is not known *a priori*, this method has almost no predictive validity.

Concentration-dependent interactions in principle provide a more satisfactory expansion of alloy configurational energies. Such interactions are most often (although not exclusively) determined within the formal construct of the CPA. As such, they are approximate themselves and their utility and reliability hinge crucially on the accuracy of the method used in their determination. In the following sections we discuss in some detail the methods that have been developed for the calculation of ECIs within either approach.

Once a set of ECIs has been obtained, it remains to specify the statistical model to be used to calculate thermodynamic and statistical properties of interest. In this regard, Kikuchi's cluster variation method (CVM) [14, 26-28], and Monte Carlo simulation techniques [29, 30] can be mentioned. Such techniques have evolved to the stage where they can incorporate both concentration and even temperature dependent interactions in the study of alloy phase stability. Unfortunately, a discussion of these methods lies outside the scope of this book. The reader interested in obtaining more details is urged to consult the references cited.

13.6 Krivoglaz-Clapp-Moss Formula

In this section, we exploit the formal analogy between the Ising model and a substitutional alloy and derive the basic equations of the mean-field approximation (MFA), or Bragg-Williams approximation. We also derive an important formula connecting the effective pair interactions (EPIs), V_{ij}, defined explicitly below with the SRO parameters. More details of the formalism presented below can be obtained in a number of published works [14, 20, 31, 33, 34].

13.6.1 Mean-field equations

We begin with an Ising-like Hamiltonian, Eq. (13.11), in the grand-canonical ensemble written in the form,

$$H(\{p_n^\alpha\}) = \frac{1}{2}\sum_{n\neq m}\sum_{\alpha\beta} V_{nm}^{\alpha\beta} p_n^\alpha p_m^\beta - \sum_{n,\alpha} \mu^\alpha p_n^\alpha. \tag{13.60}$$

Here, we use occupation numbers rather than spin variables in a notation that is more commonly used in connection with substitutional alloys. We also restrict the ECIs to those between pairs of sites. We assume that the thermodynamic behavior of the alloy can be obtained from the Hamiltonian in the last expression, in complete analogy with the Ising model. Also, when no confusion can arise, we will not use tildes to distinguish between quantities in the GC and the C ensemble.

We provide a treatment of the alloy Hamiltonian within the mean-field approximation (MFA). We assume that all sites but one, say site n, are characterized by an effective chemical potential, or *effective field*,

$$\mu_n^{\mathrm{eff},\alpha} = \mu^\alpha - \sum_{\beta,m\neq n} V_{nm}^{\alpha\beta}\bar{p}_m^\beta, \tag{13.61}$$

where $\bar{p}_m^\beta$ is an average occupation number characterizing the effective medium. The Hamiltonian of that medium is written in the form,

$$H_n^{\mathrm{eff}} = -\sum_\alpha \mu_n^{\mathrm{eff},\alpha} p_n^\alpha. \tag{13.62}$$

We note that the effective field, $\mu_n^{\mathrm{eff},\alpha}$, is the sum of external fields (or chemical potentials) μ^α and a term describing the influence of the medium on the atom at site n. Now, the single-site density matrix, ρ_n, is approximated by

$$\rho_n = \frac{\exp\left[\beta H_n^{\mathrm{eff}}\right]}{\sum_\alpha \exp\left[\beta\mu_n^{\mathrm{eff},\alpha}\right]}. \tag{13.63}$$

In this approximation, the average of p_n^α is written in the form, [see Eq. (13.16)],

$$\begin{aligned} c_n^\alpha &= \langle p_n^\alpha\rangle = \mathrm{Tr}\rho_n p_n^\alpha \\ &= \frac{\exp\left[\beta\mu_n^{\mathrm{eff},\alpha}\right]}{\sum_\gamma \exp\left[\beta\mu_n^{\mathrm{eff},\gamma}\right]}, \end{aligned} \tag{13.64}$$

with

$$\mu_n^{\mathrm{eff},\alpha} = \mu^\alpha - \sum_{\beta,m\neq n} V_{nm}^{\alpha\beta} c_m^\beta. \tag{13.65}$$

These expressions clearly preserve the identity $\sum_\alpha c_n^\alpha = 1$.

Let us illustrate the ideas introduced above in terms of both an alloy and an Ising (spin) model. In the case of binary alloys, we use the one independent occupation number, $p_n = p_n^A = 1 - p_n^B$, and $\mu = \mu^A$ to cast the alloy Hamiltonian, Eq. (13.60), in the form

$$H(\{p_n\}) = \frac{1}{2}\sum_{m\neq n} V_{nm} p_n p_m - \mu \sum_n p_n, \tag{13.66}$$

which involves the effective pair interactions,

$$V_{nm} = V_{nm}^{AA} + V_{nm}^{BB} - 2V_{nm}^{AB}. \tag{13.67}$$

At a fixed concentration the quantities V_{nm} can be interpreted as *interchange* energies associated with the exchange of an atom of type A with one of type B. Also it follows from its definition that a positive or negative value of V_{nm} indicates a tendency toward ordering (unlike-atom neighbors are favored over those of like atoms), or phase separation (like atoms prefer to cluster together). Now, Eq. (13.64) becomes

$$c_n = \left[1 + \exp\left(-\beta\mu_n^{\text{eff}}\right)\right]^{-1}, \quad \mu_n^{\text{eff}} = \mu - \sum_m V_{nm} c_m. \tag{13.68}$$

The equations derived above can be also expressed in terms of spin variables, σ_n. With the identifications

$$p_n \leftrightarrow \frac{1}{2}(1+\sigma_n), \quad V_{nm} \leftrightarrow -4J_{nm}, \quad \frac{1}{2}\mu - \frac{1}{4}\sum_m V_{nm} \leftrightarrow h, \tag{13.69}$$

we obtain

$$H(\{\sigma_n\}) = -\frac{1}{2}\sum_{m\neq n} J_{nm}\sigma_n\sigma_m - h\sum_n \sigma_n \tag{13.70}$$

and

$$\langle\sigma_n\rangle = \tanh\beta h_n^{\text{eff}}, \quad h_n^{\text{eff}} = h + \sum_m J_{m\neq n}\langle\sigma_n\rangle. \tag{13.71}$$

Equations (13.64), (13.68), and (13.71) constitute the mean-field approximation (MFA). The MFA is also referred to as the *Bragg-Williams approximation.* This approximation and many of its properties and consequences are discussed in detail by Ducastelle [14].

13.6.2 SRO and the KCM formula

The Krivoglaz-Clapp-Moss (KCM) formula connects the Fourier transform of the SRO parameter, $\alpha(\mathbf{k})$, in a binary alloy to the Fourier transform of the pairwise interactions, V_{nm}, occurring in the MFA. We begin with a study of multicomponent alloys [14] and specialize to the binary case subsequently.

From the identity in Eq. (13.32) we have

$$\chi_{nm}^{\alpha\beta} = -\frac{\partial^2 F}{\partial \mu_n^\alpha \partial \mu_n^\beta} = \frac{\partial c_n^\alpha}{\partial \mu_m^\beta} = \frac{\partial c_m^\beta}{\partial \mu_n^\alpha} = \langle p_n^\alpha p_m^\beta \rangle / k_B T. \tag{13.72}$$

Within the MFA, Eq. (13.64), we now obtain

$$\mathrm{d}c_n^\alpha = \beta c_n^\alpha \left(\mathrm{d}\mu_n^{\mathrm{eff},\alpha} - \sum_\beta c_n^\beta \mathrm{d}\mu^{\mathrm{eff},\beta} \right), \tag{13.73}$$

where

$$\mathrm{d}\mu_n^{\mathrm{eff},\alpha} = \mathrm{d}\mu_n^\alpha - \sum_{\beta, m \neq n} V_{nm}^{\alpha\beta} \mathrm{d}c_m^\beta. \tag{13.74}$$

We apply these equations to the disordered state and set $c_n^\alpha = c^\alpha$ *after* derivatives have been taken. The resulting equations can be solved in reciprocal space using the definitions in Eq. (13.45) and the expression

$$V^{\alpha\beta}(\mathbf{k}) = \sum_n \mathrm{e}^{\mathrm{i}\mathbf{k}\cdot\mathbf{R}_n} V^{\alpha\beta}(\mathbf{R}_n), \quad V^{\alpha\beta}(\mathbf{R}_n) = V_{0n}^{\alpha\beta}, \tag{13.75}$$

and with $\mu_\mathbf{k}^\alpha$ defined in a manner analogous to that in which $c_\mathbf{k}^\alpha$ is defined. From Eq. (13.73), we now obtain

$$\begin{aligned} \mathrm{d}c_\mathbf{k}^\alpha &= \sum_\gamma \beta \left(c^\alpha \delta_{\alpha\gamma} - c^\alpha c^\gamma \right) \mathrm{d}\mu_\mathbf{k}^{\mathrm{eff},\gamma} \\ \mathrm{d}\mu_\mathbf{k}^{\mathrm{eff},\gamma} &= \mathrm{d}\mu_\mathbf{k}^\gamma - \sum_\beta V^{\gamma\beta}(\mathbf{k}) \mathrm{d}c_\mathbf{k}^\beta. \end{aligned} \tag{13.76}$$

Now, the Fourier transform of $\chi_{nm}^{\alpha\beta}$ is given in analogy to Eq. (13.75),

$$\chi^{\alpha\beta}(\mathbf{k}) = \left\{ [1 + \chi_0(\mathbf{k}) V(\mathbf{k})]^{-1} \chi_0(\mathbf{k}) \right\}^{\alpha\beta}, \tag{13.77}$$

where for s-component alloys, $\chi_0(\mathbf{k})$ and $V(\mathbf{k})$ are s-dimensional matrices. The matrix elements of $\chi_0(\mathbf{k})$ are given by the expression

$$\chi_0(\mathbf{k})^{\alpha\beta} = \left(c^\alpha \delta_{\alpha\beta} - c^\alpha c^\beta \right) / k_B T, \tag{13.78}$$

and represent the susceptibility of non-interacting atoms.

As one might expect, these formulae are symmetric with respect to the concentrations, c^α. Also, since $\sum_\alpha c^\alpha = 1$, we have $\sum_\beta \chi_0^{\alpha\beta} = 0$ and hence

$\sum_\beta \chi^{\alpha\beta} = 0$. This gives $\frac{1}{2}s(s-1)$ independent susceptibilities (or SRO) parameters for the interacting s-component system. In the case of binary alloys, $s = 2$, we obtain, Eqs. (13.49) and (13.50),

$$\chi(\mathbf{k}) = \frac{c(1-c)\alpha(\mathbf{k})}{k_\mathrm{B}T} \tag{13.79}$$

so that

$$\alpha(\mathbf{k}) = \left[1 + \frac{c(1-c)V(\mathbf{k})}{k_\mathrm{B}T}\right]^{-1} . \tag{13.80}$$

This expression was originally derived by Krivoglaz and Smirnov [35], and in slightly different form by Clapp and Moss [36]. It is commonly referred to as the Krivoglaz-Clapp-Moss (KCM) formula. This result has been derived also by other authors [20, 37] in connection with different physical systems.

After this general discussion showing how the ECIs enter various expressions in the study of alloys, we turn to methods for determining ECI's from electronic structure calculations.

13.7 Connolly-Williams Method

Connolly and Williams [25] have proposed a scheme for the determination of the EPIs within a GC ensemble[13]. The method is based on the expansion of the configurational energy in terms of orthonormal cluster functions,

$$E(\mathrm{J}) = \sum_\alpha K_\alpha \Phi_\alpha(\mathrm{J}). \tag{13.81}$$

This expression defines a set of linear, inhomogeneous equations over the space of configurations in a grand canonical ensemble and, since the $\Phi_\alpha(\mathrm{J})$ are orthonormal and complete, the coefficients K_α are unique. Now, there are two distinct ways in which these coefficients can be determined: Either solve the system of equations, and obtain all K_αs at once, or determine each coefficient separately through an averaging process. Works based on both approaches have been reported.

The CWM is based on the direct solution of the system of equations in Eq. (13.81). However, it is clear that an exact solution of these equations is an impossible task, so that approximation schemes must be brought to bare upon the problem. If one assumes that only a few of the ECIs (the K_α) are significant, and further that the significant ECIs are associated with rather short distances, e.g., near-neighbor pairs, near-neighbor equilateral

[13]Mathematically, this carries the implicit assumption that the system is of finite size and that the interactions obtained are approximations to those of an infinite system at a concentration for which $\mu = 0.0$

triangles, and so on, than one need not solve the entire system of equations. It would now suffice to invert a *finite* system of equations, equal in number to the number of non-vanishing ECIs. Therefore, in Eq. (13.81) each $E(\mathrm{J})$ denotes the energy of a specific, ordered alloy configuration in which case the correlation functions, being averages over a single configuration, become identical with the cluster functions and can be found essentially by inspection. One now has to solve the *finite* system of equations

$$E(\mathrm{J}) = \sum_{\alpha} K_{\alpha}\xi_{\alpha}(\mathrm{J}). \tag{13.82}$$

for the unknown coefficients K_{α}.

The CWM has been found to yield results in acceptable agreements with experiments in a number of applications [38-41]. However, the method is beset with a number of conceptual and computational problems. First, because one inverts an incomplete system of equations the resulting coefficients are not unique. In fact, these coefficients can depend [41] strongly on the set of ordered structures used in their determination. Of course, one can test to see whether or not the parameters determined with a specific choice of ordered structures can produce the energies of configurations not used in the determining step. Such tests are necessary, and occasionally successful, although not always [42]. But, the extra configurations used for testing tend to be at concentrations in the range of those used in determining the ECIs, which greatly diminishes their significance and validity. Second, it is not altogether clear that the expansion in Eq. (13.82) converges sufficiently rapidly to justify its use, or that it converges at all. Third, the CWM cannot take directly into account configurations at very low concentration, (which can result in unmanageably large unit cells), and thus an important part of physical reality (dilute alloys) lies beyond its reach. Fourth, the parameters K_{α} lack direct physical significance and thus cannot reveal the underlying mechanisms responsible for alloy transformations. Finally, these parameters can only be used to fit and/or reproduce the energies of alloy configurations, and contain essentially no other piece of useful information. Thus, they cannot be used reliably to study effects that are demonstrably dependent on concentration, e.g. SRO effects.

Possibly the most serious shortcoming of the CW method is the neglect of the chemical potential in averaging over a grand-canonical ensemble. By setting the chemical potential equal to zero, the calculations of the cluster interactions correspond formally to a concentration of 0.5 (for a binary system), and do not in any way represent the alloy across the concentration range. Becuase of the lack of rigor in the thermodynamics, the CWM lacks the formal validity and predictive power sometimes attributed to it.

In addition to CWM, one can also solve the set of Eqs. (13.82) by direct configurational averaging (DCA) [24]. At least in principle, this approach allows closer monitoring of the convergence characteristics of the ECIs than is possible within the CWM. However, the mode in which the DCA is usually implemented washes away any benefits that could be derived along these lines. Whereas the CWM is based on a fully self-consistent determination of the band structure within LDA, the DCA is based on the use of empirical or phenomenological TB Hamiltonians. Indeed, a fully self-consistent approach seems to be inaccessible at present, since it would involve a rather large number of calculations, each of which would be characterized by very large unit cells. Furthermore, the DCA parameters are determined through averages that, because of computational considerations, involve only a small number of configurations. Recalling that the set of configurations in a GC or C ensemble has the power of the continuum, averages that only involve a small, finite number of configurations are statistically insignificant. Thus, even if the ECIs show a tendency toward convergence, it is not clear that this convergence is unique. Apart from these differences, the DCA possesses the same unsettling features, especially with respect to the chemical potential, that characterize the CWM.

13.8 CPA-Based Methods

In this section, we begin our discussion of CPA-based methods for determining ECIs that are dependent on concentration. For the sake of clarity, we first develop the formalism within a TB, single-band framework. The generalization of the results obtained in the following discussion to the KKR-CPA is given in subsection 13.8.3 and in Section 13.12.

13.8.1 MFA and the CPA

Within a GC ensemble, the phase characteristics of a system are determined by the minima in the free-energy,

$$F = E - TS - \mu_e N_e, \tag{13.83}$$

with E being the internal energy, T the absolute temperature, S the entropy, μ_e the chemical potential for the electrons, and N_e the number of electrons. At $T = 0$, we have $\mu_e = E_F$, where E_F is the Fermi energy and also the entropy term is absent. As a preliminary part of the discussion leading to the determination of ECIs, we show how the internal energy of an alloy, calculated within the formal construct of the CPA, can be used within the general framework of a mean-field approximation such as the Bragg-Williams approximation discussed above.

Strictly speaking, the internal energy of an alloy corresponds to its total energy within the LDA, and total energy expressions are discussed in

Section 13.13. However, much insight and formal experience can be gained by first considering only that part of the energy arising from the sum of orbital energies, the so-called *band energy*, which at zero temperature takes the form

$$E_{\mathrm{b}} = \sum_{\alpha}^{\mathrm{occ}} \epsilon_{\alpha} = \int_{-\infty}^{E_{\mathrm{F}}} E n(E) \mathrm{d}E. \tag{13.84}$$

Here, α denotes an (occupied) orbital, and $n(E)$ is the electron density of states. We now proceed, closely following the development in Ducastelle [14].

We recall, Eq. (6.64), that the number of electrons is given by the expression,

$$N_{\mathrm{e}} = \int_{-\infty}^{E_{\mathrm{F}}} n(E) \mathrm{d}E, \tag{13.85}$$

so that integrating the previous expression by parts we obtain

$$E_{\mathrm{b}} - N_{\mathrm{e}} E_{\mathrm{F}} = - \int_{-\infty}^{E_{\mathrm{F}}} N(E') \mathrm{d}E', \tag{13.86}$$

where

$$N(E) = \int_{-\infty}^{E} n(E') \mathrm{d}E'. \tag{13.87}$$

The quantity

$$\Omega_{\mathrm{e,b}} = E_{\mathrm{b}} - N_{\mathrm{e}} E_{\mathrm{F}} \tag{13.88}$$

is called the *grand potential* and is particularly well-suited for study. For the sake of generality we work at finite temperatures, in which case we have

$$\Omega_{\mathrm{e,b}} = - \int_{-\infty}^{\infty} \mathrm{d}E f(E) N(E), \tag{13.89}$$

where

$$f(E) = \left[1 + \exp(E - \mu_{\mathrm{e}})/k_{\mathrm{B}}T\right]^{-1}, \tag{13.90}$$

is the Fermi distribution function. The expression in Eq. (13.89) follows from standard thermodynamic considerations [14, 21] for particles that obey Fermi-Dirac statistics. It now follows from the definition, see Eq. (2.23),

$$\begin{aligned} n(E) &= -\frac{1}{\pi} \Im G(E^{+}) \\ &= -\frac{1}{\pi} \Im \left(E + \mathrm{i}\eta - H\right)^{-1}, \end{aligned} \tag{13.91}$$

that

$$\Omega_{\mathrm{e,b}} = -\frac{1}{\pi} \Im \int \mathrm{d}E f(E) \mathrm{Tr} \ln G(E^{+}). \tag{13.92}$$

The last expression is valid for any configuration of the system, ordered or disordered. In the latter case we are interested in the configurational average

$$\langle \Omega_{e,b} \rangle = -\frac{1}{\pi} \Im \int dE f(E) \langle \mathrm{Tr} \ln G(E^+) \rangle. \qquad (13.93)$$

Using Eq. (M.4) in Appendix M, and assuming a site-diagonal self-energy, we find [14]

$$\begin{aligned} \langle \mathrm{Tr} \ln G(E^+) \rangle &= \mathrm{Tr} \ln \langle G \rangle - \langle \mathrm{Tr} \ln \left[1 - (V - \Sigma) \langle G \rangle \right] \rangle \\ &= \mathrm{Tr} \ln \langle G \rangle - \sum_n \langle \ln \left[1 - (\epsilon_n - \sigma) F \right] \rangle \\ &- \langle \mathrm{Tr} \ln \left[1 - \hat{t} \langle \hat{G} \rangle \right] \rangle, \end{aligned} \qquad (13.94)$$

where $F = \bar{G}_{nn}$ and $\hat{G}$, respectively, denote the strictly site-diagonal and site off-diagonal part of the Green function. Expanding the logarithm in the last term, we obtain

$$\langle \mathrm{Tr} \ln \left[1 - \hat{t} \langle \hat{G} \rangle \right] \rangle = \sum_{n=1}^{\infty} \mathrm{Tr} \langle \left[\hat{t} \langle \hat{G} \rangle \right]^n \rangle / n. \qquad (13.95)$$

Within the CPA, averages of products such as those on the right side of the equals sign in the last expression can be replaced by products of averages and, because of the CPA condition $\langle \hat{t} \rangle = 0$, vanish identically. This gives the approximate result,

$$\begin{aligned} \langle \mathrm{Tr} \ln G \rangle &\simeq \mathrm{Tr} \ln \langle G \rangle - \sum_n \langle \ln \left[1 - (\epsilon_n - \sigma) F \right] \rangle \\ &= \Phi_{\mathrm{CPA}}(\sigma). \end{aligned} \qquad (13.96)$$

Algebraic manipulations similar to those leading to Eq. (M.8), and the replacement of $\langle G \rangle$ by $\bar{G}$ show that $\Phi_{\mathrm{CPA}}(\sigma)$ is stationary with respect to σ,

$$\frac{\delta \Phi_{\mathrm{CPA}}(\sigma)}{\delta \sigma} = \frac{\delta}{\delta \sigma} \left[\mathrm{Tr} \ln \bar{G} - \sum_n \langle \ln \left[1 - (\epsilon_n - \sigma) F \right] \rangle \right] = 0. \qquad (13.97)$$

To proceed further, we generalize the grand potential, Eq. (13.93), so that it is applicable to inhomogeneous systems, characterized by site energies ϵ_n^α, site-dependent concentrations $c_n^\alpha = \langle p_n^\alpha \rangle$, and site-diagonal self-energies, σ_n, treated within an *inhomogeneous CPA*,

$$\begin{aligned} \langle \Omega_{e,b} \rangle &= -\frac{1}{\pi} \Im \int dE f(E) \mathrm{Tr} \ln \bar{G} \\ &+ \sum_{n\alpha} c_n^\alpha \frac{1}{\pi} \Im \int dE f(E) \mathrm{Tr} \ln \left[1 - (\epsilon_n^\alpha - \sigma_n) F_n \right]. \end{aligned} \qquad (13.98)$$

The Grand potential, $\langle\Omega\rangle$, which depends on *both* the electronic (μ^e) and atomic (μ^α) chemical potentials[14] is obtained upon the addition of the entropy term. Within the framework of a single-site theory, and neglecting the contribution of lattice vibrations, we have

$$\begin{aligned}\langle\Omega\rangle = \bar{\Omega}_e(c) &+ \sum_{n\alpha} c_n^\alpha \frac{1}{\pi}\Im \int dE f(E) \mathrm{Tr} \ln\left[1 - \left(\epsilon_n^\alpha - \sigma_n\right) F_n\right] \\ &+ k_B T \sum_{n\alpha} c_n^\alpha \ln c_n^\alpha - \sum_{n\alpha} \mu^\alpha c_n^\alpha, \end{aligned} \tag{13.99}$$

where

$$\bar{\Omega}_e(c) = -\frac{1}{\pi} \int dE f(E) \mathrm{Tr} \ln \bar{G}. \tag{13.100}$$

is the configuration-independent but concentration-dependent contribution of the averaged medium, and is defined by comparison with Eq. (13.98).

We now derive a set of coupled equations within a general scheme that includes the MFA and the CPA. These equations are obtained by demanding that the grand potential be stationary under variations with respect to appropriate variables. Varying $\langle\Omega\rangle$ with respect to σ_n, we find

$$\begin{aligned}\delta_\sigma\langle\Omega\rangle &= -\frac{1}{\pi} \int dE f(E) \sum_{n\alpha} c_n^\alpha \\ &\times \left[F_n \delta\sigma_n - F_n\left(1 + t_n^\alpha F_n\right)\delta\sigma_n + t_n^\alpha \delta F_n\right], \end{aligned} \tag{13.101}$$

where t_n^α is the t-matrix given by Eq. (11.27),

$$t_n^\alpha = \left(\epsilon^\alpha - \sigma_n\right)\left[1 - F_n\left(\epsilon^\alpha - \sigma_n\right)\right]^{-1}. \tag{13.102}$$

Since

$$\delta F_n = \sum_m \bar{G}_{nm}^2 \delta\sigma_m, \tag{13.103}$$

the condition $\delta_\sigma\langle\Omega\rangle = 0$ implies the CPA condition

$$\sum_\alpha c_n^\alpha t_n^\alpha = 0, \quad \text{for all } n, \tag{13.104}$$

so that the following discussion is consistent with the CPA.

We now vary $\langle\Omega\rangle$ with respect to c_n^α. We use a Langrange multiplier, λ_n, to incorporate the constraint $\sum_\alpha c_n^\alpha = 1$ and we find

$$\begin{aligned}\delta\langle\Omega\rangle/\delta c_n^\alpha - \lambda_n &= \frac{1}{\pi} \int dE f(E) \mathrm{Tr} \ln\left[1 - \left(\epsilon^\alpha - \sigma_n\right) F_n\right] \\ &+ k_B T \ln c_n^\alpha - \mu^\alpha - \lambda_n = 0. \end{aligned} \tag{13.105}$$

[14]The presence of μ^e, which is concentration dependent, is another reason dictating the use of concentration-dependent interactions.

With the definition of an effective chemical potential,

$$\mu_n^{\text{eff},\alpha} = \mu^\alpha - \frac{1}{\pi}\Im \int \mathrm{d}E f(E) \mathrm{Tr} \ln \left[1 - (\epsilon^\alpha - \sigma_n) F_n\right], \tag{13.106}$$

we find

$$c_n^\alpha = \exp\left[\beta\mu_n^{\text{eff},\alpha}\right] / \sum_\gamma \exp\left[\beta\mu_n^{\text{eff},\gamma}\right]. \tag{13.107}$$

This expression is identical to that in Eq. (13.64) which was derived within the MFA. Thus, by setting appropriate variations of the grand potential equal to zero, both the CPA and the MFA can be obtained within a unified framework.

13.8.2 KCM formula

We are now ready to derive the KCM formula, Eq. (13.80), within the unified formalism presented above. We realize that the mean-field equations, Eqs. (13.104) and (13.107), admit the disordered solution where $c_n^\alpha = c^\alpha$ and $\sigma_n = \sigma$. We consider an expansion of $\langle\Omega\rangle$ to second order in terms of small fluctuations [43], $\delta c_n^\alpha = c_n^\alpha - c^\alpha$, and $\delta\sigma = \sigma_n - \sigma$. It follows from Eqs. (13.101), (13.103), and (13.107) that

$$\begin{aligned} 2\delta\langle\Omega\rangle &= -\frac{1}{\pi}\Im \int \mathrm{d}E f(E) \sum_{n,m\neq n} \bar{G}_{nm} t^\alpha \delta c_n^\alpha \delta\sigma_m \\ &+ k_{\mathrm{B}}T \sum_{n\alpha} \left(\delta c_n^\alpha\right)^2 / c^\alpha. \end{aligned} \tag{13.108}$$

From the CPA condition, Eq. (13.104), we have

$$\sum_\alpha t^\alpha \delta c_n^\alpha + \sum_\alpha c^\alpha \delta t_n^\alpha = 0. \tag{13.109}$$

Now, since

$$\delta t_n^\alpha = \left(t^\alpha\right)^2 \delta F_n - \left(1 + t^\alpha F\right)\delta\sigma_n, \tag{13.110}$$

we find

$$\delta\sigma_n - \sum_{n,m\neq n} \bar{G}_{nm}^2 \delta\sigma_m = \sum_\alpha t^\alpha \delta c_n^\alpha. \tag{13.111}$$

These linear equations can be solved through the Fourier transforms

$$\begin{aligned} \delta c_{\mathbf{k}}^\alpha &= \frac{1}{N}\sum_n \mathrm{e}^{-\mathrm{i}\mathbf{k}\cdot\mathbf{R}_n} \delta c_n^\alpha, \quad \delta c_n^\alpha = \sum_{\mathbf{k}} \mathrm{e}^{\mathrm{i}\mathbf{k}\cdot\mathbf{R}_n} \delta c_{\mathbf{k}}^\alpha, \\ \delta\sigma_{\mathbf{k}} &= \frac{1}{N}\sum_n \mathrm{e}^{-\mathrm{i}\mathbf{k}\cdot\mathbf{R}_n} \delta\sigma_n, \quad \delta\sigma_n = \sum_{\mathbf{k}} \mathrm{e}^{\mathrm{i}\mathbf{k}\cdot\mathbf{R}_n} \delta\sigma_{\mathbf{k}}, \end{aligned} \tag{13.112}$$

leading to the result

$$\delta\sigma_{\mathbf{k}} = \frac{1}{1-\langle t^2\rangle \mathrm{A}'_{\mathbf{k}}} \sum_{\alpha} t^{\alpha}\delta c^{a}\alpha(\mathbf{k}), \tag{13.113}$$

with

$$\begin{aligned} \mathrm{A}'_{\mathbf{k}} &= \sum_{n\neq 0} \bar{G}^2_{0n} e^{i\mathbf{k}\cdot\mathbf{R}_n} \\ &= \sum_{\mathbf{q}} \bar{G}(\mathbf{k}+\mathbf{q})\bar{G}(\mathbf{q}) - \bar{G}^2_{00}. \end{aligned} \tag{13.114}$$

In the case of binary alloys, the CPA condition assumes the form, $ct^A + (1-c)t^B = 0$ so that

$$\sum_{\alpha} t^{\alpha}\delta c^{\alpha} = \delta c \Delta t, \quad \Delta t = t^A - t^B, \tag{13.115}$$

and

$$\langle t^2\rangle = c(1-c)\left(\Delta t\right)^2. \tag{13.116}$$

Inserting Eq. (13.113) into Eq. (13.108), and passing to a reciprocal-space representation, we find [14]

$$\begin{aligned} \delta\langle\Omega\rangle &= \frac{N}{2}\sum_{\mathbf{k}}\left\{-\frac{1}{\pi}\Im\int \mathrm{d}E f(E)\frac{\mathrm{A}'_{\mathbf{k}}\left(\Delta t\right)^2}{1-\langle t^2\rangle \mathrm{A}'_{\mathbf{k}}} + \frac{k_{\mathrm{B}}T}{c(1-c)}\right\}|\delta c_{\mathbf{k}}|^2 \\ &= \frac{N}{2}\sum_{\mathbf{k}}\left\{e(\mathbf{k}) + \frac{k_{\mathrm{B}}T}{c(1-c)}\right\}|\delta c_{\mathbf{k}}|^2. \end{aligned} \tag{13.117}$$

The coefficient of $|\delta c_{\mathbf{k}}|^2$ is just the susceptibility [14], $\chi(\mathbf{k}) = \frac{\partial c_{\mathbf{k}}}{\partial \mu_{\mathbf{k}}}$, so that

$$\chi(\mathbf{k}) = \frac{1}{e(\mathbf{k}) + \frac{k_{\mathrm{B}}T}{c(1-c)}}, \tag{13.118}$$

with $e(\mathbf{k})$ defined by Eq. (13.117)

$$e(\mathbf{k}) = -\frac{1}{\pi}\Im\int \mathrm{d}E f(E)\frac{\mathrm{A}'_{\mathbf{k}}\left(\Delta t\right)^2}{1-\langle t^2\rangle \mathrm{A}'_{\mathbf{k}}}. \tag{13.119}$$

We note that the susceptibility in Eq. (13.118) has the KCM form, given by Eq. (13.80), with $e(\mathbf{k})$ playing the role of the Fourier transform of the EPI's. In fact, an inverse transformation shows that within a constant, the grand potential can be written in the form,

$$\langle\Omega\rangle = \frac{1}{2}\sum_{nm} e_{nm}\delta c_n \delta c_m, \tag{13.120}$$

which is the form obtained within the pairwise Ising model. Therefore, the MFA-CPA formalism presented above is equivalent to the Bragg-Williams approximation with effective pair interactions given in **k**-space by Eq. (13.119). By considering higher-order derivatives of the grand potential, the formal approach of this section can be generalized to yield effective cluster interactions of arbitrary order.

13.8.3 Concentration waves and the KKR-CPA

The generalized mean-field theory of the previous section can be extended to the KKR-CPA [44]. The corresponding expressions can be obtained in a number of ways; for example, by direct analysis of correlation functions [44]. They can also be obtained by transcribing the expressions obtained in the previous subsection within the formalism of the KKR-CPA.

In extending the MFA-CPA formalism of the previous section develope-d for single-band systems to the KKR-CPA, we must pay attention to the matrix nature of the various quantities in L-space. In this regard, it is convenient to introduce the tensor product $A \otimes B$ of two operators (matrices) A and B, which is defined in terms of its action on a third operator, C,

$$(A \otimes B)\, C = ACB. \tag{13.121}$$

We begin again with the expression for the grand potential written in terms of only the band energy, which in the CPA can be written in the form,

$$\bar{\Omega}(c) = -2\beta \int_{-\infty}^{\infty} \mathrm{d}E \bar{N}(E) f(E). \tag{13.122}$$

Using the property that $\bar{N} = \sum_\alpha c^\alpha N^\alpha$ (α denotes a chemical species in the alloy) is stationary with respect to variations in the (inverse) t-matrix $\underline{\bar{m}}$, and following a procedure formally identical to that which leads to Eq. (13.113), we obtain the result

$$\delta \underline{\bar{m}}_{\mathbf{k}} = \left[1 - \langle \underline{X}^2 \rangle \underline{\mathrm{A}}'_{\mathbf{k}}\right]^{-1} \sum_\alpha \underline{X}^\alpha \delta c^\alpha_{\mathbf{k}}, \tag{13.123}$$

where $\underline{X}^\alpha$ is given by Eq. (12.17), the quantities $\delta c^\alpha_{\mathbf{k}}$ include the effects of short-range order (the $\alpha(\mathbf{k})$), and where

$$\begin{aligned} \underline{\mathrm{A}}'_{\mathbf{k}} &= \sum_{\mathbf{R}\neq 0} \left[\underline{\bar{\tau}}^{0\mathbf{R}} \otimes \underline{\bar{\tau}}^{\mathbf{R}0}\right] \mathrm{e}^{\mathrm{i}\mathbf{k}\cdot\mathbf{R}} \\ &= \sum_{\mathbf{q}} \underline{\bar{\tau}}(\mathbf{k}+\mathbf{q}) \otimes \underline{\bar{\tau}}(\mathbf{q}) - \underline{\bar{\tau}}^{00} \otimes \underline{\bar{\tau}}^{00}. \end{aligned} \tag{13.124}$$

In the case of binary alloys, $\sum_\alpha \delta c^\alpha_{\mathbf{k}} \underline{X}^\alpha = \delta c_{\mathbf{k}} \Delta \underline{X}$, where $\Delta \underline{X} = \underline{X}^A - \underline{X}^B$, and we obtain

$$\begin{aligned}\delta\langle\Omega\rangle &= \frac{N}{2}\mathrm{Tr}\sum_{\mathbf{k}}\left\{-\frac{1}{\pi}\Im\int \mathrm{d}E f(E)\left[1-\langle\underline{X}^2\rangle \underline{\mathrm{A}}'_{\mathbf{k}}\right]^{-1}\underline{\mathrm{A}}'_{\mathbf{k}}(\Delta\underline{X})^2\right. \\ &+ \left.\frac{k_\mathrm{B}T}{c(1-c)}\right\}|\delta c_{\mathbf{k}}|^2. \end{aligned} \quad (13.125)$$

With the term inside the brackets denoted by $S^{(2)}(\mathbf{k})$[15], and the relations between the susceptibility and the SRO parameters, we regain the KCM formula,

$$\alpha(\mathbf{k}) = \left[1 - \beta c(1-c) S^{(2)}(\mathbf{k})\right]^{-1}. \quad (13.126)$$

As in the TB case, the Fourier transform of $S^{(2)}(\mathbf{k})$ leads to effective pair interaction in complete analogy with Eq. (13.120).

13.9 Applications

We now illustrate the formal results derived in previous sections by using them to analyze briefly some experimental results. We begin by an example that illustrates the significance of concentration waves and long range order (LRO). More details pertinent to the following discussion than is possible to give here can be found in the literature [45].

13.9.1 Concentration waves

Let us consider what happens to a binary alloy, A_cB_{1-c}, as the temperature is lowered toward zero. At high enough temperatures, each site in the lattice[16] is occupied randomly by an atom of type A or B with probability c or $1-c$, respectively. As the temperature is lowered below some critical value, T_c, the system will either phase separate into its individual elemental constituents or order in some particular configuration. Two well-known examples of compositional ordering are afforded by the alloys Cu-Zn and Cu-Au.

In the case of the $Cu_{0.5}Zn_{0.5}$ alloy, the lattice is bcc and upon ordering the Cu atoms are positioned preferentially at the corners of the unit cube with the Zn atoms at the center. Now, the system can be viewed as consisting of two interpenetrating sublattices, labeled I and II, formed, respectively, by the cube corners and cube centers of the unit cube of the original lattice. In the disordered alloy, above T_c, the concentration of Cu

[15] Because of this notation, the method is often referred to as the S2 approach.

[16] We assume that as long as it remains a solid, the alloy is characterized by the presence of a well-defined lattice regardless of temperature.

(and Zn) is independent of the sublattice. Below T_c, the concentration $c_{\rm I}^{\rm Cu}$ of Cu on sublattice I becomes different from that on sublattice II, $c_{\rm II}^{\rm Cu}$. Similar considerations apply to the concentration of Zn atoms, $c_{\rm I}^{\rm Zn} \neq c_{\rm II}^{\rm Zn}$. Clearly, the sublattice concentrations depend on the temperature. Therefore, below the critical temperature the symmetry of the system is lower than that of the completely disordered material that was characterized by the symmetry of the underlying lattice. The resulting two-sublattice structure is described as a state of *broken symmetry.*

Now, for each sublattice we have the obvious sum rules,

$$c_{\rm I}^{\rm Cu} + c_{\rm I}^{\rm Zn} = 1, \quad c_{\rm II}^{\rm Cu} + c_{\rm II}^{\rm Zn} = 1. \tag{13.127}$$

The difference

$$\eta(T) = c_{\rm I}^{\rm Cu} - c_{\rm II}^{\rm Cu} \tag{13.128}$$

between the concentrations on the two sublattices is called the *order parameter.* The order parameter signifies the presence of LRO in the alloy and it is a function of temperature; it is equal to zero above the critical temperature and increases to one as the temperature approaches zero. A schematic diagram of the behavior of the order parameter as a function of temperature for the case of the $\rm Cu_{0.5}Zn_{0.5}$ alloy is given in part (a) of Fig. 13.1. In this case, the ordering transition is continuous.

Results analogous to those for $\rm Cu_{0.5}Zn_{0.5}$ but appropriate to the case of the $\rm Cu_cAu_{1-c}$ alloy based on an fcc lattice are shown in part (b) of the figure. As the temperature is lowered, near $c = 0.75$, the alloy undergoes a transition to the $\rm L1_2$ state, in which the corners of the unit cube are occupied by Au and the centers of the faces by Cu atoms. In this case we distinguish four sublattices, labeled I, II, III, and IV. The order parameter is again given by $\eta(T) = c_{\rm I} - c_{\rm II}$. However, unlike the case of Cu-Zn alloys, the order parameter drops discontinuously to zero at T_c. The two different types of behavior exhibited by the LRO parameters of the CuAu and CuZn systems are indicative of *first-* and *second-order* phase transitions, respectively.

The concentration variation in a system characterized by LRO can be described by means of concentration waves, Eq. (13.45). For example, the $\rm L1_2$ structure of the Cu-Au alloy can be described by

$$\begin{aligned} c_n &= c + \frac{1}{2}\sum_\nu \left(c_\nu e^{i\mathbf{k}_1\cdot\mathbf{R}_n} + c_\nu^* e^{-i\mathbf{k}_1\cdot\mathbf{R}_n}\right) \\ &= c + \frac{1}{4}\eta\left(e^{i\mathbf{k}_1\cdot\mathbf{R}_n} + e^{i\mathbf{k}_2\cdot\mathbf{R}_n} + e^{i\mathbf{k}_3\cdot\mathbf{R}_n}\right), \end{aligned} \tag{13.129}$$

where c is the average concentration, and the reciprocal lattice vectors are given by $\mathbf{k}_1 = (2\pi/a)(1,0,0)$, $\mathbf{k}_2 = (2\pi/a)(0,1,0)$, and $\mathbf{k}_3 = (2\pi/a)(0,0,1)$, with a being the lattice parameter. It is easily seen that the last expression

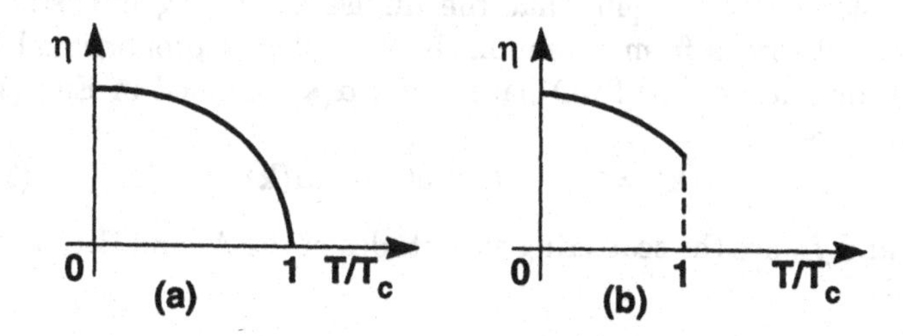

Figure 13.1: Schematic diagram of the order parameter as a function of temperature for a continuous (second-order) transition, (a), and a discontinuous (first-order) transition, (b).

gives $c_{\rm I} = c - \eta/4$ and $c_{\rm II} = c_{\rm III} = c_{\rm IV} = c + \eta/4$. For further discussion on concentration waves and order parameters in multicomponent alloys, the reader is directed to the book by Ducastelle [14].

13.9.2 Diffraction experiments

Concentration waves can be observed directly in X-ray, electron, and neutron diffraction experiments [46]. If we let $\{\mathbf{k}_i\}$ denote the reciprocal lattice vectors (recips) of the underlying parent lattice, then above T_c diffraction intensities peak at the Bragg peaks at $\{\mathbf{k}_i\}$. These peaks are called the *fundamentals*. Below T_c, where the symmetry is broken, extra peaks appear at $\{\mathbf{k}_i + \mathbf{k}_\nu\}$, the so-called ***superlattice spots***. The intensity of peaks at these spots measures the amplitudes c_ν of the concentration wave describing the superlattice (the ordered structure).

It can also be shown [46] that the diffuse scattering intensity, $I(\mathbf{k})$, of X-rays or electrons from a random binary alloy is proportional to the correlation function or the SRO parameter, $\alpha(\mathbf{k})$, defined in Eq. (13.48). In fact, we have

$$I(\mathbf{k}) = (f_A - f_B)^2\, c(1-c)\alpha(\mathbf{k}), \tag{13.130}$$

where f_A and f_B are the scattering amplitudes of the A- and B-type atoms, respectively.

13.9.3 Fermi-surface driven ordering

If a binary alloy has a tendency toward ordering at low temperatures, say in the $L1_2$ structure, then above the critical temperature the scattering intensity $I(\mathbf{k})$ shows diffuse scattering peaks at the superlattice Bragg peaks. In the specific case of the $L1_2$ structure, these peaks are at the face and the edge centers of the unit cell in the fcc reciprocal space as is shown in Fig. 13.2. Experimental verification of the existence of superlattice spots is shown in Fig. 13.3. The figure contains electron diffraction patterns [47] for the Cu_cPd_{1-c} alloy, for $c = 0.126$, 0.250, and 0.332, in panels (a), (b), and (c), respectively. The figures clearly show the four superlattice spots, whose relative positions change rapidly with concentration. A large number of alloy systems behave in similar manner and have been the subject of study [48, 49] for a long time.

The superlattice spots for the $L1_2$ structure correspond to ordering by spontaneous appearance of concentration waves with wave vectors $\mathbf{q}_1 = (1,0,0)$, $\mathbf{q}_2 = (0,1,0)$, and $\mathbf{q}_3 = (0,0,1)$, as discussed above. Since the intensity $I(\mathbf{k})$ is related to $\alpha(\mathbf{k})$, which in term is given in terms of $S^{(2)}(\mathbf{k})$, we must look for peaks in $S^{(2)}(\mathbf{k})$ along the wave vectors $\mathbf{q}_1$, $\mathbf{q}_2$, and $\mathbf{q}_3$. It follows from Eq. (13.124) that $S^{(2)}(\mathbf{k})$ can become very large (even unbounded) if there exist flat pieces of the Fermi surface that are connected

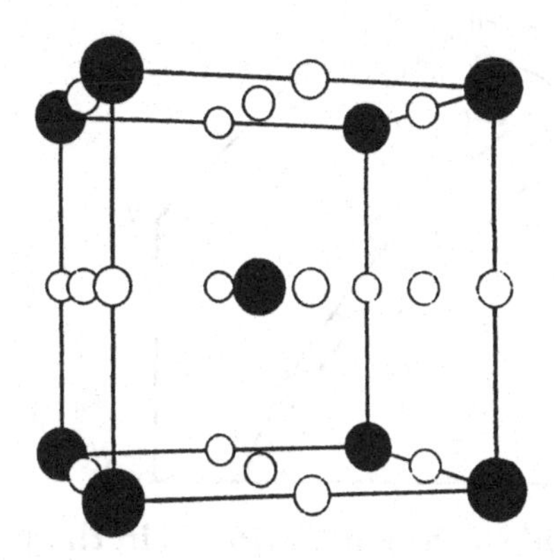

Figure 13.2: **The large solid circles are the Bragg spots for the bcc reciprocal space corresponding to an fcc lattice. The small open circles denote the positions of the diffuse scattering maxima as the ordering transition into the $L1_2$ state is approached from the high-temperature side.**

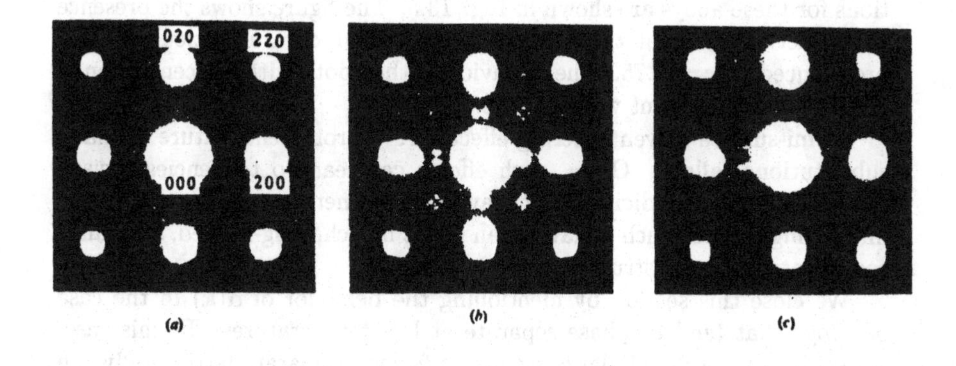

Figure 13.3: **Electron diffraction patterns from Cu_cPd_{1-c} alloys with $c = 0.126$, panel (a), $c = 0.250$, panel(b), and $c = 0.332$, panel (c) [47].**

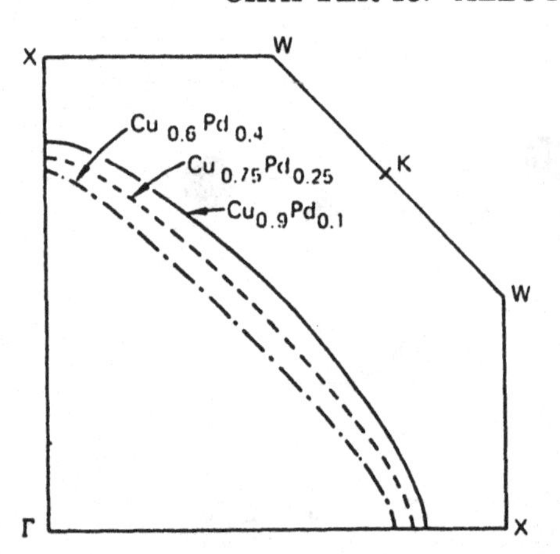

Figure 13.4: **The Fermi surface of Cu_cPd_{1-c} in the ΓXW plane [44].**

(spanned) by the vectors $\mathbf{q}_1$, $\mathbf{q}_2$, and $\mathbf{q}_3$. Such regions make a large contribution to the integral determining $A'(\mathbf{k})$, and can even cause the integral to diverge [45]. Figure 13.4 shows the shape of the Fermi surface of the Cu-Pd alloy system on the ΓXW plane, calculated within the KKR-CPA, [see Eq. (12.70] for the Bloch spectral function). The figure indicates the evolution of Fermi surface with concentration. At $c = 0.75$, the Fermi surface develops an essentially flat sheet that is perpendicular to the vector from the Γ to the K point [the (110) direction]. The corresponding correlation functions for these alloys are shown in Fig. 13.5. The figure shows the presence of diffraction peaks at the superlattice spots, which are particularly well pronounced at $c = 0.75$. The behavior of the spots with concentration is also in good agreement with experiment [45].

Fermi-surface driven ordering effects are a prominent feature of many substitutional alloys. Often, such effects can lead to tendencies toward structures with periodicities that may be incommensurate with that of the underlying lattice. Such behavior can result in rich, long-ranged, and fairly complicated ordered structures.

We close this section by mentioning the behavior of $\alpha(\mathbf{k})$ in the case of alloys that tend to phase separate at low temperatures. In this case, since each of the individual components forms a separate lattice, only the fundamental peaks survive in the diffuse scattering intensity. As an illustration, consider the system Pd_cRh_{1-c}. The experimental phase diagram [50] in Fig. 13.6 shows a single coexistence curve along which regions of fcc solid solutions of Pd and Rh with concentration at c_1 coexist with macroscopically large regions at concentration c_2. In this system the transition to phase separation is of first order (discontinuous) at all concentrations except at $c = 0.5$ where it is continuous.

The figure shows the phase boundary for Pd-Rh alloys as calculated

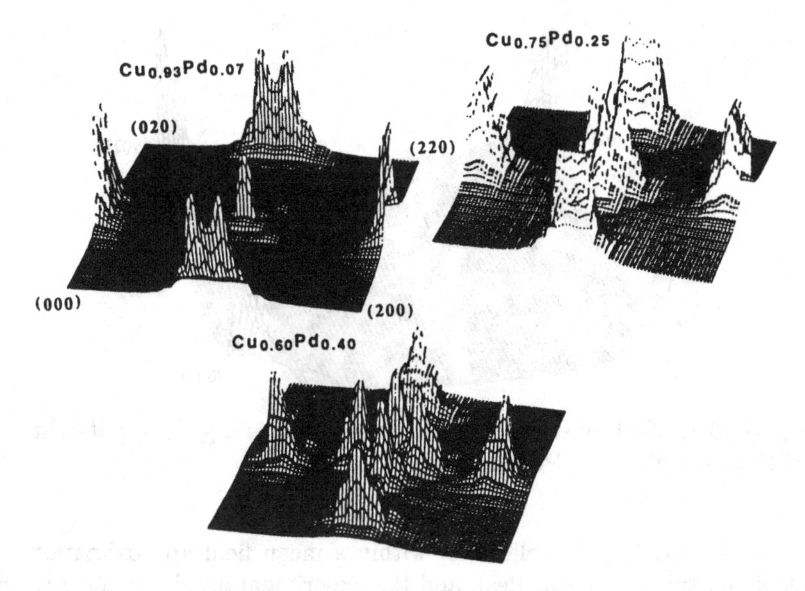

Figure 13.5: **The calculated correlation functions, $S^{(2)}(\mathbf{k})$, for three Cu-Pd alloys [45].**

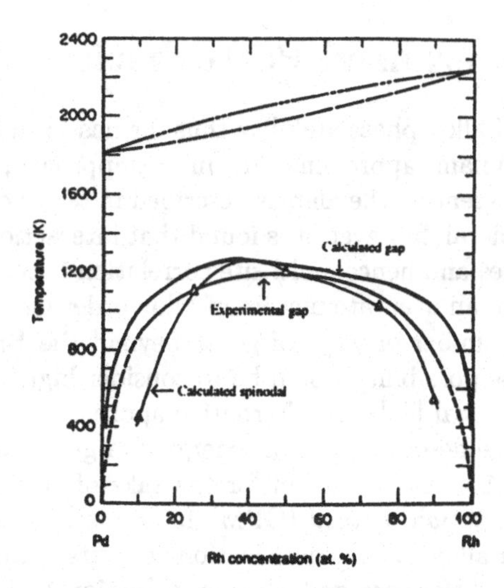

Figure 13.6: **Comparison of calculated clustering spinodal line and phase boundary with the experimentally determined phase diagram for fcc-based Pd-Rh alloys.**

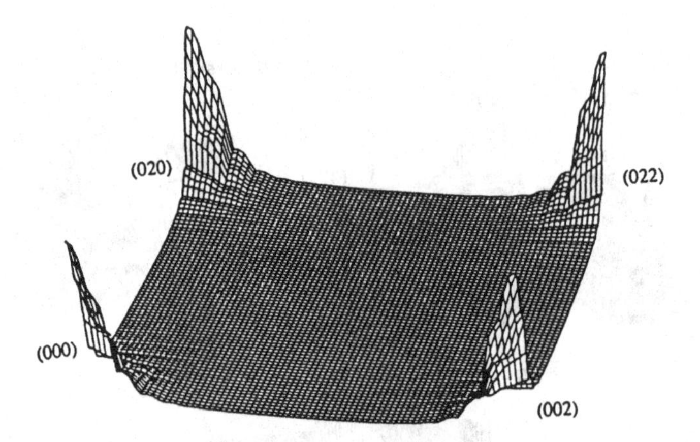

Figure 13.7: **Calculated SRO diffuse scattering intensity in the (001) plane for the $PdRh_3$ alloy.**

within the KKR-CPA (solid line) within a mean field approximation, the calculated spinodal (triangles), and the experimentally determined coexistence [50]. Consistent with this behavior, the calculated intensity, shown in Fig. 13.7, peaks only at the fundamentals.

13.10 Generalized Perturbation Method

Our discussion of alloy phase stability thus far has been based on a mean-field or Bragg-Williams approximation. In this approach, the effective Ising Hamiltonian representing the alloy is described in terms of effective pair interactions. In general, however, it is found that interactions involving more than a pair of sites and hence multi-site correlation functions are necessary in order to obtain an accurate picture of alloy ordering characteristics.

There are a number of ways of going beyond the Bragg-Williams approximation. One possibility would be to consider higher order derivatives of the grand potential [45]. An alternative approach [51-58] is to attempt a perturbation-like description of the energy of a given configuration starting from the CPA solution. Again for the sake of clarity, we present the *generalized perturbation method* (GPM) in terms of a TB, single-band[17] description of an alloy. The generalization to a first principles KKR-CPA formalism is straightforward and is given in Section 13.12.

We recall that a specific alloy configuration can be characterized by the set of occupation numbers $\{p_n^\alpha\}$. Considering each configuration as consisting of finite concentration (occupation) fluctuations away from the

[17]However, we shall keep the explicit designation of taking the trace as it simplifies the passage to multiband systems.

CPA medium and taking into account only the band contribution to the energy[18] we can write the following expression for the thermodynamic potential [55, 56],

$$\Omega(\{p_n^\alpha\}) = \bar{\Omega}(c) + \Omega'(\{p_n^\alpha\}). \tag{13.131}$$

The quantity $\bar{\Omega}(c)$ was introduced in Eq. (13.100) and is the concentration-dependent but configuration-independent contribution of the CPA medium, while the second term depends on both concentration and configuration. As follows from Eq. (13.94) this second term can be written as the generalized perturbation expansion

$$\begin{aligned} \Omega'(\{p_n^\alpha\}) &= \frac{2}{\pi N}\Im\left[\int_{-\infty}^{E_F} \mathrm{Tr}\ln\left(1 - T\hat{\bar{G}}\right)\right] \mathrm{d}E \\ &= \frac{2}{\pi N}\Im\left[\sum_{n=2}^{\infty}\frac{1}{n}\int_{-\infty}^{E_F} \mathrm{Tr}\left(T\hat{\bar{G}}\right)^n\right] \mathrm{d}E. \end{aligned} \tag{13.132}$$

Here, the scattering matrix operator is site-diagonal and is given by the expression,

$$T = \sum_n p_n^\alpha t_n^\alpha |n\rangle\langle n|, \tag{13.133}$$

the quantity $\hat{\bar{G}}$ is the site off-diagonal part of the CPA Green function, and the factor of 2 accounts for the degeneracy associated with the spin of the electron. In the case of binary alloys, it follows from Eq. (13.115) that we can write

$$T = \sum_n \delta c_n \delta t |n\rangle\langle n|, \quad \delta c_n = p_n - c. \tag{13.134}$$

Upon using Eq. (13.131) in Eq. (13.132), one obtains a cluster expansion of the configurational energy in terms of the occupation numbers,

$$\begin{aligned} \Omega'(\{p_n^\alpha\}) &= \sum_{s=2}^{\infty}\Omega'^{(s)}(\{p_n^\alpha\}) \\ &= \frac{1}{2}\sum_{nm}\sum_{\alpha\beta} v_{nm}^{\alpha\beta} p_n^\alpha p_m^\beta + \frac{1}{3}\sum_{nml}\sum_{\alpha\beta\gamma} v_{nml}^{\alpha\beta\gamma} p_n^\alpha p_m^\beta p_l^\gamma \\ &+ \frac{1}{4}\sum_{nmlk}\sum_{\alpha\beta\gamma\delta} v_{nmlk}^{\alpha\beta\gamma\delta} p_n^\alpha p_m^\beta p_l^\gamma p_k^\delta + \cdots, \end{aligned} \tag{13.135}$$

where it is to be kept in mind that no two consecutive site indices can be

[18]Total energies are considered in Section 13.13.

the same. The cluster interactions, $v_{ijk\cdots}^{\alpha\beta\gamma\cdots}$, are given by the expressions

$$v_{i_0 i_1 i_2 \cdots i_{n-1}}^{\alpha_0 \alpha_1 \alpha_2 \cdots \alpha_{n-1}} = -\frac{2}{\pi} \Im \mathrm{Tr} \left[\int_{-\infty}^{E_F} \left(t_{i_0}^{\alpha_0} \bar{G}_{i_0 i_1} t_{i_1}^{\alpha_1} \bar{G}_{i_1 i_2} \cdots t_{i_{n-1}}^{\alpha_{n-1}} \bar{G}_{i_{n-1} i_0} \right) \mathrm{d}E \right]. \tag{13.136}$$

We note that the term with $n = 1$ vanishes throughout since $\mathrm{Tr}(T\hat{\bar{G}}) = 0$. The last two expressions provide an approximation to the configurational energy of an alloy[19]. Therefore, given the νs, one can calculate the energy of an ordered alloy configuration based on Eq. (13.135).

In some cases Eqs. (13.135) and (13.136) can be simplified considerably. In the case of binary alloys, we can use Eqs. (13.134) to obtain

$$\begin{aligned} \Omega'(\{p_n^\alpha\}) &= \sum_{s=2}^{\infty} \Omega'^{(s)}(\{p_n^\alpha\}) \\ &= \frac{1}{2} \sum_{nm} v_{nm}^{(2)} \delta c_n \delta c_m + \frac{1}{3} \sum_{nml} v_{nml}^{(3)} \delta c_n \delta c_m \delta c_l \\ &+ \frac{1}{4} \sum_{nmlk} v_{nmlk}^{(4)} \delta c_n \delta c_m \delta c_l \delta c_k + \cdots, \end{aligned} \tag{13.137}$$

where the nth order cluster interactions are given by the expression,

$$v_{i_0 i_1 i_2 \cdots i_{n-1}}^{(n)} = -\frac{2}{\pi} \Im \mathrm{Tr} \left[\int_{-\infty}^{E_F} \left(\delta t \bar{G}_{i_0 i_1} \delta t \bar{G}_{i_1 i_2} \cdots \delta t \bar{G}_{i_{n-1} i_0} \right) \mathrm{d}E \right], \tag{13.138}$$

with $\delta t = t^A - t^B$, and where any two consecutive indices, i_n, i_m, must be different.

It is to be noted that Eq. (13.137) is *formally* identical to expansions that would be obtained through continuous differentiation of the grand potential, but that the coefficients which appear in the two expressions are different. One difference is that the coefficients which arise within the Bragg-Williams approximation contain the response to statistical fluctuations of the effective CPA medium throughout the material, [through the denominator in Eq. (13.113)], whereas the effective interactions defined in Eq. (13.138) are confined to specific clusters of sites. At the same time, the MFA results are derived for the case of infinitesimal concentration fluctuations, whereas the expressions derived above are applicable to finite concentration deviations from the CPA medium. In spite of such formal differences, however, numerical calculations indicate [58] that within the

[19]There are two approximations usually involved here: the use of the CPA, which yields the Green function that appears in these expressions, and the restriction to band energies. In principle, both of these approximations can be removed to various degrees.

same order of the expansion, e.g., pairs, the two methods yield very similar results, particularly in the case of metallic alloys.

As it stands, Eq. (13.137) is somewhat awkward for computational purposes. The main difficulty is that higher-order terms, such as v_{ijkl}, may contain contributions from clusters of lower order. For example in v_{ijkl} we may have $i = k$ and $j = l$, a term that should be included in the pair contribution to the expansion in Eq. (13.137). A computationally more useful expression can be obtained in terms of *irreducible contributions* associated with clusters of sites that contain no lower-order terms. These *effective cluster interactions*[20] (ECIs) are defined in connection with indices corresponding to distinct cluster sites. As is shown below, the configurationally-dependent part of the grand potential now takes the form,

$$\begin{aligned} \Omega'(\{p_n\}) &= \sum_n V_n \delta c_n + \frac{1}{2} \sum_{nm}'' V_{nm} \delta c_n \delta c_m \\ &+ \frac{1}{3} \sum_{nml}'' V_{nml} \delta c_n \delta c_m \delta c_l + \cdots, \end{aligned} \tag{13.139}$$

where the double primes indicate that no two site indices in a term can refer to the same site. The term [14, 58] $\bar{\Omega}(c)$ is commonly referred to as the *disorder energy* and is a quantity that plays an important role in the construction of alloy phase diagrams. As a simple example, we consider the effective pair interactions. In this case, Eq. (13.139) leads to the result

$$V_{nm} = v_{nm}^{(2)} + \frac{1}{2}(1-2c)^2 v_{nm}^{(4)} + c(1-c) \sum_k v_{nmk}^{(4)} + \cdots,. \tag{13.140}$$

As will be shown in the following discussion, this summation can be evaluated by considering a pair of sites embedded in a CPA medium, and it leads to the expression

$$V_{nm} = \bar{\Omega}'_{nm}(AA) + \bar{\Omega}'_{nm}(BB) - \bar{\Omega}'_{nm}(AB) - \bar{\Omega}'_{nm}(BA), \tag{13.141}$$

where $\bar{\Omega}'_{nm}(\alpha\beta)$ denotes the contribution to Ω' associated with two sites, n and m, occupied by atoms of type α and β, respectively, embedded in the CPA medium.

We note that the GPM expansions derived in this section, Eq. (13.139), are controlled by the "small" parameter $\delta t \bar{G}$ that can be shown to be less than unity almost everywhere inside the band of a metallic alloy. Thus, the GPM expansion is essentially guaranteed to yield converged results. This small quantity may in fact become large near band edges [14], but in a realistic description of the Hamiltonian (where sharp band edges are essentially

[20] Occasionally also called renormalized effective cluster interactions (RECIs).

absent) such difficulties would be confined mostly near the bottom of the band, and in general would cause no convergence difficulties.

As mentioned above, the GPM expressions can become essentially exact if the properly averaged medium Green function is used, and if the total energy, rather than the band energy is taken into account. Usually the propagation between sites in Eq. (13.137) is calculated within the CPA. In that case, the ECIs contain the contributions of clusters embedded in the CPA medium. The connection of the ECIs derived within the GPM and the embedded cluster method is made explicit in the following section.

13.11 ECIs in the ECM

In this section we show that the formalism of the embedded-cluster method [58-62] allows the summation of GPM expansions and the determination of ECIs, in a particularly useful and transparent form.

13.11.1 GPM expansions — pair interactions

Because of the significance of the role played by pair interactions in the discussion of ordering phenomena, we shall consider the case of pairs separately.

The pair interaction that results from Eq. (13.138) has a simple physical interpretation: It is the contribution to the configurational energy that includes all possible scattering events from a pair of atoms embedded in a CPA medium. Summation of all terms in Eq. (13.132) that include only a pair of sites in a specific configuration leads to the result

$$\Omega'(p_n^\alpha p_m^\beta) = \frac{2}{\pi}\Im\mathrm{Tr}\left[\int_{-\infty}^{E_F} \ln Q^{\alpha\beta}\mathrm{d}E\right], \tag{13.142}$$

representing the summation to *all* orders of the corresponding GPM expansion. Here, we have defined the matrix

$$\left(Q^{\alpha\beta}\right)_{nm} = \delta_{nm} - t_n\bar{G}_{nm}(1-\delta_{nm}), \quad t_n \equiv t_n^\alpha, \tag{13.143}$$

which is simply related to the scattering matrix $T^{\alpha\beta}$ for the two sites by the expression,

$$T^{\alpha\beta} = \delta D\left(Q^{\alpha\beta}\right)^{-1}, \tag{13.144}$$

with δ and D being site-diagonal matrices with elements $\delta_n^\alpha = \epsilon_n^\alpha - \sigma_n$ and $(D^{-1})_{nn}^\alpha = (1-\delta_n^\alpha\bar{G}_{nn})$, respectively.[21] Further, direct comparison shows

[21] Note that in the case of the homogeneous CPA, we have $\sigma_n = \sigma$, and $\bar{G}_{nn} = \bar{G}_{00}$.

that the renormalized pair interaction in Eq. (13.141) can be written in the form

$$
\begin{aligned}
V_{kl} &= -\frac{2}{\pi}\Im\mathrm{Tr}\left\{\int_{-\infty}^{E_F} \ln\left[\left(Q^{AA}Q^{BB}\right)\left(Q^{AB}Q^{BA}\right)^{-1}\right] \mathrm{d}E\right\} \\
&= \sum_n \frac{1}{n}\left[(1-c)^n - (-c)^n\right]^2 v_{kl}^{(n)}. \qquad (13.145)
\end{aligned}
$$

The last summation contains all terms in Eq. (13.140) that involve sites k and l.

13.11.2 GPM expansions — multisite interactions

As follows from Eq. (13.93), the grand potential (including only band energies) can be written in then form,

$$
\Omega = \frac{2}{\pi N_c}\Im\mathrm{Tr}\left[\int_{-\infty}^{E_F} \left(\ln\bar{G} - \ln D - \ln Q\right) \mathrm{d}E\right], \qquad (13.146)
$$

where the quantity Q is such that the Green function, $G(E)$, for a cluster of N_c sites embedded in a CPA medium is given by the expression

$$
G(E) = \bar{G}\left(D^{-1}Q^{-1}\right). \qquad (13.147)
$$

Since the matrix D is site diagonal, Eq. (13.144), the configurationally dependent contribution to the grand potential from a cluster of sites embedded in the CPA medium is given by the last term on the right of the equals sign in Eq. (13.146). Consider now a binary alloy, A_cB_{1-c}. The interchange energy for the two species is given by the expression

$$
V_1^{(1)} = -\int_{-\infty}^{E_F} \left(N_A - N_B\right) \mathrm{d}E, \qquad (13.148)
$$

which defines a "single-body" interaction resulting from replacing an A atom with a B atom in an alloy, all other sites being effective medium-like sites. Now, EPIs can be obtained as the difference in the single-body interactions at site n, when site m, with $m \neq n$, is occupied by an A or a B atom. Thus, consistent with the definition of the integrated density of states for different configurations, we find,

$$
\begin{aligned}
V_{nm}^{(2)} &\equiv \left(V_n^{(1)}\right)_m^A - \left(V_n^{(1)}\right)_m^B \\
&= -\int_{-\infty}^{E_F} \left\{ \left[N_{nm}^{AA} - N_{nm}^{BA}\right] - \left[N_{nm}^{AB} - N_{nm}^{BB}\right] \right\} \mathrm{d}E.
\end{aligned}
\qquad (13.149)
$$

According to Eq. (13.146), this expression can be written in the form of Eq. (13.145). The process just indicated can be extended to arbitrary numbers of sites. We find that n-site effective cluster interactions or n-body irreducible contributions to the energy (at $T = 0$) are given by the expression

$$\begin{aligned} V^{(n)}_{i_0 i_1 \cdots i_{n-1}} &= \left(V^{(n-1}_{i_0 i_1 \cdots i_{n-2}}\right)^{A}_{i_{n-1}} - \left(V^{(n-1}_{i_0 i_1 \cdots i_{n-2}}\right)^{B}_{i_{n-1}} \\ &= -\frac{2}{\pi N_c} \Im \left\{ \int_{-\infty}^{E_F} \mathrm{Tr} \ln \left[\Pi_J Q^{J(\mathrm{even})}\right] \left[\Pi_J Q^{J(\mathrm{odd})}\right]^{-1} \mathrm{d}E \right\}, \end{aligned} \tag{13.150}$$

where J(even) and J(odd) denote cluster configurations with even and odd numbers of B atoms, respectively.

The renormalized effective cluster interactions (RECIs) defined by the last expression represent an irreducible contribution to the configurational energy of an alloy. Thus, they do not contain contributions from smaller clusters. This is because any such terms occur with equal frequency in the numerator and the denominator of the logarithm in Eq. (13.150) and, therefore, cancel identically. Furthermore, these interactions can be expanded to yield the GPM expansions derived above, an exercise which shows that their convergence is controlled by the small parameter $\delta t\bar{G}$. Finally, it is possible to augment both the GPM series and the interactions defined by Eq. (13.150) to include contributions from the effective medium surrounding the cluster [58]. In any case, whether such additional corrections are included or not, it follows from the discussion in Appendix N [58] that the configurational energy, E_J, of an alloy can be written in the form

$$\begin{aligned} E_J &= E_{\mathrm{CPA}} + \sum_i V_i^{(1)} \delta c_i + \frac{1}{2} \sum_{ij}{}^{\prime\prime} V_{ij}^{(2)} \delta c_i \delta c_j \\ &+ \frac{1}{3} \sum_{ijk}{}^{\prime\prime} V_{ijk}^{(3)} \delta c_i \delta c_j \delta c_k + \cdots, \end{aligned} \tag{13.151}$$

where E_{CPA} is the contribution to the energy of the CPA medium, and RECIs are defined by Eq. (13.150). This expression is the central result of this section. It shows that *the configurational energy of an alloy can be calculated within the ECM as a sum of irreducible, renormalized interactions associated with clusters of sites embedded in the CPA*. In the case of monatomic lattices (one atom per unit cell), the single-body contribution, $V_1^{(1)}$, is independent of site index and makes no contribution to the configurational energy since $\sum_n \delta c_n = 0$.

13.12 GPM and ECM in the KKR-CPA

We now extend the formalism of the previous section to the KKR-CPA [63].

13.12.1 Generalized perturbation method

We continue to consider only the band contribution to the total energy, so that the grand potential (simply the energy at T=0 K) takes the form, Eq. (13.89).

$$\Omega(E_{\rm F}) = -\int_{-\infty}^{E_{\rm F}} {\rm d}EN(E), \tag{13.152}$$

where $N(E)$ is the integrated density of states. As we saw above, [refer to Eq. (13.94)], the grand potential can be written as the sum of two terms: (i) the concentration-dependent but configuration-independent contribution of the CPA effective medium, and (ii) a term that depends on configuration. In analogy with Eq. (13.132) this term can be written in the form,

$$\Omega'(\{p_n^\alpha\}, E_{\rm F}) = -\frac{1}{\pi N}\int_{-\infty}^{E_{\rm F}} {\rm Tr}\ln\left(1 - \underline{X}\hat{\underline{\tau}}\right), \tag{13.153}$$

where the trace is taken over both site and angular momentum indices, N denotes the number of sites, and $\hat{\underline{\tau}}$ is the site off-diagonal part of the scattering path operator determined in the KKR-CPA, Eq. (12.5). Upon expanding the logarithm, we obtain[22]

$$\Omega'(\{p_n^\alpha\}, E_{\rm F}) = -\frac{1}{\pi N}\Im\sum_{k=2}^{\infty}\int_{-\infty}^{E_{\rm F}} {\rm d}E{\rm Tr}\frac{\left(\underline{X}\hat{\underline{\tau}}\right)^k}{k}. \tag{13.154}$$

Through the use of CPA algebra, and for the specific case of binary alloys, we obtain the analogue of Eq. (13.137)

$$\Omega'(\{p_n^\alpha\}, E_{\rm F}) = \sum_{k=2}^{\infty}\frac{1}{k}\sum_{n_1 n_2\cdots n_k} V^{(k)}_{n_1 n_2\cdots n_k}\delta c_{n_1}\delta c_{n_2}\cdots\delta c_{n_k}. \tag{13.155}$$

Here, the quantity $V^{(k)}_{n_1 n_2\cdots n_k}$ is the n-th order ECI involving the sites appearing as subscripts, and is given by the expression,

$$V^{(k)}_{n_1 n_2\cdots n_k} = -\frac{2}{\pi}\Im\int_{-\infty}^{E_{\rm F}} {\rm d}E{\rm Tr}\left(\delta\underline{X}_{n_1}\underline{\tau}^{n_1 n_2}\delta\underline{X}_{n_2}\cdots\delta\underline{X}_{n_k}\underline{\tau}^{n_k n_1}\right), \tag{13.156}$$

where

$$\delta\underline{X}_{n_k} = \underline{X}^A_{n_k} - \underline{X}^B_{n_k}, \tag{13.157}$$

[22]For complex systems, the $k = 1$ term should be retained. This term vanishes for systems with one atom per unit cell.

the trace is taken over angular momentum indices, and the prefactor of 2 accounts for the electron spin. Thus, to lowest order in the small parameter, $|\Delta \underline{X}\hat{\underline{\tau}}|$, where $\Delta \underline{X} = \underline{X}^A - \underline{X}^B$, the configurational contribution to the grand potential is given by the expression,

$$\Omega'(\{p_n^\alpha\}, E_F) = \frac{1}{2} \sum_{n \neq 0} V_{0n}^{(2)} \delta c_0 \delta c_n, \tag{13.158}$$

where $V_{0n}^{(2)}$ is the effective pair interaction between sites 0 and n,

$$V_{0n}^{(2)} = V_{0n}^{AA} + V_{0n}^{BB} - V_{0n}^{AB} - V_{0n}^{BA}. \tag{13.159}$$

This expression is formally identical to that in Eq. (13.141).

Whether derived within a phenomenological TB or a first-principles formalism, the ECIs within the GPM have the same physical interpretation. For example, a positive (negative) value of the pair interaction is indicative of tendencies toward ordering (phase separation).

13.12.2 Embedded cluster method

A complete summation of selected terms in the GPM expansion leads to results that can be readily obtained from the ECM. As in the TB case, we find that the grand potential can be written in the form of Eq. (13.155), where the ECIs can be expressed as interchange energies,

$$\begin{aligned} V_{n_1 n_2 \cdots n_k}^{(k)} &= \left[V_{n_1 n_2 \cdots n_{k-1}}^{(k-1)}\right]_k^A - \left[V_{n_1 n_2 \cdots n_{k-1}}^{(k-1)}\right]_k^B \\ &= -\frac{2}{\pi} \Im \int_{-\infty}^{E_F} \mathrm{Tr} \ln \left[\Pi_J \underline{Q}^{J,\mathrm{even}}\right] \left[\Pi_J \underline{Q}^{J,\mathrm{odd}}\right]^{-1}. \end{aligned} \tag{13.160}$$

Here, as in Eq. (13.150), J (even) and J (odd) denote cluster configurations with even and odd numbers of B-type atoms, respectively, and $\underline{Q} = (1 - \underline{X}\hat{\underline{\tau}})$.

Upon expanding the quantity $\mathrm{Tr} \ln \underline{Q}$ in the last expression, one readily obtains the GPM expansion, Eq. (13.156). In fact, the first term of this expansion often yields results [58] that are indistinguishable from the fully summed ECM expression in Eq. (13.160). Thus, the GPM, due to the presence of the small parameter $|\Delta \underline{X}\hat{\underline{\tau}}|$, can yield rapidly convergent expressions for the ECIs.

13.12.3 Applications

We now provide a brief review of calculations based on the GPM and ECM formalisms described above, carried out within the KKR-CPA framework. A more extensive review is given in reference [64, 65].

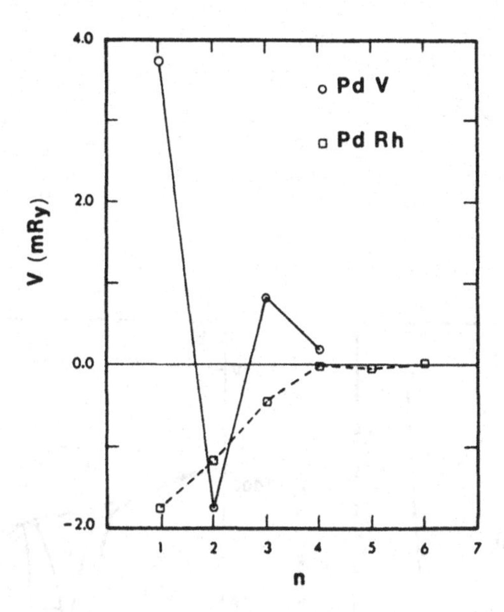

Figure 13.8: **Variation with neighbor distance of the effective pair interactions for Pd_3Rh and Pd_3V alloys.**

Effective pair interactions [63] for Pd-Rh and Pd-V alloys are shown in Fig. 13.8. In the case of Pd-Rh alloys, all the interactions are negative, which is consistent with the tendency of the alloy to phase separate at low temperatures, as is indicated in Fig. 13.6. On the other hand, the alternating sign of the pair interactions for the Pd-V system leads to much richer behavior. Taking account of pair interactions up to and including fourth neighbors, one can predict a stable DO_{22} ordered structure at $c = 0.75$, in agreement with experimental observations. Considering only first and second neighbors yields an incorrect $L1_2$ ground state. Thus, the convergence of the GPM expansion (or any other expansion) can be an important factor in the study of alloy phase stability.

The concentration-temperature phase diagram of the Cu-Zn alloy system is shown in Fig. 13.9. The diagram was constructed [65] on the basis of effective pair interactions calculated within the GPM, and the configurational entropy of the alloy was treated within the cluster variation method [26, 27, 28]. This study was limited to two underlying lattices, fcc and bcc[23]. Also, the vibrational contribution to the entropy was included with-

[23]In the experimental phase diagram, Zn is hcp, and above 50% Zn concentration there exist comlex structures, γ, δ, ϵ, which are distorted versions of the bcc lattice, and are due to the vacancy-induced formation of these alloys in this concentration range. Such phenomena form a fascinating subject of study in alloy physics but lie outside the scope

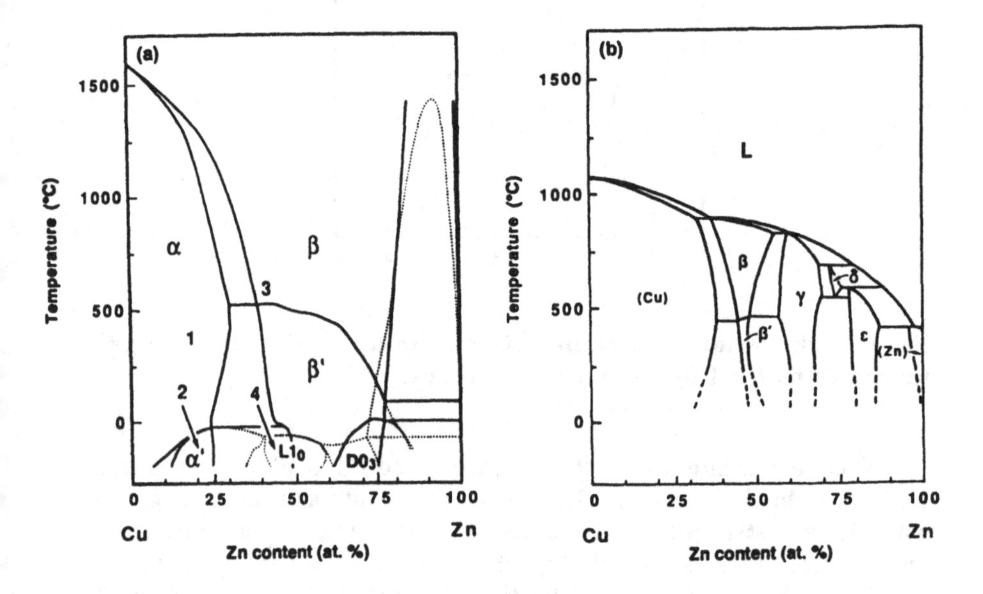

Figure 13.9: **Calculated (a) and experimentally assessed (b) Cu-Zn alloy phase diagrams. Dotted lines indicate the presence of metastable phase equilibria [66].**

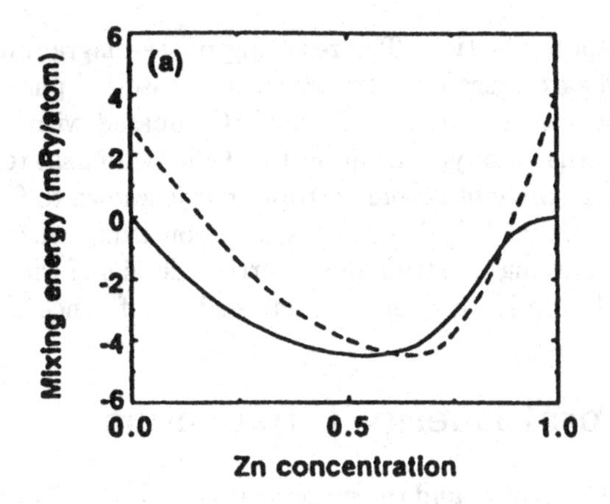

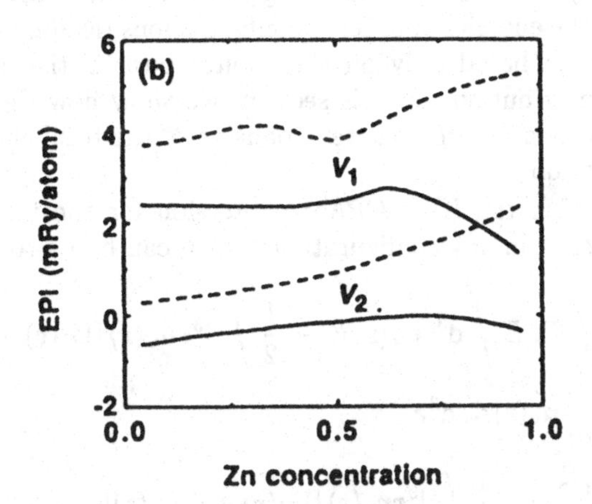

Figure 13.10: **The variation of the mixing energy and of first- and second-neighbor effective interactions of the fcc-based (solid line) and the bcc-based (dashed line) CuZn alloy with Zn concentration [66].**

in the Debye approximation. The resulting phase diagram in Fig. 13.9(b) is in very good agreement with experimentally assessed phase diagrams for this system, at least in the Cu-rich part. Compared with part (a) of the figure, it shows that the typical curvature of the two-phase region $(\alpha+\beta)$ is essentially due to the vibrational entropy that is known to favor the stability of β-based systems [66]. This behavior is consistent with the calculated SRO diffuse-scattering spectrum [65]. Further details of the phase stability properties of the Cu-Zn system are contained in reference [65].

13.13 Total Energy Expansions

The CPA-based methods and the numerical approaches discussed above are characterized by one common disadvantage. Although one can calculate self-consistently determined total energies within the single-site CPA, the calculation of the energies of ordered configurations (leading to the *ordering energy*) has been based only on the orbital part of the energy, the so-called band contribution. In this section, we show how the GPM can be generalized to yield a perturbative expansion of the total energy of a given alloy configuration.

We begin with the KKR-CPA[24] expression for the total energy of a solid, Eq. (1.211), in any configuration, which can be written in the form,

$$\begin{aligned} E &= \int E\mathrm{d}E \int \mathrm{d}^3 r n_\mathrm{v}(\mathbf{r},E) - \frac{1}{2}\int \mathrm{d}^3 r n_\mathrm{v}(\mathbf{r})\left[V_\mathrm{C}(\mathbf{r}) + 2\mu_\mathrm{xc}(\mathbf{r})\right] \\ &+ \int_\Omega n_\mathrm{v}(\mathbf{r})\epsilon_\mathrm{xc}\mathrm{d}^3 r \\ &+ \sum_i^{\mathrm{core}} \epsilon_i - \frac{1}{2}\int \mathrm{d}^3 r n_\mathrm{c}(\mathbf{r})\left[V_\mathrm{C}(\mathbf{r}) + 2\mu_\mathrm{xc}(\mathbf{r})\right] \\ &+ \int_\Omega n_\mathrm{c}(\mathbf{r})\epsilon_\mathrm{xc}\mathrm{d}^3 r \\ &- \frac{1}{2}\sum_\nu Z_\nu \left\{\int \frac{n(\mathbf{r})\mathrm{d}^3 r}{|\mathbf{r}-\mathbf{R}_\nu|} - \sum_{\alpha\neq\nu}\frac{Z_\alpha}{|\mathbf{R}_\alpha - \mathbf{R}_\nu|}\right\}, \end{aligned} \tag{13.161}$$

where n_v and n_c denote the valence and core contributions to the electron density, ϵ_xc and μ_xc are the exchange-correlation energy and potential, respectively, Z_ν is the nuclear charge at $\mathbf{R}_\nu$, and $V_\mathrm{C}(\mathbf{r})$ is the electrostatic

of our discussion in this book.

[24]It should be kept in mind that the CPA, as well as any manipulations based on it, should be carried out in connection with neutral cells in order to minimize charge-transfer effects.

(Hartree) potential at $\mathbf{r}$,

$$V_{\mathrm{C}}(\mathbf{r}) = \int \frac{n(\mathbf{r}')\mathrm{d}^3r'}{|\mathbf{r}-\mathbf{r}'|} - \sum_{\alpha} \frac{Z_\alpha}{|\mathbf{r}-\mathbf{R}_\alpha|}. \tag{13.162}$$

Let us isolate that part of the energy that is proportional to the valence charge and write it in the form,

$$\begin{aligned} E_{\mathrm{v}} &= \int \mathrm{d}E \int \mathrm{d}^3r n_{\mathrm{v}}(\mathbf{r},E)\left\{E - \frac{1}{2}\left[V_{\mathrm{C}}(\mathbf{r}) + 2\mu_{\mathrm{xc}}(\mathbf{r}) - 2\epsilon_{\mathrm{xc}}\right]\right\} \\ &= \int \mathrm{d}E \int \mathrm{d}^3r n_{\mathrm{v}}(\mathbf{r},E) Y(\mathbf{r},E). \end{aligned} \tag{13.163}$$

In addition, we need the expression for the V_n^c in terms of cluster energies. As was shown in the previous section, at least in the case of binary alloys, these coefficients are *interchange* energies and satisfy the exact relations,

$$V_n^c \equiv V_{i_1 i_2 \cdots i_n}^{(n)} = \left[V_{i_1 i_2 \cdots i_{n-1}}^{(n-1)}\right]_{i_n}^{A} - \left[V_{i_1 i_2 \cdots i_{n-1}}^{(n-1)}\right]_{i_n}^{B}, \tag{13.164}$$

so that these coefficients can be determined within an iterative process beginning with the coefficient of the lowest order, $V_i^{(1)}$, Eqs. (13.148) and (13.149). Since these iterative relations hold in the case of total energies, we have to start the iteration process based on a total energy expansion of $V_1^{(1)}$.

To a very good approximation, it is sufficient to work only with E_{v}. The core part, which is everything not included in E_{v}, does indeed contribute to $V^{(1)}$ but, at least for monatomic systems, this term does not occur in a perturbative expansion because of the property $\sum_\alpha \delta c_\alpha = 0$. Therefore, we proceed with the treatment of only the valence part of the energy, within the framework of the KKR-CPA[25].

From the formal developments of MST presented in Chapter 3, we have

$$n_{\mathrm{v}}^{\alpha}(\mathbf{r},E) = \sum_{LL'} Z_L^{\alpha}(\mathbf{r}) \tilde{\tau}_{LL'}^{\alpha}(E) Z_{L'}^{\alpha}(\mathbf{r}), \tag{13.165}$$

where n_{v}^{α} is the local electron density at E associated with an atom of type α in the alloy, $\tilde{\underline{\tau}}^{\alpha}$ is the imaginary part of the corresponding scattering-path operator (the transition matrix), and Z^α is the regular solution of the Schrödinger equation for the potential in a cell of type α that satisfies

[25] At least in this initial stage, we assume that no charge transfer takes place between cells of differing chemical composition. The generalization of the present discussion to account for charge transfer effects can be obtained within the configurational-matrix approach of neutral cells. See references 14 and 15 in the previous chapter.

well-defined matching conditions on the surface of a sphere bounding the cell.

We now define the vectors $|Z^\alpha(\mathbf{r},E)\rangle$, whose elements are the functions $Z_L^\alpha(\mathbf{r},E)$, and the matrix

$$\begin{aligned} \underline{f}^\alpha(E) &= \int d^3r |Z^\alpha(\mathbf{r},E)\rangle\langle Z^\alpha(\mathbf{r},E)| \\ &= \int d^3r \underline{f}^\alpha(\mathbf{r},E). \end{aligned} \tag{13.166}$$

In terms of these matrices, we can write

$$n_v^\alpha(E) = \mathrm{Tr}\underline{n}^\alpha(E), \tag{13.167}$$

where

$$\underline{n}^\alpha(E) = \underline{\tilde{\tau}}^\alpha(E)\underline{f}^\alpha(E). \tag{13.168}$$

In order to study disordered alloys, the energy expression must be averaged over all configurations of a C ensemble. It is consistent with the spirit of the CPA and the perturbative expansions considered here to replace the disordered material with an effective medium, determined within the CPA, and to consider all sites, occupied by different chemical species, as being embedded in such a medium. We now have, (with $E_v = E$),

$$\begin{aligned} E &= \mathrm{Tr} \int dE \underline{\tilde{\tau}}^\alpha(E)\underline{f}^\alpha(\mathbf{r},E)Y^\alpha(\mathbf{r},E) \\ &\equiv \mathrm{Tr} \int dE \underline{\tilde{\tau}}^\alpha(E)\underline{F}^\alpha(E). \end{aligned} \tag{13.169}$$

Here, $\underline{\tilde{\tau}}^\alpha(E)$ is the imaginary part of the scattering-path operator associated with a cell (atom) of type α embedded in the CPA medium, Z^α is the regular solution of the Schrödinger equation for the potential of such a cell, and $Y^\alpha(\mathbf{r},E)$ is the potential, including exchange and correlation, acting on a cell of type α. For reference purposes, we note explicitly the elements of the matrix $\underline{F}^\alpha$,

$$F_{LL'}^\alpha(E) = \int d^3r Z_L^\alpha(\mathbf{r},E) Z_{L'}^\alpha(\mathbf{r},E) Y^\alpha(\mathbf{r},E), \tag{13.170}$$

where within the CPA framework we take the "potential" $Y^\alpha(\mathbf{r},E)$ to depend only on the chemical occupation of cell α, and to have an additive dependence on the energy. Using these definitions, we can write

$$\begin{aligned} E &= \mathrm{Tr} \int dE \underline{n}^\alpha(E)\underline{f}^{\alpha-1}(E)\underline{F}^\alpha(E) \\ &= \mathrm{Tr} \int dE \underline{n}^\alpha(E)\underline{G}^\alpha(E). \end{aligned} \tag{13.171}$$

Integration of the last expression by parts leads to the result,

$$\begin{aligned} E &= \mathrm{Tr}\left[\underline{N}^{\alpha}(E)\underline{G}^{\alpha}(E)\right]\Big|_{-\infty}^{E_F} \\ &- \mathrm{Tr}\int_{-\infty}^{E_F} \underline{N}(E)\frac{\mathrm{d}G(E)}{\mathrm{d}E}\mathrm{d}E. \end{aligned} \tag{13.172}$$

If, as is often done in calculations based on band energies, we neglect configurational fluctuations in the integrated term above, we obtain the following expression for the configurational energy (also called the grand potential),

$$\Omega = -\mathrm{Tr}\int_{-\infty}^{E_F} \underline{N}(E)\underline{g}(E)\mathrm{d}E, \tag{13.173}$$

where $\underline{g}(E) = \frac{\mathrm{d}G(E)}{\mathrm{d}E}$. This expression is in a particularly convenient form for the calculation of the effective cluster interactions.

From the general expression in Eq. (13.164), and beginning with $V^{(1)}$, we find

$$V_i^{(1)} \equiv V^{(1)} = -\mathrm{Tr}\int_{-\infty}^{E_F} \left[\underline{N}_i^A(E)\underline{g}^A(E) - \underline{N}_i^B(E)\underline{g}^B(E)\right]\mathrm{d}E, \tag{13.174}$$

for the one-body contribution, and

$$\begin{aligned} V_{ij}^{(2)} &\equiv \left[V_i^{(1)}\right]_j^A - \left[V_i^{(1)}\right]_j^B \\ &= -\mathrm{Tr}\int \left\{\left[N_{ij}^{AA}(E) - N_{ij}^{BA}(E)\right]\underline{g}^A(E)\right. \\ &- \left.\left[N_{ij}^{AB}(E) - N_{ij}^{BB}(E)\right]\underline{g}^B(E)\right\}\mathrm{d}E, \end{aligned} \tag{13.175}$$

for the effective pair interaction. Higher-order ECIs can be evaluated in a fairly straightforward way.

It can also be shown that the expressions in terms of integrated densities of states can be written concisely in terms of cluster matrices. Through the definition of the scattering matrix in Eq. (12.17) and the matrix $\underline{Q}$ defined in Eq. (13.147), we can write

$$\begin{aligned} V_{i_1 i_2 \cdots i_n}^{(n)} &= -\frac{2}{\pi}\mathrm{Tr}\int \mathrm{d}E\left\{\ln\left[\left[\Pi_J \underline{Q}^{J,(\mathrm{even}),A}\right]\left[\Pi_J \underline{Q}^{J,(\mathrm{odd}),A}\right]^{-1}\right]\underline{g}^A(E)\right. \\ &- \left.\ln\left[\left[\Pi_J \underline{Q}^{J,(\mathrm{even}),B}\right]\left[\Pi_J \underline{Q}^{J,(\mathrm{odd}),B}\right]^{-1}\right]\underline{g}^B(E)\right\}, \end{aligned} \tag{13.176}$$

where $\underline{Q}^{J,(\mathrm{even}),A}$ is a cluster matrix for configuration J that has an even number of B-like atoms and contains an atom of type A at the origin (center) of the cluster. This expression yields the irreducible ECIs for a given cluster of sites within the CPA formalism.

The expression for the ECIs just derived is a perturbative expansion of the total energy whose convergence is governed by the presence of the small parameter, $|\Delta \underline{t}\hat{\underline{\tau}}|$. As such it is similar to the GPM (and also ECM) expansions that have been derived in connection with band energies. Clearly, it reduces to those expansions in a trivial way when only the band contribution to the quantity $Y^{\alpha}(\mathbf{r})$ is considered. Furthermore, since the potential $Y^{\alpha}(\mathbf{r})$ depends on the energy only in an additive way, we may expect that the kinetic energy should in general make the largest contribution to the perturbative expansions of the energy away from the CPA values.

Bibliography

[1] W. Hume-Rothery and G. V. Granger, *The Structure of Metals and Alloys*, The Institute of Metals, London (1962).

[2] S. K. Sinha, Prog. Mat. Sci. **15**, 1 (1972).

[3] K. Girgis, *Physical Metallurgy*, 3rd Edition, R. W. Cahn and P. Haasen (eds.), North-Holland, Amsterdam (1983), p. 219.

[4] L. Pauling, *The Nature of the Chemical Bond*, Cornell University Press, Ithaca (1960).

[5] F. R. de Boer, R. Boom, W. C. M. Mattens, A. R. Miedema, and A. K. Niessen, *Cohesion in Metals; Transition Metal Alloys, Cohesion and Structure*, Vol. **1**, F. R. de Boer and D. G. Pettifor (eds.) North-Holland, Amsterdam, (1989).

[6] J. C. Philips, Rev. Mod. Phys. **42**, 317 (1970).

[7] J. C. Philips, *Bonds and Bands in Semiconductors*, Academic Press, New York (1973).

[8] W. B. Pearson, *The Crystal Chemistry and Physics of Metals and Alloys*, Willey, New York (1972).

[9] M. O'Keeffe and A. Navrotsky (eds.), *Structure and Bonding*, Vol. I and II, Academic Press, New York (1982).

[10] D. G. Pettifor, J. Phys. C**19**, 285 (1986).

[11] D. G. Pettifor, New Scientist **100**, 48 (1986).

[12] D. G. Pettifor, Mater. Sci. and Technol. **4**, 675 (1988).

[13] P. Villars, K. Mathis, and F. Hulliger, "Environmental Classification and Structural Stability Maps" in *The Structure of Binary Compounds, Cohesion and Structure*, Vol. **2**, F. R. de Boer and D. G. Pettifor (eds.), North-Holland, Amsterdam (1989), p. 1.

[14] F. Ducastelle, *Order and Phase Stability in Alloys*, North-Holland, Amsterdam (1992).

[15] E. Ising, Z. Phys. **31**, 253 (1925).

[16] R. Kubo, Hiroshi Ichimura, Tsunemaru Usui, and Natsuki Hashizumi, *Statistical Physics*, North-Holland, Amsterdam (1965).

[17] F. Reif, *Fundamentals of Statistical and Thermal Physics*, McGraw-Hill Book Co. New York (1965).

[18] R. Kubo, J. Phys. Soc. Japan **12**, 570 (1957).

[19] R. Kubo, M. Toda, and N. Hashitsume, *Statistical Physics* II, Springer-Verlag, Berlin (1985).

[20] R. Brout, *Phase Transitions*, Benjamin, New York, (1965).

[21] L. D. Landau and E. M. Lifshitz, *Statistical Physics*, 2nd Ed. Pergamon Press, Oxford (1969), and subsequent editions.

[22] See any publication of the CALPHAD (Calculations of Alloy Phase Diagrams) group.

[23] Prabhakar P. Singh and A. Gonis, Phys. Rev. Lett. **71**, 1605 (1993).

[24] M. Asta, C. Wolverton, D. de Fontaine, and H. Dreysse, Phys. Rev. B**44**, 4907 (1991-I).

[25] J. W. Connolly and A. R. Williams, Phys. Rev. B**27**, 5169 (1983).

[26] R. Kikuchi, Phys. Rev. **81**, 998 (1951).

[27] D. de Fontaine, Solid St. Phys. **34**, 73 (1979).

[28] A. Finel, in *Statics and Dynamics of Alloy Phase Transformations*, P. E. A. Turchi and A. Gonis (eds.), NATO ASI series B, Vol. 319, Physics, Plenum, New York (1993), p. 495, and references therein.

[29] K. Binder, J. L. Lebowitz, M. H. Phani, and M. H. Kalos, Acta Metall. **29**, 1655 (1981).

[30] K. Binder, in *Monte Carlo Methods in Statistical Physics*, Topics in Current Physics, Vol. **7**, K. Binder (ed.), Springer-Verlag, Heidelberg, (1986).

[31] H. E. Stanley, *Introduction to Phase Transitions and Critical Phenomena*, Clarendon Press, Oxford (1971).

[32] A. G. Khachaturyan, Prog. Mat. Sci. **22**, 1 (1978).

[33] A. G. Khachaturyan, *The Theory of Structural Transformations in Solids*, Wiley, New York (1983).

[34] F. Gautier, *Solid State Transformations in Metals and Alloys*, Les Editions de Physique, Orsay (1980), p.459.

[35] M. A. Krivoglaz and A. A. Smirnov, *The Theory of Order-Disorder in Alloys*, McDonald, London (1964).

[36] P. C. Clapp and S. C. Moss, Phys. Rev. **142**, 418 (1964).

[37] P. G. de Gennes and J. Friedel, J. Phys. and Chem. Sol. **4**, 71 (1958).

[38] S. Takizawa, K. Terakura, and T. Mohri, Phys. Rev. B**39**, 5792 (1989).

[39] M. Sluiter and P. E. A. Turchi, Phys. Rev. B**40**, 11215 (1989).

[40] L. G. Ferreira, S. Wei, and A. Zunger, Phys. Rev. B**40**, 3197 (1989).

[41] M. Sluiter and P. E. A. Turchi, in *Alloy Phase Stability*, G. M. Stocks and A. Gonis (eds.), NATO-ASI Series E, Vol. 163, Kluwer Academic Publishers, Dordrecht (1988), p. 521.

[42] J. Mikalopas, Thesis, Department of Applied Science Engineering, UC Davis (1993), unpublished.

[43] F. Gautier, F. Ducastelle, and F. Giner, Phil. Mag. **31**, 1373 (1985).

[44] B. L. Györffy and G. M. Stocks, Phys. Rev. Lett. **50**, 374 (1983).

[45] B. L. Györffy, D. D. Johnson, F. J. Pinski, D. M. Nicholson, and G. M. Stocks, in *Alloy Phase Stability*, G. M. Stocks and A. Gonis (eds.), NATO-ASI Series E, Vol. 163, Kluwer Academic Publishers, Dordrecht (1988), p. 421.

[46] M. A. Krivoglaz, *Theory of X-ray and Thermal Neutron Scattering by Real Crystals*, Plenum Publications, New York (1969).

[47] K. Oshima and D. Watanabe, Acta Cryst. A**33**, 520 (1977).

[48] H. Sato and R. S. Toth, in *Alloying behavior and Effects in Concentrated Solid Solutions*, T. B. Massalski (ed.), Gordon and Breach, New York (1965).

[49] H. Sato and R. S. Toth, Phys. Rev. **127**, 469 (1961).

[50] J. E. Shield and R. K. Williams, Scripta Met. **21**, 1475 (1987).

[51] F. Ducastelle and F. Gautier, J. Phys. F **6**, 2039 (1976).

[52] G. Tréglia, F. Ducastelle, and F. Gautier, J. Phys. F**8**, 1437 (1978).

[53] A. Bieber, F. Gautier, G. Tréglia, and F. Ducastelle, Solid State Commun. **39**, 149 (1981).

[54] A. Bieber, F. Ducastelle, F. Gautier, G. Tréglia, and P. Turchi, Solid State Commun. **45**, 585 (1983).

[55] A. Bieber and F. Gautier, J. Phys. Soc. Jap. **53**, 2061 (1984).

[56] A. Bieber and F. Gautier, Z. Phys. B**57**, 335 (1984).

[57] F. Ducastelle, in *Alloy Phase Stability*, G. M. Stocks and A. Gonis (ed.), NATO-ASI Series E, Vol. **163**, Kluwer Academic Publishers, Dordrecht (1989), p. 293.

[58] A. Gonis, X.-G. Zhang, A. J. Freeman, P. Turchi, G. M. Stocks, and D. M. Nicholson, Phys. Rev. B**36**, 4630 (1987).

[59] A. Gonis and J. W. Garland, Phys. Rev. B**16**, 2424 (1977).

[60] C. W. Miles and J. B. Dow, Phys. Rev. Lett. **42**, 254 (1979); Phys. Rev. B**19**, 4939 (1979).

[61] A. Gonis, G. M. Stocks, W. H. Butler, and H. Winter, Phys. Rev. Lett. **50**, 1482 (1982).

[62] A. Gonis, G. M. Stocks, W. H. Butler, and H. Winter, Phys. Rev. B**29**, 555 (1984)

[63] P. E. A. Turchi, G. M. Stocks, W. H. Butler, D. M. Nicholson, and A. Gonis, Phys. Rev. B**37**, 5982 (1988).

[64] P. E. A. Turchi, in *Intermetallic Compounds: Vol. 1, Principles*, J. H. Westbrook and R. L. Fleischer (eds.), John Wiley and Sons, New York (1994), p. 21.

[65] P. E. A. Turchi, M. Sluiter, F. J. Pinski, D. D. Johnson, D. M. Nicholson, G. M. Stocks, and J. B. Staunton, Phys. Rev. Lett. **67**, 1779 (1991).). Erratum, Phys. Rev. Lett. **68**, 418 (1992).

[66] J. Friedel, J. Phys. (Paris) Lett. **35**, L59 (1974).

Chapter 14

Forces and Interatomic Potentials

14.1 General Comments

The concept of force was discussed on general grounds in Chapter 1. In this chapter, we take a closer look at the determination of forces acting on the nuclei in a material and the form of interatomic potentials from which such forces can be determined.

In our discussion of the ground-state properties of the interacting electron-nucleon system, the nuclei have been assumed held fixed in their equilibrium positions. In this treatment, based on the Born-Oppenheimer approximation, the forces on the nuclei must vanish for the system to remain in its equilibrium configuration in the ground state. In principle, the vanishing of the force on a nucleus could in fact be used as the condition determining the ground-state configuration. On the other hand, the resulting procedure is not nearly as reliable as that based on the total energy; unlike the total energy, the force does not obey a variational principle that would guide an iterative process to convergence.

However, there are good reasons for attempting to calculate the restoring force on nuclei displaced along a particular direction from equilibrium. First, knowledge of the force is a necessary ingredient in the study of some of the dynamic properties of the system, e.g., the motion of the nuclei. Second, in spite of the lack of a variational principle, the possibility should be explored for determining the equilibrium configuration of the nuclei through the alternative criterion that the total force on a nucleus vanishes. This could provide an independent test of the convergence of both kinds of calculation, that of the total energy as well as of the force itself. In this chapter, we develop formal methods to determine the force along a given direction acting on a nucleus using as a basis the Hellmann-Feynman theorem (HFT) discussed in Chapter 1.

The study of the dynamics of nuclear motion can be carried out much more efficiently if the forces could be determined through the knowledge of potential functions. In this approach, one relies on the fact that the electrons, which are much lighter than the nuclei, can follow nuclear movement in essentially instantaneous fashion. This allows the possibility to express the total energy of the ground state of the system in terms of many-body contributions involving pairs, triplets, etc. of distinct sites in the system. The aim of these decompositions is to describe the motion of the nuclei in classical terms through interactions that arise because of the presence of the electron gas that must necessarily be treated quantum mechanically. It follows that it would be useful to have a prescription for deriving such internuclear potentials as directly as possible from the results of an electronic structure calculation of the ground state of the system. A prescription for doing this within MST is described in Section 14.4.

For the sake of convenience of implementation, we would like to avoid the determination of forces and potentials through processes that depend on physical differentiation. Here, one would carry out fully self-consistent calculations for various nuclear positions and calculate the force as the rate of change of the energy with respect to displacement. By contrast, we are seeking formulae that yield the force through mathematical differentiation of the self-consistently determined potentials in a ground-state calculation. As is discussed below, the HFT provides a sound mathematical basis for the direct calculation of the force on a nucleus. We also discuss the various technical aspects of such an approach. In addition, we show that MST allows a straightforward calculation of many-particle interactions in a system including not only pair potentials, but potentials of any order involving in principle an arbitrary number of atoms (triplets, quadruplets, etc.), although for practical purposes one hopes that the expansion effectively terminates after a small number of terms. These system-specific potentials can be calculated for bulk systems, surfaces and interfaces, substitutionally ordered and disordered alloys, and for alloy surfaces and interfaces.

14.2 Forces

In discussing the implications of the Hellmann-Feynman theorem, we follow the development in Abraham et al. [1]. From Eq. (1.176), the HFT leads directly to an expression for the force acting on the nucleus at $\mathbf{R}_n$, (occasionally, the dependence of the density on the external potential is indicated throught the explicit appearance of nuclear coordinates in n),

$$\begin{aligned}\mathbf{F}^n &= -\frac{\partial E[n(\mathbf{r},\mathbf{R}_n),\mathbf{R}_n]}{\partial \mathbf{R}_n} \\ &= Z^n\left\{2\int \mathrm{d}^3r' \frac{n(\mathbf{r}')}{|\mathbf{R}_n-\mathbf{r}'|^3}(\mathbf{R}_n-\mathbf{r}')\right. \\ &- \left.2\sum_{n'}\frac{Z^{n'}}{|\mathbf{R}_n-\mathbf{R}_{n'}|^3}(\mathbf{R}_n-\mathbf{R}_{n'})\right\} \\ &= -Z^n\frac{\partial}{\partial \mathbf{R}_n}V_{\mathrm{M}}(\mathbf{R}_n), \qquad (14.1)\end{aligned}$$

where $n(\mathbf{r},\mathbf{R}_n)$ denotes the ground-state electron density for a given set of nuclear positions, and $V_{\mathrm{M}}(\mathbf{R}_n)$ is the Madelung potential at $\mathbf{R}_n$,

$$V_{\mathrm{M}}(\mathbf{R}_n) = 2\int \mathrm{d}^3r' \frac{n(\mathbf{r}')}{|\mathbf{R}_n-\mathbf{r}'|} - 2\sum_{n'}\frac{Z^{n'}}{|\mathbf{R}_n-\mathbf{R}_{n'}|}. \qquad (14.2)$$

We see that the force can be represented as the product of the electric field acting on the nucleus times the nuclear charge. This is a classical result even though the electron density, $n(\mathbf{r})$, requires a quantum-mechanical treatment. But the result is far from trivial. It is clear from Eq. (14.1) that the electron-electron interaction and exchange and correlation do not affect the force on the nuclei, even though they must be taken into account in calculating the density. This implies that approximations directed at the treatment of these effects, such as the Hartree and Hartree-Fock, and the LDA, do not affect the validity of the Hellmann-Feynman theorem. An explicit proof that the theorem remains valid in the LDA was given in Chapter 1. However, it is also important to note that the theorem holds for the wave functions and particle densities that *exactly* satisfy the conditions of a given approximation scheme. As is discussed below, errors in the density can lead to great inaccuracies in the calculated force. The main reason for this effect is that the force, unlike the total energy, is not protected through the extremal conditions imposed by a variational principle.

In order to illustrate the computational characteristics of the HFT, we now derive an explicit expression for the force in terms of the electron density. This will also allow us to point to various difficulties associated with force calculations, and possible ways to overcome them. For the sake of clarity of presentation[1], we use the atomic sphere approximation and assume that the potential in a cell (in the atomic sphere) is spherically symmetric within a sphere of radius R_{WS}, the Wigner-Seitz radius, and that

[1] And running the risk of considerable loss in accuracy.

it vanishes identically outside this sphere. From the solution of Poisson's equation,

$$V(\mathbf{r}) = \int d^3r' \frac{n(\mathbf{r}')}{|\mathbf{r}-\mathbf{r}'|}, \tag{14.3}$$

and the expression, $Y_{00}(\hat{r}) = 1/\sqrt{4\pi}$, we obtain,

$$\begin{aligned} V_M(\mathbf{R}_n) &= 2\sqrt{4\pi}\int_0^{R_{WS}} d^3r' r' n_{\ell=0}^{(n)}(r')\left[1-\left(\frac{r'}{R_{WS}}\right)\right] \\ &+ V_{\ell=0}(R_{WS}) + \frac{2Z^n}{R_{WS}}, \end{aligned} \tag{14.4}$$

where $n_{\ell=0}^{(n)}(r')$ is the particle density in the atomic sphere around the nucleus at $\mathbf{R}_n$, and where because of spherical symmetry only the $\ell = 0$ terms enter the discussion. The first term in the last expression gives the cellcontribution to the Madelung potential, while the second, $V_{\ell=0}(R_{WS})$, represents the effect of all other charges in the system. Because this effective potential, as determined in the course of an electronic structure calculation, also includes the effect of the nucleus in the cell, that contribution, $-2Z^n/R_{WS}$, must be subtracted explicitly in the last expression.

To obtain the Hellmann-Feynman force (HFF), we consider the derivative of the electrostatic potential. This can be expressed in terms of the *dipole* terms, $\ell = 1$, in the density and the Coulomb potential. Noting that

$$-\nabla_{\mathbf{R}} V_M(\mathbf{R}_n) = \nabla_{\mathbf{r}} V(\mathbf{r})|_{\mathbf{r}=0}, \tag{14.5}$$

and developing $V(\mathbf{r})$ in spherical harmonics, we obtain

$$\begin{aligned} \nabla_{\mathbf{r}} V(\mathbf{r})|_{\mathbf{r}=0} &= \nabla_{\mathbf{r}} \sum_L V_L(r) Y_L(\hat{r}) \\ &= \nabla_{\mathbf{r}} \sum_L r^\ell \phi_L(r) Y_L(\hat{r}). \end{aligned} \tag{14.6}$$

Therefore, for the component of the force along the αth direction, ($\alpha = x$, y, or z) we obtain

$$\begin{aligned} \nabla_{r_\alpha} V(\mathbf{r})|_{\mathbf{r}=0} &= \sum_L \left\{ \left[\nabla_{r_\alpha} \frac{V_L(r)}{r^\ell}\right] r^\ell Y_L(\hat{r}) + \frac{V_L(r)}{r^\ell}\left[\nabla_{r_\alpha} r^\ell Y_L(\hat{r})\right]\right\} \\ &= \sum_L \frac{V_L(r)}{r^\ell}\left[\nabla_{r_\alpha} r^\ell Y_L(\hat{r})\right]|_{r=0}, \end{aligned} \tag{14.7}$$

since the first term vanishes because $\frac{V_L(r)}{r^\ell}|_{r=0}$ is a constant. With

$$r = \left(x^2+y^2+z^2\right)^{1/2}, \tag{14.8}$$

and

$$Y_{1x} = \sqrt{\frac{3}{4\pi}}\frac{x}{r}, \quad Y_{1y} = \sqrt{\frac{3}{4\pi}}\frac{y}{r}, \quad Y_{1z} = \sqrt{\frac{3}{4\pi}}\frac{z}{r}, \tag{14.9}$$

we have

$$F^{n,\alpha} = Z^n \sqrt{\frac{3}{4\pi}}\phi_{1\alpha}(0), \tag{14.10}$$

where the potential $\phi_{1\alpha}(0)$ is obtained as the limit

$$\phi_{1\alpha}(0) = \lim_{r\to 0} \frac{V_{1\alpha}}{r}. \tag{14.11}$$

Now, the Lth component of the potential at the center of a sphere of radius $R_{\rm WS}$ can be written in the form,

$$\begin{aligned} \lim_{r\to 0} \frac{V_L(r)}{r^\ell} &= \frac{8\pi}{2\ell+1}\int_0^{R_{\rm WS}} dr'(r')^{-\ell+1} n_L(r') \left\{1 - \left(\frac{r'}{R_{\rm WS}}\right)^{2\ell+1}\right\} \\ &+ \frac{\tilde{V}_L^n(R_{\rm WS})}{R_{\rm WS}}, \end{aligned} \tag{14.12}$$

where

$$\tilde{V}_L^n(r) = V_L^n(r) + \frac{2Z^n}{r}. \tag{14.13}$$

Setting $L = 0$ in Eq. (14.12) yields the Madelung potential, Eq. (14.4), while setting $L = 1$ leads to the final expression for the HFF,

$$F^{n\alpha} = 2\sqrt{\frac{4\pi}{3}} Z^n \left\{\int_0^{R_{\rm WS}} dr' n_{1\alpha}(r') \left[1 - \left(\frac{r'}{R_{\rm WS}}\right)^3\right] + \frac{3}{4\pi}\frac{\tilde{V}_{1\alpha}(R_{\rm WS})}{R_{\rm WS}}\right\}. \tag{14.14}$$

14.3 Limits of the HF Force

Even though the Hellmann-Feynman theorem leads to a well-defined expression for the force on a nucleus, its practical implementation is somewhat problematic. The main difficulty lies in the inaccuracies that usually arise in calculating the density, $n(\mathbf{r})$, particularly near the nucleus of an atom. Unlike the ground-state density, which is governed by a variational principle, the force has no extremal properties associated with its calculation. Therefore, errors in the density, $\Delta n(\mathbf{r})$, produce errors of $[\Delta n(\mathbf{r})]^2$ in the energy but result in errors of order $\Delta n(\mathbf{r})$ in the force. Additional difficulties are associated with the dependence of the force on the $\ell = 1$ component of the density, $n_{1m}(\mathbf{r})$, in contrast with the energy that depends on the entire, and more robust, quantity $n(\mathbf{r})$. Let us now attempt to clarify these notions.

The ground-state energy, $E[n(\mathbf{r})]$, is an extremum in a variational scheme based on the grand-canonical energy functional, Eq. (1.188),

$$\tilde{E}[n(\mathbf{r})] = E[n(\mathbf{r})] - E_{\mathrm{F}} \left[\int \mathrm{d}^3 r n(\mathbf{r}) - N \right]. \tag{14.15}$$

The variational condition

$$\delta \tilde{E}[n(\mathbf{r})] = \int \mathrm{d}^3 r \left(\frac{\delta E}{\delta n} - E_{\mathrm{F}} \right) \delta n(\mathbf{r}) = 0 \tag{14.16}$$

guarantees that variations, $\Delta n = n - n_0$, away from the ground-state e-quilibrium density, n_0, lead to errors $\Delta \tilde{E}$ that are of the second order in Δn,

$$\Delta \tilde{E} \simeq O\left\{ (\Delta n)^2 \right\}. \tag{14.17}$$

Therefore, because of its variational properties, the total energy is an order of magnitude more accurate than the particle density.

On the other hand, the force on a nucleus is given as the difference between two terms, both of large magnitude, the attractive force due to electrons and the repulsive one due to the other nuclei in the system. The magnitudes of these terms are much larger than their difference and small errors in the density can result in gross inaccuracies. To see this, consider the energy derivative,

$$\frac{\mathrm{d}E}{\mathrm{d}\lambda} = \frac{\partial E}{\partial \lambda}|_{n_0(\mathbf{r})=\mathrm{const}} + \int \mathrm{d}\mathbf{r} \frac{\partial E}{\partial n} \frac{\partial n}{\partial \lambda}, \tag{14.18}$$

where λ is a parameter such as the position of a nucleus, (as discussed in connection with the Hellmann-Feynman theorem in Section 1.6). We now obtain

$$\frac{\mathrm{d}E}{\mathrm{d}\lambda} = \frac{\partial E}{\partial \lambda}|_{n_0(\mathbf{r})=\mathrm{const}} + E_{\mathrm{F}} \frac{\partial N(\lambda)}{\partial \lambda}, \tag{14.19}$$

where the condition

$$N(\lambda) = \int \mathrm{d}^3 r n(\mathbf{r}, \lambda) \tag{14.20}$$

has been used. Provided that the particle number remains fixed, we have,

$$\frac{\mathrm{d}E}{\mathrm{d}\lambda} = \frac{\partial E}{\partial \lambda}|_{n_0(\mathbf{r})=\mathrm{const}}. \tag{14.21}$$

The second term in Eq. (14.19) vanishes since the exact density is stationary with respect to λ. Therefore, for the exact solution of the Kohn-Sham equations, only the dependence of the total energy on λ comes into play. Setting $\lambda = Z^n$ leads to energy differences, while $\lambda = \mathbf{R}_n$ provides an expression for the force and leads to Eq. (14.1).

However, when errors occur in the solution of the Kohn-Sham equations for $n(\mathbf{r})$, the relation,

$$\int \mathrm{d}^3 r \frac{\delta E}{\delta n}\frac{\partial n}{\partial \lambda} = E_{\mathrm{F}}\frac{\partial N}{\partial \lambda} = 0 \tag{14.22}$$

no longer holds. Instead, we write

$$\mathbf{F}^n = -\frac{\mathrm{d}E}{\mathrm{d}\mathbf{R}_n} = -\frac{\partial E}{\partial \mathbf{R}_n}|_{n(\mathbf{r})=\mathrm{const}} - \int \mathrm{d}^3 r \frac{\delta E}{\delta n}\frac{\partial n}{\partial \mathbf{R}_n} = \mathbf{F}_{\mathrm{HF}} + \mathbf{F}_{\mathrm{C}}, \tag{14.23}$$

where $\mathbf{F}_{\mathrm{C}}$ is a correction term to the Hellmann-Feynman force, $\mathbf{F}_{\mathrm{HF}}$, associated with the errors in the density. This term vanishes for the *exact* solution, $n(\mathbf{r})$, but this exact solution is often very difficult to attain. Numerical procedures invariably lead to uncertainties that in the case of the force cannot be accounted for within a variational formalism. The result is that $\mathbf{F}_{\mathrm{C}}$ depends linearly on Δn,

$$\mathbf{F}_{\mathrm{C}} \simeq O\{(\Delta n)\}. \tag{14.24}$$

Perhaps the most important difficulty in the evaluation of the exact density is connected with the treatment of the core states, near the nucleus. In contrast to "chemical" properties that are calculated in terms of the valence charge density, the force on the nucleus, calculated through the HFT, depends on the density of states near the core. Usually, one treats the core density as being spherically symmetric whereas deviations from spherical symmetry exist particularly for p ($\ell = 1$) states. These deviations are indeed of small magnitude and, because they are screened at large distances, they do not influence the valence charge. However, they play a decisive role in the calculation of the force, leading to the need for the development of a modified formalism to account for them.

14.3.1 Modified HF force

The quick answer to the difficulties just mentioned appears to be a more accurate calculation of the charge density near the nucleus that takes the presence of non-spherical terms into account. However, not only is this a very difficult numerical problem, but the deviations from spherical symmetry are usually so small that it is useful to attempt to calculate corrections to the force that result from spherical densities. This is done by means of evaluating the difference in the force caused by the difference in the potential due to the non-spherical terms.

We begin with Eq. (14.23) in which the force on nucleus Z^n is written as the sum of the Hellmann-Feynman contribution and a correction term. Since we expect the corrections to be associated with errors in the core electron density, we break the total density up into a valence and a core contribution,

$$n(\mathbf{r}) = n_c(\mathbf{r}) + n_v(\mathbf{r}), \tag{14.25}$$

where the valence density is taken to be the exact solution of the Kohn-Sham equations so that

$$\frac{\delta E}{\delta n_v(\mathbf{r})} = E_F. \tag{14.26}$$

To derive an expression for the correction term, we assume again that n_v is an exact solution of the Kohn-Sham equations, and obtain

$$\begin{aligned} \mathbf{F}_C^n &= -\int d^3r \frac{\delta E}{\delta n_c(\mathbf{r})} \frac{\partial n_c(\mathbf{r})}{\partial \mathbf{R}_n} - \int d^3r \frac{\delta E}{\delta n_v(\mathbf{r})} \frac{\partial n_v(\mathbf{r})}{\partial \mathbf{R}_n} \\ &= -\int d^3r \frac{\delta E}{\delta n_c(\mathbf{r})} \frac{\partial n_c(\mathbf{r})}{\partial \mathbf{R}_n}, \end{aligned} \tag{14.27}$$

which depends only on the density of core electrons.

In proceeding further, we assume that the changes in core density are contained inside the cell itself, and that they are strongly screened. For the core density in the vicinity of $\mathbf{R}_n$, we assume an expression of the form,

$$n_c(\mathbf{r}) \simeq \sum_\nu |\psi_\nu(\mathbf{r} - \mathbf{R}_n)|^2, \tag{14.28}$$

where the $\psi_\nu(\mathbf{r})$ satisfy the Schrödinger equation,

$$H_0 \psi_\nu = \epsilon_\nu \psi_\nu, \tag{14.29}$$

with H_0 being the isotropic part of the total Hamiltonian, H. Now, the total Hamiltonian consists of this isotropic part and a perturbation, ΔV, which includes all the non-spherical contributions to the potential, $\nu_{\text{eff}}(\mathbf{r})$. Therefore, we can set $n_c(\mathbf{r} - \mathbf{R}_n) = n_c(|\mathbf{r} - \mathbf{R}_n|)$. Now, from Eq. (1.173), we have (where $\Re$ denotes the real part)

$$\mathbf{F}_C^n = -2\Re \sum_\nu \langle \frac{\partial \psi_\nu}{\partial \mathbf{R}_n} | H - \epsilon_\nu | \psi_\nu \rangle, \tag{14.30}$$

a term that would clearly vanish if the ψ_ν were the eigenfunctions corresponding to H. In the present case, Eq. (14.29) allows us to write,

$$\begin{aligned} \mathbf{F}_C^n &= -2\Re \int d^3r \frac{\partial \psi_\nu^*(\mathbf{r} - \mathbf{R}_n)}{\partial \mathbf{R}_n} \Delta V(\mathbf{r} - \mathbf{R}_n) \psi_\nu(\mathbf{r} - \mathbf{R}_n) \\ &= -\int d^3r \left[\frac{\partial}{\partial \mathbf{R}_n} n_c(\mathbf{r} - \mathbf{R}_n) \right] \Delta V(\mathbf{r} - \mathbf{R}_n) \\ &= \int d^3r' \left[\frac{\partial}{\partial \mathbf{r}'} n_c(\mathbf{r}') \right] \Delta V(\mathbf{r}') \\ &= \int d^3r' \left[\frac{\partial}{\partial r'} n_c(\mathbf{r}') \right] \frac{\mathbf{r}'}{r'} \Delta V(\mathbf{r}'). \end{aligned} \tag{14.31}$$

We can now expand the potential, $\Delta V(\mathbf{r})$, in terms of spherical harmonics and obtain the α-component of the force,

$$F_{\mathrm{C}}^{n,\alpha} = \int_0^{R_{\mathrm{WS}}} r^2 \mathrm{d}\Omega \frac{\partial}{\partial r} \left[n_{\mathrm{c}}(\mathbf{r}) \right] \sum_L \Delta V_L(r) Y_L(\hat{r}) \frac{r_\alpha}{r}, \tag{14.32}$$

where, because of the highly localized nature of the core density, the integral can be extended to the WS radius. Furthermore, since r_α / r can be written in terms of $Y_{1m}(\hat{r})$, we can integrate over the cell volume Ω and obtain

$$\begin{aligned} F_{\mathrm{C}}^{n,\alpha} &= \sqrt{\frac{4\pi}{3}} \int_0^{R_{\mathrm{WS}}} r^2 \mathrm{d}r \frac{\partial}{\partial r} n_{\mathrm{c}}(\mathbf{r}) \Delta V_{1\alpha}(r) \\ &= -\sqrt{\frac{4\pi}{3}} \int_0^{R_{\mathrm{WS}}} n_{\mathrm{c}}(\mathbf{r}) \left\{ \frac{\partial}{\partial r} \left[r^2 \Delta V_{1\alpha}(r) \right] \right\}. \end{aligned} \tag{14.33}$$

We now set,

$$\tilde{n}_{\mathrm{c}}(r) = \int \mathrm{d}\Omega r^2 n_{\mathrm{c}}(r) = 4\pi r^2 n_{\mathrm{c}}(r), \tag{14.34}$$

and realize that because the nucleus contributes only to the $\ell = 0$ term of the effective potential, the quantity $\Delta V_{1\alpha}(r)$ is simply equal to $\nu_{\mathrm{eff},1\alpha}(r)$, and we can write Eq. (14.33) in the form,

$$F_{\mathrm{C}}^{n,\alpha} = \sqrt{\frac{4\pi}{3}} \int_0^{R_{\mathrm{WS}}} \mathrm{d}r \frac{\tilde{n}_{\mathrm{c}}(r)}{4\pi} \left[\frac{2}{r} \nu_{\mathrm{eff},1\alpha}(r) + \frac{\partial}{\partial r} \nu_{\mathrm{eff},1\alpha}(r) \right]. \tag{14.35}$$

Combining this expression with that for the Hellmann-Feynman force, we obtain a generalized expression for the force on the nucleus at $\mathbf{R}_n$,

$$\begin{aligned} \mathbf{F}^n &= \mathbf{F}_{\mathrm{HF}}^n + \mathbf{F}_{\mathrm{C}}^n = Z^n \nabla_{\mathbf{r}} V(\mathbf{r})|_{\mathbf{r}=0} \\ &- \int \mathrm{d}\mathbf{r} n_{\mathrm{c}}(\mathbf{r}) \frac{\partial}{\partial \mathbf{r}} \left[\nu_{\mathrm{eff}}(\mathbf{r}) \right]_{\ell=1}, \end{aligned} \tag{14.36}$$

or, explicitly in terms of components,

$$\begin{aligned} F^{n,\alpha} &= 2\sqrt{\frac{4\pi}{3}} Z^n \left\{ \int_0^{R_{\mathrm{WS}}} \mathrm{d}r n_{1\alpha} \left[1 - \left(\frac{r}{R_{\mathrm{WS}}} \right)^3 \right] + \frac{3}{4} \frac{\tilde{V}_{1\alpha}(R_{\mathrm{WS}})}{R_{\mathrm{WS}}} \right\} \\ &- \sqrt{\frac{4\pi}{3}} \int_0^{R_{\mathrm{WS}}} \mathrm{d}r \frac{\tilde{n}_{\mathrm{c}}}{4\pi} \left[\frac{2}{r} \nu_{\mathrm{eff},1\alpha}(r) + \frac{\partial}{\partial r} \nu_{\mathrm{eff},1\alpha}(r) \right]. \end{aligned} \tag{14.37}$$

We see from these expressions that the HF contribution consists only of the electrostatic contribution to the force. The extra term provides the corrections associated with the inaccuracies in determining the core charge densities.

The procedure just described provides only one way of assessing the effect on the force of the inaccuracies that usually accompany the calculation of the core density. Alternative methods have also been suggested, such as those based on the ***frozen potential approximation*** [2], and the reader is referred to the literature for details.

14.3.2 Corrections for extremal properties

The previous discussion has indicated that it is necessary to modify the Hellmann-Feynman expression for the force in order to correct for possible inaccuracies in the determination of the density of core electrons. This allows one at least to account for such errors that are linear in variations in the density.

Unlike the total energy, the force does not satisfy an extremal principle that would lead to errors of second order in Δn. In this section, we extend our discussion so that it incorporates the extremal properties of the valence electrons. We begin with the generalized energy functional, $\tilde{E}[n(\mathbf{r})]$, defined implicitly in Eq. (1.188). The kinetic energy functional, $T_s[n(\mathbf{r})]$, of non-interacting electrons can be written in the form,

$$\begin{aligned} T_s[n(\mathbf{r})] &= 2\sum_{\nu} \epsilon_\nu - \int d^3r n(\mathbf{r}) \nu_{\text{eff}}(\mathbf{r}) \\ &= \int^{E_F} dE n(E) - \int d^3r n(\mathbf{r}) \nu_{\text{eff}}(\mathbf{r}) \\ &= E_F N(E_F) + \int [E - E_F]\, n(E) dE \\ &- \int d^3r n(\mathbf{r}) \nu_{\text{eff}}(\mathbf{r}). \end{aligned} \tag{14.38}$$

We now divide the generalized energy functional into single particle (sp) and double counting (dc) contributions

$$\tilde{E}[n(\mathbf{r})] = E_{\text{sp}} + E_{\text{dc}}, \tag{14.39}$$

as follows,

$$E_{\text{sp}} = T_s[n(\mathbf{r})] + \int d^3r n(\mathbf{r}) \nu_{\text{eff}}(\mathbf{r}) - E_F\left[N(E_F) - N\right], \tag{14.40}$$

and

$$E_{\text{dc}} = -\int d^3r n(\mathbf{r}) \nu_{\text{eff}}(\mathbf{r}) + U[n(\mathbf{r})] + E_{\text{xc}}[n(\mathbf{r})]. \tag{14.41}$$

Here, $N(E_F)$ denotes the density of states at the Fermi level, and N is the total number of particles. We note that expressing the kinetic energy as a sum over single-particle states causes both E_{sp} and E_{dc} to depend explicitly on $\nu_{\mathrm{eff}}(\mathbf{r})$. However, their sum is independent [3] of explicit variations in $\nu_{\mathrm{eff}}(\mathbf{r})$. Using the expressions above, and displaying explicitly the external potential in which the electrons move, we can write

$$\begin{aligned}\tilde{E}[n] &= E_F N - \int^{EF} N(E)\mathrm{d}E - \int \mathrm{d}^3 r n(\mathbf{r})\nu_{\mathrm{eff}}(\mathbf{r}) \\ &+ \int \mathrm{d}^3 r n(\mathbf{r})\nu_{\mathrm{ext}}(\mathbf{r}) + \int \mathrm{d}^3 r \int \mathrm{d}^3 r' \frac{n(\mathbf{r})n(\mathbf{r}')}{|\mathbf{r}-\mathbf{r}'|} \\ &+ \int \mathrm{d}^3 r \epsilon_{\mathrm{xc}}[n(\mathbf{r})]n(\mathbf{r}),\end{aligned} \tag{14.42}$$

and

$$\begin{aligned}\frac{\delta\{U[n]+E_{\mathrm{xc}}[n]\}}{\delta n(\mathbf{r})} &\equiv \nu_{\mathrm{eff}}(\mathbf{r}) = 2\int \mathrm{d}^3 r' \frac{n(\mathbf{r})n(\mathbf{r}')}{|\mathbf{r}-\mathbf{r}'|} \\ &+ \nu_{\mathrm{ext}}(\mathbf{r}) + \mu_{\mathrm{xc}}(\mathbf{r}).\end{aligned} \tag{14.43}$$

In carrying out numerical calculations within a Green function formalism, we let G correspond not to the exact potential, $\nu_{\mathrm{eff}}(\mathbf{r})$, but to a slightly different potential, $\nu_{\mathrm{eff}}(\mathbf{r})+\delta\tilde{\nu}_{\mathrm{eff}}(\mathbf{r})$. Then, the change in the Green function to first order in $\delta\tilde{\nu}_{\mathrm{eff}}$ takes the form,

$$\delta G(\mathbf{r},\mathbf{r}';E) = \int \mathrm{d}^3 r'' G(\mathbf{r},\mathbf{r}'';E)\delta\tilde{\nu}_{\mathrm{eff}}(\mathbf{r}'')G(\mathbf{r}'',\mathbf{r}';E), \tag{14.44}$$

which leads to a corresponding expression for the change in the density through the relation,

$$\delta n(\mathbf{r}) = -\frac{2}{\pi}\Im \int^{E_F} \mathrm{d}E\delta G(\mathbf{r},\mathbf{r};E), \tag{14.45}$$

and

$$\delta n(E) = -\frac{2}{\pi}\Im \int \mathrm{d}^3 r \delta G(\mathbf{r},\mathbf{r};E). \tag{14.46}$$

Therefore, for the change, δE_{sp}, we obtain the expression,

$$\delta E_{sp} = -\int \mathrm{d}^3 r \hat{n}(\mathbf{r})\delta\tilde{\nu}_{\mathrm{eff}}(\mathbf{r}), \tag{14.47}$$

where

$$\hat{n}(\mathbf{r}) = \frac{2}{\pi}\Im \int^{E_F} \mathrm{d}E \int^{E} \mathrm{d}E' \int \mathrm{d}^3 r' G(\mathbf{r}',\mathbf{r};E')G(\mathbf{r},\mathbf{r}';E'). \tag{14.48}$$

For the change in $E_{\rm dc}$, we find

$$\begin{aligned} \delta E_{\rm dc} &= \int \mathrm{d}^3 r n(\mathbf{r}) \delta \nu_{\rm ext}(\mathbf{r}) - \int \mathrm{d}^3 n(\mathbf{r}) \delta \tilde{\nu}_{\rm eff}(\mathbf{r}) \\ &+ \int \mathrm{d}^3 r \nu_{\rm eff}[\mathbf{r}, n(\mathbf{r})] \delta n(\mathbf{r}) - \int \mathrm{d}^3 r \tilde{\nu}_{\rm eff}(\mathbf{r}) \delta n(\mathbf{r}), \end{aligned} \tag{14.49}$$

where $\nu_{\rm eff}[\mathbf{r}, n(\mathbf{r})]$ denotes an effective potential corresponding to a trial charge density, $n(\mathbf{r})$. Therefore, this potential corresponds to neither the correct charge density nor the correct effective potential of the ground state. We now see that we can identify three different contributions to $\delta\tilde{E}[n]$ arising, respectively, from the change in the external potential, the trial potential, and the trial density,

$$\begin{aligned} \delta \tilde{E} &= \int \mathrm{d}^3 r n(\mathbf{r}) \delta \nu_{\rm ext}(\mathbf{r}) \\ &+ \int \mathrm{d}^3 r \left\{ \nu_{\rm eff}[\mathbf{r}, n(\mathbf{r})] - \tilde{\nu}_{\rm eff}(\mathbf{r}) \right\} \delta n(\mathbf{r}) \\ &+ \int \mathrm{d}^3 r \left[\hat{n}(\mathbf{r}) - n(\mathbf{r}) \right] \delta \tilde{\nu}_{\rm eff}(\mathbf{r}). \end{aligned} \tag{14.50}$$

The first line above corresponds to the Hellmann-Feynman theorem. Differentiation (variation) with respect to nuclear position leads to the force on the nucleus. The second term measures the effects of the lack of self-consistency in the potential, say between the input and output potentials at a particular stage of the iteration process to self-consistency. This term disappears in a fully converged self-consistent calculation in which $\tilde{\nu}_{\rm eff} \equiv \nu_{\rm eff}$. Otherwise, this term requires the calculation of the change in density, $\delta n(\mathbf{r})$, associated with a displaced nucleus. Finally, the third and last term is associated with inaccuracies in the calculation of the Green function itself, and vanishes when $\hat{n}(\mathbf{r})$ coincides with $n(\mathbf{r})$. This is the case when the calculated Green function satisfies the formal identity,

$$-\nabla_E G(E) = -\nabla_E \frac{1}{E + \mathrm{i}\epsilon - H} = \left(\frac{1}{E + \mathrm{i}\epsilon - H} \right)^2 = G(E) G(E). \tag{14.51}$$

When this is not the case, one must calculate explicitly the trial potential, $\delta\tilde{\nu}_{\rm eff}(\mathbf{r})$, for the displaced nucleus.

In general, the last two correction terms require the calculation of both the change in the potential *and* in the density associated with a displaced nucleus. Thus, they correspond to the calculation of the force obtained through *physical* differentiation, in which one calculates ΔE by subtracting the results of self-consistent calculations at different configurations of the system, such as those corresponding to the undisplaced and the infinitesimally displaced nuclei.

14.3.3 Corrections for incomplete basis

In carrying out a self-consistent calculation of the electronic structure, it is invariably necessary to employ expansions in basis functions, χ_α, such as plane waves or angular momentum states. For example, we have

$$n(\mathbf{r}) = \sum_\nu |\phi_\nu(\mathbf{r})|^2, \quad \text{where} \quad \phi_\nu(\mathbf{r}) = \sum_\alpha C_{\nu\alpha}\chi_\alpha(\mathbf{r}). \tag{14.52}$$

The expansions of the wave function ϕ_ν over the set of the χ_α can rarely be carried to formal convergence. Consequently, the resulting expressions for the density, the energy, and the force contain inaccuracies associated, effectively, with the use of an incomplete basis set (IBS). In this section, we obtain expressions describing the effects of an IBS in the calculated forces. In the literature, these effects are referred to as Pulay corrections [4, 5].

We examine the correction term,

$$\mathbf{F}_\mathrm{C} = -2\Re \sum_\nu \langle \frac{\partial \phi_\nu}{\partial \mathbf{R}_n} | H - \epsilon_\nu | \phi_\nu \rangle, \tag{14.53}$$

where ϵ_ν guarantees the normalizations of the ϕ_ν,

$$\frac{\partial}{\partial \mathbf{R}_n} \langle \phi_\nu | \phi_\nu \rangle = 0. \tag{14.54}$$

Using the expansion of the ϕ_ν in terms of the basis functions, χ_α, in Eq. (14.52), we have

$$\begin{aligned} \mathbf{F}_\mathrm{C} &= -2\Re \sum_{\nu\alpha\alpha'} \frac{\partial C^*_{\nu\alpha}}{\partial \mathbf{R}_n} C_{\nu\alpha'} \langle \chi_\alpha | H - \epsilon_\nu | \chi_{\alpha'} \rangle \\ &- 2\Re \sum_{\nu\alpha\alpha'} C^*_{\nu\alpha} C_{\nu\alpha'} \langle \frac{\partial \chi_\alpha}{\partial \mathbf{R}_n} | H - \epsilon_\nu | \chi_{\alpha'} \rangle \\ &= \mathbf{F}_\mathrm{NSC} + \mathbf{F}_\mathrm{IBS}. \end{aligned} \tag{14.55}$$

The first term on the right side of the equals sign vanishes when the Euler-Lagrange equations,

$$\sum_\alpha [\langle \chi_\alpha | H | \chi_{\alpha'} \rangle - \epsilon_\nu \langle \chi_\alpha | \chi_\alpha \rangle] = 0, \tag{14.56}$$

are fulfilled in the space of the basis functions, $\chi_\alpha(\mathbf{r})$. Therefore, this non-self-consistent (NSC) term is associated with the lack of self-consistency in the solution of the equations of DFT. The second term,

$$\mathbf{F}_\mathrm{IBS} = -2\Re \sum_{\nu\alpha\alpha'} C^*_{\nu\alpha} C_{\nu\alpha'} \langle \frac{\partial \chi_\alpha}{\partial \mathbf{R}_n} | H - \epsilon_\nu | \chi_{\alpha'} \rangle, \tag{14.57}$$

corresponds to the use of an incomplete expansion for the ϕ_ν. This correction term has no classical interpretation and corresponds to errors in the solution of the variational principle, $\delta E[n] = 0$. It disappears when the eigenvalue problem

$$(H - \epsilon_\nu)\,\phi_\nu = 0 \tag{14.58}$$

is solved exactly, as is required for the Hellmann-Feynman theorem to hold. Finally, using for $\mathbf{F}_{\mathrm{NSC}}$ the expression in the second term of Eq. (14.50), we find [5, 6]

$$\begin{aligned}
\mathbf{F}^n_{\mathrm{C}} &= \mathbf{F}_{\mathrm{NSC}} + \mathbf{F}_{\mathrm{IBS}} \\
&= -\int \mathrm{d}^3r\,(\nu_{\mathrm{eff}} - \tilde{\nu}_{\mathrm{eff}})\,\frac{\partial n(\mathbf{r})}{\partial \mathbf{R}_n} \\
&- 2\Re \sum_{\nu\alpha} C^*_{\nu\alpha} \int \mathrm{d}^3r \frac{\partial \chi^*_\alpha}{\partial \mathbf{R}_n}\,[H - \epsilon_\nu]\,\phi_\nu(\mathbf{r}).
\end{aligned} \tag{14.59}$$

14.3.4 Numerical results

The formalism of this section has been used [1] to calculate the forces on host atoms in the neighborhood of 3d and 4d impurities embedded in Cu within MST. The results are summarized in Fig. 14.1. In this figure, the full curve and the left scale refer to full-potential calculations, while the dashed curve and the right scale refer to calculations performed within the ASA. A characteristic trend of these calculations is that the forces are extremely sensitive to any approximation for the valence charge density. From the different scales in the figure, we see that the ASA can result in forces larger by more than an order of magnitude compared with the results of the full-potential calculation. At the same time, they do give the same trends as the full-cell results. This is in agreement with the general experience with non-self-consistent calculations: They drastically overshoot but often in the right direction.

14.4 Many-Body Potentials

In the previous sections, we developed the formalism for the application of the Hellmann-Feynman theorem to calculate the force on a nucleus. The aim in studying these developments is eventually to be able to understand the mechanisms governing nuclear motion or atomic displacement and propagation in metals and other materials. If the forces can be accurately calculated, then it is possible to simulate nuclear motion and atomic rearrangements by means of computer simulation methods, such as that of molecular dynamics. Such simulations have been carried out for a number

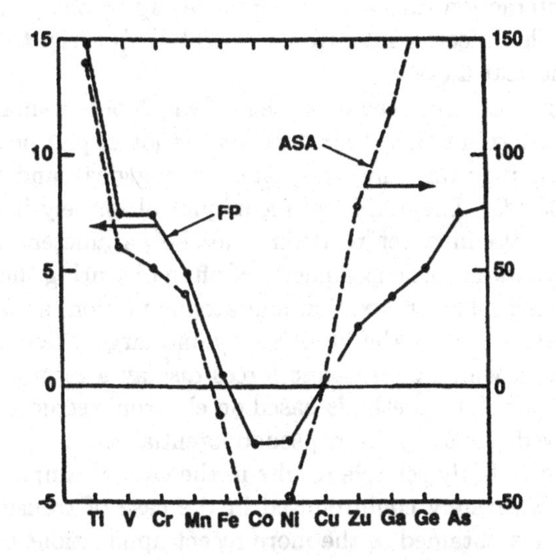

Figure 14.1: **The force F^x in the x-direction (vertical axes) in mRy/a_B, where a_B is the Bohr radius, exerted on a nearest-neighbor Cu atom by 3d and 4d impurities in Cu. The full curve and the left scale refer to full-potential calculations, while the dashed curve and the right scale refer to calculations performed within the ASA [1].**

of physical systems on the basis of interaction potentials between pairs of atoms. The main idea behind these simulations is to solve Newton's equations of motion based on the forces acting on the atoms that are given by the gradient of an interatomic potential[2].

The derivation of interatomic potentials ultimately rests on the construct of light electrons equilibrating rapidly to the changed nuclear positions. One may then consider the total energy of a system of electrons and nuclei as consisting of the kinetic energy of the nuclei and terms that mimic their pairwise interaction through a medium set up by the presence of the electron cloud. This effective interatomic potential gives rise to a force that in turn guides nuclear motion.

Various techniques have been proposed (of which only a small, representative sample is referred to here) for the construction of pair potential functions [8-30], including semi-empirical, phenomenological, and those based on *ab initio* methods. The empirical techniques ultimately involve fitting to experimental data in order to determine the parameters of a theory, and this has the undesirable consequence of often obscuring the underlying mechanisms driving lattice relaxation and atomic motion, as well as being sensitive to the data used in the fit. Also, by and large, these methods fail to treat systems in which interatomic forces display a strong angular dependence. First-principles methods based on electronic structure have also been used, derived primarily from pseudopotential theory [10-12, 28-30]. These methods give fairly reliable results in the case of simple metals, but their reliability is severely compromised in the case of transition metals. Also, the potentials obtained in the more recent applications of pseudopotential theory [28-30] display a rather slow convergence with respect to distance or number of particles (atoms).

As shown below, multiple scattering theory provides a natural framework for the expansion of the density in many-body interactions, and one such approach has been reported [30] within the rigid muffin tin approximation. In this approach, on which essentially the present treatment is based, one expands the energy in terms of irreducible, many-body terms, in the form

$$E = \sum_i E_i^{(1)} + \frac{1}{2!}\sum_{ij} E_{ij}^{(2)} + \frac{1}{3!}\sum_{ijk} E_{ijk}^{(3)} + \cdots, \tag{14.60}$$

where the various n-site terms, $E^{(n)}_{j_1 j_2 j_3 \cdots j_n}$, are irreducible in the sense that they do not contain contributions from clusters of m sites, with $m < n$.

[2]Obtaining the spatial dependence of the forces and potentials through electronic structure calculations based on a rigid lattice is by no means straightforward. It often involves conjectures and assumptions about the functional dependence of the potential on interparticle separation that are difficult to justify.

Thus, the two-body potentials, $E_{ij}^{(2)}$, do not contain single-body contributions, the three-body terms, $E_{ijk}^{(3)}$, contain no two- or one-body contributions, and so on. In the most common application of many-body potentials, only the two-body terms are retained with forces being derived from them through spatial differentiation. However, it is rarely the case that the expansion of the energy terminates with two-body contributions. Usually, higher order terms are found to be important, and must be taken into account for an accurate representation of the energy.

Multiple scattering theory has the desirable feature of leading to uniquely defined expansions of the energy in terms of multisite interactions through the analogous expansion of the density. However, in carrying out these expansions, it is important to keep in mind that the energies treated pertain to a particular volume of the system, or a particular ionic density, and that they are the result of a self-consistent calculation. Thus, the extraction of volume-dependent forces is difficult (and ill-defined) at best. Furthermore, attempts to apply the expansion in Eq. (14.60) to combinations of atomic charge densities are prone to lead to extremely slow and unreliable convergence features. Applying the method to the results of a self-consistent calculation tends to impede fluctuation effects and to facilitate convergence to the proper limit (the result of the initial calculation).

14.4.1 Energy Expansions

We begin with an expression for the total energy of a solid, in any configuration, which can be written in the form,

$$\begin{aligned} E &= \int E\mathrm{d}E \int \mathrm{d}^3 r n_{\mathrm{v}}(\mathbf{r},E) - \frac{1}{2}\int \mathrm{d}^3 r n_{\mathrm{v}}(\mathbf{r})\left[V_{\mathrm{C}}(\mathbf{r}) + 2\mu_{\mathrm{xc}}(\mathbf{r})\right] \\ &+ \int_{\Omega} n_{\mathrm{v}}(\mathbf{r})\epsilon_{\mathrm{xc}}\mathrm{d}^3 r \\ &+ \sum_i^{\mathrm{core}} \epsilon_i - \frac{1}{2}\int \mathrm{d}^3 r n_{\mathrm{c}}(\mathbf{r})\left[V_{\mathrm{C}}(\mathbf{r}) + 2\mu_{\mathrm{xc}}(\mathbf{r})\right] \\ &+ \int_{\Omega} n_{\mathrm{c}}(\mathbf{r})\epsilon_{\mathrm{xc}}\mathrm{d}^3 r \\ &- \frac{1}{2}\sum_{\nu} Z_{\nu}\left\{\int \frac{n(\mathbf{r})\mathrm{d}^3 r}{|\mathbf{r}-\mathbf{R}_{\nu}|} - \sum_{\alpha\neq\nu}\frac{Z_{\alpha}}{|\mathbf{R}_{\alpha}-\mathbf{R}_{\nu}|}\right\}, \end{aligned} \tag{14.61}$$

where n_{v} and n_{c} denote the valence and core contribution to the electron density, ϵ_{xc} and μ_{xc} are the exchange-correlation energy and potential, respectively, Z_{ν} is the nuclear charge at $\mathbf{R}_{\nu}$, and $V_{\mathrm{C}}(\mathbf{r})$ is the electrostatic

(Hartree) potential at $\mathbf{r}$,

$$V_C(\mathbf{r}) = \int \frac{n(\mathbf{r}')d^3r'}{|\mathbf{r}-\mathbf{r}'|} - \sum_\alpha \frac{Z_\alpha}{|\mathbf{r}-\mathbf{R}_\alpha|}. \tag{14.62}$$

Let us isolate that part of the energy that is proportional to the valence charge and write it in the form,

$$\begin{aligned} E_v &= \int dE \int d^3r n_v(\mathbf{r},E) \left\{ E - \frac{1}{2}\left[V_C(\mathbf{r}) + 2\mu_{xc}(\mathbf{r}) - 2\epsilon_{xc}\right]\right\} \\ &= \int dE \int d^3r n_v(\mathbf{r},E) Y(\mathbf{r},E). \end{aligned} \tag{14.63}$$

This expression should be compared to the formally identical one derived in the previous chapter for the case of alloys (see Section 13.13).

In general, we assume that we can identify a one-body contribution, E_i, associated with each site, i, in the system[3]. We set the corresponding irreducible part, $E_i^{(1)}$ equal to E_i,

$$E_i^{(1)} = E_i. \tag{14.64}$$

Clearly, $E_i^{(1)}$ is independent of site index for all equivalent sites. Now, given a two-body contribution, E_{ij}, associated with the pair of sites, i, j, we define the corresponding irreducible two-body contribution by means of the expression,

$$E_{ij}^{(2)} = E_{ij} - E_i^{(1)} - E_j^{(1)}. \tag{14.65}$$

We note that the pair energy, E_{ij}, can be written as the sum of one- and two-body irreducible contributions,

$$E_{ij} = E_{ij}^{(2)} + E_i^{(1)} + E_j^{(1)}. \tag{14.66}$$

Proceeding in this manner, we can identify a three-body irreducible contribution associated with sites i, j, k by means of the relation,

$$E_{ijk}^{(3)} = E_{ijk} - \left(E_{ij}^{(2)} + E_{ik}^{(2)} + E_{jk}^{(2)}\right) - \left(E_i^{(1)} + E_j^{(1)} + E_k^{(1)}\right). \tag{14.67}$$

We note that the three-site energy, E_{ijk}, is the sum of one-, two-, and three-body irreducible contributions.

Clearly the definitions given above can be extended to any number of sites so that the energy of an arbitrarily large (even an infinite) cluster can be written as the sum of irreducible contributions to all orders, as is indicated in Eq. (14.60)

[3]Occasionally, a zero body contribution can also be defined.

We now prescribe ways in which both the cluster energies, $E_{i_1 i_2 \cdots, i_n}$, and the corresponding irreducible parts, $E^{(n)}_{i_1 i_2 \cdots}$, can be obtained from the results of a calculation based on MST. The various contributions to the energy arise from appropriate expansions of the density, which in turn is obtained from the form of the scattering-path operator. Thus, we write for the one-body contribution to the valence density,

$$n_{i,v}(\mathbf{r}) = -\frac{1}{\pi}\Im\langle Z^i(\mathbf{r})|\underline{t}_i|Z^i(\mathbf{r})\rangle, \tag{14.68}$$

where $\underline{t}_i$ is the scattering matrix associated with the potential in cell i, and $|Z^i\rangle$ are the corresponding regular solutions of the Schrödinger equation for the cell. We assume the use of real energies, so that no single-site term, $\langle Z^i|S^i\rangle$, with $|S^i\rangle$ being the irregular solution, enters the last expression. When the expression in Eq. (14.68) is used to replace the valence density in Eq. (14.61) and the indicated integrals are carried out, we obtain the energy E_i associated with site i, which is also equal to the irreducible part, $E_i^{(1)}$.

In order to obtain the two-site contribution to the energy associated with sites i and j, we consider the corresponding expression for the density contributed by these sites (taking the wavefunctions to be real),

$$\begin{aligned} n_{ij}(\mathbf{r}) &= -\frac{1}{2!\pi}\begin{pmatrix} \langle Z^i| & 0 \\ 0 & \langle Z^j| \end{pmatrix} \mathrm{Tr}\Im \begin{pmatrix} \underline{t}_i^{-1} & -\underline{G}_{ij} \\ -\underline{G}_{ji} & \underline{t}_j^{-1} \end{pmatrix}^{-1} \\ &\times \begin{pmatrix} |Z^i\rangle & 0 \\ 0 & |Z^j\rangle \end{pmatrix}, \end{aligned} \tag{14.69}$$

where the $\underline{G}_{ij}$ are the real-space structure constants connecting sites i and j. With this expression for the valence density in Eq. (14.61), we obtain the pair contribution to the energy associated with this particular pair of sites. Then the irreducible contribution is obtained by means of Eq. (14.65). The process just indicated can be extended to define irreducible contributions to all orders.

The last expression provides at least one way of obtaining a spatial dependence for E_{ij}, albeit in highly approximate fashion. Here, one uses $n_{ij}(\mathbf{r})$ in the form of the last expression but with the argument of the structure constants allowed to vary away from its equilibrium value. As no self-consistency is used with these displaced stomic positions, this procedure lacks formal justification in spite of its possible intuitive appeal.

Bibliography

[1] P. H. Dederichs, B. Drittler, and R. Zeller, in *Applications of multiple scattering theory to materials science*, W. H. Butler, P. H. Dederichs, A. Gonis, and R. L. Weaver (eds.), Materials Research Society Symposium Proceedings, Materials Research Society, Pittsburgh (1992), Vol.**253**, p. 185. See also, K. Abraham, B. Drittler, R. Zeller, and P. H. Dederichs, Report 2451, KFA Jülich, unpublished.

[2] J. Harris, R. O. Jones, and J. E. Müller, J. Chem. Phys. 75, 3904 (1981).

[3] J. Deutz, Dissertation, RWTH Aachen (1982) unpublished. Also, Report No. 1805, KFA Jülich, unpublished.

[4] P. Pulay, Mol. Phys. **17**, 197 (1969).

[5] C. Satoko, Chem. Phys. Lett. **83**, 111 (1981).

[6] P. Bendt and A. Zunger, Phys. Rev. Lett. **50**, 1684 (1983).

[7] R. J. M. Cotterill and M. Doyama, *Lattice defects and their interaction*, R. R. Hasiguti (ed.), Gordon and Breach, New York (1967).

[8] L. Dagens, J. Phys. F: Met. Phys. **6**, 1801 (1976).

[9] J. M. Wills and W. A. Harrison, Phys. Rev. B **28**, 4363 (1983).

[10] W. A. Harrison and S. Froyen, Phys. Rev. B **21**, 3214 (1980).

[11] W. A. Harrison, Phys. Rev. B **41**, 6008 (1990).

[12] M. W. Finnis and J. M. Sinclair, Philos. Mag. A **50**, 45 (1984).

[13] R. Rebonato, D. O. Welch, R. D. Hatcher, and J. C. Bilello, Philos. Mag. A **55**, 655 (1987).

[14] G. J. Ackland and R. Thetford, Philos. Mag. A**56**, 15 (1987).

[15] A. E. Carlsson, in *Solid State Physics: Advances in Research and Applications*, H. Ehrenreich and D. Turnbull (eds.), Academic Press, San Diego (1990), Vol. **43**, pp. 42-53.

[16] A. E. Carlsson, P. A. Fedders, and C. W. Myles, Phys. Rev. B **41**, 1247 (1990).

[17] A. E. Carlsson, Phys. Rev. B 44, 6590 (1991).

[18] For a review of some methods based on the tight-binding approximation to the Hamiltonian, see *The Recursion Method and Its Applications*, D. G. Pettifor and D. L.Weaire (eds.), Springer Verlag, Ney York (1985).

[19] S. M. Foiles, Phys. Rev. B **48**, 4287 (1993).

[20] X. Wu and J. B. Adams, Surface Science **301**, 371 (1994).

[21] M. S. Daw and M. I. Baskes, Phys. Rev. Lett. **50**, 1285 (1983).

[22] M. S. Daw and M. I. Baskes, Phys. Rev. B **29**, 6443 (1984).

[23] R. A. Johnson and D. J. Oh, J. Mater. Research **4**, 1195 (1989).

[24] A. M. Guellil and J. B. Adams, J. Mater. Research **7**, 639 (1992).

[25] J. B. Adams and S. M. Foiles, Phys. Rev. B**41**, 7441 (1992).

[26] M. I. Baskes, Phys. Rev. B **46**, 2727 (1992).

[27] J. A. Moriarty, Phys. Rev. B **5**, 2066 (1972).

[28] J. A. Moriarty, Phys. Rev. B **38**, 3199 (1988).

[29] J. A. Moriarty, Phys. Rev. B **42**, 1609 (1988).

[30] Peter Jewsbury, J. Phys. F: Metal Phys. **13**, 805 (1983).

Part IV

Surfaces and Interfaces

Chapter 15

Structure and Properties of Surfaces

15.1 General Comments

It is convenient to think of an *interface* as the region where two different materials come into contact. A *surface* can then be defined as an interface when one of the two materials is vacuum[1]. Geometrical structures that conform to the general description of an interface include grain boundaries, domain walls, stacking faults, and others. The difference between the materials forming an interface can take various forms, including changes in chemical composition, changes in structure, changes in physical properties, or combinations of these. In chemically different systems, the materials on either side of the interface region and sufficiently far away from it are characterized by different species of atoms or different concentrations of a given species. When the difference is one of structure, or crystal orientation, the arrangement of atoms in space on one side of the material cannot be obtained from that on the other by a simple translation. Therefore, the presence of an interface breaks the translational symmetry of a system in directions perpendicular to the interface itself. In addition, changes in physical properties, such as atomic or electronic density, or ferromagnetic or paramagnetic behavior may be used to define an interface. Clearly, these various features are not mutually exclusive, and interfaces formed by heterogeneous solids with different crystal structures and properties are very much in evidence in many technologically important materials.

[1]The notion of vacuum should be invoked with caution in this case. The space in the vicinity of a real surface is hardly empty, containing, for example, vapors and other emitted matter, including spilled out electrons.

The changes in the characteristics and properties of a system across an interface can occur with various degrees of abruptness. Atomically sharp interfaces can arise in layered materials formed of solids that are immiscible with respect to one another. On the other extreme lies the configuration at the critical point of a liquid in contact with its own vapor at high temperature and pressure. In this case, the thickness of the interface approaches infinity as the critical point is approached. Finally, an intermediate case arises when mutually miscible materials are brought into contact. Now, a substitutional alloy several planes thick may form, possibly accompanied by structural changes as well. Such alloy layers may greatly influence the physical properties of a material.

Both in principle as well as in practice, there are two methods of forming an interface. As a specific example, consider the formation of a surface. One can imagine a *cleavage plane*, often taken to be planar, passing through the material, thus separating the original structure into two distinct parts. Experimentally, this process corresponds to cleaving, crushing, or grinding a crystal into smaller pieces with an accompanying increase in surface area. Alternatively, one may think of joining two materials together along their surfaces to form a larger system. This process of *synthesis* can be realized experimentally through the growth of a crystal from vapor, the formation of thin films and multilayers through sputtering, or direct bicrystal formation from two separate single crystals.

In this part of the book, we are concerned primarily with the description of the electronic structure of surfaces and interfaces based on a Green-function, multiple-scattering approach. Space limitations prevent us from delving deeply into the physics of surfaces and interfaces, or even providing a complete discussion of electronic properties. Fortunately, there exist a number of excellent text books on this subject [1, 2], which the interested reader is urged to consult.

This part of the book is broadly structured as follows. After a brief exposition of the historical development of surface science, we mention experimental techniques for studying the atomic and electronic structure of surfaces. We introduce some useful parameters, such as the work function, and also derive some characteristic features of surface electronic structure based on simple quantum-mechanical arguments. In the following chapter, we show how the equation of motion for the Green function within a TB description of the Hamiltonian can be solved in the presence of a surface or an interface. We also show how the CPA along with its perturbative extensions, such as the generalized perturbation method and the embedded cluster method, can be extended to the study of alloy surfaces and interfaces. In the third chapter, we discuss the extension of the first-principles MST formalism to semi-infinite systems, including samples of applications to real systems. We also discuss the construction of self-consistent charge densities for such systems, as well as the treatment of the potential barrier near a surface.

15.2 Historical Development of Surface Science

A surface is the region where a material comes into contact with its environment. Consequently, surfaces play an important role in determining materials properties and have been of scientific interest for a very long time. Surface science can lay claim to beginnings that date to the times of the Babylonian civilization. Cuneiform writings [3] tell of a form of divination called *lecanomancy* which was connected with the behavior of oil poured into a bowl of water. In more modern times, Benjamin Franklin [4] wrote rather descriptively of the calming effect of a few drops of oil on the waters of a pond (even in the presence of winds). Even today, night fishermen will often sprinkle oil on the surface of the ocean near the shore that allows them to see clearly and harpoon fish on the ocean floor.

The beginning of surface science as commonly understood today can be traced to three important results obtained in the nineteenth century. The elements of catalytic behavior were discovered by Michael Faraday, who in 1833 realized that in the presence of platinum the reaction of hydrogen and oxygen could proceed well below the nominal combustion temperature [5]. The qualitative theory of catalytic action (so termed by Berzelius in 1936) proposed by Faraday on the basis of further scientific work remains valid to this day.

The second discovery of importance in the development of surface science is associated with the dramatic influence that can be exerted by the presence of interfaces on electronic transport. In 1874, Karl Ferdinand Braun [6] observed deviations from Ohm's law in the resistivity of a Cu/FeS sandwich. This phenomenon of *rectification* was attributed by Braun to the presence of thin surface layers at the interface.

These experimental discoveries were followed in 1877 by the publication by J. Willard Gibbs of his monumental memoir, "The Equilibrium of Heterogeneous Substances", in the *Transactions of the Connecticut Academy.* In this work, Gibbs provides a complete description of the thermodynamics of surface phases. Quite deservedly, this work is considered one of the crowning scientific achievements of the 19th century [7].

Following these developments came the work of Irving Langmuir in the early years of this century. While still a student under Nerst in Göttingen [8], Langmuir studied the dissociation of various gases produced by a hot platinum wire. Langmuir's remarkable scientific achievements in the research laboratory of General Electric include the invention of the nitrogen-filled tungsten incandescent lamp, the introduction of concepts such as the adsorption chemical bond, the surface adsorption lattice, the accommodation coefficient, and adsorption precursors, as well as studies of the work functions of metals, heterogeneous catalysis and adsorption kinetics, the introduction of a detailed model of thermionic emission and, in collaboration with Katherine Blodgett, the study of monomolecular films. For these achievements Langmuir was awarded the Nobel prize in 1932 for "outstanding discoveries and inventions within the field of surface chemistry". The field of surface science now came to be recognized in its own right.

A great deal of experimental and theoretical activity followed that contributed to the growth of this scientific endeavor. Einstein provided an explanation of the photoelectric effect, and Davisson and Germer conducted electron diffraction experiments that solidly confirmed the wave nature of matter. Both of these achievements were awarded with Nobel prizes, Einstein in 1921 and Davisson in 1937. Further work includes that of Tamm [9], Maue [10], Goodwin [11], and Shockley [12] on the nature of electronic states at crystal surfaces.

Work performed in the late 1930s and in the years during and following World War II [13-18] was concerned with the physics of adsorbates, metal/semiconductor interfaces, and rectifying junctions. This work culminated in the invention of the transistor [19], which irrevocably changed the course of technological developments.

Present-day surface science consists of well-established experimental techniques for the study of the atomic and electronic structure of surfaces. These techniques are greatly aided by the development of technology allowing the production and maintenance of high-vacuum chambers where samples can be kept clean from contamination for sufficiently long periods. This allows the performance of experiments on well-characterized surfaces whose results can be reliably compared with the predictions of theoretical investigations. As in every field of science, the interplay of theory and experiment can be expected to yield new insights and a deeper understanding of the physics of surfaces in the years to come.

15.3 Atomic Structure of Surfaces

The spatial arrangement of atoms in a material plays a crucial role in determining its physical, chemical, and mechanical properties. In attempting to understand materials properties, one is compelled to know as accurately as possible the atomic structure of the system under study. This requires the development of methods to determine the structure of a system as well as models to describe it. In the case of bulk solids, this description makes use of the so-called Bravais lattices. In the case of surfaces, one can easily conceive of a two-dimensional equivalent of a Bravais lattice, and indeed surfaces that can be so described form an important part of surface science. At the same time, such *ideal* surfaces make up only a small fraction of surfaces encountered in real systems. Real surfaces can be characterized by features that require specific description.

In this section, we briefly consider ways to describe the atomic structure of surfaces, and comment on the method of low-energy electron diffraction (LEED), which can be developed within the framework of MST, to determine this structure. The method of LEED has been discussed in great detail in the literature [20, 21] and the reader is referred there for detail descriptions of the method.

15.3.1 TLK model

Let us consider the simplest possible kind of surface, formed when a cleavage plane passes through a single Bravais lattice separating it into two distinct halves. In a so-called *ideal surface*, it is assumed that the arrangement of atoms in the surface and the bulk is unaffected by the formation of the surface. Thus, the atoms bear the same relative positions with respect to one another as they did in the bulk crystal. In this case, the surface structure can be described in terms of a two-dimensional lattice, in a manner analogous to the description of bulk systems in terms of three-dimensional Bravais lattices. A *real surface*, in contrast, may exhibit effects of *reconstruction* whereby atomic arrangements at or near the surface (or interface) region may differ drastically from those of an ideal configuration.

The structure of an ideal or real surface in general can be described in terms of the so-called *terrace-ledge-kink*, or TLK, model [23]. Within this model a surface is classified as *singular*, *vicinal*, or *rough*. Singular surfaces are *smooth* on an atomic scale and correspond to surface orientations lying along low-index planes of the bulk crystal such as the (100), (110), or (111) planes of a cubic structure. Such a surface can be represented as an arrangement of cubes (or spheres), each containing a single atom, as indicated in Fig. 15.1(a) for the (100) surface of a simple cubic crystal.

When a surface is formed along a high index plane or along a direction misoriented with respect to a singular orientation, the smoothness of the profile associated with a singular surface is often lost. A vicinal surface may contain flat terraces of atomic thickness of the closest singular orientations arranged so as to account for the surface orientation with respect to a singular orientation. Such an arrangement is illustrated in panel (b) of Fig. 15.1 for the case of a (100) vicinal surface of a single-cubic crystal misoriented along one direction. This panel clearly shows the terrace structure of the surface. Panel (c) shows a vicinal surface for the same crystal structure misoriented along two directions relative to a singular orientation.

Finally, a surface that is so disordered that cannot be classified as singu-

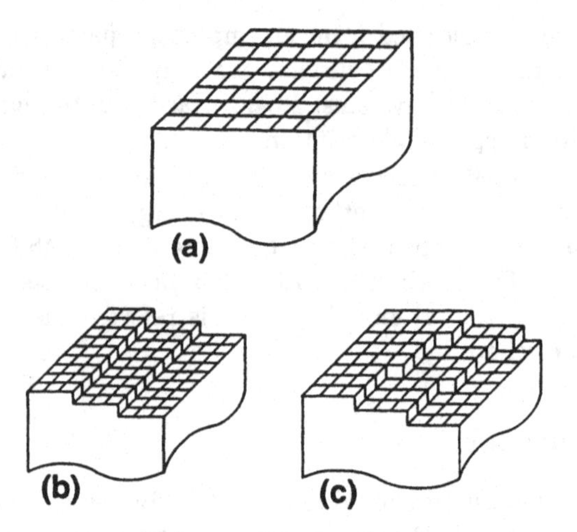

Figure 15.1: **TLK model for a singular surface, panel (a), and two vicinal surfaces, panels (b) and (c).**

lar or vicinal within the TLK model it is said to be *rough*. Thermodynamic arguments indicate that rough surfaces are unlikely to form as equilibrium structures at low temperature, and experimental data do not seem to support their existence.

The TLK model can be used to obtain a pictorial representation of ideal as well as real surfaces that have undergone a certain amount of reconstruction. These descriptions can be compared with structures observed experimentally, or predicted on the basis of theoretical arguments. Among experimental techniques for determining surface structure (such as field ion microscopy), possibly the most commonly used procedure is that associated with LEED. The physical basis of a LEED experiment is briefly mentioned in the following subsection. A more detailed discussion of LEED crystallography and the formal aspects of the analysis of LEED spectra is given later in this part of the book.

15.3.2 Low energy electron diffraction

As is well known, the atomic structure of crystalline solids can be studied through x-ray diffraction. The analysis of the interference patterns produced by the scattered x-rays provides unique information about the structure of the material producing the scattering. The atomic structure of surfaces can also be studied by diffraction techniques, but x-rays are not suitable for such a study. The penetration depth of x-ray beams, typically with energies in the 20 to 40 keV range, is of the order of 10^5 nm and therefore is not sensitive to surface structure. On the other hand, low-energy electrons, with energy of about 100 to 300 KeV, penetrate no more than a few nm inside a material before being scattered. For this reason LEED can provide structural information associated with the outermost surface layers and is commonly used to determine of surface structure.

15.3.3 Surface crystallography

Scattering experiments such as LEED indicate that surfaces by and large exhibit lattice structures that are amenable to crystallographic analysis and classification. Ideally, a surface can be described as a sequence of parallel planes each characterized by perfect two-dimensional periodicity and extending indefinitely in a direction perpendicular to the surface. An atom on such a plane, (say labeled by *is*), can be reached from the atom at the origin (on the same plane) by means of translation vector of the form, $T = \mathbf{a}_1^{(i)} n + \mathbf{a}_2^{(i)} m$, where n and m are integers. For the sake of simplicity, we assume that the planes have identical structures but may be displaced (shifted) with respect to one another. If the vector τ^i connects the origin of the surface plane to that of plane i , then the lattice positions on plane i are located with respect to the origin at the surface by $T = \mathbf{a}_1^{(i)} n + \mathbf{a}_2^{(i)} m + \tau^i$. If for a value of i the vector τ^i turns out to be equal to a lattice vector, $\mathbf{a}_1 n + \mathbf{a}_2 m$, then the system consists of the periodic repetition along one direction of a stack of i planes.

The primitive two-dimensional vectors, $\mathbf{a}_1$ and $\mathbf{a}_2$, define a *unit mesh* or a *surface mesh.* There are five possible nets [22] in two dimensions (Fig. 15.2). The positions of the atoms on the nets must conform to the requirements of rotational and reflection symmetry. There are 10 point groups, consisting of mirror reflections across a line and rotations through an angle of $2\pi/p$, where $p = 1, 2, 3, 4, 5, 6$, that are consistent with the five nets and leave one point unmoved. These point groups combine with the surface nets to lead to 13 space groups. Adding a glide plane leads to 17 two-dimensional space groups.

It is customary to characterize a surface by the ideal bulk structure formed on the cleavage plane or the plane of termination. Thus, we speak of a Ni(111) or a Cu(110) surface. Here, the periodicity and orientation of the surface derives directly from that of the bulk material; these are called 1x1 structures. If the primitive vectors of a surface are multipliers of those of an ideal 1x1 surface, e.g., $\hat{\mathbf{a}}_1 = N\mathbf{a}_1$ and $\hat{\mathbf{a}}_2 = M\mathbf{a}_2$, as may happen in a reconstructed surface (see Chapter 17), then the surface is described by an additional index, R(hkl)NxM, e.g., Au(110)2x1 or Si(111)7x7. Finally, if the surface net is rotated by an angle ϕ with respect to the ideal structure, we append the angle, i.e., R(hkl)NxM-ϕ.

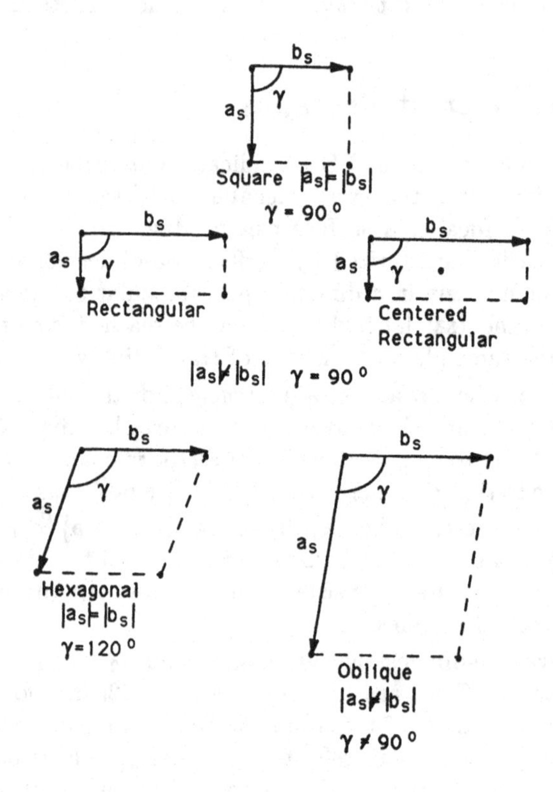

Figure 15.2: **Illustrative depiction of surface nets.**

15.4 Basics of Surface Electronic Structure

It is fairly evident that an understanding of the physical, chemical, and mechanical properties of surfaces and interfaces requires a knowledge of the corresponding electronic structure. Phenomena such as oxidation, heterogeneous catalysis, crystal growth, electronic transport, and brittle fracture are intimately connected with the behavior of electrons near a surface or an interface.

The investigation of the electronic properties of surfaces and interfaces hinges on the consideration of a number of basic, often interrelated questions. What is the electron density near a surface or an interface? How do electron states near a surface or interface differ from those in the bulk material? Is chemical bonding affected by the presence of a surface? What are the electrostatic and transport properties of an interface? The attempt to provide answers to questions such as these is possibly the greatest impetus behind research in surface science.

On the experimental side, photoelectron spectroscopy can be used to obtain data directly related to the electronic structure of a surface. Spectroscopic measurements are usually not difficult to perform, but data analysis is based on comparisons with the predictions of theoretical models. Spectroscopy is one area of scientific research where experimental results can be immediately compared to those of theory. Since such comparisons are based on the ability to predict the behavior of electrons near a surface or interface, the knowledge of the electronic structure is an indispensable ingredient of surface science. Consequently, the following two chapters contain a detailed exposition of formal methods based on MST that allow the calculation of the Green function for electrons in surface or interface regions.

In preparation for the discussion of these methods, in this chapter we consider a number of simple results obtained on the basis of a free-electron (jelium) model. Although hardly applicable in a quantitative sense to real surfaces, the results obtained through the use of the model provide a qualitative understanding of the effects of a surface or interface on the electronic structure. They form the foundation that supports the conceptual development and practical application of more sophisticated methods.

15.4.1 Nature of the problem

In broad outline, the elements of the study of surface electronic structure are the same as those involved in the case of bulk materials. Let us assume that the atomic arrangements of a surface or interface system are known[2]. Denoting the positions of the atoms by $\mathbf{R}$, and invoking the Born-Oppenheimer approximation, i.e., nuclei fixed in position, we can easily

[2]Possibly through the application of LEED or other experimental methods.

write down the Hamiltonian of a semi-infinite (or, in the case of interfaces, doubly semi-infinite) system of N electrons of momenta $\mathbf{p}_i$ and mass m,

$$H = \sum_{i=1}^{N} \frac{p_i^2}{2m} - \sum_{\mathbf{R}} \sum_{i=1}^{N} \frac{Z_{\mathbf{R}} e^2}{|\mathbf{r}_i - \mathbf{R}|} + \frac{1}{2} \sum_{i,j \neq i}^{N} \frac{e^2}{|\mathbf{r}_i - \mathbf{r}_j|}. \tag{15.1}$$

We recognize this as the Hamiltonian of Eq. (1.94) describing an electron gas in the presence of nuclei of charge $Z_{\mathbf{R}}e$. The three terms on the right of the equals sign are associated, respectively, with the kinetic energy of the electrons, the interaction of the electrons with the nuclei, and the interaction of the electrons among themselves.

As in the case of bulk systems, we seek solutions of the Schrödinger equation

$$H\Psi = E\Psi, \tag{15.2}$$

associated with the Hamiltonian in Eq. (15.1) and the appropriate boundary conditions imposed by the presence of a surface or an interface. As in the case of the bulk solid, the presence of the electron-electron interaction precludes an exact solution of this equation, and necessitates the use of approximate treatments. The methods currently in use for the study of the electronic structure of surfaces and interfaces are invariably based on *density functional theory*[3] (DFT) and in particular the *local density approximation* (LDA) of DFT. We recall that within DFT, the ground-state energy of a complex system containing electrons and nuclei is given as a *unique* functional of the ground-state electron density, $n(\mathbf{r})$,

$$\begin{aligned} E[n(\mathbf{r})] &= T[n(\mathbf{r})] - \sum_{\mathbf{R}} Z_{\mathbf{R}} e \int \frac{n(\mathbf{r})}{|\mathbf{R} - \mathbf{r}|} d^3r \\ &+ \frac{1}{2} \int\int \frac{n(\mathbf{r})n(\mathbf{r}')}{|\mathbf{r} - \mathbf{r}'|} d^3r d^3r' + E_{xc}[n(\mathbf{r})], \end{aligned} \tag{15.3}$$

where $T[n(\mathbf{r})]$ is the kinetic-energy functional of a non-interacting electron gas characterized by the charge density $n(\mathbf{r})$, and $E_{xc}[n(\mathbf{r})]$ is the *exchange and correlation* part of the energy that formally contains all many-body effects. The two middle terms on the right side of the last expression represent the electron-nuclear interaction and the mutual repulsion of the electrons, respectively. The expression for the ground-state energy obtained within DFT is *exact* and is stationary with respect to variations in the electron density,

$$\frac{\delta E[n(\mathbf{r})]}{\delta n(\mathbf{r})} = 0. \tag{15.4}$$

[3] See Chapter 1 for a discussion and references.

One remarkable feature of DFT, as expressed within the Kohn-Sham formalism, is that the density can be written in the form, Eq. (1.194),

$$n(\mathbf{r}) = \sum_i |\psi_i(\mathbf{r})|^2, \tag{15.5}$$

where the orbitals (more precisely spin-orbitals) $\psi_i(\mathbf{r})$ can be found as the solutions of a set of coupled *ordinary* differential equations,

$$\left[-\frac{1}{2}\nabla^2 + v_{\text{eff}}(\mathbf{r})\right]\psi_i(\mathbf{r}) = \epsilon_i\psi_i(\mathbf{r}), \tag{15.6}$$

where

$$v_{\text{eff}}(\mathbf{r}) = -\mathrm{e}\sum_{\mathbf{R}}\frac{Z_{\mathbf{R}}}{|\mathbf{r}-\mathbf{R}|} + \int\frac{n(\mathbf{r}')}{|\mathbf{r}-\mathbf{r}'|}\mathrm{d}^3r' + v_{\text{xc}}(\mathbf{r}), \tag{15.7}$$

is an effective *one-electron potential* that includes the effects of the electron-electron interaction in the exchange-correlation potential,

$$v_{\text{xc}}(\mathbf{r}) = \frac{\delta E_{\text{xc}}[n(\mathbf{r})]}{\delta n(\mathbf{r})}. \tag{15.8}$$

Obviously, a practical implementation of DFT requires a good approximation for $v_{\text{xc}}(\mathbf{r})$.

As discussed in Chapter 1, it is extremely useful to treat $v_{\text{xc}}(\mathbf{r})$ within the local density approximation. In this approximation, $v_{\text{xc}}(\mathbf{r})$ at position $\mathbf{r}$ is set equal to the exchange correlation energy density of an interacting but *homogeneous* electron gas with the same density as the density of the physical, interacting system at $\mathbf{r}$. This approximation is easy to implement because the exchange-correlation energy of free electrons is known [24]. Computational experience shows that the use of the LDA can reliably yield a number of electronic properties for surfaces and interfaces such as ground state energies and charge densities, as it has been found to do in the case of bulk materials. As an approximation, however, it fails to capture some of the physics of complex systems, particularly that which is connected to the excited states of the electron gas. Its most notable failures in the case of surface and interface electronic structure are mentioned below[4].

At this point, a few words about the solutions, $\psi_i(\mathbf{r})$, and the energies, ϵ_i, appearing in Eq. (15.6) are in order. Strictly speaking, these quantities have no physical meaning. Their purpose is to allow an evaluation of the

[4]Technically, the implementation of the LDA to the study of the electronic structure of surfaces and interfaces is considerably more complicated than that to bulk materials. For example, the abrupt change in geometry near such two-dimensional defects leads to charge rearrangements and the formation of dipole layers and image charges whose treatment is far from simple. The interested reader is urged to consult cited references for more detail.

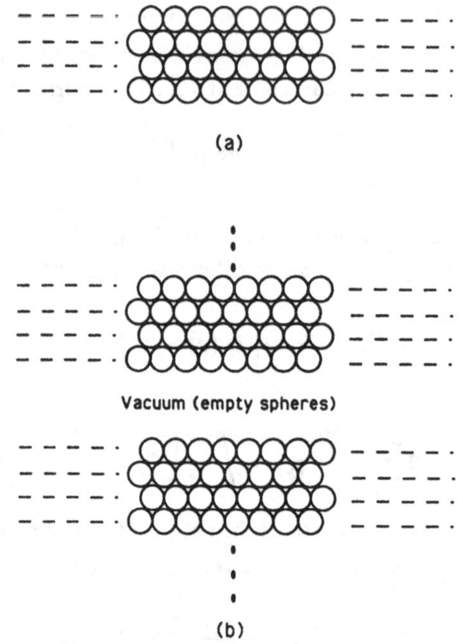

Figure 15.3: **Modeling of a surface by means of a free-standing slab, panel (a), or a repeating slab, panel (b).**

physical observable, $n(\mathbf{r})$. However, they are frequently and successfully interpreted as single-particle eigenstates and eigenenergies, respectively [25]. As we have already seen, in the case of periodic solids the $\psi_i(\mathbf{r})$ become the Bloch wave functions, $\psi_{\mathbf{k}}(\mathbf{r})$ (suppressing spin variables), and the ϵ_i correspond to the band energies $E_\nu(\mathbf{k})$ associated with band index ν and wave vector $\mathbf{k}$ in the first Brillouin zone of the reciprocal lattice.

In the case of semi-infinite systems, i.e., systems containing surfaces or interfaces, the construction of Bloch states becomes problematic because these states require the existence of three-dimensional periodicity. One approach is to model [26] a surface by means of free-standing or repeating slabs, as is indicated schematically in Fig. 15.3(a). In the free-standing slab scheme, a surface (or interface) can be approximated with a two-dimensional periodic lattice and an appropriately chosen unit cell. In a repeating-slab scheme, panel (b), the system is three-dimensional, and the repeating cell (supercell) extends across the vacuum layers separating the slabs. The distance between the slabs in the repeating-slab scheme must be chosen large enough so as to minimize the interaction between adjacent surfaces (material/vacuum interfaces). At the same time, the slabs themselves must contain enough layers so that the electronic charge density at the innermost layer, the "center" of the slab, resembles that of the three-dimensional solid. Because electronic wave functions decay exponentially into the vacuum, one finds that at least in metallic systems a separation of 3 to 5 layers replaced by vacuum, accompanied by a slab thickness of 5 to 9 layers can yield [26] accurate representations of a number of surface electronic properties such as charge density and ground-state energy. S-lab calculations of both the free-standing and repeating varieties have been used successfully [27] to calculate work functions (defined below) and surface reconstruction.

However, in spite of successful applications, the slab approach to the calculation of surface electronic structure is not fully satisfactory. For example, although a free-standing slab approach can yield reliable results for integrated quantities such as charge densities and energies, it can lead to a very poor representation of **k**-dependent spectral functions. Illustrations of this failure are given in the a following chapter in connection with the development of methods for the treatment of semi-infinite solids within MST. The poor convergence of **k**-dependent quantities with slab thickness can adversely affect the calculation of such physical properties as electrical conductivity and phonon spectra.

Another unsatisfactory aspect of free-standing or repeating slabs is the fact that they approximate the *structure* of surface or interface system. Such approximations are not well controlled and are based on boundary conditions of the Schrödinger equation that can be very different from those associated with the semi-infinite geometry of a surface or interface. With increasing distance, the wave function far from a surface (or interface) should more closely resemble the wave function of the bulk materials forming the interface. Thus, it should approach a free wave far into the vacuum side of a surface, and a Bloch wave deep into the material side. No slab geometry can be fully consistent with these requirements. Thus, although such approximation schemes can yield very useful and often accurate information, a fully satisfactory treatment of an interface requires the development of methods that take into account the correct boundary conditions of the problem. We will see how MST accomplishes this non-trivial task.

15.4.2 Use of the Green function

We have had several opportunities to mention some of the advantages of working directly with the Green function rather than the wave function of an extended system. These advantages are clear, for example, in the case of substitutionally disordered alloys discussed in the previous part of the book. It is not surprising that the Green function is also of great utility in the study of semi-infinite systems, such as surfaces or interfaces, or of solids containing two-dimensional defects such as stacking faults. This utility is based on the ability to directly calculate the charge density of a system from the Green function, without relying on the knowledge of the wave function that can be difficult to evaluate in certain cases. Therefore, even though we will use some arguments based on the properties and behavior of wave functions to gain an intuitive understanding of certain basic properties of surface and interface electronic structure, the thrust of our development will be based on the calculation of the Green function of semi-infinite solids within multiple scattering theory.

15.5 Jellium Model

Before turning to our discussion of the electronic structure of real surfaces in the next chapter, it is instructive to consider a simplified treatment based on the *jellium model.* We saw in Chapter 1 that the jellium model consists of a cloud of non-interacting electrons moving inside a box of side L containing a background of constant, positive charge with density n_+ that ensures the neutrality of the entire system. The corresponding electron states are plane waves (in the limit as $L \to \infty$), Eq. (1.36) leading to a constant charge density, Eq. (1.34). Let us now consider the effect on the wave functions and the charge density in the jellium model brought about bythe presence of a surface that we represent by a potential barrier at $z = 0$.

15.5.1 Infinite surface barrier

The simplest model of a surface potential consists of an infinitely high, square potential barrier at $x = 0$ so that the potential inside the box takes the form

$$V = \begin{cases} 0 & x \leq 0 \\ \infty & x > 0 \end{cases} \tag{15.9}$$

Therefore, the potential inside the box is a constant (that can be chosen equal zero) everywhere in the half space, $x \leq 0$, and is infinitely high for $x > 0$.

Because of the presence of the barrier, the electron wave function must vanish everywhere on the $z-y$ plane (where $x = 0$) so that instead of plane waves we have

$$\psi_{\mathbf{k}} = \left(\frac{2}{L^3}\right)^{1/2} \exp\left[\mathrm{i}(k_z z + k_y y)\right] \sin k_x x. \tag{15.10}$$

In this case the allowed values of k_x are

$$k_x = \frac{n_x \pi}{L}, \quad n_x = 1, 2, \cdots, \tag{15.11}$$

with k_z and k_y remaining unchanged from those of the three-dimensional case, Eq. (1.37).

We now calculate the charge density, n_-, associated with the wave functions of Eq. (15.10),

$$\begin{aligned} n_- &= -e\sum_{\mathbf{k}} |\psi_{\mathbf{k}}|^2 \\ &= -e\sum_{\mathbf{k}} \left[\left(\frac{4}{L^3}\right)\sin^2(k_x x)\right] g(x), \end{aligned} \tag{15.12}$$

where $g(x)$ denotes the number of states with a given k_z but different values of k_x and k_y. The sum in the last expression can be evaluated to yield,

$$n_- = -n_+ e\left[1 + \frac{3\cos\chi}{2\chi} - \frac{3\sin\chi}{3\chi}\right], \tag{15.13}$$

where $\chi = 2k_F x$. We now consider the implications of this result.

The electronic charge density must vanish at the impenetrable wall at $x = 0$ and the wave function consequently goes to zero there. However, as it does so it creates a net positive charge adjacent to the barrier, since now there are not enough electrons in the vicinity of the barrier to neutralize the positive background in the region. In order to compensate for this effect and conserve charge, electrons pile up in regions slightly away from the barrier, producing a layer of negative charge. Combined with the positive charge next to the barrier, this produces a *dipole* near the surface. This effect is also present in real materials. Proceeding even further inside the material, the charge oscillates with ever decreasing amplitude until it reaches the constant value of n_+ deep inside the material. These are the characteristic *Friedel oscillations* associated with screening of impurities in metals, and are found in most surfaces and interfaces, as well as in the screening of impurity charges in general. Friedel oscillations are characterized by a wave vector that equals $2k_F$. These oscillations are shown schematically in Fig. 15.4. We see the vanishing of the charge density at $z = 0$, and the formation of a dipole layer near the surface. We also note that since the electronic charge density becomes equal to that of the positive background away from the surface, it is necessary to terminate the positive charge at some distance from the barrier in order to preserve the neutrality of the system.

15.5.2 Finite surface barrier

Let us now consider a jellium model that is somewhat more realistic than the one above. In this case, we take the surface barrier to have finite height that we define in terms of the *work function*, W. Let $v(\infty)$ denote the potential in vacuum macroscopically far (infinitely far) from the surface.

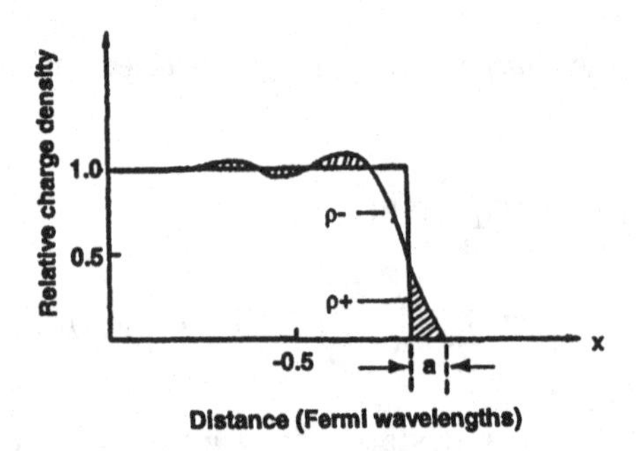

Figure 15.4: **Friedel oscillations for a jellium model with an infinite surface barrier at $x = 0$. Distances are given in terms of Fermi wave length, $\lambda_F = 2\pi/k_F$. Note that the constant, positive charge density is moved back a small distance from the wall in order to facilitate conservation of charge. After [29].**

One can define the work function as the energy needed to remove an electron from inside the material to a point outside far awayfrom the surface,

$$\begin{aligned} W &= v(\infty) + E_{N-1} - E_N \\ &= v(\infty) - \mu, \end{aligned} \tag{15.14}$$

where μ is the chemical potential referenced, along with the electrostatic potential, to the mean electrostatic potential, $v(-\infty)$, deep inside the bulk. Often, $v(\infty)$ is set equal to zero, in which case we have

$$W = -\mu. \tag{15.15}$$

The work function can also be defined alternatively as the difference in potential between an electron *just outside* the surface and the chemical potential *just inside* the material,

$$W = -\mathrm{e}V - \mu. \tag{15.16}$$

Defining the zero of potential as the potential just outside the material[5], we have,

$$W = -\mu. \tag{15.17}$$

[5] Just outside means typically a distance of the order of 10^{-4} cm. Because of the presence of the dipole layer and the fact that electron charge penetrates somewhat into the vacuum side, the potential of an electron may be different close to the surface than far away. Also, the presence of the so-called image potential can have a large effect which in fact is not treated properly within the LDA. Consequently, the value of the reference potential may change with surface orientation, an effect which is observed experimentally. This point is discussed further in the section on inner potential.

Having chosen a reference potential, we can now define μ as the energy change of an electron when it is moved from a position of rest "just outside" the surface to "just inside" with kinetic energy equal to E_F. This energy change involves three distinct contributions:

- The potential energy change, $e\Delta V$, due to the dipole layer.
- The energy change associated with exchange and correlation effects, $E_{xc}(k_F)$, at the Fermi energy.
- The change in kinetic energy, E_F.

We now obtain,

$$W = -\mu = e\Delta V - E_{xc}(k_F) - \left(\frac{\hbar^2}{2m}\right) k_F^2. \tag{15.18}$$

Although this expression involves energies that individually can be very large, of the order of 1 Ry, the work functions of most metals are found to be in the range of 2-5 eV.

The change in the electrostatic energy due to the dipole layer can be characterized by the dipole moment, P, per unit area, which is given in terms of the total charge distribution across the layer(s),

$$\Delta V = -4\pi P = -4\pi \int_{-\infty}^{\infty} z n(z) \mathrm{d}z, \tag{15.19}$$

where the *total* charge, $n(z)$, has the form

$$\begin{aligned} n(z) &= n_+ + n_-(z), \quad z < 0 \\ n(z) &= n_-(z) \qquad\qquad z > 0, \end{aligned} \tag{15.20}$$

with n_- taken to be negative. We see that although the positive background charge stops at the surface, the (negative) electron density extends outward into the vacuum region. The charge distribution of a jellium model with finite surface barrier exhibits the formation of a dipole layer and the characteristic Friedel oscillations encountered in the case of the infinite step. A schematic diagram of this behavior is shown in Fig. 15.5 for two different values of the electron density ($4\pi r_s^3/3 = 1/n$).

Real surfaces display many of the features exhibited by the simple jellium model. One usually finds a dipole layer with the electronic charge extending into the vacuum side a distance of 0.1 to 0.3 nm. Friedel oscillations are also observed, extending into the material a distance of about 5 to 7 nm. Finally, the work function is found to depend on surface orientation, an effect which because of the neglect of the ions, is missed in the jellium model. The rule of thumb is that the work function is greatest for densely packed surfaces. This can be understood based on the fact that more open surfaces can have ion cores protruding out into the dipole layer, as

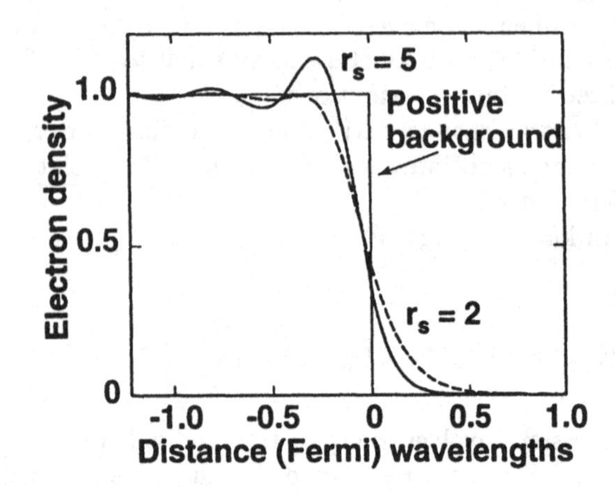

Figure 15.5: **Charge distribution near the region of a finite surface barrier at $z = 0$, for two values of the electron density, $r_s = 2$ (dashed curve), and $r_s = 5$ (solid curve) [30].**

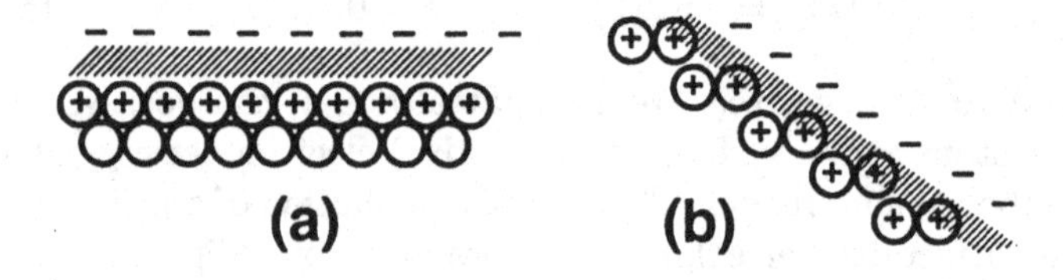

Figure 15.6: **Schematic representation of the surface electron charge densities for a close-packed surface, panel (a), characterized by a high value of the work function, and an "open" surface, panel (b), typically characterized by smaller values of the work function.**

shown schematically in Fig. 15.6. The positively charged ion cores reduce the dipole moment, leading to a lower value for ΔV and of W for open, rougher surfaces.

15.5.3 Image charge and potential

Among the features of the effective potential that are missed in the jellium model is one that is also missed in sophisticated, LDA-based calculations. It follows on physical grounds that an electron far into the vacuum should experience a *classical* image potential. In fact, if the *exact* exchange-correlation potential were known, we would find

$$\lim_{z\to\infty} v_{\mathrm{xc}}(z) = -4\frac{\mathrm{e}}{|z-d_{\perp}|}, \tag{15.21}$$

where $d_{\perp}$ is the centroid of the charge distortion, $\delta n(z)$, induced by the presence of the charge outside the surface [1],

$$d_{\perp} = \int_{-\infty}^{\infty} \mathrm{d}z\; z\, \delta n(z) / \int_{-\infty}^{\infty} \mathrm{d}z \delta n(z). \tag{15.22}$$

Typical values of $d_{\perp}$ are of the order of $1\mathring{A}$.

The inability to reproduce the image potential is one important failure of the LDA. At the same time, this drawback is not found to greatly affect the calculation of charge distributions or total energies of surfaces and interfaces.

15.5.4 Inner potential

A discussion of surface potentials would not be complete without at least a cursory examination of what is commonly referred to as the *inner potential.* This term is very important in the analysis of LEED experiments. The inner potential is defined as the difference in the potential energy of an electron with a given kinetic energy in a metal relative to the energy of an electron at a large distance from the surface into the vacuum. Denoting by $\Delta V'$ the difference in the electrostatic potential inside the metal relative to $v(\infty)$, we can write

$$\Phi_{\mathrm{inner}}(\mathbf{k}) = -\mathrm{e}\Delta V' + \epsilon_{\mathrm{xc}}(\mathbf{k}). \tag{15.23}$$

In this expression, $\epsilon_{\mathrm{xc}}(\mathbf{k})$ is calculated for the value of $\mathbf{k}$ appropriate to the electron in question, and consequently the inner potential is $\mathbf{k}$-dependent. The inner potential is related to the work function by the expression,

$$\Phi_{\mathrm{inner}}(\mathbf{k}) = -W - \mathrm{e}\left(\Delta V' - \Delta V\right) + \left[\epsilon_{\mathrm{xc}}(\mathbf{k}) - \epsilon_{\mathrm{xc}}(k_{\mathrm{F}})\right] - \left(\frac{\hbar^2}{2m}\right) k_{\mathrm{F}}^2. \tag{15.24}$$

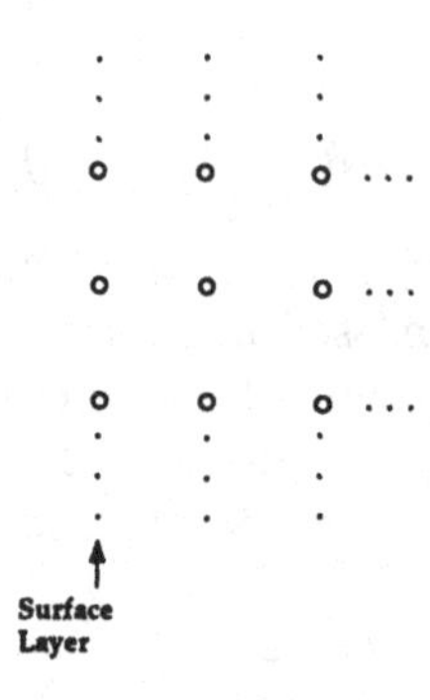

Figure 15.7: **A surface along (100) of a simple cubic crystal.**

15.6 Surface Band Structure

Before leaving this chapter, we briefly discuss the related concepts of reciprocal lattice and of surface electronic band structure.

15.6.1 Reciprocal lattice

To fix ideas, we consider a simple cubic crystal cleaved along the (100) direction. As illustrated in Fig. 15.7, the resulting semi-infinite system may be viewed as a collection of semi-infinite lines of atoms arranged perpendicularly to the surface plane on the sites of a two-dimensional square lattice. In other words, each site in this lattice is occupied by a molecule with an infinite number of atoms arranged in a half line.

The two-dimensional periodic lattice, described by unit translation vectors $\mathbf{a}_1$ and $\mathbf{a}_2$, gives rise to a corresponding two-dimensional reciprocal space characterized by the wave vectors $\mathbf{k}_{\|}$ that are good quantum numbers. This description is analogous to that of the three-dimensional, periodic lattice described by unit translation vectors, $\mathbf{a}_1$, $\mathbf{a}_2$, and $\mathbf{a}_3$, whose reciprocal space can be described as the collection of vectors $\mathbf{k}$ that are linear combinations of the unit vectors

$$\mathbf{b}_1 = \frac{2\pi(\mathbf{a}_2 \times \mathbf{a}_3)}{\mathbf{a}_1 \cdot (\mathbf{a}_2 \times \mathbf{a}_3)}$$

$$\mathbf{b}_2 = \frac{2\pi(\mathbf{a}_3 \times \mathbf{a}_1)}{\mathbf{a}_1 \cdot (\mathbf{a}_2 \times \mathbf{a}_3)}$$

$$\mathbf{b}_3 = \frac{2\pi(\mathbf{a}_1 \times \mathbf{a}_2)}{\mathbf{a}_1 \cdot (\mathbf{a}_2 \times \mathbf{a}_3)}. \tag{15.25}$$

For example, in the case of the sc structure the vectors $\mathbf{a}_1$, $\mathbf{a}_2$, and $\mathbf{a}_3$ can be taken to point along the positive x, y, and z directions of a right-handed Cartesian coordinate system. In this case, unit vectors in reciprocal space, Eq. (15.25), also form a simple cube. To derive the corresponding relations for a surface, we may proceed as follows.

We consider the system obtained when the length of one of the unit vectors in real space, say $\mathbf{a}_1$, is allowed to increase beyond all bounds. The resulting system consists of independent two-dimensional lattices separated by an infinite distance one from another. The corresponding reciprocal space associated with these two-dimensional layers is characterized by the primitive vectors,

$$\begin{aligned} \mathbf{b}_1 &= \frac{2\pi(\mathbf{a}_3 \times \mathbf{n})}{\mathbf{n} \cdot (\mathbf{a}_2 \times \mathbf{n})} \\ \mathbf{b}_2 &= \frac{2\pi(\mathbf{n} \times \mathbf{a}_2)}{\mathbf{n} \cdot (\mathbf{a}_2 \times \mathbf{n})}, \end{aligned} \tag{15.26}$$

where $\mathbf{n}$ is any vector perpendicular to the surface. In the case of a sc structure cleaved along the (100) direction, the reciprocal primitive vectors are orthonormal and the first Brillouin zone of the reciprocal lattice is a square of side $2\pi/a$, where a is the lattice constant of the direct lattice.

15.6.2 Surface states

The discussion of the jellium model has revealed that the presence of a surface (or interface) can produce features in the electronic charge distribution, such as the dipole layer, which are generally absent in bulk materials (although they are similar to screening effects around impurities). These features are present even in a model that provides a very approximate one-dimensional description of a surface, ignoring completely such features as the presence of the ion cores, and the electron-electron interaction. When the presence of the ions is taken into account, new features begin to emerge in the electronic structure of a surface. These features can be traced to the new boundary conditions imposed on the solution of the Schrödinger equation associated with a surface (or interface). The requirement that the wave function be continuous with a continuous derivative everywhere, especially across a surface or an interface, can cause electronic states to appear at energies that would be excluded in a bulk material. These states are called *surface states* and their presence can have a profound influence on a number of technologically important materials such as semi-conductor devices. It can also strongly influence mechanical properties, e.g., bonding across an interface.

In order to gain a preliminary understanding of surface states, we consider a simplified, one-dimensional model which, while neglecting the electron-electron interaction and self-consistency in the charge and potential, does introduce some of the effects of the periodically arranged ion cores. Our discussion in this subsection will be brief, aimed only at providing an indication of the emergence of surface states due to the changed boundary conditions at the surface. A lengthy exposition of surface states can be found in the literature [1, 2]. The treatment of surfaces and interfaces within MST, including the performance of self-consistent calculations within the LDA, is presented in the following chapter.

The Schrödinger equation associated with an effective one-particle potential, which only includes the contribution of the ion cores arranged periodically at distances a along a line in the z direction, takes the form,

$$\left[-\frac{\mathrm{d}^2}{\mathrm{d}z} + V(z)\right]\psi(z) = E\psi(z). \tag{15.27}$$

The periodic potential due to the ions is modeled by the form,

$$V(z) = -V_0 + 2V_g \cos gz, \tag{15.28}$$

where $g = 2\pi/a$ is the shortest reciprocal lattice vector of the linear chain.

The solution of Eq. (15.27) for the complete line is well known [1, 28]. The feature relevant to our discussion can be captured by a trial wave function of the form,

$$\psi_k = \alpha \mathrm{e}^{ikz} + \beta \mathrm{e}^{\mathrm{i}(k-g)z}. \tag{15.29}$$

Substitution of this expression for $\psi_k(z)$ into the Schrödinger Eq. (15.27), provides the secular equation for the coefficients[6] α and β,

$$\begin{bmatrix} k^2 - V_0 - E & V_g \\ V_g & (k-g)^2 - V_0 - E \end{bmatrix} \begin{bmatrix} \alpha \\ \beta \end{bmatrix} = 0. \tag{15.30}$$

The solutions of this equation yield the wave function for the full chain

$$\psi_k(z) = \mathrm{e}^{ikz}\cos(\frac{1}{2}gz + \delta), \tag{15.31}$$

and the eigenvalues

$$E_\kappa = -V_0 + \left(\frac{1}{2}g\right)^2 + \kappa^2 \pm \left(g^2\kappa^2 + V_g^2\right)^{1/2}, \tag{15.32}$$

[6] We note the similarity with the secular equation of MST derived in Chapter 3, which is a special case of the present general construction. There, as here, a secular equation is used to determine the coefficients of linear combination of basis functions that determines the wave function of the system.

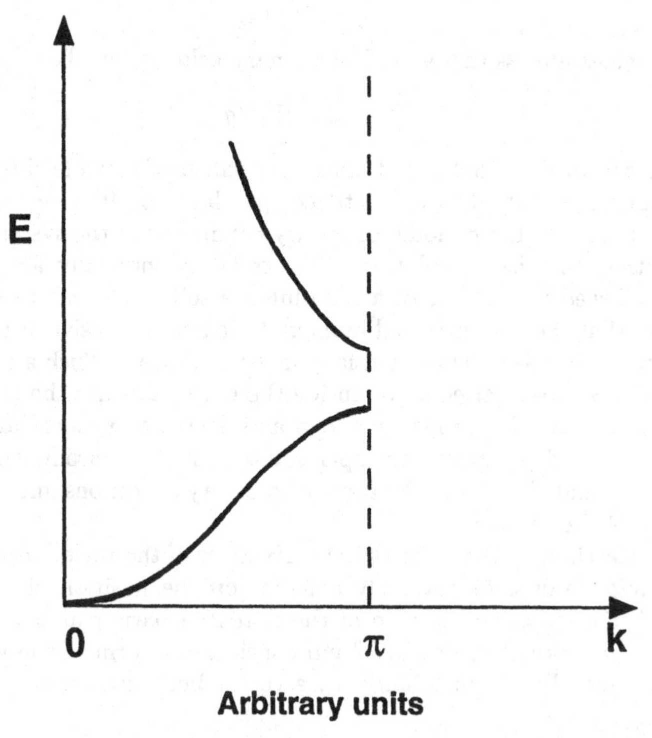

Figure 15.8: **Schematic representation of the band structure of a one-dimensional chain of atoms, showing the existence of a gap in the energy spectrum.**

where the phase shift, δ, is defined by means of the relation,

$$e^{2i\delta} = (E - k^2)/V_g, \tag{15.33}$$

and the wave vector, κ, has been defined in terms of its deviation from the Brillouin zone boundary,

$$\kappa = k - g/2. \tag{15.34}$$

These expressions show that the lowest energy solution is even (odd) with respect to reflection about $z = 0$ depending on whether $V_g < 0$ (> 0).

Figure 15.8 shows a schematic representation of the band structure, $E(\kappa^2)$, of the one-dimensional chain. The figure displays the band gap at the Brillouin zone boundary ($\kappa = \pi$) (in arbitrary units). As we will see shortly, states in the gap would grow exponentially with distance in the bulk solid (the full line of atoms) and do not correspond to acceptable solutions. At the same time, Eq. (15.32) yields a *continuous* function of κ^2 provided that κ^2 is allowed to take on negative values. The corresponding energy values are associated with imaginary values of κ when

$$0 < |\kappa| < |V_g|/g. \tag{15.35}$$

It is clear from Eq. (15.31) that imaginary values of κ can lead to solutions that grow exponentially with distance, (for large positive or negative z), and fail to satisfy the periodic boundary conditions of the system.

On the other hand, solutions that grow exponentially for positive z can be allowed in the case of a semi-infinite solid (half-line in this case), provided that they are matched in magnitude and derivative at the surface onto a wave that decays exponentially into the vacuum. Such a wave would satisfy the Schrödinger equation inside the material, since the functions in Eq. (15.31) have this property, and would decrease exponentially outside so that it would constitute a proper solution of the Schrödinger equation everywhere and also satisfy the proper boundary conditions imposed by the presence of the surface.

The matching of the logarithmic derivatives of the wave function (magnitude and derivative) obviously depends on the position of the surface [1, 2]. Let us position the step of the potential barrier at $z = a/2$ away from the ion core at the surface, and consider the form of the wave function corresponding to imaginary wave vector both inside and outside the material,

$$\begin{aligned} \psi(z) &= e^{\kappa z}\cos\left(\frac{1}{2}gz + \delta\right) & z < a/2 \\ \psi(z) &= e^{-qz}, & z \geq a/2, \end{aligned} \tag{15.36}$$

with $q^2 = V_0 - E$. We note that this wave function decays exponentially away from the surface both inside and outside the surface. Such wave functions are called *evanescent* waves and give rise to surface states[7] that are localized in the vicinity of the surface. These states arise solely due to the change in the boundary conditions at a surface and are commonly referred to as *Shockley states* [12]. In addition to Shockley states, surface states called *Tamm states* [9], can also arise because of changes in the potential near a surface, an effect that can have dramatic implications in semiconductor devices.

15.6.3 Surface band structure

Our discussion thus far has demonstrated the strong influence that the presence of a surface can have on the electronic structure. It now remains to

[7] Surface states for arbitrary positions of the vacuum potential can also be found by graphical means. One draws the solutions associated with ever increasing energies in the gap, for either positive or negative values of V_g, until a solution is found whose logarithmic derivative matches to an exponentially decaying solution outside the surface.

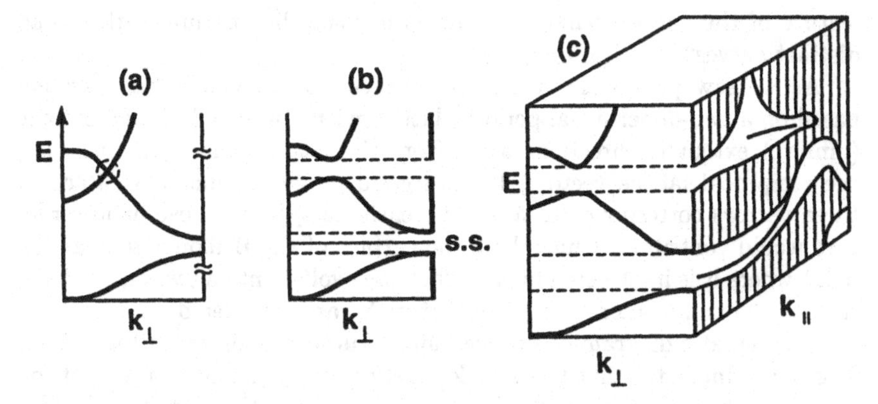

Figure 15.9: **Schematic representation of typical band structure of one-dimensional lattice. Hybridization causes the appearance of a gap in a region where two bands would otherwise cross (dashed circle). A surface state appears in the gap at the zone boundary. The surface dispersion relation in this case consists of the projection of the bulk band structure, (thick lines), onto the surface of the Brillouin zone which in this case is a point. Panel (c) shows the corresponding effect for the case of a three-dimensional band structure and its projection onto a plane defined by a constant value of the perpendicular component of k.**

provide a description of the energy dispersion relation, or the surface band structure, $E(\mathbf{k})$. In the case of a bulk solid, $E(\mathbf{k})$ can be represented as a function of the wave vector **k**, (and band index), along different directions inside the first Brillouin zone. The two-dimensional nature of the surface introduces a new feature into the description of the surface band structure.

Let us revisit the one-dimensional case discussed in the previous subsection. The reciprocal space is linear with a Brillouin zone extending from $-\pi/2a$ to $\pi/2a$. A typical band structure as a function of k inside the first Brillouin zone is shown in Fig. 15.9. The surface of the direct lattice is a point, say at $z = a/2$, and corresponds to a Brillouin zone whose surface is the two end points, $-\pi/2a$ and $\pi/2a$. The projection of the bulk band structure onto one of these points, say at $\pi/2a$, gives rise to the thick lines shown in the figure, panel (b).

First, we note that in the bulk solid, hybridization effects cause a gap to appear in a region where two bands would otherwise cross. This gap, along with the one at a Brillouin zone boundary appear as gaps on the projection of the bulk band structure, thick solid lines, onto the surface plane of the zone at $k = \pi/2a$. We also note the presence of a surface state (ss) in the Brillouin zone gap in the surface band structure. Perhaps the most novel feature of the surface band structure is its band-like nature, rather than distinct curves[8].

As we saw previously, a cleaved surface can be viewed as a system based on a two-dimensional periodic lattice with a unit cell of macroscopic (infinite) extent in directions away from the lattice plane. In this case, the reciprocal lattice vectors, $\mathbf{k}_{\|}$, are good quantum numbers. Each of these vectors corresponds to a one-dimensional system whose band structure would contribute a line along the surface of the Brillouin zone at the point where it is intersected by $\mathbf{k}_{\|}$. This line evolves into a two-dimensional surface or band as more and more vectors $\mathbf{k}_{\|}$ are considered.

A schematic diagram of surface band structure is shown in Fig. 15.10. The figure indicates that to every $\mathbf{k}_{\|}$ there corresponds a $k_{\perp}$ rod that extends into the bulk three-dimensional Brillouin zone. Also shown are surface states, one of which at some point mixes with the bulk conduction states. This mixing is a result of *surface resonance* in which a propagating bulk state has the same energy and can be matched onto a state that decays exponentially into the vacuum. Finally, the (111) surface band structure for Cu is shown in Fig. 15.10.

Bibliography

[1] Andrew Zangwill, *Physics at Surfaces*, Cambridge University Press, New York (1990).

[2] F. Garcia-Moliner and F. Flores, *Introduction to the Physics of Solid Surfaces*, Cambridge Monographs on Physics, Cambridge University Press, Cambridge (1979).

[3] D. Tabor, J. Colloid Interface Sci. **75**, 240 (1980).

[4] R. J. Seeger, *Benjamin Franklin: New World Physicist*, Pergamon, Oxford (1973).

[5] L. P. Williams, *Michael Faraday*, Chapman and Hall, London (1965).

[6] C. Susskind, in *Advances in Electronic and Electron Physics*, Marton and Marton (eds.), vol. **50**, pp. 241-260, Academic Press, New York (1980).

[7] J. Rice, in *Commentary on the Scientific Writings of J. Willard Gibbs*, Donnan and Haas (eds.), vol. **1**, pp. 505-708, Yale University Press, New Haven CT, (1936).

[8] This novelty is mostly a matter of perception. In three-dimensions, the band structure is a collection of two-dimensional surfaces whose intersections with chosen planes in the Brillouin zone are graphed in the form of continuous curves. At the same time, each point on the two-dimensional periodic lattice of a surface is occupied by a basis with an infinite number of atoms so that the distinct curves in the band structure of a bulk solid merge into continuous distributions.

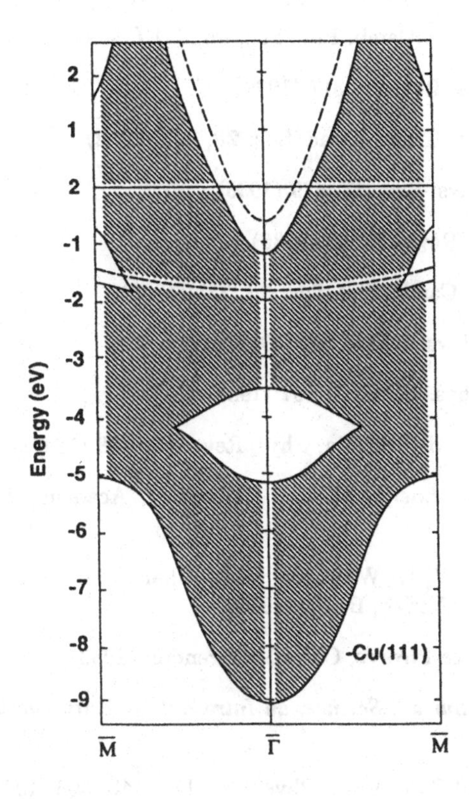

Figure 15.10: **Surface band structure for Cu.**

[8] A. Rosenfeld, *The Collected Works of Irving Langmuir*, Suits (ed.), vol. **12**, pp. 5-229, Pergamon Press, New York (1962).

[9] I. Tamm, Phys. Z. Soviet Union, **1**, 733 (1932).

[10] A. W. Maue, Z. Physik, **94**, 717 (1935).

[11] E. T. Goodwin, Proc. Camb. Phil. Soc. **35**, 221 (1939).

[12] W. Shockley, Phys. Rev. **56**, 317 (1939).

[13] J. E. Lenard-Jones, Trans. Farad. Soc. **28**, 333 (1932).

[14] R. W. Gurney, Phys. Rev. **47**, 479 (1935).

[15] J. Bardeen, Phys. Rev. **49**, 653 (1936).

[16] N. F. Mott, Proc. Camb. Phil. Soc. **34**, 221 (1938).

[17] W. Schottky, Z. Physik, **113**, 367 (1939).

[18] B. Davydov, J. Phys. USSR, **1**, 167 (1939).

[19] J. Bardeen and W. H. Brattain, Phys. Rev. **75**, 1208 (1949).

[20] J. B. Pendry, *Low Energy Electron Diffraction*, Academic Press, London (1974).

[21] M. A. Van Hove, W. H. Weinberg, C.-M. Chan, *Low Energy Electron D-iffraction*, Springer Verlag, Berlin (1986).

[22] M. Prutton, Surface Physics, Oxford, Clarendon (1983).

[23] John B. Hudson, *Surface Science, an Introduction*, Butterworth and Heinemann, Boston (1992).

[24] D. M. Ceperly and B. J. Alder, Phys. Rev. Lett. **45**, 566 (1980).

[25] D. D. Koelling, Rep. Prog. Phys. **44**, 139 (1981).

[26] H. Krakauer, M. Posternak, and A. J. Freeman, Phys. Rev. B**19**, 1706 (1979).

[27] C. L. Fu, A. J. Freeman, E. Wimmer, and M. Weinert, Phys. Rev. Lett. **54**, 2261 (1985).

[28] C. Kittel, *Introduction to Solid State Physics*, Wiley and Sons, New York (1966).

[29] W. Swiatechi, Proc. Phys. Soc. (L), A**64**, 227 (1951).

[30] N. D. Lang and W. Kohn, Phys. Rev. B**1**, 4555 (1970).

Chapter 16

TB Description of Surfaces

16.1 General Comments

Having given a brief overview of some basic elements of surface atomic and electronic structure in the previous chapter, we now turn to the main goal of this part of the book: developing Green-function methods to calculate the electronic structure of surfaces and interfaces. Beginning in this chapter, we discuss methods based on a tight-binding (TB) description of the Hamiltonian of a semi-infinite system. First-principles, self-consistent applications of MST to the electronic structure of surfaces and interfaces are presented in Chapter 17.

Both TB and first-principles descriptions are useful as computational tools, and often complement one another. Tight-binding methods, particularly those based on a non-self-consistent, phenomenological determination of the Hamiltonian matrix elements, are usually easier to apply computationally than self-consistent, first-principles methods, and can quickly lead to a qualitative understanding of physical phenomena. The more recently developed first-principles tight-binding methods [1-3] allow for incorporation of the LDA and can yield self-consistently determined charge densities and potentials.

16.2 TB Green Functions for Surfaces

Our treatment of surfaces and interfaces within a TB formalism is characterized by certain commonly made assumptions and approximations. In general the behavior of electrons in the solid depends on their position relative to a surface or interface, and can be assumed to approach the value in the bulk material as the distance of the atom from the surface increases. At least for metallic systems, this assumption is justified by the results of realistic calculations of the electronic structure of films. These calculations [4, 5] show indeed that the DOS in the central layer of a 5- or 7-layer film can be practically identical to that of the corresponding bulk material. We will also assume that all planes parallel to the surface are characterized by two-dimensional periodicity, and that there exists a basic stack of planes, called the *repeating unit*, whose periodic repetition along a given direction generates the semi-infinite solid. The layers at or near the surface region may have different structure and be characterized by different potentials than those of the repeating unit that generates the ideal surface system. These deviations are usually associated with *surface reconstruction effects*, a feature that can be accommodated, at least in some approximate fashion, within the formalism of the following pages[1]. In our initial development we also explicitly consider monatomic materials characterized by a single (s) s-tate. More complex cases can be treated within appropriate generalizations of the formalism.

It is consistent with the spirit of the TB model to assume that the hopping terms of the Hamiltonian are of finite range. This allows an easy and unified treatment of surfaces and interfaces of ordered as well as substitutionally disordered materials. Some of these treatments are considered explicitly in the following pages.

A number of works [6-18] on surface electronic structure within a TB-like description of the Hamiltonian can be found in the bibliography section[2]. These works encompass various approaches to the treatment of surfaces within a TB formalism, including methods based on the propagator and the locator picture. In our development, we follow a locator/renormalized interactor approach (see Chapter 11). This allowsan easy extension of many formal aspects of a TB formalism to a first-principles approach based on MST, as discussed in the next chapter.

We begin our discussion with the simple example of a single-band, one-dimensional lattice, characterized by near-neighbor hopping. The generalization to three-dimensional systems is presented in the following subsection.

---

[1] A total energy approach, along the lines that can be followed within a first-principles TB method or the formalism of the next chapter, can often yield reliable information about the relative stability of competing surface structures.

[2] It is not possible to provide a complete list of references to the large body of work on surfaces and interfaces. The references given are restricted to a small number that are more or less related to the formalism in the following pages.

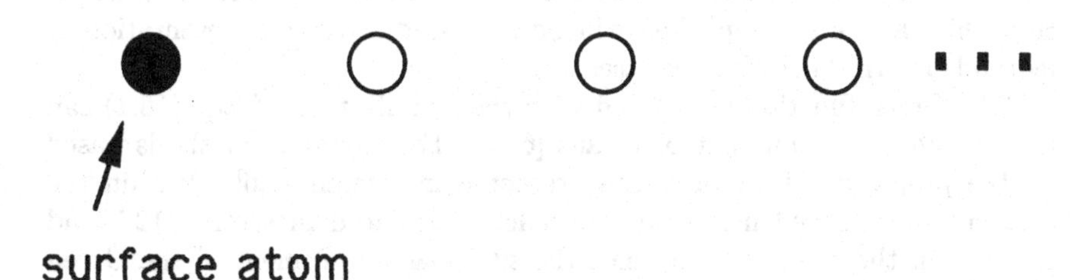

Figure 16.1: **A semi-infinite line of atoms as a prototype of a one-dimensional surface.**

16.2.1 One-dimensional surfaces

A one-dimensional surface is depicted schematically in Fig. 16.1. We consider a system consisting of a semi-infinite line of atoms beginning at the site labeled 0, the surface atom, and repeating with period a along a given direction. In this initial model system, we neglect the presence of a surface barrier. We are interested in evaluating the elements G_{nm} of the Green function associated with the sites n and m in the semi-infinite line.

The Hamiltonian of the system takes the general form of Eq. (10.2) which, for the case of a single orbital reduces to the form,

$$H = \sum_i |i\rangle\epsilon_i\langle i| + \sum_{i,j>i} |i\rangle W_{ij}\langle j|. \tag{16.1}$$

The associated Green function is given by the inverse

$$G(z) = \begin{pmatrix} z-\epsilon_0 & -W_{01} & -W_{02} & \cdots & -W_{0n} & \cdots \\ -W_{10} & z-\epsilon_1 & -W_{12} & \cdots & -W_{1n} & \cdots \\ \cdot & \cdot & \cdot & \cdots & \cdot & \\ \cdot & \cdot & \cdot & \cdots & \cdot & \\ -W_{n0} & -W_{n1} & -W_{n2} & \cdots & z-\epsilon_n & \cdots \\ \cdot & \cdot & \cdot & \cdots & \cdots & \end{pmatrix}^{-1} \tag{16.2}$$

whose site matrix elements satisfy the equation of motion, Eq. (10.14),

$$G_{ij} = g_i\delta_{ij} + g_i \sum_{k\neq i} W_{ik}G_{kj}, \tag{16.3}$$

with $g_i = (z-\epsilon_i)^{-1}$ as the bare locator associated with site i.

The matrix elements, ϵ_i and W_{ij}, that enter Eq. (16.1) are generally functions of distance from the surface (the end of the line). Thus, both Shockley and Tamm states can be described within the present formalism. How this can be accomplished will be discussed after the presentation of methods for treating ideal surfaces.

The Green function associated with the Hamiltonian of Eq. (16.1) can be evaluated by a number of means [6, 7]. For example, methods based on the propagator formalism are presented in reference [6]. Within the locator/renormalized interactor approach introduced in section 12.7 and followed in the present treatment, the site-diagonal element of the Green function for site n takes the form,

$$G_{nn} = (z - \epsilon_n - \Delta_n)^{-1}, \tag{16.4}$$

where the renormalized interactor Δ_n represents the interaction of site n with the surrounding medium and has the general form given in Eq. (10.58). Specializing to the case of the surface atom, $n = 0$, we have

$$G_{00} = (z - \epsilon_0 - \Delta_0)^{-1}, \tag{16.5}$$

where

$$\begin{aligned} \Delta_0 &= \sum_{nm} W_{0n}\Gamma^{(0)}_{nm}W_{m0} \\ &= W_{01}G_{00}W_{10} \end{aligned} \tag{16.6}$$

with 1 labeling the site on the chain just next to the surface atom. The second line in the last expression follows from the fact that in the case of surface systems with only nearest-neighbor hopping, and in the absence of reconstruction effects, the defective-medium Green function, $\Gamma^{(0)}_{11}$, which is the Green function of the surface atom when site 0 is removed, is identical to G_{00},

$$\Gamma^{(0)}_{11} = G_{00}. \tag{16.7}$$

This equality is a simple manifestation of the principle of *removal invariance* that characterizes systems with semi-infinite periodicity. This principle is defined and used extensively in the developments presented in the next chapter. Now, with $W_{01} = W_{10} = W$, Eqs. (16.5) and (16.6) yield the expression

$$\Delta_0 = \frac{z - \epsilon_0 \pm \sqrt{(z-\epsilon_0)^2 - 4W^2}}{2}, \tag{16.8}$$

for the renormalized interactor. In the upper complex-energy half plane, $\Im z > 0$, only the negative sign leads to $\Im \Delta_0 < 0$, and to physical values for G_{00}. These values are such that, for example, the density of states,

$$n_0 = -\frac{1}{\pi}\Im G_{00}(E), \tag{16.9}$$

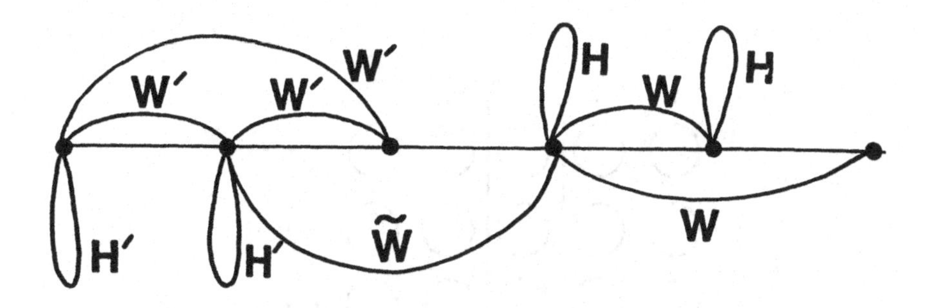

Figure 16.2: **This one-dimensional surface impurity cluster is characterized by values of the Hamiltonian matrix elements that are different from those of the bulk material.**

is always non-negative.

The knowledge of Δ_0 allows one to determine any of the elements, G_{nm}, of the Green function, and also to take into account any changes in the Hamiltonian matrix elements, ϵ_i and W_{ij}, that are brought about by the introduction of the surface. The relevant geometry is depicted in Fig. 16.2. The figure shows a cluster of n sites on the surface of a semi-infinite chain which is characterized by values of ϵ_i and W_{ij} that are generally different from the corresponding bulk values, ϵ and W. It is assumed that the Hamiltonian of the system beyond site n is characterized by the bulk values of the Hamiltonian, as indicated in the figure. In this case the interaction of site n with the material away from the surface, or of any site further into the material, is given by the renormalized interactor, Δ, obtained from the general expression in Eq. (16.8) with ϵ_0 set equal to ϵ. Now, the matrix elements of the Green function for any two sites, m and n, are obtained from the equation of motion, Eq. (16.3), as the inverse of a matrix in site space. With $n > m$, this matrix contains all sites from 0 to n, and we have

$$G_{ij} = \left\{ \begin{bmatrix} z-\epsilon_0 & -W_{01} & 0 & \cdots \\ -W_{10} & z-\epsilon_1 & -W_{12} & \cdots \\ 0 & -W_{12} & \cdots & \cdots \\ \cdot & \cdot & \cdot & \cdots \\ \cdot & \cdot & \cdot & z-\epsilon_n-\Delta \end{bmatrix}^{-1} \right\}_{ij} . \tag{16.10}$$

It is fairly obvious that both Shockley and Tamm states can be accounted for by the last expression. Also the treatment of impurities at a surface or an interface, in this case formed by two half lines, can be easily effected within the present formalism. Finally, the presence of a polyatomic basis or

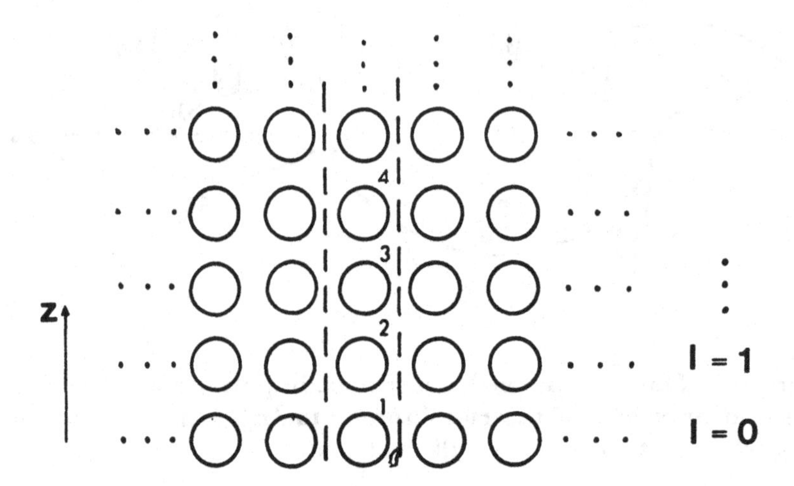

Figure 16.3: **A semi-infinite system viewed as a collection of linear clusters of semi-infinite extent.**

of multiple bands and extended hopping terms can be taken into account by a generalization of scalar quantities to matrices. Some of these generalizations are discussed explicitly in later sections of this chapter and in the next chapter.

16.2.2 Higher-dimensional surfaces

We now extend the formalism of the previous subsection to the treatment of three-dimensional surfaces. In doing so, it is convenient to view a prototype of a semi-infinite material as consisting of clusters, C, of semi-infinite extent along a direction perpendicular to the surface, arranged on the sites of a two-dimensional periodic lattice. A schematic diagram of this construction is shown in Fig. 16.3. Here, the clusters are labeled by the symbol $C = \bar{2},\ \bar{1},\ 0,\ 1,\ 2, \ldots$, etc., while the sites within each cluster are labeled by the positive integers, i, starting with 1 at the surface atom. In the simple example shown here, the repeating unit consists of only one plane parallel to the surface. In this case, the index i also labels the planes away from the surface. More complicated structures, in which the atoms of one plane can be brought into coincidence with those of another by means of translations as well as rotations, can be treated by considering repeating units consisting of stacks of planes. The various geometric figures next to the sites in Fig. 16.3 indicate possible differences in potential (Hamiltonian matrix elements) near the surface, and indicate the convergence of these quantities to the bulk values as one proceeds deep into the material. We assume that the values of ϵ and W defining the Hamiltonian of the system are known, both for the bulk and for the planes near the surface. We also assume that the bulk values are attained after a finite number n of planes away from the surface. We seek methods to determine the matrix elements of the Green function, G_{ij}, associated with sites i and j of the system.

Our starting point is again the equation of motion, Eq. (16.3). This equation can be expressed in terms of the clusters C, described above,

$$\underline{G}_{CC'} = \underline{g}_C \delta_{CC'} + \underline{g}_C \sum_{C'' \neq C} \underline{W}_{CC''} \underline{G}_{C''C'}, \tag{16.11}$$

where $\underline{g}_C$ is the bare locator of a cluster of semi-infinite extent. Thus, we have

$$\underline{g}_C = (\underline{z} - \underline{H}_C)^{-1}. \tag{16.12}$$

In a site representation, $\underline{H}_C$ is a semi-infinite matrix with elements,

$$[\underline{H}_C]_{ij} = \epsilon_i \delta_{ij} + W_{ij}(1 - \delta_{ij}), \quad i,\ j \in C. \tag{16.13}$$

Since the semi-infinite solid possesses two-dimensional periodicity along the surface, Eq. (16.11) can be solved by means of a two-dimensional Fourier transformation. Letting $\mathbf{k}_\parallel$ denote a two-dimensional vector in the first Brillouin zone of the reciprocal lattice defined by the periodicity of the surface, we use Fourier transforms with respect to cluster indices to obtain the expression,

$$\underline{G}(\mathbf{k}_\parallel) = \left[\underline{g}_C^{-1} - \underline{W}(\mathbf{k}_\parallel)\right]^{-1}, \tag{16.14}$$

where each quantity is a semi-infinite matrix in site space in the direction perpendicular to the surface. In particular, we have,

$$[\underline{W}(\mathbf{k}_\parallel)]_{ij} = \sum_{C'} [\underline{W}_{0C'}]_{ij}\, e^{i\mathbf{k}_\parallel (\mathbf{R}_0 - \mathbf{R}_{C'})}, \quad i \in C,\ j \in C', \tag{16.15}$$

where the indices i and j correspond to sites along one of the semi-infinite clusters in the material. It may clarify our discussion to explicitly exhibit the form of the Green function. Corresponding to Eq. (16.2), and with the notation $H_{ii} = z - \epsilon_i - W_{ii}(\mathbf{k}_\parallel)$, we can write

$$\underline{G}(\mathbf{k}_\parallel) = \begin{bmatrix} H_{11}(\mathbf{k}_\parallel) & -W_{12}(\mathbf{k}_\parallel) & -W_{13}(\mathbf{k}_\parallel) & \cdots & \cdots \\ -W_{21}(\mathbf{k}_\parallel) & H_{22}(\mathbf{k}_\parallel) & -W_{23}(\mathbf{k}_\parallel) & \cdots & \cdots \\ \cdot & \cdot & \cdot & \cdot & \cdot \\ -W_{n1}(\mathbf{k}_\parallel) & -W_{n2}(\mathbf{k}_\parallel) & \cdots & H_{nn}(\mathbf{k}_\parallel) & \cdots \\ \cdot & \cdot & \cdot & \cdot & \cdot \end{bmatrix}^{-1} \tag{16.16}$$

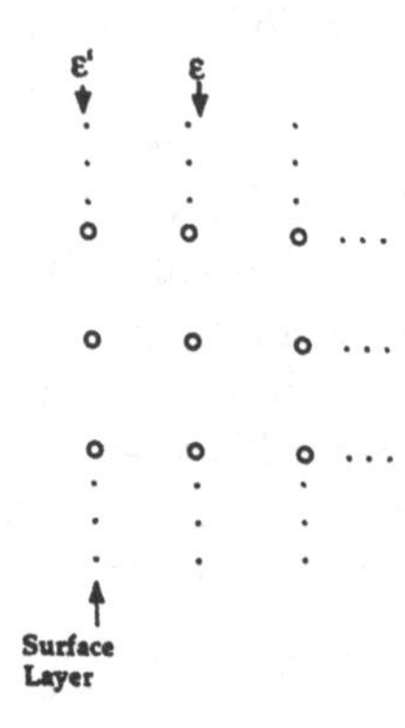

Figure 16.4: **Possible matrix element configuration near the surface of a three-dimensional material. Primes indicate quantities whose values differ from the corresponding bulk values of the Hamiltonian matrix elements.**

where the intracluster hopping terms W_{ij} for sites i and j belonging to the same cluster are incorporated into the quantity $W_{ij}(\mathbf{k}_{\|})$. It is to be noted that in general the quantity $\underline{G}(\mathbf{k}_{\|})$ is a semi-infinite matrix with site matrix elements between the sites of a linear semi-infinite cluster. We now recognize that for each value of $\mathbf{k}_{\|}$, the quantity $\underline{G}(\mathbf{k}_{\|})$ satisfies an equation of motion that is formally identical to that of a single semi-infinite line. Therefore, it can be treated by the methods of the previous section.

As was the case with the single semi-infinite line, the site matrix elements $H_{ii}(\mathbf{k}_{\|})$ and $W_{ij}(\mathbf{k}_{\|})$ may depend on the position of the sites i and j relative to the surface. We assume that the Hamiltonian matrix elements of the cleaved material attain their bulk values after n planes away from the surface. For the sake of generality, we now also allow the hopping matrix elements to extend to m neighbors. Figure 16.4 displays a schematic representation of the special case $m = 2$ and $n = 3$. Accordingly, we set the Hamiltonian matrix elements, H_{ij}, equal to the corresponding bulk values whenever site i or j has index greater than 3. For values less than 3, the Hamiltonian matrix elements may differ from those of the bulk, as indicated by the primes in the figure. It is now convenient to think of the semi-infinite system corresponding to each $\mathbf{k}_{\|}$ point in terms of a repeating unit containing enough planes, n_u, that reduce the interaction between units to nearest neighboring ones. A convenient choice is often dictated by the extent of the hopping, so that in the present case we can choose $n_u = 2$. Let us denote the repeating unit, or cluster, by the symbol C_u. Then, for the case of an ideal surface, the Green function corresponding to the unit at the surface, $C_u = 0$, is given by the expression

$$\underline{G}^{00}(\mathbf{k}_\parallel) = \left[\underline{g}_0^{-1}(\mathbf{k}_\parallel) - \Delta_0(\mathbf{k}_\parallel)\right]^{-1}. \tag{16.17}$$

Provided that two-dimensional periodicity is preserved, this expression can be easily generalized along the lines leading to Eq. (16.10) to allow the treatment of impurity layers characterized by Hamiltonian matrix elements that differ from their bulk values (Fig. 16.4). Such a matrix inversion also allows the determination of the matrix elements $\underline{G}_{ij}(\mathbf{k}_\parallel)$ for sites i and j arbitrarily far into the material.

Once the Green function has been determined in reciprocal space, its real-space matrix elements can be obtained by a Fourier transformation. For example, for site i in cluster C parallel to the surface and site j in cluster C', we have

$$G_{ij}^{CC'} = \frac{1}{\Omega_{\mathrm{SBZ}}} \int \mathrm{d}^2 k_\parallel \underline{G}_{ij}(\mathbf{k}_\parallel) \mathrm{e}^{\mathrm{i}\mathbf{k}_\parallel \cdot \mathbf{R}_{CC'}}, \tag{16.18}$$

where $\mathbf{R}_{CC'}$ is a direct vector between clusters C and C' of the two-dimensional surface lattice, and Ω_{SBZ} denotes the volume of the Brillouin zone defined by the periodicity of the surface. Quantities related to the Green function, such as the density of states, can be now obtained through well-known expressions.

16.2.3 Model calculations

In this subsection, we present the results of calculations of the surface electronic structure of model three-dimensional systems.

In general, we have seen that there are two different types of states that contribute to the density of states. First, there are bulk-like states that extend throughout the semi-infinite crystal and lie entirely inside the allowed energy region associated with the bulk material. Second, there are states that are localized primarily in the region of a surface (or interface) and decay exponentially with increasing distance from that region. The total DOS, given by the imaginary part of the Green function, Eq. (16.9), is the sum of these two distinct contributions.

Figure 16.5 shows the local DOSs as a function of the distance from the surface of a simple cubic, s-band TB material, with site-diagonal energies ϵ, cleaved along the (100) direction [6]. This figure clearly indicates the difference between the surface ($n = 0$) and bulk DOSs, and that the DOS approaches that of the bulk with increasing distance from the surface. Although the presence of the surface can be felt for a rather large distance into the material, especially near the band edges, the overall form of the DOS two planes below the surface, $n = 2$, closely approximates that of the bulk material (dotted line).

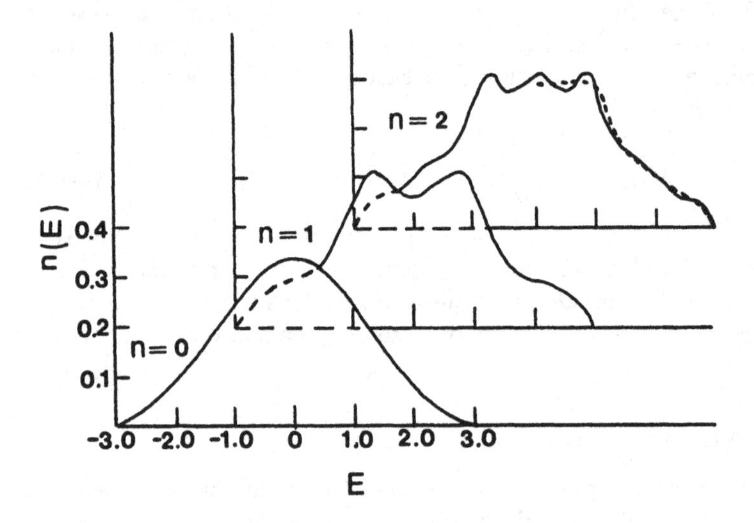

Figure 16.5: **Local DOSs for sites 0 (surface), 1, and 2, for a cubic material cleaved along the (100) direction. Here, the potential at the surface (the matrix elements of the Hamiltonian) are taken to have the same value as in the bulk material [6].**

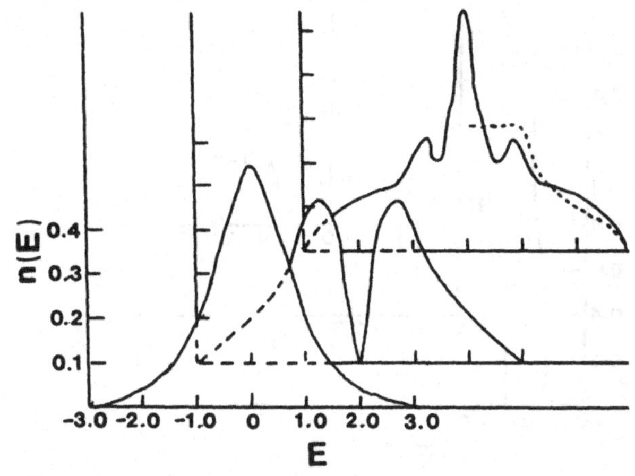

Figure 16.6: **Results analogous to those depicted in Fig. 16.5, but for a (111) surface [6].**

This situation can change dramatically with changes in surface orientation and/or changes, ΔV, of the potential at the surface, due to the possible onset of surface states. Such states can greatly distort the DOS curves, as is indicated in the next two figures [6]. Figure 16.6 shows the DOS at the surface layer and at the two subsequent layers below it for a (111) surface of a simple cubic crystal characterized by a single-band, TB Hamiltonian, with $\Delta V = 0$. The difference from the results obtained for the (100) surface, Fig. 16.5, is evident. Now, the presence of the surface can be seen clearly at $n = 2$, where the DOS is quite different from the bulk DOS.

The effects of the surface can have strong consequences when the potential ΔV is also allowed to vary. Figure 16.7 shows results analogous to those of Fig. 16.6 but with a finite change in the potential, $\Delta V = 0.5\epsilon$, at the surface. The total surface DOS is the sum of two very distorted contributions associated with the change in bulk states, of the kind shown at $n = 1$, and with the presence of surface states, $n = 0$.

16.3 Alloy Surfaces

A generalization of the previous discussion to the case of surfaces and interfaces of substitutionally disordered alloys[3] is fairly straightforward. Again we follow our usual procedure and attempt to replace the semi-infinite disordered material with an effective medium that is translationally invariant in directions parallel to the surface. When perpendicularl to the surface,

[3]Ordered alloys can be treated within the formalism presented above.

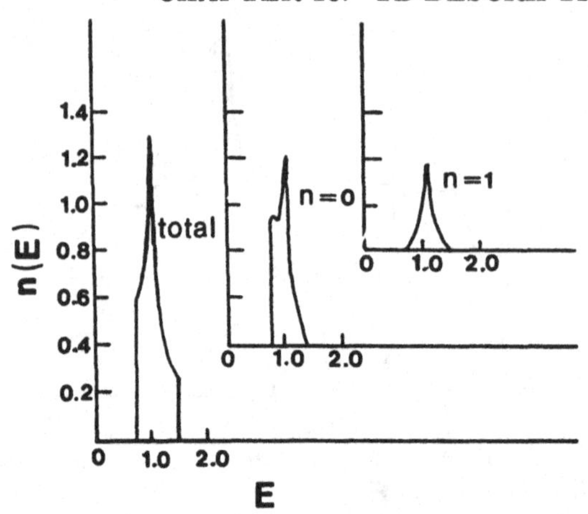

Figure 16.7: **Results analogous to those depicted in Fig. 16.6 but including a change in the surface diagonal matrix element of the Hamiltonian, $\Delta V = 0.5\epsilon$ [6].**

each plane i is characterized by effective potentials (self-energies), σ_i[4]. We can again make the reasonable assumption that, at least in the case of metallic systems, these self energies attain the same asymptotic bulk value, σ. Near the surface, the self energies depend on distance from the surface and on the composition of particular planes, which may indeed be different from the overall bulk composition.

16.3.1 Surface CPA

We begin by examining the case in which the composition profile at the surface is given. We assume that each plane is characterized by an effective concentration, c_i, and that these concentrations approach the bulk concentration c after n planes, (and for practical purposes can be set equal to it). This system can be treated with a straightforward extension of the single-site CPA.

Within this extension, the self-energy σ_i can be determined through the relation

$$\langle G_{ii} \rangle = \bar{G}_{ii}, \tag{16.19}$$

where the Green function associated with a site in plane I, and is given by the general form

$$G_{ii} = (z - \epsilon_i - \Delta_i)^{-1}. \tag{16.20}$$

We note that because of interlayer coupling, the self-consistency condition, Eq. (16.19), represents an infinite system of coupled equations. These

[4]For monatomic materials, the plane index and the site index in directions perpendicular to the surface can coincide.

equations can be truncated at a convenient point determined by the convergence of the self-energy as a function of distance from the surface (or interface).

Because of possible surface reconstruction effects, the site-diagonal elements of the Hamiltonian, ϵ_i, may depend on i at least for a finite number of layers. The same is true of the hopping terms, but this dependence is neglected withinthe present simple treatment. The effective medium Green function, $\bar{G}_{ii}$, has the form

$$\bar{G}_{ii} = (z - \sigma_i - \Delta_i)^{-1}, \tag{16.21}$$

reflecting the fact that it also depends on position relative to the surface (or interface). The asymptotic expression of the Green function, achieved after n layers, is given when σ_i is replaced by σ, the self-energy of the bulk material, and the renormalized interactor is assigned the value appropriate to the nth layer. Clearly, the *value* of the Green function approaches its asymptotic bulk value much more slowly than the self-energy. This is because the latter quantity is local in nature, whereas the Green function, being a global quantity, is characterized by the system as a whole.

The values of the Green functions that appear in Eq. (16.19) can be obtained using expressions such as that in Eq. (16.10). In the general case of a three-dimensional surface, a matrix containing all sites that are characterized by Hamiltonian matrix elements different from those of the bulk is constructed, and is terminated by a renormalized interactor, Δ, that describes the interaction of the cluster of surface planes with the semi-infinite substrate.

Results of numerical tests of the formalism associated with the extension of the CPA to surfaces and interfaces have been given in previous publications [19]. These results clearly indicate that the self-energy can approach its asymptotic value at a much more rapid rate than the Green function with increasing distance from the surface. In particular, the self-energy for a two-dimensional semi-infinite alloy based on a square lattice reaches its asymptotic value after about two layers, whereas the Green function at the ninth layer below the surface can still be distinguished from its bulk counterpart. This behavior is exhibited in Figs. 16.8 and 16.9, respectively. We see that the self-energy at the second plane beneath the surface is very similar to that of the bulk alloy (last panel on the right). We also see that the Green function converges to its bulk value but somewhat more slowly than the self-energy.

16.3.2 Perturbative corrections to the CPA

The various extensions of the CPA discussed with respect to bulk systems in Chapters 11 and 12, such as the generalized perturbation methods and

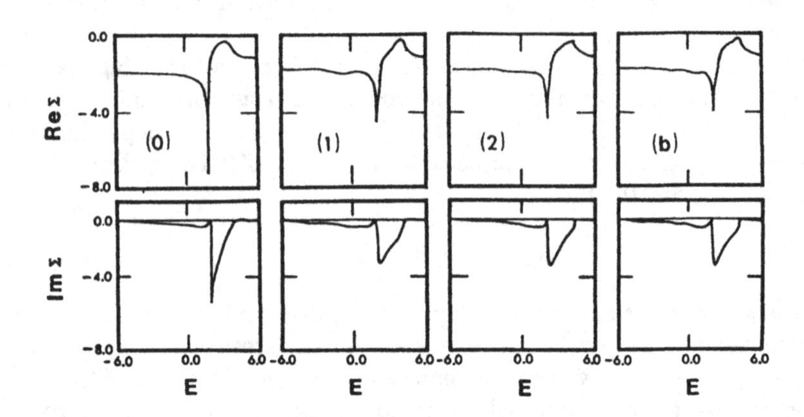

Figure 16.8: **Real and Imaginary parts of the self-energy associated with the surface (0), first (1), and second (2) layers below the surface, and with the bulk material for a two-dimensional alloy, $A_{0.8}B_{0.2}$, based on a square lattice.**

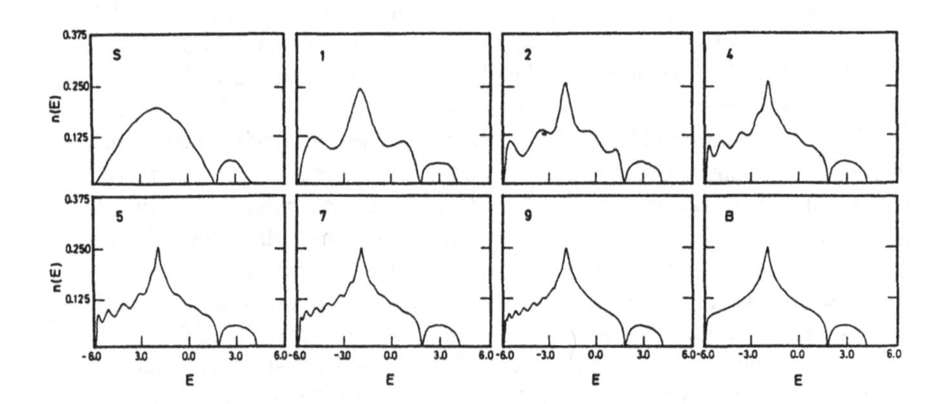

Figure 16.9: **Densities of states at various planes below the surface of the same alloy as in Fig. 16.8, compared with the bulk (B) value.**

the embedded cluster method, can easily be extended to the applications of the CPA to surfaces and interfaces. Results of model calculations based on an application of the ECM to alloy surfaces are quoted in reference [19]. These extensions can be applied to the study of disordered interfaces in a manner that is fairly straightforward. Finally, the GPM has also been extended to surfaces within the first-principles TB methods, and the reader is referred to the literature for details [1, 20].

Bibliography

[1] J. Kudrnovsky, S. K. Bose, and V. Drchal, *Phys. Rev. Lett.* **69**, 308 (1992).

[2] J. Kudrnovsky, V. Drchal, and P. Weinberger, *Phys. Rev. B* **44**, 6410 (1991).

[3] J. Kudrnovsky, I. Turek, V. Drchal, P. Weinberger, S. K. Bose, and A. Pasturel, *Phys. Rev. B* **47**, 16525 (1993)

[4] H. Krakauer, M. Posternak, and A. J. Freeman, *Phys. Rev. B* **19**, 1706 (1979).

[5] M. Posternak, H. Krakauer, A. J. Freeman, and D. D. Koelling, *Phys. Rev.B* **21**, 5601 (1980).

[6] D. Kalkstein and P. Soven, Surf. Sci. **26**, 85 (1971).

[7] F. Cyrot-Lackmann, M. C. Desjonqures, and J. P. Gaspard, J. Phys. C**7**, 925 (1974).

[8] E. Ni Foo and D. J. Giangiulio, Physica **48**B, 167 (1976)

[9] E. J. Mele and J. D. Joannopoulos, Phys. Rev. B**17**, 1816 (1978).

[10] J. Pollmann and S. T. Pantelides, Phys. Rev. B**18**, 5524 (1978).

[11] M. J. Kelly, Solid State Phys. Vol. **35**, 295 (1980), and references therein.

[12] D. H. Lee and J. D. Joannopoulos, Phys. Rev. B**23**, 4988 (1981); ibid. 4997 (1981).

[13] M. Schmeitz, A. Mazur, and J. Pollman, Phys. Rev. B**27**, 5012 (1983).

[14] C. Mailhiot, C. B. Duke, and D. J. Chadi, Phys. Rev. B**31**, 2213 (1985).

[15] N. F. Berk, Surf. Sci. **48**, 289 (1975).

[16] J. L. Moran-Lopez and K. H. Bennemann, Phys. Rev. B**15**, 4790 (1977).

[17] C. Pisni, Phys. Rev. B**30**, 6841 (1984).

[18] X.-G. Zhang, M. Hwang, A. Gonis, and A. J. Freeman, Phys. Rev. B**34**, 5169 (1986).

[19] A. Gonis, *Green Functions for Ordered and Disordered Materials*, North Holland, Amsterdam (1993).

[20] V. Drchal, J. Kudrnovsky, L. Udvardi, P. Weinberger, and A. Pasturel, *Phys. Rev. B* **45**, 14328 (1992).

Chapter 17

Electronic Structure Properties

17.1 General Comments

In this chapter, we present methods for the study of electronic structure of surfaces, interfaces, and other layered structures, based on a first-principles approach. As in previous chapters, our aim is to provide a means to determine the single-particle Green function, and hence the charge density and ground state energy of a system. Each of the system's parts is characterized by a structure that repeats along a given direction over a given half-space. Surfaces and interfaces form two prominent and physically important examples of such systems. Other examples include stacking faults, domain walls, etc., whose treatment can usually be carried out along formal lines similar to those developed in the following pages.

Most of our development will be given in terms of MST and the corresponding Green functions. This allows a proper handling of the boundary conditions associated with the semi-infinite nature of a surface or interface. However, a full-cell implementation of MST has not been widely used. At the same time, excellent results of surface properties have been obtained using alternative procedures in which the semi-infinite nature of the systems under discussion is simulated through the use of slabs or repeating slabs. This has allowed the computations to proceed very effectively and to yield many interesting results. We will quote some of them to illustrate the usefulness of electronic structure methods based on DFT and the LDA in the study of surface electronic properties.

Even though a surface or an interface can represent a small part of a system[1], they may profoundly influence the chemical, physical, and mechanical properties of materials. One important physical property that is strongly dependent on surface atomic and electronic structure is catalytic behavior [1-4]. Stacking faults, and grain boundaries play a dominant role in determining deformation behavior [5, 6], corrosion resistance, and electrical resistivity [7-9]. For example, the high stacking-fault energies generally observed in Al-based alloys tend to prevent slip and lead to poor ductility. On the other hand, impurity segregation at grain boundaries may strongly influence and even completely alleviate this effect [10, 11]. The importance of the electronic origins of these properties is attested to by the large number of works directed toward the development of methods to calculate the electronic structure of semi-infinite systems. References to work based on a phenomenological tight-binding description of the Hamiltonian are given in the previous chapter. In the following pages, we augment that list with a few more references of works carried out within first-principles approaches, including MST.

First-principles treatments of surface and interface electronic structure have been provided within a number of formal approaches. These treatments by and large avoid the approximations inherent in determining the Hamiltonian matrix element in a phenomenological TB description, but often at the expense of imposing approximations to the geometry of the system being studied. In order to treat the lack of translational invariance in directions perpendicular to a surface or interface, many methods [12-16] employ a slab or film construction. Both free-standing and periodically repeating films, as well as supercell structures have been used. Although a first-principles treatment based on such geometric constructs can yield physically accurate results [17-19] the artificial boundary conditions associated with such structures may induce unphysical interactions between the surfaces of the films, or the boundaries of the supercells. Similarly unsatisfactory formal features characterize the supercell approach to the calculation of the electronic structure of surfaces [20-22]. The boundary conditions imposed in such calculations may cause surface or interface states, which otherwise would decay rapidly with distance, to maintain essentially a constant amplitude ($\simeq 90\%$) even at a distance of six or seven layers away from the surface [23]. Therefore, the need arises for developing a method that allows for the proper formal treatment of the semi-infinite nature of surfaces and interfaces in electronic structure calculations. Such a treatment is developed here based on MST.

Among the first treatments of the electronic structure of a real surface was that of Applebaum and Haman [24], who matched surface and bulk wave functions using a transfer matrix approach. A rather similar for-

[1] In some cases, such as in a monolayer or a thin film, the surface may indeed constitute the greater or even the entire part of a system.

mal method has been developed by Holzwarth and Lee [25, 26]. Although conceptually sound, these techniques suffer from the difficulties associated with the projection of an essentially infinite number of bulk states onto a two-dimensional surface band structure [27].

Matching methods based on the Green function, rather than the wave function, have been proposed by a number of authors [28-36]. In more recent applications, Green-function techniques based on the so-called screened linear-muffin-tin orbital (LMTO) [or tight-binding (TB) LMTO] method have been used to study the electronic structure of surfaces. This methodology strikes a balance between the accuracy of fully *ab initio* methods that can be rather sluggish in execution and the speed of the phenomenological TB approach. Complete details of this methodology are included in the book by Turek et al. [37]. These and other Green-function-based methods indicate that the Green function is in general a much more flexible quantity to manipulate than the wave function in treating semi-infinite solids (as is also often the case for bulk materials).

Our development in the following pages is directed at determining the surface or interface Green function within the framework of MST. Following the historical order of development, we describe two different approaches to the calculation of the electronic structure of systems with reduced symmetry. The first method is specifically suited for the treatment of strictly two-dimensional, layered structures and is commonly referred to as the layer Korringa-Kohn-Rostoker (LKKR) method [5, 6, 35, 38, 39]. The second approach allows the solution of the equation of motion for the scattering-path operator entirely in real space [with no recourse to reciprocal (**k**) space] and allows the study of structures with very reduced symmetry, such as edges and corners. It is known as the real-space, multiple-scattering theory (RS-MST) [39-42]. Not surprisingly, these two methods are not entirely unrelated and tend to complement each other in terms of applicability and ease of implementation. The advantages and disadvantages of each method are presented below along with their formal development and examples of calculations based on them.

17.2 Layer Korringa-Kohn-Rostoker Method

What has come to be known as the layer Korringa-Kohn-Rostoker (LKKR) method has its roots in the formalism introduced [44, 45] for the analysis of LEED spectra. In this section, we present the methods as has been developed for the calculation of the electronic structure of semi-infinite (or doubly semi-infinite) solids.

17.2.1 Aim of the calculation

As in the case of periodic solids, we wish to calculate the Green function for a surface or interface system. In operator language, this quantity takes the usual form,

$$G = G_0 + G_0 T G_0, \tag{17.1}$$

where G_0 is the free-particle propagator and the system t-matrix, T, can be written formally as the sum of all scattering paths that connect all pairs of sites in the system

$$T = \sum_{nm} \tau^{nm}. \tag{17.2}$$

In the angular momentum representation, and setting $r > r'$ with the cell vectors $\mathbf{r}$ in cell n and $\mathbf{r}'$ in cell m, Eq. (17.1) can be writtenas

$$G(\mathbf{r},\mathbf{r}') = \langle Z^n(\mathbf{r})|\underline{\tau}^{nm}|Z^m(\mathbf{r})\rangle - \langle Z^n(\mathbf{r})|S^n(\mathbf{r}')\rangle\delta_{nm}, \tag{17.3}$$

where the wave functions $|Z^m(\mathbf{r})\rangle$ and $|S^n(\mathbf{r})\rangle$[2] are regular and irregular solutions of the Schrödinger equation associated with cell n that satisfy the boundary conditions specified in connection with Eqs. (3.79) and (3.80). The quantity $\underline{\tau}^{nm}$ is the scattering-path operator[3] connecting sites n and m. The SPO satisfies the equation of motion, Eq. (3.53),

$$\underline{\tau}^{nm} = \underline{t}^n \left\{ \delta_{nm} + \sum_{\ell \neq n} \underline{G}(\mathbf{R}_{n\ell})\underline{\tau}^{\ell m} \right\}, \tag{17.4}$$

or, equivalently,

$$\underline{\tau}^{nm} = \left\{ \delta_{nm} + \sum_{\ell \neq n} \underline{G}(\mathbf{R}_{n\ell})\underline{\tau}^{\ell m} \right\} \underline{t}^m. \tag{17.5}$$

Once the Green function, $G(\mathbf{r},\mathbf{r}')$, has been determined, the charge density and the density of states are obtained, respectively, by means of the well-known expressions,

$$n(\mathbf{r}) = -\frac{1}{\pi}\Im \int G(\mathbf{r},\mathbf{r};E)\mathrm{d}E, \tag{17.6}$$

and

$$n(E) = -\frac{1}{\pi}\Im \int G(\mathbf{r},\mathbf{r};E)\mathrm{d}^3 r. \tag{17.7}$$

[2] The irregular solution that matches onto $|J(\mathbf{r})\rangle$ on the surface of the bounding sphere has also been denoted by $|\tilde{J}(\mathbf{r})\rangle$ in previous discussion.

[3] As already mentioned, scattering-path matrix may be a more appropriate name for this quantity.

In developing the formalism for the evaluation of the Green function for a layered material, we will assume the presence of space-filling cells. This, of course, brings up the question of internal summations in angular momentum space, and of the associated convergence. As discussed in Section 5.9, these summations can always be performed in converged fashion and we will proceed on that basis. In any practical application of the ensuing formalism, the convergence of internal sums should be monitored closely on a case by case basis.

17.2.2 Evaluation of the scattering matrix

It is clear from Eq. (17.1) that the knowledge of G depends on the knowledge of $\underline{T}$, or in the angular momentum representation, on the knowledge of the SPO. In order to clarify some important features of the solution of the equation of motion for layered systems, we consider anew a number of basic scattering processes[4]. In our development of the LKKR, we follow the lines in references [35] and [39]. Similar derivations have also been given by Pendry [38].

We begin by considering the multiple scattering processes associated with two spatially-bounded scatterers, t^1 and t^2. We phrase our initial discussion in general operator language and subsequently convert to the angular momentum representation[5]. Denoting by T^1 and T^2 the sums of all scattering paths that terminate at scatterers 1 and 2, respectively, we have,

$$T^1 = t^1 + t^1 G_0^{12} t^2 + t^1 G_0^{12} t^2 G_0^{21} t^1 + t^1 G_0^{12} t^2 G_0^{21} t^1 G_0^{12} t^2 \cdots, \tag{17.8}$$

and

$$T^2 = t^2 + t^2 G_0^{21} t^1 + t^2 G_0^{21} t^1 G_0^{12} t^2 + t^2 G_0^{21} t^1 G_0^{12} t^2 G_0^{21} t^1 \cdots, \tag{17.9}$$

where the individual scattering events in each term are to be perceived as proceeding chronologically from right to left. As is so often the case with the equations of MST, these operator expressions remain valid in the angular momentum representation, under the replacement of operator quantities by the corresponding matrices. For example, under such a replacement, t^i, with $i = 1, 2$, becomes the matrix, $\underline{t}^i_{LL'}$, and G_0^{ij} is replaced by the real-space structure constant, $G_{LL'}(\mathbf{R}_{ij})$. We can write the last two equations

[4]The formalism developed in Chapter 3 suffices to give a condensed derivation of the basic equations of LLKR. However, such a derivation might obscure some interesting and important details of the method that will be stated explicitly in the present derivation.

[5]Much of this initial material should be familiar from our discussion in Chapter 3. It is repeated here for the sake of completeness of the presentation, and in order to alleviate the need for the reader to search for it in previous pages.

in the compact form,

$$\begin{aligned} T^1 &= t^1 + t^1 G_0^{12} T^2 \\ T^2 &= t^2 + t^2 G_0^{21} T^1. \end{aligned} \tag{17.10}$$

This set of equations can be solved by means of a matrix inversion,

$$\begin{pmatrix} T^1 \\ T^2 \end{pmatrix} = \begin{pmatrix} I & -t^1 G_0^{12} \\ -t^2 G_0^{21} & I \end{pmatrix}^{-1} \begin{pmatrix} t^1 \\ t^2 \end{pmatrix}. \tag{17.11}$$

The generalization of the last expression to the case of N scatterers is straightforward and takes the obvious form,

$$\begin{pmatrix} T^1 \\ T^2 \\ \cdot \\ \cdot \\ \cdot \\ T^N \end{pmatrix} = \begin{pmatrix} I & -t^1 G_0^{12} & -t^1 G_0^{13} & \cdots & -t^1 G_0^{1N} \\ -t^2 G_0^{21} & I & -t^2 G_0^{23} & \cdots & -t^1 G_0^{2N} \\ \cdot & \cdot & \cdot & \cdots & \cdot \\ \cdot & \cdot & \cdot & \cdots & \cdot \\ \cdot & \cdot & \cdot & \cdots & \cdot \\ -t^N G_0^{N1} & -t^N G_0^{N2} & -t^N G_0^{N3} & \cdots & I \end{pmatrix}^{-1} \begin{pmatrix} t^1 \\ t^2 \\ \cdot \\ \cdot \\ \cdot \\ t^N \end{pmatrix}. \tag{17.1}$$

The various terms, T^i with $i = 1,\ 2,\ \cdots N$, can be combined to yield the scattering matrix of any group of scatterers viewed as a single scattering unit. In particular, the t-matrix for the entire system can be written as the sum, [see Eq. (3.3)],

$$T = \sum_{i=1}^{N} T^i. \tag{17.13}$$

It is also convenient to define scattering matrices, T^{ij}, describing the sum of all scattering paths that start at site j and end at site i. This quantity is simply the ijth element of the inverse in Eq. (17.12) multiplied by t^j. We then have,

$$T = \sum_{ij} T^{ij}. \tag{17.14}$$

Furthermore, all these expressions can be converted into the angular momentum representation. For example, in this representation, T^{ij} becomes the SPO appearing in Eq. (17.4) and the last expression takes the form of Eq. (3.64).

17.2.3 Two-dimensional lattice

Let us now consider that the N sites (scatterers) of the system occupy the points of two-dimensional periodic lattice (with $N \to \infty$). In this case, we have $T^i = T^j = T$ for all j and i and we can write (in operator form),

$$T^i \equiv T = t + t\left[\sum_j G^{ij}\right]T, \qquad (17.15)$$

where the sum extends over all sites other than i, and the explicit use of the invariance of T with respect to site index allows us to factor it out of the sum. Passing to the angular momentum representation, and using the *two-dimensional* lattice Fourier transform of the *three-dimensionals* structure constants, we have

$$\begin{aligned} G_{LL'}(\mathbf{k}_\parallel) &\equiv G^{ii}_{LL'}(\mathbf{k}_\parallel) \\ &= -4\pi\mathrm{i}\sqrt{E}\sum_{j\neq i}\sum_{L_1}\mathrm{i}^{\ell_1}C\left(L_1L'L_1\right)H_{L_1}(k_\parallel\mathbf{R}_{ij}), \end{aligned} \qquad (17.16)$$

or

$$\underline{G}(k_\parallel) = \sum_{j\neq i}\underline{G}(\mathbf{R}_j)\mathrm{e}^{\mathrm{i}\mathbf{k}_\parallel\cdot\mathbf{R}_j}. \qquad (17.17)$$

In both of the expressions above, the point $\mathbf{R}_{ii} \equiv 0$ is excluded from the sums. The quantities $G_{LL'}(\mathbf{k}_\parallel)$ are the in-plane structure constants in two-dimensional reciprocal space, a fact that is designated explicitly by the symbol $\mathbf{k}_\parallel$. These quantities are the formal analogues of the TB quantities, $\underline{W}(\mathbf{k}_\parallel)$, introduced in Eq. (16.15). Methods for the evaluation of these "layer" structure constants have been given by Kambe [46, 47]. We can now write Eq. (17.15) in the form,

$$\begin{aligned} \underline{T} &= \underline{\tau}(\mathbf{k}_\parallel)\left[1 - \underline{t}\underline{G}(\mathbf{k}_\parallel)\right]^{-1}\underline{t} \\ &= \underline{t}\left[1 - \underline{G}(\mathbf{k}_\parallel)\underline{t}\right]^{-1} \\ &= \left[\underline{m} - \underline{G}(\mathbf{k}_\parallel)\right]^{-1}. \end{aligned} \qquad (17.18)$$

We recognize the formal similarity between the structure constants $\underline{G}(\mathbf{k}_\parallel)$ and the TB quantities $\underline{W}_{ii}(\mathbf{k}_\parallel)$ introduced in the last chapter. The last expression for $\underline{\tau}(\mathbf{k}_\parallel)$ is formally identical to that for the TB Green function in Eq. (16.14).

17.2.4 Several periodic layers

We now extend the formalism presented above to pairs and larger numbers of periodic planes of scatterers. We assume that the planes are characterized by common periodicity, and are arranged parallel to one another along a given direction, usually taken to be the positive z axis. We explicitly treat the case of monoatomic unit cells, but the generalization to polyatomic cells is not conceptually difficult (although the computational complications may indeed become prohibitive in specific cases).

In developing the formalism, it is convenient to introduce a different system of labeling atomic sites in a semi-infinite material from that used in the last chapter. We let capital letters, I, $J, \cdots$, denote entire planes, and use lower case letters, i, $j, \cdots$, to denote sites within each plane[6].

First, consider two planes, labeled 1 and 2, characterized by individual cell scattering matrices t^1 and t^2, respectively. Let T^1 and T^2 represent all multiple-scattering paths that terminate at an atom of plane 1 and 2, respectively, and τ^1 and τ^2 the sum of all such paths that are confined entirely in the corresponding planes. This means that T^1, for example, consists of the scattering events in τ^1, and all scattering contributions that terminate at any site i in plane 2, and thus are given by T^2, then propagate back to and terminate on plane 1, giving τ^1. With a similar reasoning for plane 2, we can write

$$\begin{aligned} T^1 &= \tau^1 + \tau^1 \sum_{i \in 2} G_0^{12(i)} T^2 \\ T^2 &= \tau^2 + \tau^2 \sum_{i \in 1} G_0^{21(i)} T^1. \end{aligned} \tag{17.19}$$

The analogy with Eq. (17.10) should be noted. The two equations are essentially identical in form and content, provided one considers each plane as a single scattering unit. The only difference between the two expressions is the presence of sums over the sites in a plane, which clearly arises because each scattering unit has more than a single scattering cell. In fact the two sets of equations can be made to look exactly the same if we define the interplanar propagators

$$G_0^{12} = \sum_{i \in 2} G_0^{12(i)}, \quad \text{and} \quad G_0^{21} = \sum_{i \in 1} G_0^{21(i)}, \tag{17.20}$$

so that the last pair of equations can be solved in the form,

$$\begin{pmatrix} T^1 \\ T^2 \end{pmatrix} = \begin{pmatrix} I & -\tau^1 G_0^{12} \\ -\tau^2 G_0^{21} & I \end{pmatrix}^{-1} \begin{pmatrix} \tau^1 \\ \tau^2 \end{pmatrix}. \tag{17.21}$$

This expression clearly indicates the treatment of entire planes as single scattering units characterized by the scattering matrices τ^i. The generalization to N planes is straightforward and the solution for T^i, $i = 1, \cdots, N$,

[6]This notation is to be compared with that of the previous chapter where the coordinates of a site were given in terms of its position in a "linear" cluster perpendicular to the surface.

takes a form analogous to that in Eq. (17.12),

$$\begin{pmatrix} T^1 \\ T^2 \\ \cdot \\ \cdot \\ \cdot \\ T^N \end{pmatrix} = \begin{pmatrix} I & -\tau^1 G_0^{12} & -\tau^1 G_0^{13} & \cdots & -\tau^1 G_0^{1N} \\ -\tau^2 G_0^{21} & I & -\tau^2 G_0^{23} & \cdots & -\tau^1 G_0^{2N} \\ \cdot & \cdot & \cdot & \cdots & \cdot \\ \cdot & \cdot & \cdot & \cdots & \cdot \\ \cdot & \cdot & \cdot & \cdots & \cdot \\ -\tau^N G_0^{N1} & -\tau^N G_0^{N2} & -\tau^N G_0^{N3} & \cdots & I \end{pmatrix}^{-1} \begin{pmatrix} \tau^1 \\ \tau^2 \\ \cdot \\ \cdot \\ \cdot \\ \tau^N \end{pmatrix}. \tag{17.22}$$

We now allow N to go to infinity and pass over to the angular momentum representation. In this case each τ^I is given by Eq. (17.18), while the Fourier transform of the interplanar ($I \neq J$) structure constants becomes

$$\underline{G}^{IJ}(\mathbf{k}_{\parallel}) = \sum_j{}' \underline{G}(\mathbf{R}_{0j} + \mathbf{R}_J)e^{-i\mathbf{k}_{\parallel}\cdot\mathbf{R}_j}. \tag{17.23}$$

Here, the prime indicates that the term with $\mathbf{R}_J = 0$ is not included in the sum. The two-dimensional lattice constants, $\underline{G}^{IJ}(\mathbf{k}_{\parallel})$, can be evaluated through methods introduced by Kambe [46, 47]. Using these structure constants, explicitly taking into account the infinite number of planes in the system, for each value of $\mathbf{k}_{\parallel}$ we can obtain from Eq. (17.22) the expression

$$\begin{pmatrix} T^1 \\ T^2 \\ \cdot \\ \cdot \\ \cdot \end{pmatrix} = \begin{pmatrix} I & -\tau^1 G_0^{12} & -\tau^1 G_0^{13} & \cdots & -\tau^1 G_0^{1N} \\ -\tau^2 G_0^{21} & I & -\tau^2 G_0^{23} & \cdots & -\tau^1 G_0^{2N} \\ \cdot & \cdot & \cdot & \cdots & \cdot \\ \cdot & \cdot & \cdot & \cdots & \cdot \\ \cdot & \cdot & \cdot & \cdots & \cdot \end{pmatrix}^{-1} \begin{pmatrix} \tau^1 \\ \tau^2 \\ \cdot \\ \cdot \\ \cdot \end{pmatrix}. \tag{17.24}$$

Each value of $\mathbf{k}_{\parallel}$ gives rise to a linear scattering system whose solution is described formally by the last equation.

Before leaving this subsection, we mention that in analogy with the discussion that follows Eq. (17.13), it is also possible (and often useful) to define interplanar scattering-path operators[7] τ^{IJ} whose site elements describe all scattering paths that start on a site in plane J and end on a site in plane I. This scattering-path operator is obtained from the IJth element of the inverse in Eq. (17.22) [or Eq. (17.24)] multiplied by τ^J.

As already stated, our aim is to obtain the scattering-path operator and corresponding Green function associated with a semi-infinite system.

[7] In either abstract space or in the angular-momentum representation

Proceeding from this point, this can be accomplished in at least two distinct ways based, respectively, on the layer Korringa-Kohn-Rostoker (LKKR) approach, and on the RS-MST method. Since the LKKR was developed prior to the introduction of RS-MST, we begin with that.

17.2.5 From L-space to k-space

Within the LKKR method, one considers the cumulative process of reflections from and transmissions through individual planes when an incident (plane) wave impinges upon a semi-infinite solid. The in-plane scattering is described in an angular-momentum (L) representation, while the propagation between planes is expressed in reciprocal ($\mathbf{k}$) space. In order to proceed with the development of the LKKR method, we need expressions for the reflection and transmission coefficients for a stack of layers (planes), and we also need to know how to pass back and forth between the L and $\mathbf{k}$ representations.

We consider an electron beam of energy $E = k^2$, expressed in a plane-wave basis as $\chi(\mathbf{r}) = \mathrm{e}^{\mathrm{i}\mathbf{k}\cdot\mathbf{r}}$, incident on a stack of planes, I, arranged in periodic fashion perpendicularly along the positive z direction of a right-handed Cartesian coordinate system. We wish to calculate the transmitted part $\psi^+(\mathbf{r})$ and reflected part $\psi^-(\mathbf{r})$ of the scattered wave. Because of the two-dimensional periodicity of the surface, any three-dimensional vector $\mathbf{k}$ is equivalent with the wave vector

$$\mathbf{K}_{\mathbf{g}}^{\pm} = \left[k_x + g_x, k_y + g_y, \pm\sqrt{E - (k_x + g_x)^2 - (k_y + g_y)^2}\right], \qquad (17.25)$$

where $\mathbf{g} = (g_x, g_y)$ is a vector in the two-dimensional reciprocal space defined by the periodicity of the planes, and the + (-) sign indicates waves propagating along the positive (negative) z direction. It follows that an incident wave function (in general a wave packet) can be expanded in the form,

$$\chi(\mathbf{r}) = \sum_{\mathbf{g}} U_{\mathbf{g}}^{\pm} \mathrm{e}^{\mathrm{i}\mathbf{K}_{\mathbf{g}}^{\pm}\cdot\mathbf{r}}. \qquad (17.26)$$

Similarly, the transmitted part of the scattered wave can be expressed as

$$\psi^+(\mathbf{r}) = \sum_{\mathbf{g}} U_{\mathbf{g}}^{+} \mathrm{e}^{\mathrm{i}\mathbf{K}_{\mathbf{g}}^{+}\cdot\mathbf{r}} + \sum_{\mathbf{g}} V_{\mathbf{g}'}^{+} \mathrm{e}^{\mathrm{i}\mathbf{K}_{\mathbf{g}'}^{+}\cdot\mathbf{r}}, \qquad (17.27)$$

while the reflected part of the wave takes the form,

$$\psi^-(\mathbf{r}) = \sum_{\mathbf{g}} V_{\mathbf{g}'}^{-} \mathrm{e}^{\mathrm{i}\mathbf{K}_{\mathbf{g}'}^{-}\cdot\mathbf{r}}. \qquad (17.28)$$

In a plane-wave representation, the reflection and transmission coefficients are related by the expression,

$$V^{\pm}_{\mathbf{g}'} = \sum_{\mathbf{g}} M^{\pm +}_{\mathbf{gg}'} U^{+}_{\mathbf{g}}. \tag{17.29}$$

More generally, considering waves incident along either the positive or negative z directions, we have

$$V^{\pm}_{\mathbf{g}'} = \sum_{\mathbf{g}} M^{\pm\pm}_{\mathbf{gg}'} U^{\pm}_{\mathbf{g}}. \tag{17.30}$$

As follows from the discussion in Appendix O, the quantities $M^{\pm\pm}_{\mathbf{gg}'}$ are related to the scattering-path operator $\underline{\tau}^{IJ}$ between planes I and J by the expression,

$$\begin{aligned} M^{\pm\pm}_{\mathbf{gg}'} &= -\frac{8\pi^2 \mathrm{i}}{AK^{\pm}_{g'_z}} \sum_{IJ} \mathrm{e}^{-\mathrm{i}\mathbf{K}^{\pm}_{\mathbf{g}'}\cdot\mathbf{C}_I} \mathrm{e}^{\mathrm{i}\mathbf{K}^{\pm}_{\mathbf{g}}\cdot\mathbf{C}_J} \\ &\times \langle Y(\mathbf{K}^{\pm}_{\mathbf{g}'}) | \underline{\tau}^{IJ}(\mathbf{k}_{\parallel}) | Y(\mathbf{K}^{\pm}_{\mathbf{g}}) \rangle, \end{aligned} \tag{17.31}$$

where $\mathbf{C}_I$ denotes the center of plane I, A denotes the area of the two-dimensional unit cell of the surface, and the bracket notation indicates summation in L-space. We note that for complex argument, we have

$$Y^{*}_{\ell m}(\hat{k}) = (-1)^m Y_{\ell m}(\hat{k}). \tag{17.32}$$

We can now define operators,

$$\begin{aligned} \Gamma^{\pm}_{I,gL} &= 4\pi \mathrm{e}^{\mathrm{i}\mathbf{K}^{\pm}_{\mathbf{g}}\cdot\mathbf{C}_I} Y^{*}_{\ell -m}(\mathbf{K}^{\pm}_{\mathbf{g}}) \\ \tilde{\Gamma}^{\pm}_{I,gL} &= \frac{2\pi}{AK^{\pm}_{g'_z}} \mathrm{e}^{\mathrm{i}\mathbf{K}^{\pm}_{\mathbf{g}}\cdot\mathbf{C}_I} Y_{\ell m}(\mathbf{K}^{\pm}_{\mathbf{g}}), \end{aligned} \tag{17.33}$$

which project an incident plane wave onto an angular momentum basis and back, respectively. We also need the plane-wave matrix elements of the free-particle propagator between the origins of two successive planes,

$$P_{I\mathbf{gg}'} = \langle \mathbf{K}^{\pm}_{\mathbf{g}} | G_0 | \mathbf{K}^{\pm}_{\mathbf{g}'} \rangle = \delta_{\mathbf{gg}'} \mathrm{e}^{\mathrm{i}\mathbf{K}^{\pm}_{\mathbf{g}}\cdot(\mathbf{C}_{I+1} - \mathbf{C}_I)}. \tag{17.34}$$

17.2.6 Layer doubling method

We now have assembled all the basic elements needed in the LKKR method for the solution of the MST equations for a layered system. The central idea is to express the total reflection and transmission coefficients for the semi-infinite system in terms of the corresponding quantities associated with a single layer (or a repeating unit).

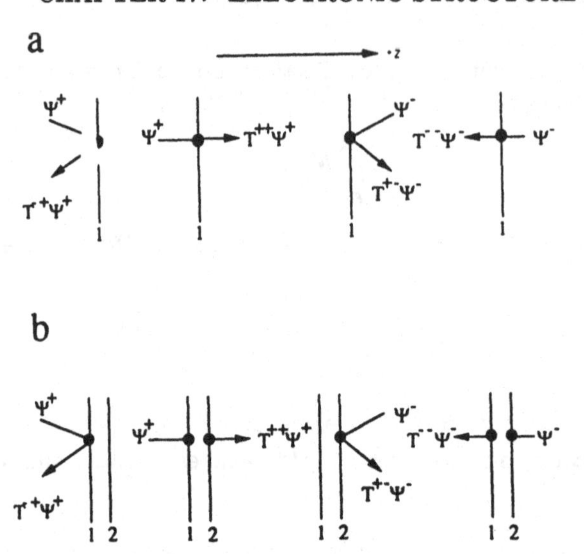

Figure 17.1: **Schematic representation of the transmission and reflection matrices for a single layer, (a), and for two layers, (b).**

Reflection and transmission at a single layer, I, can be represented by the four operators $T_I^{\pm\pm}$ which in view of Eqs. (17.33) and (17.34) are formally defined as follows,

$$\begin{aligned} T_I^{-+} &= \Gamma^- T_I \tilde{\Gamma}^+ \quad T_I^{++} = 1 + \Gamma^+ T_I \tilde{\Gamma}^+ \\ T_I^{+-} &= \Gamma^+ T_I \tilde{\Gamma}^- \quad T_I^{--} = 1 + \Gamma^- T_I \tilde{\Gamma}^-, \end{aligned} \tag{17.35}$$

where the explicit indication of matrix elements in reciprocal and L space is suppressed for ease of presentation.

The structure of these equations allows a straightforward interpretation. An incident wave in the plane-wave representation, proceeding along the (positive) negative z direction is reflected or transmitted by layer I as a result of the scattering processes within the layer. Consider explicitly the first of the four expressions above. The quantity $\tilde{\Gamma}^+$ converts the plane wave into an angular momentum representation, T_I describes the sum of all scattering events inside the layer, and Γ^- converts the outgoing wave back into reciprocal space. The presence of unity in the second and fourth expressions indicates the fact that the incident and transmitted waves are proceeding along the same direction. The matrices $T^{\pm\pm}$ can be represented schematically as shown in Fig. 17.1(a).

Let us now consider transmission and reflection through two layers, 1 and 2. We denote the quantities analogous to those defined for a single layer, Eq. (17.35), by the symbol $T_{12}^{\pm\pm}$. These matrices describe the cumulative effect of the combined reflection and transmission from the two layers considered as a single unit. For example, the overall transmission matrix $T_{12}^{\pm+}$ is the result of transmission through layer 1, described by T_2^{++} and through layer 2, described by T_1^{++}, with all possible reflections between the two layers, T_2^{+-} and T_1^{-+}, with all possible free-particle propagation in between. Considered in this manner, and based on the general arguments of Appendix P, these combined coefficients can be expressed in the form [35] [see also Fig. 17.1(b)],

$$\begin{aligned}
T_{12}^{++} &= T_1^{++}\left(1 - P_1^+ T_2^{+-} P_1^- T_1^{-+}\right)^{-1} P_1^+ T_2^{++} \\
T_{12}^{--} &= T_2^{++}\left(1 - P_1^+ T_1^{+-} P_1^- T_2^{-+}\right)^{-1} P_1^+ T_1^{++} \\
T_{12}^{+-} &= T_1^{+-} + T_1^{++} P_1^+ T_2^{+-} \\
&\times \left(1 - P_1^- T_1^{-+} P_1^+ T_2^{+-}\right)^{-1} P_1^- T_1^{--} \\
T_{12}^{-+} &= T_2^{-+} + T_2^{--} P_1^- T_1^{+-+} \\
&\times \left(1 - P_1^+ T_2^{+-} P_1^- T_1^{-+}\right)^{-1} P_1^+ T_2^{++}.
\end{aligned} \tag{17.36}$$

Here, $P_I^{\pm}$ is the **k**-space representation of free-particle propagation between two successive layers, with the $T^{\pm}$ describing intra-layer scattering in the angular momentum representation.

The last set of equations suggests an iterative process. If $T_{1,2}^{\pm}$ on the right-hand side is replaced by $T_{12}^{\pm\pm}$ and $P_1^{\pm}$ is replaced by $P_2^{\pm}$, the equations yield the transmission and reflection coefficients for a pair of two layers, or four layers total. These four layers can be treated as a unit, and with the appropriate replacements again, the equations yield the reflection and transmission coefficients for eight layers. This process can be repeated as many times as necessary until the reflection and transmission coefficients converge within a preset tolerance.

The iterative scheme just described leads to the reflection and transmission coefficients for a semi-infinite solid by means of combining layers into single scattering units, taking the units two at a time. It is commonly referred to as the ***layer doubling method***, and has been described extensively in a number of texbooks [38, 39] and review articles [35].

As we have pointed out, the layer doubling approach uses a mixed L-space/**k**-space representation. The scattering inside the layers is described in the angular-momentum representation, while interlayer propagation is expressed in reciprocal space (plane waves). Because of the form of Eq. (17.25), the quantity $\mathbf{K}_{\mathbf{g}}^{\pm}$ acquires an increasingly larger imaginary part with increasing magnitude of **g** which damps out the amplitude of the basis plane-wave states at successive layers. Consequently the expansion in reciprocal space usually requires a small number of reciprocal-lattice vectors to reach convergence. In actual applications, it is found [35] that 10 to 20 reciprocal lattice vectors are sufficient to yield converged results.

However, the number of reciprocal lattice vectors (also referred to as beams) needed in a given expansion becomes increasingly large as the successive layers get closer together. This can occur in the case of surfaces and interfaces described by low Miller indices (large angles) and are characterized by small interplanar spacing and large areas of two-dimensional unit cells. In such cases, the convergence of the layer doubling process can become sluggish, or fail altogether. As discussed in the next section, alternative techniques have been developed which allow the treatment of semi-infinite systems characterized by essentially arbitrary interplanar spacing.

17.2.7 SPO for a semi-infinite solid

To complete the description of the electronic structure of a semi-infinite system, we must now proceed with the evaluation of the scattering-path operator based on the knowledge of the total reflection and transmission coefficients determined from an application of the layer doubling process.

Let T^I be the scattering-path operator[8] representing all scattering paths confined *entirely* inside plane I. An explicit expression for T^I in the angular momentum representation is given in Eq. (17.18) (denoted there by $\underline{T}$). Now, imagine an infinite system consisting of identical layers extending periodically from $-\infty$ to ∞ along the z axis. We wish to determine the scattering-path operator, τ^{II}, which represents the sum of all scattering paths throughout the infinite system that start and end on layer I. Clearly, this quantity is the sum of T^I (strictly intralayer scattering) and of all scattering paths that emanate from layer I and arrive back there after traveling throughout the semi-infinite spaces on either side of layer I. As is shown in Appendix P, τ^{II} has the general form,

$$\tau^{II} = T^I + T^I R^I T^I, \tag{17.37}$$

where R^I is an effective reflectivity of the solid on either side of layer I. This expression represents the solution of the problem of embedding an impurity layer (or a stack of such layers) into an unperturbed host lattice. Explicit expressions for this effective reflectivity in the angular momentum representation in terms of the reflection and transmission coefficients derived above are given in Appendix P, Eq. (P.6), and in the literature [35]. It is perhaps evident that in the case of a strictly semi-infinite system, Eq. (17.37) retains its general form, but with R^I denoting the reflectivity of only the system on one side of layer I. Expressions for the interplanar scattering-path operator can also be obtained in a straightforward way.

[8] For ease of presentation, we employ abstract operator space, but the expressions remain essentially identical in the angular momentum representation.

Knowledge of the scattering-path operator allows the calculation of the Green function for the system, and hence of all single-particle properties. Thus, we have now completed the description of the electronic structure of a semi-infinite or doubly semi-infinite solid within the LKKR formalism. For the sake of completeness, we quote an expression for the Green function.

With the vector $\mathbf{r}$ in cell Ω_i in plane (layer) I and vector $\mathbf{r}'$ in cell j in plane J, and setting $r' > r$, we have,

$$G(\mathbf{r},\mathbf{r}') = \langle Z^{I_i}(\mathbf{r})|\underline{\tau}^{I_iJ_j}|Z^{J_j}(\mathbf{r}')\rangle - \langle Z^{I_i}(\mathbf{r})|S^{I_i}(\mathbf{r}')\rangle\delta_{I_i J_j}. \tag{17.38}$$

The meaning of the various quantities that appear in this expression has been specified in previous discussion, e.g., Eqs. (3.79) and (3.80).

17.3 Real-Space MST

In this section, we describe an alternative method for the calculation of the electronic structure of layered and bulk systrems[9]. The method described below allows the solution of the equation of motion of the SPO entirely within the L representation and in direct space. Thus, in contrast to the LKKR, it makes no use of expansions in reciprocal space. Consequently, this approach is labeled [40, 48, 49] real-space multiple scattering theory (RS-MST). We begin our discussion by recalling a few basic notions of multiple scattering theory, and by pointing out a few central points of comparison and contrast between the LKKR and RS-MST methods.

A central step within any method based on MST is the solution of the equation of motion of the SPO associated with the particular structure under consideration. Having the SPO, τ, the scattering matrix of the system as a whole, i.e., the scattering matrix describing the entire system as a single scatterer, is given by the formal relation, Eqs. (17.13) and (17.14),

$$T = \sum_i \tau^i = \sum_{ij} \tau^{ij}. \tag{17.39}$$

As we saw in Eq. (3.64), in the angular momentum representation the last expression takes the form,

$$\begin{aligned} \underline{T} &= \langle \underline{g}|\underline{\tau}|\underline{g}\rangle \\ &= \sum_{ij} \underline{g}(-\mathbf{R}_{i0})\underline{\tau}^{ij}\underline{g}^{\dagger}(-\mathbf{R}_{j0}). \end{aligned} \tag{17.40}$$

We recall that the quantities $g_{LL'}(\mathbf{R})$ form the angular momentum representation of the translation operator, and that the quantity $\underline{\underline{\tau}}$ is the inverse

[9] Bulk systems can also be viewed as consisting of periodically repeating layers and are thus amenable to treatment by the methods of this chapter.

of the matrix $\underline{\underline{M}}$ whose elements in site space are given by the expression,

$$\left[\underline{\underline{M}}\right]_{ij} = \underline{m}^i \delta_{ij} - \underline{G}(\mathbf{R}_{ij})(1 - \delta_{ij}). \tag{17.41}$$

The quantities $\underline{G}(\mathbf{R}_{ij})$ are the real-space structure constants defined in Eq. (F.11).

In the previous section, we showed that the LKKR method allows one to obtain $\underline{T}$ in Eq. (17.40) by coupling together layers in the layer-doubling process. In the LKKR method, the scattering within a layer is described fully within the angular momentum representation, while propagation between the layers is expressed in a $\mathbf{k}$ representation. The method to be developed below uses only the angular-momentum representation. In addition, instead of the layer-doubling process, the RS-MST method leads to a set of self-consistent equations for the scattering matrix of a semi-infinite, periodic system. The replacement of the layer-doubling process (based on expansions in plane-waves) by a set of consistency conditions, expressed entirely in the angular-momentum representation and in real space, is the main conceptual difference between the LKKR and the RS-MST methods.

In principle, either of these two methods can be used to solve the equations of MST for a layered structure. However, the conceptual differences between the two methods have computational ramifications as well. The use of $\mathbf{k}$-space usually implies that the LKKR converges faster than the RS-MST in the case of systems where both are applicable. However, in many cases, the RS-MST can lead to converged results when the layer-doubling procedure may diverge. As we pointed out earlier, this may happen in the case of closely spaced layers [50]. At the same time, the question of convergence of sums over L is much more pronounced in the RS-MST than in the LKKR, and requires close monitoring on a case-by-case basis. Furthermore, the increase in matrix size that usually accompanies applications of RS-MST to closely spaced layers can lead to sluggish convergence, although even in fairly extreme cases convergence can be usually achieved with values of ℓ of the order of 6 or 8.

Finally, the RS-MST, since it can be developed entirely within real (rather than reciprocal) space, allows the solution of the MST equations even in the case of systems with severely reduced translational symmetry, such as edges and corners in two- and three-dimensional space. Applications of RS-MST involve the continued inversion of matrices and are consequently computationally cumbersome. The calculation of the electronic structure of even simple elements, e.g. Cu, is essentially as difficult as that involved in the treatment of reduced symmetry structures. This is a central disadvantage of the use of RS-MST. On the other hand,this difficulty does not increase significantly with increased reduction in symmetry, and the RS-MST can provide a proper treatment of the boundary conditions associated with specific structures with reduced symmetry, which is not always possible within methods based on slabs or supercells. Some of these features of the RS-MST are addressed further in the material that follows.

We now turn to a detailed description of RS-MST, beginning with the concept of *removal invariance.*

17.3.1 Removal invariance

Real-space MST consists of the solution of the equations of multiple-scattering theory for an infinite system entirely in real space and not, as may be initially thought, of a real-space inversion of the MST matrices for a finite cluster of cells. The theory is based on the concept of removal invariance that characterizes systems with semi-infinite periodicity. We define *semi-infinite periodicity* (SIP) as the periodic repetition of a basic unit (an atom, a plane, or stack of planes of atoms) along a given direction, with a well-defined starting position in space. A semi-infinite, periodic, linear array of atoms and an ideal surface are prototypes of systems with SIP. Clearly, many structures, such as bulk, periodic solids, can be viewed as being formed by parts, each of which is characterized by SIP. Furthermore, the existence of SIP in three dimensions does not imply the existence of two-dimensional periodicity. Two identical, periodic half lines of atoms arranged perpendicularly with a common starting point and repeated a-long a direction that bisects the angle between them form a repeating unit that generates a quarter of a square monolayer. The resulting structure is certainly characterized by SIP but does not possess full two-dimensional periodicity. Similar structures, such as three-dimensional corners and edges can be built on the basis of a repeating unit. Because of the possible absence of full periodicity in structures with SIP, and thus the inability to use Fourier transforms in reciprocal space, we must look for ways to solve the equations of MST entirely in real space.

This solution can be accomplished through the use of the principle of *removal invariance* that characterizes systems with SIP. This principle s-tates that *all observable properties of a system with SIP remain invariant when an integral number of repeating units is removed from the free end of the system.* Alternatively, this principle is a special case of the invariance of any system under translations or rotations in free space. When this principle is applied to the scattering properties of an arrangement of scattering potential cells, it states that these properties remain unchanged (within possibly a trivial phase factor) when an integral number of repeating units is removed from the free end of the system with SIP. It is precisely this property of removal invariance that allows the solution of the MST equations for systems with SIP entirely in real space, and leads to useful expressions for the Green function and related properties. Since this concept forms the cornerstone of a development of RS-MST, we examine it closely, beginning

Figure 17.2: **A half line of periodically repeating scattering cells, each characterized by the same scattering matrix.**

with the case of a semi-infinite periodic line of atoms.

17.3.2 Semi-infinite line

We consider a one-dimensional, semi-infinite, periodic array of scattering cells (atoms), is illustrated schematically in Fig. 17.2. Each atom, beginning with the one at the free end of the line which we label by 0, is characterized by a scattering matrix, $\underline{t}$. In order to arrive directly at computable expressions, we work within the angular momentum representation. Let $\underline{T}_1$ denote the scattering matrix of this half line considered as a single scatterer. With the origin of coordinates chosen at the center of the cell labeled by 0, Eq. (17.40) takes the form,

$$\underline{T}_1 = \sum_{ij} \underline{g}(-\mathbf{R}_{0i})\underline{\tau}^{ij}\underline{g}(-\mathbf{R}_{j0}), \tag{17.42}$$

where $\mathbf{R}_{ij} = \mathbf{R}_j - \mathbf{R}_i$ is a vector connecting the centers of the cells labeled i and j. The quantity $\underline{\tau}^{ij}$ is ijth matrix element of the scattering-path operator associated with the half line and given by the inverse of the matrix $\underline{\underline{M}}$ defined in Eq. (17.41).

Recalling our discussion of removal invariance, we realize that $\underline{T}_1$ is also the scattering matrix of the system that remains when any number of scattering cells, say cell 0, are removed from the free end of the line. Therefore, we can view the original system as a two-scatterer assembly[10] consisting of a "bare" scatterer, cell 0, described by a scattering matrix $\underline{t}$,

[10] We assume the presence of ideal surfaces. Surface reconstruction can be treated by attaching a stack of layers representing the reconstructed part onto the ideal surface.

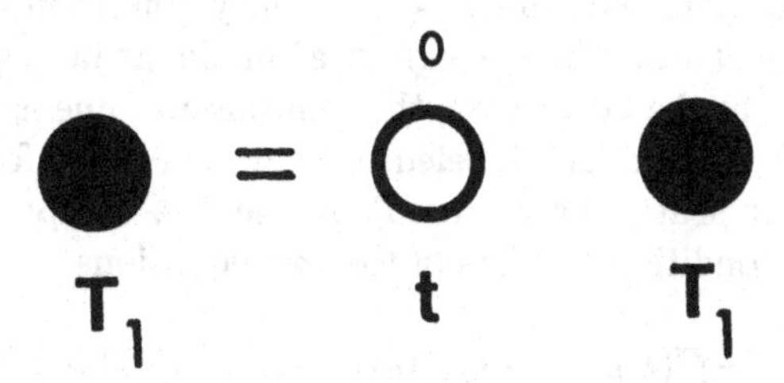

Figure 17.3: **The scattering matrix of a semi-infinite system is equal to that of a two-scatterer assembly composed of a bare scatterer and a renormalized scatterer representing the half line.**

and a "renormalized" scatterer characterized by $\underline{T}_1$, placed in the position following that of cell 0 on the line. This construction effectively reduces the infinite system of scatterers on a half line to a two-scatterer assembly, as indicated schematically in Fig. 17.3.

Now, the scattering-path operator associated with the two-scatterer assembly is given by the inverse,

$$\underline{\underline{\tau}} = \begin{pmatrix} \underline{t}^{-1} & -\underline{G}(\mathbf{R}_{01}) \\ -\underline{G}(\mathbf{R}_{10}) & \underline{T}_1^{-1} \end{pmatrix}^{-1} . \tag{17.43}$$

This two-scatterer assembly, treated as a single unit, is also characterized by $\underline{T}_1$ so that from Eq. (17.40), we obtain the following self-consistency condition,

$$\underline{T}_1 = \sum_{ij} \underline{g}(-\mathbf{R}_{0i}) \underline{\tau}^{ij}(\underline{T}_1) \underline{g}^{\dagger}(-\mathbf{R}_{j0}), \tag{17.44}$$

where the dependence of the SPO on the total scattering matrix is explicitly indicated. It is clear that the last two expressions allow an iterative determination of $\underline{T}_1$, and of $\underline{\underline{\tau}}$. One begins with an approximate guess[11] for $\underline{T}_1$, say , $\underline{T}_1 = \underline{t}$, and evaluates[12] $\underline{\underline{\tau}}$ from Eq. (17.43). Then Eq. (17.44) is used

[11]It is usually the case that site-diagonal matrix elements the SPO converge faster than the system scattering matrix, and the criterion for convergence could be associated with those elements.

[12]Other choices for the starting value of $\underline{T}_1$ may prove more efficient in terms of

to obtain a new value of $\underline{T}_1$. This process can be continued until $\underline{T}_1$ has converged within a preset tolerance. Once the SPO has been determined, one can obtain the Green function and related quantities using well-known expressions.

It is important to note that the self-consistency condition determining the SPO can be applied either in a "single-site" mode, as just described, or in a "cluster mode". In the latter case, the semi-infinite line is replaced by a finite cluster of N "bare cells", labeled 0, 1, 2, ..., $N-1$, followed by a renormalized cell with scattering matrix $\underline{T}_1$ in the $N+1$th position. Now, the self consistency condition consists of the two equations,

$$\underline{\tau} = \begin{pmatrix} \underline{t}^{-1} & -\underline{G}(\mathbf{R}_{01}) & -\underline{G}(\mathbf{R}_{02}) & \cdots & -\underline{G}(\mathbf{R}_{0N}) \\ -\underline{G}(\mathbf{R}_{10}) & \underline{t}_1^{-1} & -\underline{G}(\mathbf{R}_{12}) & \cdots & -\underline{G}(\mathbf{R}_{1N}) \\ \cdot & \cdot & \cdot & \cdot & \cdot \\ \cdot & \cdot & \cdot & \cdot & \cdot \\ \cdot & \cdot & \cdot & \cdot & \cdot \\ -\underline{G}(\mathbf{R}_{N0}) & -\underline{G}(\mathbf{R}_{N1}) & -\underline{G}(\mathbf{R}_{N2}) & \cdots & \underline{T}_1^{-1} \end{pmatrix}^{-1} \tag{17.45}$$

and

$$\underline{T}_1 = \sum_{ij}^{N+1} \underline{g}(-\mathbf{R}_{0i})\underline{\tau}^{ij}(\underline{T}_1)\underline{g}^{\dagger}(-\mathbf{R}_{j0}). \tag{17.46}$$

One important advantage of the cluster mode is that it can greatly enhance the rate of convergence of the angular momentum expansions, by trading off expansions in L space for expansions in real (site) space. The presence, and importance, of sums over L is evident in Eqs. (17.44) and (17.46). Numerical work [49, 51] indicates that the rate of convergence of these sums increases with increasing size of cluster of bare cells. Thus, for a given dimensionality of the matrix $\underline{\tau}$ in Eq. (17.45), the overall converge is usually faster the larger the cluster used in the calculation. However, the problem of L convergence is indeed a real one, and should be addressed on a case-by-case basis. Within RS-MST, the monitoring of convergence can be checked in a unique way by increasing the maximum value of the angular momentum index used in the various expansions.

17.3.3 Full line

The scattering matrix, $\underline{T}_l$, describing an infinite chain of cells can be constructed from the knowledge of the scattering matrices, $\underline{T}_{\mathrm{L}}$ and $\underline{T}_{\mathrm{R}}$, associated with semi-infinite chains extending to the "left" and to the "right", respectively. Assume that $\underline{T}_{\mathrm{R}}$ has been determined by a process of the

convergence. For example, $\underline{T}_1$ could be set equal to the scattering matrix for a cluster of cells, rather than a single cell.

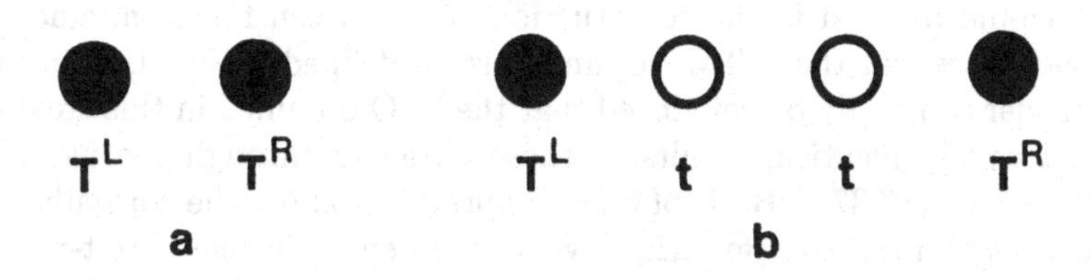

Figure 17.4: **Two alternative methods for constructing the t-matrix of an infinite line.**

type described above[13]. The t-matrix for the mirror image line, $\underline{T}_{\mathrm{L}}$, can be determined either by a similar process or, more efficiently, by pre- and post-multiplying $\underline{T}_{\mathrm{R}}$ by the matrix $\underline{J}$ representing the inversion operator in angular-momentum space,

$$\underline{T}_{\mathrm{L}} = \underline{J}\underline{T}_{\mathrm{R}}\underline{J}. \tag{17.47}$$

Then, a full line can be represented by a two-scatterer system consisting of two renormalized scatterers $\underline{T}_{\mathrm{R}}$ and $\underline{T}_{\mathrm{L}}$ placed a distance a apart, where a is the lattice constant of the line. This construction is shown in panel (a) of Fig. 17.4. Alternatively, panel (b), a cluster of $N-2$ bare scatterers, with scattering matrix $\underline{t}$, could be "dressed" at either end by the renormalized t-matrices, $\underline{T}_{\mathrm{L}}$ and $\underline{T}_{\mathrm{R}}$. This cluster construction affords relatively easy convergence, and allows the treatment of impurities embedded in the line. In this general case, the SPO for the line is given by the relation (see Fig. 17.4 for labeling system)

$$\underline{\tau} = \begin{pmatrix} \underline{T}_{\mathrm{L}}^{-1} & -\underline{G}(\mathbf{R}_{12}) & -\underline{G}(\mathbf{R}_{13}) & \cdots & -\underline{G}(\mathbf{R}_{1N}) \\ -\underline{G}(\mathbf{R}_{21}) & \underline{t}_{1}^{-1} & -\underline{G}(\mathbf{R}_{23}) & \cdots & -\underline{G}(\mathbf{R}_{2N}) \\ \cdot & \cdot & \cdot & \cdot & \cdot \\ \cdot & \cdot & \cdot & \cdot & \cdot \\ \cdot & \cdot & \cdot & \cdot & \cdot \\ -\underline{G}(\mathbf{R}_{N1}) & -\underline{G}(\mathbf{R}_{N2}) & -\underline{G}(\mathbf{R}_{N3}) & \cdots & \underline{T}_{\mathrm{R}}^{-1} \end{pmatrix}^{-1} \tag{17.48}$$

[13]We assume that convergence has been reached in the sense that increasing the cut-off value of the angular-momentum index or the number of bare cells does not change the results in a significant way.

and

$$\underline{T}_l = \sum_{ij}^{N} \underline{g}(-\mathbf{R}_{1i})\underline{\tau}^{ij}\underline{g}^{\dagger}(-\mathbf{R}_{j1}). \tag{17.49}$$

We should point out that only elements of the SPO associated with bare sites should be used in the construction of the Green function, since only for such sites can the cell wave functions be defined. With this in mind, the reader can easily be convinced that the SPO obtained in this treatment gives formally identical results to those obtained through the use of the LKKR, Eq. (17.37). Both of these approaches allow the formally exact treatment of a stack of impurity layers, represented by the bare t-matrices in Eq. (17.48), embedded in an otherwise pure host material. Differences between the two approaches do exist, however, in terms of computational ease and convergence, as we have already pointed out.

17.3.4 Higher-dimensional systems

The RS-MST method can be used to solve the equations of MST for general three-dimensional systems that consist of parts, each of which is characterized by SIP. Each of these parts is built individually through an application of the RS-MST method, and the parts are combined using the algebra of MST to form the composite system. Figure 17.5 may help illustrate the general lines of the process involved. Panel (a) shows a semi-infinite, periodic arrangement of scattering units, denoted by open circles, and characterized by a scattering matrix $\underline{t}$ (in the angular-momentum representation). The total t-matrix, $\underline{T}^{(1)}$, of this system can be determined, for example, in the manner described in connection with Eqs. (17.45) and (17.46). Using the cluster mode, the infinite number of scatterers in the original system is replaced by a finite number, $N_b + 1$, of scatterers, including N_b bare cells and a renormalized cell with scattering matrix $\underline{T}^{(1)}$ at the boundary, panel (b). As we saw above, this cell represents the entire half line, with $\underline{T}^{(1)}$ being determined through Eqs. (17.45) and (17.46).

Once $\underline{T}^{(1)}$ has been determined (within certain convergence limits), the scattering matrix, $\underline{T}^{(2)}$ of a semi-infinite line along a direction perpendicular to the first, say along the positive y direction, panel (c), can be obtained through the use of rotation matrices, $\underline{D}(R)$, expressed in the angular-momentum representation. In the case at hand, panel (c), $\underline{D}(R)$ is the matrix representing a rotation through a positive angle of 90° about the z axis. Similarly, the t-matrix of a half line of atoms extending in a direction opposite that of the original line can be obtained through the use of the inversion operator, Eq. (17.47). Once the scattering matrix of the semi-infinite lines have been obtained, those of full lines, panels (d) and (e),

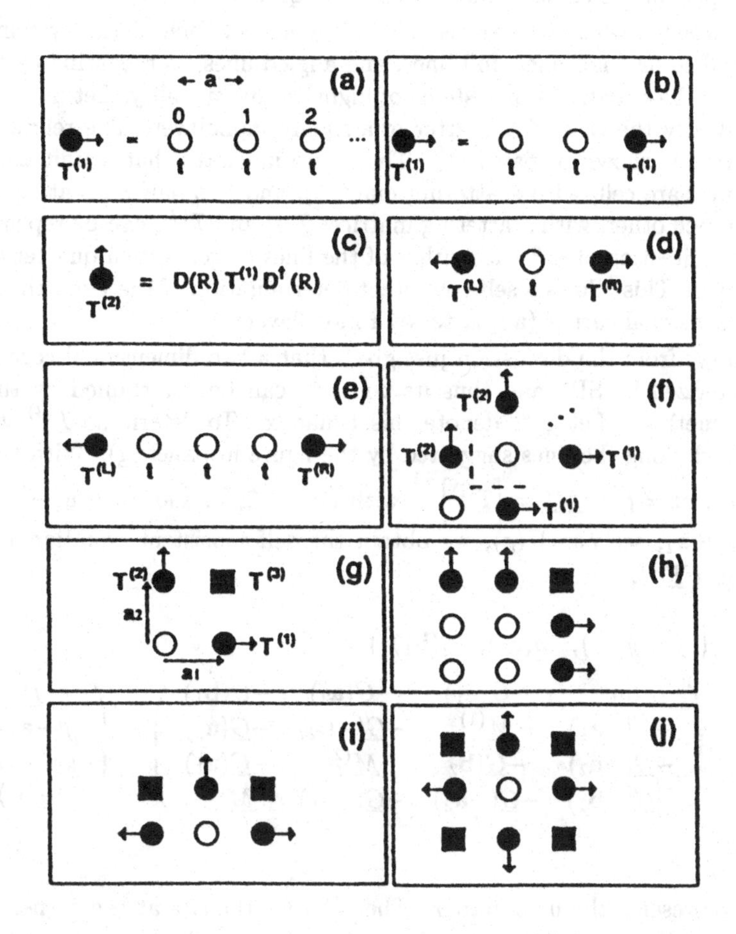

Figure 17.5: **Representation of the RS-MST method.**

of angled lines, panel (f), and the corresponding SPOs can be determined through appropriate combinations of half-line quantities.

To proceed further, we can use the t-matrices of each of the systems determined so far, half lines, full lines, and angled lines, as repeating units to generate the t-matrices of systems of higher dimensionality. Let us consider explicitly the case of a quarter of a square monolayer. The relevant construction is shown in panel (f). The figure indicates that a unit consisting of a bare cell, with scattering matrix $\underline{t}$, and two half lines at right angles to each other, with scattering matrices $\underline{T}^{(1)}$ and $\underline{T}^{(2)}$, can be repeated along a direction of 45° with either of the lines to generate a quarter of a monolayer. This allows a self-consistent determination of the t-matrix of a two-dimensional corner (a quarter of a monolayer).

It follows from the discussion just given that a two-dimensional corner is characterized by SIP, and thus its t-matrix can be determined by the RS-MST method. Let $\underline{T}^{(3)}$ denote this t-matrix. To determine $\underline{T}^{(3)}$ we can proceed along the lines suggested by the figure in panel (g). With the definitions, $\underline{m} = \underline{t}^{-1}$, $\underline{M}^{i} = \left[\underline{T}^{(i)}\right]^{-1}$ with $i = 1,\ 2,\ 3$, and $\mathbf{b}_1 = \mathbf{a}_1 + \mathbf{a}_2$, $\mathbf{b}_2 = \mathbf{a}_1 + \mathbf{a}_2$, see panel (g), we obtain the self-consistent equation for determining $\underline{T}^{(3)}$,

$$\begin{aligned}\underline{T}^{(3)} &= \left(\ \underline{I},\ \ \underline{g}(\mathbf{a}_1),\ \ \underline{g}(\mathbf{a}_2),\ \ \underline{g}(\mathbf{b}_1)\ \right)\\ &\times \begin{pmatrix} \underline{m} & -\underline{G}(\mathbf{a}_1) & -\underline{G}(\mathbf{a}_2) & -\underline{G}(\mathbf{b}_1) \\ -\underline{G}(-\mathbf{a}_1) & \underline{M}^{(1)} & -\underline{G}(-\mathbf{b}_2) & -\underline{G}(\mathbf{a}_2) \\ -\underline{G}(-\mathbf{a}_2) & -\underline{G}(\mathbf{b}_2 & \underline{M}^{(2)} & -\underline{G}(\mathbf{a}_1) \\ -\underline{G}(-\mathbf{b}_1) & -\underline{G}(-\mathbf{a}_2) & -\underline{G}(-\mathbf{a}_1) & \underline{M}^{(3)} \end{pmatrix}^{-1} \begin{pmatrix} \underline{I} \\ \underline{g}(-\mathbf{a}_1) \\ \underline{g}(-\mathbf{a}_2) \\ \underline{g}(-\mathbf{b}_1) \end{pmatrix},\end{aligned} \tag{17.50}$$

where $\underline{I}$ represents the unit matrix. The SPO for the site at the corner is given by the $(1,1)$th matrix element of the inverse in the last expression. Furthermore, we mention that the convergence of the entire process may benefit through an application of the RS-MST in the cluster mode, as indicated in panel (h).

The t-matrices representing half lines, full lines, and corners can be combined to form other two-dimensional systems such as half-planes, panel (i), and full planes, panel (j). Those structures, in turn, can serve as repeating units to generate three-dimensional systems such as corners, edges, surfaces, interfaces, and bulk systems. The reader may profit by providing specific descriptions for setting up the corresponding processes for a number of these structures. The reader is also referred to the literature for model and realistic applications of RS-MST, and the study of its convergence properties [41, 43].

17.3.5 Green function

The RS-MST method allows the determination of the scattering-path operator for systems with SIP, and hence the determination of the single-particle propagator through Eq. (17.3). All related properties, such as charge densities and densities of states, can now be obtained through well-known expressions. Thus, we have completed the path that allows the solution of the Schrödinger equation for a multi-cell assembly entirely in real space. Sample numerical results of the application of the RS-MST as well as the LKKR method are given in the following section.

17.3.6 Alternative forms of the RS-MST equations (*)

The single-site and the cluster mode of the RS-MST method have one-element in common: both of them reduce a system of semi-infinite extent to a finite cluster of bare and renormalized sites. Although convergence can always be reached by increasing the angular-momentum cut-off and/or the number of bare sites, the approach to convergence can be sluggish and is often achieved after a threshold number of bare cells have been used [41, 43]. The need to stabilize convergence has led to a number of alternative expressions for the equations of the RS-MST method.

One approach [49] represents a semi-infinite system by means of a cluster of renormalized cells rather than a single cell. For example, one could view a semi-infinite linear chain as consisting of n-cell molecules spaced a distance na apart, where a is the lattice constant of the chain. One could then use the single-cell or cluster mode of the RS-MST to determine a renormalized t-matrix of an n-cell molecule to represent the half line. This method does indeed speed the convergence in L, but at the expense of using larger clusters of sites.

One can arrive at a very stable form of the RS-MST equations based on the following two-observations. First, the self-consistency condition ought to be formulated in terms of the scattering-path operator, rather than the system t-matrix, since it is indeed the SPO that explicitly enters the construction for the Green function. Second, the convergence problems encountered in the RS-MST are strongly connected with summations over the elements of the translation operator, $g_{LL'}(\mathbf{R})$. It is in fact possible to express removal invariance in terms of τ and to bypass completely the truncation problems associated with $g_{LL'}(\mathbf{R})$[14].

Let us consider a semi-infinite line of scatterers represented by two different finite clusters, as depicted schematically in Fig. 17.6. In this figure,

[14] It is not to be expected that *all* convergence problems can be alleviated by the methods described here, but only those that are associated with sums over the elements of the translation operator. For example, the genuine need to use high values of L to represent a certain structure is not addressed by the methods of this section.

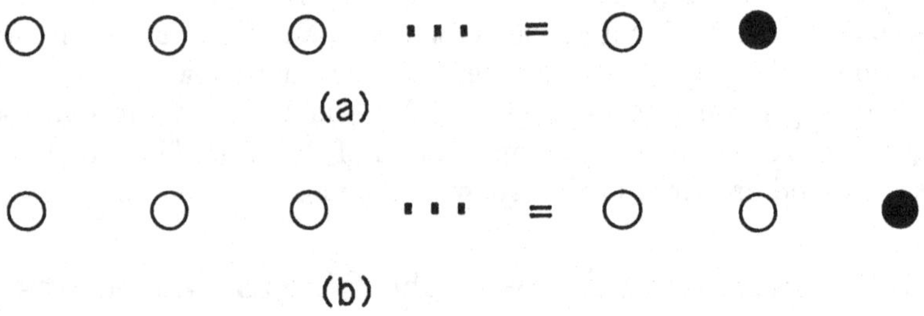

Figure 17.6: **A semi-infinite line of scattering cells represented by two different cluster configurations.**

the semi-infinite line is represented by a single renormalized cell attached to the end of two different clusters of bare cells. Because these two configurations have identical scattering properties, the site-diagonal elements of the SPO associated with the site at the free end, cell 0, determined within either cluster must be equal. With the positions of the cluster sites labeled by 0, 1, 2,..., this leads to the following expression for the consistency condition of the RS-MST, (with $\underline{m} = \underline{t}^{-1}$ as usual)

$$\left\{ \begin{array}{cc} \underline{m} & -\underline{G}(\mathbf{R}_{01}) \\ -\underline{G}(\mathbf{R}_{10}) & \underline{T}^{-1} \end{array} \right\}_{00} = \left\{ \begin{array}{ccc} \underline{m} & -\underline{G}(\mathbf{R}_{01}) & -\underline{G}(\mathbf{R}_{02}) \\ -\underline{G}(\mathbf{R}_{10}) & \underline{m} & -\underline{G}(\mathbf{R}_{12}) \\ -\underline{G}(\mathbf{R}_{20}) & -\underline{G}(\mathbf{R}_{21}) & \underline{T}^{-1} \end{array} \right\}_{00},$$

or

$$\begin{aligned} \underline{T} &= [\underline{I}, \underline{G}^{-1}(\mathbf{R}_{01})\underline{G}(\mathbf{R}_{02})] \left[\begin{array}{cc} \underline{m} & -\underline{G}(\mathbf{R}_{12}) \\ -\underline{G}(\mathbf{R}_{21}) & \underline{T}^{-1} \end{array} \right]^{-1} \\ &\times \left[\begin{array}{c} \underline{I} \\ \underline{G}(\mathbf{R}_{21})\underline{G}^{-1}(\mathbf{R}_{10}) \end{array} \right]. \end{aligned} \qquad (17.51)$$

We note that in this expression the quantities $\underline{g}(\mathbf{R})$ are replaced by the ratio of two structure constants. Clearly, the ratio $\underline{G}^{-1}(\mathbf{R}_{01})\underline{G}(\mathbf{R}_{02})$ approaches the quantity $\underline{g}(\mathbf{R}_{02} - \mathbf{R}_{01})$ in the limit $\ell \to \infty$, and in that limit the last equation becomes identical to Eq. (17.44). However, for finite values of L, Eq. (17.51) does not suffer from difficulties in the convergence of L sums associated with $\underline{g}$. And if the value of L used in the calculation is sufficient to represent the structure under consideration, the problems encountered in applications of Eq. (17.44) are completely alleviated.

Further improvements that solidify the convergence of the RS-MST expressions can be made. We mention the so-called *folding mode* [49] in which the self-consistency condition is expressed explicitly in terms of the SPO, and which has been found to be the most efficient form of the consistency condition. The reader interested in further details is referred to the literature [41, 52].

17.3.7 Hybrid mode of the RS-MST

The RS-MST method allows the solution of the equation of motion for the scattering-path operator without recourse to reciprocal space. At the same time, this method requires the inversion of matrices whose size can become large enough to make self-consistent calculations, in which matrix inversions would have to be repeated many times, exceedingly difficult, or even impractical. On the other hand, the RS-MST can also be used in a *hybrid* mode in which it takes advantage of any two-dimensional periodicity, while continuing to treat the dimension in which translational invariance is broken in real space [52].

Consider an ideal surface. At any point $\mathbf{k}$ of the associated two-dimensional Brillouin zone, the semi-infinite solid can be viewed as a half line of $\mathbf{k}$-dependent scatterers, each described by the scattering matrix associated with a single layer[15]

$$\underline{m}^I(\mathbf{k}) = (\underline{t}^I)^{-1} - \underline{G}_{II}(\mathbf{k}), \tag{17.52}$$

where $\underline{t}^I$ is the single site t-matrix associated with every site in layer I, and $\underline{G}_{II}(\mathbf{k})$ is the interlayer structure constant defined in Eq. (17.23). The corresponding equation of motion takes the one-dimensional form,

$$\underline{\tau}^{IJ}(\mathbf{k}) = \underline{t}^I(\mathbf{k}) \left\{ \delta_{IJ} + \sum_{K \neq I} \underline{G}_{IK}(\mathbf{k}) \underline{\tau}^{KJ}(\mathbf{k}) \right\}. \tag{17.53}$$

The quantities $\underline{G}_{IK}(\mathbf{k})$ are the interlayer structure constants defined by the

[15]Strictly speaking we should use the symbol $\mathbf{k}_{\|}$ to denote points in the two-dimensional Brillouin zone. However, just $\mathbf{k}$ is simpler and its use should cause no confusion since in the present discussion we do not need to explicitly consider a component of $\mathbf{k}$ perpendicular to the surface.

expression[16]

$$\underline{G}_{IJ}(\mathbf{k}) = \sum_j \underline{G}(\mathbf{R}_{I_0 J_j}) e^{i\mathbf{k}\cdot\mathbf{R}_{0j}}. \tag{17.54}$$

Here, the vector $\mathbf{R}_{I_0 J_j}$ connects the "centers" of cell 0 in plane I to that of cell j in layer J, while $\mathbf{R}_{0j}$ is the intercell vector between cells 0 and j in the same plane. This equation can be solved using RS-MST, in any of its implementations (see previous subsection) as the reader is urged to verify. For example, explicit expressions for the SPO within the folding mode have been presented in the literature [52].

Having obtained the scattering-path operator $\underline{\tau}^{IJ}(\mathbf{k})$ for an ideal, semi-infinite solid as a function of $\mathbf{k}$ in the two-dimensional Brillouin zone, the site matrix elements $\underline{\tau}^{I_i J_j}$ can be obtained as the integral

$$\underline{\tau}^{I_i J_j} = \frac{1}{\Omega_{2\mathrm{DBZ}}} \int_{\Omega_{2\mathrm{DBZ}}} d^2k \underline{\tau}^{IJ}(\mathbf{k}) e^{-i\mathbf{k}\cdot\mathbf{R}_{ij}}, \tag{17.55}$$

where $\Omega_{2\mathrm{DBZ}}$ denotes the volume of the two-dimensional Brillouin zone. The surface Green function, in turn, is given by Eq. (17.38), in terms of the site matrix elements of the SPO and the corresponding cell wave functions (or basis functions). Finally, it can be readily shown that the SPO for an interface, including the presence of impurities[17] can be treated in a straightforward way by placing renormalized cells representing the $\mathbf{k}$-dependent half lines on either side of the impurity layers and integrating over $\mathbf{k}$ to obtain the SPO in real space. It is not difficult to show that in both $\mathbf{k}$ and real space the SPO and the Green function take the *same* form as already discussed in connection with the LKKR, although the numerical aspects of the two methods may differ substantially in specific cases.

17.4 KKR-CPA for Semi-Infinite Systems

The formalism of the KKR-CPA discussed in Chapter 12 can be generalized in a straightforward way to the case of substitutionally-disordered systems with SIP, such as surfaces and interfaces. This can be done in a straightforward way using the same conceptual approach that was used in the TB description of surfaces in Section 16.3.

To fix ideas, we begin by considering a simple model for an ideal disordered surface, namely the planes of atoms on either side of a cleavage plane

[16] We are still using the assumption of monoatomic layers that can be brought into coincidence by simple translation along a given direction. More complicated cases, such as polyatomic unit cells or layers that require the use of rotations or secondary translation for coincidence, can also be treated through a generalization of the methods presented here.

[17] It is assumed that both semi-infinite solids forming an interface, as well the impurity layers, are characterized by the same two-dimensional periodicity.

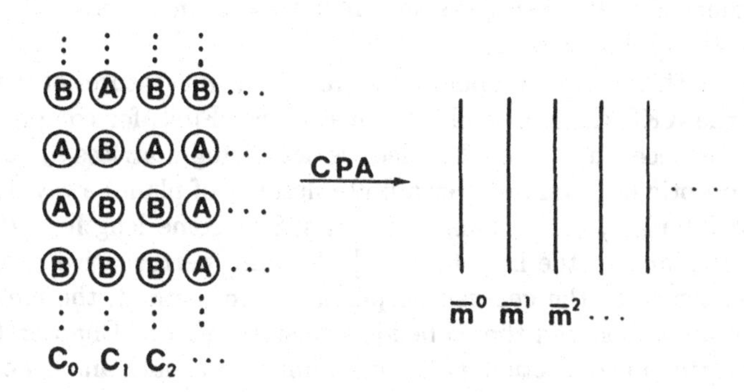

Figure 17.7: **Schematic representation of a substitutionally-disordered semi-infinite material and a medium of parallel ordered planes whose effective t-matrices are determined within the CPA.**

passing through a homogeneous substitutionally-disordered alloy. The ideal surfaces generated by this construction can be treated within either the LKKR or RS-MST methods simply by replacing the atomic (cell) scattering matrices by those determined within an application of the CPA. The procedure for performing the required averages is described below. At the same time, it is intuitively obvious that in a real surface (or interface), the potential and concentration can vary as a function of distance from the surface, and these variations must be taken into account. The formalism presented below can be applied to alloy surfaces with known potential and composition profiles, or can be used to predict them.

In a straightforward extension of the spirit of the CPA as applied to bulk alloys, we seek to replace the true, disordered semi-infinite material (or doubly semi-infinite material) by a medium consisting of "ordered" planes, I, each of which is characterized by an effective (inverse) scattering matrix, $\underline{m}^I$, in the manner indicated in Fig. 17.7. These effective t-matrices are given as the solutions of the following, in principle infinite, set of equations,

$$\langle \underline{\tau}^{II} \rangle_I = \bar{\underline{\tau}}^{II}, \tag{17.56}$$

associated with each plane I. As indicated in this expression, the average is taken over the concentration of a given plane, but the SPO for plane I represents scattering throughout the disordered material. This quantity can be evaluated either within the LKKR method, or the RS-MST method used exclusively within real space or in its hybrid mode. The reader may wish to generalize the relevant equations, derived in previous sections for the treatment of ordered surfaces and interfaces, to incorporate the effects of substitutional disorder [49].

In closing this section, we point out two important features of the application of the KKR-CPA to semi-infinite systems. First, for computational reasons, it is necessary to set the effective scattering matrices $\underline{\bar{m}}^I$ equal to their asymptotic bulk values after a finite number of planes away from the surface or interface, in the manner discussed at some length in the previous chapter and in the literature [49]. Second, perturbative extensions of the CPA, such as the generalized perturbation method, the method of concentration waves, and the embedded cluster method, Chapter 13, can be implemented in connection with semi-infinite materials, and conceivably be used to predict surface and interface composition profiles.

17.5 Self Consistency

In the previous sections, we described two alternative ways of determining the scattering properties of systems with semi-infinite periodicity, assuming the cell potentials are given functions of coordinates. A realistic study of two-dimensional defects in solids invariably involves a self-consistent determination of the potential and the charge distribution within the LDA, or LSDA. In carrying out such calculations, the MST formalism presented in this chapter can be used at each iteration step to obtain the solutions of the Schrödinger equation for the effective single-particle potential. The surface energies resulting from such calculations can be used to compare the relative stability of competing structures in a manner quite analogous to that on which similar comparisons for bulk materials are based. This approach can be used in connection with surface and interface regions in pure or ordered materials or substitutionally-disordered alloys. The latter systems may be studied within extensions of the KKR-CPA to surfaces along the lines indicated above.

It should also be kept in mind that the application of MST to the study of two-dimensional defects should conceivably be carried out within a full-cell framework. This would allow the treatment of the reduced symmetry near such regions as well as the non-cubic arrangement of charge (leading to the formation of dipole moments near a surface or interface).

Finally, the treatment of charge oscillations and dipole layers near a surface or interface requires very special care. The reader interested in further details of methods for the fully self-consistent treatments of the electronic structure of surfaces and interfaces is referred to the literature [1-3, 16, 28].

17.6 Direct Applications of MST

The methods described above, especially the LKKR and the RS-MST, have found a number applications with respect to realistic systems. In this section, we provide only a few examples that emphasize the close relationship and complimentary nature of the two approaches. The first example also illustrates the need to treat the semi-infinite extent of a surface or interface system, at least when **k**-dependent spectral densities are the subject of calculation.

We can define **k**-resolved densities of states, $n(E, \mathbf{k})$, from the imaginary part of the corresponding expression for the Green function,

$$G(\mathbf{r}, \mathbf{r}'; \mathbf{k}) = \langle Z(\mathbf{r}) | \underline{\tau}(E, \mathbf{k}) | Z(\mathbf{r}) \rangle - \langle Z(\mathbf{r}) | S(\mathbf{r}') \rangle \delta_{\mathbf{k}0}, \tag{17.57}$$

where the scattering-path operator is given by the solution of Eq. (17.53) for any wave vector **k** in the surface Brillouin zone. It can also be calculated simply by truncating the number of layers in Eq. (17.53) and inverting the resulting matrix. This clearly corresponds to approximating the semi-infinite system by a slab of finite thickness.

17.6.1 Layered structures

Figure 17.8 shows [52] the DOS at the Γ ($\mathbf{k} = 0$) point calculated for a slab of three layers, panel (a), and of nine layers, panel (b), representing the (100) surface of Cu elemental solid. The potentials used in this calculation were converged, self-consistently determined bulk potentials, and an imaginary part of 1 mRy was assigned to the energy variable. The figure clearly demonstrates the possible slow convergence of the k-resolved DOS with respect to slab thickness, so that such calculations should always be viewed with great caution. On the other hand, either the LKKR or the RS-MST yield converged results as is illustrated in Fig. 17.9. The curve in panel (a) shows the converged DOS for the semi-infinite system obtained within the LKKR method with 25 plane waves used in the expansion of the interplanar propagator. Panel (b) shows the corresponding DOS obtained within two different modes of application of the RS-MST. In one application, the shaded curve, a single plane was used as the repeating unit, while in the other, unshaded curve, the repeating unit consisted of two layers. Comparing with the previous figure, we see that the proper treatment of the semi-infinite nature of the system results in smooth, converged results for **k**-resolved spectral functions, in contrast to the greatly structured curves obtained in a slab calculation. In addition, the RS-MST used in connection with a larger repeating unit provides a faster approach to convergence for the reasons already discussed. Similarly satisfactory convergence is obtained [52] when two semi-infinite **k**-dependent lines are brought together to simulate a bulk system.

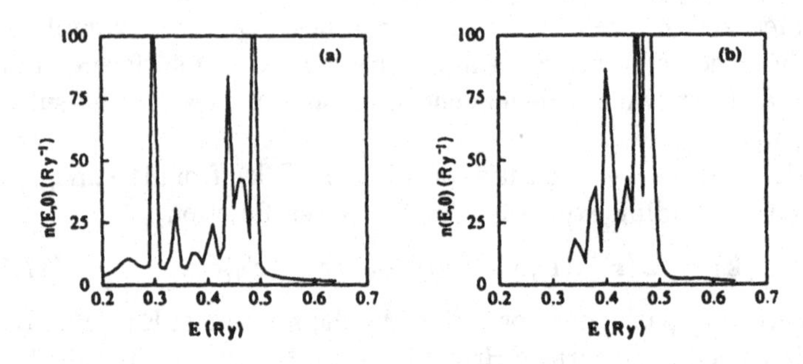

Figure 17.8: **k-resolved DOSs for Cu (100) slabs of (a) three layers, and (b) nine layers.**

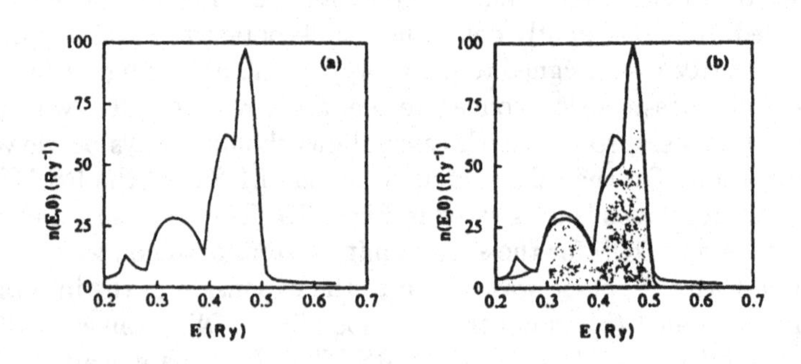

Figure 17.9: **k-resolved DOSs for Cu (100) calculated within the LKKR, panel (a), and the RS-MST, panel (b). In the latter calculation, one and two layers were used as the repeating unit with the results shown, respectively, by the shaded and unshaded curves.**

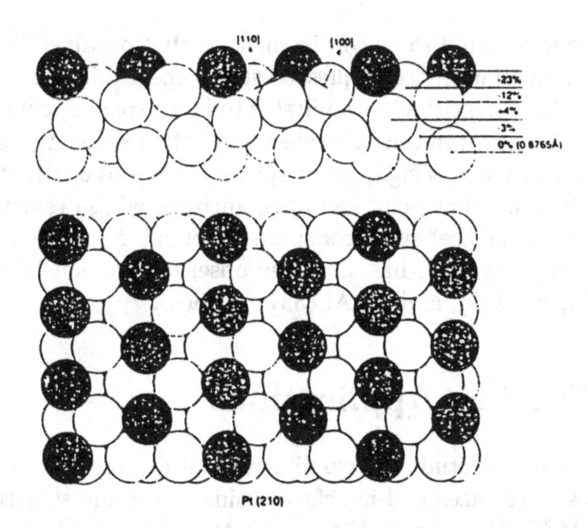

Figure 17.10: **Side view (top panel) and top view (bottom panel) of the Pt (210) surface, with outermost surface atoms shown in gray. Interlayerspacing relaxations found by LEED are shown as percentages of bulk interplanar spacings.**

17.6.2 LEED analysis

The techniques discussed above can be used in the analysis of LEED spectra. The present study involves the multilayer relaxation of the stepped and kinked (210) surface of Pt metal and gives rise to a number of computational difficulties. These difficulties are directly connected with the open structure, Fig. 17.10, of the (210) surface of an fcc lattice.

A stepped surface is characterized by the relatively large area, A, of its two-dimensional unit cell, and the small interplanar spacing, d, between atomic layers parallel to the surface. This causes two major difficulties in the analysis of the LEED data. First, the size of the plane-wave basis set necessary to represent the electronic wave functions at a given energy scales roughly as A/d, which rapidly becomes prohibitively large for stepped surfaces. Second, in the traditional approach, one uses layer stacking methods, such as layer doubling, to obtain the intensities of the reflected electron beams. This process can also fail to converge for small interplanar spacings. Improvements can result from using a stack of planes as the basic unit in implementing [52-54] the layer-doubling procedure, but this technique only delays the divergence without curing it.

In the present case the layer doubling procedure was indeed found to diverge, and stable solutions were obtained [50] only within the hybrid mode of the RS-MST. The resulting structure of the Pt (210) surface is shown in

Fig. 17.10, and is found to be consistent with that obtained in applications of numerical simulation techniques to study the equilibrium structure of the surface. Not surprisingly, the first interlayer spacing exhibits a rather large contraction, a trend usually observed with respect to open surfaces. The relative constancy of sign, (most spacings seem to contract), is at some variance with the behavior of less open surfaces whose relaxation usually exhibits "oscillatory" behavior from plane to plane. However, such behavior as found in the case of Pt has also been observed previously in the case of bcc Fe (111), Fe (210), and fcc Al (331) surfaces.

17.7 Further Applications

We now turn to the study of two-dimensional defects and other electronic properties of surfaces and interfaces using electronic structure methods other than MST. Once again, the hope is to emphasize the underlying importance of the electronic structure itself rather than a particular method used in its determination.

17.7.1 Surface reconstruction

Compared to the atoms in the bulk, surface atoms usually have a significantly smaller coordination number. The resulting nearest-neighbor configuration may vary dramatically from equilibrium. As a case in point, consider the (111), (100), and (110) faces of an fcc crystal. On the surface each atom has, respectively, 9, 8, and 7 nearest neighbors, compared to 12 for bulk atoms. This deviation from equilibrium generally results to atomic rearrangements on the surface.

Surface reconstruction is directly related to the electronic structure. We may get an idea as to how a surface may tend to relax by comparing the interatomic distance in the bulk of a material to that of a diatomic molecule. In general, the latter are smaller than the former, so that we may expect that surface atoms will contract inwardly. Indeed, many metallic surfaces exhibit this behavior. Smoluchowski [56] and Finnis and Heine [57] have discussed the mechanisms for this contraction: The sp electrons move toward the holes between surface atoms, smoothing out the density and thus reducing their kinetic energy. This has the tendency of increasing the electrostatic energy, which in turn is reduced by the inward motion of the surface atoms. This effect depends on surface orientation and structure. For example, for a fcc crystal, it is weakest for the close-packed (111) surface and most pronounced for the open (110) face. Planes below the surface can also relax by moving inward or outward (see discussion of the Pt surface given in the previous section).

Alternating expansions and contractions with diminishing amplitude toward the bulk have been observed in a number of cases. Pseudopotential calculations based on the LDA have shown [58] that the first three interlayer distances in the Al (110) surfaces exhibit a contraction-expansion-contraction pattern, whose numerical values are -6.8±0.5%, +3.5±0.5%, and -2.0±0.5%, respectively. The corresponding experimental results are +8.5±1.0%, +5.5±1.5%, and about 0.0. The calculated work function of 4.32 eV compares very well with the experimental value of 4.28 eV.

Reconstruction effects can be very complicated as illustrated by the Au (110) surface in which alternating rows in the first atomic plane are missing. Similar effects are observed on Ir and Pt (110) surfaces. By contrast, the corresponding 4d metals, Rh, Pd, and Ag do not exhibit this missing row effect.

First-principles, LDA-based electronic structure calculations shed light on the difference in behavior between materials containing electrons in 4d and the 5d states. This difference in behavior indicates strongly that the d electrons play a significant role in surface reconstruction. The lattice spacings of Au and Ag are almost identical, although the 5d wave functions are much less localized than the 4d orbitals. Thus, the sp charge in Au is significantly compressed and the missing rows on the Au (110) surface permit it to spread out and thus reduce its kinetic energy.

The sp charge in the 4d metals is much less compressed and there is little to be gained by spreading further. On the other hand, missing-row behavior is indeed exhibited by the (110) faces of these metals under small amounts of adsorbed alkali atoms. The alkali atoms are known to be partially ionized and the donated charge increases the sp electron density in the surface region. Missing rows have indeed been observed [59] on the (110) surface of Ag, Pd, and Rh, under submonolayer coverage by alkali atoms.

17.7.2 Semiconductor surfaces

Because of its great technological importance to the semiconductor industry, the Si (111) surface has been studied extensively. In the calculations reported here [60], the Si surface was simulated by a 12-layer Si(111) slab and a vacuum region. The calculations were based on a surface LMTO (SLMTO) code within the atomic sphere approximation (ASA), in which the combination of planes and vacuum was repeated periodically in the (111) direction. Arrays of empty spheres were introduced in order to treat the vacuum region.

Figure 17.11 shows the site-projected (local) DOSs for an atom at the center of the slab (bottom), to the first empty sphere on the vacuum side, (top). The calculations reveal the presence of a large peak in the semiconductor gap for atoms at and near the surface. This peak is due to the

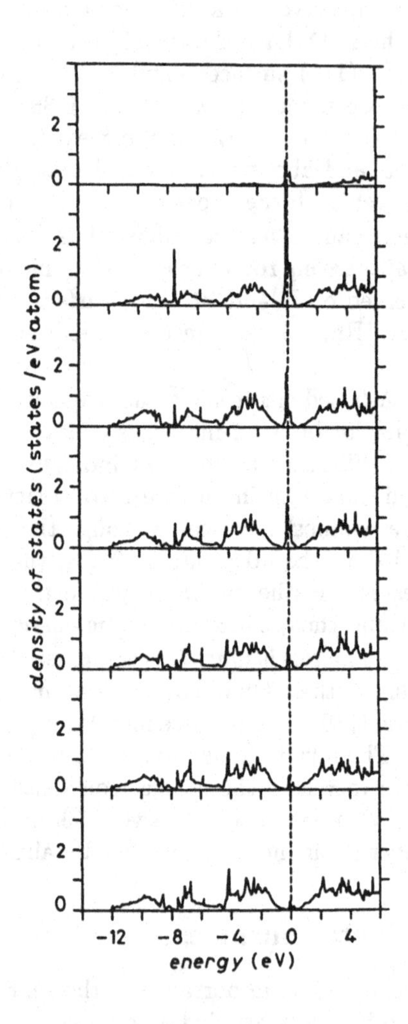

Figure 17.11: **Local DOSs for a Si (111) surface, beginning with an atom at the center of a 12-layer slab (bottom) and proceeding to that of an empty sphere on the vacuum side (top panel).**

so-called dangling-bond surface state which is only half filled and results in metallic behavior of the Si (111) surface. Other differences between the DOSs for surface and bulk-like atoms can be ascribed to various surface s-tates. One prominent example is the peak near -8.0 eV in the hybridization gap for Si.

The surface reconstruction of the Si (111) surface plays host to number of complex phenomena. According to Kahn [61]:

The cleaved surface of Si (111) exhibits a 2x1 metastable reconstruction which, upon annealing at 380° C, transforms irreversibly into a 7x7 structure. The 7x7 structure can also be obtained from a chemically polished and sputtered surface annealed at about 1100° C. Upon annealing the cleaved surface at 900° C, an unstable 1x1 structure appears which can be quenched or impurity-stabilized to room temperature. Laser annealing of the (111) surface also produces a 1x1 structure.

This remarkable behavior can be traced to the electronic and chemical properties of the surface. The surface state exhibited in Fig. 17.11 is rather weak because the dangling bonds interact at second nearest-neighbor distances. However, a shear distortion of the top two layers brings the bonds into nearest-neighbor positions The energy of the resulting 2x1 structure is quite lower than that of the ideal surface and leads to occupied and unoccupied surface states corresponding to the bonding and anti-bonding π-orbitals. Further discussion with references of the reconstructed Si (111) surface can be found in the literature [62].

17.8 Magnetism at Surfaces

We have seen that the electronic structure at a surface can differ substantially from the structure in the bulk of a material, and that this can have important consequences for the physical behavior of a surface. This is particularly true for magnetism at the surface, especially in connection with the Stoner criterion (discussed in Chapter 8). Model calculations based on TB Hamiltonians suggest that the surfaces of the 3d transition metals, Co, Fe, and Ni remain magnetic. In addition, the model predicts that paramagnetic vanadium and antiferromagnetic chromium order ferromagnetically, with manganese as a marginal case.

In the remainder of this section, we concentrate on surface magnetism of non-magnetic and of magnetic systems.

17.8.1 Non-magnetic systems

A typical case of a non-magnetic bulk system whose surface becomes magnetic is that of elemental vanadium. Vanadium is a bcc metal and is known experimentally to be non-magnetic. On the other hand, a vanadium atom in its ground state has a magnetic moment of $3.0\mu_B$. On this basis, we may expect vanadium solid to become magnetic at some (hypothetical) volume that would be larger than the equilibrium volume. Fully self-consistent, spin-polarized calculations [63] based on the augmented plane-wave (AP-W) method show both the lack of magnetism at the equilibrium volume, corresponding to a lattice parameter $a = 2.62$Å, and a fairly sharp transition to a ferromagnetic state at $a = 3.70$Å, with a moment of $2.22\mu_B$ per atom.

This behavior is clearly reflected in the energy bands of the metal [63] shown in Fig. 17.12, for the equilibrium lattice volume (left), and for the magnetic case (right). The figure on the right clearly shows the contraction of the bands at the expanded lattice constant and the different behavior of the majority and the minority spins there. In particular, the 3d bands narrow substantially in the regions of the Fermi energy and produce the high peak in the density of states that is necessary for the setting in of magnetic behavior in the Stoner model. The calculated results are in good qualitative agreement with the appearance of magnetic behavior in V alloys in which the distance between V atoms is of the order of 3.10Å.

Let us now turn to the magnetic behavior of vanadium surfaces. A great amount of interest exists in the V (001) surface because of the possibility of inducing ferromagnetism on the surface layers while the bulk material beneath the surface remains paramagnetic [66-69]. This possibility is supported by experimental findings of magnetic behavior of hyperfine particles (100 to 1000Å) of vanadium [68].

The full-potential, linearized APW (FLAPW) calculations of Ohnishi, Fu, and Freeman show that indeed the (001) surface of vanadium does *not* exhibit magnetic behavior. This is illustrated in the DOSshown in Fig. 17.13. As shown here, the sharp peak at the surface DOS, which is responsible for the magnetic behavior of Fe and Ni, lies 0.3eV above E_F and, consequently, the surface is non-magnetic[18].

The experimental results of reference [68] can be understood by considering the presence of oxygen in the system, which is found to exhibit a V (001) 5x1 reconstruction [69, 70]. The proposed configuration results in the decrease of coordination for V atoms leading, essentially, to isolated-atom behavior and, hence, to magnetism.

17.8.2 Magnetic systems

We explicitly consider Fe and Ni whose bulk magnetic moments are $2.12\mu_B$ and 0.56_B, respectively. These metals are particularly interesting because of the initial observation [71] of *dead layers*, i.e., surface layers in Fe and Ni that are *paramagnetic*. It is now known that the surface increases the

[18] It is now established that a V *monolayer* is magnetic with a moment of $3.09\mu_B$.

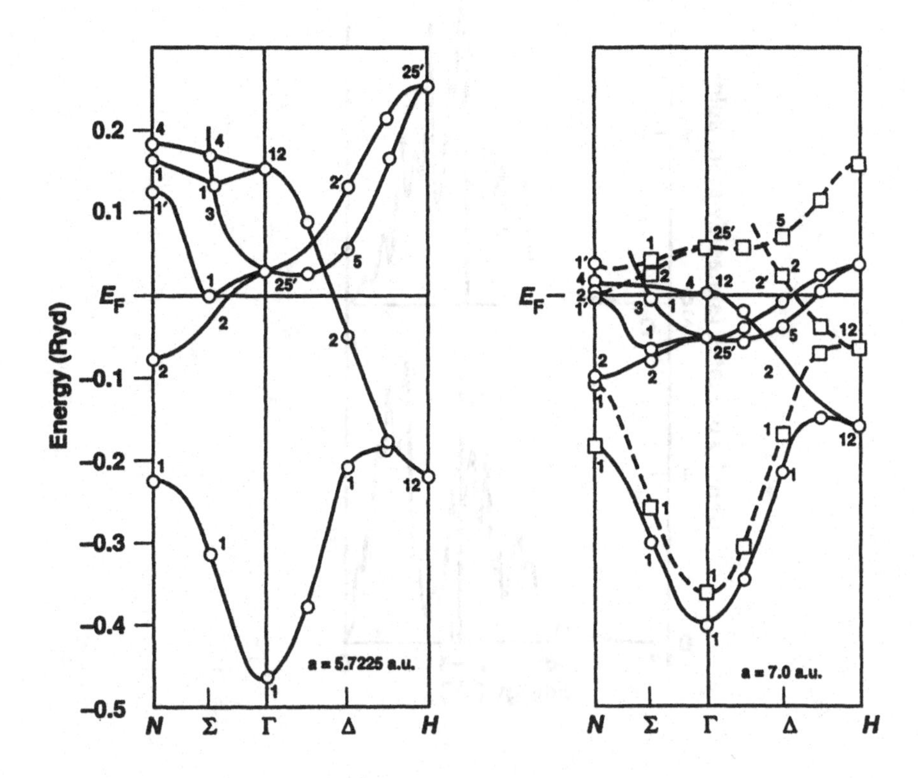

Figure 17.12: **Energy bands for V along two symmetry directions in the Brillouin zone. The left panel corresponds to the non-magnetic metal, $a = 2.62$Å, and the right to the expanded volume with $a = 3.70$Å. The solid lines in the right panel indicate the majority and the dashed lines the minority spin bands.**

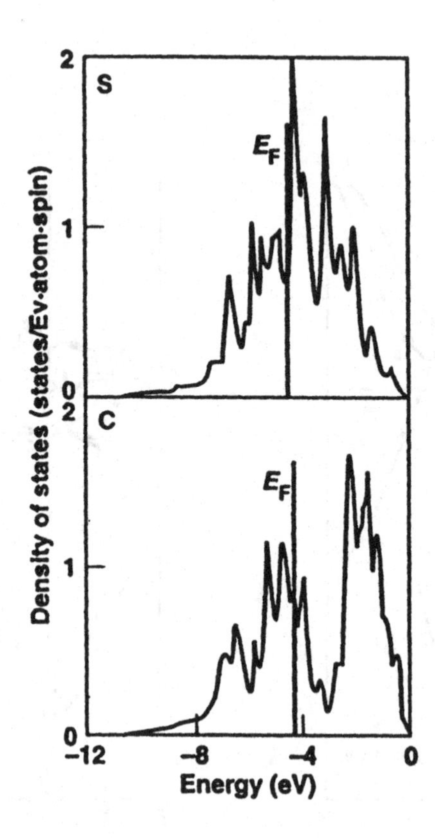

Figure 17.13: **The local (site-projected) DOS at the surface (S) and the center (C) layers of a 7-layer slab calculation based on the FLAPW method.**

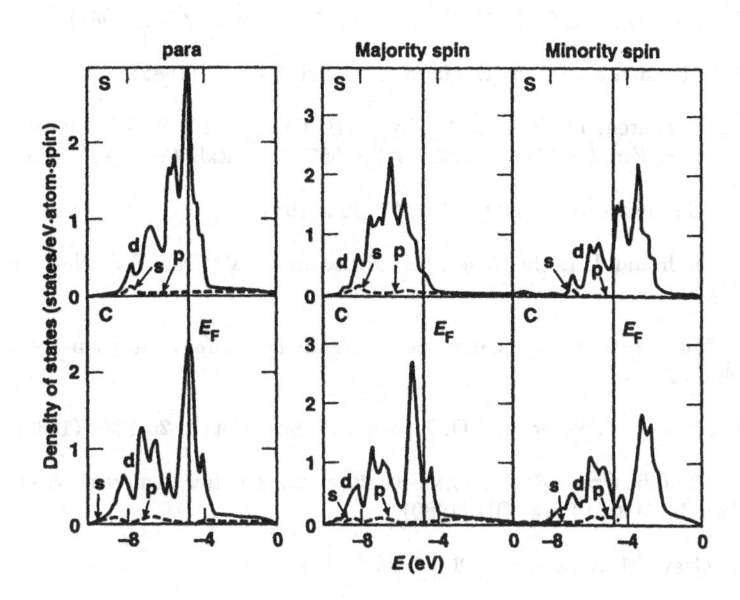

Figure 17.14: **Site and symmetry (angular-momentum) resolved DOSs for Fe in the paramagnetic (two left panels) and ferromagnetic (four right panels) state at the (001) surface (S) and the center (C) layers of a 5-layer slab calculation using the FLAPW method.**

moment of Ni and Fe, and that the experimental observations are consistent with the presence of impurities on the surface.

The DOSs for the center layer and the surface of a 5-layer slab of Fe (001) obtained [67] in the FLAPW method are shown in Fig. 17.14. The enhancement of the magnetization at the Fe (001) surface is evident. The surface DOS is drastically changed compared to the bulk, with the majority d band becoming almost completely filled. The minority DOS exhibits a surface structure around the Fermi level. The narrowing of the d band due to the surface and the occurrence of surface states lead to an enhancement of the moment from its bulk value of $2.27\mu_B$ to that at the surface of $2.96\mu_B$.

Similar effects are found [72] in the case of Ni. For example, the moment on a Ni (001) monolayer is found to be equal to $0.95\mu_B$, enhanced dramatically from its value in the bulk, $0.56\mu_B$. The magnetic moment at the surface is $0.65\mu_B$. The interested reader is referred to the literature [72] for more details.

Bibliography

[1] P. J. Feibelman and D. R. Hamann, Phys. Rev. Lett. **52**, 61 (1984)

[2] P. J. Feibelman and D. R. Hamann, Surf. Sci. **149**, 48 (1985).

[3] J. M. MacLaren, D. D. Vveddensky, J. B. Pendry, and R. W. Joyner, *J. Chem. Soc. Faraday Trans.* 1 **83**, 1945 (1987); J. Catal. **110**, 243 (1988).

[4] D. W. Goodman, Acc. Chem. Res. **17**, 194 (1984).

[5] J. P. Hirth and J. Lothe, *Theory of Dislocations*, McGraw-Hill, New York (1968).

[6] L. E. Murr, *Interfacial Phenomena in Metals and Alloys*, Addison-Wesley, London (1975).

[7] S. P. Chen, A. F. Voter, and D. J. Srolovitz, Scr. Metall. **20**,1389 (1986).

[8] R. W. Ballufi, *Grain Boundaries Structure and Kinetics*, American Society for Metals, Metals Park, OH (1980).

[9] G. T. Gray, III, Acta Metall. **36**, 1745 (1988).

[10] J.-H. Xu, T. Oguchi, and A. J. Freeman, Phys. Rev. B **36**, 4186 (1987).

[11] T. Hong and A. J. Freeman, Mater. Res. Soc. Symp. Proc. **133**, 75 (1989).

[12] J. E. Inglesfield, Rep. Prog. Phys. **45**, 223 (1982), and references therein.

[13] J. E. Inglesfield, in *Equilibrium Structure and properties of Surfaces and interfaces*, NATO ASI Conf. Proc. No. 300, A. Gonis and G. M. Stocks (eds.), Plenum, New York (1992).

[14] W. Kohn, Phys. Rev. B **11**, 3756 (1975).

[15] N. Kar and P. Soven, Phys. Rev. B **11**, 3671 (1975).

[16] H. Krakauer, M. Posternak, and A. J. Freeman, Phys. Rev. B **19**, 1706 (1979).

[17] H. Krakauer, M. Posternak, and A. J. Freeman, Phys. Rev. Lett. **41**, 1072 (1979).

[18] D. Wang, A. J. Freeman, H. Krakauer, and M. Posternak, Phys. Rev. B **23**, 1685 (1981).

[19] E. Wimmer, C. L. Fu, and A. J. Freeman, Phys. Rev. Lett. **55**, 2618 (1981).

[20] M. Posternak, H. Krakauer, A. J. Freeman, and D. D. Koelling, Phys. Rev. B **21**, 5601 (1980).

[21] D. P. Di Vincenzo, O. L. Alerhand, M. Schlüter. and J. W. Wilkins, Phys. Rev. Lett. **56**, 1925 (1986).

[22] M. Y. Chou, M. L. Cohen, and S. G. Louie, Phys. Rev. B **32**, 7279 (1986).

[23] J. R. Chelikowsky, M. Schlüter, S. G. Louie, and M. L. Cohen, Solid State Commun. **17**, 1103 (1975).

[24] J. A. Applebaum and D. R. Hamann, Phys. Rev. B **6**, 2166 (1972).

[25] N. A. W. Holzwarth and M. J. G. Lee, Phys. Rev. B **18** 5350, (1978).

[26] M. J. G. Lee and N. A. W. Holzwarth, Phys. Rev. B **18**, 5365 (1978).

[27] J. E. Inglesfield and G. Benesh, Phys. Rev. B **37**, 6682 (1988).

[28] F. Garcia-Moliner and V. R. Velasco, Prog. Surf. Sci. **21**, 93 (1986).

[29] J. E. Inglesfield, J. Phys. C **14**, 3795 (1981).

[30] J. E. Inglesfield, Surf. Sci. **76**, 355 (1978).

[31] G. Benesh and J. E. Inglesfield, J. Phys. C **17**, 1595 (1984).

[32] W. R. Lambrecht and O. K. Andersen, Surf. Sci. **178**, 256 (1986).

[33] A. R. Williams, P. J. Feibelman, and N. D. Lang, Phys. Rev. B **18**, 616 (1978).

[34] P. J. Feibelman, Phys. Rev. Lett. **54**, 2627 (1985).

[35] J. M. MacLaren, S. Crampin, D. D. Vveddensky, and J. B. Pendry, Phys. Rev. B **40**, 12164 (1989).

[36] J. M. MacLaren, S. Crampin, D. D. Vveddensky, R. C. Albers, and J. B. Pendry, Comp. Phys. Commun. **60**, 365 (1990).

[37] I. Turek, V. Drchal, J. Kudrnovský, M. Šob, and P. Weinberger,, *Electronic Structure of Disordered Alloys, Surfaces and Interfaces*, Kluwer Academic, New York (1997).

[38] J. B. Pendry, *Low Energy Electron Diffraction*, Academic Press, London (1974).

[39] M. A. Van Hove, W. H. Weinberg, C.-M. Chan, *Low Energy Electron Diffraction*, Springer Verlag, Berlin (1986).

[40] A. Gonis, Phys. Rev. B **34**, 8313 (1985).

[41] X.-G. Zhang, A. Gonis, and James M. MacLaren, Phys. Rev. **40**, 3694 (1989).

[42] A. Gonis, X.-G. Zhang, J. M. MacLaren, and S. Crampin, Phys. Rev. **42**, 3798 (1990).

[43] X.-G. Zhang, in *Applications of Multiple Scattering Theory to Materials Science*, Materials Research Society Conf. Proc. No. 253, W. H. Butler, P. H. Dederichs, A. Gonis, and R. L. Weaver (eds.), Materials Research Society, Pittsburgh (1992)

[44] D. W. Jepsen, P. M. Marcus, and F. Jona, Phys. Rev. **12**, 3933 (1972).

[45] A. P. Shen, Phys. Rev. B **12**, 4200 (1972).

[46] K. Kambe, Z. Naturforsch. **22a**, 322 (1967); **22a**, 422 (1967); **23a**, 1280 (1968).

[47] Certain corrections to Kambe's original formulation can be found in K. Kambe, Z. Naturforsch. **24c**, 1432 (1969).

[48] A. Gonis, X.-G. Zhang, and D. M. Nicholson, Phys. Rev. B **38**, 3564 (1988).

[49] A. Gonis, *Green Functions for Ordered and Disordered Systems*, North Holland, Amsterdam (1992).

[50] X.-G. Zhang, M. A. Van Hove, G. A. Samorjai, P. J. Rous, D. Tobin, A. Gonis, J. M. MacLaren, K. Heinz, M. Michl, H. Lindner, K. Müller, M. Ehsasi, and J. H. Block, Phys. Rev. Lett. **67**, 1298 (1991).

[51] S. Crampin, D. D. Vveddensky, J. M. MacLaren, and M. E. Eberhart, Mater. Res. Soc. Symp. Proc. **141**, 373 (1989).

[52] J. M. MacLaren, X.-G. Zhang, A. Gonis, and S. Crampin, Phys. Rev. B **40**, 9955 (1989).

[53] D. W. J epsen, Phys. Rev. B **22**, 5701 (1980).

[54] P. J. Rous and J. B. Pendry, Surf. Sci. **173**, 1 (1986).

[55] J. M. MacLaren, S. Crampin, and D. D. Vveddensky, Phys. Rev B **40**, 12176 (1989).

[56] R. Smoluchowski, Phys. Rev. **60**, 661 (1941).

[57] M. W. Finnis and V. Heine, J. Phys. F **4**, L37 (1974).

[58] K. M. Ho and K. P. Bohnen, Phys. Rev Lett. **56**, 934 (11986); Phys. Rev. B**32**, 3446 (1985).

[59] J. W.-M. Frenken, R. L. Krans, J. F. van der Veen, E. Holub-Krappe, and K. Horn, Phys. Rev. Lett **59**, 2307 (1987).

[60] S. Ossicini and O. Bisi, Surf. Sci. **211/212**, 572 (1989).

[61] A. Kahn, Surf Sci. Rep. **3**, 193 (1983).

[62] Andrew Zangwill, *Physics at Surfaces*, Cambridge University Press, New York (1978).

[63] T. M. Hattox, J. B. Conklin, J. C. Slater, and S. B. Trickey, J. Phys. Chem. Solids **34**, 1627 (1973).

[64] G. Allan, Phys. Rev. B**19**, 4774 (1979).

[65] D. R. Grempel and S. C. Ying, Phys. Rev. Lett. **45**, 1018 (1980).

[66] G. Yokoyama, H. Hirashita, T. Oguchi, T. Kambaya, and K. I. Gondaira, J. Phys. F **11**, 1463 (1981).

[67] S. Ohnishi, C. L. Fu, and A. J. Freeman, J. Mag. Magn. Mater. **50**, 161 (1985).

[68] H. Akoh and A. Tasahi, J. Phys. Soc. Jpn. **42**, (791) (1977).

[69] J. S. Ford, A. P. C. Reed, and R. M. Lambert, Surf. Sci. **129**, 79 (1983).

[70] V. Jensen, J. N. Andersen, H. B. Nielsen, and D. L. Adams, Surf. Sci. **116**, 66 (1984).

[71] L. N. Lieberman, J. Clinton, P. M. Edwards, and J. Mathon, Phys. Rev. Lett. **25**, 232 (1970).

[72] E. Wimmer, A. J. Freeman, and H. Krakauer, Phys. Rev. B **30**, 3113 (1984).

Part V

Transport

Chapter 18

Classical Theory of Transport

18.1 General Comments

Our discussion up to this point has been confined almost exclusively to equilibrium properties of materials, and their connection to the underlying electronic structure. By and large, equilibrium properties depend only on *average* values of the parameters characterizing the state of a system, such as the positions and velocities of the particles in it. On the other hand, ***transport phenomena*** are associated with the response of a system to an externally applied field and in general require the knowledge of ***instantaneous*** values of parameters. Therefore, a proper discussion of transport must be based on ***distribution functions*** which describe the number of particles in the system as a function of position, velocity, and time.

Two important and frequently studied transport phenomena are those of electrical and thermal conduction. These two terms can be taken to include a large number of specific physical effects, such as electronic transport in metals and semiconductors, magnetoresistance, heat conduction, thermoelectric effects, and many others. In this chapter, we begin our discussion of transport properties of materials, concentrating first on classical theories of conduction. Quantum mechanical approaches to transport and the use of Green functions within the framework of multiple scattering theory in the study of conduction are discussed in the following two chapters.

18.2 Drude Theory of Conduction

There is no evidence that electric phenomena *per se* were known to the ancient Greeks, although the extraordinary properties of amber and lodestone had been observed as early as 800 BC. The first attempt to explain the passage of electricity through a metal was not made until about 100 years after Coulomb made his first observations of the forces that exist between charged particles. In 1875, Weber [1] visualized metals as consisting of relatively unstable molecules whose individual, charged particles could jump from molecule to molecule thus mediating transport. After the discovery of the electron by J. J. Thompson in 1897, a number of somewhat more sophisticated theories evolved, such as that of Riecke [2] in 1898. Further work at this early stage culminated in the publication of Drude's [3] classic paper in 1890 which, among other accomplishments, included a derivation of the Wiedemann-Franz law [4, 5], stating that the ratio of electrical to thermal conductivity at a given temperature is the same for all metals.

In Drude's theory, the electrons in a metal are taken to move randomly through the material with some average speed. The electrons can be deflected from their paths through elastic collisions with the ions which, due to their comparatively much larger mass, can be considered as remaining stationary. Under these random scattering conditions, the average electron velocity (along a given direction) vanishes. However, the situation changes in the presence of an electric field, **E**, which induces a net current, or net electron motion, along the direction of the field.

Let e be the charge of the electron (in electrostatic units) and m its mass, and let 2τ be the average time the electron spends between collisions with the ions. During this "free motion" the electron experiences an acceleration equal to -eE/m, with E being the magnitude of the applied field along the field direction. If u_0 is the component of the electron velocity along the field direction immediately after the collision, then the value of u_0 at the next collision is on average equal to $u_0 - \mathrm{e}E\tau/m$. The *electric current density* can be defined as a sum over all electrons in the system,

$$j = \sum -\mathrm{e}\left(u_0 - \mathrm{e}\tau E/m\right). \tag{18.1}$$

Since $\sum u_0 = 0$, we can write

$$j = n\mathrm{e}^2\tau E/m, \tag{18.2}$$

where n is the net (average) number of electrons per unit volume. The *electric conductivity*, defined as theratio of current to field, takes the form,

$$\sigma = n\mathrm{e}^2\tau/m. \tag{18.3}$$

Drude's theory can also be used to calculate the electronic thermal conductivity that is a response to a temperature gradient, $\frac{\partial T}{\partial x}$. In order to treat the effects of a temperature gradient, we must assume that the average velocity, v_0, and the energy, E, of an electron are functions of position. Consider an electron whose velocity makes an angle θ with the x axis reaching unit area in the plane $x = x_0$ and colliding with an ion there. This electron on the average will have made its previous collision with an ion in the plane $x = x_0 - 2\tau v_o \cos\theta$ and will have had an energy, as a function of position, $E(x_0 - 2\tau v_0 \cos\theta)$ there. The number of free electrons per unit volume whose directions of motion fall within a solid angle $\mathrm{d}\Omega$ is $n\mathrm{d}\Omega/4\pi$ so that the number with velocities lying between θ and $\theta + \mathrm{d}\theta$ with respect to the x axis is equal to $\frac{1}{2}n\sin\theta\mathrm{d}\theta$. We also need the number of electrons with velocities between θ and $\theta + \mathrm{d}\theta$ crossing unit area in plane x_0 in a time interval $\mathrm{d}t$. This number is equal to the number inside a cylinder with unit cross section in plane x_0 and with perpendicular height $v_0 \cos\theta\mathrm{d}t$. The volume of the cylinder is then $v_0\cos\theta\mathrm{d}t$ and the energy flux across unit area in the plane x_0 in time $\mathrm{d}t$ is given by

$$\begin{aligned}
\mathrm{d}\Phi &= \mathrm{d}t\int_0^\pi E(x_0 - 2\tau v_0\cos\theta)\frac{1}{2}nv_0\cos\theta\sin\theta\mathrm{d}\theta \\
&= \mathrm{d}t\int_0^\pi \left[E(x_0) - 2\tau v_0\cos\theta\left(\frac{\partial E}{\partial x}\right)_{x=x_0}\right]\frac{1}{2}nv_0\cos\theta\sin\theta\mathrm{d}\theta \\
&= -\frac{2}{3}\mathrm{d}tn\tau v_0^2\left(\frac{\partial E}{\partial x}\right)_{x=x_0}. \qquad (18.4)
\end{aligned}$$

If we assume that the energy, E, can only depend on x through the temperature, T, as may be expected in the case of a system with uniform composition, we have $\frac{\partial E}{\partial x} = \left(\frac{\mathrm{d}E}{\mathrm{d}T}\right)\frac{\partial T}{\partial x}$. The current density, w, defined as the ratio $\mathrm{d}\Phi/\mathrm{d}t$ now becomes,

$$w = -\frac{2}{3}n\tau v_0^2\frac{\mathrm{d}E}{\mathrm{d}T}\frac{\partial T}{\partial x}, \qquad (18.5)$$

and the *thermal conductivity*, $K = w/(-\partial T/\partial x)$, takes the form,

$$K = \frac{2}{3}n\tau v_0^2\mathrm{d}E/\mathrm{d}T = \frac{2}{3}\tau v_0^2 C_v, \qquad (18.6)$$

where

$$C_v = n\frac{\mathrm{d}E}{\mathrm{d}T} \qquad (18.7)$$

is the specific heat per unit volume of the electron gas at constant volume.

There is a clear analogy between the definitions of the electrical and thermal conductivities, Eqs. (18.3) and (18.6), which is probably not lost on the reader. Both conductivities are defined as ratios of a current (response) to a field (stimulus), where the field in the case of thermal conduction is the gradient of the temperature. It is interesting to form the ratio of the two conductivities,

$$\frac{K}{\sigma} = \frac{2mv_0^2}{3\mathrm{e}^2}\frac{\mathrm{d}E}{\mathrm{d}T}. \tag{18.8}$$

Assuming that the energy (and the average velocity) of the electrons is the same for all metals, then one arrives at the *Wiedemann-Franz* law mentioned above. In fact assuming that formulae derived within a dynamic theory of a classical gas are applicable to electrons as well, we have $E \equiv \frac{1}{2}mv_0^2 = \frac{3}{2}\mathrm{k_B}T$, with $\mathrm{k_B}$ as Boltzmann's constant. In this case we obtain from Eq. (18.8) the expression

$$\frac{K}{\sigma} = \frac{3T\mathrm{k_B^2}}{\mathrm{e}^2}. \tag{18.9}$$

This shows that the Wiedemann-Franz ratio is proportional to the temperature, a result established by Lorentz [6] in 1872. Finally, the Lorentz number, $L = K/\sigma T$, becomes

$$L = 3\left(\frac{\mathrm{k_B}}{\mathrm{e}}\right)^2. \tag{18.10}$$

In electrostatic units, this ratio has a numerical value of 2.48×10^{-13} e.s.u., which agrees well with the value 2.72×10^{-13} e.s.u. obtained as the average of the observed Lorentz numbers of several metals at 18°C.

In spite of such agreements, the validity of Drude's theory was quickly called into question. It was realized that the calculated electrical and thermal conductivities *separately* often did not agree with experimental observations. First, no electronic contribution to the specific heat equal to $3n\mathrm{k_B}$ was ever observed. Second, experimental observations indicate that at ordinary (room) temperatures σ varies as $1/T$, so that $n\tau$ should also have the same functional dependence on temperature. At the same time, the mean free path, ℓ, of an electron, i.e., the average distance traveled between collisions, is equal to $2\tau v_0$ and should be of the order of the interatomic distance, and hence temperature independent. Since $v \simeq \sqrt{T}$, it follows that the electron density n should vary as $1/\sqrt{T}$. It is indeed difficult to see how the number of electrons could decrease with increasing temperature.

It is, of course, now well known that the ideal gas laws cannot be applied to the electrons in a metal. The "success" of Drude's theory (apart from multiplicative factors) results from a cancellation of errors. For example, at room temperature, the electronic contribution to the specific heat is 100 times smaller than the classical predictions, but the mean square electron speed is about 100 times larger, a combination that conspires to yield a surprisingly accurate Lorentz number.

18.3 Boltzmann Equation

Drude's theory suffers from a number of shortcomings because it is based on assumptions that are inconsistent with experimental observations of gas dynamics. Lorentz [7] and Bohr [8] provided further refinements in the hope of arriving at values for the conductivity that would be in closer accord with experiment. Lorentz's theory uses Maxwell-Boltzmann statistics and is based on the following five assumptions [5]:

- The atoms in a metal are rigid spheres that behave like perfectly elastic particles in collisions with free electrons[1].
- Electron-electron collisions can be neglected.
- The number of electrons per unit volume, n, in a given metal is a function of temperature, and satisfies a principle of dynamic balance: The number of electrons leaving a given region of space (atom) equals the number that enter it.
- The atoms occupy a small volume of a metal.
- Temperature gradients and other inhomogeneities are small over a distance of the order of the electron mean free path, ℓ. Similarly, the presence of external fields produces changes in velocities over ℓ that are small compared to the mean free speed of the electrons in the material. Wilson [5] provides a detailed description of the line of argument developed by Lorentz based on these assumptions.

With these assumptions in mind, we examine the transport properties of electrons in a metal. Unlike the theory introduced by Drude, it is fairly evident that mean velocities and energies cannot be used to describe the response of an electron gas to fields and temperatures that may vary from point to point. It is now necessary to consider the instantaneous values of a *distribution function*, $f_{\mathbf{k}}(\mathbf{r}, t)$, with $\mathbf{v} = \hbar\mathbf{k}/m$, describing the number of electrons per unit volume in real and momentum space at time t. Furthermore, this distribution is an explicit function of the interactions between particles and fields and cannot be determined once for all. Transport is now described in terms of an integrodifferential equation, known as the *Boltzmann equation*, satisfied by the distribution function, $f_{\mathbf{k}}(\mathbf{r}, t)$. The Boltzmann equation can be derived along the following lines.

Consider the number of electrons at time t

$$f_{\mathbf{k}}(\mathbf{r}, t)\mathrm{d}^3r\mathrm{d}^3k, \tag{18.11}$$

in the element of volume of phase space, $\mathrm{d}^3r\mathrm{d}^3k$, at the real space point $\mathbf{r}$ that move with momentum $\mathbf{k}$. If there are no collisions, an electron which at time t is in the phase-space point $(\mathbf{r}, \mathbf{k})$ will, at time $t + \mathrm{d}t$, be in the position $\mathbf{r}' = \mathbf{r} + \mathbf{v}\mathrm{d}t$, $\mathbf{v}' = \mathbf{v} + \mathbf{a}\mathrm{d}t$, where $\mathbf{a}$ is the acceleration caused by

[1]This assumption is not absolutely necessary and can be ignored for most of the formal development of the theory.

the action of the fields. Because of the fifth assumption above, a can be taken to be a constant. Now, let

$$\left[\frac{\partial f}{\partial t}\right]_{\text{coll}} \mathrm{d}t\mathrm{d}^3r\mathrm{d}^3k \tag{18.12}$$

denote the number of electrons forced into $\mathrm{d}^3r\mathrm{d}^3k$ because of collisions suffered in the time interval $\mathrm{d}t$. Then based on Louiville's theorem about the invariance of the volume occupied in phase space, we demand that the total number of electrons in the infinitesimal volume $\mathrm{d}^3r\mathrm{d}^3k$ at $(\mathbf{r}, \mathbf{k})$ be equal to the number inside the "same" volume at $\mathbf{r} + \mathbf{v}\mathrm{d}t$, $\mathbf{v} + \mathbf{a}\mathrm{d}t$. This leads to the relation

$$f_{\mathbf{k}+\dot{\mathbf{k}}\mathrm{d}t}\left(\mathbf{r} + \mathbf{v}\mathrm{d}t, t + \mathrm{d}t\right) = f_{\mathbf{k}}\left(\mathbf{r}, t\right) + \left[\frac{\partial f}{\partial t}\right]_{\text{coll}} \mathrm{d}t. \tag{18.13}$$

Expanding the left-hand side in a Taylor series, and keeping only first-order terms, we obtain the celebrated *Boltzmann equation*,

$$\frac{\partial f_{\mathbf{k}}}{\partial t} - \dot{\mathbf{k}} \cdot \nabla_{\mathbf{k}} f_{\mathbf{k}} - \mathbf{v}_{\mathbf{k}} \cdot \nabla_{\mathbf{r}} f_{\mathbf{k}} = \left[\frac{\partial f_{\mathbf{k}}}{\partial t}\right]_{\text{coll}} \mathrm{d}t. \tag{18.14}$$

We note that in a steady state we have $\partial f_{\mathbf{k}}/\partial t = 0$.

The form of the last equation can be simplified considerably through the following process. We identify the term $\mathbf{v}_{\mathbf{k}} \cdot \nabla_{\mathbf{r}} f_{\mathbf{k}}$ with the change in the distribution function resulting from the presence of diffusion and we write it in the form $\left[\frac{\partial f_{\mathbf{k}}}{\partial t}\right]_{\text{diff}} \mathrm{d}t$. Similarly, the term $\dot{\mathbf{k}} \cdot \nabla_{\mathbf{k}} f_{\mathbf{k}}$ is the change due to the presence of the field and can be written in the form $\left[\frac{\partial f_{\mathbf{k}}}{\partial t}\right]_{\text{field}} \mathrm{d}t$. Then the steady-state Boltzmann equation (with the time rate of change of $f_{\mathbf{k}}$ set equal to zero) can be transformed into the simple-looking expression,

$$\left[\frac{\partial f_{\mathbf{k}}}{\partial t}\right]_{\text{diff}} + \left[\frac{\partial f_{\mathbf{k}}}{\partial t}\right]_{\text{field}} + \left[\frac{\partial f_{\mathbf{k}}}{\partial t}\right]_{\text{coll}} = 0. \tag{18.15}$$

It is also of interest to derive a somewhat more explicit form for the scattering (collision) contribution. We note that the process of scattering from state $\mathbf{k}$ to state $\mathbf{k}'$ causes a decrease of $f_{\mathbf{k}}(\mathbf{r}, t)$. The chance for this process to take place is proportional to a product one of whose factors is $f_{\mathbf{k}}(\mathbf{r}, t)$, representing the number of carriers in the original state, and the other $1 - f_{\mathbf{k}'}(\mathbf{r}, t)$, representing the number of available final states. Also taking into account the inverse process, that of scattering from state $\mathbf{k}'$ into state $\mathbf{k}$, which increases the value of $f_{\mathbf{k}}(\mathbf{r}, t)$, denoting the transition probability from an occupied state $\mathbf{k}$ to an unoccupied state $\mathbf{k}'$ by $Q(\mathbf{k}, \mathbf{k}')$, and integrating over all values of $\mathbf{k}'$, we can write,

$$\begin{aligned}\left[\frac{\partial f_{\mathbf{k}}(\mathbf{r},t)}{\partial t}\right]_{\text{coll}} &= \int \{f_{\mathbf{k}'}(1-f_{\mathbf{k}}) - f_{\mathbf{k}}(1-f_{\mathbf{k}'})\}\, Q(\mathbf{k},\mathbf{k}')\mathrm{d}^3k' \\ &= \int (f_{\mathbf{k}'} - f_{\mathbf{k}})\, Q(\mathbf{k},\mathbf{k}')\mathrm{d}^3k'. \qquad (18.16)\end{aligned}$$

In deriving this expression we have made implicit use of the principle of *microscopic reversibility* [5], which states that the transition rate from state $\mathbf{k}$ to state $\mathbf{k}'$ is equal to that from $\mathbf{k}'$ to $\mathbf{k}$. This allows us to set $Q(\mathbf{k},\mathbf{k}') = Q(\mathbf{k}',\mathbf{k})$ and to move the common factor $Q(\mathbf{k},\mathbf{k}')$ outside the brackets in the last expression. Now, combining the last two expressions, we can write the Boltzmann equation in the form,

$$\left[\frac{\partial f_{\mathbf{k}}}{\partial t}\right]_{\text{diff}} + \left[\frac{\partial f_{\mathbf{k}}}{\partial t}\right]_{\text{field}} = \int (f_{\mathbf{k}'} - f_{\mathbf{k}})\, Q(\mathbf{k},\mathbf{k}')\mathrm{d}^3k'. \qquad (18.17)$$

18.3.1 Fermi distribution function

It is important to realize that the steady state is not the same as the *equilibrium* state, denoted by the distribution function $f^0_{\mathbf{k}}$, which is characterized by the absence of fields and temperature gradients. On the other hand, the results obtained in an application of the formal Boltzmann equation depend strongly on the form used for this equilibrium function, $f^0_{\mathbf{k}}$. In Lorentz's theory, this was taken to be the Maxwell distribution,

$$f^0_{\mathbf{k}} = n\left(\frac{m}{2\pi k_B T}\right)^{3/2} e^{E_{\mathbf{k}}/k_B T}. \qquad (18.18)$$

Although the use of this function allowed great improvement on Drude's theory, many discrepancies with experiment remained [5]. For example, the theory yields a Lorentz number that is too small by a factor of 2/3, and the prediction that σ should vary as $T^{-1/2}$ rather than as T^{-1}. It also led to contradictions with the law of Dulong and Petit in predicting that the specific heat of a metal should be equal to about twice the limiting value of $3Nk_B$ attained at high temperature. Further discrepancies between the predictions of the Drude-Lorentz theory and experimental results arose with respect to the Hall coefficient, and finally a reconciliation of theory and experiment was not proved possible.

The situation was not clarified until the advent of quantum mechanics and the introduction of Fermi-Dirac statistics [9]. In this formalism, the average occupation number of particles obeying the exclusion principle in a quantum state with energy $E_{\mathbf{k}}$ is given by the *Fermi* function,

$$f^0_{\mathbf{k}} = \frac{1}{e^{(E_{\mathbf{k}}-\zeta)/k_B T} + 1}, \qquad (18.19)$$

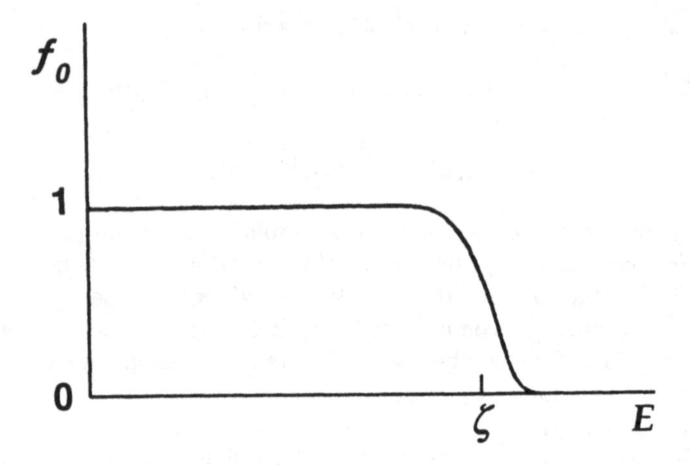

Figure 18.1: **The Fermi function.**

where ζ is a parameter determined by the total number of particles[2]. In the presence of spin, the average occupation number is twice that given by the last expression. Finally, as ζ/k_BT becomes very large compared to 1, the Fermi distribution function passes over into the Maxwell distribution, Eq. (18.18).

The Fermi distribution function approaches 1 as $E_\mathbf{k}$ becomes increasingly smaller than ζ, it is equal to 1/2 at $E_\mathbf{k} = \zeta$, and for large values of $E_\mathbf{k}$ falls exponentially to zero at the rate $e^{-(E_\mathbf{k}-\zeta)/k_BT}$. The Fermi function as a function of $E \equiv E_\mathbf{k}$ is given in Fig. 18.1.

18.4 Linearized Boltzmann Equation

The Boltzmann Eq. (18.15) implies that in a steady state the net change in $f_\mathbf{k}(\mathbf{r})$ vanishes at all points $\mathbf{r}$ and all momenta $\mathbf{k}$. At equilibrium, in the absence of external fields and temperature gradients, the distribution function, $f^0_\mathbf{k}(\mathbf{r})$, has the form of Eq. (18.19). Let us assume that the steady state form of the distribution function, which is achieved in the presence of fields and temperature gradients, differs only slightly from the equilibrium form,

$$f_\mathbf{k}(\mathbf{r}) = f^0_\mathbf{k}(\mathbf{r}) + g_\mathbf{k}(\mathbf{r}). \tag{18.20}$$

[2]This is the chemical potential, which is often also denoted by μ.

It is to be expected that $f_{\mathbf{k}}(\mathbf{r})$ should reflect the positional dependence of temperature, so that in general we can write,

$$f_{\mathbf{k}}(\mathbf{r}) = \frac{1}{\mathrm{e}^{(E_{\mathbf{k}}-\zeta)/k_{\mathrm{B}}T(\mathbf{r})}+1}. \tag{18.21}$$

Since the form of $T(\mathbf{r})$ may not be easy to discover, a subsidiary condition can be used to guarantee conservation of particles. It is often assumed that $g_{\mathbf{k}}(\mathbf{r})$ is such that

$$\int \mathrm{d}^3 k g_{\mathbf{k}}(\mathbf{r}) = 0, \tag{18.22}$$

at all points $\mathbf{r}$.

With these considerations in mind, we can now derive an approximate but more tractable form of the Boltzmann equation.

First, we recall that the rate of change of $f_{\mathbf{k}}(\mathbf{r})$ due to diffusion is given by the expression,

$$\left[\frac{\partial f_{\mathbf{k}}}{\partial t}\right]_{\mathrm{diff}} = -\mathbf{v}_{\mathbf{k}} \cdot \nabla f_{\mathbf{k}}. \tag{18.23}$$

This follows from our previous arguments showing that $f_{\mathbf{k}}(\mathbf{r},t) = f_{\mathbf{k}}(\mathbf{r} - \mathbf{v}_{\mathbf{k}}t, 0)$. We can also write an explicit expression for $\dot{\mathbf{k}}$, the acceleration of an electron due to the presence of electric and magnetic fields,

$$\dot{\mathbf{k}} = \frac{\mathrm{e}}{\hbar}\left[\mathbf{E} + \frac{1}{\mathrm{c}}\mathbf{v}_{\mathbf{k}} \times \mathbf{H}\right], \tag{18.24}$$

so that from the condition $f_{\mathbf{k}}(\mathbf{r},t) = f_{\mathbf{k}-\dot{\mathbf{k}}t}(\mathbf{r},0)$, we obtain,

$$\begin{aligned}\left[\frac{\partial f_{\mathbf{k}}}{\partial t}\right]_{\mathrm{field}} &= -\dot{\mathbf{k}} \cdot \nabla_{\mathbf{k}} f_{\mathbf{k}} \\ &= -\frac{\mathrm{e}}{\hbar}\left[\mathbf{E} + \frac{1}{\mathrm{c}}\mathbf{v}_{\mathbf{k}} \times \mathbf{H}\right] \cdot \nabla_{\mathbf{k}} f_{\mathbf{k}}.\end{aligned} \tag{18.25}$$

(We recall in passing the kinematic principle that expresses the velocity of an electron as the gradient of $E_{\mathbf{k}}$,

$$\mathbf{v}_{\mathbf{k}} = \nabla_{\mathbf{k}} E_{\mathbf{k}}.) \tag{18.26}$$

These expressions allow us to cast the Boltzmann equation in the form,

$$-\mathbf{v}_{\mathbf{k}} \cdot \nabla f_{\mathbf{k}} - \frac{\mathrm{e}}{\hbar}\left[\mathbf{E} + \frac{1}{\mathrm{c}}\mathbf{v}_{\mathbf{k}} \times \mathbf{H}\right] \cdot \nabla_{\mathbf{k}} f_{\mathbf{k}} = \left[\frac{\partial f_{\mathbf{k}}}{\partial t}\right]_{\mathrm{scatt}}. \tag{18.27}$$

Now, using Eq. (18.20) and the chain rule for differentiation, we obtain the expression,

$$- \quad \mathbf{v}_{\mathbf{k}} \cdot \frac{\partial f^0_{\mathbf{k}}}{\partial T}\nabla T - \frac{\mathrm{e}}{\hbar}\left[\mathbf{E} + \frac{1}{\mathrm{c}}\mathbf{v}_{\mathbf{k}} \times \mathbf{H}\right] \cdot \nabla_{\mathbf{k}} f_{\mathbf{k}}$$

$$= \left[\frac{\partial f_{\mathbf{k}}}{\partial t}\right]_{\text{scatt}} + \mathbf{v}_{\mathbf{k}} \cdot \nabla_{\mathbf{k}} g_{\mathbf{k}} + \frac{\mathrm{e}}{\hbar}\left[\mathbf{E} + \frac{1}{\mathrm{c}}\mathbf{v}_{\mathbf{k}} \times \mathbf{H}\right] \cdot \nabla_{\mathbf{k}} g_{\mathbf{k}}(\mathbf{r}). \tag{18.28}$$

In view of the explicit form of the Fermi-Dirac distribution function, Eq. (18.19), and dropping terms such as $\mathbf{E} \cdot \nabla_{\mathbf{k}} g_{\mathbf{k}}(\mathbf{r})$ which would be of order E^2 and lead to deviations from Ohm's law, we obtain [10] the *linearized Boltzmann* equation,

$$\begin{aligned} &\left(-\frac{\partial f_{\mathbf{k}}^0}{\partial E_{\mathbf{k}}}\right) \mathbf{v}_{\mathbf{k}} \cdot \left\{-\frac{E_{\mathbf{k}} - \zeta}{T}\nabla T + \mathrm{e}\left[\mathbf{E} - \frac{1}{\mathrm{e}}\nabla\zeta\right]\right\} \\ = &\left(\frac{\partial f_{\mathbf{k}}}{\partial t}\right)_{\text{scatt}} + \mathbf{v}_{\mathbf{k}} \cdot \nabla_{\mathbf{k}} g_{\mathbf{k}} + \frac{\mathrm{e}}{\hbar c}(\mathbf{v}_{\mathbf{k}} \times \mathbf{H}) \cdot \nabla_{\mathbf{k}} g_{\mathbf{k}}(\mathbf{r}). \end{aligned} \tag{18.29}$$

We note that the magnetic field does not appear on the left-hand side of the last equation since the term $\mathbf{v}_{\mathbf{k}} \cdot \mathbf{v}_{\mathbf{k}} \times \mathbf{H}$ vanishes identically. The nature of the Boltzmann equation as an integrodifferential equation for $g_{\mathbf{k}}(\mathbf{r})$ is revealed when Eq. (18.16) is put into the last expression. This form shows the similar roles played by the temperature and potential gradients as the driving forces for transport. In the remainder of this chapter, we are concerned with methods for obtaining approximate solutions of the Boltzmann equation.

18.5 Electrical Conductivity

In order to derive an expression for the electrical conductivity, we consider the Boltzmann Eq. (18.16) for an infinite system in the presence of an electric field, $\mathbf{E}$. We do not consider temperature effects, keeping the temperature constant throughout the system. In this case, the linearized form of Eq. (18.16) becomes,

$$\begin{aligned} \left[\frac{\partial f_{\mathbf{k}}}{\partial t}\right]_{\text{coll}} &= \left(\frac{\partial f_{\mathbf{k}}^0}{\partial E}\right) \mathbf{v}_{\mathbf{k}} \cdot \mathrm{e}\mathbf{E} \\ &= \int (g_{\mathbf{k}'} - g_{\mathbf{k}})\, Q(\mathbf{k}, \mathbf{k}') \mathrm{d}^3 k'. \end{aligned} \tag{18.30}$$

This is an integral equation for the function $g_{\mathbf{k}}(\mathbf{r})$ whose solution can be fairly complicated in the case of general fields, even for interaction functions $Q(\mathbf{k}, \mathbf{k}')$ of fairly simple form. However, under certain conditions [5, 10] on the external fields and the interaction function, it is possible to make the simplifying assumption that

$$\left[\frac{\partial f_{\mathbf{k}}}{\partial t}\right]_{\text{coll}} = -\frac{1}{\tau} g_{\mathbf{k}}. \tag{18.31}$$

The quantity τ is referred to as the *time of relaxation.* It has the physical significance that under the sudden removal of the field, the distribution function would approach equilibrium according to the law,

$$\frac{\partial g_{\mathbf{k}}}{\partial t} = -\frac{1}{\tau} g_{\mathbf{k}}, \tag{18.32}$$

so that

$$g_{\mathbf{k}}(t) = g_{\mathbf{k}}(0)\mathrm{e}^{-t/\tau}, \tag{18.33}$$

or

$$\left(f_{\mathbf{k}} - f_{\mathbf{k}}^{0}\right)_t = \left(f_{\mathbf{k}} - f_{\mathbf{k}}^{0}\right)_{t=0} \mathrm{e}^{-t/\tau}. \tag{18.34}$$

The introduction of a relaxation time greatly simplifies the discussion of all transport phenomena, but the assumption of the existence of a time of relaxation needs justification. One can show [5] that a relaxation time is a meaningful concept provided that the ions in a metal are treated as fixed spheres of finite radii that scatter electrons only elastically. This is the first and most specialized assumption in Lorentz's theory of conduction mentioned previously. The use of the relaxation time approximation in solving the Boltzmann equation is illustrated in the following discussion.

Combining Eqs. (18.32) and (18.30), we obtain the expression,

$$g_{\mathbf{k}} = \left(-\frac{\partial f_{\mathbf{k}}^{0}}{\partial E}\right) \tau \mathbf{v}_{\mathbf{k}} \cdot \mathrm{e}\mathbf{E}. \tag{18.35}$$

To calculate the electrical conductivity, we need an expression for the current density. Using the fact that in equilibrium the average value of the electron velocity vanishes,

$$\int \mathbf{v}_{\mathbf{k}} f_{\mathbf{k}}^{0} \mathrm{d}^3 k = 0, \tag{18.36}$$

we obtain the following expression for the vector current density

$$\begin{aligned} \mathbf{J} &= 2\int \mathrm{e}\mathbf{v}_{\mathbf{k}} f_{\mathbf{k}} \mathrm{d}^3 k \\ &= 2\int \mathrm{e}\mathbf{v}_{\mathbf{k}} g_{\mathbf{k}} \mathrm{d}^3 k, \end{aligned} \tag{18.37}$$

where the prefactor of 2 accounts for the presence of electron spin. Using Eq. (18.26) and the fact that an integral over $\mathbf{k}$-space is equivalent to integrals over surfaces, S_E, at energy E, integrated over all energies, we have

$$\mathbf{J} = \frac{1}{4\pi^3} \int \mathrm{e}^2 \tau \mathbf{v}_{\mathbf{k}} \left(\mathbf{v}_{\mathbf{k}} \cdot \mathbf{E}\right) \left(-\frac{\partial f_{\mathbf{k}}^{0}}{\partial E}\right) \frac{\mathrm{d}S_E}{v_{\mathbf{k}}} \mathrm{d}E, \tag{18.38}$$

where $v_{\mathbf{k}}$ denotes the magnitude of the velocity $\mathbf{v_k}$. Making further progress requires an assumption about the form of the distribution function, $f^0_{\mathbf{k}}$. In Lorentz's theory of conduction, $f^0_{\mathbf{k}}$ is taken to be the Maxwell distribution function, Eq. (18.18). As already mentioned, the use of this function seems to improve upon Drude's simple theory, but still fails to provide a fully satisfactory account of conduction [5].

A form of $f_{\mathbf{k}}$ more appropriate for the description of electrons in metals is that of the Fermi distribution, Eq. (18.19). Considering the low temperature case, we can replace $-\partial f^0_{\mathbf{k}}/\partial E$ by a delta function at the Fermi energy (more appropriately a Lorentzian with narrow half width), which allows us to integrate over E and obtain

$$\mathbf{J} = \frac{\mathrm{e}^2\tau}{4\pi^3}\int \frac{\mathbf{v_k v_k}}{v_{\mathbf{k}}}\mathrm{d}S_{\mathrm{F}} \cdot \mathbf{E}, \tag{18.39}$$

where $\mathrm{d}S_{\mathrm{F}}$ is an element of the Fermi surface. From the standard relation

$$\mathbf{J} = \overleftrightarrow{\sigma} \cdot \mathbf{E}, \tag{18.40}$$

we obtain the following general expression for the conductivity tensor (dyadic)

$$\overleftrightarrow{\sigma} = \frac{\mathrm{e}^2\tau}{4\pi^3}\int \frac{\mathbf{v_k v_k}}{v_{\mathbf{k}}}\mathrm{d}S_{\mathrm{F}}. \tag{18.41}$$

In many cases Eq. (18.41) can be simplified considerably. For example, when the material is characterized by cubic symmetry, the conductivity tensor reduces to a scalar. Taking the field $\mathbf{E}$ and the current $\mathbf{J}$ to point along the (positive) x direction, we have

$$\begin{aligned} \mathbf{v_k v_k} \cdot \mathbf{E} &= v_x^2 E \\ &= \frac{1}{3}v^2 E, \end{aligned} \tag{18.42}$$

where v_x^2 is set equal to $v^2/3$ for isotropic materials. Now, we can write,

$$\begin{aligned} \overleftrightarrow{\sigma} &= \frac{\mathrm{e}^2\tau}{12\pi^3}\int v\mathrm{d}S_{\mathrm{F}} \\ &= \frac{\mathrm{e}^2}{12\pi^3}\int \Lambda\mathrm{d}S_{\mathrm{F}}, \end{aligned} \tag{18.43}$$

where the *mean free path* Λ is defined by the relation

$$\Lambda = \tau v. \tag{18.44}$$

Rather loosely speaking, Λ is the average distance traveled by an electron between collisions.

The use of Fermi-Dirac statistics produces both improved agreement with experimental results as well as physical insight into the mechanisms of electrical conduction. For example, it leads [5] to a Lorentz number that agrees remarkably well with the observed average value, 2.71x10^{-13} e.s.u. vs. the experimental value of 2.72x10^{-13} e.s.u. at room temperature. It also clearly shows that conduction is due to a small number of electrons near the Fermi energy moving with large speeds (corresponding to the energy at the Fermi level), rather than to all so-called free electrons drifting slowly under the action of a field. This feature is also consistent with the values of the electronic specific heat in metals that are considerably smaller than the classical predictions based on the motion of the entire electron cloud.

A simple example may serve as an illustration of these comments. In the free-electron approximation, Eq. (18.43) takes the classical form, Eq. (18.3),

$$\sigma = n\frac{e^2\tau}{m}. \tag{18.45}$$

Consistent with our discussion above, the quantity n should be interpreted as the density of states at the Fermi level $n = n(E_F)$.

However, even this improved theory possesses various unsatisfactory features. The most important among these is the use of a mean free path that cannot be calculated within the theory. Provided there is adequate justification for invoking the relaxation-time approximation, one can attempt to determine Λ through comparison with experimental results. Such fitting procedures can lead to entirely non-physical consequences, for instance yielding a mean free path of the order of 100 interatomic distances (5.2x10^{-6}) in Ag. Such values are inexplicable on the basis of classical mechanics, upon which the previous formalism is based (the use of the Fermi function not withstanding). We will see in our continuing discussion how the theory of conduction can be further improved through a proper description of the electron states in a metal.

18.6 Relaxation Time

Let us take a closer look at Eq. (18.30). For elastic scattering, we set $E_{\mathbf{k}} = E_{\mathbf{k}'}$ (so that $f^0_{\mathbf{k}} = f^0_{\mathbf{k}'}$), and using the linear response condition, $g_{\mathbf{k}} \approx \mathbf{E}$, we can write,

$$\begin{aligned} g_{\mathbf{k}} &= e\mathbf{E}\cdot\mathbf{\Lambda}(\mathbf{k})\left(-\frac{\partial f^0_{\mathbf{k}}}{\partial E}\right) \\ &= e\mathbf{E}\cdot\mathbf{\Lambda}(\mathbf{k})\delta\left(E_{\mathbf{k}} - E_F\right). \end{aligned} \tag{18.46}$$

The vector quantity $\mathbf{\Lambda}$ indicates that the mean free path in general varies in magnitude and direction over the Fermi surface. The Boltzmann Eq.

(18.30) can now be cast into an integral form for $\mathbf{\Lambda}$,

$$\mathbf{\Lambda} = \tau(\mathbf{k}) \left[\mathbf{v_k} + \int \mathrm{d}^3k' Q(\mathbf{k}, \mathbf{k}') \mathbf{\Lambda} \right], \tag{18.47}$$

where we have introduced the **k**-vector dependent relaxation time [11]

$$\tau^{-1}(\mathbf{k}) = \int \mathrm{d}^3k' Q(\mathbf{k}, \mathbf{k}'). \tag{18.48}$$

It is straightforward to show that the conductivity tensor, Eq. (18.43), can also be written in the form,

$$\overleftrightarrow{\sigma} = \frac{2e^2}{(2\pi)^2} \int \frac{\mathbf{v_k} \mathbf{\Lambda}(\mathbf{k})}{v_\mathbf{k}} \mathrm{d}S_\mathrm{F}. \tag{18.49}$$

In the case of systems with cubic symmetry, this expression reduces to Eq. (18.43)

It is now seen that to determine $\overleftrightarrow{\sigma}$ in general one must solve Eq. (18.47) for a given host band structure, which determines $\mathbf{v_k}$ and the morphology of the Fermi surface, and to employ a given transition probability $Q(\mathbf{k}, \mathbf{k}')$, which also depends on $E_\mathbf{k}$. As is indicated in further discussion, the solution can be carried out by means of an iterative procedure, or more sophisticated methods.

18.7 Impurity Resistivity

Bloch's theorem states that the electrons in a perfectly periodic system (for our purposes a metal) exist in eigenstates with an infinite lifetime, and hence infinite mean free path[3]. Thus, such systems are characterized by perfect conductivity. Finite resistivity arises only in the presence of imperfections, such as atomic displacements and impurities of a foreign kind embedded in the host material. The resistance due to lattice distortions is commonly referred to as *ideal resistance* and becomes more and more important with increasing temperature. The treatment of this contribution to the resistance of a metal is rather difficult and somewhat outside the scope of this book. The reader is referred to the literature [10] for further details on this topic.

As the temperature is lowered, lattice excitations (phonons) become less and less significant and one approaches the regime of *residual resistivity* caused by the presence of impurities distributed at random in an otherwise

[3]We see now why fitting to experiment can yield the very large electronic mean free paths mentioned earlier.

periodic material. We begin our discussion of residual resistivity by obtaining an expression for the contribution to resistivity due to the scattering by impurities embedded in a free-electron gas. We base our discussion on the relaxation-time approximation. This discussion also provides a contact with the theory of electron scattering in solids discussed in previous parts of the book.

From the simple expression for the conductivity, Eq. (18.45), we obtain

$$\rho = \frac{1}{\sigma} = \frac{m}{ne^2}\frac{1}{\tau} \tag{18.50}$$

for the resistivity of a free-electron gas. It remains only to determine the relaxation time, τ. (We note that in view of our previous discussion the quantity n in this expression should be interpreted as the electron density at the Fermi level.)

An expression for τ can be obtained from Eq. (18.47) which, in view of Eq. (18.48), can be cast into the integral form,

$$\mathbf{v}_{\mathbf{k}} = \int \mathrm{d}^3k' Q(\mathbf{k},\mathbf{k}')\left[\mathbf{\Lambda}_{\mathbf{k}} - \mathbf{\Lambda}_{\mathbf{k}'}\right], \tag{18.51}$$

where we continue to use the condition of elastic scattering, $E_{\mathbf{k}} = E_{\mathbf{k}'}$. In agreement with the relaxation-time approximation, we set$\mathbf{\Lambda}_{\mathbf{k}} = \tau \mathbf{v}_{\mathbf{k}}$, and write the last expression in the form,

$$\frac{1}{\tau} = \int \mathrm{d}^3k' Q(\mathbf{k},\mathbf{k}')\left(1 - \frac{\mathbf{v}_{\mathbf{k}}\cdot\mathbf{v}_{\mathbf{k}'}}{v_{\mathbf{k}}v_{\mathbf{k}'}}\right). \tag{18.52}$$

We consider the isotropic case so that the right side is independent of $\mathbf{k}$ since $\mathbf{v}_{\mathbf{k}} = \mathbf{v}_{\mathbf{k}'} = v_{\mathrm{F}} = \mathrm{const}$ and, as shown below, $Q(\mathbf{k},\mathbf{k}')$ depends only on the angle θ between $\mathbf{k}$ and $\mathbf{k}'$. Thus, the last term in Eq. (18.52) can be replaced by $\cos\theta$ and we obtain for the resistivity the expression,

$$\rho = \frac{m}{ne^2}\int \mathrm{d}^3k' Q(\mathbf{k},\mathbf{k}')(1-\cos\theta). \tag{18.53}$$

We now examine the form of $Q(\mathbf{k},\mathbf{k}')$ for the case of impurity scattering in an otherwise free electron gas.

The quantity $Q(\mathbf{k},\mathbf{k}')$ has been defined as the transition probability from the state $\mathbf{k}$ to the state $\mathbf{k}'$ (and vice versa) arising from the interparticle interactions in the system. In the present case, Fermi's golden rule yields

$$Q(\mathbf{k},\mathbf{k}') = N^i\frac{2\pi}{\hbar}\left|t_{\mathbf{k},\mathbf{k}'}\right|^2 \delta(E_{\mathbf{k}} - E_{\mathbf{k}'}), \tag{18.54}$$

where N^i is the number of impurities given by cN in terms of the impurity concentration c and the total number of sites N in the system. In the last expression, we must set $E_{\mathbf{k}} = E_{\mathbf{k}'} = E_{\mathrm{F}}$ and the delta function guarantees the restriction to elastic scattering. Therefore the transition rate per site is obtained from Eq. (18.54) upon the replacement of N^i by c.

Let us now consider the case of a spherically symmetric potential at the impurity site. Then, from Eq. (D.47) we have [with $f(\theta,\phi) \simeq t_{\mathbf{k},\mathbf{k}'}$, and with various factors ($\hbar$, m), etc., restored]

$$t_{\mathbf{k},\mathbf{k}'} = -\frac{4\pi}{v}\frac{\hbar^2}{2mk_{\mathrm{F}}}\sum_{\ell}(2\ell+1)\sin\eta_\ell \mathrm{e}^{\mathrm{i}\eta_\ell}P_\ell(\cos\theta), \tag{18.55}$$

where v denotes the volume of the impurity cell, k_{F} is the magnitude of the (free-electron) Fermi momentum, η_ℓ denotes the phase shift of the ℓth partial wave, and where we have set $\eta_\ell(E_{\mathrm{F}}) = \eta_\ell$. We now obtain for the residual resistivity per site the expression [11]

$$\rho = \frac{2\mathrm{ch}}{Zk_{\mathrm{F}}\mathrm{e}^2}\sum_{\ell\ell'}(2\ell+1)(2\ell'+1)\sin\eta_\ell\sin\eta_{\ell'}\mathrm{e}^{\mathrm{i}(\eta_\ell-\eta_{\ell'})}I_{\ell\ell'}, \tag{18.56}$$

where Z is the total electron charge on the impurity and we have defined the quantity,

$$I_{\ell\ell'} = \frac{1}{2}\int_{-1}^{1}\mathrm{d}xP_\ell(x)P_{\ell'}(x)(1-x). \tag{18.57}$$

Using the properties of the Legendre polynomials (see Appendix C)

$$\int_{-1}^{1}P_\ell(x)P_{\ell'}(x) = \frac{2}{2\ell+1}\delta_{\ell\ell'} \tag{18.58}$$

and

$$(2\ell+1)xP_\ell(x) = (\ell+1)P_{\ell+1}(x) + \ell P_{\ell-1}(x), \tag{18.59}$$

we find

$$I_{\ell\ell'} = \frac{1}{2\ell+1}\left[\delta_{\ell\ell'} - \frac{\ell'+1}{2\ell'+1}\delta_{\ell,\ell'+1} - \frac{\ell'}{2\ell'+1}\delta_{\ell,\ell'-1}\right], \tag{18.60}$$

so that the double sum in Eq. (18.56) can be simplified to yield the expression [11]

$$\rho = \frac{2\mathrm{ch}}{Z\mathrm{e}^2k_{\mathrm{F}}}\sum_{\ell}(\ell+1)\sin^2(\eta_{\ell+1}-\eta_\ell). \tag{18.61}$$

We see that in this simple model, the residual resistivity is characterized by a coupling of neighboring partial waves which, in turn, arises from the presence of the term $\cos\theta$ in Eq. (18.53). The presence of this factor shows that large-angle scattering makes the greatest contribution to resistivity since it corresponds to large momentum transfer.

The discussion just presented was based on the assumption of non-interacting, free electrons. A more realistic treatment should account for the presence of a lattice and the interaction of the electrons with the ions in the system, and with one another. Models for improved treatments of the Boltzmann equation, such as the spherical-band model, have been reported in the literature [12-14]. The resulting expression for the resistivity in, for example, the spherical-band model has essentially the same form as that of the last expression above, but with a number of expected changes. It properly depends on the electron velocity and density at the Fermi level, and the phase shifts are replaced by quantities that measure the scattering of a Bloch wave due to the presence of the additional potential introduced by the impurity (see the discussion on two-potential scattering in Chapter 9).

The solutions of the Boltzmann equation discussed so far are based on the relaxation time approximation. In the general case, when the mean free path is *not* proportional to the velocity, the solution becomes much more complicated. One approach [15] is to solve the equation iteratively starting from an initial guess of the form

$$\mathbf{\Lambda}_{\mathbf{k}}^{0} = \tau_{\mathbf{k}} \mathbf{v}_{\mathbf{k}}. \tag{18.62}$$

Substitution of this expression into the right side of Eq. (18.47) yields a new value, $\mathbf{\Lambda}_{\mathbf{k}}^{1}$, and the procedure is repeated till convergence is reached. The so-called *degenerate kernel* method has also been used with some success [16-20], and the reader is referred to the corresponding literature for details.

18.8 Transport Coefficients

Within the framework of the Boltzmann equation, a discussion of the conductivity of a material hinges on the behavior of the distribution function in the presence of potential and/or temperature gradients. We now derive a set of equations that allow us to define general coefficients associated with the response of a system to such gradients. In our discussion, we follow Ziman [10], where the reader is referred for more details.

In addition to an electric field, if a temperature gradient also exists in the material, Eq. (18.29) leads to the expression

$$\begin{aligned} &\left(-\frac{\partial f_{\mathbf{k}}^{0}}{\partial E_{\mathbf{k}}}\right) \mathbf{v}_{\mathbf{k}} \cdot \left\{-\frac{E_{\mathbf{k}} - \zeta}{T}(-\nabla T) + \mathrm{e}\left[\mathbf{E} - \frac{1}{\mathrm{e}}\nabla\zeta\right]\right\} \\ = &\left[\frac{\partial f_{\mathbf{k}}}{\partial t}\right]_{\mathrm{scatt}}. \end{aligned} \tag{18.63}$$

In the relaxation-time approximation, this equation can be solved in the form

$$\begin{aligned} g_{\mathbf{k}} &= f_{\mathbf{k}} - f_{\mathbf{k}}^0 \\ &= \tau \left(-\frac{\partial f_{\mathbf{k}}^0}{\partial E_{\mathbf{k}}} \right) \mathbf{v}_{\mathbf{k}} \cdot \left\{ -\frac{E_{\mathbf{k}} - \zeta}{T} (-\nabla T) + \mathrm{e} \left[\mathbf{E} - \frac{1}{\mathrm{e}} \nabla \zeta \right] \right\}. \end{aligned} \tag{18.64}$$

An electric current can be calculated from an equation of the form of Eq. (18.38),

$$\begin{aligned} \mathbf{J} &= \frac{\mathrm{e}^2 \tau}{4\pi^3 \hbar} \int \mathbf{v}_{\mathbf{k}} \mathbf{v}_{\mathbf{k}} \left(-\frac{\partial f_{\mathbf{k}}^0}{\partial E_{\mathbf{k}}} \right) \frac{\mathrm{d}S_E}{v_{\mathbf{k}}} \mathrm{d}E \cdot \left(\mathbf{E} - \frac{1}{\mathrm{e}} \nabla \zeta \right) \\ &+ \frac{\mathrm{e}^2 \tau}{4\pi^3 \hbar} \int \mathbf{v}_{\mathbf{k}} \mathbf{v}_{\mathbf{k}} \left(\frac{E - \zeta}{T} \right) \left(-\frac{\partial f_{\mathbf{k}}^0}{\partial E_{\mathbf{k}}} \right) \frac{\mathrm{d}S_E}{v_{\mathbf{k}}} \mathrm{d}E \cdot (-\nabla T). \end{aligned} \tag{18.65}$$

This expression indicates the possible presence of *thermoelectric effects*, that is to say, the production of an electric current due to the presence of thermal gradients acting in the absence of an electric field. We note the existence of a term proportional to the gradient of ζ. This term arises due to the difference in chemical potential caused by the temperature gradients in the material. In the following discussion, we assume that this termis absorbed in the observed electric fields, and we will not consider it explicitly.

We are now in a position to give an operational definition of a *heat current*, $\mathbf{U}$. In basic thermodynamics, heat is defined as the internal energy, $E_{\mathbf{k}}$, minus the free energy, ζ, so that we define the heat current by the expression,

$$\mathbf{U} = 2 \int f_{\mathbf{k}} \left[E_{\mathbf{k}} - \zeta \right] \mathbf{v}_{\mathbf{k}} \mathrm{d}^3 k. \tag{18.66}$$

Combining the last three expressions, we can write the general transport equations,

$$\mathbf{J} = \mathrm{e}^2 \overleftrightarrow{\mathbf{K}}_0 \cdot \mathbf{E} + \frac{\mathrm{e}}{T} \overleftrightarrow{\mathbf{K}}_1 \cdot (-\nabla T), \tag{18.67}$$

and

$$\mathbf{U} = \mathrm{e} \overleftrightarrow{\mathbf{K}}_1 \cdot \mathbf{E} + \frac{1}{T} \overleftrightarrow{\mathbf{K}}_2 \cdot (-\nabla T), \tag{18.68}$$

where the tensor transport coefficients $\overleftrightarrow{\mathbf{K}}_n$, with $n = 0,\ 1,\ 2$, are defined by the expression

$$\overleftrightarrow{\mathbf{K}}_n \equiv \frac{\tau}{4\pi^3 \hbar} \int \int \mathbf{v}\mathbf{v} \left(E_{\mathbf{k}} - \zeta \right)^n \left(-\frac{\partial f^0}{\partial E} \right) \frac{\mathrm{d}S}{v} \mathrm{d}E. \tag{18.69}$$

To evaluate these coefficients, we use the fact that the derivative of the Fermi function is a Lorentzian at finite temperature (a delta function at zero temperature) and write [10] to second order in T,

$$\int \Phi(E)\left(-\frac{\partial f^0}{\partial E}\right) \mathrm{d}E \approx \Phi(\zeta) + \frac{1}{6}\pi^2(\mathrm{k_B}T)^2 \left[\frac{\partial^2 \Phi(E)}{\partial E^2}\right]_{E=\zeta} + \cdots, \tag{18.70}$$

where $\Phi(E)$ represents an integral over the constant energy surface $E_{\mathbf{k}} = \zeta$. Now, the coefficient $\overleftrightarrow{\mathbf{K}}_0$ takes the form

$$\overleftrightarrow{\mathbf{K}}_0 = \frac{\tau}{4\pi^3\hbar}\int \frac{\mathbf{v_k v_k}}{v_{\mathbf{k}}} \mathrm{d}S_{\mathrm{F}}, \tag{18.71}$$

which leads immediately to Eq. (18.39) for the conductivity. To evaluate $\overleftrightarrow{\mathbf{K}}_2$, we consider $\Phi(E) = (E-\zeta)^2\overleftrightarrow{\mathbf{K}}_0(E)$ which vanishes at $E = \zeta$ and leads to the expression,

$$\overleftrightarrow{\mathbf{K}}_2 = \frac{1}{3}\pi^2(\mathrm{k_B}T)^2\overleftrightarrow{\mathbf{K}}_0, \tag{18.72}$$

arising from the second term in Eq. (18.70). Finally, this second term also leads to the expression

$$\overleftrightarrow{\mathbf{K}}_1 = \frac{1}{3}\pi^2(\mathrm{k_B}T)^2\left[\frac{\partial \overleftrightarrow{\mathbf{K}}_0}{\partial E}\right]_{E=\zeta}. \tag{18.73}$$

Under the appropriate conditions, these transport coefficients reduce to familiar expressions. For example, when temperature gradients vanish, we obtain from Eq. (18.67) the result

$$\overleftrightarrow{\sigma} = \mathrm{e}^2\overleftrightarrow{\mathbf{K}}_0. \tag{18.74}$$

This is just Eq. (18.41).

18.9 Thermal Conductivity

Thermal conductivity is determined by the flow of heat as a response to a temperature gradient. If the system consists of charged particles, then thermal conductivity is defined in the absence of an electric field. Before turning to a discussion of heat transport based on the discussion of transport coefficients within the framework of the Boltzmann equation, let us attempt to gain some intuitive understanding of the processes associated with the conduction of heat through a metal.

18.9.1 Heat transport

First, consider a classical gas of neutral, non-interacting particles in the presence of a temperature gradient. The energy of the particles in the gas is entirely kinetic and hence proportional to temperature. The particles move faster in regions of higher temperature and due to collisions impart kinetic energy to other particles. This process can be viewed as a transfer of heat (kinetic energy) from regions of high to low temperature.

In a system consisting of electrons and nuclei, e.g., a metal, heat transport cannot be described in such simple terms. Now, both the electrons and the nuclei contribute to heat transport in interrelated ways. It is clear that the electrons play an important role since it is well-known that high electrical conductivity usually implies high thermal conductivity (although exceptions even in metallic systems exist, e.g., Bi). The experimental verification of this relation in the case of metals is embodied in the Wiedemann-Frantz law, Eq. (18.9). The relation between thermal and electrical conductivity becomes more complex in semiconductors and insulators, and leads to interesting phenomena in so-called thermoelectric materials, e.g., $OsSi_2$ and $FeSi_2$. We now inquire as to the mechanisms giving rise to heat transport in a solid, particularly a metal.

It is clear at the outset that electron-electron collisions can be neglected. The effects of temperature gradients are to be sought in the behavior of the ions, and the interactions (collisions) of the electrons with them. Elevated temperatures cause ions to be displaced from their equilibrium positions and thus provide scattering centers for the electron gas. This has the effect of altering the Fermi distribution function so that it becomes "sharper" in regions of lower temperature and "broader" in regions of higher temperature. Thermal conduction occurs without a net transfer of charge, i.e., no electrical conduction, because "hot" and "cold" electrons, characterized by appropriate distribution functions, travel in equal numbers in opposite directions.

It is also of interest to inquire about the nature of the mechanisms responsible for relaxation, i.e., the elimination of temperature gradients. Consider an elastic scattering event in which a hot electron with energy E (slightly above E_{F}) reverses its direction. We can think of this *horizontal* process [10] as depleting the ranks of "hot" electrons while augmenting those of the "cold" ones. Such processes are also responsible for reducing electrical resistance, and for them the Wiedemann-Frantz law holds. As we saw in the previous section, such processes depend on the angle of scattering through a factor $1 - \cos\theta$, showing that large-angle scattering is more effective in reducing conduction than scattering at small angles.

However, in addition to horizontal processes, inelastic scattering plays a significant role in reducing thermal conductivity. In a so-called *vertical* process, an electron loses all its extra energy and falls below the Fermi level. Such a process does not affect electrical conduction, but can be as effective in reducing heat conduction as scattering at large angles. Clearly such inelastic processes are associated with the creation or absorption of phonons in the lattice and their analysis, ultimately involving the electron-phonon interaction, is rather complicated. Some features of this process, however, are worth a closer look.

At high enough temperatures, electron-phonon interactions are not expected to have much of an effect on the heat current. The maximum energy that an electron could gain or lose through scattering by a phonon is of the order $k_B\Theta$, where Θ is the Debye temperature (see subsection 21.3.2). When $T > \Theta$, this energy loss or gain is smaller than the spread of the Fermi function which is of the order k_BT, allowing electrons to exchange energy with the phonons while maintaining their position with respect to the Fermi level. Thus, the Wiedemann-Franz law can be expected to hold also for the "ideal" resistance of a metal at high enough temperatures, and this more or less turns out to be the case. At low temperatures, however, the energy of a phonon is also of the order k_BT and an electron could fall below the Fermi energy through its interaction with a phonon. Now, *all* scattering processes are equally effective in reducing thermal conduction and the factor $1 - \cos\theta$ can be omitted from the calculations. A somewhat simplified analysis [10] shows that the Lorentz number is now of the form,

$$L \approx \left(\frac{T}{\Theta}\right)^2, \tag{18.75}$$

instead of a constant.

Electron-phonon scattering, or *phonon drag*, plays a very important role in determining the transport properties of many technologically import ant materials such as thermoelectrics. However, it is complex enough to defy a rigorous treatment within the formal framework underlying our discussion in this book and will not be considered further. In the following, we concentrate on the effects of elastic scattering events on the thermal conductivity of a solid.

18.9.2 Thermal conduction

Analogous to electrical conductivity, thermal conductivity can be defined as the proportionality tensor connecting the heat current to the temperature gradient in the material. From Eq. (18.68), this coefficient appears to be equal to $\overleftrightarrow{\mathbf{K}}_2/T$. However, we should exercise some care in making this identification.

We recall that the electrical conductivity was defined in the absence of a temperature gradient. Similarly, we should define the thermal conductivity in the absence of an electric field. However, it is not always feasible to carry out experiments so that **E** vanishes (recall that it is the local or microscopic field that concerns us here), whereas it is a simple matter to place the specimen in an open circuit, so that **J**=0. This means that there will be an electric field set up along the specimen, which from Eq. (18.67), is equal to

$$\mathbf{E} = \frac{1}{\mathrm{e}} \left(\overleftrightarrow{\mathbf{K}}_0\right)^{-1} \overleftrightarrow{\mathbf{K}}_1 \frac{-\nabla T}{T}. \tag{18.76}$$

With this expression in Eq. (18.68), we obtain

$$\begin{aligned} \mathbf{U} &= \frac{1}{T}\left(\overleftrightarrow{\mathbf{K}}_2 - \overleftrightarrow{\mathbf{K}}_1 \overleftrightarrow{\mathbf{K}}_0^{-1} \overleftrightarrow{\mathbf{K}}_1\right)(-\nabla T) \\ &= \overleftrightarrow{\kappa}\,(-\nabla T), \end{aligned} \tag{18.77}$$

for the heat current. This expression defines the thermal conductivity, $\overleftrightarrow{\kappa}$, under open circuit conditions (vanishing current). In the case of a metal, the second term inside the brackets is usually small compared to the first term, (metals in general have large electrical conductivities), and can be neglected. In this case we obtain the expression

$$\overleftrightarrow{\kappa} = \frac{1}{T}\overleftrightarrow{\mathbf{K}}_2, \tag{18.78}$$

for the thermal conductivity. We note that according to Eqs. (18.72) and (18.74) this can be written in the form of the Wiedemann-Franz law,

$$\overleftrightarrow{\kappa} = \frac{\pi^2 \mathrm{k}_\mathrm{B}^2}{3\mathrm{e}^2} T \overleftrightarrow{\sigma}. \tag{18.79}$$

The ratio $\overleftrightarrow{\kappa}\overleftrightarrow{\sigma}^{-1}/T$ is the Lorentz number encountered in Eq. (18.10).

When compared to Eq. (18.10), the expression for the Lorentz number derived within Drude's theory, the two expressions differ in the coefficient multiplying the ratio $(\mathrm{k_B}/\mathrm{e})^2$. This ratio reflects the fact that electrons have charge e, and when acted upon by a force e**E**, contribute a current per unit field equal to e^2. Similarly, in thermal conductivity, each electron carries thermal energy $\mathrm{k_B}T$, and when acted upon by a thermal force, $\mathrm{k_B}\nabla T$, contributes a thermal current per unit temperature gradient equal to $\mathrm{k_B^2}$. The ratio of the two currents is equal to $(\mathrm{k_B}/\mathrm{e})^2$. The coefficient of 3 in Drude's theory is replaced by the number $\pi^2/3$ in Eq. (18.79) because of Fermi-Dirac statistics used to represent the electron distribution.

18.9.3 Thermoelectric effects

Equations (18.67) and (18.68) provide a clear indication of the interconnected effects of electric potential and temperature gradients on the electrical and thermal currents. With $\mathbf{J} = 0$ in Eq. (18.67), we obtain Eq. (18.76), which we write in the form,

$$\mathbf{E} = \overleftrightarrow{\mathbf{Q}} \nabla T. \tag{18.80}$$

This expression reveals the presence of an e.m.f. (which is amenable to measurement) resulting from the presence of a temperature gradient. The presence of an electric current following the introduction of a temperature gradient is known as the *Seebeck effect*, and the coefficient $\overleftrightarrow{\mathbf{Q}}$ is referred to as the *thermopower*. From Eqs. (18.73) and (18.74), we can obtain [10]

$$\overleftrightarrow{\mathbf{Q}} = \frac{\pi^2 \mathrm{k}_\mathrm{B}^2}{3\mathrm{e}} T \frac{\partial \ln \overleftrightarrow{\sigma}(E)}{\partial E} \Big|_{E=E_\mathrm{F}}, \tag{18.81}$$

where $\partial \ln \overleftrightarrow{\sigma}(E)$ can be thought of the conductivity of a "hypothetical metal" with a Fermi level equal to E.

In deriving the expression for $\overleftrightarrow{\mathbf{Q}}$ in Eq. (18.81) we neglected the effects associated with lattice distortions, the so-called phonon drag. This expression for $\overleftrightarrow{\mathbf{Q}}$ is commonly referred to as *diffusion thermopower*, and it is a good approximation for the thermopower at high enough temperatures, above liquid nitrogen temperature, where phonon drag effects can be neglected [10].

In addition to the Seebeck effect, other thermoelectric effects include the *Peltier effect*, connecting the heat flux to the electric current, and the *Thompson effect*, observed when electric fields and temperature gradients are both present in a circuit. The associated coefficients for these effects can be worked out in fairly straightforward fashion from Eqs. (18.67) and (18.68). The interested reader can obtain a more detailed discussion of thermoelectric effects by referring to the standard literature [10].

18.10 Magnetic Field Effects

Up to this point we have examined the response of an electron gas to gradients of electric potential and temperature. We have left out those effects associated with the presence of magnetic fields in a system. Such effects can have very important physical and technological consequences, as the discovery of giant magnetoresistance indicates, and at least a brief discussion of the fundamental issues involved is warranted.

We start with Eq. (18.29) and assume the existence of both electric and magnetic fields, but not temperature gradients. In the relaxation-time approximation, we obtain

$$\left(-\frac{\partial f_\mathbf{k}^0}{\partial E_\mathbf{k}}\right) \mathrm{e}\mathbf{v}_\mathbf{k} \cdot \mathbf{E} = \frac{g_\mathbf{k}}{\tau} + \frac{\mathrm{e}}{\hbar c} (\mathbf{v}_\mathbf{k} \times \mathbf{H}) \cdot \nabla_\mathbf{k} g_\mathbf{k}(\mathbf{r}) \tag{18.82}$$

in place of Eqs. (18.30) and (18.31). This is a differential equation for $g_{\mathbf{k}}$ which is not easy to solve in general. However, interesting qualitative information can be obtained by considering the case of the non-interacting, free electron gas.

18.10.1 Hall effect

In this case we have,

$$\mathbf{k} = \frac{m\mathbf{v}}{\hbar}, \tag{18.83}$$

so that states can be labeled by a velocity as well as a $\mathbf{k}$ index. As a trial solution of Eq. (18.82), we choose a form analogous to that in Eq. (18.35)

$$g_{\mathbf{k}} = \left(-\frac{\partial f^0}{\partial E}\right) \tau \mathbf{v_k} \cdot e\mathbf{A}, \tag{18.84}$$

where the vector $\mathbf{A}$ replaces the electric field, $\mathbf{E}$, and remains to be specified. Using Eqs. (18.84) and (18.82), we obtain the expression

$$\mathbf{v} \cdot \mathbf{E} = \mathbf{v} \cdot \mathbf{A} + \frac{e\tau}{mc} (\mathbf{v} \times \mathbf{H}) \cdot \mathbf{A}, \tag{18.85}$$

from which we obtain the result

$$\mathbf{E} = \mathbf{A} + \frac{e\tau}{mc} (\mathbf{H} \times \mathbf{A}). \tag{18.86}$$

It is of some interest to note that this expression can now be inverted to solve for $\mathbf{A}$ for given values of the vectors $\mathbf{E}$ and $\mathbf{H}$. The solution is elementary [10], Fig. 18.2,

$$\mathbf{A} = \frac{\mathbf{E} - \frac{e\tau}{mc}\mathbf{H} \times \mathbf{E}}{1 + \left(\frac{e\tau}{mc}\right)^2 H^2}. \tag{18.87}$$

However, to study transport, we can work directly with Eq. (18.86). In the relaxation-time approximation, Eq. (18.84), we have

$$\mathbf{J} = \sigma_0 \mathbf{A}, \tag{18.88}$$

where σ_0 is the electrical conductivity of a material, with its tensor character suppressed, in the absence of a magnetic field. Combining Eqs. (18.86) and (18.88), we get

$$\begin{aligned} \mathbf{E} &= \frac{1}{\sigma_0}\mathbf{J} + \frac{e\tau}{mc}\mathbf{H} \times \frac{1}{\sigma_0}\mathbf{J} \\ &= \rho_0 \mathbf{J} + \frac{e\tau}{mc}\rho_0 \mathbf{H} \times \mathbf{J}, \end{aligned} \tag{18.89}$$

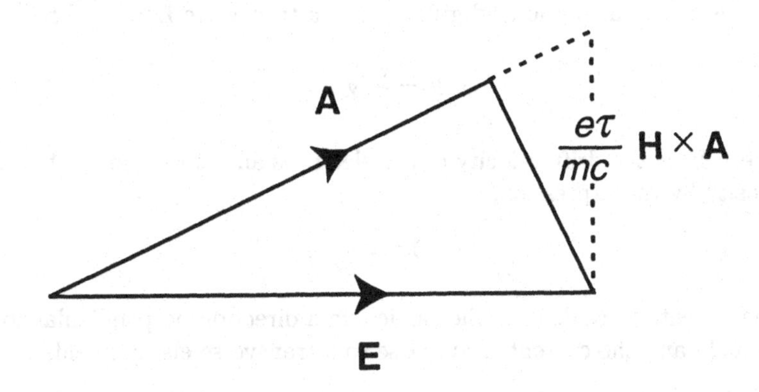

Figure 18.2: **Geometric construction for the solution of the vector A defined in the text.**

where $\rho_0 = 1/\sigma_0$ is the ordinary resistivity of a material[4].

We now examine Eq. (18.89). For currents parallel to any external magnetic fields, we have

$$E_{\parallel} = \rho_0 J, \tag{18.90}$$

and the field does not affect electrical transport. In this case there is no *magnetoresistance*, i.e., effects on the electrical transport properties of the system associated with the presence of a magnetic field. Magnetoresistance also vanishes when **H** is perpendicular to **J**. Then a transverse electric field is set up, as is indicated by the second term in Eq. (18.89), of magnitude,

$$E_H = \frac{e\tau}{mc}\rho_0 H J. \tag{18.91}$$

The presence of this field is the celebrated *Hall effect.* The proportionality factor E_H/HJ is called the *Hall coefficient*, and for free electrons it has the form,

$$\begin{aligned} R &= \frac{e\tau}{mc}\rho_0 \\ &= \frac{1}{nec}, \end{aligned} \tag{18.92}$$

where Eq. (18.45) has been used to eliminate the relaxation time from the second line.

[4]In general, the resistivity is also a tensor that reduces to a scalar for isotropic systems. This reduction, which is employed here, considerably simplifies the discussion of transport and related phenomena.

The Hall effect and the dependence of the Hall coefficient on the inverse of carrier density can be understood through elementary arguments. The presence of a magnetic field gives rise to a transverse Lorentz force

$$\mathbf{F} = \frac{e}{c}\delta\mathbf{v} \times \mathbf{H}, \tag{18.93}$$

where $\delta\mathbf{v}$ is the drift velocity of the electrons and is related to the current density by the expression,

$$\delta\mathbf{v} = \frac{1}{ne}\mathbf{J}. \tag{18.94}$$

The Lorentz force deflects the carriers in a direction perpendicular to both the field and the current, giving rise to a transverse electric field,

$$\begin{aligned} E_H &= \frac{1}{c}\mathbf{H} \times \delta\mathbf{v} \\ &= \frac{1}{c}\mathbf{H} \times \frac{1}{ne}\mathbf{J}. \end{aligned} \tag{18.95}$$

This is the field needed to compensate for the pile up of the charge that results from the sideways deflection of the electrons. It is proportional to $1/n$ because, for a given current, the smaller the density of carriers, the faster they must travel and the greater the deflection that each suffers in the presence of the magnetic field.

We see that in this case there is no magnetoresistance. The compensating electric fields exactly cancel the effects of the sideways deflection of the carriers so that the longitudinal current under the influence of $E_{\|}$ remains unaffected.

One more feature of the Hall effect is worthy of note. The sign of the Hall coefficient can change depending on whether the current is carried predominantly by negative charges (electrons), or positive ones (holes). This result can be proved directly for a spherical Fermi surface [10], and is in close accord with experimental observations for many metallic systems. However, the formal results derived within the free-electron model may not be quite appropriate in the case of real materials. Even when transport is mediated by carriers in a single band, the relaxation time may depend strongly on energy and wave vector. Also, the use of the Boltzmann equation may be insufficient for the treatment of transport in disordered materials, such as substitutionally disordered alloys, where $\mathbf{k}$ is no longer a good quantum number. In this case, a more fundamental rather than phenomenological approach, such as one based on the Kubo formula, may be needed. A discussion of transport bas ed on the Kubo formula is presented in the next chapter.

18.10.2 Ordinary magnetoresistance

The discussion given above has revealed that when a single type of carrier is responsible for current propagation, the resistivity of a material remains unaffected by the presence of a magnetic field. This situation can change dramatically when more than one type of carrier, say electrons and holes, or electrons moving along different paths on the Fermi surface, are present. It may not be possible to set up a transverse electric field that will exactly cancel the deflections of the two different carriers, and the *longitudinal* resistance of the materials, i.e., the resistance in the direction of the field, may indeed change.

To see in a fairly simple way how this effect can materialize, we consider the presence of two types of carriers, each characterized by its own conductivity, associated current, and Hall coefficient. Under the action of an electric and a magnetic field, Eq. (18.96) leads to the expression

$$\mathbf{E} = \frac{1}{\sigma_i}\mathbf{J}_i + \beta_i \mathbf{H} \times \frac{1}{\sigma_i}\mathbf{J}_i, \quad i = 1, 2, \tag{18.96}$$

for each type of carrier, i, where,

$$\beta_i = \frac{e\tau_i}{m_i c}. \tag{18.97}$$

Of course, the total current is given by the vector sum,

$$\mathbf{J} = \mathbf{J}_1 + \mathbf{J}_2. \tag{18.98}$$

Equations (18.96) can be solved for $\mathbf{J}_i$ using a procedure similar to that used to obtain Eq. (18.87), leading to the result [10],

$$\mathbf{J} = (\alpha_1 + \alpha_2)\mathbf{E} - (\alpha_1\beta_1 - \alpha_2\beta_2)\mathbf{H} \times \mathbf{E}, \tag{18.99}$$

where

$$\alpha_i = \frac{\sigma_i}{1 + \beta_i^2 H^2}. \tag{18.100}$$

The Hall coefficient follows from Eq. (18.99). Its determination requires an inversion to express $\mathbf{E}$ in terms of $\mathbf{J}$ and $\mathbf{H} \times \mathbf{J}$, which can be accomplished using the method leading to Eq. (18.87). For small magnetic fields, we have,

$$R = \frac{\sigma_1^2 R_1 + \sigma_2^2 R_2}{(\sigma_1 + \sigma_2)^2}, \tag{18.101}$$

where R_1 and R_2 are the Hall coefficients associated with each type of carrier separately. Clearly, if these two coefficients have opposite signs, the value of R is a compromise between the individual responses of the two types of carriers.

We may also inquire about the effects of a magnetic field on the resistivity of a material characterized by two different carriers. First, we establish [10] the relation between the electric field and the current in the field direction for this case,

$$\begin{aligned} \rho &= (\mathbf{J} \cdot \mathbf{E}) / J^2 \\ &= \frac{\alpha_1 + \alpha_2}{(\alpha_1 + \alpha_2)^2 + (\alpha_1 \beta_1 + \alpha_2 \beta_2)^2 H^2}. \end{aligned} \tag{18.102}$$

After some algebra, this expression can be written in the form

$$\frac{\Delta\rho}{\rho_0} \equiv \frac{\rho - \rho_0}{\rho_0} = \frac{\sigma_1 \sigma_2 (\beta_1 - \beta_2)^2 H^2}{(\sigma_1 + \sigma_2)^2 + (\sigma_1 \beta_1 + \sigma_2 \beta_2)^2 H^2}, \tag{18.103}$$

where

$$\rho_0 = \frac{1}{\sigma} = \frac{1}{\sigma_1 + \sigma_2} \tag{18.104}$$

is the resistance in the absence of a magnetic field. A non-vanishing value of $\Delta\rho/\rho_0$ is referred to as *ordinary magnetoresistance.* We see that ordinary magnetoresistance varies essentially as H^2. Figure 18.3 shows magnetoresistance curves for a number of cadmium wires at various temperatures.

Ordinary magnetoresistance is usually positive, and its existence requires the presence of at least two distinct types of carriers. The carriers may differ in mass, charge, or in the mean distance traveled between collisions. As long as $\beta_1 \neq \beta_2$, it is not possible to find an electric field that will keep the various contributions to the current going in the same direction. Since the net current is the vector sum of these contributions, Eq. (18.98), the net current is reduced.

The development presented above is appropriate for the case in which the magnetic field is perpendicular to the current, and Eq. (18.103) describes *transverse magnetoresistance. Longitudinal magnetoresistance* can also arise when the field is parallel to the current, but the effect vanishes identically in spherically symmetric bands (free electron case). The onset of longitudinal magnetoresistance requires the presence of non-spherical Fermi surfaces, and its analysis is more complicated than that of transverse MR. The magnetoresistance of real materials is often a complicated function of the orientation of the field and the morphology of the Fermi surface

In addition to ordinary magnetoresistance, which depends directly on the applied external field, effects associated with the internal magnetization of the sample are also observed. The phenomenon of *anisotropic magnetoresistance* [21] is caused by the presence of the spin-orbit interaction. The more recently discovered [22] effect of *giant magnetoresistance* (GMR) has opened new and exciting possibilities for the construction of magnetic recording heads and other technologies.

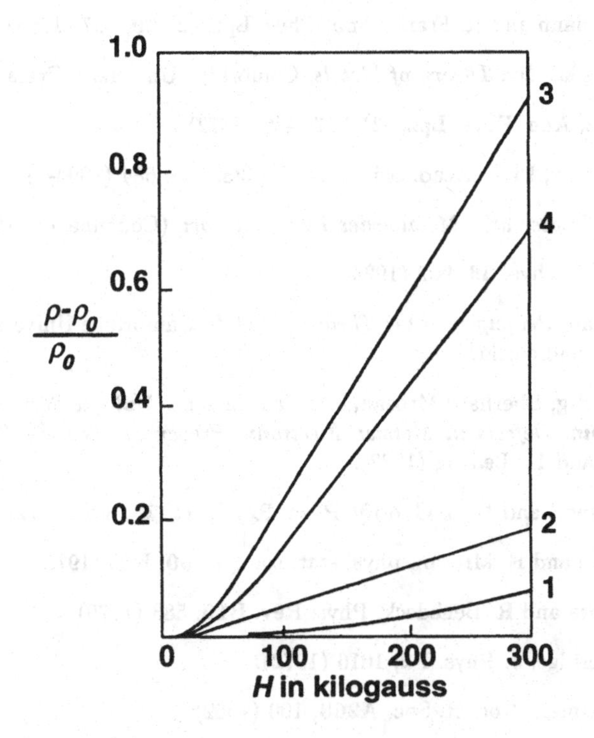

Figure 18.3: Magnetoresistance (MR) curves for cadmium wires. Curves 1, 2, 3, show transverse MR at T = 290, 190, and 88 °K. Curve 4 shows longitudinal MR at T =88 °K.

Bibliography

[1] W. Weber, Ann. Phys. Lpz. (2) **156**, 1 (1875).

[2] E. Riecke, Ann. Phys. Lpz. (3) **66**, 353 (1898).

[3] P. Drude, Ann. Phys. Lpz. (4) **1**, 566 (1890).

[4] G. Wiedemann and R. Franz, Ann. Phys. Lpz. (2) **89**, 497 (1853).

[5] A. H. Wilson, *The Theory of Metals*, Cambridge University Press (1965).

[6] L. Lorentz, Ann. Phys. Lpz. (2) **147**, 429 (1872).

[7] H. A. Lorentz, Proc. Acad. Sci. Amst. **7**, 438, 585, 684 (1904-5).

[8] N. Bohr, *Studier über Metallernes Elektrontheori*, (Copenhagen, 1911).

[9] E. Fermi, Z. Phys. **36**, 902 (1926).

[10] J. M. Ziman, *Principles of the Theory of Solids*, Cambridge University Press (1989), second edition.

[11] Ingrid Mertig, Eberhard Mrosan, and Paul Ziesche, *Multiple Scattering Theory of Point Defects in Metals: Electronic Properties*, Teubner Texte zur Physik, Band 11, Leipzig (1987).

[12] G. D. Gaspari and B. L. Györffy, Phys. Rev. Lett. **28**, 801 (1972).

[13] R. Schöpke and E. Mrosan, phys. stat. sol. (b), **90**, K95 (1978).

[14] R. P. Gupta and R. Benedeck, Phys. Rev. B**19**, 583 (1979).

[15] P. T. Coleridge, J. Phys. F**2**, 1016 (1972).

[16] D. Sondheimer, Proc. R. Soc. A**268**, 100 (1962).

[17] A. Seeger, E. Mann, and K. Clausecker, phys. stat. sol. (b) **24**, 721 (1967).

[18] K. Clausecker, E. Mann, and A. Seeger, Phys. Kondens. Materie **9**, 73 (1969).

[19] P. T. Coleridge, N. A. Holtzwarth, and M. J. G. Lee, Phys. Rev. B**10**, 1213 (1974).

[20] I. Mertig, E. Mrosan, and R. Schöpke, J. Phys. F**12**, 1689 (1982a).

[21] T. R. McGuire and R. I. Potter, IEEE Transactions on Magnetics, Vol. MAG-11, No. **4**, 1018 (1975), and references therein.

[22] M. N. Baibich, J. M. Broto, A. Fert, F. Nguyen van Dau, F. Petroff, P. Etienne, G. Creuzet, A. Friederich, and J. Chazeles, Phys. Rev. Lett. **61**, 2472 (1988).

Chapter 19

Kubo Formula

19.1 General Comments

The phenomenological treatment of transport discussed in the previous chapter concerns itself only with the kinematics of the time evolution of the distribution function. It provides no insight into the dynamics of the response of a system to an external perturbation. A rigorous treatment of transport phenomena should contain the microscopic origin of the distribution function and the dynamics characterizing its behavior in the presence of externally imposed fields.

The formalism of this chapter addresses this issue within the context of *linear response* theory. Simply stated, linear response implies that a system's reaction to an external perturbation is linearly proportional to the perturbation itself.

Within linear response theory, the expression for the response to a perturbation has the form of a *Kubo formula*, [1] who provided the original derivation. There is a Kubo formula for every transport coefficient, but we focus on electronic transport and electrical conductivity. We also derive an expression for the magnetic susceptibility. For a discussion of other transport properties within linear response, the reader is referred to the literature [2, 3].

In the following discussion, we first derive a general result within linear response theory that relates the response of an interacting system characterized by a Hamiltonian, $\hat{H}$, to an external perturbation described by $\hat{H}^{\text{ext}}$. This yields the basic expression for the change induced in the (ground state) expectation value of an operator, $\hat{O}$, by the action of $\hat{H}^{\text{ext}}$. Then, we identify $\hat{O}$ with the current operator, and $\hat{H}^{\text{ext}}$ with the Hamiltonian describing an external electric field, and derive the famous Kubo formula for the electrical conductivity. This formula is subsequently expressed in terms of the single-particle Green function of the electron cloud. We then derive expressions for the ac and dc conductivities, i.e., the quantities measured in a conductivity experiment, and point out some of the fundamental relations connecting the conductivity of a solid to its optical properties. The magnetic susceptibility is discussed subsequently, and is used as the basis to derive an expression of the fluctuation-dissipation theorem.

19.2 Linear Response

When a system is perturbed by an external stimulus whose magnitude can be characterized as "small"[1], then the response of the system is linearly proportional to the perturbation. In order to derive the basic equation of linear response theory for an interacting (non-relativistic) quantum mechanical system, we start with the general form of the time-dependent Schrödinger equation, (we set $\hbar = 1$ throughout)

$$\mathrm{i}\frac{\partial | \Psi_{\mathrm{S}}(t) \rangle}{\partial t} = \hat{H} | \Psi_{\mathrm{S}}(t) \rangle . \tag{19.1}$$

The Hamiltonian, $\hat{H}$, of the interacting many-particle system is taken to be time-independent, so that we can write

$$| \Psi_{\mathrm{S}}(t) \rangle = \mathrm{e}^{\mathrm{i}\hat{H}t} | \Psi_{\mathrm{S}}(0) \rangle . \tag{19.2}$$

We wish to examine the behavior of an observable, represented by the operator $\hat{O}_{\mathrm{S}}(t)$, under the action of an external perturbation[2,3]. In this discussion we follow closely the lines of development in Fetter and Walecka [4].

Let the external perturbation be represented by the Hamiltonian $\hat{H}^{\mathrm{ext}}$, and to vanish identically for all times t prior to a time t_0 when the perturbation is "switched on". Then, for $t > t_0$, the Schrödinger equation, Eq. (19.1), takes the form,

$$\mathrm{i}\frac{\partial | \bar{\Psi}_{\mathrm{S}}(t) \rangle}{\partial t} = \left[\hat{H} + \hat{H}^{\mathrm{ext}}\right] | \bar{\Psi}_{\mathrm{S}}(t) \rangle , \tag{19.3}$$

[1] If $\hat{H}^{\mathrm{ext}}$ represents the perturbation, then "small" implies that the expectation value $\langle\psi|\hat{H}^{\mathrm{ext}}|\psi\rangle$, where ψ is an eigenstate of the unperturbed system, is numerically small in comparison with the difference between the eigenvalues corresponding to ψ.

[2] The wave functions and operators associated with a quantum mechanical system can be expressed in anyone of the Schrödinger, Heisenberg, or interaction pictures. We consider a system characterized by an unperturbed Hamiltonian, $\hat{H}^0$, and a total Hamiltonian $\hat{H} = \hat{H}^0 + \hat{H}'$. Denoting the quantities in the Schrödinger picture by a subscript S, those in the Heisenberg picture by a subscript H, and those in the interaction picture by a subscript I, we have the following relations for the wave functions: $|\Psi_{\mathrm{H}}(t)\rangle = \mathrm{e}^{\mathrm{i}\hat{H}t}|\Psi_{\mathrm{S}}(t)\rangle$ and $|\Psi_{\mathrm{I}}(t)\rangle = \mathrm{e}^{\mathrm{i}\hat{H}_0 t}|\Psi_{\mathrm{S}}(t)\rangle$.

[3] In the three pictures the following relations hold for operators: $\hat{O}_{\mathrm{H}}(t) = \mathrm{e}^{\mathrm{i}\hat{H}t}\hat{O}_{\mathrm{S}}\mathrm{e}^{-\mathrm{i}\hat{H}t}$ and $\hat{O}_{\mathrm{I}}(t) = \mathrm{e}^{\mathrm{i}\hat{H}^0 t}\hat{O}_{\mathrm{S}}\mathrm{e}^{-\mathrm{i}\hat{H}^0 t}$. We note that the three pictures coincide at $t = 0$.

where the bar over Ψ denotes the wave function of the perturbed system.

In analogy with Eq. (19.2), we attempt a solution in the form

$$|\,\bar{\Psi}_{\rm S}(t)\,\rangle = {\rm e}^{-{\rm i}\hat{H}t}\hat{A}(t)\,|\,\Psi_{\rm S}(0)\,\rangle\,, \tag{19.4}$$

where the operator $\hat{A}(t)$ satisfies the causal initial condition,

$$\hat{A}(t) = 1, \quad \text{for } t \le t_0. \tag{19.5}$$

Equations (19.3) and (19.4) yield an equation for $\hat{A}(t)$,

$$\begin{aligned} {\rm i}\frac{\partial \hat{A}(t)}{\partial t} &= {\rm e}^{{\rm i}\hat{H}t}\hat{H}^{\rm ext}(t){\rm e}^{-{\rm i}\hat{H}t}\hat{A}(t) \\ &= \hat{H}_{\rm I}^{\rm ext}(t)\hat{A}(t), \end{aligned} \tag{19.6}$$

where

$$\hat{H}_{\rm I}^{\rm ext}(t) = {\rm e}^{{\rm i}\hat{H}t}\hat{H}^{\rm ext}(t){\rm e}^{-{\rm i}\hat{H}t} \tag{19.7}$$

is the interaction picture representation[4] of $\hat{H}^{\rm ext}$.

One way to solve Eq. (19.6) is by means of an iteration in terms of $\hat{H}_{\rm I}^{\rm ext}(t)$. Thus, for times $t \ge t_0$, we can write

$$\hat{A}(t) = 1 - {\rm i}\int_{t_0}^{t} {\rm d}t'\,\hat{H}_{\rm I}^{\rm ext}(t') + \cdots, \tag{19.8}$$

which satisfies the initial condition, Eq. (19.5), since $\hat{H}_{\rm I}^{\rm ext}(t) = 0$ for $t \le t_0$. The state vector, $|\,\bar{\Psi}(t)\,\rangle$, in Eq. (19.4) is now given by the expression,

$$|\,\bar{\Psi}_{\rm S}(t)\,\rangle = {\rm e}^{{\rm i}\hat{H}t}\,|\,\Psi_{\rm S}(0)\,\rangle - {\rm i}\int_{t_0}^{t} {\rm d}t\hat{H}_{\rm I}^{\rm ext}(t')\,|\,\Psi_{\rm S}(0)\,\rangle + \cdots. \tag{19.9}$$

Let us now consider an observable represented by the operator, $\hat{O}_{\rm S}(t)$. All physical information corresponding to $\hat{O}_{\rm S}(t)$ is contained in the general matrix elements,

$$\begin{aligned} \left\langle \hat{O}(t) \right\rangle_{\rm ext} &= \left\langle \bar{\Psi}'_{\rm S}(t) \middle| \hat{O}_{\rm S} \middle| \bar{\Psi}_{\rm S}(t) \right\rangle \\ &= \langle\, \Psi'_{\rm S}(0)\,| \left[1 + {\rm i}\int_{t_0}^{t} {\rm d}t'\,\hat{H}_{\rm I}^{\rm ext}(t') + \cdots\right] {\rm e}^{{\rm i}\hat{H}t}\hat{O}_{\rm S}(t){\rm e}^{-{\rm i}\hat{H}t} \\ &\times \left[1 - {\rm i}\int_{t_0}^{t} {\rm d}t'\,\hat{H}_{\rm I}^{\rm ext}(t') + \cdots\right] |\,\Psi_{\rm S}(0)\,\rangle \\ &= \left\langle \Psi'_{\rm I}(0) \middle| \hat{O}_{\rm I}(t) \middle| \Psi_{\rm I}(0) \right\rangle \\ &+ {\rm i}\left\langle \Psi'_{\rm I}(0) \middle| \int_{t_0}^{t} {\rm d}t' \left[\hat{H}_{\rm I}^{\rm ext}(t'), \hat{O}_{\rm I}(t)\right] \middle| \Psi_{\rm I}(0) \right\rangle + \cdots. \end{aligned} \tag{19.10}$$

[4] We consider $\hat{H}^{\rm ext}$ as being the perturbation imposed on an otherwise unperturbed system described by $\hat{H}$.

In writing the last expression, we have used the fact that the wave functions in the Schrödinger, Heisenberg, and interaction pictures coincide at $t = 0$.

In the last expression, we have retained only the terms linear in $\hat{H}^{\text{ext}}$, and have defined the interaction picture form of the wave function,

$$|\,\Psi_{\text{I}}(t)\,\rangle = \text{e}^{\text{i}\hat{H}t}\,|\,\Psi_{\text{S}}(t)\,\rangle\,. \tag{19.11}$$

If we let $|\,\Psi_{\text{S}}(t)\,\rangle$ and $|\,\Psi'_{\text{S}}(t)\,\rangle$ both denote the normalized ground state, $|\,\Psi_0\,\rangle$, the linear response of the ground-state expectation value of an operator is given by the first-order change in the matrix element,

$$\begin{aligned} \delta\langle\,\hat{O}(t)\,\rangle &\equiv \langle\,\hat{O}(t)\,\rangle_{\text{ext}} - \langle\,\hat{O}(t)\,\rangle \\ &= \text{i}\int_{t_0}^{t}\text{d}t'\,\langle\,\Psi_0\left|\left[\hat{H}_{\text{I}}^{\text{ext}}(t'),\hat{O}_{\text{I}}(t)\right]\right|\Psi_0\,\rangle\,. \end{aligned} \tag{19.12}$$

This is the basic equation of linear response theory. It involves the difference in the matrix elements of an operator between the externally perturbed and the unperturbed states of the interaction picture representation for a total Hamiltonian, $\hat{H} + \hat{H}^{\text{ext}}$. We note that it is often the case for the quantity $\langle\,\hat{O}(t)\,\rangle$ to vanish, as is for example true when the operator $\hat{O}$ represents the current in the system. In this case, the usual ground state of a system is an equilibrium state and cannot support any current.

Equation (19.12) is the starting point for the discussion of transport properties within linear response theory. The Kubo formulae for particular transport properties, e.g., the electrical conductivity, can be obtained through a proper identification of $\hat{H}^{\text{ext}}$ and the operator $\hat{O}_{\text{I}}(t)$ with relevant quantities of the system at hand.

19.3 Kubo Formula for the Conductivity

In order to apply the formalism of linear response theory, Eq. (19.12), to the treatment of electrical conduction in a solid, we identify $\hat{H}^{\text{ext}}$ with the Hamiltonian associated with an externally applied electric field, and $\hat{O}$ with the operator representing the electric current in the system. In proceeding further, we first recall a few basic relations between fields and currents.

19.3.1 Relations between fields and currents

By definition, the electric current is the summation, over all particles, of the product of particle velocity times the corresponding charge,

$$\begin{aligned} \mathbf{j}(\mathbf{r}) &= \frac{1}{2}\sum_i \text{e}_i\left[\mathbf{v}_i(\mathbf{r})\delta(\mathbf{r}-\mathbf{r}') + \delta(\mathbf{r}-\mathbf{r}')\mathbf{v}_i(\mathbf{r})\right] \\ &= \frac{1}{2}\sum_i \frac{\text{e}_i}{m_i}\left[\mathbf{p}_i(\mathbf{r})\delta(\mathbf{r}-\mathbf{r}') + \delta(\mathbf{r}-\mathbf{r}')\mathbf{p}_i(\mathbf{r})\right]. \end{aligned} \tag{19.13}$$

This expression is valid in the absence of any external potentials and fields. Therefore, in this form, it cannot be used to define the conductivity of a material which, within linear response, is the proportionality tensor between the current *induced* by an external field and the *total* field in the system. To clarify the concepts involved in defining the electronic conductivity of a material, we consider an external field of the form,

$$\mathbf{E}^{\text{ext}}(\mathbf{r},t) = \mathbf{E}_0^{\text{ext}} e^{i(\mathbf{q}\cdot\mathbf{r}-\omega t)}. \tag{19.14}$$

Linear response leads to an induced current which is proportional to the external field,

$$\mathbf{J}(\mathbf{r},t) = \overleftrightarrow{\sigma}'(\underline{f}q,\omega) \cdot \mathbf{E}^{\text{ext}}(\mathbf{r},t). \tag{19.15}$$

However, the quantity $\overleftrightarrow{\sigma}'(\mathbf{q},\omega)$ is not the conductivity we seek. Rather, we are after an expression between $\mathbf{J}(\mathbf{r},t)$ and the total field, which in analogy with Eq. (19.14) we take to be of the form,

$$\mathbf{E}(\mathbf{r},t) = \mathbf{E}_0 e^{i(\mathbf{q}\cdot\mathbf{r}-\omega t)}. \tag{19.16}$$

The conductivity is defined by means of the relation,

$$\mathbf{J}(\mathbf{r},t) = \overleftrightarrow{\sigma}(\mathbf{q},\omega) \cdot \mathbf{E}(\mathbf{r},t), \tag{19.17}$$

an expression which bears further scrutiny.

First, we note that the current operator and the field are evaluated at the same point in space, a feature that characterizes homogeneous materials. In a non-homogeneous but periodic solid, Eq. (19.17) must be generalized to read

$$\mathbf{J}(\mathbf{r},t) = \int d^3r' \int_{-\infty}^{t} dt' \overleftrightarrow{\sigma}(\mathbf{r}-\mathbf{r}';t-t') \cdot \mathbf{E}(\mathbf{r}',t'). \tag{19.18}$$

With $\mathbf{q}$ being a vector in the Brillouin zone of the reciprocal lattice, Eq. (19.18) is just the Fourier transform of Eq. (19.17). However, even in periodic systems, the dependence of the conductivity only on the distance $\mathbf{r}-\mathbf{r}'$ may not be valid at very small distances, say within atomic core states, or even between atomic positions. A general and rigorous treatment requires that the conductivity be a function of $\mathbf{r}$ and $\mathbf{r}'$ independently,

$$\mathbf{J}(\mathbf{r},t) = \int d^3r' \int_{-\infty}^{t} dt' \overleftrightarrow{\sigma}(\mathbf{r},\mathbf{r}';t-t') \cdot \mathbf{E}(\mathbf{r}',t'). \tag{19.19}$$

In calculating the electrical conductivity of a periodic solid, one can use Eq. (19.18) provided that the current is averaged over many unit cells. Alternatively, Eq. (19.17) is valid in the long wave length limit, $\mathbf{q} \to 0$.

Second, it is important to examine the behavior of the conductivity tensor in the limits $\mathbf{q} \to 0$ followed by $\omega \to 0$. This is the limit that leads to the dc conductivity, a quantity that is often of great interest. In obtaining the dc conductivity, one must take the limits in the order just stated, or incorrect results can be obtained. Reversing the order, $\omega \to 0$ followed by $\mathbf{q} \to 0$, implies the presence of a static electric field, which in turn can lead to a static rearrangement of charge that balances the external field so that the current vanishes in the steady state.

We now turn our attention to the total electric field appearing in Eq. (19.17). This field is induced by and hence is related to the external field, $\mathbf{E}^{\text{ext}}(\mathbf{r}, t)$, but the exact nature of that relation is not of particular relevance here[5]. The relation that we need is that which connects [5] the electric field, $\mathbf{E}$, to a scalar potential, $V(\mathbf{r})$, and a vector potential, $\mathbf{A}(\mathbf{r}, t)$,

$$\mathbf{E}(\mathbf{r}, t) = -\nabla V(\mathbf{r}) - \frac{1}{\mathrm{c}} \frac{\partial \mathbf{A}(\mathbf{r}, t)}{\partial t}. \tag{19.20}$$

It is often convenient to work in the Coulomb gauge in which

$$\nabla \cdot \mathbf{A} = 0. \tag{19.21}$$

This condition implies that the commutator of $\mathbf{A}$ and the particle momentum vanishes[6],

$$[\mathbf{A}, \mathbf{p}] = \mathbf{A} \cdot \mathbf{p} - \mathbf{p} \cdot \mathbf{A} = 0. \tag{19.22}$$

In the Coulomb gauge, the scalar potential is given by the solution of the Poisson equation,

$$\nabla^2 V(\mathbf{r}) = -4\pi\rho(\mathbf{r}), \tag{19.23}$$

or

$$V(\mathbf{r}) = \int \mathrm{d}^3 r \frac{\rho(\mathbf{r}')}{|\mathbf{r} - \mathbf{r}'|}. \tag{19.24}$$

In this gauge, the vector potential is obtained as the solution of the equation [2, 5],

$$\nabla^2 \mathbf{A} - \frac{1}{\mathrm{c}^2} \frac{\partial^2 \mathbf{A}(\mathbf{r}, t)}{\partial t^2} = \frac{4\pi}{\mathrm{c}} \nabla \times \left[\nabla \times \int \mathrm{d}^3 r' \frac{\mathbf{j}(\mathbf{r}')}{|\mathbf{r} - \mathbf{r}'|} \right]. \tag{19.25}$$

This expression shows that the vector potential is a transverse quantity, i.e., it responds only to the transverse part of the current, a feature that is consistent with Eq. (19.22). Consequently, our discussion of transport is

[5] For example, in the case of a static, longitudinal electric field, the two fields are related by the dielectric vector, $\mathbf{D}$.

[6] To see this, we consider the operator nature of the vector potential. Then, because of the Coulomb gauge, we have $\int \psi^*(\mathbf{r})[\nabla \cdot \mathbf{A}\psi(\mathbf{r})]\mathrm{d}^3 r = \int \psi^*(\mathbf{r})[\nabla \cdot \mathbf{A}]\psi(\mathbf{r})\mathrm{d}^3 r + \int \psi^*(\mathbf{r})\mathbf{A} \cdot [\nabla\psi(\mathbf{r})]\mathrm{d}^3 r = \int \psi^*(\mathbf{r})[\mathbf{A} \cdot \nabla]\psi(\mathbf{r})\mathrm{d}^3 r$.

applicable to transverse quantities. Alternatively, the longitudinal part of the current does not affect the vector potential. Here, the terms transverse and longitudinal are to be understood in terms of the direction of motion of the charge giving rise to the current, at least in the case of homogeneous media. In heterogeneous materials, the meaning of these terms is not always straightforward, even for cubic (periodic) systems. We confine our attention to systems that are homogeneous, at least in an average sense, and we write,

$$\nabla^2 \mathbf{A} - \frac{1}{c^2}\frac{\partial \mathbf{A}(\mathbf{r},t)}{\partial t} = -\frac{4\pi}{c}\mathbf{j}_t, \tag{19.26}$$

where $\mathbf{j}_t$ is defined by the expression on the right-hand side of Eq. (19.25), and the subscript t denotes the transverse component of the current. In the same gauge, we also have [5],

$$\nabla\frac{\partial V}{\partial t} = 4\pi\mathbf{j}_l, \tag{19.27}$$

with $\mathbf{j}_l$ denoting the longitudinal component of the current.

The Coulomb (or transverse gauge) lends itself conveniently to calculations in the case in which no potentials are present, so that $V(\mathbf{r}) = 0$. In this case, we have

$$\mathbf{E} = -\frac{1}{c}\frac{\partial \mathbf{A}}{\partial t}. \tag{19.28}$$

Using Eq. (19.16), and integrating the last expression over the time variable, we find[7]

$$\frac{1}{c}\mathbf{A}(\mathbf{r},t) = \frac{-i}{\omega}\mathbf{E}(\mathbf{r},t). \tag{19.29}$$

We are now in a position to derive an expression for the Hamiltonian, $\hat{H}^{\text{ext}}$, associated with an external electric field. Because we are interested in microscopic, induced quantities, such as the total induced current in a material, it is convenient to express $\hat{H}^{\text{ext}}$ in terms of the vector potential, $\mathbf{A}(\mathbf{r},t)$.

19.3.2 External Hamiltonian

Let us consider the Hamiltonian of a single electron of mass m and momentum $\mathbf{p}$, moving in an external scalar potential, $V(\mathbf{r})$,

$$\hat{H}^0 = \frac{p^2}{2m} + V(\mathbf{r}). \tag{19.30}$$

[7]The integral is straightforward provided the limit $t = -\infty$ is assumed to vanish. This is consistent with the absence of fields at the remote past. Formally, one way of guaranteeing this behavior is to assign to ω an infinitesimal positive imaginary part that is set to zero after the limit $t \to -\infty$ has been taken.

In the presence of a vector potential, $\mathbf{A}(\mathbf{r})$, and in the Coulomb gauge, we have (recall that $\mathbf{p}$ and $\mathbf{A}$ commute)

$$\begin{aligned} \hat{H} &= \frac{\left(\mathbf{p}-\frac{e}{c}\mathbf{A}\right)^2}{2m}+V(\mathbf{r}) \\ &= \hat{H}^0-\frac{e}{mc}\mathbf{A}\cdot\mathbf{p}+\frac{e^2}{2m}A^2 \\ &= \hat{H}^0-\mathbf{A}\cdot\left[\frac{e}{mc}\mathbf{p}-\frac{e^2}{2m}\mathbf{A}\right]. \end{aligned} \tag{19.31}$$

Our treatment of a many-electron system is given in terms of an effective single-particle picture in which each electron is taken to move in a field set up by the nuclei and the other electrons in the system. Correspondingly, the single-electron Hamiltonian associated with an external electric field can be written in the form,

$$\hat{H}^{\mathrm{ext}}=-\frac{e}{mc}\sum_i \mathbf{A}(\mathbf{r}_i)\cdot\mathbf{p}_i, \tag{19.32}$$

where the summation is over all particles in the system, and only terms linear in the vector potential have been retained. We can take $\hat{H}^0$ to include the effective single-particle potential acting on an electron in the absence of an external field. A convenient form for $\hat{H}^{\mathrm{ext}}$ can be obtained as the integral,

$$\hat{H}^{\mathrm{ext}}=-\frac{1}{c}\int d^3r\mathbf{A}(\mathbf{r})\cdot\mathbf{j}(\mathbf{r}). \tag{19.33}$$

Integrating over $\mathbf{r}$, we obtain an expression in terms of the Fourier component of the current operator. In view of Eqs. (19.16) and (19.13), we find

$$\hat{H}^{\mathrm{ext}}=\frac{i}{\omega}\mathbf{j}(\mathbf{q})\cdot\mathbf{E}_0e^{-i\omega t}, \tag{19.34}$$

where

$$\mathbf{j}(\mathbf{q})=\frac{1}{2m}\sum_i e_i\left(\mathbf{p}_ie^{i\mathbf{q}\cdot\mathbf{r}_i}+e^{i\mathbf{q}\cdot\mathbf{r}_i}\mathbf{p}_i\right). \tag{19.35}$$

It is important to recall the distinction between the current operator $\mathbf{j}$, which enters the Hamiltonian in Eq. (19.34), and the induced current, $\mathbf{J}$, which is what is measured in an experiment. Conductivity, a measured quantity, must be defined with respect to this induced current. To do so, we need to establish the connection between the two kinds of current operators, which we can accomplish along the following lines. We note that $\mathbf{J}$, the measured, average current, is given by a summation over all particle velocities in the presence of an external field, divided by the volume, Ω, of the material,

$$\begin{aligned} \mathbf{J}(\mathbf{r},t) &= \mathrm{e}\sum_i \mathbf{v}_i \delta(\mathbf{r}-\mathbf{r}_i) \\ &= \frac{\mathrm{e}}{\Omega}\sum_i \langle \mathbf{v}_i \rangle . \end{aligned} \tag{19.36}$$

Now, the velocity in the presence of a vector field is given by

$$\mathbf{v}_i = \frac{1}{m}\left[\mathbf{p}_i - \frac{\mathrm{e}}{\mathrm{c}}\mathbf{A}(\mathbf{r}_i,t)\right], \tag{19.37}$$

so that

$$\mathbf{J}(\mathbf{r},t) = \frac{\mathrm{e}}{m\Omega}\sum_i \langle \mathbf{p}_i \rangle - \frac{\mathrm{e}^2}{mc\Omega}\sum_i \mathbf{A}(\mathbf{r}_i,t). \tag{19.38}$$

Next, in view of Eq. (19.29), we can write

$$\mathbf{J}(\mathbf{r},t) = \langle \mathbf{j}(\mathbf{r}) \rangle + \mathrm{i}\frac{n_0 \mathrm{e}^2}{m\omega}\mathbf{E}(\mathbf{r},t), \tag{19.39}$$

where in the second term we have replaced the summation over particles divided by the volume with the particle density, n_0.

We see that the measured current contains two terms, one proportional to the electric field, and one given by the expectation value of the current operator, $\langle \mathbf{j}(\mathbf{r}) \rangle$. The term proportional to $\mathbf{E}$ gives rise to a contribution to the conductivity of the form,

$$\overleftrightarrow{\sigma}^1(\mathbf{q},\omega) = \frac{n_0 \mathrm{e}^2}{m\omega}\overleftrightarrow{1}, \tag{19.40}$$

where $\overleftrightarrow{1}$ is the three-dimensional unit matrix. We now show that within linear response, the average of the current operator, $\langle \mathbf{j}(\mathbf{r}) \rangle$, is also proportional to the electric field, and that the constant of proportionality is given by the Kubo formula.

19.3.3 Kubo formula for transverse currents

We use Eq. (19.12) with the identification $\hat{O} \equiv \mathbf{J}$ and with $\hat{H}^{\mathrm{ext}}$ as given by Eq. (19.33), [or Eq. (19.34)]. We also note that no current flows in the ground state of the system in the absence of an external electric field so that[8]

$$\langle \Psi_0 | \mathbf{j} | \Psi_0 \rangle = 0. \tag{19.41}$$

[8]The state $|\Psi_0\rangle$ can represent either the true, many-particle ground state or an approximation to it, such as a properly antisymmetrized product of single-particle states. In any case, the condition of vanishing current must be taken to hold.

This implies that in Eq. (19.12) the quantity $\langle \hat{O}(t) \rangle$ can be set equal to zero, and we can write

$$\begin{aligned}\langle \mathrm{j}_\alpha(\mathbf{r},t)\rangle &\equiv J'_\alpha(\mathbf{r},t) \\ &= -\mathrm{i}\int_{-\infty}^{t} \mathrm{d}t' \langle \Psi_0 \left| \left[j_\alpha(\mathbf{r},t), \hat{H}^{\text{ext}}(\mathbf{r},t') \right] \right| \Psi_0 \rangle, \end{aligned} \quad (19.42)$$

where α denotes one of the space coordinates, x, y, or z. To derive an expression for the conductivity, we must be able to remove a factor of $\mathbf{E}(\mathbf{r},t)$ from the right-hand side of this expression. With $\hat{H}^{\text{ext}}$ given by Eq. (19.35), and with the definitions in the interaction picture

$$\hat{H}^{\text{ext}}(t) = \mathrm{e}^{\mathrm{i}\hat{H}t} \hat{H}^{\text{ext}} \mathrm{e}^{-\mathrm{i}\hat{H}t}, \quad (19.43)$$

and

$$\mathrm{j}(t) = \mathrm{e}^{\mathrm{i}\hat{H}t} \mathrm{j} \mathrm{e}^{-\mathrm{i}\hat{H}t}, \quad (19.44)$$

we have

$$\begin{aligned}\left[j_\alpha(\mathbf{r},t), \hat{H}^{\text{ext}}(\mathbf{r},t') \right] &= \frac{\mathrm{i}}{\omega} \sum_\beta E_{0\beta} \mathrm{e}^{-\mathrm{i}\omega t} \left[j_\alpha(\mathbf{r},t), j_\beta(\mathbf{q},t') \right] \\ = \quad \frac{\mathrm{i}}{\omega} \sum_\beta E_{0\beta}(\mathbf{r},t) \mathrm{e}^{\mathrm{i}\mathbf{q}\cdot\mathbf{r}} \mathrm{e}^{\mathrm{i}\omega(t-t')} &\left[j_\alpha(\mathbf{r},t), j_\beta(\mathbf{q},t') \right]. \end{aligned} \quad (19.45)$$

This can be written in the form,

$$\begin{aligned} J'_\alpha(\mathbf{r},t) &= \frac{1}{\omega} \sum_\beta E_\beta(\mathbf{r},t) \mathrm{e}^{-\mathrm{i}\mathbf{q}\cdot\mathbf{r}} \\ &\times \int_{-\infty}^{t} \mathrm{d}t' \mathrm{e}^{\mathrm{i}\omega(t-t')} \langle \Psi_0 \left| \left[j_\alpha(\mathbf{r},t), j_\beta(\mathbf{q},t') \right] \right| \Psi_0 \rangle. \end{aligned} \quad (19.46)$$

This expression clearly indicates the proportionality of $\mathbf{J}'$ to the electric field. Extracting the proportionality coefficient and including the term that directly comes from the presence of the field, Eq. (19.40), we obtain the important preliminary result for the conductivity,

$$\begin{aligned} \sigma_{\alpha\beta}(\mathbf{q},\omega) &= \frac{1}{\omega} \mathrm{e}^{\mathrm{i}\mathbf{q}\cdot\mathbf{r}} \int_{-\infty}^{t} \mathrm{d}t' \mathrm{e}^{\mathrm{i}\omega(t-t')} \langle \Psi_0 \left| \left[j_\alpha(\mathbf{r},t), j_\beta(\mathbf{q},t') \right] \right| \Psi_0 \rangle \\ &+ \frac{n_0 \mathrm{e}^2}{m\omega} \mathrm{i}\delta_{\alpha\beta}. \end{aligned} \quad (19.47)$$

To complete the derivation of the Kubo formula, we integrate the last expression over $\mathbf{r}$ and divide by the volume in order to eliminate (average over) small-distance fluctuations. Using the result,

$$\frac{1}{\Omega} \int \mathrm{d}^3 r \mathrm{e}^{-\mathrm{i}\mathbf{q}\cdot\mathbf{r}} j_\alpha(\mathbf{r},t) = j_\alpha(-\mathbf{q},t) = j_\alpha^\dagger(\mathbf{q},t), \quad (19.48)$$

we arrive at the expression

$$\begin{aligned}\sigma_{\alpha\beta}(\mathbf{q},\omega) &= \frac{1}{\omega}\int_{-\infty}^{t} dt' e^{i\omega(t-t')}\langle\Psi_0 \left| \left[j_\alpha^\dagger(\mathbf{q},t), j_\beta(\mathbf{q},t')\right] \right| \Psi_0\rangle \\ &+ \frac{n_0 e^2}{m\omega} i\delta_{\alpha\beta}. \end{aligned} \tag{19.49}$$

This is the celebrated Kubo formula for the electronic conductivity of a material[9].

19.4 Green-Function Expressions

In this section, we derive expressions for the Kubo formula in terms of the single-particle electronic Green function of a system. This provides a direct way for computing the conductivity[10] of a material through the performance of electronic structure calculations. The following discussion follows closely the lines suggested by MacLaren [6].

We set $\mathbf{q} = 0$ in Eq. (19.49), and for ease of formal presentation, also set $t = 0$. When $\mathbf{q} = 0$, we have $j_\alpha^\dagger = j_\alpha$ and we can write

$$\begin{aligned}\sigma_{\alpha\beta}(\omega) &= \frac{1}{\omega}\int_{-\infty}^{0} dt' e^{-i\omega t'}\langle\Psi_0 \left| \left[j_\alpha^\dagger j_\beta(t') - j_\beta(t') j_\alpha\right] \right| \Psi_0\rangle \\ &+ \frac{n_0 e^2}{m\omega} i\delta_{\alpha\beta}. \end{aligned} \tag{19.50}$$

We concentrate on the first term[11] on the right side of the equals sign, denoting it by $\sigma'_{\alpha\beta}(\omega)$. Interpreting the ground-state expectation value appearing in Eq. (19.50) at $T = 0$ as the trace over all occupied states, $|n\rangle$, (with energies below the Fermi energy), and inserting a *complete* set of states between products of operators and summing over them, we obtain the expression,

$$\begin{aligned}\sigma'_{\alpha\beta}(\omega) &= \frac{1}{\omega}\int_{-\infty}^{0} dt' e^{-i\omega t'} \sum_{n_<,m} \left[\langle n|j_\alpha|m\rangle\langle m|j_\beta(t')|n\rangle\right. \\ &- \left.\langle n|j_\beta(t')|m\rangle\langle m|j_\alpha|n\rangle\right], \end{aligned} \tag{19.51}$$

where $n_<$ denotes a restriction to occupied states. We now use the interaction-picture representation of $j_\alpha(t)$, Eq. (19.44), and the fact that the states $|n\rangle$

[9] It is customary to set $t = 0$ in applying this formula.

[10] Conductivity expressions based on the two-particle Green function are given in Chapter 24.

[11] The second term will often be omitted, even without warning, when its omission will not cause confusion or affect formal developments.

and $|m\rangle$ are eigenstates of the system Hamiltonian, $\hat{H}$, to write,

$$\begin{aligned}\sigma'_{\alpha\beta}(\omega) &= \frac{1}{\omega}\int_{-\infty}^{0} \mathrm{d}t' \mathrm{e}^{-\mathrm{i}\omega(t')} \sum_{n_<,m} \Big[\langle n|j_\alpha|m\rangle\langle m|j_\beta(t')|n\rangle \mathrm{e}^{\mathrm{i}(E_m-E_n)t'} \\ &- \langle n|j_\beta(t')|m\rangle\langle m|j_\alpha|n\rangle \mathrm{e}^{\mathrm{i}(E_n-E_m)t'}\Big]. \end{aligned} \tag{19.52}$$

In performing the integrals over t', it is convenient to provide ω with a positive infinitesimal imaginary part, iη, and take the limit $\eta \to 0$ after the integration is completed. This procedure guarantees that the definite integral is well-behaved at $t = -\infty$, and yields the result

$$\sigma'_{\alpha\beta}(\omega) = \frac{\mathrm{i}}{\omega}\lim_{\eta\to 0}\sum_{n<,m}\left[\frac{\langle n|j_\alpha|m\rangle\langle m|j_\beta|n\rangle}{(\omega+E_n-E_m+\mathrm{i}\eta)} - \frac{\langle n|j_\beta|m\rangle\langle m|j_\alpha|n\rangle}{(\omega+E_m-E_n+\mathrm{i}\eta)}\right]. \tag{19.53}$$

Now, we recall that the current operator has no matrix elements between occupied states[12], Eq. (19.41), and cast the last expression in the form,

$$\sigma'_{\alpha\beta}(\omega) = \frac{\mathrm{i}}{\omega}\lim_{\eta\to 0}\sum_{n_<,m_>}\left[\frac{\langle n|j_\alpha|m\rangle\langle m|j_\beta|n\rangle}{(\omega+E_n-E_m+\mathrm{i}\eta)} - \frac{\langle n|j_\beta|m\rangle\langle m|j_\alpha|n\rangle}{(\omega+E_m-E_n+\mathrm{i}\eta)}\right], \tag{19.54}$$

where the summation index m runs only over unoccupied states. Suppressing the limit and interchanging the dummy summation indices, n and m, we have[13]

$$\sigma'_{\alpha\beta}(\omega) = \frac{\mathrm{i}}{\omega}\left[\sum_{n_<,m}\frac{\langle n|j_\alpha|m\rangle\langle m|j_\beta|n\rangle}{(\omega+E_n-E_m+\mathrm{i}\eta)} - \sum_{m_<,n}\frac{\langle n|j_\alpha|m\rangle\langle m|j_\beta|n\rangle}{(\omega+E_n-E_m+\mathrm{i}\eta)}\right]. \tag{19.55}$$

This result can now be expressed in terms of the single-particle Green functions of the fully interacting (but unperturbed) system[14]. First, we note that we can write

$$\frac{1}{(\omega+E_n-E_m+2\mathrm{i}\gamma)} = \frac{1}{2\pi\mathrm{i}}\oint_C \frac{\mathrm{d}E}{(E-E_n-\mathrm{i}\gamma)\,(\omega+E-E_m+\mathrm{i}\gamma)}, \tag{19.56}$$

where the contour of integration, C, extends from $-\infty$ to the Fermi level and doubles back, as indicated by the solid line in Fig. 19.1. Thus, the first

[12]We take this to be true in any approximation used to represent the wave function or Green function of the system in its ground state.

[13]Note that even though the summands now become identical, the expression does not vanish since the indices in the two sums extend over different sets of states.

[14]At this point, we introduce an approximation in describing the single-particle Green function of an interacting system in terms of the Green function of a single particle moving in an effective external field. This description is useful in deriving computable expressions within the LDA of DFT.

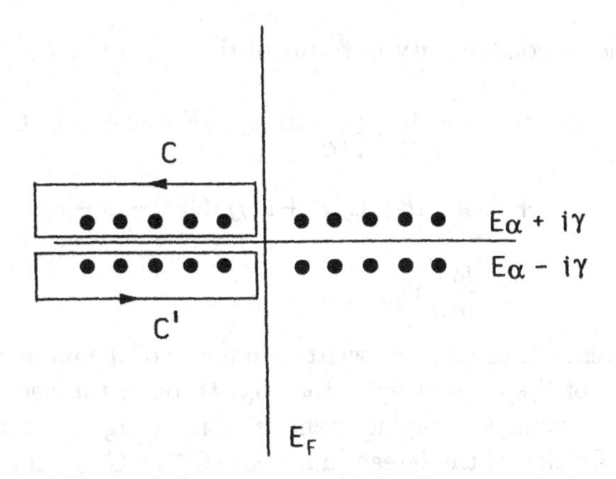

Figure 19.1: **Contours of energy integrations.**

term inside the brackets in Eq. (19.55) can be written in the form,

$$\sum_{n_<,m} \frac{\langle n|j_\alpha|m\rangle\langle m|j_\beta|n\rangle}{(\omega + E_n - E_m + \mathrm{i}\eta)} = \frac{1}{2\pi \mathrm{i}} \oint_{C_1} \mathrm{d}E \sum_{n,m} \frac{\langle n|j_\alpha|m\rangle\langle m|j_\beta|n\rangle}{(E - E_n - \mathrm{i}\gamma)(\omega + E - E_m + \mathrm{i}\gamma)}, \tag{19.57}$$

where the sums over n and m are now unrestricted since the domain of integration keeps track of the sets of states involved in the calculation. Interchanging sums and integrals in the last expression, and using the spectral representation for the Green function, see Eq. (2.17),

$$G(E + \mathrm{i}\eta) = \sum_n \frac{|n\rangle\langle n|}{E - E_n + \mathrm{i}\eta}, \tag{19.58}$$

we obtain

$$\sum_{n_<,m} \frac{j_\alpha|m\rangle\langle m|j_\beta|n\rangle\langle n|}{(\omega + E_n - E_m + \mathrm{i}\eta)} = \frac{1}{2\pi \mathrm{i}} \oint_C \mathrm{d}E j_\alpha G(E + \omega + \mathrm{i}\gamma) j_\beta G(E - \mathrm{i}\gamma). \tag{19.59}$$

Similarly, using contour C' in Fig. 19.1, we can write the second term inside the brackets in Eq. (19.55) in the form

$$\begin{aligned}\sum_{n,m_<} \frac{j_\alpha|m\rangle\langle m|j_\beta|n\rangle\langle n|}{(\omega + E_n - E_m + \mathrm{i}\eta)} &= \frac{1}{2\pi \mathrm{i}} \oint_{C'} \mathrm{d}E \sum_{n,m} \frac{j_\alpha|m\rangle\langle m|j_\beta|n\rangle\langle n|}{(-E + \omega + E_n + \mathrm{i}\gamma)(E - E_m + \mathrm{i}\gamma)} \\ &= -\frac{1}{2\pi \mathrm{i}} \oint_{C'} \mathrm{d}E j_\alpha G(E + \mathrm{i}\gamma) j_\beta G(E - \omega - \mathrm{i}\gamma).\end{aligned} \tag{19.60}$$

Combining the last two expressions with Eq. (19.58), we obtain an expression for the ac conductivity in terms of the single-particle Green function

$$\begin{aligned}\sigma_{\alpha\beta}(\omega) &= \frac{1}{2\pi\omega}\lim_{\gamma\to 0}\left[\oint_{C_1} \mathrm{d}E j_\alpha G(E+\omega+\mathrm{i}\gamma) j_\beta G(E-\mathrm{i}\gamma)\right. \\ &+ \left.\oint_{C'} \mathrm{d}E j_\alpha G(E+\mathrm{i}\gamma) j_\beta G(E-\omega-\mathrm{i}\gamma)\right] \\ &+ \frac{n_0 \mathrm{e}^2}{m\omega}\mathrm{i}\delta_{\alpha\beta}. \end{aligned} \quad (19.61)$$

These expressions can also be written in terms of the momentum operators, p_α, instead of the current operators, j_α, through the use of the relations, $j_\alpha = \frac{\mathrm{e}}{m}p_\alpha$. Making this replacement, incorporating the limiting procedures into the definition of the Green functions (G^+ or G^-), and integrating over the identity in the form $\int \mathrm{d}^3r|\mathbf{r}\rangle\langle\mathbf{r}|$ around the Green functions, we obtain the real-space ($\mathbf{r}$) representation of the ac conductivity at $T = 0^0 K$,

$$\begin{aligned}\sigma_{\alpha\beta}(\omega) &= -\frac{\mathrm{e}}{m\pi\omega}\int\int \mathrm{d}^3r\mathrm{d}^3r' \\ &\left[\oint_C \mathrm{d}E\nabla_\alpha G^+(\mathbf{r},\mathbf{r}';E+\omega)\nabla_\beta G^-(\mathbf{r}',\mathbf{r};E)\right. \\ &+ \left.\oint_{C'} \mathrm{d}E\nabla_\alpha G^+(\mathbf{r},\mathbf{r}';E)\nabla_\beta G^-(\mathbf{r}',\mathbf{r};E-\omega)\right] + \frac{n_0\mathrm{e}^2}{m\omega}\mathrm{i}\delta_{\alpha\beta}, \end{aligned} \quad (19.62)$$

where a factor of 2 has been included to account for the spin of the electrons. In performing the energy integrations indicated in this expression, it is usually convenient to choose semicircular or triangular contours into the complex plane whose base on the real energy axis extends from just below the lower part of the valence band in a metal to the Fermi energy. The Green functions occurring in this expression can be calculated using the methods discussed in previous chapters leading to computable formulae for the conductivity. Further illustration of the use of this result is given in the next chapter.

The expressions derived thus far are general results of linear response theory, relating the response of a system to an externally applied perturbation. Clearly, these are complex quantities. The conductivity measured in an experiment, however, is a real quantity, since real fields induce real currents. As is evident from the last expression above, such measured quantities are obtained as the imaginary part of the Kubo formula.

We now wish to generalize the results derived so far to the calculation of general transport coefficients. In the process, we will also indicate the connection of these general results to measured quantities. To do this, we consider the *complex admittance* characterizing the response of a system to an external perturbation.

19.5 Complex Admittance

The formalism developed below contains the explicit presence of the Fermi-Dirac distribution function, $f(E)$, at finite temperature. The limit $T = 0$ can be obtained through the corresponding limiting behavior of $f(E)$.

The usual definition of conductivity (or a transport coefficient) involves a proportionality factor, in general a tensor quantity, connecting the response to the perturbation. Let the external perturbation be given in terms of an operator (observable) $\hat{C}_1$ in the form,

$$\hat{H}^{\text{ext}}(\omega) = \mathrm{e}^{-\mathrm{i}(\omega+\mathrm{i}\delta)t}\hat{C}_1, \tag{19.63}$$

where ω is real. The operator $\hat{C}_1$ is taken to be independent of time. The restriction of $\hat{H}^{\text{ext}}$ to a single Fourier component (harmonic) is not particularly limiting; within linear response theory, each harmonic component of a perturbation acts independently of the others, and the net result is the sum of these independent contributions.

We wish to write the change, $\delta\langle\hat{C}_2(t)\rangle$, of an observable (operator) $\hat{C}_2$ under the action of the external perturbation, $\hat{H}^{\text{ext}}$, in the form,

$$\delta\langle\hat{C}_2(t)\rangle \equiv \chi_{\hat{C}_2\hat{C}_1}(\omega+\mathrm{i}\delta)\mathrm{F}\mathrm{e}^{-\mathrm{i}(\omega+\mathrm{i}\delta)t}. \tag{19.64}$$

This form defines [7] the microscopic *complex admittance*, $\chi_{\hat{C}_2\hat{C}_1}(\omega+\mathrm{i}\delta)$, as a proportionality factor between the response of a system to a perturbation and the perturbation itself. Our purpose in this section is to obtain an expression for this quantity starting from the general result of linear response theory, Eq. (19.12).

19.5.1 Derivation(*)

We continue to work within an operator (invariant or representation-free) formalism. Specific representations and expansions in the corresponding basis sets will be introduced explicitly when we derive formulae for practical applications. Also, it is convenient to slightly generalize the form of Eq. (19.12) by introducing the Fermi distribution function, $f(E)$, at finite temperatures in the performance of statistical averaging. The $T = 0$ limit can easily be taken by using the proper form of the $f(E)$.

Now, interpreting the ground state expectation value, $\langle\Psi_0|\cdots|\Psi_0\rangle$, Eq. (19.12), as the trace over a complete set of states in the presence of $f(\hat{H})$, we can write Eq. (19.12) in the form [7]

$$\chi_{\hat{C}_2\hat{C}_1}(z) = -\mathrm{i}\int_0^{\infty} \mathrm{d}t\mathrm{e}^{\mathrm{i}zt}\mathrm{Tr}\left[\left(\mathrm{e}^{\mathrm{i}\hat{H}t}\hat{C}_1\mathrm{e}^{-\mathrm{i}\hat{H}t},\hat{C}_2\right)f(\hat{H})\right], \tag{19.65}$$

where z is a complex frequency with $\Im z \geq 0$. We consider operators $\hat{C}_1$ and $\hat{C}_2$, which can be represented by the general form,

$$\hat{C} = \frac{1}{2}\left[\gamma(\hat{H})\hat{C}' + \hat{C}'\gamma(\hat{H})\right]. \tag{19.66}$$

This class of operators is broad enough to include the current operator, $e\hat{\mathbf{p}}/m$, the total energy, $\hat{H}$, and the total energy flux, $w^\alpha = \left(\hat{p}^\alpha\hat{H} + \hat{H}\hat{p}^\alpha\right)/2m$, where w^α, with $\alpha = x,\ y,\ z$ is the heat current along the αth direction. We now replace the Hamiltonians in the exponents by λ and η, introduce integrations over these variables with the factors $\delta(\lambda - \hat{H})$ and $\delta(\eta - \hat{H})$ inserted between products of operators, and carry out the time integration in Eq. (19.65) to obtain the expression,

$$\begin{aligned}\chi_{\hat{C}_2\hat{C}_1}(z) &= \frac{1}{4}\int\int \mathrm{d}\lambda\mathrm{d}\eta\frac{f(\lambda)-f(\eta)}{z-\lambda-\eta}\left[\gamma_1(\eta)+\gamma_1(\lambda)\right]\left[\gamma_2(\eta)+\gamma_2(\lambda)\right] \\ &\times\ I_{C_2',C_1'}(\lambda,\eta),\end{aligned} \tag{19.67}$$

where

$$I_{C_2',C_1'}(\lambda,\eta) = \mathrm{Tr}\hat{C}_2'\delta(\lambda-\hat{H})\hat{C}_1'\delta(\eta-\hat{H}). \tag{19.68}$$

The common factor $I_{C_2',C_1'}(\lambda,\eta)$ results from the cyclic permutation properties of operators under the trace.

We now recall that the quantity $\delta(\lambda - \hat{H})$ is the imaginary part of the Green function, Eq. (2.19),

$$\hat{G}(z) = \left(z - \hat{H}\right)^{-1}, \tag{19.69}$$

so that in the limit $\delta \to 0$ we have

$$\delta(\lambda - \hat{H}) = (2\pi\mathrm{i})^{-1}\left[\hat{G}(\lambda - \mathrm{i}\delta) - \hat{G}(\lambda + \mathrm{i}\delta)\right]. \tag{19.70}$$

Consequently, the quantity $I_{C_2',C_1'}(\lambda,\eta)$ in Eq. (19.68) can be written in terms of the (unperturbed) system Green function[15],

$$\begin{aligned}I_{C_2',C_1'}(\lambda,\eta) &= \left(4\pi^2\right)^{-1}\mathrm{Tr}\left\{\hat{C}_2'\left[K(\lambda^+,\hat{C}_1',\eta^-)\right.\right. \\ &+\left.\left. K(\lambda^-,\hat{C}_1',\eta^+) - K(\lambda^+,\hat{C}_1',\eta^+) - K(\lambda^-,\hat{C}_1',\eta^-)\right]\right\},\end{aligned} \tag{19.71}$$

where $\lambda^\pm = \lambda \pm \mathrm{i}\delta$, etc., and the quantities K are defined by the expression,

$$K\left(z_1,\hat{C},z_2\right) = \hat{G}(z_1)\hat{C}\hat{G}(z_2). \tag{19.72}$$

[15]Because only the unperturbed Hamiltonian and Green function enter the definition of the complex admittance, this quantity represents an internal property of the system.

With Eq. (19.71) inserted into Eq. (19.67), we obtain an expression for the complex admittance (the Kubo formula) explicitly in terms of the system Green function.

The presence of the Green function in the Kubo formula presents a number of advantages. The often simple properties and analytic structure of Green functions allow the study of the corresponding properties and structures of the complex admittance, as is illustrated in the next subsection. They also allow the introduction of approximation schemes and the treatment of systems, e.g., substitutionally disordered alloys, for which the wave function may not be a very useful concept. The usefulness of expressions based on Green functions will be further demonstrated in the next chapter, where we discuss electronic transport in the case of substitutionally disordered alloys.

19.5.2 Exact properties(*)

The complex admittance, $\chi_{\hat{C}_2,\hat{C}_1}(z)$, satisfies a number of exact mathematical properties, of which we mention the most prominent. The satisfaction of these properties is an important criterion for judging the validity of any approximation scheme, e.g. the CPA, in specific applications of linear response theory.

We examine the behavior of the function $\chi_{\hat{C}_2,\hat{C}_1}(z)$ in the complex frequency plane [1, 7, 8]. Much of this behavior canbe easily derived by inspection of Eqs. (19.67) and (19.68), and the various expressions involving the Green function. The function $\chi_{\hat{C}_2,\hat{C}_1}(z)$ is analytic in both complex half planes and satisfies the *crossing* relation,

$$\chi_{\hat{C}_2,\hat{C}_1}(z) = \chi^*_{\hat{C}_2,\hat{C}_1}(-z^*). \tag{19.73}$$

In the limit $z \to \infty$, we have

$$\chi_{\hat{C}_2,\hat{C}_1}(z) = z^{-1}\mathrm{Tr}\left\{\left[\hat{C}_2,\hat{C}_1\right] f(\hat{H})\right\} + O(z^{-2}), \tag{19.74}$$

so that the sum rule

$$\pm \mathrm{i}\frac{1}{\pi}\lim_{\delta\to 0}\int_{-\infty}^{\infty}\chi_{\hat{C}_2,\hat{C}_1}(E+\mathrm{i}\delta)\mathrm{d}E = \mathrm{Tr}\left\{\left[\hat{C}_2,\hat{C}_1\right] f(\hat{H})\right\}, \tag{19.75}$$

is satisfied. Finally, the values of $\chi_{\hat{C}_2,\hat{C}_1}(z)$ in the upper and lower half planes are connected by the relation,

$$\chi_{\hat{C}_2,\hat{C}_1}(z) = \chi^*_{\hat{C}_2,\hat{C}_1}(z^*). \tag{19.76}$$

In addition to these properties, $\chi_{\hat{C}_2,\hat{C}_1}(z)$ must also preserve the number of particles in the system, and because energy is not dissipated in a linear approximation (Joule heat is a quadratic effect), $\chi_{\hat{C}_2,\hat{C}_1}(z)$ must also be consistent with energy conservation. Particle conservation is expressed by

$$\chi_{1,\hat{C}_1} = 0, \tag{19.77}$$

while energy conservation implies

$$\chi_{\hat{H},\hat{C}_1} = 0. \tag{19.78}$$

The quantity $K(z_1, \hat{C}, z_2)$ defined in Eq. (19.72) satisfies the following properties that are sufficient for the validity of Eqs. (19.73) to (19.78):

$$K(z_1, \hat{C}, z_2) = K^\dagger(z_2^*, \hat{C}, z_1^*), \tag{19.79}$$

$$\begin{aligned} \mathrm{Tr}\left\{\left[\hat{C}_2, \hat{C}_1\right] f(\hat{H})\right\} &= \frac{1}{(4\pi)^2} \int\int \mathrm{d}\lambda \mathrm{d}\eta \frac{f(\lambda) - f(\eta)}{z - \lambda - \eta} \left[\gamma_1(\eta) + \gamma_1(\lambda)\right] \\ &\times \left[\gamma_2(\eta) + \gamma_2(\lambda)\right] I_{C_2', C_1'}(\lambda, \eta), \end{aligned} \tag{19.80}$$

$$\mathrm{Tr}\left\{\hat{C}_2 K(z_1, \hat{C}_1, z_2)\right\} = \mathrm{Tr}\left\{\hat{C}_1 K(z_2, \hat{C}_2, z_1)\right\}, \tag{19.81}$$

and

$$K(z_1, 1, z_2) = -(z_1 - z_2)^{-1} \left[G(z_1) - G(z_2)\right]. \tag{19.82}$$

In addition, under the action of any system symmetry operator, $\hat{S}$, we have

$$\hat{S} K(z_1, \hat{C}, z_2) \hat{S}^\dagger = K(z_1, \hat{S}\hat{C}\hat{S}^\dagger, z_2), \tag{19.83}$$

which implies that

$$\chi_{\hat{S}\hat{C}_2\hat{S}^\dagger, \hat{S}\hat{C}_1\hat{S}^\dagger} = \chi_{\hat{C}_2,\hat{C}_1}. \tag{19.84}$$

The reader can verify that Eq. (19.79) implies Eq. (19.73), Eq. (19.80) implies the validity of Eqs. (19.74) and (19.75), Eq. (19.81) implies Eq. (19.76), Eq. (19.83) that of Eq. (19.84) and, finally, Eq. (19.82) implies the validity of Eqs. (19.77) and (19.78). Equation (19.82) is an example of a Ward identity [8] expressing the response of K to changes of the origin of energies. This relation will be encountered again in the next chapter in connection with the so-called vertex corrections characterizing the discussion of transport in disordered systems.

19.5.3 Conductivity relations

In addition to the general analytic properties derived above, it is useful to consider [9] some specific properties of the conductivity tensor. For ease of presentation in deriving these properties, we explicitly treat the case of cubic systems for which the conductivity tensor reduces to a scalar.

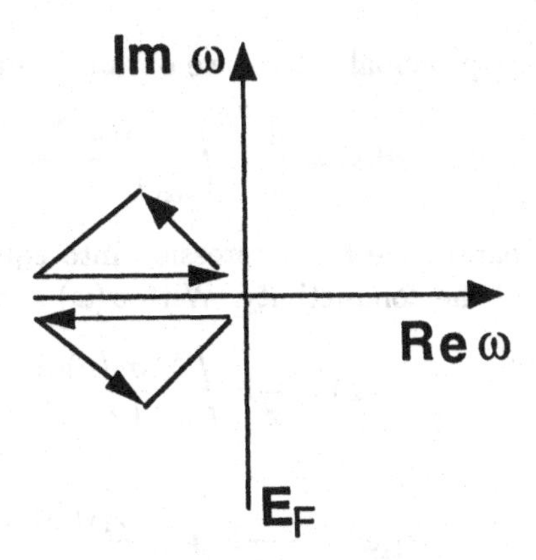

Figure 19.2: **Contours for integrations in the complex frequency plane.**

Essentially identical results, with scalar quantities replaced by their tensor counterparts, hold in the general case.

It is reasonable to assume that the current cannot start before an electric field has been applied, so that $\sigma(\mathbf{r}, t)$ can be set equal to zero for negative values of t. This is the requirement of causality, Eq. (19.5). It implies that the Fourier transform, $\sigma(\omega)$, (suppressing the space variable) is an analytic function in the upper half of the complex ω plane. Thus, in the expression,

$$\sigma(t) = \frac{1}{2\pi \mathrm{i}} \int_{-\infty}^{\infty} \sigma(\omega) \mathrm{e}^{-\mathrm{i}\omega t} \mathrm{d}\omega, \tag{19.85}$$

the contour of integration should be closed by a large semicircle in the upper half plane, as indicated by the contour C_1 in Fig. 19.2. For ω_0 a frequency with a positive imaginary part, we have

$$\sigma(\omega_0) = \frac{1}{2\pi \mathrm{i}} \oint_{C_2} \frac{\sigma(\omega)\mathrm{d}\omega}{(\omega - \omega_0)}, \tag{19.86}$$

where C_2 is any contour surrounding the point ω_0, such as the contour indicated in Fig. 19.2. If this contour is expanded to C_1, and if we assume that $\sigma(\omega) \to 0$ as $\omega \to \infty$, then only the contribution from the integral along the real axis survives, and we have

$$\sigma(\omega) = \frac{1}{2\pi \mathrm{i}} \int_{-\infty}^{\infty} \frac{\sigma(\nu)\mathrm{d}\nu}{(\nu - \omega_0)}, \tag{19.87}$$

where ν is the real part of ω. Taking ω_0 to lie close to the real axis, $\omega_0 = \omega_0 + \mathrm{i}\epsilon$, ω real, and using the identity

$$\frac{1}{(\nu - \omega - \mathrm{i}\epsilon)} = \mathrm{P}\frac{1}{(\nu - \omega)} + \mathrm{i}\pi\delta(\nu - \omega), \tag{19.88}$$

(where P indicates principal value), we obtain from Eq. (19.87) the result

$$\sigma(\omega) = \frac{1}{\pi \mathrm{i}} \mathrm{P} \int_{-\infty}^{\infty} \frac{\sigma(\nu)\mathrm{d}\nu}{(\nu - \omega)}. \tag{19.89}$$

It is useful to separate the last expression into equations for the real and imaginary parts of the conductivity. With $\sigma(\omega) = \sigma_1(\omega) + \mathrm{i}\sigma_2(\omega)$, we find

$$\sigma_1(\omega) = \frac{1}{\pi} \mathrm{P} \int_{-\infty}^{\infty} \frac{\sigma_2(\nu)\mathrm{d}\nu}{(\nu - \omega)}, \tag{19.90}$$

and

$$\sigma_2(\omega) = -\frac{1}{\pi} \mathrm{P} \int_{-\infty}^{\infty} \frac{\sigma_1(\nu)\mathrm{d}\nu}{(\nu - \omega)}. \tag{19.91}$$

Thus, the real and imaginary parts of $\sigma(\omega)$ are Hilbert transforms of one another. These integrals exist provided that the functions σ_1 and σ_2 vanish sufficiently rapidly as $\nu \to \infty$, which is the case if these functions are square integrable.

Further analytic properties follow from the requirement that (for real frequencies) the conductivity must be real, since real fields produce real (measurable) currents. This requires the condition,

$$\sigma^*(\nu) = \sigma(-\nu), \tag{19.92}$$

so that

$$\sigma_1(\nu) = \sigma_1(-\nu), \tag{19.93}$$

$$\sigma_2(\nu) = -\sigma_2(-\nu). \tag{19.94}$$

Therefore, the real part of the conductivity is an even function, and the imaginary part an odd function of frequency.

One more set of relations can be derived by using Eqs. (19.93) and (19.94) to eliminate the negative frequency part in Eqs. (19.90) and (19.91). From Eq. (19.90), we obtain

$$\begin{aligned} \sigma_1(\omega) &= \frac{\mathrm{P}}{\pi}\left[\int_{-\infty}^{0} \frac{\sigma_2 \nu \mathrm{d}\nu}{(\nu - \omega)} + \int_{0}^{\infty} \frac{\sigma_2 \nu \mathrm{d}\nu}{(\nu - \omega)}\right] \\ &= \frac{\mathrm{P}}{\pi} \int_0^{\infty} \sigma(\nu) \left[\frac{1}{(\nu - \omega)} + \frac{1}{(\nu + \omega)}\right] \mathrm{d}\nu \\ &= \frac{2\mathrm{P}}{\pi} \int_0^{\infty} \frac{\nu \sigma_2(\nu)\mathrm{d}\nu}{(\nu^2 - \omega^2)}, \end{aligned} \tag{19.95}$$

where Eq. (19.93) has been used to arrive at the second line. Similarly, we find

$$\sigma_2(\omega) = -\frac{2\omega \mathrm{P}}{\pi} \int_0^{\infty} \frac{\sigma_1(\nu)\mathrm{d}\nu}{(\nu^2 - \omega^2)}. \tag{19.96}$$

The last two expressions constitute the Kramers-Kronig [10, 11, 12] relations for the conductivity.

This completes the purely formal discussion of linear response theory and the way in which various Kubo formulae associated with different transport coefficients can be derived. In the next section, we provide a detailed derivation of the ac and dc electronic conductivities.

19.6 The ac and dc Conductivity

We now turn our attention to the derivation of formulae for measurable quantities, specifically the ac and dc conductivity of a solid. These are macroscopic properties of the system and are given as averages of their corresponding microscopic counterparts. Such averages will be introduced as appropriate in the following discussion.

As we mentioned above, the ac conductivity is obtained as the imaginary part of the general Kubo formula, and the dc conductivity as the $\omega \rightarrow 0$ limit of that imaginary part. We assume that $\mathbf{q} = 0$ throughout the following discussion.

19.6.1 The ac conductivity

In the case of electrical conduction, we set $\gamma(\hat{H}) \equiv 1$ so that Eq. (19.67) takes the form,

$$\chi_{\alpha\beta} = \frac{\mathrm{e}^2}{m^2\Omega}\int\int \mathrm{d}\lambda \mathrm{d}\eta \frac{f(\lambda) - f(\eta)}{z - \lambda + \eta} I_{\alpha\beta}(\lambda, \eta), \tag{19.97}$$

where we have written,

$$I_{\alpha\beta}(\lambda, \eta) = \frac{1}{\Omega}\mathrm{Tr}\langle p_\alpha \delta(\lambda - \hat{H}) p_\beta \delta(\lambda - \hat{H})\rangle, \tag{19.98}$$

in which the angular brackets indicate an average over the sample. Now, with $|n\rangle$ and E_n denoting the eigenfunctions and eigenenergies of $\hat{H}$, we can write,

$$I_{\alpha\beta} = \frac{1}{\Omega}\sum_{nm}\langle p_\alpha^{nm} \delta(\lambda - E_m) p_\beta^{mn} \delta(\lambda - E_n)\rangle, \tag{19.99}$$

where $p_\alpha^{nm} = \langle n|p_\alpha|m\rangle$. Inserting this expression into Eq. (19.97), and integrating over λ and η leads to the expression (with $z = \omega + \mathrm{i}\epsilon$)

$$\chi_{\alpha\beta}(z) = \frac{2\mathrm{e}^2}{m\Omega}\sum_{nm}\langle \frac{f(E_n) - f(E_m)}{(z - E_n + E_m)} p_\alpha^{nm} p_\beta^{mn}\rangle. \tag{19.100}$$

The quantity that is measured in an experiment is given by the imaginary part of the last expression, (neglecting the diagonal part), divided by ω,

$$\begin{aligned}\sigma_{\alpha\beta}(\omega) &= \Im\chi_{\alpha\beta}(z)/\omega \\ &= \frac{2\pi e^2}{m^2\Omega\omega}\sum_{nm}\langle[f(E_n)-f(E_m)]\,p_\alpha^{nm}p_\beta^{mn}\delta(\omega+E_n-E_m)\rangle,\end{aligned} \tag{19.101}$$

which can also be written in the form,

$$\begin{aligned}\sigma_{\alpha\beta}(\omega) &= \frac{2\pi e^2}{m^2\Omega\omega}\sum_{nm}\int \mathrm{d}E\,[f(E_n)-f(E_m)] \\ &\times \langle p_\alpha^{nm}\delta(E-\omega-E_n)p_\beta^{mn}\delta(E-E_m)\rangle \\ &= \frac{2\pi e^2}{m^2\Omega\omega}\sum_{nm}\int \mathrm{d}E\,[f(E_n)-f(E_m)] \\ &\times \langle p_\alpha^{nm}\Im G(E-\omega-E_n)p_\beta^{mn}\Im G(E-E_m)\rangle.\end{aligned} \tag{19.102}$$

Because of the properties of the delta functions, this form is exactly equivalent to the one before it. In both expressions, the factor of 2 accounts for the presence of electron spin.

19.6.2 The dc conductivity

An expression for the dc conductivity can be obtained from the imaginary part of Eq. (19.97) in the limit $\omega \to 0$. We begin by writing,

$$\chi_{\alpha\beta}(z) = B\int\int \mathrm{d}\lambda\mathrm{d}\eta\frac{f(\lambda)-f(\eta)}{(z-\lambda+\eta)}\sum_{nm}\langle p_\alpha^{nm}\delta(\lambda-E_n)p_\beta^{nm}\delta(\lambda-E_m)\rangle, \tag{19.103}$$

where $B = \frac{2e^2}{m\Omega}$. Introducing the factor ω^{-1}, we now consider the quantity

$$\chi_{nm}(z) = \frac{B}{\omega}\int\int \mathrm{d}\lambda\mathrm{d}\eta\left[\frac{f(\lambda)}{(z-\lambda+\eta)} - \frac{f(\eta)}{(z-\lambda+\eta)}\right]\delta(\lambda-E_n)\delta(\eta-E_m). \tag{19.104}$$

Interchanging the variables λ and η, as well as the indices n and m in the second term inside the integral, and integrating over η, we obtain

$$\chi_{nm}(z) = \frac{B}{\omega}\int \mathrm{d}\lambda\left[\frac{f(\lambda)}{(z-\lambda+E_m)} - \frac{f(\lambda)}{(z-E_m+\lambda)}\right]\delta(\lambda-E_n). \tag{19.105}$$

Taking the imaginary part of the last expression, we have

$$\Im\chi_{nm}(z) = -\frac{B}{\omega}\int \mathrm{d}\lambda f(\lambda)\,[\pi\delta(\omega-\lambda+E_m)-\pi\delta(\omega+\lambda-E_m)]\,\delta(\lambda-E_n), \tag{19.106}$$

and take the limit $\omega \to 0$. Since the term inside the brackets, as well as ω, approach zero in that limit, we differentiate numerator and denominator of the ratio to obtain,

$$\lim_{\omega\to 0}\frac{\pi}{\omega}\left[\delta(\omega-\lambda+E_m)-\delta(\omega-E_m+\lambda)\right]=\pi\frac{\mathrm{d}}{\mathrm{d}\lambda}\delta(\lambda-E_m). \quad (19.107)$$

This expression follows upon a change of variable, and upon using the fact that the derivative of a delta function is odd under a change in sign of its argument. We now note that for any A_{nm} and B_{nm} independent of λ, we have

$$\sum_{nm}A_{nm}\delta'(\lambda-E_m)B_{mn}\delta(\lambda-E_n)=\frac{1}{2}\frac{\mathrm{d}}{\mathrm{d}\lambda}\sum_{nm}A_{nm}\delta(\lambda-E_m)B_{mn}\delta(\lambda-E_n), \quad (19.108)$$

and upon substitution into Eq. (19.103), in the limit $\omega \to 0$, we obtain the preliminary result,

$$\sigma_{\alpha\beta}=\pi B\int_{-\infty}^{\infty}\mathrm{d}\lambda f(\lambda)\frac{\mathrm{d}}{\mathrm{d}\lambda}\sum_{nm}p_\alpha^{nm}\delta(\lambda-E_m)p_\beta^{mn}\delta(\lambda-E_n). \quad (19.109)$$

Finally, integrating by parts, and noting that the delta functions vanish at $\lambda=-\infty$, since there are no states there, and that $f(\lambda)$ vanishes at $\lambda=\infty$, we obtain

$$\begin{aligned}\sigma_{\alpha\beta} &= \frac{2\pi e^2}{m^2\Omega}\int_{-\infty}^{\infty}\mathrm{d}\lambda\left(-\frac{\partial f(\lambda)}{\partial\lambda}\right)\sum_{nm}\langle p_\alpha^{nm}\delta(\lambda-E_m)p_\beta^{mn}\delta(\lambda-E_n)\rangle \\ &= \frac{2\pi e^2}{m^2\Omega}\int_{-\infty}^{\infty}\mathrm{d}\lambda\left(-\frac{\partial f(\lambda)}{\partial\lambda}\right)\sum_{nm}\langle p_\alpha^{nm}\Im G(\lambda-E_m)p_\beta^{mn}\Im G(\lambda-E_n)\rangle,\end{aligned} \quad (19.110)$$

where once again a factor of 2 accounts for the electron spin. This is the Kubo-Greenwood formula [13] for the dc conductivity of a metal.

Before leaving this section, we point out a few important features of the formulae just derived. It follows from Eq. (19.102) that the ac conductivity depends strongly on the value of the *joint* density of states,

$$D(\omega)=\int_{\mu-\omega}^{\mu}\mathrm{d}E\sum_{nm}\delta(E-E_m)\delta(E-\omega-E_n). \quad (19.111)$$

This quantity is the contribution arising from the integration over the allowed range for the dipole transitions described by the matrix element $\langle n|p_\alpha|m\rangle\langle m|p_\beta|n\rangle$ (see Section 19.8). These matrix elements can make dominant contributions to the ac conductivity when the energy difference

$|E_n - E_m|$ is large. Therefore, the ac conductivity is determined by the strength of the dipole transitions as well as the availability of states for these transitions to take place.

Finally, in the expression for the dc conductivity, Eq. (19.110), we see that the electrons responsible for conduction are energetically confined to the vicinity of the Fermi level. In the limit $T = 0$, only electrons at E_{F} contribute to the current.

19.7 Representations

Up to this point, we have developed the formalism of linear response theory within an operator (representation-free) framework. However, a practical application of these results requires the introduction of a particular representation and the corresponding expressions for the single-particle Green functions and other operators that occur in the formalism.

In this section, we briefly discuss the forms taken by these operators in a TB description of the Hamiltonian, as well as within the angular momentum representation of first-principles multiple scattering theory.

19.7.1 Tight-binding formalism

For the sake of clarity, we consider a simple, single-band TB Hamiltonian of the form[16], Eq. (10.3),

$$\hat{H} = \sum_i \epsilon_i a_i^\dagger a_i + \sum_{ij} W_{ij} a_i^\dagger a_j, \tag{19.112}$$

where the $a_i^\dagger$ (a_i) are electron creation (annihilation) operators associated with site i, ϵ_i is the single-site particle energy, and W_{ij} denotes the matrix elements describing electron hopping from site i to site j. In order to obtain an expression for the current operator, it is convenient to use the relation,

$$\mathrm{j} = \frac{\partial \mathbf{P}}{\partial t} = \mathrm{i}\left[\hat{H}, \mathbf{P}\right], \tag{19.113}$$

where $\mathbf{P}$ is the polarization vector (also the position operator). In the tight-binding model, the polarization vector has the form [2]

$$\mathbf{P} = \sum_i \mathbf{R}_i n_i = \sum_i \mathbf{R}_i a_i^\dagger a_i. \tag{19.114}$$

[16]The extension to more general forms is straightforward. Here, we replace the states $|n\rangle$ with the equivalent expression $a_n^\dagger|0\rangle$ where $|0\rangle$ represents the vacuum state, and $a_n^\dagger$ is an operator creating an electron in state $|n\rangle$. The commutation relations, $[a_n^\dagger, a_m] = \delta_{nm}$, and $[a_n^\dagger, a_m^\dagger] = [a_n, a_m] = 0$ are well-known results of elementary quantum mechanics.

Now, it follows from the commutation properties of the creation and annihilation operators that

$$\mathbf{j} = -\mathrm{i}\sum_{ij} W_{ij}(\mathbf{R}_j - \mathbf{R}_i)a_i^{\dagger}a_j. \tag{19.115}$$

A TB description of transport properties is most appropriate in calculations involving largely localized electrons. Thus, it can be used to treat solids characterized by rather narrow bands, as might occur in ionic solids and organic materials.

The form taken by the Green function,

$$\hat{G}(z) = \left(z - \hat{H}\right)^{-1}, \tag{19.116}$$

in a TB model depends on the system under study, and the approximation used in its construction. Explicit expressions for the specific case of substitutionally disordered alloys are given in the next chapter.

19.7.2 First-principles MST

In this case, the current operator takes the form, (with $\hbar = 1$)

$$j_\alpha = -\mathrm{i}\frac{\mathrm{e}}{m}\frac{\partial}{\partial r_\alpha}. \tag{19.117}$$

In the angular momentum representation and for vectors $\mathbf{r}$ and $\mathbf{r}'$ in cells n and m, respectively, the Green function in the single-particle picture used here is given by Eq. (3.82), which we write in the form

$$G(z;\mathbf{r},\mathbf{r}') = \langle Z^n(\mathbf{r})|\underline{\tau}^{nm}|Z^m(\mathbf{r}')\rangle, \tag{19.118}$$

where the single-scatterer term, which is real for real energies, is omitted, $|Z^n(\mathbf{r})\rangle$ is the regular solution of the Schrödinger equation for the potential in cell n, and $\underline{\tau}^{nm}$ is the scattering path matrix connecting sites n and m. Once again, the specific form taken by this matrix depends on the system being studied, and on the method (approximation) used in its evaluation. A detailed discussion of the conductivity of substitutionally disordered alloys within the coherent potential approximation based on the last expression is given in the next chapter.

19.8 Connection to Optical Properties

The optical properties of a material are closely related to its conductivity. When a photon impinges on a solid, the action of the electric field can cause electronic transitions between states. Transitions between band states are prominent in the energy range of 0.1 to 20 eV. At lower energies, lattice vibrations become important [9], and at higher energies atomic level excitations begin to predominate.

A photon with energy of 20 eV has a wavelength of 600 Å. This is sufficiently larger than a typical lattice constant, say 6 Å, so that spatial variations of the electric field over a unit cell can be neglected. The optical properties of a material can now be described by a conductivity and dielectric function that depend only on frequency, rather than frequency and wave vector.

Electromagnetic theory yields relations between the dielectric function, $\kappa(\omega)$, and the conductivity. With $\kappa(\omega) = \kappa_1(\omega) + i\kappa_2(\omega)$, and denoting the permitivity of free space by ϵ_0, we have

$$\kappa_1(\omega) = 1 - [\sigma_2(\omega)/\epsilon_0\omega] \tag{19.119}$$

and

$$\kappa_2(\omega) = \sigma_1(\omega)/\epsilon_0\omega \tag{19.120}$$

so that using the Kramers-Kronig relations, Eqs. (19.95) and (19.96), we find [9]

$$\kappa_1(\omega) = 1 + \frac{2\mathrm{P}}{\pi}\int_0^\infty \frac{\nu\kappa_2(\nu)\mathrm{d}\nu}{(\nu^2 - \omega^2)}, \tag{19.121}$$

and

$$\kappa_2(\omega) = \frac{2\mathrm{P}}{\pi\omega}\int_0^\infty \frac{\nu^2[1 - \kappa_1(\nu)]\mathrm{d}\nu}{(\nu^2 - \omega^2)}. \tag{19.122}$$

It is useful to note that the quantities σ_1, σ_2, κ_1, and κ_2 can be measured independently so that the relations quoted above can be checked experimentally. However, these relations are derived on very broad and general assumptions and are customarily accepted as valid and used to analyze experimental data. For example, if $\kappa_1(\omega)$ is measured over a sufficiently broad (and properly dense) range of frequency, the Kramers-Kronig relations can be used to calculate $\kappa_2(\omega)$. On the other hand, when measured quantities fail to satisfy these relations they provide strong evidence that the limits characterizing the derivations of the relations have been exceeded. This in turn indicates a possible breakdown of linear response theory and the need for more sophisticated and difficult treatments.

19.9 Magnetic Susceptibility

Linear response theory can be used to discuss the response of a system to a magnetic disturbance. In this section, we derive a general expression for the *dynamic susceptibility* that characterizes this response. The development follows the lines of that in March and Jones [14].

19.9.1 Susceptibility

We consider the response of a system to a weak, externally-applied magnetic field, which for the moment we take to have the form of a localized impulse,

$$\mathbf{H}(\mathbf{r},t) = \mathbf{H}\delta(\mathbf{r}-\mathbf{r}')\delta(t). \tag{19.123}$$

Within linear response and given the form of the external field, we assume that the magnetic moment along the βth Cartesian direction at point $\mathbf{r}$ at time t is given by

$$\Delta M_\beta(\mathbf{r}',t') = \sum_\alpha \chi(\mathbf{r},\mathbf{r}';t'-t)H_\alpha. \tag{19.124}$$

The function $\chi_{\alpha\beta}(\mathbf{r},\mathbf{r}';t-t')$ is called the *susceptibility*. Because there can be no induced moment before the application of the field, we must have,

$$\chi(\mathbf{r},\mathbf{r}';t) = 0, \quad t<0. \tag{19.125}$$

For general, time-dependent fields, we write in place of Eq. (19.124),

$$\Delta M_\beta(\mathbf{r},t) = \sum_\alpha \int \mathrm{d}^3r' \int_{-\infty}^{t} \chi(\mathbf{r},\mathbf{r}';t'-t)H_\alpha(\mathbf{r}',t')\mathrm{d}t'. \tag{19.126}$$

Passing to a momentum representation in the usual way, we can write,

$$\Delta M_\beta(\mathbf{k},t) = \sum_\alpha \int_{-\infty}^{t} \chi(\mathbf{k},\mathbf{k}';t-t')H_\alpha(\mathbf{k}',t')\mathrm{d}t'. \tag{19.127}$$

This expression can be considerably simplified in the case of a periodic system. In this case, the response at $\mathbf{r}+\mathbf{R}$ to a field such as that given by Eq. (19.123) is the same as the response at $\mathbf{r}$ to the field $\mathbf{H}(\mathbf{r}-\mathbf{r}'+\mathbf{R})\delta(t)$, so that,

$$\chi_{\alpha\beta}(\mathbf{r}+\mathbf{R},\mathbf{r}',t) = \chi_{\alpha\beta}(\mathbf{r},\mathbf{r}'-\mathbf{R},t). \tag{19.128}$$

Starting with Eq. (19.127), we now have,

$$\begin{aligned}
\chi_{\alpha\beta}(\mathbf{k},\mathbf{k}',t) &= \int \mathrm{d}^3r \sum_{\mathbf{R}} \mathrm{e}^{\mathrm{i}\mathbf{k}'\cdot\mathbf{R}} \int_\Omega \mathrm{d}^3\mathbf{r}'\chi_{\alpha\beta}(\mathbf{r},\mathbf{r}',t)\mathrm{e}^{\mathrm{i}\mathbf{k}'\cdot\mathbf{r}'-\mathrm{i}\mathbf{k}\cdot\mathbf{r}} \\
&= \sum_{\mathbf{R}} \mathrm{e}^{\mathrm{i}\mathbf{k}\cdot\mathbf{R}} \int_\Omega \chi_{\alpha\beta}(\mathbf{r}-\mathbf{R},\mathbf{r}',t)\mathrm{e}^{\mathrm{i}\mathbf{k}'\cdot\mathbf{r}',t}\mathrm{e}^{\mathrm{i}\mathbf{k}'\cdot\mathbf{r}'-\mathrm{i}\mathbf{k}\cdot\mathbf{r}} \\
&= \sum_{\mathbf{R}} \mathrm{e}^{\mathrm{i}(\mathbf{k}'-\mathbf{k})\cdot\mathbf{R}} \int_\Omega \mathrm{d}^3r'\mathrm{e}^{\mathrm{i}\mathbf{k}'\cdot\mathbf{r}'} \int \mathrm{d}^3r\chi_{\alpha\beta}(\mathbf{r},\mathbf{r}',t)\mathrm{e}^{-\mathrm{i}\mathbf{k}\cdot\mathbf{r}} \\
&= \Omega_{\mathrm{BZ}} \sum_{\mathbf{K}} \delta(\mathbf{k}'-\mathbf{k}+\mathbf{K}) \int_\Omega \mathrm{d}^3r'\mathrm{e}^{\mathrm{i}\mathbf{k}'\cdot\mathbf{r}'} \int \mathrm{d}^3r\chi_{\alpha\beta}(\mathbf{r},\mathbf{r}',t)\mathrm{e}^{-\mathrm{i}\mathbf{k}\cdot\mathbf{r}},
\end{aligned} \tag{19.129}$$

or

$$\chi_{\alpha\beta}(\mathbf{k},\mathbf{k}',t) = \sum_{\mathbf{K}} \chi_{\alpha\beta}(\mathbf{k},\mathbf{k}+\mathbf{K},t)\delta_{\mathbf{k}',\mathbf{k}+\mathbf{K}}. \tag{19.130}$$

Because of this property, we will write the susceptibility in the form $\chi(\mathbf{k},t)$, keeping in mind that in general $\mathbf{k}$ can be replace by $\mathbf{k},\mathbf{k}'$. If we take the applied field to have a harmonic dependence on time, $\mathbf{H}(\mathbf{k},t) = \mathbf{H}(\mathbf{k})\mathrm{e}^{\mathrm{i}\omega t}$, we have,

$$\Delta M_\beta(\mathbf{k},t) = \mathrm{e}^{\mathrm{i}\omega t}\sum_{\alpha\beta} \chi_{\alpha\beta}(\mathbf{k},\omega)H_\alpha(\mathbf{k}), \tag{19.131}$$

where

$$\chi_{\alpha\beta}(\mathbf{k},\omega) = \int_0^\omega \mathrm{e}^{-\mathrm{i}\omega t}\chi_{\alpha\beta}(\mathbf{k},t)\mathrm{d}t \tag{19.132}$$

is called the ***dynamical susceptibility***, and is in general complex and energy dependent.

19.9.2 Correlation functions

It is useful to express the susceptibility in terms of commutators and correlation functions of the electronic magnetic-moment density, $\boldsymbol{\mu}(\mathbf{r})$, and thus make formal contact with corresponding expressions for the electrical conductivity. We assume that the perturbing part of the Hamiltonian has the form,

$$H^{\mathrm{ext}} = \int \boldsymbol{\mu}(\mathbf{r}')\cdot\mathbf{H}(\mathbf{r}',t)\mathrm{d}^3r'. \tag{19.133}$$

The electronic magnetic-moment density can be written as

$$\boldsymbol{\mu}(\mathbf{r}) = \sum_j \delta(\mathbf{r}-\mathbf{r}_j)\boldsymbol{\mu}_j, \tag{19.134}$$

where $\boldsymbol{\mu}_j$ is the magnetic-moment operator of the jth electron. Following a development along the lines of subsection 19.3.3, with the formal replacement of the current operator with the magnetic-density operator, we obtain the result,

$$\chi_{\alpha\beta}(\mathbf{r},\mathbf{r}';t) = \begin{cases} \frac{\mathrm{i}}{\hbar}\langle\Psi_0|[\mu_\alpha(\mathbf{r}),\mu_\beta(\mathbf{r}',t)]|\Psi_0\rangle & \text{if } t>0 \\ 0 & \text{if } t<0, \end{cases} \tag{19.135}$$

and, by definition, (with $\hbar = 1$)

$$\chi_{\alpha\beta}(\mathbf{r},\mathbf{r}';t) = \begin{cases} \mathrm{i}\langle\Psi_0|[\mu_\alpha(-\mathbf{k}),\mu_\beta(\mathbf{k}',t)]|\Psi_0\rangle & \text{if } t>0 \\ 0 & \text{if } t<0. \end{cases} \tag{19.136}$$

A Fourier transform with respect to time leads to the dynamical susceptibility, $\chi(\mathbf{k},\mathbf{k}';\omega)$, in a form analogous to Eq. (19.47) for the conductivity.

The discussion of frequency-dependent properties can be facilitated through the introduction of the Fourier transform,

$$S_{\alpha\beta}(\mathbf{k},\mathbf{k}';\omega) = \frac{1}{2\pi}\int_{-\infty}^{\infty} \mathrm{e}^{-\mathrm{i}\omega t}\langle\Psi_0|[\mu_\alpha(-\mathbf{k}')\mu_\beta(\mathbf{k},t)]|\Psi_0\rangle, \tag{19.137}$$

where the integral contains the correlation function of the magnetic-moment densities. As in the case of Eq. (19.130), we have

$$S_{\alpha\beta}(\mathbf{k},\mathbf{k}';\omega) = \sum_{\alpha\beta} S_{\alpha\beta}(\mathbf{k},\mathbf{k}+\mathbf{K};\omega)\delta_{\mathbf{k}',\mathbf{k}+\mathbf{K}}, \tag{19.138}$$

and we'll write $S_{\alpha\beta}(\mathbf{k},\omega)$. This quantity is sometimes referred to as the *structure factor* or *power factor*. It can be shown [14] that it obeys the relation (with $\beta = 1/k_{\mathrm{B}}T$),

$$S_{\alpha\beta}(-\mathbf{k}-\omega) = \mathrm{e}^{-\beta\omega}S_{\alpha\beta}(\mathbf{k},\omega). \tag{19.139}$$

So that by expressing the commutators in Eq. (19.136) in terms of correlation functions, we have,

$$\begin{aligned}\chi_{\alpha\beta} &= \frac{1}{\mathrm{i}}\int_{-\infty}^{\infty}[S_{\alpha\beta}(\mathbf{k},\omega) - S_{\alpha\beta}(-\mathbf{k},-\omega)]\,\mathrm{e}^{\mathrm{i}\omega t}\mathrm{d}\omega \\ &= \frac{1}{\mathrm{i}}\int_{-\infty}^{\infty} S_{\alpha\beta}(1-\mathrm{e}^{-\beta\omega})\mathrm{e}^{\mathrm{i}\omega t}\mathrm{d}\omega. \end{aligned} \tag{19.140}$$

19.10 Fluctuation-Dissipation Theorem

When an acting external field, $\mathbf{H}(\mathbf{k})$, is switched off, the system relaxes toward its steady state in a manner that can be described by a *relaxation factor*, R, as follows,

$$\Delta M_\beta(\mathbf{k},t) = \sum_\alpha H_\alpha(\mathbf{k})R_{\alpha\beta}(\mathbf{k},t). \tag{19.141}$$

It is not surprising that R is related to the susceptibility, χ. Indeed from Eq. (19.127), we have,

$$\Delta M_\beta(\mathbf{k},t) = \sum_\alpha \int_{-\infty}^{0} H_\alpha(\mathbf{k})\chi_{\alpha\beta}(\mathbf{k},t-t')\mathrm{d}t', \tag{19.142}$$

so that

$$R_{\alpha\beta}(\mathbf{k},t) = \int_t^\infty \chi_{\alpha\beta}(\mathbf{k},t')\mathrm{d}t', \quad t>0. \tag{19.143}$$

We now express the relaxation factor in terms of the correlation function, $S_{\alpha\beta}(\mathbf{k},\omega)$, defined in the previous subsection.

We see immediately that Eq. (19.143) can be written in the form,

$$R_{\alpha\beta}(\mathbf{k},t) = \frac{1}{\mathrm{i}}\int_t^{\infty} \mathrm{d}t' \int_{-\infty}^{\infty} S_{\alpha\beta}(\mathbf{k},\omega)(1-\mathrm{e}^{-\beta\omega})\mathrm{e}^{\mathrm{i}\omega t}\mathrm{d}\omega. \tag{19.144}$$

We can perform the integral over t' by replacing the upper limit by τ, and taking the limit $\tau \to \infty$. Then we obtain,

$$\begin{aligned} R_{\alpha\beta} &= \int_{-\infty}^{\infty} S_{\alpha\beta}(\mathbf{k},\omega)\frac{(1-\mathrm{e}^{-\beta\omega})}{\omega}\mathrm{e}^{-\mathrm{i}\omega t}\mathrm{d}\omega \\ &- \lim_{\tau\to} \int_{-\infty}^{\infty} S_{\alpha\beta}(\mathbf{k},\omega)\frac{(1-\mathrm{e}^{-\beta\omega})}{\omega}\mathrm{d}\omega. \end{aligned} \tag{19.145}$$

Now, in the limit $\tau \to \infty$, the second term in the last expression becomes significant about $\omega \to 0$ only, where $\frac{(1-\mathrm{e}^{-\beta\omega})}{\omega} \to \beta$. So that we have,

$$\lim_{\tau\to} \int_{-\infty}^{\infty} S_{\alpha\beta}(\mathbf{k},\omega)\frac{(1-\mathrm{e}^{-\beta\omega})}{\omega}\mathrm{d}\omega = \beta S_{\alpha\beta}(\mathbf{k},\omega)|_{\tau=\infty}. \tag{19.146}$$

However, in the limit of large times, the quantities $\mu_\alpha(-\mathbf{k})$ and $\mu_\beta(\mathbf{k},t)$ are uncorrelated so that we can write,

$$S_{\alpha\beta}(\mathbf{k},\omega)|_{\tau=\infty} = \langle\mu_\alpha(-\mathbf{k})\mu_\beta(\mathbf{k},\infty)\rangle = \langle\mu_\alpha(-\mathbf{k})\rangle\langle\mu_\beta(\mathbf{k})\rangle. \tag{19.147}$$

If we define,

$$S'_{\alpha\beta}(\mathbf{k},\omega) = S_{\alpha\beta}(\mathbf{k},\omega) - \delta(\omega)\langle\mu_\alpha(-\mathbf{k})\rangle\langle\mu_\beta(\mathbf{k})\rangle, \tag{19.148}$$

we can write Eq. (19.145) in the form

$$R_{\alpha\beta}(\mathbf{k},t) = \int_{-\infty}^{\infty} S'(\mathbf{k},\omega)\frac{(1-\mathrm{e}^{-\beta\omega})}{\omega}\mathrm{e}^{\mathrm{i}\omega t}\mathrm{d}\omega. \tag{19.149}$$

Because the mean values $\langle\mu_\alpha(-\mathbf{k})\mu_\beta(\mathbf{k})\rangle$ are subtracted out, $S'_{\alpha\beta}(\mathbf{k},\omega)$ refers to a spectrum of spontaneous fluctuations. The last expression for $R_{\alpha\beta}(\mathbf{k},t)$ is one form of the fluctuation-dissipation theorem.

Bibliography

[1] R. Kubo, J. Phys. Soc. Japan **12**, 570 (1957).

[2] Gerald D. Mahan, *Many-Particle Physics*, Plenum Press, New York (1981).

[3] P. L. Rossiter, *The electrical Resistivity of Metals and Alloys*, Cambridge Solid State Science Series, Cambridge University Press, New York (1987).

[4] Alexander L. Fetter and John Dirk Walecka, *Quantum Theory of Many Particle Systems*, McGraw-Hill, New York (1971).

[5] J. D. Jackson, *Classical Electrodynamics*, John Wiley and Sons, New York (1996).

[6] James M. MacLaren, private communication.

Wesley, 5Reading, Massachusetts (1965).

[7] B. Velicky, Phys. Rev. **184**, 614 (1969).

[8] P. Nozieres, *Theory of Interacting Fermi Systems*, W. A. Benjamin, Inc., New York (1964).

[9] Joseph Callaway, *Quantum Theory of the Solid State*, Academic Press, New York (1974).

[10] H. A. Kramers, Estratto dagli Atti del Congr. Int. Fis. **2**, 545 (1927).

[11] R. de L. Kronig, J. Opt. Soc. Amer. **12**, 547 (1926).

[12] R. de L. Kronig, Ned. Tijdschn. Natuurk **9**, 402 (1942).

[13] D. A. Greenwood, Proc. Phys. Soc. **71**, 585 (1958).

[14] William Jones and Norman H. March, *Theoretical Solid State Physics*, Dover Publications, New York 1973).

Chapter 20

Transport in Random Alloys

20.1 General Comments

A perfect metallic crystal exhibits perfect conductivity. That is, in a translationally invariant material containing no impurities or structural imperfections, the electron states are characterized by crystal momentum wave vectors that are good quantum numbers. Consequently, these states have infinite lifetimes. Under the action of an external electric field, the electrons in these states will move unimpeded, qualitatively in a way electrons in vacuum would move in the direction of the field[1]. It is only in the presence of impurities, structural imperfections, and other defects that a finite resistivity is observed. As was discussed in Chapter 18, the presence of impurities gives rise to the so-called residual resistivity at $T = 0$. In this chapter, we use the Kubo formula to study the effects of random substitutional impurities on the conductivity of a material. We explicitly treat the transverse conductivity and also work within the single-particle picture and the LDA of DFT to obtain the single-particle Green function. The following discussion addresses the case of finite impurity concentration, thus

[1]There are certainly important differences between the motion of free electrons and electrons in a periodic solid. For example, free electrons are characterized by a momentum (wave vector) that can vary continuously in magnitude and direction, whereas crystal momentum is quantized according to crystal periodicity. Furthermore, electrons in a crystal behave as if they had an effective mass that can be different from the free electron mass depending on the particular state occupied by the electron (the shape of the Fermi surface). However, provided that motion is allowed under the action of an electric field, in a perfect crystal this motion proceeds unimpeded so that the resistivity of a defect-free metallic material vanishes.

extending the single-impurity results presented in Chapter 18[2].

The study of transport in disordered materials has a long history. It has been studied by many authors from a great variety of viewpoints and on the basis of many different formal frameworks. Our discussion is based on the coherent potential approximation (CPA) and the corresp onding form taken by the Kubo formula. The CPA was discussed in Chapter 12 in connection with the calculation of the single-particle Green function of disordered substitutional alloys.

The study of transport properties is based on the two-particle Green function. In the approach taken here[3], this is represented by the product of two single-particle propagators. In the following pages, we examine the form taken by this two-resolvent formalism within the CPA and derive computable expressions for the electrical conductivity of substitutionally random materials.

The first application of the CPA to the study of electronic transport was made by Velicky [1] within the context of a tight-binding model, and has been followed by the works of a number of authors [2-21]. Velicky's work was extended by Levin et al. [3], Brouers et al. [4], and Chen et al. [2]. The last authors studied the effects of thermal disorder and explained some unusual features of metallic alloys that exhibit negative temperature coefficients of the electronic resistivity. Niizeki et al. [5, 6] formulated a general theory to include the effects of off-diagonal disorder (ODD) in the additive limit. Blackman [7] used the formalism of Blackman et al. [8] to treat transport in alloys with general ODD within the CPA. Schwartz [9]

[2]It is interesting to examine the following argument purportedly showing that the resistivity of any system, ordered or disordered, vanishes identically. Consider a given material, say a substitutionally disordered, random alloy. This material exists in one particular configuration that is characterized by real potentials (cell potentials) so that the corresponding Hamiltonian is Hermitean. The eigensolutions of the associated Schrödinger equation for such a system have real eigenvalues and must have infinite lifetimes. This situation appears to be identical to that in a pure, translationally invariant system, which seems to indicate that the resistivity of the random system must also vanish. This argument, however, is incomplete. One must think of the system in the presence of an electric field that we can take to be independent of both position and time. In the case of an ordered system, periodicity is preserved and the Hamiltonian in the presence of the field commutes with the momentum operator so that momentum continues to be a good quantum number. On the other hand, in a disordered material momentum is not conserved (with or without the field), and the perturbation caused by the presence of the field will induce transitions between states characterized by different values of the crystal momentum giving rise to a finite resistivity. We also note that this argument indicates that in a disordered system changes in momentum will cause the recoil of atoms and can lead to energy dissipation and the production of Joule heat. Joule heat is in fact required by energy conservation to dissipate the energy supplied in bringing the system in the region of the field. On the other hand, the Kubo formula, which is based on linear response, does not allow for Joule heat that is a second (and higher) order effect.

[3]Expressions based on the two-particle Green function are given in Chapter 24.

discussed electronic transport within the CPA for a multiband model, while Wysokinski et al. [10] investigated thermal and ODD effects on the ac and dc conductivity of binary and ternary alloys. Czycholl et al. [11] calculated transport quantities by applying the molecular CPA (see Chapter 12) to the calculation of the two-particle Green function. This last work, based on a cluster theory, allows the calculation of so-called *vertex corrections*, i.e., the effects associated with the correlated motion of two electrons interacting simultaneously with the same alloy configuration. Vertex corrections can also be calculated within the embedded cluster method (ECM) [12]. Vertex corrections, which are discussed in some detail in the body of this chapter, effectively measure the difference between the average of the product of two single-particle Green functions and the product of their configurational averages. Although they can be shown to vanish identically within the CPA in the case of certain TB model Hamiltonians, they are in general non zero and can have a profound effect on the calculated values of transport coefficients. A realistic application of the Kubo formula within a TB framework to calculate the absorption coefficient of amorphous Si has been reported by Pickett, et al [13].

The CPA-based works mentioned above were all concerned with TB models of substitutionally disordered alloys. Treatments of transport properties within the CPA based on a first-principles application of multiple-scattering theory have been also reported [14-21]. In their initial work, Butler et al. [14-16] made two approximations beyond linear response: they assumed the existence of well-defined energy bands in a substitutionally disordered alloy, and neglected vertex corrections (which in the present case do *not* vanish within the CPA). Both of these shortcomings were removed in subsequent work [17] that reported quite a satisfactory agreement between theoretical and experimental results for the dc conductivity of a number of Cu- and Ag-based alloys. Since then, a number of works have appeared [18-21], either reporting an additional application or extending the theory to include relativistic effects [21].

Before turning to the main discussion in this chapter, it is useful to emphasize the meaning and significance of the vertex corrections. As was mentioned above, the study of electronic transport is based on the study of the two-particle Green function [22], or, as presented here, on products of two single-particle propagators. In the case of disordered alloys, it is necessary to average such products over all alloy configurations. Each Green function is associated with the motion of a single electron, which within density functional theory, moves in the effective field of the nuclei and the other electrons. However, since the two electrons represented by the two propagators in the product of Green functions are affected by the *same* alloy configuration, their motion becomes statistically correlated. This means that one cannot replace the average of the product, $\langle GG \rangle$, by the product of the averages, $\langle G\rangle\langle G\rangle$. It is essentially the non-vanishing difference between these two quantities that gives rise to vertex corrections[4]. We will show that in some cases, these vertex corrections vanish identically within the CPA[5]. In general, however, vertex corrections must be taken into account within some approximation, and we shall describe methods that have been proposed for doing this.

We begin our discussion of electronic transport in random substitutional alloys based on the CPA by studying the ac and dc conductivity of a single band, diagonally disordered substitutional alloy within a TB model Hamiltonian. In this discussion, we follow closely the lines in the paper by Velicky [1]. We derive a number of important relations for the application of the theory, such as the satisfaction of particle conservation requirements (Ward identity) and the vanishing of vertex corrections. We then extend the discussion to systems with ODD. Our presentation based on TB models concludes with a discussion of cluster methods, especially the ECM, in the calculation of vertex corrections.

The presentation based on the TB model is followed by the treatment of electronic transport within a first-principles formulation of multiple scattering theory in the angular momentum representation and the KKR-CPA. We point out some salient features of the formalism and also the similarities and differences with the treatment based on the TB model.

20.2 CPA and Transport

At this point, it would be useful if the reader were to review the discussion of the CPA given in Chapter 12. It may also be useful to give a brief description of the CPA, emphasizing the features that are of most importance in the discussion of transport phenomena. Such a selective review is provided in the following subsection.

20.2.1 Review of the CPA

We consider a binary disordered material, A_cB_{1-c}, described by a single-band, TB Hamiltonian of the form, Eq. (19.112),

$$\hat{H} = \sum_n \epsilon_n a_n^\dagger a_n + \sum_{nm} W_{nm} a_n^\dagger a_m. \tag{20.1}$$

[4]Strictly speaking, vertex corrections denote the difference $\langle GCG\rangle - \langle G\rangle C\langle G\rangle$, where $\langle G\rangle$ is calculated within some (usually approximate) scheme, such as the LDA, and C is an operator that is often independent of configuration.

[5]This is no great advantage of the theory since vertex corrections are generally non zero.

The atomic (site) energies, ϵ_n, assume values ϵ_A with probability c and ϵ_B with probability $1-c$ associated with the two species forming the alloy, while the hopping terms, W_{nm}, are taken to depend only on the distance between sites n and m and not on the species of atom occupying those sites. The single-particle Green function associated with the Hamiltonian of Eq. (20.1) has the general form[6],

$$G(z) = (z - H)^{-1} = (z - \epsilon - W)^{-1}. \tag{20.2}$$

In the CPA, the true disordered material is replaced by an effective medium characterized by a site-diagonal, energy-dependent self-energy of the form[7]

$$\Sigma_{nm}(z) = \sigma(z)\delta_{nm}, \tag{20.3}$$

so that the self-energy can be written as (with explicit dependence on energy suppressed)

$$\Sigma = \sum_n \sigma_n, \tag{20.4}$$

with $\sigma_n = \sigma$. The Green function associated with the CPA translationally invariant medium is given by the general expression,

$$\bar{G}(z) = (z - \sigma - W)^{-1}, \tag{20.5}$$

and can also be written as an integral over the first Brillouin zone of the reciprocal lattice. In particular, the site-diagonal element has the form,

$$\bar{G}(z)_{00} = \frac{1}{\Omega_{\text{BZ}}} \int_{\text{BZ}} \mathrm{d}^3k \, [z - \sigma(z) - W(\mathbf{k})]^{-1}. \tag{20.6}$$

Here, the Fourier transform of the hopping matrix is given by the usual expression,

$$W(\mathbf{k}) = \frac{1}{N} \sum_n \mathrm{e}^{\mathrm{i}\mathbf{k}\cdot\mathbf{R}_{0n}} W_{0n}. \tag{20.7}$$

When the effective potential, σ, at a site is replaced by a "real" potential, ϵ_A or ϵ_B, the resulting system can be thought of as consisting of an impurity (the real atom) embedded in an otherwise pure, translationally invariant host material. The additional scattering of a wave resulting from

[6]For the sake of simplicity we do not explicitly exhibit the matrix form of the various quantities, and also dispense with operator notation involving hats over symbols. We trust that this will not cause confusion, since the material presented here has been discussed in previous pages of this book.

[7]We consider the case of monoatomic unit cells so that all sites in the CPA medium are equivalent.

the presence of the impurity is described by the single-site t-matrix, (with the explicit dependence on complex energy suppressed)

$$\begin{aligned} t_n &= (\epsilon_n - \sigma)\left[1 + \bar{G}t_n\right] \\ &= (\epsilon_n - \sigma)\left[1 - \bar{G}(\epsilon_n - \sigma)\right]^{-1}. \end{aligned} \tag{20.8}$$

The full t-matrix (operator) associated with the disordered alloy can be expressed in terms of the site t-matrices, [see Eq. (11.14)]

$$T = \sum_n t_n + \sum_n \sum_{m \neq n} t_n \bar{G} t_m + \cdots. \tag{20.9}$$

We note the exclusion in the summations in the second (and higher order) terms that prevent the electrons from scattering consecutively from the same site. The last expression can be replaced by a closed set of equations in two equivalent ways. We can write either

$$T = \sum_n Q_n \tag{20.10}$$

$$Q_n = t_n \left[1 + \bar{G} \sum_{m \neq n} Q_m\right] \tag{20.11}$$

or

$$T = \sum_n \tilde{Q}_n \tag{20.12}$$

$$\tilde{Q}_n = \left[1 + \sum_{m \neq n} Q_m \bar{G}\right] t_n. \tag{20.13}$$

Because $T(z^*) = T^\dagger(z)$ [see Eq. (20.8)], we have

$$\tilde{Q}_n(z^*) = Q_n^\dagger(z). \tag{20.14}$$

The CPA is based on the assumption that in a statistical average over alloy configurations, say the average of the t-matrix in Eq. (20.10), the correlation between different sites can be neglected. Thus, from Eq. (20.10) and (20.11) we obtain,

$$\langle T \rangle = \sum_n \langle Q_n \rangle, \tag{20.15}$$

$$\langle Q_n \rangle = \langle t_n \rangle \left(1 + \bar{G} \sum_{m \neq n} \langle Q_m \rangle\right), \tag{20.16}$$

with a corresponding expression holding with respect to $\tilde{Q}_n$. Based on these assumptions, the CPA self-consistency condition determining the self-energy $\sigma(z)$ takes the form,

$$\langle T \rangle = \langle t_n \rangle = 0. \tag{20.17}$$

As was shown in Chapter 12, this condition can also be written in terms of the Green function,

$$\langle G_{00} \rangle = \bar{G}_{00}, \tag{20.18}$$

where in the last two expressions the brackets $\langle \cdots \rangle$ denote an average over the occupation of a single site. In the case of a binary alloy, we can obtain the following expression for the self-energy that is entirely equivalent to the last two conditions,

$$\sigma = \langle \epsilon \rangle - (\epsilon_A - \sigma)\bar{G}_{00}(\epsilon_B - \sigma). \tag{20.19}$$

We also quote the following general relations between Green functions and t-matrices that are valid in the case of multicomponent alloys,

$$\begin{aligned} G &= \bar{G} + \bar{G}(\epsilon - \sigma)G \\ &= \bar{G} + \bar{G}T\bar{G}, \end{aligned} \tag{20.20}$$

where the matrix character of the various quantities in site space is suppressed but is to be understood throughout.

20.2.2 Transport quantities in the CPA

We now apply the CPA formalism to the evaluation of the Kubo formula. In particular, we are interested in obtaining an approximate expression for the operator, K, Eq. (19.72), involving the product of two single-particle Green functions. Inserting Eq. (20.20) into Eq. (19.72), and using the CPA condition we can write

$$\begin{aligned} K &= \bar{G}C\bar{G} + \bar{G}\langle T\bar{G}C\bar{G}T \rangle \bar{G} \\ &= \bar{G}(C + \Gamma)\bar{G}. \end{aligned} \tag{20.21}$$

In writing down these expressions, we have used the fact that the operator C (here the momentum or current operator) is configuration-independent, and have interpreted the brackets in Eq. (19.72) as an average over alloy configurations, as well as a spatial average over the sample. The quantity Γ which appears in the second line above is defined explicitly by the quantity in the brackets in the first line, and denotes the *vertex corrections* to the operator C. As has been mentioned, vertex corrections arise because of the correlation in the motion of two particles moving in and interacting with the same alloy configuration. We now turn our attention to the use of the CPA in determining the vertex corrections Γ and hence K.

From Eqs. (19.72), (20.11), and (20.21), we get

$$\Gamma = \sum_n \sum_m \langle Q_n \bar{G} C \bar{G} \tilde{Q}_m \rangle, \tag{20.22}$$

where

$$\langle Q_n \bar{G} C \bar{G} \tilde{Q}_m \rangle = \langle t_n [1 + \bar{G} \sum_{l \neq n} Q_l] \bar{G} C \bar{G} [1 + \sum_{k \neq m} \tilde{Q}_k \bar{G}] t_m \rangle. \tag{20.23}$$

Consistent with the spirit of the CPA, we neglect statistical correlations between sites and approximate Eq. (20.23) in the form,

$$\langle Q_n \bar{G} C \bar{G} \tilde{Q}_m \rangle = \langle t_n \langle [1 + \bar{G} \sum_{l \neq n} Q_l] \bar{G} C \bar{G} [1 + \sum_{k \neq m} \tilde{Q}_k \bar{G}] \rangle t_m \rangle. \tag{20.24}$$

For $n \neq m$, the t-matrices t_n and t_m are statistically independent and can be averaged separately. Because of the CPA condition, Eq. (20.17), this average vanishes, so that a non-zero value for the vertex corrections is obtained only when $n = m$. Therefore, we now define quantities Γ_n through the relation

$$\langle Q_n \bar{G} C \bar{G} \tilde{Q}_m \rangle = \Gamma_n \delta_{nm}. \tag{20.25}$$

This result can be used to simplify the internal average in Eq. (20.24). Since only the diagonal terms, $n = m$, are non-zero, we have

$$\Gamma_n = \langle t_n \bar{G} \left(C + \sum_{l \neq n} \Gamma_l \right) \bar{G} t_n \rangle, \tag{20.26}$$

and from Eqs. (20.22) and (20.25) it follows that

$$\Gamma = \sum_n \Gamma_n. \tag{20.27}$$

In view of Eq. (20.21), the last expression leads to the result,

$$K = \bar{G} C \bar{G} + \bar{G} \sum_n \Gamma_n \bar{G}, \tag{20.28}$$

where

$$\Gamma_n = \langle t_n K t_n \rangle - \langle t_n \bar{G} \Gamma_n \bar{G} t_n \rangle. \tag{20.29}$$

The last two expressions represent the application of the CPA to the determination of the quantity K, and through Eq. (19.67) and (19.71) also allow that of the complex admittance, $\chi_{\hat{C}_1\hat{C}_2}$. It is important to emphasize that the site-diagonality of the vertex corrections is a direct consequence of the assumptions underlying the CPA. Namely, in random systems the motion of the two electrons remains (essentially) uncorellated as long as they find themselves in the vicinity of different sites; such sites are averaged separately and due to the CPA condition the average of the corresponding t-matrices vanishes. On the other hand, when the electrons are in the vicinity of the same site, then the CPA condition is no longer sufficient for the average of the product of *two* t-matrices to vanish and leads to a non-zero contribution to the vertex corrections. Therefore, in the CPA the vertex corrections become asum of single-site contributions, consistent with the single-site nature of this approximation. The situation becomes much more complex in the presence of SRO in which different sites cannot be averaged independently.

We note that Eq. (20.26) can be solved through iteration to yield the following formal expression for Γ

$$\begin{aligned}\Gamma &= \sum_n \langle t_n \bar{G} C \bar{G} t_n \rangle \\ &+ \sum_n \sum_{m \neq n} \langle t_n \bar{G} \langle t_m \bar{G} C \bar{G} t_m \rangle \bar{G} t_n \rangle + \cdots . \end{aligned} \tag{20.30}$$

Like Eq. (20.9), this expression contains summation exclusions preventing consecutive scatterings from the same site. In addition to this iterative solution, vertex corrections within the CPA can be determined in essentially closed form, as is demonstrated further on in this chapter. Also, they can be calculated with ever increasing accuracy directly by means of cluster theories.

20.2.3 Complex admittance in the CPA

As with the single-particle Green function, it is important to inquire as to the extent to which the exact properties of various transport quantities, depending on the two-particle propagator, are preserved in the CPA. In particular, we examine the properties of the complex admittance embodied in Eqs. (19.73) to (19.82).

From Eq. (20.30), the formal solution for K, Eq. (20.28), takes the form,

$$\begin{aligned} K &= \bar{G} C \bar{G} + \frac{c}{1-c} \sum_n \bar{G} t_n^A \bar{G} C \bar{G} t_n^A \bar{G} \\ &+ \left(\frac{c}{1-c}\right)^2 \sum_n \sum_{m \neq n} \bar{G} t_n^A \bar{G} \bar{G} t_m^A \bar{G} C \bar{G} t_m^A \bar{G} \bar{G} t_n^A \bar{G} + \cdots , \end{aligned} \tag{20.31}$$

where all scattering is expressed in terms of the t-matrix for atoms (cells) of species A in the alloy through the CPA condition,

$$ct_n^A + (1-c)t_n^B = 0. \tag{20.32}$$

The Green functions $\bar{G}$ are those associated with the effective medium determined in the CPA.

Term by term consideration of the last expression shows that the identities in Eq. (19.79), (19.81), and (19.83) are satisfied by the approximate K. An important requirement of any approximate theory is the satisfaction of Eq. (19.82), when the exact Green functions[8], G, in that expression are replaced by approximate Green functions, $\bar{G}$, say those of the CPA. To verify [1] that Eq. (19.82) holds in the CPA, we note that $\bar{G}(z) \to z^{-1}$ in the limit $z \to \infty$. Therefore, the "free" part of $\bar{G}C\bar{G}$ (not involving vertex corrections) has the asymptotic behavior $z_1^{-1}C\bar{G}(z_2)$. If we can show that $\Gamma_n \to 0$ in the same limit, then Eq. (19.82) will be verified. But, as $z \to \infty$ we have $t_n(z) \to \epsilon_n - \langle\epsilon_n\rangle$, since $\sigma_n(z) \to \langle\epsilon_n\rangle$ as follows from Eq. (20.26). This leads to the behavior $\Gamma_n \to z_1^{-1}$ as $z_1 \to \infty$, which completes the proof of validity of the identity (19.82) in the CPA. It follows that we can write,

$$K(z_1, C, z_2) \to z_1^{-1}C\bar{G}(z_2) + O(z_1^{-2}), \quad \text{as} \quad z_1 \to \infty. \tag{20.33}$$

One important consequence of the validity of this identity is that it allows us to evaluate averages of concentration-independent operators, such as those in Eq. (19.80) through the use of single-particle Green functions. In particular, we can write,

$$\begin{aligned}\langle\delta(\lambda - H)\rangle &= \sum_n |n\rangle\langle n|\delta(\lambda - E_n) \\ &= \frac{1}{2\pi i}\left[\bar{G}(\lambda^-) - \bar{G}(\lambda^+)\right],\end{aligned} \tag{20.34}$$

with $\lambda^{\pm} = \lambda \pm i\epsilon$, with ϵ being a positive infinitesimal and $\bar{G}$ the averaged Green function.

The final formal property that we consider is the identity in Eq. (19.82). This equation is an example of a *Ward's identity* that expresses the response, K, of the electrons to changes in the origin of energies[9] by means of G. Equivalently, Eq. (19.82) relates the exact vertex part, Γ, to the exact self-energy, Σ,

$$\Gamma(z_1, 1, z_2) = -(z_1 - z_2)^{-1}\left[\Sigma(z_1) - \Sigma(z_2)\right]. \tag{20.35}$$

[8]The term exact here refers to the single-particle Green function, evaluated within, say, the LDA, averaged exactly over alloy configurations rather than the exact many-particle Green function.

[9]In a many-body treatment, this would correspond to changes in the chemical potential.

We wish to show that Eqs. (19.82) and (20.35) remain valid when they are expressed in terms of the Green functions and self-energies obtained in the CPA. That is to say, we wish to show that the relations,

$$K(z_1, 1, z_2) = -(z_1 - z_2)^{-1}\left[\bar{G}(z_1) - \bar{G}(z_2)\right] \tag{20.36}$$

and

$$\Gamma(z_1, 1, z_2) = -(z_1 - z_2)^{-1}\left[\sigma(z_1) - \sigma(z_2)\right] \tag{20.37}$$

hold. The satisfaction of the Ward's identity has important implications for the CPA [1]. This identity provides the link between K and $\bar{G}$ and implies particle-number conservation. Since its validity depends on the detailed structure of the CPA equation, it presents important evidence for the internal consistency of this approximation. We now proceed with the proof of the validity of the last two equations.

From the site-diagonal nature of the self-energy and the vertex corrections in the CPA, e.g., Eq. (20.27), and from Eqs. (20.5) and (20.8), we can cast the general expression for Γ, Eq. (20.26), in the form

$$\begin{aligned} \Gamma_n(z_1, 1, z_2) &= \langle t_n(z-1)\bar{G}(z-1) \\ &\times \left\{1 - \sum_{m\neq n}(z_1 - z_2)^{-1}\left[\sigma_m(z_1) - \sigma_m(z_2)\right]\right\} \\ &\times \bar{G}(z_2)t_n(z_2)\rangle \\ &= -(z_1 - z_2)^{-1}\langle t_n(z_1)\bar{G}(z_1)\left[\epsilon_n - \sigma_n(z_1)\right] - \left[\epsilon_n - \sigma_n(z_2)\right] \\ &\times \bar{G}(z_2)t_n(z_2)\rangle. \end{aligned} \tag{20.38}$$

It is instructive to note the role played by the summation exclusions in the multiple scattering expressions, Eq. (20.9). In the absence of such exclusions, instead of Eq. (20.38) one obtains

$$\Gamma(z_1, 1, z_2) = -(z_1 - z_2)^{-1}\langle t_n(z_1)\left[\bar{G}(z_1) - \bar{G}(z_2)\right]t_n(z_2)\rangle. \tag{20.39}$$

Equation (20.38) can now be simplified through the use of the CPA self-consistency condition, $\langle t_n\rangle = 0$, and Eq. (20.8) to yield Eq. (20.37). Thus, Ward's identity remains valid and particle-number conservation is preserved in the CPA.

20.2.4 Vanishing of vertex corrections in the CPA

Within the single-band, TB model discussed above, the vertex corrections vanish identically in the CPA. To see this, we consider the expression,

$$\begin{aligned} [\bar{G}p_\alpha\bar{G}] &= \sum_l \bar{G}_{0n}p_\alpha^{nm}\bar{G}_{m0} \\ &= \frac{1}{N}\sum_{\mathbf{k}} \bar{G}(\mathbf{k})p_\alpha(\mathbf{k})\bar{G}(\mathbf{k}). \end{aligned} \tag{20.40}$$

Now, the CPA Green function has the form of Eqs. (20.5) and (20.6), and since $W(\mathbf{k}) = W(-\mathbf{k})$ it is an even function of $\mathbf{k}$. On the other hand, it follows from Eq. (19.115) that $j_\alpha(\mathbf{k}) = -\mathbf{j}_\alpha(-\mathbf{k})$ so that $p_\alpha(\mathbf{k})$ is an odd function of $\mathbf{k}$ and the quantity in Eq. (20.40) vanishes identically. Therefore, in the CPA one has

$$K(z_1, p_\alpha, z_2) = \bar{G}(z_1) p_\alpha \bar{G}(z_2), \tag{20.41}$$

which has non-vanishing matrix elements K_{nm} only for $n \neq m$.

Before leaving this discussion, it is important to point out that the vertex corrections do *not* always vanish within the CPA. For example, vertex corrections do not vanish if the band structure of a material possesses reversal symmetry. This is often the case [9] in multiband alloys in which the current operator contains site-diagonal elements. An explicit example of non-vanishing vertex corrections is provided by the application of first-principles MST and the KKR-CPA to transport properties discussed in a following section.

20.2.5 Alloys with ODD

It is fairly straightforward to extend the TB model results obtained above for diagonally disordered alloys to alloys exhibiting off-diagonal disorder (ODD). This is done through the formalism of Blackman et al. [8], which allows the implementation of the CPA to alloys with ODD. One simply replaces scalar quantities, occurring in the treatment of diagonally disordered systems, by matrices in the configurational space defined by the possible occupations of a single site in the alloy. Thus, an n-component alloy gives rise to n-dimensional matrices. This matrix replacement[10] is to be introduced in all quantities, such as hopping terms, self energies, or Green functions, which enter the discussion of transport properties.

As an example, we consider the case of binary alloys. One can use the Green functions defined in Eq. (11.60) and (11.65) to define a two-dimensional t-matrix as indicated in Eq. (11.70). Then, the two-particle response function, K, and the vertex corrections, Γ_n, follow along lines familiar from the treatment of diagonally disordered treatments, but with the appropriate replacement of scalar quantities by two dimensional matrices.

We forego a detailed discussion of the transport properties of alloys with ODD because the associated formalism is essentially identical to that which is used in the following section in a first-principles application of the KKR-CPA to transport properties. However, the reader may find it useful to derive some of the relevant expressions at this point.

[10] As cautioned in Chapter 12, care should be taken not to invert any singular matrices that may arise within this formalism.

20.2.6 ECM and vertex corrections

We saw above that vertex corrections result because of the correlated motion of two electrons interacting with the same alloy configuration. This suggests that a cluster theory could be useful in the study of transport properties of disordered materials because it may allow a direct and systematic study of the effects of vertex corrections. Such a theory is afforded by the embedded cluster method (ECM)[11].

Let us consider the Kubo formula for the observed ac conductivity, Eq. (19.101),

$$\sigma_{\alpha\beta}(\omega) = \frac{2\pi e^2}{m^2\omega\Omega}\sum_{nm}\sum_{pk}\int \mathrm{d}E[f(E_n) - f(E_m)]$$
$$\times \quad \langle p_\alpha^{nm}\Im G^{mk}(E - E_n)p_\beta^{kp}\Im G^{pn}(E - E_m)\rangle, \tag{20.42}$$

where $E_m = \omega + E_n$, and the angular brackets denote configurational averaging. In the case of the CPA, each Green function in the last expression is replaced by its corresponding value in the CPA effective medium. In the ECM, one may use the Green functions associated with clusters of a finite (usually small) number of sites embedded in the CPA medium. Taking the momenta p_α to be configuration independent, and denoting by P_J the probability of occurrence of cluster configuration, J, we have

$$\langle p_\alpha^{nm}\Im G^{mk}(E - E_n)p_\beta^{kp}\Im G^{pn}(E - E_m)\rangle = \frac{1}{N_J}\sum_J P_J$$
$$\times \quad \left[p_\alpha^{nm}\Im G_J^{mk}(E - E_n)p_\beta^{kp}\Im G_J^{pn}(E - E_m)\right], \tag{20.43}$$

where N_J is the number of cluster configurations, and the sites n, m, k, and p, are restricted to lie in the cluster. Because the Green functions, G_J, are configuration dependent, and both Green functions in the product are calculated at the same configuration, one obtains a proper description of the correlated electron motion and, hence, of vertex corrections. In fact, we note that if a cluster theory such as the ECM is used, it is not necessary to separately calculate corrections to the product of two averaged Green functions. Rather, a direct calculation of K is obtained (which approaches the exact value as the size of the cluster increase).

[11] A self-consistent cluster theory such as the MCPA, discussed in Chapter 11, can also be used in the study of transport properties, as is indicated in some of the references to this chapter. However, the inherent computational difficulties of implementing the MCPA compared to the ECM tends to discourage practical applications. Other self-consistent cluster theories can also be used with due attention being paid to such matters as analyticity and the satisfaction of fundamental sum rules.

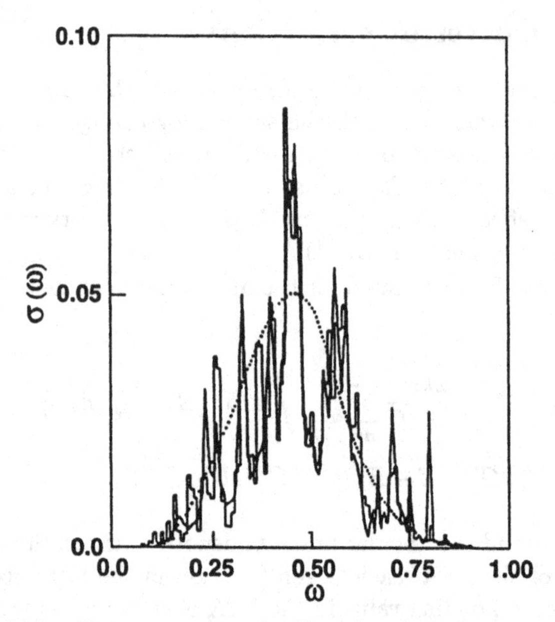

Figure 20.1: **Exact, numerically simulated ac conductivity (histogram) for a one-dimensional, single-band binary alloy, A_cB_{1-c} with $\epsilon_A = -\epsilon_B = 2.0$, near-neighbor hopping, $W = 1.0$, and $c = 0.5$, compared with the results obtained in the CPA (dotted line), and in a seven-site application of the ECM (solid line). The frequency ω is given in terms of the bandwidth.**

20.2.7 Numerical results

We now present numerical results [23] of the CPA-based formalism for electronic transport in alloys obtained above. Figure 20.1 shows a comparison between the ac conductivity of a one-dimensional, single band TB binary alloy obtained in an exact numerical simulation based on large linear segments (histogram) in the CPA, and in a seven-site application of the ECM. The Fermi energy was chosen to be $\mu = 0.5$, corresponding to an energy $E = 0.0$. This energy falls in the gap formed by the subbands of the two alloy species.

Not unexpectedly, the CPA yields a smooth conductivity curve in sharp contrast to the exact results, which exhibit a great deal of structure. As in the case of the single-particle DOS, the CPA provides a smooth interpolation curve through the exact results. Numerical integration also shows that within numerical accuracy the CPA does indeed satisfy the exact properties discussed above, Eqs. (19.79) to (20.35).

On the other hand the conductivity calculated within the CPA does have a flaw, which is masked in this figure. It fails in general to reproduce the exact result that in strictly one-dimensional systems the dc conductivity, $\sigma(0)$, vanishes identically. This is clearly the case since in such a system even the slightest amount of disorder destroys the single path through the system that would mediate conduction. The fact that the CPA results shown here *do* vanish at $\omega = 0$ is a consequence of the fact that at $\mu = 0.5$ both the DOS and the joint DOS vanish, and *not* an indication of proper behavior of the CPA.

In contrast to the single-site CPA results, those obtained in a seven-site application of the ECM contain a great deal of structure that resolves practically all peaks in the exact histogram. In fact, a seven-site ECM calculation of the conductivity appears to yield comparatively more accurate results than it does for the DOS (compare with Fig. 11.4). In particular, it represents quite accurately the edges of the spectrum, in contrast to the case of the DOS where both the CPA and the ECM yield analytically incorrect results. This result is not hard to understand. The ac conductivity involves the convolution of two DOSs and consequently contains less structure than either of them. On the other hand, a seven-site ECM calculation retains the dominant peaks in the DOS, i.e., those that can be expected to survive the convolution process.

It is also interesting to examine the nearly flawless reproduction of the edges in the spectrum of $\sigma(\omega)$ by the ECM results. This is at first surprising because a calculation based on a small cluster yields accurate results for quantities which in principle can be influenced by long-distance fluctuations. This feature arises because increasing disorder tends to increase the localization of states and transport properties become rather insensitive to fluctuations beyond a moderate distance. On the other hand, no calculation based on a finite-size cluster can be expected to yield accurate results for the transport properties of (nearly) ordered materials, or materials with a large degree of short-range order. For such systems, a much more judicious choice of an effective medium [23] may be necessary, before accurate results can be obtained.

The application of the ECM to the calculation of transport properties has a number of desirable features. First, it allows the incorporation from the start of the effects of local fluctuations on the conductivity, and avoids the separate calculation of vertex corrections at a later stage. Second, it becomes exact in the limit of a large cluster (and an appropriately chosen effective medium). Third, it allows the investigation of the effects of short-range order since, as in calculations of the DOS, it enables one to identify peaks in the conductivity spectra associated with specific cluster configurations. Fourth, the method is computationally simple and can be applied to disordered materials in any dimension and to Hamiltonians of more general type than that of the simple TB model considered here. In fact, since realistic, three-dimensional systems can be expected to possess less structure in their DOS spectra than one-dimensional TB systems, the ECM can conceivably be used in connection with fairly small clusters (perhaps encompassing no more than two or three sites). Larger clusters and media other than those obtained in CPA-based calculations may be necessary in order to account for SRO effects in real materials.

20.3 First-Principles MST and Transport

In this section, we present the formal and computational aspects of the application of multiple scattering theory to the study of transport in random alloys. In a manner analogous to the case of the TB model, we base our discussion on the Korringa-Kohn-Rostoker coherent potential approximation, beginning with a brief review of the method itself. For more details of the CPA formalism, the reader is referred to Chapter 12.

20.3.1 Review of the KKR-CPA

For reasons that will probably become clear below [see also discussion following Eq. (12.49)], our review of the KKR-CPA will be based[12] on the matrix formalism of BEB. For the sake of clarity, we confine our exposition to the case of a binary alloy, A_cB_{1-c}; the extension to multicomponent alloys is straightforward.

Within the KKR-CPA, the effective medium that replaces the true, disordered alloy material is determined through a self-consistency condition which can be written in the general form, Eq. (12.33),

$$\langle \underline{G}_{00} \rangle = \bar{\underline{G}}_{00}. \tag{20.44}$$

Here, $\underline{G}_{00}$ and $\bar{\underline{G}}_{00}$ denote single-site matrix elements of the Green function associated with the disordered material and the effective medium, respectively, and $\langle \cdots \rangle$ denotes an average over the occupation of a single site. In the matrix version of the KKR-CPA, the single-particle Green function, $\bar{\underline{G}}(\mathbf{r}, \mathbf{r}'; E)$, when $\mathbf{r}$ and $\mathbf{r}'$ are in no other cells than Ω_n and Ω_m, takes the form,

$$\underline{G}(\mathbf{r}, \mathbf{r}'; E) = \langle \underline{Z}(\mathbf{r}; E) | \underline{\tau}^{nm}(E) | \underline{Z}(\mathbf{r}'; E) \rangle, \tag{20.45}$$

[12] Although most of the formal development and applications of the KKR-CPA has been carried out within the muffin-tin approximation to the cell potentials of a solid, we will not make the MT assumption explicitly in our discussion. Given the various conditions of convergence mentioned in Chapter 3, we will use the MT forms of the various equations but will consider them applicable to generally shaped full-cell potentials under appropriate modifications of various quantities such as the structure constants.

where the single-scatterer term is omitted being real at real energies and not contributing to the charge density[13]. Care should be taken by recalling the matrix nature of the variousquantities that appear in the last expression. A single underline deno tes a matrix in configuration space, that is to say in AB space for a binary alloy, while a double underline denotes a matrix in both site and configuration space. We do not explicitly exhibit the vector or matrix nature of various quantities in angular momentum (L) space trusting that it will be apparent from the content of the discussion. In this notation, the wave function "matrices" $|\underline{Z}\rangle$ are given by the expressions

$$\underline{Z}(\mathbf{r};E) = \begin{pmatrix} |Z^A(\mathbf{r};E)\rangle & 0 \\ 0 & |Z^B(\mathbf{r};E)\rangle \end{pmatrix}, \tag{20.46}$$

where $|Z^\alpha(\mathbf{r};E)\rangle$, is the regular solution of the Schrödinger equation associated with the potential in a cell of type α, with $\alpha = A$ or B, and the ket notation denotes a vector indexed by L. A similar notation describes the bra vector, $\langle\underline{Z}(\mathbf{r};E)|$. It is obvious from Eq. (20.46) that the vector wave function is independent of the occupation of a given site. The Green function, $\underline{G}(\mathbf{r},\mathbf{r}';E)$, for $\mathbf{r}$ ($\mathbf{r}'$) in no other cells than Ω_n (Ω_m), has the form

$$\underline{G}(\mathbf{r},\mathbf{r}';E) = \begin{pmatrix} x_n G^{AA}(\mathbf{r},\mathbf{r}';E)x_m & x_n G^{AB}(\mathbf{r},\mathbf{r}';E)y_m \\ y_n G^{BA}(\mathbf{r},\mathbf{r}';E)x_m & y_n G^{BB}(\mathbf{r},\mathbf{r}';E)y_m \end{pmatrix}, \tag{20.47}$$

with x_n and $y_n = 1 - x_n$ as occupation numbers that equal 1 and 0 when site n is occupied by an atom of type A, or 0 and 1 for an atom of type B. Each individual component, $G^{\alpha\beta}$, in the last matrix is to be interpreted as the corresponding quantity determined within the ordinary (scalar) version of MST. Finally, the scattering-path matrix (also called the scattering-path operator), $\underline{\tau}^{nm}(E)$, is given by an expression analogous to that of the Green function,

$$\underline{\tau}^{nm}(E) = \begin{pmatrix} x_n \tau^{nm;AA}(E)x_m & x_n \tau^{nm;AB}(E)y_m \\ y_n \tau^{nm;BA}(E)x_m & y_n \tau^{nm;BB}(E)y_m \end{pmatrix}, \tag{20.48}$$

where each element, $\tau^{nm;\alpha\beta}$, is a matrix in L-space.

To complete the description of the KKR-CPA self-consistency condition in the matrix form, we note that the effective-medium wave functions, Green functions, and scattering-path matrices are given, respectively, by Eq. (20.46) and by the expressions

$$\bar{\underline{G}}(\mathbf{r},\mathbf{r}';E) = \begin{bmatrix} G^{AA}(\mathbf{r},\mathbf{r}';E) & G^{AB}(\mathbf{r},\mathbf{r}';E) \\ G^{BA}(\mathbf{r},\mathbf{r}';E) & G^{BB}(\mathbf{r},\mathbf{r}';E) \end{bmatrix}, \tag{20.49}$$

[13]This term must be taken into account at complex energies, in which case it can be incorporated into the formalism without undue effort.

and

$$\bar{\underline{\tau}}^{nm}(E) = \left[\begin{array}{cc} \bar{\tau}^{nm;AA}(E) & \bar{\tau}^{nm;AB}(E) \\ \bar{\tau}^{nm;BA}(E) & \bar{\tau}^{nm;BB}(E) \end{array} \right]. \tag{20.50}$$

Here the bars denote effective medium quantities obtained within the KKR-CPA. For example, because of the translational invariance of the CPA effective medium, the scattering-path matrix is given by an integral over the first Brillouin zone,

$$\bar{\underline{\tau}}^{nm} = \frac{1}{\Omega_{\mathrm{BZ}}} \int_{\mathrm{BZ}} \mathrm{d}^3k \, [\bar{\underline{m}} - \underline{G}(\mathbf{k})]^{-1} \, \mathrm{e}^{\mathrm{i}\mathbf{k}\cdot\mathbf{R}_{nm}}, \tag{20.51}$$

where $\bar{\underline{m}} = \bar{\underline{t}}^{-1}$ is the inverse of the effective cell t-matrix which in configuration space is given by the expression,

$$\bar{\underline{t}}(E) = \left[\begin{array}{cc} \bar{t}^{AA}(E) & \bar{t}^{AB}(E) \\ \bar{t}^{BA}(E) & \bar{t}^{BB}(E) \end{array} \right], \tag{20.52}$$

and $\underline{G}(\mathbf{k})$ is the configurational matrix expression for the KKR structure constants

$$\underline{G}(E) = \left[\begin{array}{cc} G(\mathbf{k}) & G(\mathbf{k}) \\ G(\mathbf{k}) & G(\mathbf{k}) \end{array} \right], \tag{20.53}$$

where $G(\mathbf{k})$ represents the familiar KKR structure constants (which are a matrix in L-space, a fact not denoted explicitly in the present discussion). The KKR-CPA self-consistency condition now involves the site-diagonal matrix elements of the alloy and effective medium Green functions defined in Eqs. (20.47) and (20.49).

Since the wave function matrices $\langle \underline{Z} |$ and $| \underline{Z} \rangle$ are configuration independent, the KKR-CPA self-consistency condition, Eq. (20.44), can also be written in the form

$$\langle \underline{\tau}^{00} \rangle = \bar{\underline{\tau}}^{00}, \tag{20.54}$$

where the site-diagonal element of the effective scattering-path matrix, $\bar{\underline{\tau}}^{00}$, is obtained from the expression in Eq. (20.51) by setting $n = m$. The corresponding quantity for the alloy has the form

$$\underline{\tau}^{00,\alpha} = \left\{ \underline{1} - \underline{t}_0^{\alpha} \left[\bar{\underline{m}} - (\bar{\underline{\tau}}^{00})^{-1} \right] \right\}^{-1} \underline{t}_0^{\alpha}, \tag{20.55}$$

where the site t-matrix is given by the expression

$$\underline{t}_n^{\alpha} = \left(\begin{array}{cc} x_n t^A & 0 \\ 0 & y_n t^B \end{array} \right). \tag{20.56}$$

Finally, it is also possible to express the KKR-CPA self-consistency condition in the form, (see discussion in Chapter 12),

$$\langle \underline{X}_{00} \rangle = 0, \tag{20.57}$$

where the "scattering matrix" $\underline{X}$, which measures the additional scattering induced by the presence of a real atom in the KKR-CPA effective medium has the form[14],

$$\underline{X}_n = \left[\underline{t} - (\underline{t}\underline{\bar{m}} - \underline{1})\,\underline{\bar{\tau}}^{00}\right]^{-1} (\underline{t}\underline{\bar{m}} - \underline{1})\,. \tag{20.58}$$

The matrix $\underline{X}$ is seen to correspond to the site t-matrix, t, encountered in the TB model.

It is a matter of some straightforward algebra to show that the scattering-path matrix for the alloy is given in terms of the corresponding quantity for the CPA medium by the expression [compare with Eq. (20.20)]

$$\underline{\tau} = \underline{\bar{\tau}} + \underline{\bar{\tau}}\underline{X}\underline{\bar{\tau}}, \tag{20.59}$$

where the full X-matrix (operator) is given in terms of individual site X-matrices by

$$\underline{X} = \sum_n \underline{X}_n + \sum_n \sum_{m \neq n} \underline{X}_n \underline{\bar{\tau}} \underline{X}_n + \cdots, \tag{20.60}$$

which corresponds to Eq. (20.10). Now, the entire formalism contained in Eqs. (20.10) to (20.20) can be transcribed intact in terms of the matrices $\underline{X}$, $\underline{\tau}$, and $\underline{\bar{\tau}}$, which play, respectively, the roles of t, G, and $\bar{G}$. As is shown in the following section, the resulting formal relations can be used to express transport coefficients within the KKR-CPA as were the analogous expressions in the TB model.

Before leaving this section, it may be useful to comment on some special features of the matrix formulation of the KKR-CPA. First, this approach allows the KKR-CPA condition to be expressed in terms of the Green function as well as the scattering-path operator and the scattering matrix, $\underline{X}$. The direct relation involving Green functions is not possible in the scalar version of the KKR-CPA because of the dependence of the cell wave functions on the type of atom occupying the cell. In a matrix version, these wave functions are arranged in configurationally independent matrices and can be "divided out" on both sides of the self-consistency condition, Eq. (20.44), to lead to the self-consistency condition in terms of the scattering-path matrices, Eq. (12.50).

Second, even though the matrix approach to KKR-CPA is very useful in the discussion of transport phenomena, as is shown below, it is not necessary to solve the matrix equations in order to obtain the effective-medium

[14]The form quoted here is designed to avoid the inversion of singular matrices such as those representing the cell t-matrix.

quantities, such as effective site t-matrices and scattering-path matrices. One may solve the scalar version of the KKR-CPA and then arrange the various quantities in a matrix for further application to transport quantities. Finally, we must keep in mind that the matrix formulation leads to quantities with elements in configuration space. Some of these quantities may have individual significance, others may not [see discussion following Eq. (11.62)]. In any case, physical quantities, such as the physical Green function (whose imaginary part yields the total particle density), are given as sums over *all* elements in configuration space. This must be born in mind in deriving formulae for measurable, i.e., physical, quantities describing transport in random alloys, and in the following discussion we will point out where such sums become necessary in order to yield physical results.

20.4 KKR-CPA and the Kubo Formula

20.4.1 General comments

The main advantages of the configurational-matrix formulation of the KKR-CPA applied to transport phenomena in random alloys is that it leads to formulae which, apart from the detailed meaning of the symbols in them, are formally identical to those arrived at within the TB model. This allows a unified exposition of the subject that emphasizes the underlying physics rather than the mechanics of a calculation. At the same time, one should not loose sight of the detailed meaning of the quantities occurring in the various expressions, since their behavior may differ for different models. For example, even though the expressions for the vertex corrections in the TB model and in the KKR-CPA are essentially identical, vertex corrections do *not* vanish in the latter case. With these caveats in mind, we now derive the form taken by the Kubo formula for the transverse conductivity within the KKR-CPA framework.

As we have seen in the previous discussion, the evaluation of most transport coefficients, C, involves expressions of the type,

$$C = \mathrm{Tr}\langle O_1 G O_2 G\rangle, \tag{20.61}$$

where G is a single-particle Green function that depends on the configuration of the system, and O_1 and O_2 are operators which are often independent of configuration (as is the current or momentum differential operator that occurs in the study of electrical transport). The brackets in the last equation denote a spatial average, which in the case of a disordered material, includes an average over all alloy configurations. We now show the relevance of Eq. (20.61) in the study of transport for the specific case of the dc conductivity.

20.4.2 dc conductivity and Green functions

We consider the dc conductivity at zero temperature[15]. From Eq. (19.110), we have

$$\sigma_{\alpha\beta} = \frac{2\pi}{N\Omega}\int_{-\infty}^{\mu} \mathrm{d}E \sum_{nm} \langle\langle n|j_\alpha|m\rangle\langle m|j_\beta|n\rangle\delta(E-E_n)\delta(E-E_m)\rangle, \quad (20.62)$$

where N denotes the number of unit cells each having volume Ω. The current operator has the form (with $\hbar = 1$)

$$j_\alpha = -\mathrm{i}\frac{\mathrm{e}}{m}\frac{\partial}{\partial r_\alpha}, \quad (20.63)$$

and is clearly configuration independent. From Eq. (20.62) we abstract the energy dependent quantity

$$\begin{aligned}\sigma_{\alpha\beta}(E) &= \frac{2\pi}{N\Omega}\sum_{nm}\langle\langle n|j_\alpha|m\rangle\langle m|j_\beta|n\rangle\delta(E-E_n)\delta(E-E_m)\rangle \\ &= \frac{2\pi}{N\Omega}\sum_{nm}\langle\langle n|j_\alpha|m\rangle\Im G(E-E_n)\langle m|j_\beta|n\rangle\Im G\rangle. \quad (20.64)\end{aligned}$$

Using the expressions relating the imaginary part of the single-particle Green functions to delta functions, Eq. (20.61), in the form

$$\begin{aligned}\sum_n |n\rangle\langle n|\delta(E-E_n) &= -\frac{1}{\pi}\Im G(E+\mathrm{i}\eta) \\ &= \frac{1}{2\pi\mathrm{i}}\lim_{\eta\to 0}\left[G(E+\mathrm{i}\eta)-G(E-\mathrm{i}\eta)\right], \quad (20.65)\end{aligned}$$

we can write [16] Eq. (20.62) in the form of Eq. (20.61)

$$\begin{aligned}\sigma_{\alpha\beta}(E) &= \frac{1}{4}\lim_{\eta\to 0}\left[\tilde{\sigma}_{\alpha\beta}(z_1,z_1)+\tilde{\sigma}_{\alpha\beta}(z_2,z_2)\right. \\ &- \left.\tilde{\sigma}_{\alpha\beta}(z_1,z_2)-\tilde{\sigma}_{\alpha\beta}(z_2,z_1)\right], \quad (20.66)\end{aligned}$$

where

$$\tilde{\sigma}_{\alpha\beta}(z_1,z_2) = -\frac{1}{\pi N\Omega}\mathrm{Tr}\langle j_\alpha G(z_1) j_\beta G(z_2)\rangle, \quad (20.67)$$

with z_1 (z_2) denoting $E+\mathrm{i}\eta$ ($E-\mathrm{i}\eta$). The brackets in the last expression include an average over all configurations of a random system. In view

[15] The finite temperature dc conductivity can be discussed along essentially identical lines but with the explicit presence of the Fermi function inside energy integrals.

of the expression for the current operator given in Eq. (20.63), the last expression has the following representation in real (coordinate) space,

$$\begin{aligned}\tilde{\sigma}_{\alpha\beta}(z_1,z_2) &= -\frac{1}{\pi N\Omega}\int \mathrm{d}^3r \int \mathrm{d}^3r' \langle\left(-\mathrm{i}\frac{\mathrm{e}}{m}\frac{\partial}{\partial r_\alpha}\right) G(\mathbf{r},\mathbf{r}';z_1) \\ &\times \left(-\mathrm{i}\frac{\mathrm{e}}{m}\frac{\partial}{\partial r'_\beta}\right) G(\mathbf{r}',\mathbf{r};z_2)\rangle. \end{aligned} \quad (20.68)$$

These expressions can be rewritten in forms more convenient for computations. In view of Eq. (20.45), we can write[16]

$$\tilde{\sigma}_{\alpha\beta}(z_1,z_2) = -\frac{4}{\pi N\Omega}\mathrm{Tr}\langle [j^m_\alpha(z_2,z_1)\tau^{nm}(z_1)j^n_\beta(z_2,z_1)\tau^{mn}(z_2)]\rangle, \quad (20.69)$$

where we have defined a matrix current operator with elements in the angular momentum representation having the form

$$[j^m_\alpha(z_1,z_2)]_{LL'} = -\frac{\mathrm{ie}}{m}\int_{\Omega_m} \mathrm{d}^3r Z^m_L(\mathbf{r};z_1)\left[\frac{\partial}{\partial r_\alpha} Z^m_{L'}(\mathbf{r};z_2)\right]. \quad (20.70)$$

Here, the vector $\mathbf{r}$ is confined to the cell Ω_m, and the integral extends correspondingly only over the domain of that cell.

At this point some observations are in order. The expressions derived thus far are exact. For example, the single-particle Green functions occurring in them are the exact single-particle Green functions[17] associated with the disordered material, and the angular brackets signify exact averages. Furthermore, an important technical point has arisen in view of the definition of the current matrix, Eq. (20.70). Because the wave functions $|Z^m_L(\mathbf{r};z)\rangle$ depend on the chemical species occupying cell Ω_m, the current operator itself also depends on the occupation of the cell. Therefore, the current operator is configurationally dependent, and care should be taken in accounting for this dependence.

It is indeed possible [17] to modify the Kubo formula so that it accounts for the configurational dependence of the current operator that arises within the KKR-CPA. When this is done, it induces formal differences between the formulae in the TB model and those that occur in the KKR-CPA. For example, it is necessary to treat the single-site terms in the conductivity formula individually, which yields in principle two different types of vertex corrections. For the sake of completeness, the discussion of transport theory within the framework of the ordinary (scalar) KKR-CPA is described

[16]We continue the practice of indicating matrices in L-space by plain (not underlined) symbols.

[17]The term exact is used in the sense of configurational averaging, rather than of many-body character.

in Appendix Q. The material there is based on the work by Butler [17] to which the reader is referred for more details and references. Another approach consists in rewriting the various formulae in a form that renders the current operator independent of configuration. This is the approach that is developed in the following lines, drawing heavily on the essentially exact analogy to the formalism of the TB model. However, we should be aware that now vertex corrections no longer vanish since the current operator has site-diagonal elements and is not odd under space inversion. The ensuing formalism allows the evaluation of these vertex corrections within the KKR-CPA. However, these are not the only corrections associated with the correlated motion of two particles interacting with the same alloy configuration. Vertex correction because of the presence of statistical fluctuations in the environment of a site are not accounted for in the ensuing formalism. A possible way of treating these effects is afforded by the cluster methods discussed in the previous section.

20.4.3 Configurational matrix approach

We begin by introducing a configuration independent current operator as the matrix[18]

$$\underline{j}^m(z,z') = \begin{pmatrix} j^{m,A}(z,z') & 0 \\ 0 & j^{m,B}(z,z') \end{pmatrix}, \tag{20.71}$$

where $j^{m,\alpha}(z,z')$ is the species dependent site-diagonal matrix defined by Eq. (20.70). Using these matrices, which are no longer dependent on site occupancy, and the matrix formulation of the SPO discussed in Chapter 12, we can write Eq. (20.69) in the form

$$\tilde{\sigma}_{\alpha\beta}(z_1,z_2) = -\frac{4}{\pi N\Omega}\tilde{\mathrm{Tr}}\langle \underline{j}^m(z_1,z_2)\underline{\tau}^{mn}(z_2)\underline{j}^n(z_2,z_1)\underline{\tau}^{mn}(z_1)\rangle, \tag{20.72}$$

where the symbol $\tilde{\mathrm{Tr}}$ denotes a trace over sites, angular momentum indices, *and* the sum over all matrix elements in AB space. We are now in a position to begin the transcription of the various formulae that arise within the TB model description of transport to the KKR-CPA.

We recall from Chapter 12 that the exact scattering-path matrix of an alloy satisfies the equation,

$$\underline{\tau} = \underline{\bar{\tau}} + \underline{\bar{\tau}}\underline{X}\underline{\bar{\tau}}, \tag{20.73}$$

where the bar over a symbol indicates effective quantities (determined here within the CPA), and $\underline{X}$ is the scattering matrix describing the scattering

[18] For n-component alloys the matrix would be n-dimensional rather than two-dimensional.

resulting from the presence of real atoms in the effective host medium. The last expression can also be written in terms of matrices in configuration space, where it takes the form,

$$\underline{\underline{\tau}} = \underline{\underline{\bar{\tau}}} + \underline{\underline{\bar{\tau} X \bar{\tau}}}, \tag{20.74}$$

where double underlines denote a matrix in site and AB-space[19]. The matrix elements of $\underline{\underline{X}}$ satisfy the equation of motion, [see Eq. (12.20)],

$$\underline{X}^{nm} = \underline{X}^{n}\delta_{nm} + \underline{X}^{n}\sum_{l}\underline{\bar{\tau}}^{nl}\underline{X}^{lm}. \tag{20.75}$$

Since the current operator is configuration-independent, in evaluating Eq. (20.69) we need only explicitly consider the matrix quantity[20],

$$\begin{aligned} \underline{\tilde{K}} &= \mathrm{Tr}\langle \underline{\underline{\tau j \tau}} \rangle \\ &= \sum_{nm} \langle \underline{\tau}^{nm} \underline{j}^{m} \underline{\tau}^{mn} \rangle. \end{aligned} \tag{20.76}$$

For the sake of clarity, we continue to use a site index with the current operator even though this operator in its matrix form is the same for all sites. A comparison of the last expression with the one in Eq. (20.21) should prove useful. With Eq. (20.74) inserted in Eq. (20.76), we can write,

$$\begin{aligned} \underline{\tilde{K}} &= \left[\underline{\underline{\bar{\tau} j \bar{\tau}}} + \underline{\underline{\bar{\tau}}} \langle \underline{\underline{X \bar{\tau} j \bar{\tau} X}} \rangle \underline{\underline{\bar{\tau}}} \right] \\ &= \left[\underline{\underline{\bar{\tau}}} \left(\underline{\underline{j}} + \underline{\underline{\Gamma}} \right) \underline{\underline{\bar{\tau}}} \right], \end{aligned} \tag{20.77}$$

which defines the vertex corrections $\underline{\underline{\Gamma}}$ in complete analogy with the formalism based on the TB model[21]. Once again, the equations derived above are exact. It remains to show how the formalism of the KKR-CPA allows one to obtain useful approximations to these expressions.

We proceed along lines familiar from the treatment of the transport within the TB model. We define quantities $\underline{Q}_n$ and $\underline{\tilde{Q}}_n$ such that

$$\underline{X} = \sum_{n} \underline{Q}_n \tag{20.78}$$

[19] We continue with the practice of designating matrices in L-space by plain symbols.

[20] At the end, we need to sum over all elements in the configuration space defined by the occupation of a single site and take the trace over L to obtain physical results.

[21] We note the important distinction that in this case we are dealing with configurational matrices in contrast to the scalar quantities we encountered in discussions based on the single band, TB model.

$$\underline{Q}_n = \underline{X}_n \left(\underline{\underline{1}} + \underline{\bar{\tau}} \sum_{m \neq n} \underline{Q}_m \right) \tag{20.79}$$

and

$$\underline{X} = \sum_n \underline{\tilde{Q}}_n \tag{20.80}$$

$$\underline{\tilde{Q}}_n = \left(\underline{\underline{1}} + \sum_{m \neq n} \underline{\tilde{Q}}_m \underline{\bar{\tau}} \right) \underline{X}_n. \tag{20.81}$$

We now obtain[22] a result corresponding to Eq. (20.22),

$$\begin{aligned} \underline{\underline{\Gamma}} &= \sum_n \sum_m \langle \underline{Q}_n \underline{\bar{\tau}} \underline{\underline{j}} \underline{\bar{\tau}} \underline{\tilde{Q}}_m \rangle \\ &= \sum_n \sum_{m \neq n} \langle \underline{X}_n \langle \left(\underline{\underline{1}} + \underline{\bar{\tau}} \sum_{l \neq n} \underline{Q}_l \right) \underline{\bar{\tau}} \underline{\underline{j}} \underline{\bar{\tau}} \left(\underline{\underline{1}} + \sum_{m \neq n} \underline{\tilde{Q}}_m \underline{\bar{\tau}} \right) \rangle \underline{X}_m \rangle. \end{aligned} \tag{20.82}$$

Since for $n \neq m$ the matrices $\underline{X}_n$ and $\underline{X}_m$ can be averaged independently, we find that the vertex corrections are site diagonal, and in analogy with Eq. (20.25), are given by the expression

$$\langle \underline{Q}_n \underline{\bar{\tau}} \underline{\underline{j}} \underline{\bar{\tau}} \underline{\tilde{Q}}_m \rangle = \underline{\underline{\Gamma}}_n \delta_{nm}. \tag{20.83}$$

We can use this result to simplify the internal summations in Eq. (20.83) to obtain

$$\underline{\underline{\Gamma}}_n = \langle \underline{X}_n \underline{\bar{\tau}} \left(\underline{\underline{j}} + \sum_{k \neq n} \underline{\underline{\Gamma}}_k \right) \underline{\bar{\tau}} \underline{X}_n \rangle, \tag{20.84}$$

a result that can be compared to Eq. (20.26). Therefore, we can write

$$\underline{\underline{\Gamma}} = \sum_n \underline{\underline{\Gamma}}_n, \tag{20.85}$$

which is analogous to Eq. (20.27). Finally, this relation and Eq. (20.77) allow us to cast Eq. (20.84) in the form

$$\underline{\underline{K}} = \underline{\bar{\tau}} \underline{\underline{j}} \underline{\bar{\tau}} + \underline{\bar{\tau}} \sum_n \underline{\underline{\Gamma}}_n \underline{\bar{\tau}}, \tag{20.86}$$

[22] It is suggested that the reader fills in the steps leading to the next equation as well as other equations that are obtained in analogy with the TB model.

$$\begin{aligned} \underline{\Gamma}_n &= \langle \underline{X}_n \underline{\underline{K}} \underline{X}_n \rangle - \langle \underline{X}_n \underline{\underline{\tau}} \underline{\Gamma}_n \underline{\underline{\tau}} \underline{X}_n \rangle \\ &= \langle \underline{X}_n \underline{\underline{\tau}} \underline{\underline{j}} \underline{\underline{\tau}} \underline{X}_n \rangle + \langle \underline{X}_n \underline{\underline{\tau}} \sum_{m \neq n} \underline{\Gamma}_m \underline{\underline{\tau}} \underline{X}_n \rangle. \end{aligned} \quad (20.87)$$

These expressions are formally identical to the corresponding ones derived within the TB model, and fully define the form of the Kubo formula within the framework of the KKR-CPA. As we have pointed out, the only technical difference between the two types of expressions is the replacement of scalar quantities in the case of the single-band, diagonally-disordered TB model by matrices in the configurational space of a single site.

20.4.4 Closed formulae

The evaluation of the transport formulae derived above requires a means for calculating the vertex corrections, Γ_n. The vertex corrections can be evaluated in a more or less straightforward way by means of repeated expansions or iterations, along the lines indicated in Eq. (20.30). Alternatively, it is possible to derive a closed form of the conductivity expressions in terms of the direct product of matrices. We illustrate how this is done for the case of the zero temperature dc conductivity. From Eqs. (20.86) and (20.87), we can write

$$\left[\underline{\Gamma}_n - \sum_{m \neq n} \underline{W}_{nm} \underline{\Gamma}_m \right] = \underline{A}_n, \quad (20.88)$$

where we have defined the quantities

$$\underline{W}_{nm} \equiv \underline{W}_n = \langle \underline{X}_{nm} \underline{\underline{\tau}} \otimes \underline{\underline{\tau}} \underline{X}_m \rangle \delta_{nm} \quad (20.89)$$

and

$$\underline{A}_n = \langle \underline{X}_n \underline{\underline{K}} \underline{X}_n \rangle, \quad (20.90)$$

and $\otimes$ denotes the direct product of matrices. Further, defining the lattice Fourier transforms

$$\underline{W}(\mathbf{q}) = \frac{1}{N} \sum_n \underline{W}_{n0} e^{i\mathbf{q}\cdot\mathbf{R}_{n0}}, \quad (20.91)$$

with similar expressions for $\underline{A}(\mathbf{q})$ and $\underline{\Gamma}(\mathbf{q})$, and the inverse transformations, we obtain from Eq. (20.88) the formal solution

$$\underline{\Gamma}(\mathbf{q}) = \left[\underline{1} - \underline{W}(\mathbf{q})\right]^{-1} \underline{A}(\mathbf{q}). \quad (20.92)$$

However, because $\underline{W}$ and $\underline{A}$ are site diagonal (as well as configuration independent), only the $\mathbf{q} = 0$ limit contributes to Eq. (20.92). Thus, taking the inverse transform, we obtain the "closed" result for the vertex corrections

$$\underline{\Gamma}_n = \left[\underline{1} - \underline{W}(0)\right]^{-1} \underline{A}(0). \quad (20.93)$$

This allows us to write Eq. (20.69) for the conductivity in the form

$$\tilde{\sigma}_{\alpha\beta}(z_1,z_2) = -\frac{4}{\pi N\Omega}\tilde{\mathrm{Tr}}\sum_{nm}\{\underline{j}^n(z_2,z_1)\underline{\tau}^{nm}(z_1) \times \left[\underline{j}^n(z_1,z_2)+\underline{\Gamma}_m(z_1,z_2)\right]\underline{\tau}^{mn}(z_2)\}, \quad (20.94)$$

with $\underline{\Gamma}_m(z_1,z_2)$ being given by Eq. (20.93), and where $\tilde{\mathrm{Tr}}$ denotes the summation of elements in configuration space and an ordinary trace over L. For the sake of easy reference, we exhibit the detailed structure of the last expression,

$$\tilde{\sigma}_{\alpha\beta}(z_1,z_2) = -\frac{4}{\pi N\Omega}\tilde{\mathrm{Tr}}\sum_{nm}\{\underline{j}^n(z_2,z_1)\underline{\tau}^{nm}(z_1)\left[\underline{j}^n(z_1,z_2) + \left[\underline{1}-\langle\underline{X}_0\underline{\underline{\tau}}(z_1)\otimes\underline{\underline{\tau}}(z_2)\underline{X}_0\rangle\right]^{-1} \times \langle\underline{X}_0\underline{\underline{\tau}}(z_1)\underline{\underline{j}}(z_1,z_2)\underline{X}_0\underline{\underline{\tau}}(z_2)\rangle\right]\underline{\underline{\tau}}(z_2)\}, \quad (20.95)$$

At this point, a number of observations are in order. First, as already pointed out, the matrix formulation of the KKR-CPA, although very convenient for the discussion of transport phenomena, need not be used in obtaining the solution of the KKR-CPA self-consistency conditions. It suffices to solve the "scalar" version of the KKR-CPA (not involving configurational matrices but only matrices in L-space), and then construct the configurational matrices from the quantities determined in the scalar version. Second, the matrix form may, in fact, not be the most convenient for numerical work, in spite of its formal simplicity. The use of configurational matrices, whose elements are themselves matrices in L space, makes this form somewhat more cumbersome to use than the ordinary form of the KKR-CPA in evaluating transport coefficients. A detailed discussion of the Kubo formula within the ordinary version of the KKR-CPA has been given in the literature [17]. The main elements of that discussion are presented in Appendix Q

20.5 Derivation of the Boltzmann Equation

Within linear response theory, the Kubo formula provides the most general expression for transport coefficients. Therefore, under the appropriate assumptions, it can be used to obtain "classical" results such as the Boltzmann equation. It can also be used to derive expressions for the scattering rates $Q(\mathbf{k},\mathbf{k}')$, discussed in connection with the Boltzmann equation in Chapter 18. Here, we quote only the final results, with a more detailed discussion presented on Appendix Q.

Under the assumption that the alloy is characterized by well-defined (sharp) bands, and neglecting vertex corrections, the dc conductivity can be written in the form, (with $\hbar$ restored)

$$\tilde{\sigma}_{\alpha\beta} = \frac{e^2}{(2\pi)^3\hbar} \int_{S_F} \frac{dS_{\mathbf{q}}}{v_{\mathbf{q}}} v_{\mathbf{q}}^{\alpha} v_{\mathbf{q}}^{\beta} \tau_{\mathbf{q}}^{B}, \tag{20.96}$$

where $v_{\mathbf{q}}^{\alpha}$ is the αth component of the electron velocity in the solid (given by the corresponding gradient of the energy as a function of wave vector), and the integral extends over the Fermi surface of the alloy (which is assumed to be well-defined in energy and wave vector). We see that the last expression is a somewhat generalized form of Eq. (18.71) because it includes the possible dependence of the Boltzmann lifetime $\tau_{\mathbf{q}}^{B}$ on $\mathbf{q}$. The details of the derivation of the last expression for the conductivity are given in Appendix Q.

In addition to allowing a derivation of the Boltzmann equation, the Kubo formula within the framework of the KKR-CPA can be used to obtain an expression for the scattering probability, Eq. (18.16). As is shown in Appendix Q, we obtain [see also Eq. (18.54)]

$$Q(\mathbf{k}, \mathbf{k}') = \frac{2\pi}{\hbar} \sum_{\alpha} c^{\alpha} \left| \underline{T}_{\mathbf{k},\mathbf{k}'}^{\alpha} \right|^2 \delta(\epsilon_{\mathbf{k}} - \epsilon_{\mathbf{k}'}), \tag{20.97}$$

where $\underline{T}_{\mathbf{k},\mathbf{k}'}^{\alpha}$ is the reciprocal-space representation of the scattering matrix associated with the presence of an atom of type α in the KKR-CPA medium. For explicit details the reader is referred to Appendix Q.

Finally, it can be shown that the equations derived within a KKR-CPA application of the Kubo formula reduce to acceptable results in various limits, such as the low concentration limit (see Chapter 18 for a discussion of the single-impurity results), and weak scattering [1].

20.6 Applications

The Kubo formula in connection with the KKR-CPA has been used [18] to study electrical transport in a number of binary alloys, and some of the results obtained in such studies are reviewed in the following lines. In the calculations reported here, the KKR-CPA was implemented in connection with muffin-tin potentials, and all angular momentum expressions were truncated at $L = 2$.

Figure 20.2 shows the dc residual resistivity, $\rho = 1/\sigma$, in units of $\mu\Omega$-cm as a function of concentration c for the alloys $Cu_{1-c}Zn_c$, $Cu_{1-c}Ga_c$, $Cu_{1-c}Ge_c$, and $Ag_{1-c}Pd_c$. The calculated residual resistivities of the various alloys represented in the figure are shown as the open triangles. For

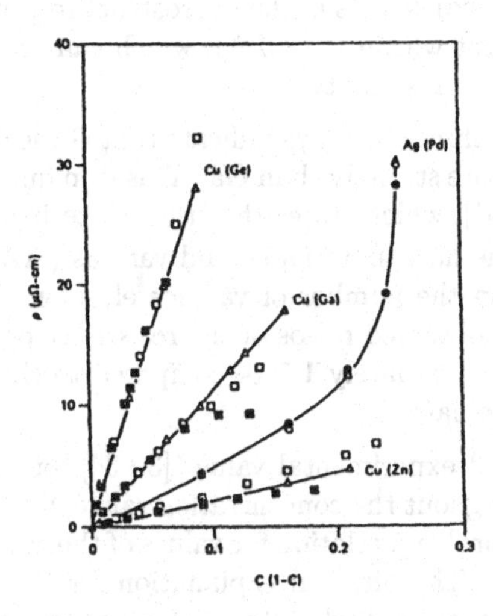

Figure 20.2: **Experimental and calculated residual resistivities for Cu and Ag alloys.**

the copper alloys, the straight lines represent the Nordheim [24] relation, $\rho \approx c(1-c)$, adjusted to go through one of the calculated points. The squares indicate experimental data [25-30] for the copper alloys, with open squares corresponding to cold-worked samples and filled squares to annealed samples. Corresponding results for the Ag alloys are shown in circles [30-34], filled for cold-worked and open for annealed alloys. For AgPd alloys, the calculated points at $c = 0.1$, 0.2, and 0.3 fall exactly on the experimental values, as do the points at $c = 0.1$ for CuZn, at $c = 0.05$ for CuGa, and at $c = 0.333$ and 0.10 for CuGe alloys.

The figure indicates that both the experimental, cold-worked as well as annealed, and calculated resistivities of the Cu alloys follow closely the Nordheim relation, and are in very good agreement with one another at least at low concentrations. The calculated results begin to deviate from the experimental ones near the boundaries of the single-phase regions, which occur at $c = 0.38$ for CuZn, at $c = 0.19$ for CuGa, and at $c = 0.085$ for CuGe, where the measured resistivities of the annealed samples are lower than the calculated ones. This discrepancy can be attributed to the presence of short range order (SRO) in the annealed samples. When the SRO is disrupted by cold-working, the resistivity increases, and in most cases, lies slightly above the calculated values. Cold-working tends to produce a more random system compared to the one in the presence of SRO, thus making comparison with the CPA results more meaningful. At the same time, this procedure introduces defects in the microstructure, such as dislocations (not taken into account within the CPA), which can scatter electrons and cause an increase in the resistivity.

The slopes for the three Cu alloys indicate that Ga scatters more strongly than Zn, and Ge more strongly than Ga. This is in qualitative agreement with Linde's "law" [35], which states that the scattering rates for different impurities in the same host material should vary as $(\Delta Z)^2$, where ΔZ denotes the difference in the number of valance electrons between host and impurity atoms. The observed ratios of the resistivity per atom of Zn, Ga, and Ge in Cu are approximately 1:5:14 compared to the 1:4:9 that would be predicted by Linde' law.

The calculated and experimental values [30-34] for AgPd alloys are in good agreement throughout the concentration range, and both deviate substantially from the Nordheim relation for values of the concentration greater than about $c = 0.20$. The physical explanation for this deviation can be traced to the presence of the Pd d-states that contribute the dominant part of the DOS at the Fermi energy for Pd concentrations greater than 0.50. Cold work appears to have a much smaller effect on AgPd alloys than on the copper alloys. Interestingly, cold-working causes an initial decrease in the electrical resistivity for Pd alloys at concentrations below 0.5.

The resistivity of the alloys reported in this figure was calculated both in the absence and presence of vertex corrections. For all alloys, vertex corrections were found to be important. In the case of copper alloys vertex corrections reduced the calculated resistivity by factors of approximately 2.51±0.01, 2.4±0.1, and 2.1±0.05, for the Zn, Ga, and Ge alloys, respectively. Under the assumption of weak scattering, the ratio of the resistivities in the absence, ρ^0, and presence, ρ^1, of vertex corrections can be expressed [36] in terms of the differences in the scattering phase shifts characterizing the host and impurity atoms,

$$\frac{\rho^0}{\rho^1} = \frac{\sum_{l=0}(2l+1)\sin^2\left(\delta_l^h - \delta_l^i\right)}{\sum_{l=0}(l+1)\sin^2\left[\left(\delta_{l+1}^h - \delta_{l+1}^i\right) - \left(\delta_l^h - \delta_l^i\right)\right]}. \tag{20.98}$$

For example, this leads to a value of 3.7±0.01 for all three concentrations of CuZn alloys, which is quite a bit larger than the values obtained in a more rigorous calculation. This is conceivably due to the neglect of concentration in the last expression, which is taken into account in the CPA. Further discussion of the various features characterizing the application of the Kubo formula to alloys can be found in published works [17] and references therein.

Bibliography

[1] B. Velicky, Phys. Rev. **184**, 614 (1969).

[2] A. B. Chen, G. Weisz, and A. Sher, Phys. Rev. B **5**, 2897 (1972).

[3] K. Levin, B. Velicky, and H. Ehrenreich, Phys. Rev. B **2**, 1771 (1970).

[4] F. Brouers and A. V. Vedyayev, Phys. Rev. B **5**, 348 (1972).

[5] K. Niizeki, J. Phys. C **10**, 3351 (1977), and references therein.

[6] K. Niizeki and M. Endo, Solid State Commun. **14**, 689 (1974).

[7] J. A. Blackman, Phys. Rev. B **12**, 3482 (1975).

[8] J. A. Blackman, D. M. Esterling, and N. F. Berk, Phys. Rev. B **4**, 2412 (1971).

[9] L. Schwartz, Phys. Rev. B **24**, 1091 (1981).

[10] K. I. Wysokinski, E. Taranko, and R. Taranko, J. Phys. C **13**, 6659 (1980), and references therein.

[11] G. Czycholl, Z. Phys. B **30**, 383 (1972); G. Czycholl and J. Zittartz, Z. Phys. B **30**, 375 (1972).

[12] Miaogy Hwang, A. Gonis, and A. J. Freeman, Phys. Rev. B **33**, 2872 (1986).

[13] W. Pickett, D. A. Papaconstantopoulos, and E. N. Economou, Phys. Rev. B**28**, 2232 (1983).

[14] G. M. Stocks and W. H. Butler, Phys. Rev. Lett. **48**, 55 (1982).

[15] W. H. Butler and G. M. Stocks, Phys. Rev. B **29**, 4217 (1984).

[16] W. H. Butler, Phys. Rev. B **29** 4224 (1984).

[17] W. H. Butler, Phys. Rev. **31**, 3260 (1985).

[18] J. C. Swihart, W. H. Butler, G. M. Stocks, D. M. Nicholson, and R. C. Ward, Phys. Rev. Lett. **67**, 1181 (1986).

[19] R. H. Brown, P. B. Allen, D. M. Nicholson, and W. H. Butler, in *Atomic Scale Calculations in Materials Science*, J. Tersoff, D. Vanderbilt, and V. Vitek (eds.), MRS Symposium proceedings No. 141, Materials Research Society (1989), p. 91.

[20] J. Barnahart, R. Bernstein, J. Voitländer, and P. Weinberger, Solid State Commun. **77**, 107 (1991).

[21] J. Barnhart, H. Ebert, P. Weinberger, and J. Voitländer, Phys. Rev. B **50**, 2104 (1994).

[22] Gerald D. Mahan, *Many-Particle Physics*, Plenum Press, New York (1981).

[23] A. Gonis and A. J. Freeman, Phys. Rev. B **29**, 4277 (1984).

[24] L. Nordheim, Ann. Phys. **9**, 664 (1931).

[25] H. A. Fairbank, Phys. Rev. **66**, 274 (1944).

[26] W. G. Henry and P. A. Schroeder, Can. J. Phys. **41**, 1076 (1963).

[27] C. Y. Ho, M. W. Ackerman, K. Y. Wu, T. N. Havill, R. H. Bogaard, R. A. Matula, S. G. Oh, and H. M. James, J. Phys. Chem. Ref. Data **12**, 183 (1983).

[28] R. W. R. S. Crisp, W. G. Henry, and Schroeder, Philos. Mag. **10**, 553 (1964).

[29] W. Köster and H.-P. Rave, Z. Metaallk. **55**, 750 (1964). The room temperature results shown here were adjusted to low temperature by the assumption that the temperature-dependent resistivity as a function of concentration reported in the following reference was not affected by cold work.

[30] W. R. G. Kamp, P. G. Klemens, A. K. Sneedhar, and G. K. White, Proc. Roy. Soc. (London), Ser. A **233**, 480 (1956).

[31] B. R. Coles and J. C. Taylor, Proc. Roy. Soc. (London) Ser. A **267**, 139 (1962).

[32] R. W. Westrelund and M. E. Nicholson, Acta Metall. **14**, 569 (1966).

[33] W. -K. Chen and M. E. Nicholson, Acta Metall. **12**, 687 (1964).

[34] W. H. Arts and A. S. Houston-MacMillan, Acta Metall. **5**, 525 1957.

[35] J. O. Linde, Ann. Phys. **15**, 219 (1936).

[36] P. T. Colleridge, J. Phys. F **9**, 473 (1979).

Part VI

Phonons and Photons

Chapter 21

Lattice Vibrations

21.1 General Comments

Up to this point, our discussion has dealt exclusively with the electrons in an interacting electron-nucleon system such as a solid. By invoking the Born-Oppenheimer approximation, we reduced the role of the stationary nuclei to supplying the external potential in which the electrons move. However, the physical properties of materials are by no means determined solely by electronic behavior. The interaction of the nuclei among themselves and with the electron cloud, and the motion of the nuclei in this fully interacting system, play a very important role in determining materials behavior. In this chapter, we provide a brief description of lattice vibrations following a Green-function formalism. In our initial discussion, we do not take into account the interaction of the electrons with the nuclei, although the presence of the electrons is manifest through their effect on producing the medium in which the ions vibrate, and the forces governing these vibrations. However, the electron-phonon interaction will not be completely ignored. An approach directed at calculating in a self-consistent way not only the ground state properties of the electron cloud but also the response of the electrons to changes in the external potential, e.g., the motion of an atom, is provided in the following chapter.

Atomic vibrations in a solid, and the interaction of the ions with the electron gas, determine not only a material's thermal properties, but also govern a broad spectrum of physical phenomena, such as diffuse X-ray scattering, neutron scattering, spin-lattice relaxation, superconductivity in many materials, and others. However, it was primarily the desire to understand the thermal behavior of solids that provided the initial motivation for the study of lattice dynamics.

The modern era in the study of lattice vibrations begins with an important empirical discovery made in 1819 by Dulong and Petit, who determined that the specific heat per gram-atom of a crystal of any element was 3R, with R as the gas constant. This "law" was found to hold for most elemental solids at room temperature and above, but to fail badly at low temperatures where the specific heat of crystals was found to be significantly lower. Such a decrease was predicted by Einstein who, in 1907, developed a quantum theory of specific heat. This theory was based on the so-called "independent oscillator model", in which each ion (nucleus) is assumed to execute harmonic motion around its equilibrium position independently of the other ions in the crystal.

However, this approach led to a decrease in the specific heat that was far more rapid than what was observed. This discrepancy was removed in 1912 by Debye and almost simultaneously by Max Born and von Kármán. Allowing for the strong coupling between ions in the crystal, Debye's theory was remarkably successful in explaining the temperature dependence of the specific heat of most crystals[1].

Specific heat measurements, being of an integral nature, cannot provide a sensitive test of Debye's theory of lattice dynamics. And even though the theory also allows a fairly accurate analysis of temperature effects on the intensity of Laue spots in X-ray diffraction studies, these effects also depend on the overall frequency spectrum of the crystal and do not allow one to deduce the details of the spectrum from such experiments. It was the availability of high intensity thermal neutron beams from reactors that made possible the study of processes involving single lattice waves, along with the determination of the frequency distribution function and the dispersion relation (the dependence of frequency on wave vector). Neutron experiments have revealed that constructing a theory of lattice dynamics is by no means as simple as once it might have seemed.

In retrospect, the complex nature of the description of lattice dynamics is not surprising and it mirrors the also immensely complex problem of the accurate description of the electron cloud. The electron-phonon interaction couples the two fields and can often have dramatic effects. This interaction influences electronic transport and cyclotron resonances, and can lead to the spectacular phenomenon of superconductivity in many materials.

In what follows, we provide an elementary discussion of the specific heat of crystals, motivating the study of the frequency spectrum of lattice vibrations. We then provide a Green-function-based description of lattice vibrations in ordered (periodic) as well as substitutionally disordered solids. In this discussion, we make use of the so-called "harmonic approximation", and also neglect the electron-phonon interaction. The harmonic

[1] A few exceptions such as graphite, bismuth, selenium, and tellurium were discovered, and it took some time before their behavior was understood and explained.

approximation[2] assumes that each displaced ion is acted upon by restoring forces which vary linearly in the displacement (or that the potential energy of a displaced atom is quadratic in the displacement). The bare ion-ion interaction, and the electron-electron and electron-phonon interactions combine to produce an effective coupling between the ions that govern their oscillatory behavior (at least for sufficiently "small" displacements). We will, however, allow the individual "oscillators" to interact and describe ways for obtaining the dispersion relation for both ordered, translationally invariant materials as well as random, substitutionally disordered alloys. The latter discussion will be given within the framework of the coherent potential approximation (CPA). We will point out the important similarities, but also the differences with electronic systems arising when the CPA is applied to dynamical systems.

In the following chapter, we attempt an estimate of the electron-phonon interaction. This is a difficult subject, which strictly speaking lies outside the scope of this book, but a discussion within MST is possible and can prove useful. The reader interested in the theoretical treatment of the electron-phonon interaction in solids is referred to a number of excellent texts [1, 2] in the literature.

21.2 Elementary Discussion

In this section, we give a brief overview of the physics that underlies the properties of lattice vibrations. We concentrate primarily on processes occurring during the interaction of phonons with photons and neutrons, and which lead to experiments that allow the determination of the phonon spectrum. All of these experiments are based on the conservation laws for momentum and energy when photons (infrared light, X-rays) or thermal neutrons are scattered by the presence of lattice vibrations [3]. Even though the concept of phonons, the quanta of lattice vibrations, is used often, our discussion is based mostly on a classical treatment of lattice vibrations. A derivation of the quantum mechanical description of lattice vibrations in terms of second quantization in given in Appendix R.

21.2.1 Quanta of lattice vibrations

Lattice vibrations are quantized, and the quantum of energy in an elastic wave is called a *phonon*. Each phonon of frequency ω ($\nu = \omega/2\pi$) has

[2]The harmonic approximation is very convenient computationally and sheds much light into the physics of lattice dynamics. At the same time, it is an approximation that is not able to address a number of important physical phenomena, such as thermal expansion, the generation of Joule heat, and heat transport, all of which are effects of order higher than the second in the lattice displacements.

an energy $\epsilon = \hbar\omega = h\nu$, where $\hbar$ is Planck's constant (divided by 2π), and we have neglected the term $\frac{1}{2}\hbar\omega$ associated with the zero-point motion of a harmonic oscillator[3]. The quantization of lattice waves is verified by the vanishing of the specific heat of a crystal at zero temperature that can only be explained if lattice vibrations are quantized. In addition, the scattering of electromagnetic radiation and neutrons by crystals can be used to yield information on the properties of individual phonons, such as the dependence of the frequency on wave vector (dispersion relation). Also, the analysis of scattering experiments often depends on the phonon density of modes (density of states), a quantity whose calculation will occupy a good part of this chapter.

Phonons in a crystal are described by their frequencies and wave vectors (as electrons are described by their energy and momentum vector). A phonon of wave vector $\mathbf{k}$ behaves as if it had a momentum $\hbar\mathbf{k}$, even though a phonon, describing oscillations of *relative* coordinates, carries *no* real momentum. Physical momentum is carried only by a phonon of zero wave vector that corresponds to the motion of the entire crystal as a whole. The wave vector $\mathbf{k}$, however, is useful in describing the interactions of phonons with other particles and fields, including phonons.

21.2.2 Phonon-photon scattering

Consider a quantum of radiation, a photon, of wave vector $\mathbf{k}$ interacting with the lattice vibrations in a crystal. For example, one way that such an interaction can come about is when the oscillating electric field vector of the photon excites the optical mode in a diatomic lattice by causing the ions of opposite charge in the basis to oscillate with opposite phases (see examples below). Assuming that such an interaction has created a single phonon of wave vector $\mathbf{K}$, momentum conservation yields the relation,

$$\mathbf{k}' + \mathbf{K} = \mathbf{k} + \mathbf{G}, \tag{21.1}$$

where $\mathbf{k}'$ is the wave vector of the photon emerging from the collision, and $\mathbf{G}$ is a reciprocal lattice vector[4]. Similarly, if a phonon is absorbed in the scattering process, momentum conservation takes the form,

$$\mathbf{k}' - \mathbf{K} = \mathbf{k} + \mathbf{G}. \tag{21.2}$$

Conservation relations involving the energy can also be easily worked out. Let the energy of the photon be given by $\hbar\omega$, where in a medium of refractive index, n, we have

$$\omega = \frac{ck}{n}, \tag{21.3}$$

[3]In principle, zero point motion must be included in the calculations of energy of the electronic ground state in a solid.

[4]In a periodic lattice, momentum is conserved modulo a reciprocal lattice vector.

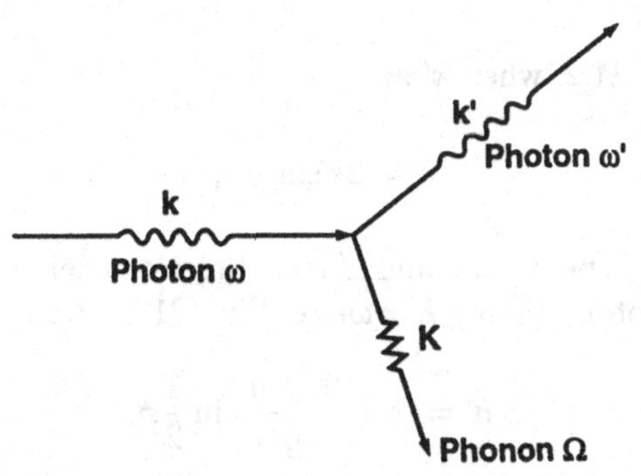

Figure 21.1: **A photon of wave vector k and energy $\hbar\omega$ is scattered inelastically by a phonon of wave vector K and energy $\hbar\Omega$, emerging with energy $\hbar\omega'$ in the direction of $\mathbf{k}'$.**

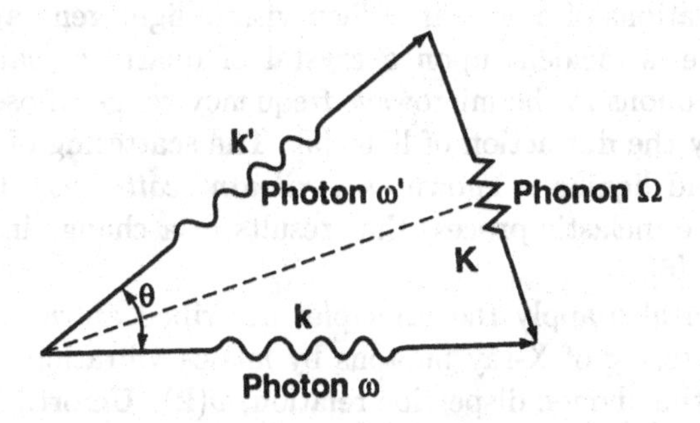

Figure 21.2: **Momentum conservation when a photon is scattered by a phonon. If $k' \approx k$, the triangle formed by k, $\mathbf{k}'$, and K is isosceles and the length of its base equals $K = 2k\sin\frac{1}{2}\phi$.**

with c being the velocity of light in vacuum. Interacting with a crystal, a photon can create or absorb (annihilate) a phonon of energy $\hbar\Omega$, emerging with energy $\hbar\omega'$ in the direction of $\mathbf{k}'$. Neglecting Bragg scattering, which would be described in terms of reciprocal lattice vectors, the requirement of conservation of energy yields the relation,

$$\hbar\omega = \hbar\omega' \pm \hbar\Omega, \tag{21.4}$$

with a + or a - denoting creation or absorption of a phonon. Figure 21.1 provides a representation of Eq. (21.4) and shows the inelastic scattering process of a photon that leads to the creation of a phonon.

In general, a phonon can only carry away a small part of the energy of a photon. Even if $K \approx k$, we have $ck/n >> v_s K$, since the velocity of light c/n in a material is very much larger than the velocity of sound, v_s. Therefore, $\omega >> \Omega$, and from Eq. (21.4) it follows that $\omega' \approx \omega$ and $k' \approx k$.

As shown in Fig. 21.2, when $k' \approx k$

$$K = 2k \sin \frac{1}{2}\phi, \tag{21.5}$$

where ϕ is the angle between **k** and **k**′ (the directions defined by the incident and scattered photon). Using $k = \omega n/c$, Eq. (21.3), we have,

$$v_s K = \Omega \approx \frac{2v_s \omega n}{c} \sin \frac{1}{2}\phi, \tag{21.6}$$

for the energy Ω of the phonons created when photons are scattered by an angle ϕ.

These basic considerations govern the interaction of radiation with the lattice vibrations of a crystal. When visible light generated by an intense laser source is incident upon a crystal of quartz or sapphire [4], it can produce phonons in the microwave frequency range whose presence can be detected by the diffraction of light [5]. The scattering of light by phonons in solids and liquids is known as *Brillouin scattering.* The *Raman effect* refers to the inelastic process that results in a change in the frequency of the photon [6].

One can also apply the principles described above to the inelastic or diffuse scattering of X-ray photons by lattice vibrations in an attempt to determine the phonon dispersion relation, $\omega(\mathbf{k})$. Unfortunately, X-rays suffer only a relatively small shift in frequency that is difficult to determine experimentally. On the other hand, neutron scattering experiments usually allow a direct measurement of energy shifts in an inelastic process.

21.2.3 Phonon-neutron scattering

A non-magnetic crystal[5] is seen by a neutron primarily through its interaction (collisions) with the nuclei of the atoms. The kinematics describing the inelastic scattering of neutrons by phonons leads to relations essentially identical to those associated with photon scattering.

If **k** and **k**′ are the wave vectors of the incident and the scattered neutron, and **K** that of a phonon, then momentum conservation yields the relation,

$$\mathbf{k} = \mathbf{k}' \pm \mathbf{K}, \tag{21.7}$$

where Bragg reflections are neglected. The momentum of a neutron with wave vector **k** is $\mathbf{p} = \hbar\mathbf{k}$ and the associated kinetic energy is given by $\hbar^2 k^2/2\mathrm{M_n}$, where $\mathrm{M_n}$ is the neutron mass. The statement of conservation

[5] In magnetic crystals, neutrons can create collective excitations of the magnetization field, called *nagnons*. The analysis of magnon scattering has many common elements with that of phonon scattering, but will not be considered further.

of energy in a scattering process in which a phonon is created or absorbed (+ or -) takes the form,

$$\frac{\hbar^2 k^2}{2\mathrm{M_n}} = \frac{\hbar^2 k'^2}{2\mathrm{M_n}} \pm \hbar\omega(\mathbf{k}), \tag{21.8}$$

where $\hbar\omega(\mathbf{k})$ is the energy of the phonon. Neutron scattering experiments can usually be carried out with sufficient accuracy[6] to determine the dispersion relation, $\omega(\mathbf{k})$.

All scattering experiments involving phonons either require prior knowledge of, or lead to, a determination of the phonon density of states, i.e., the distribution of phonon states in frequency. Phonon densities of states are needed in the definition of scattering cross sections of phonons by phonons or other excitations. Thus, the mode density of states, and along with it the more basic information embodied in the dispersion relation, play a central role in the study of lattice vibrations. The importance of this quantity becomes apparent when we consider the thermal properties of crystals associated with the presence of lattice vibrations.

21.3 Thermal Properties of Crystals

The thermal properties of a material are determined by the response of both the electrons and the nuclei when heat is supplied externally to the system. Thus, metals, insulators, and superconductors exhibit different heat capacities.The electronic specific heat in a solid depends directly on the density of states at the Fermi level, and at temperatures small campared to the Fermi energy, varies linearly with temperature (see Chapter 7). In this section, we consider the heat capacity associated with lattice vibrations. This leads directly to a description of the phonon spectrum and the associated density of modes.

The heat capacity at constant volume[7] is defined by the relation

$$C_\mathrm{V} \equiv T\left(\frac{\partial S}{\partial T}\right)_\mathrm{V} = \left(\frac{\partial E}{\partial T}\right)_\mathrm{V}. \tag{21.9}$$

We use this relation to calculate the heat capacity of the lattice vibrations of a crystal in the Einstein and the Debye models.

[6]In contrast to radiation scattering, the interacting particles in this case, neutrons and nuclei, are of comparable mass.

[7]The heat capacities at constant pressure, C_P, and at constant volume, C_V, are connected by the relation $C_\mathrm{P} - C_\mathrm{V} = 9\alpha^2 BVT$, where α is the coefficient of linear expansion, V the volume, and B the bulk modulus of the material. The quantity C_P is more convenient to work with since it is easier to carry out experiments at constant pressure than at constant volume.

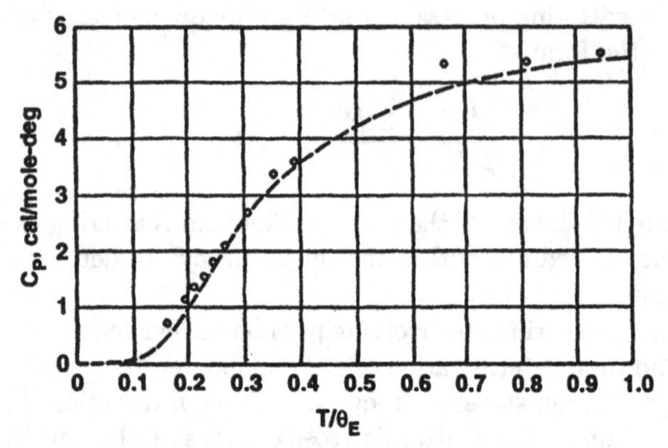

Figure 21.3: **Comparison of experimental (open circles) and calculated values of the specific heat of diamond. Calculated results are based on the Einstein model and were obtained using a characteristic temperature of $\theta_E = \hbar\omega/k_B = 1320^o K$.**

21.3.1 Einstein model

In the Einstein model, each atom (we consider monoatomic crystals), executes harmonic motion independently of the motion of other atoms. Thus, the energy of a vibrating lattice depends on the frequency of vibration, ω, and the phonon occupancy, n, at that frequency. In thermal equilibrium at temperature T, the occupancy is given by the Planck distribution function,

$$\langle n \rangle = \frac{1}{e^{\beta\hbar\omega} - 1}, \tag{21.10}$$

where $\langle n \rangle$ denotes an average at thermal equilibrium, and $\beta = 1/k_B T$, with k_B as Boltzmann's constant. Therefore, the average energy of an oscillator of frequency ω is $\langle n \rangle \hbar\omega$. The energy of N oscillators in one dimension[8] all having frequency ω is equal to

$$E = N\langle n \rangle \hbar\omega = \frac{N\hbar\omega}{e^{\beta\hbar\omega} - 1}. \tag{21.11}$$

The heat capacity of the N oscillators follows from Eq. (21.9),

$$C_V = \left(\frac{\partial E}{\partial T}\right)_V = Nk_B \left(\beta\hbar\omega\right)^2 \frac{e^{\beta\hbar\omega}}{\left(e^{\beta\hbar\omega} - 1\right)^2}. \tag{21.12}$$

A comparison of the results of the Einstein model with experimental heat capacities for diamond [7] are shown in Fig. 21.3. The agreement

[8]In three dimensions, N is replaced by $3N$ to account for the presence of an independent oscillator along each coordinate direction.

between the two types of results is very close at higher temperatures, where both approach the limiting value of 3R in accordance with the law of Dulong and Petit, but as $T \to 0$, the experimental curve behaves as T^3 in contrast to the theoretical prediction of an exponential decay of the form $e^{-\beta\hbar\omega}$. This discrepancy is explained in the following discussion where the Debye model is considered. Einstein's model fails at low temperatures because it assumes that all oscillators have the same frequency. This assumption is removed in the Debye model, which leads to much better agreement between theory and experiment at low temperatures. At the same time, the Einstein model does demonstrate that quantization is necessary to explain the vanishing value of the specific heat of a crystal at zero temperature.

21.3.2 Debye model

The Einstein model can be improved by allowing the crystal to vibrate at different frequencies. The energy of a system consisting of oscillators at various frequencies, $\omega(\mathbf{k})$, is given by

$$E = sum_{\mathbf{k}} n_{\mathbf{k}} \omega(\mathbf{k}), \tag{21.13}$$

where $n_{\mathbf{k}}$ is the number of oscillations with frequency $\omega(\mathbf{k})$ and is given by the Planck distribution function, Eq. (21.10). The summation in the last equation can also be replaced as an integral over frequency,

$$E = \int \mathrm{d}\omega D(\omega) n(\omega, T) \hbar\omega, \tag{21.14}$$

where $D(\omega)$ is the *density of modes* of vibration[9], i.e., the number of vibrational modes in the frequency range ω and $\omega + \mathrm{d}\omega$. Like its electronic counterpart, $n(E)$, the quantity $D(\omega)$ is by definition non-negative. It plays an important role in determining the dynamic properties of a lattice and describing and analyzing experimental results. This and the related quantity $D(\omega^2)$ will be the central focus of our formal discussion in this section. For example, the heat capacity is obtained from the last expression upon differentiation with respect to temperature.

In order to carry out the integral in Eq. (21.14), and derive an expression for C_V, we must set the limits of integration. Let us consider periodic boundary conditions in a system with N^3 cells (atoms) contained inside a cubic box of side L. Periodic boundary conditions imply that in reciprocal space there is one allowed value of the $\mathbf{k}$ vector in a volume[10] $\left(\frac{L}{2\pi}\right)^3 = \frac{V}{8\pi^3}$.

[9]Corresponding to the density of states, $n(E)$, in the electron case.

[10]This follows from the relation, $e^{i(xk_x+yk_y+zk_z)} = e^{i[(x+L)k_x+(y+L)k_y+(z+L)k_z]}$, which implies that k_x, k_y, k_z can take the values $\pm n\pi/L$, $n = 0, 2, 4, \cdots$ (choosing N to be even).

In the Debye (or continuum) approximation, the speed of sound, v_s, is taken to be a constant and $\omega(\mathbf{k}) = v_s\mathbf{k}$. The total number of modes, $N(\omega)$, with wave vector less than $\mathbf{k}$ is the product of $\left(\frac{L}{2\pi}\right)^3$ times the volume of a sphere of radius k, so that

$$\begin{aligned} N(\omega) &= \left(\frac{L}{2\pi}\right)^3 \frac{4\pi}{3} k^3 = \frac{V}{8\pi^3}\frac{4\pi}{3}k^3 \\ &= \frac{V}{8\pi^3}\frac{4\pi\omega}{3v_s^3} \\ &= \frac{V\omega^3}{6\pi v_s^3}, \end{aligned} \tag{21.15}$$

along each direction (or each polarization, a concept that is discussed in more detail in Appendix R). The quantity v_s defined implicitly in the last expression denotes the speed of sound in the material and describes the speed of propagation of a lattice (vibrational) wave in the system. Differentiating with respect to ω, we have

$$D(\omega) = \frac{\mathrm{d}N(\omega)}{\mathrm{d}\omega} = \frac{V\omega^2}{2\pi v_s^3}, \tag{21.16}$$

for each polarization type. If there are N primitive cells, the total number of (acoustic) phonon modes is N, and from Eq. (21.15) we obtain the cutoff frequency, ω_D,

$$\omega_D^3 = \frac{6\pi^2 v_s^3 N}{V}. \tag{21.17}$$

The frequency ω_D is also called the *Debye frequency* of the crystal. The length K_D of the corresponding vector is given by

$$K_D = \frac{\omega_D}{v_s} = \left(\frac{6\pi^2 N}{V}\right)^{\frac{1}{3}}. \tag{21.18}$$

It follows that the integral in Eq. (21.14) extends from 0 to ω_D, and we have,

$$E = \int_0^{\omega_D} \mathrm{d}\omega \left(\frac{V\omega^3}{2\pi^2 v_s^3}\right)\left(\frac{\hbar\omega}{e^{\beta\hbar\omega}-1}\right), \tag{21.19}$$

for each polarization type. Assuming further that the velocity of sound, v_s, is independent of polarization, we multiply the last integral by 3 and, defining the quantities,

$$x_D \equiv \frac{\hbar\omega_D}{k_B T} = \frac{\theta_D}{T}, \tag{21.20}$$

we obtain

$$E = 9Nk_B T\left(\frac{T}{\theta_D}\right)^3 \int_0^{x_D} \mathrm{d}x \frac{x^3}{e^x - 1}. \tag{21.21}$$

The quantity θ_D is called the *Debye temperature*. Differentiating the last equation with respect to temperature yields the expression

$$C_V = 9Nk_B\left(\frac{T}{\theta_D}\right)^3\int_0^{x_D} dx\frac{x^4 e^x}{(e^x-1)^2}, \tag{21.22}$$

for the specific heat at constant volume.

The Debye T^3 law can now be derived. At low enough temperatures, we may let x_D approach infinity obtaining

$$\int_0^\infty dx\frac{x^3}{e^x-1} = \int_0^\infty dx x^3\sum_{n=1}^{\infty} e^{-nx} = 6\sum_{1}^{\infty}\frac{1}{n^4} = \frac{\pi^4}{15}. \tag{21.23}$$

Therefore, in the limit $T << \theta_D$, we have $E \approx 3\pi^4 Nk_B T^4/5{\theta_D}^3$, and

$$C_V \approx \frac{12\pi^4}{5}Nk_B\left(\frac{T}{\theta_D}\right)^3. \tag{21.24}$$

This is the Debye T^3 law for the specific heat of a crystal at low temperatures.

The Debye approximation reinforces the conclusions drawn from the Einstein model about the quantization of lattice vibrations. It also improves upon that model and can be expected to be quite accurate at low temperatures.

21.4 Harmonic Approximation

In this section we describe the dynamics of lattice vibrations within the *harmonic approximation*. Our aim is to derive the equation of motion for the ions moving in an effective field provided by the presence of the other ions and the electrons, and also derive the secular equation, which in the case of periodic systems, determines the variation of frequency with wave vector. Following that, we discuss the density of modes functions, and show how lattice vibrations can be studied in terms of Green functions. This in principle allows a unified approach to the phonon structure of ordered and disordered alloys. The interested reader can find elementary as well as more advanced treatments of lattice vibrations than presented here in a number of published works [3, 8-10]. In our discussion in this section, we follow closely the lines of development in Ghatak and Kothari [9].

21.4.1 Equation of motion

For the sake of simplicity of exposition, we consider a monoatomic perfect crystal[11] in which atomic positions, $\mathbf{R}_l$, are defined with respect to any (arbitrarily chosen) one atom according to the well-known relation,

$$\mathbf{R}_l = l_1\mathbf{a}_1 + l_2\mathbf{a}_2 + l_3\mathbf{a}_3, \tag{21.25}$$

where the $\mathbf{a}_i$, with $i = 1, 2, 3$, are the primitive vectors of the lattice, and the l_i are integers, positive, negative, or zero. These three integers will be denoted collectively by l.

The atoms in the lattice can execute small oscillations about their equilibrium positions. We denote the instantaneous displacement along the αth direction of the atom at $\mathbf{R}_l$ by $u_\alpha(l)$, where $\alpha = 1, 2, 3$, corresponding to the x-, y-, and z-directions. If M denotes the mass of an atom, then its kinetic energy is given by

$$\frac{1}{2}M\dot{\mathbf{u}}\cdot\dot{\mathbf{u}} = \frac{1}{2}\sum_{\alpha=1}^{3}\dot{u}_\alpha^2(l), \tag{21.26}$$

where a dot over a symbol denotes differentiation with respect to time. The total kinetic energy of the lattice is given by a sum over all atomic positions,

$$T = \frac{1}{2}M\sum_{l}\sum_{\alpha=1}^{3}\dot{u}_\alpha^2(l), \tag{21.27}$$

where a sum over l denotes summations over all values of l_1, l_2, and l_3.

The oscillations of the atoms take place in the field produced by the other ions and electrons in the system. We can obtain an expression for the (restoring) forces acting on a displaced atom by expanding the potential energy, V, of the lattice in a Taylor series in terms of the atomic displacements, $\mathbf{u}(l)$. [A process for obtaining the quantities $V_{\alpha\beta}(ll')$ as well as higher order potentials from first principles is discussed in Chapter 14.] Thus, we obtain

$$V = V_0 + \sum_{l}\sum_{\alpha}V_\alpha(l)u_\alpha(l) + \frac{1}{2}\sum_{ll'}\sum_{\alpha\beta}V_{\alpha\beta}(ll')u_\alpha(l)u_\beta(l') + \cdots, \tag{21.28}$$

where V_0 is the potential energy of the lattice corresponding to the equilibrium (undisplaced) configuration of the atoms and without loss of generality can be set equal to zero. The term $V_\alpha(l)$ is given by

$$V_\alpha(l) = \left.\frac{\partial V}{\partial u_\alpha(l)}\right|_0, \tag{21.29}$$

[11] In a following subsection, we work out an example of a diatomic lattice in order to illustrate some important features.

where a subscript of 0 indicates that the derivative is evaluated at the equilibrium atomic positions. We see that $V_\alpha(l)$ is the negative of the force acting on the atom at $\mathbf{R}_l$ in its equilibrium position and must vanish by definition,

$$V_\alpha(l) = 0. \tag{21.30}$$

The second-order term,

$$V_{\alpha\beta}(ll') = \left.\frac{\partial^2 V}{\partial u_\alpha(l) u_\beta(l')}\right|_0 = V_{\beta\alpha}(l'l), \tag{21.31}$$

is the first non-vanishing term in Eq. (21.28) and it is the only term that we retain in the following discussion. The truncation of the potential energy at the term of second order constitutes the *harmonic approximation*, in which the potential energy due to the atomic displacements is taken to be of the quadratic form,

$$V = \frac{1}{2}\sum_{ll'}\sum_{\alpha\beta} V_{\alpha\beta}(ll') u_\alpha(l) u_\beta(l'). \tag{21.32}$$

This approximation neglects terms of order higher than quadratic leading to a particularly useful formal development of the dynamics of lattice vibrations. At the same time, the terms that are neglected are responsible for a number of important physical effects such as thermal expansion, and phonon-phonon interactions that lead to finite phonon lifetimes. We confine ourselves to the harmonic approximation, in which the Hamiltonian describing the moving atoms takes the form,

$$\begin{aligned} H &= T + V \\ &= \frac{1}{2M}\sum_{l\alpha} P_\alpha^2(l) + \frac{1}{2}\sum_{ll'}\sum_{\alpha\beta} V_{\alpha\beta}(ll') u_\alpha(l) u_\beta(l'), \end{aligned} \tag{21.33}$$

where

$$P_\alpha(l) = M\dot{u}_\alpha(l), \tag{21.34}$$

is the momentum along the αth direction of the atom (nucleus) at $\mathbf{R}_l$.

In order to derive the equation of motion for the atoms, we construct the Lagrangian, L, of the system,

$$\begin{aligned} \mathtt{L} &= T - V \\ &= \frac{1}{2}M\sum_{l\alpha} \dot{u}_\alpha^2(l) - \frac{1}{2}\sum_{ll'}\sum_{\alpha\beta} V_{\alpha\beta}(ll') u_\alpha(l) u_\beta(l'), \end{aligned} \tag{21.35}$$

and from the relation

$$\frac{\partial \mathtt{L}}{\partial u_\alpha(l)} = -\sum_{l'\beta} V_{\alpha\beta}(ll') u_\beta(l'), \tag{21.36}$$

and the generalized equation of motion [11],

$$\frac{\mathrm{d}}{\mathrm{d}t}\left(\frac{\partial \mathrm{L}}{\partial \dot{u}_\alpha(l)}\right) = \frac{\partial \mathrm{L}}{\partial u_\alpha(l)}, \tag{21.37}$$

we obtain the equation of motion of the atom at $\mathbf{R}_l$,

$$M\ddot{u}_\alpha(l) = -\sum_{l'\beta} V_{\alpha\beta}(ll')u_\beta(l'). \tag{21.38}$$

In this expression, $-V_{\alpha\beta}(ll')u_\beta(l')$ represents the force in the α direction on the atom at $\mathbf{R}_l$ due to the displacement of the atom at $\mathbf{R}_{l'}$ along the β direction. The linear dependence of the restoring force on displacement is the characteristic signature of oscillatory motion. We expect that each atom will execute harmonic motion around its equilibrium position in accordance with Eq. (21.38). The last expression can also be written in the form,

$$M\ddot{u}_\alpha(l) = \sum_{l'} F_\alpha(ll'), \tag{21.39}$$

where

$$F_\alpha(ll') = -\sum_{\beta} V_{\alpha\beta}(ll')u_\beta(l'), \tag{21.40}$$

is the αth component of the force acting on the particle at $\mathbf{R}_l$ arising from the displacement of the particle at $\mathbf{R}_{l'}$ along the β direction, with the effects of displacement along different directions being superposed in linear fashion.

The expansion coefficients, $V_{\alpha\beta}$, satisfy a particularly interesting relation. We note that no force acts along the αth direction on the atom at $\mathbf{R}_l$ when the crystal as *a whole* is shifted along the β direction by an infinitesimal amount. Since the displacement is the same for *all* atoms, the vanishing of the force is expressed by the equation,

$$\sum_{ll'}\sum_{\alpha\beta} V_{\alpha\beta}(ll') = 0. \tag{21.41}$$

Introducing the *force constant matrix*, $\underline{\Phi}_{ll'}$, with elements[12],

$$[\underline{\Phi}_{ll'}]_{\alpha\beta} = V_{\alpha\beta}(ll'), \tag{21.42}$$

we obtain the important result,

$$\underline{\Phi}_{ll} = -\sum_{l'\neq l} \underline{\Phi}_{ll'}. \tag{21.43}$$

As discussed below, this relation has important consequences in attempts to treat substitutionally disordered systems using methods such as the coherent potential approximation (CPA) that may have been well-established in the study of disorder with respect to the electronic structure of a material.

[12]Often, the force constant matrix is represented in dyadic notation.

21.4.2 Secular equation

We are now in a position to attempt a solution to the equations of motion, Eq. (21.38). These constitute an infinite set of equations for the lattice displacements, $\mathbf{u}(l)$. To solve them, we realize that in a translationally invariant crystal, the quantities $V_{\alpha\beta}(ll')$ can only depend on the difference $\mathbf{R}_l - \mathbf{R}_{l'}$, so that

$$V_{\alpha\beta}(ll') = V_\alpha(l-l') = V_{\alpha\beta}(l'l). \tag{21.44}$$

We are, therefore, motivated to seek solutions in terms of lattice Fourier transforms.

Since we expect a harmonic behavior, we assume a solution of the form

$$u_\alpha(l) = A_\alpha \mathrm{e}^{-\mathrm{i}(\omega t - \mathbf{q}\cdot\mathbf{R}_l)}, \tag{21.45}$$

where A_α is the amplitude of the oscillations along the α direction, which because of translational invariance, is independent of l. The last equation describes a plane wave propagating through the lattice in the direction of the wave vector $\mathbf{q}$. Substituting the last expression for $u_\alpha(l)$ into Eq. (21.38), we obtain the result,

$$\omega^2 A_\alpha = \sum_\beta D_{\alpha\beta}(\mathbf{q}) A_\beta, \tag{21.46}$$

where

$$D_{\alpha\beta}(\mathbf{q}) = \frac{1}{M}\sum_{l'} V_{\alpha\beta}(l-l')\mathrm{e}^{-\mathrm{i}\mathbf{q}\cdot[\mathbf{R}_{l'}-\mathbf{R}_l]}, \tag{21.47}$$

and where we have used the fact that

$$\ddot{u}_\alpha(l) = -\omega^2 u_\alpha(l). \tag{21.48}$$

Because $V_{\alpha\beta}(l-l')$ depends only on the difference $\mathbf{R}_l - \mathbf{R}_{l'}$, the sum in Eq. (21.47) extends only over all values of l' for any fixed value of l, and is thus independent of l. Setting $l = 0$, and denoting $V_{\alpha\beta}(0l)$ simply by $V_{\alpha\beta}(l)$, we have

$$D_{\alpha\beta}(\mathbf{q}) = \frac{1}{M}\sum_{l'} V_{\alpha\beta}(l')\mathrm{e}^{\mathrm{i}\mathbf{q}\cdot\mathbf{R}_{l'}}. \tag{21.49}$$

Comparison of Eq. (21.46) with Eq. (21.38) reveals that the problem of solving an infinite set of differential equations has been reduced to that of solving a set of three linear homogeneous equations in the three unknowns, A_α. Obviously, this simplification is possible only because we are considering an infinite periodic lattice. Rewriting Eq. (21.46) in the form,

$$\sum_\beta \left[D_{\alpha\beta}(\mathbf{q}) - \omega^2\delta_{\alpha\beta}\right] A_\beta = 0, \tag{21.50}$$

we find that non-trivial solutions, A_α, exist provided that the determinant of the coefficient matrix vanishes,

$$\det\left[D_{\alpha\beta}(\mathbf{q}) - \omega^2\delta_{\alpha\beta}\right] = 0. \tag{21.51}$$

This is the ***secular equation*** that determines the solutions $\omega^2(\mathbf{q})$, often also called the ***phonic band structure.*** For the case of monoatomic systems, the secular matrix is three dimensional, and Eq. (21.51) yields a cubic equation for $\omega^2(\mathbf{q})$. The corresponding three solutions can be denoted by $\omega_j(\mathbf{q})$, with $j = 1,\ 2,\ 3$, and are referred to as the ***acoustical branches*** of the phonon spectrum. In the case of a lattice with a basis of r atoms, the secular equation leads to $3r$ solutions (branches), three of which are the acoustical branches, and $3r - 3$ are the ***optical*** branches. An explicit example of a linear diatomic chain is presented in the following section.

We close the present discussion by writing the secular equation in a form involving the force-constant matrix. From Eq. (21.42), we have

$$\det\left[\underline{\Phi}(\mathbf{q}) - \omega^2\underline{I}\right] = 0, \tag{21.52}$$

where $\underline{I}$ is the three-dimensional unit matrix. One cannot help but notice the formal similarity between the last equation and the secular equation determining the electronic structure of a solid.

21.4.3 One-dimensional examples

We now illustrate the considerations presented above by means of one-dimensional examples. The difference in behavior between $D(\omega^2)$ and $D(\omega)$ as functions of ω^2 and ω, respectively, is discussed. Also, we work out [3] the case of a diatomic lattice and show the manner in which optical modes arise in lattice vibrations.

Monoatomic lattice

Consider an infinite chain of atoms[13] of mass M arranged a distance a apart along the length of the chain. In the harmonic approximation, the force acting on atom n due to the displacement of atom $n+m$ is proportional to the net displacement, $u_{n+m} - u_n$, so that the total force on atom n is given by the expression, Eq. (21.40),

$$F_n = \sum_m C_{nm}(u_{n+m} - u_n), \tag{21.53}$$

where m runs over all positive and negative integers.

---

[13] Or planes arranged periodically along one dimension and assumed to vibrate so that the atoms in them remain co-planar.

The force constants, C_{nm}, are given by the second derivative of the potential energy of atom n with respect to displacements of atoms at n and at $n+m$, and are in general different for transverse and longitudinal waves (i.e., waves formed by oscillations perpendicular or parallel to the line). It follows that the equation of motion of atom n is given by

$$M\frac{\mathrm{d}^2u_n}{\mathrm{d}t^2} = \sum_m C_{nm}(u_{n+m} - u_n). \tag{21.54}$$

As suggested by Eq. (21.45), we look for solutions of the form,

$$u_n = A\mathrm{e}^{\mathrm{i}(nak-\omega t)}, \tag{21.55}$$

where k is the wave vector. Now, Eq. (21.54) reduces to the form,

$$\omega^2 M = -\sum_m C_{nm}\left(\mathrm{e}^{\mathrm{i}mka} - 1\right). \tag{21.56}$$

It follows from symmetry considerations that $C_{n+m} = C_{n-m}$, so that we have

$$\omega^2 M = -\sum_{m>0} C_{n+m}\left(\mathrm{e}^{\mathrm{i}mka} + \mathrm{e}^{-\mathrm{i}mka} - 2\right), \tag{21.57}$$

or

$$\omega^2 = \frac{2}{M}\sum_{m>0} C_{n+m}(1 - \cos mka). \tag{21.58}$$

It is important to note that the slope of ω^2 vs k, $\mathrm{d}\omega^2/\mathrm{d}k$, vanishes at $k = \pm n\pi/a$. This has consequences for the behavior of the density of modes (density of states) $D(\omega^2)$, which are discussed below.

Now, if we assume that C_{n+m} vanishes unless the atoms n and m are nearest neighbors, and denote the associated force constant by C_1, we have

$$\omega^2 = \frac{2}{M}C_1\left(1 - \cos ka\right). \tag{21.59}$$

We can also write this expression in the form,

$$\omega^2 = \frac{4C_1}{M}\sin^2\frac{1}{2}ka, \tag{21.60}$$

from which it follows that

$$\omega = \sqrt{\frac{4C_1}{M}}\left|\sin\frac{1}{2}ka\right|. \tag{21.61}$$

We choose the root of Eq. (21.60) so that the frequency is always positive.

The functions $\omega^2(\mathbf{k})$ and $\omega(\mathbf{k})$ are plotted as functions of $\mathbf{k}$ in Fig. 21.4,

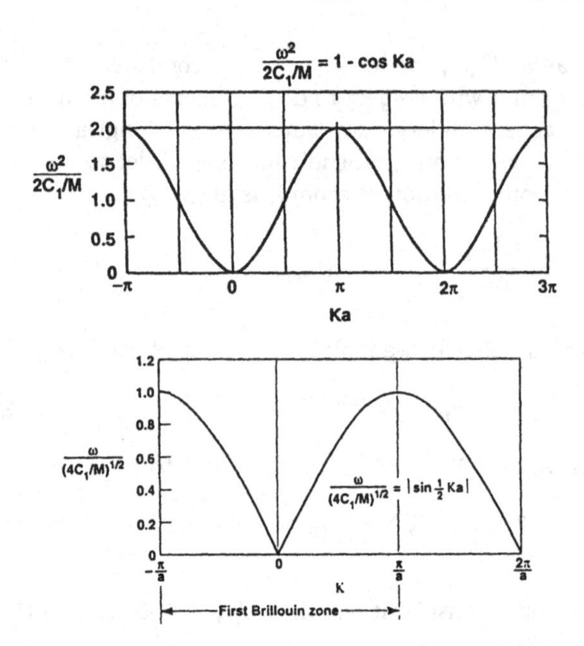

Figure 21.4: **The dispersion relations $\omega^2(\mathbf{k})$ and $\omega(\mathbf{k})$ for the monoatomic linear chain discussed in the text [3].**

upper and lower panels, respectively. We note that as k ranges between $-\pi/a$ to π/a, it covers all independent values of the exponential e^{ika}. The k-values $-\pi/a \leq k < \pi/a$ form the first Brillouin zone of the reciprocal lattice for this one-dimensional system. From physical optics (and in analogy with the electron case) the group velocity of a phonon wave packet is given by the expression,

$$\begin{aligned} v_g &= \frac{d\omega}{dk} \\ &= \left(\frac{C_1 a}{M}\right)^{\frac{1}{2}} \cos \frac{1}{2} ka, \end{aligned} \tag{21.62}$$

where Eq. (21.61) has been used.

Diatomic lattice

Consider an infinite linear chain formed by atoms of alternating masses M_1 and M_2 a distance a apart. We assume that only adjacent atoms interact and that the force constant is C, independent of the masses of the atoms. We let u_n and v_m denote the displacements of masses M_1 and M_2, respectively. In a manner analogous to that leading to Eq. (21.54), we obtain the coupled equations,

$$M_1 \frac{d^2 u_n}{dt^2} = C \left(v_n + v_{n-1} - 2u_n \right) \tag{21.63}$$

and

$$M_2 \frac{d^2 v_n}{dt^2} = C \left(u_{n+1} + u_n - 2v_n \right). \tag{21.64}$$

Assuming solutions of oscillatory form, e.g., $u_n = u\mathrm{e}^{\mathrm{i}(kna-\omega t)}$, and $v_m = v\mathrm{e}^{\mathrm{i}(kma-\omega t)}$, we obtain the following two homogeneous equations for the coefficients (amplitudes) u and v,

$$\begin{aligned} -\omega^2 M_1 u &= Cv\left(1+\mathrm{e}^{-\mathrm{i}ka}\right) - 2Cu \\ -\omega^2 M_2 v &= Cu\left(\mathrm{e}^{\mathrm{i}ka}+1\right) - 2Cv. \end{aligned} \tag{21.65}$$

These equations have non-trivial solutions only when the determinant of the coefficients vanishes,

$$\begin{vmatrix} 2C - M_1\omega^2 & -C\left(1+\mathrm{e}^{-\mathrm{i}ka}\right) \\ -C\left(\mathrm{e}^{\mathrm{i}ka}+1\right) & 2C - M_2\omega^2 \end{vmatrix} = 0, \tag{21.66}$$

or

$$M_1 M_2 \omega^4 - 2C\left(M_1 + M_2\right)\omega^2 - 2C^2\left(1 - \cos ka\right) = 0. \tag{21.67}$$

This is a quadratic equation in ω^2 whose solutions are given by

$$\omega^2 = \frac{2C\left(M_1+M_2\right) \pm \sqrt{4C^2\left(M_1+M_2\right)^2 + 8M_1M_2C^2(1-\cos ka)}}{2M_1M_2}. \tag{21.68}$$

This expression can be analyzed as a whole, but it is instructive to consider the limits $ka << 1$ and $ka = \pm\pi$ (at the zone boundary). For small ka, we have $\cos ka \approx 1 - k^2a^2/2$ and the two roots of Eq. (21.68) become,

$$\omega^2 \approx 2C\left(\frac{1}{M_1} + \frac{1}{M_2}\right), \quad \text{(optical branch)} \tag{21.69}$$

and

$$\omega^2 \approx \frac{\frac{1}{2}C}{M_1 + M_2} k^2 a^2, \quad \text{(acoustical branch)}. \tag{21.70}$$

At the zone boundary, $ka = \pm\pi$, we have

$$\omega^2 = \frac{2C}{M_1}; \quad \text{and} \quad \omega^2 = \frac{2C}{M_2}. \tag{21.71}$$

The dependence of ω^2 on k, the dispersion relation, for the diatomic linear chain with $M_1 > M_2$ is shown in Fig. 21.5. Experimental dispersion curves,

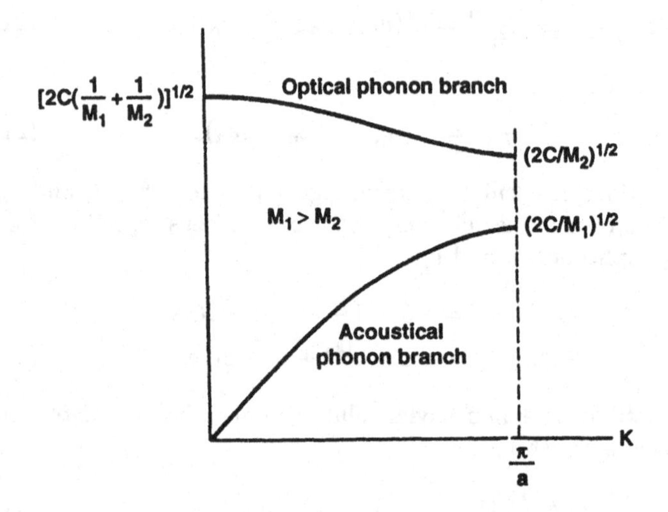

Figure 21.5: **Optical and acoustical branches of the dispersion relation for the diatomic linear chain discussed in the text, with $M_1 > M_2$.**

ω vs **k**, for diamond [12] in the (001) and (111) directions in reciprocal space are shown in Fig. 21.6, with k in units of π/a. We note the existence of the optical branches in diamond associated with the two atoms per unit cell of the diamond lattice even though the two atoms are identical. The two different branches in the spectrum reflect the fact that the two atoms in the unit cell can oscillate in phase (acoustical branch), or with a phase difference of π (optical branch).

We also note in Fig. 21.5 that there exist no solutions for frequencies ω such that $\sqrt{2C/M_1} < \omega < \sqrt{2C/M_2}$, so that there is a *frequency gap* at the zone boundary. Such frequency gaps are a common feature of the phonon dispersion relations for systems with polyatomic unit cells.

21.4.4 Connection to optical properties

The optical branch in a diatomic system corresponds to a mode of vibration in which the atoms vibrate "against" one another while their center of mass remains fixed. When the two atoms of a basis have different electric charges, say in an ionic compound, an optical mode can be excited by the passage of radiation through the material. The oscillatory electric field of

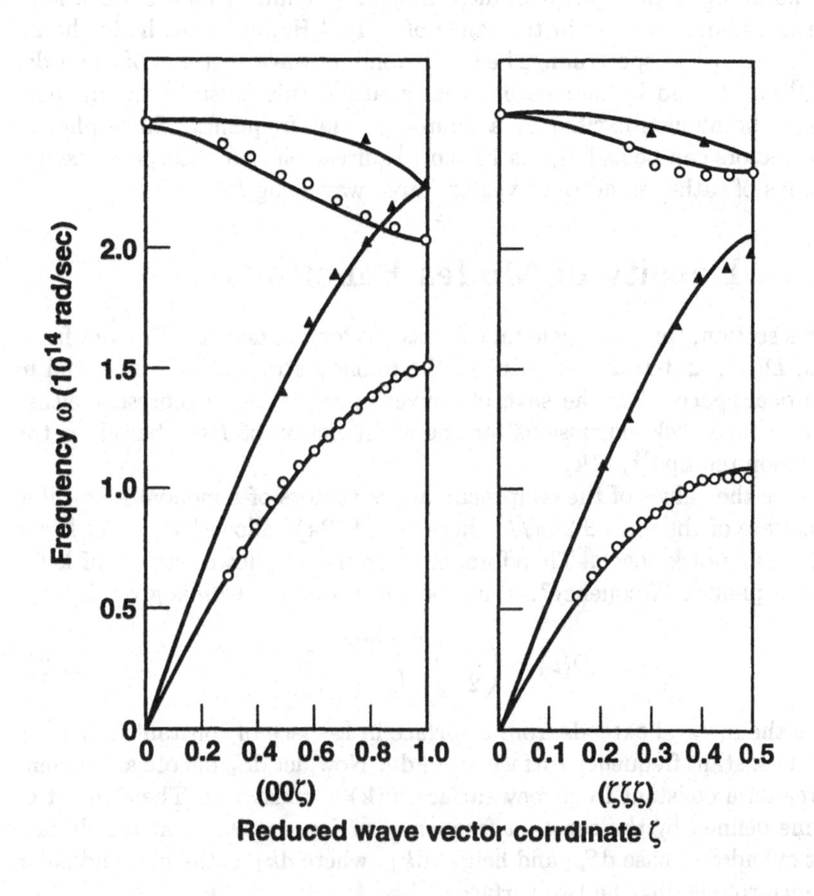

Figure 21.6: **Measured (symbols) and calculated (curves) frequencies for diamond along the (001) and (111) directions in reciprocal space [31].**

the radiation can cause[14] the atoms to oscillate out of phase, thus coupling the radiation to the optical branch of the spectrum of lattice vibrations. Relevant frequencies are in the range of $\simeq 10^{13}$ Hz, and thus lie in the infrared part of the spectrum. The corresponding wave vector is of the order of 10^3 cm^{-1}, and by momentum conservation, this must be the momentum of the phonon excited by a photon of that frequency. Since phonon wave vectors can be as large as 10^8 cm^{-1}, direct photon absorption excites phonons of rather small wave vector (large wave length).

21.5 Density of Modes Functions

In this section, we derive general expressions for the density of modes functions, $D(\omega^2)$ and $D(\omega)$, in terms of the phonon structure of a material in reciprocal space. For the sake of convenience, we let ν represent either ω^2 or ω, and seek expressions for the construction of $D(\nu)$ based on the dispersion relation[15], $\nu(\mathbf{k})$.

Since the values of the reciprocal lattice vectors of a monoatomic cubic system are of the form $\pm 2n\pi/L$, there are $(L/2\pi)^3$ allowed values of $\mathbf{k}$ per unit volume of $\mathbf{k}$-space. Therefore, the number of allowed values of $\mathbf{k}$ for which a phonon "frequency", ν, lies between ν and $\nu + \mathrm{d}\nu$ is given by

$$D(\nu) = \left(\frac{L}{2\pi}\right)^3 \int_{\nu}^{\nu+\mathrm{d}\nu} \mathrm{d}^3k, \tag{21.72}$$

where the integral extends from a surface in $\mathbf{k}$-space of constant frequency ν, to a constant frequency surface at $\nu+\mathrm{d}\nu$. Now, let $\mathrm{d}S_\nu$ denote an element of area on a constant frequency surface, $\nu(\mathbf{k}) = \mathrm{const} = \nu$. The element of volume defined by the constant frequency surfaces at ν and at $\nu + \mathrm{d}\nu$ is a right cylinder of base $\mathrm{d}S_\nu$, and height $\mathrm{d}k_\perp$, where $\mathrm{d}k_\perp$ is the perpendicular distance connecting the two surfaces.

Clearly, $\mathrm{d}k_\perp$ is a function of position in $\mathbf{k}$-space. Furthermore, by the definition,

$$|\nabla_{\mathbf{k}}\nu(\mathbf{k})|\, \mathrm{d}k_\perp = \mathrm{d}\nu, \tag{21.73}$$

so that

$$\mathrm{d}S_\nu \mathrm{d}k_\perp = \mathrm{d}S_\nu \frac{\mathrm{d}\nu}{|\nabla_{\mathbf{k}}\nu(\mathbf{k})|}. \tag{21.74}$$

Therefore, we have

$$D(\nu)\mathrm{d}\nu = \left(\frac{L}{2\pi}\right)^3 \int \mathrm{d}\nu \frac{\mathrm{d}S_\nu}{|\nabla_{\mathbf{k}}\nu(\mathbf{k})|}, \tag{21.75}$$

[14]Provided that selection rules, such as momentum and energy, are satisfied, and that the resulting oscillations are allowed.

[15]Dispersion relations for $D(\omega^2)$ and for $D(\omega)$ for the case of the linear chain were derived in the previous section.

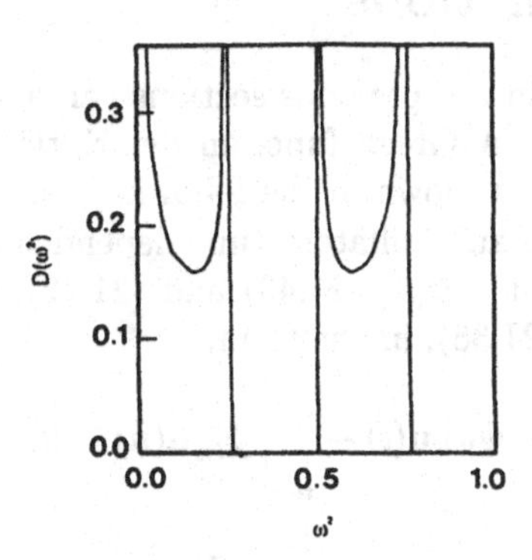

Figure 21.7: **Phonon density of states, $D(\omega^2)$, for an ordered binary alloy with mass ratio equal to two.**

so that the density of modes is given by the expression

$$D(\nu) = \left(\frac{L}{2\pi}\right)^3 \int \frac{dS_\nu}{|\nabla_{\mathbf{k}}\nu(\mathbf{k})|}. \tag{21.76}$$

The integral extends over the surface $\nu(\mathbf{k}) = \text{const} = \nu$ in **k**-space. This expression should be compared with the corresponding expression for electrons, Eq. (6.17). We note that for $\nu = \omega$, we have $\nabla_{\mathbf{k}}\omega(\mathbf{k}) = v_g(\mathbf{k})$, (where v_g is the group velocity), so that we can write,

$$D(\omega) = \left(\frac{V}{8\pi^3}\right) \int \frac{dS_\omega}{v_g(\mathbf{k})}. \tag{21.77}$$

It is useful at this point to take a closer look at the behavior of the two different kinds of density of modes that are represented by $D(\nu)$ for the case of a linear lattice. It follows from Eqs. (21.59) and (21.61) that the function $D(\omega^2)$ becomes unboundedly large at $\omega = 0$, since at zero frequency $\nabla_{\mathbf{k}}\omega(\mathbf{k}) \to 0$, and the integral in Eq. (21.76) becomes singular. No such singularity occurs for the $D(\omega)$ function, since $\omega(\mathbf{k})$ has a finite slope at $\omega = 0$ (see Fig. 21.4). Such behavior is usually absent in two and three dimensions because it is unlikely that the gradient of $\omega(\mathbf{k})$ will vanish along all three directions at the Brillouin zone boundary. Finally, from the example of the one-dimensional diatomic lattice, we see that $D(\omega^2)$ exhibits the presence of the acoustical and optical branches, of which the acoustical branch diverges at $\omega = 0$ (Fig. 21.7).

21.6 Green Functions

The discussion presented in the previous sections can also be expressed in terms of Green functions. A Green function description is also useful in treating random alloys, as is shown in the following section.

With the assignment of an oscillatory time dependence to the displacement vector, $\mathbf{u}(l)$, and using Eqs. (21.43) and (21.42), we can write the equation of motion, Eq. (21.38), in the form,

$$\left[\omega^2 \underline{M} - \underline{\Phi}_{ll}\right] \mathbf{u}(l) - \sum_{n \neq l} \underline{\Phi}_{ln} \mathbf{u}(n) = 0, \tag{21.78}$$

where $\underline{M}$ is a diagonal matrix with elements[16]

$$[\underline{M}]_{\alpha\beta} = M_\alpha \delta_{\alpha\beta}. \tag{21.79}$$

We now make the identifications,

$$\underline{\epsilon}_l = \frac{\underline{\Phi}_{ll}}{\underline{M}}, \quad (\underline{M} \text{ is diagonal}), \tag{21.80}$$

and

$$\underline{W}_{ij} = (\underline{M})^{-1/2} \underline{\Phi}_{ij} (\underline{M})^{-1/2}, \tag{21.81}$$

and write Eq. (21.78) in the form of a matrix equation whose elements are defined in both site space (over the sites of the lattice) and in the three-dimensional space defined by the directions in space,

$$\left[\omega^2 \underline{\underline{I}} - \underline{\underline{\epsilon}} - \underline{\underline{W}}\right] \mathbf{u} = 0, \tag{21.82}$$

where a double underline denotes matrices in site and direction space, and the vector $\mathbf{u}$ has elements

$$[\mathbf{u}]_{l\alpha} = u_\alpha(l). \tag{21.83}$$

Now, in complete analogy with the tight-binding description of electron systems, we define the Green function,

$$\underline{\underline{G}}(\omega^2) = \left[\omega^2 \underline{\underline{I}} - \underline{\underline{\epsilon}} - \underline{\underline{W}}\right]^{-1}. \tag{21.84}$$

Many of the considerations associated with the electronic TB model remain valid in the case of the lattice vibrational Green function just defined (with some minor and fairly obvious technical modifications). For example, in the case of translationally invariant systems, a Fourier transformation can

[16]This definition allows for the case in which the atom (ion) moves along different directions with different effective masses, M_α.

be used to arrive at the secular Eq. (21.51), as the reader can easily verify. More specifically, the density of phonon modes is given by the expression,

$$\begin{aligned} D(\omega^2) &= -\frac{1}{\pi N}\Im \mathrm{Tr}\underline{MG}(\omega^2) \\ &= -\frac{1}{\pi}\Im \underline{MG}_{00}(\omega^2). \end{aligned} \tag{21.85}$$

It should be pointed out, however, that in spite of many formal similarities between electronic and phonic systems, the underlying physical principles governing each system are different. In addition to the replacement of the energy parameter, E ($E \pm i\eta$), by the strictly positive quantity, ω^2 ($\omega^2 + i\eta$), many other distinctions arise. Some of these differences will be discussed in the next section, where the Green function treatment of lattice vibrations in random, disordered alloys is presented.

21.7 Substitutional Disorder and the CPA

We now turn our attention to the lattice vibrations of random systems. Randomness in lattice vibrations can imply either strictly ***diagonal disorder***, where the mass varies from site to site but where the force constants are configuration independent, or ***off-diagonal disorder*** (ODD), in which the force constants can vary as a function of configuration, or most commonly, combinations of the two.

The Green function formalism is ideally suited for the study of lattice dynamics in random systems. In the following pages we formulate the problem of determining the density of phonon modes in random alloys, and the extent to which well-developed approximation schemes in the study of electronic systems can be used in the present case.

21.7.1 Mass disorder

We consider a random, substitutionally disorder binary alloy, A_cB_{1-c}, consisting of masses M_A and M_B distributed over the sites of a lattice with concentration c and $1-c$, respectively[17]. From Eq. (21.84), and assigning a site index to the mass tensor, we can write,

$$\omega^2\underline{M}_i\underline{G}_{ij} = \delta_{ij} + \underline{\Phi}_{ii}\underline{G}_{ij} + \sum_{k\neq i}\underline{\Phi}_{ik}\underline{G}_{kj}. \tag{21.86}$$

Multiplying this expression from the left by $\underline{M}_i^{-1}$, and subsequently from the left and from the right by $\underline{M}_i^{1/2}$ and $\underline{M}_j^{1/2}$, respectively, and defining

[17] We continue to explicitly treat the case of monoatomic systems. The extension to lattices with a basis is not conceptually complicated.

the quantities,

$$\underline{G}'_{ij} = \underline{M}_i^{1/2}\underline{G}_{ij}\underline{M}_j^{1/2}, \tag{21.87}$$

$$\underline{W}_{ij} = \underline{M}_i^{-1/2}\underline{\Phi}_{ij}\underline{M}_j^{-1/2}, \tag{21.88}$$

and

$$\underline{g}_i = \left(\omega^2\underline{I} - \underline{W}_{ii}\right)^{-1}, \tag{21.89}$$

we can cast Eq. (21.86) in the form,

$$\underline{G}'_{ij} = \underline{g}_i\left[\delta_{ij} + \sum_{k\neq i}\underline{W}_{ik}\underline{G}'_{kj}\right]. \tag{21.90}$$

We recognize this last expression as the equation of motion for the Green function within a TB formalism, as developed for the case of electrons.

It is important to point out that because of the factors $\underline{M}_i^{-1/2}$, the "hopping" matrix $\underline{W}_{ij}$ depends on the species of atom occupying sites i and j, even if the force constant matrix is configuration independent. Therefore, the TB-like Hamiltonian associated with the lattice vibrations of a mass (diagonally) disordered alloy is characterized by the presence of ODD. In this case, the ODD is of the geometric type and can be treated within a scalar (rather than a configurational-matrix) framework. The situation becomes more complicated in the case of general ODD, i.e., when the force constant matrix is configuration dependent (the force constants, $\underline{\Phi}_{ij}$, depend on the species of atoms occupying sites i and j). We will see in the following discussion that these features have serious implications in attempting to construct approximate methods for the determination of the ensemble average of the system Green function.

However, the density of modes, $D(\omega^2)$, for a random alloy, with either diagonal or off-diagonal disorder, can always be determined numerically. The methods at one's disposal for doing this are essentially identical to those used to calculate the density of states of random, substitutionally disordered electronic systems [13]. The history of the numerical simulation of the lattice spectrum is very long [14-21], and it includes model calculations as well as applications to real systems [17].

The following figures show phonon density of states, $D(\omega^2)$, curves as functions of ω^2 for one-dimensional, substitutionally disordered binary alloys. As specified in the captions, Fig. 21.8 corresponds to alloys with only mass disorder, whereas Fig. 21.9 shows the density of modes for an alloy with both mass and force constant disorder. These results were obtained using the negative eigenvalue counting theorem [13], whose application to phonon systems is identical (apart from technical modifications) to its application to electronic systems.

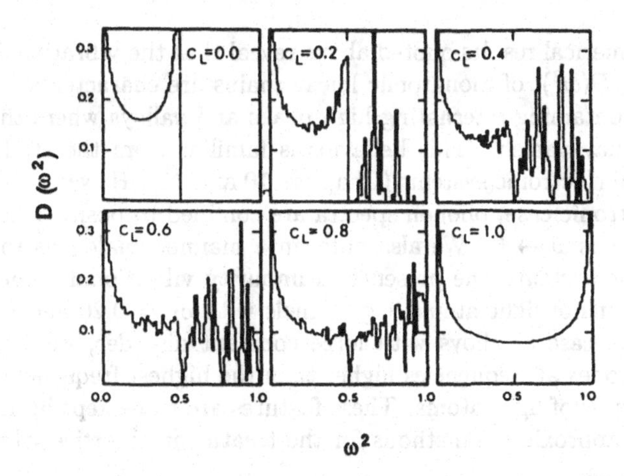

Figure 21.8: **Squared frequency spectra for two-component disordered chains of 5,000 atoms. The alloys are characterized by a mass ratio of 2:1, and the dynamic matrix (force constants) extends only to nearest neighbors. The parameter c_L denotes the concentration of the light atoms. The abscissa is given in terms of the maximum squared frequency of the monoatomic light chain as a unit: thus, 1.0 corresponds to the squared frequency $\omega^2_{\max}$(light), and 0.5 to $\omega^2_{\max}$(heavy).**

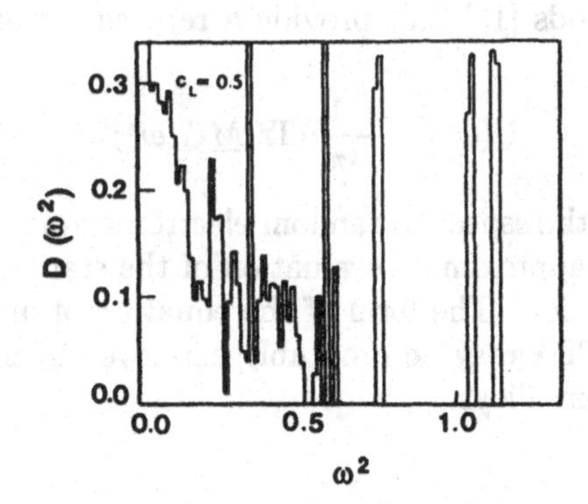

Figure 21.9: **Results analogous to those of the previous figure, but for an alloy with both mass and force-constant disorder. Here, the mass ratio is 5:1, and $\Phi^{AA} = 5.0$, $\Phi^{AB} = \Phi^{BA} = 2.0$, and $\Phi^{BB} = 1.0$. Note the presence of impurity levels at frequencies higher than the maximum frequency for a chain of light atoms.**

The numerical results quoted above reveal that the vibrational frequency spectra, $D(\omega^2)$, of monatomic linear chains are characterized by a rich structure containing alternating high peaks and valleys where the density of modes may vanish. This behavior is familiar from the study of one-dimensional electronic systems (Chapters 10 and 11). However, in contrast to the electronic case, phonon spectra are confined to positive values of ω, and diverge as $\omega \to 0$. We also note, in a manner analogous to the case of electronic systems, the presence of impurity vibrational modes for low concentrations of light atoms, e.g., panels with $c_L = 0.20$ and $c_L = 0.30$. Also, in the case of alloys with force constant disorder, we can identify impurity modes at frequencies higher than the highest frequency exhibited by pure chains of light atoms. These features are to be kept in mind when we discuss approximate methods for the treatment of lattice vibrations in random alloys.

Frequency distribution curves for two- and three-dimensional systems [13] are shown in Figs. 21.10 and 21.11. The higher-dimensional spectra show considerably less structure than their one-dimensional counterparts, but share a number of common features with them as well. For example, the presence of impurity bands is clearly noticeable in all cases.

21.8 CPA for phonons

The frequency spectra shown in the figures mentioned above were obtained by numerical methods [13] that provide a representation of the ensemble average

$$D(\omega^2) = -\frac{1}{\pi}\Im\mathrm{Tr}\langle \underline{MG}(\omega^2)\rangle. \tag{21.91}$$

As was the case with respect to random electronic systems, we seek methods that allow the approximate evaluation of the statistical average of the phonon Green function. The form of the equation of motion, Eq. (21.90) suggests that the CPA may be profitably employed in the study of lattice vibrations of random alloys.

21.8.1 Mass disordered alloys

Indeed, in the case of alloys with only mass disorder, the TB model can be used to construct a Green function that corresponds to an alloy with both diagonal and off-diagonal disorder. This can be treated using the method of Blackman et al. [21], developed for the application of the CPA to random electronic systems with ODD. The method of BEB has been discussed in Chapter 11. In fact, the ODD that arises from the presence of mass disorder

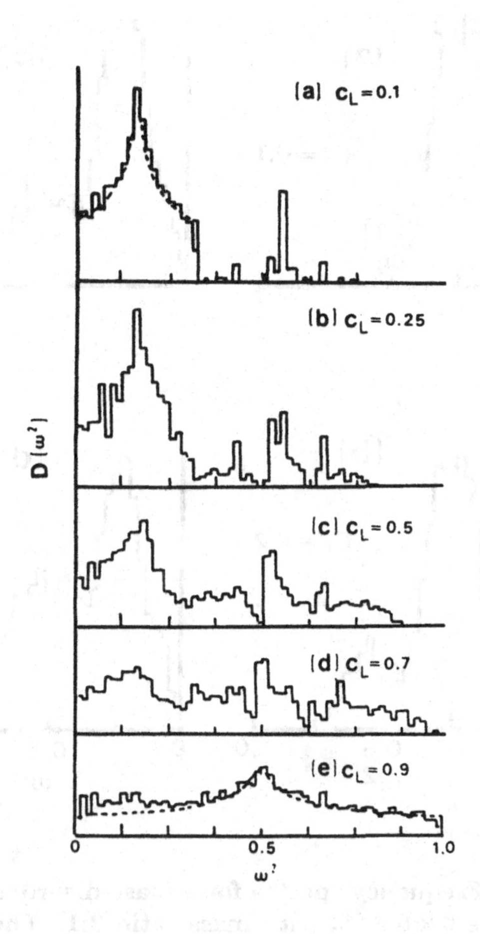

Figure 21.10: Frequency spectra for a mass-disordered square lattice of size 56 × 16 sites with atomic mass ratio 3:1. The broken lines in panel (a) indicate the spectrum of an infinite, pure system of heavy atoms, and in panel (e) that of an infinite, pure system of light atoms [13].

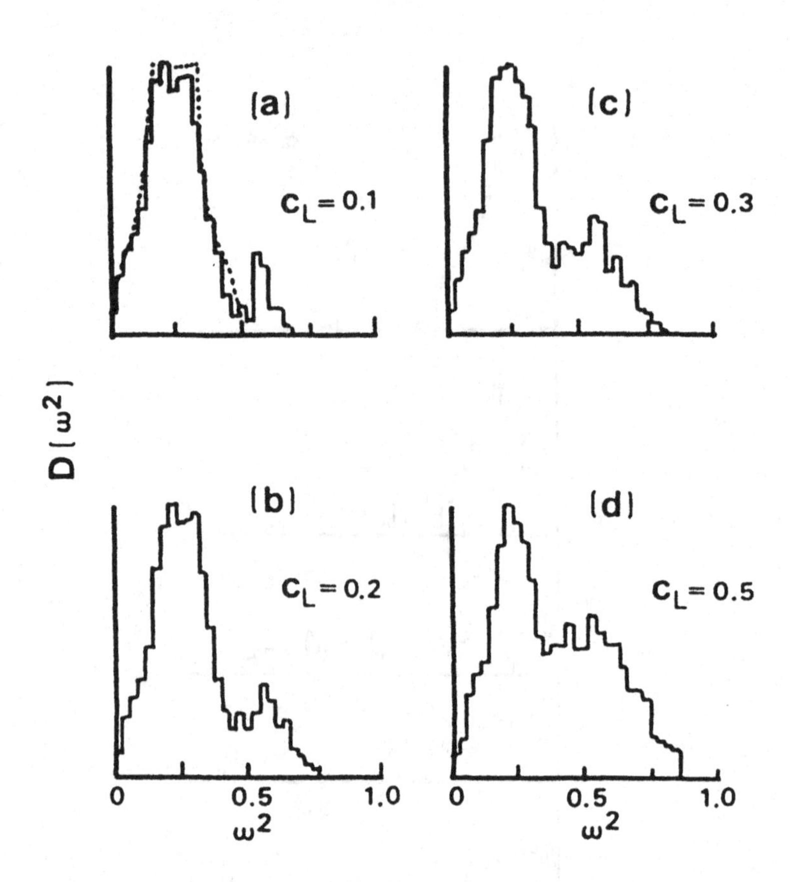

Figure 21.11: **Frequency spectra for a mass-disordered simple cubic lattice of size 6 × 6 × 25, with mass ratio 2:1. The broken line in panel (a) indicates the exact spectrum for the infinite monatomic cubic lattice of heavy atoms [13].**

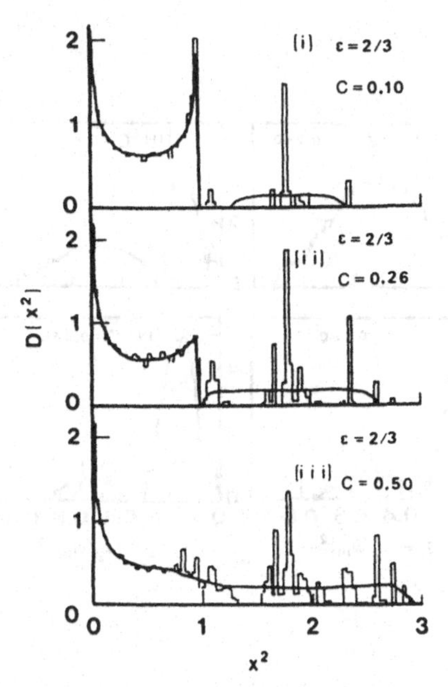

Figure 21.12: **Phonon density of modes, $D(x^2)$, vs. x^2, where $x^2 = \omega^2/\omega_{\max}^2$, for mass-disordered linear chains with mass ratio 3:1. The value of c denotes the concentration of light atoms [23].**

is of the geometric type,

$$W^{AB^2} = W^{AA}W^{BB}, \tag{21.92}$$

and the matrix formalism of BEB can be reduced to a scalar representation. Such a scalar application of the CPA to mass disordered alloys was given by Taylor [22]. Phonon spectra for alloys with only mass disorder calculated [22] numerically (histograms) and in the CPA are shown in Figs. 21.12 and 21.13 for one- and three-dimensional systems, respectively. The overall agreement of the CPA results with those obtained numerically is quite satisfactory, and resembles the behavior of the CPA in connection with electronic systems. As for electronic systems, the CPA is more accurate than non-self-consistent single-site theories such as the ATA, and it reproduces more accurately the structure in the interior of the band than at the band edges. The neglect of long-distance statistical fluctuations causes the CPA to introduce gaps somewhat too easily [see panel (i) in Fig. 21.12], and to lead to the wrong analytic form of the spectrum at band edges. The accuracy of the CPA increases with increasing dimensionality (Fig. 21.13).

In addition to the single-site CPA, cluster theories developed for the treatment of random electronic systems can be used in the study of the lat-

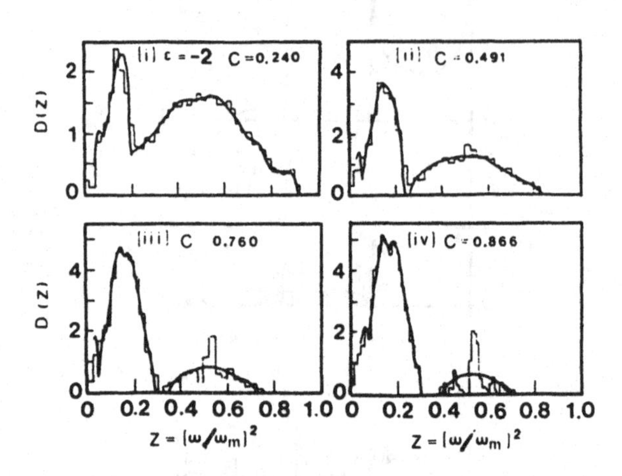

Figure 21.13: **Results analogous to those of the previous figure but for simple cubic, mass-disordered alloys with mass ratio 3:1.**

tice vibrations of substitutionally disordered alloys. The application of self-consistent as well as non-self-consistent methods is possible, each method being characterized by essentially the same features that distinguish it in applications to electronic systems. An application of the embedded cluster method [23] (ECM) to mass-disordered alloys [24] is shown in Fig. 21.14, panels (a) to (d), with the CPA results shown in panel (e). As for electronic systems, the ECM allows a fairly accurate reproduction of the exact spectra when clusters of moderately large size are employed. It also allows the i-dentification of configurations that lead to specific structure in the spectra. In three-dimensional systems, smaller cluster sizes than necessary in one dimension may suffice for an accurate reproduction of $D(\omega^2)$.

21.8.2 Force-constant disorder

Serious difficulties appear in applications of the CPA to alloys with both mass and force-constant disorder. The culprit responsible for these difficulties is the site-diagonal force constant, $\underline{\Phi}_{ll}$. Because of Eq. (21.43), the site-diagonal term, $\underline{\epsilon}_i$, in Eq. (21.80) can be a function of the occupancy of sites surrounding site i. Theextent of the region around site i which influences $\underline{\epsilon}_i$ depends on the range of the force constant matrix, $\underline{\Phi}_{ll'}$. It is precisely this correlation of a site to its environment that cannot be treated

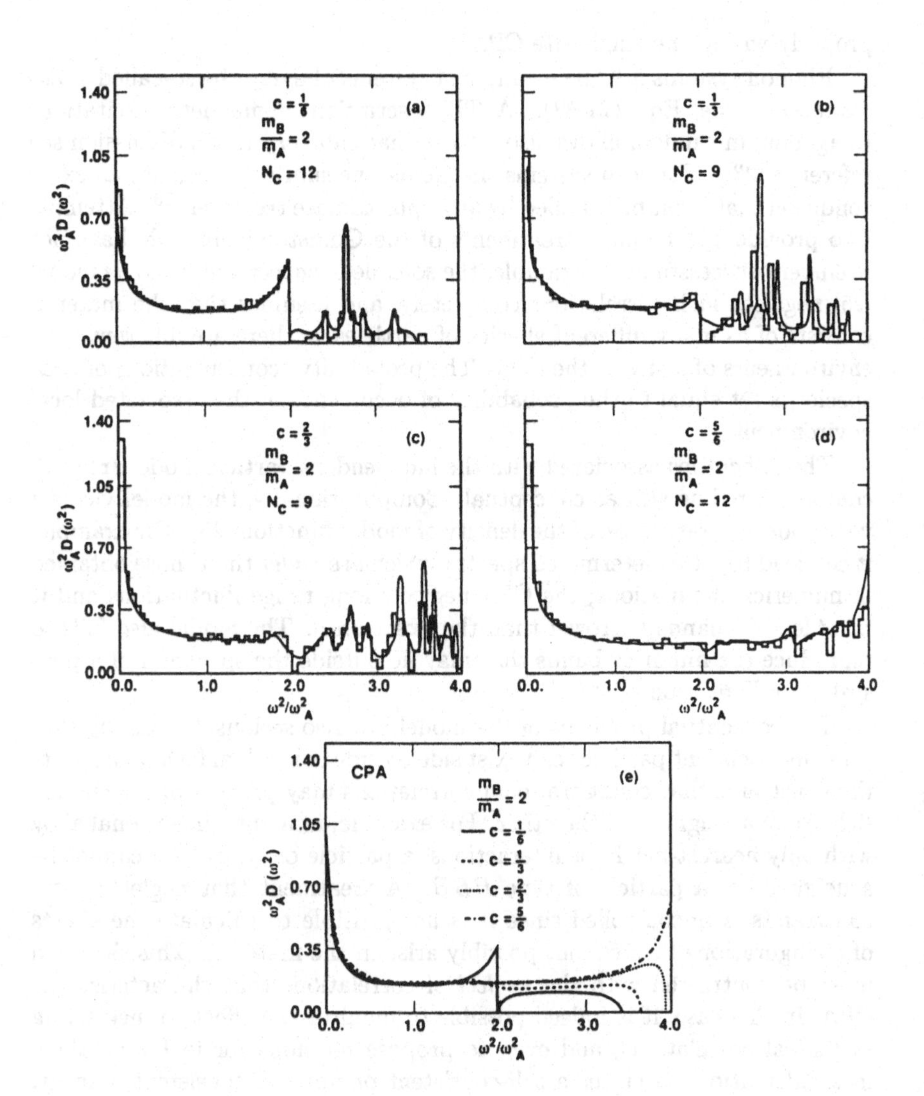

Figure 21.14: **Vibrational DOS of mass-disordered one-dimensional alloys with mass ratio 1:2 as a function of the concentration c of light atoms (various panels). The histograms represent the results of numerical simulations, and the solid curves those of theoretical methods, here the ECM, panels (a) to (d), and the CPA, panel (e). The number of sites, N_c, in the cluster used in applying the ECM is indicated in the panels [25].**

properly within the single-site CPA.

Phonon systems are one example of systems obeying the so-called *Goldstone sum rule*, Eq. (21.43). A TB description of magnetic excitations (magnons) in random alloys also falls in that category (for a discussion see reference [23]). For such systems the Goldstone sum rule becomes an extra condition that must be satisfied by any approximate treatment[18]. Attempts to provide approximate treatments of the Goldstone sum rule have not been very successful. For example, the so-called *independent particle* model [25] neglects such correlations completely, and assumes that the material consists of as many different species of particles as there are different local environments of a site in the alloy. The probability (concentration) of each species is set equal to the probability of occurrence of the associated local environment.

The difficulties associated with the independent particle model are both computational as well as conceptual. Computationally, the model yields a very poor representation of the density of modes function [25]. For example, it can lead to CPA-determined spectra which are *wider* than those obtained in numerical simulations; the CPA neglects long range fluctuations and it *must* lead to bands narrower than the exact ones. The model also fails to reproduce the impurity bands that may lie outside the spectrum of a pure system of light atoms.

The conceptual problems of the model are also serious. Assuming that these independent particles can exist side by side in random fashion neglects the constraint that configurational correlations may prevent one such particle to be a neighbor of the other. For example, in a one-dimensional alloy with only nearest-neighbor interactions, a particle of type AAA cannot be a neighbor of a particle of type BBB. A treatment that neglects these constraints is uncontrolled since it is not possible to calculate the effects of configurations that cannot possibly arise in the material. This situation must be contrasted with the *neglect* of correlations that characterize the CPA. In this case, it is indeed possible to monitor the effects of neglecting statistical correlations, and even to propose methods for including them in a calculation, in either a self-consistent or non-self-consistent manner. Attempts [27-29] to incorporate the Goldstone sum rule into a CPA-like treatment of lattice dynamics have until presently failed to produce fully satisfactory results.

The considerations presented above are also relevant to certain types of electronic disorder, especially to systems that exhibit so-called "charge transfer" effects. The independent particle model has been invoked [29] in attempts to treat such environmentally dependent effects. In the case of

[18] We recall that in the case of lattice vibrations, the Goldstone sum rule is a reflection of the fact that no force is acting on the atoms under uniform translation of the crystal as a whole. This is clearly a condition that must be observed in an approximate theory.

realistic systems, say a binary alloy based on an fcc lattice, considering only effects extending to nearest neighbors and neglecting differences caused by specific arrangements of atoms, this model results in an alloy of 26 components. The CPA can be carried out on this multicomponent system, and results indicate some improvement over an ordinary treatment in which charge transfer effects are neglected. Unfortunately, this "improvement" is largely fortuitous and cannot be justified on rigorous grounds. The difficulties the model faces with respect to electronic systems are conceptually equivalent to those encountered in the treatment of lattice vibrations, and as discussed above, lead to uncontrolled approximation schemes.

Bibliography

[1] Alexander L. Fetter and John Dirk Walecka, *Quantum Theory of Many-Particle Systems*, McGraw-Hill, Inc., San Francisco, CA (1971.)

[2] Gerald D. Mahan, *Many-Particle Systems*, Plenum Press, NY (1981).

[3] C. Kittel, *Introduction to Solid State Physics*, John Wiley and Sons, New York (1971), Fourth Edition.

[4] R. Y. Chiao, C. H. Towneg, and B. P. Stoicheff, Phys. Rev. Lett. **12**, 592 (1964).

[5] K. N. Baranskii, Soviet Phys. Doklady **2**, 237 (1957).

[6] R. Loudon, Adv. in Physics **13**, 424 (1964).

[7] A. Einstein, Ann. Physik **22**, 180 (1907).

[8] Neil W. Ashcroft and N. David Mermin, *Solid State Physics*, Saunders College, Philadelphia, PA (1976).

[9] A. K. Ghatak and L. S. Kothari, *An Introduction to Lattice Dynamics*, Addison-Wesley Publishing Company, Reading, MA (1972).

[10] C. Kittel, *Quantum Theory of Solids*, John-Wiley and Sons, New York, NY (1963).

[11] Herbert Goldstein, *Classical Mechanics* Addison Wesley, Reading, MA (1950).

[12] J. L. Warren, R. G. Wenzel, and J. L. Yarnell, *Inelastic Scattering of Neutrons*, IAEA, Vienna (1965).

[13] P. Dean, Rev. Mod. Phys. **44**, 127 (1972).

[14] E. W. Kellermann, Phil. Trans. A**238**, 513 (1940).

[15] M. Blackman, *Specific Heats of Solids*, Encycl. of Phys. **7**, 325, Springer Verlag, Heidelberg, Berlin (1955).

[16] B. Dayal and B. Sharan, Proc. Roy. Soc. A**262**, 136 (1961).

[17] K. Krebs, Phys. Rev. A**143**, 138 (1965).

[18] E. W. Montroll, J. Chem. Phys. **10**, 218 (1942).

[19] E. W. Montroll, J. Chem. Phys. **11**, 481 (1943).

[20] K. G. Aggarwal, J. Mahanty, and V. K. Tewary, Proc. Phys. Soc. (London) **86**, 1225 (1965).

[21] J. A. Blackman, D. M. Esterling, and N. F. Berk, Phys. Rev. B**4**, 2412 (1971).

[22] D. W. Taylor, Phys. Rev. **156**, 1017 (1967).

[23] A. Gonis, *Green Functions for Ordered and Disordered Systems*, North Holland, Amsterdam, New York, NY (1992), and references therein.

[24] C. W. Myles and J. D. Dow, Phys. Rev. B **19**, 4939 (1979).

[25] D. J. Whitelaw, J. Phys. C **14**,2871 (1980).

[26] T. Kaplan and M. Mostoller, Phys. Rev. B **9**, 1783 (1974).

[27] R. Bass, J. Phys. C **10**, L325 (1977).

[28] H. W. Diel and P. Leath, Phys. Rev. B **19**, 596 (1976).

[29] D. D. Johnson and F. J. Pinski, Phys. Rev. B**48**, 11553 (1993).

[30] J. L. Warren et al., Phys. rev. **158**, 806 (1967).

Chapter 22

The Electron-Phonon Interaction

22.1 General Comments

The electron-phonon (e-ph) interaction is responsible for a number of dramatic effects on the properties of materials. This interaction scatters electrons by lattice distortions, and because of these atomic displacements, this interaction goes beyond the Born-Oppenheimer approximation. The main effect is to degrade an electron's energy and momentum imparted by the presence of applied fields or temperature gradients. This obviously affects the transport properties of a material. For example, it provides the dominant contribution to the resistivity of reasonably pure metals except at very low temperatures, where other effects such as electron-electron scattering may become prominent.

The e-ph interaction is also responsible for the so-called *electron-phonon mass enhancement* for the modification of phonon frequencies and life times, and affects the phonon self-energy and the thermal conductivity of a material. As we have seen already, it is the main reason for the existence of superconductivity in a broad class of metals[1].

In this chapter, we review a number of formal elements that enter the study of the e-ph interaction. In the following discussion, it is useful to keep in mind a visual illustration of the various different ways in which this interaction manifests itself for some of the more basic processes (Fig. 22.1). Here, a filled dot represents the interaction and it is called a *vertex*. Straight

[1] Certain ceramics, with superconducting transition temperature in excess of 100°K, may be subject to different mechanisms that cause condensation into the superconducting state.

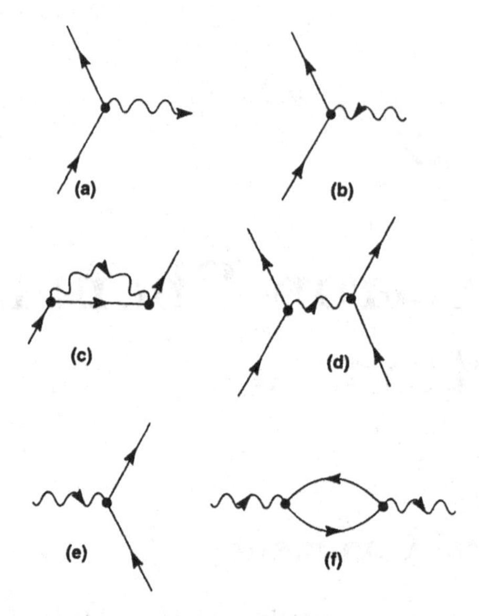

Figure 22.1: **Diagrammatic representation of various process associated with the electron-phonon interaction.**

lines coming out of a vertex, as indicated by the arrows, represent electrons, whiles wavy lines represent phonons. For example, Fig. 22.1a represents the process by which an electron changes its momentum and energy through the *emission* of a phonon, while phonon *absorption* is represented by Fig. 22.1b. We will refer to this figure as our discussion of the processes proceeds.

Because of the intricate coupling in the behavior of electrons and lattice vibrations, the study of the e-ph interaction in a material must be based on the knowledge of three distinct but interrelated ingredients:

1. the electronic structure of a material is needed to construct the electron propagators, or Green functions (straight lines in Fig. 22.1);

2. phonon frequencies and polarizations are used to construct the phonon propagators (wavy lines in Fig. 22.1); and

3. electron-phonon matrix elements yield the strength of the interaction [the vertices (dots) in the figure where two electron lines and a phonon line meet].

Much of our previous discussion in this book has been concerned with the determination of the electronic structure of a material, which directly addresses the first of the three ingredients just mentioned. We have seen, for example, how to determine the electronic structure of ordered, periodic solids, as well as random substitutionally disordered alloys. Realistic phonon frequencies can be calculated reliably for free-electron-like metals

[1], with rapid progress made in the case of transition metals and intermetallic alloys and compounds [2-6]. In addition, the availability of accurate neutron-scattering data [7] allows us to determine accurate phonon frequencies for most elements and many compounds. Thus, the second ingredient is fairly well in hand, at least for the purpose of possessing realistic values of phonon frequencies to use in a calculation. The third element needed in a discussion of the e-ph interaction, the e-ph matrix elements, represented by dots (vertices) in Fig. 22.1, can in principle be obtained through a first-principles determination of the electronic and phonic structure. However, given the complexity of the problem, at present such c alculations entail a number of approximation schemes and educated prescriptions. At the same time, some of the approximations appear to be sufficiently accurate to allow for a meaningful comparison between experimental and theoretical results.

22.2 Electronic Transport

The electron-phonon scattering process depicted in Figs. 22.1a and 22.1b leads to a dissipation of electron momentum and energy and thus directly affects the electrical conductivity of a metal. In this section, we discuss this effect within the context of the Boltzmann equation. The present formalism may help to shed some light into the nature of the various quantities that were introduced in our discussion of the Boltzmann equation in Chapter 18. The discussion in this chapter closely follows that given by Butler [9].

We saw in Chapter 18 that the Boltzmann equation describing transport in the presence of an electric field, $\mathbf{E}$, and a temperature gradient, ∇T, takes the form [8],

$$[e\mathbf{E}\cdot\mathbf{v}(\mathbf{k}) + (E_{\mathbf{k}}/T)\nabla T\cdot\mathbf{v}(\mathbf{k})]\frac{\partial f_{\mathbf{k}}}{\partial E_{\mathbf{k}}} = \sum_{\mathbf{k}'} Q_{\mathbf{k}\mathbf{k}'}g_{\mathbf{k}'}. \tag{22.1}$$

Here, $\mathbf{v}(\mathbf{k})$ is the group velocity of the electrons, $\mathbf{v}(\mathbf{k}) = \hbar\nabla_{\mathbf{k}}E_{\mathbf{k}}$, $f_{\mathbf{k}}$ is the equilibrium Fermi-Dirac distribution function, and $g_{\mathbf{k}}$ is the deviation function that describes the departure from equilibrium of the electron distribution function, $F_{\mathbf{k}}$,

$$F_{\mathbf{k}} = f_{\mathbf{k}} - \left(\frac{\partial f_{\mathbf{k}}}{\partial E_{\mathbf{k}}}\right)g_{\mathbf{k}}. \tag{22.2}$$

We assume that the phonons remain in equilibrium, which is justified at not too low temperatures, especially for systems with complicated Fermi surfaces.

The treatment of the Boltzmann equation hinges on assumptions made about the *scattering operator*, $Q_{\mathbf{k}\mathbf{k}'}$, and methods for solving for the deviation function, $g_{\mathbf{k}}$. The scattering operator has both *scattering-in*, $Q^{\mathrm{in}}_{\mathbf{k}\mathbf{k}'}$, and *scattering-out*, $Q^{\text{out}}_{\mathbf{kk}'}$, terms describing, respectively, the loss and gain of electrons in state **k** because of scattering between this and the other states. We write,

$$Q_{\mathbf{kk}'} = Q^{\text{out}}_{\mathbf{kk}'} - Q^{\text{in}}_{\mathbf{kk}'} \tag{22.3}$$

$$Q^{\text{out}}_{\mathbf{kk}'} = \frac{\delta_{\mathbf{kk}'}}{k_{\text{B}}T} \sum_{\mathbf{k}''} P_{\mathbf{kk}''} \tag{22.4}$$

$$Q^{\text{in}}_{\mathbf{kk}'} = \frac{1}{k_{\text{B}}T} P_{\mathbf{kk}'}, \tag{22.5}$$

where $P_{\mathbf{kk}'}$ is the probability at equilibrium (transition probability) to s-catter from state **k** to state **k**$'$. The electrical or thermal current density can be calculated in terms of $g_{\mathbf{k}}$,

$$\mathrm{j_e} = \frac{-2e}{V_a} \sum_{\mathbf{k}} \mathbf{v}(\mathbf{k}) g_{\mathbf{k}} \left(-\partial f_{\mathbf{k}} / \partial E_{\mathbf{k}}\right), \tag{22.6}$$

and

$$\mathrm{j_Q} = \frac{-2}{V_a} \sum_{\mathbf{k}} E_{\mathbf{k}} \mathbf{v}(\mathbf{k}) g_{\mathbf{k}} \left(-\partial f_{\mathbf{k}} / \partial E_{\mathbf{k}}\right), \tag{22.7}$$

where V_a denotes the volume per atom. The deviation function, $g_{\mathbf{k}}$, is obtained formally as the solution of the Boltzmann equation,

$$g_{\mathbf{k}'} = \sum_{\mathbf{k}} \left[Q^{-1}\right]_{\mathbf{k}'\mathbf{k}} \left[e\mathbf{E} \cdot \mathbf{v}(\mathbf{k}) + (E_{\mathbf{k}}/T)\nabla T \cdot \mathbf{v}(\mathbf{k})\right] \frac{\partial f_{\mathbf{k}}}{\partial E_{\mathbf{k}}}. \tag{22.8}$$

However, the inversion of the scattering operator is a very difficult task and methods other than that indicated by the last expression must be sought. One approach utilizes a variational procedure based on a theorem first proved by Kohler [10]. In a simple implementation of this variational theorem, $g_{\mathbf{k}}$ is set proportional to the velocity along the direction of the applied field, which in turn is taken to point along the positive x-direction, and write

$$g_{\mathbf{k}} = -e\mathbf{E}_x v_x(\mathbf{k})\tau, \tag{22.9}$$

where τ is a proportionality constant. It will turn out that τ plays the role of a relaxation time, describing the interval between successive collisions of an electron. The conductivity is now given in terms of τ by substituting the last expression into Eq. (22.6),

$$\sigma = \frac{2e^2 n(0)\langle v_x^2\rangle \tau}{V_a} = e^2 (n/m)_{\text{eff}}\tau, \tag{22.10}$$

where $n(0)$ is the single-spin electron density of states per atom at the Fermi level, and $\langle v_x^2\rangle$ is defined by the expression,

$$2n(0)\langle v_x^2\rangle = 2\sum_{\mathbf{k}} v^2(\mathbf{k}) \left(-\partial f_{\mathbf{k}} / \partial E_{\mathbf{k}}\right). \tag{22.11}$$

Here the quantity $2n(0)\langle v_x^2\rangle$ plays the role of $e^2(n/m)_{\text{eff}}$ in the classical formula [last expression in Eq. (22.10)]. At not too high temperature, for most metals the electronic structure does not vary appreciably on a scale of $4k_BT$, which is approximately the width of $\partial f_{\mathbf{k}}/\partial E_{\mathbf{k}}$. In the sum over $\mathbf{k}$ in Eq. (22.11), the derivative of the Fermi function can be replaced by a delta function, $-\partial f_{\mathbf{k}}/\partial E \approx \delta(E)$, and $\langle v_x^2\rangle$ can be calculated as the average over the Fermi surface of the square of the x-component of the velocity there (Fermi velocity).

The relaxation time, τ, can be obtained through a substitution of Eq. (22.9) into Eq. (22.1), followed by an operation with $\sum_{\mathbf{k}} v_x(\mathbf{k})$ on both sides of the expression, leading to,

$$n(0)\langle v_x^2\rangle = \tau \sum_{\mathbf{k}\mathbf{k}'} Q_{\mathbf{k}\mathbf{k}'} v_x(\mathbf{k}) v_x(\mathbf{k}'). \tag{22.12}$$

From Eqs. (22.3) to (22.5), and the symmetry of the transition probability, $P_{\mathbf{k}\mathbf{k}'} = P_{\mathbf{k}'\mathbf{k}}$, it follows that

$$2k_BTn(0)\langle v_x^2\rangle/\tau = \sum_{\mathbf{k}\mathbf{k}'} P_{\mathbf{k}\mathbf{k}'}\left[v_x(\mathbf{k}) - v_x(\mathbf{k}')\right]^2. \tag{22.13}$$

The foregoing discussion has been general in the sense that no specific mechanism has been connected to the scattering events described by $P_{\mathbf{k}'\mathbf{k}}$. We now proceed by tying the transition probabilities to the e-ph interaction. To do this we make use of the *golden rule* of perturbation theory. We consider an electron, initially in state $|\mathbf{k}\rangle$, making a transition to a state $|\mathbf{k}'\rangle$ by emitting a phonon of momentum $\mathbf{q} = \mathbf{k}' - \mathbf{k}$ and energy $\hbar\omega^j_{\mathbf{k}'-\mathbf{k}} = E_{\mathbf{k}'} - E_{\mathbf{k}}$, where j represents the polarization of the phonon. A schematic representation of this process is shown in Fig. 22.1a. The probability, $P^{\mathrm{e}}_{\mathbf{k}\mathbf{k}'}$, describing the emission of a phonon can be written in the form,

$$\begin{aligned} P^{\mathrm{e}}_{\mathbf{k}\mathbf{k}'} &= \frac{2\pi}{\hbar}\sum_j |M^j_{\mathbf{k}\mathbf{k}'}|^2 f_{\mathbf{k}}(1-f_{\mathbf{k}'}) \\ &\times \left[n\left(\omega^j_{\mathbf{k}'-\mathbf{k}}\right)+1\right]\delta\left(E_{\mathbf{k}} - E_{\mathbf{k}'} - \hbar\omega^j_{\mathbf{k}'-\mathbf{k}}\right), \end{aligned} \tag{22.14}$$

where $M^j_{\mathbf{k}\mathbf{k}'}$ denotes the electron-phonon matrix element for transition between electronic states $\mathbf{k}$ and $\mathbf{k}'$ mediated by the emission of a phonon of polarization j. Here, $f_{\mathbf{k}}(E)$ and $n(\omega)$ are the Fermi-Dirac and Bose-Einstein distribution functions, respectively, and the factor $f_{\mathbf{k}}(1-f_{\mathbf{k}'})$ ensures that the initial state is occupied and the final state empty so the process can take place. The terms in the phonon factor, $[n(\omega)+1]$, arise because of the presence of stimulated, $n(\omega)$, and spontaneous emission, 1, transition probabilities [11]. The delta function describes energy conservation in the emission process. Similarly, the probability for absorption of a phonon (Fig. 22.1b) is given by

$$\begin{aligned} P^{\mathrm{a}}_{\mathbf{kk}'} &= \frac{2\pi}{\hbar}\sum_j |M^j_{\mathbf{kk}'}|^2 f_{\mathbf{k}}(1-f_{\mathbf{k}'}) \\ &\times\; n\left(\omega^j_{\mathbf{k}'-\mathbf{k}}\right)\delta\left(E_{\mathbf{k}}-E_{\mathbf{k}'}+\hbar\omega^j_{\mathbf{k}'-\mathbf{k}}\right). \end{aligned} \qquad (22.15)$$

The total transition probability is given by the sum $P_{\mathbf{kk}'} = P^{\mathrm{e}}_{\mathbf{kk}'} + P^{\mathrm{a}}_{\mathbf{kk}'}$, and using it in Eq. (22.13) yields an expression for τ.

Further insight into Eqs. (22.13) to (22.15) can be obtained by using the fact that the e-ph matrix elements $M^j_{\mathbf{kk}'}$ (discussed in more detail below) and the group velocities, $v_x(\mathbf{k})$, change with energy on the scale[2] of eV, while the temperature and phonon frequencies are of the order of meV. This difference in energy scales allows us to separate the $\mathbf{k}$ dependence of the sum in Eq. (22.13) into an energy dependence and a variation over surfaces in $\mathbf{k}$ space. Accordingly, we replace the sums over $\mathbf{k}$ and $\mathbf{k}'$ by integrals over constant-energy surfaces,

$$\sum_{\mathbf{k}} = \int \mathrm{d}E \sum_{\mathbf{k}} \delta(E-E_{\mathbf{k}}) = \int \mathrm{d}E \frac{V_{\mathrm{a}}}{(2\pi)^3}\int \frac{\mathrm{d}S_{\mathbf{k}}(E)}{\hbar v_{\mathbf{k}}}, \qquad (22.16)$$

where $\mathrm{d}S_{\mathbf{k}}(E)$ is an element of the constant-energy surface at E. It is useful to insert an integral over ω, so that Eq. (22.13) takes the form,

$$\begin{aligned} \frac{k_{\mathrm{B}}Tn(0)\langle v_x^2\rangle}{\tau\pi} &= \int \mathrm{d}E \int \mathrm{d}E' \int \mathrm{d}\omega \\ &\times \left\{ \frac{V_{\mathrm{a}}^2}{(2\pi)^6}\int \frac{\mathrm{d}S_{\mathbf{k}}(E)}{\hbar v(\mathbf{k})}\int \frac{\mathrm{d}S_{\mathbf{k}'}(E')}{\hbar v(\mathbf{k}')}\sum_j |M_{\mathbf{kk}'}|^2 \right. \\ &\times \left. [v_x(\mathbf{k})-v_x(\mathbf{k}')]^2\,\delta(\hbar\omega-\hbar\omega^j_{\mathbf{q}})\right\} \\ &\times\; f(E)[1-f(E')]\left\{[n(\omega)+1]\right. \\ &\times\; \left.\delta(E-E'-\hbar\omega)+n(\omega)\delta(E-E'+\hbar\omega)\right\}. \end{aligned}$$

(22.17)

This expression can be greatly simplified by evaluating the surface integrals at the Fermi surface. Although not exact, this evaluation makes only a small error for materials with non-pathological topologies of the Fermi surface, at least at moderate temperatures, and allows us to replace the expressions within the first set of curly braces by a dimensionless function

[2]Special considerations may be needed in the cases of some high-T_c, A-15 materials characterized with very narrow bands near the Fermi energy.

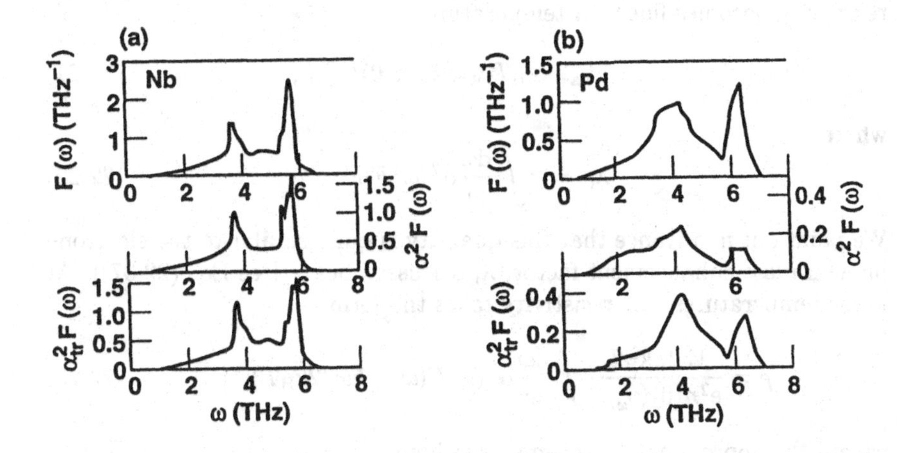

Figure 22.2: **The functions $F(\omega)$ and $\alpha_{\rm tr}^2(\omega)F(\omega)$ for the elemental solids Nb and Pd.**

of ω, which we designate by $\alpha_{\rm tr}^2(\omega)F(\omega)$. This function was introduced by Allen [12] and tends to be similar in shape to the phonon density of states, $F(\omega)$. The functions $F(\omega)$ and $\alpha_{\rm tr}^2(\omega)F(\omega)$ for the elemental solids Nb and Pd are shown in Fig. 22.2. The figure also shows the corresponding function, $\alpha^2(\omega)F(\omega)$, which occurs in the theory of the electron-phonon mass enhancement as well as the theory of the superconducting transition temperature.

Now, Eq. (22.17) can be written in the form,

$$\hbar/\tau = 4\pi k_B T \int \frac{d\omega}{\omega} \alpha_{\rm tr}^2(\omega)F(\omega)I(\hbar\omega/2k_BT), \tag{22.18}$$

where the function $I(\hbar\omega/2k_BT)$ is defined by

$$I(\hbar\omega/2k_BT) = \left(\frac{\hbar\omega}{2K_B^2T^2}\right)\int dE\int dE' f(E)[1-f(E')] \\ \{[n(\omega)+1]\delta(E-E'-\hbar\omega)+n(\omega)\delta(E-E'+\hbar\omega)\}, \tag{22.19}$$

and can be shown to have the simple form,

$$I(x) = (x/\sinh x)^2 . \tag{22.20}$$

At sufficiently high temperatures, so that $k_B T >> \hbar\omega_{max}$, where ω_{max} is the maximum phonon frequency, $I(x)$ tends to unity and the electrical resistivity becomes linear in temperature,

$$\rho = V_a 2\pi k_B T \lambda_{tr} / 2e^2 n(0) \langle v_x^2 \rangle \hbar, \tag{22.21}$$

where

$$\lambda_{tr} = 2 \int \frac{d\omega}{\omega} \alpha^2(\omega) F(\omega). \tag{22.22}$$

We point out in advance that this quantity is quite similar to the electron–phonon mass enhancement factor, λ, discussed below [see Eq. (22.37)]. At lower temperatures, the resistivity takes the form,

$$\rho = \frac{V_a 2\pi k_B T}{e^2 n(0) \langle v_x^2 \rangle \hbar} \int \frac{d\omega}{\omega} \alpha^2(\omega) F(\omega) I(\hbar\omega / 2k_B T), \tag{22.23}$$

within the approximation scheme used here.

This expression can be thought of as a modern version of the Bloch-Grüneisen theory [13] and can be used to derive a number of well-known results. For example, the Bloch-Grüneisen formula is obtained by replacing the function $\alpha^2(\omega)F(\omega)$ by $C\omega^4$, where C is an adjustable constant, and placing an upper limit of ω_D on the frequency integral, where ω_D is also an adjustable parameter. In fact, it can be shown that $\alpha^2(\omega)F(\omega)$ does vary as ω^4 as $\omega \to 0$, so that Eq. (22.23) yields a T^5 law for the resistivity at sufficiently low temperatures. On the other hand, it should be pointed out that the approximation used here becomes unreliable in that limit.

Figure 22.3 shows the results of calculations [9] for the resistivity of elements Nb and Pd compared to experimental results [14, 15]. Given that no adjustable parameters were used in the calculations, the agreement with the experiment is quite good.

Further refinements to the method can be introduced based on an energy- and **k**-dependent distribution function obtained from an (approximate) inversion of the scattering operator in Eq. (22.8). These improvements are very important in the region below 50°K, but they would be hardly noticeable on Fig. 22.3

22.3 Mass Enhancement

One manifestation of the e-ph interaction in a metal is that it tends to *slow down* the electrons near the Fermi energy in a metal. This effect does *not* contribute to the resistivity because it does not cause energy or momentum dissipation. The interaction does not affect the Fermi momentum k_F, so that the slowing down of an electron can be understood in terms of a larger

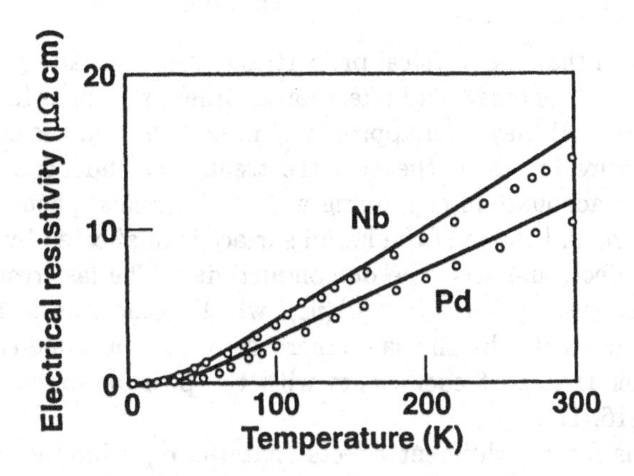

Figure 22.3: **Calculated resistivities of Nb and Pd compared to experimental results.**

effective mass. For a metal described to a good approximation by a free-electron dispersion relation, $E \approx k^2$, the Fermi velocity can be written in the form, $v_F = \hbar k_F/m$, where m is the effective mass of the electron in the absence of the e-ph interaction. A reduction in the Fermi velocity $v_F^* < v_F$ caused by the e-ph interaction implies a larger effective mass for the electron, $m^* = m(1+\lambda)$, where λ is called the *electron-phonon mass enhancement parameter*. This quantity can vary from about 0.1 for the alkali metals, to 1.5 for Pd.

The process leading to the *electron-phonon mass enhancement* is the *emissions* and *reabsorption* by the electrons of "virtual" phonons (Fig. 22.1c). The word "virtual" implies that energy need not be conserved during the short time intervening between the emission and the reabsorption of the phonon. Physically, a n electron slows down because its charge polarizes the lattice in its vicinity (the ions are displaced toward it), and can move more slowly through the region in which the ions exert an increased force on it. This effect on the velocity can be observed in de Haas-van Alphen and cyclotron resonance experiments. Also, the low-temperature specific heat is enhanced by this effect since it is proportional to the density of states at the Fermi level, and as Eq. (22.16) indicates, this quantity varies as the inverse of the velocity,

$$N(E) = \int \mathrm{d}E \frac{V_{\mathrm{a}}}{(2\pi)^3} \int_E \frac{\mathrm{d}S_{\mathbf{k}}(E)}{\hbar v_{\mathbf{k}}}. \tag{22.24}$$

It is thus seen that the physical properties of electrons and phonons are inextricably bound together, and referring to either of them as independent entities (or particles) may be inappropriate in general. On the other hand, the simple picture in which the two are treated as independent of one another is often adequate in explaining a host of physical phenomena. For example, the size and shape of the Fermi surface is unaffected by the mass-enhancement effect, and so is the dc conductivity. The last result follows because the decrease by $1 + \lambda$ in $n(0)\langle v_x^2 \rangle$, which would lead to a decrease in σ, is canceled exactly by the same increase in τ. The same can be said about the other transport coefficients with the possible exception of the thermopower [16, 17].

The reasons for the different effects that the e-ph interaction has on electronic properties can be related to the so-called *self-energy* of the electron, which describes the effect on an electron produced by the ions and the other electrons in the system. In the following discussion, we give a simplified discussion of the contribution made by the e-ph interaction to the electron self-energy.

We consider the behavior of an electron in a material as described by the single-particle Schrödinger equation,

$$\left[-\frac{\hbar}{2m} \nabla^2 + \sum_n v_{\mathrm{eff}}(\mathbf{r} - \mathbf{R}_n) \right] \psi_{\mathbf{k}}(\mathbf{r}) = E_{\mathbf{k}} \psi_{\mathbf{k}}(\mathbf{r}). \tag{22.25}$$

Here, $v_{\mathrm{eff}}(\mathbf{r} - \mathbf{R}_n)$ is an effective, one-electron potential describing the self-consistent field seen by an electron set up by the ions and the rest of the electrons. This potential includes exchange and correlation (and can be non-local), but it is required to be real and independent of energy. In this description, the quasi-particle spectrum is not given by $E_{\mathbf{k}}$, regardless of how ingeniously v_{eff} has been chosen. This spectrum is obtained by considering the change in energy of the *system* resulting from the addition of an extra electron of momentum $\mathbf{k}$. This energy is equal to $E_{\mathbf{k}}$ plus an additional term,

$$E_{N+1}(\mathbf{k}) - E_N = E_{\mathbf{k}} + \Sigma(\mathbf{k}, E). \tag{22.26}$$

This additional term, $\Sigma(\mathbf{k}, E)$, is the self-energy which by definition contains all relaxation effects in the system caused by the addition of the extra electron. One such effect arises from the e-ph interaction. This contribution to the self-energy, however, is extremely difficult to calculate exactly and is not at all well-known for most metals, with the alkalis as a possible exception [18]. Conceivably, it can be made rather small near the Fermi energy through an appropriate choice of v_{eff}, but it can remain sizable for strongly paramagnetic metals, such as Pd.

We now discuss at a rather simple level the contribution to the self-energy made by the e-ph interaction, a quantity which we shall denote by Σ_{ph}. This part in general is known to vary rapidly with energy near the Fermi surface. At low temperatures, this part of the self-energy is given by [19]

$$\Sigma_{\text{ep}}(\mathbf{k}, E) = \sum_{\mathbf{k}'j} |M^j_{\mathbf{k}\mathbf{k}'}|^2 \left[\frac{(1 - f_{\mathbf{k}'})}{E - E_{\mathbf{k}'} - \hbar\omega^j_{\mathbf{k}'-\mathbf{k}}} - \frac{f_{\mathbf{k}'}}{E_{\mathbf{k}'} - E - \hbar\omega^j_{\mathbf{k}'-\mathbf{k}}} \right]. \tag{22.27}$$

This expression can be understood as follows. The first term inside the brackets is connected to an energy shift, arising in second-order Brillouin-Wigner perturbation theory from transitions between the state in which the added electron has energy E to one in which it has emitted a phonon and is in state $E_{\mathbf{k}'}$. This transition can occur only if the state $|\mathbf{k}'\rangle$ is unoccupied, hence the factor $(1 - f_{\mathbf{k}'})$. The second term arises because prior to the introduction of the added electron, the energy of the system was reduced by transitions in which an electron within the Fermi sea emits a phonon and goes over to state $|\mathbf{k}'\rangle$. These transitions are now blocked because of the added electron in that state.

In evaluating the e-ph self-energy, it is useful to introduce an auxiliary quantity, the so-called electron-phonon spectral function. This function is k-dependent and is given by the expression,

$$\begin{aligned} \alpha^2_{\mathbf{k}}(\omega)F(\omega) &= \frac{V_{\text{a}}}{(2\pi)^3} \int_{\text{FS}} \frac{\mathrm{d}S_{\mathbf{k}'}}{\hbar v_{\mathbf{k}'}} \\ &\times \sum_j |M^j_{\mathbf{k}\mathbf{k}'}|^2 \delta\left(\omega - \omega^j_{\mathbf{k}'-\mathbf{k}}\right), \end{aligned} \tag{22.28}$$

where FS denotes an integral over the Fermi surface.

We note the formal similarity between the spectral function defined by the last expression and that defined by Eq. (22.18). Now, Eq. (22.27) can be simplified in a manner analogous to that used to simplify Eq. (22.13). Replacing sums over $\mathbf{k}'$ by an integral over energy and integration over constant energy surfaces, as indicated in Eq. (22.16), we find

$$\begin{aligned} \Sigma_{\text{ep}}(\mathbf{k}, E) &= \int \mathrm{d}\omega \int \mathrm{d}E \frac{V_{\text{a}}}{(2\pi)^3} \int_{E'} \frac{\mathrm{d}S_{\mathbf{k}'}}{\hbar v_{\mathbf{k}'}} \sum_j |M^j_{\mathbf{k}\mathbf{k}'}|^2 \delta\left(\omega - \omega^j_{\mathbf{k}'-\mathbf{k}}\right) \\ &\times \left[\frac{(1 - f_{E'})}{E - E' - \hbar\omega^j_{\mathbf{k}'-\mathbf{k}}} - \frac{f_{E'}}{E - E' + \hbar\omega^j_{\mathbf{k}'-\mathbf{k}}} \right]. \end{aligned} \tag{22.29}$$

For $\omega < \omega_{\max}$, the major contributions to the integral over E' come from the region around $E \pm \hbar\omega$. For energies a few times $\hbar\omega_{\max}$ around the Fermi surface, the Fermi velocity and the e-ph matrix elements can be treated as being essentially constant[3]. Under this approximation, we obtain the expression,

$$\Sigma_{\mathrm{ep}}(\mathbf{k},E) = \hbar \int \mathrm{d}\omega\alpha_{\mathbf{k}}^2(\omega)F(\omega) \int \mathrm{d}E' \left[\frac{(1-f_{E'})}{E-E'-\hbar\omega} - \frac{f_{E'}}{E-E'+\hbar\omega}\right], \tag{22.30}$$

which can be integrated to yield,

$$\Sigma_{\mathrm{ep}}(\mathbf{k},E) = \hbar \int \mathrm{d}\omega\alpha_{\mathbf{k}}^2(\omega)F(\omega) \ln\left|\frac{E-\hbar\omega}{E+\hbar\omega}\right|, \tag{22.31}$$

a result that is rigorously valid at $T = 0°\mathrm{K}$. Expressions for finite temperatures have been derived by Allen [20] and Grimvall [21]. An immediate consequence of the last expresion is that Σ_{ep} vanishes at the Fermi energy, $\Sigma_{\mathrm{ep}}(\mathbf{k},0) = 0$. If energies are measured with respect to the Fermi energy, the vanishing of $\Sigma(\mathbf{k},0)$ implies that the size and shape of the Fermi surface are not affected by the e-ph interaction.

On the other hand, the Fermi velocity *is* affected. This can be seen by considering the gradient of the dispersion relation, obtained as the roots of the equation, $E - E_{\mathbf{k}} - \Sigma_{\mathrm{ep}}(\mathbf{k},0) = 0$. We have,

$$\hbar v_{\mathbf{k}}^* = \nabla_{\mathbf{k}}E = \nabla_{\mathbf{k}}E_{\mathbf{k}} + \nabla_{\mathbf{k}}\Sigma_{\mathrm{ep}}(\mathbf{k},E). \tag{22.32}$$

Although the explicit $\mathbf{k}$ derivative of Σ_{ep} vanishes at $E = 0$, according to Eq. (22.31), an implicit depende nce on $\mathbf{k}$ arises because of the dependence of $\Sigma_{\mathrm{ep}}(\mathbf{k},E)$ on energy. Solving for $\nabla_{\mathbf{k}}E$ at $E = 0$ in Eq. (22.32), we find

$$\hbar v_{\mathbf{k}}^* = \nabla_{\mathbf{k}}E_{\mathbf{k}}/(1+\lambda) = \hbar v_{\mathbf{k}}/(1+\lambda), \tag{22.33}$$

where

$$\lambda_{\mathbf{k}} = -\left[\frac{-\partial\Sigma_{\mathrm{ep}}(\mathbf{k},E)}{\partial E}\right]_{E=0}. \tag{22.34}$$

Equation (22.31) can be used to evaluate the last expression yielding the result,

$$\begin{aligned} \lambda_{\mathbf{k}} &= 2\int \frac{\mathrm{d}\omega}{\omega}\alpha^2(\omega)F(\omega) \\ &= \left[\frac{V_{\mathrm{a}}}{(2\pi)^3}\right] \int_{\mathrm{FS}} \frac{\mathrm{D}S_{\mathbf{k}}}{\hbar v_{\mathbf{k}}} \sum_{\mathbf{k}'j} \left|M_{\mathbf{k}\mathbf{k}'}^j\right|^2 / \hbar\omega_{\mathbf{k}-\mathbf{k}'}^j. \end{aligned} \tag{22.35}$$

[3]Metals with rapidly varying electronic structures near the Fermi energy may require more specialized treatment.

Averaging $\lambda_{\mathbf{k}}$ over the Fermi surface yields a quantity λ that determines the enhanced density of states at the Fermi energy, $n^*(0)$, in agreement with experimental measurements of specific heat at low temperatures. To see how the enhancement of the density of states comes about, recall that $n(0)$ within band theory is obtained as an integral of the inverse velocity over the Fermi surface. In the presence of the e-ph interaction, we find,

$$\begin{aligned} n^*(0) &= \sum_{\mathbf{k}} \delta[E(\mathbf{k})] = \left[\frac{V_a}{(2\pi)^3}\right] \int_{FS} \frac{dS_{\mathbf{k}}}{\hbar v_{\mathbf{k}}} \\ &= n(0)(1+\lambda). \end{aligned} \tag{22.36}$$

The quantity λ can now be written in terms of an average over the Fermi surface of the spectral function, $\alpha_{\mathbf{k}}^2(\omega)F(\omega)$, a quantity which we denote by $\alpha^2(\omega)F(\omega)$,

$$\lambda = 2\int \frac{d\omega}{\omega}\alpha^2(\omega)F(\omega). \tag{22.37}$$

As is discussed in the following section, it is primarily the quantity $\alpha^2(\omega)F(\omega)$ that determines the superconducting transition temperature.

22.4 Transition Temperature

The e-ph interaction leads to an effective electron-electron attractive interaction, which in the cases of many metals at low enough (but finite) temperatures, can overcome the Coulomb repulsion and cause superconducting behavior (Fig. 22.1d).

Let us imagine a lattice of positively charged ions immersed in a sea of electrons. A single electron, moving through the lattice (with a Fermi velocity of order of 10^8 cm/sec), will attract the ions in nearby positions that will respond by moving closer to the electron. But given the much larger mass of the ions compared to the electrons, their maximum response will occur in a time $\tau \approx 1/\omega_{ph} \approx 10^{-13}$ sec after the electron has passed, where ω_{ph} is a typical phonon frequency in the material. Thus, the electron leaves a tail of positively charged lattice polarization in its wake, whose maximum strength occurs at a distance of $\ell \approx v_F\tau$ behind the electron and amounts to several hundred lattice spacings. A second electron approaching the region feels the attraction of the polarized area, and consequently an attractive interaction between the electrons is introduced. The bare Coulomb repulsion between electrons is therefore reduced considerably by the slow response of the ions, which allows the interacting electrons to be well separated in space.

The normal to superconducting transition is due to the formation of Cooper pairs [22, 23], which are bound states of two electrons of opposite spins and momenta. The total energy of the pair, $E_{\rm p}$, is smaller than the sum of the energies of the individual electron states, $E_{\rm p} < E_{\mathbf{k}\uparrow}+E_{\mathbf{k}\downarrow} = 2E_{\mathbf{k}}$. The transition to the superconducting state occurs at a temperature $T_{\rm c}$, which depends strongly on the magnitude of the e-ph interaction. In the following discussion, we present a fairly basic theory of $T_{\rm c}$, while attempting to emphasize the underlying physics. Then, we delve a bit more deeply into the formal aspects of determining the superconducting transition temperature than can be gleaned from elementary treatments usually based on a coupling constant (g^2).

Let us exploit the simple picture of phonon exchange (Fig. 22.1d). Consider an electron at position $\mathbf{x}$ in the crystal at time t. This electron experiences a force arising from the potential fields of the ions at positions $\mathbf{R}_\ell$, a quantity that we shall denote by $V(\mathbf{x}, \{\mathbf{R}_\ell\})$. Because of translational invariance, we can write, $V(\mathbf{x}+\mathbf{d}, \{\mathbf{R}_\ell\}) = V(\mathbf{x}, \{\mathbf{R}_\ell - \mathbf{d}\})$. This is a useful result because it allows us to express the total force exerted by the electrons on the crystal as a sum of contributions from each lattice site. Denoting this total force along a direction α by $F_\alpha^{\rm T}$ we have

$$F_\alpha^{\rm T} = \nabla_\alpha V(\mathbf{x}, \{\mathbf{R}_\ell\}) = -\sum_\ell \frac{\delta V(\mathbf{x}, \{\mathbf{R}_\ell\})}{\delta R_{\ell\alpha}}, \tag{22.38}$$

so that the force on an atom at $\mathbf{R}_\ell$ in direction α exerted by an electron at $(\mathbf{x}, t)$, i.e., $F_\alpha^{\rm T}$, is given by $-\delta V/\delta R_{\ell\alpha}$, where $R_{\ell\alpha}$ is the component of $\mathbf{R}_\ell$ along the αth direction.

Now a force exerted at lattice site $\mathbf{R}_\ell$ at time t will propagate through the system causing other ions to be displaced as well. This effect can be calculated in the harmonic approximation by standard methods [24]. The calculation relies on a function, $D_{\beta\alpha}(\ell'\ell; t'-t)$, which describes the displacement in direction β on the ion at $\mathbf{R}_{\ell'}$ at time t' due to the force exerted on the ion at $\mathbf{R}_\ell$ in the direction α at time t. This quantity is called the *lattice Green function* or propagator, and was introduced in our discussion of the lattice vibrations in the preceding chapter where it was denoted by the symbol $\underline{G}$. It yields the derived displacements in the form,

$$u_{\ell'\beta}(t') = \sum_{\ell\alpha} \int {\rm d}t D_{\beta\alpha}(\ell'\ell; t'-t) F_{\ell\alpha}(t). \tag{22.39}$$

The propagator can be written in terms of the phonon frequency, $\omega_{\mathbf{q}}^j$, and the corresponding polarization vectors, $\hat{\epsilon}^j(\mathbf{q})$,

$$D_{\beta\alpha}(\ell'\ell; t'-t) = -\Theta(t'-t) \sum_{\mathbf{q}j} \frac{\epsilon_\beta^j(\mathbf{q})\epsilon_\beta^j(\mathbf{q})}{M_{\rm N}\omega_{\mathbf{q}}^j} \sin\left[\omega_{\mathbf{q}}^j(t'-t)\right] {\rm e}^{i\mathbf{q}\cdot(\mathbf{R}_\ell - \mathbf{R}_{\ell'})}, \tag{22.40}$$

where M_N denotes the nuclear mass (for a monoatomic system), and $\Theta(x) = 1$ for $x \geq 0$ and 0 otherwise.

Given the displacement of the atom at $\mathbf{R}_{\ell'}$, we can calculate the effect on the electrons in its vicinity. The change in the energy of an electron at $(\mathbf{x}', t')$ caused by the displacement $\mathbf{u}_{\ell'}(t')$ is equal to the change in potential energy and is given by the expression,

$$\delta V = \sum_{\ell'\beta} [\delta V(\mathbf{x}', \{\mathbf{R}_\ell\})/\delta R_{\ell'\beta}]\, u_{\ell'\beta}(t'). \tag{22.41}$$

It follows that the change in energy of an electron at $(\mathbf{x}', t')$ due to the existence of an electron at $(\mathbf{x}, t)$ for a short time Δt is given by the expression,

$$v(\mathbf{x}', t'; \mathbf{x}, t) = -\sum_{\ell'\beta;\ell\alpha} \frac{\delta V(\mathbf{x}')}{\delta R_{\ell'\beta}} D_{\beta\alpha}(\ell'\ell; t'-t) \frac{\delta V(\mathbf{x})}{\delta R_{\ell\alpha}} \Delta t. \tag{22.42}$$

We can convert this space-time form of the interaction into a form in terms of momentum and energy by noting that an electron with wave function $\psi_{\mathbf{k}}(\mathbf{x})\mathrm{e}^{\mathrm{i}E_{\mathbf{k}}t/\hbar}$ is annihilated at the vertex of Fig. 22.1b, while an electron with wave function $\psi_{\mathbf{k}'}(\mathbf{x})\mathrm{e}^{\mathrm{i}E_{\mathbf{k}'}t/\hbar}$ is created there. A similar process takes place at $(\mathbf{x}', t')$, except that the states are time reversed, so that $\mathbf{k} \to \mathbf{k}'$, $t \to -t'$, etc. Neglecting electron spin, since it is not affected by the e-ph interaction, we obtain the expression,

$$\begin{aligned} & v\left(\mathbf{k}, \mathbf{k}'; E_{\mathbf{k}} - E_{\mathbf{k}'}\right) \\ = & -\left(\frac{1}{V_{\mathrm{a}}^2}\right) \int \mathrm{d}^3x' \int \mathrm{d}^3x \int \mathrm{d}\tau \sum_{\ell'\ell\beta\alpha} \psi^*_{-\mathbf{k}'}(\mathbf{x}') \frac{\delta V(\mathbf{x}')}{\delta R_{\ell'\beta}} \psi_{-\mathbf{k}}(\mathbf{x}') \\ \times & \;\; \psi^*_{\mathbf{k}'}(\mathbf{x}') \frac{\delta V(\mathbf{x})}{\delta R_{\ell\alpha}} \psi_{\mathbf{k}}(\mathbf{x}) \mathrm{e}^{\mathrm{i}(E_{\mathbf{k}} - E_{\mathbf{k}'})\tau/\hbar}, \end{aligned} \tag{22.43}$$

for the electron-electron interaction mediated by the emission and absorption of a phonon. This expression can be simplified based on the property of the Bloch functions under translations by lattice vectors, $\psi_{\mathbf{k}}(\mathbf{x}+\mathbf{R}) = \mathrm{e}^{\mathrm{i}\mathbf{k}\cdot\mathbf{R}}\psi_{\mathbf{k}}(\mathbf{x})$, time-reversed symmetry, $\psi_{\mathbf{k}}(\mathbf{x}) = \psi^*_{-\mathbf{k}}(\mathbf{x})$, and the translational invariance of the potential function, $V(\mathbf{x}, \{\mathbf{R}_\ell\})$. We can write,

$$\int \mathrm{d}^3x \psi_{\mathbf{k}'}(\mathbf{x}) \frac{\delta V(\mathbf{x})}{\delta R_{\ell\alpha}} \psi_{\mathbf{k}}(\mathbf{x}) = \mathrm{e}^{\mathrm{i}(\mathbf{k}-\mathbf{k}')\cdot\mathbf{R}_\ell} V_{\mathrm{a}} I^{\alpha}_{\mathbf{k}\mathbf{k}'}, \tag{22.44}$$

where $I^{\alpha}_{\mathbf{k}\mathbf{k}'}$ is the electronic part of the e-ph matrix element. Explicitly, we have

$$I^{\alpha}_{\mathbf{k}\mathbf{k}'} = \frac{1}{V_{\mathrm{a}}} \int \mathrm{d}^3x \psi_{\mathbf{k}'}(\mathbf{x}) \frac{\delta V(\mathbf{x})}{\delta R_{\ell\alpha}} \psi_{\mathbf{k}}(\mathbf{x}). \tag{22.45}$$

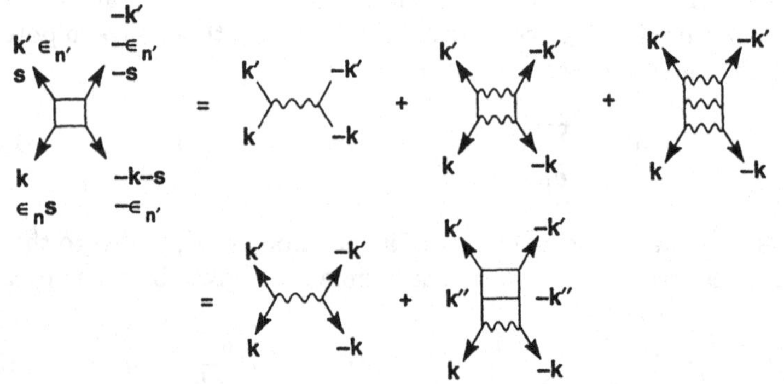

Figure 22.4: **Repeated phonon exchange contributing to the attractive electron-electron interaction.**

We note that because of translational invariance, the region of integration in the last expression is confined to the interior of an atomic unit cell. Using Eqs. (22.40), (22.43) and (22.44), we obtain the expression,

$$v(\mathbf{k},\mathbf{k}';E_{\mathbf{k}'}-E_{\mathbf{k}}) = \sum_j \left|M^j_{\mathbf{kk}'}\right|^2 \frac{2\hbar\omega_{\mathbf{k}'-\mathbf{k}}}{\left(E_{\mathbf{k}'}-E_{\mathbf{k}}\right)^2-\left(\omega_{\mathbf{k}'-\mathbf{k}}\right)^2}, \tag{22.46}$$

where the square of the *electron-phonon matrix element* is given by the expression,

$$\left|M^j_{\mathbf{kk}'}\right|^2 = \hbar \sum_{\alpha\beta}\sum_{\alpha\beta} \frac{\epsilon^j_\alpha \epsilon^j_\beta I^\alpha_{\mathbf{kk}'} I^\beta_{\mathbf{kk}'}}{\left(2M_{\mathrm{N}}\omega_{\mathbf{k}'-\mathbf{k}}\right)^2}. \tag{22.47}$$

We'll show later that this is the same matrix element that appears in Eqs. (22.17) and (22.29), the expressions, respectively, for the resistivity and for mass enhancement.

We now turn to the role played by the attractive interaction, v, in the determination of the superconducting transition temperature, T_c. The bound Cooper-pair state is obtained when we consider repeated phonon exchange (Fig. 22.4). If we let the function $\Gamma(\mathbf{k}',\mathbf{k};E_{\mathbf{k}'}-E_{\mathbf{k}})$ denote the total scattering amplitude for electrons in states $\mathbf{k}$ and $-\mathbf{k}$ to interact and scatter into states $\mathbf{k}'$ and $-\mathbf{k}'$, we can write

$$\Gamma = v + vKv + vKvKv + \cdots, \tag{22.48}$$

where Γ, the *vertex function*, is represented by the box on the left side of the diagrammatic expression. In this approximation, the phonon mediated electron-electron interaction is represented by wavy lines with attached dots. Here, K represents the lowest-order approximation to the two-particle propagator and is depicted in terms of two straight lines connecting two successive interaction (wavy lines). The last expression can be summed formally to yield

$$\Gamma = v + vK\Gamma \tag{22.49}$$

The formation of a bound state between a pair of electrons is associated with the divergence[4] in the scattering amplitude, Γ. In the present case, the transition from the normal to the superconducting state occurs when the temperature becomes low enough for Γ to become unbounded or, equivalently, that a non-trivial solution of the equation, $\Gamma = vK\Gamma$, exists. This last expression has the explicit form,

$$\Gamma(\mathbf{k}', \mathbf{k}; E', E) = \sum_{\mathbf{k}''} \int dE'' v(\mathbf{k}', \mathbf{k}''; E' - E'') K(\mathbf{k}'', E'') \Gamma(\mathbf{k}'', \mathbf{k}; E'', E). \tag{22.50}$$

It is useful to solve the last equation in an approximate manner in order to illustrate the determination of the transition temperature. We introduce the lowest-order two-particle propagator as the product of two single-particle propagators,

$$K(\mathbf{k}, E) = \frac{1}{(E - E_{\mathbf{k}})} \frac{1}{(-E - E_{-\mathbf{k}})}. \tag{22.51}$$

We must now also introduce the temperature into our zero-temperature formalism. This can be done by means of the formulation introduced by Matsubara [25] in which energies are replaced by imaginary integral quantities, $E \to \mathrm{i}E_n = \mathrm{i}(2n+1)\pi k_B T$. From the last expression, we obtain,

$$\begin{aligned} \Gamma(\mathbf{k}', \mathbf{k}; E_n, E_{n'}) &= k_B T \sum_{n''} \sum_{\mathbf{k}'' j} |M^j_{\mathbf{k}\mathbf{k}'}|^2 \frac{2\hbar\omega^j_{\mathbf{k}-\mathbf{k}''}}{(E_n - E_{n''})^2 + \left(\hbar\omega^j_{\mathbf{k}-\mathbf{k}''}\right)^2} \\ &\times \frac{1}{E^2_{n''} + E^2_{\mathbf{k}''}} \Gamma(\mathbf{k}'', \mathbf{k}'; E_{n''}, E_{n'}). \end{aligned} \tag{22.52}$$

As was done in connection with the self-energy of Eq. (22.26), we replace the sum over $\mathbf{k}''$ by an integral over $E_{\mathbf{k}''}$, and an integral over constant energy surfaces. We suppress the arguments $\mathbf{k}'$ and $E_{n'}$ of Γ and assume again that the band structure does not change significantly over an integral

[4] We recall that the formation of a bound state in "single-particle" scattering theory is signaled by the divergence of the corresponding t-matrix that satisfies the equation, $t = V + VG_0t$, with V being the scattering potential and G_0 the free-particle propagator.

about the Fermi energy equal to a typical phonon energy. This assumption allows us to evaluate the surface integrals at the Fermi energy, and to perform the integral over E'' analytically. Using the definition,

$$\lambda_{\mathbf{kk}'}(m) = n(0)\sum_j |M^j_{\mathbf{kk}'}|^2 \frac{2\hbar\omega^j_{\mathbf{k}-\mathbf{k}''}}{(2\pi mks_{\mathrm{B}}T)^2 + \left(\hbar\omega^j_{\mathbf{k}-\mathbf{k}''}\right)^2}, \tag{22.53}$$

of a set of temperature-dependent coupling functions, we can write

$$\Gamma(\mathbf{k}, n) = \frac{n^{-1}(0)V_{\mathrm{a}}}{(2\pi)^3}\int \frac{\mathrm{d}S'_{\mathbf{k}}}{\hbar v_{\mathbf{k}'}}\sum_{n'}\frac{\lambda_{\mathbf{kk}'}(n-n')}{|2n'+1|}\Gamma(\mathbf{k}', n'). \tag{22.54}$$

The quantity $\Gamma(\mathbf{k}, n)$ is sometimes referred to as the *gap function*, and Eq. (22.54) may be called the *linearized gap equation.* This expression is valid near T_{c}; below the transition temperature the equation determining $\Gamma(\mathbf{k}, n)$ becomes non-linear.

The variation of $\Gamma(\mathbf{k}, n)$ along the Fermi surface is known as *gap anisotropy.* This anisotropy is weak for many materials, in which case we can write, $\Gamma(\mathbf{k}, n) = \bar{\Gamma}(n) + \gamma_{\mathbf{k}}(n)$, where the average of $\gamma_{\mathbf{k}}(n)$ over the Fermi surface vanishes, and $\gamma_{\mathbf{k}}(n) << \bar{\Gamma}(n)$. Neglecting $\gamma_{\mathbf{k}}(n)$, Eq. (22.54) can be reduced to a rather simple form,

$$\bar{\Gamma}(n) = \sum_{n'=-\infty}^{\infty} [\lambda(n-n')/|2n'+1|]\,\bar{\Gamma}(n'), \tag{22.55}$$

where $\lambda(m)$ is a temperature-dependent generalization of the isotropic mass-enhancement parameter,

$$\lambda(m) = \langle\langle\lambda_{\mathbf{kk}'}\rangle_{\mathbf{k}}\rangle_{\mathbf{k}'} = 2\int \frac{\alpha^2(\omega)\Gamma(\omega)\omega\mathrm{d}\omega}{\omega^2 + (2\pi mks_{\mathrm{B}}T)^2}. \tag{22.56}$$

Here, $\langle\cdots\rangle_{\mathbf{k}}$ indicates a point by point average over the Fermi surface.

The highest temperature for which Eq. (22.55) has non-trivial solutions is T_{c}. Since T_{c} depends primarily on the overall magnitude of $\lambda(m)$ rather than on its detailed dependence on m, we can attempt to find an approximate solution of Eq. (22.55) by assuming that $\lambda(n-n')$ has the simple (and rather unrealistic) form,

$$\lambda(n-n') = \lambda\theta(N-|n|)\theta(N-|n'|), \tag{22.57}$$

where $\theta(x)$ is taken to vanish for non-negative values of x [so that it has the opposite behavior from $\Theta(x)$], and where the cutoff Matsubara frequency is set by the scale of the phonon frequencies $(2N+1)\pi k_{\mathrm{B}}T = \hbar\omega_0$. Under these assumptions, Eq. (22.55) is satisfied by $\bar{\Gamma}(n) = \bar{\Gamma}\theta(N - |n|)$ and T_c is determined by

$$\frac{1}{\lambda} = 2\sum_{n=0}^{N-1} \frac{1}{2n+1} = \gamma + 2\ln 2 + \psi\left(N + \frac{1}{2}\right), \tag{22.58}$$

where $\gamma = 0.5772...$ is Euler's constant, and $\psi(z)$ is the Digamma function [26] which for large values of its argument may be approximated by $\psi(z) \approx \ln z$. Now, T_c is given by the expression,

$$k_B T_c = \left(\frac{2e^{\gamma}}{\pi}\right) \hbar\omega_0 e^{-1/\lambda} = 1.13\hbar\omega_0 e^{-1/\lambda}. \tag{22.59}$$

This is indeed the result obtained by Bardeen et al. [23] in the BCS theory. Even though Eq. (22.57) is far from realistic, the last result has essentially the correct form if $\alpha^2(\omega)F(\omega)$ is a physically reasonable function.

At this point, it should be mentioned that Eq. (22.55) leaves out two important effects. One is the electron-phonon mass enhancement that should have been included in the two-particle propagator, K. Its effect is essentially to replace $1/\lambda$ in Eq. (22.59) by $(1+\lambda)/\lambda$. The second effect is the Coulomb repulsion between the electrons. Although the magnitude of this effect is difficult to calculate, it has been argued that because the Coulomb repulsion acts instantaneously on the scale of lattice relaxation time, it may be replaced by a frequency-independent pseudopotential, μ^*, over the frequency range where the e-ph interaction is attractive. The approximate effect of the Coulomb repulsion is to replace the exponent $(1+\lambda)/\lambda$ by $(1+\lambda)/(\lambda - \mu^*)$. Values for μ^* and $\alpha^2(\omega)F(\omega)$ can be obtained [27] through the analysis of superconducting tunneling experiments. For simple metals, one typically obtains $\mu^* \approx 0.1$. This small value is due to the short range of the screened Coulomb interaction in a metal and to the slow response of the ions, which allows an attractive interaction between electrons that are well-separated in space.

Solutions of Eq. (22.55) for realistic forms of $\lambda(m)$ have been given by Allen and Dynes [28]. Their results can be summarized fairly accurately by the following analytic expression for T_c,

$$T_c = \frac{\langle\omega\rangle_{\log}}{1.2} \exp\left[-\frac{1.04(1+\lambda)}{\lambda - \mu^* - 0.62\lambda\mu^*}\right], \tag{22.60}$$

where

$$\langle\omega\rangle_{\log} = \exp\left[\frac{1}{\lambda}\int \frac{d\omega}{\omega}\alpha^2(\omega)F(\omega)\ln\omega\right]. \tag{22.61}$$

As already mentioned, Eqs. (22.55) and (22.60) correspond to a material with an isotropic gap function. This is occasionally referred to as the "dirty limit", because the presence of impurities tends to smooth out[5] anisotropies in $\Gamma(\mathbf{k}, n)$. For very pure materials, this anisotropy should be taken into account. In the weak coupling, weak anisotropy limit, it can be shown that the effect of anisotropy is to enhance T_c by an amount,

$$\frac{\delta T_c}{T_c} = \langle \lambda_{\mathbf{k}}^2 - \lambda^2 \rangle_{\mathbf{k}} / \lambda^2, \tag{22.62}$$

where $\langle \cdots \rangle_{\mathbf{k}}$ indicates an average over the Fermi surface. The reader interested in further details of the role played by the e-ph interaction in the determination of T_c is urged to consult the literature [29, 30].

22.5 Phonon Lifetimes and Energy Shifts

In addition to its effect on the electrons, the e-ph interaction also affects the phonon spectrum. For example, consider the process depicted in Fig. 22.1b from the point of view of the phonon. As is illustrated in part e of the figure, a phonon can decay into an *electron-hole* pair. In this nomenclature, a hole coming out of a vertex is equivalent to an electron coming in, so that the process can be viewed as a phonon being absorbed by an electron in the Fermi sea that is thereby promoted out of the Fermi sea. This process limits phonon lifetimes to finite values, τ_{ph}, and produces a width in the spectrum, $\gamma \approx \hbar/\tau_{\text{ph}}$. Such widths can be observed in neutron scattering experiments.

Another effect of the e-ph interaction is the modification of phonon frequencies. A process such as that depicted in part f of Fig. 22.1 describes the emission and reabsorption of virtual electron-hole pairs. Physically, phonon frequencies are reduced because of the motion of electrons to screen out interionic forces between displaced nuclei. In this section, we discuss the effects of the e-ph interaction on phonon lifetimes and frequencies.

22.5.1 Lifetimes

We begin by an expression for the transition rate governing the number of phonons, $n_j(\mathbf{q})$, with polarization j and momentum, $\mathbf{q}$. This rate, $\tau_j^{-1}(\mathbf{q}) = -n_j^{-1}(\mathbf{q}) dn_j(\mathbf{q})/dt$, is obtained from Fermi's "golden rule" in the form

$$\begin{aligned} \tau_j^{-1}(\mathbf{q}) &= \frac{2\pi}{\hbar} \sum_{\mathbf{k}\mathbf{k}'} |M_{\mathbf{k}\mathbf{k}'}^j|^2 \delta\left(E_{\mathbf{k}} - E_{\mathbf{k}'} + \hbar\omega_{\mathbf{q}}^j\right) \\ &\times \left[f_{\mathbf{k}}\left(1 - f_{\mathbf{k}'}\right) - f_{\mathbf{k}'}\left(1 - f_{\mathbf{k}}\right)\right] \delta(\mathbf{k} - \mathbf{k}' + \mathbf{q}). \end{aligned} \tag{22.63}$$

[5]A good example of the smoothing effect of impurities is provided by the DOS of random alloys which are usually much less structured than those for pure, periodic solids.

The first and second terms in this expression correspond to diagrams e and a, respectively, in Fig. 22.1. Now, phonon and thermal energies are small on the energy scale of electron bands, so that $f_{\mathbf{k}} - f_{\mathbf{k}'}$ may be approximated by $\delta(E_{\mathbf{k}})\hbar\omega_{\mathbf{q}}^j$ and Eq. (22.63) can be written in the form,

$$\begin{aligned} \hbar/\tau_j(\mathbf{q}) &= 2\pi\hbar\omega_{\mathbf{q}}^j \frac{V_{\mathrm{a}}^2}{(2\pi)^6} \int_{\mathrm{FS}} \frac{\mathrm{d}S_{\mathbf{k}}}{\hbar v_{\mathbf{k}}} \int_{\mathrm{FS}} \frac{\mathrm{d}S_{\mathbf{k}'}}{\hbar v_{\mathbf{k}'}} \\ &\times \; |M_{\mathbf{k}\mathbf{k}'}^j|^2 \delta(\mathbf{k} - \mathbf{k}' + \mathbf{q}). \end{aligned} \tag{22.64}$$

The lifetime given by this expression may be used to calculate the contribution of the e-ph interaction to the thermal resistivity of the lattice [31],

$$\left(W_{\mathrm{ep}}^{\mathrm{P}}\right)^{-1} = \frac{1}{V_{\mathrm{a}}} \sum_{j\mathbf{q}} v_{jx}^2(\mathbf{q}) \tau_j(\mathbf{q}) I\left(\hbar\omega_{\mathbf{q}}^j / 2k_{\mathrm{B}}T\right), \tag{22.65}$$

where $I(x)$ is given by Eq. (22.20). Here, $x = \hbar\omega_{\mathbf{q}}^j/2k_{\mathrm{B}}T$ is the specific heat of the phonon mode $(\mathbf{q}, j)$ and $v_{jx}^2(\mathbf{q})$ is the corresponding group velocity. The lifetime can also be used to define a phonon linewidth

$$\gamma_j(\mathbf{q}) = \frac{1}{2} I_j^{-1}(\mathbf{q}), \tag{22.66}$$

which can be observed in inelastic neutron scattering experiments of sufficient resolution. We note the factor of 1/2 in the last expression. This arises because $1/\tau_j(\mathbf{q})$ defined by Eq. (22.63) is a decay rate for the phonon *probability*, which decays twice as fast the phonon *amplitude.*

The phonon linewidth can be related directly to the e-ph interaction, so that its value (obtained from an experiment) leads to an estimate of the interaction. We recall that the mass enhancement, λ, and the spectral function, $\alpha^2(\omega)F(\omega)$, can both be expressed as simple averages over the phonon linewidth,

$$\lambda = \frac{\hbar}{\pi n(0)} \sum_{\mathbf{q}j} \gamma_j(\mathbf{q}) / \left(\hbar\omega_{\mathbf{q}}^j\right)^2, \tag{22.67}$$

$$\alpha^2(\omega)F(\omega) = \frac{1}{2\pi n(0)\hbar\omega} \sum_{\mathbf{q}j} \gamma_j(\mathbf{q}) \delta\left(\omega - \omega_{\mathbf{q}}^j\right). \tag{22.68}$$

The quantity $\gamma_j(\mathbf{q})/(\hbar\omega_{\mathbf{q}}^j)^2$ entering the expression for λ is a measure of the contribution of the phonons in mode $(\mathbf{q}, j)$ to the overall strength of the phonon contribution to the electron-electron interaction. Phonon linewidths due to the electron-electron interaction are shown [32-34] in Fig. 22.5 for the elemental solids Nb and Pd. These linewidths were used to calculate the spectral functions shown in Figs. 22.2 and the resistivities in Fig. 22.3.

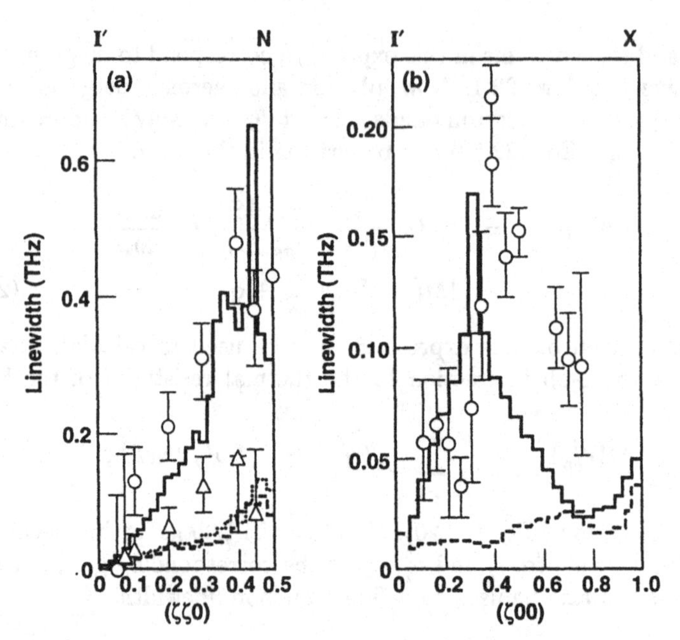

Figure 22.5: **Phonon linewidths for the elemental solids Nb and Pd.**

22.5.2 Energy shifts

The electron-phonon interaction modifies the phonon spectrum just as it does that of the electron [35]. The emission and reabsorption of virtual electron-hole pairs leads to a complex self-energy for the phonons so that the dispersion relation is modified from its non-interacting form into the expression,

$$\left(\omega_{\mathbf{q}}^{j}\right)^{2} - 2\mathrm{i}\gamma_{j}(\mathbf{q})\omega_{\mathbf{q}}^{j} = \left(\Omega_{\mathbf{q}}^{j}\right)^{2} + 2\Omega_{\mathbf{q}}^{j}\Pi_{j}\left(\mathbf{q},\omega_{\mathbf{q}}^{j}\right). \tag{22.69}$$

Here, $\Omega_{\mathbf{q}}^{j}$ is the "bare" phonon frequency (in the absence of the e-ph interaction), and $\Pi_{j}(\mathbf{q},\omega_{\mathbf{q}}^{j})$ is the phonon self-energy. These two quantities correspond, respectively, to the electron energy, $E_{\mathbf{k}}$, and the self-energy, $\Sigma(\mathbf{k},E)$ defined in Eq. (22.26). Expressing the phonon self-energy in terms of its real and imaginary parts,

$$\Pi_{j}\left(\mathbf{q},\omega_{\mathbf{q}}^{j}\right) = \Delta_{j}\left(\mathbf{q},\omega_{\mathbf{q}}^{j}\right) - \mathrm{i}\Gamma_{j}\left(\mathbf{q},\omega_{\mathbf{q}}^{j}\right), \tag{22.70}$$

we can identify $\Delta_{j}(\mathbf{q},\omega_{\mathbf{q}}^{j})$ as the cause of the shift in the phonon spectra, and $\Gamma_{j}(\mathbf{q},\omega_{\mathbf{q}}^{j})$ as the cause of the exponential decay of the phonon wave function with respect to time. We write, $\phi_{j}(\mathbf{q}) \approx \exp(\mathrm{i}\omega_{\mathbf{q}}^{j}t)\exp(-\gamma_{j}(\mathbf{q})t)$, where the factor $\gamma_{j}(\mathbf{q})$ is the linewidth discussed in the previous subsection and is given in terms of Γ_j by the relation,

$$\gamma_j(\mathbf{q}) = \Gamma_j\left(\mathbf{q}, \omega_{\mathbf{q}}^j\right) \Omega_{\mathbf{q}}^j / \omega_{\mathbf{q}}^j. \tag{22.71}$$

It is interesting to note that, unlike the linewidth, the frequency shift, $\Delta_j(\mathbf{q}, \omega_{\mathbf{q}}^j)$, is not uniquely defined, depending on the choice of the bare frequencies, $\Omega_{\mathbf{q}}^j$. These in turn depend on the different possible ways in which the force-constant matrix, $\Phi_{\alpha\beta}(\ell\ell')$, defined in the preceding chapter as the negative of the force along direction α on the nucleus at $\mathbf{R}_\ell$ per unit displacement along β of the atom at $\mathbf{R}_{\ell'}$, may be separated into "electronic" and "ionic" bare contributions.

22.5.3 Force constants and the HFT

We now discuss the interatomic force constants in a solid based on the Hellmann-Feynman theorem (HFT) (see Chapter 1). According to this theorem, the force $F_{\ell\alpha}$, exerted on the nucleus at $\mathbf{R}_\ell$ along the αth direction, can be expressed in terms of the internuclear potential, V_{NN}, the electron-nuclear interaction, V_{eN}, and the electronic charge, $\rho(\mathbf{r})$, by means of the relations (where only monoatomic materials are considered),

$$V_{\mathrm{NN}} = \frac{1}{2} \sum_{\ell\ell'} \frac{Z^2 \mathrm{e}^2}{|\mathbf{R}_\ell - \mathbf{R}_{\ell'}|}, \tag{22.72}$$

$$V_{\mathrm{eN}} = \sum_{\ell} \frac{Z\mathrm{e}}{|\mathbf{r} - \mathbf{R}_\ell|}, \tag{22.73}$$

$$F_{\ell\alpha} = \frac{\delta V_{\mathrm{NN}}(\{\mathbf{R}_\ell\})}{\delta R_{\ell\alpha}} - \int \mathrm{d}^3 r \frac{\delta V_{\mathrm{eN}}(\mathbf{r}, \{\mathbf{R}_\ell\})}{\delta R_{\ell\alpha}} \rho(\mathbf{r}). \tag{22.74}$$

By definition (recall the simple expression $F = -kx$), the force constant is obtained from the gradient of $F_{\ell\alpha}$ with respect to $R_{\ell'\beta}$, with $\ell' \neq \ell$,

$$\Phi_{\alpha\beta}(\ell\ell') = \frac{\delta^2 V_{\mathrm{NN}}(\{\mathbf{R}_\ell\})}{\delta R_{\ell\alpha} \delta R_{\ell'\beta}} + \int \mathrm{d}^3 r \frac{\delta V_{\mathrm{eN}}(\mathbf{r}, \{\mathbf{R}_\ell\})}{\delta R_{\ell\alpha}} \frac{\delta \rho(\mathbf{r})}{\delta R_{\ell'\beta}}. \tag{22.75}$$

We can obtain an expression for $\delta\rho$, the perturbation on the charge density caused by the displaced ions, in terms of the "bare", or lowest-order, susceptibility,

$$\delta\rho(\mathbf{r}) = \int \mathrm{d}^3 r' \chi_0(\mathbf{r}, \mathbf{r}') \delta V(\mathbf{r}'), \tag{22.76}$$

where $\chi_0(\mathbf{r}, \mathbf{r}')$ converts a change in potential to a change in charge density and is given explicitly by the expression,

$$\chi_0(\mathbf{r}, \mathbf{r}') = \frac{\mathrm{e}}{V_{\mathrm{a}}^2} \sum_{\mathbf{k}\mathbf{k}'} \frac{\psi_{\mathbf{k}}^*(\mathbf{r}) \psi_{\mathbf{k}'} \psi_{\mathbf{k}}(\mathbf{r}') \psi_{\mathbf{k}'}^*(\mathbf{r}')}{E_{\mathbf{k}} - E_{\mathbf{k}'}} (f_{\mathbf{k}} - f_{\mathbf{k}'}). \tag{22.77}$$

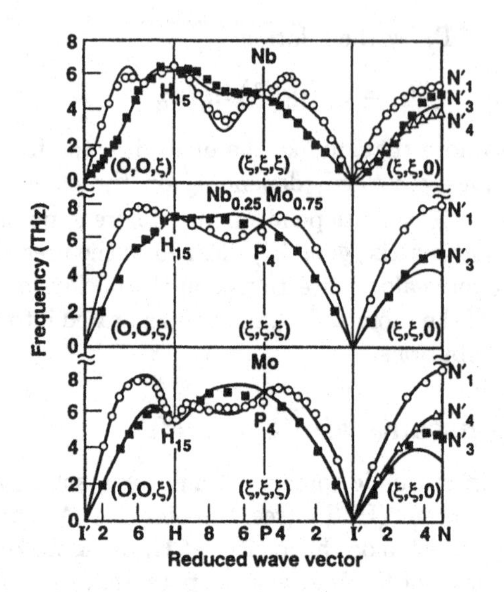

Figure 22.6: **Phonon dispersion curves for Nb (solid lines) compared with observed longitudinal mode frequencies (circles) and transverse mode frequencies (squares and triangles). Experiments [49], calculations [4].**

Using Eqs. (22.77) and (22.76) in the second term of Eq. (22.75) leads to the expression,

$$\Phi^{e}_{\alpha\beta}(\ell\ell') = \sum_{\mathbf{kk}';j} \frac{I^{\alpha 0}_{\mathbf{kk}'} I^{\alpha}_{\mathbf{kk}'} (f_{\mathbf{k}} - f_{\mathbf{k}'})}{E_{\mathbf{k}} - E_{\mathbf{k}'}} e^{i(\mathbf{k}-\mathbf{k}')\cdot(\mathbf{R}_{\ell}-\mathbf{R}'_{\ell})}, \tag{22.78}$$

where $I^{\alpha}_{\mathbf{kk}'}$ denotes the electronic part of the e-ph matrix element defined in Eq. (22.45), and $I^{\alpha 0}_{\mathbf{kk}'}$ denotes the matrix element of the "bare" gradient of the nuclear potential. Phonon dispersion curves calculated within a "modified tight-binding" approximation schemes are shown in Fig. 22.6.

It is now important to note that the formal considerations just given mask a serious difficulty associated with the use of Eq. (22.75). The quantity $\delta\rho/\delta R_{\ell\alpha}$ is extremely difficult to calculate (refer to discussion of forces in Chapter 14), while it must be calculated very accurately because of the cancellation of the two large terms contained in it. A number of approximation schemes and formal approaches have been introduced for the calculation of force constants, and hence of the phonon dispersion relations, including tight-binding schemes [4] as well as first-principles methods based on DFT.

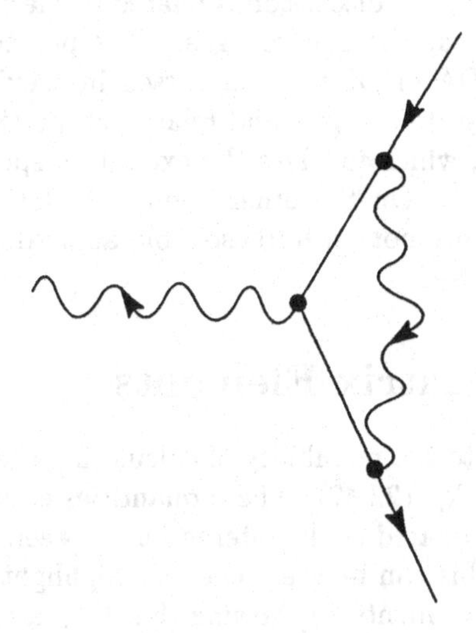

Figure 22.7: **Diagrammatic representation of vertex corrections.**

22.6 Vertex Corrections

Our discussion of the e-ph interaction thus far was based on approximation schemes that neglect a number of "higher-order" physical processes. This allowed us to derive [9] simple and revealing expressions for a number of physical properties, such as:

- the resistivity, Eqs. (22.18) and (22.23);
- the e-ph mass enhancement, Eq. (22.35);
- the superconducting transition temperature, Eq. (22.60);
- phonon lifetimes, Eqs. (22.64) and (22.66);
- lattice thermal resistivity, Eq. (22.65); and
- interatomic force constants, Eqs. (22.75) and (22.78).

These various results must be reconsidered by taking into account the so-called *vertex correction* terms, such as the one shown in Fig. 22.7. This figure shows a phonon connecting the members of an electron pair that has just emitted a phonon, leading to further contributions to the self-energy. Migdal [36] has shown that these terms are smaller than the lowest-order terms by at least a factor of order $\hbar\omega/E_{\mathrm{B}}$, where ω is a typical phonon frequency (of the order of 10 meV), and E_{B} is a characteristic band energy (of the order of eV). This difference in scale is connected to the large difference between the electronic and nuclear masses. This allows the electrons to respond to nuclear motion in essentially adiabatic fashion, without giving rise to transitions among different electronic levels (states).

Another approximation in our discussion is related to the single-particle picture that was used to treat the electron gas. This procedure is fairly well justified within the LDA of DFT, as discussed in previous chapters in connection with ground-state energies and related properties. Its use in calculating transition rates, which involves the excitation spectrum of the system, is less well justified[6]. On the other hand, calculations [38] tend to indicate that it is probably not an unreasonable approximation in the study of the e-ph interaction.

22.7 The e-ph Matrix Elements

We now turn our attention to the possibility of calculating the e-ph matrix elements, $M^j_{\mathbf{kk'}}$, defined in Eq. (22.47). These quantities enter the discussion of every physical effect related to the interaction between the electrons and phonons in a metal. This can be seen when we highlight the physical significance of these matrix elements by showing that they are given by the expression,

$$M^j_{\mathbf{kk'}} = \langle\psi_{\mathbf{k}}|\delta V|\psi_{\mathbf{k'}}\rangle, \tag{22.79}$$

where δV is the change in the electronic Hamiltonian caused by a phonon of polarization j, and momentum $\mathbf{q} = \mathbf{k'} - \mathbf{k}$. Now, the change in a crystal potential due to a phonon is given by

$$\delta V = \sum_{\ell} \nabla_{\mathbf{R}_\ell} V \cdot \mathbf{u}^j_\ell, \tag{22.80}$$

where $\mathbf{u}^j_\ell$ is the phonon displacement field, [see Eq. (21.26)] for the equation of motion),

$$\mathbf{u}^j_\ell = \sum_{\mathbf{q}} \hat{\epsilon}^j_{\mathbf{q}} \sqrt{\hbar/2M_{\mathrm{N}}\omega^j_{\mathbf{q}}} e^{i\mathbf{q}\cdot\mathbf{R}_\ell}. \tag{22.81}$$

When the last two expressions are substituted into Eq. (22.79), and use is made of the Bloch condition on the wave function (for translationally invariant materials) and of the translational symmetry of the potential, V, we find that the matrix elements can be written in the form,

$$M^j_{\mathbf{kk'}} = \sum_{\alpha} \epsilon^j_\alpha(\mathbf{k'} - \mathbf{k}) I^\alpha_{\mathbf{k'k}} \sqrt{\hbar/2M_{\mathrm{N}}\omega^j_{\mathbf{k'}-\mathbf{k}}}, \tag{22.82}$$

where $I^\alpha_{\mathbf{k'k}}$ is given by Eq. (22.45). The relevance of the matrix elements in treating the various effects of the e-ph interaction follows.

[6] For the treatment of excited states within DFT, the reader is referred to the literature [37].

The calculation of the matrix elements is particularly delicate because it requires the evaluation of $\delta V/\delta R_\alpha$ which appears in $I^\alpha_{\mathbf{k}'\mathbf{k}}$. This quantity, as discussed above, describes the change in total crystal potential due to an infinitesimal displacement of the atom at the origin along the direction α, with all other atoms remaining fixed in position. The difficulty in calculating this quantity arises because of the dynamic response of the electrons that would tend to readjust themselves to compensate the resultant electrostatic field and to screen the perturbation caused by the displacements. One can express the change in total crystal potential as the sum of the change in bare potential, δV_b, plus the compensating induced potential, δV_ind,

$$\delta V/\delta R_\alpha = \delta V_\mathrm{b}/\delta R_\alpha + \delta V_\mathrm{ind}/\delta R_\alpha. \tag{22.83}$$

The change in the bare potential, which neglects the electronic response to the change in nuclear positions, can be easily calculated. The second term on the right side of the last equation can be written in the form,

$$\delta V_\mathrm{ind}/\delta R_\alpha = \int d^3r'\bar{v}(\mathbf{r},\mathbf{r}')\delta\rho(\mathbf{r}')/\delta R_\alpha, \tag{22.84}$$

where the kernel, $\bar{v}(\mathbf{r},\mathbf{r}')$, which converts a change in charge to a change in potential, is given as a sum of a "Hartree" term and an exchange-correlation term,

$$\bar{v}(\mathbf{r},\mathbf{r}') = \frac{1}{|\mathbf{r}-\mathbf{r}'|} + \frac{e\delta^2 E_\mathrm{xc}}{\delta\rho(\mathbf{r})\delta\rho(\mathbf{r}')V_\mathrm{a}}, \tag{22.85}$$

where E_xc is the exchange-correlation energy of the interacting electron gas. Of course, as we have seen, this quantity is far from being known exactly. In many calculations, it is replaced with the corresponding functional obtained within the LDA of DFT.

A number of approximation schemes have been introduced for calculating $\delta V/\delta R_{\ell\alpha}$. In the *weak pseudopotential* model, the *screened pseudopotential* is calculated by means of the relation,

$$V_\mathrm{sc}(\mathbf{r}) = V_\mathrm{b}(\mathbf{r}) + \int d^3r'\chi(\mathbf{r}-\mathbf{r}')\int d^3r''\bar{v}(\mathbf{r}-\mathbf{r}')V_\mathrm{b}, \tag{22.86}$$

where $\chi(\mathbf{r}-\mathbf{r}')$ is the *interacting* susceptibility given in terms of the *bare* susceptibility, χ_0, defined in Eq. (22.77), by means of the relation.

$$\chi = \left(\chi_0^{-1} - \bar{v}\right)^{-1}. \tag{22.87}$$

In spite of apparent simplicity, the inversions indicated in the last expresion are far from trivial to perform for the case of the inhomogeneous electron gas. In any case, the potential V_sc of Eq. (22.86) also determines the band structure of the material. The band structure potential is given by

$V_{bs} = \sum_{\ell} V_{sc}(\mathbf{r} - \mathbf{R}_{\ell})$, which expresses the simple result that the potential perturbation that enters the discussion of the e-ph interaction matrix elements results from the displacement of one of the potentials whose sum is the total crystal potential. This result is consistent with the more general result of Eq. (22.38), which in turn means that the sum over all sites of the quantity $-\delta V/\delta \mathbf{R}_{\ell}$ is the force on an electron due to the change in the crystal potential, i.e., is the gradient of the potential used to calculate the electronic (band) structure of the material,

$$\sum_{\ell} \delta V(\mathbf{r})/\delta R_{\ell\alpha} = -\nabla_{\alpha} V(\mathbf{r}). \tag{22.88}$$

This is an important sum rule because most quantities entering the discussion can be written in terms of sums of the form [see Eq. (22.42] $\sum_{\ell\ell'} (\delta V(\mathbf{r})/\delta \mathbf{R}_{\ell})(\delta V(\mathbf{r})/\delta \mathbf{R}_{\ell'}) F(\ell\ell')$. However, this sum rule does not determine the e-ph matrix elements uniquely because there is more than one way to write the total potential as a sum of atomic contributions. The reader wishing more detailed information, especially about the calculation of $M^{j}_{\mathbf{kk}'}$ and its comparison with experiment is urged to peruse the relevant literature [39-45] and references cited therein.

This is a good point to turn our attention to the very important and rather difficult problem of calculating the electronic structure of a material in the *presence* of lattice vibrations. We concern ourselves with a formal way to account (approximately) for the effects of phonons on the electronic structure in a self-consistent way within the formalism of MST.

22.8 Effects on the Electronic Structure

This may be a convenient point to address, albeit in very brief fashion, the response of the electron gas when atoms (ions) move away from their equilibrium positions.

Under the Born-Oppenheimer approximation, the positive nuclei provide the external (and static) potential in which the electrons move. However, as already mentioned, this approximation cannot account for a number of important physical phenomena that are associated with nuclear motion. Therefore, we must have a way calculate the combined effects of nuclear motion and the electronic response to it. Clearly, when the nuclei move away from their equilibrium positions, the surrounding electron gas, especially in a metal, can be expected to change its configuration so as to screen out any charge imbalance caused by this displacement. In this section, we derive expressions describing the rearrangement of electronic charge when a single atom is displaced from equilibrium. In the following section, we set forth a series of steps designed to lead to a self-consistent determination of the ground state of an electron-nucleon system not constrained by the Born-Oppenheimer approximation, thus accounting for nuclear motion. In our discussion, we follow the development given by Butler [46].

22.8.1 Rigid muffin-tin approximation

We saw above that the calculation of the electron-phonon interaction involves the matrix element,

$$I = \langle \Psi | \nabla_{\mathbf{R}} V | \Psi \rangle, \tag{22.89}$$

where $|\Psi\rangle$ is an eigenfunction of the many-body, interacting electron system in the field of the (stationary) nuclei, V is the total potential felt by the electron cloud, and $\nabla_{\mathbf{R}} V$ denotes the change induced in this potential when a nucleus at $\mathbf{R}$ is displaced by an infinitesimal amount. One may attempt an evaluation of this matrix element within various approximation schemes.

In the so-called rigid muffin-tin approximation (RMTA), the self-consistent change in the potential due to the displacement of an atom (say the one at the origin) is replaced by the gradient of the LDA, single-electron potential inside the Wigner-Seitz cell [47] [7]. When this approximation is used [46, 47] to calculate the resistivity of a pure metal such as Pd and Nb, it results in discrepancies of about 10% compared to experimental results at T=300°K. This discrepancy increases at lower temperatures, amounting to a factor of two for Nb at T=10°K. It seems likely that the RMTA is an important cause of this inaccuracy.

The RMTA was first put forward by Gaspari and Gyorffy [49], and the method was successfully applied to realistic systems by W. H. Butler, et al [50], and D. A. Papaconstantopoulos, et al [51]. In order to improve on the original RMTA, the change in the potential and the electronic charge density caused by the displacement of a single atom from equilibrium must be evaluated self-consistently. An approach toward such a calculation is outlined below.

22.8.2 Self-consistent treatment(*)

In attempting to evaluate the matrix element in Eq. (22.89), we approximate the wave function, $|\Psi\rangle$, by a Bloch eigenfunction of the unperturbed crystal and consider the integral

$$I^{\alpha}_{\mathbf{kk}'} = \langle \Psi_{\mathbf{k}} | \frac{\delta V}{\delta \mathbf{R}_{\alpha}} | \Psi_{\mathbf{k}'} \rangle, \tag{22.90}$$

[7] In practical applications, the approximation has usually been implemented with respect to the muffin-tin approximation to the potential, whence it derives its name.

where $\mathbf{R} \equiv \mathbf{R}_0$ denotes the position of the atom at the origin. In the RMTA, one sets

$$\frac{\delta V}{\delta \mathbf{R}_\alpha} = \left(-\frac{\partial V(\mathbf{r})}{\partial \mathbf{r}_\alpha}\right) \theta(\mathbf{r}), \tag{22.91}$$

where $\theta(\mathbf{r})$ is defined to be unity within the Wigner-Seitz cell at the origin and zero elsewhere. Here, $V(\mathbf{r})$ is the self-consistently determined cell potential obtained in the LDA.

To proceed beyond the RMTA, one must attempt a rigorous evaluation of the integral in Eq. (22.90). In general, this represents a very difficult calculational task. In what follows, we use the framework of MST to calculate self-consistently the change in both the charge density and the potential caused by the displacement of a single atom. This is done by considering the RMTA part as "known" and deriving expressions for the difference, or the unknown, parts. The discussion that follows has many common elements with the formalism presented in Chapter 14 for the calculation of forces and interatomic potentials in solids.

We write the changes in the electron density, $n(\mathbf{r})$, and the potential, $V(\mathbf{r})$, caused by the displacement of the atom at the origin as the sum of RMTA parts and remainders. This leads to a pair of coupled equations for the remainders that can be solved simultaneously within MST. Denoting these remainder parts for the density and the potential, respectively, by $\delta\tilde{n}(\mathbf{r})$ and $\delta\tilde{V}(\mathbf{r})$, we obtain the basic relations,

$$\delta n(\mathbf{r})/\delta R_\alpha = -\frac{\partial n(\mathbf{r})}{\partial r_\alpha}\theta_0(\mathbf{r}) + \delta\tilde{n}(\mathbf{r}), \tag{22.92}$$

and

$$\delta V(\mathbf{r})/\delta R_\alpha = -\frac{\partial V(\mathbf{r})}{\partial r_\alpha}\theta_0(\mathbf{r}) + \delta\tilde{V}(\mathbf{r}). \tag{22.93}$$

We now recall the form of the potential acting on the electron cloud obtained within the LDA (see Chapter 1),

$$V(\mathbf{r}) = \sum_n \frac{-2Z}{|\mathbf{r} - \mathbf{R}_n|} + 2\int \mathrm{d}^3r' \frac{n(\mathbf{r}')}{|\mathbf{r} - \mathbf{r}'|} + V_{\mathrm{xc}}[n(\mathbf{r})]. \tag{22.94}$$

Based on this expression, the change in the potential due to the displacement of the atom at the origin along the αth direction takes the form,

$$\frac{\delta V(\mathbf{r})}{\delta R_\alpha} = -\frac{2Zr_\alpha}{r^3} + 2\int \mathrm{d}^3r' \frac{\delta n(\mathbf{r}')/\delta R_\alpha}{|\mathbf{r} - \mathbf{r}'|} + \int \mathrm{d}^3r' \frac{\delta V_{\mathrm{xc}}[n(\mathbf{r}')]\delta n(\mathbf{r}')}{\delta n(\mathbf{r}')\delta R_\alpha}, \tag{22.95}$$

where Z is the nuclear charge. The functional $V_{\mathrm{xc}}[n(\mathbf{r}')]$ is generally unknown, but in the LDA it is approximated to be a function of the local electron density. Then, the last term in the equation above takes the form of a derivative, $(\mathrm{d}V_{\mathrm{xc}}/\mathrm{d}n)\,\delta n(\mathbf{r})$, with the derivative being evaluated at the local electron density for the perfect crystal. Now, substituting Eq. (22.92) into Eq. (22.95) yields the expression,

$$\begin{aligned}\frac{\delta V(\mathbf{r})}{\delta R_\alpha} &= -\frac{2Zr_\alpha}{r^3} - 2\int_0 \mathrm{d}^3r' \frac{\partial n(\mathbf{r}')/\partial r'_\alpha}{|\mathbf{r}-\mathbf{r}'|} - \left(\frac{\mathrm{d}V_{\mathrm{xc}}}{\mathrm{d}n}\right)\left(\frac{\mathrm{d}n(\mathbf{r})}{\partial r_\alpha}\right)\theta_0(\mathbf{r}) \\ &+ \int \mathrm{d}^3r' U(\mathbf{r}-\mathbf{r}')\delta\tilde{n}(\mathbf{r}'), \end{aligned} \tag{22.96}$$

where the zero subscript in the second term indicates that the integral is confined to the Wigner-Seitz cell at the origin, and where we have defined the quantity,

$$U(\mathbf{r}-\mathbf{r}') = \frac{2}{|\mathbf{r}-\mathbf{r}'|} + \left(\frac{\mathrm{d}V_{\mathrm{xc}}}{\mathrm{d}n}\right)\delta(\mathbf{r}-\mathbf{r}'). \tag{22.97}$$

One may think of $U(\mathbf{r}-\mathbf{r}')$ as a Green function that converts a change in density to a change in potential. If we subtract from Eq. (22.96) the RMTA contribution, we obtain an expression for the remainder term, $\delta\tilde{V}(\mathbf{r})$,

$$\delta\tilde{V}(\mathbf{r}) = \delta V_0(\mathbf{r}) + \int \mathrm{d}^3r' U(\mathbf{r}-\mathbf{r}')\delta\tilde{n}(\mathbf{r}'), \tag{22.98}$$

where

$$\begin{aligned}\delta V_0(\mathbf{r}) &= -2\int_0 \mathrm{d}^3r' \left[\frac{\partial n(\mathbf{r}')/\partial r'_\alpha}{|\mathbf{r}-\mathbf{r}'|} + n(\mathbf{r}')\frac{\partial}{\partial r'_\alpha}\frac{1}{|\mathbf{r}-\mathbf{r}'|}\right] \\ &+ \delta V^{\mathrm{LF}}(\mathbf{r}), \end{aligned} \tag{22.99}$$

with

$$\begin{aligned}\delta V^{\mathrm{LF}}(\mathbf{r}) &= -2\frac{\partial}{\partial r_\alpha}\left\{\sum_{n\neq 0}\left[\frac{Z}{|\mathbf{r}-\mathbf{R}_n|} - \int_n \mathrm{d}^3r' \frac{n(\mathbf{r}')}{|\mathbf{r}-\mathbf{r}'|}\right]\right\}\theta_0(\mathbf{r}) \\ &+ 2\frac{\partial}{\partial r_\alpha}\left[\frac{Z}{r} - \int_0 \mathrm{d}^3r' \frac{n(\mathbf{r}')}{|\mathbf{r}-\mathbf{r}'|}\right][1-\theta_0(\mathbf{r})]. \end{aligned} \tag{22.100}$$

In Eq. (22.98), the unknown part of the potential change is expressed as the sum of a part, $\delta V_0(\mathbf{r})$, which can be calculated from the charge density of the perfect crystal, and a part due to the "unknown" part of change in electron density, $\delta\tilde{n}(\mathbf{r})$. This term, $\delta V_0(\mathbf{r})$, itself consists of two terms: the first has the form of a dipolar interaction, while the second, local field (LF), is of considerably shorter range. For $\mathbf{r}$ *inside* the cell at the origin, $\delta V^{\mathrm{LF}}(\mathbf{r})$ is the gradient of the local Coulomb field due to the nuclei and electrons *outside* that cell, while for $\mathbf{r}$ *outside* the cell at the origin, $\delta V^{\mathrm{LF}}(\mathbf{r})$ is the gradient of the total Coulomb potential due to the nucleus and electrons *inside* the cell[8].

The first term in Eq. (22.99) can be converted into an integral over the surface of the cell at the origin,

$$-2\int_0 \mathrm{d}S_\alpha \frac{n(\mathbf{r})}{|\mathbf{r}-\mathbf{r}'|}, \tag{22.101}$$

and, to a very good approximation, the total charge density, $n(\mathbf{r})$, can be replaced by $n^{\mathrm{v}}(\mathbf{r})$, the valence charge density, since the core electron density can be expected to be negligible at the cell boundary.

We now turn our attention to the equation for $\delta\tilde{n}(\mathbf{r})$, Eq. (22.92). Assuming that the core electron density, $n^{\mathrm{c}}(\mathbf{r})$, moves rigidly with the nucleus[9], we can write,

$$\frac{\delta n(\mathbf{r})}{\delta R_\alpha} = -\frac{\partial n^{\mathrm{c}}(\mathbf{r})}{\partial r_\alpha}\theta_0(\mathbf{r}) + \int \mathrm{d}^3r'\chi_0^{\mathrm{v}}(\mathbf{r},\mathbf{r}')\frac{\delta V(\mathbf{r}')}{\delta R_\alpha}, \tag{22.102}$$

where $\chi_0^{\mathrm{v}}(\mathbf{r},\mathbf{r}')$ is the ***susceptibility*** (or polarizability) of the valence electrons. This quantity converts a change in potential to a change in charge density, and an approach to its evaluation within MST is provided in the next section. Substituting Eqs. (22.92) and (22.93) into Eq. (22.102) yields the expression,

$$\delta\tilde{n}(\mathbf{r}) = \delta n_1(\mathbf{r}) + \int \mathrm{d}^3r'\chi_0^{\mathrm{v}}(\mathbf{r},\mathbf{r}')\delta\tilde{V}(\mathbf{r}'), \tag{22.103}$$

where

$$\delta n_1(\mathbf{r}) = \frac{\partial n^{\mathrm{v}}(\mathbf{r})}{\partial r_\alpha}\theta_0(\mathbf{r}) + \int \mathrm{d}^3r'\chi_0^{\mathrm{v}}(\mathbf{r},\mathbf{r}')\left[-\frac{\partial}{\partial r_\alpha}V(\mathbf{r}')\right]. \tag{22.104}$$

The "unknown" part of the charge density, $\delta\tilde{n}$, is the sum of an inhomogeneous term, δn_1, and a term that represents the response of the valence electrons to the unknown part, $\delta\tilde{V}$. The inhomogeneous term is the difference between the rigidly shifted valence charge density and the response of the valence electrons to the rigid muffin-tin part of the change in the potential.

Equations (22.98) and (22.103) form a closed set of equations for the "unknowns", $\delta\tilde{V}(\mathbf{r})$ and $\delta\tilde{n}(\mathbf{r})$. In attempting a solution of these equations, it may be advantageous to eliminate the inhomogeneous term in Eq.

---

[8]The reader is encouraged to fill in the steps leading to these equations, as well as to those that follow.

[9]This approximation can be the cause of great inaccuracies, as our discussion of forces has revealed.

(22.103) by defining $\delta\bar{n}(\mathbf{r}) = \delta\tilde{n}(\mathbf{r}) - \delta n_1(\mathbf{r})$. Then, Eqs. (22.97) and (22.103) become,

$$\delta\bar{n}(\mathbf{r}) = \int d^3r' \chi_0^{\mathrm{v}}(\mathbf{r},\mathbf{r}')\delta\tilde{V}(\mathbf{r}'), \tag{22.105}$$

and

$$\delta\tilde{V}(\mathbf{r}) = \delta V_1(\mathbf{r}) + \int d^3r' U(\mathbf{r}-\mathbf{r}')\delta\bar{n}(\mathbf{r}'), \tag{22.106}$$

where

$$\begin{aligned}\delta V_1(\mathbf{r}) &= \delta V_0(\mathbf{r}) + \int d^3r' U(\mathbf{r}-\mathbf{r}')\delta n_1(\mathbf{r}') \\ &= V^{\mathrm{LF}}(\mathbf{r}) + \int d^3r' \int_0 d^3r'' U(\mathbf{r}-\mathbf{r}')\chi_0^{\mathrm{v}}(\mathbf{r}',\mathbf{r}'') \left[-\frac{\partial V(\mathbf{r}'')}{\partial r''_\alpha}\right] \\ &+ \frac{\partial}{\partial r_\alpha}\int_0 d^3r' U(\mathbf{r}-\mathbf{r}')n^{\mathrm{v}}(\mathbf{r}'). \end{aligned} \tag{22.107}$$

The physical meaning of the second term in this expression is the change in potential caused by the response of the valence electrons in the RMTA, and the third term is the change in potential caused by rigidly shifting the potential arising from the valence electrons. The last expression provides some insight into the success enjoyed by the RMTA in applications to tightly-bound electrons. In this case, the local field term, V^{LF}, is expected to be small, while the second and third terms may approximately cancel since in such systems the valence charge density moves essentially rigidly with the nucleus.

We are now faced with the non-trivial problem of simultaneously solving Eqs. (22.105) and (22.106). A series of steps within the framework of MST designed to lead to this solution will be described. A somewhat more detailed description of a fully self-consistent evaluation of the potential, total energy, and susceptibility of a material is provided in the next section.

The first step toward the solution of Eqs. (22.105) and (22.106) is the calculation of the change in electron density due to an arbitrary, infinitesimal change in crystal potential. Within MST, the density is given by the imaginary part of the Green function, which for vectors $\mathbf{r}_n$ and $\mathbf{r}'_n$ inside cell Ω_n, is given by (see Chapter 3),

$$G(\mathbf{r}_n,\mathbf{r}'_n;E) = \langle Z^n(\mathbf{r}_n;E)\,|\underline{\tau}^{nn}(E)|\,Z^n(\mathbf{r}'_n;E)\rangle - \langle Z^n(\mathbf{r}_n;E)|J^n(\mathbf{r}'_n;E)\rangle, \tag{22.108}$$

where $\mathbf{r}_n = \mathbf{r} - \mathbf{R}_n$, and the meaning of the various other symbols and terms in this expression should be familiar from previous discussion. We recall that for real energies the solutions $|Z^n\rangle$ and $|J^n\rangle$ are real and the charge density is given by the expression,

$$\rho(\mathbf{r};E) = -e\frac{1}{\pi}\Im G(\mathbf{r},\mathbf{r};E)$$

$$= \langle Z^n(\mathbf{r}_n;E)\left|-\frac{1}{\pi}\Im\underline{\tau}^{nn}(E)\right|Z^n(\mathbf{r}_n;E)\rangle. \quad (22.109)$$

Now, if the crystal potential is perturbed slightly, the charge density will change according to

$$\begin{aligned} \delta\tilde{\rho}(\mathbf{r}_n;E) &= \langle\delta\left[Z^n(\mathbf{r}_n;E)\,|\otimes|\,Z^n(\mathbf{r};E)\right]\rangle\left[-\frac{1}{\pi}\Im\underline{\tau}^{nn}(E)\right] \\ &+ \langle Z^n(\mathbf{r}_n;E)\left|\delta\left[-\frac{1}{\pi}\Im\underline{\tau}^{nn}(E)\right]\right|Z^n(\mathbf{r}_n;E)\rangle, \quad (22.110) \end{aligned}$$

where $\otimes$ denotes the cross (direct) product of vectors. From the general expression of the equation of motion for the scattering-path matrix (or operator) Eq. (9.89), and keeping terms only linear in $\delta\underline{m}$, we have

$$\delta\underline{\tau}^{nn} = \sum_k\langle\underline{\tau}^{nk}\,|\delta\underline{m}^k|\,\underline{\tau}^{kn}\rangle, \quad (22.111)$$

where $\delta\underline{m}^k$ is the change in the inverse cell scattering matrix caused by the perturbation of the moving atom. In order to derive an expression for $\delta\underline{m}^k$ we begin with Eq. (9.7). We consider V to be an infinitesimal perturbation to U so that $\Psi(U+V)\simeq\Phi(U)$, and note that $\delta m = -m\delta t m$. Since $m\Phi^+ = mJ - ikH = Z$, outside the range of the potential, we have,

$$\delta\underline{m}^k = -\int_k \mathrm{d}^3r_k|Z^k(\mathbf{r}_k,E)\rangle\delta\tilde{V}(\mathbf{r}_k)\langle Z^k(\mathbf{r}_k;E)|. \quad (22.112)$$

The changes in the wave functions, $\langle Z^n(\mathbf{r}_n;E)|$, brought about by the perturbation can be evaluated in terms of the Green function for a single cell, $G^n(\mathbf{r},\mathbf{r}';E)$, as is indicated[10] in Eqs. (2.82) and (9.32),

$$\delta|Z^n(\mathbf{r}_n;E)\rangle = \int_n \mathrm{d}^3r'_n G^n(\mathbf{r}_n,\mathbf{r}'_n;E)\delta\tilde{V}_n(\mathbf{r})_n|Z^n(\mathbf{r}'_n;E)\rangle. \quad (22.113)$$

Using Eqs. (22.111) and (22.113) to evaluate Eq. (22.110), the perturbation in electron density can be written in the form,

$$\delta\tilde{n}(\mathbf{r}) = \sum_n\theta_n(\mathbf{r}_n)\int_0^{E_F}\mathrm{d}E\tilde{\rho}(\mathbf{r};E). \quad (22.114)$$

The next step is to convert $\delta\tilde{n}(\mathbf{r})$ in Eq. (22.106) into a perturbation of the potential. The exchange-correlation part can be evaluated in the

[10] We set $\delta|Z^n\rangle \equiv |\Psi^\pm\rangle - |\Phi^\pm\rangle$, $\delta V \equiv V$, $G^n \equiv G$, and $|Z^n\rangle \equiv |\Phi^\pm\rangle$.

LDA, while the Coulomb part can be handled by means of an expansion in spherical harmonics[11],

$$\frac{1}{|\mathbf{r}-\mathbf{r}'+\mathbf{R}_{nm}|} = \langle J(\mathbf{r}_n) \left| \underline{S}(\mathbf{R}_{nm}) \right| J(\mathbf{r}'_m) \rangle. \tag{22.115}$$

Here, $J(\mathbf{r})$ denotes the product of a spherical Bessel function of order ℓ and a spherical harmonic of order $L = (\ell, m)$, and $\underline{S}(\mathbf{R}_{nm})$ is the zero-energy structure constant of the lattice (see Chapter 4). For $\mathbf{R}_{nm} = 0$, the integral in Eq. (22.106) must be calculated directly.

Now, Eq. (22.106) can be evaluated for a given $\delta n(\mathbf{r})$. As the final step, an iterative procedure could be used to solve the set of coupled Eqs. (22.105) and (22.106).

Even though the method outlined above consists of several steps, each of which is individually straightforward, the resulting process is long and numerically tedious. The hope remains, however, that increasing computational power will lead to realistic calculations in the not too distant future.

22.9 Fully Self-Consistent MST(*)

In previous chapters and sections we have discussed methods for determining the ground-state properties, such as charge densities and total energies, of an interacting electron cloud in the field of the stationary nuclei. We have also seen how to self-consistently determine the response of the electrons to small shifts of a nucleus from its equilibrium position. In this section, we consider the possibility of carrying out a fully self-consistent calculation of the electron-nucleon system that takes into account nuclear displacements. As the formalism within which such a scheme could be implemented, we choose the Green-function approach based [48] on MST.

A fully self-consistent implementation of MST that would include the electron-phonon interaction could follow two steps.

Step 1:

Use the methods of MST to solve the single-particle Schrödinger equation in the LDA for either an ordered (periodic) or substitutionally disordered solid. Treatments for ordered solids were outlined in Chapters 1 and 6, while disordered alloys could be studied within the formal construct of the coherent potential approximation (CPA) and its possible extensions, discussed in Chapters 10 and 11. Ordered materials can be studied by a number of formal methods, many based on the determination of the ground-state wave function. However, a unified formalism applicable to both ordered and disordered alloys that takes proper account of the

[11]This expansion can be shown to converge absolutely only in the case of non-overlapping muffin-tin charges. For general cell charges, such an expansion can be effected through the construction of a shifted cell or shifted center discussed earlier.

boundary conditions associated with the geometry of the system can be constructed only on the basis of the Green function, and MST provides a particularly convenient framework for such a development.

Within this approach, the potential and the Green function of an interacting electron-nucleon system, with the nuclei held fixed in position, are obtained by means of an iterative procedure:

- the initial charge densities are usually constructed as either a superposition of overlapping atomic charges, or as the results of some other, related, self-consistent calculation; and
- based on those initial charges, the corresponding single-particle potential within the LDA is constructed, the associated Schrödinger equation for the solid is solved, and a new charge density and potential (through the use of the Green function) is obtained.

The process is repeated until the changes in calculated quantities between successive iterations fall within preset acceptance criteria. Upon completion, this procedure leads to the self-consistent determination of charge densities and total energies for the system with the nuclei held at their equilibrium positions.

Step 2:

This part of a fully self-consistent application of MST involves the calculation of the response of the charge density to a change in the potential,

$$\delta n(\mathbf{r}) = \int d^3r' \chi_0(\mathbf{r},\mathbf{r}')\delta v(\mathbf{r}'). \tag{22.116}$$

The case in which the change in potential was associated with the displacement of a nucleus was discussed in the previous section, and led to the definition of the ***static susceptibility*** of the system, χ_0.

Since the charge density is given in terms of the imaginary part of the Green function, we first consider the general expression for the change in the Green function due to changes in the potential (see relevant formalism in Chapter 9). Denoting the Green function associated with the unperturbed (but fully interacting) system by G, we have in general operation notation,

$$G(z) = (z - H)^{-1}. \tag{22.117}$$

With G' denoting the Green function for the system in the presence of the perturbation, we find

$$\begin{aligned} G'(z) &= (z - H - \delta v)^{-1} \\ &= G + G\delta v G + \cdots, \end{aligned} \tag{22.118}$$

so that in the limit $\delta v \to 0$, we have

$$\delta G(\mathbf{r},\mathbf{r}') = \int d^3r'' G(\mathbf{r},\mathbf{r}''; E)\delta v(\mathbf{r}'')G(\mathbf{r}'',\mathbf{r}'). \tag{22.119}$$

This is the result of linear response theory for changes in the Green function of the system. This result allows us to calculate changes in the electron density due to a change in the potential,

$$\begin{aligned}\delta n(\mathbf{r}) &= -\frac{1}{\pi}\Im\int_{-\infty}^{E_F}\delta G(\mathbf{r},\mathbf{r}';E)\mathrm{d}E\\ &= -\frac{1}{\pi}\Im\int_{-\infty}^{E_F}\mathrm{d}E\int \mathrm{d}^3r' G(\mathbf{r},\mathbf{r}';E)\delta v(\mathbf{r}')G(\mathbf{r}',\mathbf{r};E).\end{aligned} \tag{22.120}$$

Therefore, comparison with Eq. (22.116) leads to the expression for the static, "bare" susceptibility function,

$$\chi_0(\mathbf{r},\mathbf{r}') = -\frac{1}{\pi}\Im\int_{-\infty}^{E_F}\mathrm{d}EG(\mathbf{r},\mathbf{r}';E)G(\mathbf{r}',\mathbf{r};E). \tag{22.121}$$

This is the result for a static perturbation. The dynamic susceptibility is only slightly more complicated, requiring integrals of the form[12],

$$\chi_0(\mathbf{r},\mathbf{r}';\omega) \approx -\frac{1}{\pi}\Im\int_{-\infty}^{E_F}\mathrm{d}EG(\mathbf{r},\mathbf{r}';E)G(\mathbf{r}',\mathbf{r};E+\omega). \tag{22.122}$$

These susceptibility functions are termed "bare" because they only allow for the direct response of the electrons to the perturbation. However, as the electrons relax, they affect the rest of the electron cloud. This "indirect" interaction can be expressed in terms of "dressed" response functions or susceptibilities. If we denote the fully (dressed) interaction by w, we can write the following formal expression in terms of the "bare" Coulomb interaction, u,

$$\begin{aligned} w &= u + u\chi_0 u + u\chi_0 u\chi_0 u + \cdots\\ &= (1-u\chi_0)^{-1}u.\end{aligned} \tag{22.123}$$

The t-matrix-like appearance of the last expression is obvious. Within MST, we can employ an expansion in angular-momentum states and write,

$$\underline{w}^{nm} = \underline{S}^{nm} + \sum_{n_1,n_2\neq n_1} \underline{S}^{nn_1}\underline{\chi}_0^{n_1n_2}(\omega)\underline{S}^{n_2m} + \cdots, \tag{22.124}$$

where underlines denote matrices in angular momentum space, and $\underline{S}^{nm}$ is the zero-energy structure constant[13] for the lattice, Eq. (22.115), and

[12] A comparison with the forms for the dc and ac conductivities discussed in Chapter 20 may prove useful.

[13] In the case of full-cell potentials, a modified expression for these structure constants may have to be used, possibly based on a shifted cell arguments, to guarantee convergence.

where the matrix $\underline{\chi}_0^{nm}(\omega)$ is defined by

$$\underline{\chi}_0^{nm}(\omega) = \int \mathrm{d}^3 r_n \int \mathrm{d}r'_m |J(\mathbf{r})\rangle \underline{\chi}_0^{nm}(\mathbf{r}_n, \mathbf{r}'_m; \omega)\langle J(\mathbf{r}'_m)|. \qquad (22.125)$$

Steps 1 and 2, taken together, allow the self-consistent determination of single-particle quantities such as charge densities, total energies, and the response to a potential change (brought about, say, by atomic displacement or motion), of the interacting electron-nucleon system. However, the numerical implementation of these two steps is far from trivial.

Bibliography

[1] L. Dagens, M. Rasolt, and R. Taylor, Phys. Rev. B**11**, 2726 (1975).

[2] S. K. Sinha and B. N. Harmon, p. 269 in *Superconductivity in d- and f-band Metals, (I)*, H. Douglas (ed.), Plenum, New York (1976).

[3] C. M. Varma and W. Weber, Phys. Rev. Lett. **39**, 1094 (1977).

[4] C. M. Varma and W. Weber, Phys. Rev B**19**, 6142 (1979)

[5] G. L. Zhao and B. N. Harmon, Phys. Rev. B**45**, 2818 (1992)

[6] J. Mizuki, Y. Chen, K. M. Ho, and C. Stassis, Phys. Rev. B**32**, 666 (1985).

[7] A. Larose and J. Vanderwall, *Scattering of Thermal Neutrons: A Bibliography (1932-1974)*, Plenum, New York (1974).

[8] J. M. Ziman, *Electrons and Phonons*, Oxford Universtity Press, London and New York (1960), Chapter 7.

[9] W. H. Butler, in Treatise on Materials Science and Technology, Vol. **21**, *Electronic Structure and Properties*, Frank Y. Fraden (ed.), Academic Press, New York (1981), p. 165.

[10] M. Kohler, Z. Phys. (Leipzig) **124**, 772 (1947).

[11] L. I. Schiff, *Quantum Mechanics*, McGraw-Hill, New York (1955), 2nd Edition, p. 399.

[12] P. B. Allen, Phys. Rev. B**3**, 305 (1971)

[13] F. Bloch, Z. Phys. **52**, 555 (1928).

[14] M. J. Laubitz and T. Matsubara, Can J. Phys. **50**, 196 (1972).

[15] J. M. Abraham and B. Deviot, J. Less -Common Met. **29**, 311 (1972).

[16] S. K. Lyo, Phys. Rev. B**17**, 2545 (1978).

[17] A. Vilenkin and P. L. Taylor, Phys. Rev. Lett. **42**, 597 1979.

[18] L. Hedin and S. Lundquist, Solid State Phys. Supplement **23**, 1 (1969).

[19] D. J. Scalapino, p. 399 in *Superconductivity*, R. D. Parks (ed.), McGraw-Hill, New York (1955), 2nd Edition.

[20] P. B. Allen, Proc. Int. Conf. Low Temp. Phys. 12th Kyoto, (1970), p. 517.

[21] G. Grimvall, Phys. Kondens. Mater. **9**, 283 (1969).

[22] L. N. Cooper, Phys. Rev. **104**, 1189 (1956).

[23] J. Bardeen, L. N. Cooper, and J. Schriefer, Phys. Rev. **105**, 1175 (1957).

[24] J. A. Reissland, *The Physics of Phonons*, Wiley, New York (1973).

[25] T. Matsubara, Prog. Theor. Phys. (Kyoto) **14**, 351 (1955).

[26] M. Abramovitz and I. A. Stegun, *Handbook of Mathematical Functions*, Dover, New York (1964).

[27] W. L. McMillan and J. M. Rowell, p. 561 in *Superconductivity*, R. D. Parks (ed.), Dekker, New York, (1969), Vol. **1**.

[28] P. B. Allen and R. C. Dynes, Phys. Rev. B**12**, 905 (1975).

[29] W. H. Butler and P. B. Allen, in *Superconductivity in d- and f-band Metals*, D. H. Douglas (ed.), Plenum, New York (1976).

[30] M. Petr, J. Ashkanazi, and M. Dacorogna, Helv. Phys. Acta **50**, 267 (1977).

[31] W. H. Butler and R. K. Williams, Ohys. Rev. B**18**, 6483 (1978).

[32] W. H. Butler, H. G. Smith, and N. Wakabayashi, Phys. Rev. Lett. **39**, 1004 (1977).

[33] F. J. Pinski and W. H. Butler, Phys. Rev. B**19**, 6010 (1979).

[34] R. Youngblood, Y. Noda, and G. Shirane, Phys. Rev. B**19**, 6016 (1979).

[35] P. B. Allen, in *Dynamical Properties of Solids*, G. M. Horton and A. A. Maradudin (eds.), North Holland, Amsterdam and New York (1980), Vol. **3**.

[36] A. B. Migdal, Zh. Eksp. Teor. Fiz. **34**, 1438 (1958). [Engl. transl. Sov. Phys. JETP **7**, 996 (1958).]

[37] R. M. Dreitzler and E. K. U. Gross, *Density Functional Theory*, Spinger-Verlag, Berlin (1990).

[38] M. Rasolt and J. F. Devlin, Phys. Rev. B**13**, 3290 (1976).

[39] V. Heine, P. Nozieres, and J. W. Wilkins, Phil. Mag. **13**, 741 (1966).

[40] R. E. Prange and A. Sachs, Phys. Rev. **158**, 672 (1967).

[41] D. L. Martin, Phys. Rev. **139**, A150 (1965).

[42] G. Gilat and R. M. Nicklow, Phys. Rev. **143**, 489 (1965).

[43] A. Birnboin and H. Gutfreund, Phys. Rev. B**9**, 139 (1974).

[44] D. A. Papaconstantopoulos and B. M. Klein, Ferroelectrics **16**, 307 (1977).

[45] J. Ashkenazi, M. Dacorogna, and M. Peter, Solid State Commun. **29**, 181 (1979).

[46] W. H. Butler, Can. J. Phys. **60**, 735, (1982).

[47] F. J. Pinski, P. B. Allen, and W. H. Butler, Phys. Rev. B **23**, 5080 (1981).

[48] W. H. Butler, private communication.

[49] G. D. Gaspari and B. L. Gyorffy, Phys. Rev. Lett **29**, 801 (1972).

[50] W. H. Butler, J. J. Olson, J. S. Faulkner, and B. L. Gyorffy, Phys. Rev. B**14**, 3823 (1976).

[51] D. A. Papaconstantopoulos, L. L. Boyer, B. M. Klein, A. R. Williams, V. L. Moruzzi, and J. F. Janak, Phys. Rev. B**15**, 4221 (1977).

Chapter 23

Light in Matter

23.1 General Comments

Whether a solid is transparent or opaque to incident radiation depends on the frequency of the radiation field. Similar dependence on frequency governs the reflectivity or absorptive behavior of a surface. This dependence determines a number of related physical properties, such as the color of a metal.

What may not be obvious, however, is the role played by microscopic properties, such as conductivity, in the propagation of electromagnetic radiation through matter. To see how the conductivity enters the picture, consider the response of the electrons in a metal to an externally applied constant electric field, **E**. In this case, the "free" charges in the metal, i.e., the valence electrons, can move over essentially arbitrary distances in the direction of the field in a manner described directly by the conductivity (tensor), σ. At the same time, electrons that are bound to a particular region, say those in core states near the nucleus, may also respond by being slightly displaced through a finite distance from their equilibrium position under the action of **E**. Such a displacement of charge would give rise to a *dipole* density on opposite surfaces of a dielectric material and is usually described in terms of a dielectric constant, ϵ. Even in this "simple", static case, one may expect both the conductivity and the dielectric constant[1] to figure prominently in determining the response of the electrons to applied external fields.

Let us consider an externally applied oscillating field, such as the fields associated with electromagnetic radiation. Both free and bound charges respond in a dynamic way, and depending on the frequency of the radia-

[1]The dielectric constant will also be referred to as the dielectric function since it is in general a function of both frequency and wave vector.

tion, their responses may be indistinguishable. At small enough frequencies, $\omega << 1/\tau$, where τ is a measure of the time between collisions of the electrons with imperfections in the material, the motion of the free charges would still be described by the dynamic conductivity, $\sigma(\omega)$, whereas that of the bound charges would be given by the dynamic dielectric function, $\epsilon_0(\omega)$. In this case, free charges would again move more or less in phase with the field, so that $\sigma(\omega)$ would be predominantly real, whereas bound charges would tend to respond sufficiently out of phase requiring a predominantly imaginary dielectric constant. However, at sufficiently higher frequencies, the motion of unbound electrons may acquire a substantial out-of-phase component, making the conductivity predominantly imaginary, while bound charges can move largely in-phase with the field that would prescribe a dielectric function with a large real component. In such cases, the separation between the features associated with the conductivity and the dielectric function become somewhat blurred, and it is more useful to describe the response to radiation by the *generalized dielectric response function*,

$$\epsilon(\omega) = \epsilon_0(\omega) + \frac{4\pi i\sigma(\omega)}{\omega}, \tag{23.1}$$

where ϵ_0 is the permitivity of free space (i.e., in the absence of a conducting medium). This combination of the conductivity and dielectric functions arises naturally in the discussion of the interaction of classical electromagnetic radiation with matter.

In this chapter, we discuss the optical properties of materials by considering the propagation of electromagnetic radiation through a solid. Since this propagation is in general frequency dependent, our discussion includes the entire electromagnetic spectrum, from long-wavelength radio waves to X-rays. As may be expected, we base our discussion on Maxwell's equations in the presence of matter. Our discussion is aimed at providing solutions to these equations, and understanding the form and significance of these solutions, both within a phenomenological approach as well as by means of a multiple-scattering, Green-function treatment of the resulting vector wave equation.

23.2 Maxwell's Equations in Matter

The starting point of our discussion is the set of Maxwell's equations in the presence of matter[2],

$$\nabla \times \mathbf{H} = \frac{4\pi \mathbf{J}}{\sigma} + \frac{1}{c}\frac{\partial \mathbf{D}}{\partial t}$$

[2]The tensor nature of various quantities, such as conductivity and dielectric function, should be kept in mind even though for ease of presentation it will not always be explicitly displayed.

$$\begin{aligned}\nabla \cdot \mathbf{D} &= 4\pi\rho \\ \nabla \times \mathbf{E} &= -\frac{1}{c}\frac{\partial \mathbf{B}}{\partial t} \\ \nabla \cdot \mathbf{H} &= \nabla \cdot \mathbf{B} = 0,\end{aligned} \tag{23.2}$$

where

$$\mathbf{D} = \epsilon\mathbf{E}, \quad \mathbf{B} = \mu\mathbf{H}, \quad \text{and} \quad \mathbf{J} = \sigma\mathbf{E}, \tag{23.3}$$

with μ being the permeability of the material. For much of our discussion we set $\mu = 1$, thus neglecting the effects of any internal magnetic properties of a material.

At this point it is important to clarify the meaning of the various quantities, such as ρ and $\mathbf{J}$, that occur in Eqs. (23.2). We first study the dielectric function of an interacting electron system. In this discussion, we follow the lines in Jones and March [1].

23.2.1 Dielectric function of a Fermi gas

We begin with the study of the static, $\omega = 0$, wave-vector dependent dielectric constant, $\epsilon(\mathbf{k})$, describing the response of an electron cloud to the presence of a static charge embedded in the gas. Our discussion is based on the Schrödinger equation associated with the self-consistent potential, $V(\mathbf{r})$, in which the electrons move under the action of the charge.

We recall the Schrödinger equation (in atomic units),

$$\left[\nabla^2 + E\right]\Psi(\mathbf{r}) = V(\mathbf{r})\Psi(\mathbf{r}), \tag{23.4}$$

or its integral form, the Lippmann-Schwinger equation[3],

$$\Psi(\mathbf{r}) = \chi(\mathbf{r}) - \int \mathrm{d}r'^3 G_0(\mathbf{r},\mathbf{r}')V(\mathbf{r}')\Psi(\mathbf{r}'), \tag{23.5}$$

where the free-particle propagator at energy $k^2 = E$,

$$G_0(\mathbf{r},\mathbf{r}') = \frac{\mathrm{e}^{ik|\mathbf{r}-\mathbf{r}'|}}{4\pi|\mathbf{r}-\mathbf{r}'|}, \tag{23.6}$$

satisfies the equation,

$$\left[\nabla^2 + E\right]G_0(\mathbf{r},\mathbf{r}') = \delta(\mathbf{r}-\mathbf{r}'). \tag{23.7}$$

Under the action of the potential, the solution $\Psi(\mathbf{r})$ evolves out of the free-particle solution $\chi(\mathbf{r})$ at the same energy, E. If we choose $\chi(\mathbf{r})$ to have the

[3]The minus sign in front of the integral sign is consistent with the definition of the Green function that follows, and is used for clarity of presentation.

form of a plane wave, $\chi_{\mathbf{k}}(\mathbf{r}) = \exp\{\mathrm{i}\mathbf{k}\cdot\mathbf{r}\}/\sqrt{\Omega}$, where Ω is the volume of the system, then $\Psi(\mathbf{r})$ in the case of a periodic solid can be taken to be a Bloch state at wave vector $\mathbf{k}$ and energy[4] E.

Let us now focus on a weak potential, so that $\Psi_{\mathbf{k}}(\mathbf{r})$ can be obtained in the first Born approximation,

$$\Psi_{\mathbf{k}}(\mathbf{r}) = \frac{1}{\sqrt{\Omega}}\left[\mathrm{e}^{\mathrm{i}\mathbf{k}\cdot\mathbf{r}} - \int \mathrm{d}^3r' G_0(\mathbf{r},\mathbf{r}')V(\mathbf{r}')\mathrm{e}^{\mathrm{i}\mathbf{k}\cdot\mathbf{r}'}\right]. \tag{23.8}$$

We are interested in obtaining the form of the charge density, $\rho(\mathbf{r})$, which is given as a sum over occupied states, Eq. (6.5),

$$\rho(\mathbf{r}) = 2\sum_{|\mathbf{k}|<\mathbf{k}_F} \Psi_{\mathbf{k}}(\mathbf{r})\Psi^*_{\mathbf{k}}(\mathbf{r}). \tag{23.9}$$

Using the form given in Eq. (23.8) for the wave function and including a factor of 2 for spin, we have

$$\begin{aligned}\rho(\mathbf{r}) &= \sum_{|\mathbf{k}|<\mathbf{k}_F} \Omega^{-1} - 2\sum_{|\mathbf{k}|<\mathbf{k}_F} \Omega^{-1}\int \mathrm{d}^3r'V(\mathbf{r}') \\ &\times \left[G_0(\mathbf{r},\mathbf{r}')\mathrm{e}^{\mathrm{i}\mathbf{k}\cdot(\mathbf{r}'-\mathbf{r})} + G_0^*(\mathbf{r},\mathbf{r}')\mathrm{e}^{-\mathrm{i}\mathbf{k}\cdot(\mathbf{r}'-\mathbf{r})}\right],\end{aligned} \tag{23.10}$$

where only terms linear in the potential have been retained. There are $\rho_0 = \Omega/(2\pi)^3$ states per unit volume in $\mathbf{k}$-space and two spin directions, so that replacing a sum over $\mathbf{k}$ by an integral, we have

$$\begin{aligned}\rho(\mathbf{r}) &= \rho_0 - \frac{1}{(2\pi)^3}\Omega^{-1}\int \mathrm{d}^3r'V(\mathbf{r}')\int \mathrm{d}^3k \\ &\times \left[G_0(\mathbf{r},\mathbf{r}')\mathrm{e}^{\mathrm{i}\mathbf{k}\cdot(\mathbf{r}'-\mathbf{r})} + G_0^*(\mathbf{r},\mathbf{r}')\mathrm{e}^{-\mathrm{i}\mathbf{k}\cdot(\mathbf{r}'-\mathbf{r})}\right].\end{aligned} \tag{23.11}$$

Integration over the angles of $\mathbf{k}$ effectively replaces the exponential, $\mathrm{e}^{\mathrm{i}\mathbf{k}\cdot(\mathbf{r}'-\mathbf{r})}$, by the $l = 0$ (s) term in the expansion of a plane wave,

$$\mathrm{e}^{\mathrm{i}\mathbf{k}\cdot(\mathbf{r}'-\mathbf{r})} \to \frac{\sin k|\mathbf{r}-\mathbf{r}'|}{k|\mathbf{r}-\mathbf{r}'|}, \tag{23.12}$$

so that combining G_0 and G_0^* based on Eq. (23.6) allows us to write,

$$\rho(\mathbf{r}) = \rho_0 - \frac{1}{(2\pi)^4}\int \mathrm{d}^3r'V(\mathbf{r}')\int_0^{k_F} \mathrm{d}k 4\pi k^2\left[\frac{\sin k|\mathbf{r}-\mathbf{r}'|}{k|\mathbf{r}-\mathbf{r}'|}\frac{2\cos k|\mathbf{r}-\mathbf{r}'|}{k|\mathbf{r}-\mathbf{r}'|}\right]. \tag{23.13}$$

[4]The explicit dependence of the wave function on energy will often be suppressed but must be understood throughout.

The integral over k can be performed to yield an expression for the "displaced" charge, $\rho(\mathbf{r}) - \rho_0$,

$$\rho(\mathbf{r}) - \rho_0 = -\frac{k_F^3}{4\pi^3}\int d^3r' V(\mathbf{r}')\frac{j_1(2k_F|\mathbf{r}-\mathbf{r}'|)}{|\mathbf{r}-\mathbf{r}'|^2}, \tag{23.14}$$

with $j_1(x)$ being the spherical Bessel function for $l = 1$, given explicitly by the expression, Eq. (C.31),

$$j_1(x) = x^{-2}\left[\sin x - x\cos x\right]. \tag{23.15}$$

Let us now discuss the form of the displaced charge given by Eq. (23.14).

The displaced charge represents the attempt of the electron cloud to shield the perturbation caused by the presence of the external charge. To gain some insight into the spatial variation of this screening, we assume that $V(\mathbf{r})$ is varying slowly enough to be treated as a constant in the integral of Eq. (23.14). Setting $V(\mathbf{r}') = V(\mathbf{r})$ we obtain,

$$\begin{aligned}\rho(\mathbf{r}) - \rho_0 &= -\frac{k_F^3}{4\pi^3}V(\mathbf{r})\int d^3r'\frac{j_1(2k_F|\mathbf{r}-\mathbf{r}'|)}{|\mathbf{r}-\mathbf{r}'|^2} \\ &= -\frac{q^2}{4\pi}V(\mathbf{r}), \quad \text{with} \quad q^2 = \frac{k_F}{\pi a_0},\end{aligned} \tag{23.16}$$

where $a_0 = 1/2e^2$ is the first Bohr radius in the units used here. This result is also obtained within a (linearized) Thomas-Fermi model, discussed in Chapter 1.

To obtain a somewhat more accurate treatment, we use the Poisson equation to connect the potential to the total charge,

$$\nabla^2 V(\mathbf{r}) = \frac{e^2 k_F^2}{\pi^2}\int d^3r' V(\mathbf{r}')\frac{j_1(2k_F|\mathbf{r}-\mathbf{r}'|)}{|\mathbf{r}-\mathbf{r}'|^2}. \tag{23.17}$$

This equation can be treated analytically for a potential associated with the presence of a point charge, Z, at $\mathbf{r} = 0$. We introduce the Fourier components of the Potential,

$$\tilde{V}(\mathbf{k}) = \int d^3r e^{i\mathbf{k}\cdot\mathbf{r}} V(\mathbf{r}), \tag{23.18}$$

and obtain [2]

$$\tilde{V}(\mathbf{k}) = -4\pi Z e^2 / \left[k^2 + \frac{k_F}{\pi a_0} g\left(\frac{k}{2k_F}\right)\right], \tag{23.19}$$

where the function $g(x)$ has the form

$$g(x) = 2 + \frac{x^2-1}{x}\ln\left|\frac{1-x}{1+x}\right|. \tag{23.20}$$

The results of the linear Thomas-Fermi approximation can now be recovered by noting that as $k \to 0$ the function $g(x) \to 4$, so that

$$\tilde{V}(\mathbf{k}) \approx \frac{4\pi Z e^2}{k^2 + q^2}, \tag{23.21}$$

whose transformation into $\mathbf{r}$-space leads to the expression

$$V(\mathbf{r}) = -\frac{Ze^2}{r} \mathrm{e}^{-qr}. \tag{23.22}$$

This is the Coulomb potential arising through the response of an otherwise free electron gas in screening out a point charge. We can now define a general, static dielectric constant, $\epsilon(\mathbf{k})$, by

$$\tilde{V}(\mathbf{k}) = -\frac{4\pi Z e^2}{k^2 \epsilon(\mathbf{k})}, \tag{23.23}$$

so that

$$\epsilon(\mathbf{k}) = \frac{k^2 + \frac{k_F}{\pi a_0} g\left(\frac{k}{2k_F}\right)}{k^2}. \tag{23.24}$$

From what has been said above, the semi-classical Thomas-Fermi result, $\epsilon_{sc}(\mathbf{k})$, takes the form

$$\epsilon_{sc}(\mathbf{k}) = \frac{k^2 + q^2}{k^2}, \tag{23.25}$$

which is consistent with the Eqs. (23.21) and (23.23).

Although the previous manipulations are formally straightforward, some qualifying remarks are in order. We note that the use of a perturbative treatment based on the Born approximation in dealing with the Coulomb potential, which is of infinite range, is questionable from the start. On the other hand, the expression defining the dielectric constant, Eq. (23.23), is meaningful and can be made the starting point of a discussion in which realistic potentials, e.g., those arising in a solid, are considered. In this case, the determination of the dielectric constant must be attempted primarily by numerical means (see also the discussion on the electron-phonon interaction in the previous chapter).

23.2.2 Dynamic dielectric constant

Let us now consider the response of an electron gas to time-varying fields, such as those associated with incident radiation. Once again we begin with Maxwell's equations, Eq. (23.2), and define a wave-vector and frequency dependent dielectric constant[5] by means of the expression,

$$\mathbf{D} = \epsilon \mathbf{E}, \tag{23.26}$$

where

$$\nabla \cdot \mathbf{D} = 4\pi \rho_t \tag{23.27}$$

and

$$\begin{aligned} \nabla \cdot \mathbf{E} &= 4\pi (\rho_t + \rho_m) \\ &= 4\pi \rho_{tot}, \end{aligned} \tag{23.28}$$

with ρ_t being the density of a test charge, ρ_m the density of the medium charge, and their sum, ρ_{tot}, the total charge. We note that the divergence of $\mathbf{E}$ is related to the total charge, and that in the absence of ρ_t the divergence of $\mathbf{D}$ vanishes. If we let ρ denote the charge density of the electrons, and ρ_0 that of a uniform, positive background, we have

$$\rho_m = \rho - \rho_0. \tag{23.29}$$

We study the response of the system to the switching-on of a test charge having a harmonic dependence on time,

$$\rho_t(\mathbf{k}; \mathbf{r}, t) = e\rho_t(\mathbf{k}) e^{i(\mathbf{k}\cdot\mathbf{r} - \omega t)} + \text{c.c.}, \tag{23.30}$$

where the complex conjugate is included since a charge must be real. From Eqs. (23.26) and (23.28), we have[6]

$$4\pi e \rho_t(\mathbf{k}) e^{-i\omega t} = i\mathbf{k} \cdot \mathbf{D}(\mathbf{k}) = i\epsilon(\mathbf{k}, \omega)\mathbf{k} \cdot \mathbf{E}(\mathbf{k}). \tag{23.31}$$

In similar fashion, we assume that ρ_m varies harmonically with time, and obtain,

$$4\pi e \left[\rho_t(\mathbf{k}) + \rho_m(\mathbf{k})\right] e^{-i\omega t} = i\mathbf{k} \cdot \mathbf{E}(\mathbf{k}), \tag{23.32}$$

where $\rho_m(\mathbf{k})$ is the Fourier component of the medium charge density corresponding to the wave vector $\mathbf{k}$. Dividing Eq. (23.31) by (23.32), we find the expression,

$$\begin{aligned} \epsilon(\mathbf{k}, \omega) &= \frac{e^{-i\omega t} \rho_t(\mathbf{k})}{e^{-i\omega t} \left[\rho_t(\mathbf{k}) + \rho_m(\mathbf{k})\right]} \\ &= \frac{\rho_t}{\rho_t + \rho_m}. \end{aligned} \tag{23.33}$$

[5]For ease of presentation, the explicit dependence of various quantities on frequency and wave vector will often be suppressed. We will also refrain from explicitly exhibiting the dyadic (tensor) nature of various quantities, such as the dielectric constant, trusting that this will cause no confusion.

[6]The dielectric function defined here is parallel to $\mathbf{k}$ and comprises only the longitudinal part of the dielectric-function tensor. A more general discussion that involves the entire tensor character of the dielectric function is given in the following section.

Thus, the longitudinal component of the dynamic dielectric constant is the ratio of the test charge to the total charge at a given wave vector and frequency of radiation.

An analysis based on linear response, or equivalently to first order in perturbation theory, shows that we can write [1],

$$\frac{1}{\epsilon(\mathbf{k},\omega)} = 1 - \frac{4\pi e^2}{k^2} \sum_n |\langle n|\rho_{\mathbf{k}}|0\rangle|^2 \left[\frac{1}{\omega - \omega_n + i\delta} + \frac{1}{\omega_n - \omega - i\delta}\right], \quad (23.34)$$

where $|n\rangle$ denotes an eigenstate of the electron system with energy E_n, $\rho_{\mathbf{k}}$ is a Fourier component of the electron charge density, $\omega_n = E_n - E_0$, and δ is a positive infinitesimal. This expression indicates that the resonances of a system occur at the roots of the equation[7] $\epsilon(\mathbf{k},\omega) = 0$.

From the general expression for the charge density in $\mathbf{k}$ space

$$\rho(\mathbf{r}) = \sum_{\mathbf{k}} \rho_{\mathbf{k}} e^{i\mathbf{k}\cdot\mathbf{r}}, \quad (23.35)$$

it follows [1] that the expectation value of the potential energy is given by the expression,

$$\langle V\rangle = 2\pi e^2 \sum_{\mathbf{k}} \left[\frac{\langle\rho_{\mathbf{k}}^{\dagger}\rho_{\mathbf{k}}\rangle}{k^2} - \frac{\rho_0}{k^2}\right], \quad (23.36)$$

which shows the cancellation of the contributions from the electron and background charge at $\mathbf{k} = 0$. The averages involving $\rho_{\mathbf{k}}$ in the last expression can be evaluated through the use of the formal identity,

$$\lim_{\eta\to 0} \frac{1}{x \pm i\eta} = \mathrm{P}\left(\frac{1}{x}\right) \mp i\delta(x), \quad (23.37)$$

where P denotes principal value, and $\delta(x)$ is the Dirac delta function. Applied to Eq. (23.34), this identity leads to the result,

$$\Im\frac{1}{\epsilon(\mathbf{k},\omega)} = \frac{4\pi e^2}{k^2} \sum_n |\langle n|\rho_{\mathbf{k}}|0\rangle|^2 \left[\delta(\omega + \omega_n) - \delta(\omega - \omega_n)\right]. \quad (23.38)$$

The evaluation of $\langle\rho_{\mathbf{k}}^{\dagger}\rho_{\mathbf{k}}\rangle \equiv |\sum_n \langle\rho_{\mathbf{k}}\rangle|^2$ requires an integral over ω from 0 to ∞, so that finally we can write

$$\langle V\rangle = -\sum_{\mathbf{k}} \left[\frac{2\pi\rho_0 e^2}{k^2} + \frac{1}{2\pi}\int_0^{\infty} \Im\frac{1}{\epsilon(\mathbf{k},\omega)} d\omega\right]. \quad (23.39)$$

[7] Although this result is derived within first order perturbation theory, it is true in general, since a resonance defines the normal modes of a system that can be excited by infinitesimally small external perturbations.

This is an important result. One of its most useful formal properties is that it allows derivation of an exact expression for the correlation energy of an interacting electron gas (in terms of the exact dielectric constant, which of course is usually not known). Such an expression can be derived using the formalism of Chapter 1, and Pauli's argument that the expectation value of the Hamiltonian $H(\lambda) = T + \lambda V$, with λ a parameter varying from 0 to 1, is given by Eq. (1.177),

$$\langle H \rangle = \langle H(0) \rangle + \int_0^1 \frac{\mathrm{d}\lambda}{\lambda} \langle \Psi_\lambda | V | \Psi_\lambda \rangle. \tag{23.40}$$

Since the average energy per unit volume of a non-interacting electron gas is given by [see Eq. (1.42), which gives the average energy per particle],

$$\langle H(0) \rangle = \langle T \rangle = \frac{3}{5} \rho_0 E_{\mathrm{F}}, \tag{23.41}$$

we find that the correlation energy of the fully interacting system has the form,

$$\frac{E_{\mathrm{total}}}{\Omega} = \frac{3}{5} \rho_0 E_{\mathrm{F}} - \frac{2\pi^2 e^2}{\Omega} \sum_{\mathbf{k}} \int_0^1 \frac{\mathrm{d}\lambda}{\lambda} \left[\frac{\lambda}{k^2} + \frac{1}{4\pi^3 e^2} \int_0^\infty \mathrm{d}\omega \Im \frac{1}{\epsilon_\lambda(\mathbf{k}, \omega)} \right]. \tag{23.42}$$

If we use the value of $\epsilon(\mathbf{k}, \omega)$ obtained in a self-consistent field calculation with e^2 replaced with λe^2, and exclude the exchange term, we find [3]

$$\frac{E_{\mathrm{total}}}{N} = \frac{2.21}{r_s^2} - \frac{0.916}{r_s} + 0.062 \ln r_s - 0.096 + O(r_s \ln r_s), \tag{23.43}$$

where r_s is the characteristic electron radius, Eq. (1.35).

The last result is exact to order $\ln r_s$ and $(r_s)^0$. The first two terms in the last expansion comprise the Hartree-Fock energy (the first being the mean Fermi energy and the second the exchange contribution), so that the remainder is the correlation energy of the interacting electron cloud. This expression was also quoted in Eq. (1.151), and the reader is encouraged to review that material.

23.2.3 Dielectric function and permeability

Maxwell's equations couple the electric and magnetic fields acting on a system, and the response functions of the system, such as dielectric function, conductivity, and permeability. Therefore, it is not obvious upon mere scrutiny what the precise meaning of the various symbols[8] that occur in

[8] We should keep in mind that the response functions are in general tensor quantities, a feature that will often be suppressed for ease of presentation.

these equations may be. For example, what does one take for the current, J, in Eq. (23.2)? At first glance, it may be thought that J denotes the current contributed by the conduction electrons in the system, the "moving charges". However, since this current is related to the applied field by the expression $\mathbf{J} = \underline{\sigma}\mathbf{E}$ ($\underline{\sigma}$ a tensor), it must be interpreted as the total or measurable current. This follows from our discussion of the conductivity in Chapter 18, where the reader is referred for details.

To clarify ideas, we discuss the three response functions, ϵ, σ, and μ associated with a uniform electron gas [1]. This discussion may point out some of the important interrelationships of these functions and elucidate their meaning in discussions still to come.

Maxwell's equations relating the magnetic and electric fields to the current in a uniform gas take the form (with the velocity of light set equal to one)

$$\epsilon\nabla \times \mathbf{B} = \mathbf{J} + \epsilon_0\frac{\partial \mathbf{E}}{\partial t}. \tag{23.44}$$

At the same time, for any medium with electric displacement $\mathbf{D}$, we can write

$$\nabla \times \mathbf{B} = \frac{\partial \mathbf{D}}{\partial t}. \tag{23.45}$$

Upon comparison of these equations, and using the relation $\mathbf{D} = \epsilon\mathbf{E}$, it follows that

$$\mathbf{J} = (\epsilon - 1)\,\epsilon_0\frac{\partial \mathbf{E}}{\partial t}. \tag{23.46}$$

If a harmonic time dependence is assigned to $\mathbf{E}$, so that $\mathbf{E} \approx \mathrm{e}^{i\omega t}$, we obtain

$$\mathbf{J} = \mathrm{i}\omega(\epsilon - 1)\epsilon_0\mathbf{E}. \tag{23.47}$$

Furthermore, if $\mathbf{E}$ varies in space according to $\mathrm{e}^{\mathrm{i}\mathbf{q}\cdot\mathbf{r}}$, we can take account of the tensor character of the dielectric function by explicitly considering its components parallel (longitudinal) and perpendicular (transverse) to the wave vector $\mathbf{q}$ and write

$$\frac{\mathbf{D}}{\epsilon_0} = \epsilon_\mathrm{l}\mathbf{E}_\mathrm{l} + \epsilon_\mathrm{t}\mathbf{E}_\mathrm{t}, \tag{23.48}$$

where $\mathbf{E}_\mathrm{l}$ and $\mathbf{E}_\mathrm{t}$ denote, respectively, the components of the electric field parallel and perpendicular to $\mathbf{q}$. Comparison with Eq. (23.31) reveals that in only the longitudinal component, ϵ_l, is involved.

It is now interesting to note that instead of using the transverse part of the dielectric function explicitly, we may ascribe part of the current to a magnetic moment density, $\mathbf{M}$,

$$\mathbf{J} = \sigma\mathbf{E} + \nabla \times \mathbf{M}. \tag{23.49}$$

Comparison with Eq. (23.47) yields the relation,

$$\mathbf{J} = \mathrm{i}\omega(\epsilon_l - 1)\epsilon_0\mathbf{E} + \nabla \times \mathbf{M}. \tag{23.50}$$

However, since ϵ is a tensor, we also have from Eq. (23.47) the relation,

$$\begin{aligned} \mathbf{J} &= \mathrm{i}\omega(\epsilon_t - 1)\epsilon_0\mathbf{E}_t + \mathrm{i}\omega(\epsilon_l - 1)\epsilon_0\mathbf{E}_l \\ &= \mathrm{i}\omega(\epsilon_t - 1)\epsilon_0\mathbf{E} + \mathrm{i}\omega(\epsilon_t - \epsilon_l)\epsilon_0\mathbf{E}_l. \end{aligned} \tag{23.51}$$

Comparing with the previous expression, we find

$$\nabla \times \mathbf{M} = \mathrm{i}\omega(\epsilon_t - \epsilon_l)\epsilon_0\mathbf{E}_t. \tag{23.52}$$

We can now connect the dielectric function and the permeability by using another one of Maxwell equations,

$$-\frac{\partial \mathbf{B}}{\partial t} = \nabla \times \mathbf{E}, \tag{23.53}$$

so that for a field with the harmonic behavior $\mathbf{E} \approx \mathrm{e}^{\mathrm{i}(\mathbf{q}\cdot\mathbf{r}-\omega t)}$, we find

$$-\mathrm{i}\omega\mathbf{B} = \mathbf{q} \times \mathbf{E}, \tag{23.54}$$

and, upon taking the curl of both sides, we have,

$$-\mathrm{i}\omega\mathbf{q} \times \mathbf{B} = -q^2\mathbf{E}_t. \tag{23.55}$$

Now, Eq. (23.52) takes the form,

$$\mathbf{q} \times \mathbf{M} = \frac{\omega^2}{q^2}(\epsilon_t - \epsilon_l)\,\epsilon_0\mathbf{q} \times \mathbf{B}, \tag{23.56}$$

and since

$$\mathbf{M} = \epsilon_0(\mathbf{B} - \mathbf{H}) = \epsilon_0\left(1 - \mu^{-1}\right)\mathbf{B}, \tag{23.57}$$

we establish the relation

$$-\mu^{-1} = \left(\frac{\omega}{q}\right)^2(\epsilon_t - \epsilon_l) \tag{23.58}$$

between the electric and magnetic susceptibilities[9].

This is an important result. It shows, for example, that at $\omega \neq 0$, when $\mathbf{M} = 0$ (or equivalently when $\mathbf{H} = \mathbf{B}$ and $\mu = 1$) the transverse and longitudinal components of the dielectric tensor are equal. This is a sensible result since the non-vanishing difference between these two quantities reveals the presence of magnetization that is connected with the transverse part of the electric field. We also see that at $\omega = 0$ and in the absence of an electric field, the current that arises due to the presence of $\mathbf{M}$ is the total current in the system, as is indicated by Eq. (23.49).

[9]Although not explicitly noted, the quantities ϵ and μ appearing in this expression are functions of $\mathbf{q}$ and ω.

23.2.4 Current

We now focus on the meaning of the current, $\mathbf{J}$, entering the Maxwell Eqs. (23.2). We note that the displacement in a material has the general form,

$$\mathbf{D} = \mathbf{E} + 4\pi\mathbf{P}, \tag{23.59}$$

where $\mathbf{P}$, the *polarization* vector, results from a charge density ρ' such that

$$\nabla \cdot \mathbf{P} = \rho'. \tag{23.60}$$

Associated with this polarization there exists a current $\mathbf{J}'$ given by

$$-\nabla \cdot \mathbf{J}' = \frac{\partial \rho'}{\partial t} \tag{23.61}$$

so that we have

$$-\mathbf{J}' = \frac{\partial \mathbf{P}}{\partial t}. \tag{23.62}$$

This current is an additional contribution to that arising from the conduction electrons, or the "free charges" in the system and can be thought of as a "bound current". Therefore, as anticipated, the current density entering the Maxwell equations is the total, and measurable, current in the material.

23.2.5 Index of refraction

There is an alternative way of seeing the connection between the microscopic transport properties of a material, such as the conductivity, and its optical properties. We can write,

$$\nabla \times \underline{f}H = \epsilon_0 \frac{\partial \mathbf{E}}{\partial t} + 4\pi\sigma\mathbf{E}, \tag{23.63}$$

and

$$\nabla \times \mathbf{E} = -\mu \frac{\partial \mathbf{H}}{\partial t}. \tag{23.64}$$

We set $\mu = 1$, neglecting magnetic effects. The difference between the propagation of radiation in free space and in matter is expressed by two (tensor) quantities. The dielectric function defines the magnitude of the displacement currents due to the time variation of the electric field, and the conductivity measures real currents created by the action of an electric field.

Eliminating the magnetic field between the equations above using the relation $\nabla \cdot \mathbf{E} = 0$, we obtain[10]

$$\nabla^2 \mathbf{E} = \epsilon_0 \frac{\partial^2 \mathbf{E}}{\partial^2 t} + 4\pi\sigma \frac{\partial \mathbf{E}}{\partial t}. \tag{23.65}$$

[10]Assuming that no free charges are present so that $\nabla \cdot \mathbf{E} = 0$.

This equation describes wave propagation with dissipation. If we choose a field of the form

$$\mathbf{E} = \mathbf{E}_0 e^{i(\mathbf{K}\cdot\mathbf{r}-\omega t)}, \tag{23.66}$$

then Eq. (23.65) requires that

$$K = \omega \left[\epsilon_0 + \frac{4\pi i\sigma}{\omega}\right]^{1/2}, \tag{23.67}$$

and the propagation constant is complex, consistent with the dissipative nature of the propagation of the wave. Since in free space[11],

$$K = \omega, \tag{23.68}$$

Eq. (23.67) can be taken to define the propagation constant of a wave with a complex velocity,

$$v_{ph} = \frac{1}{N}, \tag{23.69}$$

where the *complex refractive index*, N, is defined by the expression,

$$\begin{aligned} N &= \left(\epsilon_0 + \frac{4\pi i\sigma}{\omega}\right)^{1/2} \\ &= \epsilon^{1/2}. \end{aligned} \tag{23.70}$$

This expression defines the dynamic dielectric constant

$$\epsilon(\omega) = \left(\epsilon_0 + \frac{4\pi i\sigma}{\omega}\right). \tag{23.71}$$

The significance of the refractive index is evident from the fact that the entire macroscopic theory of the optical properties of the material can be formulated in terms of this index. As an example, consider radiation in the form of a plane wave of frequency ω traveling along the z-direction through a material with refractive index N. Writing,

$$N = n + i\kappa, \tag{23.72}$$

we have

$$K = n\omega + i\kappa\omega, \tag{23.73}$$

so that Eq. (23.66) takes the form,

$$\mathbf{E} = \mathbf{E}_0 e^{i\omega(nz-t)} e^{-\frac{\kappa\omega z}{c}}. \tag{23.74}$$

[11] With the velocity of light explicitly exhibited, $K = \omega/c$.

This describes a wave with velocity $1/n$ decaying in the direction of propagation by a fraction $2\pi\kappa/n$ per wavelength. The constant κ is often referred to as the *extinction coefficient.*

Now, the damping of the wave implies the loss of energy that must appear as Joule heat in the material. To calculate this effect, we first find the current [4] from the first equation in the set of Eqs. (23.2),

$$\begin{aligned} \mathbf{J} &= (-\mathrm{i}\omega\epsilon + 4\pi\sigma)\,\mathbf{E} \\ &= -\mathrm{i}\omega N^2\mathbf{E}, \end{aligned} \tag{23.75}$$

where Eq. (23.70) has been used. The rate of production of Joule heat is the real part of the product,

$$\mathbf{J}\cdot\mathbf{E} = -\mathrm{i}\omega N^2 E^2. \tag{23.76}$$

Therefore, the ***absorption coefficient***, i.e., the fraction of radiation energy absorbed by unit thickness of the material, is given by,

$$\eta = \frac{\Re\,(\mathbf{j}\cdot\mathbf{E})}{nE^2} = \frac{2\kappa\omega}{\mathrm{c}}. \tag{23.77}$$

Finally, it is elementary to show that the ***reflection coefficient***, i.e., the ratio of the complex amplitudes of the waves incident and reflected by a surface, is given by

$$R = \left|\frac{1-N}{1+N}\right|^2 = \frac{(n-1)^2+\kappa^2}{(n+1)^2+\kappa^2}. \tag{23.78}$$

Before leaving this section, we note that the refractive index is a tensor quantity, as are the dielectric function and conductivity, so that Eq. (23.70) takes the form,

$$(N)^2_{\alpha\beta} = \epsilon_{\alpha\beta} + \frac{4\pi\mathrm{i}}{\omega}\sigma_{\alpha\beta}, \tag{23.79}$$

reflecting the fact that in an inhomogeneous medium light propagates with different velocities along different directions. Also, in the case of the free-electron gas, we can set $\epsilon_0 = 1$ and obtain,

$$(N)^2_{\alpha\beta} = \delta_{\alpha\beta} + \frac{4\pi\mathrm{i}}{\omega}\sigma_{\alpha\beta}. \tag{23.80}$$

Thus, the index of refraction consists of its free-space value, 1, and corrections that arise from the presence of the medium (electrons) and are expressed in terms of the conductivity, $\underline{\sigma}$. Further discussion of the optical properties of a material and its conductivity is provided in Chapter 19.

23.3 Vector Wave Equation

We now generalize the phenomenological treatment of the passage of radiation through matter given in the previous section to a microscopic theory. The following formalism will allow us to address such issues as the formation of phonic bands in solids, the existence of phonic band gaps corresponding to frequencies at which radiation cannot propagate through a medium, and the treatment of radiation passing through random, substitutionally disordered alloys. The basis of this approach is the *vector wave equation* which we now derive.

We start with Eqs. (23.2), the Maxwell equations in the presence of matter. It is important to keep in mind that the medium quantities ϵ, μ, and σ are tensors reflecting the fact that the medium of propagation may not be uniform or isotropic.

We assume fields of (circular) frequency ω, and write

$$\mathbf{E}(t) = \mathbf{E}\mathrm{e}^{i\omega t} + \mathbf{E}^{*}\mathrm{e}^{-i\omega t}, \tag{23.81}$$

with similar expressions for the other fields. Then, the equations involving the vector product in Eq. (23.2) take the form

$$\nabla \times \mathbf{H} = -\mathrm{i}kN^2\mathbf{E}, \quad \nabla \times \mathbf{E} = \mathrm{i}k\mathbf{B}, \tag{23.82}$$

where [see Eq. (23.70)],

$$k = \omega = \frac{2\pi}{\lambda_0}, \quad \text{and} \quad N^2 = \epsilon_0 + \frac{4\pi\mathrm{i}}{\omega}\sigma, \tag{23.83}$$

with λ_0 being the wavelength of the radiation in vacuum.

Applying the curl operator in Eq. (23.82) leads to the expression,

$$\nabla \times (\nabla \times \mathbf{E}) - k^2\mathbf{E} = k^2 (N-1)\,\mathbf{E} + \nabla \times \left[\left(1-\mu^{-1}\right)\nabla \times \mathbf{E}\right], \tag{23.84}$$

where the index of refraction, dielectric function, conductivity, and permeability must be understood as being tensor quantities in general. We note that in this case the double curl, $\nabla \times (\nabla \times \mathbf{E})$, does not reduce to $\nabla^2\mathbf{E}$ because in general $\nabla \cdot \mathbf{E} \neq 0$.

The last expression, Eq. (23.84), constitutes the *vector wave equation*. Our aim is to develop methods, based on Green functions and multiple scattering theory, for finding solutions to this equation both in the case of translationally invariant materials as well as substitutionally disordered alloys.

23.4 Vector Spherical Harmonics

As in the case of the "scalar" wave equation, e.g., the Poisson or the Schrödinger equations, we wish to study the vector wave equation, Eq.

(23.84), by means of a partial-wave analysis. This requires an extension of the scalar spherical harmonic functions, $Y_{\ell m}(\hat{r})$, to vector quantities [5].

Let $\hat{x}$, $\hat{y}$, and $\hat{z}$ denote unit vectors along the axes of a right-handed coordinate system, and define the *complex* vector quantities,

$$\hat{\chi}_1 \equiv \frac{1}{\sqrt{2}}(\hat{y} - i\hat{x}), \quad \hat{\chi}_0 \equiv i\hat{z}, \quad \hat{\chi}_{-1} \equiv \frac{1}{\sqrt{2}}(\hat{y} + i\hat{x}). \tag{23.85}$$

Clearly, we have,

$$\sum_{\mu=-1,01} \hat{\chi}_\mu \hat{\chi}^*_\mu = \underline{1} \quad \text{and} \quad \hat{\chi}^*_\mu \cdot \hat{\chi}_{\mu'} = \delta_{\mu\mu'}, \tag{23.86}$$

where $\underline{1}$ denotes the three-dimensional unit matrix (tensor), and an asterisk denotes complex conjugation. The vector functions $\hat{\chi}_\mu$ can be seen to satisfy the phase relation,

$$\hat{\chi}^*_\mu = (-1)^{\mu+1}\,\hat{\chi}_\mu. \tag{23.87}$$

Now, let the *spin operator*, **S**, be defined by its action on any three-dimensional vector function, **f**,

$$S_x\mathbf{f} \equiv i\hat{x} \times \mathbf{f}, \quad S_y\mathbf{f} \equiv i\hat{y} \times \mathbf{f}, \quad \text{and} \quad S_z\mathbf{f} \equiv i\hat{z} \times \mathbf{f}. \tag{23.88}$$

We see that[12],

$$S_z\hat{\chi}_\mu = \mu\hat{\chi}_\mu \tag{23.89}$$

$$S^2\hat{\chi}_\mu = 2\hat{\chi}_\mu. \tag{23.90}$$

It follows from the properties of the ordinary, scalar, spherical harmonics, and the properties of the vector functions defined above that the direct products[13],

$$\mathrm{Y}_{\ell m\mu}(\hat{r}) = Y_{\ell m}(\hat{r})\chi_\mu, \tag{23.91}$$

are the eigenfunctions of the angular momentum and spin operators, so that

$$\begin{aligned} L^2\mathrm{Y}_{\ell m\mu} &= \ell(\ell+1)\,\mathrm{Y}_{\ell m\mu} \\ L_z\mathrm{Y}_{\ell m\mu} &= m\mathrm{Y}_{\ell m\mu} \\ S_z\mathrm{Y}_{\ell m\mu} &= \mu\mathrm{Y}_{\ell m\mu}. \end{aligned} \tag{23.92}$$

The total angular momentum, **J**, is obtained as the vector sum of **L** and **S**, $\mathbf{J} = \mathbf{L} + \mathbf{S}$, and since $S = 1$, the modulus of **J** can take the values,

[12] Note that considering μ to be the projection along the z-axis of an angular momentum, $\ell = 1$, we have, $S^2\hat{\chi}_\mu = \ell(\ell+1)\,\hat{\chi}_\mu = 2\hat{\chi}_\mu$.

[13] For ease of presentation, we will no longer designate the vector character of the $\hat{\chi}_\mu$ with a hat symbol.

$j = \ell - 1,\ \ell,\ \ell + 1$. Using Clebsch-Gordan coefficients (see Appendix C), we can combine the functions $\mathrm{Y}_{\ell m \mu}$ into eigenfunctions of $\mathbf{J}$,

$$\begin{aligned}\mathrm{Y}_{J\ell M}(\hat{r}) &\equiv \sum_{m\mu} C\left(\ell 1 J, m\mu M\right) \mathrm{Y}_{\ell m\mu}(\hat{r}) \\ &= \sum_{\mu} C\left[\ell 1 J, (M-\mu)\mu M\right] \mathrm{Y}_{\ell (M-\mu)\mu}(\hat{r}). \end{aligned} \quad (23.93)$$

These functions satisfy the relations,

$$\begin{aligned} J^2 \mathrm{Y}_{J\ell M}(\hat{r}) &= (\mathbf{L}+\mathbf{S})^2\, \mathrm{Y}_{J\ell M}(\hat{r}) = J(J+1)\, \mathrm{Y}_{J\ell M}(\hat{r}), \\ L^2 \mathrm{Y}_{J\ell M}(\hat{r}) &= \ell(\ell+1)\, \mathrm{Y}_{J\ell M}(\hat{r}), \\ J_z \mathrm{Y}_{J\ell M}(\hat{r}) &= (L_z + S_z)\, \mathrm{Y}_{J\ell M}(\hat{r}) = M \mathrm{Y}_{J\ell M}(\hat{r}), \end{aligned} \quad (23.94)$$

and they are orthonormal,

$$\int d\Omega_r \mathrm{Y}_{J\ell M}(\hat{r}) \cdot \mathrm{Y}_{J'\ell' M'}(\hat{r}) = \delta_{JJ'}\delta_{\ell\ell'}\delta_{MM'}, \quad (23.95)$$

and complete

$$\sum_{J\ell M} \mathrm{Y}_{J\ell M}(\hat{r}) \mathrm{Y}^*_{J\ell M}(\hat{r}') = \underline{1}\delta_\Omega(\hat{r}, \hat{r}'). \quad (23.96)$$

We note that for $J = 0$, only Y_{010} is non-vanishing.

The *vector* spherical harmonics, $\mathrm{Y}_{J\ell M}(\hat{r})$, can be formed into various convenient combinations. For example, transverse and longitudinal vector spherical harmonics can be defined as[14]:

$$\begin{aligned} \mathrm{Y}^{(e)}_{JM} &\equiv \left(\frac{J+1}{2J+1}\right)^{1/2} \mathrm{Y}_{J,J-1,M} - \left(\frac{J}{2J+1}\right)^{1/2} \mathrm{Y}_{J,J+1,M} \\ \mathrm{Y}^{(m)}_{JM} &\equiv \mathrm{Y}_{JJM} \\ \mathrm{Y}^{(0)}_{JM} &\equiv \left(\frac{J}{2J+1}\right)^{1/2} \mathrm{Y}_{J,J-1,M} + \left(\frac{J+1}{2J+1}\right)^{1/2} \mathrm{Y}_{J,J+1,M}. \end{aligned} \quad (23.97)$$

The first two combinations result in transverse quantities in the sense that

$$\mathbf{r} \cdot \mathrm{Y}^{(e)}_{JM} = \mathbf{r} \cdot \mathrm{Y}^{(m)}_{JM} = 0, \quad (23.98)$$

in contrast to the longitudinal character of the third combination,

$$\mathbf{r} \times \mathrm{Y}^{(0)}_{JM} = 0. \quad (23.99)$$

[14] The superscripts e and m signify the special use of the corresponding functions in expanding electric and magnetic fields.

These new functions preserve the vector orthogonality condition in terms of the index $\lambda = \mathrm{e},\ \mathrm{m},\ 0$,

$$\mathrm{Y}_{JM}^{(\lambda)*}(\hat{r}) \cdot \mathrm{Y}_{JM}^{(\lambda')}(\hat{r}) = 0, \quad \text{unless} \quad \lambda = \lambda', \tag{23.100}$$

and satisfy the phase relations,

$$\begin{aligned} \mathrm{Y}_{JM}^{(\mathrm{e})}(-\hat{r}) &= (-1)^{J+1}\,\mathrm{Y}_{JM}^{(\mathrm{e})}(\hat{r}) \\ \mathrm{Y}_{JM}^{(\mathrm{m})}(-\hat{r}) &= (-1)^{J}\,\mathrm{Y}_{JM}^{(\mathrm{m})}(\hat{r}) \\ \mathrm{Y}_{JM}^{(0)}(-\hat{r}) &= (-1)^{J+1}\,\mathrm{Y}_{JM}^{(0)}(\hat{r}), \end{aligned} \tag{23.101}$$

and

$$\begin{aligned} P\mathrm{Y}_{JM}^{(\mathrm{e})}(\hat{r}) &= (-1)^{J}\,\mathrm{Y}_{JM}^{(\mathrm{e})}(\hat{r}) \\ P\mathrm{Y}_{JM}^{(\mathrm{m})}(\hat{r}) &= (-1)^{J+1}\,\mathrm{Y}_{JM}^{(\mathrm{m})}(\hat{r}) \\ P\mathrm{Y}_{JM}^{(0)}(\hat{r}) &= (-1)^{J}\,\mathrm{Y}_{JM}^{(0)}(\hat{r}), \end{aligned} \tag{23.102}$$

where P is the parity operator. The last three relations follow from the observation that under a reflection of coordinates, we have,

$$PY(\mathbf{r}) = -Y(-\mathbf{r}). \tag{23.103}$$

We also quote a set of interrelations among these longitudinal and transverse vector spherical harmonics that makes their association with electric and magnetic fields somewhat easier to see,

$$\begin{aligned} \mathrm{Y}_{JM}^{(\mathrm{e})}(\hat{r}) &= [J(J+1)]^{-1/2}\,r\nabla \mathrm{Y}_J^M \\ \mathrm{Y}_{JM}^{(\mathrm{m})}(\hat{r}) &= \hat{r} \times \mathrm{Y}_{JM}^{(\mathrm{e})}(\hat{r}) = r\nabla \times \mathrm{Y}_{JM}^{(\mathrm{e})}(\hat{r}) \\ \mathrm{Y}_{JM}^{(0)}(\hat{r}) &= -r^2\,[J(J+1)]^{-1/2}\,\nabla \times \left[\nabla \times \mathrm{Y}_{JM}^{(\mathrm{e})}(\hat{r})\right]. \end{aligned} \tag{23.104}$$

Like the functions $\mathrm{Y}_{J\ell M}(\hat{r})$, these vector spherical harmonics are complete and orthonormal,

$$\sum_{\lambda JM} \mathrm{Y}_{JM}^{(\lambda)}(\hat{r})\mathrm{Y}_{JM}^{(\lambda)}(\hat{r}') = \underline{1}\delta(\hat{r},\hat{r}') \tag{23.105}$$

$$\int d\Omega_r \mathrm{Y}_{JM}^{(\lambda)}(\hat{r}) \cdot \mathrm{Y}_{J'M'}^{(\lambda')}(\hat{r}) = \delta_{JJ'}\delta_{MM'}\delta_{\ell\ell'}. \tag{23.106}$$

Either set of the vector spherical harmonics defined above can be used to expand any *vector* function of position. For example, for the electric field, $\mathbf{E}(\mathbf{r})$, we may write either of the following expressions,

$$\begin{aligned} \mathbf{E}(\mathbf{r}) &= \sum_{J\ell M} \underline{E}_{J\ell M}(r)\mathrm{Y}_{J\ell M}(\hat{r}) \\ &= \sum_{\lambda JM} \underline{E}_{JM}^{(\lambda)}(r)\mathrm{Y}_{JM}^{(\lambda)}(\hat{r}). \end{aligned} \tag{23.107}$$

In Eq. (23.104), the terms labeled by e, (J, and M) correspond to electric 2^J-dipole radiation, and those labeled m, (and J, M) correspond to magnetic 2^J-dipole radiation [6]. In the last expression in Eq. (23.107), λ represents e or m.

An expansion for the magnetic field in terms of vector spherical harmonics can be obtained from Eq. (23.107) by using the relation,

$$\nabla \times \mathbf{E} = -\frac{1}{\mathrm{c}}\frac{\partial \mathbf{B}}{\partial t}. \tag{23.108}$$

For a harmonic time dependence, $\mathbf{B} \approx \mathrm{e}^{i\omega t}$, we have

$$\mathbf{B} = -ik^{-1}\nabla \times \mathbf{E}, \tag{23.109}$$

where $k = \omega$ (or $k = \omega/\mathrm{c}$ with c restored). Explicit expressions for the magnetic field expanded in vector spherical harmonics can now be obtained from Eq. (23.107) through the application of the curl operator on the vector functions $\mathrm{Y}^{(\lambda)}_{JM}(\hat{r})$, (or $\mathrm{Y}_{J\ell M}(\hat{r})$) and the coefficients, $\underline{E}^{(\lambda)}_{JM}(r)$ (or $\underline{E}_{J\ell M}(r)$).

23.4.1 Plane wave expansions

A partial wave analysis is based on the expansion of a plane wave into spherical functions. In the present case, we are interested in expanding the tensor quantity, $\underline{1}\mathrm{e}^{i\mathbf{k}\cdot\mathbf{r}}$ in terms of any one of the sets of vector spherical harmonics defined above.

We begin with the addition theorem, Eq. (C.19), which we write in the form

$$(2\ell+1)\,P_\ell(\hat{r}_1\cdot\hat{r}_2) = 4\pi\sum_{m=-\ell}^{\ell} Y_{\ell m}(\hat{r}_1)Y^*_{\ell m}(\hat{r}_2). \tag{23.110}$$

Using Eq. (23.86) and the definitions of the vector spherical harmonics, $\mathrm{Y}_{J\ell M}(\hat{r})$, we obtain

$$(2\ell+1)\,\underline{1}P_\ell(\hat{r}_1\cdot\hat{r}_2) = 4\pi\sum_{JM} \mathrm{Y}_{J\ell M}(\hat{r}_1)\mathrm{Y}^*_{J\ell M}(\hat{r}_2), \tag{23.111}$$

a result following from the relation,

$$Y_{\ell m}(\hat{r})\chi_\mu = \sum_{JM} C\,(\ell 1J, m\mu M)\,\mathrm{Y}_{J\ell M}(\hat{r}), \tag{23.112}$$

which is the inverse of the transformation in Eq. (23.93), and can be obtained from the properties of the Clebsch-Gordan coefficients,

$$\sum_{JM} C\,(\ell sJ, m\mu M)\,C\,(\ell sJ, m'\mu' M) = \delta_{mm'}\delta_{\mu\mu'}, \tag{23.113}$$

and

$$\sum_{m\mu} C(\ell sJ, m\mu M)\, C(\ell sJ', m\mu M') = \delta_{JJ'}\delta_{MM'}. \tag{23.114}$$

Now, from the relation,

$$e^{i\mathbf{k}\cdot\mathbf{r}} = 4\pi \sum_{\ell} i^{\ell} j_{\ell}(kr)\,(2\ell+1) \sum_{m=-\ell}^{\ell} Y_{\ell m}(\hat{r}) Y^{*}_{\ell m}(\hat{k}), \tag{23.115}$$

we readily obtain the relation that we have been seeking,

$$\underline{1} e^{i\mathbf{k}\cdot\mathbf{r}} = 4\pi \sum_{\ell JM} i^{\ell} j_{\ell}(kr) \mathbf{Y}_{J\ell M}(\hat{r}) \mathbf{Y}^{*}_{J\ell M}(\hat{k}), \tag{23.116}$$

where the dyadic nature of the products $\mathbf{YY}^*$ is to be noted.

23.4.2 Free-particle propagator

The final basic ingredient in a partial-wave analysis of the vector wave Eq. (23.84) is the expansion of the free-particle propagator in vector spherical harmonics. For the vector wave equation, the free-particle propagator must be a tensor, and we define the tensor Green function, $\underline{G}_0(\mathbf{r}, \mathbf{r}')$, by the relation,

$$\nabla \times [\nabla \times \underline{G}_0(\mathbf{r}, \mathbf{r}')] - k^2 \underline{G}_0(\mathbf{r}, \mathbf{r}') = \underline{1}\delta(\mathbf{r} - \mathbf{r}'). \tag{23.117}$$

For $\mathbf{r} \neq \mathbf{r}'$, $\underline{G}_0(\mathbf{r}, \mathbf{r}')$ satisfies the relation,

$$\nabla \cdot \underline{G}_0(\mathbf{r}, \mathbf{r}') = 0, \quad \mathbf{r} \neq \mathbf{r}'. \tag{23.118}$$

Now, through the direct application of the operator $\nabla \times \nabla \times -k^2$, it can be shown that the quantity,

$$\underline{G}_0(\mathbf{r}, \mathbf{r}') = \left[\underline{1} + \frac{1}{k^2}\nabla\nabla\right] \frac{e^{ik|\mathbf{r}-\mathbf{r}'|}}{4\pi |\mathbf{r} - \mathbf{r}'|}, \tag{23.119}$$

satisfies the defining equation for the propagator, Eq. (23.117), as well as the subsidiary condition, Eq. (23.118). This Green function also satisfies the scattering boundary condition, and therefore is the proper Green function to use in the scattering treatment of the vector wave equation.

23.4.3 Partial-wave expansions of the Green function

Having obtained a form of the free-particle propagator in r-space, we seek its expansion in vector spherical harmonics. As in the case of the scalar Green function, we define vector spherical functions as the products of the spherical Bessel, Neumann, and Hankel functions with vector spherical harmonics, [with $h_\ell(kr)$ denoting either of the functions $h_\ell^{\pm}(kr)$],

$$\begin{aligned} \mathrm{J}_{J\ell M}(\mathbf{r}) &= j_\ell(kr)\mathrm{Y}_{J\ell M}(\hat{r}) \\ \mathrm{N}_{J\ell M}(\mathbf{r}) &= n_\ell(kr)\mathrm{Y}_{J\ell M}(\hat{r}) \\ \mathrm{H}_{J\ell M}(\mathbf{r}) &= h_\ell(kr)\mathrm{Y}_{J\ell M}(\hat{r}). \end{aligned} \tag{23.120}$$

We note that these solid vector spherical harmonics have the property

$$\mathrm{Z}^{(\mathrm{e})}_{JM}(\mathbf{r}) = -\frac{\mathrm{i}}{k}\nabla \times Z^{(\mathrm{m})}_{JM}(\mathbf{r}), \tag{23.121}$$

where Z stands for either J or H. Often, it will also be convenient to absorb a factor $-ik$ into the definition of H. Thus, the expansion of the tensor plane wave, Eq. (23.116), takes the form,

$$\underline{1}e^{\mathrm{i}\mathbf{k}\cdot\mathbf{r}} = 4\pi \sum_{\ell JM} \mathrm{i}^\ell \mathrm{J}_{J\ell M}(\mathbf{r})\mathrm{Y}^*_{J\ell M}(\hat{k}). \tag{23.122}$$

Now, consider the result,

$$\underline{1}e^{\mathrm{i}\mathbf{k}\cdot(\mathbf{a}+\mathbf{b})} = \underline{1}e^{\mathrm{i}\mathbf{k}\cdot\mathbf{a}}\underline{1}e^{\mathrm{i}\mathbf{k}\cdot\mathbf{b}}, \tag{23.123}$$

which can be expanded in vector spherical functions to yield the relation,

$$\begin{aligned} 4\pi \sum_{J\ell M} \mathrm{i}^\ell \mathrm{J}_{J\ell M}(\mathbf{a}+\mathbf{b})\mathrm{Y}^*_{J\ell M}(\hat{k}) &= (4\pi)^2 \sum_{J_1\ell_1 M_1} \mathrm{i}^{\ell_1} \mathrm{J}_{J_1\ell_1 M_1}(\mathbf{a})\mathrm{Y}^*_{J_1\ell_1 M_1}(\hat{k}) \\ &\times \sum_{J_2\ell_2 M_2} \mathrm{i}^{\ell_2} \mathrm{J}_{J_2\ell_2 M_2}(\mathbf{b})\mathrm{Y}^*_{J_2\ell_2 M_2}(\hat{k}). \end{aligned} \tag{23.124}$$

Multiplying both sides by $\mathrm{Y}_{J_0\ell_0 M_0}(\hat{k})$ and integrating over the angles of $\hat{k}$, we obtain

$$\mathrm{J}_{J_0\ell_0 M_0}(\mathbf{a}+\mathbf{b}) = \sum_{J_1\ell_1 M_1} g_{J_0\ell_0 M_0, J_1\ell_1 M_1}(\mathbf{a})\mathrm{J}_{J_1\ell_1 M_1}(\mathbf{b}), \quad \text{all } \mathbf{a}, \mathbf{b}, \tag{23.125}$$

where

$$g_{J_0\ell_0 M_0, J_1\ell_1 M_1}(\mathbf{a}) = 4\pi \sum_{J_2\ell_2 M_2} F(J_1\ell_1 M_1; J_0\ell_0 M_0; J_2\ell_2 M_2)\, \mathrm{J}_{J_2\ell_2 M_2}(\mathbf{a}), \tag{23.126}$$

is the matrix element that connects two regular spherical functions evaluated at two different points in space. In the last expression, the quantities

F are vector Gaunt numbers given as integrals over three vector spherical functions,

$$F(J_1\ell_1M_1; J_0\ell_0M_0; J_2\ell_2M_2) = \int d\Omega_r \mathrm{Y}^*_{J_1\ell_1M_1}(\hat{r})\mathrm{Y}^*_{J_2\ell_2M_2}(\hat{r})\mathrm{Y}_{J_0\ell_0M_0}(\hat{r}). \tag{23.127}$$

We recognize Eqs. (23.125) and (23.127) as the vector generalization of the corresponding scalar relations for ordinary spherical harmonics given in Appendix C

In proceeding further, it is convenient to denote the set of indices, $J\ell M$, by a combined index, Q. Then Eq. (23.126) takes the form,

$$\mathrm{J}_Q(\mathbf{a}+\mathbf{b}) = \sum_{Q'} g_{QQ'}(\mathbf{a})\mathrm{J}_{Q'}(\mathbf{b}). \tag{23.128}$$

It is a matter of some algebra to rewrite the expansions exhibited above in terms of the transverse and longitudinal vector harmonics, $\mathrm{Y}^{(\lambda)}_{JM}(\hat{r})$. The foregoing formalism remains essentially unchanged, and all relations, e.g., Eq. (23.128), remain valid with appropriate redefinitions of parameters. Thus, in Eq. (23.128), the index Q would denote the three indices, λ, J, and M.

In addition to the expansions of a plane wave, we also require a corresponding expansion of the free-particle propagator. We start with the well-known expansion of the scalar free-particle Green function in terms of spherical functions,

$$G_0(\mathbf{r},\mathbf{r}') = -\frac{e^{ik|\mathbf{r}-\mathbf{r}'|}}{4\pi|\mathbf{r}-\mathbf{r}'|} = \sum_L J_L(\mathbf{r})H_L(\mathbf{r}'), \quad \text{with} \quad r' > r. \tag{23.129}$$

To obtain an expression for the tensor $\underline{G}_0$, we operate on the last result with $\left(\underline{1} + \frac{\nabla\nabla}{k^2}\right)$. Using the recurrence relations for the spherical Bessel and Hankel functions, Eqs. (C.32) and (C.33), we can write,

$$\begin{aligned} \underline{G}_0(k;\mathbf{r},\mathbf{r}') &= \sum_{JM\lambda} \mathrm{J}_{JM\lambda}(\mathbf{r})\mathrm{H}^*_{JM\lambda}(\mathbf{r}') \\ &= \sum_Q \mathrm{J}_Q(\mathbf{r})\mathrm{H}^*_Q(\mathbf{r}), \quad r' > r. \end{aligned} \tag{23.130}$$

Once again, the transverse and longitudinal forms of the vector spherical harmonics, $\mathrm{Z}^{(\lambda)}_{JM}(\mathbf{r})$, can also be used in the expansion of the Green function.

The irregular solutions of the free-particle vector wave equations, like the regular ones, can be expanded about a shifted origin. The coefficients of that expansion are the *vector* real-space structure constants. Here, we quote only some final results, leaving the details to the reader.

Considering the functions, $\mathtt{H}^{(\lambda)}_{JM}$, we find,

$$\begin{aligned} \mathtt{H}_Q(\mathbf{r}+\mathbf{R}) &\equiv \mathtt{H}^{(\lambda)}_{JM}(\mathbf{r}+\mathbf{R}) \\ &= \sum_{\lambda' J' M'} \underline{G}^{\lambda\lambda'}_{JM,J'M'}(\mathbf{R})\mathtt{J}^{(\lambda')}_{J'M'}(\mathbf{r}) \\ &= \sum_{Q'} \underline{G}_{QQ'}(\mathbf{R})\mathtt{J}_{Q'}(\mathbf{r}), \quad R > r. \end{aligned} \tag{23.131}$$

The real-space vector structure constants, $G_{QQ'}$, are given explicitly by the expressions

$$\underline{G}^{\lambda\lambda'}_{JM,J'M'}(\mathbf{R}) = \begin{cases} \sum_\mu C\,(J1J, M-\mu, \mu) \\ \times G_{JM-\mu;J'M'-\mu}(\mathbf{R}) C\,(J'1J', M'-\mu, \mu) & \text{if } \lambda = \lambda' \\ \\ \left[\frac{2J'+1}{J'+1}\right]^{1/2} \sum_\mu C\,(J1J, M-\mu, \mu) \\ \times G_{JM-\mu;J'-1,M'-\mu}(\mathbf{R}) \\ \times C\,(J'-1, 1J', M'-\mu, \mu) & \text{if } \lambda = \mathrm{m}, \\ & \lambda' = \mathrm{e} \\ -\left[\frac{2J'+1}{J'+1}\right]^{1/2} \sum_\mu C\,(J1J, M-\mu, \mu) \\ \times G_{JM-\mu;J'-1,M'-\mu}(\mathbf{R}) \\ \times C\,(J'-1, 1J', M'-\mu, \mu) & \text{if } \lambda = \mathrm{e}, \\ & \lambda' = \mathrm{m}. \end{cases} \tag{23.132}$$

This completes the development of the formal basis upon which the partial wave analysis of the vector wave equation, Eq. (23.84), is based. We now turn to solving that equation using the general formalism of multiple scattering theory.

23.5 Single-Cell Scattering

The vector wave equation lends itself well to analysis through the general formalism of scattering theory. One can define the vector or tensor generalizations of scattering states, Møller wave operators, and other formal concepts of scattering theory (see Appendix E), and also obtain appropriate generalizations of important results such as the optical theorem. The reader interested in the formal aspects of the application of scattering theory to the vector wave equation is referred to excellent presentations in the literature [5]. In our discussion, we will use the formal construct of MST to solve the vector wave equation in the presence of matter by drawing on the close formal analogy that exists between the scalar and vector forms of the equation.

The first step in an implementation of MST is the solution of the wave equation associated with a single cell (scatterer). This leads to expressions for the single-cell scattering matrix[15] and wave function (basis functions) that are then used to obtain the solution for an assembly of scatterers.

We begin with Eq. (23.84) and assume that the quantity $\epsilon(\mathbf{r})-1$ vanishes identically[16] outside a finite region of space, Ω. Also, we neglect magnetic effects and set $\mu = 1$. We write the differential Eq. (23.84) in the form of the integral Lippmann-Schwinger equation,

$$\begin{aligned}\mathbf{E}(\mathbf{r}) &= \mathbf{E}_0(\mathbf{r}) + \int_\Omega \mathrm{d}^3r' \underline{G}_0(\mathbf{r},\mathbf{r}')k^2\left[\epsilon(\mathbf{r}')-1\right]\mathbf{E}(\mathbf{r}') \\ &= \mathbf{E}_0(\mathbf{r}) + \int_\Omega \mathrm{d}^3r' \underline{G}_0(\mathbf{r},\mathbf{r}')\underline{f}(\mathbf{r}')\mathbf{E}(\mathbf{r}'),\end{aligned} \tag{23.133}$$

where $\underline{G}_0$ is the free-particle Green function defined in Eq. (23.119), and we have defined the tensor (matrix) function $\underline{f}(\mathbf{r}) = k^2\left[\underline{\epsilon}(\mathbf{r}) - \underline{1}\right]$. This form of the Lippmann-Schwinger equation is essentially identical to the form- s treated previously, the only difference being the rather straightforward generalization of scalar quantities to vectors and/or matrices (tensors). We follow the same steps as in the treatment of the scalar wave equation, Appendix D, which the reader may wish to review briefly at this point.

To solve for $\mathbf{E}(\mathbf{r})$, we introduce the basis functions,

$$\tilde{\mathrm{P}}_Q(\mathbf{r}) = \sum_{Q'}\left[\underline{c}_{QQ'}(r)\mathrm{J}_{Q'}(\mathbf{r}) + \underline{s}_{QQ'}(r)\mathrm{H}_{Q'}(\mathbf{r})\right]. \tag{23.134}$$

With the expansion,

$$\mathbf{E}_0(\mathbf{r}) = \sum_Q \underline{\bar{C}}_{QQ'}\mathrm{J}_{Q'}(\mathbf{r}), \tag{23.135}$$

the vector phase functions, $\underline{c}_{QQ'}(r)$, and $\underline{s}_{QQ'}(r)$ are given by the expressions,

$$\underline{c}_{QQ'}(r) = \underline{\bar{C}}_{QQ'} - k\int^{r'>r} \mathrm{d}^3r'\mathrm{H}_Q(\mathbf{r}')\underline{f}(\mathbf{r}')\tilde{\mathrm{P}}^*_Q(\mathbf{r}'), \tag{23.136}$$

and

$$\underline{s}_{QQ'}(r) = k\int_0^r \mathrm{d}^3r'\mathrm{J}_Q(\mathbf{r}')\underline{f}(\mathbf{r}')\tilde{\mathrm{P}}^*_Q(\mathbf{r}'). \tag{23.137}$$

[15] It is assumed throughout that we are dealing with space-filling, non-muffin-tin scattering cells. The various special considerations that were presented with respect to the scalar wave equation applied to such potentials find straightforward and fairly obvious generalizations in this case.

[16] In free space, $\epsilon = 1$.

In these expressions, the combined index Q denotes either the triplet of indices $J\ell M$, or λJM, whichever is employed in the discussion. Because of the curl relations between the electric and magnetic fields that enter the Maxwell equations, the set λJM, which emphasizes the transverse relations between these fields, is often more convenient for the treatment of electromagnetic radiation than the set $J\ell M$.

The formal analogy of the expressions derived above with those given in Appendix D should be noted. However, once again we must emphasize a special feature related to the vector/matrix nature of the various quantities involved in these expressions. We recall that the underlined quantities are matrices in x, y, z space, and that the various functions, Z, (i.e., vector spherical harmonics or solid harmonics) are vectors in this space. Therefore, expressions of the form $\mathrm{Z}_1\underline{A}\mathrm{Z}_2^*$ must be interpreted as involving the dyadic product of the vector functions Z_1 and Z_2^*. This is indicated by the use of the asterisk that denotes the complex-conjugate transpose of the (generally) complex vector functions.

We can now proceed in fairly familiar fashion with the construction of the solution of the vector equation. We realize that outside a sphere bounding the "potential" region Ω, the phase functions $\underline{c}_{QQ'}(r)$ and $\underline{s}_{QQ'}(r)$ reach their asymptotic constant values, which in analogy with Eqs. (D.91) and (D.92), are given by the expressions,

$$\underline{C}_{QQ'} = \delta_{QQ'} - k\int_{\Omega} \mathrm{d}^3r' \mathrm{H}_Q(\mathbf{r}')\underline{f}(\mathbf{r}')\tilde{\mathrm{P}}_Q^*(\mathbf{r}'). \tag{23.138}$$

and

$$\underline{S}_{QQ'} = k\int_{\Omega} \mathrm{d}^3r' \mathrm{J}_Q(\mathbf{r}')\underline{f}(\mathbf{r}')\tilde{\mathrm{P}}_Q^*(\mathbf{r}'). \tag{23.139}$$

The basis functions, $\tilde{\mathrm{P}}_Q(\mathbf{r})$, behave as $\mathrm{J}_Q(\mathbf{r})$ at the origin. It is also convenient to define functions $\mathrm{P}_Q(\mathbf{r})$ which outside a sphere bounding the potential behave as $\mathrm{J}_Q(\mathbf{r}) + \sum_{Q'} \underline{t}_{QQ'}\mathrm{H}_{Q'}(\mathbf{r})$. These functions are given as the linear combinations

$$\mathrm{P}_Q(\mathbf{r}) = \sum_{QQ'} \underline{A}_{QQ'}\tilde{\mathrm{P}}_{Q'}(\mathbf{r}), \tag{23.140}$$

with

$$\underline{A} = [\underline{C} - \mathrm{i}\underline{S}]^{-1}, \tag{23.141}$$

where the underlined quantities appearing in the last expression are to be understood as being matrices in x, y, z (real), as well as in Q (angular momentum) space[17]. The cell t-matrix $\underline{t}_{QQ'}$ is given by the expression,

$$\underline{t} = [\underline{C} - \mathrm{i}\underline{S}]^{-1}\,\underline{S}, \tag{23.142}$$

[17] For the sake of ease of notation, we avoid an explicit designation of supermatrices.

which is to be compared with its "scalar" counterpart, Eq. (D.107). This cell t-matrix is also given by the expression,

$$\underline{t}_{QQ'} = \int_{\Omega} \mathrm{d}^3 r \, \mathtt{J}_Q(\mathbf{r}) \underline{f}(\mathbf{r}) \mathtt{P}^*_{Q}(\mathbf{r}), \tag{23.143}$$

which is analogous to Eq. (23.139), but in terms of the functions defined following that equation. Explicit expressions for the t-matrices and phase functions in some special cases, such as that of a sphere of radius characterized by a constant dielectric function, are given in the literature [5, 7].

As was the case with the scalar wave equation, the phase functions $\underline{C}$ and $\underline{S}$ can also be written [8] in terms of surface (rather than volume) integrals over any surface bounding the volume Ω. In Eqs. (23.138), (23.139), and (23.143), we replace $\underline{f}(\mathbf{r})$ by $\nabla \times \left[\nabla \times -k^2\right] \mathbf{E}(\mathbf{r})$, integrate by parts, and use the fact that the vector solid harmonics satisfy the free-particle vector wave equation to obtain the expressions,

$$\underline{C}_{QQ'} = k \left[\mathtt{H}_Q, \mathtt{P}^*_{Q'}\right]_S, \tag{23.144}$$

$$\underline{S}_{QQ'} = -k \left[\mathtt{J}_Q, \mathtt{P}^*_{Q'}\right]_S, \tag{23.145}$$

and

$$\underline{t}_{QQ'} = k \left[\mathtt{H}_Q, \mathtt{P}^*_{Q'}\right]_S, \tag{23.146}$$

where the Wronskian-type integrals, $[\mathtt{A}, \mathtt{B}^*]_S$, are defined by the expression,

$$[\mathtt{A}, \mathtt{B}^*]_S = \int_S \mathrm{d}\mathbf{S} \cdot [\mathtt{A} \times (\nabla \times \mathtt{B}^*) + (\nabla \times \mathtt{A}) \times \mathtt{B}^*]. \tag{23.147}$$

The reader may wish to verify the following relations,

$$\left[\mathtt{Z}_Q, \mathtt{Z}^*_{Q'}\right]_S = 0, \quad \text{when} \quad Z = J \text{ or } H, \tag{23.148}$$

$$\left[\mathtt{J}_Q, \mathtt{H}^*_{Q'}\right]_S = \underline{1} \delta_{QQ'}, \tag{23.149}$$

$$\left[\mathtt{H}_Q, \mathtt{J}^*_{Q'}\right]_S = \underline{1} - \delta_{QQ'}, \tag{23.150}$$

for any closed surface S.

23.6 Multiple Scattering Expressions

The multiple scattering expressions describing the propagation of electromagnetic waves through matter can be obtained in essentially identical fashion with the case of the scalar wave equation. With the generalization of scalar quantities to matrices, the various expressions assume identical forms in the scalar and vector case.

We consider the total solution, $\mathbf{E}(\mathbf{r})$, associated with a collection of "scattering" cells, Ω_n, each of which is characterized by a "potential" $\underline{f}^n(\mathbf{r})$. A solution that satisfies the proper boundary conditions at infinity can be written in the form,

$$\mathbf{E}(\mathbf{r}) = \sum_Q \underline{A}_Q^n \mathrm{P}_Q^n(\mathbf{r}), \tag{23.151}$$

for the vector $\mathbf{r}$ in cell Ω_n. The coefficients, $\underline{A}_Q^n$, can be obtained as the solutions of a secular equation that can be derived through the same process used in the derivation of Eqs. (3.55) or (3.61). In a notation that should be obvious, the secular equation can be written in the form,

$$\det\left|(\underline{t}^n)^{-1}\delta_{nn'} - \underline{G}(\mathbf{R}_{nn'})\right| = 0, \tag{23.152}$$

where the structure constants $\underline{G}(\mathbf{R}_{nn'})$ are defined in Eq. (23.132). The system Green function for vectors $\mathbf{r}$ and $\mathbf{r}'$ in no other cells than Ω_n and $\Omega_{n'}$, respectively, is given by the expression,

$$\begin{aligned} \underline{G}(\mathbf{r},\mathbf{r}';E) &= \sum_{QQ'} \mathrm{Z}_Q^n(\mathbf{r})\underline{\tau}_{QQ'}^{nn'}(E)\mathrm{Z}^{n'*}(\mathbf{r}') - \sum_Q \mathrm{Z}_Q^n(\mathbf{r})\mathrm{S}_Q^{n*}(\mathbf{r}')\delta_{nn'} \\ &= \langle \mathrm{Z}^n(\mathbf{r})\left|\underline{\tau}^{nn'}(E)\right|\mathrm{Z}^{n'*}(\mathbf{r})\rangle - \langle \mathrm{Z}^n(\mathbf{r})|\mathrm{S}^{n*}(\mathbf{r}')\rangle\delta_{nn'}. \end{aligned} \tag{23.153}$$

Here, the functions $\mathrm{Z}_Q^n(\mathbf{r})$ are those regular solutions of the vector wave equation associated with cell Ω_n, which at the surface of a sphere bounding the cell join smoothly to the function $\mathrm{J}_Q(\mathbf{r})+\sum_{Q'}\underline{t}_{QQ'}^n\mathrm{H}_{Q'}^*(\mathbf{r})$. The irregular solutions, $\mathrm{S}_Q^n(\mathbf{r})$, join smoothly to $\mathrm{J}_Q^n(\mathbf{r}')$ on the surface of a bounding sphere. Finally, the scattering-path matrix, $\underline{\tau}$, is given as the inverse of the matrix

$$\underline{M}_{nn'} = (\underline{t}^n)^{-1} - \underline{G}(\mathbf{R}_{nn'}), \tag{23.154}$$

in exact analogy with the case of the scalar wave equation.

For a periodic structure, the solutions of the secular Eq. (23.152) in reciprocal space determine the *phonic* band structure of a material, that is to say the dispersion relation, $\omega(\mathbf{k})$, governing the propagation of radiation through matter. The results of such a calculation [8] for the case of touching vacuum spheres arranged on a diamond lattice in a material with a constant dielectric function equal to 12.96 are shown in Fig. 23.1.

23.7 Disorder and the CPA

It is not difficult to see how the formalism of MST describing the propagation of radiation through a periodic solid outlined above can be generalized

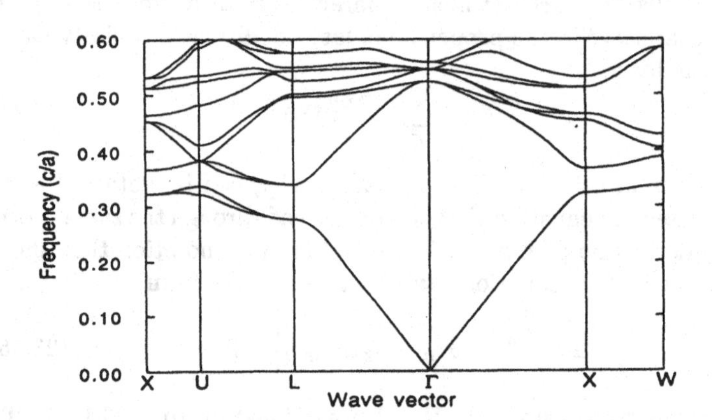

Figure 23.1: **Calculated phonic band structure for diamond lattice as described in the text. The frequency is given in units c/a, where a is the lattice constant of the diamond structure, and c is the speed of light in vacuum. See reference [8].**

to treat the case of substitutionally disordered alloys. The various expressions involved in the CPA formalism discussed in Chapter 12 have exact counterparts in the case of electromagnetic waves, when scalar quantities are appropriately interpreted as matrices, as the reader is urged to verify.

Bibliography

[1] William Jones and Norman H. March, *Theoretical Solid State Physics*, Dover Publications, Inc., New York (1985).

[2] J. Lindhart, Kgl. Danske Mat. Fys. Medd. **28**, 8 105 (1954).

[3] M. Gell-Mann and K. A. Bruckner, Phys. Rev. **106**, 369 (1957).

[4] J. M. Ziman, *Principles of the Theory of Solids*, Cambridge University Press, Cambridge, New York and Port Chester (1979).

[5] Roger G. Newton, *Scattering Theory of Waves and Particles*, Springer Verlag, New York (1982).

[6] John David Jackson, *Classical Electrodynamics*, John Wiley and Sons, New York (1966).

[7] Karamjeet Arya, Zhao-Bin Su, and Joseph Birman, in *Scattering and Localization of Classical Waves in Random Media*, P. Sheng, (ed.)World Scientific, Singapore (1990).

[8] Xindong Wang, X.-G. Zhang, Qingliang Yu, and B. N. Harmon, Phys. Rev. B **47**, 4161 1993-II.

Part VII

Formal Green-Function Theory

Chapter 24

Formal Theory of Green Functions

24.1 General Comments

The material presented in previous parts of this book was aimed at the derivation of computable formulae for the calculation of various properties of solids. The requirement of computability was met by working almost exclusively within the specific representation of angular momentum eigenstates (L–representation). Within this representation, we used the framework of multiple scattering theory to develop formalism that allows the determination of the single-particle Green function appropriate for the calculation of various materials properties.

The emphasis given to the single-particle Green function, and the use of a specific formal scheme, such as MST, and of a given representation, however, may obscure some of the more general features and properties of Green functions, and also hide much of their potential power and usefulness. For example, in a fully interacting many-particle system, an interconnected set of many-particle Green functions can be defined. The resulting equations form the basis for various attempts to account for the effects of interparticle interactions in a many-body system.

Having presented several specific examples of the use of single-particle Green functions, we now turn to a review of some important features of the general theory. It is hoped that the general discussion thats follows may be more accessible after the long discussion of specific applications than if it had been given at the beginning. At the same time, we will continue to confine our discussion to the non-relativistic case and to zero temperature. Reference to time-temperature Green functions is made but only in passing.

Also, initially we work in the time domain in which we derive a number of important relations involving the Green functions of an interacting system. The passage to the energy domain, within which the formalism in previous pages was developed, is indicated subsequently.

The discussion of Green functions is arranged as follows. We begin with a general definition of Green functions and derive the equations satisfied by the many-particle Green functions of an interacting system within the time domain. Then, we specifically discuss some of the important characteristics and properties of single-particle and two-particle Green functions and their spectral representations. We show how the energy of a system can be written in terms of these Green functions, and indicate the main approximations that form the basis for some well-known methods introduced into the study of interacting many-particle systems. Subsequently, we illustrate the use of Fourier transforms that allow us to pass into the energy domain and derive explicit expressions for the single-particle and two-particle Green functions in free space. These energy Green functions form the starting point of the developments that were presented in the previous parts of this work.

In most of our discussion, we follow the lines of development in Kato et al. [1], which, in turn, is based on the original work of Feynman [2, 3] and Schwinger [4]. This development relies for the most part on the concept of functional differentiation discussed in Appendix A, which the reader may wish to review at this point.

24.2 Definitions of Green Functions

Possibly the simplest example of a Green function is provided by the elementary solution of the equation,

$$Du = 0, \tag{24.1}$$

where D is a known operator, such as the differential operator, ∇^2. In *classical* field theories, a Green function denotes the contribution to the field strength at a given point from a unit source at another point. Let us consider a static electrostatic charge density, $\rho(\mathbf{r})$, and the corresponding potential, $V(\mathbf{r})$, related to it by the Poisson equation,

$$\nabla^2 V(\mathbf{r}) = 4\pi\rho(\mathbf{r}). \tag{24.2}$$

As shown in Eqs. (A.15) and (A.16), the corresponding *Green function*,

$$G(\mathbf{r}-\mathbf{r}') = \frac{1}{|\mathbf{r}-\mathbf{r}'|}, \tag{24.3}$$

is the potential at point $\mathbf{r}$ produced by a unit point charge at $\mathbf{r}'$. Therefore, it is the solution of the corresponding Poisson equation,

$$\nabla^2 G(\mathbf{r}-\mathbf{r}') = -4\pi\delta(\mathbf{r}-\mathbf{r}'). \tag{24.4}$$

In analogy with Eq. (24.2), the Green function $G(\mathbf{r}-\mathbf{r}')$ is the potential due to the charge distribution $\delta(\mathbf{r}-\mathbf{r}')$. Also, because of the homogeneity of free space, the Green function depends only on the distance $|\mathbf{r}-\mathbf{r}'|$ and consequently we have,

$$(\nabla + \nabla')\, G(\mathbf{r}-\mathbf{r}') = 0, \tag{24.5}$$

where a prime denotes differentiation with respect to $\mathbf{r}'$.

24.2.1 Classical field theories

Somewhat more generally, let us consider a scalar field, $\phi(x)$, where x represents the space-time point $(\mathbf{r}, t)$. We suppose that the field is associated with the presence of a source function, $j(x)$, to which it is connected by means of the field equation,

$$D_x \phi(x) = j(x), \tag{24.6}$$

where D_x is a linear operator. Under the condition that $\phi(x)$ vanishes as $j(x)$ tends to zero, this equation has the particular solution[1],

$$\phi(x) = \int G(x, x') j(x') \mathrm{d}^4 x', \tag{24.7}$$

where $\mathrm{d}^4x = \mathrm{d}^3 r \mathrm{d}t$. The kernel, $G(x, x')$, of this integral equation is commonly referred to as the *Green function* or the *propagator*. Because of the linearity of D_x, the Green function is independent of the source, $j(x)$, and in analogy with Eq. (A.11) can be written as the functional derivative,

$$G(x, x') = \frac{\delta \phi(x)}{\delta j(x')}. \tag{24.8}$$

Using Eqs. (A.6), (24.6), and (24.8), we obtain the defining equation for $G(x, x')$,

$$D_x G(x, x') = \delta^{(4)}(x - x'), \tag{24.9}$$

where $\delta^{(4)}(x - x') = \delta(\mathbf{r} - \mathbf{r}')\delta(t - t')$.

The last equation should be compared with Eq. (24.4). It follows that $G(x, x')$ can be interpreted as the field at x contributed by a unit source at x'. Therefore, the original Eq. (24.6) has been reduced to solving Eq. (24.9) under the appropriate boundary conditions[2].

---

[1] Otherwise the solution contains a term $\phi_0(x)$.

[2] For example, in the present case, the boundary conditions are such that the field vanishes as the source term goes to zero. In the specific case of the Poisson equation, this is realized by demanding that the potential goes to zero at infinity where it is stipulated that no sources are present.

24.2.2 Non-linear operators

Green functions can also be defined in connection with non-linear field theories by means of a limiting process. Considering the case in which $\phi(x)$ vanishes as $j(x)$ tends to zero, we can write (provided that the various derivatives are well behaved),

$$\begin{aligned}\phi(x) &= \int \left[\frac{\delta\phi(x)}{\delta j(x')}\right] \mathrm{d}^4x' \\ &+ \frac{1}{2!}\int\int \left[\frac{\delta^2\phi(x)}{\delta j(x')j(x'')}\right] \mathrm{d}^4x'\mathrm{d}^4x'' + \cdots .\end{aligned} \tag{24.10}$$

Therefore, a simple linear relation such as that in Eq. (24.7) between source and field holds only for an infinitesimally small source, $\delta j(x)$, so that we can write,

$$\phi(x) = \int G(x,x')\delta j(x')\mathrm{d}^4x', \tag{24.11}$$

where we have defined the Green function as the limit,

$$G(x,x') = \lim_{j\to 0} \frac{\delta\phi(x)}{\delta j(x')}. \tag{24.12}$$

Generally, the Green function obeys the complicated equation,

$$\hat{D}_x G(x,x') = \delta^{(4)}(x-x'). \tag{24.13}$$

The operator $\hat{D}_x$, which is not necessarily equal to D_x, is defined by the expression,

$$\hat{D}_x G(x,x') = \lim_{j\to 0} \frac{\delta}{\delta j(x')} D_x[\phi] = \int \left[\frac{\delta D_x[\phi]}{\delta\phi(x'')}\right] G(x'',x')\mathrm{d}^4\mathrm{x}''. \tag{24.14}$$

This complicated expression indicates that the operator $\hat{D}_x$ and the equation for G cannot be constructed until the full solution $\phi(x)$ is known. This necessitates the use of approximate, iterative solutions for determining the Green function.

24.2.3 Quantum fields

We now return to linear operators[3]. In addition to classical field theories, Green functions can also be introduced into second-quantized theories of

[3]On occasion, results from the non-linear case will be quoted without specific comment.

many-particle systems. We consider fields described by the operator function, $\phi(x)$, obeying the equation,

$$D_x\phi(x) = j(x), \tag{24.15}$$

where D_x is a linear operator and $j(x)$ is a c-number function representing an external source distribution. The notion of second quantization and field operators is introduced in Appendix S, which the reader may wish to review before proceeding further. In analogy with classical theories, we wish to obtain the solution of Eq. (24.15) by means of a c-number Green function. This requires the evaluation of the (c-number) expectation values, $\langle\phi(x)\rangle$, of field operators such as $\phi(x)$. To accomplish this, we consider the expectation value of an operator, Q,

$$\langle Q\rangle = \langle m|Q|m\rangle, \tag{24.16}$$

where $|m\rangle$ denotes the state of the medium. When the medium is in a mixed state described by a density matrix, ρ, we must replace the last expression with,

$$\langle Q\rangle = \mathrm{Tr}\{\rho Q\}/\mathrm{Tr}\{\rho\}, \tag{24.17}$$

where

$$\rho = \sum_m |m\rangle w_m\langle m|, \tag{24.18}$$

with w_m being the weight factor of eigenstate $|m\rangle$ in the mixed state of the medium[4].

We now consider the expectation value $\langle\phi(x)\rangle$, based on the definition in Eqs. (24.16) or (24.17). Clearly, $\phi(x)$ is non-vanishing only in the presence of the source, $j(x)$, because when $j(x) = 0$, the operator $\phi(x)$ decreases the number of particles by one so that $\langle m|\phi(x)|m\rangle = 0$. We may then attempt an expansion of the function $\langle\phi(x)\rangle$ in a series such as that occurring in Eq. (24.10),

$$\begin{aligned}\langle\phi(x)\rangle &= \int\left[\frac{\delta\phi(x)}{\delta j(x')}\right]\mathrm{d}^4x' \\ &+ \frac{1}{2!}\int\int\left[\frac{\delta^2\phi(x)}{\delta j(x')j(x'')}\right]\mathrm{d}^4x'\mathrm{d}^4x'' + \cdots.\end{aligned} \tag{24.19}$$

[4]In the case of time-dependent sources and fields, the description in terms of stationary eigenstates becomes much less meaningful. In this case, we can design the external sources so that they are independent of time at the beginning and end of the process. Denoting by $|m_\mathrm{I}\rangle$ and by $|m_\mathrm{F}\rangle$ the corresponding eigenstates of the total Hamiltonian, and assuming that the two types of states coincide as the external source tends to zero, we obtain an expression for the expectation value $\langle Q\rangle$ that has the usual form involving a density matrix, but in which the density matrix is given by $\rho = \sum_m |m_\mathrm{I}\rangle w_m\langle m_\mathrm{F}|$.

In the linear case, only the first term on the right of the equals sign is non-vanishing. In general, we can write,

$$\langle\phi(x)\rangle = \int G(x,x')\delta j(x')\mathrm{d}^4x', \tag{24.20}$$

for infinitesimally small fields, $\delta j(x)$, where the green function is defined as the limit,

$$G(x,x') \equiv \lim_{j\to 0}\frac{\delta\langle\phi(x)\rangle}{\delta j(x')}. \tag{24.21}$$

The last expression provides the basic definition of the ***one-particle Green function*** or ***propagator*** describing single-particle (or hole) propagation in the medium characterizing the interacting system. Higher-order propagation involving two, three, . . ., particles can also be described in terms of corresponding Green functions, as is discussed in the following subsection. In general, these Green functions satisfy sets of interconnected equations that couple propagators of a given order to those of order immediately above and below it. Like Eq. (24.14), these complicated equations in principle may not be constructed until the solution to the entire problem is fully known. Therefore, the study of Green functions usually proceeds on the basis of approximations and iterative procedures that often provide great insight into the physical properties of a system.

24.2.4 Bosons and fermions

We now specialize the general discussion presented above to the distinct cases of the two kinds of particles that characterize the interacting systems of interest to us, bosons and fermions. Denoting the boson field by $\phi(x)$ and the fermion field by $\psi_\alpha(x)$, where α is a spin index[5], and letting $j(x)$ and $\eta_\alpha(x)$ denote the external sources for bosons and fermions, respectively, we can define the one-boson Green function, $K(x,x')$, and the one-fermion Green function, $G_{\alpha\beta}(x,x')$, by means of the expressions[6],

$$K(x,x') \equiv \lim_{j,\eta\to 0}\frac{\delta\langle\phi(x)\rangle}{\delta j(x')}, \tag{24.22}$$

and

$$G_{\alpha\beta}(x,x') \equiv \lim_{j,\eta\to 0}\frac{\delta\langle\psi_\alpha(x)\rangle}{\delta\eta_\beta(x')}. \tag{24.23}$$

[5]For the sake of simplicity of notation, spin indices will often be suppressed without warning.

[6]The external sources can be introduced through the addition of the term

$$-\phi(x)j(x) - \sum_\alpha \psi_\alpha^*(x)\eta_\alpha(x) - \sum_\alpha \eta_\alpha^*(x)\psi_\alpha(x)$$

to the Lagrangian density of the system. See also Appendix T.

Now, using Eq. (T.13), the functional derivatives in these expressions take the forms,

$$\frac{\delta\langle\phi(x)\rangle}{\delta j(x')} = \frac{1}{\mathrm{i}\hbar}\{\langle T\,[\phi(x)\phi(x')]\rangle - \langle\phi(x)\rangle\langle\phi(x')\rangle\}, \tag{24.24}$$

and

$$\frac{\delta\langle\psi_\alpha(x)\rangle}{\delta\eta_\beta(x')} = \frac{1}{\mathrm{i}\hbar}\{\langle T\,[\psi_\alpha(x)\psi_\beta(x')]\rangle - \langle\psi_\alpha(x)\rangle\langle\psi_\beta(x')\rangle\}, \tag{24.25}$$

from which follow the explicit definitions,

$$K(x,x') = \frac{1}{\mathrm{i}\hbar}\langle T\,[\phi(x)\phi(x')]\rangle, \tag{24.26}$$

and

$$G_{\alpha\beta}(x,x') = \frac{1}{\mathrm{i}\hbar}\langle T\,[\psi_\alpha(x)\psi^*_\beta(x')]\rangle, \tag{24.27}$$

where T denotes the time-ordering operator that arranges time arguments from earlier to later proceeding to the left in a product of operators. This guarantees the proper sequence of action of operators on states since operators at earlier times must necessarily act before those at later times.

Let us consider somewhat more closely the significance of the time-ordered products of operators. We recall that the symbol $\langle\cdots\rangle$ must be interpreted in the sense of Eq. (24.16) or (24.17) and consider the case of the one-fermion Green function[7] with reference to a pure state $|m\rangle$,

$$G(x,x') = \frac{1}{\mathrm{i}\hbar}\langle m|T\,(\psi(x)\psi^*(x'))\,|m\rangle. \tag{24.28}$$

When $t > t'$, (i.e., when t' is an earlier time than t), t' appears to the right of t in the operator product so that the expression,

$$G(x,x') = \frac{1}{\mathrm{i}\hbar}\langle m|\psi(\mathbf{r},t)\psi^*(\mathbf{r}',t')|m\rangle. \tag{24.29}$$

can be interpreted as the amplitude of propagation of a particle from point $(\mathbf{r}',t')$, (created there by $\psi^*(\mathbf{r}',t')$), to point $(\mathbf{r},t)$, where it is annihilated by $\psi(\mathbf{r},t)$. On the other hand, when $t < t'$, the expression,

$$G(x,x') = -\frac{1}{\mathrm{i}\hbar}\langle m|\psi^*(\mathbf{r}',t')\psi(\mathbf{r},t)|m\rangle, \tag{24.30}$$

may be interpreted as describing the propagation of a hole from $\psi(\mathbf{r},t)$ to $(\mathbf{r}',t')$. Note the minus sign appearing in the last expression; it is the result of the anticommuting properties of fermion operators.

[7]With spin indices suppressed.

It must now be pointed out that in an interacting system, the notion of individual particles may be ill-defined so that the interpretations just given are characterized by a certain degree of crudeness. More appropriate interpretations can be given in terms of the amplitude of adding and removing a particle at x and x', respectively, without the constraint that a particle itself propagates through the medium. With this understanding, we may define higher-order Green functions for two, three, $\cdots$, n particles through the expressions,

$$\begin{aligned} G^{(2)}(x_1, x_2; x_1', x_2^prime) &\equiv \lim_{j,\eta\to 0} \frac{\delta^2 \langle T[\psi(x_1)\psi(x_2)]\rangle}{\delta\eta(x_1')\delta\eta(x_2')} \\ &= (\mathrm{i}\hbar)^{-2}\langle T\left[\psi(x_1)\psi(x_2)\psi^*(x_2')\psi^*(x_1')\right]\rangle, \end{aligned} \tag{24.31}$$

$$\begin{aligned} G^{(3)}(x_1, x_2, x_3; x_1', x_2^prime, x_3') &\equiv \lim_{j,\eta\to 0} \frac{\delta^3 \langle T[\psi(x_1)\psi(x_2)\psi(x_3)]\rangle}{\delta\eta(x_1')\delta\eta(x_2')\delta\eta(x_3')} \\ &= (\mathrm{i}\hbar)^{-3}\langle T\left[\psi(x_1)\psi(x_2)\psi(x_3)\psi^*(x_3')\psi^*(x_2')\psi^*(x_1')\right]\rangle, \end{aligned} \tag{24.32}$$

and

$$\begin{aligned} &G^{(n)}(x_1, x_2, \cdots, x_n; x_1', x_2^prime, \cdots, x_n') \\ &= (\mathrm{i}\hbar)^{-n}\langle T\left[\psi(x_1)\psi(x_2)\cdots\psi(x_n)\psi^*(x_n')\cdots\psi^*(x_2')\psi^*(x_1')\right]\rangle, \end{aligned} \tag{24.33}$$

for fermions, and

$$\begin{aligned} &K^{(n)}(x_1, x_2, \cdots, x_n; x_1', x_2^prime, \cdots, x_n') \\ &= (\mathrm{i}\hbar)^{-n}\langle T\left[\phi(x_1)\phi(x_2)\cdots\phi(x_n)\phi^*(x_n')\cdots\phi^*(x_2')\phi^*(x_1')\right]\rangle, \end{aligned} \tag{24.34}$$

for bosons. Various relationships among these Green functions of different orders are derived in the following section.

The definitions just presented allow us to make certain statements about the form of the dependence of the Green function on spatial or temporal variables depending on the degree of uniformity of the medium of propagation. Differentiation with respect to $\mathbf{r}$ yields the expression,

$$\nabla G(\mathbf{r}, t; \mathbf{r}', t') = -\nabla' G(\mathbf{r}, t; \mathbf{r}', t') + \hbar^{-2}\langle T\left[\psi(\mathbf{r}, t)\psi^*(\mathbf{r}', t')\right], P\rangle \tag{24.35}$$

where we have used the equation

$$\nabla\psi(\mathbf{r}, t) = \frac{\mathrm{i}}{\hbar}\left[\mathbf{P}, \psi(\mathbf{r}, \mathbf{t})\right], \tag{24.36}$$

with $\mathbf{P}$ being the total momentum operator,

$$\mathbf{P}(t) = \int \psi^*(\mathbf{r},t)\frac{\hbar}{\mathrm{i}}\nabla\psi(\mathbf{r},t)\mathrm{d}^3r. \tag{24.37}$$

If the medium is uniform in space or, equivalently, if $|m\rangle$ is an eigenstate of $\mathbf{P}$ so that the density matrix, ρ, commutes with $\mathbf{P}$, then the last term on the right in Eq. (24.35) vanishes and we obtain,

$$(\nabla + \nabla')G(\mathbf{r},t;\mathbf{r}',t') = 0. \tag{24.38}$$

This expression implies that in the presence of a spatially uniform medium the Green function depends on the variables $\mathbf{r}$ and $\mathbf{r}'$ only through the difference $\mathbf{r} - \mathbf{r}'$. An analogous result holds for the case of media uniform with respect to time. In general, however, the medium is not uniform, e.g., a solid, so that the Green function depends on individual coordinates separately. Any partial symmetry, such as the lattice translational invariance of a periodic solid will be reflected on the form of the spatial dependence of the Green function, as was indeed found to be the case throughout our discussion in previous parts of this book.

24.3 Green Function Equations

The Green functions associated with an interacting many-particle system satisfy a set of equations connecting Green functions of different orders. In this section, these equations are derived based on the fundamental definition of Green functions given in Eqs. (24.29) to (24.34).

24.3.1 Basic equations

We consider an interacting many-particle system characterized by an interparticle interaction, $U(x,x')$, which is assumed to be symmetric in the coordinates x and x', and to have the form, $U(x-x') = V(\mathbf{r}-\mathbf{r}')\delta(t-t')$. In the presence of an external (c-number) field, $\phi(x)$, the Hamiltonian of the system takes the form (see Appendix S),

$$\begin{aligned} H &= \int \psi^*(x)\left[-\frac{\hbar^2}{2m}\nabla^2 + \phi(x)\right]\psi(x)\mathrm{d}^3r \\ &+ \frac{1}{2}\int\int \psi^*(x)\psi^*(x')U(x-x')\psi(x')\psi(x)\mathrm{d}^4x'\mathrm{d}^3r, \end{aligned} \tag{24.39}$$

from which it is seen that in general the Hamiltonian is time dependent. The c-number function (external potential) $\phi(x)$ is a formal aid that allows a number of useful algebraic manipulations as is discussed below. Final expressions after such manipulations are obtained by setting $\phi(x)$ equal to zero.

It now follows from Heisenberg's equation of motion for an operator, A,

$$i\hbar\frac{\partial A}{\partial t} = [H, A], \tag{24.40}$$

that $\psi(x)$ obeys the equation

$$i\hbar\frac{\partial\psi(x)}{\partial t} = \left[-\frac{\hbar^2}{2m}\nabla^2 + \phi(x) + \int \psi^*(x')U(x-x')\psi(x')\mathrm{d}^4x'\right]\psi(x). \tag{24.41}$$

From Appendix S, we use the formula

$$i\hbar\frac{\delta\langle Q\rangle}{\delta\phi(x)} = \langle T[Q\psi^*(x)\psi(x)]\rangle - \langle Q\rangle\langle\psi^*(x)\psi(x)\rangle \tag{24.42}$$

to differentiate Eq. (24.39) with respect to t and obtain (with spin indices suppressed)

$$\frac{\partial G(x,x')}{\partial t} = (i\hbar)^{-1}\delta^{(4)}(x-x') + (i\hbar)^{-1}\langle T\left[\frac{\partial\psi(x)}{\partial t}\psi^*(x')\right]\rangle \tag{24.43}$$

where the delta function arises from the jump of G at its discontinuous point $t = t'$. Substituting Eq. (24.41) into the last expression, we obtain the equation,

$$\begin{aligned}\left[i\hbar\frac{\partial}{\partial t} + \frac{\hbar^2}{2m}\nabla^2 - \phi(x)\right]G(x,x') &= \delta^{(4)}(x-x') \\ &+ (i\hbar)^{-1}\int \mathrm{d}^4x''U(x-x'')\langle T\left[\psi^*(x'')\psi(x'')\psi(x)\psi^*(x')\right]\rangle,\end{aligned} \tag{24.44}$$

satisfied by the one-particle Green function. But from Eq. (24.31), the last term on the right member of this equation can be written in terms of the two-particle Green function, so that we obtain the expression,

$$\begin{aligned}\left[i\hbar\frac{\partial}{\partial t} + \frac{\hbar^2}{2m}\nabla^2 - \phi(x)\right]G(x,x') &= \delta^{(4)}(x-x') \\ &+ (i\hbar)^{-1}\int \mathrm{d}^4x''U(x-x'')\langle G^{(2)}(x,x'';x',x''),\end{aligned} \tag{24.45}$$

connecting the one-particle Green function to the two-particle Green function. Similarly, the equation for $G^{(2)}$ involves $G^{(1)}$ and $G^{(3)}$. In general, the equation for the N-particle Green function contains both the $N+1$– and the $N-1$–particle Green functions and, for $N \geq 2$, takes the form,

$$\begin{aligned}
&\left[i\hbar\frac{\partial}{\partial t_1} + \frac{\hbar^2}{2m}\nabla_1^2 - \phi(x_1)\right] G^{(N)}(x_1, x_2, \cdots, x_N; x_1', x_2', \cdots, x_N') \\
+ &\; i\hbar \int d^4x_{N+1} U(x - x_{N+1}) G^{(N+1)}(x_1, \cdots, x_{N+1}; x_1', \cdots, x_{N+1}') \\
= &\sum_{i=1}^{N} \delta^{(4)}(x_1 - x_i')(-1)^{i-1} G_{N-1}(x_2, \cdots, x_N; x_1', \cdots, x_{i-1}', x_{i+1}', \cdots, x_N').
\end{aligned} \tag{24.46}$$

Through functional differentiation with respect to $\phi(x)$, we can eliminate all higher order Green functions and obtain expressions in terms of $G^{(1)}$ alone. We now turn our attention to various equations satisfied by the one-particle and the two-particle Green functions.

24.3.2 One-particle Green functions

Let us denote the one-particle Green function in the presence of $\phi(x)$ by G_ϕ. Using Eq. (24.42), we can rewrite the last term in the right member of Eq. (24.44) as follows,

$$\begin{aligned}
&(i\hbar)^{-1} \int d^4x'' U(x - x'') \langle T\left[\psi^*(x'')\psi(x'')\psi(x)\psi^*(x')\right]\rangle \\
= &\; \bar{V}_\phi(x) G_\phi(x, x') + i\hbar \int d^4x'' U(x - x'') \frac{\delta}{\delta\phi(x'')} G_\phi(x, x'),
\end{aligned} \tag{24.47}$$

where we have defined the expression,

$$\begin{aligned}
\bar{V}_\phi(x) &\equiv \int d^4x'' U(x - x'') \langle \psi^*(x'')\psi(x'')\rangle \\
&= -i\hbar \int d^4x'' U(x - x'') \mathrm{tr} G_\phi(\mathbf{r}'', t''; \mathbf{r}'', t'' + 0),
\end{aligned} \tag{24.48}$$

in which tr denotes a trace over spinor indices, and $t'' + 0$ indicates a time infinitesimally larger (later) than t''. Now, the equation for G_ϕ takes the form,

$$\begin{aligned}
&\left[i\hbar\frac{\partial}{\partial t} + \frac{\hbar^2}{2m}\nabla^2 - \phi(x) - \bar{V}(x)\right. \\
- &\left. i\hbar \int d^4x'' U(x - x'') \frac{\delta}{\delta\phi(x'')}\right] G(x, x') = \delta^{(4)}(x - x'),
\end{aligned} \tag{24.49}$$

which includes only the one-particle Green function. Recalling that final expressions are to be obtained by setting $\phi(x)$ equal to zero, we define the expressions,

$$G(x,x') \equiv \lim_{\phi \to 0} G_\phi(x,x') \tag{24.50}$$

and

$$\begin{aligned} \bar{V}(x) &\equiv \lim_{\phi \to 0} = \int \mathrm{d}^3 r'' \langle m|\psi^*(\mathbf{r}'',t)\psi(\mathbf{r}'',t)|m\rangle V(\mathbf{r}-\mathbf{r}'') \\ &= \int \mathrm{d}^3 r'' \frac{\mathrm{Tr}[\rho\psi^*(\mathbf{r}'',t)\psi(\mathbf{r}'',t)]}{\mathrm{Tr}\rho} V(\mathbf{r}-\mathbf{r}''). \end{aligned} \tag{24.51}$$

The last expression can be interpreted as the *average, static potential* for a particle moving in the medium containing all other particles. If we take the state $|m\rangle$ to be an eigenstate of H, the function $\bar{V}(x) = \bar{V}(\mathbf{r})$ is independent of time. Furthermore, if G is uniform in space or, equivalently, the state $|m\rangle$ is an eigenstate of the momentum operator, $\bar{V}(\mathbf{r})$ is reduced to a constant independent of $\mathbf{r}$. The deviation of $\bar{V}(\mathbf{r})$ from a constant independent of $\mathbf{r}$ is a reflection of the lack of uniformity of the medium. We may recall the structure of the effective field, $v_{\mathrm{eff}}(\mathbf{r})$, obtained in LDA-based methods for calculating the electronic structure of solids. This effective potential is an approximate description of $\bar{V}(\mathbf{r})$ and in the case of a translationally invariant material possesses the spatial periodicity of the underlying lattice.

In the limit $\phi \to 0$, the term involving the functional derivative in Eq. (24.49) represents the *exchange effects* between the particle in consideratio-nand the medium particles with which it interacts. It can also be interpreted as representing pair excitation effects (since it stands in the place of the two-particle Green function), of the medium caused by the motion of the particle because the application of the operator $\delta/\delta\phi(x)$ implies the creation of a particle-hole pair at x. Thus, functional derivatives with respect to $\phi(x)$ allow us formally to eliminate the many-particle Green functions from the set of equations where they represent all possible excitations of the medium.

It is now useful to define the following quantities,

$$\int \mathrm{d}^4 x'' G_\phi^{-1}(x,x'') G_\phi(x'',x') = \delta^{(4)}(x-x') \tag{24.52}$$

$$\int \mathrm{d}^4 x'' G_\phi(x,x'') G_\phi^{-1}(x'',x') = \delta^{(4)}(x-x') \tag{24.53}$$

$$\langle\phi(x)\rangle \equiv \phi(x) + \bar{V}_\phi(x) - \bar{V}(x) \tag{24.54}$$

$$\Delta_\phi(x,x') \equiv \int \mathrm{d}^4 x'' U(x-x'') \frac{\delta\langle\phi(x')\rangle}{\delta\phi(x'')}, \tag{24.55}$$

in terms of which Eq. (24.49) takes the form,

$$\begin{aligned}&\left[i\hbar\frac{\partial}{\partial t}+\frac{\hbar^2}{2m}\nabla^2-\langle\phi(x)\rangle-\bar{V}(x)\right.\\ &-\left. i\hbar\int d^4x''\Delta(x,x'')\frac{\delta}{\delta\phi(x'')}\right]G(x,x')=\delta^{(4)}(x-x').\end{aligned} \tag{24.56}$$

The quantity G_ϕ^{-1} is the inverse Green function which applied to G_ϕ gives unity (in a delta-function sense). The functions $\langle\phi(x)\rangle$ and $\Delta(x,x')$ can, respectively, be interpreted as the external field and the two-particle potential modified by the interaction with the medium. The quantity $\delta\langle\phi(x')\rangle/\delta\phi(x'')$ denotes what is commonly referred to as the *inverse dielectric function* of the system of interacting particles.

Differentiation of either of the first two equations in Eq. (24.55) with respect to $\langle\phi(x)\rangle$ yields the expression,

$$\begin{aligned}\frac{\delta G_\phi(x,x')}{\delta\langle\phi(x')\rangle} &= -\int d^4\zeta\int d^4\xi G_\phi(x,\xi)\frac{\delta G_\phi^{-1}(\xi,\zeta)}{\delta\langle\phi(x'')\rangle}G_\phi(\zeta,x')\\ &= \int d^4\zeta\int d^4\xi G_\phi(x,\xi)\Gamma(\xi,zeta;x'')G_\phi(\zeta,x'),\end{aligned} \tag{24.57}$$

where we have defined the *vertex function*,

$$\Gamma(\xi,zeta;x'')=-\frac{\delta G_\phi^{-1}(\xi,zeta)}{\delta\langle\phi(x'')\rangle}. \tag{24.58}$$

Using Eq. (24.57) to replace the functional derivative in Eq. (24.56), we obtain the following form for the inverse Green function,

$$\begin{aligned}G_\phi^{-1}(x,x') &= \left[i\hbar\frac{\partial}{\partial t}+\frac{\hbar^2}{2m}\nabla^2-\langle\phi(x)\rangle-\bar{V}(x)\right]\delta^{(4)}(x-x')\\ &- \Pi_\phi(x,x'),\end{aligned} \tag{24.59}$$

with the definition

$$\Pi_\phi(x,x')\equiv i\hbar\int d^4x''\int d^4\xi\Delta_\phi(x,x'')G_\phi(x,\xi)\Gamma_\phi(\xi,x';x''). \tag{24.60}$$

Defining the *self-energy* $\Sigma_\phi(x,x')$ through the expression,

$$\Sigma_\phi(x,x')\equiv\bar{V}(x)\delta^{(4)}(x-x')+\Pi_\phi(x,x'), \tag{24.61}$$

we can rewrite $G_\phi^{-1}(x,x')$ in the form,

$$G_\phi^{-1}(x,x') = \left[i\hbar\frac{\partial}{\partial t} + \frac{\hbar^2}{2m}\nabla^2 - \langle\phi(x)\rangle \right] \delta^{(4)}(x-x') - \Sigma_\phi(x,x'). \quad (24.62)$$

From this, we obtain the relation,

$$\Gamma_\phi(\xi,\zeta;x'') = \delta^{(4)}(\xi-\zeta)\delta^{(4)}(\xi-x'') + \frac{\delta\Sigma_\phi(\xi,\zeta)}{\delta\langle\phi(x'')\rangle}. \quad (24.63)$$

Also, from Eqs. (24.48) and (24.55), we obtain the expressions,

$$\begin{aligned} \Delta_\phi(x,x') &= U(x-x') - (\mathrm{i})\hbar \mathrm{tr} \int\int\int\int \mathrm{d}^4x''\mathrm{d}^4\xi'\mathrm{d}^4\xi''\mathrm{d}^4\zeta' \\ &\times \Gamma_\phi(\xi',\xi'';\zeta')G_\phi(\xi'')G_\phi(x'',\xi')U(x''-x'), \end{aligned} \quad (24.64)$$

and

$$\begin{aligned} \langle\phi(x)\rangle &= \phi(x) - i\hbar \int \mathrm{d}^4x''U(x-x'') \\ &\times \mathrm{tr}\left[G_\phi(\mathbf{r}'',t'';\mathbf{r}'',t''+0) - G(\mathbf{r}'',t'';\mathbf{r}'',t''+0)\right], \end{aligned} \quad (24.65)$$

where the presence of G_ϕ and G inside the brackets in the last expression is to be noted. Furthemore, introducing the *polarization* term,

$$P_\phi(\zeta',x'') \equiv (i\hbar)\mathrm{tr}\int\int \mathrm{d}^4\xi''\mathrm{d}^4\xi'\Gamma_\phi(\xi',\xi'';\zeta')G_\phi(\xi'',x'')G_\phi(x'',\xi'), \quad (24.66)$$

we can cast Eq. (24.64) into the form,

$$\Delta_\phi(x,x') = U(x-x') - \int\int \mathrm{d}^4\zeta'\mathrm{d}^4x''\Delta(x,\zeta')P_\phi(\zeta',x'')U(x''-x'). \quad (24.67)$$

We have now formulated a set of six equations in the six unknowns, G_ϕ, Δ_ϕ, Σ_ϕ, Γ_ϕ, P_ϕ, and $\langle\phi\rangle$, which can be written in the matrix form,

$$\left[i\hbar\frac{\partial}{\partial t} + \frac{\hbar^2}{2m}\nabla^2 - \langle\phi\rangle - \Sigma_\phi \right] G_\phi = 1, \quad (24.68)$$

$$\Sigma_\phi = -i\hbar UG + i\hbar\Delta_\phi G_\phi \Gamma_\phi \quad (24.69)$$

$$\Delta_\phi = i\hbar U - \Delta_\phi P_\phi U, \quad (24.70)$$

$$P_\phi = i\hbar\Gamma_\phi G_\phi G_\phi, \quad (24.71)$$

$$\Gamma_\phi = 1 + \frac{\delta\Sigma}{\delta\langle\phi\rangle}, \tag{24.72}$$

and

$$\langle\phi\rangle = \phi - \mathrm{i}\hbar U(G_\phi - G). \tag{24.73}$$

All physical quantities needed to describe an interacting particle system are obtained from their ϕ-dependent counterparts in the limit $\phi \to 0$. Denoting any function (or operator) at that limit by the same symbol but without a subscript, e.g., Eq. (24.50), we can write the following expression for the single-particle Green function,

$$\left[\mathrm{i}\hbar\frac{\partial}{\partial t} + \frac{\hbar^2}{2m}\nabla^2\right] G(x,x') - \int \mathrm{d}^4x''\Sigma(x,x'')G(x'',x') = \delta^{(4)}(x-x'), \tag{24.74}$$

where

$$\Sigma(x,x') = \bar{V}(x)\delta^{(4)}(x-x') + \Pi(x,x'), \tag{24.75}$$

and

$$\Pi(x,x') = \mathrm{i}\hbar\int\int \mathrm{d}^4x''\mathrm{d}^4\xi\Delta(x,x')G(x,\xi)\Gamma(\xi,x';x''). \tag{24.76}$$

Now, denoting the inverse of $\left[\mathrm{i}\hbar\frac{\partial}{\partial t} + \frac{\hbar^2}{2m}\nabla^2\right]$ by G_0, we can write the integral form of Eq. (24.74) in the matrix form,

$$G = G_0 + G_0\Sigma G. \tag{24.77}$$

It is to be noted that Eq. (24.74) is exact, leading to the exact single-particle Green function for the interacting many-particle system. However, because of its time dependence (energy dependence in an energy picture), the determination of the exact self-energy, $\Sigma(x,x')$, presents an immensely complex problem that allows only approximate solutions in most cases.

24.3.3 Uniform medium

The case of the uniform medium deserves special consideration. In this case, all quantities introduced above are independent of the origin of the coordinate system, depending on the differences among space-time variables. For example, we have,

$$\begin{aligned} G(x,x') &= G(x-x') \quad \Delta(x,x') = \Delta(x-x') \\ \Sigma(x,x') &= \Sigma(x-x') \quad \Gamma(\xi,\zeta;x'') = \Gamma(\xi-\zeta,\xi-x''), \end{aligned} \tag{24.78}$$

and so on. In this case, Eqs. (24.68) to (24.73) can be simplified through Fourier transformation into the momentum representation.

Because we wish to treat space and time variables in equivalent fashion, we define a momentum four vector $\hbar p = (\mathbf{p}, p_0)$ such that the scalar product px takes the form,

$$\begin{aligned} px &= (\mathbf{p}\cdot\mathbf{r} - p_0 t)/\hbar \\ &= (\mathbf{P}\cdot\mathbf{r} - Et)/\hbar. \end{aligned} \tag{24.79}$$

From the general definition,

$$G(p,p') = \int\int \mathrm{d}^4x\mathrm{d}^4x' \mathrm{e}^{-\mathrm{i}px} G(x,x') \mathrm{e}^{\mathrm{i}p'x'}, \tag{24.80}$$

and the uniformity of the medium we obtain,

$$G(p,p') = (2\pi)^4 G(p)\delta^{(4)}(p-p'), \tag{24.81}$$

where

$$G(p) = \int \mathrm{d}^4(x-x')\mathrm{e}^{-\mathrm{i}p(x-x')} G(x-x') \tag{24.82}$$

Similarly, we have,

$$\Delta(p) = \int \mathrm{d}^4(x-x')\mathrm{e}^{-\mathrm{i}p(x-x')} \Delta(x-x') \tag{24.83}$$

$$\Sigma(p) = \int \mathrm{d}^4(x-x')\mathrm{e}^{-\mathrm{i}p(x-x')} \Sigma(x-x') \tag{24.84}$$

and

$$\begin{aligned} \Gamma(p',p;k) &= \int\int\int \mathrm{d}^4\xi\mathrm{d}^4\zeta\mathrm{d}^4x'' \mathrm{e}^{-\mathrm{i}p'\xi + \mathrm{i}p\zeta - \mathrm{i}kx''} \Gamma(\xi - zeta, \xi - x'') \\ &\times\ \mathrm{e}^{\mathrm{i}p'x'} = (2\pi)^4 \Gamma(p,k)\delta^{(4)}(p-p'+k) \end{aligned} \tag{24.85}$$

where

$$\Gamma(p,k) = \int\int \mathrm{d}^4X\mathrm{d}^4Y \mathrm{e}^{-\mathrm{i}pX + \mathrm{i}kY} \Gamma(X;Y). \tag{24.86}$$

The inverse transformations can be used to obtain the space-time dependence of various functions from the momentum-energy quantities. For example,

$$G(x-x') = \int \mathrm{d}^4p \mathrm{e}^{\mathrm{i}p(x-x')} G(p), \tag{24.87}$$

and

$$\Gamma(X,Y) \equiv \Gamma(\xi-\zeta;\xi-x'') = \int\int \mathrm{d}^4p\mathrm{d}^4k \mathrm{e}^{\mathrm{i}pX - \mathrm{i}kY} \Gamma(X;Y). \tag{24.88}$$

We can now express the set of six equations, Eqs. (24.68) to (24.73), in the momentum representation. The equation for the single-particle Green function takes the form, ($\hbar\omega = p_0$),

$$\left[\omega - \frac{P^2}{2m} - \Sigma(p)\right] G(p) = 1, \tag{24.89}$$

which can be solved formally to yield,

$$G(p) = \left[\omega - \frac{P^2}{2m} - \Sigma(p)\right]^{-1}. \tag{24.90}$$

The self-energy has the form,

$$\Sigma(p) = \bar{V} + \Pi(p), \tag{24.91}$$

where, because of the uniformity of the medium, $\bar{V}$ is a constant and

$$\Pi(p) = \frac{i\hbar}{(2\pi)^4} \int d^4k \Delta(k) G(p-k) \Gamma(p;k). \tag{24.92}$$

It follows from Eqs. (24.67) and (24.66) that $\Delta(k)$ obeys the equation,

$$\Delta(k) = U(k) - \Delta(k) P(k) U(k), \tag{24.93}$$

where[8]

$$P(k) = \frac{i\hbar}{(2\pi)^4} \mathrm{tr} \int d^4p \Gamma(p;k) G(p) G(p-k). \tag{24.94}$$

From Eq. (24.93), we obtain the formal solution for $\Delta(k)$,

$$\Delta(k) = U(k) \left[1 + P(k) U(k)\right]^{-1}. \tag{24.95}$$

It follows from these expressions that all the functions, G, Δ, Σ, Π, and P can be determined once $\Gamma(p;k)$ is known. However, Γ cannot be determined unless we solve the functional-derivative equation

$$\Gamma(p;k) = \delta^{(4)}(p - p' + k) + \lim_{\phi \to 0} \frac{\delta \Pi_\phi(p', p)}{\delta \langle \phi(k) \rangle}, \tag{24.96}$$

which again involves ϕ-dependent functions. Some approximate procedures to the evaluation of the single-particle Green function are briefly mentioned below.

[8]It is hoped that the use of the same symbol P to denote the polarization and the magnitude of the space part of the momentum four-vector will not cause confusion.

24.3.4 Two-particle Green functions(*)

The two-particle Green function, $G^{(2)}(x_1, x_2; x_1', x_2')$, also satisfies an equation from which higher-order Green functions can be eliminated through functional differentiation, and this equation is derived in the following lines. We specialize the discussion to the case of fermions. For ease of presentation, we denote the space-time point $(\mathbf{r}_n, t_n)$ by n, so that we have the definition,

$$\begin{aligned} G(2)(x_1, x_2; x_1', x_2') &\equiv G(2)(1, 2; 1', 2') \\ &= (\mathrm{i}\hbar)^{-2}\langle T\left[\psi(1)\psi(2)\psi^*(2')\psi^*(1')\right]\rangle. \end{aligned} \tag{24.97}$$

Differentiating this expression with respect to t_1 yields the expression

$$\begin{aligned} \mathrm{i}\hbar G_\phi^{(2)}(1, 2; 1', 2') &= \delta(1, 1')G_\phi(2, 2') - \delta(1, 2')G_\phi(2, 1') \\ &+ (\mathrm{i}\hbar)^{-2}\langle T\left[\frac{\partial\psi(1)}{\partial t_1}\psi(2)\psi^*(2')\psi^*(1')\right]\rangle, \end{aligned} \tag{24.98}$$

which, in view of Eq. (24.41) and (24.42) yields the equation,

$$F_1 G_\phi^{(2)}(1, 2; 1', 2') = \delta(1, 1')G_\phi(2, 2') - \delta(1, 2')G_\phi(2, 1'). \tag{24.99}$$

where

$$F_1 \equiv \mathrm{i}\hbar\frac{\partial}{\partial t_1} + \frac{\hbar^2}{2m}\nabla_1^2 - \bar{V} - \langle\phi(1)\rangle - D(1), \tag{24.100}$$

with $D(1)$ being an integrodifferential operator,

$$D(1) \equiv \mathrm{i}\hbar\int \mathrm{d}(3)\Delta_\phi(1, 3)\frac{\delta}{\delta\langle\phi(3)\rangle}, \tag{24.101}$$

and where we have introduced the notation $\mathrm{d}(n) = \mathrm{d}^4x_n$. The explicit presence of the external source field, $\phi(x)$, is to be noted. We recall that final expressions are to be obtained in the limit $\phi(x) \to 0$ at the end of formal arguments. From Eqs. (24.56) and (24.99), there results the following equation for the two-particle Green function,

$$F_1 F_2 G_\phi(2)(1, 2; 1', 2') = \delta(1, 1')\delta(2, 2') - \delta(1, 2')\delta(2, 1'), \tag{24.102}$$

where F_2 is given by an expression identical to that in Eq. (24.100) but refers to particle 2. The last equation certainly involves only $G^{(2)}$ but also contains products of functional derivatives.

A more transparent and more useful expression can be obtained by separating out the single-particle components of $G^{(2)}$. We seek an expression of the form[9] (with ϕ subscripts suppressed for convenience of notation),

$$\begin{aligned} &\int\int \mathrm{d}(1'')\mathrm{d}(2'')\left[G^{-1}(1,1'')G^{-1}(2,2'') - W(1,2;1'',2'')\right] \\ \times\ & G(2)(1'',2'';1',2') = \delta(1,1')\delta(2,2') - \delta(1,2')\delta(2,1'), \end{aligned} \tag{24.103}$$

or, in matrix form,

$$\left[G^{-1}(1,1')G^{-1}(2,2') - W\right]G(2) = 1_{12}. \tag{24.104}$$

Here, $G^{-1}(1,1')$ is the inverse of the exact single-particle Green function given by Eq. (24.59), and 1_{12} stands for the matrix element,

$$\langle 12|1_{12}|1'\rangle 2' = \delta(1,1')\delta(2,2') - \delta(1,2')\delta(2,1'). \tag{24.105}$$

24.3.5 Function $W(*)$

We now examine somewhat more closely the function W that occurs in the equation for $G^{(2)}$. Since G represents the exact single-particle Green function, the function W represents the remaining interaction of a pair of particles with the medium, that is to say, the part that is not included in the single-particle Green functions. The integral form of Eq. (24.103) is,

$$\begin{aligned} G^{(2)}(1,2;1',2') &= G(1,1')G(2,2') - G(1,2')G(2,1') \\ &+ \int\int\int\int \mathrm{d}(3)\mathrm{d}(4)\mathrm{d}(3')\mathrm{d}(4')G(1,3)G(2,4) \\ &\times\ W(3,4;3',4')G^{(2)}(3',4';1',2'), \end{aligned} \tag{24.106}$$

or, in matrix form,

$$G^{(2)} = G_1G_2 1_{12} + G_1G_2WG^{(2)}, \tag{24.107}$$

in a notation which is hopefully obvious. This equation for $G^{(2)}$ involves no higher-order Green functions and is analogous to Eq. (24.77) for the single-particle Green function.

Further scrutiny of the function W proves revealing. Applying the operator, F_1, Eq. (24.100), to Eq. (24.106) yields the result,

$$F_1G^{(2)}(1,2;1',2') = \delta(1,1')G(2,2') - \delta(1,2')G(2,1')$$

[9]The resulting expressions are, of course, valid immediately in the limit of vanishing $\phi(x)$.

$$
\begin{aligned}
+ & \int G(2,2'') \Big\{ \iint W(1,2'';3',4') G^{(2)}(3',4';1',2') \mathrm{d}(3')\mathrm{d}(4') \\
- & \mathrm{i}\hbar \iint \Delta(1,6)\Gamma(2'',5';6) \left[G(1,1')G(5',2') - G(1,2')G(5',1') \right. \\
+ & \iiiint G(1,3)G(5',4)W(3,4;3',4') \\
\times & \left. G^{(2)}(3',4';1',2')\mathrm{d}(3)\mathrm{d}(4)\mathrm{d}(3')\mathrm{d}(4') \right] \mathrm{d}(5')\mathrm{d}(6) \\
- & \mathrm{i}\hbar \iiiint G(1,3)\Delta(1,6) \frac{\delta}{\delta\langle\phi(6)\rangle} \left[W(3,2'';3',4')G^{(2)}(3',4';1',2') \right] \\
\times & \mathrm{d}(3)\mathrm{d}(6)\mathrm{d}(3')\mathrm{d}(4') \Big\} \mathrm{d}(2''). \qquad (24.108)
\end{aligned}
$$

Comparing this expression with that in Eq. (24.99), we see that the extra term on the right must vanish leading to the equation,

$$
\begin{aligned}
& \iint W(1,2;3',4') G^{(2)}(3',4';1',2') \mathrm{d}(3')\mathrm{d}(4') \\
= & \mathrm{i}\mathrm{h} \iint \Delta(1,6)\Gamma(2,5';6) G^{(2)}(1,5';',2') \\
+ & \mathrm{i}\hbar \iiiint G(1,3)\Delta(1,6) \frac{\delta}{\delta\langle\phi(6)\rangle} \left[W(3,2;3',4')G^{(2)}(3',4';1',2') \right] \\
\times & \mathrm{d}(3)\mathrm{d}(6)\mathrm{d}(3')\mathrm{d}(4'), \qquad (24.109)
\end{aligned}
$$

which can also be written in the matrix form

$$
WG^{(2)} = \mathrm{i}\hbar\Delta_{12}\Gamma_2 G^{(2)} + \mathrm{i}\hbar G_1 \Delta_{12} \frac{\delta}{\langle\delta\phi_2\rangle} \left[WG^{(2)} \right]. \qquad (24.110)
$$

If we denote the first term on the right-hand side by $WG^{(2)}$, we have

$$
\begin{aligned}
W(1,2;3',4') & \equiv \mathrm{i}\hbar \iint \mathrm{d}(5')\mathrm{d}(6)\Delta(1,6)\Gamma(2,5';6)\langle 1,5'|1_{12}|3',4'\rangle \\
& = \mathrm{i}\hbar \int \mathrm{d}(6) \left[\Delta(1,6)\Gamma(2,4;6)\delta(1,3') \right. \\
& - \left. \Delta(1,6)\Gamma(2,3';6)\delta(1,4') \right]. \qquad (24.111)
\end{aligned}
$$

The two terms in this expression represent exchange effects.

24.3.6 Non-zero external field

The form of the various Green function equations derived above can be readily extended to the case in which an (c-number) external field, $\phi(x)$, is acting on the system. In this case, it may be convenient to exhibit the presence of the external source explicitly in the equations. The single-particle Green function satisfies the equation,

$$\begin{aligned}\left[i\hbar\frac{\partial}{\partial t} \right. &+ \left. \frac{\hbar^2}{2m}\nabla^2 - \phi(x)\right] G(x,x') = \delta^{(4)}(x-x') \\ &+ (i\hbar)^{-1}\int d^4x'' U(x-x'')\langle G^{(2)}(x,x'';x',x''),\end{aligned} \tag{24.112}$$

which explicitly contains the two-particle Green function. We have also seen that the two-particle Green function can be eliminated in favor of a self-energy in terms of which the single-particle Green function satisfies the following equation,

$$\begin{aligned}\left[i\hbar\frac{\partial}{\partial t} + \frac{\hbar^2}{2m}\nabla^2 - \phi(x)\right] G(x,x') &- \int d^4x''\Sigma(x,x'')G(x'',x') \\ &= \delta^{(4)}(x-x').\end{aligned} \tag{24.113}$$

The study of the interacting electron gas in a solid almost invariably proceeds on the basis of Eq. (24.113), and consists of attempts to obtain an approximate determination of the self-energy. For example, the LDA-based methods of electronic structure calculations, discussed at length in previous parts of this book, are an example of this procedure. Below, we devote some space to the discussion of perturbation methods in the evaluation of the single-particle Green function.

24.3.7 Total energy

The expectation values of one-particle and two-particle operators can be obtained from the corresponding Green functions by taking appropriate limits at equal times. We can use such relations to obtain expressions for the ground-state energy of the system. Denoting the one-particle part of the Hamiltonian of an electron gas in the presence of the nuclear potential by $h(\mathbf{r})$,

$$h(\mathbf{r}) = -\frac{\hbar^2}{2m}\nabla^2 - \sum_n Z_n U(\mathbf{r}-\mathbf{R}_n), \tag{24.114}$$

we obtain the expression[10],

$$E = -i\int d^3r\,\{h(\mathbf{r})G(\mathbf{r},t;\mathbf{r}_1,t+0)\}_{\mathbf{r}_1\to\mathbf{r}}$$

[10]Where spatial integrations can be taken to include summations over spin variables as well.

$$- \frac{1}{2}\int d^3r \int d^3r' U(\mathbf{r}-\mathbf{r}')G^{(2)}(\mathbf{r},t;\mathbf{r}',t';\mathbf{r},t+0,\mathbf{r}',t+0)$$
$$+ \frac{1}{2}\sum_{nm}{}' Z_n Z_m U(\mathbf{R}_n - \mathbf{R}_m). \qquad (24.115)$$

Here, Z_n is the nuclear charge at $\mathbf{R}_n$, and a prime over a summation symbol indicates the exclusion of the term with $n = m$.

We can also obtain an expression for the total energy in terms of the single-particle Green function. To accomplish this, we multiply the equation of motion for the operator $\psi(x)$, Eq. (24.41), by $\psi^*(x)$, integrate over spatial variables (and sum over spin where appropriate), and finally let the external field, $\phi(x)$, vanish. We then obtain the following relation between the Coulomb energy and the single-particle Green function,

$$\begin{aligned}
& \tfrac{1}{2} \int d^3r \int d^3r' U(\mathbf{r}-\mathbf{r}')\langle T\left[\psi^*(\mathbf{r},t)\psi^*(\mathbf{r}',t)\psi(\mathbf{r}',t)\psi(\mathbf{r},t))\right]\rangle \\
= & \frac{1}{2}\langle\psi^*(\mathbf{r},t)\left[i\hbar\frac{\partial}{\partial t} - h(\mathbf{r})\right]\psi(\mathbf{r},t)\rangle \\
= & -\frac{i}{2}\int d^3r \left[i\hbar\frac{\partial}{\partial t} - h(\mathbf{r})\right] G(\mathbf{r},t\mathbf{r}',t'), \quad \mathbf{r}',t' \to \mathbf{r},t. \qquad (24.116)
\end{aligned}$$

This allows us to write the total energy in Eq. (24.115) in the form,

$$E = -\frac{i}{2}\int d^3r \left[i\hbar\frac{\partial}{\partial t} - h(\mathbf{r})\right] G(\mathbf{r},t\mathbf{r}',t')_{\mathbf{r}',t'\to\mathbf{r},t} + V_{\text{nucl}}, \qquad (24.117)$$

where V_{nucl} denotes the internuclear repulsion. Thus, the total energy of the system can be written explicitly in terms of the single-particle Green function.

24.4 Perturbation Theory

One approach to the evaluation of the single-particle Green function is provided by perturbation theory. Let the Hamiltonian of the system be divided into two parts,

$$H = H_0 + H_{\mathrm{I}} \qquad (24.118)$$

where H_0 and H_{I} are, respectively, the Hamiltonians of an unperturbed system, whose Green function, G_0, is assumed known, and the perturbation. We write H_0 in the form

$$H_0 \equiv \int \psi^*(x)\left(-\frac{\hbar^2}{2m}\nabla^2\right)\psi(x)d^3r + \int\int \psi^*(x)\Xi(x,x')\psi(x')d^3r d^4x', \qquad (24.119)$$

where $\Xi(x, x')$ is an appropriately chosen non-local potential. The unperturbed single-particle Green function, G_0, corresponding to H_0, satisfies the equation,

$$\left[i\hbar\frac{\partial}{\partial t} + \frac{\hbar^2}{2m}\nabla^2\right] G_0(x, x') - \int d^4x''\Xi(x, x'')G_0(x'', x') = \delta^{(4)}(x - x'). \tag{24.120}$$

To use G_0 as the basis for a perturbation treatment of G, it is convenient to rewrite G in the form of a Dyson equation,

$$\begin{aligned} G(x, x') &= G_0(x, x') + \int\int d^4x''d^4x'''G_0(x, x'') \\ &\times [\Sigma(x'', x''') - \Xi(x'', x''')] G_0(x''', x'), \end{aligned} \tag{24.121}$$

or, in matrix notation

$$G = G_0 + G_0 [\Sigma - \Xi] G. \tag{24.122}$$

Clearly, the convergence properties and the accuracy of any perturbation series based on this expresson depends on the choice of Ξ.

To proceed further, we recall the expression for the self-energy given in Eqs. (24.75). We note that the function $\Pi(x, x')$, Eq. (24.76), includes a term

$$\Pi^{ex}(x, x') = i\hbar U(x - x')G(x - x'), \tag{24.123}$$

as can be easily seen from the expressions for Δ and Γ given in Eqs. (24.70) and (24.72). This term can be interpreted as the exchange effect between two particles corresponding to the term $\bar{V}$ in Σ. To see this, we use Eq. (24.48) to write

$$\bar{V}(x)G(x, x') = -i\hbar \int d^4\xi U(x - \xi)G(\xi, \xi)G(x, x'), \tag{24.124}$$

and compare this expression with

$$\int d^4\xi\Pi^{ex}(x, \xi)G(\xi, x') = i\hbar \int d^4\xi U(x - \xi)G(x, \xi)G(\xi, x'). \tag{24.125}$$

The second member of the last expression is obtained from the corresponding member of the previous one upon an exchange of the particle coordinates ξ and x. Let us now divide Σ into two parts,

$$\Sigma^s(x, x') \equiv \bar{V}(x)\delta^{(4)}(x - x') + \Pi^{ex}(x, x') \tag{24.126}$$

$$\Sigma^F(x, x') \equiv \Sigma s(x, x') - \Pi^{ex}(x, x'). \tag{24.127}$$

Replacing Σ by Σ^s is often referred to as the static approximation.

We now return to the perturbation treatment of G. As a first choice for Ξ we may consider,

$$\Xi = \Sigma^s[G_0], \tag{24.128}$$

which is a simple application of the Hartree-Fock method to the problem at hand. We can now attempt to solve for G_0 in the symbolic form,

$$\left[p_0 - \frac{P^2}{2m} - \Sigma^s[G_0]\right] G_0 = 1, \tag{24.129}$$

an expression that ultimately must be treated by an iterative procedure. Other choices for Ξ are also possible, such as the one suggested by the formal approach of Brueckner [5, 6].

The perturbation series for the Green function is obtained by successive iteration. This, in turn, requires the series for Σ, Δ, P, and Γ. The initial terms in the corresponding expressions take the forms:

$$\begin{aligned} \Sigma^{\mathrm{F}}(x,x') &= -(\mathrm{i}\hbar)^2 \int\int \mathrm{d}^4x''\mathrm{d}^4\zeta' U(x'-x'')G_0(x'',\zeta')G_0(\zeta',x'') \\ &\times \; U(x-\zeta')G_0(x,x') \\ &+ \; (\mathrm{i}\hbar)^2 \int\int \mathrm{d}^4x''\mathrm{d}^4\xi' \prime U(x-x'')G_0(x,\xi)U(\xi-x') \\ &\times \; G_0(\xi,x'')G_0(x'',x') + \cdots, \end{aligned} \tag{24.130}$$

$$\begin{aligned} \Delta(x,x') &= U(x-x') - (\mathrm{i}\hbar) \int\int \mathrm{d}^4x''\mathrm{d}^4\zeta' U(x'-x'') \\ &\times \; \mathrm{tr}\left[G_0(x'',\zeta')G_0(\zeta',x'')\right] U(x-\zeta') + \cdots, \end{aligned} \tag{24.131}$$

in which the lowest order contribution to P,

$$P(\zeta',x'') = \mathrm{i}\hbar \mathrm{tr}\left[G_0(x'',\zeta')G_0(\zeta',x'')\right] + \cdots, \tag{24.132}$$

has been used, and

$$\Gamma(\xi,\zeta;x'') = \delta^{(4)}(\xi-\zeta)\delta^{(4)}(\xi-x'') + \mathrm{i}\hbar U(\xi-\zeta)G_0(\xi,x'')G_0(x'',\zeta) + \cdots. \tag{24.133}$$

24.4.1 GW approximation

It follows from the set of Eqs. (24.68) to (24.73) that the various quantities appearing there can be determined in iterative fashion once Γ is known. In the so-called *GW approximation*, we set $\Gamma = 1$. We now have,

$$\Delta = U\,(1 + PU)^{-1}, \tag{24.134}$$

and, in the limit $\phi \to 0$,

$$\Sigma = WG, \tag{24.135}$$

where W, the *screened Coulomb interaction*, is given by the expression,

$$W = -UG + U(1 + PU)^{-1}. \tag{24.136}$$

The expression for the self energy can now be inserted into Eq. (24.68) to obtain an estimate for G. This, in turn, can yield new values for the other parameters in the set of Eqs. (24.68) to (24.73). Proceeding in such an iterative fashion, the quantities G, Σ, Δ, and P can be determined in a self-consistent way.

It is customary to define the *inverse dielectric constant* by the expression, Eq. (24.55),

$$\epsilon^{-1} = \frac{\delta\langle\phi(x')\rangle}{\delta\phi(x'')}. \tag{24.137}$$

The *irreducible polarization propagator* is given in related to ϵ^{-1} by the expression,

$$\epsilon(1,2) = \delta(1,2) - \int \mathrm{d}(3)P(3,2)U(1,3), \tag{24.138}$$

leading to an expression for the *exact, fully screened* Coulomb interaction (in matrix notation),

$$\begin{aligned} W &= U + WPU \\ &= U(1 - PU)^{-1}. \end{aligned} \tag{24.139}$$

The GW approximation leads to perturbative expressions that include the first term in an expansion in terms of the screened interaction.

24.5 Spectral Representations

The *spectral representation*, developed originally by Landau [7], Galitskii and Migdal [8], and Martin and Schwinger [9], and subsequently by many other authors, provides a particularly useful approach to the study of Green functions. In this section, we begin by developing the formalism for the case of uniform media and in the absence of external fields. Therefore, the Green function, $G(x - x')$, depends only on the difference of its space-time coordinates. The effects of an external field are considered in the following subsection in connection with electronic transport.

For the sake of generality, we consider, in addition to the zero-temperature Green function of Eqs. (24.27), the Green function describing particles moving in a medium in thermal equilibrium at temperature $1/\beta$. The quantity,

$$G(x,x') = (\mathrm{i}\hbar)^{-1}\mathrm{Tr}\left\{\mathrm{e}^{-\beta[H-\mu N]}T\left[\psi(x)\psi^*(x')\right]\right\}/\mathrm{Tr}\left\{\mathrm{e}^{-\beta[H-\mu N]}\right\} \tag{24.140}$$

is called the *time-temperature* Green function. If t and t' in this expression are replaced, respectively, by $\mathrm{i}\beta$ and $\mathrm{i}\beta'$, we obtain the *temperature* Green function introduced by Matsubara [10]. In this expression, μ denotes the chemical potential and N represents the operator corresponding to the number of particles.

Now, in a representation, $|n\rangle$, which diagonalizes the total Hamiltonian, the space-time dependence of the matrix elements of the operator $\psi(x)$ takes the form,

$$\psi_{nm}(\mathbf{r},t) = \psi_{nm}^{(0)} \exp\{\mathrm{i}\mathbf{k}_{nm}\cdot\mathbf{r} - \omega_{nm}t\}, \qquad (24.141)$$

where

$$\hbar\mathbf{k}_{nm} = \mathbf{P}_n - \mathbf{P}_m \quad \hbar\omega_{nm} = E_n - E_m = E_{nm}, \qquad (24.142)$$

where n and m identify eigenstates of H, and E_n, $\mathbf{P}_n$, and N_n are the eigenvalues of the total energy, momentum, and number of particles, respectively. In terms of these matrix elements, the single-particle Green function defined by Eqs. (24.27) or (24.140) can be expressed as follows,

$$\begin{aligned} G(\mathbf{r},t) &= (\mathrm{i}\hbar)^{-1}\sum_{nm}\exp\left[-\beta\left(E_n - \mu N - \Omega\right)\right] \\ &\times \;|\psi_{nm}^{(0)}|^2 \exp\left[\mathrm{i}\left(\mathbf{k}_{nm}\cdot\mathbf{r} - \omega_{nm}t\right)\right], \\ &\qquad \text{for } t > 0, \\ &= \pm(\mathrm{i}\hbar)^{-1}\sum_{nm}\exp\left[-\beta\left(E_n - \mu N - \Omega\right)\right] \\ &\times \;|\psi_{nm}^{(0)}|^2 \exp\left[\mathrm{i}\left(\mathbf{k}_{nm}\cdot\mathbf{r} - \omega_{nm}t\right)\right], \\ &\qquad \text{for } t > 0, \\ &= \pm\mathrm{i}\hbar)^{-1}\sum_{nm}|\psi_{nm}^{(0)}|^2\exp\left[-\beta\left(E_n - \mu N - \Omega\right)\right] \\ &\times \;\exp\left[\beta\left(E_{nm} + \mu\right)\right]\exp\left[\mathrm{i}\left(\mathbf{k}_{nm}\cdot\mathbf{r} - \omega_{nm}t\right)\right], \\ &\qquad \text{for } t < 0, \end{aligned} \qquad (24.143)$$

where Ω denotes the Gibbs free energy (thermodynamic potential) of the system, and the last expression follows from the fact that ψ_{nm} is nonzero only when $N_m = N_n + 1$. The + and - signs correspond to bosons and fermions, respectively. In the case of a non-interacting system, Eq. (24.142) reduces to the expression,

$$\begin{aligned} G_0(\mathbf{r},t) &= (\mathrm{i}\hbar)^{-1}\int \mathrm{d}^3k\langle 1 - n_{\mathbf{k}}\rangle \exp\left[\mathrm{i}\left(\mathbf{k}\cdot\mathbf{r} - \omega t\right)\right] \quad t > 0, \\ &= (\mathrm{i}\hbar)^{-1}\int \mathrm{d}^3k\langle n_{\mathbf{k}}\rangle \exp\left[\mathrm{i}\left(\mathbf{k}\cdot\mathbf{r} - \omega t\right)\right] \quad t < 0, \end{aligned} \qquad (24.144)$$

where $\langle n_{\mathbf{k}}\rangle = \langle a^*_{\mathbf{k}} a_{\mathbf{k}}\rangle$ is the expectation value of the number operator associated with the state $\mathbf{k}$. In the momentum representation, the Green function becomes,

$$G(\mathbf{k},t) = \int G(\mathbf{r},t)\mathrm{e}^{-\mathrm{i}\mathbf{k}\cdot\mathbf{r}}\mathrm{d}^3 r, \tag{24.145}$$

so that for the non-interacting case we have,

$$\begin{aligned} G_0(\mathbf{k},t) &= (\mathrm{i}\hbar)^{-1}(2\pi)^{-3}\langle 1-n_{\mathbf{k}}\rangle \mathrm{e}^{-\omega t} \quad t>0, \\ &= \pm(\mathrm{i}\hbar)^{-1}(2\pi)^{-3}\langle n_{\mathbf{k}}\rangle \mathrm{e}^{-\omega t} \quad t<0. \end{aligned} \tag{24.146}$$

In general, the Fourier transform of the Green function defined by Eq. (24.80) is given by the expression,

$$\begin{aligned} G(\mathbf{k},E) &= -(2\pi)^{-3}\Bigg\{\pm\sum_{nm}|\psi^{(0)}_{nm}|^2 \exp\left[-\beta\left(E_m-\mu N_m-\Omega\right)\right] \\ &\times \frac{1}{E-E_{nm}-\mathrm{i}\delta} \\ &+ \sum_{nm}|\psi^{(0)}_{nm}|^2 \exp\left[-\beta\left(E_n-\mu N_n-\Omega\right)\right]\frac{1}{E-E_{nm}-\mathrm{i}\delta}\Bigg\} \\ &\times \delta^{(3)}(\mathbf{k}-\mathbf{k}_{nm}), \end{aligned} \tag{24.147}$$

where the δ in the denominators denotes a positive infinitesimal quantity.

In the limiting case of large systems with a given density, or in the limit of continuous quantum numbers, Eq. (24.147) can be written in the form,

$$G(\mathbf{k},E) = \int_0^\infty \mathrm{d}\omega\left[fracA(\mathbf{k},\omega)E-\omega-\mathrm{i}\delta + \frac{B(\mathbf{k},\omega)}{E-\omega+\mathrm{i}\delta}\right], \tag{24.148}$$

where the *spectral functions*, $A(\mathbf{k},\omega)$, and $B(\mathbf{k},\omega)$ are defined as the limits just described of the corresponding quantities in Eq. (24.147),

$$\begin{aligned} A(\mathbf{k},\omega)\mathrm{d}\omega &= \lim(\mp)(2\pi)^3\sum_{nm}|\psi^{(0)}_{nm}|^2 \exp\left[-\beta\left(E_n-\mu N_n-\Omega\right)\right] \\ &\times \delta^{(3)}(\mathbf{k}_{nm}-\mathbf{k}), \\ &\text{for} \quad E<E_{nm}<E+\mathrm{d}E \end{aligned} \tag{24.149}$$

$$\begin{aligned} B(\mathbf{k},\omega)\mathrm{d}\omega &= \lim(-)(2\pi)^3\sum_{nm}|\psi^{(0)}_{nm}|^2 \exp\left[-\beta\left(E_n-\mu N_n-\Omega\right)\right] \\ &\times \delta^{(3)}(\mathbf{k}_{nm}-\mathbf{k}), \\ &\text{for} \quad E<E_{nm}<E+\mathrm{d}E. \end{aligned} \tag{24.150}$$

The expressions in these equations are essentially identical to the spectral representations introduced by Lehmann into quantum field theory.

Using the well-known formula,

$$\frac{1}{\mathrm{i}}\int_0^\infty \mathrm{e}^{\pm \mathrm{i}\alpha x}\mathrm{d}x = \mathrm{i}\pi\delta(\alpha) \pm P\left(\frac{1}{\alpha}\right), \tag{24.151}$$

where P denotes Cauchy principal value, and with the use of Eqs. (24.142), Eq. (24.147) can be cast in the form,

$$G(\mathbf{k}, E) = (2\pi)^3 \sum_{nm} |\psi_{nm}^{(0)}|^2 \exp\left[-\beta\left(E_n - \mu N_n - \Omega\right)\right]\delta^{(3)}(\mathbf{k}_{nm} - \mathbf{k})$$
$$\times \left[P\left(\frac{1}{E_{nm} - E}\right)\left(1 \pm \mathrm{e}^{\beta(E_{nm}+\mu)}\right) + \mathrm{i}\pi\delta(E_{nm} - E)\left(1 \pm \mathrm{e}^{\beta(E_{nm}+\mu)}\right)\right]. \tag{24.152}$$

From this expression, there follows a relation between the real and imaginary parts of the Green function. For the case of the Bose field, we find,

$$\Re G(\mathbf{k}, E) = \frac{1}{\pi} P \int \tanh\frac{\beta(z+\mu)}{2}\frac{\Im G(\mathbf{k}, z)}{z - E}\mathrm{d}z, \tag{24.153}$$

while for the Fermi field, we have,

$$\Re G(\mathbf{k}, E) = \frac{1}{\pi} P \int \coth\frac{\beta(z+\mu)}{2}\frac{\Im G(\mathbf{k}, z)}{z - E}\mathrm{d}z. \tag{24.154}$$

24.5.1 Two-particle Green functions and transport

We now derive an expression for the electrical conductivity of a material in terms of the two-particle Green function. As in our previous study of transport expressed in terms of the single-particle Green function, we consider the response of the system to an external field represented by the vector potential, $\mathbf{A}(\mathbf{r}, t)$. The external Hamiltonian can be written in the form,

$$H^{\mathrm{ext}} = -\frac{1}{\mathrm{c}}\int \mathrm{d}^3r\,\mathbf{A}(\mathbf{r}, t)\cdot\mathbf{j}^{(0)}(\mathbf{r}, t) - \frac{1}{2\mathrm{c}}\int \mathrm{d}^3r\,\mathbf{A}(\mathbf{r}, t)\cdot\mathbf{j}^{(1)}(\mathbf{r}, t), \tag{24.155}$$

where

$$\mathbf{j}^{(0)}(\mathbf{r}, t) = \frac{\mathrm{e}}{2m}\left[-\frac{\hbar}{\mathrm{i}}\left(\nabla\psi^*(\mathbf{r}, t)\right)\psi(\mathbf{r}, t) + \psi^*(\mathbf{r}, t)\frac{\hbar}{\mathrm{i}}\left(\nabla\psi(\mathbf{r}, t)\right)\right] \tag{24.156}$$

and

$$\mathbf{j}^{(1)}(\mathbf{r}, t) = \frac{\mathrm{e}^2}{2m}\psi^*(\mathbf{r}, t)\psi(\mathbf{r}, t)\mathbf{A}(\mathbf{r}, t). \tag{24.157}$$

To first order in the vector potential, we find

$$\mathbf{j}^{(0)}(\mathbf{r},t) = \mathbf{j}_0(\mathbf{r},t) - \frac{1}{\mathrm{i}c\hbar}\int_0^t \mathrm{d}t' \int \mathrm{d}^3r' \, [\mathbf{j}_0(\mathbf{r},t), \mathbf{j}_0(\mathbf{r}',t')]\,\mathbf{A}(\mathbf{r}',t'), \quad t > t', \tag{24.158}$$

where $\mathbf{j}_0(\mathbf{r},t)$ is the current operator in the absence of external fields. We now obtain the relation in terms of the two-particle Green function

$$\begin{aligned}
\langle \mathbf{j}_0(\mathbf{r},t)\mathbf{j}_0(\mathbf{r}',t')\rangle &= \left(\frac{e\hbar}{2im}\right)^2 \lim_{\substack{\mathbf{r}_1\to\mathbf{r}_2=\mathbf{r} \\ \mathbf{r}_1'\to\mathbf{r}_2'=\mathbf{r}'}} \lim_{\substack{t_1\to t_2=t \\ t_1'\to t_2'=t'}} (\nabla_1 - \nabla_2)(\nabla_1' - \nabla_2') \\
&\times \langle \psi^*(\mathbf{r}_1,t_1)\psi(\mathbf{r}_2,t_2)\psi^*(\mathbf{r}_1',t_1')\psi(\mathbf{r}_2',t_2')\rangle \\
&= \left(\frac{e\hbar^2}{2im}\right)^2 \lim_{\mathbf{r},t} (\nabla_1 - \nabla_2)(\nabla_1' - \nabla_2') \\
&\times G^{(2)}(\mathbf{r}_2,t_2,\mathbf{r}_2',t_2';\mathbf{r}_1,t_1,\mathbf{r}_1',t_1'),
\end{aligned} \tag{24.159}$$

where the symbol $\lim_{\mathbf{r},t}$ is defined implicitly in the first line of the equation. It follows that the current can be expressed in terms of the two-particle Green function,

$$\begin{aligned}
\langle \mathbf{j}(\mathbf{r},t)\rangle &= \frac{e}{mc}\langle \psi^*(\mathbf{r},t)\psi(\mathbf{r},t)\rangle \mathbf{A}(\mathbf{r},t) \\
&+ \frac{1}{\mathrm{i}c\hbar}\left(\frac{e\hbar^2}{2im}\right)^2 \int_0^t \mathrm{d}t' \int \mathrm{d}^3r' \lim_{\mathbf{r},t} (\nabla_1 - \nabla_2)(\nabla_1' - \nabla_2') \\
&\times G^{(2)}(\mathbf{r}_2,t_2,\mathbf{r}_2',t_2';\mathbf{r}_1,t_1,\mathbf{r}_1',t_1')\mathbf{A}(\mathbf{r},t) + \text{c.c.},
\end{aligned} \tag{24.160}$$

where c.c. denotes complex conjugate. Therefore, the two-particle Green function contains sufficient information to yield the current density to first order in $\mathbf{A}(\mathbf{r},t)$. *This is the basic result of linear response theory expressed in terms of the two-particle Green function and should be compared with the expressions derived previously in terms of the single-particle Green function.*

In order to obtain an explicit expression for the conductivity, we consider the Fourier transform

$$\begin{aligned}
F_{\alpha\beta}(\mathbf{k}\omega,\mathbf{k}'\omega') &= -\frac{1}{(2\pi)^8} \\
&\times \int\!\!\int \mathrm{d}^3k_1\mathrm{d}^3k_1'\mathrm{d}\omega\mathrm{d}\omega'(\mathbf{k}+2\mathbf{k}_1)_\alpha(\mathbf{k}'+2\mathbf{k}_2')_\beta \\
&\times G^{(2)}(\mathbf{k}+\mathbf{k}_1,\omega+\omega_1,\mathbf{k}'+\mathbf{k}_1',\omega'+\omega_1',\mathbf{k}_1,\omega_1,\mathbf{k}_1',\omega_1'),
\end{aligned} \tag{24.161}$$

where α and β represent spatial directions, and the spatial and reciprocal space representations of the Green function are connected by the relation,

$$\begin{aligned} G^{(2)}(\mathbf{r}_1t_1, \mathbf{r}_2t_2; \mathbf{r}_1't_1', \mathbf{r}_2't_2') &= \frac{1}{(2\pi)^{16}} \int\int\int \mathrm{d}^3k_1\mathrm{d}\omega_1\mathrm{d}^3k_1'\mathrm{d}\omega_1' \\ \times \quad & \mathrm{d}^3k_2\mathrm{d}\omega_2\mathrm{d}^3k_2'\mathrm{d}\omega_2' \\ \times \quad & \exp\{\mathrm{i}(\mathbf{k}_1\cdot\mathbf{r}_1 + \mathbf{k}_2\cdot\mathbf{r}_2 - \mathbf{k}_1'\cdot\mathbf{r}_1' - \mathbf{k}_2'\cdot\mathbf{r}_2') \\ - \quad & \mathrm{i}(\omega_1t_1 + \omega_2t_2 - \omega_1't_1' - \omega_2't_2')\} \\ \times \quad & G^{(2)}(\mathbf{k}_1\omega_1, \mathbf{k}_2\omega_2; \mathbf{k}_1'\omega_1', \mathbf{k}_2'\omega_2'). \end{aligned} \tag{24.162}$$

Similarly, we have for the correlation function

$$\begin{aligned} F_{\alpha\beta}(\mathbf{r}t;\mathbf{r}'t') &= \frac{1}{(2\pi)^3}\int\int \exp\{\mathrm{i}(\mathbf{k}\cdot\mathbf{r} - \mathbf{k}'\cdot\mathbf{r}') - \mathrm{i}(\omega t - \omega' t')\} \\ & \times \quad F_{\alpha\beta}(\mathbf{k}\omega;\mathbf{k}'\omega')\mathrm{d}^3k\mathrm{d}\omega\mathrm{d}^3k'\mathrm{d}\omega'. \end{aligned} \tag{24.163}$$

It is more convenient to proceed with the analysis on the basis of the correlation function rather than the Green function.

The second term of Eq. (24.158) can also be expressed in terms of current-current correlation functions in the form,

$$-\frac{1}{\mathrm{i}c\hbar}\int_0^t\int \mathrm{d}^3r'\langle j_0(\mathbf{r}',t')j_0(\mathbf{r},t+\mathrm{i}\beta\hbar) - j_0(\mathbf{r},t)j_0(\mathbf{r},t+\mathrm{i}\beta\hbar)\rangle\mathbf{A}(\mathbf{r},t). \tag{24.164}$$

Introducing the symmetrized product,

$$\begin{aligned} S_{\alpha\beta} &= \langle\{j_\alpha(\mathbf{r},t), j_\beta(\mathbf{r}',t')\}\rangle \\ &= \langle j_\alpha(\mathbf{r},t)j_\beta(\mathbf{r}',t') + j_\beta(\mathbf{r}',t')j_\alpha(\mathbf{r},t)\rangle \\ &= F_{\alpha\beta}(\mathbf{r}t;\mathbf{r}'t') + F_{\alpha\beta}(\mathbf{r}'t';\mathbf{r}t), \end{aligned} \tag{24.165}$$

and its Fourier transform

$$S_{\alpha\beta}(\mathbf{r},\mathbf{r}';\omega) = \int S_{\alpha\beta}(\mathbf{r},\mathbf{r}';t-t')\mathrm{e}^{\{\mathrm{i}\omega(t-t')\}}\mathrm{d}(t-t'), \tag{24.166}$$

we can write the expectation value of the commutator of $j_\alpha(\mathbf{r},t)$ and $j_\beta(\mathbf{r}',t')$ in the form,

$$\langle[j_\alpha(\mathbf{r},t), j_\beta(\mathbf{r}',t')]\rangle = \int_{-\infty}^{\infty}\frac{\mathrm{d}\omega}{2\pi}\mathrm{e}^{\{-\mathrm{i}\omega(t-t')\}}\tanh\frac{\beta\hbar\omega}{2}S_{\alpha\beta}(\mathbf{r},\mathbf{r}';\omega), \tag{24.167}$$

where the invariance property of the trace under cyclic permutation of operators has been used.

We now specialize the discussion to the case of a uniform medium in which the momentum representation of $S_{\alpha\beta}(\mathbf{r},\mathbf{r}';\omega)$ takes the form,

$$S_{\alpha\beta}(\mathbf{k},\omega) = \int \mathrm{d}^3|\mathbf{r}-\mathbf{r}'|\mathrm{e}^{-\mathrm{i}\mathbf{k}\cdot(\mathbf{r}-\mathbf{r}')} S_{\alpha\beta}(\mathbf{r}-\mathbf{r}',\omega). \tag{24.168}$$

Separating the tensor elements of $S_{\alpha\beta}(\mathbf{k},\omega)$,

$$S_{\alpha\beta}(\mathbf{k},\omega) = \delta_{\alpha\beta}S_1 + \left(k_\alpha k_\beta - k^2\right)\delta_{\alpha\beta}S_2, \tag{24.169}$$

we cast Eq. (24.161) in the form,

$$\begin{aligned}\langle[j_\alpha(\mathbf{r},t), j_\beta(\mathbf{r}',t')]\rangle &= \frac{1}{(2\pi)^4}\left[\delta_{\alpha\beta}\frac{\partial}{\partial t'}\int_{-\infty}^{\infty}\mathrm{d}\omega\int \mathrm{d}^3k\right.\\ \times\ & \mathrm{e}^{-\mathrm{i}(t-t')+\mathrm{i}\mathbf{k}\cdot(\mathbf{r}-\mathbf{r}')}[S_1]\frac{1}{\mathrm{i}\omega}\tanh\frac{\beta\hbar\omega}{2}\\ +\ & \left(\nabla'_\alpha\nabla'_\beta - \nabla^2_\alpha\delta_{\ alpha\beta}\right)\int_{-\infty}^{\infty}\mathrm{d}\omega\int \mathrm{d}^3k\mathrm{e}^{-\mathrm{i}(t-t')+\mathrm{i}\mathbf{k}\cdot(\mathbf{r}-\mathbf{r}')}[S_2]\\ \times\ & \left.\frac{1}{\mathrm{i}\omega}\tanh\frac{\beta\hbar\omega}{2}\right].\end{aligned} \tag{24.170}$$

For the problem at hand, the electric field, $\mathbf{E}(\mathbf{r},t)$, and the magnetic field, $\mathbf{H}(\mathbf{r},t)$, are determined from the vector potential by means of the relations,

$$\mathbf{E}(\mathbf{r},t) = -\frac{1}{\mathrm{c}},\quad \nabla\times\mathbf{H}(\mathbf{r},t) = \left(\nabla\nabla - \nabla^2\right)\mathbf{A}(\mathbf{r},t). \tag{24.171}$$

After a certain amount of algebraic manipulation, we obtain the following expression for the expectation value of the current operator[11],

$$\begin{aligned}\langle\mathbf{j}(\mathbf{r},t)\rangle &= \int \mathrm{d}^3r'\int_0^t \mathrm{d}t'\left[\sigma(\mathbf{r}-\mathbf{r}',t-t')\mathbf{E}(\mathbf{r}',t')\right.\\ &+ \left.\mathrm{c}\frac{\partial}{\partial t}\gamma(\mathbf{r}-\mathbf{r}',t-t')\nabla\times\mathbf{H}(\mathbf{r}',t')\right],\end{aligned} \tag{24.172}$$

where

$$\sigma(\mathbf{r}-\mathbf{r}',t-t') = \frac{1}{(2\pi)^4\hbar}\int_{-\infty}^{\infty}\mathrm{d}\omega\int \mathrm{d}^3k\mathrm{e}^{-\mathrm{i}(t-t')+\mathrm{i}\mathbf{k}\cdot(\mathbf{r}-\mathbf{r}')}[S_1]$$

[11] In deriving the expressions for the conductivity, there arises a non-gauge-invariant term of the form $\int_{-\infty}^{\infty}\frac{\mathrm{d}\omega}{2\pi}\frac{\tanh(\beta\hbar\omega/2)}{\omega}S_1 - \frac{ne^2}{m}$, where n denotes the particle density. Because of charge conservation, this term vanishes, leading to the sum rule $\int_{-\infty}^{\infty}\frac{\mathrm{d}\omega}{2\pi}\frac{\tanh(\beta\hbar\omega/2)}{\omega}S_1 = \frac{ne^2}{m}$ for the spectral distribution function, S_1.

$$\times \quad \frac{1}{i\omega}\tanh\frac{\beta\hbar\omega}{2}$$

$$\gamma(\mathbf{r}-\mathbf{r}',t-t') = \frac{1}{(2\pi)^4\hbar c62}\int_{-\infty}^{\infty}d\omega\int d^3k e^{-i(t-t')+i\mathbf{k}\cdot(\mathbf{r}-\mathbf{r}')}[S_2]$$
$$\times \quad \frac{1}{i\omega}\tanh\frac{\beta\hbar\omega}{2}. \tag{24.173}$$

By virtue of Fourier transformations, these expressions lead directly to expressions for the electric conductivity, polarizabilty, and magnetic susceptibility. It can be shown that the real and imaginary parts of the Fourier transforms of σ and γ satisfy the well-known Kramers-Kronig relations.

24.6 Energy Domain

Having given a summary of fundamental Green function relations in the time representation, we now indicate how the time variable can be eliminated in favor of its conjugate variable, the energy. Energy expressions are very convenient for the derivation of computable formulae, as the discussion in previous parts of the book, carried out exclusively in the energy domain, illustrates.

In the following, we indicate the conversion from time to energy explicitly for the case of free-particle propagators, and also exhibit the form of the corresponding spectral functions. This will bring us to a point from which we can begin the development of Green-function based computational materials science which is the aim of this book. Thus, we close the loop between the points of purely formal development of Green functions and their computational application to the study of materials properties.

The time-dependent Green function or propagator is defined as the solution of the equation (with $\hbar = 1$),

$$\left[i\frac{\partial}{\partial t} - H_0\right] G_0(\mathbf{r},t;\mathbf{r}',t') = \delta(\mathbf{r}-\mathbf{r}')\delta(t-t'). \tag{24.174}$$

Through the definitions, $\mathbf{R} = \mathbf{r} - \mathbf{r}'$ and $\tau = t - t'$, we can write this equation in the form,

$$\left[i\frac{\partial}{\partial \tau} + \nabla^2_{\mathbf{R}}\right] G_0(\mathbf{R},\tau) = \delta(\mathbf{R})\delta(\tau). \tag{24.175}$$

It is convenient to introduce the quantity $\tilde{G}_0(\mathbf{k},\omega)$, the Fourier transform of $G_0(\mathbf{r},t)$, through the relation,

$$G_0(\mathbf{R},t) = (2\pi)^{-4}\int e^{i\mathbf{k}\cdot\mathbf{R}}e^{-i\omega\tau}\tilde{G}_0(\mathbf{k},\omega)d^3k d\omega. \tag{24.176}$$

Upon substituting Eq. (24.176) and the relation,

$$\delta(\mathbf{R})\delta(t) = (2\pi)^{-4}\int \mathrm{e}^{\mathrm{i}\mathbf{k}\cdot\mathbf{R}}\mathrm{e}^{-\mathrm{i}\omega\tau}\mathrm{d}^3k\mathrm{d}\omega \tag{24.177}$$

into Eq. (24.175) we obtain the expression,

$$(2\pi)^{-4}\int \left[(\omega - k^2)\tilde{G}_0(\mathbf{k},\omega) - 1\right] \exp \mathrm{i}\mathbf{k}\cdot\mathbf{R} - \omega\tau\mathrm{d}^3k\mathrm{d}\omega = 0, \tag{24.178}$$

so that,

$$\tilde{G}_0(\mathbf{k},\omega) = \frac{1}{\omega - k^2}. \tag{24.179}$$

It now follows that $G_0(\mathbf{R},\tau)$ has the integral representation,

$$G_0(\mathbf{R},\tau) = (2\pi)^{-4}\int \frac{\exp\{\mathrm{i}(\mathbf{k}\cdot\mathbf{R} - \omega\tau)\}}{\omega - k^2}\mathrm{d}^3k\mathrm{d}\omega. \tag{24.180}$$

In order to evaluate this integral, we first consider the integral over ω,

$$I = \int_{\infty}^{-\infty} \frac{\mathrm{e}^{-\mathrm{i}\omega\tau}}{\omega - \omega_0}\mathrm{d}\omega, \tag{24.181}$$

where $\omega_0 = k^2$. The integrand has a pole at $\omega = \omega_0$ and we need a prescription on how to avoid this singularity by choosing an appropriate path of integration in the complex plane. The criterion for selecting this path is the physical requirement of causality, i.e., the Green function must lead to physically meaningful behavior. As is well known, proper paths for the integration of Eq. (24.181) are obtained by considering the limiting process,

$$I^{\pm} = \lim_{\epsilon\to 0+}\int_{-\infty}^{\infty} \frac{\mathrm{e}^{-\mathrm{i}\omega\tau}}{\omega - \omega_0 \pm \mathrm{i}\epsilon}\mathrm{d}\omega. \tag{24.182}$$

Clearly, the introduction of the term $\pm\mathrm{i}\epsilon$ has the effect of shifting the poles of the integrand below (+) or above (-) the real ω axis by a small positive quantity. Since

$$\exp(-\mathrm{i}\omega\tau) = \exp(-\mathrm{i}\tau\Re\omega)\exp(\tau\Im\omega), \tag{24.183}$$

for negative values of τ ($\tau < 0$ or $t < t'$) we should close the contour in the upper half plane. Choosing the semicircle C_1 indicated in Fig. 24.1, and letting the radius of C_1 tend to infinity, we obtain,

$$I^{+}(\tau) = 0, \qquad \tau < 0, \tag{24.184}$$

and

$$I^{-}(\tau) = 2\pi\mathrm{i}\exp(-\mathrm{i}\omega_0\tau)\exp(\epsilon\tau), \quad \tau < 0, \tag{24.185}$$

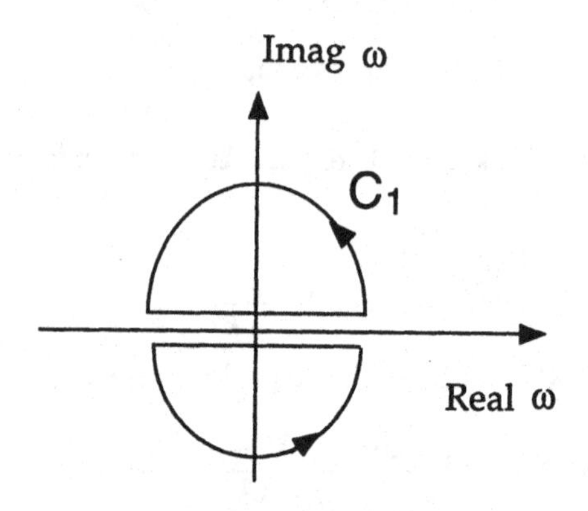

Figure 24.1: **Contours for Green function integrations.**

where the limiting process $\epsilon \to 0$ is to be understood. Similarly, for contour C_2, we have

$$I^{+}(\tau) = -2\pi \mathrm{i} \exp(-\mathrm{i}\omega_0\tau)\exp(-\epsilon\tau), \quad \tau > 0, \tag{24.186}$$

while

$$I^{-}(\tau) = 0, \qquad \tau > 0. \tag{24.187}$$

We can rewrite Eqs. (24.184) through (24.187) in more condensed form by introducing the step function $\Theta(\tau)$ such that

$$\Theta(\tau) = \begin{cases} 1 & \text{if } \tau > 0 \\ 0 & \text{if } \tau < 0 \end{cases}. \tag{24.188}$$

We now obtain,

$$I^{+}(\tau) = -2\pi \mathrm{i} \exp(-\mathrm{i}\omega_0\tau)\exp(-\epsilon\tau)\Theta(\tau), \tag{24.189}$$

and

$$I^{-}(\tau) = 2\pi \mathrm{i} \exp(-\mathrm{i}\omega_0\tau)\exp(-\epsilon\tau)\Theta(-\tau), \tag{24.190}$$

These equations can be inserted into the integral representation for G_0, Eq. (24.180), to yield the expression,

$$G_0^{+}(\mathbf{R},\tau) = -(2\pi)^{-3}\mathrm{i}\exp(-\epsilon\tau)\Theta(\tau)\int \exp(\mathrm{i}\mathbf{k}\cdot\mathbf{R})\exp(-\mathrm{i}k^2\tau)\mathrm{d}^3k, \tag{24.191}$$

corresponding to I^+, and

$$G_0^-(\mathbf{R},\tau) = (2\pi)^{-3}\mathrm{i}\exp(-\epsilon\tau)\Theta(-\tau)\int\exp(\mathrm{i}\mathbf{k}\cdot\mathbf{R})\exp(-\mathrm{i}k^2\tau)\mathrm{d}^3k, \tag{24.192}$$

corresponding to I^-. Note that the Green function G_0^+ vanishes for past times ($t' < t$), while G_0^- vanishes into the future, ($t' > t$). The function G_0^+ is called the *causal* or *retarded* Green function or propagator, and G_0^- the *advanced* Green function.

It now remains to evaluate the k-space integrals in Eqs. (24.180). Using Cartesian coordinates such that $\mathbf{R} = (X, Y, Z)$ and $\mathbf{k} = (k_x, k_y, k_z)$, we note that we must evaluate integrals of the kind,

$$\begin{aligned}\int_{-\infty}^{\infty} & \exp(\mathrm{i}k_xX)\exp(\mathrm{i}k^2\tau)\mathrm{d}k_x \\ = & \exp\left(\mathrm{i}\frac{X^2}{\tau}\right)\int_{-\infty}^{\infty}\exp\left[-\mathrm{i}\tau\left(k_x-\frac{X}{2\tau}\right)^2\right]\mathrm{d}k_x \\ = & \exp\left(\mathrm{i}\frac{X^2}{\tau}\right)\left(\frac{1}{\tau}\right)^{1/2}\int_{-\infty}^{\infty}\exp(-\mathrm{i}q_x^2)\mathrm{d}q_x,\end{aligned} \tag{24.193}$$

where $q_x = \sqrt{\tau}(k_x - X/2\tau)$. Although not strictly convergent, the last integral can be evaluated by including an integration factor, $\exp(-\alpha q_x^2)$, for α real and positive, and taking the limit $\alpha \to 0$ at the end of the calculation. Since,

$$\int_{-\infty}^{\infty}\exp[-(\alpha+\mathrm{i})q_x^2]\mathrm{d}q_x = \left(\frac{\pi}{\alpha+\mathrm{i}}\right)^{1/2}, \tag{24.194}$$

we find,

$$\int_{-\infty}^{\infty}\exp(\mathrm{i}k_xX)\exp(-\mathrm{i}k_x^2\tau)\mathrm{d}k_x = \left(\frac{\pi}{\mathrm{i}\tau}\right)^{1/2}\exp\left(\mathrm{i}\frac{X^2}{4\tau}\right). \tag{24.195}$$

With similar results for the integrals over k_y and k_z we obtain,

$$G_0^+(\mathbf{R},\tau) = -\mathrm{i}(2\pi)^{-3}\exp(-\epsilon\tau)\Theta(\tau)\left(\frac{\pi}{\mathrm{i}\tau}\right)^{1/2}\exp\left(\mathrm{i}\frac{R^2}{4\tau}\right). \tag{24.196}$$

Returning to the original variables, $(\mathbf{r}, t; \mathbf{r}', t')$, and taking the limit $\epsilon \to 0$, we can write the retarded Green function in the form,

$$G_0^+(\mathbf{r},t;\mathbf{r}',t') = -\mathrm{i}\left[\frac{1}{4\pi\mathrm{i}(t-t')}\right]^{3/2}\Theta(t-t')\exp\left[\mathrm{i}\frac{|\mathbf{r}-\mathbf{r}'|^2}{4(t-t')}\right], \tag{24.197}$$

with a similar expression for the advanced propagator,

$$G_0^-(\mathbf{r},t;\mathbf{r}',t') = \mathrm{i}\left[\frac{1}{4\pi\mathrm{i}(t-t')}\right]^{3/2}\Theta(t'-t)\exp\left[\mathrm{i}\frac{|\mathbf{r}-\mathbf{r}'|^2}{4(t-t')}\right]. \tag{24.198}$$

24.6.1 Time-independent Green functions

We now turn our attention to the determination of the free-particle propagator, $G_0^+(\mathbf{r}-\mathbf{r}')$. The spectral representation of this propagator takes the form,

$$G_0^+(\mathbf{r}-\mathbf{r}') = (2\pi)^{-3/2} \lim_{\epsilon\to 0} \int \mathrm{d}^3k' \frac{\mathrm{e}^{\mathrm{i}\mathbf{k}'\cdot(\mathbf{r}-\mathbf{r}')}}{(k'^2 - k_i^2 - \mathrm{i}\epsilon)}. \tag{24.199}$$

First, through a change of variables, $\mathbf{R} = \mathbf{r} - \mathbf{r}'$, we can write Eq. (24.199) in the form,

$$G_0^+(R) = -(2\pi)^{-3} \int_0^\infty \mathrm{d}k' k'^2 \int_0^\pi \mathrm{d}\theta' \sin\theta' \int_0^{2\pi} \mathrm{d}\phi' \frac{\exp(\mathrm{i}kR\cos\theta')}{k' - k^2 - \mathrm{i}\epsilon}, \tag{24.200}$$

where we have chosen polar coordinates so that the z-axis is along the vector $\mathbf{R}$. Upon performing the angular integration in Eq. (24.200), we obtain

$$G_0^+(R) = -(4\pi^2 R)^{-1} \int_{-\infty}^{\infty} \frac{k' \sin k'R}{k' - k^2 - \mathrm{i}\epsilon} \mathrm{d}k'. \tag{24.201}$$

Since the poles of the integrand are displaced from the real axis into the upper half plane, the contour of integration can be closed in the upper plane yielding the result,

$$\begin{aligned} G_0^+(|\mathbf{r}-\mathbf{r}'|) &= -\frac{1}{4\pi}\frac{\mathrm{e}^{\mathrm{i}kR}}{R} \\ &= -\frac{1}{4\pi}\frac{\mathrm{e}^{\mathrm{i}k|\mathbf{r}-\mathbf{r}'|}}{|\mathbf{r}-\mathbf{r}'|}. \end{aligned} \tag{24.202}$$

The reader is urged to derive the corresponding expression for G_0^-. The derivation just given provides a justification for the retention of the terms retarded and advanced when referring to the propagator in the energy rather than the time representation. The expression for the free-particle Green function given in the last equation, and its expansion in terms of angular-momentu states, have figured prominently in our previous discussion.

24.6.2 Spectral representations

The spectral representation of the free single-particle propagator is given by Eq. (24.180). A Fourier transformation over the time variable leads to,

$$G_0(\mathbf{r},\mathbf{r}',E) = \int \mathrm{d}^3k \frac{\exp \mathrm{i}\,[\mathbf{k}\cdot(\mathbf{r}-\mathbf{r}')]}{E - k^2}. \tag{24.203}$$

This can also be written in the form,

$$G_0(\mathbf{r},\mathbf{r}',E) = \int \mathrm{d}^3k \frac{u_{\mathbf{k}}(\mathbf{r})u^*_{\mathbf{k}}(\mathbf{r}')}{E-k^2}, \tag{24.204}$$

where $u_{\mathbf{k}}(\mathbf{r}) = \exp \mathrm{i}\,(\mathbf{k}\cdot\mathbf{r})$ denotes a plane wave [normalized to $(1\pi)^3$]. Similarly, for the two-particle Green function, we obtain

$$G_0^{(2)}(\mathbf{r}_1,\mathbf{r}_2;\mathbf{r}'_1,\mathbf{r}'_2;E) = \int \mathrm{d}^3k_1 \int \mathrm{d}^3k_2 \frac{U_{\mathbf{k}_1\mathbf{k}_2}(\mathbf{r}_1,\mathbf{r}_2)\cdot U^*_{\mathbf{k}_1\mathbf{k}_2}(\mathbf{r}_1,\mathbf{r}_2)}{E-k_1^2-k_2^2}, \tag{24.205}$$

where the + sign corresponds to bosons and the − to fermions[12] and $U_{\mathbf{k}_1\mathbf{k}_2}(\mathbf{r}_1,\mathbf{r}_2) = [u_{\mathbf{k}_1}(\mathbf{r}_1)u_{\mathbf{k}_2}(\mathbf{r}_2) \pm u_{\mathbf{k}_1}(\mathbf{r}_2)u_{\mathbf{k}_2}(\mathbf{r}_1)]$. These expressions can be made the staring points for the study of interacting many-particle systems, such as the electrons and nuclei comprising a solid material.

24.7 Superconductivity

As a final application of the formalism of many-particle Green functions given above, we discuss briefly the BCS theory of superconductivity. Since temperature plays a vital role in the phenomenon of superconductivity, our previous expositions of Green functions, developed exclusively for the $T = 0$ case, must be generalized to finite temperatures. The basic elements of such a generalization are provided in the next subsection. A Green function formulation of the theory of BCS is given in the following subsection. For detailed discussions of temperature Green functions, the reader is urged to consult the literature, e.g., the book by Mahan [11], which forms the basis for the discussion in the next section.

24.7.1 Temperature Green functions

The concept of temperature enters the discussion through the assumption that a particle (electron, phonon, spin) is in contact with a reservoir consisting of other particles with an assigned average energy. This energy can be expressed in terms of a temperature.

Defining the parameter $\beta = 1/k_{\mathrm{B}}T$, where k_{B} is Boltzmann's constant, we begin by examining the quantity,

$$\frac{\mathrm{Tr}[\mathrm{e}^{-\beta H}C_{\mathbf{p}}(t)C^\dagger_{\mathbf{p}}(t')]}{\mathrm{Tre}^{-\beta H}}. \tag{24.206}$$

[12] We recall that spatial coordinates include spin indices and that integrals contain summations over spin variables.

This expression obviously contains a thermodynamic average and is a good candidate for the Green function of an electron at finite temperature. Here, the trace denotes a sum over a complete set of states,

$$\mathrm{Tr} \equiv \sum_n \langle n| \cdots |n\rangle, \tag{24.207}$$

and the creation and destruction operators are given in the Heisenberg picture by the expression,

$$C_{\mathbf{p}}(t) = \mathrm{e}^{\mathrm{i}Ht} C_{\mathbf{p}} \mathrm{e}^{-\mathrm{i}Ht}. \tag{24.208}$$

We now note that the Hamiltonian, $H = H_0 + V$, occurs in two different places in Eq. (24.206): once in the exponential, $\mathrm{e}^{-\beta H}$, and also in the definition of the Heisenberg operators. In a formal analysis, say through a perturbation expansion, the perturbing part of the Hamiltonian, V, would make its appearance once in the expansion of the time exponentials and once in the expansion of the thermodynamic factors. This complication can be avoided through a formalism introduced by Matsubara [12].

Let us motivate the Matsubara formalism by examining the thermal occupation numbers for both fermions, $n_{\mathrm{F}}(\xi_{\mathbf{p}}) = (\mathrm{e}^{\beta\xi_{\mathbf{p}}}+1)^{-1}$, and for bosons, $n(\omega_{\mathbf{q}}) = (\mathrm{e}^{\beta\omega_{\mathbf{q}}} - 1)^{-1}$. Using the well-known theorem of complex analysis, stating that a meromorphic function can be expanded as a summation over its poles and the residues there, we have,

$$n_{\mathrm{F}}(\xi_{\mathbf{p}}) = (\mathrm{e}^{\beta\xi_{\mathbf{q}}} + 1)^{-1} = \frac{1}{2} + \frac{1}{\beta} \sum_{n=-\infty}^{\infty} \frac{1}{(2n+1)\mathrm{i}\pi/\beta - \xi_{\mathbf{p}}} \tag{24.209}$$

$$n_{\mathrm{B}}(\omega_{\mathbf{q}}) = (\mathrm{e}^{\beta\omega_{\mathbf{q}}} - 1)^{-1} = \frac{1}{2} + \frac{1}{\beta} \sum_{n=-\infty}^{\infty} \frac{1}{2n\mathrm{i}\pi/\beta - \omega_{\mathbf{p}}}, \tag{24.210}$$

because the poles of n_{F} are at the points $\xi_{\mathbf{p}} = (2n+1)\mathrm{i}\pi/\beta$ and those of n_{B} at $\omega_{\mathbf{q}} = 2n\mathrm{i}\pi/\beta$. At these poles, we define the corresponding frequencies,

$$\omega_n = \frac{(2n+1)\pi}{\beta}, \quad \text{for fermions}, \tag{24.211}$$

$$\omega_n = \frac{2n\pi}{\beta}, \quad \text{for bosons}. \tag{24.212}$$

We see that fermions (bosons) have poles at odd (even) multiples of π/β. Now, the summations in Eqs. (24.209) and (24.210) can be written in the forms,

$$\sum_n \frac{1}{\mathrm{i}\omega_n - \xi_{\mathbf{p}}} \quad \text{or} \quad \sum_n \frac{1}{\mathrm{i}\omega_n - \omega_{\mathbf{q}}}, \tag{24.213}$$

where n runs over odd integers for the case of fermions and over even integers for that of bosons. We note that in both cases, the factors,

$$\frac{1}{i\omega_n - \xi_{\mathbf{p}}} \text{ or } \frac{1}{i\omega_n - \omega_{\mathbf{q}}}, \tag{24.214}$$

have the forms of Green functions and are, as we see below, the unperturbed Green functions for fermions and bosons, respectively, in the Matsubara formalism.

We can now take advantage of the interplay between temperature and frequency in either of two ways. We can either think of $\beta = 1/k_{\mathrm{B}}T$ as a complex time, or think of time as a complex temperature. In his formalism, Matsubara followed the latter approach. Now, time becomes a complex quantity, usually denoted by $\tau = it$. The corresponding Green functions are functions of τ which is defined in the domain,

$$-\beta \leq \tau \leq \beta. \tag{24.215}$$

The interplay between temperature (complex time) and frequency (energy) can be strengthened through considerations of the corresponding Fourier transformations. Let $f(\tau)$ be a function of τ over the domain, $-\beta \leq \tau \leq \beta$. We can write the general results,

$$f(\tau) = \frac{1}{2}a_0 + \sum_{n=1}^{\infty}\left[a_n \cos\left(\frac{n\pi\tau}{\beta}\right) + b_n \sin\left(\frac{n\pi\tau}{\beta}\right)\right] \tag{24.216}$$

where

$$a_n = \frac{1}{\beta}\int_{-\beta}^{\beta} \mathrm{d}\tau\tau f(\tau) \cos\left(\frac{n\pi\tau}{\beta}\right) \tag{24.217}$$

$$b_n = \frac{1}{\beta}\int_{-\beta}^{\beta} \mathrm{d}f(\tau) \sin\left(\frac{n\pi\tau}{\beta}\right). \tag{24.218}$$

Defining the quantity

$$f(i\omega_n) = \frac{\beta(a_n + ib_n)}{2}, \tag{24.219}$$

we can write,

$$f(\tau) = \frac{1}{\beta}\sum_{n=-\infty}^{\infty} \mathrm{e}^{-in\pi\tau/\beta} f(i\omega_n) \tag{24.220}$$

and

$$f(i\omega_n) = \frac{1}{2}\int_{-\beta}^{\beta} \mathrm{d}\tau f(\tau)\mathrm{e}^{in\pi\tau/\beta}. \tag{24.221}$$

These expressions can be simplified further. Using the fact that for bosons[13],

$$f(\tau+\beta) = f(\tau), \quad \text{when} \quad -\beta < \tau < 0 \quad (\text{or } 0 < \tau + \beta < \beta), \tag{24.222}$$

we divide the integral in Eq. (24.221) into negative and positive parts of its domain of integration,

$$f(\mathrm{i}\omega_n) = \frac{1}{2}\left[\int_0^\beta \mathrm{d}\tau f(\tau)\mathrm{e}^{\mathrm{i}n\pi\tau/\beta} + \int_{-\beta}^0 \mathrm{d}\tau f(\tau)\mathrm{e}^{\mathrm{i}n\pi\tau/\beta}\right], \tag{24.223}$$

and change the variable of integration from τ to $\tau+\beta$ in the second term in the brackets to obtain,

$$f(\mathrm{i}\omega_n) = \frac{1}{2}\int_0^\beta \mathrm{d}\tau f(\tau)\mathrm{e}^{\mathrm{i}n\pi\tau/\beta}. \tag{24.224}$$

It is seen that $f(\mathrm{i}\omega_n) = 0$ when n is odd. Therefore, for bosons, we have,

$$f(\mathrm{i}\omega_n) = \int_0^\beta \mathrm{d}\tau f(\tau)\mathrm{e}^{\mathrm{i}\omega_n\tau}, \tag{24.225}$$

$$f(\tau) = \frac{1}{\beta}\sum_n f(\mathrm{i}\omega_n)\mathrm{e}^{-\mathrm{i}\omega_n\tau}, \tag{24.226}$$

$$\omega_n = \frac{2n\pi}{\beta}. \tag{24.227}$$

These are the standard forms of Fourier transformations between a time and an energy domain, and are consistent with the fact that boson frequencies contain only even integers.

For the case of fermions, we have,

$$f(\tau+\beta) = -f(\tau), \quad \text{when} \quad -\beta < \tau < 0, \tag{24.228}$$

and in place of Eqs. (24.225) to (24.227), we obtain the expressions,

$$f(\mathrm{i}\omega_n) = \int_0^\beta \mathrm{d}\tau f(\tau)\mathrm{e}^{\mathrm{i}\omega_n\tau}, \tag{24.229}$$

$$f(\tau) = \frac{1}{\beta}\sum_n f(\mathrm{i}\omega_n)\mathrm{e}^{-\mathrm{i}\omega_n\tau}, \tag{24.230}$$

$$\omega_n = \frac{(2n+1)\pi}{\beta}, \tag{24.231}$$

where the fermion frequencies occur at odd multiples of π/β.

[13]The reader may wish to prove this result.

24.7.2 Matsubara Green functions

Because finite-temperature Green functions eventually involve thermodynamic averages, it is convenient to work within a grand-canonical ensemble that allows for a variable number of particles. In this formulation, the Hamiltonian is replaced by $K = H - \mu N$, where μ is the chemical potential for the particles under consideration. The formulation also includes the thermodynamic potential, Ω, as the normalization factor in a thermodynamic average,

$$\mathrm{e}^{-\beta\Omega} = \mathrm{Tr}\left(\mathrm{e}^{-\beta(H-\mu N)}\right) = \mathrm{Tr}\mathrm{e}^{-\beta K}. \tag{24.232}$$

We now define the electron Green function by means of the expression,

$$\mathcal{G}(\mathbf{p}, \tau - \tau') = -\langle T_\tau C_{\mathbf{p}}(t) C_{\mathbf{p}}^\dagger(t')\rangle \tag{24.233}$$

$$\mathcal{G}(\mathbf{p}, \tau - \tau') = \mathrm{Tr}\left[\mathrm{e}^{-\beta(K-\Omega)} T_\tau \mathrm{e}^{\tau K} C_{\mathbf{p}} \mathrm{e}^{-K(\tau-\tau')} C_{\mathbf{p}}^\dagger \mathrm{e}^{-\tau K}\right], \tag{24.234}$$

where T_τ is a τ ordering operator that arranges operators with earliest τ (closest to $-\beta$) to the right. We use the symbol $\mathcal{G}$ to denote a Green function of complex time and complex frequency.

We now show that $\mathcal{G}$ is a function only of the difference $\tau - \tau'$ as indicated in the last two expressions. Separating the cases $\tau > \tau'$ and $\tau < \tau'$, we have

$$\begin{aligned}\mathcal{G}(\mathbf{p}, \tau - \tau') = & - \Theta(\tau - \tau')\mathrm{Tr}\left[\mathrm{e}^{-\beta(K-\Omega)} \mathrm{e}^{\tau K} C_{\mathbf{p}} \mathrm{e}^{-K(\tau-\tau')} C_{\mathbf{p}}^\dagger \mathrm{e}^{-\tau' K}\right] \\ & + \Theta(\tau' - \tau)\mathrm{Tr}\left[\mathrm{e}^{-\beta(K-\Omega)} \mathrm{e}^{\tau' K} C_{\mathbf{p}}^\dagger \mathrm{e}^{-K(\tau'-\tau)} C_{\mathbf{p}} \mathrm{e}^{-\tau K}\right],\end{aligned} \tag{24.235}$$

where the sign difference between the two terms appears because of the interchange of two fermion operators. At this point we can use the invariance of the trace under cyclic interchange of operators to move the factors $\mathrm{e}^{-\tau' K}$ and $\mathrm{e}^{-\beta(K-\Omega)}$ to the beginning of the corresponding products, and commute the exponential operators in the form,

$$\mathrm{e}^{-\tau' K} \mathrm{e}^{-\beta(K-\Omega)} = \mathrm{e}^{-\beta(K-\Omega)} \mathrm{e}^{-\tau' K}, \tag{24.236}$$

an operation which is justified because Ω is only a scalar function of β and μ, and not an operator. This leads to the expression,

$$\begin{aligned}\mathcal{G}(\mathbf{p}, \tau - \tau') = & - \Theta(\tau - \tau')\mathrm{Tr}\left[\mathrm{e}^{-\beta(K-\Omega)} \mathrm{e}^{K(\tau-\tau')} C_{\mathbf{p}} \mathrm{e}^{-K(\tau-\tau')} C_{\mathbf{p}}^\dagger\right] \\ & + \Theta(\tau' - \tau)\mathrm{Tr}\left[\mathrm{e}^{-\beta(K-\Omega)} \mathrm{e}^{K(\tau-\tau')} C_{\mathbf{p}}^\dagger \mathrm{e}^{-K(\tau'-\tau)} C_{\mathbf{p}}\right],\end{aligned} \tag{24.237}$$

which is indeed a function of the difference $\tau - \tau'$. We can also eliminate one of the time variables since it only fixes an inessential origin of the time axis, so that we can write,

$$\begin{aligned} \mathcal{G}(\mathbf{p},\tau) &= -\langle T_\tau C_\mathbf{p}(\tau) C_\mathbf{p}(0)\rangle \\ &= -\mathrm{Tr}\left[\mathrm{e}^{-\beta(K-\Omega)} T_\tau C_\mathbf{p} \mathrm{e}^{-\tau K} C_\mathbf{p}^\dagger\right]. \end{aligned} \tag{24.238}$$

It can be shown [11] that $\mathcal{G}(\mathbf{p},\tau)$ satisfies the property of Eq. (24.228), namely $\mathcal{G}(\mathbf{p},\tau+\beta) = -\mathcal{G}(\mathbf{p},\tau)$. It follows that the Green function can be expanded in a Fourier series according to Eqs. (24.229) and (24.230),

$$\mathcal{G}(\mathbf{p}, \mathrm{i}\omega_n) = \int_0^\beta \mathrm{d}\tau \mathcal{G}(\mathbf{p},\tau) \mathrm{e}^{\mathrm{i}\tau\omega_n} \tag{24.239}$$

and

$$\mathcal{G}(\mathbf{p},\tau) = \frac{1}{\beta}\sum_n \mathcal{G}(\mathbf{p}, \mathrm{i}\omega_n) \mathrm{e}^{-\mathrm{i}\tau\omega_n}. \tag{24.240}$$

Finally, we define expressions for the unperturbed or free-particle Green function. For the unperturbed Hamiltonian, H_0, we take the form,

$$H_0 = \sum_\mathbf{p} \epsilon_\mathbf{p} C_\mathbf{p}^\dagger C_\mathbf{p}, \tag{24.241}$$

so that

$$K_0 = H_0 - \mu N = \sum_\mathbf{p} (\epsilon_\mathbf{p} - \mu) C_\mathbf{p}^\dagger C_\mathbf{p} = \sum_\mathbf{p} \xi_\mathbf{p} C_\mathbf{p}^\dagger C_\mathbf{p}. \tag{24.242}$$

From the time evolution[14] of the operators $C_\mathbf{p}$ and $C_\mathbf{p}^\dagger$,

$$C_\mathbf{p}(\tau) = \mathrm{e}^{\tau K_0} C_\mathbf{p} \mathrm{e}^{-\tau K_0} = \mathrm{e}^{-\tau\xi_\mathbf{p}} C_\mathbf{p} \tag{24.243}$$

$$C_\mathbf{p}^\dagger(\tau) = \mathrm{e}^{\tau K_0} C_\mathbf{p}^\dagger \mathrm{e}^{-\tau K_0} = \mathrm{e}^{\tau\xi_\mathbf{p}} C_\mathbf{p}^\dagger, \tag{24.244}$$

we obtain the expression,

$$\mathcal{G}_0(\mathbf{p},\tau) = -\Theta(\tau)\mathrm{e}^{-\tau\xi_\mathbf{p}} \langle C_\mathbf{p} C_\mathbf{p}^\dagger\rangle + \Theta(-\tau)\mathrm{e}^{-\tau\xi_\mathbf{p}} \langle C_\mathbf{p}^\dagger C_\mathbf{p}\rangle. \tag{24.245}$$

This can also be put in the form,

$$\begin{aligned} \mathcal{G}_0(\mathbf{p},\tau) &= -\mathrm{e}^{-\tau\xi_\mathbf{p}}\{\Theta(\tau)[1 - n_\mathbf{p}(\xi_\mathbf{p})] - \Theta(-\tau) n_\mathbf{p}(\xi_\mathbf{p})\} \\ &= -\mathrm{e}^{-\tau\xi_\mathbf{p}}[\Theta(\tau) - n_\mathrm{F}(\xi_\mathbf{p})], \end{aligned} \tag{24.246}$$

[14] This follows from the theorem $\mathrm{e}^A C \mathrm{e}^{-A} = C + [A,C] + \frac{1}{2!}[A,[A,C]] + \cdots$.

where $n_F(\xi_\mathbf{p})$ is the expectation values of the number operator for fermions,

$$n_F(\xi_\mathbf{p}) = \langle C_\mathbf{p} C_\mathbf{p}^\dagger \rangle, \tag{24.247}$$

which for non interacting particles has the form,

$$n_F(\xi_\mathbf{p}) = \frac{1}{e^{\beta \xi_\mathbf{p}} + 1}. \tag{24.248}$$

The Green function in the frequency domain is obtained by means of a Fourier transform,

$$\begin{aligned} \mathcal{G}_0(\mathbf{p}, i\omega_n) &= \int_0^\beta d\tau e^{i\omega_n \tau} \mathcal{G}_0(\mathbf{p}, \tau) = -(1 - n_F) \int_0^\beta d\tau e^{\tau(i\omega_n - \xi_\mathbf{p})} \\ &= -\frac{(1 - n_F)(e^{\beta(i\omega_n - \xi_\mathbf{p})} - 1)}{i\omega_n - \xi_\mathbf{p}}. \end{aligned} \tag{24.249}$$

Recalling that,

$$i\beta\omega_n = i(2n+1)\pi \tag{24.250}$$

$$e^{\beta i\omega_n} = -1, \tag{24.251}$$

we find,

$$\begin{aligned} \mathcal{G}_0(\mathbf{p}, i\omega_n) &= \frac{(1 - n_F)(e^{\beta \xi_\mathbf{p}} + 1)}{i\omega_n - \xi_\mathbf{p}} \\ &= \frac{1}{i\omega_n - \xi_\mathbf{p}}, \end{aligned} \tag{24.252}$$

where the last equality follows from Eq. (24.248). This is indeed the expression for the frequency Green function for free particles quoted in Eq. (24.214).

Temperature Green functions for phonons and photons can also be derived [11] following procedures analogous to those indicated above for the case of fermions. In particular, for the case of phonons, we have [11],

$$\mathcal{D}(\mathbf{q}, \tau - \tau') = -\langle T_\tau A(\mathbf{q}, \tau) A(-\mathbf{q}, \tau') \rangle \tag{24.253}$$

where

$$A(\mathbf{q}, \tau) = e^{\tau H} \left(a_\mathbf{q} + a_{-\mathbf{q}}^\dagger \right) e^{-\tau H}, \tag{24.254}$$

and where $a_\mathbf{q}$ ($a_{-\mathbf{q}}^\dagger$) is an annihilation (creation) operator for a phonon of wave vector $\mathbf{q}$, with the brackets denoting thermodynamic average,

$$\langle \cdots \rangle = \mathrm{Tr} \left(e^{-\beta(H - \Omega)} \ldots \right). \tag{24.255}$$

The free-particle propagator has the form,

$$\mathcal{D}_0(\mathbf{q}, \mathrm{i}\omega_n) = -\frac{2\omega_\mathbf{q}}{\omega_n^2 + \omega_\mathbf{q}^2}. \tag{24.256}$$

Finally, photon propagators can be defined according to the expression,

$$\mathcal{D}^{\mu\nu}(\mathbf{k}, \tau) = -\langle \mathrm{T}_\tau A_\mu(\mathbf{q}, \tau) A_\nu(-\mathbf{k}, 0)\rangle, \tag{24.257}$$

where,

$$A_\mu(\mathbf{k}, \tau) = \mathrm{e}^{\tau H} \zeta_{\mathbf{k}\mu} \left(\frac{2\pi}{\omega_\mathbf{k}}\right)^{1/2} (a_{\mathbf{k}\mu} + a_{-\mathbf{k}\mu}) \mathrm{e}^{-\tau H} \tag{24.258}$$

with A_μ being the vector potential operator associated with the μ direction. The ζ are unit polarization vectors and the $a_{\mathbf{k}\mu}$ ($a^\dagger_{-\mathbf{k}\mu}$) annihilation (creation) operators for the associated fields. The free-particle propagator has the form [11],

$$\mathcal{D}^{\mu\nu}(\mathbf{k}, \mathrm{i}\omega_n) = \frac{4\pi(\delta_{\mu\nu} - k_\mu k_\nu/k^2)}{\omega_n^2 + \omega_\mathbf{k}^2}. \tag{24.259}$$

Matsubara Green functions are particularly useful in the study of interacting quantum systems. One of their desirable properties is that they lead directly to physical results. For example, the retarded (advanced) Green function is obtained by the analytic continuation $\mathrm{i}\omega_n \to \omega + \mathrm{i}\delta$ ($\mathrm{i}\omega_n \to \omega - \mathrm{i}\delta$). Therefore, the Matsubara formalism yields quantities that are directly comparable to experiment. A more detailed exposition of the Matsubara formalism can be found in the book by Mahan [11].

24.7.3 Superconductivity

We are now ready to consider some of the basic elements of a Green-function formulation [11, 13] of the BCS theory of superconductivity. As we saw, this theory is based on the pairing of two electrons of opposite momenta and spins, $(\mathbf{k}, \uparrow)$ and $(-\mathbf{k}, \downarrow)$. Because of this, it is necessary to consider, in addition to Green functions, two so-called correlation functions, which are Green-function like quantities but for Cooper pairs. Following Abrikosov, et al. [14], we define

$$\begin{aligned}
\mathcal{G}(\mathbf{p}, \tau - \tau') &= -\langle \mathrm{T}_\tau C_{\mathbf{p}\sigma}(\tau) C^\dagger_{\mathbf{p}\sigma}(\tau')\rangle \\
\mathcal{F}(\mathbf{p}, \tau - \tau') &= \langle \mathrm{T}_\tau C_{-\mathbf{p}\downarrow}(\tau) C_{\mathbf{p}\uparrow}(\tau')\rangle \\
\mathcal{F}^\dagger(\mathbf{p}, \tau - \tau') &= \langle \mathrm{T}_\tau C^\dagger_{\mathbf{p}\uparrow}(\tau) C^\dagger_{-\mathbf{p}\downarrow}(\tau')\rangle.
\end{aligned} \tag{24.260}$$

The Green function, $\mathcal{G}$, is defined in the usual way, as discussed above. But the two correlation functions, $\mathcal{F}$ and $\mathcal{F}^\dagger$, are distinguished from $\mathcal{G}$ by some very important features. First, we note that both vanish in the normal state because no correlation exists between the spins there. Although they are non-zero in the superconducting state, the averages contained in them connect states that differ from one another by two particles. For example, in $\mathcal{F}$, a ket $|\rangle$ is operated upon by two destruction operators so that the bra, $\langle|$, which closes the expression must represent a state with two particles less than that denoted by $|\rangle$. Particle non-conservation is a basic feature of BCS theory and the grand canonical ensemble, which allows for variable numbers of particles, is ideally suited for the discussion.

We seek self-consistent equations for $\mathcal{G}$, $\mathcal{F}$, and $\mathcal{F}^\dagger$. It is clear on physical grounds that at high enough temperatures, $\mathcal{F} = \mathcal{F}^\dagger = 0$, but non-zero solutions can become possible at low temperatures. The discussion proceeds on the basis of a model phenomenological Hamiltonian of the form,

$$H = \sum_{\mathbf{p}\sigma} \xi_{\mathbf{p}} C^\dagger_{\mathbf{p}\sigma} C_{\mathbf{p}\sigma} + \frac{1}{2\Omega} \sum_{\mathbf{q}\mathbf{p}\mathbf{p}'} \sum_{\sigma\sigma'} V(\mathbf{q}) C^\dagger_{\mathbf{p}+\mathbf{q},\sigma} C^\dagger_{\mathbf{p}'-\mathbf{q},\sigma'} C_{\mathbf{p}'\sigma'} C_{\mathbf{p}\sigma} \quad (24.261)$$

where Ω denotes the volume of the system. The first term represents the energy of the electron gas, whereas the second contains an effective two-particle interaction. For the sake of simplicity of exposition, we take this interaction to be a constant, $V(\mathbf{q}) = -V_0$, over a range of energies within the Debye energy of the Fermi surface. More realistic expressions for $V(\mathbf{q})$ are possible [11], but this form is sufficient to reveal some of the most salient features of the theory of BCS.

In order to derive the self-consistent equations coupling the Green function and the two correlation functions, we consider the equation of motion of various operators. For example,

$$\frac{\partial}{\partial\tau} C_{\mathbf{p}\sigma}(\tau) = [H, C_{\mathbf{p}\sigma}] = -\xi_{\mathbf{p}} C_{\mathbf{p}\sigma} - \frac{1}{\Omega} \sum_{\mathbf{p}'\mathbf{q}\sigma'} V(\mathbf{q}) C^\dagger_{\mathbf{p}'-\mathbf{q},\sigma'} C_{\mathbf{p}'\sigma'} C_{\mathbf{p}-\mathbf{q},\sigma}. \quad (24.262)$$

The definition of the Green function as a time-ordered product leads to the expression,

$$\begin{aligned}
\frac{\partial}{\partial\tau} \mathcal{G}(\mathbf{p}, \tau - \tau') &= -\frac{\partial}{\partial\tau} \left[\Theta(\tau - \tau') \langle C_{\mathbf{p}\sigma}(\tau) C^\dagger_{\mathbf{p}\sigma}(\tau')\rangle \right. \\
&- \left. \Theta(\tau' - \tau) \langle C^\dagger_{\mathbf{p}\sigma}(\tau') C_{\mathbf{p}\sigma}(\tau)\rangle\right] \\
&= -\delta(\tau - \tau') \langle [C_{\mathbf{p}\sigma}, C^\dagger_{\mathbf{p}\sigma}]_+\rangle \\
&- \langle \mathrm{T}_\tau \left[\frac{\partial}{\partial\tau} C_{\mathbf{p}\sigma}\right] C^\dagger_{\mathbf{p}\sigma}(\tau')\rangle \\
&= -\delta(\tau - \tau') - \langle \mathrm{T}_\tau \left[\frac{\partial}{\partial\tau} C_{\mathbf{p}\sigma}(\tau)\right] \\
C & {}^\dagger_{\mathbf{p}\sigma}(\tau')\rangle.
\end{aligned} \quad (24.263)$$

From the equation of motion, Eq. (24.262), we have

$$\left(-\frac{\partial}{\partial \tau} - \xi_{\mathbf{p}}\right)\mathcal{G}(\mathbf{p},\tau-\tau') + \frac{1}{\Omega}\sum_{\mathbf{p}'\mathbf{q}\sigma} V(\mathbf{q})$$
$$\times \langle \mathrm{T}_\tau C^\dagger_{\mathbf{p}'-\mathbf{q},\sigma}(\tau)C_{\mathbf{p}'\sigma'}(\tau)C_{\mathbf{p}-\mathbf{q},\sigma}(\tau)C^\dagger_{\mathbf{p}\sigma}(\tau')\rangle = \delta(\tau-\tau'). \tag{24.264}$$

We now must find ways to evaluate the product of the four operators that occurs in the last expression. Clearly, combinations such as $\langle CC^\dagger\rangle\langle CC^\dagger\rangle$ will lead to the appearance of the Green function. It is also possible to decouple in the form $\langle CC\rangle\langle C^\dagger C^\dagger\rangle$ for which the correlation functions $\mathcal{F}$ and $\mathcal{F}^\dagger$ become relevant. It is usually a reasonable approximation to assume that in the long wavelength limit, $\mathbf{q}\to 0$, phonons do not contribute to the effective electron-electron interaction, and to neglect pairing when $\mathbf{q}=0$. In the case of a normal metal, we would have, $\mathbf{p}=\mathbf{p}'$ and $\sigma=\sigma'$, so that the only pairing would be of the form, $\delta_{\mathbf{p}\mathbf{p}'}\delta_{\sigma\sigma'}n_{\mathbf{p}-\mathbf{q}}\mathcal{G}(\mathbf{p},\tau-\tau')$. Such pairings also occurs in the superconducting state, but other pairings occur as well. These other pairings lead to correlation functions, $\mathcal{F}$ and $\mathcal{F}^\dagger$, and in evaluating them special attention must be paid to the spin variables. Choosing $\sigma=-\sigma'=\uparrow$ leads to the expression

$$\begin{aligned} &- \langle C_{\mathbf{p}'\downarrow}(\tau)C_{\mathbf{p}-\mathbf{q}\uparrow}(\tau)\rangle\langle \mathrm{T}_\tau C^\dagger_{\mathbf{p}\uparrow}(\tau')C^\dagger_{\mathbf{p}'-\mathbf{q}\downarrow}(\tau')\rangle \\ &= -\delta_{\sigma,-\sigma'}\delta_{\mathbf{p}'+\mathbf{p}-\mathbf{q}}\mathcal{F}(\mathbf{p}-\mathbf{q},0)\mathcal{F}^\dagger(\mathbf{p},\tau-\tau'). \end{aligned} \tag{24.265}$$

Similarly, choosing $\sigma=-\sigma'=\downarrow$, we have

$$\begin{aligned} &- \langle C_{\mathbf{p}-\mathbf{q},\downarrow}(\tau)C_{\mathbf{p}'\uparrow}(\tau)\rangle\langle \mathrm{T}_\tau C^\dagger_{\mathbf{p}'-\mathbf{q},\uparrow}(\tau)C^\dagger_{\mathbf{p}\downarrow}(\tau')\rangle \\ &= -\delta_{\sigma,-\sigma'}\delta_{\mathbf{p}'-\mathbf{p}+\mathbf{q}}\mathcal{F}(-\mathbf{p}+\mathbf{q},0)\mathcal{F}^\dagger(-\mathbf{p},\tau-\tau'). \end{aligned} \tag{24.266}$$

It is shown below that the correlation functions, $\mathcal{F}$ and $\mathcal{F}^\dagger$, are independent of the signs of their arguments, either momentum or τ, so that the last two expressions are identical. Now, the last term in Eq. (24.264) takes the form,

$$\begin{aligned} &\frac{1}{\Omega}\sum_{\mathbf{p}\mathbf{q}\sigma} V(\mathbf{q})\langle \mathrm{T}_\tau C_{\mathbf{p}'-\mathbf{q}\sigma'}(\tau)C_{\mathbf{p}'\sigma'}(\tau)C_{\mathbf{p}-\mathbf{q},\sigma}(\tau)C^\dagger_{\mathbf{p}\sigma}(\tau')\rangle \\ &= \frac{1}{\Omega}\sum_{\mathbf{q}} V(\tau)\left[\mathcal{G}(\mathbf{p},\tau-\tau')n_{\mathbf{p}-\mathbf{q}} - \mathcal{F}(\mathbf{p}-\mathbf{q},0)\mathcal{F}^\dagger(\mathbf{p},\tau-\tau')\right]. \end{aligned} \tag{24.267}$$

The first term inside the brackets represents the usual contribution to the self-energy of the electron arising from the interaction with the other electrons. This term can be shown to be essentially insensitive to the onset of superconductivity. It is the term mainly responsible for the mass enhancement of the electron as discussed in Chapter 22. In strong superconductors, such as lead, the mass enhancement is large enough to warrant inclusion of this term throughout the discussion. In what follows, we consider only the weak-coupling limit in which this term is usually neglected. The second term inside the brackets in Eq. (24.267) contains a factor,

$$\Delta(\mathbf{p}) = -\frac{1}{\Omega}\sum_{\mathbf{q}} V(\mathbf{q})\mathcal{F}(\mathbf{p}-\mathbf{q},0). \tag{24.268}$$

This is the well-known gap function of BCS and plays a central role in determining the properties of superconductors. When $V(\mathbf{p})$ is negative, $\Delta(\mathbf{p})$ is positive. With the definition of the gap function, Eq. (24.264) takes the form,

$$\left(-\frac{\partial}{\partial\tau} - \xi_{\mathbf{p}}\right)\mathcal{G}(\mathbf{p},\tau-\tau') + \Delta(\mathbf{p})\mathcal{F}'(\mathbf{p},\tau-\tau') = \delta(\tau-\tau'). \tag{24.269}$$

This equation contains two unknowns, $\mathcal{G}$ and $\mathcal{F}^\dagger$, so that another equation is needed for a solution. To obtain it, we consider the equation of motion of $\mathcal{F}^\dagger$, which gives,

$$\frac{\partial}{\partial\tau}\mathcal{F}^\dagger(\mathbf{p},\tau-\tau') = \langle \mathrm{T}_\tau \left[\frac{\partial}{\partial\tau}C^\dagger_{\mathbf{p}\uparrow}(\tau)\right] C_{-\mathbf{p}\downarrow}(\tau')\rangle. \tag{24.270}$$

We note the absence of a $\delta(\tau-\tau')$ compared to the corresponding equation for the Green function, which is due to the anticommutation property of creation and annihilation operators for fermions. From the time development of $C^\dagger_{\mathbf{p}\sigma}$,

$$\frac{\partial}{\partial\tau}C^\dagger_{\mathbf{p}\sigma}(\tau) = [H, C^\dagger_{\mathbf{p}\sigma}] = \xi_{\mathbf{p}}C^\dagger_{\mathbf{p}\sigma} + \frac{1}{\Omega}\sum_{\mathbf{p}'\mathbf{q}\sigma'} V(\mathbf{q})C^\dagger_{\mathbf{p}'-\mathbf{q},\sigma}C_{\mathbf{p}'+\mathbf{q},\sigma'}C_{\mathbf{p}\sigma'}, \tag{24.271}$$

we obtain the expression,

$$\begin{aligned}\left(-\frac{\partial}{\partial\tau}\right. &+ \left.\xi_{\mathbf{p}}\right)\mathcal{F}^\dagger(\mathbf{p},\tau-\tau') + \frac{1}{\Omega}\sum_{\mathbf{p}'\mathbf{q}\sigma'} V(\mathbf{q}) \\ &\times \langle \mathrm{T}_\tau C^\dagger_{\mathbf{p}'-\mathbf{q}\uparrow}(\tau)C^\dagger_{\mathbf{p}'+\mathbf{q},\sigma'}(\tau)C_{\mathbf{p}'\sigma'}(\tau)C^\dagger_{-\mathbf{p}\downarrow}(\tau')\rangle = 0.\end{aligned} \tag{24.272}$$

Evaluating the product of four operators provides a somewhat different challenge than in the case of the Green function. Note in particular that three of the operators act at time τ, and only one, $C^\dagger_{-\mathbf{p}\downarrow}(\tau')$, at time

τ'. There are three different possible ways of pairing $C^\dagger_{-\mathbf{p}\downarrow}(\tau')$ with the remaining operators, leading to the results

$$\begin{aligned} C^\dagger_{-\mathbf{p}\downarrow}(\tau')C_{\mathbf{p}'\sigma'}(\tau) &\rightarrow \delta_{sigma'\downarrow}\delta_{\mathbf{p}',-\mathbf{p}}\Delta^\dagger(\mathbf{p})\mathcal{G}(-\mathbf{p},\tau-\tau') \\ C^\dagger_{-\mathbf{p}\downarrow}(\tau')C_{\mathbf{p}-\mathbf{q}\uparrow}(\tau) &\rightarrow \delta_{\mathbf{q}0}n_{\mathbf{p}'}\mathcal{F}^\dagger(\mathbf{p},\tau-\tau') \\ C^\dagger_{-\mathbf{p}\downarrow}(\tau')C_{\mathbf{p}'+\mathbf{q},\sigma'}(\tau) &\rightarrow -\delta_{\sigma'\uparrow}\delta_{\mathbf{p}',\mathbf{p}-\mathbf{q}}n_{\mathbf{p}-\mathbf{q}}\mathcal{F}^\dagger(\mathbf{p},\tau-\tau') \end{aligned} \tag{24.273}$$

The second term requires $\mathbf{q}=0$ and corresponds to phonons of long (infinite) wavelength whose effects we are neglecting. The last term leads again to the usual exchange self-energy, $\Sigma_x = \sum_{\mathbf{q}} V(\mathbf{q})n_{\mathbf{p}-\mathbf{q}}$, which we also neglect. The first term is retained and leads to the equation,

$$\left(-\frac{\partial}{\partial\tau}+\xi_{\mathbf{p}}\right)\mathcal{F}^\dagger(\mathbf{p},\tau-\tau')+\Delta(\mathbf{p})\mathcal{G}^\dagger(\mathbf{p},\tau-\tau')=0. \tag{24.274}$$

This equation follows from the assumed reality of the gap function, $\Delta^\dagger(\mathbf{p}) = \Delta(\mathbf{p})$, an assumption which is justified below. Using the Fourier transforms of the Green function and the correlation functions, as indicated in Eq. (24.240), we can rewrite Eqs. (24.272) and (24.274) in the form,

$$\begin{aligned} (ip_n-\xi_{\mathbf{p}})\,\mathcal{G}(\mathbf{p},ip_n)+\Delta(\mathbf{p})\mathcal{F}^\dagger(\mathbf{p},ip_n) &= 1 \\ (ip_n+\xi_{\mathbf{p}})\,\mathcal{F}(\mathbf{p},ip_n)+\Delta(\mathbf{p})\mathcal{G}(\mathbf{p},ip_n) &= 0. \end{aligned} \tag{24.275}$$

The solutions of these algebraic equations are elementary and we find,

$$\begin{aligned} \mathcal{G}(\mathbf{p},ip_n) &= -\frac{ip_n+\xi_{\mathbf{p}}}{p_n^2+\xi_{\mathbf{p}}^2+\Delta^2(\mathbf{p})} \\ \mathcal{F}(\mathbf{p},ip_n) &= \mathcal{F}^\dagger(\mathbf{p},ip_n)=\frac{\Delta(\mathbf{p})}{p_n^2+\xi_{\mathbf{p}}^2+\Delta^2(\mathbf{p})}. \end{aligned} \tag{24.276}$$

At this point, the equivalence of $\mathcal{F}$ and $\mathcal{F}^\dagger$ can be shown explicitly by deriving for $\mathcal{F}$ the equation analogous to Eq. (24.274). These results can be used to show the validity of some of the assumptions used in their derivation. For example, considering the Fourier transforms of $\mathcal{F}(\mathbf{p},-\tau)$ and changing $p_n \rightarrow -p_n$, we easily show that $\mathcal{F}(\mathbf{p},-\tau)=\mathcal{F}(\mathbf{p},\tau)$, since $\mathcal{F}(\mathbf{p},-ip_n) = \mathcal{F}(\mathbf{p},ip_n)$ as follows from Eq. (24.226). We also have,

$$\mathcal{F}(\mathbf{p},0)=\frac{1}{\beta}\sum_n\mathcal{F}(\mathbf{p},ip_n)=\frac{1}{\beta}\sum_n\mathcal{F}^\dagger(\mathbf{p},ip_n)=\mathcal{F}^\dagger(\mathbf{p},0), \tag{24.277}$$

so that the gap function is given by the expressions,

$$\Delta=-\frac{1}{\Omega}\sum_{\mathbf{q}}V(\mathbf{q})\mathcal{F}(\mathbf{p}-\mathbf{q},0)=-\frac{1}{\Omega}\sum_{\mathbf{q}}V(\mathbf{q})\mathcal{F}^\dagger(\mathbf{p}-\mathbf{q},0), \tag{24.278}$$

and is, consequently, real. We note that when the gap function is assumed to vanish, the Green function reduces to the form describing a normal (non-superconducting) material,

$$\mathcal{G}(\mathbf{p}, ip_n) = \frac{1}{ip_n - \xi_{\mathbf{p}}}, \quad \Delta = 0, \tag{24.279}$$

and both correlation functions, $\mathcal{F}$ and $\mathcal{F}^{\dagger}$, vanish.

It is now seen that $\mathcal{G}$, $\mathcal{F}$, and $\mathcal{F}^{\dagger}$ have poles at the energies $\pm E_{\mathbf{p}}$, where,

$$E_{\mathbf{p}} = \left[\xi_{\mathbf{p}}^2 + \Delta^2(\mathbf{p})\right]^{1/2}, \tag{24.280}$$

where $E_{\mathbf{p}}$ is called the *excitation energy* of the superconductor. Some analysis of the implications of this form may be useful. At a fixed temperature, we can treat $\Delta(\mathbf{p})$ simply as a constant independent of both momentum and energy. [In strong-coupling theories an energy dependence, $\Delta(\omega)$, must be taken into account.] It is seen that the minimum excitation energy is Δ. Because it is not possible to excite just one member of a Cooper pair with excitation energy $E_{\mathbf{p}} = \left[\xi_{\mathbf{p}}^2 + \Delta^2(\mathbf{p})\right]^{1/2}$, both particles must be raised to the excitation band. Thus, to break a pair, one must supply a minimum excitation energy of $E_{\mathbf{p}} + E_{\mathbf{p}'} > 2\Delta$, and the superconducting energy gap equals 2Δ.

Further elaboration [11] allows one to show that the energy gap has the form,

$$E_q = 2\Delta = 4\omega_{\mathrm{D}} \mathrm{e}^{-1/N_{\mathrm{F}} V_0}. \tag{24.281}$$

The energy gap decreases with increasing temperature. The superconducting transition temperature has the form,

$$K_{\mathrm{B}} T_{\mathrm{c}} = 1.14\omega_{\mathrm{D}} \mathrm{e}^{-1/N_{\mathrm{F}} V_0}. \tag{24.282}$$

These results can be tested experimentally and, at least in the case of weak superconductors, they attest fully to the validity of the BCS theory.

Bibliography

[1] Tomokazu Kato, Tetsuro Kobayashi, and Mikio Namiki, Supplement of the Progress of Theoretical Physics, No. **15** (1960).

[2] R. P. Feynman, rev. Mod. Phys. **20**, 367 (1948).

[3] R. P. Feynman, rev. Mod. Phys. **80**, 440 (1950).

[4] J. Schwinger, Proc. Nat. Acad. Sci. **37**, 452, 455 (1951).

[5] K. A. Brueckner, Phys. Rev. **100**, 36 (1956).

[6] K. A. Brueckner and L. J. Gammel, Phys. Rev. **110**, 431 (1958), and references therein.

[7] L. D. Landau, JETP (USSR) **34**, 262 (1958); Soviet Phys. **7**, 183 (1958).

[8] V. M. Galitskii and A. B. Migdal, JETP (USSR) **34**, 139 (1958); Soviet Phys. **7**, 96 (1958).

[9] P. C. Martin and J. Schwinger, Phys. Rev. **115**, 1342 (1959).

[10] T. Matsubara, Prog. Theor. Phys. **14**, 351 (1955).

[11] Gerald D. Mahan, *Many-Particle Physics*, Plenum Press, New York (1981).

[12] T. Matsubara, Prog. Theor. Physics (Kyoto) **14**, 351 (1955).

[13] C. Kittel, *The Quantum Theory of Solids*, John Willey and Sons, Inc., New York (1963).

[14] A. A. Abrikosov, L. P. Gorkov, and I. E. Dzyaloshinski, *Methods of Quantum Field Theory in Statistical Physics,* Prentice Hall, Englewood Cliffs, NJ (1963).

Appendix A

Functional Derivatives

Functionals are a very powerful analytical tool of mathematical physics. For example, they provide a particularly straightforward approach to the theory of Green functions, as is illustrated by part of the formalism in Chapter 2.

A functional is a prescription for assigning a (complex) number, $F[\phi]$, to a function, $\phi(x)$. As simple examples of functionals, we may consider,

$$F[\phi] = \phi(a), \tag{A.1}$$

which assigns to a function ϕ its value at a fixed point a, and

$$F_D[\phi] = \int_D u(x)\phi(x)\mathrm{d}x, \tag{A.2}$$

which assigns to ϕ the value of an integral of the product of ϕ with a function u over some domain D.

Functional differentiation of a functional with respect to its argument function can be defined in a way that is a natural extension of the usual concept of differentiation in calculus. In general, we define the functional derivative, $\delta F[\phi]/\delta\phi(x)$, by means of the expression,

$$\int \frac{\delta F[\phi]}{\delta\phi(x)} f(x)\mathrm{d}x = \lim_{\epsilon\to 0}\frac{1}{\epsilon}\{F[\phi+\epsilon f] - F[\phi]\}, \tag{A.3}$$

where $f(x)$ denotes an arbitrary (but) smooth integrable function. It follows from this definition that functional differentiation has the properties of the usual differential operation of calculus, that is,

$$\frac{\delta a}{\delta\phi(x)} = 0,$$

$$\frac{\delta}{\delta\phi(x)}\{aF[\phi]+bG[\phi]\} = a\frac{\delta F[\phi]}{\delta\phi(x)} + b\frac{\delta G[\phi]}{\delta\phi(x)},$$
$$\frac{\delta}{\delta\phi(x)}\{F[\phi]G[\phi]\} = \frac{\delta F[\phi]}{\delta\phi(x)}G(x) + F(x)\frac{\delta G[\phi]}{\delta\phi(x)}, \tag{A.4}$$

for arbitrary functionals F and G, and arbitrary functions a and b which are independent of ϕ.

As a way of illustrating Eq. (A.3), we consider the functional derivatives of the functionals defined in Eqs. (A.1) and (A.2). For the functional $F[\phi] = \phi(a)$, the definition (A.3) yields

$$\int \frac{\delta F[\phi]}{\delta\phi(x)} f(x)\mathrm{d}x = \lim_{\epsilon\to 0}\frac{1}{\epsilon}[\phi(a)+\epsilon f(a) - \phi(a)] = f(a), \tag{A.5}$$

so that we can identify,

$$\frac{\delta\phi(a)}{\delta\phi(x)} = \delta(x-a), \tag{A.6}$$

consistent with the properties of δ-functions. For the functional of Eq. (A.2), we obtain,

$$\begin{aligned} &\int \frac{\delta F[\phi]}{\delta\phi(x)} f(x)\mathrm{d}x \\ = \; &\lim_{\epsilon\to 0}\left\{\int u(x)[\phi(x)+\epsilon f(x)]\mathrm{d}x - \int u(x)f(x)\mathrm{d}x\right\} \\ = \; &\int u(x)f(x)\mathrm{d}x. \end{aligned} \tag{A.7}$$

Because this relation holds for arbitrary (but sufficiently well-behaved) functions, $f(x)$, we can identify,

$$\frac{\delta}{\delta\phi(x)}\int u(x)\phi(x)\mathrm{d}x = u(x). \tag{A.8}$$

As a somewhat more general example, consider a simple scalar field, $V(x)$, where $x = (\mathbf{x}, i)$ represents a point in space time, characterized by the field equation,

$$\mathrm{D}_x V(x) = \rho(x), \tag{A.9}$$

where D_x is a linear operator, and $\rho(x)$ is the source function of the field. Under the conditions that V vanishes as ρ tends to zero, Eq. (A.9) has the particular solution,

$$V(x) = \int G(x,x')\rho(x')\mathrm{d}x'. \tag{A.10}$$

Because D_x is a linear operator, we can write,

$$G(x,x') = \frac{\delta V(x)}{\delta\rho(x')}. \tag{A.11}$$

Using Eqs. (A.6) and (A.9) we obtain the equation for the Green function,

$$\frac{\delta}{\delta\rho(x')}[D_x V(x)] = D_x G(x,x') = \delta(x-x'). \tag{A.12}$$

Thus, the Green function can be interpreted as the contribution to $V(x)$ from a unit (point) source at x'. Therefore, through the introduction of the Green function all linear differential equations can be reduced to solving an equation of the type given in Eq. (A.12) under appropriate boundary conditions.

As a simple example, we consider the Poisson equation,

$$\nabla^2 V(\mathbf{r}) = -4\pi\rho(\mathbf{r}), \tag{A.13}$$

whose solution is given by the integral

$$V(\mathbf{r}) = \int \frac{\rho(\mathbf{r}')}{|\mathbf{r}-\mathbf{r}'|} d^3r'. \tag{A.14}$$

The Green function for the operator ∇^2 is found through the use of Eqs. (A.8) and (A.11)

$$G(\mathbf{r}-\mathbf{r}') = \frac{\delta V(\mathbf{r})}{\delta\rho(\mathbf{r}')} = -\frac{1}{4\pi}\frac{1}{|\mathbf{r}-\mathbf{r}'|}. \tag{A.15}$$

From Eq. (A.12) we have

$$\nabla_{\mathbf{r}}^2 \frac{1}{|\mathbf{r}-\mathbf{r}'|} = -4\pi\delta(\mathbf{r}-\mathbf{r}'). \tag{A.16}$$

Appendix B

Vector Spaces and Linear Operators

B.1 Basic Concepts

A vector space, V, is defined to be a set of elements, $\mathbf{x}_1$, $\mathbf{x}_2$, $\cdots$, called *vectors*, which possess the following properties:

1. For any two vectors, $\mathbf{x}_1$ and $\mathbf{x}_2$, in V the sum $\mathbf{x}_1 + \mathbf{x}_2$ is a uniquely defined vector in V.
2. If S denotes the set of scalar numbers, real or complex, then for any scalar c, and any vector $\mathbf{x}$ in V, the product $c\mathbf{x}$ is a uniquely defined vector in V.
3. In addition, the elements of V satisfy the axioms:
   (a) $\mathbf{x}_1 + \mathbf{x}_2 = \mathbf{x}_2 + \mathbf{x}_1$, (commutative law of addition).
   (b) $\mathbf{x}_1 + (\mathbf{x}_2 + \mathbf{x}_3) = (\mathbf{x}_1 + \mathbf{x}_2) + \mathbf{x}_3$, (associative law of addition).
   (c) There exists a vector, $\mathbf{0}$, such that for all $\mathbf{x}$ in V, $\mathbf{0} + \mathbf{x} = \mathbf{x}$.
   (d) $c(\mathbf{x}_1 + \mathbf{x}_2) = c\mathbf{x}_1 + c\mathbf{x}_2$, (scalar multiplication distributive over vector addition).
   $(c + d)\mathbf{x} = c\mathbf{x} + d\mathbf{x}$, (vector multiplication distributive over scalar addition).
   (e) $c(d\mathbf{x}) = cd\mathbf{x}$.
   $1\mathbf{x} = \mathbf{x}$.

It is easy to see that the set of three-dimensional geometric vectors satisfy axioms 1 to 3 above and thus forms a vector space.

A set of vectors $\{\mathbf{e}_1, \mathbf{e}_2, \cdots\}$ is a *basis* in V if every vector in V can be written as a *linear combination* of the vectors in the set,

$$\mathbf{x} = \sum_i \mathbf{e}_i x_i, \tag{B.1}$$

where the x_i are called the *components* of $\mathbf{x}$ in the basis $\{\mathbf{e}_i\}$, and can be complex. The vectors in a basis are *linearly independent*, i.e., one vector in the set cannot be expressed as a linear combination of the others. If the basis set contains a finite number n of vectors, the space is called *finite dimensional*, or *n-dimensional*. Otherwise, the space is *infinite dimensional*. In any case, the vectors in a basis can be chosen to be orthogonal and of unit length (see below), comprising an *orthonormal basis*. We say that the basis vectors *span* the space V.

Analogous to the notion of length in three-dimensional Euclidean space, we define the length or *modulus* of a vector $\mathbf{x}$ in an n-dimensional space through the relation,

$$\begin{aligned} x^2 &= \sum_{i=1}^{n} x_i^* x_i \\ &= \sum_{i}^{n} |x_i|^2, \end{aligned} \tag{B.2}$$

in terms of the components of $\mathbf{x}$ in an orthonormal basis. Here and subsequently, an asterisk denotes the complex conjugate of a quantity. Clearly, the length of a vector is a real number, and equals zero if and only if $x_i = 0$ for all i.

The concept vector products can be generalized to n-dimensional complex vector spaces through the introduction of the *dual* space $V^\dagger$ to a given vector space, V. First, we define the dual basis $\{\mathbf{e}_i^\dagger\}$ such that

$$\mathbf{e}_i^\dagger \mathbf{e}_j = \delta_{ij}, \tag{B.3}$$

where δ_{ij} is the Kronecker delta. (As an example, consider a real vector space. If the basis is represented by column vectors, then the dual basis is represented by the corresponding row vectors.) For every vector $\mathbf{x}$ in V there is a vector $\mathbf{x}^\dagger$ in $V^\dagger$ given by the expression,

$$\mathbf{x}^\dagger = \sum_{i=1}^{n} x_i^* \mathbf{e}_i^\dagger. \tag{B.4}$$

A basis for which Eq. (B.3) holds is said to be biorthonormal. If $\mathbf{e}_i^\dagger = \mathbf{e}_i$ the basis is called *unitary*. Assuming the distributive law and axiom 3d, we can write the square of the modulus of a vector in the form,

$$\begin{aligned} x^2 &= \mathbf{x}^\dagger \mathbf{x} \\ &= \sum_{i=1}^{n} \sum_{j=1}^{n} (x_i^* \mathbf{e}_i^\dagger)(x_j^* \mathbf{e}_j) \\ &= \sum_{i=1}^{n} |x_i|^2, \end{aligned} \tag{B.5}$$

where Eq. (B.3) has been used. The scalar product of two vectors is then defined by the expression

$$\begin{aligned} \mathbf{x}^{\dagger}\mathbf{y} &= \sum_{i=1}^{n}\sum_{j=1}^{n}(x_i^*\mathbf{e}_i^{\dagger})(y_j^*\mathbf{e}_j) \\ &= \sum_{i=1}^{n} x_i^* y_i. \end{aligned} \tag{B.6}$$

We note that the product of vectors is *not* commutative. Reversing the order in the last expression leads to the product $\mathbf{y}\mathbf{x}^{\dagger}$, which is a *dyadic* (or a matrix) in n-dimensional space. Dyadics in three-dimensional space are discussed in more detail in Appendix J; the basic notions introduced there find immediate generalization to n dimensions.

Using Eq. (B.1) and the properties of the scalar product of two vectors, we can derive an important expansion relation. From Eq. (B.1) we have,

$$\mathbf{e}_j^{\dagger}\mathbf{x} = \sum_{i=1}^{n} \mathbf{e}_j^{\dagger} x_i \mathbf{e}_i = x_i, \tag{B.7}$$

a result that can be substituted back into Eq. (B.1) to yield,

$$\mathbf{x} = \sum_{i=1}^{n} (\mathbf{e}_i\mathbf{e}_i^{\dagger})\mathbf{x}. \tag{B.8}$$

It now follows from axiom 3e that

$$\sum_{i=1}^{n} (\mathbf{e}_i\mathbf{e}_i^{\dagger}) = \underline{I}_n, \tag{B.9}$$

where $\underline{I}_n$ denotes the n-dimensional unit matrix.

A set of basis vectors satisfying Eq. (B.9) is said to be *complete* and Eq. (B.9) is called a *completeness* relation. A complete basis set (which may or may not be orthonormal) can be used to expand any vector in V as a linear combination of the basis vectors.

B.1.1 Change of basis

It is convenient to phrase our discussion in terms of Dirac's notation for vectors in terms of bras and kets, as outlined in Appendix J. For example, denoting the basis vectors by $|i\rangle$, the completeness relation, Eq. (B.9) takes the form,

$$\sum_i |i\rangle\langle i| = \underline{I}_n, \tag{B.10}$$

where $\langle i|$ denotes the elements of the dual basis.

Let $\{|i\rangle\}$ and $\{|\alpha\rangle\}$ denote two biorthonormal, complete basis sets in an n-dimensional Euclidean space. Each element of either set can be expanded in terms of the elements of the other set. Multiplying the completeness relation, Eq. (B.10), by $|\alpha\rangle$ on the right we obtain,

$$|\alpha\rangle = \sum_{i=1}^{n} |i\rangle\langle i|\alpha\rangle. \tag{B.11}$$

Similarly, we have,

$$|i\rangle = \sum_{i=1}^{n} |\alpha\rangle\langle\alpha|i\rangle. \tag{B.12}$$

The quantities $\langle i|\alpha\rangle$ in Eq. (B.11) are the components of the basis vector $|\alpha\rangle$ in the $|i\rangle$ basis. They can be conveniently arranged in a matrix, $\underline{S}$, with matrix elements, $S_{i\alpha} = \langle i|\alpha\rangle$, called the ***transformation matrix*** from the $|\alpha\rangle$ to the $|i\rangle$ basis or ***representation***. Similarly, the matrix $\underline{T}$ with elements $T_{\alpha i} = \langle\alpha|i\rangle$ gives the transformation form the $|i\rangle$ to the $|\alpha\rangle$ representation. From the definition of the scalar product in complex vector spaces, it follows that,

$$\langle i|\alpha\rangle = \langle\alpha|i\rangle^*, \tag{B.13}$$

so that,

$$\underline{T} = \underline{S}^\dagger \tag{B.14}$$

and,

$$\underline{S} = \underline{T}^\dagger. \tag{B.15}$$

Here, a dagger denotes the Hermitean conjugate of a matrix defined as the complex conjugate of the transpose,

$$\underline{A}^\dagger = \left(\underline{A}^{\mathrm{T}}\right)^*. \tag{B.16}$$

It follows from the orthonormality of the basis sets that

$$\sum_i \langle\alpha|i\rangle\langle i|\beta\rangle = \delta_{\alpha\beta}, \tag{B.17}$$

or,

$$\sum_i T_{\alpha i} S_{i\alpha} = \delta_{\alpha\beta}, \tag{B.18}$$

and the matrices $\underline{S}$ and $\underline{T}$ are unitary, i.e.,

$$\underline{S}\underline{S}^\dagger = \underline{T}\underline{T}^\dagger = \underline{I}_n. \tag{B.19}$$

Any vector $|y\rangle$ given in terms of its components in any biorthonormal basis, can be expressed in terms of its components in another biorthonormal basis. Let

$$|y\rangle = \sum_i |i\rangle\langle i|y\rangle. \tag{B.20}$$

Taking the scalar product with $\langle\alpha|$, and using Eq. (B.10), we have,

$$\langle\alpha|y\rangle = \sum_i \langle\alpha|i\rangle\langle i|y\rangle = \sum_i T_{i\alpha}\langle i|y\rangle, \tag{B.21}$$

which expresses the components of $|y\rangle$ in the $\{|\alpha\rangle\}$ basis in terms of those in the $\{|i\rangle\}$ basis.

B.1.2 Some important properties of matrices

Some basic properties of matrices were used in Appendix J. Here, we provide a brief summary of matrix definitions and relations that are useful in the algebra of operators discussed in the next section.

1. $\underline{S}^\dagger = (\underline{S}^{\mathrm{T}})^*$ is called the *Hermitean conjugate* of the matrix S.
2. If $\underline{S}^\dagger = \underline{S}$, then $\underline{S}$ is said to be *Hermitean* or *self-adjoin.*
3. If $\underline{S}^\dagger\underline{S} = \underline{S}\underline{S}^\dagger = \underline{I}_n$, then $\underline{S}$ is said to be unitary.
4. If $\underline{T}\underline{S} = \underline{S}\underline{T} = \underline{I}_n$, $\underline{T}$ is called the *inverse* of $\underline{S}$ and is written as $\underline{S}^{-1}$.

A necessary and sufficient condition that a given n-dimensional matrix has an inverse is that its determinant is nonzero. Denoting the elements of $\underline{S}$ by s_{ij}, the elements of $\underline{S}^{-1}$ are given by the expression,

$$(\underline{S}^{-1})_{ij} = (-1)^{i+j}\frac{S_{ji}}{\det|\underline{S}|}, \tag{B.22}$$

where S_{ji} denotes the cofactor of the element s_{ij}. (The cofactor of an element of a matrix is the determinant of the matrix that remains when the row and the column containing the element are removed from the matrix.)

5. The trace of a matrix is defined as the sum of its diagonal elements,

$$\mathrm{Tr}\,\underline{S} = \sum_i s_{ii}. \tag{B.23}$$

6. The trace of a matrix remains invariant under a similarity transformation, $\underline{T}' = \underline{S}\underline{T}\underline{S}^{-1}$ (see Appendix J for a proof).

$$\mathrm{Tr}\,\underline{T}' = \mathrm{Tr}\,\underline{T}. \tag{B.24}$$

7. Let a function $f(x)$ have a convergent Taylor series expansion about $x = a$,

$$f(x) = \sum_{n=0}^{\infty}\frac{(x-a)^n}{n!}\left(\frac{\mathrm{d}^n f(x)}{\mathrm{d}x^n}\right)\bigg|_{x=a}. \tag{B.25}$$

Then, the function $f(\underline{A})$ of a matrix $\underline{A}$ can be defined by the series,

$$f(\underline{A}) = \sum_{m=0}^{\infty} \frac{(\underline{A} - a\underline{I}_n)^m}{m!} \left(\frac{\mathrm{d}^m f(x)}{\mathrm{d}x^m} \right) \bigg|_{x=a}. \tag{B.26}$$

8. If, for a given matrix $\underline{A}$, there exists a (column) vector $|x\rangle$ and a scalar λ_x such that,

$$\underline{A}|x\rangle = \lambda_x |x\rangle, \tag{B.27}$$

then $|x\rangle$ is called an *eigenvector* of $\underline{A}$ and λ_x the corresponding *eigenvalue*. An n-dimensional matrix has n eigenvectors and n eigenvalues, not necessarily all distinct. The matrix $\underline{S}$ whose columns are formed by the eigenvectors of $\underline{A}$ diagonalizes $\underline{A}$ by a similarity transformation,

$$\underline{A}_\mathrm{d} = \underline{S}^{-1}\underline{A}\underline{S}. \tag{B.28}$$

The elements of the diagonal matrix, $\underline{A}_\mathrm{d}$, are the eigenvalues of $\underline{A}$. The set of eigenvalues is called the *spectrum* of $\underline{A}$.

9. The eigenvalues of a Hermitean matrix are real.

10. A matrix $\underline{A}$ that satisfies the relation,

$$\underline{A}^\dagger \underline{A} = \underline{A}\underline{A}^\dagger, \tag{B.29}$$

is called *normal.* The eigenvectors of a normal matrix form a biorthonormal basis set. Clearly, Hermitean and unitary matrices are normal, but the converse is not necessarily true.

11. Using (7) we can evaluate the action of the function $f(\underline{A})$ on an eigenvector $|y\rangle$ of $\underline{A}$,

$$f(\underline{A})|y\rangle = \sum_{n=0}^{\infty} \left(\frac{\mathrm{d}^n f(x)}{\mathrm{d}x^n} \right) \bigg|_{x=a} \frac{(\lambda_y - a)^n |y\rangle}{n!}, \tag{B.30}$$

or,

$$f(\underline{A})|y\rangle = f(\lambda_y)|y\rangle. \tag{B.31}$$

More generally,

$$f(\underline{A})\underline{S} = \underline{S}f(\underline{A}_\mathrm{d}), \tag{B.32}$$

or,

$$f(\underline{A}) = \underline{S}f(\underline{A}_\mathrm{d})\underline{S}^{-1}. \tag{B.33}$$

The last expression is called the *spectral representation* of the operator function $f(\underline{A})$.

B.2 Linear Operators

An *operator* $\hat{L}$ in a vector space V is a prescription for relating a vector $|x\rangle$ in V to a vector $|y\rangle$ in V (not necessarily different from $|x\rangle$). We write,

$$\hat{L}|x\rangle = |y\rangle. \tag{B.34}$$

If for any vectors $|x\rangle$ and $|x\rangle$, and any scalar α, the relations,

$$\hat{L}\left[|x\rangle + |y\rangle\right] = \hat{L}|x\rangle + \hat{L}|y\rangle, \tag{B.35}$$

and,

$$\hat{L}\left[\alpha|x\rangle\right] = \alpha\hat{L}|x\rangle, \tag{B.36}$$

hold, then $\hat{L}$ is called a *linear operator*. A necessary and sufficient condition for the determination of a linear operator is knowledge of its effect on every vector of a complete basis set $\{|i\rangle\}$. In particular, we have

$$\hat{L}|i\rangle = \sum_j |j\rangle\langle j|\hat{L}|i\rangle = \sum_j |j\rangle l_{ji}, \tag{B.37}$$

so that the action of $\hat{L}$ is represented by the *matrix* l_{ji}. If we form the dyadic product of Eq. (B.37) with $\langle i|$ and sum over all elements i of the basis, we obtain,

$$\begin{aligned} \hat{L} &\equiv \sum_i \sum_j |j\rangle\langle j|\hat{L}|i\rangle\langle i| \\ &= \sum_i \sum_j |j\rangle l_{ji}\langle i|, \end{aligned} \tag{B.38}$$

so that the matrix $\underline{l}$ provides a complete description of the operator $\hat{L}$.

An important operator is the *identity* (we no longer explicitly exhibit the dimensionality of the space by means of a subscript),

$$\hat{I}|x\rangle = |x\rangle, \tag{B.39}$$

represented by the unit matrix, $\underline{I}_{ij} = \delta_{ij}$. If the matrix elements of $\underline{l}$ are given in some basis, i.e., the $l_{\alpha\beta}$ are known, then the matrix $\underline{l}$ in a different basis can be written in the form,

$$\hat{L} = \sum_\alpha \sum_\beta \sum_i \sum_j |i\rangle\langle i|\alpha\rangle l_{\alpha\beta}\langle\beta|j\rangle\langle j|, \tag{B.40}$$

where Eqs. (B.20) and (B.38) have been used.

The Hermitean conjugate, $\hat{L}^\dagger$ of a linear operator $\hat{L}$ can be defined in a natural way,

$$\hat{L}^\dagger = \sum_{ij} |i\rangle \left(l^\dagger\right)_{ij} \langle j| = \sum_{ij} |i\rangle l^*_{ji} \langle j|. \tag{B.41}$$

The inverse, $\hat{L}^{-1}$, of an operator, $\hat{L}$, (if it exists) is defined by the relation,

$$\hat{L}\hat{L}^{-1} = \hat{L}^{-1}\hat{L} = \hat{I}. \tag{B.42}$$

At least for finite-dimensional spaces, the properties of matrices summarized above can also be ascribed to operators. Thus, the eigenvectors and eigenvalues of an operator are those that satisfy the relations under (8) above with respect to the matrix representing the operator in a particular basis. In particular, denoting the eigenvalues of $\hat{L}$ by λ_i, we can write,

$$f(\underline{L}) = \sum_i |i\rangle f(\lambda_i) \langle i|, \tag{B.43}$$

for the spectral representation of the operator $f(\hat{L})$.

B.2.1 Infinite dimensions

Quantum mechanical applications of the concepts considered above usually involve vector spaces with the dimensionality of the continuum. For example, if $\hat{X}$ denotes the position operator, we may inquire as to the possible eigenvectors $|x\rangle$ and eigenvalues x satisfying the relation,

$$\hat{X}|x\rangle = x|x\rangle. \tag{B.44}$$

Let $|n\rangle$ be any vector in the space. Then, formally, we could write,

$$|n\rangle = \sum_x |x\rangle\langle x|n\rangle, \tag{B.45}$$

provided that our previous discussion remains valid. However, the set of eigenvectors of $\hat{X}$ is continuous, $-\infty < x < \infty$, and it is not clear *a priori* that Eq. (B.45) is meaningful. In fact since x is a continuous variable, the summation in Eq. (B.45) should be replaced by an integral. Thus, for any vector $|\phi\rangle$ we can write,

$$|\phi\rangle = \int \mathrm{d}x |x\rangle\langle x|\phi\rangle = \int \mathrm{d}x |x\rangle \phi(x). \tag{B.46}$$

The vectors $|x\rangle$ determined in Eq. (B.44) are called *generalized eigenvectors*, and the set,

$$|x\rangle, \quad -\infty < x < \infty \tag{B.47}$$

is called a *generalized basis.* By contrast, we call the eigenvectors corresponding to a discrete spectrum *proper* eigenvectors. Taking the "scalar product" of $|\phi\rangle$ in Eq. (B.46) with an arbitrary element, $|x'\rangle$, of the generalized basis (more precisely the functional $|x'\rangle$ at the element ϕ), we obtain,

$$\begin{aligned} \langle x'|\phi\rangle &\equiv \phi(x') \\ &= \int \mathrm{d}x \langle x'|x\rangle\langle x|\phi\rangle \\ &= \int \mathrm{d}x \langle x'|x\rangle \phi(x). \end{aligned} \tag{B.48}$$

Thus, $\langle x'|x\rangle$ by integration maps the "well behaved" function $\phi(x) \equiv \langle x|\phi\rangle$ onto its value at x'. The quantity $\langle x'|x\rangle$ is an example of a *generalized function* or *distribution*, and is known as the Dirac delta function,

$$\langle x'|x\rangle = \delta(x' - x). \tag{B.49}$$

With this notation, Eq. (B.48) becomes

$$\phi(x') = \int \mathrm{d}x \delta(x' - x)\phi(x). \tag{B.50}$$

Thus, in the transition from a finite and discrete basis to an infinite-dimensional continuous basis the summation over basis vectors is replaced by an integral,

$$\sum_x \langle x'|x\rangle \rightarrow \int \mathrm{d}x \delta(x' - x). \tag{B.51}$$

We note that in a continuous basis, the analog of the completeness relation, Eq. (B.10), takes the form,

$$\int \mathrm{d}x |x\rangle\langle x| = 1, \tag{B.52}$$

where 1 stands for the infinite-dimensional unit matrix. Applying the operator $\hat{X}$ to an arbitrary vector $\phi\rangle$ we obtain,

$$\hat{X}|\phi\rangle = \int \mathrm{d}x |x\rangle x \langle x|\phi\rangle, \tag{B.53}$$

so that omitting the arbitrarily chosen ϕ, we find the spectral representation of the operator $\hat{X}$,

$$\hat{X} = \int \mathrm{d}x |x\rangle x \langle x|. \tag{B.54}$$

B.2.2 Representations

To express a state vector or an operator in a particular representation means to explicitly exhibit the components of the vector or the matrix elements of the operator along each of the basis vectors forming the representation. For example, let us consider the states of a harmonic oscillator of mass m described by the Hamiltonian,

$$\hat{H} = \frac{1}{2m}\hat{P}^2 + \frac{m\omega^2}{2}\hat{X}^2. \tag{B.55}$$

The eigenstates of the system satisfy the relation,

$$\hat{H}|n\rangle = E_n|n\rangle, \tag{B.56}$$

with the eigenenergies given by,

$$E_n = \hbar\omega(n + \frac{1}{2}), \quad n = 0, 1, 2, \cdots. \tag{B.57}$$

To obtain the coordinate representation of $|n\rangle$, we form the inner product $\langle x|n\rangle$, where $\langle x|$ is a (generalized) eigenvector of the operator $\hat{X}$. As is well known, the functions $\phi_n(x) = \langle x|n\rangle$ have the form,

$$\phi_n(x) = \left(\frac{m\omega}{\pi\hbar}\right)^{\frac{1}{4}} \frac{1}{\sqrt{2^n n!}} H_n\left(\sqrt{\frac{m\omega}{\hbar}}x\right) e^{-\left(\frac{m\omega}{2\hbar}\right)x^2}, \tag{B.58}$$

where $H_n(x)$ is the Hermite polynomial of degree n.

The harmonic oscillator states can also be expressed in the momentum representation,

$$\begin{aligned}\phi_n(p) &\equiv \langle p|n\rangle \\ &= \left(\frac{1}{\pi m\omega\hbar}\right) \frac{1}{\sqrt{2^n n!}} H_n\left(\frac{p}{\sqrt{\hbar m\omega}}\right) e^{-\frac{p^2}{2m\omega\hbar}}.\end{aligned} \tag{B.59}$$

We can pass from one representation to the other by means of the inner product between elements of the two basis sets, as indicated in Eq. (B.13). In this case, we have

$$\langle x|p\rangle = \langle p|x\rangle^* = \frac{1}{\sqrt{2\pi\hbar}} e^{\frac{ixp}{\hbar}}. \tag{B.60}$$

It is easily shown that

$$\phi_n(p) = \int dx \langle p|x\rangle \phi_n(x), \tag{B.61}$$

which is simply the Fourier transformation between direct and reciprocal space.

B.2.3 Plane-wave states

In three-dimensional systems, the momentum states $|p\rangle$ become the plane-wave states $|\mathbf{k}\rangle$. These vectors are the eigenstates of the momentum operator and their inner product with the eigenstates of the position operator can be normalized in the form,

$$\langle x|\mathbf{k}\rangle = \frac{1}{(2\pi)^{3/2}} e^{i\mathbf{k}\cdot\mathbf{x}}. \tag{B.62}$$

These wave functions satisfy the relations of orthonormality

$$\langle \mathbf{k}'|\mathbf{k}\rangle = \frac{1}{(2\pi)^{3/2}} \int d^3x e^{i(\mathbf{k}'-\mathbf{k})\cdot\mathbf{x}} = \delta(\mathbf{k}' - \mathbf{k}), \tag{B.63}$$

and completeness,

$$\frac{1}{(2\pi)^{3/2}} \int d^3k e^{i(\mathbf{x}'-\mathbf{x})\cdot\mathbf{k}} = \delta(\mathbf{x}' - \mathbf{x}). \tag{B.64}$$

B.2.4 Angular momentum representation

The simultaneous eigenstates of the square of the angular momentum operator, $\hat{L}^2$, and of the component of $\hat{L}$ along the z-direction, $\hat{L}_z$, are commonly denoted by $|\ell, m\rangle$, where ℓ is the *orbital* quantum number and m the *azimuthal* (magnetic) quantum number. These vectors satisfy the relations,

$$\hat{L}^2|\ell, m\rangle = \ell(\ell+1)|\ell, m\rangle, \tag{B.65}$$

and,

$$\hat{L}_z|\ell, m\rangle = m|\ell, m\rangle, \tag{B.66}$$

where $\ell = 0, 1, 2, \cdots$, and $-\ell < m < \ell$. In a coordinate representation in which θ and ϕ denote the angular coordinates of a point on the unit sphere, the eigenvectors become the familiar spherical harmonics,

$$\langle \theta, \phi|\ell, m\rangle = Y_{\ell,m}(\theta, \phi). \tag{B.67}$$

These state vectors form a biorthonormal basis satisfying the conditions of orthonormality,

$$\langle \ell, m|\ell', m'\rangle = \int\int \sin\theta d\theta d\phi Y^*_{\ell',m'}(\theta, \phi) Y_{\ell,m}(\theta, \phi) = \delta_{\ell'\ell}\delta_{m'm}, \tag{B.68}$$

and completeness,

$$\sum_{\ell}\sum_{m} Y_{\ell,m}(\theta, \phi) Y_{\ell,m}(\theta', \phi') = \delta(\theta' - \theta)\delta(\phi' - \phi)/\sin\theta. \tag{B.69}$$

Some basic relations satisfied by the spherical harmonics are presented in Appendix C. The angular momentum representation is used almost exclusively in the formal developments throughout this book.

Appendix C

Spherical Functions

C.1 Spherical Harmonics

The solutions of the generalized Legendre equation,

$$\frac{\mathrm{d}}{\mathrm{d}x}\left[(1-x^2)\frac{\mathrm{d}P}{\mathrm{d}x}\right]+\left[\ell(\ell+1)-\frac{m^2}{1-x^2}\right]P=0, \tag{C.1}$$

are the associated Legendre polynomials, P_ℓ^m, which for m positive are defined by the formula,

$$P_\ell^m(x)=(-1)^m(1-x^2)^{m/2}\frac{\mathrm{d}^m}{\mathrm{d}x^m}P_\ell(x), \tag{C.2}$$

where the $P_\ell(x) = P_\ell^0(x)$ are the Legendre polynomials. These are given by *Rodrigues' formula*,

$$P_\ell(x)=\frac{1}{2^\ell \ell!}\frac{\mathrm{d}^\ell}{\mathrm{d}x}(x^2-1)^\ell, \tag{C.3}$$

which in combination with Eq. (C.3) yields a formula for the associated Legendre polynomials that is valid for both positive and negative m, $m = -\ell, -\ell+1, \cdots \ell-1, \ell$,

$$P_\ell^m(x)=\frac{(-1)^m}{2^\ell \ell!}(1-x^2)^{m/2}\frac{\mathrm{d}^{\ell+m}}{\mathrm{d}x^{\ell+m}}(x^2-1)^\ell. \tag{C.4}$$

It can be shown that,

$$P_\ell^{-m}(x)=(-1)^m\frac{(\ell-m)!}{(\ell+m)!}P_\ell^m(x). \tag{C.5}$$

The associated Legendre polynomials satisfy the orthogonality relation,

$$\int_{-1}^{1} P_\ell^{-m}(x)P_{\ell'}^{-m}(x)\mathrm{d}x = \frac{2}{2\ell+1}\frac{(\ell+m)!}{(\ell-m)!}\delta_{\ell\ell'}. \tag{C.6}$$

As is well known, the functions $Q_m(\phi) = \mathrm{e}^{\mathrm{i}m\phi}$ form a complete set of orthogonal functions in the index m on the interval $0 \le \phi \le 2\pi$. From the orthogonality of the $P_\ell^m(\cos\theta)$ in the interval $-1 \le \cos\theta \le -1$ it follows that the product $P_\ell^m Q_m$ forms a complete orthogonal set of functions in the indices ℓ, m on the surface of the unit sphere. The *spherical harmonics* $Y_{\ell m}(\theta,\phi)$ are suitably normalized functions on the surface of the unit sphere, and are given explicitly by the expression,

$$Y_{\ell m}(\theta,\phi) = \sqrt{\frac{2\ell+1}{4\pi}\frac{(\ell-m)!}{(\ell+m)!}}P_\ell^m(\cos\theta)\mathrm{e}^{\mathrm{i}m\phi}. \tag{C.7}$$

From Eq. (C.5) we see that,

$$Y_{\ell,-m}(\theta,\phi) = (-1)^m Y_{\ell m}^*(\theta,\phi), \tag{C.8}$$

with A^* denoting the complex conjugate of A. Using L as a combined index, $L = (\ell, m)$, we can also write,

$$Y_{\ell m}(\theta,\phi) \equiv Y_L(\theta,\phi) \equiv Y_{\ell m}(\hat{r}), \tag{C.9}$$

where $\hat{r}$ is a unit vector in the direction (θ,ϕ).

The spherical harmonics are eigenfunctions of the square of the angular momentum operator, $\hat{L}$,

$$\hat{L}^2 Y_{\ell m}(\theta,\phi) = \ell(\ell+1)Y_{\ell m}(\theta,\phi), \tag{C.10}$$

and of the component of $\hat{L}$ along the z-axis,

$$L_z Y_{\ell m}(\theta,\phi) = mY_{\ell m}(\theta,\phi). \tag{C.11}$$

In the interval $0 \le \theta \le \pi$, $0 \le \phi \le 2\pi$ the $Y_{\ell m}(\theta,\phi)$ satisfy, respectively, the orthogonality and completeness relations,

$$\int_0^{2\pi} \mathrm{d}\phi \int_0^{\pi} \sin\theta \mathrm{d}\theta Y_{\ell' m'}^*(\theta,\phi)Y_{\ell m}(\theta,\phi) = \delta_{\ell\ell'}\delta_{mm'} \tag{C.12}$$

and

$$\sum_{\ell=0}^{\infty}\sum_{-\ell}^{\ell} Y_{\ell m}^*(\theta',\phi')Y_{\ell m}(\theta,\phi) = \delta(\phi-\phi')\delta(\cos\theta-\cos\theta'). \tag{C.13}$$

The first few spherical harmonics are listed below.
$\ell = 0$:

$$Y_{00} = \frac{1}{\sqrt{4\pi}} \tag{C.14}$$

$\ell = 1$:

$$\begin{aligned} Y_{11} &= -\sqrt{\frac{3}{8\pi}} \sin\theta e^{i\phi} \\ Y_{10} &= \sqrt{\frac{3}{4\pi}} \cos\theta \end{aligned} \tag{C.15}$$

$\ell = 2$:

$$\begin{aligned} Y_{22} &= \frac{1}{4}\sqrt{\frac{15}{4\pi}} \sin^2\theta e^{2i\phi} \\ Y_{21} &= -\sqrt{\frac{15}{8\pi}} \sin\theta\cos\theta e^{i\phi} \\ Y_{20} &= \sqrt{\frac{5}{4\pi}} \left(\frac{3}{2}\cos^2\theta - \frac{1}{2}\right) \end{aligned} \tag{C.16}$$

$\ell = 3$:

$$\begin{aligned} Y_{33} &= -\frac{1}{4}\sqrt{\frac{35}{4\pi}} \sin^3\theta e^{3i\phi} \\ Y_{32} &= \frac{1}{4}\sqrt{\frac{105}{2\pi}} \sin^2\theta\cos\theta e^{2i\phi} \\ Y_{31} &= -\frac{1}{4}\sqrt{\frac{21}{4\pi}} \sin\theta(5\cos^2\theta - 1)e^{i\phi} \\ Y_{30} &= \sqrt{\frac{7}{4\pi}} \left(\frac{5}{2}\cos^3\theta - \frac{3}{2}\cos\theta\right) \end{aligned} \tag{C.17}$$

We see that for $m = 0$

$$Y_{\ell 0} = \sqrt{\frac{2\ell+1}{4\pi}} P_\ell(\cos\theta). \tag{C.18}$$

A detailed discussion of the properties of the functions $Y_{\ell m}(\theta, \phi)$ is given in Condon and Shortley [1]. Of particular interest is the relation known as *the addition theorem for spherical harmonics.* If two unit vectors, $\hat{r}$ and

$\hat{r}'$, are defined by the angles (θ, ϕ) and (θ', ϕ'), and make between them an angle Θ, then,

$$P_\ell(\Theta) = \frac{4\pi}{2\ell+1} \sum_{m=-\ell}^{\ell} Y^*_{\ell m}(\theta', \phi') Y_{\ell m}(\theta, \phi). \tag{C.19}$$

This relation can also be expressed in terms of the associated Legendre polynomials,

$$\begin{aligned} P_\ell(\Theta) &= P_\ell(\theta) P_\ell(\theta') \\ &+ 2 \sum_{m=-\ell}^{\ell} \frac{(\ell-1)!}{(\ell+m)!} P_\ell^m(\cos\theta) P_\ell^m(\cos\theta') \cos\left[m(\phi - \phi')\right]. \end{aligned} \tag{C.20}$$

The addition theorem is important in obtaining many useful expressions, such as the expansions of the free-particle propagator discussed in Appendix D.

It is often convenient to define linear combinations of spherical harmonics that are real. Thus, we have,

$$\begin{aligned} \overline{Y}_{\ell 0}(\hat{r}) &= Y_{\ell 0}(\hat{r}) \\ \overline{Y}_{\ell,m}(\hat{r}) &= \frac{1}{\sqrt{2}} \left(Y_{\ell,m}(\hat{r}) + Y_{\ell,-m}(\hat{r})\right), \quad m > 0 \\ \overline{Y}_{\ell,m}(\hat{r}) &= \frac{\mathrm{i}}{\sqrt{2}} \left(Y_{\ell,m}(\hat{r}) - Y_{\ell,-m}(\hat{r})\right), \quad m < 0. \end{aligned} \tag{C.21}$$

These functions satisfy the proper orthonormality and completeness relations and form a complete basis set.

C.2 Bessel, Neumann, and Hankel Functions

The Bessel functions of order ν are the solutions of the Bessel equation,

$$\frac{\mathrm{d}^2 R}{\mathrm{d}x^2} + \frac{1}{x}\frac{\mathrm{d}R}{\mathrm{d}x} + \left(1 - \frac{\nu^2}{x^2}\right) R = 0. \tag{C.22}$$

There are two solutions for each value of ν, given in terms of the gamma function,

$$J_\nu(x) = \left(\frac{x}{2}\right)^\nu \sum_{j=0}^{\infty} \frac{(-1)^j}{j!\Gamma(j+\nu+1)} \left(\frac{x}{2}\right)^{2j}, \tag{C.23}$$

and

$$J_{-\nu}(x) = \left(\frac{x}{2}\right)^{-\nu} \sum_{j=0}^{\infty} \frac{(-1)^j}{j!\Gamma(j-\nu+1)} \left(\frac{x}{2}\right)^{2j}, \tag{C.24}$$

called Bessel functions of the first kind of order $\pm\nu$. If ν is an integer, these two solutions are linearly dependent,

$$J_{-\nu}(x) = (-1)^{\nu} J_{\nu}(x), \tag{C.25}$$

in which case a second, linearly independent, solution is given by the Bessel function of the second kind, or *Neumann function*,

$$N_{\nu} = \frac{J_{\nu}(x)\cos\nu\pi - J_{-\nu}}{\sin\nu\pi}. \tag{C.26}$$

The Bessel functions of the third kind, or *Hankel functions*, are defined as linear combinations of J_{ν} and N_{ν},

$$\begin{aligned} H_{\nu}^{(+)}(x) &= J_{\nu}(x) + \mathrm{i}N_{\nu}(x) \\ H_{\nu}^{(-)}(x) &= J_{\nu}(x) - \mathrm{i}N_{\nu}(x). \end{aligned} \tag{C.27}$$

Each of the two combinations, $\{J_{\nu}, N_{\nu}\}$ and $\{H_{\nu}^{(+)}, H_{\nu}^{(-)}\}$ forms a linearly independent set of solutions to Bessel's equation, Eq. (C.22).

When ν is not an integer, the functions $\{J_{\nu}, J_{-\nu}\}$ form a linearly independent set of solutions to Bessel's equation. For $\nu = \ell + \frac{1}{2}$, with ℓ being an integer, it is customary to define spherical Bessel, Neumann, and Hankel functions through the relations,

$$\begin{aligned} j_{\ell}(x) &= \sqrt{\frac{\pi}{2x}} J_{\ell+\frac{1}{2}}(x) \\ n_{\ell}(x) &= \sqrt{\frac{\pi}{2x}} N_{\ell+\frac{1}{2}}(x) \\ h_{\ell}^{(\pm)}(x) &= \sqrt{\frac{\pi}{2x}} H_{\ell+\frac{1}{2}}^{(\pm)}(x). \end{aligned} \tag{C.28}$$

For x real, we have,

$$h_{\ell}^{(\pm)}(x) = h_{\ell}^{(\mp)^*}(x). \tag{C.29}$$

It follows from the series representations, Eqs. (C.23) and (C.24), that,

$$\begin{aligned} j_{\ell}(x) &= (-x)^{\ell} \left(\frac{1}{x}\frac{\mathrm{d}}{\mathrm{d}x}\right)^{\ell} \left(\frac{\sin x}{x}\right) \\ n_{\ell}(x) &= -(-x)^{\ell} \left(\frac{1}{x}\frac{\mathrm{d}}{\mathrm{d}x}\right)^{\ell} \left(\frac{\cos x}{x}\right). \end{aligned} \tag{C.30}$$

For the first few values of ℓ, the functions j_{ℓ}, n_{ℓ}, and $h_{\ell}^{(+)}$ are given by the expressions,

$$j_0(x) = \frac{\sin x}{x}; \quad j_1(x) = \frac{\sin x}{x^2} - \frac{\cos x}{x}$$

$$\begin{aligned}
& j_2(x) = \left(\frac{3}{x^3} - \frac{1}{x}\right)\sin x - 3\frac{\cos x}{x} \\
n_0(x) &= -\frac{\cos x}{x}; \quad n_1(x) = -\frac{\cos x}{x^2} - \frac{\sin x}{x} \\
& n_2(x) = -\left(\frac{3}{x^3} - \frac{1}{x}\right)\cos x - 3\frac{\sin x}{x} \\
h_0^{(+)}(x) &= \frac{\mathrm{e}^{\mathrm{i}x}}{\mathrm{i}x}; \quad h_1^{(+)}(x) = -\frac{\mathrm{e}^{\mathrm{i}x}}{x}\left(1 + \frac{\mathrm{i}}{x}\right) \\
& h_2^{(+)}(x) = \frac{\mathrm{i}\mathrm{e}^{\mathrm{i}x}}{x}\left(1 + \frac{3\mathrm{i}}{x} - \frac{3}{x^2}\right).
\end{aligned} \tag{C.31}$$

The spherical Bessel, Neumann, and Hankel functions, denoted collectively by z_ℓ, satisfy the recursion formulae,

$$\frac{2\ell+1}{x} z_\ell(x) = z_{\ell-1}(x) + z_{\ell+1}(x) \tag{C.32}$$

and

$$z'_\ell(x) = \frac{1}{2\ell+1}\left[\ell z_{\ell-1}(x) - (\ell+1) z_{\ell+1}(x)\right]. \tag{C.33}$$

Finally, in the limit of small argument, $x << \ell$, we have the asymptotic forms,

$$\begin{aligned}
j_\ell(x) &\to \frac{x^\ell}{(2\ell+1)!!} \\
n_\ell(x) &\to -\frac{(2\ell-1)!!}{x^{\ell+1}}
\end{aligned} \tag{C.34}$$

where $(2\ell+1)!! = (2\ell+1)(2\ell-1)(2\ell-3)\cdots(5)(3)(1)$. In the limit of large argument, $x >> \ell$, we have,

$$\begin{aligned}
j_\ell(x) &\to \frac{1}{x}\sin\left(x - \frac{\ell\pi}{2}\right) \\
n_\ell(x) &\to -\frac{1}{x}\cos\left(x - \frac{\ell\pi}{2}\right) \\
h_\ell^{(+)}(x) &\to (-\mathrm{i})^{\ell+1}\frac{\mathrm{e}^{\mathrm{i}x}}{x}.
\end{aligned} \tag{C.35}$$

Finally, the spherical functions satisfy a set of useful Wronskian relations. With

$$W[f(x), g(x)] = fg' - gf' \qquad f' = \frac{\mathrm{d}f}{\mathrm{d}x} \tag{C.36}$$

we have

$$W[j_\ell(x), n_\ell(x)] = \frac{1}{i} W[j_\ell(x), h_\ell(x)] = -W[n_\ell(x), h_\ell(x)] = \frac{1}{x^2}. \tag{C.37}$$

As an illustration of the use of the spherical harmonics, we present the solutions of the force-free Schrödinger equation, the Helmholtz equation, in one, two and three dimensions.

C.3 Helmholtz Equation

We now give the multipolar solutions to the Helmholtz equation in one, two, and three dimensions as well as some useful identities. We use a notation and normalization for the regular and irregular solutions such that most of the formulae in the text are valid in one, two, or three dimensions. $J_L(\mathbf{r})$ is used to denote the regular solution that is proportional to r^ℓ near the origin. It is useful to define two irregular solutions. $N_L(\mathbf{r})$ represents the irregular solution appropriate to standing-wave boundary conditions, and $H_L(\mathbf{r})$ represents the irregular solution compatible with outgoing waves at large distances. $H_L(\mathbf{r})$ is also the irregular solution that vanishes at infinity when analytically continued to negative energy.

The Helmholtz equation, $[\nabla^2 + E]\psi(x) = 0$, has solutions of the form $\psi(x) = \exp(i\mathbf{k}\cdot\mathbf{r})$ where $k^2 = E$. We desire, however, solutions in the form of radial functions multiplying angular functions, e.g. $J_L(\mathbf{r}) = f_L(r)Y_L(\hat{r})$. To achieve this end we make use of the angular harmonics, $Y_L(\mathbf{r})$, which are orthonormal and complete over the angular variables,

$$\int d\hat{r}\; Y_L(\hat{r})Y_{L'}^{(*)}(\hat{r}) = \delta_{LL'}, \tag{C.38}$$

$$\sum_L Y_L(\hat{r})Y_L^{(*)}(\hat{r}') = \delta(\hat{r},\hat{r}'), \tag{C.39}$$

In one dimension there are only two angular harmonics corresponding to functions that are either symmetric or anti-symmetric about the origin,

$$Y_0(\hat{r}) = \frac{1}{\sqrt{2}} \quad , Y_1(\hat{r}) = \frac{x}{r\sqrt{2}} \qquad (d=1). \tag{C.40}$$

In two dimensions the angular harmonics may be defined as,

$$Y_L(\hat{r}) = \frac{\exp(iL\theta)}{\sqrt{2\pi}}, \qquad (L = 0, \pm 1, \pm 2, \pm 3, \ldots), \qquad (d=2) \tag{C.41}$$

but we can equally well take them to be real,

$$Y_L(\hat{r}) = \begin{cases} \frac{1}{\sqrt{\pi}} \cos L\theta & L > 0\,; \\ \frac{1}{\sqrt{2\pi}} & L = 0; \\ \frac{1}{\sqrt{\pi}} \sin L\theta & L < 0. \end{cases} \qquad (d=2) \tag{C.42}$$

In three dimensions the angular harmonics are the spherical harmonics,

$$Y_L(\hat{r}) = N_\ell^m P_\ell^{|m|}(\cos\theta)\exp(im\phi), \qquad (d=3) \tag{C.43}$$

$$L = \{\ell, m\} \quad \ell = 0,1,2,3,\ldots, \quad m = (-\ell, -\ell+1, \ldots, \ell-1, \ell) \tag{C.44}$$

where $P_\ell^m(x)$ is an associated Legendre polynomial,

$$P_\ell^m(x) = (1-x^2)^{m/2}\frac{1}{d^{\ell+m}}dx^{\ell+m}(x^2-1)^\ell, \tag{C.45}$$

and N_ℓ^m is a normalization factor,

$$N_\ell^m = \left(-\frac{m}{|m|}\right)^m \sqrt{\frac{2\ell+1}{4\pi}}\sqrt{\frac{(\ell-|m|)!}{(\ell+|m|)!}}. \tag{C.46}$$

There is substantial arbitrariness in the phases of the spherical harmonics and it is often convenient to define them so that they are real,

$$Y_L(\hat{r}) = N_\ell^m P_\ell^{|m|}(\cos\theta)\begin{cases}\sqrt{2}\cos m\phi & (m>0),\\ 1 & (m=0),\\ \sqrt{2}\sin m\phi & (m<0).\end{cases} \qquad (d=3) \tag{C.47}$$

Functions, $J_L(\mathbf{r})$, of the desired form that are regular (finite, continuous, and smooth) everywhere can be defined as the coefficients in an expansion of $\exp(i\mathbf{k}\cdot\mathbf{r})$,

$$\exp(i\mathbf{k}\cdot\mathbf{r}) = \begin{cases} 2\sum_{L=0,1} i^L J_L(\mathbf{r})Y_L(\hat{k}), & (d=1),\\ 2\pi\sum_{L=-\infty}^{\infty} i^{|L|}J_{|L|}(\mathbf{r})Y_L(\hat{k}), & (d=2),\\ 4\pi\sum_{\ell=0}^{\infty}\sum_{m=-\ell}^{\ell} i^\ell J_L(\mathbf{r})Y_L(\hat{k}), & (d=3),\end{cases} \tag{C.48}$$

or

$$J_L(\mathbf{r}) = \int d\hat{k}\exp(i\mathbf{k}\cdot\mathbf{r})Y_L(\hat{k})\begin{cases}\frac{i^{-L}}{2} & (d=1),\\ \frac{i^{-|L|}}{2\pi} & (d=2),\\ \frac{i^{-\ell}}{4\pi} & (d=3).\end{cases} \tag{C.49}$$

These regular, multipolar solutions are easily evaluated in one, two, or three dimensions,

$$J_L(\mathbf{r}) = Y_L(\hat{r})\begin{cases}\cos(kr - l\pi/2) & (d=1),\\ \mathcal{J}_{|L|}(kr) & (d=2),\\ j_\ell(kr) & (d=3).\end{cases} \tag{C.50}$$

Here we have used $\mathcal{J}_{|L|}(z)$ to represent the cylindrical Bessel function and $j_\ell(z)$ to represent the spherical Bessel function.

It may be helpful to view Eq. (C.48) as the *generating* function for the Bessel functions. The generating function for the cylindrical Bessel coefficients is given by the expression,

$$\exp\left[z\left(t-\frac{1}{t}\right)\right] = \sum_{L=-\infty}^{\infty} t^L \mathcal{J}_L(z). \tag{C.51}$$

This can be converted into Eq. (C.48) by the substitution $t \to \exp i(\theta + \pi/2)$. Similarly an expression equivalent to the $d = 3$ version of Eq. (C.48),

$$\exp(izx) = \sum_{\ell=0}^{\infty}(2\ell+1)P_\ell(x)i^\ell j_\ell(z), \tag{C.52}$$

where $P_\ell(x)$ is a Legendre polynomial, may be taken as a definition of the spherical Bessel functions.

The cylindrical and spherical Bessel functions have the following limiting forms for small arguments,

$$\begin{aligned} \mathcal{J}_{|L|}(z) &\to \frac{(z/2)^{|L|}}{|L|!}[1+O(z^2)] \\ j_\ell(z) &\to \frac{z^\ell}{(2\ell+1)!!}, \end{aligned} \tag{C.53}$$

and vary at large arguments as,

$$\begin{aligned} \mathcal{J}_{|L|}(z) &\to \sqrt{\frac{2}{\pi z}}\cos(z-|L|\pi/2-\pi/4) \\ j_\ell(z) &\to \sin(z-\ell\pi/2)/z. \end{aligned} \tag{C.54}$$

In addition to these regular solutions to the Helmholtz equation, irregular solutions are also needed in order to construct the Green function that satisfies the inhomogeneous Helmholtz equation,

$$[\nabla^2 + E]G_0(\mathbf{r},\mathbf{r}') = \delta(\mathbf{r}-\mathbf{r}'). \tag{C.55}$$

It is easy to verify that this equation is satisfied by the following integral expression for G_0,

$$G_0(\mathbf{r},\mathbf{r}') = (\frac{1}{2\pi})^d \int_0^\infty k^{d-1}\mathrm{d}k \int \mathrm{d}\hat{k}\frac{\exp[i\mathbf{k}\cdot(\mathbf{r}-\mathbf{r}')]}{E-k^2}. \tag{C.56}$$

In addition to satisfying the inhomogeneous Helmholtz equation, Eq. (C.55), the Green function specifies the boundary condition satisfied by the wave functions. We customarily use outgoing-wave boundary conditions because they are the easiest to understand physically. They correspond to a causal or retarded Green function; i.e., a perturbation at the origin at time t propagates outward as time increases. Sometimes, however, it is convenient to use stationary-wave boundary conditions that correspond to the average of the advanced and retarded Green function. With standing-wave boundary conditions a disturbance initiates both incoming and outgoing waves of equal amplitude that set up a standing-wave pattern.

Mathematically, the two types of Green function arise from different ways of handling the singularities at $k = \pm k$ in Eq. (C.56) when the integral over k is converted into a contour integral. The stationary-wave Green function arises from averaging the residues from both poles while the outgoing-wave Green function arises from assuming that the energy has a small positive imaginary part and closing contours in such a way that the integral remains finite. The outgoing-wave Green function has the advantage that its analytic continuation to negative energies vanishes at large distances so it is appropriate for describing bound-state wave functions.

The outgoing-wave or causal Green function in one, two, or three dimensions can be calculated from Eqs. (C.56) and (C.48) together with the following identities,

$$\int_0^\infty \frac{\mathrm{d}k}{E-k^2}\left\{\begin{array}{c}\cos(kr)\\ k\mathcal{J}_0(kr)\\ k^2 j_0(kr)\end{array}\right\} = \left\{\begin{array}{c}-(1/2)\pi i E^{-(1/2)}\exp(ikr)\\ -(1/2)\pi i\mathcal{H}_0(kr)\\ -(1/2)\pi ikrh_\ell(kr)\end{array}\right\}. \tag{C.57}$$

The resultant expression for the Green function is $G_0(\mathbf{r},\mathbf{r}') = H_0(\mathbf{r}-\mathbf{r}')$, where H_0 is an $L = 0$ irregular solution to the Helmholtz equation given by,

$$H_0(\mathbf{r}-\mathbf{r}') = Y_L(\widehat{r}-r')\left\{\begin{array}{cc}-iE^{-(1/2)}\exp(ikr|\mathbf{r}-\mathbf{r}'|) & (d=1)\\ -(1/4)i\mathcal{H}_0(kr|\mathbf{r}-\mathbf{r}'|) & (d=2)\\ -ikrh_0(kr|\mathbf{r}-\mathbf{r}'|) & (d=3)\end{array}\right\}. \tag{C.58}$$

In the main text we make extensive use of an expansion for the Green function as a sum of terms, each of which is separable in $\mathbf{r}$ and $\mathbf{r}'$. Using Eqs. (C.56) and (C.48) again, but substituting an expansion in terms of angular harmonics for both $\exp(i\mathbf{k}\cdot\mathbf{r})$ and $\exp(-i\mathbf{k}\cdot\mathbf{r}')$ and using the orthonormality of the angular harmonics, we have,

$$G_0(\mathbf{r},\mathbf{r}') = \frac{\Omega^2}{(2\pi)^3}\int_0^\infty \frac{k^{d-1}\mathrm{d}k}{E-k^2}\sum_L J_L(\mathbf{r};k)J_{L'}^{(*)}(\mathbf{r}';k), \tag{C.59}$$

where Ω is the angular phase space, equal to 2, 2π, and 4π in one, two, and three dimensions respectively. The dependence of the Helmholtz equation solutions on the energy parameter is shown explicitly for clarity. The integral gives,

$$G_0(\mathbf{r},\mathbf{r}') = \sum_L J_L(\mathbf{r};k)H_L^{(*)}(\mathbf{r}';k)\Theta(r'-r) + H_L(\mathbf{r};k)J_L^{(*)}(\mathbf{r}';k)\Theta(r-r'), \tag{C.60}$$

where $\Theta(x)$ is the Heaviside step function that is unity if its argument is greater than zero and vanishes if its argument is less than zero, and where the irregular solution, $H_L(\mathbf{r})$ is given by

$$H_L(\mathbf{r}) = Y_L(\mathbf{r}) \begin{cases} -iE^{-(1/2)}\exp[i(kr-\ell\pi/2)] & (d=1) \\ -(1/4)i\mathcal{H}_{|L|}(kr) & (d=2) \\ -ikrh_\ell(kr) & (d=3) \end{cases} . \tag{C.61}$$

The notation $H_L^{(*)}$ indicates complex conjugation of the angular harmonic part of the function only. If real angular harmonics are used all of the $(*)$s may be omitted. $\mathcal{H}_L = \mathcal{J}_L + i\mathcal{Y}_L$ is the cylindrical Hankel function that is given approximately for small values of its argument by,

$$\mathcal{H}_{|L|}(z) \rightarrow \begin{cases} \frac{2}{\pi}(\ln(z/2)+\gamma)[1+O(z^2)] & (L=0) \\ -\frac{(|L|-1)!}{\pi(z/2)^{|L|}}[1+O(z^2)] & (|L|>0) \end{cases} \tag{C.62}$$

where γ is Euler's constant, $\gamma = 0.5772156649\ldots$. For large values of its argument, $\mathcal{H}_L(z)$ varies as,

$$\mathcal{H}_{|L|}(z) \rightarrow \sqrt{\frac{2}{\pi z}}\exp i(z-|L|\pi/2-\pi/4). \tag{C.63}$$

Similarly the three dimensional irregular solutions are proportional to spherical Hankel functions $h_\ell = j_\ell + in_\ell$. At small values of their arguments these functions are approximately,

$$h_\ell(z) \rightarrow \frac{(2\ell-1)!!}{z^{\ell+1}}, \tag{C.64}$$

and at large values they vary as,

$$h_\ell(z) \rightarrow \exp i(z-\ell\pi/2)/z. \tag{C.65}$$

It is important that the regular and irregular solutions satisfy the following Wronskian-like relation,

$$\int_S d\mathbf{S}\cdot[J_L(\mathbf{r})\nabla H_{L'}(\mathbf{r}) - \nabla J_L(\mathbf{r})H_{L'}(\mathbf{r})] = \begin{cases} \delta_{LL'} & \text{if origin in } S \\ 0 & \text{otherwise.} \end{cases} \tag{C.66}$$

This relation is equivalent to the requirement that the Green function be correctly normalized since if $[\nabla^2+E]G(\mathbf{r},\mathbf{r}') = \delta(\mathbf{r}-\mathbf{r}')$ and $[\nabla^2+E]J_L(x) = 0$ it follows from Green's theorem that,

$$\int d\mathbf{S}\cdot[J_L(\mathbf{r})\nabla G(\mathbf{r},\mathbf{r}') - \nabla J_L(\mathbf{r})G(\mathbf{r},\mathbf{r}')] = J_L(\mathbf{r}'), \tag{C.67}$$

provided $\mathbf{r}'$ is within the interval bounded by the surface of integration. Expanding the Green function in the usual way yields,

$$\int \mathrm{d}\mathbf{S} \cdot \sum_{L'} [J_L(\mathbf{r}) \nabla H_{L'}(\mathbf{r}) - \nabla J_L(\mathbf{r}) H_{L'}(\mathbf{r}')] J_L(\mathbf{r}') = J_L(\mathbf{r}'), \quad \text{(C.68)}$$

which requires the Wronskian condition (C.66).

Both the regular and irregular functions can be expanded about a shifted origin (see Appendix D),

$$J_L(\mathbf{r} - \mathbf{R}) = \sum_{L'} g_{LL'}(\mathbf{R}) J_{L'}(\mathbf{r}) \quad \text{(C.69)}$$

$$H_L(\mathbf{r} - \mathbf{R}) = \sum_{L'} G_{LL'}(\mathbf{R}) J_{L'}(\mathbf{r}). \quad \text{(C.70)}$$

The coefficients, $g_{LL'}(\mathbf{R})$ for the expansion of $J_L(\mathbf{r} - \mathbf{R})$ play the role of a translation operator. The coefficients, $G_{LL''}(\mathbf{R})$, for the expansion of the irregular solutions in terms of the regular solutions play a central part in multiple scattering theory. They are given by,

$$G_{LL'}(\mathbf{R}) = \begin{cases} 2\sum_{L''} i^{L'-L-L''} C(L,L',L'') H_{L''}^{(*)}(\mathbf{R}) & (d=1), \\ 2\pi \sum_{L''} i^{|L'|-|L|-|L''|} C(L,L',L'') H_{|L''|}^{(*)}(\mathbf{R}) & (d=2), \\ 4\pi \sum_{L''} i^{\ell'-\ell-\ell''} C(L,L',L'') H_{L''}^{(*)}(\mathbf{R}) & (d=3). \end{cases} \quad \text{(C.71)}$$

Here the gaunt numbers, $C(L, L', L'')$ are defined as integrals over three angular harmonics,

$$C(L, L', L'') = \int \mathrm{d}\hat{r} Y_L^{(*)}(\hat{r}) Y_{L'}(\hat{r}) Y_{L''}^{(*)}(\hat{r}). \quad \text{(C.72)}$$

The expression for $g_{LL'}(\mathbf{R})$ is identical to that for $G_{LL'}(\mathbf{R})$ except that $H_{L''}^{(*)}(\mathbf{R})$ is replaced by $J_{L''}^{(*)}(\mathbf{R})$.

In one dimension, simple analytic expressions for the structure constants both in real and reciprocal space are available,

$$H_L(\mathbf{r} - \mathbf{R}) = \sum_{L'=0,1} G_{LL'}(\mathbf{R}) J_{L'}(\mathbf{r}) = \sum_{L'=0,1} J_{L'}(\mathbf{r}) G_{L'L}(-\mathbf{R}) \quad \text{(C.73)}$$

$$G_{LL'}(\mathbf{R}) = \frac{\exp(ikr)}{ik} \begin{pmatrix} 1 & \mathrm{i}^{-1}\mathbf{R}/R \\ \mathrm{i}\mathbf{R}/R & 1 \end{pmatrix} \quad \text{(C.74)}$$

$$\tilde{G}_{LL'}(k) = \sum_{n=-\infty}^{\infty} G_{LL'}(na)$$

$$
\begin{aligned}
&= \frac{1}{ik}\begin{pmatrix} -1+\frac{i\sin\phi}{\cos\theta-\cos\phi} & \frac{\sin\theta}{\cos\theta-\cos\phi} \\ \frac{-\sin\theta}{\cos\theta-\cos\phi} & -1+\frac{i\sin\phi}{\cos\theta-\cos\phi} \end{pmatrix} \\
\phi &= kra \quad \theta = ka.
\end{aligned}
\tag{C.75}
$$

For $E < 0$ it is convenient to define the regular solutions to be,

$$
J_L(\mathbf{r}) = Y_L(\hat{r}) \begin{cases} i^{-L}\cos(i\sqrt{-E}r - \ell\pi/2) & (d=1) \\ , i^{-|L|}\mathcal{J}_{|L|}(i\sqrt{-E}r) & (d=2) \\ , i^{-\ell}j_\ell(i\sqrt{-E}r) & (d=3), \end{cases} \tag{C.76}
$$

and the irregular solutions to be

$$
H_L(\mathbf{r}) = Y_L(\hat{r}) \begin{cases} -[-E]^{-(1/2)}\exp(-\sqrt{-E}r) & (d=1) \\ , (1/4)i^{|L|-1}\mathcal{H}_L(i\sqrt{-E}r) & (d=2) \\ , i^{\ell}\sqrt{-E}h_\ell(i\sqrt{-E}r) & (d=3). \end{cases} \tag{C.77}
$$

These definitions yield real values for J_L and H_L at real negative energies provided that the angular harmonics are chosen to be real.

Bibliography

[1] E. U. Condon and G. H. Shortley, *The Theory of Atomic Spectra* Cambridge (1967).

Appendix D

Partial Waves

D.1 General Considerations

In this appendix, we give a summary of the theory of partial-wave analysis of the scattering of a single particle by a spatially-bounded potential, $V(\mathbf{r})$. For comparative purposes, and in order to ease into the discussion of non-spherical potentials, we begin with a summary of the partial wave analysis of the scattering from a central field of force. In this part of the discussion we closely follow the exposition in standard texts [1, 2]. Then, we proceed with a partial-wave analysis of the wave function for generally-shaped, spatially-bounded potentials, $V(\mathbf{r})$. We discuss the differential and integral equations satisfied by the wave function, as well as the form taken by the wave function in various regions of space, such as in the so-called "moon" regions, i.e., the parts of free space that lie inside a sphere circumscribing a non-spherical cell. Also, we point out the many similarities that exist between the cases of central and non-central force fields. At the same time, very important considerations arise in connection with expansions in angular-momentum eigenfunctions, and questions of convergence that such expansions ultimately entail. These questions are addressed as they are brought forth in the formal development.

We begin by considering the scattering of a non-relativistic, spinless particle of mass m and momentum $\mathbf{p}_i$ interacting with a time-independent potential, V($\mathbf{r}$.) In order to simplify our notation, we describe the particle in terms of its wave vector, $\mathbf{k}_i = \mathbf{p}_i/\hbar$, and choose units such that $\hbar = 1$, and $2m = 1$. In this form, the equations also describe the motion of a wave front of incident wave vector $\mathbf{k}_i$.

The motion of the particle (wave) is governed by the time-dependent

Schrödinger equation, which in the present case we write in the form,

$$[-\nabla_{\mathbf{r}}^2 + V(\mathbf{r})]\Psi(\mathbf{r},t) = \mathrm{i}\frac{\partial}{\partial t}\Psi(\mathbf{r},t). \tag{D.1}$$

We are interested in the stationary solutions of this equation of the type,

$$\Psi(\mathbf{r},t) = \psi_{\mathbf{k}_i}(\mathbf{r})\exp(-\mathrm{i}E_i t), \tag{D.2}$$

where $E = \mathbf{k}_i{}^2$. Substitution of this expression into Eq. (D.1) yields the time-independent Schrödinger equation,

$$[-\nabla_{\mathbf{r}}^2 + V(\mathbf{r})]\psi_{\mathbf{k}_i}(\mathbf{r}) = E_i\psi_{\mathbf{k}_i}(\mathbf{r}). \tag{D.3}$$

For potentials of the type under consideration here (in general, for potentials that vanish faster than r^{-1} as $r \to \infty$), the wave function[1] at regions outside the field of force, i.e., in the limit $r \to \infty$, can be written in the form [2],

$$\psi_{\mathbf{k}_i}^{(+)}(\mathbf{r}) \to A\left[\exp(-\mathrm{i}\mathbf{k}_i\cdot\mathbf{r}) + f(\theta,\phi)\frac{\exp(ikr)}{r}\right], \quad as \quad r \to \infty. \tag{D.4}$$

This asymptotic boundary condition clearly exhibits the form of the wave function at large r. There, this so-called "scattering wave" consists of the incident wave (free-particle solution), represented by the first term on the right-hand side, and a scattered contribution (wave). (Note the presence of the causal Green function in this equation and the corresponding use of a (+) superscript in the wave function.) In the last equation, the constant A is independent of r, and the angles θ and ϕ of the outgoing (scattered) wave vector, $\mathbf{k}_s$, are measured in a coordinate system in which the z-axis coincides with the direction of the incident wave vector, $\mathbf{k}_i$. It is important to note that in general the particular solution given by Eq. (D.4) satisfies Eq. (D.3) only asymptotically, including terms of order $1/r$ in the region in which the potential can be neglected [provided that V($\mathbf{r}$) vanishes faster than $1/r$ for large r].

The quantity $f(\theta,\phi) = f(\mathbf{k}_i,\mathbf{k}_s)$ in Eq. (D.4) is the so-called "scattering amplitude", a quantity of central importance in scattering theory in general, and in multiple scattering theory in particular. We now proceed with an evaluation of the explicit forms of the wave functions and the scattering amplitudes associated with spherically-symmetric potentials. The generalization to non-central potentials is given in a subsequent section.

[1] Occasionally, we shall refer to solutions of the Schrödinger equation as "wave functions" even though they may satisfy only the differential equation and a *single* boundary condition at the origin. Such solutions should more properly be referred to as "basis functions", in terms of which one can expand the wave function. The distinction between basis functions and wave functions will be emphasized as necessary in the following discussion.

D.2 Spherically-Symmetric Potentials

The solution of the Schrödinger equation, Eq. (D.3), is considerably simplified when the potential, $V(\mathbf{r})$, is spherically symmetric. We choose a right-handed coordinate system with the z-axis along the wave vector $\mathbf{k}_i$, and with the origin coinciding with that of the vector $\mathbf{r}$. In spherical coordinates, the Hamiltonian operator, $\hat{H} = -\nabla^2 + \hat{V}$ reads,

$$H = -\left[\frac{1}{r^2}\frac{\partial}{\partial r}\left(r^2\frac{\partial}{\partial r}\right) + \frac{1}{r^2\sin\theta}\frac{\partial}{\partial\theta}\left(\sin\theta\frac{\partial}{\partial\theta}\right) + \frac{1}{r^2\sin^2\theta}\frac{\partial^2}{\partial\phi^2}\right] + V(r), \tag{D.5}$$

where the dependence of V only on the magnitude of $\mathbf{r}$ is explicitly indicated. Now, the Schrödinger equation, Eq. (D.1), for the stationary scattering wave function, $\psi^{(+)}_{\mathbf{k}_i}(\mathbf{r})$, can be written in the form,

$$\left[\frac{1}{r^2}\frac{\partial}{\partial r}\left(r^2\frac{\partial}{\partial r}\right) + \frac{1}{r^2\sin\theta}\frac{\partial}{\partial\theta}\left(\sin\theta\frac{\partial}{\partial\theta}\right) + \frac{1}{r^2\sin^2\theta}\frac{\partial^2}{\partial\phi^2}\right]\psi^{(+)}_{\mathbf{k}_i}(\mathbf{r}) + V(\mathbf{r})\psi^{(+)}_{\mathbf{k}_i}(\mathbf{r}) = E_i\psi^{(+)}_{\mathbf{k}_i}(\mathbf{r}). \tag{D.6}$$

To solve Eq. (D.6) we make use of the observation that the square of the angular momentum,

$$\mathbf{L} = \mathbf{r}\times\mathbf{p}, \tag{D.7}$$

of a system with spherical symmetry, and its projection, L_z, along the z-axis are constants of the motion. This follows immediately from the fact that they satisfy the relations,

$$[H, L^2] = [H, L_z] = 0, \tag{D.8}$$

where $[A,B] = AB - BA$ denotes the commutator of two operators, A and B. Using the correspondence $\mathbf{p} \to -i\nabla_{\mathbf{r}}$ in the definition of angular momentum, Eq. (D.7), and the expressions of the Cartesian coordinates, (x,y,z), in terms of polar coordinates, (r,θ,ϕ), we can show that,

$$L^2 = -\left[\frac{1}{\sin\theta}\frac{\partial}{\partial\theta}\left(\sin\theta\frac{\partial}{\partial\theta}\right) + \frac{1}{\sin^2\theta}\frac{\partial^2}{\partial\phi^2}\right]. \tag{D.9}$$

Because L^2 commutes with its components, $[L^2, L_i] = 0, \quad i = x,y,z$, it can be diagonalized simultaneously with any one of them by the same set of eigenfunctions. Choosing that component to lie along the z-axis, the corresponding eigenfunctions are the spherical harmonics [3], $Y_{lm}(\theta,\phi)$, such that,

$$L^2 Y_{lm}(\theta,\phi) = l(l+1)Y_{lm}(\theta,\phi) \tag{D.10}$$

and

$$L_z Y_{lm}(\theta,\phi) = m Y_{lm}(\theta,\phi). \tag{D.11}$$

Here, l and m are called, respectively, the ***orbital angular momentum quantum number***, and the ***magnetic quantum number***. A brief description of the spherical harmonics and some of their properties is given in Appendix C.

It follows from Eq. (D.9) that the Hamiltonian operator, Eq. (D.5), can be expressed in the form,

$$H = -\left[\frac{1}{r^2}\frac{\partial}{\partial r}\left(r^2\frac{\partial}{\partial r}\right) - \frac{L^2}{r^2}\right] + V(r). \tag{D.12}$$

Because of the commutation relations, Eq. (D.7), we look for eigenfunctions that are common to H, L^2, and L_z. Thus, we expand the scattering wave function, $\psi^{(+)}_{\mathbf{k}_i}(\mathbf{r})$, in ***partial waves*** corresponding to given values of the quantum numbers l and m, in the form,

$$\begin{aligned} \psi^{(+)}(\mathbf{k}_i,\mathbf{r}) &= \sum_{l=0}^{\infty}\sum_{m=-l}^{l} c_{lm}(k)R_{lm}(k,r)Y_{lm}(\theta,\phi) \\ &= \sum_{L} c_L R_L(k,r)Y_L(\hat{\mathbf{r}}), \end{aligned} \tag{D.13}$$

where $k = \sqrt{E}$, and we have introduced the combined index, $L = (l,m)$. (It is not immediately evident that the functions $R_L(k,r)Y_L(\hat{\mathbf{r}})$ form an appropriate basis for the expansion of $\psi^{(+)}(\mathbf{k}_i,\mathbf{r})$. That this is the case is shown explicitly in reference [5].) The radial wave functions, $R_L(k,r)$, are examples of basis functions, defined as solutions of the Schrödinger equation that satisfy only a single boundary condition, the one at the origin, but not at infinity. These basis functions could be chosen so that the wave function, which is given as a linear combination of such basis functions, satisfies the proper boundary conditions at infinity.

It is easily verified directly that because of azimuthal symmetry, the ***radial wave functions***, $R_{lm}(k,r)$, are independent of the magnetic quantum number, m, and that each satisfies the ***radial*** Schrödinger equation,

$$-\left[\frac{1}{r^2}\frac{\mathrm{d}}{\mathrm{d}r}\left(r^2\frac{\mathrm{d}}{\mathrm{d}r}\right) - \frac{l(l+1)}{r^2}\right]R_l(k,r) + V(r)R_l(k,r) = ER_l(k,r). \tag{D.14}$$

It is convenient to introduce the functions,

$$u_l(k,r) = rR_l(k,r), \tag{D.15}$$

which can easily be shown to satisfy the equation,

$$\left[\frac{\mathrm{d}^2}{\mathrm{d}r^2} + k^2 - \frac{l(l+1)}{r^2} - V(r)\right]u_l(k,r) = 0. \tag{D.16}$$

Two features of Eqs. (D.14) and (D.16) are worth emphasizing. First, they are both ordinary differential equations and, second, the functions R_l and u_l can be chosen to be real. This follows because the real and imaginary parts of the complex functions would separately satisfy the differential equations.

Equation (D.16) is particularly convenient for study. However, before considering the solutions to this equation in the presence of a potential, it is instructive to determine the solutions to the corresponding free-particle, i.e., $V = 0$, equation.

D.2.1 Free-particle solutions

Since the spherical harmonics are known functions, the determination of the scattering wave function in Eq. (D.13) requires the knowledge of the radial functions, R_l, and the coefficients, c_{lm}. In this section, we obtain these quantities for the case in which the potential, $V(r)$, vanishes identically. In this case, we obtain from Eq. (D.16) the radial equation for free motion,

$$\left[\frac{\mathrm{d}^2}{\mathrm{d}r^2} + k^2 - \frac{l(l+1)}{r^2}\right] y_l(k,r) = 0. \tag{D.17}$$

Making a further change of variables to $\rho = kr$ and defining the function,

$$f_l(\rho) = y_l/\rho, \tag{D.18}$$

we can rewrite Eq. (D.17) in the form,

$$\left[\frac{\mathrm{d}^2}{\mathrm{d}\rho^2} + \frac{2}{\rho}\frac{\mathrm{d}}{\mathrm{d}\rho^2} + \left(1 - \frac{l(l+1)}{\rho^2}\right)\right] f_l(\rho) = 0. \tag{D.19}$$

This is the well-known *spherical Bessel differential equation.* The particular solutions to this equation that are often used in scattering theory are the spherical Bessel functions, $j_l(\rho)$, the spherical Neumann functions, $n_l(\rho)$, and the spherical Hankel functions, $h_l^{(\pm)}(\rho)$, of the first, (+), and the second, (-), kind, respectively. The Hankel functions are given in terms of the Bessel and Neumann functions by,

$$h_l^{(\pm)}(\rho) = j_l(\rho) \pm i n_l(\rho). \tag{D.20}$$

(The definitions and some important properties of these functions are included in Appendix C.) Because each of the pairs (j_l, n_l) and $(h_l^{(+)}, h_l^{(-)})$ contains two linearly independent solutions to Eq. (D.17), the general solution to that equation can be written in either of the forms,

$$y_l(kr) = kr\left[C_l^{(1)}(k) j_l(kr) + C_l^{(2)}(k) n_l(kr)\right] \tag{D.21}$$

or

$$\tilde{y}_l(kr) = kr\left[D_l^{(1)}(k)h_l^{(+)}(kr) + D_l^{(2)}(k)h_l^{(-)}(kr)\right], \tag{D.22}$$

where the integration constants $[C_l^{(1)}(k), C_l^{(2)}(k)]$ and $[D_l^{(1)}(k), D_l^{(2)}(k)]$ may still depend on the energy, k.

As shown in Appendix C, the Bessel function, j_l, vanishes as r^l as $r \to 0$ and is said to be *regular* at the origin. All other functions, $n_l, h_l^{(\pm)}$, fail to vanish at the origin, in fact they diverge there as $r^{-(l+1)}$, and are said to be *irregular* solutions to the spherical Bessel equation. It now follows that the regular solution, y_l, to Eq. (D.17) vanishes at the origin,

$$y_l(k,0) = krj_l(0) = 0, \tag{D.23}$$

behaving there as

$$y_l(k,r) \to r^{l+1}, \qquad \text{as } r \to 0. \tag{D.24}$$

We are now ready to obtain the coefficients c_{lm} that enter Eq. (D.13) for the case of free motion. This is done by comparing the partial wave decomposition in Eq. (D.21) or Eq. (D.22) with that of the solutions to the corresponding wave equation in coordinate space. It can readily be verified that the Schrödinger equation for free motion,

$$-\nabla^2\psi(\mathbf{r}) = E\psi(\mathbf{r}), \tag{D.25}$$

has the plane-wave solutions, $\exp(\mathrm{i}\mathbf{k}_i\cdot\mathbf{r})$, which are also the eigenfunctions of the linear momentum operator. Since the eigenfunctions $j_l(kr)Y_{lm}(\hat{r})$ form a complete set, we may expand a plane wave in terms of them (Bauer's identity) to obtain,

$$\exp(\mathrm{i}\mathbf{k}_i\cdot\mathbf{r}) = 4\pi\sum_{l=0}^{\infty}\sum_{m=-l}^{l}\mathrm{i}^l j_l(kr)Y_{lm}^*(\hat{\mathbf{k}}_i)Y_{lm}(\hat{\mathbf{r}}). \tag{D.26}$$

Clearly, in the case of free motion the radial wave functions, R_l, are the spherical Bessel functions, j_l. Further comparison of Eqs. (D.13) and (D.26) allows us to identify the coefficients c_{lm}, which in the present case we denote by c_{lm}^0,

$$c_{lm}^0 = \mathrm{i}^l\left[4\pi(2l+1)\right]^{1/2}\delta_{m,0}. \tag{D.27}$$

Here, we have used the fact that the wave vector $\mathbf{k}_i$ points along the z-axis, $(\theta = 0)$, and the relation,

$$Y_{l0}(\theta) = \sqrt{\frac{2l+1}{4\pi}}P_l(\cos\theta), \tag{D.28}$$

where the $P_l(\cos\theta)$ are the Legendre polynomials. (Appendix C includes the definitions and some of the properties of these functions.)

This completes our discussion of the partial wave analysis of free motion. We now turn to the solution to the radial equation, Eq. (D.14) or Eq. (D.16), in the presence of a potential.

D.2.2 Radial equation for central potentials

To obtain the solution to the radial equation in the presence of a potential, we examine the behavior of the wave function in regions far removed from the range of force (in the case of spatially bounded potentials this asymptotic behavior is achieved outside a sphere bounding the potential region). Outside the range of the potential, the particle (or wave) executes essentially free motion and, according to our discussion above, we may write,

$$u_l(kr) = kr\left[C_l^{(1)}(k)j_l(kr) + C_l^{(2)}(k)n_l(kr)\right], \qquad \text{(D.29)}$$

for $r >> a$, where a provides a measure of the extend of the potential region (e.g., the radius of a bounding sphere). That this representation of the wave function is reasonable can be justified on physical grounds, as argued above, or by means of rigorous arguments based on the method of variation of parameters, as is done in a later section, for solving inhomogeneous second-order differential equations.

Now, from the fact that for large x (see Appendix C),

$$j_l(x) \to \frac{1}{x}\sin(x - \frac{l\pi}{2}), \qquad \text{(D.30)}$$

$$n_l(x) \to \frac{1}{x}\cos(x - \frac{l\pi}{2}), \qquad \text{(D.31)}$$

$$h_l^{(+)}(x) \to -\mathrm{i}\frac{\exp\left[\mathrm{i}(x - \frac{l\pi}{2})\right]}{x}, \qquad \text{(D.32)}$$

$$h_l^{(-)}(x) \to \mathrm{i}\frac{\exp\left[-\mathrm{i}(x - \frac{l\pi}{2})\right]}{x}, \qquad \text{(D.33)}$$

we can write

$$u_l(k,r) \to A_l(k)\sin\left[kr - \frac{l\pi}{2} + \delta_l(k)\right], \qquad \text{(D.34)}$$

with

$$A_l(k) = \left\{\left[C_l^{(1)}(k)\right]^2 + \left[C_l^{(2)}(k)\right]^2\right\}^{1/2}, \qquad \text{(D.35)}$$

and

$$\tan\delta_l(k) = -\frac{C_l^{(2)}(k)}{C_l^{(1)}(k)}. \qquad \text{(D.36)}$$

From Eq. (D.15) we may also write,

$$R_l(k,r) \stackrel{r\to\infty}{\longrightarrow} A_l'(k)\left[j_l(kr) - \tan\delta_l n_l(kr)\right], \tag{D.37}$$

where the constant $A_l'(k)$ is independent of r.

The quantities $\delta_l(k)$ introduced in Eq. (D.34) are called the *phase shifts* for the lth partial wave, and contain the effect of the potential on the scattered wave. Note that in the absence of the interaction, the phase shifts vanish and the radial functions reduce to the spherical Bessel functions. The trigonometric functions $\cos\delta_l(k)$ and $\sin\delta_l(k)$ are called the *phase* functions, and they can be made the basis of a complete analysis of scattering by central fields of force [4]. These functions can be continued inside the range of the potential where they become r-dependent corresponding to the potential being abruptly truncated at a given value of r. We encounter this use of a generalized version of the phase functions in our discussion of non- spherical potentials given in Section D.6.

The boundary condition at large r can also be expressed in terms of radially incoming $[\exp(-ikr)]$ and outgoing $[\exp(ikr)]$ waves. Equation (D.37) can also be written in the form,

$$u_l(k,r) \stackrel{r\to\infty}{\longrightarrow} A_l''(k)\left[-(-)^l e^{-ikr} + S_{nl}(k)e^{ikr}\right], \tag{D.38}$$

where

$$A_l''(k) = A_l i^l \exp[-i\delta_l](-)^l/2i \tag{D.39}$$

and

$$S_l(k) = \exp[2i\delta_l(k)]. \tag{D.40}$$

The coefficient of the outgoing wave, $S_l(k)$, is called an *S-matrix element.*

D.2.3 Scattering amplitude

We are now in a position to derive an expression for the scattering amplitude. This is done by comparing the asymptotic expressions of the partial wave expansion, Eq. (D.13), and of the scattering wave, Eq. (D.4). First, using Bauer's identity, Eqs. (D.26) and (D.30), we can rewrite Eq. (D.4) in the form,

$$\psi_{\mathbf{k}_i}^{(+)}(k,r) \stackrel{r\to\infty}{\longrightarrow} A(k)\left[\sum_{l=0}^{\infty}(2l+1)i^l \frac{\sin(kr-\frac{l\pi}{2})}{kr} P_l(\cos\theta) + f(k,\theta,\phi)\frac{e^{ikr}}{r}\right]. \tag{D.41}$$

Upon using Eq. (D.28) and the properties of Legendre polynomials we also have,

$$\psi_{\mathbf{k}_i}^{(+)}(k,r) \stackrel{r\to\infty}{\longrightarrow} A(k)\left\{\sum_{l=0}^{\infty}\sum_{m=-l}^{l}[4\pi(2l+1)]^{1/2}i^l\right.$$

$$\begin{aligned} \times & \quad \frac{exp\left[\mathrm{i}(kr - \frac{l\pi}{2})\right] - \exp\left[-\mathrm{i}(kr - \frac{l\pi}{2})\right]}{2\mathrm{i}kr} Y_{lm}(\hat{\mathbf{r}})\delta_{m,0} \\ + & \quad f(k,\theta,\phi)\frac{\mathrm{e}^{\mathrm{i}kr}}{r}\Bigg\}. \end{aligned} \tag{D.42}$$

At the same time, we can consider the asymptotic behavior of the partial wave expansion, Eq. (D.13), and use the relation in Eq. (D.15) to write,

$$\begin{aligned} \psi^{(+)}_{\mathbf{k}_i}(k,r) & \overset{r\to\infty}{\Longrightarrow} \sum_{l=0}^{\infty}\sum_{m=-l}^{l} c_{lm}(k)A_l(k)\frac{1}{2\mathrm{i}r}\left\{\exp\left[\mathrm{i}(kr - \frac{l\pi}{2} + \delta_l)\right]\right. \\ & - \left.\exp\left[\mathrm{i}(kr - \frac{l\pi}{2} + \delta_l)\right]\right\} Y_{lm}(\hat{\mathbf{r}}). \end{aligned} \tag{D.43}$$

Comparing the coefficients of the incoming spherical waves in Eqs. (D.42) and (D.43) we find,

$$c_{lm}(k) = \frac{A(k)}{kA_l(k)}\left[4\pi(2l+1)\right]^{1/2}\mathrm{i}_l\exp\{\mathrm{i}\delta_l\}\delta_{m,0}. \tag{D.44}$$

Therefore, we may rewrite the partial-wave expansion, Eq. (D.13), in the forms,

$$\psi^{(+)}_{\mathbf{k}_i}(k,r) = A(k)\sum_{l=0}^{\infty}\frac{(2l+1)}{kA_l(k)}\mathrm{i}_l\exp\{\mathrm{i}\delta_l\}R_l(k,r)P_l(\cos\theta), \tag{D.45}$$

or

$$\psi^{(+)}_{\mathbf{k}_i}(k,r) = A(k)\sum_{l=0}^{\infty}\frac{\sqrt{4\pi(2l+1)}}{kA_l(k)}\mathrm{i}_l\exp\{\mathrm{i}\delta_l\}R_l(k,r)Y_{l,0}(\theta). \tag{D.46}$$

Finally, by matching the coefficients of the outgoing spherical waves, and using Eq. (D.44) we obtain the following expression for the scattering amplitude,

$$\begin{aligned} f(k,\theta) & = \frac{1}{2\mathrm{i}k}\sum_{l=0}^{\infty}(2l+1)\left[\exp\{2\mathrm{i}\delta_l(k) - 1\}\right]P_l(\cos\theta) \\ & = \frac{1}{k}\sum_{l=0}^{\infty}(2l+1)\sin\delta_l(k)\mathrm{e}^{\mathrm{i}\delta_l(k)}P_l(\cos\theta). \end{aligned} \tag{D.47}$$

This can also be written in the form,

$$f(k,\theta) = \sum_{l=0}^{\infty}(2l+1)a_l(k)P_l(\cos\theta), \tag{D.48}$$

where the *partial wave amplitudes* $a_l(k)$ are such that,

$$a_l(k) = \frac{1}{2ik}\left[\exp\{2i\delta_l(k) - 1\}\right] = \frac{1}{2ik}\left[S_l(k) - 1\right]. \tag{D.49}$$

It follows that for spherically symmetric potentials the scattering amplitudes are independent of the angles θ and ϕ and are determined completely from a knowledge of the phase shifts. In view of Eq. (D.40), the partial amplitudes $a_l(k)$ can also be written in the form,

$$a_l(k) = \frac{1}{k}\exp\{i\delta_l(k)\}\sin\delta_l(k). \tag{D.50}$$

This form of the partial amplitudes is particularly convenient for use in connection with multiple scattering theory.

D.3 Argand diagrams

The partial scattering amplitudes defined in Eq. (D.50), can be represented graphically in an *Argand diagram.* With $f_\ell = ka_\ell(k)$, we have (where $\Re$ denotes the real and $\Im$ the imaginary part of a quantity)

$$f_\ell - \frac{i}{2} = -\frac{i}{2}e^{2i\delta_\ell}, \tag{D.51}$$

or,

$$\Re f_\ell + i\left(\Im f_\ell - \frac{1}{2}\right) = -\frac{i}{2}e^{2i\delta_\ell}. \tag{D.52}$$

Multiplying each side of the last equality by its complex conjugate, we obtain the expression,

$$(\Re f_\ell)^2 + \left(\Im f_\ell - \frac{1}{2}\right)^2 = -\frac{1}{4}. \tag{D.53}$$

In a coordinate system with horizontal axis $\Re f_\ell$ and vertical axis $\Im f_\ell$, Eq. (D.53) represents a circle with center at $(0, \frac{1}{2})$, and with radius equal to $\frac{1}{2}$. Thus, the quantity $f_\ell - \frac{i}{2}$ is a complex number lying on this circle, called the *unitary* or the *unitarity circle.* A schematic representation of the unitarity circle is shown in Fig. D.1. The figure shows the *Argand* plot or diagram exhibiting the connection between the "vector" f and the phase shift δ. As δ varies over the interval $0 \leq \delta \leq \pi$ (modulo) π, the tip of f varies over the circumference of the unitarity circle. The entire circle is described if δ varies over the entire region $[0, \pi]$. We note that the imaginary part of the scattering amplitude is never negative, a result that is consistent with the *optical theorem* of scattering theory [see Eq. (2.95)].

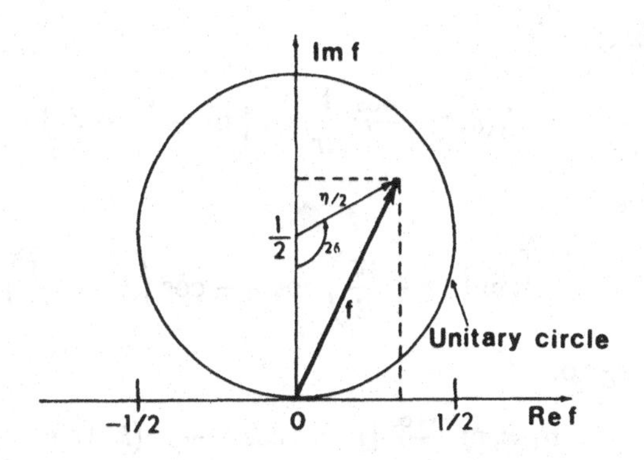

Figure D.1: **The Argand plot for the partial amplitude f associated with elastic scattering.**

In the presence of inelastic scattering, or absorptive processes, accompanied by loss of energy, we have,

$$(\Re f_\ell)^2 + \left(\Im f_\ell - \frac{1}{2}\right)^2 = -\frac{1}{4}\eta_\ell^2, \tag{D.54}$$

where $\eta_\ell < 1$. Thus, the partial amplitude lies entirely inside the unitarity circle. The exact form of the diagram depends on the interaction between particle and target and is particularly telling when that interaction is *resonant.* In that case, resonances manifest themselves in the form of closed or nearly closed loops inside the unitary circle.

D.4 Normalization of the Wave Function

Up to this point, we have not explicitly considered the role played by the multiplicative constants, $A_\ell(k)$. Even though these constants serve mainly to fix the normalization of the wave function[2], their specific choice may have important computational ramifications. This is an especially important feature in the case of non-spherical potentials that is considered below.

[2]More appropriately, of the basis function.

Some commonly used normalizations and the corresponding forms of the scattering wave function are listed below [2].

$$\text{(i)} \quad u_l(k,r) \overset{r\to\infty}{\longrightarrow} \frac{1}{k}\sin\left(kr - \frac{l\pi}{2} + \delta_l\right) \tag{D.55}$$

corresponds to

$$R_l(k,r) \overset{r\to\infty}{\longrightarrow} \frac{1}{kr}\sin\left(kr - \frac{l\pi}{2} + \delta_l\right). \tag{D.56}$$

We also have,

$$u_l(k,r) \overset{r\to\infty}{\longrightarrow} \frac{1}{k}\left[\sin\left(kr - \frac{l\pi}{2}\right)\cos\delta_l - \cos\left(kr - \frac{l\pi}{2}\right)\sin\delta_l\right], \tag{D.57}$$

corresponding to,

$$R_l(k,r) \overset{r\to\infty}{\longrightarrow} \left[j_l(kr)\cos\delta_l - n_l(kr)\sin\delta_l\right]. \tag{D.58}$$

The choice,

$$\text{(ii)} \quad u_l(k,r) \overset{r\to\infty}{\longrightarrow} \frac{1}{k}\left[\sin\left(kr - \frac{l\pi}{2}\right) - \cos\left(kr - \frac{l\pi}{2}\right)\tan\delta_l\right] \tag{D.59}$$

corresponds to the expression,

$$R_l(k,r) \overset{r\to\infty}{\longrightarrow} \left[j_l(kr) - n_l(kr)\tan\delta_l\right]. \tag{D.60}$$

Using the relations in Eqs. (D.20) and Eq. (D.50), we can write,

$$R_l(k,r) \overset{r\to\infty}{\longrightarrow} \left[j_l(kr) - t_l(k)h_l(kr)\right], \tag{D.61}$$

where the *transition matrix*, or t-matrix, t_l, is defined by the expression,

$$t_l(k) = -a_l(k) = -\frac{1}{k}e^{i\delta_l(k)}\sin\delta_l(k). \tag{D.62}$$

Finally, the choice,

$$\text{(iii)} \quad u_l(k,r) \overset{r\to\infty}{\longrightarrow} \frac{1}{k}\left[\cot\delta_l\sin\left(kr - \frac{l\pi}{2}\right) - \cos\left(kr - \frac{l\pi}{2}\right)\right], \tag{D.63}$$

corresponds to the expression,

$$R_l(k,r) \overset{r\to\infty}{\longrightarrow} \left[\cot\delta_l j_l(kr) - n_l(kr)\right], \tag{D.64}$$

for the radial wave function. Choosing the Hankel function rather that the Neumann function as the irregular solution of the Schrödinger equation, we can write,

$$R_l(k,r) \overset{r\to\infty}{\longrightarrow} \left[m_l j_l(kr) - h_l(kr)\right], \tag{D.65}$$

where $m_l = t_l^{-1}$ is the inverse of the t-matrix defined in Eq. (D.62). In the language of multiple scattering theory, the transition matrix or t-matrix, t_l, is commonly referred to as the scattering matrix, and when there is no danger of confusion, we shall follow this practice. In the form given by Eq. (D.62), the t-matrix prominently enters the expressions of multiple scattering theory.

Having fixed the asymptotic behavior of the radial wave function, we can also derive expressions for the normalization factor, $A_\ell(k)$. For example, the explicit expression for the radial wave function of Eq. (D.58) is

$$R_\ell(k,r) = \mathrm{e}^{\mathrm{i}\delta_\ell}\left[j_l(kr)\cos\delta_l - n_l(kr)\sin\delta_l\right]. \tag{D.66}$$

Other expressions can be found by simple algebraic means.

D.5 Integral Equations for the Phase Shifts

It can be shown [2] that various trigonometric functions of the phase shifts can be obtained by means of integral expressions involving the radial wave functions, R_l. The specific trigonometric functions given by these relations depend on the choice of normalization of the radial wave functions. For example, with the wave function normalized as in Eq. (D.60), we have,

$$\tan\delta_l = -k\int_0^\infty j_l(kr)V(r)R_l(k,r)r^2\mathrm{d}r. \tag{D.67}$$

On the other hand, with the normalization indicated in Eq. (D.61), one obtains an integral expression for the t-matrix,

$$t_l(k) = -\frac{\mathrm{e}^{\mathrm{i}\delta_l(k)}\sin\delta_l(k)}{k} = \int_0^\infty j_l(kr)V(r)R_l(k,r)r^2\mathrm{d}r. \tag{D.68}$$

D.5.1 Properties of the t-matrix

The cell t-matrix defined in Eq. (D.68) guarantees that the wave function and its radial derivative are continuous across the surface of the MT sphere. This continuity amounts to the continuity of the logarithmic derivative. Using Eq. (D.61), we obtain

$$t_\ell = -\frac{\gamma_\ell^j - \gamma_\ell}{\gamma_\ell^h - \gamma_\ell}\frac{j_\ell(kr_n)}{-\mathrm{i}kh_\ell(kr_n)}, \tag{D.69}$$

where r_n is the radius of the MT sphere, $k = \sqrt{E}$, and

$$\gamma_\ell = \frac{\mathrm{d}}{\mathrm{d}r}\left[\ln R_\ell(r,E)\right]_{r_n},\ \gamma_\ell^j = \frac{\mathrm{d}}{\mathrm{d}r}\left[\ln j_\ell(kr)\right]_{r_n},\ \gamma_\ell^h = \frac{\mathrm{d}}{\mathrm{d}r}\left[\ln h_\ell(kr)\right]_{r_n}. \tag{D.70}$$

The asymptotic behavior of the t-matrix at large ℓ follows easily from Eq. (D.69),

$$t_\ell \simeq \frac{(kr_n)^{2\ell+1}}{(2\ell+1)!!(2\ell-1)!!}, \tag{D.71}$$

where for m odd, $m!! = 1 \cdot 3 \cdot 5 \cdots m$. As kr_n is of order unity in typical applications, the t-matrix decreases rapidly with increasing angular momentum index. This feature, which is also present in the case of non-spherical potentials, is very helpful in increasing the rate of convergence of angular momentum expansions in MST.

D.6 Partial Waves for Non-Spherical Potentials

In this section, we discuss the solutions to the Schrödinger equation or the Lippmann-Schwinger equation for the case of a single, spatially-bounded potential cell. We place no restriction on the shape of the cell or on the potential other than that the potential does not support bound states at positive energies. (We also exclude pathological potentials that may contain non-integrable singularities.) In contrast to spherically symmetric potentials, angular momentum is no longer conserved and we expect coupling between different values of L (channels). We are interested in obtaining the regular, i.e., everywhere finite, or, more precisely, square integrable solutions of the Schrödinger equation (under proper boundary conditions at infinity),

$$(\nabla^2 + k^2)\,|\psi(k,\mathbf{r})\rangle = V(\mathbf{r})\,|\psi(k,\mathbf{r})\rangle\,, \tag{D.72}$$

associated with the potential, $V(\mathbf{r})$, at a fixed energy, $E = k^2$. Here, we use a bra and ket notation to denote quantities that are indexed by a single angular-momentum eigenstate. An explanation of this and the corresponding matrix notation is given in Appendix F[3].

When the potential vanishes, a regular solution to Eq. (D.72) at energy E can be obtained as the linear combination [5]

$$\begin{aligned}\psi(k,\mathbf{r}) &= \sum_L a_L J_L(k\mathbf{r}) \\ &= \sum_L a_L\,[j_l(kr)Y_L(\hat{r})] \\ &= \langle a|J(\mathbf{r})\rangle.\end{aligned} \tag{D.73}$$

[3]In order to simplify our notation and because we will almost invariably use retarded Green functions and associated wave functions, we will not differentiate between retarded and advanced quantities.

Here, the functions $J_L(k\mathbf{r})$ are solutions of the Helmholtz equation,

$$(\nabla^2 + E)\psi = 0, \tag{D.74}$$

and satisfy the orthonormality condition,

$$\int J_L(k\mathbf{r})J_L(k'\mathbf{r})d^3r = \frac{\pi}{2k^2}\delta(k-k')\delta_{LL'}. \tag{D.75}$$

In these expressions and subsequently, we use capital letters to denote the products of the spherical Bessel, Neumann, and Hankel functions with the spherical harmonics, i.e., $J_L = j_l Y_L, N_L = n_l Y_L, \text{and} H_L^{(\pm)} = h_l^{(\pm)} Y_L$ (see Appendix C). The coefficients a_L in Eq. (D.75) are to be chosen so that the function $\psi(k,\mathbf{r})$ satisfies a given set of boundary conditions, such as prescribed values on the surface of a sphere circumscribing the cell.

In the presence of a potential, the regular solutions to Eq. (D.72) can be written in the form,

$$\psi_\alpha(k\mathbf{r}) = \sum_L \psi_{\alpha L}(k,r)Y_L(\hat{r}), \tag{D.76}$$

where the functions $\psi_{\alpha L}(k,r)$ are to be determined and the index α, to be specified below, keeps track of the regular solutions. Upon substituting Eq. (D.76) into Eq. (D.72) and projecting on a sphere of radius r, we obtain a system of coupled differential equations for the $\psi_{\alpha L}$,

$$\left[\frac{1}{r^2}\frac{\mathrm{d}}{\mathrm{d}r}r^2\frac{\mathrm{d}}{\mathrm{d}r} + k^2 - \frac{\ell(\ell+1)}{r^2}\right]\psi_{\alpha L}(k,r) = \sum_{L'} V_{LL'}(r)\psi_{\alpha L'}(k,r), \tag{D.77}$$

where,

$$V_{LL'}(r) = \int Y_L^*(\hat{r})V(\mathbf{r})Y_{L'}(\hat{r})d\Omega_r. \tag{D.78}$$

Now, in the spirit of the well-known method of variation of parameters we express the solutions to Eq. (D.77) as a linear combination of the regular and irregular solutions of the free-particle equation. Thus, we obtain,

$$\psi_{\alpha L}(k,r) = c_{\alpha L}(k,r)j_\ell(kr) - s_{\alpha L}(k,r)n_\ell(kr), \tag{D.79}$$

where the parameters $c_{\alpha L}$ and $s_{\alpha L}$ are subject to the initial conditions,

$$c_{\alpha L}(k,0) \neq 0 \quad \text{and} \quad s_{\alpha L}(k,0) = 0, \tag{D.80}$$

in order to guaranty the regularity of the solutions, $\psi_{\alpha L}$, at the origin. We also have the condition, required by the variation of parameters method,

$$\dot{c}_{\alpha L}(k,r)j_\ell(kr) - \dot{s}_{\alpha L}(k,r)n_\ell(kr) = 0, \tag{D.81}$$

where a dot denotes differentiation with respect to r. This last condition preserves the logarithmic continuation of the $\psi_{\alpha L}(k,\mathbf{r})$ (and of ψ_α) across every spherical surface. In particular, for spatially-bounded potentials, the logarithmic derivative of $\psi_{\alpha L}$ is continuous across the surface of a sphere circumscribing the potential region. We now choose the index α from the set of angular momentum states, L, and with the use of Eq. (D.79) write the solution to the Schrödinger equation in the form,

$$\psi_L(k,\mathbf{r}) = \sum_{L'} \left[c_{LL'}(r)J_{L'}(k\mathbf{r}) - s_{LL'}(r)N_{L'}(k\mathbf{r})\right], \tag{D.82}$$

where the energy (k) dependence of the *phase functions* $c_{LL'}$ and $s_{LL'}$ is no longer shown explicitly. We note that the phase functions $\underline{c}$ and $\underline{s}$ are the generalizations to non-spherical potentials of the functions $\cos\delta_l(r)$ and $\sin\delta_l(r)$ that occur in the discussion of spherically symmetric potentials, corresponding to the asymptotic form of R_l given in Eq. (D.58). Here and subsequently, an underbar denotes a matrix in angular momentum space. The phase functions $\underline{c}$ and $\underline{s}$ fully describe the scattering resulting from the potential inside a sphere of radius r. Since any linear combination of the form,

$$\psi(k,\mathbf{r}) = \sum_L a_L \psi_L(k,\mathbf{r}), \tag{D.83}$$

with arbitrary coefficients, a_L, also satisfies Eq. (D.72), the functions ψ_L play the same role with respect to the inhomogeneous Eq. (D.73) that the J_L play with respect to the homogeneous equation ($V = 0$). In fact the two solutions can be made to coalesce in the limits $V \to 0$ or $r \to 0$. By making an appropriate choice of the coefficients, α_L, the solutions $\psi(k,\mathbf{r})$ can be made to satisfy various sets of conditions, such as a desired asymptotic behavior at infinity. Thus, the $\psi_L(k,\mathbf{r})$ are examples of basis functions. The proof that Eq. (D.83) provides a proper representation of the wave function in terms of these basis functions is given in reference [5].

The phase functions $c_{LL'}(r)$ and $s_{LL'}(r)$ satisfy sets of integral and differential equations that can be obtained from the Lippmann-Scwinger Eq. (2.81), which we write in the form,

$$\psi_L(k,\mathbf{r}) = \chi_L(k,\mathbf{r}) + \int G_0(\mathbf{r}-\mathbf{r}')V(\mathbf{r}')\psi_L(k,\mathbf{r}')d^3r'. \tag{D.84}$$

Here, $\psi_L(k,\mathbf{r})$ is given by Eq. (D.82) and $\chi_L(k,\mathbf{r})$ is taken to be a linear superposition of regular solutions of the free-particle Schrödinger equation,

$$\chi_L(k,\mathbf{r}) = \sum_{L'} \bar{C}_{LL'} J_{L'}(k\mathbf{r}). \tag{D.85}$$

In general, the coefficients $\bar{C}_{LL'}$ depend on the potential, V, and different choices for these coefficients lead to various forms of the wave function and to varying behavior with respect to convergence of expansions in angular momentum space. We now expand the free-particle propagator in terms of the regular and irregular solutions of the Helmholtz equation, Eq. (D.74), and compare the coefficients of these solutions to obtain the set of integral equations,

$$c_{LL'}(r) = \bar{C}_{LL'} - k\int_{r'>r} N_L(k\mathbf{r}')V(\mathbf{r}')\psi_{L'}(k,\mathbf{r}')d^3r' \tag{D.86}$$

and

$$s_{LL'}(r) = k\int_{r'<r} J_L(k\mathbf{r}')V(\mathbf{r}')\psi_{L'}(k,\mathbf{r}')d^3r'. \tag{D.87}$$

It is to be noted that the $\psi_L(k,\mathbf{r})$ contain an infinite sum over the phase functions, $\underline{c}$ and $\underline{s}$, and thus the determination of these functions for any given values of L and L' requires knowledge of them over the *entire* (infinite) range of angular momentum states. Upon using the explicit expression for $\psi_L(k,\mathbf{r})$, Eq. (D.83), integrating over angles and differentiating Eqs. (D.86) and (D.87) with respect to r, we can also write a set of differential equations for the phase functions,

$$\dot{c}_{LL'}(r) = -kr^2 n_\ell(kr)\sum_{L''} V_{LL''}(r)\psi_{L''L'}(k,r), \tag{D.88}$$

and

$$\dot{s}_{LL'}(r) = kr^2 j_\ell(kr)\sum_{L''} V_{LL''}(r)\psi_{L''L'}(k,r), \tag{D.89}$$

with $\psi_{LL'}$ being given by Eq. (D.79) when the index α in that equation refers to an angular momentum state.

At this point, two comments are warranted about these equations. First, their solution depends on the choice of boundary conditions, or equivalently, on the choice of initial conditions at the origin, $r = 0$, which are reflected in the coefficients $\underline{\bar{C}}$ in Eq. (D.85). Second, it is easy to convert Eqs. (D.88) and (D.89) to integral equations. Depending on the choice of initial conditions made in this conversion, some of the resulting integrals may appear to violate the conditions of the expansion of the free-particle propagator, leading to divergent expansions. This feature, which has caused a certain amount of confusion in past attempts to generalize MST to space-filling cells will be resolved in detail in the following discussion.

We now proceed with an examination of some general characteristics of the solutions to the Schrödinger equation for the case of non-spherical potentials, and a comparison with the corresponding quantities for the case of central potentials. We begin by noting that outside a sphere bounding the cell, the phase functions, $\underline{c}$ and $\underline{s}$, assume their asymptotic constant values, $\underline{C}$ and $\underline{S}$, respectively, with $\underline{S}$ being given by the expression,

$$S_{LL'}(k) = k \int J_L(k\mathbf{r}')V(\mathbf{r}')\psi_{L'}(k,\mathbf{r}')d^3r'. \tag{D.90}$$

As discussed in more detail below and as is the case with spherical potentials, $\underline{S}$ can represent various quantities depending on the choice of $\bar{\underline{C}}$ and whether the Neumann, n_l, or Hankel, h_l, functions are employed as the irregular solutions to the free-particle Schrödinger equation. Under an appropriate choice of the initial conditions and with the use of Neumann functions we can also write,

$$C_{LL'}(k) = \delta_{LL'} - k \int N_L(k\mathbf{r}')V(\mathbf{r}')\psi_{L'}(k,\mathbf{r}')d^3r'. \tag{D.91}$$

Through the use of Green's theorem and the fact that the functions J_L and N_L are solutions to the free-particle Schrödinger equation, Eqs. (D.90) and (D.91) can be expressed as the surface integrals,

$$\begin{aligned} S_{LL'}(k) &= k \int_S \left[J_L(k\mathbf{r})\nabla\psi_{L'}(k,\mathbf{r}) - \psi_{L'}(k,\mathbf{r})\nabla J_L(k\mathbf{r})\right] \mathrm{d}^2r \\ &= k \int \mathrm{d}^2r \left[J_L(k,\mathbf{r}), \psi_{L'}(k,\mathbf{r})\right], \end{aligned} \tag{D.92}$$

and

$$\begin{aligned} C_{LL'}(k) &= -k \int_S \left[N_L(k\mathbf{r})\nabla\psi_{L'}(k,\mathbf{r}) - \psi_{L'}(k,\mathbf{r})\nabla N_L(k\mathbf{r})\right] \mathrm{d}^2r, \\ &= -k \int \mathrm{d}^2r \left[N_L(k,\mathbf{r}), \psi_{L'}(k,\mathbf{r})\right], \end{aligned} \tag{D.93}$$

where the brackets denote the Wronskian of the functions inside them. Here, the surface S can be taken as the surface of the cell itself or any other surface that encloses the cell such as that of a circumscribing sphere. Everywhere outside that sphere, the ψ_L can be written in the form,

$$\psi_L(k,\mathbf{r}) = \sum_{L'} \left[C_{LL'}(k)J_{L'}(k\mathbf{r}) - S_{LL'}(k)N_{L'}(k\mathbf{r})\right]. \tag{D.94}$$

or,

$$|\psi\rangle = \underline{C}(k)\,|J\rangle - \underline{S}(k)\,|N\rangle\,, \tag{D.95}$$

where the last equation is written in terms of the vector and matrix notation of Appendix F. Expressions corresponding to those in Eqs. (D.91) through (D.95) can be written in terms of the Hankel functions, H_L, rather than the Neumann functions, N_L, corresponding to the alternate choice of the irregular solution to the homogeneous Schrödinger equation.

The different choices of initial conditions, which as we see below correspond to different boundary conditions at infinity, coupled with the use either of Neumann or Hankel functions, lead to different forms of the solution to the Schrödinger equation that may be characterized by different numerical behavior and convergence properties. It may be helpful in clarifying the effects of these various alternatives to compare them to the corresponding cases of spherically symmetric potentials.

D.6.1 Correspondence with spherical potentials

Through a proper choice of initial (boundary) conditions, the explicit form given above, Eq. (D.95), can be made to correspond to the solution for a spherically symmetric potential that has the asymptotic behavior, Eq. (D.58),

$$\psi_\ell(k,r) = \cos\delta_\ell j_\ell(kr) - \sin\delta_\ell n_\ell(kr). \tag{D.96}$$

Treating $\underline{C}$ as an invertible matrix, we can also write,

$$\underline{C}^{-1}\left|\psi\right\rangle = \left|J\right\rangle - \underline{C}^{-1}\underline{S}\left|N\right\rangle , \tag{D.97}$$

which corresponds to the form[4],

$$\psi_\ell(k,r) = j_\ell(kr) - \tan\delta_\ell n_\ell(kr), \tag{D.98}$$

for potentials with spherical symmetry, Eq. (D.60). In this case, the integral in Eq. (D.90) directly yields the tangent of the phase shift, or more precisely, the generalization of that quantity to the case of non-spherical potentials. Note that the chosen normalization is such that outside a bounding sphere the wave function behaves as J_L augmented by the scattered part involving the tangent of the phase-shift. Similarly, if the phase- function matrix $\underline{S}$ can be treated as being invertible, we can write,

$$\underline{S}^{-1}\left|\psi\right\rangle = \underline{S}^{-1}\underline{C}\left|J\right\rangle - \left|N\right\rangle , \tag{D.99}$$

corresponding to the form,

$$\psi_\ell(k,r) = \cot\delta_\ell j_\ell(kr) - n_\ell(kr), \tag{D.100}$$

for spherical potentials, Eq. (D.64).

The discussion just given was based on the choice of the Neumann functions as the irregular solutions to the homogeneous Schrödinger equation.

[4]We hope that no confusion will arise from the use of the same generic symbol to denote different forms of the wave function.

Analogous results are obtained[5] when the Hankel functions are used instead. Upon using the relation, Eq. (D.20),

$$H_\ell^{\pm} = J_\ell \pm \mathrm{i}N_\ell, \tag{D.101}$$

we can write Eq. (D.95) in the form, (with $h \equiv h^+$),

$$|\psi\rangle = [\underline{C} - \mathrm{i}\underline{S}]\,|J\rangle + \mathrm{i}\underline{S}\,|H\rangle\,. \tag{D.102}$$

Equation (D.102) can also be written as (with ik absorbed into the Hankel function),

$$\begin{aligned}[\underline{C} - \mathrm{i}\underline{S}]^{-1}\,|\psi\rangle &= |J\rangle + [\underline{C} - \mathrm{i}\underline{S}]^{-1}\,\underline{S}\,|H\rangle \\ &= |J\rangle + \underline{t}\,|H\rangle\,,\end{aligned} \tag{D.103}$$

which can be recognized as a generalization of the form ,

$$\psi_\ell(k,r) = j_\ell(kr) - \mathrm{i}kt_\ell h_\ell(kr), \tag{D.104}$$

appropriate to spherically symmetric potentials, Eq. (D.61). In this instant, the integral in Eq. (D.90) *directly* yields the t-matrix for the cell potential, V. Finally, treating $\underline{S}$ as an invertible matrix, we obtain,

$$|\psi\rangle = \underline{m}|J\rangle + |H\rangle, \tag{D.105}$$

which for potentials with spherical symmetry corresponds to Eq. (D.65).

We recall that in the case of spherically symmetric potentials the t-matrix has the explicit representation, Eq. (D.62),

$$t_\ell(k) = -\frac{1}{k}\mathrm{e}^{-\mathrm{i}\delta_\ell}\sin\delta_\ell. \tag{D.106}$$

Consequently, we identify the quantity,

$$\underline{t}(k) = -\left[\underline{S}^{-1}\underline{C} + \mathrm{i}\right]^{-1}, \tag{D.107}$$

with the t-matrix for a non-spherical, generally shaped potential cell.

D.7 Single-Scatterer Green Function

We saw in Chapter 2 that the Green function associated with the presence of a potential can be obtained, at least formally, by means of a series expansion, Eqs. (2.75) and (2.76). Here, we derive explicit expressions for the single-scatterer Green function in the angular momentum representation.

[5]Note that as $r \to \infty$, the scattered wave assumes its asymptotic form proportional to $1/r$.

The Green function corresponding to the operator $-\nabla^2 + v_n(\mathbf{r})$, where $v_n(\mathbf{r})$ is spatially bounded and vanishes outside a sphere of radius r_n, is the solution to the inhomogeneous equation, Eq. (2.12),

$$\left[\nabla^2 - v_n(\mathbf{r}) + E\right] G_n(\mathbf{r},\mathbf{r}') = \delta(\mathbf{r}-\mathbf{r}'). \tag{D.108}$$

It follows from the Dyson equation, Eq. (2.61), that $G_n(\mathbf{r},\mathbf{r}')$ satisfies the integral equation,

$$G_n(\mathbf{r},\mathbf{r}') = G_0(\mathbf{r},\mathbf{r}') + \int \mathrm{d}^3r'' G_0(\mathbf{r},\mathbf{r}'') v_n(\mathbf{r}'') G_n(\mathbf{r}'',\mathbf{r}'). \tag{D.109}$$

Furthermore, using the relation given in Eq. (2.89), we can write,

$$G_n(\mathbf{r},\mathbf{r}') = G_0(\mathbf{r},\mathbf{r}') + \int \mathrm{d}^3r_1 \int \mathrm{d}^3r_2 G_0(\mathbf{r},\mathbf{r}_1) t^n(\mathbf{r}_1,\mathbf{r}_2) G_0(\mathbf{r}_2,\mathbf{r}'), \tag{D.110}$$

where the t-matrix $t^n(\mathbf{r}_1,\mathbf{r}_2)$ is given by the expression in Eq. (2.91), with v_n replacing V. If one of the arguments of $G_n(\mathbf{r},\mathbf{r}')$, say r, is larger than r_n, the radius of the sphere bounding the potential $v_n(\mathbf{r})$, we can use the angular momentum expansion of $G_0(\mathbf{r},\mathbf{r}')$, Eq. (2.68), with $r > r'$ and $r > r_n$, to write,

$$\begin{aligned}
G_n(\mathbf{r},\mathbf{r}') &= \sum_L J_L(\mathbf{r}') H_L(\mathbf{r}) \\
&+ \sum_L \int \mathrm{d}^3r_1 \int \mathrm{d}^3r_2 G_0(\mathbf{r}',\mathbf{r}_1) t^n(\mathbf{r}_1,\mathbf{r}_2) J_L(\mathbf{r}_2) H_L(\mathbf{r}) \\
&= \sum_L \left[J_L(\mathbf{r}') + \int \mathrm{d}^3r_1 \int \mathrm{d}^3r_2 G_0(\mathbf{r}',\mathbf{r}_1) t^n(\mathbf{r}_1,\mathbf{r}_2) J_L(\mathbf{r}_2) \right] H_L(\mathbf{r}) \\
&= \sum_L \phi_L^n(\mathbf{r}') H_L(\mathbf{r}), \quad (r > r', \ r > r_n).
\end{aligned} \tag{D.111}$$

The last line follows since the terms inside the square brackets give the coordinate representation of the wave function, $\phi_L^n(\mathbf{r})$, in Eq. (2.86). Now, for points $r < r_n$ but still $r > r'$, the irregular solution to the free-particle equation must join smoothly and continuously to the irregular solution to the Schrödinger equation in the presence of the potential. Denoting that solution by $F_L^n(\mathbf{r})$, we can express the Green function for the single scatterer in the form,

$$G_n(\mathbf{r},\mathbf{r}') = \sum_L \phi_L^n(\mathbf{r}') F_L^n(\mathbf{r}), \quad (r > r'), \tag{D.112}$$

an expression that is valid both inside and outside the potential region. We note that away from the origin, the irregular solution satisfies the Schrödinger equation,

$$\left[\nabla^2 + E - v_n(\mathbf{r})\right] F_L^n(\mathbf{r}) = 0. \tag{D.113}$$

D.8 Density of States for a Spherical Scatterer

The density of states per unit volume and at a given energy in the presence of a single, spherically symmetric scatterer can be obtained as the imaginary part of the Green function in Eq. (D.112),

$$n_n(\mathbf{r};E) = -\frac{1}{\pi}\Im G_n(\mathbf{r},\mathbf{r};E) = -\frac{1}{\pi}\Im\sum_L \phi_L^n(\mathbf{r})F_L^n(\mathbf{r}). \tag{D.114}$$

For points $\mathbf{r}$ outside the range of the potential, $r > r_n$, we can write,

$$\phi_L(\mathbf{r})F_L(\mathbf{r}') = r_\ell(r,E)f_\ell(r',E)Y_L(\hat{r})Y_L(\hat{r}'), \tag{D.115}$$

where (for points outside the bounding sphere),

$$R_\ell(r,E) \to j_\ell(kr)\cos\delta_\ell - n_\ell(kr)\sin\delta_\ell, \tag{D.116}$$

and

$$\begin{aligned} f_\ell(r',E) &\to -\mathrm{i}kh_\ell(kr)\mathrm{e}^{\mathrm{i}\delta_\ell} \\ &\to k\left[(j_\ell(kr)\sin\delta_\ell + n_\ell(kr)\cos\delta_\ell)\right. \\ &- \left.\mathrm{i}\left(j_\ell(kr)\cos\delta_\ell - n_\ell(kr)\sin\delta_\ell\right)\right], \end{aligned} \tag{D.117}$$

so that

$$n_n(\mathbf{r},E) = \frac{k}{4\pi}\sum_\ell (2\ell+1)R_\ell^2(r,E). \tag{D.118}$$

These results can be used to calculate the change in the density of states due to the introduction of a single, physically symmetric potential into otherwise free space. Combining the last equation and the corresponding expression for free particles, Eq. (2.69), and integrating over all space, we obtain,

$$\begin{aligned} \Delta n(E) &= \int \mathrm{d}^3r\left[n_n(\mathbf{r},E) - n(\mathbf{r},E)\right] \\ &= \frac{1}{\pi}\int \mathrm{d}^3r\Im\left[G_n(\mathbf{r},\mathbf{r},E) - G_0(\mathbf{r},\mathbf{r},E)\right] \\ &= \frac{k}{4\pi^2}4\pi\int r^2\mathrm{d}r\sum_\ell(2\ell+1)\left[R_\ell^2(r,E) - j_\ell^2(r,E)\right]. \end{aligned} \tag{D.119}$$

The volume integrals can be evaluated after being converted into surface integrals. To accomplish this, we consider the Schrödinger equation at energy E,

$$\left[\nabla^2 + E - V(\mathbf{r})\right]\psi(\mathbf{r}) = 0, \tag{D.120}$$

and at $E + \delta E$,

$$\nabla^2 \psi^*(\mathbf{r}, E + \delta E) + [E + \delta E - V(\mathbf{r})] \psi^*(\mathbf{r}, E + \delta E) = 0. \tag{D.121}$$

We now multiply the first equation by $\psi^*(\mathbf{r}, E+\delta E)$ and subtract from the product the second equation multiplied by $\psi(\mathbf{r}, E)$, to obtain

$$\psi(\mathbf{r}, E)\psi^*(\mathbf{r}, E+\delta E)\delta E = \psi(\mathbf{r}, E)\nabla^2\psi^*(\mathbf{r}, E+\delta E) - \psi^*(\mathbf{r}, E+\delta E)\nabla^2\psi(\mathbf{r}, E). \tag{D.122}$$

Taking the limit $\delta \to 0$, integrating the last expression over the range of the potential, and using Green's theorem, Eq. (J.33), we obtain[6],

$$\int \mathrm{d}^3r|\psi(\mathbf{r})|^2 = \int \mathrm{d}\mathbf{S} \cdot \left[\frac{\partial\psi(\mathbf{r})}{\partial E}\nabla - \nabla\frac{\partial\psi(\mathbf{r})}{\partial E}\right]\psi^*(\mathbf{r}). \tag{D.123}$$

Now, in terms of the radial wave functions R_ℓ and j_ℓ, we readily obtain,

$$\Delta n(E) = \frac{1}{\pi}\sum_\ell (2\ell+1)\frac{\mathrm{d}\delta_\ell(E)}{\mathrm{d}E}. \tag{D.124}$$

Integration of this formula leads to the remarkable result that the change in the *total* number of states, $\delta n(E)$, below energy E, i.e., the change in the *integrated density of states*, induced by the potential is given by,

$$\delta n(E) = \frac{1}{\pi}\sum_\ell (2\ell+1)\delta_\ell(E). \tag{D.125}$$

This relation is known as the *Friedel sum rule* [6].

D.8.1 Non-spherical scatterer—the Lloyd formula

In the case of a non-spherical scatterer, the change in the integrated density of states is given by *Lloyd's formula* [7],

$$\delta n(E) = \frac{1}{\pi}\Im \mathrm{Tr} \ln t^n. \tag{D.126}$$

We will not prove this result at this point because it is a special case of the more general expression for Lloyd's formula in connection with a collection of scatterers that is derived in the body of the book.

[6] One should be careful when using this formula, especially for bound states. Because the energy derivative of the Schrödinger equation can be taken only when no boundary conditions are specified, the boundary of the integration must be placed so that the boundary conditions are matched to avoid including discontinuities of the wave function into the integrant.

D.9 Wave Function Inside the Bounding Sphere

In this section, we establish that the t-matrix associated with a non-spherical potential cell suffices to describe the wave function everywhere outside the cell, even inside a sphere bounding the cell. The following discussion introduces an important formal element—a divergent sum may be summed through its replacement by a conditionally convergent double sum. We use this approach in developing multiple scattering theory for space-filling cells.

That the t-matrix suffices to represent the wave function[7] everywhere outside a non-spherical potential cell is no longer obvious when scattering theory is cast in the angular momentum representation. For example, the multipole expansion in terms of the t-matrix, Eq. (D.103), is only valid *outside* a sphere bounding the cell. In fact, if that expression is used at points inside the sphere serious divergences arise in the product $\sum_{L'} t_{LL'} H_{L'}(\mathbf{r})$. The divergence becomes more pronounced as the point $\mathbf{r}$ moves closer to the surface of the cell. This behavior may lead one to suspect that in the angular-momentum representation the wave function in the moon region surrounding a cell, i.e., outside the cell but inside a bounding sphere, is not described completely in terms of the cell t-matrix. This has led to the further conjecture that the scattering from two adjacent, generally-shaped cells whose bounding spheres overlap cannot be described in terms of the cell scattering matrices. Instead, multiple scattering theory should be adjusted to take account of the near-field corrections (NFCs) [8-10] arising from the presence of a non-vanishing potential inside the moon region of a given cell. Therefore, within the angular momentum representation, we are led to asking this question: *What quantity, in addition to the t-matrix, is needed in order to properly describe the wave function in the potential-free region inside a bounding sphere circumscribing a potential cell?*

To begin with, it is not difficult to anticipate that the wave function in the moon region is determined completely in terms of the t-matrix. After all, we may view the cell (to any desired accuracy) as a collection of non-overlapping spheres (of varying radii). Then the points in the moon region lie outside *all* spherical surfaces and the wave disturbance there can be written as the superposition of the asymptotic forms of the waves scattered by the potentials inside the spheres. The resulting (multiple scattering) expression for the t-matrix or for the wave function, in the angular momentum representation, yields a summation that by definition becomes equal to the t-matrix of the cell as the number of spheres increases, with their radii chosen so that the spheres completely cover the volume of the cell (except, perhaps, for sets of measure zero that do not contribute to the scattering.)

We now rigorously show that the cell t-matrix alone suffices to complete-

[7]More precisely, the basis functions in terms of which the wave function can be constructed.

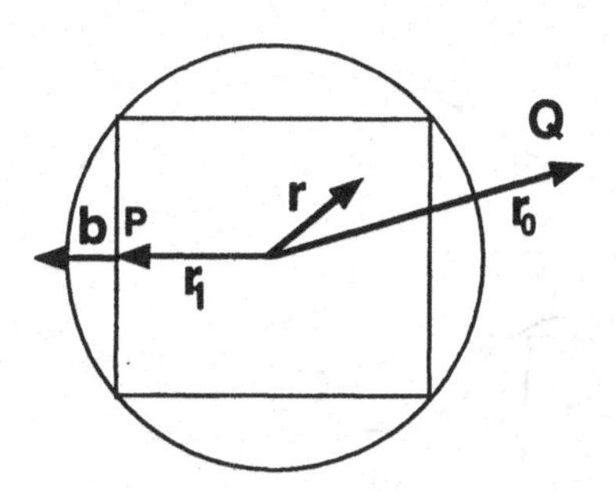

Figure D.2: **A cell with a bounding sphere, a cell vector, r, an observation vector, $\mathbf{r}_0$, and a vector, b, which moves points in the moon region outside the bounding sphere.**

ly determine the wave function everywhere outside a cell, even inside the moon region. This is true regardless of the shape of the cell, i.e., whether it is convex or concave. In fact, there are at least two different ways of showing this based, respectively, on a shifted-center and shifted-cell approach. Both of these formal arguments are discussed in this section. They both project the nature of the problem as one associated strictly with geometry alone, being quite independent of the nature of the potential in the cell.

The first line of argument presented below, that of the shifted-center, is much simpler than that of the shifted-cell, and intuitively clearer. It is, however, of somewhat limited computational, and even formal value. For example, it is only valid in the case of cells with convex, polyhedral shapes. On the other hand, as shown in Appendix Gthe shifted-cell argument applies to arbitrarily-shaped, even concave cells.

D.9.1 Displaced-center approach—convex cells

We begin with the Lippmann-Schwinger equation written in the form (with energy arguments suppressed),

$$\psi_L(\mathbf{r}_0) = J_L(\mathbf{r}_0) + \int G_0(\mathbf{r} - \mathbf{r}_0) V(\mathbf{r}) \psi_L(\mathbf{r}) \mathrm{d}^3 r. \tag{D.127}$$

For points, Q, that lie outside a sphere bounding a cell, Fig. D.2, the vector $\mathbf{r}_0$ is larger than any intracell vector, $\mathbf{r}$, and the free-particle propagator can be expanded in the form, Eq. (F.6),

$$G_0(\mathbf{r} - \mathbf{r}_0) = -\mathrm{i}k \sum_L H_L(\mathbf{r}_0) J_L(\mathbf{r}). \tag{D.128}$$

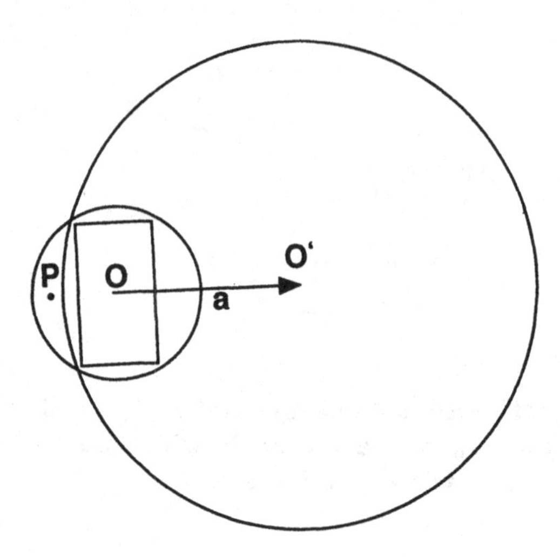

Figure D.3: **A point, P, inside the moon region of the sphere centered at point O, lies outside a sphere centered at point O'.**

When this expression is substituted into Eq. (D.127) and the integral over $\mathbf{r}$ is carried out, we obtain the expression,

$$\psi_L(\mathbf{r}_0) = J_L(\mathbf{r}_0) - ik \sum_{L'} H_{L'}(\mathbf{r}_0) t_{L'L}, \tag{D.129}$$

which is Eq. (D.103) explicitly displayed. Evidently, for $\mathbf{r}_0$ inside the bounding sphere such as point P in Fig. D.2, Eq. (D.128) may represent a divergent expression; now $\mathbf{r}_0$ may be smaller than some cell vector $\mathbf{r}$ and Eq. (D.129) is no longer justified.

In spite of these difficulties, a convergent expression for the wave function at P in terms of the cell t-matrix can be obtained along the following lines. First, we note that the point P that lies *inside* a sphere centered at O, taken to be the geometric center of the cell shown in Fig. D.3, lies *outside* a sphere centered at a point O', displaced from O by a vector $\mathbf{a}$. Denoting the cell t-matrix obtained in terms of angular momentum expansions about O' by $\underline{t}(\mathbf{a})$, we can write the wave function at P in the form,

$$\psi_L(\mathbf{r}'_0) = J_L(\mathbf{r}'_0) - ik \sum_{L'} H_{L'}(\mathbf{r}'_0) t_{L'L}(\mathbf{a}) \tag{D.130}$$

where $\mathbf{r}'_0 = -\mathbf{a} + \mathbf{r}_0$ is the radius vector to P measured from the shifted center at O'. But, as will be shown in a later section (see Appendix F), we have,

$$\underline{t}(\mathbf{a}) = \underline{g}(\mathbf{a})\underline{t}\underline{g}(-\mathbf{a}), \tag{D.131}$$

where $\underline{t}$ is the cell t-matrix evaluated about the original center at O. The last expression illustrates the existence of a similarity transformation, depending only on the amount of shift, which relates the t-matrices expanded about shifted centers. Now, Eq. (D.130) takes the form,

$$\psi_L(\mathbf{r}_0') = J_L(\mathbf{r}_0') - ik\sum_{L'} H_{L'}(\mathbf{r}_0')\left[\sum_{L_1}\sum_{L_2} g_{L'L_1}(\mathbf{a})t_{L_1L_2}g_{L_2L}(-\mathbf{a})\right], \tag{D.132}$$

which explicitly expresses the wave function at a point inside the moon region of a cell in terms of the cell t-matrix. Here, the quantities $g_{LL'}(\mathbf{a})$ are the matrix elements of the translation operator in the angular momentum representation and are defined explicitly in Appendix F. It is important to realize that the sums inside the brackets in Eq. (D.132) must be performed before those outside in order to lead to converged results. In particular, the sum over L_1 must precede that over L'. If the order of these two sums is reversed, the resulting expression reduces to Eq. (D.129), which in the present case may diverge.

In the construction just described, we can take the point P to lie arbitrarily close to the cell boundary by allowing the vector $\mathbf{a}$ to become very large (essentially approaching infinity as the point moves right against the face of the cell). Therefore, at least formally, the t-matrix in the angular momentum representation yields a complete description of the wave function *everywhere* outside a convex, spatially bounded potential cell. We point out that the basic feature characterizing the shifted-center construction just described is the replacement of a *single*, divergent sum, that over L' in Eq. (D.129), by a *double*, conditionally convergent sum, that over L_1 and L' in Eq. (D.132). Note that this replacement is based *purely* on the underlying geometry, i.e., ratios of vectors and the shape of a potential cell, not involving the potential in the cell itself.

It is clear from the discussion so far, that the shifted-center argument can only be used with potential cells of convex shape. Also, its numerical implementation becomes increasingly cumbersome as the point P approaches the face of a cell, and the length of the vector $\mathbf{a}$ becomes very large. The method described in the next subsection completely removes the first limitation, and also computationally yields a much more convenient expression.

D.9.2 Displaced-cell approach—convex cells

Our starting point is again the Lippmann-Schwinger equation in the form of Eq. (D.127). We seek a multipole, convergent expression of the wave function in the moon region, point P in Fig. D.3, in terms of the cell t-matrix. Such an expression can be obtained as follows. We note that for all points confined to the moon region adjacent to a face of a convex polyhedral cell there exist vectors $\mathbf{b}$ that satisfy the inequalities,

$$|\mathbf{b}| < |\mathbf{r}_0 - \mathbf{r} + \mathbf{b}|, \tag{D.133}$$

and

$$|\mathbf{r}| < |\mathbf{r}_0 + \mathbf{b}|, \tag{D.134}$$

for all cell vectors $\mathbf{r}$. Any vector $\mathbf{b}$ that is perpendicular to the face of the cell, directed away from the face of the cell and larger than the largest vertical distance from the bounding sphere to the face satisfies the inequalities above.

We now add and subtract such a vector $\mathbf{b}$ in the argument of G_0, Eq. (D.127), and use the first inequality above to obtain the expression,

$$\psi_L(\mathbf{r}_0) = J_L(\mathbf{r}_0) - ik\sum_{L'} J_{L'}(\mathbf{b}) \left[\int H_{L'}(\mathbf{r} - \mathbf{r}_0 + \mathbf{b})V(\mathbf{r})\psi_L(\mathbf{r})\mathrm{d}^3r\right]. \tag{D.135}$$

In view of the second inequality, the Hankel function in the last equation can be further expanded to yield the result,

$$\begin{aligned} \psi_L(\mathbf{r}_0) &= J_L(\mathbf{r}_0) + \sum_{L'} J_{L'}(\mathbf{b}) \left[\sum_{L''} G_{L'L''}(\mathbf{r}_0 + \mathbf{b}) \right. \\ &\times \left. \int J_{l''}(\mathbf{r})V(\mathbf{r})\psi_L(\mathbf{r})d^3r\right], \\ &= J_L(\mathbf{r}_0) + \sum_{L'} J_{L'}(\mathbf{b}) \left[\sum_{L''} G_{L'L''}(\mathbf{r}_0 + \mathbf{b})\underline{t}_{L''L}\right]. \end{aligned} \tag{D.136}$$

The quantities $G_{LL'}(\mathbf{R})$ denote the real-space structure constants and are the elements of the free-particle propagator in the angular momentum representation. Explicit expressions for the structure constants are given in Appendix F. The expression involving the sum over $J_{L'}$ defines a set of *modified structure constants*; they can be used in place of the canonical structure constants of KKR theory to guarantee convergence of expansions in angular-momentum eigenstates, provided that the sum over L' properly follows those over both L'' and L. Thus, calculations carried out in terms of these quantities should be closely monitored for convergence as a function of L' cut off.

Evidently, the last equation yields an expression for the wave function at a point inside the moon region explicitly in terms of the cell t-matrix. It should be kept in mind that the brackets in Eq. (D.136) indicate the proper order in which the sums should be performed in order to yield converged results. In particular, the sum over L'', corresponding to the *second* expansion must be carried out before the sum over L' that arose as a result of the *first* expansion.

There are obvious similarities between Eqs. (D.132) and (D.136). In both a divergent sum has been replaced by a conditionally convergent double sum in a construction that highlights the geometric nature of the problem. However, the expression in Eq. (D.136) holds a number of advantages over that in Eq. (D.132). First, it is of much greater conceptual value since it can be generalized to the case of concave cells. Second, it is also of greater computational value because it only involves sums over values of L that are not excessively large and vectors of finite and usually small length in contrast to the large values of L and the infinite vectors that could occur in Eq. (D.132). The results of calculations [5] indicate that converged expressions can be obtained over a broad range of the outside sum, L', when the internal sum, L_1, is taken to values about twice as large as those of the outer sum, L'. Also, the value of b for which optimal convergence rates can be expected was found numerically to be of the order of the radius of the circumscribing sphere. The convergence properties of the double sum in Eq. (D.136) have been discussed at length in connection with the three-dimensional Poisson equation [8].

In addition, Eq. (D.136) can be interpreted along formally distinct lines from Eq. (D.132). The effect of the double expansion there was to shift the position of the cell center. In Eq. (D.136), the points inside the moon region are shifted to positions outside the bounding sphere through the addition of the vector **b**. Then, the wave function is calculated at these shifted positions in terms of the t-matrices and brought back inside the moon regionby means of the outer sum involving $J_L(\mathbf{b})$. Alternatively, we may think of the points in the moon region as being kept in place but with the entire cell being shifted by $-\mathbf{b}$. In contrast to the shifted-center approach of the previous subsection, during the shifting operations in the present argument the center of the cell remains fixed with respect to the cell boundary leading to expressions in terms of the cell t-matrix from the beginning.

Finally, it is interesting to note that Eq. (D.136) provides an alternative but equivalent expression to Eq. (D.130) for the wave function in the asymptotic region. Although Eq. (D.136) is more difficult than Eq. (D.130) to handle computationally, it is of great formal value. As we have shown, it is valid everywhere outside the boundary of a convex cell, as well as in the asymptotic region. Therefore, its use bypasses convergence questions associated with the wave function given by Eq. (D.130).

Bibliography

[1] Leonard I. Schiff, *Quantum Mechanics*, McGraw Hill Book Company, New York (1955).

[2] Charles J. Joachain, *Quantum Collision Theory*, North Holland, New York and Amsterdam (1983).

[3] E. U. Condon and G. H. Shortley, *The Theory of Atomic Spectra*, Cambridge University Press (1976).

[4] F. Calogero, *Variable Phase Approach to Potential Scattering*, Academic, New York and London (1967).

[5] A. Gonis and W. H. Butler, *Multiple Scattering Theory in Solids*, Springer Verlag, Berlin (1999).

[6] J. Friedel, Nuovo Cimento, Suppl. **7**, Series X, 287 (1958).

[7] P. Lloyd, Proc. Phys. Soc. London, **90**, 207 (1967).
b**53**, 287

[8] A. Gonis, Erik C. Sowa, and P. A. Sterne, Phys. Rev. Lett. **66**, 2207 (1991).

Appendix E

Scattering Theory

E.1 General Comments

In this appendix, we provide a summary of the mathematical formalism that leads up to the time-independent Lippmann-Schwinger equation. We will rigorously justify some of the results quoted in Appendix D, and clarify the role of the boundary conditions (initial conditions) of scattering theory. In the process, we explicitly indicate how a time- independent picture is obtained from the time-dependent one. Our discussion follows closely that of Schmid and Ziegelman [1].

Two-particle scattering theory, of interest to us here, is concerned with the motion of two particles under the influence of an interaction potential. The Hamiltonian operator describing this motion has the form,

$$H = \frac{k_1^2}{2m_1} + \frac{k_2^2}{2m_2} + V, \tag{E.1}$$

where $\mathbf{k}_1$ and $\mathbf{k}_2$ are the momenta and m_1 and m_2 the masses of the two particles. In classical as well as in quantum mechanics, it is convenient to treat separately the motion of the center of mass of the system from the relative motion of the two particles. Defining a center-of-mass momentum, $\mathbf{k}$, and a relative momentum, $\mathbf{p}$,

$$\mathbf{k} = \mathbf{k}_1 + \mathbf{k}_2 \tag{E.2}$$

and

$$\mathbf{p} = \frac{m_2\mathbf{k}_1 - m_1\mathbf{k}_2}{m_1 + m_2}, \tag{E.3}$$

along with the reduced mass,

$$m = \frac{m_1 m_2}{m_1 + m_2} \tag{E.4}$$

and relative coordinate, $\mathbf{r} = \mathbf{r}_1 - \mathbf{r}_2$, where $\mathbf{r}_1$ and $\mathbf{r}_2$ are the coordinates of the two particles in the laboratory, we can express the Hamiltonian in the form,

$$H = \frac{p^2}{2m} + V \equiv H_0 + V. \tag{E.5}$$

The Hamiltonian H_0 corresponds to the motion of a particle of mass m in free space, i.e., with the interaction, V, switched off. In the form of Eq. (E.5), the Hamiltonian operator can also be used to describe the motion of a particle incident upon a stationary center of force (e.g., an electron scattered by a much more massive nucleus), and is thus appropriate for a discussion of the electron states in solids within the single-particle picture, and within the Born-Oppenheimer approximation.

E.2 Initial Conditions and the Møller Operators

Within a time-dependent picture, the scattering of two particles, or of a single particle by a stationary field, is described by the time-dependent Schrödinger equation, (with $\hbar = 1$),

$$\mathrm{i}\frac{\partial}{\partial t}\Psi_\alpha^+(t) = H\Psi_\alpha^+(t). \tag{E.6}$$

The superscript (+) denotes a state that with increasing time evolves out of a prepared state under the influence of the potential, V. Corresponding states, designated by a superscript (−), are also possible to formally define and are in fact a necessary ingredient of scattering theory. These states, however, would correspond to a collimated beam *emerging* as a result of a scattering experiment, and thus are not experimentally verifiable. As in Appendix D, the index α denotes all observables necessary to completely define the initial state.

The formal solution to Eq. (E.6) is given by,

$$\Psi_\alpha^+(t) = \mathrm{e}^{-\mathrm{i}Ht}\Psi_\alpha^+, \tag{E.7}$$

where Ψ_α^+ is a time-independent wave-packet.

As is discussed in Appendix D, the initial (or boundary) condition governing the motion of the states $\Psi_\alpha^+(t)$ can be expressed as follows: The scattering state, $\Psi_\alpha^+(t)$ has to develop from a state that is free in the infinite past and whose properties are fully characterized by α. The term free implies that the amplitude of the in-state vanishes in the region of interaction, i.e., where $V \neq 0$.

The initial condition stated above can be expressed in mathematical terms along the following lines: At $t = 0$ a reference wave-packet, Φ_α, is introduced, i.e., an experimentally prepared in-state, whose time evolution is described by the Hamiltonian, H_0, of free space,

$$\Phi_\alpha(t) = \mathrm{e}^{-\mathrm{i}H_0 t}\Phi_\alpha. \tag{E.8}$$

We wish to impose the condition that Ψ_α^+ and Φ_α coincide in the limit of the infinite past. However, a pointwise agreement between these two functions would not be sensible because wave packets disperse in the course of time. Consequently, we demand instead that the norm of the difference of these two states as $t \to -\infty$ vanishes,

$$\lim_{t\to-\infty} ||\mathrm{e}^{-\mathrm{i}Ht}\Psi_\alpha^+ - \mathrm{e}^{-\mathrm{i}H_0 t}\Phi_\alpha|| = 0. \tag{E.9}$$

This is a meaningful condition because the norm implies an integration over all space. The limit of the norm is called a ***strong limit*** and will subsequently be denoted by "s- limit". Equation (E.9) removes one important difficulty that was implicit in the expression of the initial condition. That description becomes ill-defined because a wave packet that is confined to the laboratory at a finite time, say at $t = 0$, becomes rather diffuse in the infinite past (unless its production is included in the formalism). The use of a reference wave packet, Φ_α, which is also confined to the laboratory at the same time as Ψ_α^+, and the s-limit in Eq. (E.9) allows us to circumvent this difficulty. It should be noted that the limit in Eq. (E.9) does not exist for every potential, but its existence has been proved for a large class of potentials that decrease with distance faster than the Coulomb potential. This class is broad enough to include most (if not all) the potential functions likely to be encountered in determining the electronic structure of solids.

Equation (E.9) can also be written in the form,

$$\Psi_\alpha^+ = s - \lim_{t\to-\infty} \mathrm{e}^{\mathrm{i}Ht}\mathrm{e}^{-\mathrm{i}H_0 t}\Phi_\alpha \equiv \Omega^+\Phi_\alpha, \tag{E.10}$$

where the s-limit implies the vanishing of the norm of the difference of the two sides of this equation.

E.2.1 Møller wave operators

Equation (E.10) contains the definition of the Møller wave operator, Ω^+, which has the property that when applied to a free state, Φ_α, produces the scattering state, Ψ_α^+.

$$\Psi_\alpha^+ = \Omega^+\Phi_\alpha. \tag{E.11}$$

In general, we can define the operators,

$$\Omega^\pm = s - \lim_{t\to\mp\infty} \mathrm{e}^{\mathrm{i}Ht}\mathrm{e}^{-\mathrm{i}H_0 t}. \tag{E.12}$$

The operator Ω^- produces the scattering state Ψ_α^-, which goes over to the free state Φ_α in the limit $t \to +\infty$. Note that because the Hamiltonian operators H and H_0 do not commute, the two exponentials in Eq. (E.12) cannot be combined into a single one, $e^{iHt}e^{-iH_0t} \neq e^{i(H-H_0)t}$. It will be convenient in our subsequent discussion to use an alternative interpretation of the Møller operators. This representation follows upon replacement of the time limit by an *Euler limit*,

$$\begin{aligned} \Omega^{\pm} &= s - \lim_{t\to\mp\infty} e^{iHt}e^{-iH_0t} \\ &= s - \lim_{\epsilon\to 0} \pm\epsilon \int_{\mp\infty}^{0} dt e^{\pm\epsilon t} e^{iHt} e^{-iH_0t}, \end{aligned} \tag{E.13}$$

which can be shown to be valid [2] under the assumption that the strong limit on the left exists.

To understand the procedure of Euler limit, we apply it to an ordinary function, $f(t)$, which goes to f_∞ as $t \to -\infty$. We easily see that,

$$\int_{-\infty}^{0} \epsilon e^{\epsilon t} dt = 1, \tag{E.14}$$

and that finite time intervals do not contribute to the integral,

$$\lim_{\epsilon\to 0} \int_{-\infty}^{0} \epsilon e^{\epsilon t} f(t) dt. \tag{E.15}$$

Upon setting,

$$\epsilon t = x, \tag{E.16}$$

we have,

$$\int_{-\infty}^{0} \epsilon e^{\epsilon t} f(t) dt = \int_{-\infty}^{0} e^{\epsilon x} f(\frac{x}{\epsilon}) dx, \tag{E.17}$$

so that as ϵ goes to zero, $f(\frac{x}{\epsilon})$ goes to f_∞ and can be put in front of the integral.

It should be pointed out that the introduction of Euler limit causes time to disappear from two-particle scattering theory. This will become clearer in the next section where we derive a time-independent equation that exactly incorporates the boundary (initial) condition for scattering discussed above. We now return to further discussion of the Møller operators.

From the definition, Eq. (E.12), it follows that for any finite time τ, we have the relation,

$$\begin{aligned} s - \lim_{t\to\mp\infty} e^{iH(t+\tau)}e^{-iH_0(t+\tau)} &= \Omega^{\pm} \\ &= e^{iH\tau}\Omega^{\pm}e^{-iH_0\tau} \end{aligned} \tag{E.18}$$

Through differentiation, we obtain,

$$\begin{aligned} 0 &= \frac{d\Omega^{\pm}}{d\tau} \\ &= e^{iH\tau}(H\Omega^{\pm} - \Omega^{\pm}H_0)e^{-iH_0\tau}, \end{aligned} \tag{E.19}$$

which leads to the *intertwining* relation,

$$H\Omega^{\pm} = \Omega^{\pm}H_0. \tag{E.20}$$

The last equation can be used to prove an important result of scattering theory—states with a well-defined value of the momentum (energy) evolve out of free states at the same energy. The proof requires application of a Møller operator on a wave packet of sharp momentum that consequently must fill all space, and therefore is not normalizable (square integrable). That this procedure is meaningful for a large class of potentials has been shown by Faddeev [3]. One finds that the domain of the Møller operator generally consists of the Hilbert space of free states that are square integrable and hence normalizable. By performing a limiting procedure toward sharp states on the wave packets, the domain can be extended to include states of sharp momentum, e.g., plane waves.

We now apply the operators, $\Omega^{\pm}$, to a momentum eigenstate, $|\mathbf{p}\rangle$, instead of a wave packet, Φ_α, thus obtaining scattering states, $|\mathbf{p}\rangle^{\pm}$. These are the states that include an "incident" plane wave and a scattered spherical wave. That the states $|\mathbf{p}\rangle^{\pm}$ correspond to the same sharp energy, $E = \frac{p^2}{2m}$, follows from the intertwining relation, Eq. (E.20),

$$\begin{aligned} H|\mathbf{p}\rangle^{\pm} &= H\Omega^{\pm}|\mathbf{p}\rangle = \Omega^{\pm}H_0|\mathbf{p}\rangle \\ &= \Omega^{\pm}\frac{p^2}{2m}|\mathbf{p}\rangle = \frac{p^2}{2m}|\mathbf{p}\rangle^{\pm} \end{aligned} \tag{E.21}$$

In general, the Møller operators map the proper, (+), and improper, (-), free Hilbert spaces onto the corresponding scattering states of H. The bound states of H are not reached by this process, and therefore the Møller operators are not unitary, but only isometric (a dagger denotes Hermitian conjugate),

$$\Omega^{\pm\dagger}\Omega^{\pm} = 1. \tag{E.22}$$

The last two expressions reveal an important property of the normalization of the scattered states. In fact, they show that these states have the same normalization as the free-particle states from which they evolve,

$$\langle\psi^{\pm}|\psi^{\pm}\rangle = \langle\mathbf{p}|\Omega^{\pm\dagger}\Omega^{\pm}|\mathbf{p}\rangle = \langle\mathbf{p}|\mathbf{p}\rangle. \tag{E.23}$$

When the Møller operators are applied to the bound states, Ψ_n, of H they give zero, as follows from the relation,

$$\begin{aligned} \langle \Omega^{\pm\dagger}\Psi_n|\mathbf{p}\rangle &= \langle \Psi_n|\Omega^{\pm}|\mathbf{p}\rangle \\ &= \langle \Psi_n|\mathbf{p}\rangle^{\pm} = 0 \end{aligned} \tag{E.24}$$

The last step in the equation above follows because the bound and scattering states are eigenstates of H with different energies. Since the expression on the far left equals zero, and the free states, $|\mathbf{p}\rangle$, form a complete set, we have,

$$\Omega^{\pm}|\Psi_n\rangle = 0. \tag{E.25}$$

It follows that the Møller operators are unitary only when the spectrum of H does not include bound states. It follows easily from the definition of the Møller operators, e.g., Eq. (E.11), and the fact that they are isometric that the solutions of the Lippmann-Schwinger equations have the same normalization as the free states from which they evolve,

$$\begin{aligned} \langle \Psi_\alpha^{\pm}|\Psi_\alpha^{\pm}\rangle &= \langle \Phi_\alpha|(\Omega^{\pm\dagger}\Omega^{\pm}|)\Phi_\alpha^{\pm}\rangle \\ &= \langle \Phi_\alpha|\Phi_\alpha\rangle. \end{aligned} \tag{E.26}$$

This statement expresses the conservation of probability (number of particles) in an event of elastic scattering. It follows from the normalization of the basis set, $\{\Phi_\alpha\}$, that the states $\{\Psi_\alpha^{+}\}$ satisfy the relation,

$$\langle \Psi_\alpha^{+}|\Psi_\beta^{+}\rangle = \delta_{\alpha\beta}, \tag{E.27}$$

and thus form an orthonormal basis set (at least in a δ- function sense). A relation similar to that in the last equation holds also for the states $\{\Psi_\alpha^{-}\}$.

E.3 Lippmann-Schwinger Equation

The Møller operators can be used to derive a time-independent equation for the scattering states that properly incorporates the condition at the infinite past (initial conditions). Suppressing the explicit indication of the s-limit, we use Eq. (E.13) to write,

$$\begin{aligned} |\mathbf{p}\rangle^{\pm} &= \lim_{\epsilon\to 0} \pm\epsilon \int_{\mp\infty}^{0} dt e^{\pm\epsilon t} e^{iHt} e^{-iH_0 t}|\mathbf{p}\rangle \\ &= \lim_{\epsilon\to 0} \pm\epsilon \int_{\mp\infty}^{0} dt e^{\pm\epsilon t} e^{iHt} e^{-iEt}|\mathbf{p}\rangle \\ &= \lim_{\epsilon\to 0} \pm\epsilon \int_{\mp\infty}^{0} dt e^{(H-E\mp\epsilon)t}|\mathbf{p}\rangle \\ &= \lim_{\epsilon\to 0} \pm i\epsilon (E\pm\epsilon - H)^{-1}|\mathbf{p}\rangle, \end{aligned} \tag{E.28}$$

where the exponentials after the first line can be combined since ϵ and E are scalars. Now, defining the Green function operator, or resolvent,

$$G(z) = (z - H)^{-1}, \tag{E.29}$$

where z is a complex number, we obtain the relation,

$$|\mathbf{p}\rangle^{\pm} = \lim_{\epsilon \to 0} \pm\epsilon G(E \pm \epsilon)|\mathbf{p}\rangle. \tag{E.30}$$

This equation between the scattering states, $|\mathbf{p}\rangle^{\pm}$, and the Green function operator marks ***the transition from the time-dependent picture of scattering to the time-independent one.*** We are now ready to derive a time- independent integral equation for the scattering states.

To this end, we note that the Green function satisfies the following two identities,

$$\begin{aligned} G(z') - G(z) &= (z - z')G(z')G(z) \\ &= (z - z')G(z)G(z'), \end{aligned} \tag{E.31}$$

and

$$G(z) = G_0(z) + G_0 V G(z). \tag{E.32}$$

The last equation follows upon the definition of the Green function for free motion,

$$G_0(z) = (z - H_0)^{-1}, \tag{E.33}$$

and the obvious relation,

$$V = H - H_0 = G_0{}^{-1}(z) - G^{-1}(z). \tag{E.34}$$

We now insert the second identity, Eq. (E.31), into Eq. (E.30) and obtain the relation,

$$|\mathbf{p}\rangle^{\pm} = \lim_{\epsilon \to 0} \pm\epsilon \left[G_0(E \pm \epsilon) + G_0(E \pm \epsilon)VG(E \pm \epsilon)\right]|\mathbf{p}\rangle. \tag{E.35}$$

It is easy to see that when $V = 0$ ($H = H_0$), we have,

$$\pm\epsilon G_0(E \pm \epsilon)|\mathbf{p}\rangle = \pm\epsilon(E \pm \epsilon - H_0)^{-1}|\mathbf{p}\rangle = |\mathbf{p}\rangle, \tag{E.36}$$

so that Eq. (E.35) takes the form,

$$|\mathbf{p}\rangle^{\pm} = |\mathbf{p}\rangle + G_0(E \pm 0)V|\mathbf{p}\rangle^{\pm}, \tag{E.37}$$

where the notation $E \pm 0$ indicates the limit $\epsilon \to 0$. This is the time-independent Lippmann-Schwinger equation for the scattering states, $|\mathbf{p}\rangle^{\pm}$, which properly reflects the time-dependent initial conditions of scattering theory. In the coordinate representation, the state $|\mathbf{p}\rangle$ describes a plane wave, and with the free-particle Green function given by Eq. (2.68), the scattering states far away from the field of force are seen to consist of a plane wave and an outgoing spherical wave. The time independent Lippmann-Schwinger equation can become the starting point for the development of multiple scattering theory, an approach taken in the main text.

Bibliography

[1] Erich W. Schmid and Horst Ziegelmann, *The Quantum Mechanical Three-Body Problem*, Viewveg Tracts in Pure and Applied Physics, Pergamon Press, Oxford (1974).

[2] J. M. Jauch, Helv. Phys. Acta **31**, 127 (1958).

[3] L. D. Faddeev, *Mathematical Aspects of the Three-Body Problem in the Quantum Scattering Theory*, Israel Program for Scientific Translations, Jerusalem (1965).

Appendix F

Translation of Spherical Functions

The angular momentum representation of scattering theory contains expansions of the free-particle propagator, $G_0(\mathbf{r}-\mathbf{r}')$, in terms of the regular and irregular solutions of the Schrödinger equation in free space. Similarly, multiple scattering theory involves the so-called structure constants that are obtained from specific expansions of G_0. In fact, many of the difficulties encountered in attempting to apply multiple scattering theory to space-filling potentials are of purely geometric origin, being connected to the conditions under which one may carry out such expansions. For reference purposes, it is convenient to summarize a number of expansions of spherical functions, which we define as the products of a spherical Bessel, Neumann, or Hankel function with a spherical harmonic. In the following discussion, such products are denoted by $A_L(\mathbf{r}) = a_l(r)Y_L(\hat{\mathbf{r}})$, where $a_l(r)$ denotes either $j_l(r)$, $n_l(r)$, or $h_l(r)$, with h_l being a Hankel function of the first or second kind (h_l^+ or h_l^-). The notation $Y_L(\hat{\mathbf{r}})$ denotes a real spherical harmonic (expressions analogous and very similar to those derived below can also be obtained with complex spherical harmonics). We derive expressions [1] connecting the values of $A_L(\mathbf{r}-\mathbf{a})$ about a displaced origin to the undisplaced values $A_L(\mathbf{r})$ explicitly for the case of the Bessel and Hankel functions. The analogous expressions for the case of the Neumann functions can be easily derived along lines essentially identical to those followed here.

We begin with the expansion of a plane wave in terms of spherical functions using the well-known expression (Bauer's identity),

$$e^{i\mathbf{k}\cdot\mathbf{r}} = 4\pi \sum_{l,m} i^l j_l(kr) Y_{l,m}(\hat{\mathbf{r}}) Y_{l,m}(\hat{\mathbf{k}})$$

$$= 4\pi \sum_{L} \mathrm{i}^{l} J_L(k\mathbf{r}) Y_L(\hat{\mathbf{r}}) Y_L(\hat{\mathbf{k}}). \tag{F.1}$$

Thus, for any vectors $\mathbf{r}$ and $\mathbf{a}$ we obtain,

$$\begin{aligned} \mathrm{e}^{\mathrm{i}\mathbf{k}\cdot(\mathbf{r}+\mathbf{a})} &= (4\pi)^2 \sum_{L_2} \sum_{L_3} \mathrm{i}^{l_2+l_3} J_{L_2}(k\mathbf{a}) J_{L_3}(k\mathbf{r}) Y_{L_2}(\hat{\mathbf{k}}) Y_{L_3}(\hat{\mathbf{k}}) \\ &= 4\pi \sum_{L_1} \mathrm{i}^{l_1} J_{L_1}[k(\mathbf{r}+\mathbf{a})] Y_{L_1}(\hat{\mathbf{k}}). \end{aligned} \tag{F.2}$$

Upon multiplying the equality of the last two expressions by $Y_L(\hat{\mathbf{k}})$ and integrating over the angles of $\mathbf{k}$, we obtain the result,

$$\begin{aligned} J_{L_1}[k(\mathbf{r}+\mathbf{a})] &= 4\pi \sum_{L_2} \sum_{L_3} i^{l_2+l_3-l_1} C(L_1 L_2 L_3) J_{L_2}(k\mathbf{a}) J_{L_3}(k\mathbf{r}) \\ &= \sum_{L_2} g_{L_1 L_2}(\mathbf{a}) J_{L_2}(k\mathbf{r}). \end{aligned} \tag{F.3}$$

Here, we have used the notation (for Y_L real),

$$C(L_1 L_2 L_3) = \int Y_{L_1}(\Omega) Y_{L_2}(\Omega) Y_{L_3}(\Omega) \mathrm{d}\Omega, \tag{F.4}$$

for the Gaunt numbers (the integral over three spherical harmonics), and the expansion coefficients, $g_{L_1 L_2}(\mathbf{a})$, are given by the expression,

$$g_{L_1 L_2}(\mathbf{a}) = 4\pi \sum_{L_3} \mathrm{i}^{l_2+l_3-l_1} C(L_1 L_2 L_3) J_{L_3}(k\mathbf{a}). \tag{F.5}$$

These coefficients are the matrix elements of the translation operator [1] in the angular momentum representation and form a unitary matrix as is shown below.

Equation (F.3) provides the desired expression for the "shifted" function $J_L(\mathbf{r}+\mathbf{a})$ in terms of the unshifted functions $J_L(\mathbf{r})$, for all vectors $\mathbf{r}$ and $\mathbf{a}$. In order to derive an analogous result for the Hankel function $H_L(\mathbf{r}+\mathbf{a})$, we recall the usual expression, Eq. (2.68), for the outgoing free-particle propagator,

$$\begin{aligned} G_0(\mathbf{r},\mathbf{r}') &= -\frac{\mathrm{e}^{\mathrm{i}k|\mathbf{r}-\mathbf{r}'|}}{|\mathbf{r}-\mathbf{r}'|} \\ &= -4\pi \mathrm{i}k \sum_{L} j_l(kr_<) h_l(kr_>) Y_L(\hat{\mathbf{r}}) Y_L(\hat{\mathbf{r}'}) \\ &= -4\pi \mathrm{i}k \sum_{L} J_L(k\mathbf{r}) H_L(k\mathbf{r}') \quad \text{for } r' > r, \end{aligned} \tag{F.6}$$

where $r_>(r_<)$ denotes the larger (smaller) of the vectors $\mathbf{r}$ and $\mathbf{r}'$. For any vectors such that $R > r$, there exist arbitrary vectors $\mathbf{a}$ such that $R > a$, $|\mathbf{R}-\mathbf{r}| > a$, and $|\mathbf{r}-\mathbf{a}| < R$, so that we have,

$$-\frac{e^{ik|\mathbf{r}-\mathbf{a}-\mathbf{R}|}}{|\mathbf{r}-\mathbf{a}-\mathbf{R}|} = -4\pi ik\sum_{L_2} H_{L_2}(k\mathbf{R})J_{L_2}[k(\mathbf{r}-\mathbf{a})]. \qquad \text{(F.7)}$$

We can also write,

$$-\frac{e^{ik|\mathbf{r}-\mathbf{a}-\mathbf{R}|}}{|\mathbf{r}-\mathbf{a}-\mathbf{R}|} = -4\pi ik\sum_{L} H_{L}[k(\mathbf{R}+\mathbf{a})]J_{L}(k\mathbf{r}), \qquad \text{(F.8)}$$

which upon comparison with Eq. (F.7) and use of Eq. (F.3) yields the expression,

$$\sum_L H_L[k(\mathbf{R}+\mathbf{a})]J_L(k\mathbf{r}) = \sum_L H_L(k\mathbf{R})\sum_{L_2} g_{L_2L}(\mathbf{r})J_L(k\mathbf{a}). \qquad \text{(F.9)}$$

It now follows that,

$$\begin{aligned} H_{L_1}[k(\mathbf{R}+\mathbf{a})] &= \sum_{L_2} H_{L_2}(k\mathbf{R})g_{L_2L_1}(\mathbf{r}) \\ &= \sum_{L_3} G_{L_1L_3}(\mathbf{R})J_{L_3}(k\mathbf{r}), \quad \text{for } R > r, \end{aligned} \qquad \text{(F.10)}$$

where we have defined the expansion coefficients,

$$G_{L_1L_2}(\mathbf{R}) = 4\pi\sum_{L_3} i^{l_2+l_3-l_1}C(L_1L_2L_3)H_{L_3}(k\mathbf{R}), \quad R \neq 0. \qquad \text{(F.11)}$$

[For values of $\mathbf{R}$ corresponding to the translation vectors of a Bravais lattice, these coefficients give rise to the real-space structure constants of the method of Korringa, Kohn, and Rostoker (KKR)].

It is straightforward to derive a set of useful relations involving the various expansion coefficients, $\underline{g}$ and $\underline{G}$. Note that for vectors such that $|\mathbf{r}-\mathbf{a}| > a$, Eq. (F.10) yields,

$$\begin{aligned} H_L(\mathbf{r}-\mathbf{a}+\mathbf{a}) &\equiv H_L(\mathbf{r}) \\ &= \sum_{L_1} G_{LL_1}(\mathbf{r}-\mathbf{a})J_{L_1}(-\mathbf{a}), \quad |\mathbf{r}-\mathbf{a}| > a. \end{aligned} \qquad \text{(F.12)}$$

Also,

$$\begin{aligned} H_L(\mathbf{r}-\mathbf{a}+\mathbf{a}) &= \sum_{L_1} G_{LL_1}(\mathbf{r})J_{L_1}(\mathbf{a}-\mathbf{a}) \quad \text{all } \mathbf{r},\mathbf{a} \\ &= \sum_{L_1}\sum_{L_2} G_{LL_1}(\mathbf{r})g_{L_1L_2}(\mathbf{a})J_{L_2}(-\mathbf{a}). \end{aligned} \qquad \text{(F.13)}$$

Upon comparison of the last two expressions, we obtain the relation,

$$G_{LL_1}(\mathbf{r}+\mathbf{a}) = \sum_{L_2} G_{LL_2}(\mathbf{r}) g_{L_2L_1}(+\mathbf{a}), \quad r > a. \tag{F.14}$$

The expressions derived so far can be conveniently summarized by means of a vector and matrix notation. Denoting by bras and kets row and column vectors indexed by L, we can write Eqs. (F.3), (F.6), and (F.10) in the forms,

$$|J(\mathbf{r}+\mathbf{a})\rangle = g(\mathbf{a})|J(\mathbf{r})\rangle, \quad \text{all } \mathbf{r} \text{ and } \mathbf{a}, \tag{F.15}$$

or

$$\langle J(\mathbf{r}+\mathbf{a})| = \langle J(\mathbf{r})|g(\mathbf{a}), \tag{F.16}$$

$$\begin{aligned} G(\mathbf{r},\mathbf{r}') &= 4\pi\langle J(\mathbf{r})|H(\mathbf{r}')\rangle, \\ &= 4\pi\langle H(\mathbf{r}')|J(\mathbf{r}), \quad r' > r \end{aligned} \tag{F.17}$$

$$|H(\mathbf{R}+\mathbf{a})\rangle = G(\mathbf{R})|J(\mathbf{a})\rangle, \quad R > a, \tag{F.18}$$

or

$$\langle H(\mathbf{R}-\mathbf{a})| = \langle J(\mathbf{a})|G(\mathbf{R}), \quad R > a \tag{F.19}$$

and

$$\begin{aligned} G(\mathbf{R}+\mathbf{a}) &= G(\mathbf{R})g(\mathbf{a}) \\ &= g(\mathbf{a})G(\mathbf{R}) \quad R > a. \end{aligned} \tag{F.20}$$

Furthermore, one can easily show that the following relations hold,

$$\begin{aligned} g(\mathbf{a}+\mathbf{b}) &= g(\mathbf{a})g(\mathbf{b}) \\ &= g(\mathbf{b})g(\mathbf{a}) \quad \text{all } \mathbf{a} \text{ and } \mathbf{b}, \end{aligned} \tag{F.21}$$

and

$$\begin{aligned} G(\mathbf{R}+\mathbf{a}-\mathbf{b}) &\equiv G(\mathbf{b}-(\mathbf{R}+\mathbf{a})) \\ &= g(-\mathbf{b})G(\mathbf{R}+\mathbf{a}) \quad |\mathbf{R}+\mathbf{a}| > b \\ &= g(-\mathbf{b})G(\mathbf{R})g(\mathbf{a}) \quad R > |\mathbf{a}-\mathbf{b}| \\ &= g(-\mathbf{a})G(\mathbf{R})g(\mathbf{b}) \quad R > |\mathbf{a}-\mathbf{b}|. \end{aligned} \tag{F.22}$$

As a special case of Eq. (F.22), we have the unitary condition,

$$g(\mathbf{a})g(-\mathbf{a}) = 1, \quad \text{or} \quad g(\mathbf{a})g^{\dagger}(\mathbf{a}) = 1, \tag{F.23}$$

which combined with Eq. (F.22) yields the result,

$$G(\mathbf{R}) = g(\mathbf{a})G(\mathbf{R}+\mathbf{a}-\mathbf{b})g(-\mathbf{b}), \quad |\mathbf{R}+\mathbf{a}-\mathbf{b}| > |\mathbf{a}-\mathbf{b}|. \tag{F.24}$$

We also note the relations,

$$\begin{aligned} G(\mathbf{R})_{L0} &= G(\mathbf{R})_{0L} \\ &= H_L(\mathbf{R}), \end{aligned} \tag{F.25}$$

which follow from Eqs. (F.11) or (F.13).

It is useful to note that $G(\mathbf{R}+\mathbf{a})$ can be obtained from $G(\mathbf{R})$ even in the case in which $a > R$, in an apparent contradiction of the conditions of validity of Eq. (F.14). This can be accomplished through an expansion around, rather than through the pole of the Green function by means of a process that is inherently convergent. We consider the vector $\mathbf{a}$ as a sum of vectors $\mathbf{a} = \sum_{\alpha=1}^{N} \mathbf{a}_\alpha$, chosen so that the inequalities,

$$\left|\mathbf{R} + \sum_{\alpha=1}^{n} \mathbf{a}_\alpha\right| > |\mathbf{a}_{n+1}|, \tag{F.26}$$

are satisfied for all $n \leq N$. We can then write,

$$\begin{aligned} G(\mathbf{R}+\mathbf{a}) &\equiv G\left(\mathbf{R} - \sum_{\alpha=1}^{N} \mathbf{a}_\alpha\right) \\ &= G\left(\mathbf{R} + \sum_{\alpha=1}^{N-1} \mathbf{a}_\alpha)g(\mathbf{a}_N\right) \\ &= \left[G\left(\mathbf{R} + \sum_{\alpha=1}^{N-2} \mathbf{a}_\alpha\right) g(\mathbf{a}_{N-1})\right] g(\mathbf{a}_N) \\ &= \cdots = \{\{[G(\mathbf{R})g(\mathbf{a}_1)]\, g(\mathbf{a}_2)\} \cdots\} g(\mathbf{a}_N), \end{aligned} \tag{F.27}$$

where the braces indicate the order of operations. We designate this process of expanding $G(\mathbf{R})$ with the symbol $\odot$, and we have

$$G(\mathbf{r}+\mathbf{a}) = G(\mathbf{R}) \odot g(\mathbf{a}), \tag{F.28}$$

for *all* vectors $\mathbf{R}$ and $\mathbf{a}$. (Note that this process may not always be possible in strictly one-dimensional cases, but the need to invoke it arises *only* in two and three dimensions where it is always possible to "walk around" the pole of the Green function.)

Bibliography

[1] M. Danos and L. C. Maximon, J. Math. Phys. **6**, 766 (1965).

Appendix G

Alternative Derivation of MST

G.1 Non-Spherical MT Potentials

We begin with the solutions, $\psi_L^n(\mathbf{r}_n)$, associated with the potential in cell Ω_n in an assembly of scatterers. Again, we attempt to construct a solution to the Schrödinger equation for a collection of potentials in terms of a linear combination of ψ_L^n's, the solutions for individual cells,

$$\psi(\mathbf{r}_n) = \sum_L \psi_L^n(\mathbf{r}_n) A_L^n, \tag{G.1}$$

for $\mathbf{r}$ in cell Ω_n. We can now write,

$$\begin{aligned}\Lambda &= \int d^2S \int d^2S' \left[\nabla\psi^*(\mathbf{r}) - \psi*(\mathbf{r})\nabla\right] \\ &\times \left[\psi(\mathbf{r}'_n)\nabla' G_0(\mathbf{r}-\mathbf{r}'_n) - G_0(\mathbf{r}-\mathbf{r}'_n)\nabla'\psi(\mathbf{r}'_n)\right],\end{aligned} \tag{G.2}$$

where the prime on ∇' indicates that the gradient is to be applied to functions of $\mathbf{r}'$. From now on we suppress the conditions on r and r' in the surface integrals, but these conditions are to be understood as being imposed throughout the discussion. In order to make further progress, we recall that for vectors $\mathbf{r}$ and $\mathbf{r}'$, confined inside non-overlapping spheres around cells Ω_n and Ω_m, we can express the free-particle propagator in the "factored" form,

$$\begin{aligned}G_0(-\mathbf{r}_n + \mathbf{R}_{nm} + \mathbf{r}_m) &= -ik\sum_L J_L(\mathbf{r}_n)H_L(\mathbf{r}_m)\delta_{nm} \\ &+ \sum_{LL'} J_L(\mathbf{r}_n)G_{LL'}(\mathbf{R}_{nm})\end{aligned}$$

$$\times \quad J_{L''}(-\mathbf{r}_m)\,(1-\delta_{nm})\,, \tag{G.3}$$

where we have set $r_m > r_n$, and in all expansions of the form $\langle J|H\rangle$ the argument of the Hankel function is the larger of r and r'. The expression in Eq. (G.3) can now be inserted into Eq. (G.2), and the integrals in that equation can be broken up over the surfaces of spheres bounding individual cells to lead to the expression,

$$\begin{aligned}
\Lambda &= \sum_n \sum_m \int_{S_n} \int_{S_m} \{\langle A^n \mid \nabla\psi^{n*}\rangle\, [\langle \nabla J \mid H\rangle \delta_{nm} \\
&+ \langle J\,|G_{nm}|\,\nabla J\rangle\,(1-\delta_{nm})]\,\langle \psi^m \mid A^m\rangle \\
&- \langle A^n \mid \nabla\psi^{n*}\rangle\, [\langle J \mid H\rangle \delta_{nm} \\
&+ \langle J\,|G_{nm}|\,J\rangle\,(1-\delta_{nm})]\,\langle \nabla\psi^m \mid A^m\rangle \\
&- \langle A^n \mid \psi^{n*}\rangle\, [\langle \nabla J \mid \nabla H\rangle \delta_{nm} \\
&+ \langle \nabla J\,|G_{nm}|\,\nabla J\rangle\,(1-\delta_{nm})]\,\langle \psi^m \mid A^m\rangle \\
&+ \langle A^n \mid \psi^{n*}\rangle\, [\langle \nabla J \mid H\rangle \delta_{nm} \\
&+ \langle \nabla J\,|G_{nm}|\,J\rangle\,(1-\delta_{nm})]\,\langle \nabla\psi^m \mid A^m\rangle\}\,,
\end{aligned} \tag{G.4}$$

where $\underline{G}_{nm} \equiv \underline{G}(\mathbf{R}_{nm})$. In this expression, we have suppressed integration variables, and have made use of a bra and ket notation (that we hope is obvious) to denote various linear combinations and summations over L. In fact, we can condense the notation even further through the introduction of the matrices $\underline{\mathbf{G}}$ with elements $[\underline{\mathbf{G}}]^{nm}_{LL'} = G_{LL'}(\mathbf{R}_{nm})$, and the "vectors" $\langle A \mid \psi\rangle$ with elements, $[\langle A \mid \psi\rangle]_n = \langle A^n \mid \psi^n\rangle$, and obtain the expression,

$$\begin{aligned}
\Lambda &= \int\!\!\int \{\, \langle A \mid \nabla\psi^*\rangle\, [\langle \nabla J \mid H\rangle + \langle J\,|\underline{\mathbf{G}}|\,\nabla J\rangle]\,\langle \psi \mid A\rangle \\
&- \langle A \mid \nabla\psi^*\rangle\, [\langle J \mid H\rangle + \langle J\,|\underline{\mathbf{G}}|\,J\rangle]\,\langle \nabla\psi \mid A\rangle \\
&- \langle A \mid \psi^*\rangle\, [\langle \nabla J \mid \nabla H\rangle + \langle \nabla J\,|\underline{\mathbf{G}}|\,\nabla J\rangle]\,\langle \psi \mid A\rangle \\
&+ \langle A \mid \psi^*\rangle\, [\langle \nabla J \mid H\rangle + \langle \nabla J\,|\underline{\mathbf{G}}|\,J\rangle]\,\langle \nabla\psi \mid A\rangle\,\} \\
&= \int\!\!\int \langle A|\,\{\,[|\,\nabla\psi^*\rangle\langle J\,| - |\,\psi^*\rangle\langle \nabla J\,|] \\
&\times\ [\,\underline{\mathbf{G}}\,(|\,\nabla J\rangle\langle \psi\,| - |\,J\rangle\langle \nabla\psi\,|) \\
&- (|\,\nabla H\rangle\langle \psi\,| - |\,H\rangle\langle \nabla\psi\,|)\,]\,\}\,|\,A\rangle.
\end{aligned} \tag{G.5}$$

We can now write,

$$\Lambda = \langle A \mid \underline{\tilde{\mathbf{S}}}\underline{\mathbf{G}}\underline{\mathbf{S}} - \underline{\tilde{\mathbf{S}}}\underline{\mathbf{C}} \mid A\rangle. \tag{G.6}$$

Note that $\underline{\mathbf{S}}$ and $\underline{\mathbf{C}}$ are cell-diagonal matrices. Note also that the last expression for Λ is made possible by the MT geometry that allows G_0 to be written in the form of Eq. (G.3), and consequently allows the independent integration of the cell variables $\mathbf{r}_n$ and $\mathbf{r}'_m$. It is precisely at this stage that difficulties may arise in attempts to apply the present formalism to space-filling cells. Two different ways of handling the non-MT geometry are discussed in the following sections.

Now, setting the variation of Λ with respect to A_L^n equal to zero leads to the set of linear homogeneous equations,

$$\left[\underline{\tilde{\mathbf{S}}\mathbf{GS}} - \underline{\tilde{\mathbf{S}}\mathbf{C}}\right] \mid A\rangle = 0, \tag{G.7}$$

which may have non-trivial solutions only when the determinant of the coefficients vanishes,

$$\det\left|\underline{\tilde{\mathbf{S}}\mathbf{GS}} - \underline{\tilde{\mathbf{S}}\mathbf{C}}\right| = 0. \tag{G.8}$$

Provided that we can treat the $\underline{\mathbf{S}}$ matrices as being invertible we can also write,

$$\det\left[\underline{\mathbf{GS}} - \underline{\mathbf{C}}\right] = 0, \tag{G.9}$$

or,

$$\det\left[\underline{\mathbf{G}} - \underline{\mathbf{CS}}^{-1}\right] = 0. \tag{G.10}$$

In the last expression, we can identify the quantity $\underline{\mathbf{CS}}^{-1}$ with the inverse of the cell t-matrix, provided the proper choice of normalization of the cell wave function has been made. Then we can write the secular equation in the familiar form,

$$\det\left[\underline{\mathbf{M}}\right] = 0, \tag{G.11}$$

where,

$$\underline{M}^{ij} = \underline{m}^i\delta_{ij} - \underline{G}(\mathbf{R}_{ij})(1 - \delta_{ij}). \tag{G.12}$$

Not unexpectedly, this has the well-known form of the KKR secular equation for MT potentials.

G.2 Space-Filling Cells of Convex Shape

We now provide a derivation of MST for non-spherical MT potentials and for space-filling cells by explicitly treating the geometric aspects of the problem. We begin by noting an important feature of the MST formalism, namely that the free- particle propagator couples a pair of cells at a time. Thus, it is convenient to begin by treating the case of two adjacent cells of convex, polyhedral shape. The incorporation of the resulting argument into the discussion of an arbitrary number of cells can be effected in a straightforward way. In the following discussion, it is formally useful to consider the cells as being separated by a thin strip of zero potential and infinitesimal width. The width of the strip will be allowed to go to zero at the end of formal arguments.

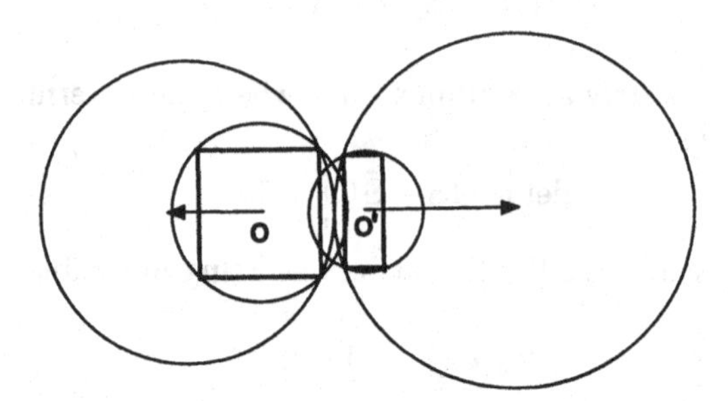

Figure G.1: **Two adjacent non-spherical cells separated by a narrow strip of zero potential. The spheres bounding the cells overlap the adjacent potential when centered at positions O and O', but not when the centers are displaced in the manner indicated.**

Now, for $\mathbf{r}_n$ and $\mathbf{r}_m$ cell vectors in cells Ω_n and Ω_m, respectively, an expansion of $G_0(-\mathbf{r}_n + \mathbf{R}_{nm} + \mathbf{r}_m)$ in the factored form, Eq. (3.40), can be obtained along the following lines. We take the intercell vector to be perpendicular to the common face of the cells, and imagine the centers of the cells as being displaced along $\mathbf{R}_{nm}$ and away from this face. As long as the distance between cells remains finite, albeit infinitesimal, there exist vectors, $\mathbf{a}_n$ and $\mathbf{a}_m$, such that spheres centered at the displaced centers, $\mathbf{R}_n + \mathbf{a}_n$ and $\mathbf{R}_m + \mathbf{a}_m$, and covering the cells, Ω_n and Ω_m, respectively, do not overlap (Fig G.1).

Therefore, in terms of vectors measured with respect to the shifted centers, we recover the MT case. With the obvious expressions,

$$\bar{\mathbf{r}}_n = \mathbf{r}_n - \mathbf{a}_n, \quad \bar{\mathbf{r}}'_n = \mathbf{r}'_n - \mathbf{a}_n, \quad \bar{\mathbf{R}}_{nm} = \mathbf{R}_{nm} - \mathbf{a}_n + \mathbf{a}_m, \tag{G.13}$$

we have the inequalities,

$$\begin{aligned} |\,\bar{\mathbf{r}}_m + \bar{\mathbf{R}}_{nm}\,| &> \bar{r}_n, \quad |\,\bar{\mathbf{R}}_{nm} - \bar{\mathbf{r}}_m\,| > \bar{r}_m \\ \bar{R}_{nm} &> \bar{r}_n, \quad \bar{R}_{nm} > \bar{r}_m, \end{aligned} \tag{G.14}$$

and we can write,

$$\begin{aligned} G_0(-\mathbf{r}_n + \mathbf{R}_{nm} + \mathbf{r}'_m) &= G_0(-\bar{\mathbf{r}}_n + \bar{\mathbf{R}}_{nm} + \bar{\mathbf{r}}'_m) \\ &= \sum_{LL'} J_L(\bar{\mathbf{r}}_n) G_{LL'}(\bar{\mathbf{R}}_{nm}) J_{L'}(-\bar{\mathbf{r}}_m) \\ &= \langle J(\bar{\mathbf{r}}_n) \mid G(\bar{\mathbf{R}}_{nm}) \mid J(-\bar{\mathbf{r}}_m) \rangle. \qquad \text{(G.15)} \end{aligned}$$

Furthermore, the functions $\langle J(\bar{\mathbf{r}}_n)|$ can be expanded (see Appendix F) in the form,

$$\begin{aligned} \langle J(\bar{\mathbf{r}}_n)| &= \langle J(\mathbf{r}_n - \mathbf{a}_n)| \\ &= \langle J(\mathbf{r}_n) \mid \underline{g}(-\mathbf{a}_n), \qquad \text{(G.16)} \end{aligned}$$

for all values of $\mathbf{r}_n$ and $\mathbf{a}_n$. A similar expression holds for $| J(-\bar{\mathbf{r}}'_m)\rangle$. We now substitute these expansions into Eq. (G.15), and obtain the expression,

$$G_0(-\bar{\mathbf{r}}_n + \bar{\mathbf{R}}_{nm} + \bar{\mathbf{r}}'_m) = \langle J(\mathbf{r}_n) | \underline{g}(-\mathbf{a}_n) || \underline{G}(\mathbf{R}_{nm} - \mathbf{a}_n + \mathbf{a}_m) || \underline{g}(\mathbf{a}_m) | J(-\mathbf{r}'_m)\rangle. \qquad \text{(G.17)}$$

Here, the double bars indicate the order of operations: They are a reminder that the multiplications of the $\underline{g}$s by the structure constant matrices, $\underline{G}$, must *follow* those between the $\underline{g}$s and the Js. Otherwise, one formally obtains the expansion in Eq. (3.40) in terms of the vectors defined with respect to undisplaced centers, which in this case may diverge for certain values of the cell vectors. With the proper order of the sums retained, Eq. (G.17) can be substituted into Eq. (G.2), and the integrals over the cell vectors $\mathbf{r}_n$ and $\mathbf{r}'_m$ can be performed in straightforward fashion leading to the expression,

$$\Lambda = \langle A \mid \underline{\tilde{\mathbf{S}}\mathbf{g}} \,||\, \underline{\mathbf{G}} \,||\, \underline{\mathbf{g}\mathbf{S}} - \underline{\tilde{\mathbf{S}}\mathbf{C}} \mid A\rangle. \qquad \text{(G.18)}$$

Note, that because the integrals over *both* cell vectors were confined to the surfaces of the cells, this equation reflects the direct, point-by-point matching of the wave function across cell boundaries. Once again, using the variational condition, Eq. (G.16), with respect to A^n_L, we obtain the secular equation,

$$\det\left[\underline{\tilde{\mathbf{S}}\mathbf{g}} \,||\, \underline{\mathbf{G}} \,||\, \underline{\mathbf{g}\mathbf{S}} - \underline{\tilde{\mathbf{S}}\mathbf{C}}\right] = 0. \qquad \text{(G.19)}$$

We point out that this equation is meaningful under the condition that the order of operations is that indicated by the double bars. However, provided that the matrix $\underline{\mathbf{S}}$ and its transpose are invertible, their determinants can be factored out and the last equation takes the form,

$$\det\left[\underline{\mathbf{g}} \,||\, \underline{\mathbf{G}} \,||\, \underline{\mathbf{g}} - \underline{\mathbf{C}\mathbf{S}^{-1}}\right] = 0. \qquad \text{(G.20)}$$

But at this point the products of the $\underline{g}$s with the structure constants *can* be carried out as the presence of the $\underline{\mathbf{S}}$ matrices which imposed the restriction on the summations has been removed. Using the property, $\underline{g}(-\mathbf{a}_n)\underline{G}(-\mathbf{a}_n+\mathbf{R}_{nm}+\mathbf{a}_m)\underline{g}(\mathbf{a}_m)=\underline{G}(\mathbf{R}_{nm})$, we obtain the secular equation,

$$\det\left[\underline{\mathbf{G}}-\underline{\mathbf{CS}}^{-1}\right]=0, \tag{G.21}$$

which is precisely Eq. (G.10), and also (G.12). In the case in which the intracell vectors are smaller than any intercell vector, Eq. (G.20) can also be converted to Eq. (G.9).

At this point, we can generalize these considerations to the case of an arbitrary number of cells. Following the lines of the discussion in the previous subsection, we treat each pair of scatterers in the manner indicated above and arrive at Eq(G.9) for an assembly of scattering potential cells.

It is of the utmost importance to realize that the passage from Eq. (G.19) to Eq. (G.20) does not involve an invalid exchange of summation indices. Note that for the proper order of operations, the (LL') element of the matrix in Eq. (G.19),

$$\sum_{L_1L_2}\sum_{L_3L_4}\{S^{\mathrm{T}}_{LL_1}g_{L_1L_2}\}\,G_{L_2L_3}\,\{g_{L_3L_4}S_{L_4L'}\}-\sum_{L_1}S^{\mathrm{T}}_{LL_1}C_{L_1L'}, \tag{G.22}$$

is well defined for *all* values of L and L'. Therefore, it is allowed to multiply this expression on the left by ${S^{\mathrm{T}}}^{-1}_{L_0L_1}$ and sum over L_1. In fact, multiplication by *any* matrix is allowed provided that the product can be shown independently to converge. The multiplication by ${S^{\mathrm{T}}}^{-1}_{L_0L_1}$ and the subsequent summation has the effect of replacing the quantity $S^{\mathrm{T}}_{LL_1}$ by $\delta_{L_0L_1}$. The matrix $\underline{\mathbf{S}}$ can also be removed in a similar manner. As the resulting expressions are manifestly convergent, the sums can be performed to yield Eq. (G.22). The procedure just described, when satisfied in detail, can be used to justify multiplication by rectangular matrices and the conversion of one form of the secular equation to another. However, in contrast to the case of square matrices, multiplications by rectangular matrices in general alter the convergence and variational properties of the equations. In particular, even though Eq. (G.19) yields energy eigenvalues that are variational with respect to the wave function, it has been shown numerically that Eq. (G.21) in general does not have this property.

It cannot be stressed too strongly that preserving the order of conditionally converging sums is crucial in the derivation of Eq. (G.21). This feature has lasting effects that are present in Eq. (G.21), even though this equation no longer involves double summations. For example, Eq. (G.21) cannot be converted to Eq. (G.6), in spite of the formal analogies between the two. In the case of space-filling cells, Eq. (G.6) can diverge. Nor can the matrix in Eq. (G.21) be expanded in terms of the cell t-matrices in a "power series" like manner, because that expansion may also diverge[1].

[1]Divergences can be avoided in the case in which intercell vectors are larger than

What we have shown amounts to the statement that the *summed* form of the inverse of the total t-matrix is the matrix in Eq. (G.21). Thus, we have succeeded in proving the formal existence of the matrix $\underline{\mathbf{M}}$, and to show that it takes the MT form even in the case of space- filling potential cells.

We now proceed with a derivation of MST that emphasizes the role of the cell shape and thus may further illuminate the importance of geometry in determining the validity of the theory.

G.3 Displaced Cell approach: Convex Cells

In this section, we present a variational derivation of the secular equation of MST using the displaced-cell formalism of Chapter 3. The advantage of this approach is that it can be applied to the case of concave cells, as is indicated in the next subsection. We examine in some detail the case of convex cells.

We begin with the observation that the equations of MST connect only two cells at a time. The free-particle propagator, $G_0(-\mathbf{r}_n + \mathbf{R}_{nm} + \mathbf{r}_m)$, connecting two points in adjacent cells, can be written in the form,

$$
\begin{aligned}
G_0\left(-\mathbf{r}_n + \mathbf{R}_{nm} + \mathbf{r}'_m\right) &= \sum_{L_1} J_{L_1}(\mathbf{b}_{nm}) \\
\times & \left[\sum_{L_2} G_{L_1L_2}(-\mathbf{r}_n + \mathbf{R}_{nm} + \mathbf{b}_{nm}) J_{L_2}(-\mathbf{r}_m)\right] \\
= & \sum_{L_1} J_{L_1}(\mathbf{b}_{nm}) \left[\sum_{L_2}\sum_{L_3} g_{L_1L_3}(-\mathbf{r}_n) G_{L_3L_2}(\mathbf{R}_{nm} + \mathbf{b}_{nm}) J_{L_2}(-\mathbf{r}_m)\right] \\
= & \sum_{L_1} J_{L_1}(\mathbf{b}_{nm}) \left[\sum_{L_2}\sum_{L_3}\sum_{L4} \mathrm{i}^{\ell_1-\ell_3+\ell_4} C\left(L_1L_3L_4\right)\right. \\
\times & \left. J_{L_4}(-\mathbf{r}_n) G_{L_3L_2}(\mathbf{R}_{nm} + \mathbf{b}_{nm}) J_{L_2}(-\mathbf{r}_m)\right].
\end{aligned}
\tag{G.23}
$$

This expression can be substituted into the secular equation and the integrals over the cell vectors can be performed, leading to the expression,

$$
\begin{aligned}
\Lambda^{nm} = & \sum_{LL'} A^n_L \left\{ \sum_{L_1} J_{L_1}(\mathbf{b}_{nm}) \left[\sum_{L_2}\sum_{L_3}\sum_{L_4} \mathrm{i}^{\ell_1-\ell_3+\ell_4} C\left(L_1L_3L_4\right)\right.\right. \\
\times & \; S^{n\mathrm{T}}_{LL_4} G_{L_3L_2}(\mathbf{R}_{nm} + \mathbf{b}_{nm}) \\
\times & \left.\left. S^m_{L_2L'}\left(1 - \delta_{nm}\right)\right] - \delta_{nm} \sum_{L_1} S^{n\mathrm{T}}_{LL_1} C^n_{L_1L'} \right\} A^m_{L'}
\end{aligned}
\tag{G.24}
$$

---

intracell vectors, through the use of modified structure constants as discussed following Eq. (G.29).

Now, the variational condition,

$$\frac{\delta \Lambda^{nm}}{\delta A_L^n} = 0$$

leads to

$$\begin{aligned}\sum_{L'} & \left\{ \sum_{L_1} J_{L_1}(\mathbf{b}_{nm}) \left[\sum_{L_2} \sum_{L_3} \sum_{L_4} \mathrm{i}^{\ell_1 - \ell_3 + \ell_4} C(L_1 L_3 L_4) \right.\right. \\ \times & \quad S_{LL_4}^{n\mathrm{T}} G_{L_3 L_2}(\mathbf{R}_{nm} + \mathbf{b}_{nm}) \\ \times & \quad \left.\left. S_{L_2 L'}^{m} (1 - \delta_{nm}) \right] - \delta_{nm} \sum_{L_1} S_{LL_1}^{n\mathrm{T}} C_{L_1 L'}^{n} \right\} A_{L'}^{m} = 0.\end{aligned} \tag{G.25}$$

This last expression yields the secular equation

$$\det \underline{\mathbf{M}} = 0, \tag{G.26}$$

where,

$$\begin{aligned}M_{LL'}^{nm} &= \sum_{L_1} J_{L_1}(\mathbf{b}_{nm}) \left[\sum_{L_2} \sum_{L_3} \sum_{L4} \mathrm{i}^{\ell_1 - \ell_3 + \ell_4} C(L_1 L_3 L_4) \right. \\ &\times \quad S_{LL_4}^{n\mathrm{T}} G_{L_3 L_2}(\mathbf{R}_{nm} + \mathbf{b}_{nm}) \\ &\times \quad \left. S_{L_2 L'}^{m} (1 - \delta_{nm}) \right] - \delta_{nm} \sum_{L_1} S_{LL_1}^{n\mathrm{T}} C_{L_1 L'}^{n}.\end{aligned} \tag{G.27}$$

As was the case with Eq. (G.19), we note that the last expression is valid for all values of the *outside* indices, LL'. Once again, multiplication by the matrix elements ${S^{n\mathrm{T}}}^{-1}_{L_0 L}$ on the left, and summation over L yields the expression,

$$\begin{aligned}M_{L_0 L'}^{nm} &= \sum_{L_1} J_{L_1}(\mathbf{b}_{nm}) \left[\sum_{L_2} \sum_{L_3} \sum_{L_4} \mathrm{i}^{\ell_1 - \ell_3 + \ell_4} C(L_1 L_3 L_4)\, \delta_{L_0 L_4} \right. \\ &\times \quad \left. G_{L_3 L_2}(\mathbf{R}_{nm} + \mathbf{b}_{nm}) S_{L_2 L'}^{m} (1 - \delta_{nm}) \right] - \delta_{nm} C_{L_0 L'}^{n}.\end{aligned} \tag{G.28}$$

Finally, the matrix elements of $\underline{\mathbf{S}}$ can be removed by a multiplication on the right, the sums over L_2 and L_1 can be performed, leading to $g_{L_0 L_3}$, and we obtain the result,

$$\begin{aligned}M_{L_0 L_5}^{nm} &= \sum_{L_3} g_{L_0 L_3}(\mathbf{b}_{nm}) G_{L_3 L_5}(\mathbf{R}_{nm} + \mathbf{b}_{nm}) (1 - \delta_{nm}) \\ &- \sum_{L1} C_{L_0 L_1}^{n} S_{L_1 L_5}^{-1} \delta_{nm} \\ &= G_{L_0 L_5}(\, \mathbf{R}_{nm}\,)(1 - \delta_{nm}) - m_{L_0 L_5}^{n} \delta_{nm}.\end{aligned} \tag{G.29}$$

This is clearly recognized as the secular equation of MST in the MT form. It should be pointed out once again that all summations have been performed in a manner that ensures convergence. Thus, the secular equation was derived in a way that explicitly matches the wave function across cell boundaries in the case of convex cells.

Furthermore, this form of the secular equation and its associated eigenvectors, A_L^n, can be used to construct a converging expansion of the wavefunction of a multiple-scattering assembly, in terms of the "modified structure constants", $\sum_{L_3} g_{L_0L_3}(\mathbf{b}_{nm}) G_{L_3L_5}(\mathbf{R}_{nm} + \mathbf{b}_{nm})$, appearing in the first line of Eq. (G.29). In such a process, one solves for the wavefunction for a given value of $\mathbf{b}_{nm}$ and of L_3 and monitors the convergence of the wavefunction as L_3 increases.

G.3.1 Displaced-cell approach: Concave cells

The application of the displaced-cell approach to concave cells is straightforward, although somewhat tedious. For the reasons discussed in Chapter 3 with respect to the single scatterer, it is now necessary to replace the single vector $\mathbf{b}_{nm}$ by series of vectors, allowing the analytic continuation of the wave function from outside the sphere bounding each of a pair of adjacent cells to points inside their respective moon regions. Because the intermediate expressions leading to the secular equation, Eq. (G.29), in this case are rather complicated, and as the reader has probably acquired the flavor of the process, we forego displaying these expressions here.

Appendix H

Green-Function Expansions

Here we show that the Green function for a non-spherical potential can be constructed in the form,

$$G(\mathbf{r},\mathbf{r}') = \sum_L [\phi_L(\mathbf{r})F_L(\mathbf{r}')\theta(r'-r) + \phi_L(\mathbf{r}')F_L(\mathbf{r})\theta(r-r')], \tag{H.1}$$

where $\phi_L(\mathbf{r})$ is a solution to the Schrödinger equation that approaches $J_L(\mathbf{r})$ as $r \to 0$. This is the only boundary condition that we apply to $\phi_L(\mathbf{r})$, but it is sufficient to determine the function uniquely within a sphere of radius R through Eqs. (D.82), (D.92), and (D.93).

The function $F_L(\mathbf{r})$ is also a solution to the Schrödinger equation that is, however, irregular at the origin. In order to determine the function F we investigate the form of the Green function. If the potential vanishes for $r > R$, we know that the Green function can be written, for $r > R$, $r' > R$, and $r' > r$ as,

$$G(\mathbf{r},\mathbf{r}') = \sum_L J_L(\mathbf{r})H_L(\mathbf{r}') + \sum_{LL'} H_L(\mathbf{r})t_{LL'}H_{L'}(\mathbf{r}'), \tag{H.2}$$

which by comparison with Eqs. (D.82) and (D.112) yields, for $r' \geq R$,

$$F_L(\mathbf{r}') = \sum_{L'} [C^{-1}(R)]_{LL'} H_{L'}(\mathbf{r}') \quad (r' \geq R), \tag{H.3}$$

where the t-matrix, $t_{LL'}$ is given by SC^{-1}. This boundary condition on F_L at R is sufficient to determine the function uniquely. It satisfies the integral equation,

$$F_L(\mathbf{r}) = \alpha_L H_L(\mathbf{r}) + \int_0^R G_0(\mathbf{r},\mathbf{r}')F_L(\mathbf{r}'), \tag{H.4}$$

where α is to be determined so that the boundary condition, Eq. (H.3), is satisfied. The result is that $F_L(\mathbf{r})$ may be written in a form analogous to Eq. (D.82),

$$F_L(\mathbf{r}) = \sum_{L'} [\bar{C}_{LL'}(r)J_{L'}(\mathbf{r}) + \bar{S}_{LL'}(r)H_{L'}(\mathbf{r})], \tag{H.5}$$

where $\bar{C}$ and $\bar{S}$ are given by

$$\bar{C}_{LL'}(r) = \int_r^R \mathrm{d}^3r' F_L(\mathbf{r}')V(\mathbf{r}')H_{L'}(\mathbf{r}') \tag{H.6}$$

$$\bar{S}_{LL'}(r) = C^{-1}_{LL'}(R) - \int_r^R \mathrm{d}^3r' F_L(\mathbf{r}')V(\mathbf{r}')J_{L'}(\mathbf{r}'). \tag{H.7}$$

For reference we also recall the analogous expressions for $C_{L'L}$ and $S_{L'L}$, which are used to represent $\phi_L(\mathbf{r})$ in Eq. (D.82),

$$C_{L'L}(r) = \delta_{L'L} - \int_0^r \mathrm{d}^3r' H_{L'}(\mathbf{r}')V(\mathbf{r}')\phi_L(\mathbf{r}') \tag{H.8}$$

$$S_{L'L}(r) = \int_0^r \mathrm{d}^3r' J_{L'}(\mathbf{r}')V(\mathbf{r}')\phi_L(\mathbf{r}'). \tag{H.9}$$

Thus if Eq. (H.1) is to be valid, the Green function for a non-spherical potential must be expressible in the form

$$\begin{aligned} G(\mathbf{r},\mathbf{r}') = \sum_{LL'} \{ & J_L(\mathbf{r}) \left[C(r)\bar{C}(r')\right]_{LL'} J_{L'}(\mathbf{r}') \\ + & J_L(\mathbf{r}) \left[C(r)\bar{S}(r')\right]_{LL'} H_{L'}(\mathbf{r}') \\ + & H_L(\mathbf{r}) \left[S(r)\bar{C}(r')\right]_{LL'} J_{L'}(\mathbf{r}') \\ + & H_L(\mathbf{r}) \left[S(r)\bar{S}(r')\right]_{LL'} H_{L'}(\mathbf{r}')\} . \end{aligned} \tag{H.10}$$

Now we know that the Green function may be expressed quite generally in the form,

$$G(\mathbf{r},\mathbf{r}') = G_0(\mathbf{r},\mathbf{r}') + \int \mathrm{d}^3r_1 \int \mathrm{d}^3r_2 \, G_0(\mathbf{r},\mathbf{r}_1)t(\mathbf{r}_1,\mathbf{r}_2)G_0(\mathbf{r}_2,\mathbf{r}'), \tag{H.11}$$

which leads directly to,

$$\begin{aligned} G(\mathbf{r},\mathbf{r}') = \sum_{LL'} & [J_L(\mathbf{r})g^{JJ}_{LL'}(r,r')J_{L'}(\mathbf{r}') + J_L(\mathbf{r})g^{JH}_{LL'}(r,r')H_{L'}(\mathbf{r}') \\ + & H_L(\mathbf{r})g^{HJ}_{LL'}(r,r')J_{L'}(\mathbf{r}') + H_L(\mathbf{r})g^{HH}_{LL'}(r,r')H_{L'}(\mathbf{r}')], \end{aligned} \tag{H.12}$$

where (for $r' > r$),

$$g_{LL'}^{JJ}(r,r') = \int_r^R \mathrm{d}^3 r_1 \int_{r'}^R \mathrm{d}^3 r_2 H_L(\mathbf{r}_1) t(\mathbf{r}_1,\mathbf{r}_2) H_{L'}(\mathbf{r}_2) \quad \text{(H.13)}$$

$$g_{LL'}^{JH}(r,r') = \int_r^R \mathrm{d}^3 r_1 \int_0^{r'} \mathrm{d}^3 r_2 H_L(\mathbf{r}_1) t(\mathbf{r}_1,\mathbf{r}_2) J_{L'}(\mathbf{r}_2) \quad \text{(H.14)}$$

$$g_{LL'}^{HJ}(r,r') = \int_0^r \mathrm{d}^3 r_1 \int_{r'}^R \mathrm{d}^3 r_2 J_L(\mathbf{r}_1) t(\mathbf{r}_1,\mathbf{r}_2) H_{L'}(\mathbf{r}_2) \quad \text{(H.15)}$$

$$g_{LL'}^{HH}(r,r') = \int_0^r \mathrm{d}^3 r_1 \int_0^{r'} \mathrm{d}^3 r_2 J_L(\mathbf{r}_1) t(\mathbf{r}_1,\mathbf{r}_2) J_{L'}(\mathbf{r}_2). \quad \text{(H.16)}$$

Thus the validity of the Green function expansion, Eq. (H.1), will be established if the following relations can be verified:

$$\left[C(r)\bar{C}(r')\right]_{LL'} = g_{LL'}^{JJ}(r,r') \quad \text{(H.17)}$$

$$\left[C(r)\bar{S}(r')\right]_{LL'} = g_{LL'}^{JH}(r,r') \quad \text{(H.18)}$$

$$\left[S(r)\bar{C}(r')\right]_{LL'} = g_{LL'}^{HJ}(r,r') \quad \text{(H.19)}$$

$$\left[S(r)\bar{S}(r')\right]_{LL'} = g_{LL'}^{HH}(r,r'). \quad \text{(H.20)}$$

The following expressions will be useful in working with $C\bar{C}$, $C\bar{S}$, $S\bar{C}$, and $S\bar{S}$:

$$F_L(\mathbf{r}) = H_L(\mathbf{r}) + \int_0^R \mathrm{d}^3 r' H_L(\mathbf{r}') V(\mathbf{r}') G(\mathbf{r}',\mathbf{r}) \quad \text{(H.21)}$$

$$Z_L(\mathbf{r}) = \sum_{L'} \phi_{L'}(\mathbf{r}) C_{L'L}^{-1}(R) = J_L(\mathbf{r}) + \int_0^R \mathrm{d}^3 r' G(\mathbf{r},\mathbf{r}') V(\mathbf{r}') J_L(\mathbf{r}'). \quad \text{(H.22)}$$

We will also use Eq. (H.1).

Since the manipulations are lengthy, we use a shorthand notation in which we suppress integration variables and the angular momentum index. We also take $r' > r$.

$$\begin{aligned} C\bar{C} &= \left(1 - \int_0^r HV\phi\right)\left(\int_{r'}^R FVH\right) \\ &= \int_{r'}^R FVH - \int_0^r \int_{r'}^R HVGVH \\ &= \int_{r'}^R HVH + \int_0^R \int_{r'}^R HVGVH - \int_0^r \int_{r'}^R HVGVH \\ &= \int_r^R \int_{r'}^R HtH = g^{JJ} \end{aligned} \quad \text{(H.23)}$$

$$
\begin{aligned}
C\bar{S} &= \left(C_R + \int_r^R HV\phi\right)\left(C_R^{-1} - \int_{r'}^R FVJ\right) \\
&= 1 + \int_r^R HVZ - \int_{r'}^R FVJ + \int_0^R \int_{r'}^R HVGVJ \\
&- \int_r^R \int_{r'}^R HVGVJ \\
&= 1 + \int_r^R HVJ - \int_{r'}^R HVJ \\
&+ [\int_r^R \int_0^R - \int_0^R \int_{r'}^R + \int_0^R \int_{r'}^R - \int_r^R \int_{r'}^R]HVGVJ \\
&= 1 + \int_r^R \int_0^{r'} HtJ = g^{JH} && \text{(H.24)}
\end{aligned}
$$

$$
\begin{aligned}
S\bar{C} &= \int_0^r JV\phi \int_{r'}^R FVH \\
&= \int_0^r \int_{r'}^R JVGVH \\
&= \int_0^r \int_{r'}^R JtH = g^{HJ} && \text{(H.25)}
\end{aligned}
$$

$$
\begin{aligned}
S\bar{S} &= \int_0^r JV\phi\left(C_R^{-1} - \int_{r'}^R FVJ\right) \\
&= \int_0^r JVZ - \int_0^r \int_{r'}^R JVGVJ \\
&= \int_0^r JVJ + \int_0^r \int_0^R JVGVJ - \int_0^r \int_{r'}^R JVGVJ \\
&= \int_0^r JVJ + \int_0^r \int_0^{r'} JVGVJ \\
&= \int_0^r \int_0^{r'} JtJ = g^{HH}. && \text{(H.26)}
\end{aligned}
$$

[step, we now verify directly that $[\nabla^2 + E - V(\mathbf{r})]$ applied to the Green function in the form of Eq. (H.1) yields $\delta(\mathbf{r} - \mathbf{r}')$. Using the fact that both ϕ_L and F_L satisfy the Schrödinger equation we have,

$$
[\nabla^2 + E - V(\mathbf{r})]G(\mathbf{r}, \mathbf{r}')
$$

$$
\begin{aligned}
&= \sum_L F_L(\mathbf{r}')\nabla\cdot[\nabla\phi_L(\mathbf{r})\theta(r'-r)+\phi_L(\mathbf{r})\nabla\theta(r'-r)] \\
&+ \sum_L \phi_L(\mathbf{r}')\nabla\cdot[\nabla F_L(\mathbf{r})\theta(r-r')+F_L(\mathbf{r})\nabla\theta(r-'r)] \\
&= \sum_L F_L(\mathbf{r}')[2\nabla\phi_L(\mathbf{r})\cdot\nabla\theta(r'-r)+\phi_L(\mathbf{r})\nabla^2\theta(r'-r)] \\
&+ \sum_L \phi_L(\mathbf{r}')[2\nabla F_L(\mathbf{r})\cdot\nabla\theta(r-r')+F_L(\mathbf{r})\nabla^2\theta(r-'r)] \\
&= \sum_L F_L(\mathbf{r}')\left[-2\frac{\partial}{\partial r}\phi_L(\mathbf{r})\delta(r-r')\right. \\
&- \left.\phi_L(\mathbf{r})\frac{2}{r}\delta(r-r')-\phi_L(\mathbf{r})\delta'(r-r')\right] \\
&+ \sum_L \phi_L(\mathbf{r}')\left[2\frac{\partial}{\partial r}F_L(\mathbf{r})\delta(r-r')+F_L(\mathbf{r})\frac{2}{r}\delta(r-r')\right. \\
&+ \left.F_L(\mathbf{r})\delta'(r-r')\right].
\end{aligned} \tag{H.27}
$$

Consider the two terms involving $\frac{2}{r}\delta(r-r')$. Substitution for ϕ_L and F_L from Eqs. (D.83) and (H.5) yields,

$$
\begin{aligned}
&\frac{2}{r}\delta(r-r')\sum_{L_1L_2} Y_{L_1}(\hat{r'})Y_{L_2}(\hat{r}) \\
&\times \{j_{\ell_1}(\kappa r)j_{\ell_2}(\kappa r)[C\bar{C}-\bar{C}^TC^T]_{L_1L_2} \\
&+ h_{\ell_1}(\kappa r)h_{\ell_2}(\kappa r)[S\bar{S}-\bar{S}^TS^T]_{L_1L_2} \\
&+ j_{\ell_1}(\kappa r)h_{\ell_2}(\kappa r)[C\bar{S}-\bar{C}^TS^T]_{L_1L_2} \\
&+ h_{\ell_1}(\kappa r)j_{\ell_2}(\kappa r)[S\bar{C}-\bar{S}^TC^T]_{L_1L_2}\}
\end{aligned} \tag{H.28}
$$

where C,$\bar{C}$, S, and $\bar{S}$, are evaluated at argument r, and we have used the superscript T to denote the transpose of a matrix. We have also incorporated a factor of $-i\kappa$ into the spherical Hankel functions $h_\ell(\kappa r)$. From the above expressions for these quantities it is clear that $C\bar{C}-\bar{C}^TC^T$ and $S\bar{S}-\bar{S}^TS^T$ vanish and that $C\bar{S}-\bar{C}^TS^T=\delta_{L_1L_2}$. Thus this term vanishes.

Now consider the term involving the derivative of the delta function. This generalized function is only meaningful when it appears as a factor in an integrant. Thus, we investigate the integral,

$$
\int dr\, r^2 f(r)\sum_L[\phi_L(\mathbf{r}')F_L(\mathbf{r})-\phi_L(\mathbf{r})F_L(\mathbf{r}')]\delta'(r-r'). \tag{H.29}
$$

Integration by parts gives,

$$
\int drr^2 f(r)\sum_L[\phi_L(\mathbf{r}')F_L(\mathbf{r})-\phi_L(\mathbf{r})F_L(\mathbf{r}')]\delta(r-r')
$$

$$
\begin{aligned}
&-\int \mathrm{d}r\frac{\partial}{\partial r}[r^2 f(r)]\sum_L[\phi_L(\mathbf{r}')F_L(\mathbf{r})-\phi_L(\mathbf{r})F_L(\mathbf{r}')]\delta(r-r') \\
&-\int \mathrm{d}r r^2 f(r)\sum_L\left[\phi_L(\mathbf{r}')\frac{\partial}{\partial r}F_L(\mathbf{r})-\frac{\partial}{\partial r}\phi_L(\mathbf{r})F_L(\mathbf{r}')\right]\delta(r-r'). \quad \text{(H.30)}
\end{aligned}
$$

The first two integrals above vanish for the same reasons as expression (H.28), with the result that the term involving $\delta'(r-r')$ in Eq. (H.27) is equivalent to $-\frac{1}{2}$ of the first term in that equation.

Thus we have,

$$
\begin{aligned}
[\nabla^2+E-V(\mathbf{r})]G(\mathbf{r},\mathbf{r}') &= \sum_L\left[\phi_L(\mathbf{r}')\frac{\partial}{\partial r}F_L(\mathbf{r})\right. \\
&\left.-\frac{\partial}{\partial r}\phi_L(\mathbf{r})F_L(\mathbf{r}')\right]\delta(r-r'). \quad \text{(H.31)}
\end{aligned}
$$

Substituting for ϕ_L and F_L in terms of C, S, $\bar{C}$, and $\bar{S}$ as was done in the evaluation of (H.28), and using again the relations, $C\bar{C}-\bar{C}^T C^T=0$, $S\bar{S}-\bar{S}^T S^T=0$, and $C\bar{S}-\bar{C}^T S^T=\delta_{L_1L_2}$, together with the Wronskian relation satisfied by the spherical Bessel and Hankel functions, we obtain,

$$
\begin{aligned}
[\nabla^2+E-V(\mathbf{r})]\sum_L\phi_L(\mathbf{r}_<)F_L(\mathbf{r}_>) &= \sum_L Y_L(\hat{r})Y_L(\hat{r'})\delta(r-r') \\
&= \delta(\mathbf{r}-\mathbf{r}'). \quad \text{(H.32)}
\end{aligned}
$$

Appendix I

Classical Strain and Stress

In this appendix, we provide a brief review of some fundamental concepts of strain and stress in classical mechanics. These concepts are useful in understanding the quantum-mechanical description of strain and stress given in Chapter 5.

I.1 Strain

The concepts of strain and stress are intimately interrelated in the mechanics of deformable bodies. Broadly speaking, *strain* provides a description of the purely geometric (or kinematic) aspects of the deformation of a non-rigid body, i.e., the change in the relative coordinates of two points in the body (elongation, contraction), while *stress* is concerned with the forces that cause the deformation. For example, the direct determination of different elastic moduli (elastic constants) of a solid follows from the response of the material to appropriately chosen forms of externally applied stress. Thus, the bulk modulus is related to the application of hydrostatic (uniform) pressure. Other elastic constants are related to different forms of stress, as is illustrated in further discussion.

On physical grounds, both strain and stress must be described by tensors, i.e., quantities that obey specific rules under coordinate transformations. (A brief description of tensors and the related concept of dyadics is given in Appendix J.) It is a common occurrence that stress along a given direction (tension), can produce not only an elongation in the same direction, but also a contraction in the perpendicular cross section of a material. Thus, strain and stress along a given direction can be associated with corresponding strains and stresses along other directions, necessitating the use of a tensor (matrix) description.

Consider a body deformed under the application of an external force,

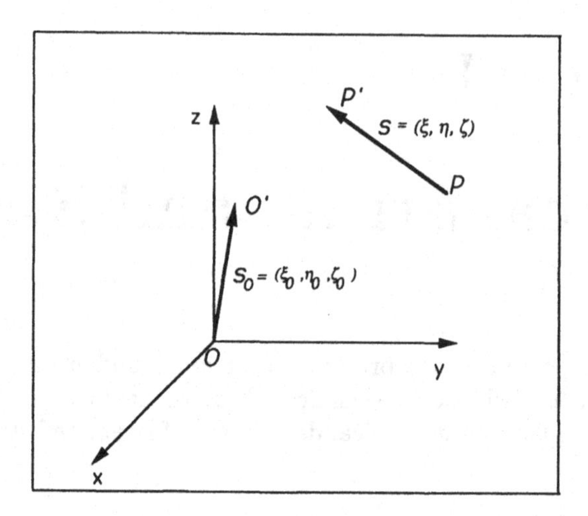

Figure I.1: **Schematic representation of strain effects on two points in a deformable body.**

with one point in the body rigidly fixed in place (a stretched rubber band with one end pinned in position). The amount of deformation will depend on the distance to the fixed point, even though the stress may well be about the same throughout the body. Thus, we must seek a description of strain and stress That is local in character, i.e., associate a strain and stress tensor[1] with every point (x, y, z) inside the body.

I.1.1 Strain tensor

Consider a point, P, in a volume element, with coordinates (x, y, z) with respect to the origin, O, of a space-fixed Cartesian coordinate system. In the presence of external forces, these points move to new positions, P' and O', as is indicated schematically in Fig. I.1. The change in position of the points can be described in terms of a *translation*, a *rotation*, and an *extension* (*contraction*) in three mutually orthogonal directions. Denote the displacement of point O by (ξ_0, η_0, ζ_0), (vector $\mathbf{s}_0$ in the figure), and that of point P by (ξ, η, ζ), (vector $\mathbf{s}$). Expanding the displacement of P in terms

[1] We will not be distinguishing between covariant and contravariant tensors, because the distinction is not relevant when Cartesian coordinates are employed.

of the space-fixed coordinates (x, y, z), we obtain,

$$\begin{aligned}\xi &= \xi_0 + \frac{\partial \xi}{\partial x}x + \frac{\partial \xi}{\partial y}y + \frac{\partial \xi}{\partial z}z + \cdots \\ \eta &= \eta_0 + \frac{\partial \eta}{\partial x}x + \frac{\partial \eta}{\partial y}y + \frac{\partial \eta}{\partial z}z + \cdots \\ \zeta &= \zeta_0 + \frac{\partial \zeta}{\partial x}x + \frac{\partial \zeta}{\partial y}y + \frac{\partial \zeta}{\partial z}z + \cdots .\end{aligned} \tag{I.1}$$

Introducing the definitions

$$\begin{aligned}a_{11} &= \frac{\partial \xi}{\partial x},\ a_{12} = \frac{\partial \xi}{\partial y},\ a_{13} = \frac{\partial \xi}{\partial z}, \\ a_{21} &= \frac{\partial \eta}{\partial x},\ a_{22} = \frac{\partial \eta}{\partial y}, \cdots ,\end{aligned} \tag{I.2}$$

and further writing the a_{ij} as sums of "odd" and "even" components

$$a_{ij} = \frac{a_{ij} - a_{ji}}{2} + \frac{a_{ij} + a_{ji}}{2}, \tag{I.3}$$

we can write the displacement vector **s** in the form,

$$\mathbf{s}(x, y, z) = \mathbf{s}_0 + \mathbf{s}_1(x, y, z) + \mathbf{s}_2(x, y, z). \tag{I.4}$$

Here, the vector,

$$\mathbf{s}_0 = \xi_0\mathbf{i} + \eta_0\mathbf{j} + \zeta_0\mathbf{k} \tag{I.5}$$

is independent of position and corresponds to a constant translation for every point P of the volume element. The vector $\mathbf{s}_1$ can be written in the form,

$$\mathbf{s}_1 = \mathbf{l} \times \mathbf{r}, \tag{I.6}$$

with

$$\mathbf{l} = \frac{a_{32} - a_{23}}{2}\mathbf{i} + \frac{a_{13} - a_{31}}{2}\mathbf{j} + \frac{a_{21} - a_{12}}{2}\mathbf{k}, \tag{I.7}$$

and corresponds to a rotation. Two such vectors would be sufficient for the description of the motion of a rigid object. The third vector, $\mathbf{s}_2(x, y, x)$ in Eq. (I.4), represents the displacement due to the distortion of the body (and vanishes if the body is perfectly rigid). Denoting its components by (ξ_2, η_2, ζ_2) we have,

$$\begin{aligned}\xi_2 &= \epsilon_{xx}x + \epsilon_{xy}y + \epsilon_{xz}z \\ \eta_2 &= \epsilon_{yx}x + \epsilon_{yy}y + \epsilon_{yz}z \\ \zeta_2 &= \epsilon_{zx}x + \epsilon_{zy}y + \epsilon_{zz}z\end{aligned} \tag{I.8}$$

where

$$\begin{aligned}
\epsilon_{xx} &= \frac{\partial \xi}{\partial x}, \quad \epsilon_{xy} = \epsilon_{yx} = \frac{1}{2}\left(\frac{\partial \xi}{\partial y} + \frac{\partial \eta}{\partial x}\right), \\
\epsilon_{yy} &= \frac{\partial \eta}{\partial y}, \quad \epsilon_{yz} = \epsilon_{zy} = \frac{1}{2}\left(\frac{\partial \eta}{\partial z} + \frac{\partial \zeta}{\partial y}\right), \\
\epsilon_{zz} &= \frac{\partial \zeta}{\partial z}, \quad \epsilon_{zx} = \epsilon_{xz} = \frac{1}{2}\left(\frac{\partial \zeta}{\partial x} + \frac{\partial \xi}{\partial z}\right).
\end{aligned} \tag{I.9}$$

The quantities $\epsilon(x, y, z)$, which clearly depend on the space-fixed coordinates, are the components of the *strain tensor* that can be written in matrix form,

$$\underline{\epsilon} = \begin{pmatrix} \epsilon_{xx} & \epsilon_{xy} & \epsilon_{xz} \\ \epsilon_{yx} & \epsilon_{yy} & \epsilon_{yz} \\ \epsilon_{zx} & \epsilon_{zy} & \epsilon_{zz} \end{pmatrix}. \tag{I.10}$$

Generally $\mathbf{s}_2$ is much smaller than $\mathbf{s}_0$ and $\mathbf{s}_1$. However, since we are interested in the internal warps and strains in a body, and not in its gross motion, we consider only $\mathbf{s}_2$, which from this point on we'll denote simply by $\mathbf{s}$.

An examination of Fig. I.1 and the ensuing discussion reveals that the displacement of point P is a function of its distance from the origin. Since the strain (and associated stress) in the body does not have this property, and is often the same throughout the extent of the material, we need to identify a quantity that reflects this feature. (In the case of a rubber band stretched within the elastic limit the fractional change in length, $\delta l/l$, is a constant independent of the length l being considered.) We will now show that the strain tensor provides such a local description [1].

At this point, it is convenient to cast the discussion above in terms of infinitesimals. Consider two points infinitesimally displaced from one another, so that we have $P_0(x, y, z)$ and $P_1(x + \mathrm{d}x, y + \mathrm{d}y, z + \mathrm{d}z)$. We seek the change in the relative position vector,

$$\mathbf{dr} = \mathbf{i}\mathrm{d}x + \mathbf{j}\mathrm{d}y + \mathbf{k}\mathrm{d}z, \tag{I.11}$$

in the presence of strain. By definition of the vector $\mathbf{s}$, point P_0 is displaced by $\mathbf{s}(x, y, z)$, and point P_1 by $\mathbf{s}(x + \mathrm{d}x, y + \mathrm{d}y, z + \mathrm{d}z)$. Thus, to first order the change in $\mathbf{dr}$ due to strain is given in accordance with Eq. (J.83) by $\mathbf{ds}$, where,

$$\begin{aligned}
\mathbf{ds} &= \mathrm{d}x\frac{\partial \mathbf{s}}{\partial x} + \mathrm{d}y\frac{\partial \mathbf{s}}{\partial y} + \mathrm{d}z\frac{\partial \mathbf{s}}{\partial z} \\
&= \mathbf{dr} \cdot \nabla \mathbf{s}.
\end{aligned} \tag{I.12}$$

To this quantity, we must also add any direct change of the vector **dr**, so that the new relative displacement vector between the points is given by,

$$\mathbf{dr}' = \mathbf{drds} = \mathbf{dr} + \mathbf{dr} \cdot \nabla \mathbf{s}. \tag{I.13}$$

We now realize that, at least formally, the ratio $\mathbf{ds}/\mathbf{dr}$ indeed describes the fractional change in length in the presence of strain. [It is as if the vector $\mathbf{s} = (\xi_2, \eta_2, \zeta_2)$ were divided by the vector $\mathbf{r} = (x, y, z)$ in Eq. (I.8).] The ratio of two vectors is an example of a *dyadic*, a quantity which under coordinate transformations transforms as indicated in Eq. (J.44) in Appendix J (roughly speaking as the product of vectors). Using the dyadic notation of Appendix J, we can write Eq. (I.13) in the form,

$$\mathbf{dr}' = \mathbf{dr}\left(\overleftrightarrow{I} + \overleftrightarrow{\epsilon}\right); \quad \overleftrightarrow{\epsilon} = \nabla \mathbf{s}. \tag{I.14}$$

The dyadic $\overleftrightarrow{\epsilon}$,

$$\overleftrightarrow{\epsilon} = \mathbf{i}\frac{\partial \mathbf{s}}{\partial x} + \mathbf{j}\frac{\partial \mathbf{s}}{\partial y} + \mathbf{k}\frac{\partial \mathbf{s}}{\partial z}, \tag{I.15}$$

measures the amount of strain at a point (x, y, z) in a space-fixed coordinate system. A number of formal manipulations can be most easily carried out and summarized in dyadic notation, and we will employ that approach in much of our discussion. For purposes of clarification, we display $\overleftrightarrow{\epsilon}$ in the notation of Eq. (J.64),

$$\overleftrightarrow{\epsilon} = \sum_{ij} \mathbf{e}_i \epsilon_{ij} \mathbf{e}_j, \tag{I.16}$$

where for simplicity in notation we denote the space-fixed coordinate axes by (x_1, x_2, x_3) rather than by (x, y, z), and the unit vectors along the axes by $\mathbf{e}_1, \mathbf{e}_2, \mathbf{e}_3$ rather than by $\mathbf{i}, \mathbf{j}, \mathbf{k}$. Here, we have defined $\epsilon_{ij} = \frac{\partial s_i}{\partial x_j}$. It is also convenient to explicitly exhibit the symmetric property[2] of the strain tensor and to write the strain dyadic in the form,

$$\begin{aligned} \overleftrightarrow{\epsilon} &= \mathbf{e}_1\epsilon_{11}\mathbf{e}_1 + \mathbf{e}_2\epsilon_{22}\mathbf{e}_2 + \mathbf{e}_3\epsilon_{33}\mathbf{e}_3 + \epsilon_{12}(\mathbf{e}_1 + \mathbf{e}_2) \\ &+ \epsilon_{13}(\mathbf{e}_1 + \mathbf{e}_3) + \epsilon_{23}(\mathbf{e}_2 + \mathbf{e}_3), \end{aligned} \tag{I.17}$$

where now we have made the definitions,

$$\epsilon_{ii} = \frac{\partial s_i}{\partial x_i}, \text{ and } \epsilon_{ij} = \frac{1}{2}\left(\frac{\partial s_i}{\partial x_j} + \frac{\partial s_j}{\partial x_i}\right). \tag{I.18}$$

This latter form corresponds exactly to the symmetric tensor in Eq. (I.10).

[2]There may also be an antisymmetric contribution to the strain, caused by twisting of the material as it is being strained, e.g., Eq. (I.7). We shall neglect this type of strain and explicitly treat the pure strain or symmetric part of the strain tensor.

The symmetric dyadic defined in the last two equations is called the *pure strain* dyadic for the point (x, y, z). Clearly, it is a function of position and provides a measure of the strain at a point (or fractional length change in the limit of vanishing length). When the pure strain dyadic vanishes at a point, there is no true strain in the medium at that point[3].

Given the tensor (matrix) nature of the strain dyadic, $\underline{\epsilon}$, in Eq. (I.10), it can be brought into diagonal form by means of a coordinate transformation. Consider the similarity transformation,

$$\underline{S}\underline{\epsilon}\underline{S}^{-1} = \underline{\epsilon}_{\mathrm{d}} \tag{I.19}$$

which diagonalizes $\underline{\epsilon}$, with

$$\underline{\epsilon}_{\mathrm{d}} = \begin{pmatrix} \epsilon_1 & 0 & 0 \\ 0 & \epsilon_2 & 0 \\ 0 & 0 & \epsilon_3 \end{pmatrix}. \tag{I.20}$$

The matrix $\underline{S}$ corresponds to a rotation that depends on position and changes the orthonormal basis $\{\mathbf{e}_i\}$ to a basis $\{\mathbf{e}'_i\}$ in terms of which we can write,

$$\overleftrightarrow{\epsilon} = \mathbf{e}'_1\epsilon_1\mathbf{e}'_1 + \mathbf{e}'_2\epsilon_2\mathbf{e}'_2 + \mathbf{e}'_3\epsilon_3\mathbf{e}'_3. \tag{I.21}$$

The three quantities ϵ_1, ϵ_2, and ϵ_3 are called the ***principal extensions*** of the medium at the point (x, y, z). The coordinate axes that are defined by the $\{\mathbf{e}'_i\}$ are called the ***principal axes***[4]. A rectangular parallelepiped of sides $\mathrm{d}\xi_1, \mathrm{d}\xi_2, \mathrm{d}\xi_3$ parallel to $\mathbf{e}'_1, \mathbf{e}'_2, \mathbf{e}'_3$, respectively, would remain a rectangular parallelepiped if the medium were strained at the point (x, y, z). However, under the strain, the length of the sides of the parallelepiped would change to $(1+\epsilon_1)\mathrm{d}\xi_1, (1+\epsilon_2)\mathrm{d}\xi_2, (1+\epsilon_3)\mathrm{d}\xi_3$. Thus, to first order, the fractional change in ***volume*** under strain is given by,

$$\theta = (\epsilon_1 + \epsilon_2 + \epsilon_3) = (\epsilon_{xx} + \epsilon_{yy} + \epsilon_{zz}) = \mathrm{Tr}\,\mathbf{s} \equiv |\overleftrightarrow{\epsilon}|. \tag{I.22}$$

This quantity is an invariant of the stress tensor, Eq. (J.49), reflecting the invariance of the trace of a matrix under similarity transformations. The quantity θ is called the ***dilation*** of the medium at the point (x, y, z).

I.1.2 Types of strain

The simplest type of strain is that which is independent of position; it is called a ***homogeneous strain***. The simplest type of homogeneous strain is

[3]In the following discussion we use either a double-pointed arrow over the symbol or an underline to denote a tensor or matrix quantity.

[4]Recall the definition of principal axes in the case of the inertia tensor in classical mechanics. Tensor quantities are also indicated by an underline, especially when they can be written as ordinary matrices.

simple expansion, which is described by a displacement vector **s** and a strain dyadic $\overleftrightarrow{\epsilon}$ of the forms,

$$\mathbf{s} = \epsilon(x\mathbf{i} + y\mathbf{j} + z\mathbf{k}); \quad \overleftrightarrow{\epsilon} = \epsilon\overleftrightarrow{I} = \epsilon[\mathbf{ii} + \mathbf{jj} + \mathbf{kk}]. \tag{I.23}$$

This type of strain is isotropic, with any set of axes being principal axes. Clearly, there is no rotation due to this type of strain. The dilation is $\theta = 3\epsilon$.

When the extension along a principal axis is equal and opposite to that along another, with the extension along the third axis being zero, we obtain another kind of homogeneous strain called *simple shear*. It is described by a displacement vector and a strain dyadic of the form,

$$\mathbf{s} = \frac{1}{2}\epsilon(\mathbf{i}x - \mathbf{j}y); \quad \overleftrightarrow{\epsilon} = \frac{1}{2}\epsilon(\mathbf{ii} - \mathbf{jj}). \tag{I.24}$$

It is easily checked that the dilation equals zero, the extension along the x-axis being canceled by the contraction along the y-axis.

A *pure shear* is described by a displacement vector and strain dyadic of the form,

$$\mathbf{s} = \frac{1}{2}\epsilon(\mathbf{j}x + \mathbf{i}y); \quad \overleftrightarrow{\epsilon} = \frac{1}{2}\epsilon(\mathbf{ij} + \mathbf{ji}). \tag{I.25}$$

A combination of pure shear and a rigid body rotation of the medium by $\frac{1}{2}\epsilon$ radians (note that ϵ is dimensionless), corresponding to the dyadic $-\frac{1}{2}\epsilon(\mathbf{ij} - \mathbf{ji})$, yields a *pure shear in the y direction*, with

$$\mathbf{s} = \epsilon y\mathbf{i}; \quad \overleftrightarrow{\epsilon} = \epsilon\mathbf{ji}. \tag{I.26}$$

A *dilationless stretch* corresponds to a homogeneous strain that leads to an extension along a single axis, say the x-axis, with a contraction along the y- and z-axes so that there is no net change in volume. It can be easily checked that the following displacement vector and strain dyadic describe a dilationless, $\theta = 0$, stretch,

$$\mathbf{s} = \epsilon\left[\mathbf{i}x - \frac{1}{2}(\mathbf{j}y + \mathbf{k}z)\right]; \quad \overleftrightarrow{\epsilon} = \epsilon\left[\mathbf{ii} - \frac{1}{2}(\mathbf{jj} + \mathbf{kk})\right]. \tag{I.27}$$

The most general type of homogeneous strain, referred to its principal axes, corresponds [2] to combinations of various amounts of simple expansion, simple shear, and dilationless stretch in three directions.

I.1.3 Strain transformations

One important use of the strain tensor is in the evaluation of functions under a strain transformation of the coordinates. Let $f(\mathbf{r})$ be a function of the position vector $\mathbf{r}$ and consider the function $f(\mathbf{r}')$, where $\mathbf{r}' = (1 + \overleftrightarrow{\epsilon})\mathbf{r}$ denotes a coordinate transformation in terms of the strain tensor $\overleftrightarrow{\epsilon}$. We have,

$$\frac{\partial f(x,y,z)}{\partial x} = \frac{\partial f(x,y,z)}{\partial x'}\frac{\partial x'}{\partial x} + \frac{\partial f(x,y,z)}{\partial y'}\frac{\partial y'}{\partial x} + \frac{\partial f(x,y,z)}{\partial z'}\frac{\partial z'}{\partial x} \tag{I.28}$$

so that,

$$\begin{aligned}
\nabla f &= \mathbf{i}\frac{\partial f}{\partial x} + \mathbf{j}\frac{\partial f}{\partial y} + \mathbf{k}\frac{\partial f}{\partial z} \\
&= \mathbf{i}\left[(1+\epsilon_{11})\frac{\partial f}{\partial x'} + \epsilon_{12}\frac{\partial f}{\partial y'} + \epsilon_{13}\frac{\partial f}{\partial z'}\right] \\
&+ \mathbf{j}\left[\epsilon_{21}\frac{\partial f}{\partial x'} + (1+\epsilon_{22})\frac{\partial f}{\partial y'} + \epsilon_{23}\frac{\partial f}{\partial z'}\right] \\
&+ \mathbf{k}\left[\epsilon_{31}\frac{\partial f}{\partial x'} + \epsilon_{32}\frac{\partial f}{\partial y'} + (1+\epsilon_{33})\frac{\partial f}{\partial z'}\right] \\
&= \overleftrightarrow{\epsilon}\cdot\nabla' f.
\end{aligned} \tag{I.29}$$

Since the function is arbitrary, we have the general result,

$$\nabla = (\overleftrightarrow{I} + \overleftrightarrow{\epsilon})\cdot\nabla'. \tag{I.30}$$

I.1.4 Some formal results involving strain

The study of stress within quantum mechanics is based on the use of wave functions (or density matrices) evaluated under the strain transformation, $\mathbf{r}' \to \{1 + \lambda\overleftrightarrow{\epsilon}\}\mathbf{r}$. The following three formal results are useful in carrying out the corresponding analysis.

First, we show that,

$$\frac{\partial}{\partial\lambda}\det(1+\lambda\overleftrightarrow{\epsilon})|_{\lambda=0} = \mathrm{Tr}\overleftrightarrow{\epsilon} = \theta. \tag{I.31}$$

To prove this result, note that it holds in the case in which the coordinate system is referred to the principal axes at a point. Now, apply a similarity transformation to both sides of the equation, corresponding to an arbitrary rotation of the coordinate system. Under this transformation, the matrix $\underline{\epsilon}$, or the dyadic $\overleftrightarrow{\epsilon}$, assumes a general, non-diagonal form, while its trace remains invariant.

Second, we note that for any function $\Phi(\mathbf{r}_1, \mathbf{r}_2, \cdots, \mathbf{r}_N)$ of the N coordinates $\{\mathbf{r}_1, \mathbf{r}_2, \cdots, \mathbf{r}_N\}$, we have

$$\frac{\partial}{\partial\lambda}\Phi(\mathbf{r}_1', \mathbf{r}_2', \cdots, \mathbf{r}_N')|_{\lambda=0} = \mathrm{i}\sum_i \mathbf{r}_i\overleftrightarrow{\epsilon}\mathbf{p}_i\Phi(\mathbf{r}_1, \mathbf{r}_2, \cdots, \mathbf{r}_N), \tag{I.32}$$

where $\mathbf{p}_i$ is the quantum mechanical momentum operator (with $\hbar = 1$). This result follows upon a Taylor series expansion of Φ about the old coordinates, and the definition of the stress tensor based on Eq. (I.30).

The third result,

$$\mathbf{p}_i \overleftrightarrow{\epsilon} \mathbf{r}_i - \mathbf{r}_i \overleftrightarrow{\epsilon} \mathbf{p}_i = -i\mathrm{Tr}\overleftrightarrow{\epsilon}, \tag{I.33}$$

follows with equal ease. Write vectors on the left (right) in row (column) form, with the momentum vector expressed in terms of spatial derivatives, and apply the resulting expression to an arbitrary function, $f(\mathbf{r})$. The result is Eq. (I.33). Finally, for future reference we note that,

$$\begin{aligned} \nabla \mathbf{r} &= (\mathbf{i}\frac{\partial}{\partial x} + \mathbf{j}\frac{\partial}{\partial y} + \mathbf{k}\frac{\partial}{\partial z})(\mathbf{i}x + \mathbf{j}y + \mathbf{k}z) \\ &= (\mathbf{ii} + \mathbf{jj} + \mathbf{kk}) \\ &= \overleftrightarrow{I}. \end{aligned} \tag{I.34}$$

I.2 Stress

Strains are produced by forces inside a deformable medium. These forces are called *stresses*. It can be shown [1, 3] that stress can be described by a symmetric tensor of the second rank, and hence also by a dyadic [2].

In order to see how stress is related to strain, let us consider a volume element in the shape of a parallelepiped with sides Δx, Δy, and Δz, and faces perpendicular to the axes of an orthogonal, right-handed coordinates system, Fig. I.2. Let this volume element be in equilibrium under any forces acting on it. Let the forces per unit area acting on the infinitesimal area perpendicular to the x-axis have components σ_{xx}, σ_{yx}, and σ_{zx}. In this notation, the first subscript denotes the direction of the force component, while the second denotes the face on which the force is acting. Thus, σ_{yx} is the force per unit area acting on the face perpendicular to the x-axis and pointing along the y-direction. A similar description applies to the other faces. The diagonal components, e.g., σ_{xx}, are also referred to as *tension* or *compression*, depending on whether they would tend to stretch or compress the body. The off-diagonal components, e.g., σ_{yx}, are tangential to the surface element and are called *shear stresses* because they would tend to produce a shear strain. Now, collecting the stresses acting on the three faces with outward normals along the positive directions of the coordinate axes, we can use the tensor matrix,

$$\underline{\sigma} = \begin{pmatrix} \sigma_{xx} & \sigma_{xy} & \sigma_{xz} \\ \sigma_{yx} & \sigma_{yy} & \sigma_{yz} \\ \sigma_{zx} & \sigma_{zy} & \sigma_{zz} \end{pmatrix}, \tag{I.35}$$

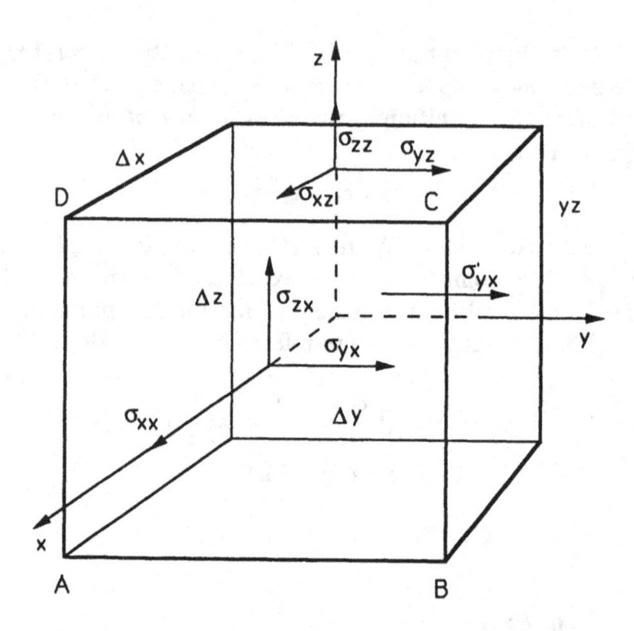

Figure I.2: **Stresses on the phase of a volume element.**

to describe the stress on the volume element. Note that we took into account only the three faces with outward normals along the positive directions of the coordinate axes. These suffice to describe the stress on the volume element, because the stress on the opposite faces is immediately given in terms of $\underline{\sigma}$. For example, consider σ'_{yx}, the shear stress in the $y-$direction acting along the surface whose normal points along the negative x-axis, Fig. I.2. Because the volume element is assumed to be in equilibrium under the forces exerted on it on all faces by the surrounding material, and the condition of equilibrium must be maintained as the volume shrinks to zero (as the width along any direction vanishes), we must have,

$$\sigma'_{yx} = -\sigma_{yx}. \tag{I.36}$$

Similar expressions hold for the other components of the stress.

We now show that the stress tensor is symmetric. The conditions of static equilibrium imposed on the volume element $\Delta x \Delta y \Delta z = \Delta \tau$ require that the moments of all forces acting on the element vanish. The force associated with the shear stress σ_{yx} is equal to $\sigma_{yx}\Delta y \Delta z$, and its moment arm about the z-axis, with the origin of coordinates at the center of the volume element, is equal to $\frac{1}{2}\Delta x$. Thus, its moment about the z-axis is equal to $\frac{1}{2}\sigma_{yx}\Delta\tau$. This moment is balanced exactly by $\frac{1}{2}\sigma_{xy}\Delta\tau$, so that[5],

$$\sigma_{yx} = \sigma_{xy}. \tag{I.37}$$

Similar considerations for the other components of the stress tensor show that $\underline{\sigma}$ is symmetric

$$\sigma_{\alpha\beta} = \sigma_{\beta\alpha}, \quad \alpha,\ \beta = x, y, z. \tag{I.38}$$

Therefore, $\underline{\sigma}$ has six independent components.

That the matrix $\underline{\sigma}$ is indeed a tensor, i.e., under coordinate transformations it follows the rule indicated in Eq. (J.44), can be shown by imposing the conditions of static equilibrium on a tetrahedron with three edges along the coordinate axes of an orthogonal Cartesian coordinate system, and a slant face with arbitrary orientation [1]. The proof is straightforward, and is given in detail in reference [1]. Thus, $\underline{\sigma}$ is indeed a symmetric tensor, referred to as the *stress tensor*.

Stresses can be readily described in terms of dyadics. The force acting on an element of area dA is given by the expression,

$$\mathbf{F} = \mathrm{d}\mathbf{A} \cdot \overleftrightarrow{T}, \tag{I.39}$$

where d$\mathbf{A}$ is the axial vector representing the surface, and $\overleftrightarrow{T}$ is the dyadic,

$$\overleftrightarrow{T} = \mathbf{F}_x\mathbf{i} + \mathbf{F}_y\mathbf{j} + \mathbf{F}_z\mathbf{k}. \tag{I.40}$$

We note that the component forces, $\mathbf{F}_x$, $\mathbf{F}_y$, and $\mathbf{F}_z$ are *vectors*, and *not* the components of a vector along the axes. In terms of the volume element in Fig. I.2, $\mathbf{F}_x$ would be the vector force on the surface with outward normal along the positive x-axis. Moment cancellation, as above, shows that the stress dyadic is symmetric,

$$(\mathbf{F}_x)_y = (\mathbf{F}_y)_x, \tag{I.41}$$

and, therefore, equal to its conjugate,

$$\overleftrightarrow{T}^* = \mathbf{i}\mathbf{F}_x + \mathbf{j}\mathbf{F}_y + \mathbf{k}\mathbf{F}_z. \tag{I.42}$$

Clearly, the components of the stress dyadic are given by $\underline{\sigma}$,

$$T_{\alpha\beta} \equiv \sigma_{\alpha\beta}, \quad \alpha, \beta = x, y, z. \tag{I.43}$$

---

[5] We note that the normal stresses do not contribute to the moment because their lines of action pass through the origin. Also, in view of Eq. (I.36), the moments set up by the forces on the faces with negatively directed outward normals cancel in pairs as well. Finally, in setting up the conditions of equilibrium we have not taken into account other forces internal to the volume element, such as gravitational forces. However such forces would be proportional to the volume of the element and become negligible much faster than stress as the volume vanishes.

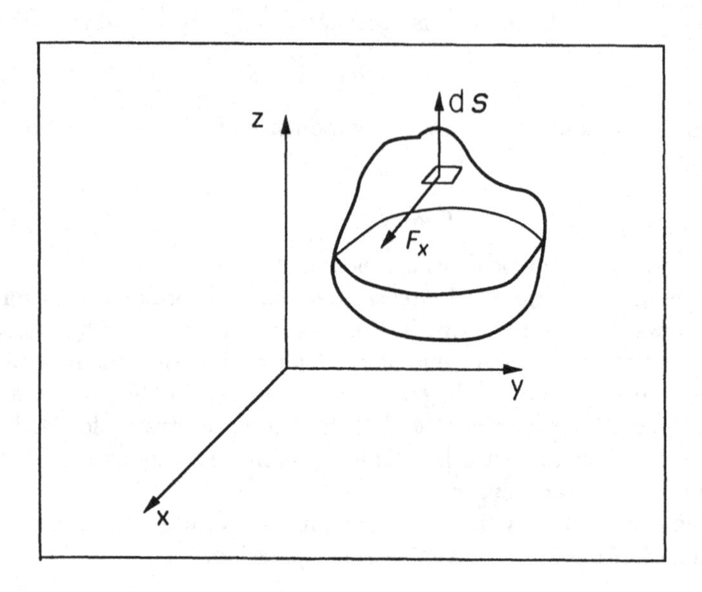

Figure I.3: **Force on a volume element.**

I.2.1 Force on a volume element

It follows from our discussion that the stress tensor can be used to determine the force acting on a volume element inside a material. Let us consider a volume V bounded by a surface S and being acted upon by internal forces, as depicted schematically in Fig. I.3. Let $\mathrm{d}F_x$ be the net force in the x direction, acting on a surface element with outward normal $\mathrm{d}\mathbf{S} = \mathrm{d}S_x\mathbf{i} + \mathrm{d}S_y\mathbf{j} + \mathrm{d}S_z\mathbf{k}$. As we saw above, this force is given by the expression,

$$\mathrm{d}F_x = \sigma_{xx}\mathrm{d}S_x + \sigma_{xy}\mathrm{d}S_y + \sigma_{xz}\mathrm{d}S_z. \tag{I.44}$$

The total force on the volume element in the $x-$direction is given by the integral,

$$\begin{aligned} F_x &= \int\!\!\int_S \sigma_{xx}\mathrm{d}S_x + \sigma_{xy}\mathrm{d}S_y + \sigma_{xz}\mathrm{d}S_z \\ &= \int\!\!\int_S \boldsymbol{\sigma}_x \cdot \mathrm{d}\mathbf{S}, \end{aligned} \tag{I.45}$$

where

$$\boldsymbol{\sigma}_x = \sigma_{xx}\mathbf{i} + \sigma_{yx}\mathbf{j} + \sigma_{zx}\mathbf{k}. \tag{I.46}$$

Similar expressions hold for σ_y and σ_z. Thus the force, **F**, on the volume element, V, is given by the integral,

$$\mathbf{F}_V = \int_V \mathbf{F} dV = \int\int_S \overleftrightarrow{\sigma} \cdot d\mathbf{S}. \tag{I.47}$$

This expression can be cast in two different forms as follows. Let the volume V shrink to an infinitesimal, $d\tau$, and note that from the last equality in Eq. (I.47) we have,

$$\mathbf{F} = \lim_{d\tau \to 0} \int\int_S \overleftrightarrow{\sigma} \cdot d\mathbf{S}. \tag{I.48}$$

Now, as shown in Eq. (J.90), the right-hand side of this expression defines the divergence of the tensor $\overleftrightarrow{\sigma}$, so that we have,

$$\mathrm{Div}\overleftrightarrow{\sigma} = \mathbf{F}. \tag{I.49}$$

It is to be emphasized that the quantity **F** is the force per unit volume acting at the point (x, y, z). The force on a volume element is given either as a volume integral of **F**, or of the divergence of the strain tensor, or as a surface integral of the stress tensor itself over the surface of the volume,

$$\mathbf{F}_V = \int_V \mathbf{F} d^3 r = \int_V d^3 r \mathrm{Div}\overleftrightarrow{\sigma} = \int\int_S \overleftrightarrow{\sigma} \cdot d\mathbf{S}. \tag{I.50}$$

We point out that Eq. (I.48) relates the divergence of the stress tensor to the force per unit volume in the material, and it is based on the conditions of static equilibrium. However, because it gives only three equations, it is not sufficient to determine the six components of the stress tensor. In fact, there is an added inherent ambiguity, since Eq. (I.49) remains valid when the curl of any tensor field is added to $\overleftrightarrow{\sigma}$[6]. This ambiguity cannot be removed by considering the conditions of moment equilibrium, because these conditions have been used in determining the symmetry of the stress field. The ambiguity is related to the question of gauge (i.e., in the choice of the tensor whose curl is added to the stress field), and persist in the quantum mechanical case. In the case in which this tensor is set equal to zero, the stress can be determined from Eq. (I.49) used in connection with the strain-stress relations [1] discussed in the following section.

I.3 Strain-Stress Relations for Isotropic Media

In this section, we discuss the connection of strain, i.e., deformation, to the applied stress causing it, and derive expressions for the elastic energy

[6]The divergence of the curl of a tensor vanishes as it does for vectors. See discussion in Appendix J.

stored in a body deformed by stress. We develop the formalism both for the case of continuous materials and for crystals.

Hooke was the first to suggest the existence of a linear relation between strain and stress, valid to first order in the stress or strain. The regime in which this law of linearity holds is called the *elastic regime.* In the following discussion, we assume that the limit of the linear law—the *elastic limit*–is not exceeded. Also, we assume that the material is *isotropic*, so that its physical properties are the same in all three directions. This assumption can be expected to be valid when large volumes of material are considered, containing a great number of microcrystals (grains). The case of macroscopic crystals is discussed in a following section.

The assumed linearity of the strain-stress relation implies that we can define *principal stresses*, σ_α, $\alpha = 1, 2, 3$, which are connected to the principal strains, ϵ_α, by relations of the form,

$$\sigma_1 = a\epsilon_1 + b\epsilon_2 + c\epsilon_3. \tag{I.51}$$

Similar expressions are obtained for the other two principal stresses under a cyclic rotation of the indices. Since the principal strain axes for directions 2 and 3 are equivalent as far as direction 1 is concerned, we must have $b = c$. The same symmetry argument applies to the other principal directions so that, within the elastic limit, a parallelepiped formed along the axes of principal strain will remain a parallelepiped under stress. Thus, the principal axes for stress coincide with those for strain. Now, adding $\pm b\epsilon_2$ on the right of the equals sign in the last equation, we can rewrite it in the form,

$$\begin{aligned}\sigma_1 &= (a-b)\epsilon_1 + b(\epsilon_1 + \epsilon_2 + \epsilon_3) \\ &= 2\mu\epsilon_1 + \lambda\theta.\end{aligned} \tag{I.52}$$

The constants μ and λ are known as *Lamé's constants*, or as *Lamé moduli.* Realizing that θ is the trace of the strain tensor, (the fractional change in volume), and therefore invariant under a transformation of coordinates, one can show [1] that in an arbitrary, orthogonal coordinate system whose axes are not parallel to the principal strain axes, we obtain,

$$\sigma_{\alpha\beta} = 2\mu\epsilon_{\alpha\beta} + \lambda\theta\delta_{\ alpha\beta}. \tag{I.53}$$

This is the general form of the strain-stress relation for an arbitrary volume element in an isotropic medium. These relations can be inverted to yield expressions for strain in terms of stress,

$$\epsilon_{\alpha\beta} = 2\mu'\sigma_{\alpha\beta} + \lambda'\Sigma\delta_{\alpha\beta}, \tag{I.54}$$

where

$$2\mu' = \frac{1}{2\mu}, \quad \lambda' = -\frac{1}{2\mu}\frac{\lambda}{2\mu + 3\lambda}, \quad \Sigma = \sigma_1 + \sigma_2 + \sigma_3. \tag{I.55}$$

It is convenient to express the strain-stress relations in terms of parameters whose physical meaning is somewhat more transparent than that of the Lamé moduli. To this end, consider a vertical bar in the shape of a prism of length l, subjected to a load P uniformly distributed over its cross-section, A. The resulting tension,

$$\sigma = \frac{P}{A}, \tag{I.56}$$

is a principal stress because no shear stress is applied. The extension per unit length resulting from the stress is a principal extension,

$$\epsilon = \frac{\Delta l}{l}, \tag{I.57}$$

where Δl is the elongation of the bar under load conditions. The ratio of stress to strain is a *constant of the material*, that is, it is independent of P, Δl, and l, as long as the elastic limit is not exceeded. This ratio is called *Young's modulus*, and is often denoted by Y,

$$Y = \frac{\sigma}{\epsilon}. \tag{I.58}$$

In addition to elongation in the direction of strain, the cross section of a loaded bar contracts uniformly under tension. Denoting this contraction by $-\epsilon'$, ($\epsilon' > 0$), we define *Poisson's ratio* by the relation,

$$\nu = -\frac{\epsilon'}{\epsilon}, \tag{I.59}$$

so that, using Eq. (I.58), the fractional contraction is given in terms of the fractional elongation in the form,

$$\epsilon' = -\frac{\nu}{Y}\sigma. \tag{I.60}$$

Young's modulus and Poisson's ratio can be used to rewrite the strain-stress relation, Eq. (I.53). Consider again the case of stresses confined to the principal axes. Because of the linearity of all occurring relations, stresses and associated strains can be added. Thus, strain ϵ_1 is caused not only by stress σ_1, but also by the stresses acting along the other two principal axes. Using the definition of Young's modulus and Poisson's ratio, Eqs. (I.58) and (I.59), we have,

$$\epsilon_1 = \frac{\sigma_1}{Y} - \frac{\nu}{Y}(\sigma_2 + \sigma_3) = \frac{1+\nu}{Y}\sigma_1 - \frac{\nu}{Y}(\sigma_1 + \sigma_2 + \sigma_3). \tag{I.61}$$

In general, we have,

$$\epsilon_i = \frac{1+\nu}{Y}\sigma_1 - \frac{\nu}{Y}\Sigma, \tag{I.62}$$

and

$$\sigma_i = \frac{Y}{1+\nu}\left(\epsilon_i + \frac{\nu}{1-2\nu}\theta\right). \tag{I.63}$$

A comparison between Eqs. (I.52) and (I.63) yields the relations,

$$2\mu = \frac{Y}{1+\nu}, \quad \lambda = \frac{\nu Y}{(1+\nu)(1-2\nu)}. \tag{I.64}$$

As a way of illustration, let us consider the case of *equal pressure in all directions*. Denoting the pressure by p, we have,

$$\sigma_1 = \sigma_2 = \sigma_3 = -p, \quad \Sigma = -3p, \quad \epsilon_1 = \epsilon_2 = \epsilon_3 = -\frac{1}{3}|\theta|. \tag{I.65}$$

Equation (I.63) then yields,

$$p = \frac{Y|\theta|}{3(1-2\nu)}. \tag{I.66}$$

In analogy with Young's modulus, Eq. (I.58), we define the *modulus of compressibility*,

$$K = \frac{p}{|\theta|}, \tag{I.67}$$

which, in view of Eq. (I.66) becomes,

$$K = \frac{Y}{3(1-2\nu)}. \tag{I.68}$$

I.3.1 Strain-stress relations in dyadic notation

The strain-stress relations established above can be readily expressed in dyadic notation. Defining the strain and stress dyadics in terms of orthogonal unit vectors[7] $\mathbf{a}_i$, $i = 1, 2, 3$, we can write

$$\overleftrightarrow{\epsilon} = \epsilon_{11}\mathbf{a}_1\mathbf{a}_1 + \cdots + \epsilon_{12}(\mathbf{a}_1\mathbf{a}_2 + \mathbf{a}_2\mathbf{a}_1) + \cdots, \tag{I.69}$$

and

$$\overleftrightarrow{\sigma} = \sigma_{11}\mathbf{a}_1\mathbf{a}_1 + \cdots + \sigma_{12}(\mathbf{a}_1\mathbf{a}_2 + \mathbf{a}_2\mathbf{a}_1) + \cdots. \tag{I.70}$$

Then, Eq. (I.53) takes the form,

$$\overleftrightarrow{\sigma} = 2\mu\overleftrightarrow{\epsilon} + \lambda|\overleftrightarrow{\sigma}|\overleftrightarrow{I}. \tag{I.71}$$

The other relations established above can be expressed in similar fashion.

[7]See reference [1] for a discussion of the transformation of these and related expressions to curvilinear coordinates.

I.4 Energetics of Strain and Stress for Isotropic Media

In this section, we consider the energetic aspects of a body in a state of stress. Our aim is to derive expressions for the *strain energy* function per unit volume, or the *elastic potential.* This, in turn, will lead directly to a definition of the elastic constants of a material. The development proceeds again in terms of space-fixed coordinates, (x, y, z), and displacement coordinates, (ξ, η, ζ), as in Eq. (I.1).

Let us consider a volume element $\Delta\tau = \Delta x\Delta y\Delta z$ in the shape of a parallelepiped, at the position (x, y, z) in a body under stress. An external force per unit volume, $\mathbf{F}$, at (x, y, z) produces an increment in the displacement,

$$\xi \rightarrow \xi + \mathrm{d}\xi, \quad \eta \rightarrow \eta + \mathrm{d}\eta, \quad \zeta \rightarrow \zeta + \mathrm{d}\zeta. \tag{I.72}$$

The work done by the stress alone on the surface $\Delta y\Delta z$ at (x, y, z) is given by the expression,

$$\mathrm{d}W_x(x, y, z) = -[\sigma_{xx}\mathrm{d}\xi + \sigma_{xy}\mathrm{d}\eta + \sigma_{xz}\mathrm{d}\zeta]\Delta y\Delta z, \tag{I.73}$$

where $-\sigma_{xx}\Delta y\Delta z$ is the force in the positive x−direction on a surface whose outward normal points along the negative x-direction, with similar expressions for the other stress components. The work done on the corresponding face at $(x + \Delta x, y, z)$ with an outward normal along the positive x−direction, is given to first order by the expression,

$$\begin{aligned}\mathrm{d}W_x(x + \mathrm{d}x, y, z) = & - \mathrm{d}W_x(x, y, z) + \frac{\partial}{\partial x}[\mathrm{d}W_x(x, y, z)]\Delta x \\ & = (\sigma_{xx}\mathrm{d}\xi + \sigma_{xy}\mathrm{d}\eta + \sigma_{xz}\mathrm{d}\zeta)\Delta y\Delta z \\ & + \frac{\partial}{\partial x}(\sigma_{xx}\mathrm{d}\xi + \sigma_{xy}\mathrm{d}\eta + \sigma_{xz}\mathrm{d}\zeta)\Delta x\Delta y\Delta z.\end{aligned} \tag{I.74}$$

The net work done by the stresses as well as by the external force along the x-direction, $\mathrm{d}W_x(x, y, z) + \mathrm{d}W_x(x + \mathrm{d}x, y, z)$, becomes,

$$\left[\frac{\partial}{\partial x}(\sigma_{xx}\mathrm{d}\xi + \sigma_{xy}\mathrm{d}\eta + \sigma_{xz}\mathrm{d}\zeta) + F_x\mathrm{d}\xi\right]\Delta x\Delta y\Delta z. \tag{I.75}$$

Similar expressions hold for the y− and z−directions. Adding the contributions from all three directions, and carrying out the indicated differentiations, the terms involving the partial derivatives of the stress tensor cancel against the force terms, in view of the conditions of equilibrium, Eq. (I.49). Furthermore, assuming that $\partial\mathrm{d}\xi = \mathrm{d}\partial\xi$, i.e., that the gradient of the incremental displacement commutes with the increment of the gradient of the displacement, and recalling the definition of the strain tensor, Eq. (I.9), we obtain for the work per unit volume on the element $\Delta\tau$ the expression,

$$dW(x,y,z) = \sum_{i,k} \sigma_{ik} d\epsilon_{ik}, \tag{I.76}$$

where $i,k = (x,y,z)$. By means of the strain-stress relations, the last expression can be written entirely either in terms of the strain or the stress tensor. For example, in view of Eq. (I.53), we obtain for the *strain energy* per unit volume,

$$dW(x,y,z) = 2\mu \sum_{i,k} \epsilon_{ik} d\epsilon_{ik} + \lambda\theta d\theta. \tag{I.77}$$

This is a *total differential* provided that the constants λ and μ are independent of the process of loading. In other words, the strain energy per unit volume is a *function of state* of the material, and not of the path (loading process) that led to that state. In this case, the expression for dW can be integrated to yield the strain energy per unit volume stored at position (x,y,z),

$$\begin{aligned} W(x,y,z) &= \mu \sum_{ik} \epsilon_{ik}^2 + \frac{\lambda}{2}\theta^2 \\ &= \frac{1}{2}\sum_{ik} \sigma_{ik}\epsilon_{ik} \\ &= \frac{1}{2}\mathrm{Tr}\underline{\sigma}\underline{\epsilon}. \end{aligned} \tag{I.78}$$

Because of the symmetry of the strain (stress) tensors, W is the trace of the product of two matrices and, therefore, is invariant under coordinate transformations. Also, even though in principle the strain energy density is a function of position, for an isotropic, uniform system W is a constant. We note also the prefactor of $\frac{1}{2}$, providing a connection with Hooke's law, i.e., the assumed linear relation between strain and stress.

As was the case for the differential quantity, dW, the expression for W can be written entirely in terms of the strain tensor, $W(\underline{\epsilon})$, or the stress tensor, $W(\underline{\sigma})$. From these expressions, one can again obtain the components of both the strain and stress tensors,

$$\epsilon_{ik} = \frac{\partial W(\underline{\sigma})}{\partial \sigma_{ik}}, \quad \sigma_{ik} = \frac{\partial W(\underline{\epsilon})}{\partial \epsilon_{ik}}. \tag{I.79}$$

If, as is common practice, one is concerned with the reaction stresses, i.e., the reaction of a system to external forces, then a minus sign must be introduced on the right side of these equations. The existence of these relations justifies the use of the term ***elastic potential*** in referring to W. Expressions analogous to these are used in calculating the elastic properties of materials, as is described more fully in the next section.

In closing the discussion of the elastic properties of an isotropic body, we should mention an important physical property of the elastic moduli. Both μ and λ are in principle temperature dependent, and so are the elastic moduli of single crystals, discussed in the next section. In addition, the elastic moduli depend on the process followed in applying the load, whether this is done isothermally or adiabatically, that is without heat exchange with individual volume elements of the body. In an adiabatic process, W is still a function of state, but the elastic moduli are different from the isothermal ones. In the general case, when the load is applied in the presence of a non-uniform temperature distribution, W is *not* a function of state but depends on the path followed. In this case, the elastic potential becomes ill-defined.

I.5 Strain-Stress Relations for Crystals

An isotropic material is a collection of a very great number of microcrystals, or grains. As discussed in the previous two sections, the elastic behavior of such an object can be described in terms of two elastic moduli. In a microcrystal, the situation becomes somewhat more complex.

Whereas isotropic bodies remain invariant under infinitesimal rotations about any given point, crystals satisfy well-defined symmetry properties under rotations. There are 32 groups of point symmetry, which make up the 7 crystal systems, i.e., ***the triclinic, monoclinic, orthorhombic, tetragonal, trigonal, hexagonal, and cubic systems.*** The state of symmetry of a physical system in a physical process is superimposed on the symmetry of the crystal. In other words, the elastic moduli must satisfy relations that reflect the point group symmetry of the crystal under consideration.

We have seen that both strain and stress are symmetric tensors, each with six independent components. If W is a function of state, then it is a quadratic form in a 6-dimensional space whose full description requires,

$$\frac{1}{2}6(6+1) = 21, \tag{I.80}$$

independent coefficients. This number can be reduced further by consideration of the symmetry properties of individual crystal types.

In the following discussion, it will be convenient to introduce quantities with a single subscript to denote the components of the strain and stress tensors. We define,

$$\epsilon_1 = \epsilon_{xx},\ \epsilon_2 = \epsilon_{yy},\ \epsilon_3 = \epsilon_{zz},\ \epsilon_4 = 2\epsilon_{yz},\ \epsilon_5 = 2\epsilon_{zx},\ \epsilon_6 = 2\epsilon_{xy}, \tag{I.81}$$

and

$$\sigma_1 = \sigma_{xx},\ \sigma_2 = \sigma_{yy},\ \sigma_3 = \sigma_{zz},\ \sigma_4 = 2\sigma_{yz},\ \sigma_5 = 2\sigma_{zx},\ \sigma_6 = 2\sigma_{xy}. \quad (I.82)$$

Now, the strain-stress relations, Eq. (I.53), take the form,

$$\sigma_k = \sum_{i=1}^{6} C_{ki}\epsilon_i, \quad (I.83)$$

where the C_{ij} are the *elastic moduli* or *elastic constants* of the crystal. In this notation, the expression for the elastic potential, Eq. (I.76), becomes

$$\mathrm{d}W = \sum_{k=1}^{6} \sigma_k \mathrm{d}\epsilon_k. \quad (I.84)$$

The last two expressions clearly illustrate how elastic constants can be calculated as the change in elastic energy associated with specific forms of applied stress.

As mentioned above, the fact that W is a function of state allows for at most 21 independent parameters, C_{ij}, so that the C_{ij} must be symmetric,

$$C_{ij} = C_{ji}, \text{ or, equivalently, } \frac{\partial \sigma_i}{\partial \epsilon_j} = \frac{\partial \sigma_j}{\partial \epsilon_i}. \quad (I.85)$$

The maximum number, 21, occurs in the case of a triclinic crystal, where the three primitive vectors have different lengths and are non-orthogonal. There are 13 independent constants for a monoclinic system, and 9 in the case of an orthorhombic structure [1]. A cubic system, characterized by the highest symmetry, requires only three elastic constants. Because we will have occasion to refer to the elastic properties of cubic crystals later in the discussion, we explicitly display the structure of the elastic constant matrix,

$$\underline{C} = \begin{pmatrix} C_{11} & C_{12} & C_{12} & 0 & 0 & 0 \\ & C_{11} & C_{12} & 0 & 0 & 0 \\ & & C_{11} & 0 & 0 & 0 \\ & & & C_{44} & 0 & 0 \\ & & & & C_{44} & 0 \\ & & & & & C_{44} \end{pmatrix}. \quad (I.86)$$

It is interesting to note that in order to adapt the present scheme to that of the strain-stress relations of an isotropic body, we need to impose the conditions, $C_{11} = 2\mu + \lambda$, $C_{12} = \lambda$, and $C_{44} = \mu$. This implies that $C_{11} = C_{12} + 2C_{44}$, which is *not* the case in cubic single crystals.

I.5.1 Elastic energy of a crystal

Through integration of Eq. (I.84), the elastic energy per unit of volume of a crystal can be written in the form,

$$W = \frac{1}{2}\sum_{i=1}^{6}\sum_{j=1}^{6} C_{ij}\epsilon_i\epsilon_j, \tag{I.87}$$

where the elastic constants are to conform to the symmetry of the structure under consideration. As an example, consider the case of a cubic crystal under uniform dilation, $\epsilon_1 = \epsilon_2 = \epsilon_3 = \frac{1}{3}|\theta|$, [see also Eq. (I.65)]. In this case, Eq. (I.87) yields,

$$W = \frac{1}{6}(C_{11} + 2C_{12})|\theta|^2. \tag{I.88}$$

The *bulk modulus* can be defined by the expression,

$$B = -V\frac{\mathrm{d}p}{\mathrm{d}V}, \tag{I.89}$$

where p denotes the pressure. Using the thermodynamic relation, $\mathrm{d}W = -p\mathrm{d}V$ ($\mathrm{d}W$ being the work done against the pressure in bringing about the change in volume), we have,

$$B = \frac{1}{3}(C_{11} + 2C_{12}). \tag{I.90}$$

It follows that the bulk modulus can also be calculated from the expression $B = V\partial^2 W/\partial V^2$, so that it is determined from the change in elastic energy under applied hydrostatic pressure. The ability to determine the same quantities e.g., B, C_{ij}, etc., by means of different computational procedures, and the interrelations of these quantities provide a very stringent test on computational methods.

Bibliography

[1] A. Sommerfeld, *Mechanics of Deformable Bodies*, Lectures on Theoretical Physics, Vol. **II**, Academic Press, New York (1964).

[2] Philip M. Morse and Herman Feshbach, *Methods of Theoretical Physics*, McGraw-Hill Book Co. New York (1953).

[3] Harry Lass, *Vector and tensor Analysis*, McGraw-Hill Book Co. New York (1950).

Appendix J

Vectors, Tensors and Dyadics

In this appendix, we provide a brief exposition of some basic concepts in vector and tensor analysis.

J.1 Vectors

In three-dimensional space, a vector may be defined in the manner in which it transforms under a rotation of the coordinates. Consider a Cartesian coordinate system with axes x_i, $i = 1, 2, 3$, and a rotation (about the origin) in which the axes assume new orientations. Let the new axes be denoted by x'_i, and let γ_{ij} be the cosine of the angle between the x'_i- and the x_j-axis. There are nine quantities, γ_{ij}, that can be arranged in matrix form,

$$\underline{\gamma} = \begin{pmatrix} \gamma_{11} & \gamma_{12} & \gamma_{13} \\ \gamma_{21} & \gamma_{22} & \gamma_{23} \\ \gamma_{31} & \gamma_{32} & \gamma_{33} \end{pmatrix}. \tag{J.1}$$

The γ_{ij} are related through the six equations,

$$\sum_j \gamma_{ij}\gamma_{jk} = \delta_{ik}, \tag{J.2}$$

three of which express the fact that both systems of coordinates, before and after the rotation, are orthogonal, and three expressing the fact that the sum of the squares of the direction cosines of a line equals unity.

Consider now the coordinates of a point in the two systems. If the coordinates in the original, unrotated system are denoted by (x_1, x_2, x_3), in the rotated system they are given by

$$x_i' = \sum_j \gamma_{ij} x_j. \tag{J.3}$$

Expressing the coordinates as single *column* matrices, we can write this equation in the form,

$$\begin{pmatrix} x_1' \\ x_2' \\ x_3' \end{pmatrix} = \begin{pmatrix} \gamma_{11} & \gamma_{12} & \gamma_{13} \\ \gamma_{21} & \gamma_{22} & \gamma_{23} \\ \gamma_{31} & \gamma_{32} & \gamma_{33} \end{pmatrix} \begin{pmatrix} x_1 \\ x_2 \\ x_3 \end{pmatrix}, \tag{J.4}$$

or,

$$\mathbf{x}' = \underline{\gamma}\mathbf{x}, \tag{J.5}$$

where bold-face symbols denote vectors. The inverse transformation is given by,

$$\begin{aligned} x_i &= \sum_j \gamma_{ji} x_j' \\ &= \sum_j \gamma_{ij}^{\mathrm{T}} x_j', \end{aligned} \tag{J.6}$$

where $\gamma_{ij}^{\mathrm{T}} = \gamma_{ji}$ is the *transpose* of the matrix $\underline{\gamma}$. It follows that $\det\underline{\gamma} = \det\underline{\gamma}^{\mathrm{T}} = 1$. In matrix notation we have,

$$\mathbf{x} = \underline{\gamma}^{\mathrm{T}}\mathbf{x}'. \tag{J.7}$$

The inverse of $\underline{\gamma}$ is a matrix defined by the property,

$$\underline{\gamma}\underline{\gamma}^{-1} = \underline{\gamma}^{-1}\underline{\gamma} = \underline{I}, \tag{J.8}$$

where $\underline{I}$ is the *identity* matrix with elements $\underline{I}_{ij} = \delta_{ij}$. Here, the transpose equals the inverse,

$$\underline{\gamma}^{\mathrm{T}} = \underline{\gamma}^{-1}. \tag{J.9}$$

With these basic notions about rotations of coordinate systems, we define a *vector* to be a set of three quantities, called the components of the vector, $\mathbf{A} = (A_1, A_2, A_3)$, which under rotations transform in the same way as the coordinates of a point,

$$A_i' = \sum_j \gamma_{ij} A_j. \tag{J.10}$$

By contrast, a *scalar* quantity remains invariant under coordinate transformations. We note that the definition of a vector given above can be generalized in a straightforward way to n-dimensional space.

J.1.1 Vector algebra

If we picture a vector as a directed line segment, then for two vectors **A** and **B** we can form the sum $\mathbf{C} = \mathbf{A} + \mathbf{B}$ following the well-known parallelogram construction. Equivalently, if $\mathbf{A} = (A_1, A_2, A_3)$ and $\mathbf{B} = (B_1, B_2, B_3)$, then $\mathbf{C} = (C_1, C_2, C_3)$, where $C_i = A_i + B_i$.

The *scalar* or *dot* product (or *inner* product) of two three-dimensional vectors, **A** and **B**, forming an angle θ is a scalar defined as,

$$\mathbf{A} \cdot \mathbf{B} = |\mathbf{A}||\mathbf{B}| \cos\theta, \tag{J.11}$$

where $|\mathbf{A}|$ is the *length* or *modulus* of **A**. It is defined as the length of a line segment connecting two points in 3-space whose coordinates along the x_i-axis differ by A_i, so that,

$$|\mathbf{A}| = \sqrt{A_1^2 + A_2^2 + A_3^2}. \tag{J.12}$$

The dot product can also be expressed in terms of the vector components,

$$\mathbf{A} \cdot \mathbf{B} = A_1 B_1 + A_2 B_2 + A_3 B_3, \tag{J.13}$$

an expression that has an immediate generalization to higher dimensional space.

The *vector* or *cross* product (or *outer* product) of two vectors **A** and **B** is a vector that is defined as,

$$\begin{aligned} \mathbf{A} \times \mathbf{B} &= \sum_{ijk} \epsilon_{ijk} \mathbf{e}_i A_j B_k \\ &= -\mathbf{B} \times \mathbf{A}, \end{aligned} \tag{J.14}$$

where ϵ_{ijk} is the permutation symbol,

$$\epsilon_{ijk} = \begin{cases} 0 & \text{if any two indices are equal} \\ 1 & \text{if } i, j, k \text{ are cyclic} \\ -1 & \text{if } i, j, k \text{ are not cyclic,} \end{cases} \tag{J.15}$$

and $\mathbf{e}_i$ is a unit vector along the positive x_i-axis. The unit vectors along the x-, y-, and z-axes are often also denoted by the symbols **i**, **j**, **k**, respectively. The magnitude of the vector product is given by,

$$|\mathbf{A} \times \mathbf{B}| = |\mathbf{A}||\mathbf{B}| \sin\theta. \tag{J.16}$$

We note that the three unit vectors, $\mathbf{e}_1$, $\mathbf{e}_2$, $\mathbf{e}_3$, obey the relations,

$$\mathbf{e}_i \cdot \mathbf{e}_j = \delta_{ij}, \quad \mathbf{e}_i \times \mathbf{e}_j = \mathbf{e}_k \epsilon_{ijk}. \tag{J.17}$$

The following list contains some useful identities of vector algebra.

$$\begin{aligned}
\mathbf{A}\cdot(\mathbf{B}\times\mathbf{C}) &= \mathbf{B}\cdot(\mathbf{C}\times\mathbf{A}) = \mathbf{C}\cdot(\mathbf{A}\times\mathbf{B}) \equiv \mathbf{ABC} \\
\mathbf{A}\times(\mathbf{B}\times\mathbf{C}) &= (\mathbf{A}\cdot\mathbf{C})\mathbf{B} - (\mathbf{A}\cdot\mathbf{B})\mathbf{C} \\
(\mathbf{A}\times\mathbf{B})\cdot(\mathbf{C}\times\mathbf{D}) &= \mathbf{A}\cdot[\mathbf{B}\times(\mathbf{C}\times\mathbf{D})] \\
&= \mathbf{A}\cdot[(\mathbf{B}\cdot\mathbf{D})\mathbf{C} - (\mathbf{B}\cdot\mathbf{C})\mathbf{D}] \\
&= (\mathbf{A}\cdot\mathbf{C})(\mathbf{B}\cdot\mathbf{D}) - (\mathbf{A}\cdot\mathbf{D})(\mathbf{B}\cdot\mathbf{C}) \\
(\mathbf{A}\times\mathbf{B})\times(\mathbf{C}\times\mathbf{D}) &= [(\mathbf{A}\times\mathbf{B})\cdot\mathbf{D}]\mathbf{C} - [(\mathbf{A}\times\mathbf{B})\cdot\mathbf{C}]\mathbf{D} \\
&= (\mathbf{ABD})\mathbf{C} - (\mathbf{ABC})\mathbf{D} \\
\mathbf{A}\times[\mathbf{B}\times(\mathbf{C}\times\mathbf{D})] &= (\mathbf{B}\cdot\mathbf{D})(\mathbf{A}\times\mathbf{C}) - (\mathbf{B}\cdot\mathbf{C})(\mathbf{A}\times\mathbf{D}) \\
(\mathbf{A}\times\mathbf{B})\cdot[(\mathbf{B}\times\mathbf{C})\times(\mathbf{C}\times\mathbf{A})] &= [\mathbf{A}\cdot(\mathbf{B}\times\mathbf{C})]^2. \qquad \text{(J.18)}
\end{aligned}$$

J.1.2 Vector differential operators

The *del* operator is defined as,

$$\nabla = \sum_i \mathbf{e}_i \frac{\partial}{\partial x_i} = \mathbf{e}_1 \frac{\partial}{\partial x_1} + \mathbf{e}_2 \frac{\partial}{\partial x_2} + \mathbf{e}_3 \frac{\partial}{\partial x_3}. \qquad \text{(J.19)}$$

The *gradient* of a scalar function, $\Phi(x, y, z) = \Phi(x_i)$, is defined by the relation,

$$\mathbf{grad}\ \Phi \equiv \nabla\Phi = \sum_i \mathbf{e}_i \frac{\partial \Phi}{\partial x_i}. \qquad \text{(J.20)}$$

The gradient operator can be used to yield the rate of change of Φ in the direction of the unit vector $\mathbf{n}$,

$$\frac{\partial \Phi}{\partial \mathbf{n}} = \mathbf{n}\cdot\nabla\Phi. \qquad \text{(J.21)}$$

A vector that depends on position through the set of coordinates $\{x_i\}$, i.e., $\mathbf{A}(x, y, z) = \mathbf{A}(\{x_i\}) = A_1(\{x_i\})\mathbf{e}_1 + A_2(\{x_i\})\mathbf{e}_2 + A_3(\{x_i\})\mathbf{e}_3$, defines a *vector field*. The action of the operator del on a vector field can be defined as follows. First, for any vector, $\mathbf{A}$, the operator $\mathbf{A}\cdot\nabla$ is defined by its action on a function Φ,

$$\begin{aligned}
(\mathbf{A}\cdot\nabla)\Phi &= \sum_i A_i \frac{\partial \Phi}{\partial x_i} \\
&= \mathbf{A}\cdot(\nabla\Phi). \qquad \text{(J.22)}
\end{aligned}$$

For a vector (field) $\mathbf{B}$, we have

$$(\mathbf{A}\cdot\nabla)\mathbf{B} = \sum_i \left(A_i \frac{\partial}{\partial x_i}\right)\mathbf{B}$$

$$= \sum_{ij} \mathbf{e}_i A_i \frac{\partial B_j}{\partial x_i}. \tag{J.23}$$

The *divergence* of a vector **A** is defined by

$$\text{div } \mathbf{A} = \nabla \cdot \mathbf{A} = \sum_i \frac{\partial A_i}{\partial x_i}. \tag{J.24}$$

The *curl* of a vector is defined by

$$\mathbf{curl}\ \mathbf{A} = \nabla \times \mathbf{A} = \sum_{ijk} \epsilon_{ijk} \mathbf{e}_i \frac{\partial A_k}{\partial x_j}. \tag{J.25}$$

The *Laplacian* of a scalar function, Φ, is defined by

$$\nabla^2 \Phi = \text{div } \mathbf{grad}\ \Phi = \nabla \cdot \nabla \Phi = \sum_i \frac{\partial^2 \Phi}{\partial x_i^2}. \tag{J.26}$$

In rectangular coordinates, we can define the Laplacian of a vector **A** as

$$\nabla^2 \mathbf{A} = (\text{div } \mathbf{grad})\ \mathbf{A} = \nabla^2 A_1 \mathbf{e}_1 + \nabla^2 A_2 \mathbf{e}_2 + \nabla^2 A_3 \mathbf{e}_3. \tag{J.27}$$

The following list contains some useful differential operations.

$$\begin{aligned}
\nabla(\Phi + \Psi) &= \nabla\Phi + \nabla\Psi \\
\nabla(\Phi\Psi) &= \Psi\nabla\Phi + \Phi\nabla\Psi \\
\nabla \cdot (\mathbf{A} + \mathbf{B}) &= \nabla \cdot \mathbf{A} + \nabla \cdot \mathbf{B} \\
\nabla \times (\mathbf{A} + \mathbf{B}) &= \nabla \times \mathbf{A} + \nabla \times \mathbf{B} \\
\nabla \cdot (\Phi \mathbf{A}) &= \mathbf{A} \cdot \nabla\Phi + \Phi\nabla \cdot \mathbf{A} \\
\nabla \times (\Phi \mathbf{A}) &= \Phi\nabla \times \mathbf{A} - \mathbf{A} \times \nabla\Phi \\
\nabla(\mathbf{A} \cdot \mathbf{B}) &= (\mathbf{A} \cdot \nabla)\mathbf{B} + (\mathbf{B} \cdot \nabla)\mathbf{A} \\
&+ \mathbf{A} \times (\nabla \times \mathbf{B}) + \mathbf{B} \times (\nabla \times \mathbf{A}) \\
\nabla \cdot (\mathbf{A} \times \mathbf{B}) &= \mathbf{B} \cdot (\nabla \times \mathbf{A}) - \mathbf{A} \cdot (\nabla \times \mathbf{B}) \\
\nabla \times (\mathbf{A} \times \mathbf{B}) &= \mathbf{A}\nabla \cdot \mathbf{B} - \mathbf{B}\nabla \cdot \mathbf{A} \\
&+ (\mathbf{B} \cdot \nabla)\mathbf{A} - (\mathbf{A} \cdot \nabla)\mathbf{B} \\
\nabla \times \nabla \times \mathbf{A} &= \nabla(\nabla \cdot \mathbf{A}) - \nabla^2 \mathbf{A} \\
\nabla \times (\nabla\Phi) &= 0 \\
\nabla \cdot (\nabla \times \mathbf{A}) &= 0.
\end{aligned} \tag{J.28}$$

The last two identities express the fact that the curl of the gradient and the divergence of the curl vanish. It follows that a vector **A** can be represented as the gradient of a scalar function iff **curl A** $= 0$,

$$\mathbf{curl}\ \mathbf{A} = 0 \leftrightarrow \mathbf{A} = \nabla\Phi. \tag{J.29}$$

J.1.3 Integral theorems

The *divergence theorem* (or *Gauss' theorem*) states that,

$$\int_V \operatorname{div} \mathbf{A}\, \mathrm{d}^3 r = \int_S \mathbf{A} \cdot \mathbf{n}\, \mathrm{d}a, \tag{J.30}$$

where V is a volume bounded by the surface S, $\mathbf{n}$ is a unit vector in the direction of the outward normal to the surface, and $\mathrm{d}a$ is an element of surface area.

Stoke's theorem states that,

$$\int_S \nabla \times \mathbf{A} \cdot \mathbf{n} \mathrm{d}a = \oint_\Gamma \mathbf{A} \cdot \mathrm{d}\mathbf{s}, \tag{J.31}$$

where the closed curve Γ bounds the open surface S. The positive sense of traversing Γ is such that the right-handed screw direction is *into* S. Proofs of the divergence and of Stoke's theorem can be found in standard treatments of vector calculus [1].

There are several theorems due to Green, all of which can be based on the equation that results when $\Psi\nabla\Phi$ is substituted for $\mathbf{A}$ into the divergence theorem,

$$\int_S \Psi \frac{\partial \Phi}{\partial \mathbf{n}} \mathrm{d}a = \int_V \left[\Psi \nabla^2 \Phi + (\nabla\Phi) \cdot (\nabla\Psi)\right] \mathrm{d}^3 r. \tag{J.32}$$

Interchanging Φ and Ψ and subtracting, we obtain,

$$\int_V \left(\Psi \nabla^2 \Phi - \Phi \nabla^2 \Psi\right) \mathrm{d}^3 r = \int_S \left(\Psi \frac{\partial \Phi}{\partial \mathbf{n}} - \Phi \frac{\partial \Psi}{\partial \mathbf{n}}\right) \mathrm{d}a. \tag{J.33}$$

Finally, setting $\Psi = 1$, we have,

$$\int_S \frac{\partial \Phi}{\partial \mathbf{n}} \mathrm{d}a = \int_V \nabla^2 \Phi \mathrm{d}^3 r. \tag{J.34}$$

The following are useful results based on Green's theorems,

$$\int_V \nabla \Phi \mathrm{d}^3 r = \int_S \Phi \mathbf{n} \mathrm{d}a, \tag{J.35}$$

$$\int_V \nabla \times \mathbf{A} \mathrm{d}^3 r = \int_S \mathbf{n} \times \mathbf{A} \mathrm{d}a, \tag{J.36}$$

$$\oint_\Gamma \mathbf{r} \times \mathrm{d}\mathbf{S} = 2 \int_S \mathbf{n} \mathrm{d}a. \tag{J.37}$$

Gauss' law has the following interesting consequence. If a vector field $\mathbf{A}$ obeys a radial inverse square law,

$$\mathbf{A} = \frac{m}{r^2} \mathbf{e}_r, \tag{J.38}$$

where $\mathbf{e}_r$ is a unit vector along $\mathbf{r}$, then,

$$\int_S \mathbf{A} \cdot \mathbf{n} da = 4\pi m, \tag{J.39}$$

where S is *any* closed surface containing the *source* of the vector field whose strength is m. Since **curl A** $= 0$, we may use Eq. (J.29) and the divergence theorem to write,

$$\int_S \nabla\Phi \cdot \mathbf{n}\, da = \int_V \nabla^2 \Phi d^3 r = 4\pi m. \tag{J.40}$$

Furthermore, if the source m has a volume distribution with density $\rho(\mathbf{r})$, then,

$$m = \int_V \rho(\mathbf{r}) d^3 r. \tag{J.41}$$

Combining with Eq. (J.40), and equating integrants, we obtain *Poisson's equation*,

$$\nabla^2 \Phi = 4\pi\rho. \tag{J.42}$$

(The minus sign that usually occurs in electrostatics is a consequence of the definition of the electric field as the negative gradient of the potential.)

J.2 Tensors

In n-dimensional space, a tensor of the mth rank is a set on n^m quantities, which under a rotation of the coordinate axes transform according to the rule,

$$T'_{abcd\cdots} = \sum_{ijkl\cdots} \gamma_{ai}\gamma_{bj}\gamma_{ck}\gamma_{dl}\cdots T_{ijkl\cdots}. \tag{J.43}$$

We will be concerned with tensors of the second rank in mostly three-dimensional space, with occasional reference to four dimensions[1]. Therefore, it will be sufficient to discuss quantities that transform according to[2],

$$T'_{ij} = \sum_{kl} \gamma_{ik}\gamma_{jl} T_{kl}, \tag{J.44}$$

where each index takes on the values 1,2,3, or 1,2,3,4, in the case of three- or four-dimensional space, respectively. For simplicity, we base our discussion on three-dimensional tensors. The extension to four dimensions, where necessary, is straightforward.

[1] See, for example, the discussion of the electromagnetic field stress tensor in Appendix K.

[2] We will not distinguish between covariant and contravariant tensor indices, because this distinction is not relevant when Cartesian coordinates are used.

A three-dimensional second rank tensor has nine *elements* or *components* that can be arranged in matrix form,

$$\overleftrightarrow{T} = \begin{pmatrix} T_{11} & T_{12} & T_{13} \\ T_{21} & T_{22} & T_{23} \\ T_{31} & T_{32} & T_{33} \end{pmatrix}. \tag{J.45}$$

Tensors (and dyadics, discussed in the next section) will be denoted with a double-sided arrow over the symbol, or an underline. The transformation rule, Eq. (J.44), can also be written in the form,

$$T'_{ij} = \sum_{kl} \gamma_{ik} T_{kl} \gamma^{\mathrm{T}}_{lj}, \tag{J.46}$$

or in matrix (dyadic) notation,

$$\overleftrightarrow{T}' = \underline{\gamma}\overleftrightarrow{T}\underline{\gamma}^{\mathrm{T}} = \underline{\gamma}\overleftrightarrow{T}\underline{\gamma}^{-1}. \tag{J.47}$$

Two tensors connected by a transformation involving the pre- and post-multiplication by an orthogonal matrix, such as $\underline{\gamma}$, and its inverse are said to be *similar*, and the transformation is called a *similarity transformation* (also see Appendix B).

A tensor is said to be *symmetric* if $T_{ij} = T_{ji}$, and antisymmetric if $T_{ij} = -T_{ji}$. In the latter case, the diagonal elements of the tensor vanish, $T_{ii} = 0$. A symmetric second rank tensor in three dimensions can have at most six independent components, and an antisymmetric tensor three. Any given tensor can always be written in terms of a symmetric and an antisymmetric part, with elements,

$$\frac{T_{ij} + T_{ji}}{2}, \quad \text{and} \quad \frac{T_{ij} - T_{ji}}{2}, \tag{J.48}$$

respectively.

The trace of a tensor is the sum of its diagonal elements,

$$\mathrm{Tr}\, \overleftrightarrow{T} = \sum_i T_{ii} = \sum_i T'_{ii}. \tag{J.49}$$

As this equation indicates, the trace of a tensor is invariant under coordinate transformations, i.e., similarity transformations, a well-known result of matrix algebra. (A derivation is provide below; also see Appendix B.)

A second invariant of a tensor is the determinant of its components,

$$\det\overleftrightarrow{T} = \begin{vmatrix} T_{11} & T_{12} & T_{13} \\ T_{21} & T_{22} & T_{23} \\ T_{31} & T_{32} & T_{33} \end{vmatrix} = \det\overleftrightarrow{T}'. \tag{J.50}$$

The proof is again elementary, based on the properties of determinant multiplication, $|AB| = |A||B|$, and the fact that $\det \underline{\gamma} = 1$ [see discussion following Eq. (J.6)].

A third tensor invariant is the set of its *characteristic values, principal values*, or *eigenvalues*, λ_j. A tensor can be written in purely diagonal form in terms of its eigenvalues,

$$T'_{ij} = \lambda_i \delta_{ij}. \tag{J.51}$$

The eigenvalues of a tensor can be determined as the solutions of a set of simultaneous linear equations,

$$\sum_l (T_{ml} - \lambda_j \delta_{ml}) \gamma_{jl} = 0. \tag{J.52}$$

This can be proved based on the law of tensor transformations. A combination of Eqs. (J.44) and (J.51) yields,

$$\lambda_j \delta_{ij} = \sum_{kl} \gamma_{ik} \gamma_{jl} T_{kl}. \tag{J.53}$$

Multiplying both sides by γ_{im}, summing over i and using the orthogonality of the cosine matrix, we obtain,

$$\lambda_j \gamma_{jm} = \sum_l \gamma_{jl} T_{ml}. \tag{J.54}$$

The left-hand side of this equation can be expressed as,

$$\lambda_j \gamma_{jm} = \sum_l \lambda_j \gamma_{jl} \delta_{ml}, \tag{J.55}$$

so that Eq. (J.54) takes the form,

$$\sum_l \lambda_j \gamma_{jl} \delta_{ml} = \sum_l \gamma_{kl} T_{ml}, \tag{J.56}$$

from which Eq. (J.52) follows. In order for a non-trivial solution to Eq. (J.52) to exist, the determinant of the coefficients must vanish. It follows that the eigenvalues λ_1, λ_2, and λ_3 are given as the roots of the so-called *secular determinant*,

$$|T_{ml} - \lambda \delta_{ml}| = 0. \tag{J.57}$$

The process of diagonalizing a tensor is called *principal axes* transformation. (An elementary application of this process is the determination of the principal axes in which the moment of inertia of a rigid body becomes diagonal.) We note that the trace of a tensor is the sum of its eigenvalues, and the determinant is the product of its eigenvalues.

In addition to the three invariant quantities, the trace, the determinant, and the eigenvalues of a tensor, other invariants can be formed as well. We note that for any value of a parameter λ,

$$\begin{aligned} D = \det(\overleftrightarrow{T}' + \lambda \overleftrightarrow{I}) &= \det[\gamma(\overleftrightarrow{T} + \lambda \overleftrightarrow{I})\gamma^{-1}] \\ &= \det(\overleftrightarrow{T} + \lambda \overleftrightarrow{I}), \end{aligned} \tag{J.58}$$

where $\overleftrightarrow{I}$ is the unit tensor (unit matrix). Defining,

$$D_\lambda = \begin{vmatrix} T_{11} + \lambda & T_{12} & T_{13} \\ T_{21} & T_{22} + \lambda & T_{33} \\ T_{31} & T_{32} & T_{33} + \lambda \end{vmatrix}, \tag{J.59}$$

we have

$$\begin{aligned} D_\lambda &= D + \lambda\Delta + \lambda^2 \mathrm{Tr}\,\overleftrightarrow{T} + \lambda^3 \\ &= D' + \lambda\Delta' + \lambda^2 \mathrm{Tr}\,\overleftrightarrow{T} + \lambda^3, \end{aligned} \tag{J.60}$$

where

$$\Delta = T_{11}T_{22} + T_{22}T_{33} + T_{33}T_{11} - T_{12}T_{21} - T_{23}T_{32} - T_{13}T_{31}, \tag{J.61}$$

and D' and Δ' correspond to $\overleftrightarrow{T}'$. Since Eq. (J.60) must hold for arbitrary values of λ, it directly yields the known equality of the determinants and the trace, and in addition shows that,

$$\Delta = \Delta'. \tag{J.62}$$

Thus, Δ is a tensor invariant. We note that for an antisymmetric tensor, $D = 0$, $\mathrm{Tr}\overleftrightarrow{T} = 0$, and that Δ reduces to the form,

$$\Delta = T_{23}^2 + T_{31}^2 + T_{12}^2. \tag{J.63}$$

J.3 Dyadics and Tensor Operations

Tensor algebra can often be facilitated through the introduction of *dyadics*. This has the desirable feature of allowing tensor manipulations to be carried out in a notation closely reminiscent of vectors.

J.3.1 Dyadics

A dyadic of the second rank in three-dimensional space is a set of nine elements or components, which under a rotation of coordinates transform according to Eq. (J.44). In other words, a dyadic is a tensor. At the same time, a dyadic can be displayed in terms of the three unit vectors, $\mathbf{e}_1$, $\mathbf{e}_2$, and $\mathbf{e}_3$ along a right-handed coordinate system of axes (ξ_1, ξ_2, ξ_3) in the form,

$$\overleftrightarrow{A} = \sum_{nm} \mathbf{e}_n A_{nm} \mathbf{e}_m. \tag{J.64}$$

The products $\mathbf{e}_n\mathbf{e}_m$ are neither vectors nor scalars. They serve rather as indicators for the proper performance of operations. For example, the inner product of $\mathbf{e}_n\mathbf{e}_m$ with a vector $\mathbf{A}$ is such that $\mathbf{e}_n\mathbf{e}_m \cdot \mathbf{A} = A_m\mathbf{e}_n$, which is a vector in the direction ξ_n of magnitude equal to the component of $\mathbf{A}$ along ξ_m. Similarly, $\mathbf{A} \cdot \mathbf{e}_n\mathbf{e}_m = A_n\mathbf{e}_m$, so that $\mathbf{A} \cdot \mathbf{e}_n\mathbf{e}_m = A_n\mathbf{e}_m \neq \mathbf{e}_n\mathbf{e}_m \cdot \mathbf{A} = A_m\mathbf{e}_n$.

Dyadics transform vectors into other vectors (as do tensors). In general, the inner product of a dyadic with a vector is defined as,

$$\begin{aligned} \overleftrightarrow{A} \cdot \mathbf{B} &= \left(\sum_{nm} \mathbf{e}_n A_{nm} \mathbf{e}_m\right)\left(\sum_{l} \mathbf{e}_l B_l\right) \\ &= \sum_{nm} \mathbf{e}_n A_{nm} B_m, \end{aligned} \tag{J.65}$$

and

$$\mathbf{B} \cdot \overleftrightarrow{A} = \sum_{nm} B_n A_{nm} \mathbf{e}_m. \tag{J.66}$$

Note that the component of the product $\overleftrightarrow{A} \cdot \mathbf{B}$ in the n direction takes the form,

$$(\overleftrightarrow{A} \cdot \mathbf{B})_n = \sum_{m} A_{nm} B_m, \tag{J.67}$$

which is the result of multiplying the matrix representing $\overleftrightarrow{A}$ with the column vector representing $\mathbf{B}$. Similar analogies exist for all operations discussed below, and are pointed out occasionally. It can readily be verified that dyadics obey the following algebraic rules,

$$\begin{aligned} \overleftrightarrow{A} + \overleftrightarrow{B} &= \sum_{mn} \mathbf{e}_m (A_{mn} + B_{mn}) \mathbf{e}_n = \overleftrightarrow{B} + \overleftrightarrow{A}, \\ \overleftrightarrow{A} \cdot \overleftrightarrow{B} &= \sum_{nm} \mathbf{e}_m \left(\sum_{i} A_{mi} B_{in}\right) \mathbf{e}_n \neq \overleftrightarrow{B} \cdot \overleftrightarrow{A} \\ c\overleftrightarrow{A} &= \sum_{mn} \mathbf{e}_m (c A_{mn}) \mathbf{e}_n = \overleftrightarrow{A} c, \quad c \text{ a scalar.} \end{aligned} \tag{J.68}$$

Thus, dyadic addition and multiplication by a scalar are commutative, but dyadic multiplication is not. We easily recognize that these expressions find direct analogies in matrix algebra. We can also define the scalar "double-dot" product,

$$\overleftrightarrow{A} : \overleftrightarrow{B} = \sum_{mn} A_{mn} B_{nm} = \overleftrightarrow{B} : \overleftrightarrow{A}, \tag{J.69}$$

which is a scalar and clearly commutative. In addition, being the trace of the product of the matrices for $\overleftrightarrow{A}$ and $\overleftrightarrow{B}$, it is invariant with respect to coordinate transformations.

In dyadic algebra we can also define a zero dyadic, $\overleftrightarrow{0}$, and a unit dyadic, $\overleftrightarrow{I}$, called the *idemfactor*, such that for any vector **F** we ,

$$\overleftrightarrow{0} \cdot \mathbf{F} = \mathbf{F} \cdot \overleftrightarrow{0} = 0, \text{ and } \overleftrightarrow{I} \cdot \mathbf{F} = \mathbf{F} \cdot \overleftrightarrow{A} = \mathbf{F}. \tag{J.70}$$

It is easily verified that the idemfactor is given by the expression,

$$\overleftrightarrow{I} = \sum_{i=1}^{3} \mathbf{e}_i \mathbf{e}_i. \tag{J.71}$$

Finally, we can define $\overleftrightarrow{A}^{-1}$, the reciprocal dyadic to $\overleftrightarrow{A}$, as the dyadic which when multiplied by $\overleftrightarrow{A}$ gives the idemfactor,

$$\overleftrightarrow{A}^{-1} \cdot \overleftrightarrow{A} = \overleftrightarrow{A} \cdot \overleftrightarrow{A}^{-1} = \overleftrightarrow{I}. \tag{J.72}$$

As is well-known from matrix algebra, the nine components of $\overleftrightarrow{A}^{-1}$ can be obtained from those of $\overleftrightarrow{A}$ by the rule

$$\left(\overleftrightarrow{A}^{-1}\right)_{nm} = (-1)^{n+m} \frac{A'_{nm}}{\det|\overleftrightarrow{A}|}, \tag{J.73}$$

where A'_{nm} is the minor of A_{mn} (note the interchange of subscripts) in the determinant $\det|\overleftrightarrow{A}|$. Also consistent with matrix algebra, we verify that,

$$\left(\overleftrightarrow{A} \cdot \overleftrightarrow{B}\right)^{-1} = \overleftrightarrow{B}^{-1} \cdot \overleftrightarrow{A}^{-1}. \tag{J.74}$$

Given a dyadic, $\overleftrightarrow{A}$, we define its conjugate $\overleftrightarrow{A}^{*}$ by the relation,

$$\left(\overleftrightarrow{A}^{*}\right)_{mn} = A_{nm}. \tag{J.75}$$

In other words, the matrix representing $\overleftrightarrow{A}^{*}$ is the transpose of the matrix representing $\overleftrightarrow{A}$. If,

$$\overleftrightarrow{A} = \sum_{i=1}^{3} \mathbf{e}_i A_i, \text{ then } \overleftrightarrow{A}^{*} = \sum_{i=1}^{3} A_i \mathbf{e}_i. \tag{J.76}$$

We see easily that,

$$\overleftrightarrow{A} \cdot \overleftrightarrow{B} = \overleftrightarrow{B} \cdot \overleftrightarrow{A}^*, \text{ and } \overleftrightarrow{B} \cdot \overleftrightarrow{A} = \overleftrightarrow{A}^* \cdot \overleftrightarrow{B}. \tag{J.77}$$

Also,

$$\left(\overleftrightarrow{A} \cdot \overleftrightarrow{B}\right)^* = \overleftrightarrow{B}^* \cdot \overleftrightarrow{A}^*. \tag{J.78}$$

A simple way of visualizing dyadic operations is to view vectors on the left (right) as rows (columns), and the dyadic as a matrix. Using Dirac's notation of bras and kets, we can write Eq. (J.64) in the form,

$$\overleftrightarrow{A} = \sum_{nm} |\mathbf{e}_n\rangle A_{nm} \langle \mathbf{e}_m|. \tag{J.79}$$

The other operations discussed above can be expressed in similar fashion.

J.3.2 Rate of change of a tensor field

Let $\mathbf{f}(x, y, z)$ define a vector field in three-dimensional space. Consider the difference in $\mathbf{f}$ between two nearby points,

$$\begin{aligned} \mathrm{d}\mathbf{f} &= \mathbf{f}(x + \mathrm{d}x, y + \mathrm{d}y, z + \mathrm{d}z) - \mathbf{f}(x, y, z) \\ &= \mathrm{d}f_x \mathbf{i} + \mathrm{d}f_y \mathbf{j} + \mathrm{d}f_z \mathbf{k}. \end{aligned} \tag{J.80}$$

An expression for $\mathrm{d}\mathbf{f}$ that is proportional to $\mathrm{d}\mathbf{r} = \mathrm{d}x\mathbf{i} + \mathrm{d}y\mathbf{j} + \mathrm{d}z\mathbf{k}$ can be found as follows. Since we have,

$$\mathrm{d}f_x = \mathrm{d}x\frac{\partial f_x}{\partial x} + \mathrm{d}y\frac{\partial f_x}{\partial y} + \mathrm{d}z\frac{\partial f_x}{\partial z}, \tag{J.81}$$

with similar expressions for the other components, we can write,

$$\begin{aligned} \mathrm{d}\mathbf{f} &= \left(\mathrm{d}x\frac{\partial}{\partial x} + \mathrm{d}y\frac{\partial}{\partial y} + \mathrm{d}z\frac{\partial}{\partial z}\right) f_x \mathbf{i} \\ &= \left(\mathrm{d}x\frac{\partial}{\partial x} + \mathrm{d}y\frac{\partial}{\partial y} + \mathrm{d}z\frac{\partial}{\partial z}\right) f_y \mathbf{j} \\ &= \left(\mathrm{d}x\frac{\partial}{\partial x} + \mathrm{d}y\frac{\partial}{\partial y} + \mathrm{d}z\frac{\partial}{\partial z}\right) f_z \mathbf{k} \\ &= (\mathrm{d}\mathbf{r} \cdot \nabla)\, \mathbf{f}. \end{aligned} \tag{J.82}$$

This can also be expressed as a dyadic relation

$$\mathrm{d}\mathbf{f} = \mathrm{d}\mathbf{r} \cdot (\nabla \mathbf{f}), \tag{J.83}$$

where

$$\nabla \mathbf{f} = \mathbf{i}\frac{\partial \mathbf{f}}{\partial x} + \mathbf{j}\frac{\partial \mathbf{f}}{\partial y} + \mathbf{k}\frac{\partial \mathbf{f}}{\partial z} \tag{J.84}$$

Finally, we can also write,

$$\nabla \mathbf{f} = \frac{\mathbf{df}}{\mathbf{dr}}. \tag{J.85}$$

This is in keeping with the definition of a dyadic, which can be interpreted as the ratio between two vectors.

J.3.3 Divergence and curl of a tensor field

The divergence and the curl of a tensor field can be defined in a straightforward way in terms of dyadic notation[3]. Using the representation of a tensor in terms of orthogonal basis vectors, Eq. (J.64), we define,

$$\begin{aligned} \text{Div}\,\overleftrightarrow{A} &= \nabla \cdot \overleftrightarrow{A} \\ &= \mathbf{i} \cdot \frac{\overleftrightarrow{A}}{\partial x} + \mathbf{j} \cdot \frac{\overleftrightarrow{A}}{\partial y} + \mathbf{k} \cdot \frac{\overleftrightarrow{A}}{\partial z} \\ &= \sum_{i,j=1}^{3} \frac{\partial A_{ij}}{\partial x_i} \mathbf{e}_j. \end{aligned} \tag{J.86}$$

This can also be written in a vector/matrix notation in the form,

$$\begin{aligned} \text{Div}\,\overleftrightarrow{A} &= \left(\mathbf{e}_1 \frac{\partial}{\partial x_1}, \mathbf{e}_2 \frac{\partial}{\partial x_2}, \mathbf{e}_3 \frac{\partial}{\partial x_3} \right) \begin{pmatrix} A_{11} & A_{12} & A_{13} \\ A_{21} & A_{22} & A_{23} \\ A_{31} & A_{32} & A_{33} \end{pmatrix} \\ &= \sum_{ij} \frac{\partial}{\partial x_i} A_{ij} \mathbf{e}_j, \end{aligned} \tag{J.87}$$

which agrees with the previous expression.

We note that the divergence of a tensor (or a dyadic) yields a vector, whereas the divergence of a vector yields a scalar Eq. (J.30). We also have the conjugate relation,

$$\overleftrightarrow{A} \cdot \nabla = \nabla \cdot \overleftrightarrow{A}^* = \mathbf{i}[\text{div}\mathbf{A}_1] + \mathbf{j}[\text{div}\mathbf{A}_2] + \mathbf{k}[\text{div}\mathbf{A}_3], \tag{J.88}$$

where $\mathbf{A}_i = \mathbf{i}A_{1i} + \mathbf{j}A_{2i} + \mathbf{k}A_{3i}$.

The divergence of a tensor can also be expressed as the limit of a surface integral, analogous to the corresponding relation for vectors, Eq. (J.30). Let a surface S with outward normal $\mathbf{n}$ enclose a volume $\delta\tau$ around the point at which the divergence of the tensor $\overleftrightarrow{A}$ is to be calculated. Then, the p-component of the divergence is given by the relation,

$$\text{Div}_p\,\overleftrightarrow{A} = \lim_{\Delta\tau \to 0} \frac{1}{\Delta\tau} \int_S A_{np} \mathrm{d}S_n. \tag{J.89}$$

[3] As already mentioned, dyadic notation is not essential in dealing with tensors, but is often helpful.

We note that A_{np} is the component of $\overleftrightarrow{A}$ along the n surface element, taken in the direction p. The proof of the theorem is a straightforward generalization of the divergence theorem for vectors [1, 2]. Equation (J.89) can also be written in the form,

$$\int_V \nabla \cdot \overleftrightarrow{A} = \int \mathbf{n} \cdot \overleftrightarrow{A} da, \tag{J.90}$$

which is analogous to the corresponding relation for vectors, Eq. (J.30).

Finally, the curl of a dyadic is also a dyadic, defined as,

$$\begin{aligned} \nabla \times \overleftrightarrow{A} &= \sum_i \mathbf{e}_i \times \frac{\partial \overleftrightarrow{A}}{\partial x_i} \\ &= \sum_i \mathbf{e}_i \left(\frac{\partial \mathbf{A}_k}{\partial x_j} - \frac{\partial \mathbf{A}_j}{\partial x_k} \right) \quad i, j, k \text{ cyclic} \\ &= \sum_i (\nabla \times \mathbf{A}_i^*) \mathbf{e}_i. \end{aligned} \tag{J.91}$$

The corresponding generalization of Stoke's theorem takes the form,

$$\oint_\Gamma \mathrm{d}\mathbf{s} \cdot \overleftrightarrow{A} = \int_S (\nabla \times \overleftrightarrow{A}) \cdot \mathbf{n}\, da. \tag{J.92}$$

Bibliography

[1] Harry Lass, *Vector and tensor Analysis*, McGraw-Hill Book Co. New York (1950).

[2] A. Sommerfeld, *Mechanics of Deformable Bodies*, Lectures on Theoretical Physics, Vol. **II**, Academic Press, New York (1964).

Appendix K

Electromagnetic Stress Tensor

First, we establish the stress qualities of the electrostatic field on somewhat intuitive grounds. Consider two particles with opposite charges, say q and $-q$, a distance $2R$ apart. The force (tension) required to increase the separation between the charges by an amount $2\Delta x$ is,

$$\Delta T = -q^2 \left(\frac{1}{(2R + \Delta x)^2} - \frac{1}{(2R)^2} \right) \simeq \frac{q^2}{(2R)^3} \Delta x, \tag{K.1}$$

to first order in Δx. We now determine the position on the perpendicular bisector plane of the line connecting the (moved) charges where the field has the same value as it does at a given position a on the bisecting plane before the charges are moved. Clearly, this position is a distance Δy closer to the axis. An elementary calculation yields[1],

$$\Delta y \simeq -\Delta x \frac{4R^2 + a^2}{3aR}, \tag{K.2}$$

to first order in infinitesimal quantities. Thus, moving the particles further apart causes a constriction of the force lines in a direction perpendicular to that of the move. Both Maxwell and Faraday viewed fields in terms of lines of force, and the behavior of these lines under tension is appropriately described by a stress tensor. We now derive a formal expression for the Maxwell stress tensor in the static case[2].

[1] This expression is meaningful provided that $\Delta y < a$. The divergence of Δy as $a \to 0$ is a manifestation of the fact that points on the line between the charges do not undergo any constriction.

[2] The case of time-varying fields where magnetic fields can be present is treated in a number of texts on electrostatics, e.g., Jackson [1].

We consider a collection of particles in a volume Ω characterized by a density $\rho(\mathbf{r})$ and interacting by purely Coulomb forces. The rate of change of the mechanical momentum of the particles in the box[3] is equal to the force acting on the particles, so that we have,

$$\frac{\mathrm{d}\mathbf{P}}{\mathrm{d}t} = \int_{\Omega} \rho(\mathbf{r})\mathbf{E}(\mathbf{r})\mathrm{d}r^3, \tag{K.3}$$

where $\mathbf{E}(\mathbf{r})$ is the electric field at $\mathbf{r}$. Now, from Maxwell's equations (also recall Poisson's equation), we have,

$$\rho = \frac{1}{4\pi}\nabla \cdot \mathbf{E}, \tag{K.4}$$

and Eq. (K.3) takes the form,

$$\frac{\mathrm{d}\mathbf{P}}{\mathrm{d}t} = \frac{1}{4\pi}\int_{\Omega} \mathbf{E}\,(\nabla \cdot \mathbf{E})\,\mathrm{d}^3r. \tag{K.5}$$

We now add zero to the integrant in the form $\nabla \times (\nabla \times \mathbf{E})$ ($\mathbf{E} = \nabla\Phi$ and the curl of a gradient vanishes), and note the existence of the relation,

$$\begin{aligned}
[\mathbf{E}\,(\nabla \cdot \mathbf{E}) - \mathbf{E} \times (\nabla \times \mathbf{E})]_1 &= E_1 \left(\frac{\partial E_1}{\partial x_1} + \frac{\partial E_2}{\partial x_2} + \frac{\partial E_3}{\partial x_3}\right) \\
&- E_2 \left(\frac{\partial E_2}{\partial x_1} - \frac{\partial E_1}{\partial x_2}\right) + E_3 \left(\frac{\partial E_1}{\partial x_3} + \frac{\partial E_3}{\partial x_1}\right) \\
&= \frac{\partial E_1^2}{\partial x_1} + \frac{\partial (E_1 E_2)}{\partial x_2} + \frac{\partial (E_1 E_3)}{\partial x_3} \\
&- \frac{1}{2}\frac{\partial}{\partial x_1}\left(E_1^2 + E_2^2 + E_3^2\right).
\end{aligned} \tag{K.6}$$

Corresponding expressions for the other Cartesian components can be written down through a cyclic permutation of the indices. In general, we have,

$$[\mathbf{E}\,(\nabla \cdot \mathbf{E}) - \mathbf{E} \times (\nabla \times \mathbf{E})]_\alpha = \sum_\beta \frac{\partial}{\partial x_\beta}\left(E_\alpha E_\beta - \frac{1}{2}\mathbf{E}\cdot\mathbf{E}\delta_{\alpha\beta}\right). \tag{K.7}$$

We now define the *Maxwell stress tensor* in the absence of a magnetic field,

$$\overleftrightarrow{M} = \frac{1}{4\pi}\left(E_\alpha E_\beta - \frac{1}{2}E^2 elta_{\alpha\beta}\right), \tag{K.8}$$

and Eq. (K.5) can be written in the form,

$$\left.\frac{\mathrm{d}\mathbf{P}}{\mathrm{d}t}\right|_\alpha = \int_\Omega \sum_\beta \frac{\partial}{\partial x_\beta} M_{\alpha\beta}(\mathbf{r})\mathrm{d}^3r. \tag{K.9}$$

[3]In the case of time-varying fields the momentum of the electromagnetic field should be added.

It is seen that the Maxwell stress tensor is a tensor field whose divergence yields the force on the particles at $\mathbf{r}$. Using the divergence theorem for tensors, Eq. (J.90), we can convert the last volume integral into an integral over the surface, S, bounding the volume,

$$\left.\frac{\mathrm{d}\mathbf{P}}{\mathrm{d}t}\right|_{\alpha} = \int_S \sum_{\beta} M_{\alpha\beta} n_{\beta} \mathrm{d}S. \tag{K.10}$$

Here, n_{α} is the component along the αth axis of the unit vector $\hat{\mathbf{n}}$, the outward normal to the surface at $\mathbf{r}$. It follows from the last expression [1] that $\sum_{\beta} M_{\alpha\beta} n_{\beta}$ is the αth component of the flow per unit area of momentum across the surface S into the volume Ω, a quantity that is clearly a function of position on the surface. *Therefore,* $\overleftrightarrow{M}$ *defines a stress field that determines the force per unit area transmitted across the surface* S *and acting on the particles in* Ω. It follows that the last equation can be used to calculate the total force on all particles in the box by enclosing the box inside a conveniently chosen surface and carrying out the indicated integral.

Bibliography

[1] J. D. Jackson, *Classical Electrodynamics*, Wiley, New York (1962).

Appendix L

Analytic Properties of the CPA

Any approximate theory of substitutionally disordered alloys should yield Green functions that possess the analytic properties of the exact ensemble-averaged Green function. In this appendix, we present a proof of the analytic behavior of the CPA using the locator-renormalized-interactor formalism. For the sake of clarity, we restrict our discussion to the case of single-band TB alloy, with only diagonal disorder.

In the locator formulation of the CPA, the Green function of the effective medium, $\bar{G}_{00}$, is given as the solution of the self-consistent Eq. (11.42),

$$\begin{aligned} \bar{G}_{00} = \langle G_{00}^{\alpha} \rangle &= \langle (z - \epsilon_{\alpha} - \bar{\Delta}_0)^{-1} \rangle \\ &= \frac{1}{\Omega_{BZ}} \int_{BZ} \left[\langle G_{00}^{\alpha} \rangle^{-1} + \bar{\Delta}_0 - W(\mathbf{k}) \right]^{-1} \mathrm{d}^3 k, \end{aligned} \quad (L.1)$$

which is a self-consistency condition for the renormalized interactor, $\bar{\Delta} \equiv \bar{\Delta}_0$. It follows easily from this expression that the self-energy (the renormalized interactor) and the Green function satisfy the conditions of reality and of limitation, i.e.,

$$\Delta(z) = \Delta^*(z^*), \quad (L.2)$$

and

$$\Im \Delta < 0 \, (> 0), \text{ when } \Im z > 0 \, (< 0), \quad (L.3)$$

where $\Im$ denotes the imaginary part of a complex quantity. It is clear that $\Delta^*(z^*)$ is a solution to Eq. (L.1) when $\Delta(z)$ is a solution, and therefore $\Delta(z)$ satisfies the requirement of reality, Eq. (L.2). Taking imaginary parts in Eq. (L.1), we obtain the result,

$$(z_2 \quad - \quad \Delta_2) \langle G_{00} G_{00}^{\dagger} \rangle$$

$$
\begin{aligned}
&= \frac{1}{\Omega_{\mathrm{BZ}}}\int_{\mathrm{BZ}} \bar{G}(\mathbf{k})^{\dagger}\bar{G}(\mathbf{k})\left[\Delta_2 + (z_2-\Delta_2)\langle G_{00}G_{00}^{\dagger}\rangle\right] \\
&\quad / \ \left(\bar{G}_{00}^{\dagger}\bar{G}_{00}\right)\mathrm{d}^3k,
\end{aligned} \tag{L.4}
$$

which can also be written in the form,

$$
\begin{aligned}
&\Delta_2\left[\frac{1}{\Omega_{\mathrm{BZ}}}\int_{\mathrm{BZ}} \bar{G}(\mathbf{k})^{\dagger}\bar{G}(\mathbf{k})\right]\mathrm{d}^3k \\
&= -(z_2-\Delta_2)\langle G_{00}G_{00}^{\dagger}\rangle \sum_{n\neq 0}\bar{G}_{0n}^{\dagger}\bar{G}_{n0}/\left(\bar{G}_{00}^{\dagger}\bar{G}_{00}\right).
\end{aligned} \tag{L.5}
$$

In deriving the last expression we have used the fact that,

$$
\frac{1}{\Omega_{\mathrm{BZ}}}\int_{\mathrm{BZ}} \bar{G}(\mathbf{k})^{\dagger}\bar{G}(\mathbf{k})\mathrm{d}^3k = \sum_n \bar{G}_{0n}^{\dagger}\bar{G}_{n0}. \tag{L.6}
$$

From these relations, we now deduce the useful inequality,

$$
\bar{G}_{00}^{\dagger}\bar{G}_{00} - \left(\langle G_{00}G_{00}^{\dagger}\rangle - \bar{G}_{00}^{\dagger}\bar{G}_{00}\right)\sum_{n\neq 0}\bar{G}_{0n}^{\dagger}\bar{G}_{n0}/\left(\bar{G}_{00}^{\dagger}\bar{G}_{00}\right) > 0. \tag{L.7}
$$

Choosing a physical solution to Eq. (L.1) for some z in the upper half-plane, i.e., a solution for which the density of states is positive, is equivalent to choosing $(z_2 - \Delta_2) > 0$. But Δ_2 and $(z_2 - \Delta_2)$ have opposite signs as follows from Eq. (L.5). Thus, the physical solution at any z in the upper half-plane, $z_2 > 0$, yields $\Delta_2 < 0$ (also for $z_2 < 0$, $\Delta_2 > 0$), so that Δ satisfies the condition of limitation, Eq. (L.3).

To complete the proof of analyticity of the CPA, we must show that a solution for the renormalized interactor that satisfies the requirement of limitation at all points in the upper half-plane can be found through the analytic continuation of the physical solution at large z_2. To prove the existence of such a solution, we first show that $\mathrm{d}\Delta/\mathrm{d}z$ is bounded for all z such that $z_2 > 0$ and $\Delta_2 < 0$. Differentiating the self-consistency condition, Eq. (L.1), we obtain the result,

$$
\begin{aligned}
\mathrm{d}\Delta/\mathrm{d}z &= \langle G_{00}^2\rangle(1-\bar{G}_{00}^{-1})\left\{\frac{1}{\Omega_{\mathrm{BZ}}}\int_{\mathrm{BZ}}\left[\bar{G}(\mathbf{k})\right]^2\mathrm{d}^3k\right\} \\
&\quad / \ \left\{\langle G_{00}^2\rangle\left[1-\langle G_{00}^2\rangle(\ldots)\right]+(\ldots)\right\},
\end{aligned} \tag{L.8}
$$

where (...) stands for the term in the braces in this equation. The derivative, $\mathrm{d}\Delta/\mathrm{d}z$, may become unboundedly large if the denominator on the right-hand side of the last equation were to vanish. That the modulus of the denominator is non-zero, however, can be seen from the inequalities,

$$
\begin{aligned}
& \left|\langle G_{00}^2\rangle \left[1 - \langle G_{00}^2\rangle (...)\right] + [...]\right| \\
\equiv & \left|\bar{G}_{00}^2 - \left(\langle G_{00}^2\rangle \bar{G}_{00}^2\right) \sum_{n\neq 0} \bar{G}_{0n}\bar{G}_{n0} / \left(\bar{G}_{00}^2\right)\right| \\
\geq & \left|\bar{G}_{00}^2\right| - \left(\langle G_{00}G_{00}^{\dagger}\rangle - \bar{G}_{00}\bar{G}_{00}^{\dagger}\right) \left(\sum_{n\neq 0} \bar{G}_{0n}\bar{G}_{n0}^{\dagger}\right) / \left(\bar{G}_{00}\bar{G}_{00}^{\dagger}\right) \\
> & \; 0, \qquad\qquad (L.9)
\end{aligned}
$$

the last step following from Eq. (L.7). It now follows that $d\Delta/dz < \infty$ for any z such that $z_2 > 0$ and $\Delta_2 < 0$. Furthermore, the implicit function theorem of complex analysis implies that Δ and hence G is an analytic function in the neighborhood of a point where the derivative, $d\Delta/dz$, is bounded. Since the upper half-plane is simply connected, such a physical, analytic solution, which can always be found uniquely at sufficiently large z, can be extended throughout the upper half-plane by analytic continuation.

Appendix M

Stationarity of the CPA

In addition to its analytic behavior, the CPA yields an integrated density of states that is stationary under (small) variations in the self energy [1, 2]. To see this, consider the expression for the integrated density of states of an alloy,

$$\begin{aligned} N(E) &= \int_{-\infty}^{E} \mathrm{d}E' n(E') \\ &= -\frac{1}{\pi}\Im \mathrm{Tr} \int_{-\infty}^{E} \mathrm{d}E' \langle (E'-H)^{-1} \rangle \\ &= -\frac{1}{\pi}\Im \mathrm{Tr} \int_{-\infty}^{E} \mathrm{d}E' \langle G(E') \rangle \\ &= -\frac{1}{\pi}\Im \mathrm{Tr} \ln (E'-H) \\ &= \frac{1}{\pi}\Im \mathrm{Tr} \ln \langle G(E') \rangle . \end{aligned} \tag{M.1}$$

Now, we note that the Green function can be written in the form of a Dyson equation, Eq. (2.72),

$$G = \langle G \rangle \left[1 - (V - \Sigma) \langle G \rangle \right], \tag{M.2}$$

and use the property[1],

$$\ln AB = \ln A + \ln B, \tag{M.3}$$

to cast Eq. (M.3) in the form,

$$\langle \mathrm{Tr} \ln G \rangle = \mathrm{Tr} \ln \langle G \rangle - \langle \mathrm{Tr} \ln \left[1 - (V - \Sigma) \langle G \rangle \right] \rangle . \tag{M.4}$$

[1] Although the logarithm of a product of non-commuting operators is not in general equal to the sum of the logarithms, the property holds when the trace is taken.

Furthermore, we consider the following functional of the self energy,

$$\Phi(\Sigma) = \mathrm{Tr}\ln\bar{G} - \langle \mathrm{Tr}\ln\left[1 - (V - \Sigma)\bar{G}\right]\rangle, \tag{M.5}$$

with

$$\bar{G} = (z - H_0 - \Sigma)^{-1}, \tag{M.6}$$

where H_0 is the translationally invariant part of the alloy Hamiltonian. Varying Σ in Eq. (M.5), we obtain,

$$\begin{aligned} \delta\Phi &= \mathrm{Tr}\bar{G}\delta\Sigma - \mathrm{Tr}\langle T\rangle\delta\bar{G} \\ &- \mathrm{Tr}\left[\bar{G} + \bar{G}\langle T\rangle\bar{G}\right]\delta\Sigma. \end{aligned} \tag{M.7}$$

If Σ is obtained through the condition $\langle T\rangle = 0$ leading to $\langle G\rangle = \bar{G}$, then $\delta\Phi = 0$. This implies also that,

$$\delta\langle \mathrm{Tr}\ln G\rangle = \delta N = 0. \tag{M.8}$$

Therefore, the integrated density of states in the CPA is stationary with respect to variations in the self energy.

Bibliography

[1] F. Ducastelle, *Order and Phase Stability in Alloys*, North Holland, Amsterdam (1991).

[2] P. Lloyd and P. R. Best, J. Phys. C: Solid State Phys. 8, 3752 (1975).

Appendix N

Exact Energy Expansions

In this appendix, we show how to obtain an exact expansion of configurational quantities, in particular the energy, in terms of exact ECIs. The method is based on the use of direct configurational averaging in determining the coefficients in an orthonormal cluster-function expansion, and is valid only within a canonical (concentration-dependent) scheme.

We begin by writing the exact thermodynamic potential of an alloy at $T = 0$K, i.e., the total energy, and at fixed concentration in the form,

$$\begin{aligned} \Omega\{p_i^\alpha\} &= \sum_{\mathrm{J}} E^{\mathrm{J}} P^{\mathrm{J}} \\ &= \sum_{\mathrm{J}} E^{\Pi_i^N \alpha_i^{\mathrm{J}}} \Pi_{i=1}^N P\left[\alpha_i^{\mathrm{J}}\right], \end{aligned} \tag{N.1}$$

where P^{J} is a configuration number that is equal to one if the alloy isin configuration J, and is zero otherwise, α_i^{J} denotes the type of atom occupying site i, $P\left[\alpha_i^{\mathrm{J}}\right]$ is an occupation number for the species at site i, and N is the number of lattice sites. In the case of binary alloys, A_cB_{1-c}, we have the relations,

$$P_i = P_i^A, \quad P_i^B = 1 - P_i, \tag{N.2}$$

and

$$\delta c_i = P_i - c, \quad 1 - P_i = (1 - c) - \delta c_i. \tag{N.3}$$

We now define the n-site *interchange energies* associated with a cluster of sites C_n,

$$E^{(n)}_{i_1 i_2 \cdots i_n}(\mathrm{J}_{C_n}) = \Pi_{j=1}^n (1 - Q_{i_j}^{AB}) E^{AA \cdots A}_{i_1 i_2 \cdots i_n}(\mathrm{J}_{C_n}). \tag{N.4}$$

Here, the operator $Q_{i_j}^{AB}$ changes an A atom into a B atom on site i_j and $E^{AA \cdots A}_{i_1 i_2 \cdots i_n}(\mathrm{J}_{C_n})$ denotes the configurational energy of a cluster all of whose sites are occupied by atoms of type A. In terms of these energies, and using Eqs. (N.2) and (N.3) we can write the grand potential of a binary alloy in the form,

$$\Omega(\{P_i\}) = E^{(0)} + \sum_{i=1}^{N} E_i^{(1)} \delta c_i + \frac{1}{2} \sum_{ij}^{\prime\prime} E_{ij}^{(2)} + \sum_{ijk}^{\prime\prime} E_{ijk}^{(3)} + \cdots, \quad \text{(N.5)}$$

where the expansion coefficients in the last expression, the fully renormalized cluster interactions (RECIs), are given as the averages of the quantities $E_{i_1 i_2 \cdots i_n}^{(n)}(\mathrm{J}_{C_n})$ over all configurations j_{C_n} of the material *surrounding* cluster C_n,

$$\begin{aligned} E_{i_1 i_2 \cdots i_n}^{(n)} &= \langle E_{i_1 i_2 \cdots i_n}^{(n)}(\mathrm{J}_{C_n}) \rangle \\ &= \sum_{\mathrm{J}_{C_n}} P_{\mathrm{J}_{C_n}} E_{i_1 i_2 \cdots i_n}^{(n)}(\mathrm{J}_{C_n}). \end{aligned} \quad \text{(N.6)}$$

The quantity $P_{\mathrm{J}_{C_n}}$ is the probability of occurrence of the configuration J_{C_n}. For example, we have,

$$\begin{aligned} E^{(0)} &= \langle E \rangle = \sum_{\mathrm{J}} P_{\mathrm{J}} E^{\mathrm{J}} \\ E_i^{(1)} &= E^{(1)} = \langle E_i^{(1)}(\mathrm{J}_i) \rangle \\ &= \sum_{\mathrm{J}_i} P_{\mathrm{J}_i} \left(E_i^A(\mathrm{j}_i) - E_i^B(\mathrm{j}_i) \right), \\ E_{ij}^{(2)} &= \langle E_{ij}(2)(\mathrm{J}_{ij}) \rangle \\ &= \sum_{\mathrm{J}_{ij}} P_{\mathrm{J}_{ij}} \left(E_{ij}^{AA}(\mathrm{J}_{ij}) - E_{ij}^{AB}(\mathrm{J}_{ij}) \right. \\ &- \left. E_{ij}^{BA}(\mathrm{J}_{ij}) + E_{ij}^{BB}(\mathrm{J}_{ij}) \right), \end{aligned} \quad \text{(N.7)}$$

and so on.

It is clear from its derivation that the expansion in Eq. (N.5) is *exact* provided that the averages indicated above are carried out in exact fashion. Also, the energies involved are the total energies of the alloy configurations. However, in general direct, exact configurational averages over large (infinite) samples and the calculation of the corresponding total energies are clearly impossible, and approximation schemes must be employed in numerical applications. One commonly used approximation considers only the band contribution to the energy, and replaces the disordered material surrounding the cluster by an effective medium, such as that determined within the CPA[1]. With this choice of effective medium, the expansion coef-

[1]Other media such as that determined within the ATA in principle could also be used. However, the CPA medium has been found to be the most satisfactory within a single-site approximation.

ficients are combinations of cluster energies obtained in the ECM, as shown in Eq. (13.150).

Appendix O

Reflectivity and Scattering Amplitudes

In this appendix we consider the connection between the reflection and transmission coefficients used in applications of the LKKR method to the t-matrices or scattering-path operators of MST. The following development relies on the exposition given by Zhang [1].

We consider an incident beam, $\chi^+ = \sum_{\mathbf{g}} U_{\mathbf{g}}^+ \mathrm{e}^{\mathrm{i}\mathbf{K}_{\mathbf{g}}^+ \cdot \mathbf{r}}$, with $\mathbf{K}_{\mathbf{g}}^+$ defined as in Eq. (17.25), scattered by a (surface) system represented by the total scattering matrix, $T(\mathbf{r}, \mathbf{r}')$. In any potential-free region, such as the interstitial region between layers, the *total* wave function can be written in the form of the Lippmann-Schwinger equation,

$$\psi(\mathbf{r}) = \sum_{\mathbf{g}} U_{\mathbf{g}}^+ \mathrm{e}^{\mathrm{i}\mathbf{K}_{\mathbf{g}}^+ \cdot \mathbf{r}} + \int \mathrm{d}^3 r' \mathrm{d}^3 r'' G_0(\mathbf{r}, \mathbf{r}') T(\mathbf{r}', \mathbf{r}'') \sum_{\mathbf{g}} \mathrm{e}^{\mathrm{i}\mathbf{K}_{\mathbf{g}}^+ \cdot \mathbf{r}''}. \tag{O.1}$$

We now write the free-particle propagator in terms of an integral in $\mathbf{k}$ space,

$$G_0(\mathbf{r}, \mathbf{r}') = \frac{1}{(2\pi)^3} \int \mathrm{d}^3 k' \frac{\mathrm{e}^{\mathrm{i}\mathbf{k}' \cdot (\mathbf{r} - \mathbf{r}')}}{E - |\mathbf{k}'| + \mathrm{i}0}, \tag{O.2}$$

and partition the double integral in Eq. (O.1) into a summation over cell integrals, thus obtaining,

$$\begin{aligned} \psi(\mathbf{r}) &= \sum_{\mathbf{g}} U_{\mathbf{g}}^+ \mathrm{e}^{\mathrm{i}\mathbf{K}_{\mathbf{g}}^+ \cdot \mathbf{r}} + \sum_{\mathbf{g}} \frac{U_{\mathbf{g}}^+}{(2\pi)^3} \sum_{ij} \int\int\int \mathrm{d}^3 r_i' \mathrm{d}^3 r_j'' \mathrm{d}^3 k' \frac{\mathrm{e}^{\mathrm{i}\mathbf{k}' \cdot (\mathbf{r} - \mathbf{r}')}}{E - |\mathbf{k}'| + \mathrm{i}0} \\ &\times \tau^{ij}(\mathbf{r}_i', \mathbf{r}_j'') \mathrm{e}^{\mathrm{i}\mathbf{K}_{\mathbf{g}}^+ \cdot \mathbf{r}''}, \end{aligned} \tag{O.3}$$

where the cell vectors $\mathbf{r}_i'$ and $\mathbf{r}_j''$ denote the same points as the vectors $\mathbf{r}'$ and $\mathbf{r}''$ but are measured from the centers of the corresponding cells. The quantity τ^{ij} is the real-space representation of the intersite matrix element of the SPO.

Because of the layered structure of the system, the summations over cells in the same layer (plane) can be grouped together, and lattice vectors, $\mathbf{R}_i$, denoting the center of cell i in plane I can be measured from the origin of the plane, $\mathbf{C}_I$. Since $\mathbf{C}_I$ can be chosen to also be the center of a cell, we have,

$$\begin{aligned}\psi(\mathbf{r}) &= \sum_{\mathbf{g}} U_{\mathbf{g}}^{+} e^{i\mathbf{K}_{\mathbf{g}}^{+}\cdot\mathbf{r}} \\ &+ \sum_{\mathbf{g}IJ} \frac{U_{\mathbf{g}}^{+}}{(2\pi)^3} \int d^3r'_I d^3r''_J d^3k' \frac{e^{i\mathbf{k}'\cdot(\mathbf{r}-\mathbf{r}'_I-\mathbf{C}_I)}}{E-|\mathbf{k}'|^2+i0} \\ &\times \sum_{ij} e^{-i\mathbf{k}'\cdot\mathbf{R}_i} \tau^{ij}(\mathbf{r}'_I,\mathbf{r}''_J) e^{i\mathbf{K}_{\mathbf{g}}^{+}\cdot\mathbf{R}_j} e^{i\mathbf{K}_{\mathbf{g}}^{+}\cdot(\mathbf{r}''_J+\mathbf{C}_J)},\end{aligned} \tag{O.4}$$

where summations over i and j are confined in planes I and J, respectively. We note that the integrals over the unit cells are the same within a given plane and can be taken outside the summations over i, j.

Now, because of the two-dimensional periodicity of the system, the summation over i and j reduces to,

$$\frac{(2\pi)^2}{A} \sum_{\mathbf{g}'} \tau(\mathbf{k}_{\|},\mathbf{r}'_I,\mathbf{r}''_J)\delta\left(\mathbf{k}'_{\|}-\mathbf{k}_{\|}-\mathbf{g}'\right), \tag{O.5}$$

where A denotes the area of the two-dimensional unit cell. Integrating out the component of $\mathbf{k}'$ parallel to the surface, we obtain,

$$\begin{aligned}\psi(\mathbf{r}) &= \sum_{\mathbf{g}} U_{\mathbf{g}}^{+} e^{i\mathbf{K}_{\mathbf{g}}^{+}\cdot\mathbf{r}} \\ &+ \sum_{\mathbf{g}\mathbf{g}'IJ} \frac{U_{\mathbf{g}}^{+}}{(2\pi)A} \int d^3r'_I d^3r''_J \int dk'_z \frac{e^{i\mathbf{k}'\cdot(\mathbf{r}-\mathbf{r}'_I-\mathbf{C}_I)}}{E-|\mathbf{k}'_{\|}|^2-{k'_z}^2+i0} \\ &\times \tau(\mathbf{k}_{\|},\mathbf{r}'_I,\mathbf{r}''_J) e^{i\mathbf{K}_{\mathbf{g}}^{+}\cdot(\mathbf{r}''_J+\mathbf{C}_J},\end{aligned} \tag{O.6}$$

with $\mathbf{k}' = (\mathbf{k}'_{\|}, k_z) = (k_x+g'_x, k_y+g'_y, k'_z)$.

We can now evaluate the integral over k'_z. We define the quantity $K_{\mathbf{g}'_z} = \sqrt{E-(k_x+g'_x)^2-(k_y+g'_y)^2}$ and use the definition of $\mathbf{K}_{\mathbf{g}'}^{\pm}$ given in Eq(17.25) to write the integral over k'_z in the form,

$$-\int dk'_z \frac{e^{i\mathbf{k}'\cdot(\mathbf{r}-\mathbf{r}'_I-\mathbf{C}_I)}}{2K_{\mathbf{g}'_z}} \left(\frac{1}{k'_z-K_{\mathbf{g}'_z}-i0} - \frac{1}{k'_z+K_{\mathbf{g}'_z}+i0}\right). \tag{O.7}$$

This integral can be evaluated along contour (a) in Fig. O.1 for $(\mathbf{r}-\mathbf{r}'_I-$

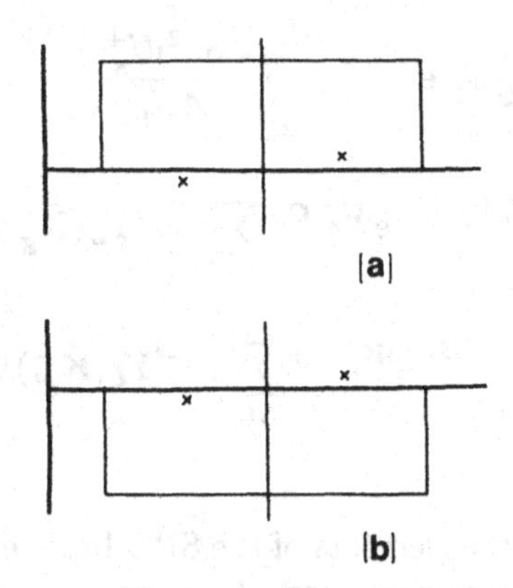

Figure O.1: **Integration contours for the integral over k'_z.**

$\mathbf{C}_I)_z > 0$, and along contour (b) for $(\mathbf{r} - \mathbf{r}'_I - \mathbf{C}_I)_z < 0$. In both cases the contribution from the part of the contour in the complex plane vanishes because of either the vanishing of the exponential factor as $\Im k'_z \to \infty$, or the vanishing of the factor $1/k'_z$ as $\Re k'_z \to \infty$. Since each contour contains only one pole, we find that the integral equals the expressions,

$$-2\pi \mathrm{i}\frac{1}{2K_{\mathbf{g}_z}}\mathrm{e}^{\mathrm{i}\mathbf{K}^+_{\mathbf{g}'}\cdot(\mathbf{r}-\mathbf{r}'_I+\mathbf{C}_I)}, \quad \text{when } (\mathbf{r} - \mathbf{r}'_I + \mathbf{C}_I)_z > 0; \tag{O.8}$$

$$-2\pi \mathrm{i}\frac{1}{2K_{\mathbf{g}_z}}\mathrm{e}^{\mathrm{i}\mathbf{K}^-_{\mathbf{g}'}\cdot(\mathbf{r}-\mathbf{r}'_I+\mathbf{C}_I)}, \quad \text{when } (\mathbf{r} - \mathbf{r}'_I + \mathbf{C}_I)_z < 0. \tag{O.9}$$

Therefore, for $\mathbf{r}$ in the interstitial region between planes N and $N+1$, the wave function can be written in the form,

$$\psi(\mathbf{r}) = \sum_{\mathbf{g}} U^+_{\mathbf{g}}\mathrm{e}^{\mathrm{i}\mathbf{K}^+_{\mathbf{g}}\cdot\mathbf{r}} - \sum_{\mathbf{g}\mathbf{g}'J}\frac{U^+_{\mathbf{g}}\mathrm{i}}{2AK_{\mathbf{g}'_z}}$$
$$\left[\sum_{I\le N}\mathrm{e}^{\mathrm{i}\mathbf{K}^+_{\mathbf{g}'}\cdot(\mathbf{r}-\mathbf{C}_I)}\mathrm{e}^{\mathrm{i}\mathbf{K}^+_{\mathbf{g}}\cdot\mathbf{C}_J}\int \mathrm{d}^3r'_I\mathrm{d}^3r''_J\mathrm{e}^{-\mathrm{i}\mathbf{K}^+_{\mathbf{g}'}\cdot\mathbf{r}'_I}\tau(\mathbf{k}_\parallel,\mathbf{r}'_I,\mathbf{r}''_J)\mathrm{e}^{\mathrm{i}\mathbf{K}^+_{\mathbf{g}}\cdot\mathbf{r}''_J}\right.$$
$$\left. + \sum_{I>N}\mathrm{e}^{\mathrm{i}\mathbf{K}^-_{\mathbf{g}'}\cdot(\mathbf{r}-\mathbf{C}_I)}\mathrm{e}^{\mathrm{i}\mathbf{K}^+_{\mathbf{g}}\cdot\mathbf{C}_I}\int \mathrm{d}^3r'_I\mathrm{d}^3r''_J\mathrm{e}^{-\mathrm{i}\mathbf{K}^-_{\mathbf{g}'}\cdot\mathbf{r}'_I}\tau(\mathbf{k}_\parallel,\mathbf{r}'_I,\mathbf{r}''_J)\mathrm{e}^{\mathrm{i}\mathbf{K}^+_{\mathbf{g}}\cdot\mathbf{r}''_J}\right]. \tag{O.10}$$

Using the expansion of a plane wave in terms of spherical harmonics, Eq. (F.1), the integrals over the unit cells can be transformed into an expansion over angular momentum states,

$$\psi(\mathbf{r}) = \sum_{\mathbf{g}} U^+_{\mathbf{g}} e^{i\mathbf{K}^+_{\mathbf{g}}\cdot\mathbf{r}} - \sum_{\mathbf{g}\mathbf{g}'J} \frac{8\pi^2 i U^+_{\mathbf{g}}}{AK_{\mathbf{g}'_z}}$$

$$\left[\sum_{I\leq N} e^{i\mathbf{K}^+_{\mathbf{g}'}\cdot(\mathbf{r}-\mathbf{C}_I)} e^{i\mathbf{K}^+_{\mathbf{g}}\cdot\mathbf{C}_J} \sum_{LL'} i^{\ell'-\ell} Y_L(\hat{\mathbf{K}}^+_{\mathbf{g}'}) Y^*_{L'}(\hat{\mathbf{K}}^+_{\mathbf{g}}) \tau^{IJ}_{LL'}(\mathbf{k}_\parallel)\right.$$

$$\left. + \sum_{I>N} e^{i\mathbf{K}^-_{\mathbf{g}'}\cdot(\mathbf{r}-\mathbf{C}_I)} e^{i\mathbf{K}^+_{\mathbf{g}}\cdot\mathbf{C}_J} \sum_{LL'} i^{\ell'-\ell} Y_L(\hat{\mathbf{K}}^-_{\mathbf{g}'}) Y^*_{L'}(\hat{\mathbf{K}}^+_{\mathbf{g}}) \tau^{IJ}_{LL'}(\mathbf{k}_\parallel)\right], \tag{O.11}$$

where $\tau^{IJ}_{LL'}(\mathbf{k}_\parallel)$ are the elements of the SPO between planes I and J in the angular-momentum representation. Also, we use a convention such that,

$$Y^*_{\ell m}(\hat{\mathbf{k}}) = (-1)^m Y_{\ell,-m}(\hat{\mathbf{k}}). \tag{O.12}$$

Now, comparing Eq. (O.11) with Eqs. (17.27) and (17.28) for the reflected and transmitted part of a wave scattered by a surface, we obtain Eq. (17.31).

Bibliography

[1] X.-G. Zhang, *Reformulation of Multiple Scattering Theory for Electronic Structure of Periodic and Semi-infinite Periodic Systems*, Ph.D. thesis (unpublished), Northwestern University, Evanston, Il, (1989).

Appendix P

Embedded Layers

In this appendix, we derive explicit expressions for the scattering matrix (transition matrix), T, corresponding to a three-dimensional layered system. For the sake of clarity, we consider only systems with one atom per unit cell. The more complex case of polyatomic unit cells can be obtained by a straightforward generalization of various scalar quantities to matrices. In our development, we follow the lines in references [1, 2], where the reader is referred for many further details.

We introduce scattering-path operators, $\underline{T}_I$, $\underline{T}_{I-1}$, and $\underline{T}_{I+1}$, corresponding to the isolated layer I, and to the two isolated semi-infinite systems consisting of the layers $-\infty, \cdots, I-1$ and $I+1, \cdots, \infty$, respectively. We also introduce the scattering-path operator $\underline{\tau}_I$ which sums all scattering events that *end* in layer I. Similarly, $\underline{\tau}_{I-1}$ and τ_{I+1} are the scattering matrices representing the sum of all scattering events that end *anywhere* in the half space to the left $(I-1)$ or to the right $(I+1)$ of plane I, respectively. The sum of all scattering paths through the three-dimensional system is clearly the total scattering matrix for the system,

$$\underline{T} = \underline{\tau}_{I-1} + \underline{\tau}_I + \underline{\tau}_{I+1}, \tag{P.1}$$

where we have written the various quantities explicitly in the angular momentum representation.

Starting from the usual equation of motion for the scattering-path operator, we can see that these scattering-path operators satisfy the relations,

$$\begin{aligned} \underline{\tau}_{I-1} &= \left(1 + \underline{\tau}_I \underline{G} + \underline{\tau}_{I+1} \underline{G}\right) \underline{T}_{I-1} \\ \underline{\tau}_I &= \left(\underline{\tau}_{I-1} \underline{G} + 1 + \underline{\tau}_{I+1} \underline{G}\right) \underline{T}_I \\ \underline{\tau}_{I+1} &= \left(\underline{\tau}_{I-1} \underline{G} + \underline{\tau}_I \underline{G} + 1\right) \underline{T}_{I+1}. \end{aligned} \tag{P.2}$$

The first equation states that all scattering paths that end in the left half space consist of all paths that are wholly contained in this subspace, plus those that end there, having previously scattered in layer I and in the other half space. In these expressions, the structure constant matrices, $\underline{G}$, have matrix elements between sites in the half spaces connected by them. These equations can be put in the form,

$$\begin{aligned} \underline{\tau}_{I-1}\left(1 - \underline{G}\underline{T}_I\underline{G}\underline{T}_{I-1}\right) &= \left(1 + \underline{\tau}_{I+1}\underline{G}\right)\left(1 + \underline{T}_I\underline{G}\right)\underline{T}_{I-1} \\ \underline{\tau}_{I+1}\left(1 - \underline{G}\underline{T}_I\underline{G}\underline{T}_{I+1}\right) &= \left(1 + \underline{\tau}_{I-1}\underline{G}\right)\left(1 + \underline{T}_I\underline{G}\right)\underline{T}_{I+1}. \end{aligned} \quad \text{(P.3)}$$

We also need the expressions for the scattering-path operators for the isolated half spaces $I-1$ (and $I+1$) with and without layer I. We let $\underline{L}_I$ and $\underline{R}_I$ denote the scattering matrices associated with the left (right) half spaces including layer I. The relations of these quantities to $\underline{T}_{I-1}$ (and $\underline{T}_{I+1}$) can be established based on the two-potential scattering matrix, Eq. (9.39). Let U denote the potential of layer I, and V that of the half space $-\infty, \cdots, I-1$ (or the half space $I+1, \cdots, \infty$). We then have,

$$\begin{aligned} \underline{L}_I &= \underline{T}_I + \left(1 + \underline{T}_I\underline{G}\right)\underline{T}_{I-1}\left(1 - \underline{G}\underline{T}_I\underline{G}\underline{T}_{I-1}\right)^{-1}\left(1 + \underline{G}\underline{T}_I\right) \\ \underline{R}_I &= \underline{T}_I + \left(1 + \underline{T}_I\underline{G}\right)\underline{T}_{I+1}\left(1 - \underline{G}\underline{T}_I\underline{G}\underline{T}_{I+1}\right)^{-1}\left(1 + \underline{G}\underline{T}_I\right). \end{aligned} \quad \text{(P.4)}$$

When these expressions are combined with those in Eqs. (P.1), to (P.3), they lead to the expression,

$$\underline{T} = \underline{T}_I + \left(1 + \underline{T}_I\underline{G}\right)\underline{R}^I_{\text{eff}}\left(1 + \underline{G}\underline{T}_I\right), \quad \text{(P.5)}$$

where we have defined the "reflectivity" of the medium surrounding layer I,

$$\begin{aligned} \underline{R}^I_{\text{eff}} &= \left[1 + \left(1 - \underline{T}_{I+1}\underline{G}\underline{T}_I\underline{G}\right)^{-1}\underline{T}_{I+1}\underline{G}\left(1 + \underline{T}_I\underline{G}\right)\right]\underline{T}_{I-1}\left(1 - \underline{G}\underline{R}_I\underline{G}\underline{T}_{I-1}\right)^{-1} \\ &+ \left[1 + \left(1 - \underline{T}_{I-1}\underline{G}\underline{T}_I\underline{G}\right)^{-1}\underline{T}_{I-1}\underline{G}\left(1 + \underline{T}_I\underline{G}\right)\right]\underline{T}_{I+1}\left(1 - \underline{G}\underline{L}_I\underline{G}\underline{T}_{I+1}\right)^{-1}. \end{aligned} \quad \text{(P.6)}$$

We note that the first term in Eq. (P.5) is the scattering matrix (scattering-path operator) associated with the isolated layer I, while the second term represents the correction due to the presence of the half spaces to the "left" and "right" of layer I. Thus, Eq. (P.5) represents the solution to the equation of motion for the case in which a layer (or stack of layers) is "embedded" within two semi-infinite materials.

Bibliography

[1] Simon Crampin, Ph.D. thesis (unpublished), Imperial College, London (1989).

[2] J. M. MacLaren, S. Crampin, D. D. Vveddensky, and J. B. Pendry, Phys. Rev. 40, 12164 (1989).

Appendix Q

KKR-CPA and the Kubo Formula

Q.1 Response Functions

In this appendix, we derive the forms taken by the expressions for the electrical conductivity in the KKR-CPA, when this approximation is expressed in its usual, "scalar", form. In our development we follow closely the paper by Butler [1], to which the reader is referred for more details. We explicitly treat the case of the dc conductivity at zero temperature. The application of the ensuing formalism to the ac conductivity (and to finite) temperature is straightforward.

Because of the configurational dependence of the current operator defined in Eq. (20.70), it is necessary to separate the expression for the dc conductivity, Eq. (20.67), into a term that contains the current operator at a single site, and terms involving the current operators of two different sites. This separation must be done in a way that accounts for the correlation between the current operators and the scattering-path matrices connected to them.

From Eq. (20.67) we can write,

$$\tilde{\sigma}_{\mu\nu}(z_1, z_2) = \tilde{\sigma}^0_{\mu\nu}(z_1, z_2) + \tilde{\sigma}^1_{\mu\nu}(z_1, z_2), \tag{Q.1}$$

where the single-site term has the form,

$$\begin{aligned} &\tilde{\sigma}^0_{\mu\nu}(z_1, z_2) \\ &= -\frac{4m^2}{\pi\hbar^3\Omega} \sum_{L_1L_2L_3L_4} \langle j^{0\mu}_{L_4L_1}(z_2, z_1)\tau^{00}_{L_1L_2}(z_1) j^{0\nu}_{L_2L_3}(z_1, z_2)\tau^{00}_{L_3L_4}(z_2)\rangle \end{aligned}$$

$$= -\frac{4m^2}{\pi\hbar^3\Omega}\mathrm{Tr}\langle \underline{j}^{0\mu}(z_2,z_1)\underline{\tau}^{00}(z_1)\underline{j}^{0\nu}(z_1,z_2)\underline{\tau}^{00}(z_2)\rangle, \tag{Q.2}$$

and

$$\begin{aligned}\tilde{\sigma}^1_{\mu\nu}(z_1,z_2) &= -\frac{4m^2}{\pi\hbar^3\Omega}\sum_{n\neq 0}\sum_{L_1L_2L_3L_4}\langle j^{0\mu}_{L_4L_1}(z_2,z_1)\tau^{0n}_{L_1L_2}(z_1)\\ &\times\ j^{n\nu}_{L_2L_3}(z_1,z_2)\tau^{n0}_{L_3L_4}(z_2)\rangle\\ &= -\frac{4m^2}{\pi\hbar^3\Omega}\sum_{n\neq 0}\mathrm{Tr}\langle \underline{j}^{0\mu}(z_2,z_1)\underline{\tau}^{0n}(z_1)\underline{j}^{n\nu}(z_1,z_2)\underline{\tau}^{n0}(z_2)\rangle,\end{aligned} \tag{Q.3}$$

where underlines designate matrices in L-space, and the trace is over L.

We attempt to evaluate these expressions by drawing as much as possible on the treatment of the TB model (or that based on the configurational matrix formulation of the KKR-CPA). Therefore, for each of the two terms above, we define response functions that involve only one current operator. Thus, we write,

$$\tilde{\sigma}^0_{\mu\nu}(z_1,z_2) = -\frac{4m^2}{\pi\hbar^3\Omega}\sum_{\mu}c_\alpha \mathrm{Tr}\left[\underline{K}^{0\alpha}_\nu(z_1,z_2)\underline{j}^{\alpha\nu}(z_2,z_1)\right] \tag{Q.4}$$

and

$$\tilde{\sigma}^1_{\mu\nu}(z_1,z_2) = -\frac{4m^2}{\pi\hbar^3\Omega}\sum_{n\neq 0}\sum_{\alpha\beta}c_\alpha c_\beta \mathrm{Tr}\left[\underline{L}^{0n;\alpha\beta}_\nu(z_1,z_2)\underline{j}_{\alpha\mu}(z_2,z_1)\right], \tag{Q.5}$$

where we have defined the response functions,

$$\underline{K}^{0\alpha}_\nu = \langle \underline{\tau}^{00}(z_1)\underline{j}^{\alpha\nu}(z_1,z_2)\underline{\tau}^{00}(z_2)\rangle_{0=\alpha} \tag{Q.6}$$

and

$$\underline{L}^{0n;\alpha\beta}_\nu = \langle \underline{\tau}^{0n}(z_1)\underline{j}^{\beta\nu}(z_1,z_2)\underline{\tau}^{n0}(z_2)\rangle_{0=\alpha,n=\beta}, \tag{Q.7}$$

with c_α being the concentration of atoms of species α in the alloy.

Obtaining computable expressions for these response functions within the KKR-CPA is a matter of some lengthy algebra that is given in detail in the paper by Butler [1]. This development leads to the definition of vertex corrections for each of the response functions, $\underline{K}$ and $\underline{L}$, so that we can write,

$$\tilde{\sigma}^0_{\mu\nu}(z_1,z_2) = -\frac{4m^2}{\pi\hbar^3\Omega}\sum_{\mu}c_\alpha \mathrm{Tr}\left[\underline{\tilde{K}}^{0\alpha}_\nu(z_1,z_2)\underline{\tilde{j}}^{\alpha\nu}(z_2,z_1)\right] \tag{Q.8}$$

and

$$\tilde{\sigma}^1_{\mu\nu}(z_1, z_2) = -\frac{4m^2}{\pi\hbar^3\Omega}\sum_{n\neq 0}\sum_{\alpha\beta} c_\alpha c_\beta \mathrm{Tr}\left[\underline{\tilde{L}}_\nu^{0n;\beta}(z_1, z_2)\underline{j}_{\alpha\mu}(z_2, z_1)\right]. \quad (Q.9)$$

The various quantities that appear in these expressions are defined as follows,

$$\underline{\tilde{j}}^{\beta\mu} = \underline{\tilde{D}}^\beta \underline{j}^{\beta\mu}\underline{D}^\beta \quad (Q.10)$$

where

$$\underline{D}^\alpha = \left[\underline{1} - \underline{\bar{\tau}}^{00}\left(\underline{\bar{t}}^{-1} - \underline{t}^{\alpha-1}\right)\right]^{-1}, \quad (Q.11)$$

and

$$\underline{\tilde{D}}^\alpha = \left[\underline{1} - \left(\underline{\bar{t}}^{-1} - \underline{t}^{\alpha-1}\right)\underline{\bar{\tau}}^{00}\right]^{-1}; \quad (Q.12)$$

$$\underline{\tilde{K}}^{0\alpha} = \underline{\bar{\tau}}^{00}\underline{j}^\alpha\underline{\bar{\tau}}^{00} + \sum_{k\neq 0}\underline{\bar{\tau}}^{0k}\underline{\gamma}_k^{0\alpha}\underline{\bar{\tau}}^{k0}, \quad (Q.13)$$

and

$$\underline{\tilde{L}}^{0n\beta} = \underline{\bar{\tau}}^{0n}\underline{\tilde{j}}^\beta\underline{\bar{\tau}}^{n0} + \sum_{k\neq 0,n}\underline{\bar{\tau}}^{0k}\underline{\gamma}_k^{0\beta(\alpha)}\underline{\bar{\tau}}^{k0}, \quad (Q.14)$$

where

$$\underline{\gamma}_k^{0\alpha} = \langle\left(\underline{Q}\underline{\bar{\tau}}\right)^{0k}\underline{j}^\alpha\left(\underline{\bar{\tau}}\underline{\tilde{Q}}\right)^{0k}\rangle_{0=\alpha}, \quad (Q.15)$$

and

$$\underline{\gamma}_k^{n\beta(\alpha)} = \langle\left(\underline{Q}^k\underline{\bar{\tau}}\right)^{kn}\underline{j}^\beta\left(\underline{\bar{\tau}}\underline{\tilde{Q}}\right)^{nk}\rangle_{0=\alpha,n=\beta}, \quad (Q.16)$$

with the matrices $\underline{Q}$ and $\underline{\tilde{Q}}$ as defined in the text, e.g., by analogy to Eqs. (20.78) to (20.81), but in terms of the ordinary, non-matrix, form of the KKR-CPA.

Closed expressions for the vertex functions, $\underline{\gamma}_k^{0\alpha}$ and $\underline{\gamma}_k^{n\beta(\alpha)}$, can be obtained [1], and using Fourier transformations to solve the resulting equations, we obtain the expressions,

$$\tilde{\sigma}^0_{\mu\nu}(z_1, z_2) = -\frac{4m^2}{\pi\hbar\Omega}\sum_\alpha c_\alpha \mathrm{Tr}\left[\underline{\tilde{j}}^{\alpha\mu}(z_1, z_2)\underline{\bar{\tau}}^{00}(z_1)\underline{j}^{\alpha\nu}(z_1, z_2)\underline{\bar{\tau}}^{00}(z_2)\right], \quad (Q.17)$$

which involves no vertex terms, and

$$\tilde{\sigma}^1_{\mu\nu}(z_1, z_2) = -\frac{4m^2}{\pi\hbar\Omega}\sum_{\alpha\beta} c_\alpha c_\beta \mathrm{Tr}\left\{\underline{\tilde{j}}^{\alpha\mu}\left[\underline{1} - \underline{\bar{\tau}}\otimes\underline{\bar{\tau}}(0)\langle\underline{X}^0\otimes\underline{X}^0\rangle\right]^{-1}\underline{\bar{\tau}}\otimes\underline{\bar{\tau}}(0)\underline{\tilde{j}}^{\beta\nu}\right\} \quad (Q.18)$$

Here, the notation (0) indicates the $\mathbf{q} = 0$ component of a quantity in reciprocal space.

Q.2 Derivation of the Boltzmann Equation

In this section, we show that under certain assumptions, Eqs. (Q.17) and (Q.18) have the same solutions as the semi-clasical Boltzmann equation. The main assumption under which this property holds is that the random alloy has well-defined bands in energy and momentum space. Further, for the sake of simplicity in our initial discussion, we neglect the presence of vertex corrections[1].

Under the assumptions just stated, the conductivity, $\tilde{\sigma}^n_{\mu\nu}$, including the contributions from Eqs. (Q.17) and (Q.17) takes the form,

$$\begin{aligned}\tilde{\sigma}^n_{\mu\nu}(z_1,z_2) &= -\frac{4m^2}{\pi\hbar^3\Omega}\sum_{\alpha\beta} c_\alpha c_\beta \mathrm{Tr}\left(\underline{\tilde{j}}^{\alpha\mu}(z_2,z_1)\right.\\ &\times \left[\frac{1}{\Omega_{\mathrm{BZ}}}\int_{\mathrm{BZ}} \mathrm{d}^3q\underline{\bar{\tau}}(\mathbf{q},z_1)\underline{\tilde{j}}^{\beta\nu}(z_1,z_2)\underline{\bar{\tau}}(\mathbf{q},z_2)\right.\\ &+ \left.\left.\underline{\bar{\tau}}^{00}(z_1)\left[\underline{j}^{\alpha\mu}(z_1,z_2)-\underline{\tilde{j}}^{\beta\nu}(z_1,z_2)\right]\underline{\bar{\tau}}^{00}(z_2)\right]\right),\end{aligned} \tag{Q.19}$$

where the sum $\sum_n \underline{\bar{\tau}}^{0n}\underline{\bar{\tau}}^{n0}$ has been represented as a Brillouin zone integral.

The single-site term in the last expression will be negligible compared to the first in the case of systems with long electron mean free paths, because the Brillouin zone integral contains singular contributions (to be discussed below) that make it larger than the single-site term. The assumption of long mean free paths is, of course, consistent with the assumption of well-defined bands made in the present development. Furthermore, the difference $\underline{j} - \underline{\tilde{j}}$ vanishes in the limit of pure systems (well-defined bands). Thus, defining the quantity,

$$\underline{\bar{j}} = \sum c_\alpha \underline{\tilde{j}}^\alpha, \tag{Q.20}$$

we can write,

$$\begin{aligned}\tilde{\sigma}^n_{\mu\nu}(z_1,z_2) &= -\frac{4m^2}{\pi\hbar^3\Omega}\mathrm{Tr}\frac{1}{\mathrm{BZ}}\int_{\mathrm{BZ}}\underline{\bar{J}}^\mu\\ &\times \frac{1}{\Omega_{\mathrm{BZ}}}\int_{\mathrm{BZ}}\mathrm{d}^3q\underline{\bar{M}}^{-1}(\mathbf{q},z_1)\underline{\bar{J}}^\nu\underline{\bar{M}}^{-1}(\mathbf{q},z_1),\end{aligned} \tag{Q.21}$$

where $\underline{\bar{M}}^{-1}(\mathbf{q},z) = \underline{\bar{\tau}}(\mathbf{q},z)$.

The integrant of the Brillouin zone integral will have singularities at those values of $\mathbf{q}$ for which the determinant of the matrix $\underline{\bar{M}}(\mathbf{q},z)$ vanishes. Because of the presence of disorder, these values of the wave vector in

[1] Vertex corrections are incorporated into the formalism below; their inclusion leads to the $(1-\cos\theta)$ factor in the Boltzmann equation.

general will be complex. However, under the assumption of well-defined bands, the singularities must occur *near* the real axis. We suppose that a singularity for energy z occurs at $\mathbf{q}_0(z)$. Then for $\mathbf{q} \approx \mathbf{q}_0$, we can write [2],

$$\left[\underline{\bar{M}}(\mathbf{q},z)\right]^{-1} = \frac{\hbar^2}{2m}\underline{C}(\mathbf{q}_0,z)\left[z-\epsilon_{\mathbf{q}_0}\right]^{-1}\underline{\tilde{C}}(\mathbf{q}_0,z), \qquad \text{(Q.22)}$$

where the matrices $\underline{C}$ and $\underline{\tilde{C}}$ consists of columns and rows of eigenvectors of $\underline{\bar{M}}$, the elements of which satisfy the relations,

$$\sum_{L'} \bar{M}_{LL'}(\mathbf{q}_0,z)\mathrm{i}^{l'}C_{L'}(\mathbf{q}_0,z) = 0, \qquad \text{(Q.23)}$$

and

$$\sum_{L'} \mathrm{i}^{-l'}\tilde{C}_{L'}(\mathbf{q}_0,z)\bar{M}_{L'L}(\mathbf{q}_0,z) = 0. \qquad \text{(Q.24)}$$

These vectors can be used to construct a "wave function"-like quantity associated with the KKR-CPA effective medium along the lines discussed in Chapter 12 (also see below).

The denominator in Eq. (Q.22) vanishes when the band energy $\epsilon_{\mathbf{q}}$ is equal to z which, under the present assumptions, happens when $\mathbf{q} \approx \mathbf{q}_0$. Thus, Eq. (Q.22) may be written as,

$$\left[\underline{\bar{M}}(\mathbf{q},z)\right]^{-1} = \frac{\hbar^2}{2m}\underline{C}(\mathbf{q}_0,z)\left[\left(\nabla_{\mathbf{q}}\epsilon_{\mathbf{q}}\right)_{\mathbf{q}_0(z)}\cdot(\mathbf{q}-\mathbf{q}_0)\right]^{-1}\underline{\tilde{C}}(\mathbf{q}_0,z). \qquad \text{(Q.25)}$$

The singular contribution to the zone integral, Eq. (Q.21), will be of the form,

$$I = \int \mathrm{d}^3q \frac{1}{[\mathbf{q}-\mathbf{q}_0(z_1)]}\frac{1}{[\mathbf{q}-\mathbf{q}_0(z_2)]}. \qquad \text{(Q.26)}$$

When $z_1 = z_2 = (E \pm \mathrm{i}\eta)$, the integral, Eq. (Q.26), will be negligibly small compared to the case in which $z_1 = z_2^*$, for then $\mathbf{q}_0(z_2) = \mathbf{q}_0^*(z_1)$, and Eq. (Q.26) can be approximated by the expression,

$$I = \int \mathrm{d}^3q \frac{1}{\left[\mathbf{q}-\mathbf{q}_0^1(z_1)\right]^2 + \gamma_{\mathbf{q}}} = \frac{\pi}{\gamma_{\mathbf{q}}}, \qquad \text{(Q.27)}$$

where $\gamma_{\mathbf{q}} = |\Im\mathbf{q}_0(z_1)|$ is essentially the inverse of the mean free path and $\mathbf{q}_0^1 = \Re\left[\mathbf{q}_0(z_1)\right]$.

As is argued in the next section, the group velocity of the electrons in the alloy, $\mathbf{v}_{\mathbf{q}} = \hbar^{-1}\nabla_{\mathbf{q}}\epsilon_{\mathbf{q}}$, is given by,

$$ev^{\mu}_{\mathbf{q}_0} = \mathrm{Tr}\left[\underline{\tilde{C}}(\mathbf{q}_0,\epsilon)\underline{j}^{\mu}\underline{C}(\mathbf{q}_0,\epsilon)\right]. \qquad \text{(Q.28)}$$

Thus, using Eqs. (Q.22), (Q.27), and (Q.28) in Eq. (Q.21), we have,

$$\tilde{\sigma}^n_{\mu\nu}(\epsilon \pm i\eta, \epsilon \mp i\eta) = -\frac{\hbar e}{(2\pi)^3} \int \frac{d\hat{q} q_0^2}{\left[\hat{q} \cdot (\nabla_{\mathbf{q}} \epsilon_{\mathbf{q}})_{\mathbf{q}0}\right]^2} \frac{v^\mu_{\mathbf{q}0} v^\nu_{\mathbf{q}0}}{\gamma_{\mathbf{q}0}}. \tag{Q.29}$$

This expression can be reduced to a more familiar form by recognizing that the quantity $\hbar / \left[\gamma_{\mathbf{q}0} \left(\hat{q} \cdot \nabla_{\mathbf{q}} \epsilon_{\mathbf{q}}\right)_{\mathbf{q}0}\right]$ is twice the Boltzmann lifetime of the state at $\mathbf{q}$,

$$\hbar / \left[\gamma_{\mathbf{q}0} \left(\hat{q} \cdot \nabla_{\mathbf{q}} \epsilon_{\mathbf{q}}\right)_{\mathbf{q}0}\right] = 2\tau^{\mathrm{B}}_{\mathbf{q}}. \tag{Q.30}$$

The factor of two arises because $\gamma_{\mathbf{q}0}$ describes an *amplitude* decay which is only one-half as large as the *density* decay that appears in the Boltzmann equation. The factor $\hat{q} \cdot \nabla_{\mathbf{q}} \epsilon_{\mathbf{q}}$ converts a width in q along the direction specified by $\hat{q}$ into a width in energy. We can now write down the final expression for the conductivity under the assumption of well-defined bands and in the absence of vertex corrections,

$$\begin{aligned} \sigma^n_{\mu\nu} &= -\frac{1}{4}\left[\tilde{\sigma}^n_{\mu\nu}(\epsilon + i\eta, \epsilon - i\eta) + \tilde{\sigma}^n_{\mu\nu}(\epsilon - i\eta, \epsilon + i\eta)\right] \\ &= \frac{e^2}{(2\pi)^3} \int_{S_F} \frac{dS_{\mathbf{q}}}{\hbar v_{\mathbf{q}}} v^\mu_{\mathbf{q}} v^\nu_{\mathbf{q}} \tau^{\mathrm{B}}_{\mathbf{q}}, \end{aligned} \tag{Q.31}$$

where the integral is over the alloy Fermi surface. This formula (which neglects the presence of electron spin) is quite analogous to that in Eq. (18.41). The present formulation also allows for the $\mathbf{q}$-dependence of the lifetime, which was not taken into account in Eq. (18.41) but was suggested in Eq. (18.49). The last expression has been used in a number of calculations of the conductivity of binary systems [3, 4, 5].

Q.3 Solution of the Boltzmann Equation

We now show that Eqs. (Q.17) and (Q.18) yield the same solutions as the Boltzmann equation in the limit of well-defined bands[2]. For an electric field in the ν-direction, the Boltzmann equation takes the form,

$$eEv^\nu_{\mathbf{q}} \frac{\partial f}{\partial \epsilon_{\mathbf{q}}} = \sum_{\mathbf{q}'} Q(\mathbf{q}, \mathbf{q}') \left(g^\nu_{\mathbf{q}} - g^\nu_{\mathbf{q}'}\right), \tag{Q.32}$$

where f is the Fermi function, $Q(\mathbf{q}, \mathbf{q}')$ is the probability for an electron to scatter between states $\mathbf{q}$ and $\mathbf{q}'$, and $g^\nu_{\mathbf{q}}$ is the "deviation" function that

[2]Note that the last expression does not "solve" the equation since the lifetime is not known.

describes the departure of the electron distribution from the Fermi function [see Eq. (18.20)]. If Eq. (Q.32) can be solved for $g^\nu_{\mathbf{q}}$, then the conductivity (per spin) may be obtained from,

$$\sigma_{\mu\nu} = -\mathrm{e}\sum_{\mathbf{q}} v^\mu_{\mathbf{q}} g^\nu_{\mathbf{q}} E^{-1}. \tag{Q.33}$$

Since $Q(\mathbf{q},\mathbf{q}')$ is the probability for an electron to scatter between states $\mathbf{q}$ and $\mathbf{q}'$, the total scattering rate, or inverse lifetime, for an electron in state $\mathbf{q}$ is,

$$\left(\tau^{\mathrm{B}}_{\mathbf{q}}\right)^{-1} = \sum_{\mathbf{q}'} Q(\mathbf{q},\mathbf{q}'). \tag{Q.34}$$

The use of the last equation in Eq. (Q.32) allows us to solve for $g^\nu_{\mathbf{q}}$,

$$g^\nu_{\mathbf{q}} = -\sum_{\mathbf{q}'} \left(1-\tau^{\mathrm{B}}Q\right)^{-1}_{\mathbf{q}\mathbf{q}'} \mathrm{e}Ev^\nu_{\mathbf{q}'}\frac{\partial f}{\partial\epsilon_{\mathbf{q}'}}\tau^{\mathrm{B}}_{\mathbf{q}'}. \tag{Q.35}$$

From Eqs. (Q.33) and (Q.35), the dc conductivity at zero temperature *including vertex corrections* is given by the expression,

$$\sigma_{\mu\nu}(\epsilon_{\mathrm{F}}) = \mathrm{e}^2\sum_{\mathbf{q}\mathbf{q}'} v^\mu_{\mathbf{q}}\left(1-\tau^{\mathrm{B}}Q\right)^{-1}_{\mathbf{q}\mathbf{q}'} v^\nu_{\mathbf{q}'}\tau^{\mathrm{B}}_{\mathbf{q}'}\delta(\epsilon_{\mathrm{F}}-\epsilon_{\mathbf{q}'}). \tag{Q.36}$$

The term involving the inverse in these expressions represents the vertex corrections. When that term is set equal to $\delta_{\mathbf{q}\mathbf{q}'}$, one recovers Eq. (Q.31).

It is also possible to compare [1] the expressions in Eqs. (Q.18) and (Q.36) term by term upon expanding the inverses (the vertex corrections terms). The equality of the first terms in such an expansion is demonstrated by Eq. (Q.31). The second terms are also equal provided that,

$$Q(\mathbf{q},\mathbf{q}') = \frac{2\pi}{\hbar}\sum_\alpha c_\alpha |T^\alpha_{\mathbf{q}\mathbf{q}'}|\delta(\epsilon_{\mathbf{q}}-\epsilon_{\mathbf{q}'}), \tag{Q.37}$$

where

$$T^\alpha_{\mathbf{q}\mathbf{q}'}(z) = \langle\tilde{C}(\mathbf{q},z)|\underline{X}^\alpha(z)|C(\mathbf{q}',z)\rangle, \tag{Q.38}$$

with $\underline{X}$ being the CPA scattering matrix defined in Chapter 12.

Q.4 CPA Wave Function

The series of steps leading to Eq. (Q.31) provide a strong indication that the equations of the KKR-CPA reduce properly to expressions in the band theory of translationally invariant materials in the limit of weak scattering or of low alloy concentration. For example, the matrices that diagonalize the matrix $\underline{\bar{M}}$ provide the coefficients for forming a "wave function", in a manner exactly analogous to the case of periodic solids, Eqs. (3.147) and (3.148). It is straightforward [1], albeit somewhat lengthy, to show that these wave functions have properties mirroring those of the wave functions of periodic solids. For example, if $\Psi_{\mathbf{q}}(\mathbf{r})$ is the wave function formed by the combination of the coefficients C_L and the local cell solutions, e.g. Eqs. (12.75) and (12.79), then one can write the Green function in the form,

$$G_{\mathbf{q}}(\mathbf{r},\mathbf{r}';z) = \frac{\Psi_{\mathbf{q}}(\mathbf{r})\Psi_{\mathbf{q}}(\mathbf{r}')}{z-\epsilon_{\mathbf{q}}}. \tag{Q.39}$$

Also, it can be shown [1] that the expression for the group velocity, Eq. (Q.28), is equivalent to the expression,

$$v_{\mathbf{q}}^{\mu} = -\frac{i\hbar}{m}\int \mathrm{d}^3 r \tilde{\Psi}_{\mathbf{q}}(\mathbf{r})\frac{\partial}{\partial r_{\mu}}\Psi_{\mathbf{q}}(\mathbf{r}), \tag{Q.40}$$

where $\tilde{\Psi}$ and Ψ are the wave functions formed by the coefficient matrices $\underline{\tilde{C}}$ and $\underline{C}$, respectively.

Bibliography

[1] W. H. Butler, Phys. Rev. **31**, 3260 (1985).

[2] W. H. Butler, J. J. Olsen, J. S. Faulkner, and B. L. Györffy, Phys. Rev. B **14**, 3823 (1976).

[3] G. M. Stocks and W. H. Butler, Phys. Rev. Lett. **48**, 55 (1982).

[4] W. H. Butler and G. M. Stocks, Phys.. Rev. B **29**, 4217 (1984).

[5] W. H. Butler, Phys. Rev. B **29**, 4224 (1984).

Appendix R

Phonons

R.1 Quantization of Lattice Vibrations

The quanta of lattice vibrations are called *phonons*. The quantum nature of lattice dynamics was established in Chapter 21, where it was shown that quantization was necessary in order to explain the behavior of lattice-specific heat as a function of temperature. In this appendix, we derive the basic relations that enter a quantum description of lattice vibrations. The following discussion is based on the work of Ghatak and Kothari [1].

R.1.1 Dynamical matrix

We begin by examining some important properties of the quantity $D_{\alpha\beta}(\mathbf{q})$, which occurs in Eq. (21.46),

$$\sum_{\beta=1,2,3} D_{\alpha\beta}(\mathbf{q}) A_\beta(\mathbf{q}) = \omega^2 A_\alpha(\mathbf{q}). \tag{R.1}$$

We show that the *dynamical matrix*, $\underline{D}(\mathbf{q})$, satisfies the properties,

$$D^*_{\beta\alpha}(\mathbf{q}) = D_{\alpha\beta}(\mathbf{q}) = D_{\beta\alpha}(\mathbf{q}), \tag{R.2}$$

so that it is Hermitean and also symmetric[1].

Because $V_{\alpha\beta}(ll')$ is real, Eq. (21.49) yields,

$$D^*_{\beta\alpha}(\mathbf{q}) = \frac{1}{M} \sum_{l'} V_{\beta\alpha}(l') \mathrm{e}^{\mathrm{i}\mathbf{q}\cdot\mathbf{R}_{l'}}. \tag{R.3}$$

[1] For a lattice with a basis the dynamical matrix is only Hermitean. See the example of the one-dimensional diatomic lattice discussed in Section 21.4.1.

Now, the index l' runs over the sites in a periodic lattice, so that $\mathbf{R}_l$ can be replaced by $-\mathbf{R}_l$, and since $V_{\alpha\beta}(l) = V_{\alpha\beta}(-l)$, we have,

$$D^*_{\beta\alpha}(\mathbf{q}) = \frac{1}{M}\sum_{l'} V_{\beta\alpha}(l')e^{-i\mathbf{q}\cdot\mathbf{R}_{l'}} = D_{\beta\alpha}(\mathbf{q}). \tag{R.4}$$

Furthermore, from Eq. (21.31), we obtain the result,

$$D^*_{\beta\alpha}(\mathbf{q}) = \frac{1}{M}\sum_{l'} V_{\alpha\beta}(l')e^{-i\mathbf{q}\cdot\mathbf{R}_{l'}} = D_{\alpha\beta}(\mathbf{q}), \tag{R.5}$$

which establishes the validity of Eq. (R.2).

Since the dynamical matrix is Hermitean, its eigenvalues are real, so that the quantity $\omega^2(\mathbf{q})$ in Eq. (R.2) must be real. Furthermore, ω^2 must be non-negative. A negative ω^2 implies that ω is imaginary and consequently the solution $u_l \approx e^{i\omega t}$ could grow exponentially large as $t \to +\infty$ or $-\infty$.

Now, for a simple (monatomic) Bravais lattice, the secular equation, Eq. (21.51), yields three eigenvalues for each vector $\mathbf{q}$, which will be denoted by $\omega_j^2(\mathbf{q})$, j =1, 2, 3. The eigenfunctions corresponding to these eigenvalues are denoted by $e_\alpha\begin{pmatrix}\mathbf{q}\\ j\end{pmatrix}$ and are defined by the relation,

$$\sum_{\beta=1,2,3} D_{\alpha\beta}(\mathbf{q})e_\beta\begin{pmatrix}\mathbf{q}\\ j\end{pmatrix} = \omega_j^2(\mathbf{q})e_\alpha\begin{pmatrix}\mathbf{q}\\ j\end{pmatrix}. \tag{R.6}$$

Because $\underline{D}$ is Hermitean, these eigenfunctions satisfy the following orthonormality and completeness properties,

$$\sum_\alpha e_\alpha\begin{pmatrix}\mathbf{q}\\ j\end{pmatrix} e_\alpha\begin{pmatrix}\mathbf{q}\\ j'\end{pmatrix} = \delta_{jj'}, \tag{R.7}$$

$$\sum_j e_\alpha\begin{pmatrix}\mathbf{q}\\ j\end{pmatrix} e_\beta\begin{pmatrix}\mathbf{q}\\ j\end{pmatrix} = \delta_{\alpha\beta}. \tag{R.8}$$

The quantities $e_\alpha\begin{pmatrix}\mathbf{q}\\ j\end{pmatrix}$ are the components of the *polarization* vector $\mathbf{e}\begin{pmatrix}\mathbf{q}\\ j\end{pmatrix}$.

Further, since

$$D_{\alpha\beta}(-\mathbf{q}) = D^*_{\alpha\beta}(\mathbf{q}), \tag{R.9}$$

we have from Eq. (R.6) the relation,

$$e_\alpha\begin{pmatrix}\mathbf{q}\\ j\end{pmatrix} = e^*_\alpha\begin{pmatrix}-\mathbf{q}\\ j\end{pmatrix}. \tag{R.10}$$

For the case of a simple Bravais lattice, the D-matrix is symmetric and hence all its elements are real. Thus, the wave functions are also real and we have,

$$e_\alpha\begin{pmatrix} -\mathbf{q} \\ j \end{pmatrix} = e_\alpha\begin{pmatrix} \mathbf{q} \\ j \end{pmatrix}. \tag{R.11}$$

In the following discussion, we make use of the more general relation, Eq. (R.10), rather than Eq. (R.11).

R.1.2 Normal coordinates

We consider a dynamical system described by the n coordinates, $\{\eta_i\}$, with $i = 1, 2, \ldots, n$, so that its kinetic and potential energies are given by the expansion,

$$T = \frac{1}{2}\sum_{ij} T_{ij}\dot{\eta}_i\dot{\eta}_j, \tag{R.12}$$

and

$$V = \frac{1}{2}\sum_{ij} V_{ij}\eta_i\eta_j, \tag{R.13}$$

with T_{ij} and V_{ij} independent of the coordinates, η_i. We now show that there exists a transformation which allows both T and V to be written as sums of squares. In other words, there exist *normal coordinates*, ζ_i, related to η_i by the transformation,

$$\eta_i = \sum_j A_{ij}\zeta_j, \tag{R.14}$$

so that

$$T = \frac{1}{2}\sum_i c_i\dot{\zeta}_i^2, \tag{R.15}$$

and

$$V = \frac{1}{2}\sum_i d_i\zeta_i^2, \tag{R.16}$$

with c_i and d_i being constants.

A transformation that accomplishes this task is generated by the expansion of the displacement, $u_\alpha(l)$, in terms of the eigenfunctions of the dynamical matrix and plane waves,

$$u_\alpha(l) = \frac{1}{\sqrt{NM}}\sum_{\mathbf{q}}\sum_j e_\alpha\begin{pmatrix} \mathbf{q} \\ j \end{pmatrix} Q\begin{pmatrix} \mathbf{q} \\ j \end{pmatrix} e^{i\mathbf{q}\cdot\mathbf{R}_l}, \tag{R.17}$$

where the summation extends over all vectors $\mathbf{q}$ in the first Brillouin zone, and $j = 1,\ 2,\ 3$. In the case of a cubic crystal of side L, there are N atoms in a box of volume L^3 and the total number of allowed values of $\mathbf{q}$ in the first Brillouin zone is also N. These values are given by the well-known relation,

$$\mathbf{q} = 2\pi\left(\frac{h_1}{c}\mathbf{b}_1 + \frac{h_2}{c}\mathbf{b}_2 + \frac{h_3}{c}\mathbf{b}_3\right), \tag{R.18}$$

where the $\mathbf{b}_i$ are reciprocal lattice vectors, and the integers h_i can take the values (with L assumed even),

$$h_i = 0,\ \pm 1,\ \pm 2,\ \cdots, \pm\left(\frac{1}{2}L - 1\right), \pm L. \tag{R.19}$$

Because of Eq. (R.10), and because $u_\alpha(l)$ is real, we have,

$$Q\begin{pmatrix} -\mathbf{q} \\ j \end{pmatrix} = Q^*\begin{pmatrix} \mathbf{q} \\ j \end{pmatrix}, \tag{R.20}$$

and the kinetic energy of the system can be written in the form,

$$\begin{aligned} T &= \frac{1}{2}M\sum_l\sum_\alpha \dot{u}_\alpha^2(l) \\ &= \frac{1}{2N}\sum_{l\alpha}\left[\sum_{\mathbf{q}j} e_\alpha\begin{pmatrix} \mathbf{q} \\ j \end{pmatrix}\dot{Q}\begin{pmatrix} \mathbf{q} \\ j \end{pmatrix}e^{i\mathbf{q}\cdot\mathbf{R}_l}\right] \\ &\times \left[\sum_{\mathbf{q}'j'} e_\alpha\begin{pmatrix} \mathbf{q}' \\ j' \end{pmatrix}\dot{Q}\begin{pmatrix} \mathbf{q}' \\ j' \end{pmatrix}e^{i\mathbf{q}\cdot\mathbf{R}_l}\right] \\ &= \frac{1}{2N}\sum_\alpha\sum_{\mathbf{q}j}\sum_{\mathbf{q}'j'} e_\alpha\begin{pmatrix} \mathbf{q} \\ j \end{pmatrix}e_\alpha\begin{pmatrix} \mathbf{q}' \\ j' \end{pmatrix}\dot{Q}\begin{pmatrix} \mathbf{q} \\ j \end{pmatrix}\dot{Q}\begin{pmatrix} \mathbf{q}' \\ j' \end{pmatrix} \\ &\times \sum_l e^{i(\mathbf{q}+\mathbf{q}')\cdot\mathbf{R}_l}. \end{aligned} \tag{R.21}$$

Because the $\mathbf{R}_l$ form the points of a periodic lattice, we have,

$$\sum_l e^{i(\mathbf{q}+\mathbf{q}')\cdot\mathbf{R}_l} = N\delta_{\mathbf{q}+\mathbf{q}',0}. \tag{R.22}$$

Using Eq. (R.7) and the last expression, we can cast Eq. (R.21) in the form,

$$T = \frac{1}{2}\sum_{\mathbf{q}j}\dot{Q}\begin{pmatrix} \mathbf{q} \\ j \end{pmatrix}\dot{Q}^*\begin{pmatrix} \mathbf{q} \\ j \end{pmatrix}. \tag{R.23}$$

Through a similar series of steps, and using Eqs. (21.47), (R.6), (R.7), (R.10), and (R.22), we can write

$$V = \frac{1}{2}\sum_{\mathbf{q}j}\omega_j^2(\mathbf{q})Q\begin{pmatrix} \mathbf{q} \\ j \end{pmatrix}Q^*\begin{pmatrix} \mathbf{q} \\ j \end{pmatrix}. \tag{R.24}$$

The last two equations express the kinetic and potential energy, respectively, as sums of squares. The quantities $Q\begin{pmatrix}\mathbf{q}\\ j\end{pmatrix}$ are called the *normal coordinates* or *normal modes* of the system. We note the greatly simplified forms that the kinetic and potential energies assume in this representation. This form corresponds to individual, uncoupled, and non-interacting harmonic oscillators described by the normal coordinates of the system.

The general solution, $u_\alpha(l)$, is given in terms of the normal coordinates by Eq. (R.17), and we also have the reverse relation,

$$Q\begin{pmatrix}\mathbf{q}\\ j\end{pmatrix} = \sqrt{\frac{M}{N}}\sum_{l\alpha} e^*_\alpha\begin{pmatrix}\mathbf{q}\\ j\end{pmatrix} u_\alpha(l)e^{-i\mathbf{q}\cdot\mathbf{R}_l}. \tag{R.25}$$

R.1.3 Quantization of normal modes

We wish to pass from the classical description given above, to a quantum-mechanical treatment of a dynamical system. To do this, we use Lagrange's canonical equations of motion and impose the appropriate quantum commutation relations between coordinates and momenta.

In terms of the normal coordinates of the system, the Hamiltonian and the Lagrangian take the forms,

$$\begin{aligned} H &= T+V \\ &= \frac{1}{2}\sum_{\mathbf{q}j}\left[\dot{Q}\begin{pmatrix}\mathbf{q}\\ j\end{pmatrix}\dot{Q}^*\begin{pmatrix}\mathbf{q}\\ j\end{pmatrix} + \omega_j^2(\mathbf{q})Q\begin{pmatrix}\mathbf{q}\\ j\end{pmatrix}Q^*\begin{pmatrix}\mathbf{q}\\ j\end{pmatrix}\right] \end{aligned} \tag{R.26}$$

and

$$\begin{aligned} \mathrm{L} &= T-V \\ &= \frac{1}{2}\sum_{\mathbf{q}j}\left[\dot{Q}\begin{pmatrix}\mathbf{q}\\ j\end{pmatrix}\dot{Q}^*\begin{pmatrix}\mathbf{q}\\ j\end{pmatrix} - \omega_j^2(\mathbf{q})Q\begin{pmatrix}\mathbf{q}\\ j\end{pmatrix}Q^*\begin{pmatrix}\mathbf{q}\\ j\end{pmatrix}\right] \end{aligned} \tag{R.27}$$

We now introduce momenta, $P\begin{pmatrix}\mathbf{q}\\ j\end{pmatrix}$, which are conjugate to the normal coordinates [2], by the usual relation,

$$\begin{aligned} P\begin{pmatrix}\mathbf{q}\\ j\end{pmatrix} &= \frac{\partial \mathrm{L}}{\partial \dot{Q}\begin{pmatrix}\mathbf{q}\\ j\end{pmatrix}} \\ &= \dot{Q}^*\begin{pmatrix}\mathbf{q}\\ j\end{pmatrix}. \end{aligned} \tag{R.28}$$

We also have,

$$P\begin{pmatrix} \mathbf{q} \\ j \end{pmatrix} = \dot{Q}^*\begin{pmatrix} \mathbf{q} \\ j \end{pmatrix} = \dot{Q}\begin{pmatrix} -\mathbf{q} \\ j \end{pmatrix} = P^*\begin{pmatrix} -\mathbf{q} \\ j \end{pmatrix}, \tag{R.29}$$

so that we can write the Hamiltonian in Eq. (R.26) in the form,

$$H = \frac{1}{2}\sum_{\mathbf{q}j}\left[P^*\begin{pmatrix} \mathbf{q} \\ j \end{pmatrix}P\begin{pmatrix} \mathbf{q} \\ j \end{pmatrix} + \omega_j^2(\mathbf{q})Q\begin{pmatrix} \mathbf{q} \\ j \end{pmatrix}Q^*\begin{pmatrix} \mathbf{q} \\ j \end{pmatrix}\right]. \tag{R.30}$$

Evidently, this expression consists of a sum over individual harmonic oscillators, each described by means of the coordinates $Q\begin{pmatrix} \mathbf{q} \\ j \end{pmatrix}$ and the momenta $P\begin{pmatrix} \mathbf{q} \\ j \end{pmatrix}$. This feature can be made clearer when we use Hamilton's equations of motion [2] to write,

$$\dot{Q}\begin{pmatrix} \mathbf{q} \\ j \end{pmatrix} = \frac{\partial H}{\partial P\begin{pmatrix} \mathbf{q} \\ j \end{pmatrix}} = P^*\begin{pmatrix} \mathbf{q} \\ j \end{pmatrix}, \tag{R.31}$$

and

$$\dot{P}\begin{pmatrix} \mathbf{q} \\ j \end{pmatrix} = -\frac{\partial H}{\partial Q\begin{pmatrix} \mathbf{q} \\ j \end{pmatrix}} = -\omega_j^2(\mathbf{q})Q^*\begin{pmatrix} \mathbf{q} \\ j \end{pmatrix}. \tag{R.32}$$

Similarly, we have

$$\dot{P}^*\begin{pmatrix} \mathbf{q} \\ j \end{pmatrix} = -\omega_j^2(\mathbf{q})Q\begin{pmatrix} \mathbf{q} \\ j \end{pmatrix}. \tag{R.33}$$

Using Eq. (R.31), we obtain

$$\ddot{Q}\begin{pmatrix} \mathbf{q} \\ j \end{pmatrix} = \dot{P}^*\begin{pmatrix} \mathbf{q} \\ j \end{pmatrix} = -\omega_j^2(\mathbf{q})Q\begin{pmatrix} \mathbf{q} \\ j \end{pmatrix} \tag{R.34}$$

so that

$$\ddot{Q}\begin{pmatrix} \mathbf{q} \\ j \end{pmatrix} + \omega_j^2(\mathbf{q})Q\begin{pmatrix} \mathbf{q} \\ j \end{pmatrix} = 0. \tag{R.35}$$

This is indeed the equation of a harmonic oscillator whose coordinates vary harmonically in time with frequency $2\pi/\omega_j(\mathbf{q})$.

Now, the passage to the quantum mechanical description is effected through the commutation relations [3] for a harmonic oscillator of coordinate Q and momentum P,

$$[Q, P] = i\hbar I, \tag{R.36}$$

with I denoting the unit matrix in a particular representation. Here, we have,

$$\left[Q\begin{pmatrix}\mathbf{q}\\ j\end{pmatrix}, P\begin{pmatrix}\mathbf{q}'\\ j'\end{pmatrix}\right] = \mathrm{i}\delta_{ij}\delta_{\mathbf{q},-\mathbf{q}'}. \tag{R.37}$$

Next, we introduce the operators,

$$a_{\mathbf{q}j} = \frac{1}{\sqrt{2\hbar\omega_j(\mathbf{q})}}\left[\omega_j(\mathbf{q})Q\begin{pmatrix}\mathbf{q}\\ j\end{pmatrix} + \mathrm{i}P\begin{pmatrix}-\mathbf{q}\\ j\end{pmatrix}\right], \tag{R.38}$$

and

$$a_{-\mathbf{q}j} = \frac{1}{\sqrt{2\hbar\omega_j(\mathbf{q})}}\left[\omega_j(\mathbf{q})Q\begin{pmatrix}-\mathbf{q}\\ j\end{pmatrix} + \mathrm{i}P\begin{pmatrix}\mathbf{q}\\ j\end{pmatrix}\right]. \tag{R.39}$$

From Eqs. (R.20) and (R.29), we have,

$$a^*_{\mathbf{q}j} = \frac{1}{\sqrt{2\hbar\omega_j(\mathbf{q})}}\left[\omega_j(\mathbf{q})Q\begin{pmatrix}-\mathbf{q}\\ j\end{pmatrix} - \mathrm{i}P\begin{pmatrix}\mathbf{q}\\ j\end{pmatrix}\right], \tag{R.40}$$

and

$$a^*_{-\mathbf{q}j} = \frac{1}{\sqrt{2\hbar\omega_j(\mathbf{q})}}\left[\omega_j(\mathbf{q})Q\begin{pmatrix}\mathbf{q}\\ j\end{pmatrix} - \mathrm{i}P\begin{pmatrix}-\mathbf{q}\\ j\end{pmatrix}\right]. \tag{R.41}$$

Conversely, we can express the normal coordinates and momenta in terms of these operators [1],

$$Q\begin{pmatrix}-\mathbf{q}\\ j\end{pmatrix} = \sqrt{\frac{\hbar}{2\omega_j(\mathbf{q})}}\left[a_{\mathbf{q}j} + a^*_{-\mathbf{q}j}\right], \tag{R.42}$$

and

$$P\begin{pmatrix}-\mathbf{q}\\ j\end{pmatrix} = \sqrt{\frac{\hbar\omega_j(\mathbf{q})}{2}}\left[a^*_{\mathbf{q}j} - a_{-\mathbf{q}j}\right]. \tag{R.43}$$

Using the last two expressions, we can write the Hamiltonian of the system, Eq. (R.30), in the form,

$$H = \sum_{\mathbf{q}j} H_{\mathbf{q}j}, \tag{R.44}$$

where

$$H_{\mathbf{q}j} = \hbar\omega_j(\mathbf{q})\left[a^*_{\mathbf{q}j}a_{\mathbf{q}j} + \frac{1}{2}\right]. \tag{R.45}$$

Once again, the Hamiltonian is written as a sum of uncoupled harmonic oscillators described in terms of the operators, $a_{\mathbf{q}j}$ and $a^*_{\mathbf{q}j}$.

The operators $a_{\mathbf{q}j}$ and $a^*_{\mathbf{q}j}$, respectively, can be viewed as creating and destroying an excitation (phonon) in the state $|\mathbf{q}j\rangle$ (with crystal momentum $\mathbf{q}$ and polarization j), and are often referred to as *creation* and *annihilation* operators [1, 3]. These operators can be used [1] to express the displacements, $u_\alpha(l)$, in the form,

$$\begin{aligned} u_\alpha(l) &= \frac{1}{2\sqrt{NM}} \sum_{\mathbf{q}j} e_\alpha \begin{pmatrix} \mathbf{q} \\ j \end{pmatrix} \left[\frac{\hbar}{2\omega_j(\mathbf{q})} \right]^{1/2} \left[\left(a_{\mathbf{q}j} + a^*_{-\mathbf{q}j} \right) e^{i\mathbf{q}\cdot\mathbf{R}_l} \right. \\ &\times \left. \left(a_{-\mathbf{q}j} + a^*_{\mathbf{q}j} \right) e^{-i\mathbf{q}\cdot\mathbf{R}_l} \right] \\ &= \left[\frac{\hbar^2}{MN} \right]^{1/2} \sum_{\mathbf{q}j} \frac{1}{\sqrt{\xi_j(\mathbf{q})}} e_\alpha \begin{pmatrix} \mathbf{q} \\ j \end{pmatrix} \left[a_{\mathbf{q}j} e^{i\mathbf{q}\cdot\mathbf{R}_l} + a^*_{\mathbf{q}j} e^{-i\mathbf{q}\cdot\mathbf{R}_l} \right], \end{aligned} \tag{R.46}$$

where the second equality follows from the first upon rearrangement of terms and the definition of the phonon energies, $\xi_j(\mathbf{q}) = \hbar\omega_j(\mathbf{q})$.

Bibliography

[1] A. K. Ghatak and L. S. Kothari, *An Introduction to Lattice Dynamics*, Addison Wesley, Reading, Massachusetts (1972).

[2] Herbert Goldstein, *Classical Mechanics*, Addison Wesley, Reading, Massachusetts (1950).

[3] Arno Bohm, *Quantum Mechanics*, Springer Verlag, Heidelberg, Berlin (1979).

Appendix S

Second Quantization

S.1 General Comments

In this appendix, we give a brief development of the concepts underlying the *occupation number representation* or *second quantization*.

Let us consider a quantum mechanical system consisting of two particles, labeled 1 and 2, located at positions x_1 and x_2, respectively. The corresponding two-particle wave function, $\psi(x_1, x_2)$, gives the probability amplitude for finding the particles at these positions in phase space. If the particles are indistinguishable, such as two electrons or two protons, the probability amplitude is given also by $\psi(x_2, x_1)$. Therefore, for indistinguishable particles, the probability amplitude for finding a particle at a and one at b is given either by $\psi(a, b)$ or $\psi(b, a)$. The corresponding probability takes the form,

$$P = |\psi(a,b)|^2 = |\psi(b,a)|^2, \tag{S.1}$$

so that $\psi(b, a)$ must differ from $\psi(a, b)$ at most by a phase factor of order unity.

Clearly, there is a redundancy in notation in specifying many-particle states that becomes exceedingly cumbersome as the number of particles increases. It would be advantageous to employ a notation that would carry the useful information, namely the number of particles occupying a certain state (in phase space), leaving out which particular particles are to be found there. Thus, for a system of n particles, the probability amplitude would be given by the single expression $\psi(n_1, n_2, n_3, \cdots)$, simply specifying that there are n_1 particles in state 1, n_2 in state 2, etc. Such a notation [1] would also be useful in considering processes in which the number of particles can change, such in the emission or absorption of photons, and in electron-positron annihilation.

The redundancy in notation can be removed by imposing the constraint that the wave function be either symmetric (for bosons) or antisymmetric (for fermions) under a single interchange between two particles. The resulting formalism is called *second quantization* or the *occupation number representation.*

S.2 Creation Operators

In establishing the formalism of second quantization, it is convenient to introduce the *vacuum state* denoted by,

$$|0\rangle, \tag{S.2}$$

which contains no particles of the type under consideration. We assume that the vacuum state is normalized according to,

$$\langle 0|0\rangle = 1. \tag{S.3}$$

It is important to emphasize that $|0\rangle$ is not necessarily empty of *all* particles (or fields) for it may indeed contain particles of kinds different from those forming the subject of a particular discussion. Therefore, if we are studying electrons only, the vacuum state could conceivably contain any number of protons, as long as they do not figure in the discussion.

Let us now consider a particle in state $|n\rangle$ whose wave function in the Schrödinger representation has the form,

$$u_n(x) = \langle x|n\rangle. \tag{S.4}$$

We can also describe this state in terms of introducing a particle at state $|n\rangle$ in the vacuum state. This can be done by means of a *creation* operator, $a_n^\dagger$, which is such that when acting on the state $|0\rangle$ gives the state $|n\rangle$,

$$|n\rangle \equiv a_n^\dagger|0\rangle. \tag{S.5}$$

For example, if $|n\rangle$ is a plane-wave state, $|n\rangle = |\mathbf{k}\rangle$, then $a_\mathbf{k}^\dagger$ creates a particle in that state. Therefore, the states of a single particle can be represented by means of a creation operator acting on the vacuum. This leads to a description of the state of a system in terms of numbers of particles in each state, rather than in terms of which specific particle is in which state. If the states $\{n\}$ constitute a complete set, e.g., plane waves, the states defined by Eq. (S.5) form such a complete set.

The bra vector corresponding to the state in Eq. (S.5) is denoted by,

$$\langle n| = \langle 0|a_n, \tag{S.6}$$

where a_n is the Hermitean conjugate of the operator $a_n^\dagger$. If the states $|n\rangle$ are orthonormal, then we have,

$$\langle 0|a_{n'}a_n^\dagger|0\rangle = \delta_{nn'}. \tag{S.7}$$

If the states $|n\rangle$ form an orthonormal basis, as, for example, do the plane-wave states, $|\mathbf{k}\rangle$, then they can be used to expand any other state as a linear combination,

$$\begin{aligned} |m\rangle &= \sum_n a_{nm}|n\rangle \\ &= \sum_n |n\rangle\langle n|m\rangle. \end{aligned} \tag{S.8}$$

We can now define a set of creation operators associated with the states $|m\rangle$,

$$\begin{aligned} |m\rangle &= a_m^\dagger|0\rangle \\ &= \sum_n a_n^\dagger|0\rangle\langle n|m\rangle, \end{aligned} \tag{S.9}$$

so that

$$a_m^\dagger = \sum_n a_n^\dagger\langle n|m\rangle. \tag{S.10}$$

Thus, $a_m^\dagger$ creates a particle in state $|m\rangle$, which is a linear combination of the basis states $|n\rangle$, as described by Eq. (S.8). For example, with $|n\rangle$ denoting plane-wave states $|\mathbf{k}\rangle$, and $|m\rangle$ denoting the eigenstates of the position operator, $|\mathbf{r}\rangle$, we have,

$$\begin{aligned} |\mathbf{r}\rangle &= (2\pi)^{-2/3}\int_{-\infty}^{\infty} \mathrm{d}^3k|\mathbf{k}\rangle\langle\mathbf{k}|\mathbf{r}\rangle \\ &= (2\pi)^{-2/3}\int_{-\infty}^{\infty} \mathrm{d}^3k|\mathbf{k}\rangle\mathrm{e}^{-\mathrm{i}\mathbf{k}\cdot\mathbf{r}}. \end{aligned} \tag{S.11}$$

Correspondingly, we have

$$a^\dagger(\mathbf{r}) = (2\pi)^{-2/3}\int_{-\infty}^{\infty} \mathrm{d}^3k a_\mathbf{k}^\dagger \mathrm{e}^{-\mathrm{i}\mathbf{k}\cdot\mathbf{r}}, \tag{S.12}$$

where $a^\dagger(\mathbf{r})$ creates a particle at position $\mathbf{r}$ in space. (It is only because of convention that $\mathbf{r}$ is put in parenthesis while $\mathbf{k}$ appears as a subscript.)

S.2.1 Two-particle states and commutation relations

There is essentially no advantage over the Schrödinger (direct) representation in introducing second quantization to describe the states of a single particle. The power of second quantized notation becomes evident only when systems containing more than one particle are considered. Let us consider a system containing two particles, one in state $|n\rangle$ and one in state $|n'\rangle$. Using the creation operators for these states, we can write,

$$|n, n'\rangle = a_n^\dagger a_{n'}^\dagger |0\rangle. \tag{S.13}$$

This expression has two creation operators acting on the vacuum state, with $a_{n'}^\dagger$ creating a particle in state $|n'\rangle$, and $a_n^\dagger$ one in state $|n\rangle$. However, the same two-particle state can be represented by

$$|n', n\rangle = a_{n'}^\dagger a_n^\dagger |0\rangle. \tag{S.14}$$

The two different expressions, Eqs. (S.13) and (S.14), represent the same physical state and can differ from one another at most by a factor. Calling this factor f, we have,

$$a_{n'}^\dagger a_n^\dagger = f a_n^\dagger a_{n'}^\dagger. \tag{S.15}$$

Using this relation, and interchanging the two operators on both sides, we obtain,

$$a_n^\dagger a_{n'}^\dagger = f a_{n'}^\dagger a_n^\dagger, \tag{S.16}$$

so that

$$f^2 = 1. \tag{S.17}$$

It follows that $f = \pm 1$, so that the creation operators $a_n^\dagger$ and $a_{n'}^\dagger$ must satisfy one of two possible commutation relations,

$$a_n^\dagger a_{n'}^\dagger = a_{n'}^\dagger a_n^\dagger, \tag{S.18}$$

or

$$a_n^\dagger a_{n'}^\dagger = -a_{n'}^\dagger a_n^\dagger. \tag{S.19}$$

Thus, two creation operators either commute, Eq. (S.18), or anticommute, Eq. (S.19). Particles whose creation operators commute are called *bosons*. Those whose operators anticommute are called *fermions*.

These simple commutation relations lead immediately to some important properties of bosons and fermions. Consider the anticommutation relation, Eq. (S.19), and set $n = n'$. We find,

$$a_n^\dagger a_n^\dagger = -a_n^\dagger a_n^\dagger = 0, \tag{S.20}$$

so that two fermions *cannot* be created in the same state. This of course is an expression of Pauli's exclusion principle. By contrast, there is no restriction to the number of bosons occupying a given state as the striking phenomena associated with Einstein condensation amply illustrates.

As in the single-particle case, we can define the Hermitean conjugate of the two-particle ket vector in Eq. (S.13),

$$\langle n, n'| = a_{n'} a_n. \tag{S.21}$$

The norm of this state is obtained as the inner product,

$$\langle 0|a_{n'} a_n a_n^\dagger a_{n'}^\dagger|0\rangle = N_{nn'}, \tag{S.22}$$

where the value of the normalization constant, $N_{nn'}$, can be chosen as specified below. More generally, we can write,

$$\langle 0|a_{n''} a_n a_n^\dagger a_{n'}^\dagger|0\rangle = N_{nn'} \delta_{n'n''}. \tag{S.23}$$

The expectation value in the last expression is the scalar product of two two-particle states. However, it can also be interpreted as the scalar product of two one-particle states, with one of these states being represented by $\langle 0|a_{n''}$, and the other by $a_n a_n^\dagger a_{n'}^\dagger|0\rangle$, if we take the three operators acting on the vacuum to have the result,

$$a_n a_n^\dagger a_{n'}^\dagger|0\rangle = N_{nn'} a_{n'}^\dagger|0\rangle. \tag{S.24}$$

Therefore, the operator a_n acting on the right on a ket vector decreases the number of particles by one (the particle created by $a_n^\dagger$ is annihilated by a_n leaving only the state $a_n^\dagger|0\rangle$). The a_n are commonly referred to as annihilation (or destruction) operators[1]. It follows that the annihilation operator a_n acting on the vacuum state must give zero, while acting on a state containing a single particle in state $|n'\rangle$ must give the vacuum state, if $n = n'$, or zero. Thus, we have

$$a_n|0\rangle = 0, \tag{S.25}$$

and

$$a_n a_{n'}^\dagger|0\rangle = \delta_{nn'}|0\rangle, \tag{S.26}$$

where the coefficient on the right of the last expression is obtained from the orthonormality relation, Eq. (S.7), for single-particle states.

The commutation relations for annihilation operators can be obtained as the Hermitean conjugates of those for creation operators, Eqs. (S.18) and (S.19). Thus, we find,

$$[a_n, a_{n'}]_\pm = (a_n a_{n'} \pm a_{n'} a_n) = 0, \tag{S.27}$$

where the (+) sign refers to fermions and the (-) to bosons.

---

[1]Note that a_n acting on the left on a bra vector *increases* the number of particles by one, while $a_n^\dagger$ *decreases* it.

We now consider the commutation relations between creation and annihilation operators. Using Eqs. (S.18), (S.19), and (S.26), we can evaluate the action of the boson commutators or the fermion anticommutators on a single-particle state,

$$(a_n a_{n'} \pm a_{n'} a_n)\, a_n^\dagger|0\rangle = \left(\pm a_n a_n^\dagger a_{n'}^\dagger \mp a_{n'}\right)|0\rangle = \pm(N_{nn'} - 1)a_{n'}^\dagger|0\rangle, \tag{S.28}$$

where the upper (lower) sign refers to bosons (fermions). We would like to choose the normalization constant, $N_{nn'}$, to make the boson commutator or the fermion anticommutator independent of the states upon which they act. Comparing the last expression with the expression obtained from Eqs. (S.25) and (S.26),

$$(a_n a_{n'} \pm a_{n'} a_n)\,|0\rangle = \delta_{nn'}|0\rangle, \tag{S.29}$$

we see that this is accomplished if we define,

$$N_{nn'} = 1 \pm \delta_{nn'}. \tag{S.30}$$

Therefore, if $n \neq n'$, we have $N_{nn'} = 1$. When $n = n'$, then $N_{nn'} = 2$ for bosons and vanishes for fermions. These results are expected in view of the exclusion principle for fermions, and the fact that more than one boson is allowed to occupy the same state.

Now, the complete set of commutation and anticommutation relations for bosons (-) and fermions (+) take the forms,

$$a_n^\dagger a_{n'}^\dagger \mp a_{n'}^\dagger a_n^\dagger = 0 = a_n a_{n'} \mp a_{n'} a_n \tag{S.31}$$

and

$$a_n a_{n'}^\dagger \mp a_{n'}^\dagger a_n = \delta_{nn'}. \tag{S.32}$$

It is easily shown that these relations remain invariant under a transformation to a new basis set, such as the one indicated in Eq. (S.9).

Finally, let us consider the consequences of the normalization shown in Eq. (S.30). From this expression, we obtain,

$$\langle 0|a_n a_n a_n^\dagger a_n^\dagger|0\rangle = 2\langle 0|a_n a_n^\dagger|0\rangle = 2 \tag{S.33}$$

and

$$\langle 0|(a_n)^m (a_n^\dagger)^m|0\rangle = m!. \tag{S.34}$$

These results look very much like those we encountered in the second-quantization treatment of the harmonic oscillator in Appendix R. In general, the normalized state vector for a state in which there are α bosons in state n_1, β bosons in state n_2, etc., has the form,

$$(\alpha!\beta!\cdots)^{-1/2}\left(a_{n_1}^\dagger\right)^\alpha\left(a_{n_2}^\dagger\right)^\beta\cdots|0\rangle. \tag{S.35}$$

For fermions, the problem of the normalization factor does not exist since there can never be more than a single particle in the same quantum state.

States of more than two particles can be constructed by appropriate application of creation operators to the vacuum state. Based on the commutation relations, Eqs. (S.31) and (S.32), these states are free of the ambiguities characterizing the configurational representation.

S.3 Number Operator

A single particle operator of particular interest is the *number operator*,

$$n_m = a_m^\dagger a_m, \tag{S.36}$$

which counts the number of particles in state $|m\rangle$. To see this, we consider the commutation relations of n_m with the creation and annihilation operators for this state. From Eqs. (S.33) and (S.34), we have

$$\left[n_m, a_{m'}^\dagger\right] = a_m^\dagger \delta_{mm'} \tag{S.37}$$

$$[n_m, a_{m'}] = -a_m \delta_{mm'} \tag{S.38}$$

and

$$\left[n_m, \left(a_m^\dagger\right)^\ell\right] = \ell \left(a_m^\dagger\right)^\ell, \tag{S.39}$$

which hold for both bosons and fermions. Because the number operator is quadratic in the creation and annihilation operators, the $(\pm)$ signs appearing in the commutation relations of these operators have disappeared. From the last expression, we obtain,

$$n_m \left(a_m^\dagger\right)^\ell |0\rangle = \ell \left(a_m^\dagger\right)^\ell |0\rangle. \tag{S.40}$$

Thus, the state with ℓ particles in state $|m\rangle$ is an eigenstate of n_m, which justifies the name number operator.

S.4 Operators in Second Quantization

In order to complete the correspondence between quantum mechanics expressed in configuration space and in second quantized notation, we must exhibit the expressions taken by operators in second quantization. Using the commutation relations, we can show [1] that a single-particle operator can be written as,

$$A = \sum_{nn'} \langle n'|A|n\rangle a_{n'}^\dagger a_n, \tag{S.41}$$

where $\langle n'|A|n\rangle$ denotes the matrix elements of A between states $|n\rangle$ and $|n'\rangle$. This expression states that the operator A can be viewed as a sum of terms that annihilate a particle at one state and create it in another. The magnitude of each such term is just the matrix element of the operator A in configuration space between these states.

Similarly, we find that a two-particle operator, B, can be written in the form,

$$B = \sum_{n_1 n_2} \sum_{n_3 n_4} \langle n_4 n_3|B|n_2 n_1\rangle a^\dagger_{n_4} a^\dagger_{n_3} a_{n_2} a_{n_1}, \tag{S.42}$$

which is a sum of terms that annihilate and create two-particle states.

S.5 Field Operators

It is often convenient to use creation and annihilation operators for particle states to obtain operators that create or annihilate a particle at point $x = (\mathbf{r}, \sigma)$. We define the *field* operators[2],

$$\psi(x) = \sum_n a_n u_n(x), \tag{S.43}$$

and

$$\psi^\dagger(x) = \sum_n a^\dagger_n u_n(x). \tag{S.44}$$

Based on the commutation relations for $a^\dagger_n$ and a_n, and the orthonormality and completeness properties of the basis states, u_n, we can show that these operators satisfy the relations,

$$[\psi(x), \psi^\dagger(x')]_\pm = \delta(x - x'), \; [\psi(x), \psi(x')]_\pm = [\psi^\dagger(x), \psi^\dagger(x')]_\pm = 0, \tag{S.45}$$

where the (+) sign (denoting an anticommutator) refers to fermions and the (-) sign (denoting a commutator) to bosons. In terms of these field operators, the Hamiltonian operator, consisting of one-particle, $h(x)$, and two-particle, $U(\mathbf{r}, \mathbf{r}')$, terms, assumes the form,

$$\begin{aligned} H &= \int \mathrm{d}x \psi^\dagger(x) h(x) \psi(x) \\ &+ \frac{1}{2} \int \mathrm{d}x \int \mathrm{d}x' \psi^\dagger(x) \psi^\dagger(x') U(\mathbf{r}, \mathbf{r}') \psi(x) \psi(x^\dagger). \end{aligned} \tag{S.46}$$

Finally, the ***average density*** in a state $|N\rangle$ of a many-particle system is given by the expression,

$$n(\mathbf{r}) = \langle N| \int \mathrm{d}\sigma \psi^\dagger(x) \psi(x) |N\rangle, \tag{S.47}$$

[2]The distinction between wave functions and field operators, when denoted by the same symbol, will be obvious from the content of the discussion.

while the *density matrix* is given by,

$$n(x, x') = \langle N|\psi^\dagger(x)\psi(x')|N\rangle. \tag{S.48}$$

Bibliography

[1] Harry J. Lipkin, *Quantum Mechanics: New Approaches to Selected Topics*, North-Holland, Amsterdam (1988).

Appendix T

Schwinger's Dynamical Principle

Schwinger's [1] dynamical principle unifies all the fundamental equations of quantum mechanics, that is, the Schrödinger (wave function) equation, the Heisenberg (operator) equation, the commutation relations, the conservation laws, and so on. The principle states that the variation of the transition probability amplitude between an initial state, $|I\rangle$, and a final state, $|F\rangle$, is given by the variation of the initial and final conditions and the Lagrangian density, L, of the system,

$$\delta\langle F|I\rangle = \frac{\mathrm{i}}{\hbar}\langle F|\delta \int L\mathrm{d}^4x|I\rangle. \tag{T.1}$$

One practical application of this principle is its use in calculating functional derivatives of amplitudes or matrix elements with respect to external sources or fields. For a brief review of functional differentiation, see Appendix A. The following discussion is follows the lines in Kato et al. [2].

We now use this principle to calculate the functional derivatives of the amplitudes and matrix elements with respect to external sources, $j(x)$ and $\eta(x)$, introduced in Chapter 24. Under a small variation, $\epsilon f(x)$, in the source function of a boson field, $\phi(x)$, the variation in the action function becomes,

$$\delta \int L\mathrm{d}^4x = -\int \phi(x)\epsilon f(x)\mathrm{d}^4x, \tag{T.2}$$

and Schwinger's principle yields the result,

$$\delta\langle F|I\rangle = \frac{1}{\mathrm{i}\hbar}\int \langle F|\phi(x)|I\rangle\epsilon f(x)\mathrm{d}^4x. \tag{T.3}$$

On the other hand, we have,

$$\delta\langle F|I\rangle = \langle F|I\rangle_{j+\epsilon f} - \langle F|I\rangle_j = \int \frac{\delta\langle F|I\rangle}{\delta j(x)}\epsilon f(x)\mathrm{d}^4x. \tag{T.4}$$

From the last two expressions, we obtain the result,

$$\frac{\delta}{\delta j(x)}\langle F|I\rangle = \frac{1}{\mathrm{i}\hbar}\langle F|\phi(x)|I\rangle. \tag{T.5}$$

Similarly, in the case of fermion fields, we obtain,

$$\frac{\delta}{\delta\eta(x)}\langle F|I\rangle = \frac{1}{\mathrm{i}\hbar}\langle F|\psi^\dagger(x)|I\rangle, \tag{T.6}$$

$$\frac{\delta}{\delta\eta^\dagger(x)}\langle F|I\rangle = \frac{1}{\mathrm{i}\hbar}\langle F|\psi(x)|I\rangle, \tag{T.7}$$

where spinor indices are suppressed.

Next, we differentiate the matrix element, $\langle F|Q(x|I\rangle)$, of an operator, $Q(x)$, with respect to $j(x)$ and $\eta(x)$. For $t > t'$, and for a time t_1 between t and t', we have,

$$\begin{aligned}\frac{\delta}{\delta j(x)}\langle F|Q(x)|I\rangle &= \frac{1}{\mathrm{i}\hbar}\sum_m \langle F|Q(x)|mt_1\rangle\frac{\delta\langle mt_1|I\rangle}{\delta j(x')} \\ &= \frac{1}{\mathrm{i}\hbar}\langle F|Q(x)\phi(x')|I\rangle,\end{aligned} \tag{T.8}$$

while for $t < t'$, we obtain,

$$\begin{aligned}\frac{\delta}{\delta j(x)}\langle F|Q(x)|I\rangle &= \frac{1}{\mathrm{i}\hbar}\sum_m \frac{\delta\langle F|mt_1\rangle}{\delta j(x')}\langle mt_1|Q(x)|mt_1\rangle \\ &= \frac{1}{\mathrm{i}\hbar}\langle F|\phi(x')Q(x)|I\rangle.\end{aligned} \tag{T.9}$$

Similar expressions hold for spinor fields. Paying attention to the anticommutativity of fermion operators, we have,

$$\begin{aligned}\delta\langle F|Q(x)|I\rangle &= \frac{1}{\mathrm{i}\hbar}\langle F|Q(x)\int \psi^\dagger(x')\delta\eta(x')\mathrm{d}^4x'|I\rangle \\ &= \frac{1}{\mathrm{i}\hbar}\langle F|\int \psi^\dagger(x')\delta\eta(x')\mathrm{d}^4xQ(x)\prime|I\rangle,\end{aligned} \tag{T.10}$$

depending on whether the non-zero point of $\delta\eta(x')$ is earlier or later than t. The right-hand side of the last expression becomes,

$$\frac{1}{\mathrm{i}\hbar}\langle F|\int \psi^\dagger(x')Q(x')\prime|I\rangle\delta\eta(x')\mathrm{d}^4x, \tag{T.11}$$

where the sign ($\pm$) results from the exchange of $Q(x)$ and $\delta\eta(x')$. Thus, we obtain,

$$\begin{aligned}
\frac{\delta}{\delta j(x')}\langle F|Q(x)|I\rangle &= \frac{1}{i\hbar}\langle F|T\,[Q(x)\phi(x')]\rangle, \\
\frac{\delta}{\delta\eta(x')}\langle F|Q(x)|I\rangle &= \frac{1}{i\hbar}\langle F|T\,[Q(x)\psi^\dagger(x')]\rangle, \\
\frac{\delta}{\delta\eta^\dagger(x')}\langle F|Q(x)|I\rangle &= \frac{1}{i\hbar}\langle F|T\,[Q(x)\psi(x')]\rangle, \qquad \text{(T.12)}
\end{aligned}$$

where T denotes time ordering. From Eqs. (T.6), (T.7), and (T.12), we obtain the formulae,

$$\begin{aligned}
\frac{\delta}{\delta j(x')}\langle Q(x)\rangle &= \frac{1}{i\hbar}\left\{\langle T\,[Q(x)\phi(x')]\rangle - \langle Q(x)\rangle\langle\phi(x')\rangle\right\}, \\
\frac{\delta}{\delta\eta(x')}\langle Q(x)\rangle &= \frac{1}{i\hbar}\left\{\langle T\,[Q(x)\psi^\dagger(x')]\rangle - \langle Q(x)\rangle\langle\psi^\dagger(x')\rangle\right\}, \\
\frac{\delta}{\delta\eta^\dagger(x')}\langle Q(x)\rangle &= \frac{1}{i\hbar}\left\{\langle T\,[Q(x)\psi(x')]\rangle - \langle Q(x)\rangle\langle\psi(x')\rangle\right\},
\end{aligned} \qquad \text{(T.13)}$$

where the symbol $\langle\cdots\rangle$ denotes expectation value in the manner described in Chapter 24.

A real scalar field, $\phi(x)$, can be introduced by adding the term $-\psi^\dagger(x)\psi(x)\phi(x)$ to the Lagrangian density. The functional derivative with respect to $\phi(x)$ can be obtained from Schwinger's dynamical principle in the form,

$$\begin{aligned}
\frac{\delta}{\delta\phi(x')}\langle Q(x)\rangle &= \frac{1}{i\hbar}\left\{\langle T\,[Q(x)\psi^\dagger(x')\psi(x')]\rangle\right. \\
&- \left.\langle Q(x)\rangle\langle\psi^\dagger(x')\psi(x')\rangle.\right\} \qquad \text{(T.14)}
\end{aligned}$$

Bibliography

[1] J. Schwinger, Proc. Nat. Acad. Sci. **37**, 452, 455 (1951).

[2] Tomokazu Kato, Tetsuro Kobayashi, and Mikio Namiki, Supplement of the Progress of Theoretical Physics, No. **15** (1960).

Glossary of Terms

APW Augmented plane wave
ASA Atomic sphere approximation
ATA Average t-matrix approximation
BEB Blackman-Esterling-Berk
BZ Brillouin zone
C Canonical
CPA Coherent potential approximation
CWM Connolly-Williams method
DCA Direct configurational average
DFT Density functional theory
DOS Density of states
ECM Embedded cluster method
FLAPW Full potential linearized augmented plane wave
GC Grand canonical
GEA Gradient expansion approximation
GGA Generalized gradient approximation
GPM Generalized perturbation method
HFF Hellmann-Feynman Force
KCM Krivoglaz-Klapp-Moss
KKR Korringa-Kohn-Rostoker
LDA Local density approximation
LEED Low-energy electron diffraction
LKKR Layer Korringa-Kohn-Rostoker
LMTO Linear muffin-tin orbital
LSDA Local spin density approximation
MFA Mean field approximation
MST Multiple scattering theory
MT Muffin tin
NFCs Near field corrections
ODD Off-diagonal disorder
RBM Rigid band model
RPA Random phase approximation
RS-MST Real space multiple scattering theory
SCBSA Self-consistent boundary site approximation
SCCSA Self-consistent central-site approximation
SCF-KKR-CPA Self-consistent field Korringa-Kohn-Rostoker coherent potential approximation
SIC Self-interaction correction
SPO Scattering path operator
SRO Short range order
SSCPA Single-site coherent potential approximation

TB Tight binding
TCA Traveling cluster approximation
TLK Terrace-ledge-kink
TTM Truncated t-matrix approximation
VCA Virtual crystal approximation

INDEX

Printed in the United States
By Bookmasters